GRAPHS

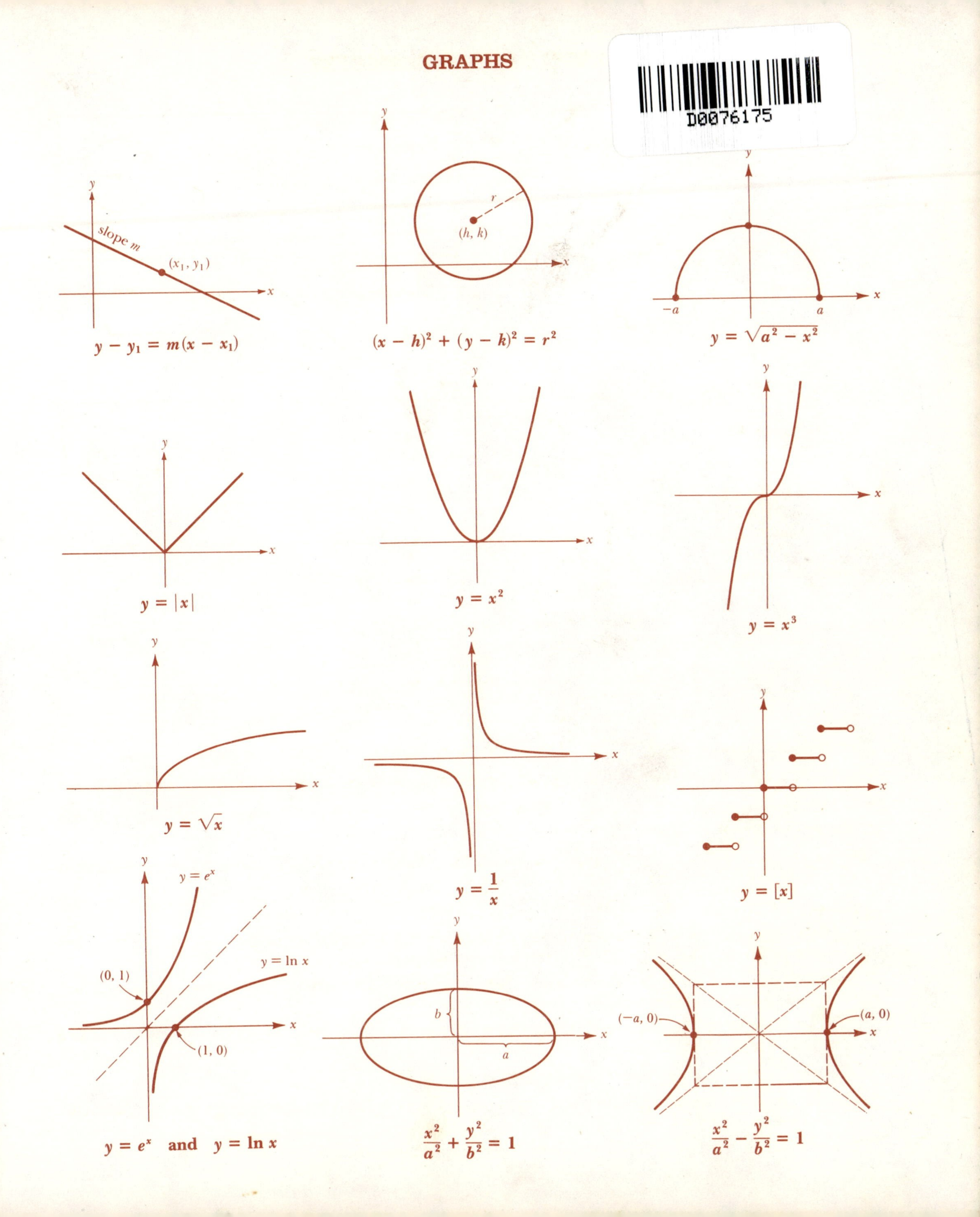

D0076175
slope m
(x_1, y_1)
$y - y_1 = m(x - x_1)$
(h, k)
r
$(x - h)^2 + (y - k)^2 = r^2$
$-a$
a
$y = \sqrt{a^2 - x^2}$
$y = |x|$
$y = x^2$
$y = x^3$
$y = \sqrt{x}$
$y = \frac{1}{x}$
$y = [x]$
$y = e^x$
$(0, 1)$
$y = \ln x$
$(1, 0)$
$y = e^x$ and $y = \ln x$
b
a
$\frac{x^2}{a^2} + \frac{y^2}{b^2} = 1$
$(-a, 0)$
$(a, 0)$
$\frac{x^2}{a^2} - \frac{y^2}{b^2} = 1$

ALGEBRA AND TRIGONOMETRY

David Cohen
Department of Mathematics
University of California
Los Angeles

West Publishing Co.
St. Paul New York Los Angeles San Francisco

For my daughter Emily

Production: Greg Hubit Bookworks
Technical illustrations: John Foster
Copy editing: Patricia Talbot Harris
Typesetting: Interactive Composition Corporation
Cover design: Delor Erickson
Indexing: Susan Cohen

Library of Congress Cataloging-in-Publication Data

Cohen, David, 1942 Sept. 28–
Algebra and trigonometry.

Includes index.
1. Algebra. 2. Trigonometry. I. Title
QA154.2.C594 1986 512.13 85-26299
ISBN 0-314-93165-1
1st Reprint—1986

Preface

This text develops the elements of college algebra and trigonometry in a straightforward manner. The presentation is student-oriented in three specific ways. First, I've tried to talk to, rather than lecture at, the student. Second, examples are consistently used to introduce, to explain, and to motivate concepts. And third, all of the initial exercises for each section are carefully correlated with the worked examples in that section.

Audience

In writing *Algebra and Trigonometry,* I have assumed that the students have been exposed to intermediate algebra, but that they have not necessarily mastered that subject. Also, for many college algebra students, there may be a gap of several years between their previous mathematics courses and the present one. For these reasons, the review material (in Chapter 1 and in parts of Chapter 2) is unusually thorough. This allows the instructor the option of covering the material in class or of assigning it to the students for study on their own.

Features

1. *Word problems and applications.* Beginning with the review material in Chapter 1, there is a constant emphasis on learning to use the *language* of algebra as a tool in problem solving. Word problems and strategies for solving them are explained and developed throughout the book. (See, for example, Section 2.3.) One complete section (Section 4.2) is devoted to setting up equations that describe or summarize given situations. Another entire section (Section 6.4) discusses applications of the exponential function.
2. *Emphasis on graphing.* Graphs and techniques for graphing are developed throughout the text, and graphs are used to explain and reinforce algebraic concepts. (See, for example, Sections 4.4 and 6.2.)
3. *Exercise sets.* The mathematical backgrounds of college algebra and trigonometry students vary widely, not only from school to school, but even within a given classroom. For this reason, it is important that the exercise sets in a mainstream text allow the instructor a great deal of flexibility in designing the course. As mentioned earlier, each exercise set in this book begins with problems that are carefully tied to the worked examples in the text. For some students, or in some classes, core problems such as these will be the principal focus. In addition to these core problems, however, each exercise set also contains a rich blend of additional problems. By selecting from these, each instructor can tailor the assignments (and thereby the course) to fit the needs of the students and the school.
4. *Calculator exercises.* Many (perhaps even most) students entering a

course in college algebra and trigonometry own, or have access to, a scientific calculator. And they are eager to use it. Calculator exercises in this text, indicated by the symbol [C], are designed to take advantage of this situation. (It should be pointed out, however, that this is not a calculator-based text.) There are four broad types of calculator exercises here:

(i) Exercises that reinforce and supplement the core material. (See for example, Exercises 63 and 64 in Section 1.3.)

(ii) Exercises with surprising results that motivate subsequent theoretical questions. (See, for example, Exercise 58 in Section 1.5 and Exercise 212 in the review exercises for Chapter 8.)

(iii) Exercises that introduce supplementary material. (See, for example, Exercise 82 in Section 1.7.)

(iv) Exercises demonstrating that the use of a calculator cannot replace thinking or the need for mathematical proofs. (See, for example, Exercise 60 in Section 1.5 and Exercise 81 in Section 6.3.)

In addition to these specially marked calculator exercises, there are many routine problems, integrated throughout the text, for which either a calculator or tables will suffice. Such problems are not preceded by the symbol [C].

5. *Starred exercises.* As a convenience for the instructor, I've often used asterisks to mark exercises that seem to be more challenging than the others. In part, of course, this is a subjective call. Indeed, in some instances, asterisks have been added or deleted based upon student comments on preliminary versions of this text.

6. *End-of-chapter material.* Each chapter concludes with a review checklist, a review exercise set, and a chapter test. The review exercises serve several key functions. First, they provide a source of routine drill-type problems in a context that is not explicitly linked to a specific section of the chapter. Thus, just as on an examination, the student must decide for himself or herself which tools to bring to bear on a problem. Second, the review exercise sets provide a source of problems at various levels, ranging from the very elementary to the challenging. Some of these approach the core material from perspectives that differ slightly from that in the exposition. This is important for students who are planning to apply their algebra or trigonometry skills in subsequent courses. The system of starring the more challenging exercises is not followed in the review exercises; the students are on their own there.

   The chapter tests, for the most part, are based on the drill-type exercises in each section that most likely would have been assigned as homework in any class.

Trigonometry

The development of most topics in this text proceeds from the specific to the general, and the trigonometry follows this pattern too. Thus, the first section on trigonometry (Section 7.1) introduces the trigonometric functions in the context of right triangles. The unit-circle approach follows shortly thereafter in Section 7.3. Between Sections 7.1 and 7.3 is an important section providing examples and drill work in the algebraic manipulations needed in trigonometry. This helps to ease the way for students when they study the more analytical portions of trigonometry in Section 7.3 and in Chapter 8.

Supplementary Materials

1. *Student's Solutions Manual.* This contains complete solutions for the odd-numbered exercises (excluding the chapter tests).
2. *Instructor's Solutions Manual.* This manual contains full solutions for every exercise and test question in *Algebra and Trigonometry.*
3. The *West MathTest.* This is a computer-generated testing program that is available to schools adopting *Algebra and Trigonometry.*
4. *Test Bank.* This contains several alternate versions for each chapter test, along with answers for those tests. This is also available to schools adopting *Algebra and Trigonometry.*

Acknowledgments

I would like to sincerely thank the following reviewers for their help and their many constructive suggestions over the course of this project.

James Brunner
Black Hawk College
Illinois

Warren Chilton
Surrey Community College
North Carolina

Douglas B. Crawford
College of San Mateo
California

Katherine Franklin
Los Angeles Pierce College
California

Frank Gunnip
Macomb Community College
Michigan

D. W. Hall
Michigan State University

B. J. Harmon
Monroe County Community College
Michigan

John Hornsby
University of New Orleans

Sarah Kennedy
Texas Tech University

Anne F. Landry
Dutchess Community College
New York

Stanley M. Lukawecki
Clemson University
South Carolina

Michael J. Mears
Manatee Junior College
Florida

Ross Rueger
College of the Sequoias
California

William Simons
West Virginia University

Gladys G. Taylor
Indiana State University

Martin E. Walter
University of Colorado

Ray Wilson
Central Piedmont Community College
North Carolina

I would also like to thank three people in particular at West Educational Publishing for their help, advice, and encouragement: Peter Marshall, Executive Editor; Mark Jacobsen, Production Editor; and Theresa O'Dell, Developmental Editor. Finally, I would like to express my appreciation to Greg Hubit of *Bookworks* for his work in transforming my manuscript into a textbook.

David Cohen
Palos Verdes, California, 1986

Contents

Chapter 1 Fundamental Concepts

Introduction This first chapter sets the foundations for our work in algebra. The central topics here are the real numbers, polynomials, and operations with polynomials. In the last section of the chapter we describe the complex number system. The notation and the results in this chapter will be used repeatedly throughout the rest of the book.

1.1 Notation and Language

Perhaps Pythagoras was a kind of magician to his followers because he taught them that nature is commanded by numbers. There is a harmony in nature, he said, a unity in her variety, and it has a language: numbers are the language of nature.

Jacob Bronowski

"Algebra is a merry science," Uncle Jakob would say. "We go hunting for a little animal whose name we don't know, so we call it x. When we bag our game we pounce on it and give it its right name."

Albert Einstein

Here, as in your previous mathematics courses, most of the numbers we deal with are *real numbers.* These are the numbers used in everyday life, in the sciences, in industry, and in business. Some examples of real numbers are

$$3 \qquad -11 \qquad \sqrt{2} \qquad 0.05 \qquad 1.666\ldots \qquad \pi$$

In algebra we often use letters to stand for numbers. This simple idea is what gives algebra much of its power and applicability in problem solving. By using letters, we can often see patterns and reach conclusions that might not be so apparent otherwise.

Arithmetic, and therefore algebra too, begins with the *operations* of addition, subtraction, multiplication, and division. Table 1 will serve to remind you that, for the most part, the notation for these operations in both arithmetic and algebra is the same. One exception: As the table indicates, the product $a \times b$ is often abbreviated to simply ab.

Operation	Arithmetic Example	Algebraic Example
Addition (+)	$5 + 7.01$	$p + q$
Subtraction (−)	$13 - \sqrt{5}$	$r - s$
Multiplication (×)	3×4	$a \times b$ or $a \cdot b$ or ab
Division (÷)	$5 \div 6$	$x \div y$ or $\frac{x}{y}$ or x/y

Table 1

There are several other notational conventions that should seem familiar to you from previous courses. A sum such as $a + a + a$ is denoted by $3a$, while a product such as $a \cdot a \cdot a$ is written a^3. In general, if n is a positive integer (i.e., one of the numbers 1, 2, 3, 4, . . .), then

$$na \text{ means } \underbrace{a + a + a + \cdots + a}_{n \text{ terms}}$$

$$a^n \text{ means } \underbrace{a \cdot a \cdot a \cdot \cdots \cdot a}_{n \text{ factors}}$$

In the expression na, n is called the *coefficient* of a. For example, in the expression $3a$ the coefficient is 3. When we write a^n we refer to a as the *base* and n as the *exponent* or *power* to which the base is raised. The process of raising a base to a power is called *exponentiation*. The notation a^2 is read "a squared" (because the area of a square with sides each of length a is $a \times a$ or a^2). Similarly, a^3 is read "a cubed," because the volume of a cube with sides each of length a is $a \times a \times a$ or a^3.

There are several important conventions regarding the order in which the operations of arithmetic and algebra are to be performed. For instance, in the absence of parentheses or other grouping symbols, multiplication takes precedence over addition. As an example of this we have $1 + 2 \times 3 = 1 + 6 = 7$. On the other hand, had we originally intended that the addition be done first, then we would have written $(1 + 2) \times 3$. In this sense the parentheses are grouping symbols, and the convention is to carry out the operations within the parentheses first. So in this case we have $(1 + 2) \times 3 = 3 \times 3 = 9$. In Table 2 we list the various conventions regarding the order of operations.

Convention	Example
1. If no grouping symbols (such as parentheses) are present:	
(a) First do the exponentiations, working from left to right;	$3 + 4^2 = 3 + 16 = 19$; $5 \times 2^3 = 5 \times 8 = 40$
(b) Next do the multiplications and divisions, working from left to right;	$9 - 2 \times 3 = 9 - 6 = 3$; $12 + 8 \div 2 = 12 + 4 = 16$ $4 + 2 \times 3^2 = 4 + 2 \times 9 = 4 + 18 = 22$
(c) Then do the additions and subtractions, again working from left to right.	$1 + 2 \times 3 - 4 = 1 + 6 - 4 = 7 - 4 = 3$ $15 + 6^2 \div 4 - 2 \times 3^2 = 15 + 36 \div 4 - 2 \times 9$ $= 15 + 9 - 18$ $= 24 - 18 = 6$
2. If grouping symbols are present:	
(a) First do the operations within the grouping symbols;	$(1 + 2) \times 5 = 3 \times 5 = 15$; $(3 + 4)^2 = 7^2 = 49$
(b) If grouping symbols appear within grouping symbols, the operations within the innermost set are to be done first.	$3[8 - 2(6 - 4)] = 3(8 - 2 \times 2)$ $= 3(8 - 4)$ $= 3 \times 4 = 12$

Table 2
The Order of Operations

The four examples that follow serve to summarize the ideas discussed in this section. The fourth example (and the exercises like it) will also help you begin to review the geometric formulas given on the inside front cover of this book.

Example 1 Rewrite each expression using exponential notation.

(a) $x \cdot x \cdot x \cdot y \cdot y \cdot y \cdot y$

(b) $(a + 3)(a + 3)(b - 1)(b - 1)(b - 1)$

Solution

(a) $x \cdot x \cdot x \cdot y \cdot y \cdot y \cdot y = x^3y^4$

(b) $(a + 3)(a + 3)(b - 1)(b - 1)(b - 1) = (a + 3)^2(b - 1)^3$

Example 2

(a) Evaluate $x - 4y^2$ when $x = 19$ and $y = 2$.

(b) Evaluate $(x - 4)y^2$ using the values for x and y given in part (a).

Solution

(a) $x - 4y^2 = 19 - 4 \times 2^2$

$= 19 - 4 \times 4 = 19 - 16 = 3$

(b) $(x - 4)y^2 = (19 - 4) \times 2^2 = 15 \times 4 = 60$

Example 3 Translate each of the following into algebraic notation:

(a) five more than twice the number x

(b) three times the sum of x and y^3

(c) the sum of three times x and y^3

Solution

(a) $2x + 5$

twice x ↑ ↑ five more

(b) $3 \times (x + y^3) = 3(x + y^3)$

three times ↑

the sum of x and y^3 ↑

(c) $3x + y^3$

↑ the sum of $3x$ and y^3

Example 4 The formula for the volume of any right circular cylinder is $V = \pi r^2h$. (See Figure 1.)

(a) Suppose that in a right circular cylinder the height is twice the radius. Write an equation relating r and h.

(b) Express the volume of the cylinder described in part (a) in terms of the radius r.

Solution

(a) We translate the English into algebra as follows:

$$\underbrace{\text{the height}}_{h} \quad \underset{=}{\text{is}} \quad \underbrace{\text{twice the radius}}_{2r}$$

That is, we have

$$h = 2r$$

(b) The general formula for the volume is $V = \pi r^2h$. Replacing h with $2r$, we obtain

$$V = \pi r^2(2r)$$

or

$$V = 2\pi r^3 \qquad \text{as required}$$

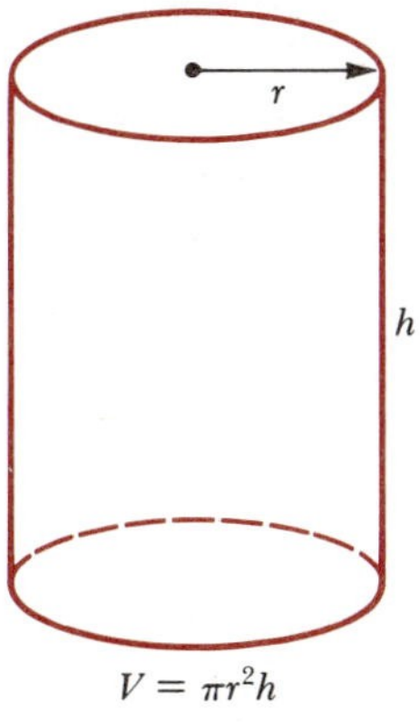

r

h

$V = \pi r^2h$

Figure 1

Exercise Set 1.1

In Exercises 1–10, rewrite the expressions using coefficients and exponents.

1. $x \cdot x \cdot x \cdot x \cdot x \cdot y \cdot y$

2. $x \cdot x \cdot y \cdot y \cdot z$

3. $(x + 1)(x + 1)(x + 1)$

4. $(a^2 + b^2)(a^2 + b^2)(a^2 + b^2)$

5. $x + x + x + x$

6. $a^5 + a^5 + a^5$

7. $a + a + a + b + b$

8. $a + a + a + b^2 + b^2$

9. $(2a + 1)(2a + 1)(2a + 1)(2b + 1)(2b + 1)$

10. $(x + y^2)(x + y^2)(x^2 + y)(x^2 + y)$

For Exercises 11–26, evaluate the expressions using the given values of the variables.

11. $x + 3y$; $x = 4$, $y = 6$

12. $a - 4b^2$; $a = 20$, $b = 2$

13. $a^2 + b^2$; $a = 3$, $b = 4$

14. $(a + b)^2$; $a = 3$, $b = 4$

15. $x^2 - 4y^2$; $x = 10$, $y = 4$

16. $x^2 - xy + y^2$; $x = 1$, $y = 1$

17. $x^2 - x + 1$; $x = 5$

18. $(x + 1)(x^2 - x + 1)$; $x = 5$

19. $2x^3 - 3y^2$; $x = 3$, $y = 2$

20. $2(x^3 - 3y^2)$; $x = 3$, $y = 2$

21. $1 \div a^2 + b^2$; $a = 1$, $b = 1$

22. $1 \div (a^2 + b^2)$; $a = 1$, $b = 1$

23. $\dfrac{1}{a^2} + \dfrac{1}{b^2} - \dfrac{1}{a^2 + b^2}$; $a = \dfrac{1}{2}$, $b = \dfrac{1}{2}$

24. $\dfrac{1}{1 - \dfrac{1}{x}}$; $x = 2$

25. $\dfrac{1000}{19 - \dfrac{12}{1 - 1/x}}$; $x = 3$

26. $[p(p^2 - 2^p)]^{p+1}$; $p = 3$

In Exercises 27–40, translate each phrase into algebraic notation.

27. Two more than x.

28. Two less than x.

29. Four times the sum of a and b^2.

30. The sum of four times a and b^2.

31. The sum of x^2 and y^2.

32. The square of the sum of x and y.

33. The sum of x and twice the square of x.

34. Twice the sum of x and the square of x.

35. The average of x, y, and z.

36. The square of the average of x, y, and z.

37. The average of the squares of x, y, and z.

38. The square of the average of the squares of x, y, and z.

39. One less than twice the product of x and y.

40. The cube of three more than the product of x^2 and y^3.

For Exercises 41–55, make use of the geometric formulas that are given on the inside front cover of this book. In Exercises 51–55, which are calculator exercises, refer to the conventions for significant digits that are explained in Section A.1 of the appendix.

41. Write expressions for the area and the circumference of a circle whose radius is $3y$.

42. Find an expression for the average of the areas of two circles whose radii are r and R, respectively.

43. The radius of a certain sphere is $3x$. Find a formula expressing the volume V of the sphere in terms of x.

44. Write an expression for the surface area of a sphere whose radius is $x/2$.

45. Suppose that in a certain right circular cylinder the height is equal to the radius.
- **(a)** Write a formula that expresses the volume of this cylinder in terms of the radius r.
- **(b)** Use the formula you found in part (a) to compute the volume of the cylinder when $r = 4$ cm.

46. Suppose that in a certain right circular cone the height is six times the radius.
- **(a)** Write a formula expressing the volume in terms of the radius r.
- **(b)** Write a formula expressing the lateral surface area in terms of the radius r.

***47.** In a certain rectangular box the length is twice the width and the height is six times the width.
- **(a)** Express the volume of the box in terms of the width w.
- **(b)** Express the surface area of the box in terms of the width w.

48. The figure at the top of the next page shows a sphere of radius r inscribed within a cylinder.
- **(a)** Find the ratio of the volume of the cylinder to the volume of the sphere.

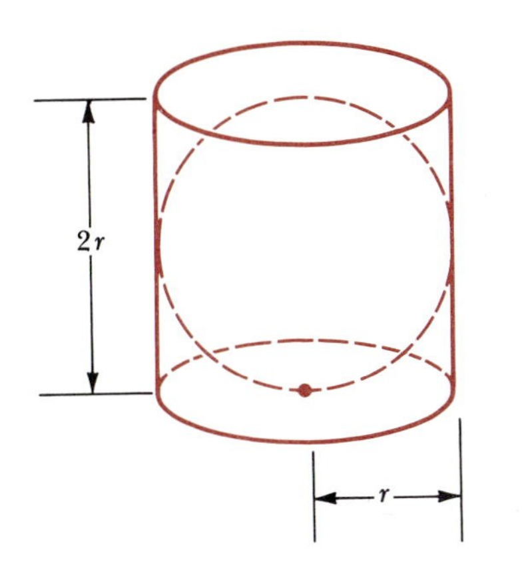

(b) Find the ratio of the total surface area of the cylinder to the surface area of the sphere.

49. The figure shows a circle inscribed within a square of side s. Express the circumference of the circle in terms of s.

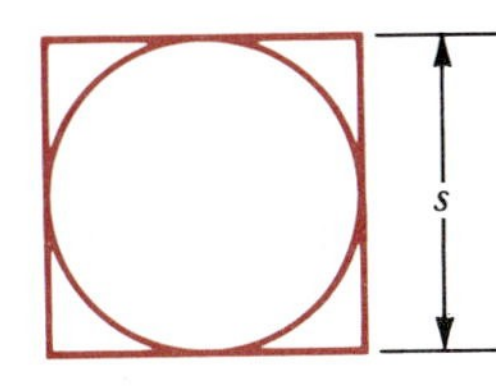

50. Suppose that the length of each side of an equilateral triangle is $2s$.
(a) Express the height of the triangle in terms of s.
(b) Express the area of the triangle in terms of s.

C **51.** Compute the volume of a box with dimensions $l = 1.46$ cm, $w = 3.50$ cm, and $h = 2.12$ cm.

C **52.** Which has the greater area, a circle with radius 5.15 cm or a square with side 9.12 cm?

C **53.** Compute the volume of a right circular cone with radius 5.05 cm and height 3.25 cm.

C **54.** The dimensions of a right circular cylinder and a right circular cone are as follows:

cylinder: $r = 3.76$ m; $h = 5.20$ m
cone: $r = 4.80$ m; $h = 6.50$ m

Which has the larger lateral surface area, the cylinder or the cone?

55. **(a)** C The dimensions of a right circular cylinder are $r = h = 4.48$ cm. Compute the ratio of the volume to the total surface area.
(b) By using more algebra and less arithmetic, you can compute the ratio in part (a) without a calculator. Do this and use the result to check your answer in part (a).

1.2 Properties of the Real Numbers

In this section we first will list the basic properties for the real number system. After that we'll summarize some of the more down-to-earth implications of those properties, such as the procedures for working with signed numbers and fractions. (Section A.2 of the appendix, at the back of this book, presents a more theoretical treatment of some of these ideas.)

The set of real numbers is *closed* with respect to the operations of addition and multiplication. This just means that when we add or multiply two real numbers, the result (that is, the *sum* or the *product*) is again a real number. Some of the other most basic properties and definitions for the real number system are listed in the following box. In the box the lowercase letters a, b, and c denote arbitrary real numbers.

Property Summary

Some Fundamental Properties of the Real Numbers

Commutative properties $a + b = b + a$ $ab = ba$

Associative properties $a + (b + c) = (a + b) + c$

$a(bc) = (ab)c$

Identity properties

1. There is a unique real number 0 (called *zero* or the *additive identity*) such that

$$a + 0 = a \quad \text{and} \quad 0 + a = a$$

2. There is a unique real number 1 (called *one* or the *multiplicative identity*) such that

$$a \cdot 1 = a \quad \text{and} \quad 1 \cdot a = a$$

Inverse properties

1. For each real number a there is a real number $-a$ (called the *additive inverse* of a or the *negative* of a) such that

$$a + (-a) = 0 \quad \text{and} \quad (-a) + a = 0$$

2. For each real number $a \neq 0$, there is a real number denoted by $\frac{1}{a}$ (or $1/a$ or a^{-1}), and called the *multiplicative inverse* or *reciprocal* of a, such that

$$a \cdot \frac{1}{a} = 1 \quad \text{and} \quad \frac{1}{a} \cdot a = 1$$

Distributive properties $a(b + c) = ab + ac$ $(b + c)a = ba + ca$

On reading this list of properties for the first time, many students ask the natural question, "Why do we even bother to list such obvious properties?" One reason is that all the other laws of arithmetic and algebra (including the "not-so-obvious" ones) can be derived from our rather short list. For example, the rule $0 \cdot a = 0$ can be proved using the distributive property, as can the rule that the product of two negative numbers is a positive number. (See Section A.2 of the appendix for these proofs.)

As you know from arithmetic, the order in which we either add two numbers or multiply two numbers does not affect the result. The commutative properties are just formal statements of these facts. The associative properties assert that the way in which we group adjacent numbers in computing a sum or product does not affect the result. So, for instance, since the two quantities $a + (b + c)$ and $(a + b) + c$ are equal, we can without ambiguity denote them both by $a + b + c$ when that is convenient. Similarly, abc denotes either of the equal expressions $a(bc)$ or $(ab)c$.

The distributive properties in a sense indicate the way in which multiplication and addition interact with each other. More specifically, each of the following examples relies on the distributive properties.

Example	*Remark*
$12 \times 103 = 12(100 + 3) = 1200 + 36 = 1236$	The distributive property can be used as an aid in mental calculations.
$3a^2(a + 2b) = 3a^2 \cdot a + 3a^2 \cdot 2b = 3a^3 + 6a^2b$	An application of the distributive property, followed by a simplification.
$A(c + d) = Ac + Ad$	A direct application of the distributive property. In the next example we are going to replace A with the quantity $(a + b)$.
$(a + b)(c + d) = (a + b)c + (a + b)d = ac + bc + ad + bd$	The distributive properties are used three times here.
$(x + 3)(x + 6) = (x + 3)x + (x + 3)6 = x^2 + 3x + 6x + 18 = x^2 + 9x + 18$	This is similar to the previous example. Examples of this type will be discussed in greater detail in Section 1.7.

The distributive and associative properties can be generalized. The *generalized distributive property* asserts that

$$a(b_1 + b_2 + \cdots + b_n) = ab_1 + ab_2 + \cdots + ab_n$$

The generalized associative property tells us that we can group the terms as we please in computing a sum or a product. For instance, we have

$$\begin{aligned} 17 + 47 + 53 + 80 &= 17 + [(47 + 53) + 80] \\ &= 17 + [100 + 80] \\ &= 17 + 180 = 197 \end{aligned}$$

Example 1 Identify the property of the real numbers that justifies each statement.

(a) $(17 \times 50) \times 2 = 17 \times (50 \times 2) = 17 \times 100 = 1700$
(b) Since π and $\sqrt{2}$ are real numbers, so is $\pi + \sqrt{2}$.
(c) $3 + y^2 = y^2 + 3$
(d) $[y + (z + 1)]x = yx + (z + 1)x$
(e) $(z + 1) + [-(z + 1)] = 0$
(f) $(a^2 + 1) \cdot \left(\dfrac{1}{a^2 + 1}\right) = 1$

Solution

(a) Associative property of multiplication.
(b) Closure property with respect to addition.
(c) Commutative property of addition.
(d) Distributive property.
(e) Additive inverse property.
(f) Multiplicative inverse property.

If you look back at the box on page 6, you'll see that the operations of subtraction and division are not mentioned. This is because addition and multiplication are actually the more fundamental operations. In fact, we can define subtraction and division in terms of addition and multiplication as follows.

Definitions of Subtraction and Division

$$a - b = a + (-b)$$

$$a \div b = a \cdot \left(\frac{1}{b}\right) = ab^{-1} \quad \text{provided that } b \neq 0$$

Division by zero is not defined. To see why this must be the case, suppose just for the moment that division by zero were defined. Then we would have $\frac{1}{0} = k$ for some real number k. But then, multiplying both sides of the equation by 0 would yield $1 = k \cdot 0$; that is, $1 = 0$, a contradiction. In summary, if we wish the rules of arithmetic and algebra to be consistent, then division by zero must remain undefined.

In basic algebra you were introduced to a number of procedures that were useful in working with algebraic expressions. In the three boxes that follow we summarize the properties of the real numbers that are behind those procedures you learned. In the statements of these properties, the lowercase letters a, b, c, and d are used to denote arbitrary real numbers. In the properties of fractions, of course, we are assuming that none of the denominators is zero.

Property Summary

Properties of Zero

Property	*Example*
1. $a \cdot 0 = 0$	$(x^3 - \pi) \cdot 0 = 0$
2. If $\frac{a}{b} = 0$, then $a = 0$.	If $\frac{x^2 + x - 1}{2x} = 0$, then $x^2 + x - 1 = 0$.
3. If $ab = 0$, then $a = 0$ or $b = 0$.	If $5x = 0$, then $x = 0$.

Property Summary

Properties of Negatives

Property	*Example*
1. $-(-a) = a$	$-[-(x + 1)] = x + 1$
2. $(-a)b = -(ab) = a(-b)$	$(-3)4 = -12 = 3(-4)$
3. $(-a)(-b) = ab$	$[-(x + y)][-(x + y)] = (x + y)^2$
4. $-a = (-1)a$	$-(x + y - z) = -1(x + y - z) = -x - y + z$

Property Summary

Properties of Fractions

(The denominators in each case are assumed to be nonzero.)

Property	*Example*
1. $\frac{a}{b} = \frac{c}{d}$ if and only if $ad = bc$	$\frac{8}{12} = \frac{2}{3}$ since $8 \cdot 3 = 12 \cdot 2$
2. $\frac{a}{b} = \frac{ac}{bc}$	$\frac{4}{7} = \frac{4 \cdot 5}{7 \cdot 5} = \frac{(4 \cdot 5)\pi}{(7 \cdot 5)\pi}$
3. $\frac{a}{b} \pm \frac{c}{b} = \frac{a \pm c}{b}$	$\frac{10}{3} + \frac{3}{3} = \frac{13}{3}; \frac{10}{3} - \frac{3}{3} = \frac{7}{3}$
4. $\frac{a}{b} \pm \frac{c}{d} = \frac{ad \pm bc}{bd}$	$\frac{2}{3} + \frac{3}{4} = \frac{2 \cdot 4 + 3 \cdot 3}{3 \cdot 4} = \frac{17}{12}$ $\frac{2}{3} - \frac{3}{4} = \frac{2 \cdot 4 - 3 \cdot 3}{3 \cdot 4} = \frac{-1}{12}$
5. $\frac{a}{b} \cdot \frac{c}{d} = \frac{ac}{bd}$	$\frac{5}{4} \cdot \frac{7}{9} = \frac{35}{36}$
6. $\frac{a}{b} \div \frac{c}{d} = \frac{a}{b} \cdot \frac{d}{c} = \frac{ad}{bc}$	$\frac{3}{5} \div \frac{2}{7} = \frac{3}{5} \cdot \frac{7}{2} = \frac{21}{10}$ $1 \div \frac{1}{10} = \frac{1}{1} \cdot \frac{10}{1} = 10$
7. $-\frac{a}{b} = \frac{-a}{b} = \frac{a}{-b}$	$-\frac{3}{4} = \frac{-3}{4} = \frac{3}{-4}$ $-\frac{x-1}{x+1} = \frac{-(x-1)}{x+1} = \frac{x-1}{-(x+1)}$
8. $\frac{a-b}{b-a} = -1$	$\frac{7-x}{x-7} = -1; \frac{x^2-y^2}{y^2-x^2} = -1$

Example 2 Simplify the following expression (that is, write it as a single fraction).

$$\frac{\frac{5}{4} + \frac{7}{3}}{\frac{1}{6} + \frac{3}{2}}$$

Solution We'll show two methods.

First Method

$$\frac{\frac{5}{4} + \frac{7}{3}}{\frac{1}{6} + \frac{3}{2}} = \frac{\frac{5}{4} \cdot \frac{3}{3} + \frac{7}{3} \cdot \frac{4}{4}}{\frac{1}{6} + \frac{3}{2} \cdot \frac{3}{3}}$$

$$= \frac{\frac{15}{12} + \frac{28}{12}}{\frac{1}{6} + \frac{9}{6}} = \frac{\frac{43}{12}}{\frac{10}{6}}$$

$$= \frac{\frac{43}{12}}{\frac{5}{3}} = \frac{43}{12} \cdot \frac{3}{5}$$

$$= \frac{43 \cdot 3}{12 \cdot 5} = \frac{43}{4 \cdot 5}$$

$$= \frac{43}{20}$$

Alternate Method The individual denominators are 4, 3, 6, and 2. Since the least common multiple* of these numbers is 12, we multiply the entire original fraction by $\frac{12}{12}$ (which is equal to 1) to obtain

$$\frac{12}{12}\cdot\frac{\frac{5}{4}+\frac{7}{3}}{\frac{1}{6}+\frac{3}{2}}=\frac{12(\frac{5}{4}+\frac{7}{3})}{12(\frac{1}{6}+\frac{3}{2})}$$

$$=\frac{15+28}{2+18}$$

$$=\frac{43}{20}\quad \text{as obtained previously}$$

Exercise Set 1.2

For Exercises 1–10, use the distributive properties to carry out the indicated multiplication.

1. $8\times 104 = 8(100+4) = ?$

2. $17\times 102 = 17(100+2) = ?$

3. $45\times 98 = 45[100+(-2)] = ?$

4. **(a)** $2a[3a+(-1)]$
(b) $2a(3a-4)$
(c) $2a(3a+b-4)$

5. **(a)** $(A+B)(A+B)$
(b) $(A+B)(A-B)$
(c) $(A-B)(A-B)$

6. **(a)** $(A+B)(C+D)$
(b) $(A+B)(2A+3B)$
(c) $(A+B)(2A-3B)$

7. **(a)** $(x+2y)(x+2y)$
(b) $(x+2y)(x-2y)$
(c) $(x-2y)(x-2y)$

8. **(a)** $(x+3)(x+3)$
(b) $(x+3)(x-3)$
(c) $(x^2+3)(x^2-3)$

9. **(a)** $A(x^2-x+1)$
(b) $(x+1)(x^2-x+1)$

10. **(a)** $(2x+3)(x-1)$
(b) $(3x+2)(x-1)$

In Exercises 11–24, identify the property or properties of the real numbers that justify each statement.

11. $x+3y = 3y+x$

12. $x(3y) = (3y)x$

13. $\frac{a^2}{b^2}+\left(-\frac{a^2}{b^2}\right)=0$

14. $6(x+8) = 6x+48$

15. $(x+1)+y = x+(1+y)$

16. $(x+y)\cdot\left(\frac{1}{x+y}\right)=1$

17. $(x+y)[(x+y)(x-y)] = (x+y)^2(x-y)$

18. $(\sqrt{2})(1)=\sqrt{2}$

19. $(5+\pi)+0 = 5+\pi$

20. $[(x+y)+(x^2+y^2)]x = (x+y)x+(x^2+y^2)x$

21. Since $\sqrt{7}$ and $\sqrt{\pi}$ are real numbers, so is $\sqrt{7}+\sqrt{\pi}$.

22. $y(y+y^2+y^3) = y^2+y^3+y^4$

23. $(x+2)(x^2+4) = (x+2)x^2+(x+2)4$

24. Since $\frac{1}{\sqrt{7}}$ and $\sqrt{5}$ are real numbers, so is $\left(\frac{1}{\sqrt{7}}\right)(\sqrt{5})$.

For Exercises 25–50, simplify the given expressions.

25. $1+2(-3)(4)$

26. $(97)(-98)(99)(0)$

27. $4-(-3+5)$

28. $(-1)^4$

29. $-[(-1)^4]$

30. -1^4

31. $1-\{1-[-(-1-1)]\}$

32. $4^2-(-4)^3$

33. $\frac{7}{5}+\frac{2}{3}$

34. $\frac{3}{4}-\frac{4}{3}$

35. $\frac{7}{x}-\frac{12}{x}$

*The *least common multiple* (l.c.m.) is the smallest positive integer that is a multiple of each of the given integers. *Examples:* The l.c.m. of 2 and 3 is 6; the l.c.m. of 3, 5, and 6 is 30.

36. $\dfrac{a+1}{a^2+1}+\dfrac{a-1}{a^2+1}$

37. $\dfrac{x^2+y^2}{x+y}-\dfrac{x^2-y^2}{x+y}$

38. $\frac{7}{12}\cdot\frac{5}{4}$

39. $\frac{7}{12}\cdot\left(\frac{-7}{3}\right)$

40. $1\div\frac{1}{3}$

41. $\frac{2}{5}\div\frac{7}{3}$

42. $1\div\frac{2}{3}+1$

43. $1\div\left(\frac{2}{3}+1\right)$

44. $\dfrac{\frac{4}{5}+\frac{1}{2}}{\frac{5}{2}}$

45. $\dfrac{\frac{4}{3}+\frac{1}{2}}{\frac{11}{6}+\frac{5}{4}}$

46. $\dfrac{3-x^4}{x^4-3}$

47. $\dfrac{1-(x+y)}{x+y-1}$

48. $\dfrac{\frac{1}{5}-\frac{1}{4}}{\frac{1}{5}+\frac{1}{4}}$

49. $\dfrac{\frac{1}{2}-\frac{1}{3}+\frac{1}{4}}{\frac{1}{2}+\frac{1}{3}-\frac{1}{4}}$

50. $1+\dfrac{1}{1-\dfrac{1}{1-\frac{1}{5}}}$

In Exercises 51–55, evaluate each expression using the given value for x.

51. $x^2-8x-4;\ x=-2$

52. $2x^3+4x^2-1;\ x=-3$

53. $8x^3-4x^2-4x;\ x=-\frac{1}{2}$

54. $\dfrac{x^2-2x+1}{x^3-3x^2+3x-1};\ x=-4$

55. $1-x-x^2-x^3-x^4;\ x=-1$

56. If $m=\dfrac{y_2-y_1}{x_2-x_1}$, find m when $y_2=3$, $y_1=-1$, $x_2=1$, and $x_1=4$.

57. If $d^2=(x_2-x_1)^2+(y_2-y_1)^2$, compute the value of d^2 when $x_2=-1$, $x_1=-2$, $y_2=6$, and $y_1=-6$.

58. In each case, give an example to show that the purported rule is invalid.

(a) $\dfrac{1}{a+b}\stackrel{?}{=}\dfrac{1}{a}+\dfrac{1}{b}$ (b) $\dfrac{a}{b}+\dfrac{c}{d}\stackrel{?}{=}\dfrac{a+c}{b+d}$

(c) $\dfrac{a+b}{b}\stackrel{?}{=}a+1$

(d) $a-(b-c)\stackrel{?}{=}a-b-c$

(e) $a^2+a^4\stackrel{?}{=}a^6$

(f) $(a+b)^2\stackrel{?}{=}a^2+b^2$

Exercises 59–62 ask you to show that certain fractions are equal. To do this, make use of property 1 in the summary box for fractions on page 9. (In each of the exercises assume that the denominators are nonzero.)

59. If $\dfrac{a}{b}=\dfrac{c}{d}$, show that $\dfrac{a}{c}=\dfrac{b}{d}$.

60. If $\dfrac{a}{b}=\dfrac{c}{d}$, show that $\dfrac{a+b}{b}=\dfrac{c+d}{d}$.

61. If $\dfrac{a}{b}=\dfrac{c}{d}$, show that $\dfrac{a-b}{b}=\dfrac{c-d}{d}$.

*62. If $\dfrac{a}{b}=\dfrac{c}{d}$, show that $\dfrac{a+b}{a-b}=\dfrac{c+d}{c-d}$.

C 63. If $y=x^3-3x^2-6x-4$, find y when $x=-0.75$.

C 64. If $y=\dfrac{x^2-4x+1}{x^3-x^2-3x-1}$, compute y when $x=-1.01$.

C 65. Let $y=\dfrac{x^2-1}{x^3}$.

(a) Complete the following table.

x	1	1.250	1.500	1.750	2.000	2.250
y						

(b) Which x-value in the table corresponds to the largest y-value?

(c) It can be shown using calculus that the (positive) x-value that produces the largest possible y-value for the equation $y=\dfrac{x^2-1}{x^3}$ is $x=\sqrt{3}$. Use a calculator to compute this largest y-value and compare your answer with the y-values obtained in part (a).

C 66. Let $y=x^4-2x^3-12x^2+13$.

(a) Complete the following table.

x	3.000	3.100	3.200	3.300	3.400	3.500
y						

(b) Which x-value in the table yields the smallest y-value?

(c) It can be shown using calculus that the x-value that produces the smallest possible y-value for the equation in part (a) is $x=\dfrac{3+\sqrt{105}}{4}$. Which x-value in the table is closest to this? Compute the y-value corresponding to $x=\dfrac{3+\sqrt{105}}{4}$.

1.3 Sets of Real Numbers. Absolute Value

Certain collections of real numbers are referred to often enough to be given special names. The *natural numbers* are the ordinary counting numbers 1, 2, 3, ... The three dots preceding the period in the previous sentence are read "and so on." They indicate in this case that the list continues endlessly in the same pattern. Two other examples of this notation are the lists

$$2, 4, 6, \ldots \quad \text{(the even numbers)}$$

and

$$1, 2, 3, \ldots, 100 \quad \text{(the natural numbers from 1 to 100)}$$

The *integers* consist of the natural numbers along with their negatives and zero. Thus we can indicate the list of integers by writing either

$$0, \pm 1, \pm 2, \pm 3, \ldots \quad \text{or} \quad \ldots, -3, -2, -1, 0, 1, 2, 3, \ldots$$

The *rational numbers,* as the name suggests, consist of all real numbers that are *ratios* of two integers (with nonzero denominators, of course). So examples of rational numbers are

$$7\left(=\frac{7}{1}\right) \quad \text{and} \quad -0.125\left(=-\frac{1}{8}\right)$$

We complete our list of categories by defining the *irrational numbers*; these are the real numbers that are not rational numbers. Some examples of irrational numbers are $\sqrt{2}$, $1+\sqrt{5}$, and π. It can be shown that any number of the form $\sqrt{n}$, where n is a natural number that's not a perfect square, is irrational. So, for instance, $\sqrt{3}$, $\sqrt{5}$, and $\sqrt{6}$ are irrational. Also, any number that is the sum or the (nonzero) product of a rational number and an irrational number is irrational. For example, $2+\sqrt{3}$ and 4π are both irrational numbers. However, as opposed to the situation with rational numbers, the sum or product of two irrationals need not be irrational. For instance, the sum of $\sqrt{5}$ and $-\sqrt{5}$ is the rational number 0; the product of $\sqrt{5}$ and $-\sqrt{5}$ is the rational number -5.

The fact that $\sqrt{2}$ is irrational (that is, not a ratio of integers) was known to Pythagoras more than two thousand years ago. Section A.3 of the appendix, at the back of this book, contains a proof of the fact that the number $\sqrt{2}$ is irrational. The proof that π is irrational is more difficult. The first person to prove that π is irrational was the Swiss mathematician J. H. Lambert (1728–1777).

There is a useful theorem that tells us whether a real number given in decimal form is rational or irrational. Before stating the theorem, we first review what is meant by a *periodic decimal expansion.* A decimal expansion such as 2.353535 . . . , in which a given sequence of digits repeats indefinitely, is said to be *periodic.* The notation $2.\overline{35}$ is used to symbolize this. The bar indicates that the sequence of digits appearing underneath it repeats indefinitely. Note that the decimal expansion

$$0.01001000100001\ldots$$

↑ 1 zero ↑ 2 zeros ↑ 3 zeros ↑ 4 zeros

is not periodic; even though there is a definite pattern, there is no fixed block of digits that repeats itself over and over. The following theorem (which we state here without a proof) tells us that the rational numbers are precisely those numbers with periodic or terminating decimal expansions. Table 1 displays several examples.

Theorem

A real number is rational if and only if its decimal expansion terminates or is periodic.

Number	Rational or Irrational	Comment
0.3	rational	The decimal expansion terminates. (Note also that 0.3 equals $\frac{3}{10}$, which is, *by definition,* a rational number.)
$2.\overline{4}$ $1.4\overline{35}$	rational	The decimal expansions are periodic.
$\sqrt{2} = 1.41421\ldots$ $\pi = 3.14159\ldots$ $0.010010001\ldots$	irrational	The decimal expansions are not periodic.

Table 1
A real number is rational if and only if its decimal expansion terminates or is periodic

According to our rule, the number $2.\overline{4}$ is rational. In other words, there is some fraction p/q such that $2.\overline{4} = p/q$. How can we find this fraction? Example 1 shows one way to do this.

Example 1 Express the repeating decimal $2.\overline{4}$ in the form p/q, where p and q are integers and $q \neq 0$.

Solution Let $x = 2.444\ldots$. Then $10x = 24.44\ldots$, and we have

$$10x - x = 24.\overline{4} - 2.\overline{4} = 22$$

or

$$9x = 22$$

$$x = \frac{22}{9}$$

We now have $2.\overline{4} = \frac{22}{9}$, as required.

Actually, a rigorous justification of each step used in Example 1 would require certain techniques from calculus. Nevertheless, you can easily verify for yourself that the answer in Example 1 is correct; just divide 22 by 9 and see what you find.

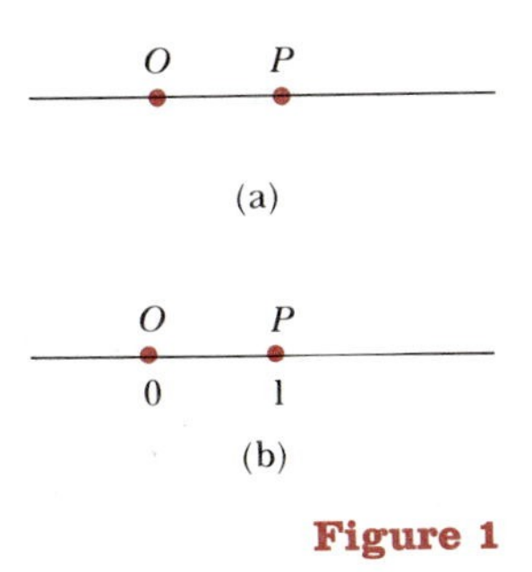

Figure 1

In previous courses you learned that the real numbers could be viewed as points on a *number line.* Because this is such an important idea, let's take a moment to review the essential features. Beginning with a straight line, we choose two points, O and P, as shown in Figure 1(a). We call O the *origin.* With the points O and P we identify the numbers 0 and 1, as shown in Figure 1(b). Now, using the distance from O to P as our unit of length, we mark off points at one-unit intervals on both sides of the origin. As indicated in Figure 2, the points that we've marked to the right of the origin are identified with the numbers 1, 2, 3, and so on; the points to the left of the origin are identified with −1, −2, −3, and so on. A fundamental fact is that

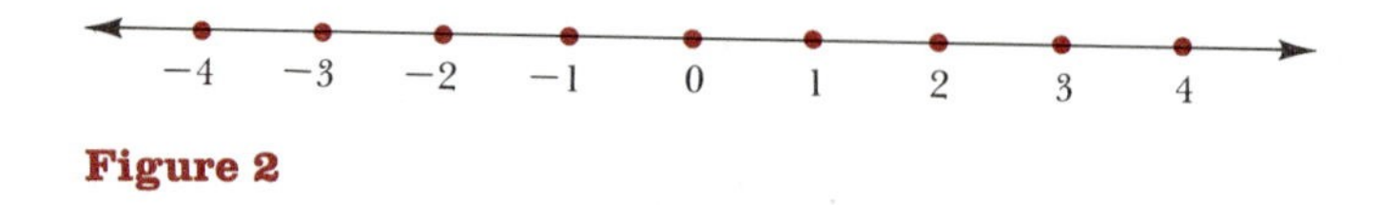

Figure 2

there is a *one-to-one correspondence* between the set of real numbers and the set of points on the line. This means that each real number is identified with exactly one point on the line; conversely, with each point on the line we identify exactly one real number. The real number associated with a given point is called the *coordinate* of the point. As a practical matter, we're usually more interested in relative locations than precise locations on a number line. For instance, since π is approximately 3.1, we show π slightly to the right of 3 in Figure 3. Similarly, since $\sqrt{2}$ is approximately 1.4, we show $\sqrt{2}$ slightly less than halfway from 1 to 2 in Figure 3.

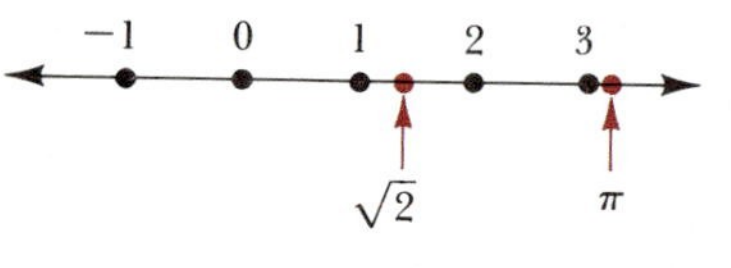

Figure 3

There are times when it is convenient to use number lines that show reference points other than the integers used in Figure 2. For instance, Figure 4(a) displays a number line with reference points that are multiples of π. In this case, of course, it is the integers that we then locate approximately. For example, in Figure 4(b) we show the approximate location of the number 1 on such a line.

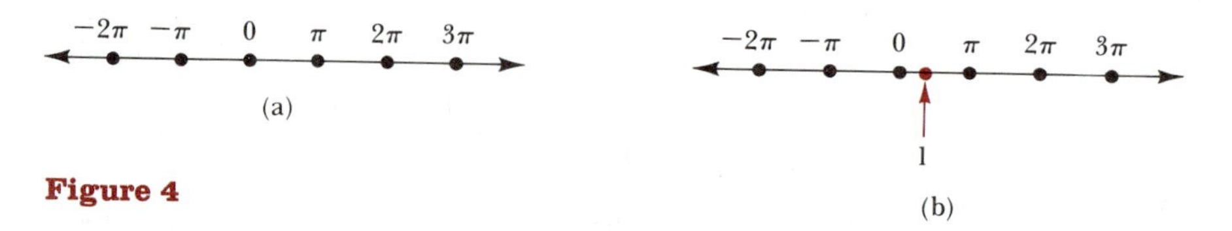

Figure 4

Given two real numbers a and b, we say that *a is less than b*, and we write $a < b$, to mean that a lies to the left of b on the number line. (We are assuming here that the number line is oriented as in Figures 2, 3, and 4.) As examples of this notation, we have

$$3 < 5 \qquad -4 < 0 \qquad \sqrt{2} < \pi$$

Another way to express the relationship $a < b$ is to write $b > a$, read *b is greater than a*. Graphically, the notation $b > a$ means that b lies to the right of a on the number line. Three examples of this notation are $5 > 3$, $0 > -4$, and $\pi > \sqrt{2}$. The notation $a \leq b$ is used to mean that either $a < b$ or $a = b$. Similarly, $b \geq a$ means that either $b > a$ or $b = a$. For example, we have $2 \leq 3$, $3 \leq 3$, $3 \geq 2$, and $3 \geq 3$.

In general, relationships involving real numbers and any of the four symbols $<$, $\leq$, $>$, and $\geq$ are called *inequalities*. One of the simplest uses of inequalities is in defining certain sets of real numbers called *intervals*. Roughly speaking, any uninterrupted portion of the number line is referred to as an *interval*. In the definitions that follow, you'll see notation such as $a < x < b$. This means that both of the inequalities $a < x$ and $x < b$ hold; in other words, the number x is between a and b.

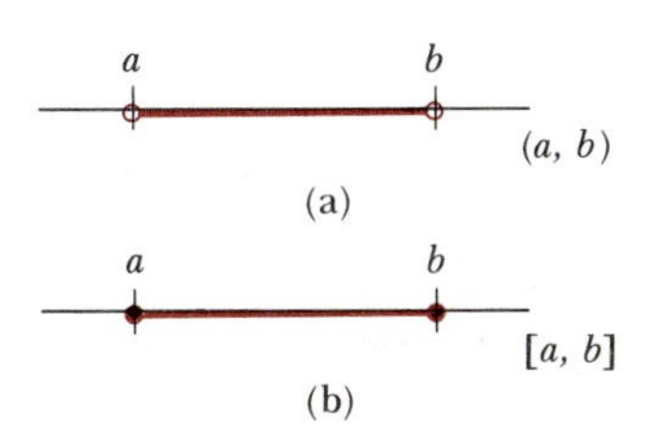

Figure 5

> **Definition: Open Intervals and Closed Intervals**
>
> The open interval (a, b) consists of all real numbers x such that $a < x < b$. See Figure 5(a).
>
> The closed interval $[a, b]$ consists of all real numbers x such that $a \leq x \leq b$. See Figure 5(b).

Notice that the filled-in circles in Figure 5(b) are used to indicate that the numbers a and b are included in the interval $[a, b]$, whereas the open circles in Figure 5(a) indicate that a and b are excluded from the interval (a, b). At times one sees notation such as $[a, b)$. This stands for the set of all real numbers x such that $a \leq x < b$. Similarly, $(a, b]$ denotes the set of all real numbers x such that $a < x \leq b$.

Example 2 Show each of the following intervals on a number line:

$$[-1, 2] \qquad (-1, 2) \qquad (-1, 2] \qquad [-1, 2)$$

Solution See Figure 6.

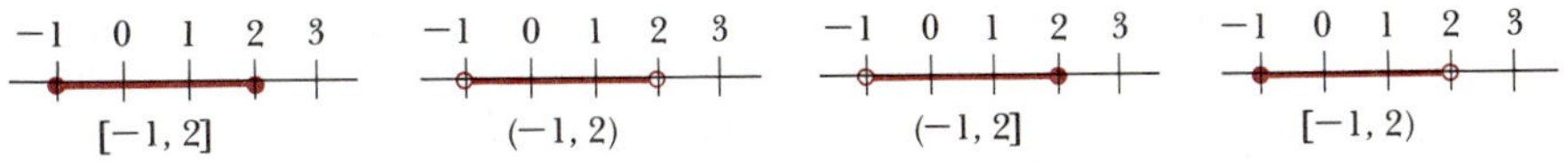

−1 0 1 2 3 [−1, 2] −1 0 1 2 3 (−1, 2) −1 0 1 2 3 (−1, 2] −1 0 1 2 3 [−1, 2)

Figure 6

Although each interval depicted in Figure 6 is of finite length, this need not always be the case. For instance, in Figures 7 and 8 we show intervals that extend indefinitely in one direction or the other. These are called *unbounded intervals.* Also, the entire real number line itself is considered to be an unbounded interval.

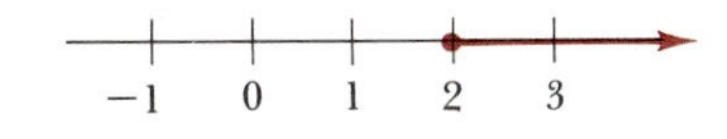

Figure 7
The set of real numbers x such that $x \geq 2$.

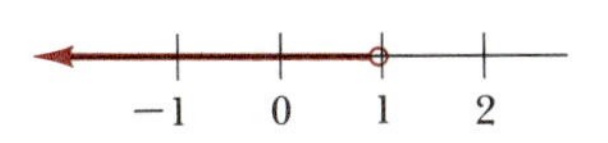

Figure 8
The set of real numbers x such that $x < 1$.

As an aid to measuring distances on the number line, we introduce the concept of absolute value. By definition, the *absolute value* of a number x, denoted by $|x|$, is the distance from x to the origin. For instance, because the numbers 5 and -5 are both five units from the origin, we have $|5| = 5$ and $|-5| = 5$. Here are three more examples:

$$|17| = 17 \qquad \left|-\frac{2}{3}\right| = \frac{2}{3} \qquad |0| = 0$$

In dealing with an expression such as $|-5 + 3|$, the convention is to compute the quantity $(-5 + 3)$ first and then to take the absolute value. We have in this case, therefore,

$$|-5 + 3| = |-2| = 2$$

There is an equivalent way of defining absolute value that is useful. According to this equivalent definition, the value of $|x|$ is x itself when $x \geq 0$, and the value of $|x|$ is $-x$ when $x < 0$. We can write this symbolically as follows:

$$|x| = \begin{cases} x & \text{when } x \geq 0 \\ -x & \text{when } x < 0 \end{cases}$$

By looking at examples with specific numbers, you should be able to convince yourself that both definitions yield the same results. This equivalent definition of absolute value is used in the solution of Example 3.

Example 3 Rewrite each expression in a form that does not contain absolute values.

(a) $|\pi - 4| + 1$

(b) $|x - 5|$ given that $x \geq 5$

(c) $|t - 5|$ given that $t < 5$

Solution **(a)** The quantity $(\pi - 4)$ is negative (since $\pi \approx 3.14$), and therefore its absolute value is equal to $-(\pi - 4)$. In view of this, we have

$$|\pi - 4| + 1 = -(\pi - 4) + 1 = -\pi + 5$$

Therefore

$$|\pi - 4| + 1 = -\pi + 5$$

(b) Since $x \geq 5$, the quantity $(x - 5)$ is nonnegative, and therefore its absolute value is equal to $(x - 5)$ itself. Thus, we have

$$|x - 5| = x - 5 \qquad \text{when } x \geq 5$$

(c) Since $t < 5$, the quantity $(t - 5)$ is negative, and therefore its absolute value is equal to $-(t - 5)$, which in turn is equal to $5 - t$. In view of this, we have

$$|t - 5| = 5 - t \qquad \text{when } t < 5$$

In the box that follows we list several basic properties of absolute value. Each of these properties can be derived from the definitions. (We shall omit the derivations.)

Property Summary

Properties of Absolute Value

1. For all real numbers x, we have

(a) $|x| \geq 0$

(b) $x \leq |x|$ and $-x \leq |x|$

2. For all real numbers a and b, we have

(a) $|a||b| = |ab|$

(b) $|a + b| \leq |a| + |b|$ (the *triangle inequality*)

Example 4 Write the expression $|-2 - x^2|$ in an equivalent form that does not contain absolute values.

Solution We have

$$\begin{aligned} |-2 - x^2| &= |-1(2 + x^2)| \\ &= |-1||2 + x^2| \qquad \text{using property 2(a)} \\ &= 2 + x^2 \end{aligned}$$

The last equality follows from the fact that x^2 is nonnegative for any real number x, and consequently the quantity $2 + x^2$ is positive.

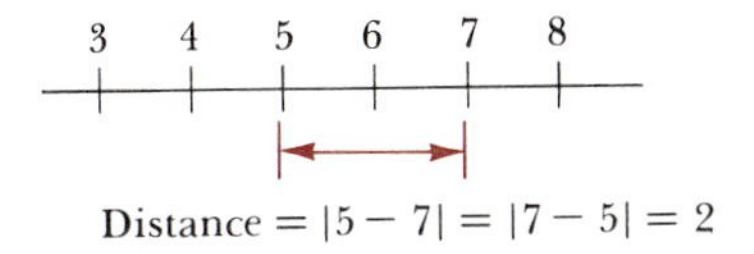

Figure 9
The distance between two points on the number line equals the absolute value of their difference.

If we think of the real numbers as points on a number line, the distance between two numbers a and b is given by the absolute value of their difference. In symbols,

$$\text{distance between } a \text{ and } b = |a - b| = |b - a|$$

For instance, as indicated in Figure 9, the distance between 5 and 7, namely 2, is given by either $|5 - 7|$ or $|7 - 5|$.

Example 5 Rewrite each of the following statements using absolute values.
(a) The distance between 12 and −5 is 17.
(b) The distance between x and 2 is 4.
(c) The distance between x and 2 is less than 4.
(d) The point t is more than five units from the origin.

Solution
(a) $|12 - (-5)| = 17$ or $|-5 - 12| = 17$
(b) $|x - 2| = 4$ or $|2 - x| = 4$
(c) $|x - 2| < 4$ or $|2 - x| < 4$
(d) $|t| > 5$

Exercise Set 1.3

For Exercises 1–12, specify the category (or categories) to which the numbers belong. For convenience in writing your answers, use the following abbreviations: nat.—natural number; int.—integer; rat.—rational number; irr.—irrational number.

Examples
(a) −27: int., rat.
(b) $\sqrt{2}$: irr.

1. 7
2. −203
3. $\frac{27}{4}$
4. $\sqrt{7}$
5. 10^6
6. 8.7419
7. $8.\overline{7419}$
8. $\sqrt{6} - 1$
9. $3\sqrt{101}$
10. $\dfrac{\sqrt{5} + 1}{4}$
11. $(3 - \sqrt{2}) + (3 + \sqrt{2})$
12. $\dfrac{0.0179}{0.0842}$

In Exercises 13–19, each of the given numbers is rational because the decimal expansion either terminates or is periodic. In each case, express the number in the form p/q, where p and q are integers and $q \neq 0$.

13. $5.\overline{4}$
14. 5.4
15. $2.\overline{8}$
16. $0.\overline{19}$ *Hint:* Let $x = 0.\overline{19}$ and consider $100x - x$.
17. $61.\overline{26}$ (See the hint given for Exercise 16.)
18. 0.3121212 . . . *Hint:* Let $x = 0.3\overline{12}$ and consider $1000x - 10x$.
19. $0.\overline{142857}$ (Adapt the hint given in Exercise 16.)
20. Draw a number line similar to the one shown in Figure 2 in the text. Then show the approximate locations of $\sqrt{5} + 3$ and $\sqrt{5} - 2$ on the line. *Note:* $\sqrt{5} = 2.24$, to three significant digits.
21. Draw a number line similar to the one shown in Figure 2 in the text. Then show the approximate locations of $\pi/2$ and $\pi/4$ on the line.
22. Draw a number line similar to the one shown in Figure 4(a) in the text. Then show the approximate locations of 6 and −3 on the line.

For Exercises 23–28, say whether the statement is true or false.

23. $-5 < -100$
24. $0 \leq -1$
25. $-1 \leq -1$
26. $\sqrt{2} + \sqrt{5} \geq 0$
27. $\frac{15}{16} > \frac{17}{18}$
28. $0.\overline{8} > 0.8$

In Exercises 29–35, show the given intervals on a number line:

29. $(-1, 3)$
30. $[4, 5]$
31. $[0, \frac{7}{2})$
32. $(-5, -1]$
33. All real numbers x such that $x > -2$.
34. All real numbers x such that $x \leq 5$.
35. All real numbers x such that $-1 < x < 1$.

For Exercises 36–42, simplify each expression.

36. $|-9|$ **37.** $|16|$

38. $|-6 + 2|$ **39.** $1 - |-2|$

40. $|-3 + 7| - |3|$ **41.** $\big||-2| - |-3|\big|$

42. $\left|\dfrac{a-b}{b-a}\right|$

In Exercises 43–50, rewrite each expression in a form that does not contain absolute values.

43. $|\sqrt{2} - 1|$ **44.** $|1 - \sqrt{2}| + 1$

45. $|x - 3|$ given that $x \geq 3$

46. $|x - 3|$ given that $x < 0$

47. $|t^2 + 1|$

48. $|x - 1| + |x - 2|$ given that $1 < x < 2$

49. $|x - 1| + |x - 2|$ given that $x < 0$

50. $|-x^2 - 5|$

For Exercises 51–60, rewrite each statement using absolute values, as in Example 5 in the text.

51. The distance between 8 and 2 is 6.

52. The distance between 8 and 2 is less than 10.

53. The distance between x and 1 is $\frac{1}{2}$.

54. The distance between x and 1 is less than $\frac{1}{2}$.

55. The distance between x and 1 is at least $\frac{1}{2}$.

56. The distance between x and 1 exceeds $\frac{1}{2}$.

57. The number y is less than three units from the origin.

58. The distance between x and -1 is less than 0.001.

59. The distance between x^2 and -2 is less than $\frac{1}{10}$.

60. The sum of the distances of a and b from the origin is greater than or equal to the distance of $a + b$ from the origin.

C **61.** Show the approximate location of the irrational number $\dfrac{\sqrt{139} - 5}{3}$ on a number line.

C **62.** Show the approximate location of the irrational number $\sqrt{\dfrac{3 - \sqrt{5}}{2}}$ on a number line.

C **63.** From grade school on, we all acquire a good deal of experience in working with rational numbers. Certainly we can tell when two rational numbers are equal, although it may require some calculation, as, for example, in the case of $\frac{129}{31} = \frac{2193}{527}$. The point of this exercise is to show you that, in working with irrational numbers, sometimes it's hard to tell (or even to show) that two numbers are equal. In each of the following cases, use a calculator to verify that the quantities on both sides of the equation agree, to six decimal places. (In each case it can be shown that the two numbers are indeed equal.)

(a) $\sqrt{6} + \sqrt{2} = 2\sqrt{2 + \sqrt{3}}$

(b) $\sqrt{3 + \sqrt{5}} + \sqrt{3 - \sqrt{5}} = \sqrt{10}$

(c) $\sqrt{\sqrt{6} + 4\sqrt{2}} = \sqrt{2} + \sqrt{2}$

(d) $\sqrt{\sqrt{\dfrac{3}{2}} + \dfrac{2}{\sqrt{2}} + \sqrt{\dfrac{3}{2}} - \dfrac{2}{\sqrt{2}}} = \sqrt{2}$

C **64.** The value of the irrational number π, correct to 10 decimal places (without rounding off), is 3.1415926535. By using a calculator, determine to how many decimal places each of the following quantities agrees with π.

(a) $\frac{22}{7}$ **(b)** $\frac{355}{113}$

(c) $\dfrac{63}{25}\left(\dfrac{17 + 15\sqrt{5}}{7 + 15\sqrt{5}}\right)$

Remark: A simple approximation that agrees with π through the first 14 decimal places is $\dfrac{355}{113}\left(1 - \dfrac{0.0003}{3533}\right)$. This approximation was discovered by the Indian mathematician Srinivasa Ramanujan (1887–1920). For a fascinating account of the history of π, see the book by Petr Beckmann, *A History of* π, 3rd Ed. (New York: St. Martin's Press, 1974).

1.4 Integer Exponents. Scientific Notation

I write a^{-1}, a^{-2}, a^{-3}, *etc., for* $\dfrac{1}{a}, \dfrac{1}{aa}, \dfrac{1}{aaa}$, *etc.*

Isaac Newton
June 13, 1676

In Section 1.1 we defined the exponential notation a^n, where a is a real number and n is a natural number. Since that definition is basic to all that follows in this and the next two sections, we repeat it here.

Definition 1

Given a real number a and a natural number n, we define a^n by

$$a^n = \underbrace{a \cdot a \cdot a \cdot \cdots \cdot a}_{n \text{ factors}}$$

In the expression a^n, a is the *base* and n is the *exponent* or *power* to which the base is raised.

Example 1 Rewrite each expression using algebraic notation:

(a) x to the fourth power, plus five
(b) the fourth power of the quantity x plus five
(c) x plus y, to the fourth power
(d) x plus y to the fourth power

Solution **(a)** $x^4 + 5$ **(b)** $(x + 5)^4$ **(c)** $(x + y)^4$ **(d)** $x + y^4$

Caution Note that (c) and (d) differ only in the use of the comma. So, for spoken purposes, the idea in (c) would be more clearly conveyed by saying "the fourth power of the quantity x plus y." In the case of (d), if you read it aloud to another student or your instructor, chances are you'll be asked, "Do you mean $x + y^4$ or do you mean $(x + y)^4$?"

Moral Algebra is a precise language; use it carefully. Learn to ask yourself if what you've written will be interpreted in the manner you intended.

In basic algebra the following four properties are developed for working with exponents that are natural numbers.

Property Summary

Properties of Exponents

Property	*Examples*
1. $a^m a^n = a^{m+n}$	$a^5 a^6 = a^{11}$; $(x + 1)(x + 1)^2 = (x + 1)^3$
2. $(a^m)^n = a^{mn}$	$(2^3)^4 = 2^{12}$; $[(x + 1)^2]^3 = (x + 1)^6$
3. $\dfrac{a^m}{a^n} = \begin{cases} a^{m-n} & \text{if } m > n \\ \dfrac{1}{a^{n-m}} & \text{if } m < n \\ 1 & \text{if } m = n \end{cases}$	$\dfrac{a^6}{a^2} = a^4$; $\dfrac{a^2}{a^6} = \dfrac{1}{a^4}$; $\dfrac{a^5}{a^5} = 1$
4. $(ab)^m = a^m b^m$	$(2x^2)^3 = 2^3 \cdot (x^2)^3 = 8x^6$
$\left(\dfrac{a}{b}\right)^m = \dfrac{a^m}{b^m}$	$\left(\dfrac{x^2}{y^3}\right)^4 = \dfrac{x^8}{y^{12}}$

Each of these properties is a direct consequence of the definition of a^n. For instance, according to the first property we have $a^2 a^3 = a^5$. To verify that this is indeed correct, we note that

$$a^2 a^3 = (aa)(aaa) = a^5$$

Now we want to extend our definition of a^n to allow for exponents that are integers but not necessarily natural numbers. We begin by defining a^0.

Definition 2	**Examples**
For any nonzero real number a,	**(a)** $2^0 = 1$
$a^0 = 1$	**(b)** $(-\pi)^0 = 1$
(0^0 is not defined)	**(c)** $\left(\frac{3}{1 + a^2 + b^2}\right)^0 = 1$

It's easy to see the motivation for defining a^0 to be 1. Assuming that the exponent zero is to have the same properties as do exponents that are natural numbers, we can write

$$a^0 a^n = a^{0+n}$$

That is,

$$a^0 a^n = a^n$$

Now divide both sides of this last equation by a^n to obtain $a^0 = 1$, which agrees with our definition.

Our next definition assigns a meaning to the expression a^{-n} when n is a natural number.

Definition 3	**Examples**
$a^{-n} = \frac{1}{a^n}$	**(a)** $2^{-1} = \frac{1}{2^1} = \frac{1}{2}$
where $a \neq 0$ and n is a natural number	**(b)** $\left(\frac{1}{10}\right)^{-1} = \frac{1}{(\frac{1}{10})^1} = 10$
	(c) $x^{-2} = \frac{1}{x^2}$
	(d) $(a^2b)^{-3} = \frac{1}{(a^2b)^3} = \frac{1}{a^6b^3}$
	(e) $\frac{1}{2^{-3}} = \frac{1}{\frac{1}{2^3}} = 2^3 = 8$

Again, it's easy to see the motivation for this definition. We have

$$a^n a^{-n} = a^{n+(-n)} = a^0 = 1$$

That is,

$$a^n a^{-n} = 1$$

Now divide both sides of this last equation by a^n to obtain $a^{-n} = 1/a^n$, in agreement with definition 3.

It can be shown that the four properties of exponents that we listed earlier continue to hold now for all integer exponents. We make use of this fact in the next three examples.

Example 2 Simplify the following expression. Write the answer in such a way that only positive exponents appear.

$$(a^2b^3)^2(a^5b)^{-1}$$

Solution *First Method*

$$(a^2b^3)^2(a^5b)^{-1} = (a^2b^3)^2 \cdot \frac{1}{a^5b}$$

$$= \frac{a^4b^6}{a^5b} = \frac{b^{6-1}}{a^{5-4}}$$

$$= \frac{b^5}{a}$$

Alternate Method

$$(a^2b^3)^2(a^5b)^{-1} = (a^4b^6)(a^{-5}b^{-1})$$

$$= a^{4-5}b^{6-1} = a^{-1}b^5$$

$$= \frac{b^5}{a} \quad \text{as obtained previously}$$

Example 3 Simplify the following expression, writing the answer so that negative exponents are not used.

$$\left(\frac{a^{-5}b^2c^0}{a^3b^{-1}}\right)^3$$

Solution

$$\left(\frac{a^{-5}b^2c^0}{a^3b^{-1}}\right)^3 = \frac{a^{-15}b^6}{a^9b^{-3}}$$

$$= \frac{b^{6-(-3)}}{a^{9-(-15)}}$$

$$= \frac{b^9}{a^{24}}$$

Example 4 Use the properties of exponents to compute the quantity $\dfrac{2^{10}\cdot 3^{13}}{27\cdot 6^{12}}$.

Solution

$$\frac{2^{10}\cdot 3^{13}}{27\cdot 6^{12}} = \frac{2^{10}\cdot 3^{13}}{(3^3)(3\cdot 2)^{12}}$$

$$= \frac{2^{10}\cdot 3^{13}}{3^3\cdot 3^{12}\cdot 2^{12}} = \frac{2^{10}\cdot 3^{13}}{3^{15}\cdot 2^{12}}$$

$$= \frac{1}{(3^{15-13})(2^{12-10})} = \frac{1}{(3^2)(2^2)}$$

$$= \frac{1}{36}$$

As an application of some of the ideas in this section, we briefly discuss *scientific notation*. Scientific notation is a convenient form for writing very large or very small numbers. Such numbers occur often in the sciences. For instance, the speed of light in a vacuum is

$$29{,}979{,}000{,}000 \text{ cm/sec}$$

As written, this number would be awkward to work with in calculating. In fact, the number as written cannot even be displayed on a hand-held calculator, since there are too many digits. To write the number 29,979,000,000 (or any positive number) in scientific notation, we express it as a number between 1 and 10, multiplied by an appropriate power of 10. That is, we write it in the form

$$a \times 10^n \quad \text{where} \quad 1 \le a < 10 \quad \text{and} \quad n \text{ is an integer}$$

According to this convention, the number 4.03×10^6 is in scientific notation, but the same quantity written as 40.3×10^5 is not in scientific notation. In order to convert a given number into scientific notation, we'll rely on the following two-step procedure.

To Express a Number Using Scientific Notation:

1. First move the decimal point until it is to the immediate right of the first nonzero digit.
2. Then multiply by 10^n or 10^{-n}, depending on whether the decimal point was moved n places to the left or to the right, respectively.

For example, to express the number 29,979,000,000 in scientific notation, first move the decimal point 10 places to the left so that it's located between the 2 and the 9. Then multiply by 10^{10}. The result is

$$29{,}979{,}000{,}000 = 2.9979000000 \times 10^{10}$$

or, more simply,

$$29{,}979{,}000{,}000 = 2.9979 \times 10^{10}$$

Incidentally, in this last form, the number can be displayed on a scientific calculator by means of keys such as [exp] or [EE]. For instance, using a Texas Instruments calculator, we would enter this number by pressing

2.9979 [EE] 10

The display would then read

[2.9979 10]

As additional examples we list the following numbers expressed in both ordinary and scientific notation. For practice you should verify each conversion for yourself using our two-step procedure.

$$55708 = 5.5708 \times 10^4$$

$$0.000099 = 9.9 \times 10^{-5}$$

$$0.0000002 = 2 \times 10^{-7}$$

Exercise Set 1.4

In Exercises 1–6, evaluate each expression using the given value of x.

1. $2x^3 - x + 4;\ x = -2$
2. $1 - x + 2x^2 - 3x^3;\ x = -1$
3. $\dfrac{1 - 2x^2}{1 + 2x^3};\ x = \dfrac{-1}{2}$
4. $\dfrac{1 - (x-1)^2}{1 + (x-1)^2};\ x = -1$
5. $\dfrac{x^2 + x^3 - x^x}{2^x + 3^x - (x+1)^2};\ x = 2$
6. $\dfrac{2^{x-1} - \dfrac{1}{2^{x-1}}}{2^{x-1} + \dfrac{1}{2^{x-1}}};\ x = 3$

For Exercises 7–14, rewrite each expression using algebraic notation.

7. The square of the quantity x plus y.
8. The square of the product of x and y.
9. Three more than the square of the quantity x plus y.
10. The square of x, minus twice the cube of y.
11. The square of half of the quantity $x^2 - 2y^3$.
12. The cube of the average of x^2 and $-2y^2$.
13. The cube of the absolute value of the quantity $x - 1$.
14. The absolute value of the fifth power of the average of $-x^2$ and y^2.

In Exercises 15–27, use the properties of exponents to simplify each expression.

15. a^3a^{12}
16. $(3^2)^3 - (2^3)^2$
17. $(x+1)(x+1)(x+1)^8$
18. $\dfrac{3^6}{3^4}$
19. $\dfrac{10^{12}}{10^{10}}$
20. $(y^3y^2)^3$
21. $\dfrac{a^{15}}{a^9}$
22. $\dfrac{(x-3)^4}{(x-3)}$
23. $\dfrac{x^6y^{15}}{x^2y^{20}}$
24. $(x^2y^3z)^4$
25. $(4x^3)^2$
26. $2(x-1)^3 - [2(x-1)]^3$
27. $\dfrac{a^{3x+y}}{a^{2x}\cdot a^{x+y}}$ (Assume that x and y are integers and $a \neq 0$.)

For Exercises 28–48, simplify each expression. Write the answers in such a way that negative exponents do not appear.

28. $(64)^0$
29. $(\frac{1}{25})^0$
30. $(\frac{2}{3})^{-1}$
31. $10^{-1} + 10^{-2}$
32. $(2^{-1} + 2^{-2})^{-1}$
33. $(5^{-1})^{-1}$
34. $(10)^{-1} + (10^{-1})^{-1}$
35. $(10^{-30})^0$
36. $(xy)^{-2}$
37. $(a^2bc^0)^{-3}$
38. $(a^3b)^3(a^2b^4)^{-1}$
39. $(a^{-2}b^{-1}c^3)^{-2}$
40. $(2^{-2} + 2^{-1} + 2^0)^{-2}$
41. $\left(\dfrac{x^3y^{-2}z}{xy^2z^{-3}}\right)^{-3}$
42. $\left(\dfrac{x^4y^{-8}z^2}{xy^2z^{-6}}\right)^0$
43. $(xy^{-5})(-xy)^{-5}$
44. $(1^{-1}\cdot 2^{-2}\cdot 3^{-3})^2$
45. $(-2x)^{-3}(-2x^{-2})^{-1}$

*46. $\dfrac{(x^{5n+1})^n}{(x^n)^{5n}}\cdot\dfrac{1}{x^{n-2}}$

47. $\dfrac{x^2}{y^{-3}} \div \dfrac{x^2}{y^3}$
48. $(2x^2)^{-3} - 2(x^2)^{-3}$

In Exercises 49–52, use the properties of exponents in computing each quantity.

49. $\dfrac{2^8\cdot 3^{15}}{9\cdot 3^{10}\cdot 12}$
50. $\dfrac{2^{12}\cdot 5^{13}}{10^{12}}$
51. $\dfrac{24^5}{32\cdot 12^4}$
52. $\left(\dfrac{144\cdot 125}{2^3\cdot 3^2}\right)^{-1}$

For Exercises 53–63, express each number in scientific notation.

53. The average distance (in miles) from Earth to the Sun:
    92,900,000
54. The average distance (in miles) from the planet Pluto to the Sun:
    3,666,000,000
55. The average orbital speed (in miles per hour) of Earth:
    66,800
56. The average orbital speed (in miles per hour) of the planet Mercury:
    107,300
57. The average distance (in miles) from the Sun to the nearest star:
    25,000,000,000,000,000,000
58. The equatorial diameter (in miles) of
    (a) Mercury: 3031
    (b) Earth: 7927
    (c) Jupiter: 88,733
    (d) the Sun: 865,000
59. The length of the "year," that is, the time to orbit the Sun once (in terms of Earth days), for the planet
    (a) Mercury: 86.688
    (b) Saturn: 10604.772
    (d) Pluto: 89424

60. The mass (in grams) of
(a) Earth:
6000000000000000000000000000
(b) Jupiter:
1900000000000000000000000000000

61. The time (in seconds) for light to travel
(a) one foot:
0.000000001
(b) across an atom:
0.0000000000000000001
(c) across the nucleus of an atom:
0.000000000000000000000001

62. The mass (in grams) of
(a) a proton:
0.00000000000000000000000167
(b) an electron:
0.000000000000000000000000000911

63. The charge of an electron (in coulombs):
−0.0000000000000000001602

64. Simplify $\dfrac{a^{4p+2q}}{a^{3p}a^{p}(a^{q})^{2}}$.

65. Simplify $\dfrac{x^{3a+2b-c}}{(x^{2a})(x^{b})} \cdot x^{3c-a-b}$.

***66.** Suppose that $p = b^x$, $q = b^y$, and $b^2 = (p^y q^x)^z$, where all the letters denote natural numbers and $b \neq 1$. Show that $xyz = 1$.

C **67.** Which number is larger, 10^9 or 9^{10}?

C **68.** **(a)** Which is larger, 11^{12} or 12^{11}?
(b) Which is larger, 18^{19} or 19^{18}?

In Exercises 69–71, assume that the numbers involved were obtained through measurements of some kind. As such, the answer that you report should not appear to be more accurate than the data you've used to obtain that answer. Use the following rule for rounding off your answers. (See Section A.1 of the appendix for the definition of significant digits.)

Suppose that we carry out a calculation involving multiplication or division using data obtained through measurements. Furthermore, suppose that there are N *significant digits in the measurement having the fewest significant digits. Then the final answer should be rounded off to* N *significant digits.*

C **69.** Compute $\dfrac{(3.21 \times 10^{4})(8.56 \times 10^{-6})}{(1.2 \times 10^{-3})}$. Express your answer in scientific notation.

C **70.** Compute (0.1456)(5.29)(275.1). Express your answer in scientific notation.

C **71.** Compute $\dfrac{(9.25 \times 10^{-8})(3.005 \times 10^{-10})}{(3.10 \times 10^{-20})}$. Express your answer in both scientific and ordinary notation.

***72.** C Use the following data to compute the time required, in minutes, for light to travel from the Sun to Earth. Round off your answer to the nearest minute.
speed of light: 2.9979×10^{10} cm/sec
distance from the Sun to Earth: 92.9×10^6 miles
1 mile = 160930 cm

C **73.** Each planet in our solar system travels around the Sun in an elliptical orbit. Let T denote the time for a planet to complete one orbit around the Sun. For example, for Earth, $T = 365$ days. (Actually, more precisely, $T = 365.26$ days.) Let a denote the length of the so-called *semimajor axis* of the ellipse. (See the figure.)

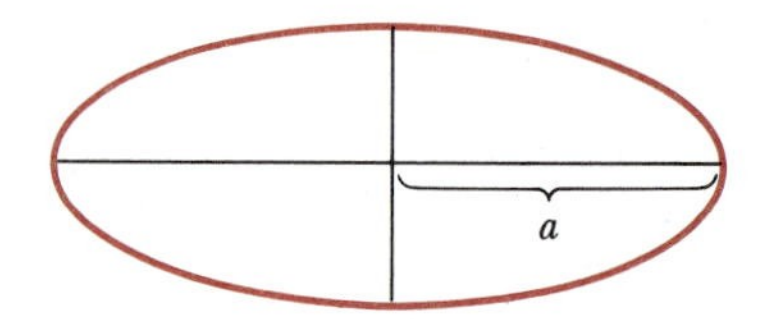

In 1618 the German astronomer Johannes Kepler announced his Third Law of Planetary Motion, which stated that the ratio $\dfrac{a^3}{T^2}$ is the same for all planets in the solar system. Verify Kepler's Third Law empirically by completing the following table. Express the required values of $\dfrac{a^3}{T^2}$ in scientific notation, rounding off to three significant digits.

Planet	a (km)	T (Earth days)	$\dfrac{a^3}{T^2}$
Mercury	5.791×10^7	87.95	
Venus	1.082×10^8	224.71	
Earth	1.496×10^8	365.26	
Mars	2.279×10^8	687.02	
Jupiter	7.783×10^8	4332.79	
Saturn	1.427×10^9	10759.72	
Uranus	2.869×10^9	30686.59	
Neptune	4.498×10^9	60191.20	
Pluto	5.900×10^9	90730.6	

1.5 Square Roots

In this section we discuss the related but distinct concepts of square root and principal square root. We begin with the definition for square root. If a and b are real numbers and

$$a^2 = b$$

then we say that a is a *square root* of b. For example, 4 is a square root of 16, because $4^2 = 16$; also, -4 is a square root of 16, since $(-4)^2 = 16$. As these two examples suggest, square roots occur in pairs, one positive and one negative. To distinguish the two square roots, we refer to the positive square root as the *principal square root*, and we use the notation $\sqrt{b}$ to denote the principal square root of the nonnegative number b. In other words,

$$\sqrt{b} = a \quad \text{if and only if } b = a^2 \quad \text{and} \quad a \geq 0$$

So, for instance, we have

$\sqrt{25} = 5$ (The principal square root of 25 is 5.)

$\sqrt{25} \neq -5$ (The principal square root of 25 is not -5.)

$-\sqrt{25} = -5$ (The negative of the principal square root of 25 is -5.)

The symbol $\sqrt{\ }$ is called a *radical sign*; the number within the radical sign is the *radicand*.*

In the box that follows we list four basic properties of principal square roots. (Each property can be derived directly from the definition of the principal square root.)

Property Summary

Properties of Square Roots

For properties 1–3, x and y denote nonnegative real numbers. For property 4, x can be any real number.

Property	*Example*
1. $(\sqrt{x})^2 = x$	$(\sqrt{7})^2 = 7$
2. $\sqrt{xy} = \sqrt{x}\sqrt{y}$	$\sqrt{(4)(3)} = \sqrt{4}\sqrt{3} = 2\sqrt{3}$
3. $\sqrt{\dfrac{x}{y}} = \dfrac{\sqrt{x}}{\sqrt{y}} \quad (y \neq 0)$	$\sqrt{\dfrac{2}{9}} = \dfrac{\sqrt{2}}{\sqrt{9}} = \dfrac{\sqrt{2}}{3}$
4. $\sqrt{x^2} = \|x\|$	$\sqrt{(-3)^2} = \|-3\| = 3$

*The radical sign ($\sqrt{\ }$) was introduced in 1525 by the German mathematician Christoff Rudolff. [Howard Eves, in *An Introduction to the History of Mathematics*, 5th ed. (Saunders College Publishing, 1983), suggests that Rudolff may have obtained this symbol by modifying the letter r, the first letter of the word *radix* (root).]

Our main use for these properties will be in simplifying expressions involving radical signs. For instance, the expression $\sqrt{72}$ is simplified as follows:

$$\begin{aligned}\sqrt{72} &= \sqrt{(36)(2)}\\ &= \sqrt{36}\,\sqrt{2}\\ &= 6\sqrt{2}\end{aligned}$$

In this procedure, we began by factoring 72 as (36)(2). Note that 36 is the largest factor of 72 that is a perfect square. If we had begun instead with a different factorization, say $72 = (9)(8)$, we could still arrive at the same answer, but it would take longer. The calculations would be as follows:

$$\begin{aligned}\sqrt{72} = \sqrt{(9)(8)} = \sqrt{9}\,\sqrt{8} = 3\sqrt{4}\,\sqrt{2} &= 3(2)\sqrt{2}\\ &= 6\sqrt{2} \qquad \text{as obtained previously}\end{aligned}$$

Example 1 Simplify: **(a)** $\sqrt{12} + \sqrt{75}$

(b) $\sqrt{\dfrac{162}{49}}$

Solution **(a)**

$$\begin{aligned}\sqrt{12} + \sqrt{75} &= \sqrt{(4)(3)} + \sqrt{(25)(3)}\\ &= \sqrt{4}\,\sqrt{3} + \sqrt{25}\,\sqrt{3}\\ &= 2\sqrt{3} + 5\sqrt{3} = (2 + 5)\sqrt{3}\\ &= 7\sqrt{3}\end{aligned}$$

(b)

$$\sqrt{\frac{162}{49}} = \frac{\sqrt{162}}{\sqrt{49}} = \frac{\sqrt{81}\,\sqrt{2}}{7} = \frac{9\sqrt{2}}{7}$$

Example 2 **(a)** Give an example to show that $\sqrt{a + b} \neq \sqrt{a} + \sqrt{b}$.
(b) Give an example in which $\sqrt{x^2} \neq x$.

Solution **(a)** Choosing $a = 16$ and $b = 9$, we have

$$\begin{aligned}\sqrt{16 + 9} &\overset{?}{=} \sqrt{16} + \sqrt{9}\\ \sqrt{25} &\overset{?}{=} 4 + 3\\ 5 &\overset{?}{=} 7 \qquad \text{No!}\end{aligned}$$

(We chose the values $a = 16$ and $b = 9$ because the computations could then be carried out without a calculator. You can construct other examples with the aid of a calculator.)

(b) The formula $\sqrt{x^2} = x$ is invalid whenever x is a negative number. For instance, when $x = -2$, we have

$$\begin{aligned}\sqrt{(-2)^2} &\overset{?}{=} -2\\ \sqrt{4} &\overset{?}{=} -2 \qquad \text{No!}\end{aligned}$$

(This example shows you why the absolute value sign is needed in property 4.)

Example 3 Simplify $\sqrt{8ab^2c^5}$, where a, b, and c are positive.

Solution

$$\begin{aligned}\sqrt{8ab^2c^5} &= \sqrt{(4b^2c^4)(2ac)}\\ &= \sqrt{4b^2c^4}\,\sqrt{2ac}\\ &= 2bc^2\,\sqrt{2ac}\end{aligned}$$

There are times in algebra and its applications when it's convenient to rewrite fractions involving radicals in alternate forms. Suppose, for example, that we want to rewrite the fraction $\frac{5}{\sqrt{3}}$ in an equivalent form not involving a radical in the denominator. This is called *rationalizing the denominator*. The procedure here is to multiply by 1 in this way:

$$\frac{5}{\sqrt{3}} = \frac{5}{\sqrt{3}} \cdot 1 = \frac{5}{\sqrt{3}} \cdot \frac{\sqrt{3}}{\sqrt{3}} = \frac{5\sqrt{3}}{3}$$

That is, $\frac{5}{\sqrt{3}} = \frac{5\sqrt{3}}{3}$, as required.

Example 4 Simplify $\frac{1}{\sqrt{2}} - 3\sqrt{50}$.

Solution First, rationalize the denominator in the fraction $\frac{1}{\sqrt{2}}$:

$$\frac{1}{\sqrt{2}} = \frac{1}{\sqrt{2}} \cdot 1 = \frac{1}{\sqrt{2}} \cdot \frac{\sqrt{2}}{\sqrt{2}} = \frac{\sqrt{2}}{2}$$

Next, simplify the expression $3\sqrt{50}$:

$$3\sqrt{50} = 3\sqrt{(25)(2)} = 3\sqrt{25}\,\sqrt{2} = (3)(5)\sqrt{2} = 15\sqrt{2}$$

Now, putting things together, we have

$$\begin{aligned}\frac{1}{\sqrt{2}} - 3\sqrt{50} &= \frac{\sqrt{2}}{2} - 15\sqrt{2} \\ &= \frac{\sqrt{2}}{2} - \frac{30\sqrt{2}}{2} = \frac{\sqrt{2} - 30\sqrt{2}}{2} = \frac{\sqrt{2}(1 - 30)}{2} \\ &= \frac{-29\sqrt{2}}{2}\end{aligned}$$

Example 5 Rationalize the denominator in each case:

(a) $\frac{4}{2 + \sqrt{3}}$

(b) $\frac{7}{\sqrt{5} - \sqrt{2}}$

Solution **(a)** We multiply by 1, writing 1 as $\frac{2 - \sqrt{3}}{2 - \sqrt{3}}$. (The quantity $2 - \sqrt{3}$ is called the *conjugate* of $2 + \sqrt{3}$.) We have

$$\begin{aligned}\frac{4}{2 + \sqrt{3}} \cdot 1 &= \frac{4}{2 + \sqrt{3}} \cdot \frac{2 - \sqrt{3}}{2 - \sqrt{3}} \\ &= \frac{4(2 - \sqrt{3})}{4 - (\sqrt{3})^2} = \frac{4(2 - \sqrt{3})}{4 - 3} = \frac{8 - 4\sqrt{3}}{1} \\ &= 8 - 4\sqrt{3}\end{aligned}$$

Note: Check for yourself that multiplying the original fraction by $\frac{2 + \sqrt{3}}{2 + \sqrt{3}}$ does *not* eliminate radicals in the denominator.

(b) We multiply by 1, writing 1 as $\frac{\sqrt{5} + \sqrt{2}}{\sqrt{5} + \sqrt{2}}$. (Again, note the sign change; the quantity $5 + \sqrt{2}$ is the *conjugate* of $5 - \sqrt{2}$.) We have

$$\begin{aligned} \frac{7}{\sqrt{5} - \sqrt{2}} \cdot 1 &= \frac{7}{\sqrt{5} - \sqrt{2}} \cdot \frac{\sqrt{5} + \sqrt{2}}{\sqrt{5} + \sqrt{2}} \\ &= \frac{7(\sqrt{5} + \sqrt{2})}{(\sqrt{5})^2 - (\sqrt{2})^2} = \frac{7(\sqrt{5} + \sqrt{2})}{5 - 2} \\ &= \frac{7\sqrt{5} + 7\sqrt{2}}{3} \quad \text{as required} \end{aligned}$$

In the next example we are asked to rationalize the numerator rather than the denominator. This is useful at times in calculus.

Example 6 Rationalize the *numerator*: $\frac{\sqrt{x} - \sqrt{3}}{x - 3}$ $(x \neq 3)$.

Solution

$$\begin{aligned} \frac{\sqrt{x} - \sqrt{3}}{x - 3} \cdot 1 &= \frac{\sqrt{x} - \sqrt{3}}{x - 3} \cdot \frac{\sqrt{x} + \sqrt{3}}{\sqrt{x} + \sqrt{3}} \\ &= \frac{(\sqrt{x})^2 - (\sqrt{3})^2}{(x - 3)(\sqrt{x} + \sqrt{3})} = \frac{x - 3}{(x - 3)(\sqrt{x} + \sqrt{3})} \\ &= \frac{1}{\sqrt{x} + \sqrt{3}} \quad \text{as required} \end{aligned}$$

To give you additional practice with square roots, as well as to review one of the key theorems from geometry, we conclude this section by discussing the Pythagorean Theorem.

Pythagorean Theorem

In any right triangle, the lengths of the three sides are related by the equation

$$a^2 + b^2 = c^2$$

where a and b are the lengths of the sides forming the right angle and c is the length of the hypotenuse (the side opposite the right angle).

Several proofs of this theorem are outlined in the exercises for this section. However, our goal for now is to learn to apply the Pythagorean Theorem. In the next two examples we'll be solving certain equations. Although this topic is not discussed in this book until Chapter 2, we'll assume that you have some experience in this area from a previous course.

Example 7 Find the lengths x and y for the sides of the right triangles shown in Figure 1.

Solution Applying the Pythagorean Theorem in Figure 1(a), we have

$$\begin{aligned} x^2 + 5^2 &= 13^2 \\ x^2 + 25 &= 169 \\ x^2 &= 169 - 25 \end{aligned}$$

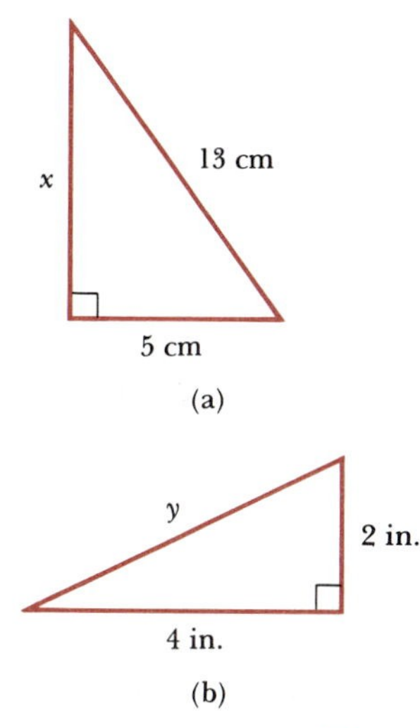

Figure 1

$$x^2 = 144$$
$$x = \sqrt{144}$$
$$x = 12 \text{ cm} \qquad \text{as required}$$

Applying the Pythagorean Theorem in Figure 1(b), we have

$$y^2 = 4^2 + 2^2$$
$$y^2 = 16 + 4$$
$$y^2 = 20$$
$$y = \sqrt{20} = \sqrt{4 \times 5}$$
$$y = \sqrt{4}\sqrt{5}$$
$$y = 2\sqrt{5} \text{ in.} \qquad \text{as required}$$

Question Is the answer, $2\sqrt{5}$ in., more or less than 6 in.? Certainly $\sqrt{5}$ is less than 3; so, twice $\sqrt{5}$ is less than twice 3. Thus, $2\sqrt{5} < 6$.

Example 8 Determine x in Figure 2.

Solution

$$x^2 = (\sqrt{2})^2 + (\sqrt{2} + 1)^2$$
$$= 2 + (2 + 2\sqrt{2} + 1)$$
$$= 5 + 2\sqrt{2}$$

Now we take the principal square root of both sides of the equation. (Clearly, x must be positive, because it represents a length.) This yields

$$x = \sqrt{5 + 2\sqrt{2}}$$

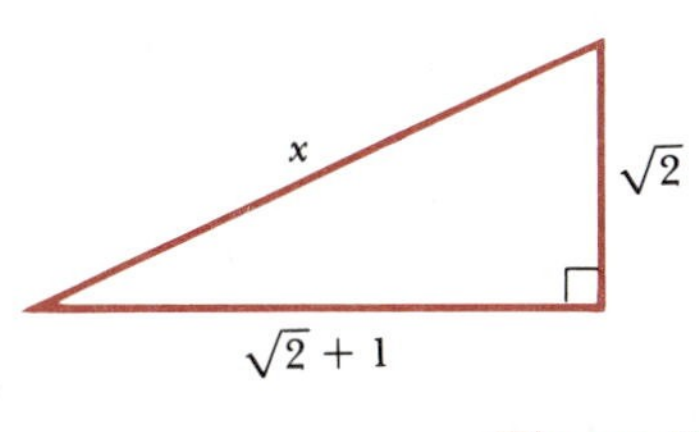

Figure 2

This is the required value for x. This particular expression, involving "nested radicals," cannot be simplified further. (Some nested radicals, however, can be simplified; see, for example, Exercises 52 and 58.) If required, a calculator can be used to obtain an approximation for x. With a calculator, we find x to be approximately 2.8.

Exercise Set 1.5

In Exercises 1–6, determine whether each statement is true or false.

1. $\sqrt{81} = -9$ **2.** $-\sqrt{100} = -10$

3. $\sqrt{7} = 49$ **4.** $\sqrt{\dfrac{4}{3}} = \dfrac{2}{\sqrt{3}}$

5. $\sqrt{10 + 6} = \sqrt{10} + \sqrt{6}$ **6.** $(\sqrt{12})^2 = 12$

For Exercises 7–16, simplify each expression.

7. $\sqrt{18}$ **8.** $\sqrt{150}$

9. $\sqrt{\dfrac{25}{4}}$ **10.** $\sqrt{2} + \sqrt{8}$

11. $4\sqrt{3} - 2\sqrt{27}$ **12.** $4\sqrt{50} - 3\sqrt{128}$

13. $\sqrt{3} - \sqrt{12} + \sqrt{48}$ **14.** $\sqrt{0.09}$

15. $\sqrt{\dfrac{49}{25}}$ **16.** $\sqrt{\dfrac{18}{121}}$

In Exercises 17–24, simplify each expression. Assume that the letters represent positive numbers.

17. $\sqrt{36x^4}$ **18.** $\sqrt{225x^4y^3}$

19. $\sqrt{ab^2}\sqrt{a^2b}$ **20.** $\sqrt{ab^3}\sqrt{a^3b}$

21. $\sqrt{72a^3b^4c^5}$ **22.** $\sqrt{\dfrac{(a+b)^5}{16a^2b^2}}$

23. $\sqrt{192a^3b^3} + \sqrt{12ab}$ **24.** $\sqrt{\sqrt{81x^4}}$

In Exercises 25–34, rationalize the denominators and simplify when possible.

25. $\frac{4}{\sqrt{7}}$

26. $\frac{3}{\sqrt{3}}$

27. $\frac{1}{\sqrt{8}}$

28. $\frac{\sqrt{2}}{\sqrt{3}}$

29. $\frac{1}{1+\sqrt{5}}$

30. $\frac{\sqrt{2}}{1-\sqrt{2}}$

31. $\frac{1+\sqrt{3}}{1-\sqrt{3}}$

32. $\frac{\sqrt{a}+\sqrt{b}}{\sqrt{a}-\sqrt{b}}$

33. $\frac{1}{\sqrt{5}} + 4\sqrt{45}$

34. $\frac{3}{\sqrt{8}} - \sqrt{450}$

For Exercises 35–38, rationalize the numerators.

35. $\frac{\sqrt{x}-\sqrt{5}}{x-5}$

36. $\frac{\sqrt{a}-\sqrt{b}}{a-b}$

37. $\frac{\sqrt{2+h}-\sqrt{2}}{h}$

38. $\frac{\sqrt{x+h}-\sqrt{x}}{h}$

In Exercises 39–47, calculate the length of the third side of a right triangle with legs a *and* b *and hypotenuse* c. *Leave your answers in terms of radicals when they appear, rather than use a calculator.*

39. $a = 5, b = 12$

40. $a = 8, c = 17$

41. $a = 1, b = 1$

42. $b = 1, c = 2$

43. $b = 2, c = \sqrt{5}$

44. $a = 20, c = 29$

45. $a = \sqrt{2} - 1, b = \sqrt{2} + 1$

46. $b = \sqrt{3} - \sqrt{2}, c = \sqrt{3} + \sqrt{2}$

47. $a = \sqrt{3}, b = 1 + \sqrt{3}$

48. Find the length a in the accompanying figure. Then find $b, c, \ldots, g$.

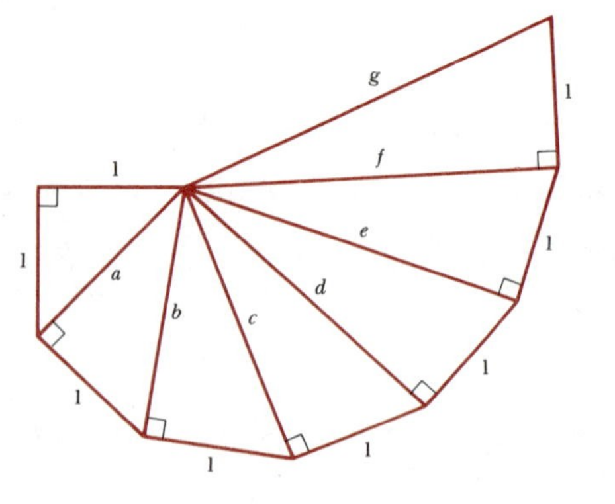

49. Verify that the principal square root of $16 - 2\sqrt{55}$ is $\sqrt{11} - \sqrt{5}$. *Hint:* That is, check that $(\sqrt{11} - \sqrt{5})^2 = 16 - 2\sqrt{55}$.

50. Verify that $\sqrt{53 - 12\sqrt{10}} = 3\sqrt{5} - 2\sqrt{2}$. (See the hint for Exercise 49.)

51. Verify that $\sqrt{4 + 2\sqrt{3}} = 1 + \sqrt{3}$.

52. Verify that $\sqrt{a + b + 2\sqrt{ab}} = \sqrt{a} + \sqrt{b}$. (Assume that a and b are nonnegative.)

***53.** Verify that

$$\sqrt{p + q - r + 2\sqrt{q(p-r)}} = \sqrt{p-r} + \sqrt{q}$$

where p, q, and r are positive and $p > r$.

***54.** Rationalize the denominator: $\frac{1}{1+\sqrt{2}+\sqrt{3}}$. *Hint:* First multiply by $\frac{1+\sqrt{2}-\sqrt{3}}{1+\sqrt{2}-\sqrt{3}}$.

***55.** Simplify $\frac{\sqrt{a}}{\sqrt{a}+\sqrt{b}} + \frac{\sqrt{b}}{\sqrt{a}-\sqrt{b}}$. *Hint:* First rationalize the denominators.

56. This problem outlines one of the shortest proofs of the Pythagorean Theorem. In the accompanying figure, we are given a right triangle ACB, with the right angle at C, and we want to prove that $a^2 + b^2 = c^2$. In the figure, CD is drawn perpendicular to AB.

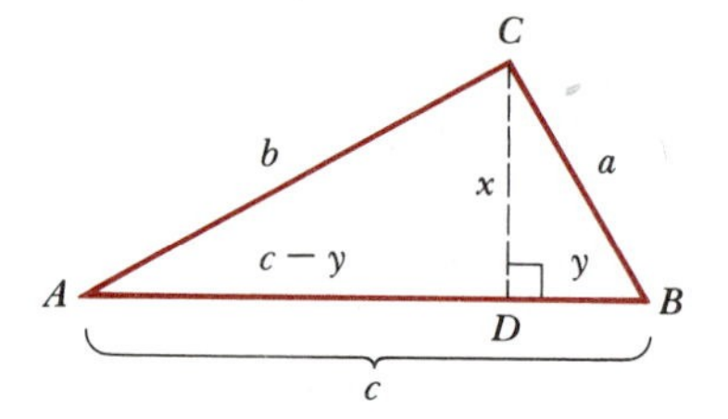

(a) Check that $\angle CAD = \angle DCB$ and that $\triangle BCD$ and $\triangle BAC$ are similar.

(b) By similar triangles, $a/y = c/a$ and therefore $a^2 = cy$.

(c) Show that $\triangle ACD$ is similar to $\triangle ABC$ and use this to deduce that $b^2 = c^2 - cy$.

(d) Combine the two equations deduced in parts (b) and (c) to arrive at $a^2 + b^2 = c^2$.

57. Here is another short proof of the Pythagorean Theorem, but one that is very different from that given in the preceding exercise. Whereas the previous proof depended on the idea of similar triangles, this proof hinges on the idea of area. The outer rectangle in the figure is *given* to be a square, each side of which has length $a + b$ as indicated.

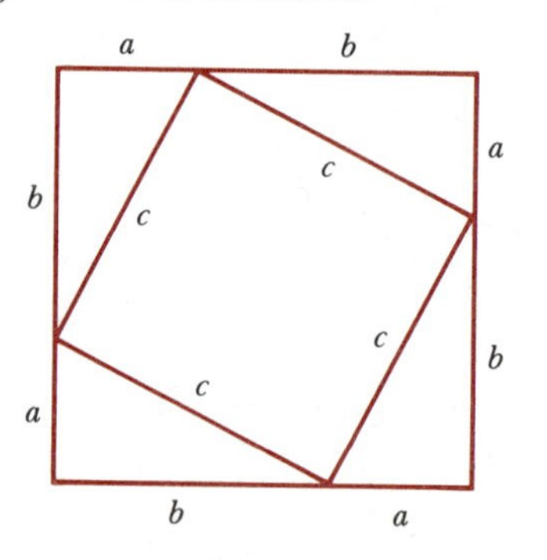

(a) Why is it correct to label all the sides of the inner figure with the same letter?

(b) Show that the inner figure is indeed a square.
(c) Why is the area of the entire figure $(a + b)^2$?
(d) Why is the area of the entire figure also equal to

$$c^2 + \frac{ab}{2} + \frac{ab}{2} + \frac{ab}{2} + \frac{ab}{2}$$

(e) Equate the expressions in parts (c) and (d) to obtain $a^2 + b^2 = c^2$.

58. **(a)** C Evaluate $\sqrt{8 - 2\sqrt{7}}$ and $\sqrt{7} - 1$. What do you observe?
(b) Prove that $\sqrt{8 - 2\sqrt{7}} = \sqrt{7} - 1$.

59. **(a)** C Evaluate the two quantities $\sqrt{5 - \sqrt{21}}$ and $\left(\sqrt{\frac{7}{2}} - \sqrt{\frac{3}{2}}\right)$.
(b) Prove that $\sqrt{5 - \sqrt{21}} = \sqrt{\frac{7}{2}} - \sqrt{\frac{3}{2}}$.

C **60.** Use a calculator to provide empirical evidence indicating that both of the following equations (at the top of the next column) may be correct:

$$\frac{2 - \sqrt{3}}{\sqrt{2} - \sqrt{2 - \sqrt{3}}} + \frac{2 + \sqrt{3}}{\sqrt{2} + \sqrt{2 + \sqrt{3}}} = \sqrt{2}$$

$$\sqrt{\sqrt{97.5 - \frac{1}{11}}} = \pi$$

The point of this exercise is to remind you that, as useful as calculators may be, there is still the need for proofs in mathematics. In fact, it can be shown that the first equation is indeed correct, but the second is not.

C **61.** Given two positive numbers a and b, we define the *geometric mean*, the *arithmetic mean*, and the *root mean square* as follows:

$$\text{G.M.} = \sqrt{ab} \qquad \text{A.M.} = \frac{a + b}{2}$$

$$\text{R.M.} = \sqrt{\frac{a^2 + b^2}{2}}$$

Complete the table.

	b	$\sqrt{ab}$ (G.M.)	$\frac{a+b}{2}$ (A.M.)	$\sqrt{\frac{a^2+b^2}{2}}$ (R.M.)	Which Is Largest, G.M., A.M., or R.M.?	Which Is Smallest, G.M., A.M., or R.M.?
1	2					
1	3					
1	4					
2	3					
3	4					
9	10					
99	100					
999	1000					

1.6 *N*th Roots and Rational Exponents

In the previous section we said that if $a^2 = b$, then a is a square root of b. Building on this idea, we now say that if $a^n = b$, then a is an *nth root* of b. Here are several examples.

2 and −2 both are fourth roots of 16
[because $2^4 = 16$ and $(-2)^4 = 16$]

3 and −3 both are sixth roots of 729
[because $3^6 = 729$ and $(-3)^6 = 729$]

2 is the cube root of 8 (because $2^3 = 8$)

−2 is the fifth root of −32 [because $(-2)^5 = -32$]

As these examples are meant to suggest, fourth roots, sixth roots, and all even roots occur in pairs, one positive and one negative, just as with square roots. In these cases we use the notation $\sqrt[n]{b}$ to denote the positive (or *principal*) nth root of b. On the other hand, cube roots, fifth roots, and all odd roots occur singly, not in pairs. In these cases we again use the notation $\sqrt[n]{b}$ for the nth root. The natural number n used in the notation $\sqrt[n]{}$ is called the *index* of the radical. For square roots, as you've already seen, we suppress the index and simply write $\sqrt{}$ rather than $\sqrt[2]{}$. The definition and examples in the box summarize our discussion so far.

Definition	Examples
1. If a and b are nonnegative real numbers, then	$\sqrt[3]{125} = 5$
$\sqrt[n]{b} = a$ if and only if $b = a^n$	$\sqrt[4]{\frac{1}{16}} = \frac{1}{2}$
2. If a and b are negative and n is an odd natural number, then	$\sqrt[3]{-8} = -2$
$\sqrt[n]{b} = a$ if and only if $b = a^n$	$\sqrt[5]{-\frac{1}{32}} = -\frac{1}{2}$

There are five properties of nth roots that are useful in simplifying certain types of expressions. The first four are entirely similar to the four properties of square roots listed in the previous section. To help you learn these new properties, we've listed them below, next to the corresponding property for square roots.

Property Summary

Properties of Nth Roots	Corresponding Properties for Square Roots
Suppose that x and y are real numbers and that m and n are natural numbers. Then each of the following properties holds, provided only that the expressions on both sides of the equation are defined (and so represent real numbers).	
1. $(\sqrt[n]{x})^n = x$	$(\sqrt{x})^2 = x$
2. $\sqrt[n]{xy} = \sqrt[n]{x}\,\sqrt[n]{y}$	$\sqrt{xy} = \sqrt{x}\,\sqrt{y}$
3. $\sqrt[n]{\frac{x}{y}} = \frac{\sqrt[n]{x}}{\sqrt[n]{y}}$	$\sqrt{\frac{x}{y}} = \frac{\sqrt{x}}{\sqrt{y}}$
4. n even: $\sqrt[n]{x^n} = \lvert x\rvert$	$\sqrt{x^2} = \lvert x\rvert$
n odd: $\sqrt[n]{x^n} = x$	
5. $\sqrt[m]{\sqrt[n]{x}} = \sqrt[mn]{x}$	

As with square roots, each of these properties can be derived from the definitions involved. As an example, we'll prove property 5, assuming that x is nonnegative. To do this, let $a = \sqrt[mn]{x}$. Then, by definition, we have $a^{mn} = x$, and consequently

$$\sqrt[m]{\sqrt[n]{x}} = \sqrt[m]{\sqrt[n]{a^{mn}}} = \sqrt[m]{\sqrt[n]{(a^m)^n}}$$

$$= \sqrt[m]{a^m} \qquad \text{We've used property 4 along with the fact that } a^m \geq 0.$$

$$= a \qquad \text{We've used property 4 again, along with the fact that } a \geq 0.$$

$$= \sqrt[mn]{x} \qquad \text{by definition of } a$$

In simplifying expressions involving nth roots, we usually try to rewrite the expression under the radical so that one factor is the largest perfect nth power that we can find. (Then we apply property 2 or 3.) Suppose, for example, that we wish to simplify $\sqrt[3]{40}$. First, what (if any) is the largest perfect cube factor of 40? Since the first few perfect cubes are

$$1^3 = 1 \qquad 2^3 = 8 \qquad 3^3 = 27 \qquad 4^3 = 64$$

we see that 8 is a perfect cube factor of 40, and we write

$$\sqrt[3]{40} = \sqrt[3]{(8)(5)} = \sqrt[3]{8}\,\sqrt[3]{5} = 2\,\sqrt[3]{5} \qquad \text{as required}$$

Example 1 Simplify $\sqrt[3]{16} + \sqrt[3]{250} - \sqrt[3]{128}$.

Solution

$$\sqrt[3]{16} + \sqrt[3]{250} - \sqrt[3]{128} = \sqrt[3]{8}\,\sqrt[3]{2} + \sqrt[3]{125}\,\sqrt[3]{2} - \sqrt[3]{64}\,\sqrt[3]{2}$$

$$= 2\,\sqrt[3]{2} + 5\,\sqrt[3]{2} - 4\,\sqrt[3]{2}$$

$$= 3\,\sqrt[3]{2}$$

Example 2 Simplify $\sqrt[4]{\dfrac{32a^6b^5}{c^8}}$, where a, b, and c are positive.

Solution

$$\sqrt[4]{\frac{32a^6b^5}{c^8}} = \frac{\sqrt[4]{32a^6b^5}}{\sqrt[4]{c^8}}$$

$$= \frac{\sqrt[4]{16a^4b^4}\,\sqrt[4]{2a^2b}}{c^2}$$

$$= \frac{2ab\,\sqrt[4]{2a^2b}}{c^2}$$

The strategy for rationalizing numerators or denominators involving nth roots is similar to that used for square roots. To rationalize a numerator or a denominator involving an nth root, we multiply that numerator or denominator by a factor that yields a product that itself is a perfect nth power. The next example displays two instances of this. Further illustrations are given in the exercises.

Example 3

(a) Rationalize the denominator: $\dfrac{6}{\sqrt[3]{7}}$

(b) Rationalize the denominator: $\dfrac{ab}{\sqrt[4]{a^2b^3}}$, where $a > 0$ and $b > 0$.

Solution (a)

$$\frac{6}{\sqrt[3]{7}} \cdot 1 = \frac{6}{\sqrt[3]{7}} \cdot \frac{\sqrt[3]{7^2}}{\sqrt[3]{7^2}}$$
$$= \frac{6\sqrt[3]{7^2}}{\sqrt[3]{7^3}}$$
$$= \frac{6\sqrt[3]{49}}{7} \quad \text{as required}$$

(b)

$$\frac{ab}{\sqrt[4]{a^2b^3}} \cdot 1 = \frac{ab}{\sqrt[4]{a^2b^3}} \cdot \frac{\sqrt[4]{a^2b}}{\sqrt[4]{a^2b}}$$
$$= \frac{ab\sqrt[4]{a^2b}}{\sqrt[4]{a^4b^4}} = \frac{ab\sqrt[4]{a^2b}}{ab}$$
$$= \sqrt[4]{a^2b}$$

We can use the concept of an nth root to give a meaning to fractional exponents that is useful and, at the same time, consistent with our earlier work. First, by way of motivation, suppose that we want to assign a value to $5^{1/3}$. Assuming that the usual properties of exponents continue to apply here, we can write

$$(5^{1/3})^3 = 5^1 = 5$$

That is,

$$(5^{1/3})^3 = 5$$

or

$$5^{1/3} = \sqrt[3]{5}$$

By repeating this reasoning, with 5 and 3 replaced with b and n, respectively, we can see that we want to define $b^{1/n}$ to mean $\sqrt[n]{b}$. Also, by thinking of $b^{m/n}$ as $(b^{1/n})^m$, we see that the definition for $b^{m/n}$ ought to be $(\sqrt[n]{b})^m$. These definitions are formalized in the box that follows.

Definition: Rational Exponents	**Examples**
1. Let b denote a real number and n a natural number. We define $b^{1/n}$ by $$b^{1/n} = \sqrt[n]{b}$$ (If n is even, we require that $b \geq 0$.)	$4^{1/2} = \sqrt{4} = 2$ $(-8)^{1/3} = \sqrt[3]{-8} = -2$
2. Let m/n be a rational number reduced to lowest terms. Assume that n is positive and that $\sqrt[n]{b}$ exists. Then, $$b^{m/n} = (\sqrt[n]{b})^m$$ or, equivalently, $$b^{m/n} = \sqrt[n]{b^m}$$	$8^{2/3} = (\sqrt[3]{8})^2 = 2^2 = 4$ or, equivalently, $8^{2/3} = \sqrt[3]{8^2} = \sqrt[3]{64} = 4$

It can be shown that the four properties of exponents listed in Section 1.4 continue to hold now for rational exponents in general. In fact, we'll take this for granted rather than follow the lengthy argument needed for its verification. We'll also assume that these properties apply to irrational exponents. So, for instance, we have

$$(2^{\sqrt{5}})^{\sqrt{5}} = 2^5 = 32$$

(The definition of irrational exponents is discussed in Section 6.1.)

Example 4 Simplify each of the following. Express the answers using positive exponents. If an expression does not represent a real number, state this.

(a) $49^{1/2}$ **(b)** $-49^{1/2}$ **(c)** $(-49)^{1/2}$ **(d)** $49^{-1/2}$

Solution

(a) $49^{1/2} = \sqrt{49} = 7$

(b) $-49^{1/2} = -(49^{1/2}) = -\sqrt{49} = -7$

(c) Does not represent a real number (because there is no real number x such that $x^2 = -49$).

(d) $49^{-1/2} = \sqrt{49^{-1}} = \sqrt{\dfrac{1}{49}} = \dfrac{\sqrt{1}}{\sqrt{49}} = \dfrac{1}{7}$

Alternately, we have

$$49^{-1/2} = (7^2)^{-1/2} = 7^{-1} = \frac{1}{7}$$

Example 5 Simplify each of the following. Write the answers using positive exponents. (Assume that $a > 0$.)

(a) $(5a^{2/3})(4a^{3/4})$

(b) $\sqrt[5]{\dfrac{16a^{1/3}}{a^{1/4}}}$

(c) $(x^2 + 1)^{1/5}(x^2 + 1)^{4/5}$

Solution

(a) $(5a^{2/3})(4a^{3/4}) = 20a^{(2/3)+(3/4)}$

$$= 20a^{17/12} \qquad \text{because } \frac{2}{3} + \frac{3}{4} = \frac{17}{12}$$

(b) $\sqrt[5]{\dfrac{16a^{1/3}}{a^{1/4}}} = \left(\dfrac{16a^{1/3}}{a^{1/4}}\right)^{1/5}$

$$= (16a^{1/12})^{1/5} \qquad \text{because } \frac{1}{3} - \frac{1}{4} = \frac{1}{12}$$

$$= 16^{1/5}a^{1/60}$$

(c) $(x^2 + 1)^{1/5}(x^2 + 1)^{4/5} = (x^2 + 1)^1 = x^2 + 1$

Example 6 Simplify: **(a)** $32^{-2/5}$ **(b)** $(-8)^{4/3}$

Solution

(a) $32^{-2/5} = (\sqrt[5]{32})^{-2} = 2^{-2} = \dfrac{1}{2^2} = \dfrac{1}{4}$

Alternately, we have

$$32^{-2/5} = (2^5)^{-2/5} = 2^{-2} = \frac{1}{2^2} = \frac{1}{4}$$

(b) $(-8)^{4/3} = (\sqrt[3]{-8})^4 = (-2)^4 = 16$

Alternately, we can write

$$(-8)^{4/3} = [(-2)^3]^{4/3} = (-2)^4 = 16$$

Example 7 Rewrite the following expression using rational exponents. (Assume that x and y are positive.)

$$\sqrt{\frac{\sqrt[3]{x}\,\sqrt[4]{y^3}}{\sqrt[5]{z^4}}}$$

Solution

$$\sqrt{\frac{\sqrt[3]{x}\,\sqrt[4]{y^3}}{\sqrt[5]{z^4}}} = \left(\frac{\sqrt[3]{x}\,\sqrt[4]{y^3}}{\sqrt[5]{z^4}}\right)^{1/2} = \left(\frac{x^{1/3}y^{3/4}}{z^{4/5}}\right)^{1/2} = \frac{x^{1/6}y^{3/8}}{z^{2/5}}$$

Example 8 Combine the radicals in the expression $\sqrt{x}\,\sqrt[3]{y^2}$. (That is, rewrite the expression using only one radical sign.) Assume that x and y are positive.

Solution

$$\sqrt{x}\,\sqrt[3]{y^2} = x^{1/2}y^{2/3}$$

$$= x^{3/6}y^{4/6}$$

We've rewritten the fractions using a common denominator.

$$= \sqrt[6]{x^3}\,\sqrt[6]{y^4}$$

$$= \sqrt[6]{x^3y^4}$$

Exercise Set 1.6

In Exercises 1–10, evaluate each of the expressions. If the expression is undefined (i.e., does not represent a real number), state this.

1. $\sqrt[3]{-64}$
2. $\sqrt[5]{32}$
3. $\sqrt[3]{\frac{1}{125}}$
4. $\sqrt[3]{-\frac{1}{1000}}$
5. $\sqrt[4]{-16}$
6. $-\sqrt[4]{16}$
7. $\sqrt[4]{\frac{81}{16}}$
8. $\sqrt[3]{-\frac{27}{64}}$
9. $\sqrt[6]{64}$
10. $\sqrt[6]{-64}$

For Exercises 11–27, simplify each expression. (Assume that all letters represent positive numbers.)

11. $\sqrt[3]{54} - \sqrt[3]{16}$
12. $\sqrt[4]{64} + \sqrt[4]{324}$
13. $\sqrt[5]{64} + \sqrt[5]{486}$
14. $\sqrt{99} - \sqrt{44} - \sqrt{176}$
15. $4\sqrt{24} - 8\sqrt{54} + 2\sqrt{6}$
16. $\sqrt[3]{192} + \sqrt[3]{-81} + \sqrt{\sqrt[3]{9}}$
17. $\sqrt{\sqrt{64}}$
18. $\sqrt[3]{\sqrt{4096}}$
19. $\sqrt[4]{16a^4b^5}$
20. $\sqrt[3]{8a^4b^6}$
21. $\sqrt{18a^3b^2}$
22. $\sqrt[5]{64a^6b^{12}}$
23. $\sqrt[3]{\frac{16a^{12}b^2}{c^9}}$
24. $\sqrt[4]{ab^3}\,\sqrt[4]{a^3b}$
25. $\sqrt[6]{\frac{5a^7}{a^{-5}b^6}}$
26. $\sqrt[3]{a^2b}\,\sqrt[3]{ab}\,\sqrt[3]{b^4}$
27. $\sqrt[4]{\frac{x^8y^6z^6}{y^2z}}\,\sqrt[4]{\frac{625}{z^{17}}}$

In Exercises 28–33, rationalize the denominators.

28. $\frac{1}{\sqrt[3]{25}}$
29. $\frac{4}{\sqrt[3]{16}}$
30. $\frac{3}{\sqrt[4]{3}}$
31. $\frac{1}{\sqrt[3]{4a^2b^8}}$
32. $\frac{1}{\sqrt[4]{2ab^5}}$
33. $\frac{\sqrt[5]{2b^2}}{\sqrt[5]{3b^3}}$

For Exercises 34–49, simplify each expression. Express the answers using positive exponents. If an expression is undefined, state this. (Assume that the letters represent positive numbers.)

34. $16^{1/2}$
35. $16^{-1/2}$
36. $(-16)^{1/2}$
37. $8^{2/3}$
38. $32^{-3/2}$
39. $(0.001)^{-1/3}$
40. $125^{2/3}$
41. $(-125)^{1/3}$
42. $(-1)^{3/5}$
43. $(27)^{4/3} + (27)^{-4/3} + 27^0$
44. $(2a^{1/3})(3a^{1/4})$
45. $\sqrt[5]{3a^4}$
46. $\sqrt[4]{\frac{64a^{2/3}}{a^{1/3}}}$
47. $(x^2+1)^{2/3}(x^2+1)^{4/3}$
48. $\frac{(x^2+1)^{3/4}}{(x^2+1)^{-1/4}}$

49. $\dfrac{(2x^2+1)^{-6/5}(2x^2+1)^{6/5}(x^2+1)^{-1/5}}{(x^2+1)^{9/5}}$

In Exercises 50–57, rewrite each expression using rational exponents rather than radicals.

50. $\sqrt[3]{(x+1)^2}$

51. $\dfrac{1}{\sqrt{x}} + \sqrt{x}$

52. $(\sqrt[5]{x+y})^2$

53. $\sqrt{\sqrt{x}}$

54. $\sqrt[3]{\sqrt{x}} + \sqrt{\sqrt[3]{x}}$

55. $\sqrt[3]{\sqrt{2}}$

56. $\sqrt{\sqrt[3]{x}\sqrt[4]{y}}$

57. $\sqrt[5]{\dfrac{\sqrt{x}\sqrt[3]{y}}{\sqrt[4]{z^2}}}$ $(z > 0)$

For Exercises 58–60, rewrite each expression using only one radical sign, as in Example 8 in the text. (Assume that x, y, a, *and* b *are positive.)*

58. $\sqrt{2}\sqrt[3]{5}$

59. $\sqrt[5]{x^3}\sqrt[7]{y^4}$

60. $\sqrt[4]{x^a}\sqrt[3]{x^b}\sqrt{x^{a/6}}$

61. Give an example showing that each statement is false.

(a) $x^{1/m} = \dfrac{1}{x^m}$ (b) $x^{-1/2} = \dfrac{1}{x^2}$

(c) $x^{m/n} = \sqrt[m]{x^n}$ (d) $\sqrt[3]{x+y} = \sqrt[3]{x} + \sqrt[3]{y}$

***62.** Simplify: $\sqrt[x]{x^{(x^2-x)}}$ $(x > 0)$

***63.** Show that $\dfrac{a-b}{a+b}\sqrt{\dfrac{a+b}{a-b}} = \left(\dfrac{a-b}{a+b}\right)^{1/2}$. (Assume that $a > b > 0$.)

64. (a) Give an example in which a rational number raised to a rational power is irrational.
(b) Give an example in which an irrational number raised to a rational power is rational.

65. Without using a calculator, decide which number is larger in each case.

(a) $2^{2/3}$ or $2^{3/2}$ (b) $5^{1/2}$ or 5^{-2}

(c) $2^{1/2}$ or $2^{1/3}$ (d) $(\frac{1}{2})^{1/2}$ or $(\frac{1}{2})^{1/3}$

(e) $10^{1/10}$ or $(\frac{1}{10})^{10}$

***66.** Given $a + b = 2c$, evaluate the following expression.

$$\left[\frac{(2^{a-b})^b(2^{b-c})^{c-a}}{(2^{c+b})^{c-b}}\right]^{1/c}$$

C **67.** Which is larger, $9^{10/9}$ or $10^{9/10}$?

C **68.** Which number is closer to 3, $13^{3/7}$ or $6560^{1/8}$?

C **69.** Given three positive numbers, a, b, and c, we define their *geometric mean* and *arithmetic mean*, respectively, as follows:

$$\text{G.M.} = (abc)^{1/3} \qquad \text{A.M.} = \frac{a+b+c}{3}$$

It can be shown that no matter how we choose the positive numbers a, b, and c, it will always be true that

$$(abc)^{1/3} \le \frac{a+b+c}{3}$$

Confirm this empirically by completing the table. (Round off to two decimal places.)

a	b	c	$(abc)^{1/3}$	$\frac{a+b+c}{3}$	Which (if Either) Is Larger, G.M. or A.M.?
1	1	6			
1	2	5			
1	3	4			
2	2	4			
2	3	3			
$\frac{8}{3}$	$\frac{8}{3}$	$\frac{8}{3}$			
12	1	1			
6	2	1			
4	3	1			
3	2	2			
$\sqrt[3]{12}$	$\sqrt[3]{12}$	$\sqrt[3]{12}$			

C **70.** Consider the expression $n^{1/n}$, where n is a natural number. In calculus it is shown that as n takes on larger and larger values, the resulting value of the expression approaches 1. Confirm this empirically by completing the table at the left. (Round off your results to four decimal places.)

n	2	5	10	100	10^3	10^4	10^5	10^6
$n^{1/n}$								

C **71.** In working with roots and radicals, it's not always apparent from the outset when two numbers are equal. (We've already seen examples of this in Exercise Set 1.3, Exercise 63.) In each of the following cases, use a calculator to verify that the quantities on both sides of the equation agree in their first six decimal places. (In each case it can be shown that the numbers are indeed equal.)

(a) $(176 + 80\sqrt{5})^{1/5} = 1 + \sqrt{5}$

(b) $(7\sqrt[3]{20} - 19)^{1/6} = \sqrt[3]{\frac{5}{3}} - \sqrt[3]{\frac{2}{3}}$

(c) $(2 + \sqrt{5})^{1/3} + (2 - \sqrt{5})^{1/3} = 1$

1.7 Polynomials

As background for our work on polynomials, we first review the terms *constant* and *variable*. By way of example, consider the familiar expression for the area of a circle of radius r, πr^2. Here π is a constant; its value never changes throughout the discussion. On the other hand, r is a variable; we can substitute any positive number for r to obtain the area of a particular circle. More generally, by a *constant* we mean either a particular number (such as π, or -17, or $\sqrt{2}$) or a letter whose value remains fixed (although perhaps unspecified) throughout a given discussion. In contrast, a *variable* is a letter for which we can substitute any number selected from a given set of numbers. The given set of numbers is called the *domain* of the variable.

Some expressions will make sense only for certain values of the variable. For instance, $\frac{1}{x - 3}$ will be undefined when x is 3 (for then the denominator is zero). So in this case we would agree that the domain of the variable x consists of all real numbers except $x = 3$. Similarly, throughout this chapter (with the exception of Section 1.10 on complex numbers) we adopt the following convention:

The Domain Convention

The domain of a variable in a given expression is the set of all real number values of the variable for which the expression is defined.

In algebra it's customary (but not mandatory) to use letters near the end of the alphabet for variables. Letters from the beginning of the alphabet are generally used for constants. So, for example, in the expression $ax + b$, x is the variable and a and b are constants.

Example 1 Specify the variable, the constants, and the domain of the variable for each expression.

(a) $3x + 4$

(b) $\dfrac{1}{(t + 1)(t - 1)}$

(c) $ay^2 + by + c$

(d) $4x + 3x^{-1}$

Solution

		Variables	Constants	Domains
(a)	$3x + 4$	x	3, 4	The set of all real numbers.
(b)	$\dfrac{1}{(t + 1)(t - 1)}$	t	1, −1	The set of all real numbers except $t = 1$ and $t = -1$. (The denominator is zero when $t = \pm 1$.)
(c)	$ay^2 + by + c$	y	a, b, c	The set of all real numbers.
(d)	$4x + 3x^{-1}$	x	4, 3	The set of all real numbers except $x = 0$. (Remember, $x^{-1} = 1/x$.)

The expressions in parts (a) and (c) of Example 1 are *polynomials*. By a *polynomial in x*, we mean an expression of the form

$$a_n x^n + a_{n-1}x^{n-1} + \cdots + a_1 x + a_0$$

where n is a nonnegative integer and $a_n \neq 0$. The individual expressions $a_k x^k$ making up the polynomial are called *terms*. In this chapter, the *coefficients* a_k will always be real numbers. For example, the terms of the polynomial $x^2 - 7x + 3$ are x^2, $-7x$, and 3; the coefficients are 1, −7, and 3. In writing a polynomial, it's customary (but not mandatory) to write the terms in order of decreasing powers of x. For instance, we'd usually write $x^2 - 7x + 3$ rather than $x^2 + 3 - 7x$. The highest power of x in a polynomial is called the *degree* of the polynomial. For example, the degree of the polynomial $x^2 - 7x + 3$ is 2. In the case of a polynomial consisting only of a nonzero constant a_0, we say that the degree is zero (because $a_0 = a_0 x^0$). No degree is defined for the polynomial whose only term is zero.

There is some additional terminology that is useful in describing polynomials that have only a few terms. A polynomial with only one term (such as $3x^2$) is a *monomial*; a polynomial with only two terms (such as $2x + 3$) is a *binomial*; and a polynomial with only three terms (such as $5x^2 - 6x + 3$) is a *trinomial*. The table that follows provides examples of the terminology.

Expression	Polynomial?	Degree (if polynomial)	Terms (if polynomial)	Coefficients (if polynomial)
$2x^3 - 3x^2 + 4x - 1$	yes	3	$2x^3, -3x^2, 4x, -1$	2, −3, 4, −1
$t^2 + 1$	yes	2	$t^2, 1$	1, 1
-12	yes	0	-12	−12
$2x^{-4} + 5$	no			
$\dfrac{1}{2x - 3}$	no			
$\sqrt{4x^2 + 1}$	no			

Just as we can combine two numbers through addition, subtraction, multiplication, or division, so can we combine polynomials. To add or subtract two polynomials, we follow the familiar process of combining like terms. For example, to add the polynomials $3x - 1$ and $x^2 + 8x - 4$, we have

$$(3x - 1) + (x^2 + 8x - 4) = \underbrace{x^2}_{\substack{\uparrow \\ 0 + x^2}} + \underbrace{11x}_{\substack{\uparrow \\ 3x + 8x}} \underbrace{- \quad 5}_{\substack{\uparrow \\ -1 + (-4)}}$$

Similarly, to subtract we write

$$(3x - 1) - (x^2 + 8x - 4) = \underbrace{-x^2}_{0 - x^2} \underbrace{- \quad 5x}_{3x - 8x} + \underbrace{3}_{-1 - (-4)}$$

Since all the terms in a polynomial ultimately stand for real numbers, we can rely on the properties of real numbers in simplifying polynomial expressions. For instance, the following rules are useful in dealing with sums and differences of polynomials.

Rules for Sums and Differences of Polynomials

Rule	*Example*
(a) If the parentheses are preceded by a plus sign, or no sign appears, then the parentheses can be omitted.	$(3x^2 + 4x) + (2x^2 - 5x)$ $= 3x^2 + 4x + 2x^2 - 5x$ $= 5x^2 - x$
(b) If the parentheses are preceded by a minus sign, then the parentheses can be omitted if the sign of every term within is reversed.	$(x^2 - 3x + 1) - (3x^2 - 8x + 4)$ $= x^2 - 3x + 1 - 3x^2 + 8x - 4$ $= -2x^2 + 5x - 3$

As we saw in Section 1.2, multiplication of polynomials can be carried out by means of the distributive properties. Example 2 displays three instances of this.

Example 2 Multiply:

(a) $3x(4x^2 - 8x + 1)$
(b) $(x^2 - 4x + 2)(x + 3)$
(c) $(2x - 3)(4x + 5)$

Solution **(a)** $3x(4x^2 - 8x + 1) = 3x(4x^2) + 3x(-8x) + 3x(1)$
$= 12x^3 - 24x^2 + 3x$

(b) $(x^2 - 4x + 2)(x + 3) = (x^2 - 4x + 2)x + (x^2 - 4x + 2)3$
$= x^3 - 4x^2 + 2x + 3x^2 - 12x + 6$
$= x^3 - x^2 - 10x + 6$

(c) $(2x - 3)(4x + 5) = (2x - 3)4x + (2x - 3)5$
$= 8x^2 - 12x + 10x - 15$
$= 8x^2 - 2x - 15$

In part (c) of the example just concluded, we showed the multiplication of two binomials. Because this type of product occurs so frequently, it's very useful to be able to carry out the work mentally. The so-called *FOIL* method allows us to accomplish this. FOIL is an acronym standing for First, Outer, Inner, Last. Here's how the procedure works in computing the product $(2x - 3)(4x + 5)$.

Step 1. Multiply the first two terms: $(2x)(4x) = 8x^2$

$$(2x - 3)(4x + 5)$$

first

Step 2. Add the products of the outer and inner terms: $(2x)(5) + (-3)(4x) = -2x$

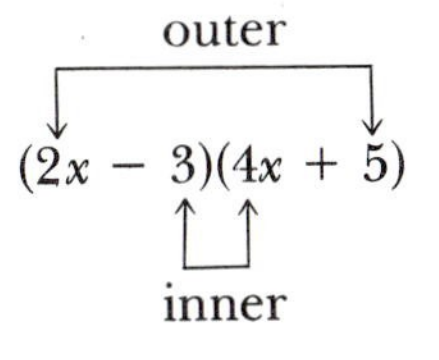

Step 3. Multiply the last two terms: $(-3)(5) = -15$

$$(2x - 3)(4x + 5)$$

last

Step 4. Add the results: $8x^2 - 2x - 15$ (as obtained previously)

Example 3 Compute each product:

(a) $(2x + 3y)(5x - y)$
(b) $(A - B)(A + B)$

Solution

(a) $(2x + 3y)(5x - y) = 10x^2 + 13xy - 3y^2$ (FOIL method)
(b) $(A - B)(A + B) = A^2 - B^2$ (The middle term, $AB - AB$, is zero.)

There are several binomial products that occur so frequently that it's convenient to memorize the results. The product $(A - B)(A + B) = A^2 - B^2$ in the example just concluded is one of these. The list that follows displays these special products. (Exercise 71 asks you to verify these results by carrying out the multiplication.) In memorizing these formulas, remember it's the *form* or *pattern* that's important, not the specific choice of letters.

Property Summary

Special Products

1. $(A - B)(A + B) = A^2 - B^2$

2. **(a)** $(A + B)^2 = A^2 + 2AB + B^2$
(b) $(A - B)^2 = A^2 - 2AB + B^2$

3. **(a)** $(A + B)^3 = A^3 + 3A^2B + 3AB^2 + B^3$
(b) $(A - B)^3 = A^3 - 3A^2B + 3AB^2 - B^3$

4. **(a)** $(A + B)(A^2 - AB + B^2) = A^3 + B^3$
(b) $(A - B)(A^2 + AB + B^2) = A^3 - B^3$

Example 4 Use the Special Product formulas to compute:

(a) $(3\sqrt{xy} - z)(3\sqrt{xy} + z)$
(b) $(4x^3 - 3)^2$
(c) $(2x + 5)^3$

Solution (a) Using Special Product 1, we have

$$(3\sqrt{xy} - z)(3\sqrt{xy} + z) = (3\sqrt{xy})^2 - z^2$$
$$= 9xy - z^2$$

(b) Using Special Product 2(b) we have

$$(4x^3 - 3)^2 = (4x^3)^2 - 2(4x^3)(3) + 3^2$$
$$= 16x^6 - 24x^3 + 9$$

(c) Using Special Product 3(a) we have

$$(2x + 5)^3 = (2x)^3 + 3(2x)^2(5) + 3(2x)(5)^2 + (5)^3$$
$$= 8x^3 + 60x^2 + 150x + 125$$

We conclude this section by describing the process of long division for polynomials. The terms *quotient, remainder, divisor,* and *dividend* are used here in the same way they're used in ordinary division of numbers. For instance, when 7 is divided by 2, the quotient is 3 and the remainder is 1. We write this as

$$\frac{7}{2} = 3 + \frac{1}{2}$$

or, equivalently,

$$7 = 2 \times 3 + 1$$

(7 ← dividend; 2 ← divisor; 3 ← quotient; 1 ← remainder)

The process of long division for polynomials follows the same four-step cycle used in ordinary long division of numbers: divide, multiply, subtract, bring down. As a first example, we divide $2x^2 - 7x + 8$ by $x - 2$. Notice that in setting up the division we write both the dividend and the divisor in decreasing powers of x.

$$\begin{array}{r} 2x - 3 \\ x - 2\overline{)2x^2 - 7x + 8} \\ \underline{2x^2 - 4x} \\ -3x + 8 \\ \underline{-3x + 6} \\ 2 \end{array}$$

1. Divide the first term of the dividend by the first term of the divisor: $\dfrac{2x^2}{x} = 2x$. The result becomes the first term of the quotient, as shown.
2. Multiply the divisor $x - 2$ by the term $2x$ obtained in the previous step. This yields the quantity $2x^2 - 4x$, which is written below the dividend, as shown.
3. From the quantity $2x^2 - 7x$ in the dividend, subtract the quantity $2x^2 - 4x$. This yields $-3x$.
4. Bring down the 8 in the dividend, as shown. The resulting quantity $-3x + 8$ is now treated as the dividend and the entire process is repeated.

We've now found that when $2x^2 - 7x + 8$ is divided by $x - 2$, the quotient is $2x - 3$ and the remainder is 2. This is summarized by writing either

$$\frac{2x^2 - 7x + 8}{x - 2} = 2x - 3 + \frac{2}{x - 2}$$

or

$$2x^2 - 7x + 8 = (x - 2)(2x - 3) + 2$$

As a second example, we divide $3x^4 - 2x^3 + 2$ by $x^2 - 1$. Notice in what follows that we have inserted the terms whose coefficients are zero in the divisor and in the dividend. These terms serve as place holders.

$$\begin{array}{r} 3x^2 - 2x + 3 \\ x^2 + 0x - 1 \overline{)\,3x^4 - 2x^3 + 0x^2 + 0x + 2} \\ \underline{3x^4 + 0x^3 - 3x^2} \\ -2x^3 + 3x^2 + 0x \\ \underline{-2x^3 + 0x^2 + 2x} \\ 3x^2 - 2x + 2 \\ \underline{3x^2 + 0x - 3} \\ -2x + 5 \end{array}$$

Thus the quotient is $3x^2 - 2x + 3$, the remainder is $-2x + 5$, and we can write

$$\frac{3x^4 - 2x^3 + 2}{x^2 - 1} = 3x^2 - 2x + 3 + \frac{-2x + 5}{x^2 - 1}$$

or

$$3x^4 - 2x^3 + 2 = (x^2 - 1)(3x^2 - 2x + 3) + (-2x + 5)$$

In the example just concluded, note that the degree of the remainder is 1, whereas the degree of the divisor is 2. So in this case we have

$$(\text{degree of remainder}) < (\text{degree of divisor})$$

In fact, it can be shown that this inequality remains valid whenever we obtain a nonzero remainder in dividing two nonconstant polynomials. (We'll return to this idea in Chapter 10.)

Exercise Set 1.7

In Exercises 1–6, specify the domain of each variable.

1. $4x^2 - 6x + 1$
2. $\dfrac{2y}{y - 1}$
3. $ax + b$
4. $4x + x^{-3}$
5. $4\sqrt{t} + t^2$
6. $\dfrac{x - 1}{(x - 2)(x - 3)}$

For Exercises 7–10, specify the degree and the (nonzero) coefficients of each polynomial.

7. $4x^3 - 2x^2 - 6x - 1$
8. $-5t^2 + 1$
9. $5x$
10. 23

In Exercises 11–40, carry out the indicated operations.

11. $(12x^2 - 4x + 2) + (8x^2 + 6x - 1)$

12. $(4x^2 - 1) + (4x^2 + 1)$

13. $(x^2 - x - 1) - (x^2 + x + 1)$

14. $(x^3 + 2x^2 + 1) - (x^2 - 1)$

15. $(2x^2 - 4x - 4) - (6x^2 + 5) - (8x - 1)$

16. $(x^3 - 1) - (x^3 - 1)$

17. $(2x^2 - 6x - 1) + (5x^2 - 5x - 1) - (3x^2 + 8x + 12)$

18. $(ax + b) - (2ax - b)$

19. $(ax^2 + bx + c) - (2ax^2 - 3bx + c)$

20. $x^3 - 4x^2 + x - 2(x^3 + 2x^2 + 1)$

21. $2x(x^2 - 4x - 5)$

22. $4ab(a^2 + 2ab + b^2)$

23. $(x - 1)(x - 2)$

24. $(y^2 + 1)(y^2 - 3)$

25. $(2x + 4)(x + 1)$

26. $(a + 2b)(a - 3b)$

27. $(x^2 + 3x)(x^2 + x)$

28. $(\sqrt{x} + 1)(\sqrt{x} + 2)$

29. $(2xy - 3)(2xy - 1)$

30. $(5xy^2 - 2)(xy^2 - 3)$

31. $\left(x + \frac{1}{x}\right)\left(x + \frac{1}{x}\right)$

32. $(2pq + r)(pq - 2r)$

33. $(\sqrt{a+b} + 1)(\sqrt{a+b} + 3)$

34. $(x^{1/2} + y^{1/2})(x^{1/2} + 2y^{1/2})$

35. $(x^{1/2} - 1)(x^{1/2} - 2)$

36. $(x - 1)(x^2 + 4x + 1)$

37. $(y + 2)(y^2 - 3y - 5)$

38. $(a + b + c)(a + b - c)$

39. $(x - 2y + z)(x + 2y - z)$

40. $a(b - c) + b(c - a) + c(a - b)$

For Exercises 41–62, use the Special Product formulas to compute the products.

41. $(x - y)(x + y)$

42. $(x - 10)(x + 10)$

43. $(a^2 - 3y)(a^2 + 3y)$

44. $[(a + b) - 1][(a + b) + 1]$

45. $[(x + y) + 2][(x + y) - 2]$

46. $(\sqrt{ab} + \sqrt{c})(\sqrt{ab} - \sqrt{c})$

47. $(x - 3)^2$

48. $(2x + 1)^2$

49. $(x^2 + 1)^2$

50. $(a^x - a^{-x})^2$

51. $(x^2 + a^2)^2$

52. $(\sqrt{a+b} - a)^2$

53. $(x + 2y)^3$

54. $(x - 3y)^3$

55. $(a + 1)^3$

56. $(x^2 - a^2)^3$

57. $(x - y)(x^2 + xy + y^2)$

58. $(x^2 - y^2)(x^4 + x^2y^2 + y^4)$

59. $(x + 1)(x^2 - x + 1)$

60. $(x + 2)(x^2 - 2x + 4)$

***61.** $(x^{1/3} - y^{1/3})(x^{2/3} + x^{1/3}y^{1/3} + y^{2/3})$

***62.** $(a^{1/3} + 5^{1/3})(a^{2/3} - a^{1/3} \cdot 5^{1/3} + 5^{2/3})$

In Exercises 63–70, use long division to compute the quotients and remainders.

63. $\frac{x^2 + 6x - 1}{x - 1}$

64. $\frac{2x^2 + x + 5}{x + 2}$

65. $\frac{2x^3 + 4x^2 - 6x + 2}{2x + 4}$

66. $\frac{x^2 + 8x - 12}{x^2 + 1}$

67. $\frac{x^6 - 64}{x - 2}$

68. $\frac{x + 4}{x - 3}$

69. $\frac{ax + b}{x + c}$

70. $\frac{8x^6 - 36x^4 + 54x^2 - 27}{2x^2 - 3}$

71. Carry out the multiplication to verify each of the Special Products listed on page 41.

For Exercises 72–75, verify each statement by carrying out the operations on the left-hand side of the equation.

***72.** $(b - c)(b + c - a) + (c - a)(c + a - b) + (a - b)(a + b - c) = 0$

***73.** $(b^{-1} + c^{-1})(b + c - a) + (c^{-1} + a^{-1})(c + a - b) + (a^{-1} + b^{-1})(a + b - c) = 6$

74. $(x^2 + 8 - 4x)(x^2 + 8 + 4x) = x^4 + 64$

75. $(\sqrt{a+x} + \sqrt{a-x})(\sqrt{a+x} - \sqrt{a-x}) = 2x$

76. By expanding both sides, verify that

$$(a^2 + b^2)(x^2 + y^2) = (ax - by)^2 + (bx + ay)^2$$

77. Simplify: $(b + c)(b - c) + (c + a)(c - a) + (a + b)(a - b)$

78. Simplify: $(a + b)(c^2 - ab) + (b + c)(a^2 - bc) + (c + a)(b^2 - ac)$

***79.** When $x^3 + kx + 6$ is divided by $x + 3$, the remainder is zero. Find k.

***80.** When $x^2 + 2px - 3q^2$ is divided by $x - p$, the remainder is zero. Show that $p^2 = q^2$.

81. **(a)** Determine the remainder when $ax^2 + bx + c$ is divided by $x - r$.

(b) Determine the remainder when $ax^3 + bx^2 + cx + d$ is divided by $x - r$.

(continued)

(c) On the basis of your answers in parts (a) and (b), guess the remainder when $ax^4 + bx^3 + cx^2 + dx + e$ is divided by $x - r$.

C 82. Notice that the expression $\frac{x^2 - 16}{x - 4}$ is undefined when $x = 4$. In this exercise we investigate the values of the expression when x is very close to 4.

(a) Complete the tables.

x	$\frac{x^2 - 16}{x - 4}$
3.9	
3.99	
3.999	
3.9999	
3.99999	

x	$\frac{x^2 - 16}{x - 4}$
4.1	
4.01	
4.001	
4.0001	
4.00001	

(b) On the basis of the tables, what value does the expression $\frac{x^2 - 16}{x - 4}$ seem to be approaching as x gets closer and closer to 4? This "target value" is referred to as the *limit* of $\frac{x^2 - 16}{x - 4}$ as x approaches 4. (The notion of a limit is made more precise in calculus.)

C 83. Notice that the expression $\frac{x^3 - 8}{x - 2}$ is undefined when $x = 2$. In this exercise we investigate the values of the expression when x is very close to 2.

(a) Complete the tables.

x	$\frac{x^3 - 8}{x - 2}$
1.9	
1.99	
1.999	
1.9999	
1.99999	

x	$\frac{x^3 - 8}{x - 2}$
2.1	
2.01	
2.001	
2.0001	
2.00001	

(b) On the basis of the tables, what "target value" does the expression $\frac{x^3 - 8}{x - 2}$ seem to be approaching as x gets closer and closer to 2?

C 84. Complete the table to compare the values of x^2 and 2^x as x grows larger and larger. Which expression yields the larger values "in the long run"?

x	1	2	3	4	5	10	20	50
x^2								
2^x								

C 85. Polynomials can be used to approximate more complicated expressions. For example, in calculus it is shown that when x is close to zero, the expression $\sqrt{1 + x}$ can be approximated by the polynomial $1 + \frac{1}{2}x$. Complete the table to see evidence of this. (Give the values of $\sqrt{1 + x}$ to six decimal places without rounding off.)

x	$1 + \frac{x}{2}$	$\sqrt{1 + x}$
0.1		
0.01		
0.001		

C 86. Using calculus, it can be shown that when x is close to zero, the expression $\frac{1}{\sqrt{1 - x^2}}$ can be approximated by the polynomial $1 + \frac{x^2}{2} + \frac{3x^4}{8}$. Complete the table to see evidence of this. (Round off your answers to six significant digits.)

x	$1 + \frac{x^2}{2} + \frac{3x^4}{8}$	$1/\sqrt{1 - x^2}$
0.1		
0.01		

1.8 Factoring

There are many cases in algebra in which the process of *factoring* simplifies the work at hand. To *factor* a polynomial means to write it as a product of two or more nonconstant polynomials. For instance, a factorization of $x^2 - 9$ is given by

$$x^2 - 9 = (x - 3)(x + 3)$$

We'll consider several techniques for factoring in this section. These techniques will be applied in Section 1.9 (Fractional Expressions), in Chapter 2 (Equations and Inequalities), and throughout the text.

There is one convention that we need to agree on at the outset. If the polynomial or expression that we wish to factor contains only integer coefficients, then the factors (if any) should involve only integer coefficients. For example, according to this convention, we will not consider the following type of factorization in this section:

$$x^2 - 2 = (x - \sqrt{2})(x + \sqrt{2})$$

because it involves coefficients that are irrational numbers. (We should point out, however, that factorizations such as this are useful at times, particularly in calculus.) As it happens, $x^2 - 2$ is an example of a polynomial that cannot be factored using integer coefficients. We say in such a case that the polynomial is *irreducible over the integers.*

There are five elementary techniques for factoring. These are summarized in the box that follows. Notice that three of the formulas in the box are just restatements of the corresponding Special Product formulas considered in the previous section. In these cases, remember it is the form or

Technique	Example or Formula	Remark
Common factor	$3x^4 + 6x^3 - 12x^2 = 3x^2(x^2 + 2x - 4)$ $4(x^2 + 1) - x(x^2 + 1) = (x^2 + 1)(4 - x)$	In any factoring problem, the first step always is to look for the common factor of highest degree.
Difference of squares	$x^2 - a^2 = (x - a)(x + a)$	There is no corresponding formula for a sum of squares; $x^2 + a^2$ is irreducible over the integers.
Trial and error	$x^2 + 2x - 3 = (x + 3)(x - 1)$	In this example, the only distinct possibilities or trials are: **(a)** $(x - 3)(x - 1)$ **(b)** $(x - 3)(x + 1)$ **(c)** $(x + 3)(x - 1)$ **(d)** $(x + 3)(x + 1)$ By inspection or by carrying out the indicated multiplications, we find that only case **(c)** checks.
Difference of cubes Sum of cubes	$x^3 - a^3 = (x - a)(x^2 + ax + a^2)$ $x^3 + a^3 = (x + a)(x^2 - ax + a^2)$	Verify these formulas for yourself by carrying out the multiplications. Then memorize the formulas.
Grouping	$x^3 - x^2 + x - 1 = (x^3 - x^2) + (x - 1)$ $= x^2(x - 1) + (x - 1)\cdot 1$ $= (x - 1)(x^2 + 1)$	This is actually an application of the common factor technique.

pattern in the formula that is important, not the specific choice of letters.

The idea in factoring is to use one or more of these techniques until each of the factors obtained is irreducible. The examples that follow show how this works in practice.

Example 1 Factor: **(a)** $x^2 - 49$ **(b)** $2x^3 - 50x$

Solution

(a) $x^2 - 49 = x^2 - 7^2$

$= (x - 7)(x + 7)$ difference of squares

(b) $2x^3 - 50x = 2x(x^2 - 25)$ common factor

$= 2x(x - 5)(x + 5)$ difference of squares

Example 2 Factor: $3x^5 - 3x$

Solution

$3x^5 - 3x = 3x(x^4 - 1)$ common factor

$= 3x[(x^2)^2 - 1^2]$

$= 3x(x^2 - 1)(x^2 + 1)$ difference of squares

$= 3x(x - 1)(x + 1)(x^2 + 1)$ difference of squares, again

Example 3 Factor: **(a)** $x^2 - 4x - 5$

(b) $x^2 - 4x + 5$

(c) $8x^4 - 24x^3 + 18x^2$

Solution

(a) $x^2 - 4x - 5 = (x - 5)(x + 1)$ trial and error

(b) Irreducible (trial and error).

(c) The first step is to check for a common factor. This is $2x^2$, and we write

$$8x^4 - 24x^3 + 18x^2 = 2x^2(4x^2 - 12x + 9)$$

Next we consider the expression in parentheses in this last equation. It can be factored by trial and error; alternately, it can be factored by recalling the Special Product formula $(A - B)^2 = A^2 - 2AB + B^2$. In either case, we find that $4x^2 - 12x + 9 = (2x - 3)^2$. Thus, our final factorization is

$$8x^4 - 24x^3 + 18x^2 = 2x^2(2x - 3)^2$$

Example 4 Factor $(t - b)^2 - (t^2 + b)^2$.

Solution This is a difference of squares, for which the basic pattern is

$$x^2 - a^2 = (x - a)(x + a)$$

In this last equation we replace x with the quantity $(t - b)$ and a with the quantity $(t^2 + b)$ to obtain

$$\begin{aligned}(t - b)^2 - (t^2 + b)^2 &= [(t - b) - (t^2 + b)][(t - b) + (t^2 + b)]\\ &= (t - b - t^2 - b)(t - b + t^2 + b)\\ &= (t - t^2 - 2b)(t + t^2)\end{aligned}$$

Finally now, notice that on the right-hand side of this last equation the common factor t can be removed from the expression $t + t^2$. Thus, the required factorization is

$$(t - b)^2 - (t^2 + b)^2 = (t - t^2 - 2b)(1 + t)t$$

Example 5 Factor $ax + ay^2 + bx + by^2$.

Solution Factor a from the first two terms and b from the second two, to obtain

$$ax + ay^2 + bx + by^2 = a(x + y^2) + b(x + y^2)$$

Now recognize the quantity $(x + y^2)$ as a common expression that can be factored out. We have then

$$a(x + y^2) + b(x + y^2) = (x + y^2)(a + b)$$

The required factorization is therefore

$$ax + ay^2 + bx + by^2 = (x + y^2)(a + b)$$

As you may wish to check for yourself, this factorization can also be obtained by trial and error.

Example 6 Use factoring to evaluate: (a) $100^2 - 99^2$
(b) $8^3 - 6^3$

Solution **(a)**

$$\begin{aligned} 100^2 - 99^2 &= (100 - 99)(100 + 99) && \text{difference of squares} \\ &= (1)(199) \\ &= 199 \end{aligned}$$

(b)

$$\begin{aligned} 8^3 - 6^3 &= (8 - 6)(8^2 + 8\cdot 6 + 6^2) && \text{difference of cubes} \\ &= 2(64 + 48 + 36) \\ &= 2[(64 + 36) + 48] = 2(148) \\ &= 296 \end{aligned}$$

Example 7 Factor $x^3 - 6ax^2 + 12a^2x - 8a^3$.

Solution Since this expression involves four terms, our first inclination is to apply the method shown in Example 5. We factor x^2 from the first two terms and $4a^2$ from the second two. This yields

$$x^3 - 6ax^2 + 12a^2x - 8a^3 = x^2(x - 6a) + 4a^2(3x - 2a)$$

But now (as opposed to the corresponding point in Example 5) there is no common expression on the right-hand side. So at this point we need to start over. Again we will use the method of Example 5; but first we rearrange our terms. We have

$$x^3 - 6ax^2 + 12a^2x - 8a^3 = \underbrace{x^3 - 8a^3}_{\substack{\text{differences} \\ \text{of cubes is} \\ \text{applicable}}} - \underbrace{6ax^2 + 12a^2x}_{\substack{\text{common term} \\ \text{is } -6ax}}$$

Thus

$$x^3 - 6ax^2 + 12a^2x - 8a^3 = (x - 2a)(x^2 + 2ax + 4a^2) - 6ax(x - 2a)$$

Now on the right-hand side of this last equation we factor out the common expression $(x - 2a)$ to obtain

$$\begin{aligned} x^3 - 6ax^2 + 12a^2x - 8a^3 &= (x - 2a)[(x^2 + 2ax + 4a^2) - 6ax] \\ &= (x - 2a)(x^2 - 4ax + 4a^2) \\ &= (x - 2a)[(x - 2a)(x - 2a)] \\ &= (x - 2a)^3 \end{aligned}$$

The required factorization is therefore

$$x^3 - 6ax^2 + 12a^2x - 8a^3 = (x - 2a)(x - 2a)(x - 2a)$$

or, more concisely,

$$x^3 - 6ax^2 + 12a^2x - 8a^3 = (x - 2a)^3$$

Note You can check this answer by recalling the Special Product formula for $(A - B)^3$. Indeed, if you're sufficiently familiar with that formula, the required factorization in this example can be obtained simply by inspection.

Example 8 Factor $x^6 - y^6$.

Solution We will show two methods; the first is reasonably straightforward, the second involves a clever manipulation.

First Method

$$\begin{aligned} x^6 - y^6 &= (x^3)^2 - (y^3)^2 \\ &= (x^3 - y^3)(x^3 + y^3) \\ &= [(x - y)(x^2 + xy + y^2)][(x + y)(x^2 - xy + y^2)] \end{aligned}$$

The required factorization is therefore

$$x^6 - y^6 = (x - y)(x + y)(x^2 + xy + y^2)(x^2 - xy + y^2)$$

Alternate Method

$$\begin{aligned} x^6 - y^6 &= (x^2)^3 - (y^2)^3 \\ &= (x^2 - y^2)[(x^2)^2 + x^2y^2 + (y^2)^2] \\ &= (x - y)(x + y)[x^4 + x^2y^2 + y^4] \end{aligned}$$

The problem now is to factor the quantity in the brackets in this last equation. (Surely it must factor, judging from the answer obtained with the first method.) The technique here is to add zero, written in the form $x^2y^2 - x^2y^2$. We have

$$\begin{aligned} x^4 + x^2y^2 + y^4 &= x^4 + x^2y^2 + y^4 + 0 \\ &= x^4 + x^2y^2 + y^4 + x^2y^2 - x^2y^2 \\ &= (x^4 + 2x^2y^2 + y^4) - x^2y^2 \\ &= (x^2 + y^2)^2 - (xy)^2 \end{aligned}$$

We now have a difference of squares, which can be factored:

$$(x^2 + y^2)^2 - (xy)^2 = [(x^2 + y^2) - xy][(x^2 + y^2) + xy]$$

or

$$(x^2 + y^2)^2 - (xy)^2 = (x^2 - xy + y^2)(x^2 + xy + y^2)$$

Putting things together, we finally obtain

$$x^6 - y^6 = (x - y)(x + y)(x^2 - xy + y^2)(x^2 + xy + y^2)$$

Except for the order of the last two factors, this agrees with the result obtained using the first method.

As an application of factoring, we describe an alternate form for writing polynomials known as *nested form*. Polynomials written in nested form are easier to evaluate, even when the x-value involves a decimal. The following example shows the steps used to convert the polynomial $6x^4 - 2x^3 + 10x^2 + 7x + 1$ to nested form.

$$\begin{aligned}
\text{start:}\quad & 6x^4 - 2x^3 + 10x^2 + 7x + 1 \\
\text{factor out } x^3\text{:}\quad & (6x - 2)x^3 + 10x^2 + 7x + 1 \\
\text{factor out } x^2\text{:}\quad & [(6x - 2)x + 10]x^2 + 7x + 1 \\
\text{factor out } x\text{:}\quad & \{[(6x - 2)x + 10]x + 7\}x + 1
\end{aligned}$$

This last expression is the required nested form. If you examine this expression carefully, you'll see why it lends itself so well to computational work. Starting with the first coefficient (namely 6), the simple two-step process

1. multiply by x
2. add the next coefficient

is used over and over again. The next example relies on this two-step process to evaluate a polynomial.

Example 9 Evaluate $2x^3 - 5x^2 + 4x - 1$ when $x = 6$, using the two-step process.

Solution

Process	*Result*
Begin with 2 (the first coefficient).	
Multiply by 6 and add -5.	$2 \cdot 6 - 5 = 7$
Multiply by 6 and add 4.	$7 \cdot 6 + 4 = 46$
Multiply by 6 and add -1.	$46 \cdot 6 - 1 = 275$

The required value is 275. You should check for yourself now that the same result is obtained when $x = 6$ is substituted directly into the given polynomial.

Exercise Set 1.8

In Exercises 1–56, factor each polynomial or expression. If a polynomial is irreducible, state this. (In Exercises 1–6, the factoring techniques are specified.)

1. (Common factor and difference of squares)
(a) $x^2 - 64$ **(b)** $7x^4 + 14x^2$
(c) $121z - z^3$ **(d)** $a^2b^2 - c^2$

2. (Common factor and difference of squares)
(a) $1 - t^4$ **(b)** $x^6 + x^5 + x^4$
(c) $u^2v^2 - 225$ **(d)** $81x^4 - x^2$

3. (Trial and error)
(a) $x^2 + 2x - 3$ **(b)** $x^2 - 2x - 3$
(c) $x^2 - 2x + 3$ **(d)** $-x^2 + 2x + 3$

4. (Trial and error)
(a) $2x^2 - 7x - 4$ **(b)** $2x^2 + 7x - 4$
(c) $2x^2 + 7x + 4$ **(d)** $-2x^2 - 7x + 4$

5. (Sum and difference of cubes)
(a) $x^3 + 1$ **(b)** $x^3 + 216$
(c) $1000 - 8x^6$ **(d)** $64a^3x^3 - 125$

6. (Grouping)
(a) $x^4 - 2x^3 + 3x - 6$
(b) $a^2x + bx - a^2z - bz$

7. $2x - 2x^3$ **8.** $3x^4 - 48x^2$

9. $100x^3 - x^5$ **10.** $2x^2 + 5x - 12$

11. $2x^4 + 3x^3 - 9x^2$ **12.** $x^2 - 6x + 1$

13. $4x^3 - 20x^2 + 25x$ **14.** $ab - bc + a^2 - ac$

15. $x^2z^2 + xzt + xyz + yt$ **16.** $x^2 + 32x + 256$

17. $a^2t^2 + b^2t^2 - cb^2 - ca^2$

18. $x^4 - 6x^2 + 9$ **19.** $x^3 - 13x^2 - 90x$

20. $a^4 - 4a^2b^2c^2 + 4b^4c^4$

21. $4x^2 - 29xy - 24y^2$

***22.** $(x + a)^4 - 2(x + a)^2(x + b)^2 + (x + b)^4$

23. $x^2 + 2x + 16$ **24.** $(x - y)^2 - z^2$

25. $1 - (x + y)^2$

26. $(a + b + 1)^2 - (a - b + 1)^2$

27. $x^8 - 1$ **28.** $x^2 - y^2 + x - y$

29. $x^3 + 3x^2 + 3x + 1$

30. $-1 + 6x - 12x^2 + 8x^3$

31. $27x^3 + 108x^2 + 144x + 64$

32. $x^6 - 1$ **33.** $64a^6 - b^6$

34. $(a + b)^3 - 1$

35. $x^4 - 25x^2 + 144$

36. $x^4 + 25x^2 + 144$

37. $x^2 + 16y^2$

38. $81x^4 - 16y^4$

39. $x^3 + 2x^2 - 255x$

40. $4(x - a)^3 - (x - a)$

41. $x^3 + a^3 + x + a$

***42.** $8(x + a)^3 - (2x - a)^3$

43. $x^2 - a^2 + y^2 - 2xy$

44. $a^4 - (b + c)^4$

45. $21x^3 + 82x^2 - 39x$

46. $x^3a^2 - 8y^3a^2 - 4x^3b^2 + 32y^3b^2$

47. $12xy + 25 - 4x^2 - 9y^2$

48. $ax^2 + (1 + ab)xy + by^2$

49. $ax^2 + (a + b)x + b$

***50.** $(5a^2 - 11a + 10)^2 - (4a^2 - 15a + 6)^2$

51. $x^4 + 64$ *Hint:* Add and subtract the term $16x^2$.

***52.** **(a)** $x^4 - 6x^2 + 9$
(b) $x^4 - 15x^2 + 9$
Hint: Add and subtract a term.

***53.** $(x + y)^2 + (x + z)^2 - (z + t)^2 - (y + t)^2$

***54.** $(a - a^2)^3 + (a^2 - 1)^3 + (1 - a)^3$

***55.** $(b - c)^3 + (c - a)^3 + (a - b)^3$

***56.** $(a + b + c)^3 - a^3 - b^3 - c^3$

For Exercises 57–60, evaluate the expressions using factoring techniques.

57. $1000^2 - 999^2$

58. $50^2 - 49^2$

59. $10^3 - 9^3$

60. $\dfrac{15^3 - 10^3}{15^2 - 10^2}$

Exercises 61–66 involve factoring expressions containing fractional exponents. For an expression such as $(x + 1)^{1/2} - (x + 1)^{3/2}$*, where the exponents are positive, the common expression to factor out is the one with the smallest exponent, in this case* $(x + 1)^{1/2}$*. In the case of expressions with negative exponents, such as* $(x + 1)^{-1/2} - (x + 1)^{-3/2}$*, the common expression to be factored out is the one whose exponent has the largest absolute value, in this case* $(x + 1)^{-3/2}$*. Use these guidelines to factor the expressions.*

61. $(x + 1)^{1/2} - (x + 1)^{3/2}$

62. $(x^2 + 1)^{3/2} + (x^2 + 1)^{7/2}$

63. $(x + 1)^{-1/2} - (x + 1)^{-3/2}$

64. $(x^2 + 1)^{-2/3} + (x^2 + 1)^{-5/3}$

65. $x^2(x - 2)^{-4} + x(x - 2)^{-3}$ Write the final answer in a form that doesn't involve negative exponents.

66. $x^2(a^2 - x^2)^{-1/2} + \sqrt{a^2 - x^2}$ Write the final answer in a form not involving negative exponents. *Hint:* Use $(a^2 - x^2)^{-1/2}$ for the common term.

In Exercises 67–72, write the polynomials in nested form.

67. $4x^3 + 5x^2 - x + 10$

68. $12x^2 - 10x + 19$

69. $5x^4 + 2x^3 - 7x^2 + 8x + 2$

70. $ax^2 + bx + c$

71. $ax^3 + bx^2 + cx + d$

72. $x^5 + x^4 + x^3 + x^2 + x + 1$

For Exercises 73–81, use the two-step process described in the text to evaluate the polynomials using the given x-values.

73. $2x^3 - 5x^2 + 10x - 15;\ x = 3$

74. $x^3 + 5x^2 - 4x - 6;\ x = 5$

75. $4x^3 - 12x^2 - x + 1;\ x = -2$

76. $x^4 - 10x^3 + 10x^2 - 5x + 4;\ x = 3$

C **77.** $4x^3 - 12x^2 + 6x - 17;\ x = 0.25$

C **78.** $2x^3 - 16x^2 - 10x + 45;\ x = 1.99$

C **79.** $12x^4 - 7x^3 - 15x^2 + 6x + 3;\ x = 35$

C **80.** $2x^4 + 3x^3 - 4x^2 - 3x + 2;\ x = 0.49$

C **81.** $6x^4 + 5x^3 - 38x^2 + 5x + 6;\ x = -\frac{1}{3}$

1.9 Fractional Expressions

But if you do use a rule involving mechanical calculation, be patient, accurate, and systematically neat in the working. It is well known to mathematical teachers that quite half the failures in algebraic exercises arise from arithmetical inaccuracy and slovenly arrangement.

George Chrystal (1893)

The rules of arithmetic for fractions were listed in Section 1.2. These same rules are used for fractions involving algebraic expressions. In algebra, as in arithmetic, we say that a fraction is *reduced to lowest terms* or *simplified* when the numerator and denominator contain no common factors (other than 1 and −1). The factoring techniques developed in the previous section are used to reduce fractions. For example, to reduce the fraction $\frac{x^2 - 9}{x^2 + 3x}$ we write

$$\frac{x^2 - 9}{x^2 + 3x} = \frac{(x - 3)(x + 3)}{x(x + 3)} = \frac{x - 3}{x}$$

In Example 1 we display two more instances in which factoring is used to reduce a fraction. After that, Example 2 indicates how these skills are used to multiply and divide fractional expressions.

Example 1 Simplify: **(a)** $\frac{x^3 - 8}{x^2 - 2x}$ **(b)** $\frac{x^2 - 6x + 8}{a(x - 2) + b(x - 2)}$

Solution **(a)** $\frac{x^3 - 8}{x^2 - 2x} = \frac{(x - 2)(x^2 + 2x + 4)}{x(x - 2)} = \frac{x^2 + 2x + 4}{x}$

(b) $\frac{x^2 - 6x + 8}{a(x - 2) + b(x - 2)} = \frac{(x - 2)(x - 4)}{(x - 2)(a + b)} = \frac{x - 4}{a + b}$

Example 2 Carry out the indicated operations and simplify:

(a) $\frac{x^3}{2x^2 + 3x} \cdot \frac{12x + 18}{4x^2 - 6x}$

(b) $\frac{2x^2 - x - 6}{x^2 + x + 1} \cdot \frac{x^3 - 1}{4 - x^2}$

(c) $\frac{x^2 - 144}{x^2 - 4} \div \frac{x + 12}{x + 2}$

Solution **(a)**
$$\frac{x^3}{2x^2 + 3x} \cdot \frac{12x + 18}{4x^2 - 6x} = \frac{x^3}{x(2x + 3)} \cdot \frac{6(2x + 3)}{2x(2x - 3)} = \frac{6x^3}{2x^2(2x - 3)} = \frac{3x}{2x - 3}$$

(b)
$$\frac{2x^2 - x - 6}{x^2 + x + 1} \cdot \frac{x^3 - 1}{4 - x^2} = \frac{(2x + 3)(x - 2)}{(x^2 + x + 1)} \cdot \frac{(x - 1)(x^2 + x + 1)}{(2 - x)(2 + x)}$$

$$= \frac{-(2x + 3)(x - 1)}{2 + x}$$

We've used the fact that $\frac{x - 2}{2 - x} = -1$.

$$= \frac{-2x^2 - x + 3}{2 + x}$$

(c)
$$\frac{x^2 - 144}{x^2 - 4} \div \frac{x + 12}{x + 2} = \frac{x^2 - 144}{x^2 - 4} \cdot \frac{x + 2}{x + 12}$$

$$= \frac{(x - 12)(x + 12)}{(x - 2)(x + 2)} \cdot \frac{x + 2}{x + 12} = \frac{x - 12}{x - 2}$$

As in arithmetic, to add or subtract two fractions the denominators must be the same. The rules in this case are

$$\frac{a}{b} + \frac{c}{b} = \frac{a + c}{b} \qquad \frac{a}{b} - \frac{c}{b} = \frac{a - c}{b}$$

For example, we have

$$\frac{4x-1}{x+1}-\frac{2x-1}{x+1}=\frac{4x-1-(2x-1)}{x+1}=\frac{4x-1-2x+1}{x+1}=\frac{2x}{x+1}$$

Fractions with unlike denominators are added or subtracted by first converting to a common denominator. For instance, to add $9/a$ and $10/a^2$ we write

$$\frac{9}{a}+\frac{10}{a^2}=\frac{9}{a}\cdot\frac{a}{a}+\frac{10}{a^2}$$

$$=\frac{9a}{a^2}+\frac{10}{a^2}=\frac{9a+10}{a^2}$$

Notice that the common denominator used was a^2. This is the *least common denominator*. In fact, other common denominators (such as a^3 or a^4) could be used here, but that would be less efficient. In general, the least common denominator for a given group of fractions is chosen as follows. Write down a product involving the irreducible factors from each denominator. The power for a given factor should be equal to (but not greater than) the highest power of that factor appearing in any of the individual denominators. For example, the least common denominator for the two fractions $\frac{1}{(x+1)^2}$ and $\frac{1}{(x+1)(x+2)}$ is $(x+1)^2(x+2)$. In the example that follows, notice that the denominators must be in factored form before the least common denominator can be chosen.

Example 3 Combine into a single fraction and simplify:

(a) $\frac{3}{4x}+\frac{7x}{10y^2}-2$ **(b)** $\frac{x}{x^2-9}-\frac{1}{x+3}$ **(c)** $\frac{15}{x^2+x-6}+\frac{x+1}{2-x}$

Solution **(a)** Denominators: $2^2\cdot x$; $2\cdot 5\cdot y^2$; 1
Least common denominator: $2^2\cdot 5xy^2=20xy^2$

$$\frac{3}{4x}+\frac{7x}{10y^2}-\frac{2}{1}=\frac{3}{4x}\cdot\frac{5y^2}{5y^2}+\frac{7x}{10y^2}\cdot\frac{2x}{2x}-\frac{2}{1}\cdot\frac{20xy^2}{20xy^2}$$

$$=\frac{15y^2+14x^2-40xy^2}{20xy^2}$$

(b) Denominators: $x^2-9=(x-3)(x+3)$; $x+3$
Least common denominator: $(x-3)(x+3)$

$$\frac{x}{x^2-9}-\frac{1}{x+3}=\frac{x}{(x-3)(x+3)}-\frac{1}{x+3}$$

$$=\frac{x}{(x-3)(x+3)}-\frac{1}{x+3}\cdot\frac{x-3}{x-3}$$

$$=\frac{x-(x-3)}{(x-3)(x+3)}=\frac{3}{(x-3)(x+3)}$$

(c) $$\frac{15}{x^2+x-6}+\frac{x+1}{2-x}=\frac{15}{(x-2)(x+3)}+\frac{x+1}{2-x}$$

$$=\frac{15}{(x-2)(x+3)}-\frac{x+1}{x-2}$$

We used the fact that $2-x=-(x-2)$.

$$= \frac{15}{(x-2)(x+3)} - \frac{x+1}{x-2} \cdot \frac{x+3}{x+3}$$

$$= \frac{15 - (x^2 + 4x + 3)}{(x-2)(x+3)} = \frac{-x^2 - 4x + 12}{(x-2)(x+3)}$$

$$= \frac{(x-2)(-x-6)}{(x-2)(x+3)} = \frac{-x-6}{x+3}$$

Notice in the last line that we were able to simplify the answer by factoring the numerator and reducing the fraction. (This is why we prefer to leave the least common denominator in factored form, rather than multiply it out, in this type of problem.)

In Examples 4 through 7 the numerators or denominators themselves contain fractions. In general, two methods of solution can be used here. Example 4 displays both methods, while Examples 5, 6, and 7 each show but one method.

Example 4 Simplify $\dfrac{\dfrac{1}{3a} - \dfrac{1}{4b}}{\dfrac{5}{6a^2} + \dfrac{1}{b}}$.

Solution *First Method*

The least common denominator for the four individual fractions is $12a^2b$. Multiplying the given expression by $\frac{12a^2b}{12a^2b}$, which equals 1, yields

$$\frac{12a^2b}{12a^2b} \cdot \frac{\dfrac{1}{3a} - \dfrac{1}{4b}}{\dfrac{5}{6a^2} + \dfrac{1}{b}} = \frac{4ab - 3a^2}{10b + 12a^2}$$

Alternate Method

We begin by working separately above and below the main fraction line.

$$\frac{\dfrac{1}{3a} - \dfrac{1}{4b}}{\dfrac{5}{6a^2} + \dfrac{1}{b}} = \frac{\dfrac{1}{3a} \cdot \dfrac{4b}{4b} - \dfrac{1}{4b} \cdot \dfrac{3a}{3a}}{\dfrac{5}{6a^2} \cdot \dfrac{b}{b} + \dfrac{1}{b} \cdot \dfrac{6a^2}{6a^2}}$$

$$= \frac{\dfrac{4b - 3a}{12ab}}{\dfrac{5b + 6a^2}{6a^2b}}$$

$$= \frac{4b - 3a}{12ab} \cdot \frac{6a^2b}{5b + 6a^2}$$

$$= \frac{4b - 3a}{2} \cdot \frac{a}{5b + 6a^2}$$

$$= \frac{4ab - 3a^2}{10b + 12a^2}$$

Example 5 Simplify $(x^{-1} + y^{-1})^{-1}$. (The answer is *not* $x + y$.)

Solution After applying the definition of negative exponents to rewrite the given expression, we'll use the first method shown in the previous example.

$$(x^{-1} + y^{-1})^{-1} = \left(\frac{1}{x} + \frac{1}{y}\right)^{-1}$$

$$= \frac{1}{\frac{1}{x} + \frac{1}{y}}.$$

$$= \frac{xy}{xy} \cdot \frac{1}{\frac{1}{x} + \frac{1}{y}} = \frac{xy}{y + x}$$

Example 6 Simplify $\dfrac{\frac{1}{x+h} - \frac{1}{x}}{h}$. (This type of expression occurs in calculus.)

Solution

$$\frac{\frac{1}{x+h} - \frac{1}{x}}{h} = \frac{(x+h)x}{(x+h)x} \cdot \frac{\frac{1}{x+h} - \frac{1}{x}}{h}$$

$$= \frac{x - (x+h)}{(x+h)xh} = \frac{-h}{(x+h)xh} = -\frac{1}{(x+h)x}$$

Example 7 Simplify $\dfrac{x - \frac{1}{x^2}}{\frac{1}{x^2} - 1}$.

Solution

$$\frac{x - \frac{1}{x^2}}{\frac{1}{x^2} - 1} = \frac{x^2}{x^2} \cdot \frac{x - \frac{1}{x^2}}{\frac{1}{x^2} - 1}$$

$$= \frac{x^3 - 1}{1 - x^2} = -\frac{x^3 - 1}{x^2 - 1}$$

$$= -\frac{(x-1)(x^2 + x + 1)}{(x-1)(x+1)} = -\frac{x^2 + x + 1}{x + 1}$$

Exercise Set 1.9

In Exercises 1–12, reduce the fractions to lowest terms.

1. $\dfrac{x^2 - 9}{x + 3}$
2. $\dfrac{25 - x^2}{x - 5}$
3. $\dfrac{x + 2}{x^4 - 16}$
4. $\dfrac{x^2 - x - 20}{2x^2 + 7x - 4}$
5. $\dfrac{x^2 + 2x + 4}{x^3 - 8}$
6. $\dfrac{a + b}{ax^2 + bx^2}$
7. $\dfrac{9ab - 12b^2}{6a^2 - 8ab}$
8. $\dfrac{a^3b^2 - 27b^5}{(ab - 3b^2)^2}$
9. $\dfrac{a^3 + a^2 + a + 1}{a^2 - 1}$
10. $\dfrac{(x - y)^2(a + b)}{(x^2 - y^2)(a^2 + 2ab + b^2)}$
11. $\dfrac{x^3 - y^3}{(x - y)^3}$
12. $\dfrac{x^4 - y^4}{(x^4y + x^2y^3 + x^3y^2 + xy^4)(x - y)^2}$

For Exercises 13–62, carry out the indicated operations and simplify where possible.

13. $\dfrac{2}{x - 2} \cdot \dfrac{x^2 - 4}{x + 2}$
14. $\dfrac{ax + 3}{2a + 1} \div \dfrac{a^2x^2 + 3ax}{4a^2 - 1}$
15. $\dfrac{x^2 - x - 2}{x^2 + x - 12} \cdot \dfrac{x^2 - 3x}{x^2 - 4x + 4}$
16. $(3t^2 + 4tx + x^2) \div \dfrac{3t^2 - 2tx - x^2}{t^2 - x^2}$
17. $\dfrac{x^3 + y^3}{x^2 - 4xy + 3y^2} \div \dfrac{(x + y)^3}{x^2 - 2xy - 3y^2}$

18. $\dfrac{a^2 - a - 42}{a^4 + 216a} \div \dfrac{a^2 - 49}{a^3 - 6a^2 + 36a}$

19. $\dfrac{x^2 + xy - 2y^2}{x^2 - 5xy + 4y^2} \cdot \dfrac{x^2 - 7xy + 12y^2}{x^2 + 5xy + 6y^2}$

20. $\dfrac{x^4y^2 - xy^5}{x^4 - 2x^2y^2 + y^4} \cdot \dfrac{x^2 + 2xy + y^2}{x^3y^2 + x^2y^3 + xy^4}$

21. $\dfrac{4}{x} - \dfrac{2}{x^2}$

22. $\dfrac{1}{3x} + \dfrac{1}{5x^2} - \dfrac{1}{30x^3}$

23. $\dfrac{6}{a} - \dfrac{a}{6}$

24. $\dfrac{1}{a} + \dfrac{1}{b} + \dfrac{1}{c}$

25. $\dfrac{1}{x + 3} + \dfrac{3}{x + 2}$

26. $\dfrac{4}{x - 4} - \dfrac{4}{x + 1}$

27. $\dfrac{3x}{x - 2} - \dfrac{6}{x^2 - 4}$

28. $1 + \dfrac{1}{x} - \dfrac{1}{x^2}$

29. $\dfrac{a}{x - 1} + \dfrac{2ax}{(x - 1)^2} + \dfrac{3ax^2}{(x - 1)^3}$

30. $\dfrac{a^2 + 5a - 4}{a^2 - 16} - \dfrac{2a}{2a^2 + 8a}$

31. $\dfrac{x}{x^2 - 9} + \dfrac{x - 1}{x^2 - 5x + 6}$

32. $\dfrac{1}{x - 1} + \dfrac{1}{1 - x}$

33. $\dfrac{4}{x - 5} - \dfrac{4}{5 - x}$

34. $\dfrac{x}{x + a} + \dfrac{a}{a - x}$

35. $\dfrac{a^2 + b^2}{a^2 - b^2} + \dfrac{a}{a + b} + \dfrac{b}{b - a}$

36. $\dfrac{3}{2x + 2} - \dfrac{5}{x^2 - 1} + \dfrac{1}{x + 1}$

37. $\dfrac{1}{x^2 + x - 20} - \dfrac{1}{x^2 - 8x + 16}$

38. $\dfrac{4}{6x^2 + 5x - 4} + \dfrac{1}{3x^2 + 4x} - \dfrac{1}{2x - 1}$

39. $\dfrac{2q + p}{2p^2 - 9pq - 5q^2} - \dfrac{p + q}{p^2 - 5pq}$

40. $\dfrac{1}{x - 1} + \dfrac{1}{x^2 - 1} + \dfrac{1}{x^3 - 1}$

***41.** $\dfrac{x + y}{2(x^2 + y^2)} - \dfrac{1}{2(x + y)} + \dfrac{x - y}{x^2 - y^2} - \dfrac{x^3 - y^3}{x^4 - y^4}$

***42.** $\dfrac{4y}{x^2 - y^2} + \dfrac{1}{x + y} + \dfrac{1}{y - x} - \dfrac{2y}{x^2 + y^2}$

43. $\dfrac{x}{(x - y)(x - z)} + \dfrac{y}{(y - z)(y - x)} + \dfrac{z}{(z - x)(z - y)}$

44. $\dfrac{3x}{x - 2} + \dfrac{4x^2}{x^2 + 2x + 1} - \dfrac{x^2}{x^2 - x - 2}$

***45.** $\dfrac{y + z}{x^2 - xy - xz + yz} - \dfrac{x + z}{xy - xz - y^2 + yz} + \dfrac{x + y}{xy - yz - xz + z^2}$

46. $\left(x + \dfrac{1}{x}\right) \cdot \left(1 - \dfrac{1}{x}\right)^2 \div \left(x - \dfrac{1}{x}\right)$

47. $\dfrac{1}{x + a} + \dfrac{x + 1}{x + a} \div \dfrac{x - a}{x - 1}$

48. $\dfrac{1}{2ax - 2a^2} - \dfrac{1}{2ax + 2a^2}$

***49.** $\dfrac{ap + q}{ax - bx - a^2 + ab} + \dfrac{bp + q}{bx - ax - b^2 + ab}$

50. $\dfrac{x^2 - 2ax + a^2}{px + q}\left[\dfrac{p}{x - a} + \dfrac{ap + q}{(x - a)^2}\right]$

***51.** $\dfrac{b - c}{a^2 - ab - ac + bc} + \dfrac{a - c}{b^2 - bc - ab + ac} + \dfrac{a - b}{c^2 - ac - bc + ab}$

52. $\dfrac{\dfrac{1}{a} + \dfrac{1}{b}}{\dfrac{1}{a} - \dfrac{1}{b}}$

53. $\dfrac{1 + \dfrac{4}{x}}{\dfrac{3}{x} - 2}$

54. $\dfrac{\dfrac{1}{a}}{1 + \dfrac{b}{c}}$

55. $\dfrac{a - \dfrac{1}{a}}{1 + \dfrac{1}{a}}$

56. $\dfrac{\dfrac{1}{x^2} - \dfrac{1}{y^2}}{\dfrac{1}{x} + \dfrac{1}{y}}$

57. $\dfrac{\dfrac{1}{2 + h} - \dfrac{1}{2}}{h}$

58. $\dfrac{\dfrac{3}{x^2 + h} - \dfrac{3}{x^2}}{h}$

59. $\dfrac{\dfrac{a}{x^2} + \dfrac{x}{a^2}}{a^2 - ax + x^2}$

60. $\dfrac{x + \dfrac{xy}{y - x}}{\dfrac{y^2}{x^2 - y^2} + 1}$

***61.** $\dfrac{\dfrac{a + b}{a - b} + \dfrac{a - b}{a + b}}{\dfrac{a - b}{a + b} - \dfrac{a + b}{a - b}} \cdot \dfrac{ab^3 - a^3b}{a^2 + b^2}$

***62.** $\dfrac{\dfrac{1}{a} - \dfrac{a - x}{a^2 + x^2}}{\dfrac{1}{x} - \dfrac{x - a}{x^2 + a^2}} + \dfrac{\dfrac{1}{a} - \dfrac{a + x}{a^2 + x^2}}{\dfrac{1}{x} - \dfrac{x + a}{x^2 + a^2}}$

***63.** Simplify: $\dfrac{\left(a + \dfrac{1}{b}\right)^a \left(a - \dfrac{1}{b}\right)^b}{\left(b + \dfrac{1}{a}\right)^a \left(b - \dfrac{1}{a}\right)^b}$

64. Simplify: $\dfrac{\sqrt{x + a}}{\sqrt{x - a}} - \dfrac{\sqrt{x - a}}{\sqrt{x + a}}$

*65. Simplify: $\dfrac{x^2 - qr}{(p - q)(p - r)} + \dfrac{x^2 - rp}{(q - r)(q - p)} + \dfrac{x^2 - pq}{(r - p)(r - q)}$

*66. Simplify: $\left(\dfrac{x^{-2} + y^{-2}}{x^{-2} - y^{-2}} - \dfrac{x^{-2} - y^{-2}}{x^{-2} + y^{-2}}\right) \div \left[\left(\dfrac{x + y}{x - y} + \dfrac{x - y}{x + y}\right)\left(\dfrac{x^2}{y^2} + \dfrac{y^2}{x^2} - 2\right)\right]^{-1}$

*67. Consider the three fractions $\dfrac{b - c}{1 + bc}$, $\dfrac{c - a}{1 + ca}$, and $\dfrac{a - b}{1 + ab}$.

(a) If $a = 1$, $b = 2$, and $c = 3$, find the sum of the three fractions. Also compute their product. What do you observe?

(b) Show that the sum and the product of the three given fractions are, in fact, always equal.

*68. Evaluate the expression

$$\frac{y - a}{x - a} - \frac{y + a}{x + a} + \frac{y - b}{x - b} - \frac{y + b}{x + b}$$

when $x = \sqrt{ab}$. *Hint:* keep the first and second terms separate from the third and fourth as long as possible.

*69. Show that $(1 + a^{x-y})^{-1} + (1 + a^{y-x})^{-1} = 1$.

*70. Simplify:

$$\frac{x^3y^{-3} - y^3x^{-3}}{(xy^{-1} - yx^{-1})(xy^{-1} + yx^{-1} - 1)} \times \frac{y^{-1} - x^{-1}}{x^{-2} + y^{-2} + x^{-1}y^{-1}}$$

1.10 The Complex Number System

When we solve equations in Chapter 2, you'll see instances in which the real number system proves to be inadequate. In particular, since the square of a real number is never negative, there is no real number x such that $x^2 = -1$. To overcome this inconvenience, mathematicians define the symbol i by the equation

$$i^2 = -1$$

For reasons that are more historical than mathematical, i is referred to as the *imaginary unit.* This name is unfortunate in a sense, because to an engineer or a mathematician i is neither less "real" nor less tangible than any real number. Having said this, however, we do have to admit that i does not belong to the real number system.

Algebraically, we operate with the symbol i as if it were any letter in a polynomial expression. However, when we see i^2, we must remember to replace it by -1. Here are four sample calculations involving i.

1. $3i + 2i = 5i$
2. $-2i^2 + 6i = -2(-1) + 6i = 2 + 6i$
3. $(-i)^2 = i^2 = -1$
4. $0i = 0$

An expression of the form $a + bi$, where a and b are real numbers, is called a *complex number.** Thus, two examples of complex numbers are $2 + 3i$ and $1 - \sqrt{2}\,i$. As a notational convenience, this latter complex number is sometimes written as $1 - i\sqrt{2}$. Given a complex number $a + bi$, we say that a is the *real part* of $a + bi$ and b is the *imaginary part* of $a + bi$. So, for example, the real part of $3 - 4i$ is 3, and the imaginary part of $3 - 4i$ is -4. Observe that the real part and the imaginary part of a complex number are themselves real numbers.

*The phrase *complex number* is attributed to Carl Friedrich Gauss (1777–1855), as is the use of i to denote $\sqrt{-1}$. The term *imaginary number* originated with René Descartes (1596–1650).

We define the notion of *equality* for complex numbers in terms of their real and imaginary parts. Two complex numbers are said to be *equal* if their corresponding real and imaginary parts are equal. We can write this definition symbolically as follows:

$$a + bi = c + di \quad \text{if and only if } a = c \quad \text{and} \quad b = d$$

Example 1 Determine real numbers c and d such that $10 + 4i = 2c + di$.

Solution Equating the real parts of the two complex numbers gives us $2c = 10$, and therefore $c = 5$. Similarly, equating the imaginary parts yields $d = 4$. These are the required values for c and d.

Next we are ready to define addition, subtraction, multiplication, and division for complex numbers. Now, technically speaking, any real number a can be viewed as a complex number, since $a = a + 0i$. Because of this, the definitions we are about to give need to be consistent with the usual rules for the arithmetic of real numbers. We shall return to this point several times later in this section. The definitions for addition and subtraction are quite straightforward. To add or subtract two complex numbers, just add or subtract the corresponding real and imaginary parts. We express this symbolically as follows:

$$(a + bi) + (c + di) = (a + c) + (b + d)i$$
$$(a + bi) - (c + di) = (a - c) + (b - d)i$$

Example 2 Simplify:

(a) $(6 - 2i) + (5 + 4i)$
(b) $(1 + i) - (1 - i)$

Solution

(a) $(6 - 2i) + (5 + 4i) = (6 + 5) + (-2 + 4)i = 11 + 2i$
(b) $(1 + i) - (1 - i) = (1 - 1) + [1 - (-1)]i = 0 + 2i = 2i$

The complex number $0 + 0i$, denoted simply by 0, plays the same role in the addition of complex numbers as it does in ordinary addition of real numbers. That is, $0 + z = z + 0 = z$, for any complex number z. To verify this last statement, let $z = a + bi$. Then we have

$$\begin{aligned} 0 + z &= (0 + 0i) + (a + bi) \\ &= (0 + a) + (0 + b)i \\ &= a + bi = z \end{aligned}$$

This shows that $0 + z = z$. In exactly the same way, you can check for yourself that $z + 0$ also equals z. This property of zero is not the only point of agreement between addition of real numbers and addition of complex numbers. In effect, all the usual rules of algebra that apply to addition (and subtraction too) carry over to the realm of complex numbers. For instance, as Exercise 52 at the end of this section asks you to verify, for all complex numbers z_1, z_2, and z_3 we have

$$z_1 + (z_2 + z_3) = (z_1 + z_2) + z_3 \qquad \text{associative property of addition}$$

and

$$z_1 + z_2 = z_2 + z_1 \qquad \text{commutative property of addition}$$

On the surface, the definition that we are now about to give for multiplication of complex numbers appears more complicated than that for addition. We define the product $(a + bi)(c + di)$ as follows:

$$(a + bi)(c + di) = (ac - bd) + (ad + bc)i$$

To see the motivation for this definition, let us compute the product $(a + bi)(c + di)$ using the usual rules of algebra along with the fact that $i^2 = -1$. We have

$$\begin{aligned}(a + bi)(c + di) &= ac + adi + bci + bdi^2 \\ &= ac + (ad + bc)i + bd(-1) \\ &= (ac - bd) + (ad + bc)i\end{aligned}$$

As you can see, this result agrees with the definition we've given. Indeed, in practice, this is the way we usually carry out multiplication for complex numbers; it is not necessary to memorize the definition.

Example 3 Multiply:
(a) $(2 + 3i)(1 - 4i)$
(b) $(1 - 4i)(2 + 3i)$

Solution **(a)**
$$\begin{aligned}(2 + 3i)(1 - 4i) &= 2(1) + 2(-4i) + 3i(1) + 3i(-4i) \\ &= 2 - 8i + 3i - 12i^2 = 2 - 5i - 12(-1) \\ &= 14 - 5i\end{aligned}$$

(b)
$$\begin{aligned}(1 - 4i)(2 + 3i) &= 1(2) + 1(3i) - 4i(2) - 4i(3i) \\ &= 2 + 3i - 8i - 12(-1) \\ &= 14 - 5i\end{aligned}$$

In the example just concluded, notice that the two answers are identical; the order of the factors does not affect the final result. As Exercise 53 at the end of this section asks you to verify, this result holds in general for multiplication of complex numbers. In fact, multiplication of complex numbers obeys the same properties as does ordinary multiplication of real numbers. For instance, the following four properties are valid in the complex number system, just as they are valid in the real number system.

$$\begin{aligned}z_1(z_2z_3) &= (z_1z_2)z_3 &&\text{associative property of multiplication} \\ z_1z_2 &= z_2z_1 &&\text{commutative property of multiplication} \\ z_1(z_2 + z_3) &= z_1z_2 + z_1z_3 &&\text{distributive property} \\ 0z &= 0 &&\text{multiplicative property of zero}\end{aligned}$$

As a consequence of these properties, we can use all the usual techniques of algebra in carrying out multiplication of complex numbers.

Before leaving the topic of multiplication for complex numbers, we point out one of the reasons why we do not define this multiplication in some seemingly less complicated manner. Basically, the reason is that we want multiplication of complex numbers to behave as much as possible like ordinary multiplication for real numbers. To illustrate this point, suppose for a moment that we had defined the multiplication this way:

$$(a + bi)(c + di) = ac + (bd)i$$

Then we would have, for example, $(2 + 0i)(0 - 3i) = 0 + 0i = 0$. But this result is contrary to our expectation or desire that the product of two nonzero numbers should be nonzero, as it is in the case for real numbers. On the other hand, it can be shown that when we define multiplication for complex numbers by the formula $(a + bi)(c + di) = (ac - bd) + (ad + bc)i$, then the product of two complex numbers is nonzero if and only if both of the factors are nonzero.

Given two complex numbers $a + bi$ and $c + di$, with $c + di \neq 0$, we define the quotient $\dfrac{a + bi}{c + di}$ as follows:

$$\frac{a + bi}{c + di} = \left(\frac{ac + bd}{c^2 + d^2}\right) + \left(\frac{bc - ad}{c^2 + d^2}\right)i$$

In order to show you the motivation for this definition, let us use the usual rules of algebra to multiply the quantity $\dfrac{a + bi}{c + di}$ by 1; however, we'll write 1 in the form $\dfrac{c - di}{c - di}$. We have

$$\begin{aligned}\frac{a + bi}{c + di} &= \frac{a + bi}{c + di} \cdot \frac{c - di}{c - di}\\ &= \frac{ac - adi + bci - bdi^2}{c^2 - d^2i^2}\\ &= \frac{(ac + bd) + (bc - ad)i}{c^2 + d^2}\\ &= \left(\frac{ac + bd}{c^2 + d^2}\right) + \left(\frac{bc - ad}{c^2 + d^2}\right)i\end{aligned}$$

As you can see, this result agrees with the initial definition. In fact, in practice, this is the method by which we divide complex numbers. As with multiplication, you need not memorize the definition. In summary then, to compute the quotient $\dfrac{a + bi}{c + di}$, multiply by $\dfrac{c - di}{c - di}$. The result can then be written in the required form $A + Bi$.

Example 4 Express each quotient in the form $a + bi$:

(a) $\dfrac{1}{2 + 3i}$

(b) $\dfrac{5 + 2i}{1 - 3i}$

Solution **(a)** The denominator is $2 + 3i$, so we multiply the given fraction by

$\frac{2 - 3i}{2 - 3i}$. This yields

$$\frac{1}{2 + 3i} = \frac{1}{2 + 3i} \cdot \frac{2 - 3i}{2 - 3i}$$

$$= \frac{2 - 3i}{4 - 9i^2} = \frac{2 - 3i}{4 + 9} = \frac{2 - 3i}{13}$$

$$= \frac{2}{13} - \frac{3}{13}i \quad \text{as required}$$

(b) The denominator is $1 - 3i$, so we multiply the given fraction by $\frac{1 + 3i}{1 + 3i}$. This yields

$$\frac{5 + 2i}{1 - 3i} = \frac{5 + 2i}{1 - 3i} \cdot \frac{1 + 3i}{1 + 3i}$$

$$= \frac{5 + 15i + 2i + 6i^2}{1 - 9i^2} = \frac{5 + 17i - 6}{1 + 9}$$

$$= \frac{-1 + 17i}{10} = \frac{-1}{10} + \frac{17}{10}i \quad \text{as required}$$

Given a complex number $c + di$, we define its *complex conjugate*, or simply *conjugate*, to be the complex number $c - di$. So, for example, the conjugate of $5 + 7i$ is $5 - 7i$, and the conjugate of $1 - i$ is $1 + i$. Using this terminology, we can summarize the method used in Example 4. To compute the quotient of two complex numbers, multiply both the numerator and the denominator by the conjugate of the denominator. If z is a complex number, the conjugate of z is denoted by $\bar{z}$. For reference, we summarize the definition and notation for conjugates as follows:

If $z = c + di$, then $\bar{z} = c - di$.

In the following box we list five of the most basic properties of conjugates.

Property Summary

Properties of Conjugates

1. $\bar{\bar{z}} = z$
2. $z = \bar{z}$ if and only if z is a real number
3. $\overline{z_1 + z_2} = \bar{z_1} + \bar{z_2}$
4. $\overline{z_1 z_2} = \bar{z_1}\,\bar{z_2}$
5. $(\bar{z})^n = \overline{(z^n)}$ for each natural number n

Each of these properties can be proved using the definition of $\bar{z}$. (The proof of the last property, however, requires the technique of mathematical induction, which is discussed in Chapter 11.

As an example of how the proofs of these properties proceed, let us prove the first property, $\bar{\bar{z}} = z$. To do this, let $z = a + bi$. Then we have

$$\bar{z} = a - bi = a + (-b)i$$

and therefore

$$\bar{\bar{z}} = a - (-b)i = a + bi = z$$

That is, $\bar{\bar{z}} = z$, as we wished to show.

The *absolute value* of a complex number $a + bi$ is defined by the equation

$$|a + bi| = \sqrt{a^2 + b^2}$$

By defining the absolute value in this fashion, we shall find that our results are consistent with the work we did in Section 1.3 dealing with the absolute value of a real number. For instance, if we write the real number -5 as $-5 + 0i$, then our new definition for absolute value yields

$$|-5| = |-5 + 0i| = \sqrt{(-5)^2 + 0^2} = \sqrt{25} = 5$$

Thus, the old and new definitions agree with each other.

Example 5 **(a)** Compute $|5 - 4i|$.

(b) Compute $|\bar{z}|$ when $z = 1 - 3i$.

Solution **(a)** $|5 - 4i| = \sqrt{5^2 + (-4)^2} = \sqrt{41}$

(b) Since $z = 1 - 3i$, we have $\bar{z} = 1 + 3i$. Thus

$$|\bar{z}| = |1 + 3i|$$
$$= \sqrt{1 + 3^2} = \sqrt{10} \qquad \text{as required}$$

We began this section by defining i by the equation $i^2 = -1$. This can be rewritten

$$i = \sqrt{-1}$$

provided that we agree to certain conventions regarding principal square roots and negative numbers. In dealing with the principal square root of a negative real number, say $\sqrt{-5}$, we shall write

$$\sqrt{-5} = \sqrt{(-1)(5)} = \sqrt{-1}\sqrt{5} = i\sqrt{5}$$

In other words, we are allowing the use of the rule $\sqrt{ab} = \sqrt{a}\sqrt{b}$ when a is -1 and b is a positive real number. However, the rule $\sqrt{ab} = \sqrt{a}\sqrt{b}$ *cannot* be used when both a and b are negative. If that were allowed, we could write

$$1 = (-1)(-1)$$

Thus

$$\sqrt{1} = \sqrt{(-1)(-1)} = \sqrt{-1}\sqrt{-1}$$

and consequently

$$1 = i^2 = -1$$

or

$$1 = -1 \qquad \text{a contradiction}$$

Again, the point here is that the rule $\sqrt{ab} = \sqrt{a}\sqrt{b}$ cannot be applied when both a and b are negative.

Example 6 Simplify:
(a) i^4 **(b)** i^{101} **(c)** $\sqrt{-12} + \sqrt{-27}$ **(d)** $\sqrt{-9}\sqrt{-4}$

Solution **(a)** We make use of the defining equation for i, which is $i^2 = -1$. Thus we have

$$i^4 = (i^2)^2 = (-1)^2 = 1$$

(The result, $i^4 = 1$, is worth remembering.)

(b) $i^{101} = i^{100}i = (i^4)^{25}i = 1^{25}i = i$

(c)
$$\begin{aligned}\sqrt{-12} + \sqrt{-27} &= \sqrt{12}\,\sqrt{-1} + \sqrt{27}\,\sqrt{-1}\\ &= \sqrt{4}\,\sqrt{3}\,i + \sqrt{9}\,\sqrt{3}\,i\\ &= 2\sqrt{3}\,i + 3\sqrt{3}\,i = 5\sqrt{3}\,i\end{aligned}$$

(d) $\sqrt{-9}\,\sqrt{-4} = (3i)(2i) = 6i^2 = -6$

Note: $\sqrt{-9}\,\sqrt{-4} \neq \sqrt{36}$; why?

Exercise Set 1.10

1. Complete the table:

i^2	i^3	i^4	i^5	i^6	i^7	i^8
-1	$-i$					

2. Specify the real part of each of the following complex numbers.
   (a) $6 - 10i$ **(b)** $\sqrt{2} + i$
   (c) $\frac{3}{2} - i$ **(d)** $16i$
3. Specify the imaginary part of each of the following complex numbers.
   (a) $3 - 6i$ **(b)** $1 + i\sqrt{2}$
   (c) $5i$ **(d)** 0
4. Determine the real numbers c and d such that
$$8 - 3i = 2c + di$$
5. Determine the real numbers a and b such that
$$27 - 64i = a^3 - b^3i$$
6. Simplify $(5 - 6i) + (3 + 9i)$.
7. Simplify $(2 + 7i) - (12 - 6i)$.
8. If $z = 1 + 4i$, compute $z - 10i$.
9. If $z = 2 - 3i$, compute $7 + 8i - 4z$.
10. Multiply $(3 - 5i)(4 + 6i)$.
11. Multiply $(1 - i\sqrt{3})(1 + i\sqrt{3})$.

For Exercises 12–31, evaluate each of the expressions using the values $z_1 = 2 + 3i$, $z_2 = 2 - 3i$, and $w = 4 + 9i$.

12. $(z_1 + z_2) + w$
13. $z_1 + (z_2 + w)$
14. z_1w
15. wz_1
16. $z_1(z_2w)$
17. $(z_1z_2)w$
18. $w(z_1 + z_2)$
19. $wz_1 + wz_2$
20. $z_1^2 - z_2^2$
21. $(z_1 - z_2)(z_1 + z_2)$
22. $(z_1w)^2$
23. $z_1^2w^2$
24. z_1^3
25. z_1^4
26. $\dfrac{z_1}{z_2}$
27. $\dfrac{z_2}{z_1}$
28. $\dfrac{z_1}{w}$
29. $\dfrac{w}{z_1}$
30. $\dfrac{z_1 + z_2}{w}$
31. $\dfrac{1}{z_1w}$

In Exercises 32–36, express each quotient in the form a + bi.

32. $\dfrac{1}{5+i}$

33. $\dfrac{3-2i}{4+5i}$

34. $\dfrac{1-i\sqrt{3}}{1+i\sqrt{3}}$

35. $\dfrac{1}{i}$

36. $\dfrac{i+i^2}{i^3+i^4}$

37. Let $z = 3 + 5i$. Compute $\bar{z}$ and $\bar{\bar{z}}$.

38. Let $z = 1 - 6i$ and $w = 5 + 7i$. Compute $\bar{z} + \bar{w}$ as well as $\overline{z+w}$. Are your answers the same?

39. Let $z = 4 + 10i$ and $w = 6 - 2i$. Compute $\bar{z}\bar{w}$ as well as $\overline{zw}$. Are your answers the same?

40. Let $z = \sqrt{3} + i$. Compute $(\overline{z^2})$ and $(\bar{z})^2$. Are your answers the same?

41. Compute the absolute value of each of the following complex numbers.
(a) $2 + 4i$ **(b)** $5 - 3i$
(c) $(2 + 4i)(5 - 3i)$

42. Suppose that $z = 7 + 8i$ and $w = 9 - 6i$. Which quantity is larger, $|z|$ or $|w|$?

Simplify each expression in Exercises 43–46.

43. $\sqrt{-9} + \sqrt{-25} - \sqrt{-81}$

44. $\sqrt{-20} + 3\sqrt{-45} - \sqrt{-80}$

45. $4\sqrt{-18} - 5\sqrt{-128}$

46. $\sqrt{-4}\sqrt{-4}$

47. Verify that $\left(\dfrac{1}{\sqrt{2}} + \dfrac{1}{\sqrt{2}}i\right)^2 = i$.

48. Let $z = \dfrac{-1+i\sqrt{3}}{2}$ and $w = \dfrac{-1-i\sqrt{3}}{2}$. Verify the following statements.
(a) $z^3 = 1$ and $w^3 = 1$ **(b)** $zw = 1$
(c) $z = w^2$ and $w = z^2$
(d) $(1 - z + z^2)(1 + z - z^2) = 4$

***49.** Let a and b be real numbers. Find the real and imaginary parts of the quantity

$$\frac{a+bi}{a-bi} + \frac{a-bi}{a+bi}$$

***50.** Find the real and imaginary parts of the quantity

$$\left(\frac{a+bi}{a-bi}\right)^2 - \left(\frac{a-bi}{a+bi}\right)^2$$

51. Show that $\left(\dfrac{-1+i\sqrt{3}}{2}\right)^2 + \left(\dfrac{-1-i\sqrt{3}}{2}\right)^2 = -1$.

52. Let z_1, z_2, and z_3 be complex numbers.
(a) Prove that $z_1 + (z_2 + z_3) = (z_1 + z_2) + z_3$. *Hint:* Let $z_1 = x_1 + iy_1$, $z_2 = x_2 + iy_2$, and $z_3 = x_3 + iy_3$.
(b) Prove that $z_1 + z_2 = z_2 + z_1$.

53. Let z_1, z_2, and z_3 be complex numbers. Prove the following statements. (See the hint for Exercise 52.)
(a) $z_1(z_2z_3) = (z_1z_2)z_3$ **(b)** $z_1z_2 = z_2z_1$
(c) $z_1(z_2 + z_3) = z_1z_2 + z_1z_3$
(d) $0z_1 = 0$ and $1z_1 = z_1$
Suggestion: Write 0 as $0 + 0i$; write 1 as $1 + 0i$.

54. Suppose that $z = \bar{z}$. Show that z is a real number.

55. Prove properties 3 and 4 in the box on page 61.

56. **(a)** Prove that $|z|^2 = z\bar{z}$.
(b) Supply reasons for each step in the following proof that $|zw| = |z|\,|w|$:

$$|zw|^2 = (zw)(\overline{zw}) = (zw)(\bar{z}\bar{w}) = z\bar{z}w\bar{w} = |z|^2\,|w|^2$$

Thus $|zw| = |z|\,|w|$.

Chapter 1 Review Checklist

- □ Coefficient, base, exponent (page 2)
- □ Order of operations (page 2)
- □ Commutative, associative, identity, inverse, and distributive properties for the real numbers (page 6)
- □ Properties of zero, of negatives, and of fractions (pages 8–9)
- □ Natural numbers, integers, rational numbers, and irrational numbers (pages 12–13)
- □ Number line and coordinates (pages 13–14)
- □ $a < b$; $a \le b$; $b > a$; $b \ge a$ (page 13)
- □ Intervals on the number line (page 15)
- □ Absolute value:
  geometric definition (page 15)
  algebraic definition (page 15)
  properties (page 16)
- □ Properties of exponents (page 19)
- □ a^0 and a^{-n} (page 20)
- □ Scientific notation (page 22)
- □ Square roots, principal square roots, and their properties (page 25)
- □ Rationalizing the denominator (pages 27, 33–34)
- □ Pythagorean Theorem (page 28)

- □ $\sqrt[n]{b}$ and properties of nth roots (pages 31–32)
- □ $b^{1/n}$ and $b^{m/n}$ (page 34)
- □ Constant, variable, domain, and the domain convention (page 38)
- □ Polynomial: terms, coefficients, and degree (page 39)
- □ Rules for sums and differences of polynomials (page 40)
- □ Special Products (page 41)
- □ Long division of polynomials (pages 42–43)
- □ Five techniques for factoring (page 46)
- □ Nested form of a polynomial (pages 49–50)
- □ $i^2 = -1$ (page 57)
- □ Complex number and its real and imaginary parts (page 57)
- □ Addition, subtraction, multiplication, and division of complex numbers (pages 58–60)
- □ Complex conjugate (page 61)
- □ Absolute value of a complex number (page 62)
- □ $i = \sqrt{-1}$ and square root computations involving i (page 62)

Chapter 1 Review Exercises

Factor each expression in Exercises 1–20.

1. $a^2 - 16b^2$
2. $x^2 + ax + yx + ay$
3. $8 - (a+1)^3$
4. $x^2 - 18x + 81$
5. $a^2x^3 + 2ax^2b + b^2x$
6. $8a^2x^2 + 16a^3$
7. $8x^2 + 6x + 1$
8. $12x^2 - 2x - 4$
9. $a^4x^4 - x^8a^8$
10. $2x^2 - 2bx + ax - ab$
11. $8 + 12a + 6a^2 + a^3$
12. $(x^2 + 2x - 8)^2 - (2x+1)^2$
13. $4x^2y^2z^3 - 3xyz^3 - z^3$
14. $(x + y - 1)^2 - (x - y + 1)^2$
15. $1 - x^6$
16. $12x^3 + 44x^2 - 16x$
17. $a^2x^2 + 2abx + b^2 - 4a^2b^2x^2$
18. $a^2 + 2ab + b^2 + a + b$
19. $a^2 - b^2 + ac - bc + a^2b - b^2a$
20. $5(a+1)^2 + 29(a+1) - 144$

Simplify each expression in Exercises 21–36. (Assume that the letters represent positive quantities.)

21. $4^{3/2}$
22. $49^{-1/2}$
23. $[(3025)^{1/2}]^0$
24. $(-125)^{2/3}$
25. $8^{-4/3}$
26. $10^{-3} + 10^3$
27. $(-243)^{-2/5}$
28. $\left(\frac{625}{16}\right)^{3/4}$
29. $(a^2b^6c^8)^{1/2}$
30. $\sqrt{25a^4b^{10}}$
31. $\sqrt{a^3b^5}\sqrt{4ab^3}$
32. $\sqrt{28a^3} + a\sqrt{63a}$
33. $\sqrt[3]{16} - \sqrt[3]{-54}$
34. $\sqrt[5]{\dfrac{-32a^{15}b^{10}}{c^5}}$
35. $\sqrt{24a^2b^3} + ba\sqrt{54b}$
36. $\sqrt{\sqrt{256x^8}}$

In Exercises 37–40, write the numbers using scientific notation.

37. 0.0014
38. 186,000
39. 12.001
40. 81×10^{12}

For Exercises 41–50, carry out the indicated operations and simplify where possible.

41. $\dfrac{x+2}{x-1} - \dfrac{x^2}{(x-1)^2}$
42. $\dfrac{3}{x} + \dfrac{1}{x+1} - \dfrac{x}{x^2-1}$
43. $\dfrac{4}{x-1} - \dfrac{3x+1}{x^2-x}$
44. $\dfrac{5}{x-1} + \dfrac{1}{x^2+2x-3}$
45. $\dfrac{x}{1-x} + \dfrac{1}{x-1} \div (x+2)$
46. $\dfrac{2}{x^2+x+1} - \dfrac{x+1}{x^3+x^2+x}$
47. $\dfrac{\dfrac{a}{x^2-a^2}+1}{\dfrac{x^2-a^2}{a}-1}$
48. $\dfrac{\dfrac{1}{x+1}-\dfrac{1}{x+2}}{\dfrac{1}{(x+1)^2}-\dfrac{1}{(x+2)^2}}$
49. $\dfrac{\dfrac{1}{x}-\dfrac{1}{x^2}+1}{\dfrac{1}{x}+\dfrac{1}{x^2}-1}$
50. $\dfrac{1-\dfrac{2}{x}+\dfrac{1}{x^2}}{1-\dfrac{3}{x}+\dfrac{2}{x^2}}$

Rationalize the denominators in Exercises 51–55.

51. $\dfrac{4}{\sqrt{2}}$
52. $\dfrac{\sqrt{6}}{\sqrt{7}}$
53. $\dfrac{1}{\sqrt{5}-\sqrt{6}}$
54. $\dfrac{1}{\sqrt{x+h}-\sqrt{x}}$
55. $\dfrac{\sqrt{a^2+x^2}+\sqrt{a^2-x^2}}{\sqrt{a^2+x^2}-\sqrt{a^2-x^2}}$

Evaluate each expression in Exercises 56–62, using the values a = -1, b = 2, c = -3, *and* x = 4.

56. $|a - b|$
57. $|a| - |c| + |a + c|$
58. $-a^b$
59. $|b^c + a|$
60. $x^2 - 8x + 1$
61. $|x^2 - 8x + 1|$
62. $x^2 - |8x + 1|$

In Exercises 63–70, carry out the indicated operations.

63. $1 - 2[3 - 4(1 - 5)]$ **64.** $x^2 - 4x - (1 - 2x^2)$

65. $1 + 3(x^2 - 5x - 4) - [1 - (15x - 3x^2)]$

66. $(ax^2 - 1)(ax^2 + 1)$

67. $(x - 1)(x + 1)(x + 2)$

68. $(x + a)(x^2 + ax + 1)$

69. $x^2 + 4 - (x - 1)(x - 2)$

70. $(x^2 - x + 1)(x^2 + x - 1) + [(x - 1)^2 - x^4]$

For Exercises 71–74, use long division to determine the quotient and remainder.

71. $(x^3 - 2x^2 + x + 4) \div (x - 1)$

72. $(x^3 + 4x^2 - 2x - 5) \div (x + 1)$

73. $(x^4 - x^2 + 1) \div (x^2 + 1)$

74. $(2x^4 + x^3 - 4x^2 - 8x - 1) \div (x + 2)$

Carry out the indicated operations in Exercises 75–88.

75. $(3 - 6i)(1 + 5i)$ **76.** $2i(1 + i) + i^2$

77. $4 - 2i + (1 - i)(1 - 2i)$

78. $(1 + i\sqrt{2})(1 - i\sqrt{2})$

79. $\dfrac{2 + 3i}{1 + i}$ **80.** $\dfrac{1}{3 - i}$

81. $\dfrac{3 - i\sqrt{3}}{3 + i\sqrt{3}}$ **82.** $|4 + 3i|$

83. $\left|\dfrac{1 + i}{1 - i}\right|$ **84.** $|(2 + 3i)^2|$

85. $(1 + i + i^2 + |3 - 4i|) \div i$

86. $\sqrt{-2}\,\sqrt{-9}$ **87.** $\sqrt{-8} - \sqrt{-72}$

88. $\dfrac{\sqrt{-4} - \sqrt{-3}\,\sqrt{-3}}{\sqrt{-100}}$

In Exercises 89–95, evaluate the expressions when z = 4 + i and w = 1 − i.

89. $\bar{z} + \bar{w}$ **90.** $\frac{1}{2}(w + \bar{w})$

91. $z\bar{w} + \bar{z}w$ **92.** $|z + w| - |z - w|$

93. $z - \bar{z} - |w|$ **94.** $z^2 - 8z + 17$

95. $\dfrac{z + 1}{z - 1}$

In Exercises 96–100, identify the properties of the real numbers that justify each statement. (The properties are listed in Table 1 in Section 1.2.)

96. $(x + y)^2 + z^2 = z^2 + (x + y)^2$

97. $(x + 1)(x + 2) = (x + 1)x + (x + 1)2 = (x^2 + x) + (2x + 2)$

98. $(ab)c = a(bc) = a(cb) = (ac)b$

99. $(x + a)y + (x + a)z = (x + a)(y + z)$

100. Since 3 and $\sqrt{\pi}$ are real numbers, $3\sqrt{\pi}$ is also a real number.

In Exercises 101–105, determine the length of the third side of the right triangle. (The letters a and b always denote the two shorter sides and c denotes the hypotenuse.)

101. $a = 10, b = 24$ **102.** $a = 7, c = \sqrt{74}$

103. $a = \sqrt{5} + 1, b = \sqrt{5} - 1$

104. $a = \sqrt{3}, c = 1 + \sqrt{3}$

105. $a = x^2 - 1, b = 2x$

For Exercises 106–110, rewrite the statements using absolute values and inequalities or equalities.

106. The distance between x and 6 is 2.

107. The distance between x and a is less than $\frac{1}{2}$.

108. The distance between a and b is 3.

109. The distance between x and -1 is 5.

110. The distance between x and 0 exceeds 10.

Rewrite each of the expressions in Exercises 111–115 in a form not containing absolute values.

111. $|\sqrt{6} - 2|$ **112.** $|1 - \sqrt{17}|$

113. $|x^4 + x^2 + 1|$ **114.** $|x - 3|$ where $x < 3$

115. $|x - 2| + |x - 3|$ where $2 < x < 3$

Determine whether the following are true or false:

116. **(a)** $-1 \leq 0$ **(b)** $\sqrt{\dfrac{3}{2}} \geq 0$
(c) $\frac{11}{10} \geq \frac{12}{11}$ **(d)** $0.\overline{9} \geq 1$

In Exercises 117–120, sketch each interval on a number line.

117. $(3, 5)$ **118.** $(3, 5]$

119. The set of all negative real numbers.

120. The set of all real numbers greater than or equal to 2.

Use the Special Products given in Section 1.7 to carry out the multiplication in Exercises 121–130.

121. $(9x - y)(9x + y)$ **122.** $(3x + 1)^2$

123. $(2ab - 3)^2$ **124.** $(3x^2 + y^2)^2$

125. $(ax + b)^2(ax - b)^2$ **126.** $(2a + 3)^3$

127. $(x^2 - 2)^3$

128. $(1 - 3a)(1 + 3a + 9a^2)$

129. $(x^n + y^n)(x^{2n} - x^n y^n + y^{2n})$

130. $(x^2 - 3x + 1)^2$

Chapter 1 Test

1. Factor $x - 81x^3$.

Carry out the indicated operations in Problems 2–5.

2. $\dfrac{1}{x-1} - \dfrac{x+1}{x^3-1}$

3. $\dfrac{x+2}{x^2+9x+20} - \dfrac{x-1}{x^2+8x+16}$

4. $\dfrac{a^3-x^3}{a^2-x^2} \div \dfrac{xa^2+x^2a+x^3}{a^2+2ax+x^2}$

5. $(x+5)(x+3)^2$

6. Rewrite using absolute values and inequality: The distance between x and 3 is less than 5.

7. Determine the hypotenuse of a right triangle in which the lengths of the two shorter sides are $\sqrt{3}-1$ and $\sqrt{3}+1$ units, respectively.

8. Express the repeating decimal $5.\overline{7}$ in the form p/q, where p and q are integers.

9. Simplify: $\sqrt[3]{16} - \sqrt[3]{-250} + \sqrt[3]{-2}$

10. Simplify: $\left(\dfrac{a^{-3}b^4c^5}{b^0ca^{-1}}\right)^{-3/2}$

11. Rationalize the denominators:

(a) $\dfrac{\sqrt{2}}{\sqrt{5}}$ **(b)** $\dfrac{1}{1+\sqrt{2}}$

12. Use long division to find the quotient and remainder when $x^3 + 8x - 1$ is divided by $x + 5$.

13. Rewrite using scientific notation: 0.000014

14. Simplify: $\dfrac{x^{-1}+y^{-1}}{x^{-2}-y^{-2}}$

15. Factor:
$(x^2+a^2)(x^3+3b^3+1) - (x^2+a^2)(2b^3+1)$

16. Indicate which property of the real numbers justifies each statement.

(a) $3xy^2 + 4x^2y^2 = (3x + 4x^2)y^2$

(b) $(a+b)(a-b) = (a-b)(a+b)$

(c) $\dfrac{1}{\sqrt{2}+1}(\sqrt{2}+1) = 1$

17. Evaluate each expression.

(a) $8^{2/3} + 4^0$ **(b)** $\left(\frac{4}{9}\right)^{-3/2}$

18. Write the expression $i^3 - i^4 + 1/i$ in the form $a + bi$, where a and b are real numbers.

For Problems 19 and 20, evaluate each expression using the values $z = 2 + 3i$ and $w = 1 - 4i$. (Write each answer in the form $a + bi$.)

19. **(a)** $z + w$ **(b)** zw **(c)** $z^2 + w^2$

20. **(a)** $\dfrac{w}{z}$ **(b)** $\bar{z} + |w|$

Chapter 2 Equations and Inequalities

Introduction

The first two sections of this chapter present the basic techniques for solving linear equations and quadratic equations. Those techniques are then used in Section 2.3 to solve a variety of applied problems. In Section 2.4, several methods are developed for solving equations that are neither linear nor quadratic. The remaining two sections of the chapter, Sections 2.5 and 2.6, deal with the solution of inequalities rather than equalities.

2.1 Linear Equations

In this section and the next we review the terminology and skills used in solving certain basic kinds of equations. Perhaps the simplest type of equation is the *linear* or *first-degree* equation in one variable. This is an equation that can be written in the form

$$ax + b = c \qquad \text{with } a, b, \text{ and } c \text{ real numbers and } a \neq 0$$

As examples of linear equations in one variable, we cite

$$2x = 10 \qquad 3m + 1 = 2 \qquad \frac{y}{2} = \frac{y}{3} + 1$$

As with any equation involving a variable, each of these equations is neither true nor false *until* we replace the variable with a number. By a *solution* or a *root* of an equation in one variable, we mean a value for x (or m or y in the foregoing examples) that makes the equation a true statement. For example, the value $x = 5$ is a solution of the equation $2x = 10$ since, with $x = 5$, the equation becomes $2(5) = 10$, which is certainly true. We also say in this case that the value $x = 5$ *satisfies* the equation.

Equations that become true statements for all values in the domain of the variable are called *identities.* Two examples of identities are

$$x^2 - 9 = (x - 3)(x + 3) \quad \text{and} \quad \frac{4x^2}{x} = 4x$$

In contrast to this, a *conditional equation* is true only for some (or perhaps even none) of the values of the variable. Two examples of conditional equations are $2x = 10$ and $x = x + 1$. The first of these is true only when

$x = 5$. The second equation has no solution (because, intuitively at least, no number can be one more than itself).

We say that two equations are *equivalent* when they have exactly the same solutions. In this section the basic method for solving an equation in one variable involves writing a sequence of equivalent equations until we finally reach an equation of the form

$$\text{variable} = \text{a number}$$

In generating equivalent equations, we rely on the following three principles. (These can be justified using the properties of real numbers discussed in Section 1.2.)

1. Adding or subtracting the same quantity on both sides of an equation produces an equivalent equation.
2. Multiplying or dividing both sides of an equation by the same nonzero quantity produces an equivalent equation.
3. Simplifying an expression on either side of an equation (using the techniques from Chapter 1) produces an equivalent equation.

The examples that follow show how these principles are applied in solving various equations.

Example 1 **(a)** Solve $4x + 5 = 33$.
(b) Solve $ax + b = c$ (a, b, and c are constants and $a \neq 0$).

Solution **(a)**

$$\begin{aligned} 4x + 5 &= 33 \\ 4x &= 28 && \text{subtracting 5 from both sides} \\ x &= 7 && \text{dividing both sides by 4} \end{aligned}$$

Check Replacing x with 7 in the original equation yields

$$\begin{aligned} 4(7) + 5 &= 33 \\ 28 + 5 &= 33 \qquad \text{True} \end{aligned}$$

(b)

$$\begin{aligned} ax + b &= c \\ ax &= c - b && \text{subtracting } b \text{ from both sides} \\ x &= \frac{c - b}{a} && \text{dividing both sides by } a \end{aligned}$$

Check Replacing x with $\frac{c - b}{a}$ in the original equation yields

$$\begin{aligned} a\left(\frac{c - b}{a}\right) + b &= c \\ c - b + b &= c \\ c &= c \qquad \text{True} \end{aligned}$$

Example 2 Solve $3[1 - 2(x + 1)] = 2 - x$.

Solution

$$3[1 - 2(x + 1)] = 2 - x$$
$$3[1 - 2x - 2] = 2 - x \quad \text{simplifying the left-hand side}$$
$$3(-1 - 2x) = 2 - x$$
$$-3 - 6x = 2 - x$$
$$-3 - 5x = 2 \quad \text{adding } x \text{ to both sides}$$
$$-5x = 5 \quad \text{adding 3 to both sides}$$
$$x = -1 \quad \text{dividing both sides by } -5$$

Check Replacing x with -1 in the original equation yields

$$3[1 - 2(0)] = 2 - (-1)$$
$$3(1) = 2 + 1 \qquad \text{True}$$

In the next two examples the equations involve fractions. The strategy in such cases is to multiply through by the least common denominator. That eliminates the need to work with fractions.

Example 3 Solve: **(a)** $\frac{x}{2} - \frac{3x}{4} = 1$ **(b)** $\frac{1}{x - 2} - \frac{4}{x + 2} = \frac{1}{x^2 - 4}$

Solution **(a)**

$$\frac{x}{2} - \frac{3x}{4} = 1$$
$$4\left(\frac{x}{2}\right) - 4\left(\frac{3x}{4}\right) = 4(1) \quad \text{multiplying both sides by the least common denominator}$$
$$2x - 3x = 4 \quad \text{simplifying}$$
$$-x = 4 \quad \text{simplifying}$$
$$x = -4 \quad \text{multiplying both sides by } -1$$

Check Replacing x with -4 in the original equation yields

$$\frac{-4}{2} - \frac{3(-4)}{4} = 1$$
$$-2 + 3 = 1 \qquad \text{True}$$

(b) By factoring the denominator on the right-hand side, we obtain

$$\frac{1}{x - 2} - \frac{4}{x + 2} = \frac{1}{(x - 2)(x + 2)}$$

From this we see that the least common denominator for the three fractions is $(x - 2)(x + 2)$. Now, multiplying both sides by this least common denominator, we have

$$\frac{(x - 2)(x + 2)}{x - 2} - \frac{4(x - 2)(x + 2)}{x + 2} = \frac{(x - 2)(x + 2)}{(x - 2)(x + 2)}$$
$$x + 2 - 4(x - 2) = 1 \quad \text{simplifying}$$
$$-3x + 10 = 1 \quad \text{simplifying}$$
$$-3x = -9 \quad \text{subtracting 10 from both sides}$$
$$x = 3 \quad \text{dividing both sides by } -3$$

Check Replacing x with 3 in the original equation yields

$$\frac{1}{3-2} - \frac{4}{3+2} = \frac{1}{(3-2)(3+2)}$$

$$1 - \frac{4}{5} = \frac{1}{5} \qquad \text{True}$$

In part (a) of the example just completed, we multiplied both sides of the original equation by 4 to produce an equivalent equation. In part (b), however, we multiplied both sides of the equation by $(x-2)(x+2)$. Since we didn't know at that point whether the quantity $(x-2)(x+2)$ was nonzero, we could not be certain that the resulting equation was actually an equivalent equation. For this reason, it is always necessary to check any solutions you may have obtained as a result of multiplying or dividing both sides of an equation by an expression involving the variable. Part (a) of the next example serves to emphasize this point.

Example 4 Solve: **(a)** $\dfrac{1}{x+5} = \dfrac{2}{x-3} + \dfrac{2x+2}{(x+5)(x-3)}$

(b) $\dfrac{1}{x} = \dfrac{1}{a} + \dfrac{1}{b}$ (a and b are positive constants)

Solution **(a)** The least common denominator for the fractions is $(x+5)(x-3)$. After multiplying both sides of the given equation by this least common denominator, we have

$$\begin{aligned} x - 3 &= 2(x+5) + 2x + 2 && \\ x - 3 &= 2x + 10 + 2x + 2 && \text{simplifying} \\ x - 3 &= 4x + 12 && \text{simplifying} \\ -3x - 3 &= 12 && \text{subtracting } 4x \text{ from both sides} \\ -3x &= 15 && \text{adding 3 to both sides} \\ x &= -5 && \text{dividing both sides by } -3 \end{aligned}$$

Check The preceding steps show that *if* the equation has a solution, then the solution is $x = -5$. With $x = -5$, however, the left-hand side of the original equation becomes $\dfrac{1}{-5+5}$, or $\dfrac{1}{0}$, which is undefined. We conclude therefore that the given equation has no solution.

(b)

$$\begin{aligned} \frac{1}{x} &= \frac{1}{a} + \frac{1}{b} && \\ (xab)\cdot\frac{1}{x} &= (xab)\cdot\frac{1}{a} + (xab)\cdot\frac{1}{b} && \text{multiplying both sides by the least common denominator, } xab \\ ab &= xb + xa && \text{simplifying} \\ ab &= x(b+a) && \text{factoring} \\ \frac{ab}{b+a} &= x && \text{dividing both sides by } b + a \end{aligned}$$

Check As Exercise 62 asks you to verify, the value $x = \dfrac{ab}{a+b}$ indeed satisfies the original equation.

The techniques used in this section can also be applied in solving equations in which absolute values appear. The next example shows how this is done.

Example 5 Solve $|2x - 1| = 7$.

Solution There are two cases to consider.

If $2x - 1 \geq 0$, the equation becomes

$$2x - 1 = 7$$
$$2x = 8$$
$$x = 4$$

If $2x - 1 < 0$, the equation becomes

$$-(2x - 1) = 7$$
$$-2x + 1 = 7$$
$$-2x = 6$$
$$x = -3$$

We've now obtained the values $x = 4$ and $x = -3$. As you should check for yourself, both of these numbers indeed satisfy the original equation.

Example 6 Solve for x in terms of the other letters: $y = \dfrac{ax + b}{cx + d}$. (Assume that $cx + d \neq 0$ and $yc - a \neq 0$.)

Solution Multiplying both sides of the given equation by the quantity $cx + d$ yields

$$\begin{aligned} y(cx + d) &= ax + b \\ ycx + yd &= ax + b \\ ycx - ax &= b - yd \\ x(yc - a) &= b - yd \\ x &= \frac{b - yd}{yc - a} \end{aligned}$$

You should check for yourself now that the expression for x on the right-hand side of this last equation indeed satisfies the original equation.

Exercise Set 2.1

Solve each equation in Exercises 1–40.

1. $2x - 3 = -5$
2. $x + 4 = 0$
3. $8x - 4 = 20$
4. $-6x + 1 = 49$
5. $1 - y = 12$
6. $3 - 4y = 8y + 3$
7. $2m - 1 + 3m + 5 = 6m - 8$
8. $1 - (2m + 5) = -3m$
9. $(x + 2)(x + 1) = x^2 + 11$
10. $t - \{4 - [t - (4 + t)]\} = 6$
11. $4(y - 1) + 5(y - 3) = 2(y + 1) - 28$
12. $2t - 3[t - 4(t + 5)] = -2 - 20t$
13. $1 - (x - 2)^2 = -x^2 + 5x + 3$
14. $(x + 2)^2 = x^2 + 4$
15. $(x - 1)^2 - (x^2 - 1) = 16$
16. $(x - 1)^2 - (x^2 - 1) = 2 - 2x$
17. $\dfrac{x}{3} + \dfrac{2x}{5} = \dfrac{-11}{5}$
18. $\dfrac{x}{2} + 1 = 0$
19. $1 - \dfrac{y}{3} = 6$
20. $y - \dfrac{y}{2} - \dfrac{y}{3} = 2$
21. $3 - \dfrac{x - 1}{4} = \dfrac{x}{9}$
22. $1 + \dfrac{x}{3} - \dfrac{x}{4} = x - \dfrac{5x}{6}$
23. $\dfrac{x - 1}{4} + \dfrac{2x + 3}{-1} = 0$
24. $\dfrac{x - 1}{2} - \dfrac{x - 2}{4} = 1 + \dfrac{x}{3}$
25. $\dfrac{1}{x - 3} - \dfrac{2}{x + 3} = \dfrac{1}{x^2 - 9}$
26. $\dfrac{1}{x - 5} + \dfrac{1}{x + 5} = \dfrac{6}{x^2 - 25}$

27. $\dfrac{1}{x-5}+\dfrac{1}{x+5}=\dfrac{2x}{x^2-25}$

28. $\dfrac{1}{x-5}+\dfrac{1}{x+5}=\dfrac{2x+1}{x^2-25}$

29. $\dfrac{4}{x+2}+\dfrac{1}{x-2}=\dfrac{4}{x^2-4}$

30. $\dfrac{1}{2x+1}+\dfrac{1}{x+1}=\dfrac{5}{2x^2+3x+1}$

31. $\dfrac{2(x+1)}{x-1}-3=\dfrac{5x-1}{x-1}$

32. $\dfrac{x-1}{x+1}=\dfrac{x+1}{x-1}$

33. $\dfrac{1}{x}=\dfrac{4}{x}-1$

34. $\dfrac{1}{y}+1=\dfrac{3}{y}-\dfrac{1}{2y}$

35. $|x-5|=1$

36. $|x+1|=9$

37. $|6x-5|=25$

38. $|5-6x|=0$

***39.** $\dfrac{|x-3|}{2}+\dfrac{|x-3|}{3}=\dfrac{-5x}{3}$ *Hint:* Consider separately the cases $x \geq 3$ and $x < 3$.

40. $\dfrac{|x|}{3}-\dfrac{2}{|x|}=\dfrac{x^2+x+1}{3|x|}$

For Exercises 41–60, solve each equation for x *in terms of the other letters.*

41. $3ax-2b=b+3$

42. $ax+b=bx-a$

43. $ax+b=bx+a$

44. $\dfrac{x}{a}+\dfrac{x}{b}=1$

45. $\dfrac{1}{x}=a+b$

46. $\dfrac{1}{ax}=\dfrac{1}{bx}-\dfrac{1}{c}$

47. $(x+b)(x+c)-a(a+b)=(x-a)(x+a)+bc$

48. $\dfrac{1}{a}-\dfrac{1}{x}=\dfrac{1}{x}-\dfrac{1}{b}$

49. $(a-x)^2=x^2+b^2$

***50.** $a^2(a-x)=b^2(b+x)-2abx$ (Assume that $a \neq b$.)

51. $\dfrac{b}{ax-1}-\dfrac{a}{bx-1}=0$ (Assume that $a \neq b$.)

52. $\dfrac{a-x}{a-b}-2=\dfrac{c-x}{b-c}$

***53.** $\dfrac{x+2p}{2q-x}+\dfrac{x-2p}{2q+x}-\dfrac{4pq}{4q^2-x^2}=0$

54. $\dfrac{x-a}{x-b}=\dfrac{b-x}{a-x}$

55. $y=mx+b$

56. $y-y_0=m(x-x_0)$

***57.** $1-\dfrac{a}{b}\left(1-\dfrac{a}{x}\right)-\dfrac{b}{a}\left(1-\dfrac{b}{x}\right)=0$

***58.** $\dfrac{q}{x+q}-\dfrac{p}{x+p}=\dfrac{q-p}{x-2pq}$

***59.** $\dfrac{1}{x-p}-\dfrac{1}{x-q}=\dfrac{p-q}{x^2-pq}$

60. $\dfrac{x+3p}{x+p+q}=\dfrac{x+p}{x+q}$

***61.** **(a)** Factor $(a+b)^2+2(a+b)+1$. [This will be needed in simplifying the answer in part (b).]

(b) Solve for x (assume that $a+b \neq -1$):
$$\frac{x-a}{b}+\frac{x-b}{a}+\frac{x-1}{ab}=\frac{2}{a}+\frac{2}{b}+2$$

62. Verify that $x=\dfrac{ab}{a+b}$ is a solution of the equation $\dfrac{1}{x}=\dfrac{1}{a}+\dfrac{1}{b}$, where a and b are positive constants.

In Exercises 63–75, solve each equation for the indicated variable.

63. $Ax+By+C=0$ for y

64. $d=\dfrac{r}{1+rt}$ for r

65. $x=\dfrac{x_1+x_2}{2}$ for x_1

66. $\dfrac{x_1x}{a^2}+\dfrac{y_1y}{b^2}=1$ for y

67. $S=2LW+2LH+2WH$ for W

68. $A=\frac{1}{2}(a+b)h$ for a

69. $S=2\pi r^2+2\pi rh$ for h

70. $s=\dfrac{n}{2}(a+l)$ for a

71. $S=\dfrac{rl-a}{r-1}$ for r

72. $A=P(1+nr)$ for n

73. $s=\frac{1}{2}(a+b+c)$ for c

74. $3y^2y'+5x^2yy'-x^2=0$ for y'

75. $A=\dfrac{b-a}{6}(f_0+4f_1+f_2)$ for f_1

***76.** Solve for x: $\dfrac{x}{x+a-b}+\dfrac{x}{x+b-c}-2=0$

For Exercises 77–81, solve each equation and round off the answers to two decimal places.

C **77.** $0.54x-0.41=3.26$

C **78.** $4.50-9.11x=1.72$

C **79.** $\dfrac{x-0.18}{x+0.23}=\dfrac{x+1.16}{x-2.24}$

C **80.** $1.28-0.40(1-x)=3.81-0.21(1+x)$

C **81.** $|0.13x-0.018|=0.56$

2.2 Quadratic Equations

Any equation that can be written in the form

$$ax^2 + bx + c = 0 \qquad \text{with } a, b, \text{ and } c \text{ real numbers and } a \neq 0$$

is called a *quadratic equation*. Examples of quadratic equations are

$$x^2 - 8x - 9 = 0 \qquad 2y^2 - 5 = 0 \qquad 3m^2 = 1 - m$$

Recall that a *solution* or *root* of an equation is a value for the unknown that makes the equation a true statement. For example, the value $x = -1$ is a solution of the quadratic equation $x^2 - 8x - 9 = 0$, because when x is replaced by -1 in this equation we have

$$\begin{aligned} (-1)^2 - 8(-1) - 9 &= 0 \\ 1 + 8 - 9 &= 0 \\ 0 &= 0 \qquad \text{True} \end{aligned}$$

When applicable, one of the simplest techniques for solving quadratic equations involves factoring. The method in this case relies on properties of the real numbers described in Section 1.2. For reference, we summarize those properties as follows.

Property Summary

$pq = 0$ if and only if $p = 0$ or $q = 0$ (or both)

For example, to solve the equation $x^2 - 2x - 3 = 0$ by factoring, we have

$$\begin{aligned} x^2 - 2x - 3 &= 0 \\ (x - 3)(x + 1) &= 0 \end{aligned}$$

$$x - 3 = 0 \;\;\big|\;\; x + 1 = 0$$
$$x = 3 \;\;\big|\;\; x = -1$$

As you can easily check, the values $x = 3$ and $x = -1$ both satisfy the given equation.

As another example of an equation that can be solved by factoring, consider $4z^2 - 9 = 0$. The steps are as follows:

$$\begin{aligned} 4z^2 - 9 &= 0 \\ (2z - 3)(2z + 3) &= 0 \end{aligned}$$

$$2z - 3 = 0 \;\;\big|\;\; 2z + 3 = 0$$
$$2z = 3 \;\;\big|\;\; 2z = -3$$
$$z = \frac{3}{2} \;\;\Big|\;\; z = \frac{-3}{2}$$

Actually, for this last example it is probably shorter to bypass the factoring in this way:

$$4z^2 - 9 = 0$$
$$4z^2 = 9$$
$$z^2 = \frac{9}{4}$$
$$z = \pm\sqrt{\frac{9}{4}} = \frac{\pm 3}{2} \qquad \text{as obtained previously}$$

For ease of reference, let's agree to refer to the method just displayed as the *square root method* (as opposed to the factoring method). The square root method can be used to solve any quadratic equation of the form $ax^2 + c = 0$.

Whether or not a simple factoring can be found to solve a given quadratic equation is not always obvious. Consider, for example, the equation $x^2 - 2x - 4 = 0$. There are only three possible factorizations with integer coefficients; as you can easily check, none yields the appropriate middle term, $-2x$, when multiplied out:

$$(x - 4)(x + 1) \qquad (x + 4)(x - 1) \qquad (x - 2)(x + 2)$$

On the other hand, it is not at all obvious that the equation $x^2 + 156x + 5963 = 0$ can be solved by factoring. (We shall come back to this equation later.) The point here is to indicate the need for a systematic approach to solving those quadratic equations that are not readily solvable through factoring. The technique of *completing the square* provides this approach. As you'll see, completing the square allows us to rewrite any quadratic equation in such a way that the square root method becomes applicable. We'll demonstrate with an example.

Consider again the equation $x^2 - 2x - 4 = 0$. First we add 4 to both sides to isolate the x and x^2 terms on the left-hand side. This yields

$$x^2 - 2x = 4$$

Now, to "complete the square," we take half of the x-coefficient $[\frac{1}{2}(-2) = -1]$, square it [to get $(-1)^2 = 1$], and then add this result (namely 1) to both sides of the equation. This yields

$$x^2 - 2x + 1 = 4 + 1$$
$$(x - 1)^2 = 5$$
$$x - 1 = \pm\sqrt{5}$$
$$x = 1 \pm \sqrt{5}$$

We've now obtained the two solutions, $1 + \sqrt{5}$ and $1 - \sqrt{5}$. (Exercise 6 asks you to verify that these values indeed satisfy the original equation.) In the box that follows we summarize the technique of completing the square, and we indicate how the process gets its name.

The description in the box explains how to complete the square for expressions of the form $x^2 + bx$. In order to complete the square in solving a quadratic equation in which the coefficient of x^2 is not 1, we can first divide both sides of the equation by the coefficient of x^2. This is done in the next example.

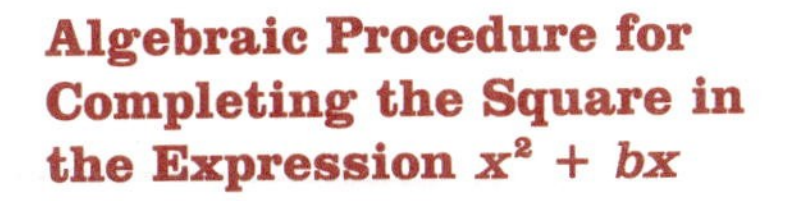

Algebraic Procedure for Completing the Square in the Expression $x^2 + bx$

Add the square of half of the x-coefficient:

$$\left(\frac{b}{2}\right)^2 = \frac{b^2}{4}$$

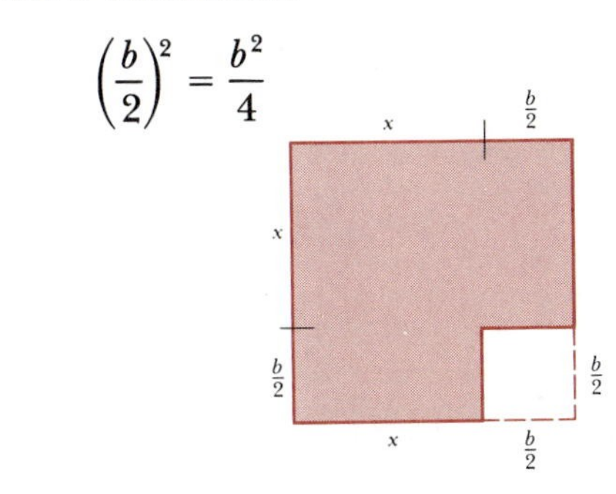

Geometric Interpretation of Completing the Square for $x^2 + bx$

The shaded region in the figure represents the quantity $x^2 + bx$ (because the area is $x^2 + \frac{b}{2}x + \frac{b}{2}x$). By adding the unshaded small square to the shaded region, we fill out or "complete" the larger square. The area of the unshaded region that completes the square is $(b/2)(b/2) = b^2/4$.

Example 1 Solve by completing the square: $9x^2 + 36x - 1 = 0$.

Solution

$$9x^2 + 36x - 1 = 0$$

$$x^2 + 4x - \frac{1}{9} = 0 \qquad \text{dividing both sides by 9}$$

$$x^2 + 4x = \frac{1}{9}$$

$$x^2 + 4x + 4 = \frac{1}{9} + 4 \qquad \text{We completed the square by adding } \left(\frac{1}{2}\cdot 4\right)^2 = 4 \text{ to both sides.}$$

$$(x + 2)^2 = \frac{37}{9}$$

$$x + 2 = \pm\sqrt{\frac{37}{9}} = \pm\frac{\sqrt{37}}{3}$$

$$x = -2 \pm \frac{\sqrt{37}}{3} = \frac{-6 \pm \sqrt{37}}{3}$$

The two solutions are therefore

$$\frac{-6 + \sqrt{37}}{3} \quad \text{and} \quad \frac{-6 - \sqrt{37}}{3}$$

The technique of completing the square can be used to derive the *quadratic formula*, a useful formula that gives the solutions to any quadratic equation. The derivation of this formula runs as follows. We start with the general quadratic equation $ax^2 + bx + c = 0 \quad (a \neq 0)$ and divide both sides by a to obtain

$$x^2 + \frac{b}{a}x + \frac{c}{a} = 0$$

Subtracting c/a from both sides yields

$$x^2 + \frac{b}{a}x = -\frac{c}{a}$$

Now, to complete the square, we add $\left[\frac{1}{2}\left(\frac{b}{a}\right)\right]^2$, or $\frac{b^2}{4a^2}$, to both sides. That gives us

$$x^2 + \frac{b}{a}x + \frac{b^2}{4a^2} = \frac{b^2}{4a^2} - \frac{c}{a}$$

$$\left(x + \frac{b}{2a}\right)^2 = \frac{b^2}{4a^2} - \frac{4ac}{4a^2}$$

$$\left(x + \frac{b}{2a}\right)^2 = \frac{b^2 - 4ac}{4a^2}$$

$$x + \frac{b}{2a} = \pm\sqrt{\frac{b^2 - 4ac}{4a^2}} = \pm\frac{\sqrt{b^2 - 4ac}}{2|a|}$$

$$= \pm\frac{\sqrt{b^2 - 4ac}}{2a}$$

This last equality follows from the fact that for any real number $a \neq 0$ the expressions $\pm 2|a|$ and $\pm 2a$ both represent the same two numbers. We conclude now that the solutions are

$$x = -\frac{b}{2a} + \frac{\sqrt{b^2 - 4ac}}{2a} \quad \text{and} \quad x = -\frac{b}{2a} - \frac{\sqrt{b^2 - 4ac}}{2a}$$

These solutions are usually written in the more compact form displayed in the box that follows.

The Quadratic Formula

The solutions of the equation $ax^2 + bx + c = 0 \quad (a \neq 0)$ are given by

$$x = \frac{-b \pm \sqrt{b^2 - 4ac}}{2a}$$

Example 2 Use the quadratic formula to solve $2x^2 = 4 - x$.

Solution First rewrite the equation as $2x^2 + x - 4 = 0$ so that it has the form $ax^2 + bx + c = 0$. By comparing these last two equations, we see that $a = 2$, $b = 1$, and $c = -4$. Therefore

$$x = \frac{-b \pm \sqrt{b^2 - 4ac}}{2a} = \frac{-1 \pm \sqrt{1^2 - 4(2)(-4)}}{2(2)} = \frac{-1 \pm \sqrt{33}}{4}$$

Thus the two solutions are $\frac{-1 + \sqrt{33}}{4}$ and $\frac{-1 - \sqrt{33}}{4}$.

Example 3 Solve $x^2 + 156x + 5963 = 0$.

Solution We use the quadratic formula and a hand-held calculator.

$$a = 1 \qquad b = 156 \qquad c = 5963$$

$$x = \frac{-b \pm \sqrt{b^2 - 4ac}}{2a} = \frac{-156 \pm \sqrt{156^2 - 4(1)(5963)}}{2(1)}$$

$$= \frac{-156 \pm 22}{2} \qquad \text{using a calculator}$$

$$= -78 \pm 11$$

The two solutions are therefore -89 and -67. Although it's conceivable that the initial equation could be solved by discovering the factorization $(x + 89)(x + 67)$, in practice it is usually more efficient to proceed directly to the formula when numbers of this magnitude are involved.

In Examples 2 and 3 we found that each quadratic equation had two real roots. As the next example indicates, however, it's also possible for quadratic equations to have only one real root, or even no real roots.

Example 4 Use the quadratic formula to solve:

(a) $4x^2 - 12x + 9 = 0$

(b) $x^2 + x + 2 = 0$

Solution **(a)** $a = 4,\ b = -12,\ c = 9$

$$x = \frac{12 \pm \sqrt{(-12)^2 - 4(4)(9)}}{2(4)} = \frac{12 \pm \sqrt{144 - 144}}{8}$$

$$= \frac{12 \pm 0}{8} = \frac{12}{8} = \frac{3}{2}$$

In this case our only solution is $\frac{3}{2}$. We refer to $\frac{3}{2}$ as a *double root. Note:* We've used the quadratic formula here only for purposes of illustration; in this case you can solve the original equation more efficiently by factoring.

(b) $a = 1,\ b = 1,\ c = 2$

$$x = \frac{-1 \pm \sqrt{(1)^2 - 4(1)(2)}}{2(1)} = \frac{-1 \pm \sqrt{-7}}{2}$$

$$= \frac{-1 \pm i\sqrt{7}}{2}$$

In this case our roots are the two complex numbers

$$\frac{-1 + i\sqrt{7}}{2} \quad \text{and} \quad \frac{-1 - i\sqrt{7}}{2}$$

The quantity $b^2 - 4ac$ that appears under the radical sign in the quadratic formula is called the *discriminant.* In Examples 2 and 3, the discriminant was positive and consequently the equations each had two real solutions. In Example 4(a), in which the discriminant was zero, we obtained but one (real) solution. In Example 4(b) the discriminant was negative, and consequently we obtained two solutions involving the imaginary unit i. These observations are generalized in the box that follows.

The Discriminant $b^2 - 4ac$

Consider the quadratic equation $ax^2 + bx + c = 0$, where a, b, and c are real numbers and $a \neq 0$. The expression $b^2 - 4ac$ is called the *discriminant.*

1. If $b^2 - 4ac > 0$, then the equation has two distinct real roots;
2. If $b^2 - 4ac = 0$, then the equation has exactly one real root;
3. If $b^2 - 4ac < 0$, then the equation has no real roots.

Example 5 (a) Use the discriminant to determine how many real solutions there are for the equation $x^2 + x - 1 = 0$.

(b) Find a value for k such that the quadratic equation $x^2 + \sqrt{2}x + k = 0$ has exactly one real solution.

Solution (a) $a = 1, b = 1, c = -1$

$$b^2 - 4ac = 1^2 - 4(1)(-1) = 5$$

Since the discriminant is positive, the equation has two distinct real solutions.

(b) For the equation to have exactly one real solution, we required that the discriminant be zero. That is,

$$\begin{aligned} b^2 - 4ac &= 0 \\ (\sqrt{2})^2 - 4(1)(k) &= 0 \\ 2 - 4k &= 0 \\ k &= \frac{1}{2} \end{aligned}$$

The required value for k is $\frac{1}{2}$.

Exercise Set 2.2

In Exercises 1–5, determine if the given value is a solution of the equation.

1. $2x^2 - 6x - 36 = 0;\ x = -3$
2. $(y - 4)(y - 5) = 0;\ y = -1$
3. $4x^2 - 1 = 0;\ x = -\frac{1}{4}$
4. $m^2 + m - \frac{5}{16} = 0;\ m = \frac{1}{4}$
5. $x^2 - 2x - 6 = 0;\ x = -1 - \sqrt{7}$
6. Verify that the numbers $1 + \sqrt{5}$ and $1 - \sqrt{5}$ both satisfy the equation $x^2 - 2x - 4 = 0$.

For Exercises 7–22, solve each equation by factoring.

7. $x^2 - 5x - 6 = 0$
8. $x^2 - 5x = -6$
9. $x^2 - 100 = 0$
10. $4y^2 + 4y + 1 = 0$
11. $25x^2 - 60x + 36 = 0$
12. $144 - t^2 = 0$
13. $10z^2 - 13z - 3 = 0$
14. $3t^2 - t - 4 = 0$
15. $(x + 1)^2 - 4 = 0$
16. $x^2 + 3x - 40 = 0$
17. $2x^2 - 15x - 8 = 0$
18. $12x^2 = 12 - 7x$
19. $x(2x - 13) = -6$
20. $x(3x - 23) = 8$
21. $x(x + 1) = 156$
22. $x^2 + (2\sqrt{5})x + 5 = 0$

Solve each equation in Exercises 23–30 by using the square root method.

23. $x^2 - 49 = 0$
24. $x^2 - \frac{15}{9} = 0$
25. $x^2 + 1 = 0$
26. $16x^2 - 25 = 0$
27. $\frac{1}{4} - x^2 = 0$
28. $2x^2 + 50 = 0$
29. $3x^2 - 192 = 0$

30. **(a)** $a^2x^2 - c^2 = 0$ $(a > 0$ and $c > 0)$
(b) $a^2x^2 - c^2 = 0$ $(a > 0$ and $c < 0)$

In Exercises 31–46, solve each equation by completing the square.

31. $x^2 - 4x - 10 = 0$ **32.** $x^2 - 2x - 7 = 0$
33. $y^2 - 5y + 1 = 0$ **34.** $y^2 + y + 3 = 0$
35. $y^2 + 8y = 0$ **36.** $y^2 + 8y + 15 = 0$
37. $2x^2 + 6x - 1 = 0$ **38.** $3x^2 - 12x + 4 = 0$
39. $4x^2 = 3 - 36x$ **40.** $x^2 = 2x$
41. $8y^2 + 22y + 5 = 0$ **42.** $9t^2 + 15t + 4 = 0$
43. $2x^2 - x - 1 = 0$ **44.** $2x^2 - x + 1 = 0$
45. $-x^2 - x + 1 = 0$ **46.** $-2x^2 + 8x + 11 = 0$

Use the quadratic formula to solve each equation in Exercises 47–62.

47. $x^2 + 10x + 9 = 0$ **48.** $x^2 - 10x + 25 = 0$
49. $x^2 + 5x - 24 = 0$ **50.** $2x^2 - 7x + 5 = 0$
51. $x^2 - x - 5 = 0$ **52.** $x^2 - x + 5 = 0$
53. $2x^2 + 3x - 4 = 0$ **54.** $4x^2 - 3x - 9 = 0$
55. $x^2 - 17 = 0$ **56.** $3x^2 + 8x = -5$
57. $12x^2 + 32x + 5 = 0$ **58.** $10x^2 - x - 1 = 0$
59. $2x^2 = x + 5$ **60.** $3 - 2x = -4x^2$
61. $-6x^2 + 12x = -1$ **62.** $-\sqrt{2}\,x^2 + x = -\sqrt{2}$

For Exercises 63–70, use the discriminant to determine how many real roots each equation has.

63. $x^2 - 12x + 16 = 0$ **64.** $2x^2 - 6x + 5 = 0$
65. $4x^2 - 5x - \frac{1}{2} = 0$ **66.** $4x^2 - 28x + 49 = 0$
67. $x^2 + \sqrt{3}\,x + \frac{3}{4} = 0$
68. $\sqrt{2}\,x^2 + \sqrt{3}\,x + 1 = 0$
69. $y^2 - \sqrt{5}\,y = -1$ **70.** $\dfrac{m^2}{4} - \dfrac{4m}{3} + \dfrac{16}{9} = 0$

In Exercises 71–74, find values for k *such that the equations have exactly one real root.*

71. $x^2 + 12x + k = 0$
72. $3x^2 + (\sqrt{2k})x + 6 = 0$
73. $x^2 + kx + 5 = 0$
74. $kx^2 + kx + 1 = 0$

For Exercises 75–100, solve the equations using any method. In Exercises 91–100, you'll need to multiply both sides of the equations by expressions involving the variable. Remember (as indicated in Section 2.1) to check your answers in these cases.

75. $25x^2 - 1 = 0$ **76.** $(3x - 2)^2 = 0$
77. $(3x - 2)^2 = 3x - 2$ **78.** $2x^2 + 11x + 12 = 0$
79. $1 - 2x^2 = x$ **80.** $5x^2 - 4x + 1 = 0$
81. $3x^2 + 4x - 3 = 0$ **82.** $13x^2 = 52$
83. $x^2 = 24$ **84.** $x^2 + 16x = 0$
85. $x(x - 1) = 1$ **86.** $x(x - 1) = -4$
87. $\frac{1}{2}x^2 - x - \frac{1}{3} = 0$ **88.** $x^2 + 34x + 288 = 0$
89. $2\sqrt{5}\,x^2 - x - 2\sqrt{5} = 0$
90. $3(x - 2)^2 - 4(x - 2) = 0$
91. $\dfrac{3}{x + 5} + \dfrac{4}{x} = 2$ **92.** $\dfrac{5}{x + 2} - \dfrac{2x - 1}{5} = 0$
93. $1 - x - \dfrac{2}{6x - 1} = 0$ **94.** $\dfrac{x^2 - 3x}{x + 1} = \dfrac{4}{x + 1}$
95. $\dfrac{3x^2 - 6x - 3}{(x + 1)(x - 2)(x - 3)} + \dfrac{5 - 2x}{x^2 - 5x + 6} = 0$
96. $\dfrac{6}{x^2 - 1} + \dfrac{x}{x + 1} = \dfrac{3}{2}$ **97.** $\dfrac{2x}{x^2 - 1} - \dfrac{1}{x + 3} = 0$
98. $\dfrac{3}{x^2 - x - 2} - \dfrac{4}{x^2 + x - 6} = \dfrac{1}{3x + 3}$
99. $\dfrac{x - 1}{x + 1} - \dfrac{x + 1}{x + 3} + \dfrac{4}{x^2 + 4x + 3} = 0$
100. $\dfrac{x}{x - 2} + \dfrac{x}{x + 2} = \dfrac{8}{x^2 - 4}$

In Exercises 101–109, solve for x *in terms of the other letters.*

101. $2y^2x^2 - 3yx + 1 = 0$ $(y > 0)$
102. $(ax + b)^2 - (bx + a)^2 = 0$ (Assume that $a > 0$, $b > 0$, and $a \neq b$.)
103. $(x - p)^2 + (x - q)^2 = p^2 + q^2$
104. $21x^2 - 2kx - 3k^2 = 0$ $(k > 0)$
105. $12x^2 = ax + 20a^2$ $(a > 0)$
106. $3Ax^2 - 2Ax - 3Bx + 2B = 0$ $(A \neq 0)$
***107.** $(a + b + c)x^2 - 2(a + b)x + (a + b - c) = 0$ $(c > 0,\ a + b + c \neq 0)$ *Hint:* Let $d = a + b$
***108.** $ab(a + b)x^2 - (a^2 + b^2)x - 2(a + b) = 0$ (Assume that a and b are nonzero and $a \neq -b$.)
***109.** $x^2 - \left(\sqrt{\dfrac{a}{b}} + \sqrt{\dfrac{b}{a}}\right)x + 1 = 0$

For Exercises 110–114, use the quadratic formula and a calculator to solve for x. *Round off each answer to two decimal places.*

C **110.** $x^2 + 0.18x - 1 = 0$
C **111.** $x^2 - 4.76x + 0.03 = 0$
C **112.** $155x^2 - 140x - 211 = 0$
C **113.** $x^2 + 3x + 2.249 = 0$
C **114.** $x^2 + 3x + 2.251 = 0$

Note: After you've completed Exercises 113 and 114, compare the two equations and the solutions. The point here is that a slight change in one of the coefficients can sometimes radically alter the nature of the solutions.

115. Here is an outline for a slightly different derivation of the quadratic formula. The advantage to this method is that fractions are avoided until the very last step. Fill in the details.

(a) Beginning with $ax^2 + bx = -c$, multiply both sides by $4a$. Then add b^2 to both sides.

(b) Now factor the resulting left-hand side and use the square root method to solve for x.

2.3 Applications

Solving problems is a practical art, like swimming, or skiing, or playing the piano: you can learn it only by imitation and practice. This book cannot offer you a magic key that opens all doors and solves all the problems, but it offers you good examples for imitation and many opportunities for practice. . . .

George Polya in his book *Mathematical Discovery* (New York: John Wiley & Sons, 1981).

The techniques developed in the previous two sections for solving linear and quadratic equations can be used in solving a wide variety of applied problems. As you approach each example and exercise in this section for the first time, you may find it useful to refer to the following suggestions for organizing your thoughts and "getting started." Eventually, of course, you'll probably want to modify this list to fit your own style.

1. Read the problem carefully. When appropriate, draw a picture or a chart to summarize the situation.
2. State in your own words, as specifically as you can, what the problem is asking for. (This usually requires rereading the problem.) Assign variables or expressions to denote the required quantities. Then, as necessary, assign auxiliary variables to any other quantities that appear relevant. Are there equations relating these auxiliary variables?
3. Set up an equation (or equations) involving the required variables. Where necessary, use the results from step 2 to obtain an equation involving but one required unknown.
4. Solve the equation that you obtained in step 3.
5. Check you answer. Relate it to the original problem. Does it satisfy the conditions of the *original* problem?

In Example 1 we solve two problems that appear to be very similar. As you'll see, however, the first problem involves a linear equation, the second a quadratic.

Example 1 The sum of two numbers is 20.

(a) Find the two numbers, assuming that their difference is 96.

(b) Find the two numbers, assuming that their product is 96.

Solution **(a)** Let

$$x = \text{the first number}$$
$$20 - x = \text{the second number}$$

(Notice that the sum of x and $20 - x$ is indeed 20.) Now, since the difference of these two numbers is 96, we have

$$\begin{aligned} x - (20 - x) &= 96 \\ 2x &= 116 \\ x &= 58 \end{aligned}$$

With $x = 58$, the second number, $20 - x$, is $20 - 58$, or -38. As you can now check for yourself, 58 and -38 are indeed the required numbers: Their sum is 20 and their difference is 96.

(b) Again we denote the two numbers by x and $20 - x$. But this time we are assuming that their product (rather than their difference) is 96. Thus we have

$$\begin{aligned} x(20 - x) &= 96 \\ 20x - x^2 &= 96 \\ x^2 - 20x + 96 &= 0 \\ (x - 8)(x - 12) &= 0 \end{aligned}$$

$$x - 8 = 0 \quad\Big|\quad x - 12 = 0$$
$$x = 8 \quad\Big|\quad x = 12$$

We've now found two possibilities for the first number, 8 or 12. If $x = 8$, then the second number is $20 - x = 20 - 8 = 12$; if $x = 12$, then the second number is $20 - 12 = 8$. So in either case we obtain the two numbers 8 and 12. As you can see, these are the required numbers: Their sum is 20 and their product is 96.

Example 2 If you have three exam scores of 91, 76, and 84, what score do you need on the fourth exam to raise your average to 85?

Solution Let

$$x = \text{the score on the fourth exam}$$

The average of the four numbers 91, 76, 84, and x is $\dfrac{91 + 76 + 84 + x}{4}$. Since this average is to be 85, we write

$$\begin{aligned} \frac{91 + 76 + 84 + x}{4} &= 85 \\ 91 + 76 + 84 + x &= 340 \qquad \text{multiplying by 4} \\ 251 + x &= 340 \\ x &= 340 - 251 = 89 \end{aligned}$$

We've now found that $x = 89$. This is the required score because, as you can check, the average of the four numbers 91, 76, 84, and 89 is in fact 85, as required.

If you look back over Examples 1 and 2, you'll see that the equations we solved were obtained by directly translating the words in the problems into algebraic notation. Example 3 is one step removed from this in that we obtain the necessary equation by means of a formula from geometry. (That formula is the Pythagorean Theorem. If you need a review, see pages 28–29.)

Example 3 Are there right triangles other than the 3-4-5 right triangle in which the lengths of the sides are three consecutive integers?

Solution Any three consecutive integers can be denoted by x, $x + 1$, and $x + 2$ (see Figure 1). If these three quantities are to be the lengths of the sides of a right triangle, then according to the Pythagorean Theorem we must have

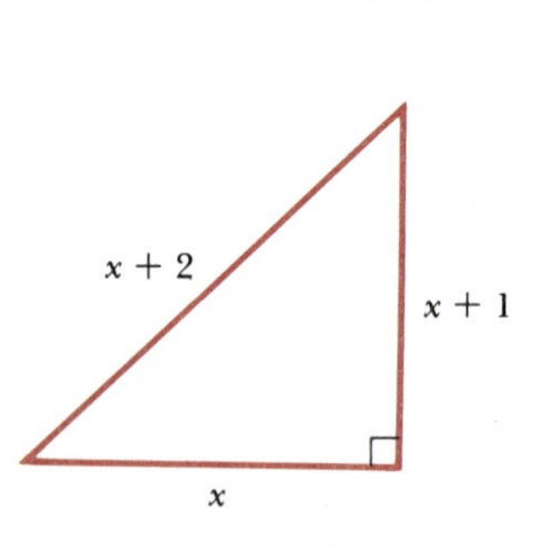

Figure 1

$$x^2 + (x + 1)^2 = (x + 2)^2$$
$$x^2 + x^2 + 2x + 1 = x^2 + 4x + 4$$
$$x^2 - 2x - 3 = 0 \quad \text{simplifying}$$
$$(x - 3)(x + 1) = 0$$

$x - 3 = 0$	$x + 1 = 0$
$x = 3$	$x = -1$

Thus, $x = 3$ or $x = -1$. We can immediately discard the value $x = -1$, for x is supposed to represent a length. On the other hand, with $x = 3$ we find that

$$x + 1 = 3 + 1 = 4 \quad \text{and} \quad x + 2 = 3 + 2 = 5$$

From this we conclude that if the lengths of the three sides in the right triangle are consecutive integers, then those integers must be 3, 4, and 5. Or, to answer the original question, there are no other right triangles in which the lengths of the sides are three consecutive integers.

The next example involves *simple interest* earned on money in a savings account. As an illustration of simple interest, suppose that you place \$100 in a savings account for one year at 10% simple interest. This means that at the end of the year, the bank contributes 10% of \$100, or \$10, to your account. Furthermore, assuming that you make no subsequent withdrawals or deposits, the bank continues to contribute \$10 to your account at the end of each year. The following general formula can be derived for computing simple interest.

> $I = Prt$
>
> where I is the simple interest earned on an initial deposit P (the *principal*) after t years at an annual rate r.

To use this formula, you need to convert the annual rate r from a percent to a decimal. For instance, to compute the simple interest earned on a principal of \$1000 invested for six years at 10%, we write

$$I = Prt = (\$1000)(0.1)(6) = \$600$$

Example 4 Suppose that you open two savings accounts. In the first you deposit \$5000 at an annual simple interest rate of 8%. In the second you deposit \$10,000 at 12% (also simple interest). After how many years will the total interest amount to \$16,000?

Solution Let

$$t = \text{the required number of years}$$

Then in t years the interest on the first account is $(5000)(.08)t$, while that on the second is $(10{,}000)(0.12)t$. Since the total interest is to be \$16,000, we have

$$\begin{aligned}(5{,}000)(0.08)t + (10{,}000)(0.12)t &= 16{,}000\\ 400t + 1200t &= 16{,}000\\ 1600t &= 16{,}000\\ t &= 10\end{aligned}$$

Thus, after 10 years the total interest will amount to \$16,000.

Check: After 10 years the interest from the first account will be (\$5000)(0.08)(10), which is \$4000. The interest from the second account after 10 years will be (\$10,000)(0.12)(10), which is \$12,000. So the total interest will be the sum of \$4000 and \$12,000, that is, \$16,000, as required.

In some problems it's helpful to set up a chart or table to see the given data at a glance. This is done in the next example.

Example 5 Suppose that a chemistry student can obtain two acid solutions from the stockroom. The first solution is 20% acid and the second is 45% acid. (The percents are by volume.) How many cubic centimeters (cm^3) of each solution should the student mix together to obtain 100 cm^3 of a 30% acid solution?

Solution Begin by assigning letters or expressions to denote the required quantities. Let x denote the number of cm^3 of the 20% solution to be used. Then

$$100 - x$$

denotes the number of cm^3 of the 45% solution to be used. All the data can now be summarized in the following table.

Type of Solution	Number of cm^3	Percent Acid	Total Acid (cm^3)
First solution (20% acid)	x	20	$(0.20)x$
Second solution (45% acid)	$100 - x$	45	$(0.45)(100 - x)$
Mixture	100	30	$(0.30)(100)$

Looking at the data in the right-hand column of the table, we can write

$$\underbrace{0.20x}_{\text{Amount of acid in } x \text{ cm}^3 \text{ of the 20\% solution}} + \underbrace{0.45(100 - x)}_{\text{Amount of acid in } (100 - x)\text{ cm}^3 \text{ of the 45\% solution}} = \underbrace{(0.30)(100)}_{\text{Amount of acid in the final mixture}}$$

So we have

$$\begin{aligned}0.20x + 0.45(100 - x) &= 30\\ 20x + 4500 - 45x &= 3000 \quad \text{multiplying by 100}\\ -25x &= -1500\\ x &= 60\end{aligned}$$

Thus $x = 60$, and therefore $100 - x = 100 - 60 = 40$. Consequently the student should use 60 cm^3 of the first solution and 40 cm^3 of the second solution.

Check: The amount of acid in 60 cm^3 of the first solution is $(60\ cm^3)(0.20)$, which is 12 cm^3. The amount of acid in 40 cm^3 of the second solution is $(40\ cm^3)(0.45)$, which is 18 cm^3. So the total amount of acid in the mixture is the sum of 12 cm^3 and 18 cm^3, which is 30 cm^3. Since the volume of the final mixture is 100 cm^3, the percent of acid there is $\frac{30}{100}(100)$; this is 30%, as required.

Example 6 The radius of a circle is 10 cm. By how many centimeters should the radius be increased so that the area increases by $5\pi\ cm^2$? Round off the answer to two decimal places.

Solution Let

$$x = \text{the number of cm by which the radius should be increased}$$

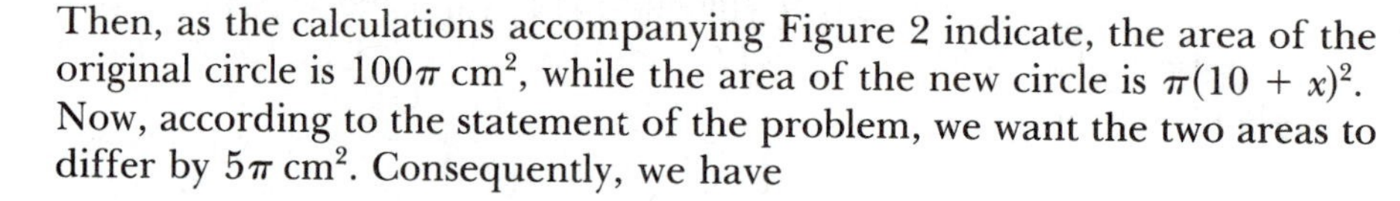

Then, as the calculations accompanying Figure 2 indicate, the area of the original circle is $100\pi\ cm^2$, while the area of the new circle is $\pi(10 + x)^2$. Now, according to the statement of the problem, we want the two areas to differ by $5\pi\ cm^2$. Consequently, we have

Original circle

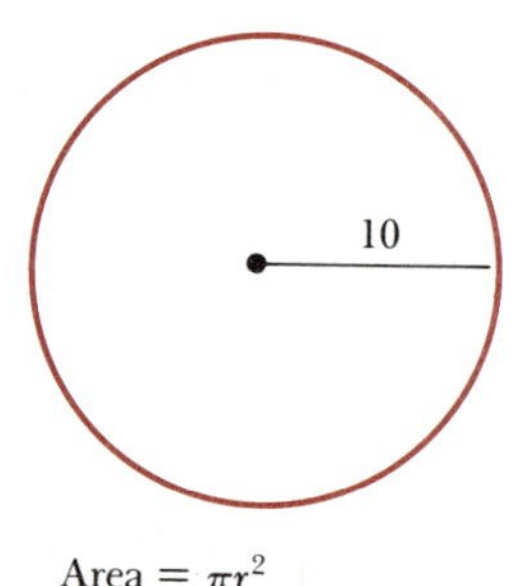

$$\text{Area} = \pi r^2 = \pi(10^2) = 100\pi$$

$$\pi(10 + x)^2 - 100\pi = 5\pi$$

$$(10 + x)^2 - 100 = 5 \quad \text{dividing by } \pi$$

$$(10 + x)^2 = 105$$

$$10 + x = \sqrt{105} \quad \text{The root } -\sqrt{105} \text{ can be disregarded here. Why is this?)}$$

$$x = \sqrt{105} - 10$$

Thus, the radius of the original circle should be increased by $(\sqrt{105} - 10)$ cm. Using a calculator and rounding off to two decimal places, we obtain $x \approx 0.25$ cm. As with all word problems, you should check for yourself here that the answer obtained satisfies the conditions of the original problem. (In checking, of course, use the exact value, $\sqrt{105} - 10$, not the calculator approximation, 0.25.)

New circle

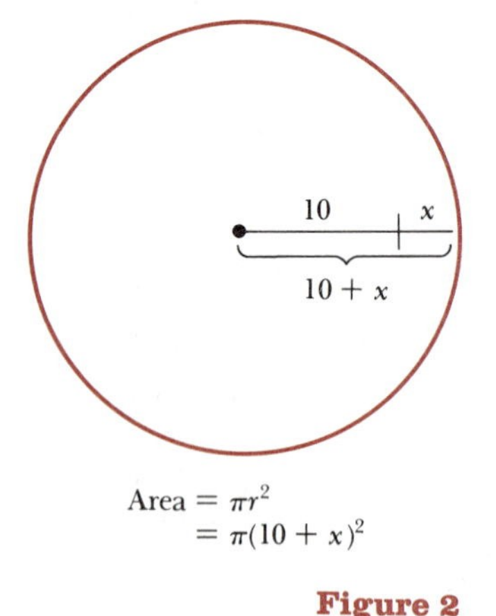

$$\text{Area} = \pi r^2 = \pi(10 + x)^2$$

Figure 2

Our final two examples for this section are motion problems. For travel at a constant speed or rate, the basic relationship is

$$\text{distance} = \text{rate} \times \text{time}$$

or, more concisely,

$$d = rt$$

For example, if you drive 55 mph for 3 hours, then the distance you cover is $d = (55)(3) = 165$ miles. If a trip involves several parts, each at a different constant speed, then the *average speed* for the trip is defined by

$$\text{average speed} = \frac{\text{total distance}}{\text{total time}}$$

Example 7 Andy and Jerry are going to run on a quarter-mile track. Andy can run at a pace of 8 minutes per mile ($= \frac{1}{8}$ mi/min), while Jerry can run at a pace of

7 minutes per mile ($=\frac{1}{7}$ mi/min). If they start together at the same time (running in the same direction), how long will it take for Jerry to lap Andy?

Solution Jerry will lap Andy on the quarter-mile track when the distance covered by Jerry is exactly $\frac{1}{4}$ mile more than that covered by Andy. After t minutes, the distance covered by each runner is as follows:

$$\text{Andy: distance} = rt = \frac{1}{8}t \qquad \text{Jerry: distance} = rt = \frac{1}{7}t$$

So the condition that Jerry be exactly $\frac{1}{4}$ mile ahead of Andy is

$$\frac{1}{7}t = \frac{1}{8}t + \frac{1}{4}$$

$$8t = 7t + 14 \qquad \text{multiplying by 56}$$

$$t = 14$$

Thus, after 14 minutes Jerry will have lapped Andy. (Check for yourself now that this answer indeed satisfies the conditions of the problem.)

Example 8 Suppose that you drive from town A to town B at 40 mph. At what (constant) speed should you make the return trip so as to average 50 mph for the entire round trip? (The answer is not 60 mph.)

Solution Let D denote the distance between the two towns, and let R denote the (required) speed on the return trip. Then we have

$$\frac{\text{total distance}}{\text{total time}} = 50$$

$$\frac{2D}{\text{total time}} = 50$$

$$\frac{D}{\text{total time}} = 25 \qquad (1)$$

The total time is the time from A to B plus the time from B back to A. Each of these can be computed by rewriting the formula $d = rt$ as $t = d/r$. Thus, the time for the trip from A to B is $D/40$, and the time for the trip from B to A is D/R. Consequently, equation (1) becomes

$$\frac{D}{D/40 + D/R} = 25$$

$$\frac{1}{1/40 + 1/R} = 25 \qquad \text{We divided the numerator and denominator on the left-hand side by } D.$$

$$\frac{40R}{R + 40} = 25 \qquad \text{We multiplied the numerator and denominator on the left-hand side by } 40R.$$

$$40R = 25R + 1000$$

$$15R = 1000$$

$$R = \frac{1000}{15} = \frac{200}{3}$$

Thus the required speed for the return trip is $\frac{200}{3}$ mph, or approximately 67 mph.

Check: Again, let D denote the distance (in miles) between A and B. The time required to go from A to B at 40 mph is $(D/40)$ hr. On the return trip, with a speed of $(200/3)$ mph, the time is $\dfrac{D}{(200/3)}$ hr, or $(3D/200)$ hr. The total time in hours for the round trip is therefore $\dfrac{D}{40} + \dfrac{3D}{200}$, which simplifies to $D/25$. Consequently, the average velocity for the round trip is $\dfrac{2D}{(D/25)}$ mph. This simplifies to 50 mph, as required.

Exercise Set 2.3

1. Find two numbers whose sum is 7 and whose difference is 17.
2. The larger of two numbers is one less than twice the smaller. The sum of the two numbers is 29. Find the two numbers.
3. Find two numbers whose sum is 17 and whose product is 52.
4. The sum of two numbers is 6 and the sum of their squares is 90. What are the two numbers?
5. The sum of a number and its double is 63. What is the number?
6. Find three consecutive integers such that twice the first plus half of the second is nine more than twice the third.
7. The product of a certain positive number and four more than that number is 96. What is the number?
8. The sum of the squares of three consecutive positive integers is 365. Find the integers.
9. What number must be added to both the numerator and the denominator of the fraction $\frac{3}{7}$ to yield a fraction that is equal to $\frac{2}{3}$?
10. Find two consecutive even integers such that when they are divided by 8 and 9, respectively, the sum of the quotients is 4.
11. The difference of two positive numbers is one third of their sum. The sum of the squares of the two numbers is 180. Find the two numbers.
12. **(a)** Is there a positive real number such that the sum of the number and its square is 1? If so, find the number.
    (b) Is there a positive real number such that the difference between the number and its square is 1? If so, find the number.
13. If you have exam scores of 70, 77, and 75, what score do you need on the fourth exam to raise your average to 75?
14. The average of three numbers is 42. The first two numbers differ by 1, and the average of the second and third numbers is 47. What are the three numbers?
15. The average of four numbers is 52. Find the fourth number, given that the first three are 80, 22, and 62.
16. The average of the reciprocals of two consecutive integers is $\frac{9}{40}$. Find the smaller of the two integers.
17. In a certain right triangle, the hypotenuse is 8 cm longer than the first leg and 1 cm longer than the second. Find the lengths of the sides of the triangle.
18. In a certain right triangle, the lengths of the two shorter sides are consecutive integers (measured in cm). Find the area of the triangle given that the sum of the lengths of the hypotenuse and longer leg is 50 cm.
19. When the sides of a square are each increased by 2 cm, the area increases by 14 cm^2. Find the length of a side in the original square.
20. In a certain rectangle, the length exceeds the width by 2 m. If the length is increased by 4 m and the width is decreased by 2 m, the area of the new rectangle is the same as that of the original rectangle. Find the dimensions of the original rectangle.
21. The perimeter and area of a rectangle are 104 cm and 640 cm^2, respectively. Find the length of the shorter side of the rectangle.
22. A rectangular picture with dimensions 5 in. by 8 in. is surrounded by a frame of uniform width. If the area of the frame (not including the space for the picture) is 114 in.^{2} greater than the area of the picture, find the width of the frame.

23. A piece of wire 12 in. long is bent to form a right triangle in which the shortest side is 2 in. Find the lengths of the other two sides.

24. A piece of wire 24 cm long is cut into two pieces. Each piece is then formed into a square. If the sum of the areas of the two squares is 20 cm^2, find the lengths of the two pieces of wire.

25. In a certain isosceles triangle, each of the two equal legs is three times as long as the base. If the perimeter of the triangle is 70 cm, find the lengths of the sides.

26. In a certain triangle, the second angle is four times the first, while the third angle is 60° more than the first angle. Find the three angles. (Use the fact that the sum of the angles in any triangle is 180°.)

27. Suppose that you open two savings accounts. In the first you deposit $4000 at an annual simple rate of 7%. In the second you deposit $5000 at 9% (also simple interest). After how many years will the total interest amount to $5840?

28. You invest $600 in a stock paying a 6% annual dividend and $800 in a stock paying an 8% annual dividend. How many years will it take for the total earnings to equal the total initial investment?

29. If you deposit $10,000 at 11% (simple interest), how long will it be until the total amount in the account is $21,000?

30. Mr. X has $5000 to deposit. He deposits part of this in an account earning 6%. The remainder is deposited in an account earning 8.5%. (Both rates are simple interest.) If the total interest earned each year is $387.50, how much was initially deposited in each account?

31. You deposit $7500 at 7.5% and $4000 at 5%, simple interest. What is the effective interest rate? (In other words, what interest rate on $11,500 would yield the same total interest at year's end?) Express your answer as a percent, rounded off to the nearest tenth of one percent.

32. You deposit $2000 at 8% and $3000 at 9%, simple interest. What additional amount would you need to invest at 12% to yield an overall return of 10% after one year?

33. Ms. X deposits $8000 in one account and $9000 in another. The interest rate on the $8000 account is one percent less than that on the $9000 account. If the $9000 account earns $165 more per year than the $8000 account, find the interest rates on each account.

34. Jenny inherits $42,900 and invests it for a year in two stocks. Stock A earns 12.5% annually and stock B earns 8.5% annually. How much was invested in each stock if stock B earned three-quarters as much as stock A?

35. Suppose that the cost of a car to an automobile dealer is $8600. At what price should the dealer list the car so that he can give the customer a discount (off the list price) of 12% and still realize a 20% profit? (Round off your answer to the nearest dollar.)

36. A publisher of mathematics books is sending a questionnaire to everyone on a certain mailing list. From experience, the publisher knows that 60% of the people receiving the questionnaire will either ignore it or throw it away. Of the remaining group, two thirds will fill out the questionnaire, but then only two thirds of those will remember to mail it back. How many names are on the mailing list if the publisher receives 2560 responses?

37. A student in a chemistry laboratory has available to her two acid solutions. The first is 10% acid and the second is 35% acid. (The percents are by volume.) How many cm^3 of each solution should she mix to obtain 200 cm^3 of a 25% acid solution?

38. One salt solution is 15% salt and another is 20% salt. How many cm^3 of each solution must be mixed to obtain 50 cm^3 of a 16% salt solution?

39. A shopkeeper has two types of coffee beans. The first type sells for $5.20 per pound, the second for $5.80 per pound. How many pounds of the first type must be mixed with 5 pounds of the second to produce a blend selling for $5.35 per pound?

40. A certain alloy contains 10% tin and 30% copper. (The percents are by weight.) How many pounds of tin and how many pounds of copper must be melted with 1000 pounds of the given alloy to yield a new alloy containing 20% tin and 35% copper?

41. Two sources of iron ore are available to a steel company. The first source of ore contains 20% iron (by weight). The second source contains 35% iron. How many tons of the second ore must be added to 16 tons of the first ore to produce a mixture that is 25% iron?

42. How much water must be evaporated from 400 pounds of a 10% salt solution to yield a solution that is 16% salt?

43. The radius of a circle is 6 cm. By how many centimeters should the radius be increased to increase the area by 288π cm^2?

44. The base and the height of a triangle are $m + 5$ and $m + 4$ units, respectively. Determine the value of m if the area of the triangle is 28 square units.

45. The two shorter sides of a right triangle differ by 7 cm. If the area is 60 cm^2, find the hypotenuse.

46. If the sides of a square are doubled, the area increases by 147 cm^2. Find the sides of the original square.

47. The area of a square with sides of length $1 - x$ is equal to the area of a rectangle with dimensions 1 by x. Find x.

48. In the figure, PT is tangent to the circle at T and $\overline{PT} = \overline{AB} = 1$. Find $\overline{PB}$. (Use the theorem from geometry stating that $\overline{PT}^2 = \overline{PA} \cdot \overline{PB}$.)

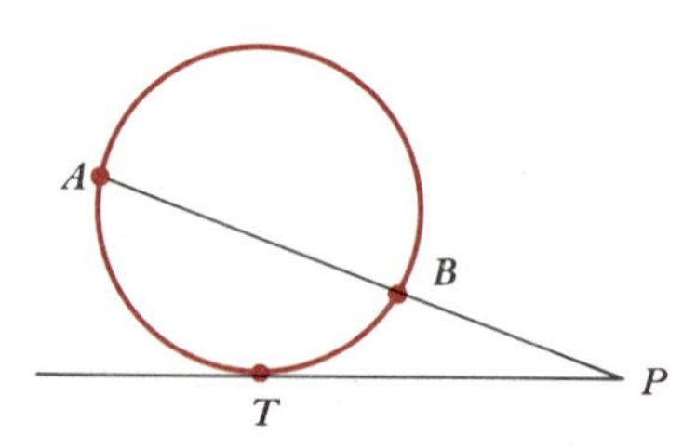

49. In the figure, $AB = 1$ and B divides AC in such a way that the ratio of the whole segment to its larger part equals the ratio of the larger part to the smaller part. Find this ratio.

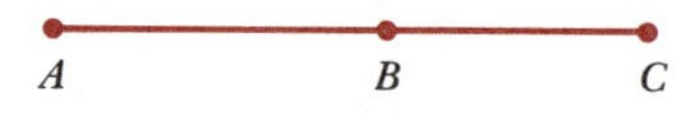

50. For a certain right circular cylinder, the height is 3 m and the total surface area is 8π m^2. Find the radius of the cylinder. (The formula for the total surface area S is $S = 2\pi r^2 + 2\pi rh$, where r is the radius and h is the height.)

51. A ball is thrown straight upward. Suppose that the height of the ball at time t is given by $h = -16t^2 + 96t$, where h is in feet and t is in seconds, with $t = 0$ corresponding to the instant that the ball is first tossed.

(a) How long does it take before the ball lands?

(b) At what time is the height 80 ft? Why are there two answers here?

52. During a flu epidemic in a small town, a public health official finds that the total number of people P who have caught the flu after t days is closely approximated by the formula

$$P = -t^2 + 26t + 106 \qquad (1 \le t \le 13)$$

(a) How many have caught the flu after 10 days?

(b) After approximately how many days will 250 people have caught the flu?

53. Emily and Ernie are going to jog on a quarter-mile track. Emily can jog at a pace of 9 minutes per mile ($= \frac{1}{9}$ mi/min), while Ernie can jog at a pace of 11 minutes per mile ($= \frac{1}{11}$ mi/min). If they start together at the same time (running in the same direction), how long will it take for Emily to lap Ernie?

***54.** If you travel from town A to town B at 35 mph, at what speed would you have to make the return trip so as to average 40 mph for the entire round trip? (Round off your answer to the nearest one mile per hour.)

55. The speed of an eastbound train is 5 mph greater than the speed of a westbound train. If the trains leave at the same time from stations 630 mi apart and pass each other after 6 hours, find the speed of the eastbound train.

56. Jim and Juan are going to drive their motorcycles from San Francisco to Los Angeles. Jim leaves at 8:00 A.M. but Juan can't leave until 8:30 A.M. If Jim drives at 50 mph and Juan at 55 mph, at what time will Juan catch up with Jim?

57. Two airplanes take off from the airport at the same time, heading in opposite directions. After 2 hours the planes are 1500 miles apart. If the speeds of the planes differ by 50 mph, find the speed of the slower plane.

***58.** Annie drives from town A to town B in 3 hours (at a fixed speed). The traffic is heavier on the return trip. As a result, Annie's speed is 16 mph less, and the return trip takes 4 hours. What were the speeds on both parts of the trip?

***59.** A charter boat captain has 2 hours in which to leave the dock, head out to sea as far as possible, and then return to the dock. The boat's speeds on the outward and return journeys are 20 mph and 12 mph, respectively. How far can the boat travel out to sea?

***60.** At noon Larry and Linda start walking in the same direction. Larry's speed is $\frac{1}{15}$ mi/min; Linda's is less than that. However, when Larry has walked 2 miles, Linda doubles her pace and catches up with Larry at 1:15 P.M. How fast was Linda walking initially?

***61.** The length of a rectangular sheet of metal is 4 in. more than the width. As indicated in the figure, 3-in. squares are cut out from each corner of the rectangular sheet, and the sides are then folded up to form a box with no top. If the volume of the box is 180 in.^{3}, determine the dimensions of the original rectangular sheet.

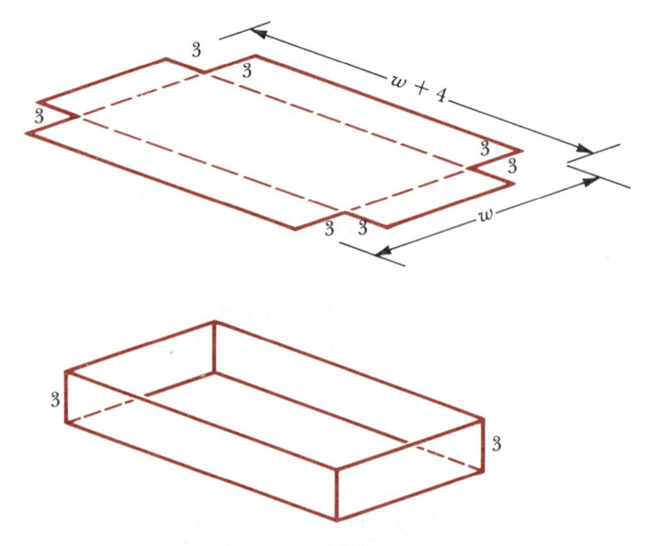

62. The units' digit in a certain two-digit number is three times the tens' digit. If the digits are reversed, the resulting number is 54 more than the original number. Find the original number. *Hint:* Let x denote the tens' digit in the original number. Then the units' digit is $3x$ and the two-digit number can be represented by the expression $10x + 3x$.

63. The sum of the digits in a certain two-digit number is 8. If the number is multiplied by the units' digit, the result is 156. Find the number.

***64.** The owner of an apartment building finds that 80 units can be rented when the rent is set at $400 per month. The owner also finds that for each $25 hike in rent (beyond the $400) two tenants will be lost. What rent should the owner charge to obtain a monthly revenue of $39,000? *Hint:* Let x denote the number of $25 price hikes that are necessary.

***65.** Suppose that you travel from town A to town B at a speed of v_1 mph and then return at a speed of v_2 mph. What is your average speed for the round trip (in terms of v_1 and v_2)?

66. If you travel for a hours at b mph and then c hours at d mph, what is your average speed for the trip?

67. In one savings account you deposit d dollars at an annual simple interest rate of $r\%$. At the same time you also deposit D dollars at $R\%$ in a second account. After how many years will the total interest amount to $d + D$ dollars?

***68.** A chemistry student has two acid solutions available to him. The first is $a\%$ acid and the second is $b\%$ acid. (The percents are by volume.) How many cm^3 of each solution should be mixed to obtain d cm^3 of a solution that is $c\%$ acid? (Assume that a, b, c, and d are positive and that $a < c < b$.)

69. The radius of a circle is r units. By how many units should the radius be increased so that the area increases by b square units?

***70.** A piece of wire a inches long is bent to form a right triangle in which the shortest side is b inches. Find the length of the hypotenuse.

71. How much water must be evaporated from b pounds of an $a\%$ salt solution to obtain a solution that is $c\%$ salt? (Assume that $c > a$.)

72. If the length of each side of a square is multiplied by a, the area increases by b^2 square units. Find the sides of the original square.

***73.** A piece of wire L inches long is cut into two pieces. Each piece is then bent to form a square. If the sum of the areas of the two squares is $5L^2/128$, how long are the two pieces of wire?

C **74.** You invest $5200 in a stock paying a 6.75% annual dividend and $8400 in a stock paying a 7.85% annual dividend. How many years will it take for the total earnings to be $2500? (Round off your answer to the nearest one-half year.)

C **75.** You deposit $2650 at 8.1% and $3300 at 9.2% simple interest. What additional amount would you need to deposit at 11.5% to yield an overall annual return of 10%? (Round off your answer to the nearest dollar.)

***76.** C A piece of wire 16 cm long is cut into two pieces, and each piece is then bent to form a circle. If the sum of the areas of the two circles is 12 cm^2, how long is the shorter piece of wire? (Round off your answer to two decimal places.)

C **77.** The height of a right circular cylinder is 1 ft. Find the radius, given that the total surface area is 660 ft^2. (Round off your answer to two decimal places.)

C **78.** In a certain right triangle, the hypotenuse is 205 cm and the two legs differ by 23 cm. Find the length of the shorter leg.

2.4 Other Types of Equations

The previous sections of this chapter dealt with linear and quadratic equations. In this section we consider several techniques that are useful in solving other types of equations. Throughout this section we'll concern ourselves only with solutions that are real numbers.

When n is a natural number, equations of the form $x^n = a$ are solved simply by rewriting the equation in terms of the appropriate nth root or roots. For example, if $x^5 = 34$, then $x = \sqrt[5]{34}$. If the exponent n is even, however, we have to remember that there are two real nth roots for each positive number. For instance, if

$$x^4 = 81$$

then

$$x = \pm\sqrt[4]{81} = \pm 3$$

Example 1 Solve: **(a)** $(x-1)^4 = 15$
(b) $3x^4 = -48$

Solution **(a)**
$$(x-1)^4 = 15$$
$$x - 1 = \pm\sqrt[4]{15}$$
$$x = 1 \pm \sqrt[4]{15}$$

As you can check for yourself now, both of the values, $1 + \sqrt[4]{15}$ and $1 - \sqrt[4]{15}$, satisfy the given equation. *Note:* You can estimate these values without a calculator. For instance, for $x = 1 + \sqrt[4]{15}$ we have

$$1 + \sqrt[4]{15} \approx 1 + \sqrt[4]{16} = 1 + 2 = 3$$

Therefore, $1 + \sqrt[4]{15}$ is a little less than 3. (In fact, a calculator shows that $1 + \sqrt[4]{15} \approx 2.97$.)

(b) For any real number x, x^4 is nonnegative. Therefore, the left-hand side of the equation is nonnegative, whereas the right-hand side is negative. Consequently, there are no real numbers satisfying the given equation.

In Example 2 we solve two equations by factoring. As with quadratic equations, the factoring method is justified by the Zero-Product property of real numbers discussed in Section 2.2.

Example 2 Solve: **(a)** $3x^3 - 12x^2 - 15x = 0$
(b) $x^4 + x^2 - 6 = 0$

Solution **(a)**
$$3x^3 - 12x^2 - 15x = 0$$
$$3x(x^2 - 4x - 5) = 0$$
$$3x(x-5)(x+1) = 0$$

Now, by setting each factor equal to zero, we obtain the three values $x = 0$, $x = 5$, and $x = -1$. As you can check, each of these numbers indeed satisfies the original equation. So we have three solutions: 0, 5, and -1. *Note:* If initially we'd divided both sides of the given equation by $3x$ to obtain $x^2 - 4x - 5 = 0$, then the solution $x = 0$ would have been overlooked.

(b)
$$x^4 + x^2 - 6 = 0$$
$$(x^2 - 2)(x^2 + 3) = 0$$

$x^2 - 2 = 0$	$x^2 + 3 = 0$
$x^2 = 2$	$x^2 = -3$
$x = \pm\sqrt{2}$	

By setting the first factor equal to zero, we've obtained the two values $x = \pm\sqrt{2}$. As you can check, these values do satisfy the given equation. When we set the second factor equal to zero, however, we obtained $x^2 = -3$, which has no real solutions (because the square of a real number must be nonnegative). We conclude therefore that the given equation in this case has just two real solutions, $\sqrt{2}$ and $-\sqrt{2}$.

The equation $x^4 + x^2 - 6 = 0$ that we considered in Example 2(b) is called an equation of *quadratic type*. This means that with a suitable substitution the resulting equation becomes quadratic. In particular here, suppose that we let

$$x^2 = t$$

and therefore

$$x^4 = t^2$$

Then the equation $x^4 + x^2 - 6 = 0$ becomes $t^2 + t - 6 = 0$, which is quadratic. This idea is exploited in the next three examples.

Example 3 Solve $x^4 - 6x^2 + 4 = 0$.

Solution This equation cannot readily be solved by factoring. However, it is of quadratic type. We let $x^2 = t$. Then $x^4 = t^2$, and the equation becomes

$$t^2 - 6t + 4 = 0$$

The quadratic formula then yields

$$t = \frac{6 \pm \sqrt{36 - 4(4)}}{2} = \frac{6 \pm \sqrt{20}}{2}$$

$$= \frac{6 \pm 2\sqrt{5}}{2} = 3 \pm \sqrt{5}$$

Consequently we have (in view of the definition of t)

$$x^2 = 3 \pm \sqrt{5}$$

and therefore

$$x = \pm\sqrt{3 \pm \sqrt{5}}$$

We've now found four values for x:

$$\sqrt{3 + \sqrt{5}} \qquad -\sqrt{3 + \sqrt{5}} \qquad \sqrt{3 - \sqrt{5}} \qquad -\sqrt{3 - \sqrt{5}}$$

As Exercise 76(a) asks you to check, each of these values indeed satisfies the original equation.

Example 4 Solve $6x^{-2} - x^{-1} - 2 = 0$.

Solution We'll show two methods. The first method depends on recognizing the equation as one of quadratic type.

First Method

Let $x^{-1} = t$. Then $x^{-2} = t^2$ and the equation becomes

$$6t^2 - t - 2 = 0$$

$$(3t - 2)(2t + 1) = 0$$

$$3t - 2 = 0 \quad \Big| \quad 2t + 1 = 0$$

$$t = \frac{2}{3} \quad \Big| \quad t = -\frac{1}{2}$$

If $t = \frac{2}{3}$, then $x^{-1} = \frac{2}{3}$. Therefore, $\frac{1}{x} = \frac{2}{3}$, and consequently, $x = \frac{3}{2}$. On the other hand, if $t = -\frac{1}{2}$, then $\frac{1}{x} = -\frac{1}{2}$, and consequently, $x = -2$. In summary, the two solutions are $\frac{3}{2}$ and -2.

Alternate Method

Multiplying both sides of the given equation by x^2 yields

$$6 - x - 2x^2 = 0$$

$$2x^2 + x - 6 = 0$$

$$(2x - 3)(x + 2) = 0$$

$$2x - 3 = 0 \quad \Big| \quad x + 2 = 0$$

$$x = \frac{3}{2} \quad \Big| \quad x = -2$$

Thus we have $x = \frac{3}{2}$ or $x = -2$, as obtained using the first method.

Example 5 Solve $4x^{4/3} + 15x^{2/3} - 4 = 0$.

Solution Let $x^{2/3} = t$. Then $x^{4/3} = t^2$, and the equation becomes

$$4t^2 + 15t - 4 = 0$$
$$(4t - 1)(t + 4) = 0$$
$$4t - 1 = 0 \quad\Big|\quad t + 4 = 0$$
$$t = \frac{1}{4} \quad\Big|\quad t = -4$$

We now have two cases to consider: $t = \frac{1}{4}$, and $t = -4$. With $t = \frac{1}{4}$ we have (in view of the definition of t)

$$x^{2/3} = \frac{1}{4}$$
$$(\sqrt[3]{x})^2 = \frac{1}{4}$$
$$\sqrt[3]{x} = \pm\frac{1}{2}$$
$$x = \left(\pm\frac{1}{2}\right)^3$$
$$= \pm\frac{1}{8}$$

Two real numbers are equal if and only if their cubes are equal.

As Exercise 76(b) asks you to verify, the values $\frac{1}{8}$ and $-\frac{1}{8}$ satisfy the original equation. On the other hand, with $t = -4$, we have

$$x^{2/3} = -4$$

or

$$(x^{1/3})^2 = -4$$

This last equation has no real solutions, since the square of a real number is never negative. In summary then, the only (real-number) solutions of the original equation are $\pm\frac{1}{8}$.

Some equations can be solved by raising both sides to the same power. In fact, if you review the solution given for Example 5, you'll see that we've already made use of this idea in cubing both sides of one of the equations there. As another example, consider the equation

$$\sqrt{x - 3} = 5$$

By squaring both sides of this equation we obtain $x - 3 = 25$ and, consequently, $x = 28$. As you can easily check, the value $x = 28$ does satisfy the original equation.

There is a complication, however, that may arise in raising both sides of an equation to the same power. Consider, for example, the equation

$$x - 2 = \sqrt{x}$$

By squaring both sides, we obtain

$$x^2 - 4x + 4 = x$$
$$x^2 - 5x + 4 = 0$$
$$(x - 4)(x - 1) = 0$$

Therefore

$$x = 4 \quad \text{or} \quad x = 1$$

The value $x = 4$ checks in the original equation, but the value $x = 1$ does not. (Verify this.) We say in this case that the value $x = 1$ is an *extraneous root* or *extraneous solution* of the original equation. [This extraneous root arises from the fact that if $a^2 = b^2$, then it need not be true that $a = b$. For instance, $(-3)^2 = 3^2$, but certainly $-3 \neq 3$.] In summary, any time that we square both sides of an equation (or raise both sides to an even integral power) there is the possibility of introducing extraneous roots. For this reason it is essential to check all candidates for solutions obtained in this manner.

Our final example for this section involves an equation containing several square root expressions. Before squaring both sides in such cases, it's usually a good idea to see if you can first isolate one of the radical expressions on one side of the equation.

Example 6 Solve $\sqrt{9+x} + \sqrt{1+x} - \sqrt{x+16} = 0$.

Solution

$$\sqrt{9+x} + \sqrt{1+x} = \sqrt{x+16}$$

We added $\sqrt{x+16}$ to both sides in preparation for squaring.

$$\begin{aligned}
(\sqrt{9+x} + \sqrt{1+x})^2 &= (\sqrt{x+16})^2 \\
9 + x + 2\sqrt{9+x}\,\sqrt{1+x} + 1 + x &= x + 16 \\
2\sqrt{9+x}\,\sqrt{1+x} &= 6 - x \\
(2\sqrt{9+x}\,\sqrt{1+x})^2 &= (6-x)^2 \\
4(9 + 10x + x^2) &= 36 - 12x + x^2 \\
3x^2 + 52x &= 0 \\
x(3x + 52) &= 0
\end{aligned}$$

$$x = 0 \quad \Big| \quad \begin{aligned} 3x + 52 &= 0 \\ x &= -\tfrac{52}{3} \end{aligned}$$

We've now found two possible solutions: 0 and $-\frac{52}{3}$. [Since these were obtained by squaring both sides of an equation, we need to check to see if either (or both) satisfy the original equation.] With $x = 0$, the original equation becomes $\sqrt{9} + \sqrt{1} - \sqrt{16} = 0$, or $3 + 1 - 4 = 0$, which is certainly true. On the other hand, with $x = -\frac{52}{3}$, the quantities underneath the radicals in the original equation are negative, and consequently the (real) square roots are undefined. So $-\frac{52}{3}$ is an extraneous solution, and $x = 0$ is the only solution of the given equation.

Exercise Set 2.4

In Exercises 1–75, find all the real solutions of each equation.

1. $2x^2 - 50x = 0$

2. $2x^3 - 50x = 0$

3. $y^3 - 1000 = 0$

4. $y^3 - y = 0$

5. $3t^4 - 48t^2 = 0$

6. $t^4 - 16 = 0$

7. $225(x-1) - x^2(x-1) = 0$

8. $x^2(x+4)^2 + 6x(x+4)^2 + 9(x+4)^2 = 0$

9. $4y^3 - 20y^2 + 25y = 0$

10. $y^4 + 27y = 0$

11. $t^4 + 2t^3 - 3t^2 = 0$

12. $2t^5 + 5t^4 - 12t^3 = 0$

13. $6x - 23x^2 - 4x^3 = 0$

14. $x^5 - 36x = 0$

15. $x^4 - x^2 - 6 = 0$

16. $x^4 - 5x^2 + 6 = 0$
17. $y^4 + 4y^2 - 5 = 0$
18. $y^4 + 6y^2 + 5 = 0$
19. $9t^4 - 3t^2 - 2 = 0$
20. $4t^4 - 5t^2 + 1 = 0$
21. $x^6 - 10x^4 + 24x^2 = 0$
22. $2x^5 - 15x^3 - 27x = 0$
23. $x^4 + x^2 - 1 = 0$
24. $x^4 - x^2 + 1 = 0$
25. $x^4 + 3x^2 - 2 = 0$
26. $2x^4 + x^2 + 2 = 0$
27. $x^6 + 7x^3 = 8$
28. $x^6 + 28x^3 + 27 = 0$
29. $t^{-2} - 7t^{-1} + 12 = 0$
30. $8t^{-2} - 17t^{-1} + 2 = 0$
31. $12y^{-2} - 23y^{-1} = -5$
32. $y^{-2} - y^{-1} = 0$
33. $4x^{-4} - 33x^{-2} - 27 = 0$
34. $x^{-4} + x^{-2} + 1 = 0$
35. $\sqrt{x-8} = 4$
36. $\sqrt{1-3x} = 2$
37. $\sqrt{x^2+5x-2} = 2$
38. $\sqrt{x^4 - 13x^2 + 37} = 1$
39. $\sqrt{y+2} = y - 4$
40. $\sqrt{2y} = y - 4$
41. $t^{2/3} = 9$
42. $t^{3/2} = 8$
43. $x^4 = 81$
44. $x^4 = -81$
45. $(y-1)^3 = 7$
46. $(2y+3)^4 = 5$
47. $(t+1)^5 = -243$
48. $(t+3)^4 = 625$
49. $9x^{4/3} - 10x^{2/3} + 1 = 0$
50. $x^{4/3} + 3x^{2/3} - 28 = 0$
51. $x^{3/2} - x^{1/2} = 0$
52. $5x + 9x^{1/2} - 4 = 0$
53. $\sqrt{1-2x} + \sqrt{x+5} = 4$
54. $\sqrt{x-5} - \sqrt{x+4} + 1 = 0$
55. $\sqrt{3+2t} + \sqrt{-1+4t} = 1$
56. $\sqrt{2t+5} - \sqrt{8t+25} + \sqrt{2t+8} = 0$
57. $\sqrt{2+y} + \sqrt{3-y} - 3 = 0$
58. $\sqrt{2y-3} - \sqrt{3y+3} + \sqrt{3y-2} = 0$
59. $\sqrt{3t+4} - \sqrt{2t-4} = 2$
***60.** $\sqrt{8+2t} + \sqrt{5+t} = \sqrt{15+3t}$
61. $\sqrt{2x+3} + \sqrt{x+2} = 2$
62. $\sqrt{2x+1} + \sqrt{x+4} = 1$
63. $\sqrt{2x+6} + \sqrt{x+4} - \sqrt{8x+9} = 0$
64. $\sqrt{x-4} - \sqrt{x+4} + \sqrt{x-1} = 0$
***65.** $\sqrt{x - \sqrt{x^2-1}} + \sqrt{x + \sqrt{x^2-1}} = \sqrt{2x^3+2}$
66. $\sqrt{a-x} + \sqrt{b-x} = \sqrt{a+b-2x} \quad (b > a > 0)$
***67.** $\sqrt{\sqrt{x}+\sqrt{a}} + \sqrt{\sqrt{x}-\sqrt{a}} = \sqrt{2\sqrt{x}+2\sqrt{b}}$
68. $x = \sqrt{3x + x^2 - 3\sqrt{3x+x^2}}$
69. $\dfrac{\sqrt{x}-a}{\sqrt{x}} - \dfrac{\sqrt{x}+a}{\sqrt{x}-b} = 0 \quad (a > 0, b > 0)$
Suggestion: Let $\sqrt{x} = t$.
70. $x - \sqrt{x^2 - x} = \sqrt{x} \quad (x > 0)$ *Suggestion:* If $x \neq 0$, then you can divide through by $\sqrt{x}$ to obtain a simpler equation.
71. $\sqrt{x^2-x-1} - \dfrac{2}{\sqrt{x^2-x-1}} = 1$ *Hint:* Let $t = x^2 - x - 1$.
***72.** $\sqrt{x^2+3x-4} - \sqrt{x^2-5x+4} = x - 1 \quad (x > 4)$
Hint: Factor the expressions underneath the radicals. Then note that $\sqrt{x-1}$ is a factor of both sides of the equation.
***73.** $\sqrt{\dfrac{x-a}{x}} + \sqrt{\dfrac{x}{x-a}} = 5 \quad (a \neq 0)$ *Hint:* Let $t = \dfrac{x-a}{x}$. Then $\dfrac{1}{t} = \dfrac{x}{x-a}$.
***74.** $\sqrt{p+4q-5t} + \sqrt{4p+q-5t} = 3\sqrt{p+q-2t}$
(Assume that p and q are constants and $q > p$.)
***75.** $-x = \sqrt{1 - \sqrt{1+x}}$ *Hint:* After squaring once and rearranging, you'll find that the quantity $\sqrt{x+1}$ is a factor of both sides.
76. **(a)** Verify that the four solutions obtained in Example 3 indeed satisfy the original equation.
(b) Verify that the values $x = \frac{1}{8}$ and $x = -\frac{1}{8}$ satisfy the equation $4x^{4/3} + 15x^{2/3} - 4 = 0$.

For Exercises 77–81, solve the equations and round off your answers to two decimal places.

C **77.** $x^4 - 2x^2 - 4 = 0$
C **78.** $(3x-5)^7 = 405$
C **79.** $2x\sqrt{x^2+4} + 2x\sqrt{x^2+1} = 3$
C ***80.** $\sqrt{x^2+x} + \dfrac{1}{\sqrt{x^2+x}} = \dfrac{5}{2}$ *Hint:* Let $t = x^2 + x$.
C **81.** The radii of three lead spheres are 10.01 cm, 15.43 cm, and 20.14 cm, respectively. Suppose that these three spheres are melted down and recast to form one large sphere. Find the radius of that sphere. *Hint:* Use the formula $V = \frac{4}{3}\pi r^3$ for the volume of a sphere.

2.5 Inequalities

The fundamental results of mathematics are often inequalities *rather than* equalities.

E. Beckenbach and R. Bellman in *An Introduction to Inequalities* (New York: Random House, 1961)

If we replace the equals sign in an equation with any one of the four symbols $<$, $\leq$, $>$, or $\geq$, we obtain an *inequality*. As with equations in one variable, a real number is a solution of an inequality if we obtain a true statement when the variable is replaced by that real number. For example, the value $x = 5$ is a solution of the inequality $2x - 3 < 8$, because when $x = 5$ we have

$$2(5) - 3 < 8$$
$$7 < 8 \qquad \text{True}$$

We also say in this case that the value $x = 5$ *satisfies* the inequality. To *solve* an inequality means to find all of the solutions. The set of all solutions of an inequality is called (naturally enough) the *solution set.*

As we solve the inequalities in this section and the next, you'll see that the solution sets here are always one or more intervals of real numbers. In Section 1.3 we defined four types of bounded intervals:

$$(a, b) \qquad [a, b] \qquad (a, b] \qquad [a, b)$$

For our present purposes it is also convenient to introduce a notation for unbounded intervals of real numbers. As an example, we shall indicate the set of all real numbers that are greater than 2 with the notation $(2, \infty)$.

Comment and Caution The symbol ∞ is read *infinity*. It is not a real number, and its use in the context $(2, \infty)$ is only to indicate that the interval has no right-hand boundary. In the box that follows we define the five types of unbounded intervals. Notice that the last interval in the table is actually the entire real number line.

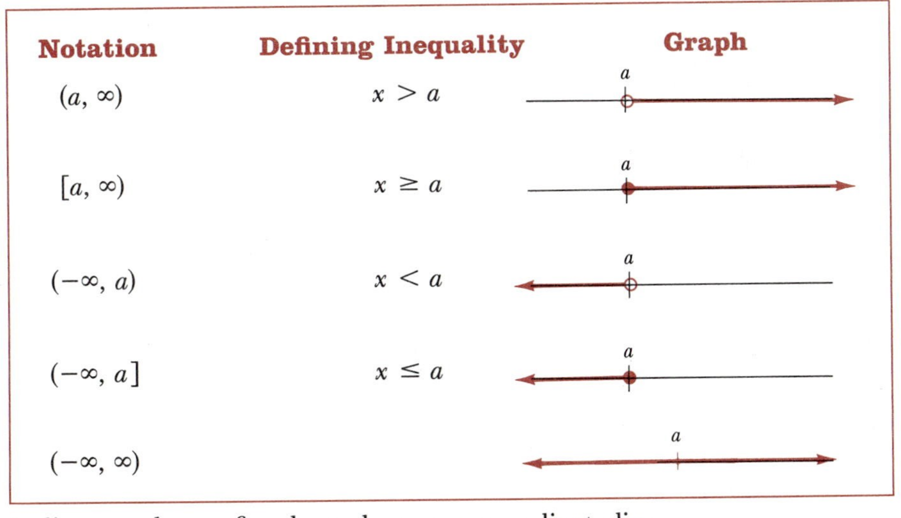

Notation	Defining Inequality	Graph
(a, ∞)	$x > a$	
$[a, \infty)$	$x \geq a$	
$(-\infty, a)$	$x < a$	
$(-\infty, a]$	$x \leq a$	
$(-\infty, \infty)$		

Example 1 Indicate each set of real numbers on a coordinate line.

(a) $(-\infty, 4]$ **(b)** $(-3, \infty)$

Solution **(a)** The interval $(-\infty, 4]$ consists of all real numbers that are less than or equal to 4. See Figure 1.

(b) The interval $(-3, \infty)$ consists of all real numbers that are greater than -3. See Figure 2.

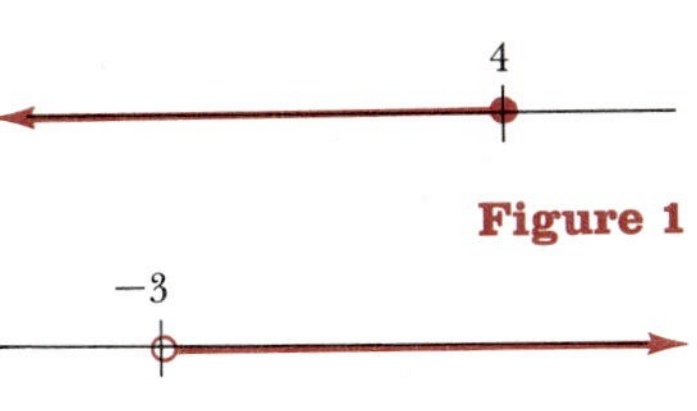

Figure 1

Figure 2

Recall that two equations are said to be equivalent if they have exactly the same solutions. Similarly, two inequalities are *equivalent* if they have the same solution set. Most of the procedures used in solving inequalities are quite similar to those for equalities. For example, adding or subtracting the same number from both sides of an inequality produces an equivalent inequality. We need to be careful, however, in multiplying or dividing both sides of an inequality by the same nonzero number. For instance, suppose that we start with the inequality $2 < 3$ and multiply both sides by 5. That yields $10 < 15$, which is certainly true. In the same way, however, had we multiplied both sides of the inequality $2 < 3$ by -5, we would have obtained $-10 < -15$, which is false. In summary, multiplying both sides of an inequality by the same positive number preserves the inequality, whereas multiplying by a negative number reverses the inequality. In the box that follows we list some of the principal properties of inequalities. In general, whenever we use properties 1 or 2 in solving an inequality, we obtain an equivalent inequality. Let us note in passing that there are obvious modifications of each property that result from the equivalence of the statements $a < b$ and $b > a$. For example, property 3 could just as well be written this way: if $b > a$ and $c > b$, then $c > a$.

Property Summary

Properties of Inequalities

Property	*Example*
1. If $a < b$, then $a + c < b + c$ and $a - c < b - c$.	If $x - 3 < 0$, then $(x - 3) + 3 < 0 + 3$, and consequently $x < 3$.
2. **(a)** If $a < b$, and c is positive, then $ac < bc$ and $a/c < b/c$.	If $\frac{1}{2}x < 4$, then $2(\frac{1}{2}x) < 2(4)$, and consequently $x < 8$.
(b) If $a < b$, and c is negative, then $ac > bc$ and $a/c > b/c$.	If $-\frac{x}{5} < 6$ then $(-5)\left(-\frac{x}{5}\right) > (-5)(6)$, and consequently $x > -30$.
3. (The transitive property) If $a < b$ and $b < c$, then $a < c$.	If $a < x$ and $x < 2$, then $a < 2$.

Example 2 Solve $2x - 3 < 5$.

Solution The work follows the same pattern you would use to solve the equation $2x - 3 = 5$. We have

$$\begin{aligned} 2x - 3 &< 5 \\ 2x &< 8 \quad \text{adding 3} \\ x &< 4 \quad \text{dividing by 2} \end{aligned}$$

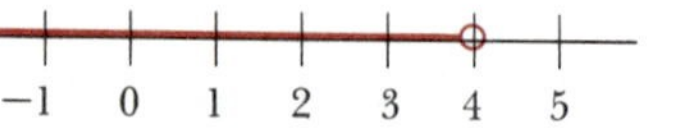

Figure 3

From this sequence of steps we see that the original inequality is equivalent to $x < 4$. The solution set is therefore $(-\infty, 4)$. Figure 3 shows this solution set on a number line.

Example 3 Solve $4t + 8 \leq 7(1 + t)$.

Solution

$$4t + 8 \le 7(1 + t)$$
$$4t + 8 \le 7 + 7t$$
$$-3t \le -1 \qquad \text{We subtracted } 7t \text{ and 8 from both sides.}$$

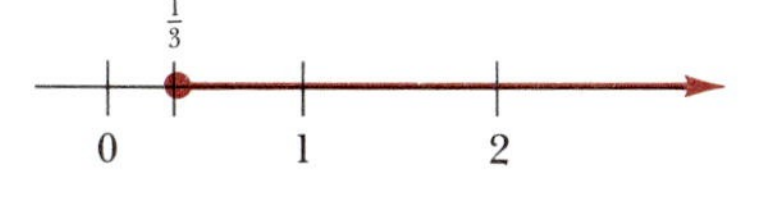

$$t \ge \frac{1}{3} \qquad \text{We multiplied by } -1/3\text{; this reverses the inequality.}$$

Figure 4

The solution set is therefore $[\frac{1}{3}, \infty)$. See Figure 4.

In the next example we solve the inequality

$$-\frac{1}{2} < \frac{3 - x}{-4} \le \frac{1}{2}$$

By definition, this is equivalent to the pair of inequalities

$$-\frac{1}{2} < \frac{3 - x}{-4} \quad \text{and} \quad \frac{3 - x}{-4} \le \frac{1}{2}$$

One way to proceed here would be to first determine the solution set for each inequality. Then the set of real numbers common to both solution sets would be the required solution set for the original inequality. However, the method shown in Example 4 is more efficient.

Example 4 Solve $-\frac{1}{2} < \frac{3 - x}{-4} \le \frac{1}{2}$.

Solution We begin by multiplying through by -4. Remember, this will reverse the inequalities:

$$2 > 3 - x \ge -2$$

Next, with a view toward isolating x, we subtract 3 to obtain

$$-1 > -x \ge -5$$

Finally, multiplying through by -1, we have

$$1 < x \le 5$$

The solution set is therefore the interval $(1, 5]$.

Example 5 Suppose that you inherit \$5000 under the stipulation that it be invested in two stocks, A and B. Stock A pays an 8% annual dividend and stock B pays a 9% annual dividend. (The dividends are computed in the same way that we computed simple interest in Section 2.3.) What is the largest amount that you can invest in stock A if the total dividends each year are to be at least \$435?

Solution First, just to get a feeling for the problem, let's see what would happen if the entire \$5000 were invested in one stock. If you invest all \$5000 in stock A, the annual dividend is (\$5000)(0.08)(1) = \$400. That falls short of the required \$435. On the other hand, if the \$5000 were invested in stock B, the annual dividend would be (\$5000)(0.09)(1), which is \$450. That exceeds the required minimum of \$435. The question the problem asks is, what's the most that can be invested in stock A if the total annual dividends are to be at least \$435? Let x denote this amount that can be invested in stock A. Then $5000 - x$ represents the amount (in dollars) to be invested in stock B. Since the combined dividends each year are to be at least \$435, we have

$$\begin{aligned} x(.08) + (5000 - x)(.09) &\geq 435 \\ 8x + (5000 - x)9 &\geq 43500 \\ 8x + 45000 - 9x &\geq 43500 \\ -x &\geq -1500 \\ x &\leq 1500 \end{aligned}$$

Thus, at most \$1500 can be invested in stock A at 8%.

In the next example reference is made to the Celsius and Fahrenheit scales* for measuring temperatures. The formula relating the temperature readings on the two scales is

$$F = \frac{9}{5}C + 32$$

Example 6 Over the temperature range $32 \leq F \leq 39.2$ on the Fahrenheit scale, water contracts (rather than expands) with increasing temperature. What is the corresponding temperature range on the Celsius scale?

Solution $32 \leq F \leq 39.2$ given

$32 \leq \frac{9}{5}C + 32 \leq 39.2$ We substituted $\frac{9}{5}C + 32$ for F.

$0 \leq \frac{9}{5}C \leq 7.2$ subtracting 32

$0 \leq C \leq \frac{5}{9}(7.2)$ multiplying by $\frac{5}{9}$

$0 \leq C \leq 4$

Thus, a range of 32°F–39.2°F on the Fahrenheit scale corresponds to 0°C–4°C on the Celsius scale.

As preparation for the final set of examples in this section, we state the following useful theorem relating absolute values and inequalities.

Theorem

Suppose that $a > 0$. Then

$$|u| < a \quad \text{if and only if } -a < u < a$$

and

$$|u| > a \quad \text{if and only if } u < -a \quad \text{or} \quad u > a$$

*The Celsius scale was devised in 1742 by the Swedish astronomer Anders Celsius. The Fahrenheit scale was first used by the German physicist Gabriel Fahrenheit in 1724.

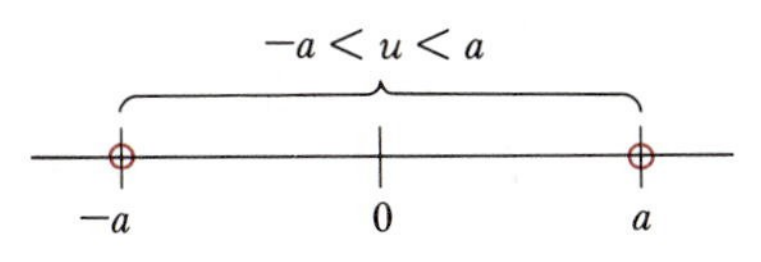

Figure 5

You can see why this theorem is valid by thinking in terms of distance and position on a number line. The condition $-a < u < a$ means that u lies between $-a$ and a, as indicated in Figure 5. But this is the same as saying that the distance from u to zero is less than a, which in turn can be written $|u| < a$. (The second part of the theorem can be justified in a similar manner.)

Example 7 Solve: **(a)** $|x| < 1$

(b) $|x| \geq 1$

Solution

(a) According to the theorem just discussed, the condition $|x| < 1$ is equivalent to

$$-1 < x < 1$$

The solution set is therefore the open interval $(-1, 1)$.

(b) In view of the second part of the theorem, the inequality $|x| \geq 1$ is satisfied when x satisfies either of the inequalities

$$x \leq -1 \quad \text{or} \quad x \geq 1$$

Now the solution sets for these last two inequalities are $(-\infty, -1]$ and $[1, \infty)$, respectively. Consequently, the solution set for the given inequality $|x| \geq 1$ consists of the two intervals $(-\infty, -1]$ and $[1, \infty)$.

In Example 7(b) we found that the solution set consisted of two intervals on the number line. There is a convenient notation for describing such sets. Given any two sets A and B, we define the set A ∪ B (read A *union* B) to be the set of all elements that are in A or in B (or in both). For example, if A = {1, 2, 3} and B = {4, 5}, then A ∪ B = {1, 2, 3, 4, 5}. As another example, the union of the two closed intervals [3, 5] and [4, 7] is given by

$$[3, 5] \cup [4, 7] = [3, 7]$$

because the numbers in the interval [3, 7] are precisely those numbers that are in [3, 5] or [4, 7] (or in both). Using this notation, we can write the solution set for Example 7(b) as

$$(-\infty, -1] \cup [1, \infty)$$

Example 8 Solve $|x - 3| < 1$.

Solution We'll show two methods.

First Method We use the theorem preceding Example 7. With $u = x - 3$ and $a = 1$, the theorem tells us that the given inequality is equivalent to

$$-1 < x - 3 < 1$$

or (by adding 3)

$$2 < x < 4$$

The solution set is therefore the open interval $(2, 4)$.

Alternate Method The given inequality tells us that x must be less than one unit away from 3 on the number line. Looking one unit to either side of 3 then, we see that x must lie strictly between 2 and 4. The solution set is therefore $(2, 4)$, as obtained using the first method.

Example 9 Solve $\left|1 - \frac{t}{2}\right| \le 5$.

Solution We use the fact that $|u| \le a$ is equivalent to $-a \le u \le a$. Letting $u = 1 - \frac{t}{2}$ and $a = 5$, it follows that the given inequality is equivalent to

$$-5 \le 1 - \frac{t}{2} \le 5$$

$$-6 \le -\frac{t}{2} \le 4 \qquad \text{subtracting 1}$$

$$12 \ge t \ge -8 \qquad \text{We multiplied by } -2\text{; remember, this reverses the inequalities.}$$

or (still equivalently)

$$-8 \le t \le 12$$

The solution set is therefore the closed interval $[-8, 12]$.

Example 10 Solve $\left|1 - \frac{t}{2}\right| > 5$.

Solution We'll show two methods.

First Method Applying the theorem with $u = 1 - \frac{t}{2}$ and $a = 5$, we have

$$1 - \frac{t}{2} < -5 \quad \text{or} \quad 1 - \frac{t}{2} > 5$$

$$-\frac{t}{2} < -6 \qquad\qquad -\frac{t}{2} > 4$$

$$t > 12 \qquad\qquad t < -8$$

This tells us that the given inequality is satisfied precisely when t is in either of the intervals $(12, \infty)$ or $(-\infty, -8)$. In other words, the solution set is $(-\infty, -8) \cup (12, \infty)$.

Alternate Method The numbers satisfying the given inequality are precisely those numbers that *do not* satisfy $\left|1 - \frac{t}{2}\right| \le 5$. But, as we've just seen in Example 9, the solution set for $\left|1 - \frac{t}{2}\right| \le 5$ is the closed interval $[-8, 12]$. Therefore, the solution set for $\left|1 - \frac{t}{2}\right| > 5$ consists of those real numbers not in $[-8, 12]$. Thus, as indicated in Figure 6, the required solution set is $(-\infty, -8) \cup (12, \infty)$.

Figure 6

Exercise Set 2.5

In Exercises 1–6, indicate each set of numbers on a coordinate line.

1. $(0, \infty)$ **2.** $(-\infty, -5)$ **3.** $[-1, \infty)$

4. $(-\infty, 1]$

5. The set of all real numbers that are not in the open interval $(0, 1)$.

6. The set of all real numbers that are not in the interval $(-\infty, 3)$.

Solve the inequalities in Exercises 7–38. (Specify the solution sets using interval notation.)

7. $x + 5 < 4$

8. $2x - 7 < 11$

9. $1 - 3x \leq 0$

10. $6 - 4x \leq 22$

11. $4x - 3 > 33$

12. $5 - 8x > -19$

13. $\frac{3}{2}x + 4 \geq -1$

14. $1 - x \geq \pi$

15. $3x - 1 < 1 + x$

16. $4x + 6 < 3(x - 1) - x$

17. $2(t - 1) - 3(t + 1) \leq -5$

18. $1 - 2(t + 3) - t \leq 1 - 2t$

19. $t - 4[1 - (t - 1)] > 7 + 10t$

20. $(t - 1)(t - 2) \geq (t - 2)(t - 3)$

21. $(t + 4)(t - 4) \geq t(t + 3) - 1$

22. $(t + a)(t + b) < t^2 \quad (a < 0 \text{ and } b < 0)$

23. $\dfrac{3x}{5} - \dfrac{x - 1}{3} < 1$

24. $\dfrac{2x + 1}{2} + \dfrac{x - 1}{3} < x + \dfrac{1}{2}$

25. $\dfrac{x - 1}{4} - \dfrac{2x + 3}{5} \leq x$

26. $\dfrac{x}{2} - \dfrac{8x}{3} + \dfrac{x}{4} > \dfrac{23}{6}$

27. $3\left(\dfrac{x}{2} + 1\right) - 2\left(\dfrac{x}{3} - 1\right) \geq 12$

28. $2(1 - x) - \dfrac{x + 3}{4} \geq 1 + \dfrac{x}{2}$

29. $-2 \leq x - 6 \leq 0$

30. $-3 \leq 2x + 1 \leq 5$

31. $-6 < \dfrac{x + 4}{3} < -2$

32. $4 < 1 - t < 5$

33. $-1 \leq \dfrac{1 - 4t}{3} \leq 1$

34. $\dfrac{2}{3} \leq \dfrac{5 - 3t}{-2} \leq \dfrac{3}{4}$

35. $0.99 < \dfrac{x}{2} - 1 < 0.999$

36. $\dfrac{9}{10} < \dfrac{3x - 1}{-2} < \dfrac{91}{100}$

***37.** $\dfrac{3}{2t + 4} > 5$ *Hint:* Consider two cases, corresponding to whether $2t + 4$ is positive or $2t + 4$ is negative.

***38.** $\dfrac{4}{2t + 4} < -2$ *Hint:* See the hint for Exercise 37.

39. Suppose that you inherit \$6000 under the stipulation that it be invested in two particular stocks, one paying a 7% dividend annually and the other paying 9%. What is the largest amount that you can invest in the stock paying the 7% dividend if you want the total yearly dividends to be at least \$500?

40. Data from the Apollo 11 moon mission in July 1969 show that temperature readings on the lunar surface vary over the interval $-183 \leq C \leq 112$ on the Celsius scale. What is the corresponding interval on the Fahrenheit scale? (Round off the numbers you obtain to the nearest integers.)

41. From the cloud tops of Venus to the planet's surface, temperature readings range over the interval $-25 \leq C \leq 475$ on the Celsius scale. What is the corresponding range on the Fahrenheit scale?

42. Measurements sent back to Earth from the Viking I spacecraft in June 1976 indicated temperature readings on the surface of Mars in the range $-139 \leq C \leq -28$. What is the corresponding range on the Fahrenheit scale? (Round off the numbers you obtain to the nearest integers.)

43. If an object is projected vertically upward with an initial velocity of v_0 ft/sec, then its velocity v (in ft/sec) after t sec is given by

$$v = -32t + v_0$$

If the initial velocity is 60 ft/sec, during what time interval will the velocity be in the range $50 \geq v \geq 40$? (Leave your answer in terms of fractions rather than decimals.)

44. The width and length of a rectangle are 5 units and x units, respectively.

(a) Show that the perimeter P is given by $P = 2x + 10$.

(b) For which interval of x-values will the perimeter be in the range $100 \leq P \leq 200$?

45. Company A charges an \$80 per month rental fee for a piano. Company B will rent the same piano for only \$50 per month, but they have an additional initial charge of \$300. For which lengths of time would it be less expensive to rent from company B? *Hint:* Let n represent the number of months. Then company A charges $80n$ dollars in that period. What does company B charge?

Solve the inequalities in Exercises 46–63. (Specify the solution sets using interval notation.)

46. $|x| < 6$

47. $|x| \leq \frac{1}{2}$

48. $|x| \geq 2$

49. $|x| > 0$

50. $|x| \geq 0$

51. $|x - 2| < 1$

52. $|x - 4| < 4$

53. $|1 - x| \leq 5$

54. $|3x + 5| < 17$

55. $|3x + 5| \geq 17$

56. $|4 - 5t| < \frac{1}{10}$

57. $|x - a| < b$

58. $\left|\dfrac{x - 2}{3}\right| < 4$

59. $\left|\dfrac{x + 1}{2} - \dfrac{x - 1}{3}\right| < 1$

60. $\left|\dfrac{3(x - 2)}{4} + \dfrac{4(x - 1)}{3}\right| \leq 2$

61. $|(x + 1)^2 - x^2| < 0.01$

62. **(a)** $|(x + h)^2 - x^2| < 3h^2 \quad (h > 0)$

(b) $|(x + h)^2 - x^2| < 3h^2 \quad (h < 0)$

63. **(a)** $|3(x+2)^2 - 3x^2| < \frac{1}{10}$
(b) $|3(x+2)^2 - 3x^2| < \epsilon \qquad (\epsilon > 0)$

***64.** Prove that each of the following inequalities is valid for all positive numbers a and b. *Hint:* Use the following property of inequalities. If x and y are positive, then the inequality $x \leq y$ is equivalent to $x^2 \leq y^2$.

(a) $\dfrac{2ab}{a+b} \leq \sqrt{ab}$ (The *harmonic mean* is less than or equal to the *geometric mean.*)

(b) $\sqrt{ab} \leq \dfrac{a+b}{2}$ (The *geometric mean* is less than or equal to the *arithmetic mean.*)

(c) $\dfrac{a+b}{2} \leq \sqrt{\dfrac{a^2+b^2}{2}}$ (The *arithmetic mean* is less than or equal to the *root mean square.*)

65. **(a)** C Complete the table, rounding off to four decimal places.

x	y	$\frac{x}{y} + \frac{y}{x}$	True or False: $\frac{x}{y} + \frac{y}{x} \geq 2$
1	1		
2	3		
3	5		
4	7		
5	9		
9	10		
49	50		
99	100		

(b) Prove that for all positive numbers x and y we have $\frac{x}{y} + \frac{y}{x} \geq 2$. *Hint:* Use the inequality given in Exercise 64(b). Take $a = \frac{x}{y}$ and $b = \frac{y}{x}$.

66. **(a)** C Complete the table, rounding off your answers to three decimal places.

n	$\sqrt{n+1} - \sqrt{n-1}$
1	
2	
3	
4	
5	
500	
1000	

(b) What is the smallest natural number n for which $\sqrt{n+1} - \sqrt{n-1} < 0.1$? *Hint:* You need to solve the given inequality and then round your answer up to the next whole number. To solve the inequality, first write it as

$$\sqrt{n+1} < 0.1 + \sqrt{n-1}$$

and then apply the hint given in Exercise 64.

67. **(a)** C Complete the table, rounding off your answers to two decimal places.

n	$\sqrt{n+1} + \sqrt{n-1}$
1	
2	
3	
4	
5	
10	
5000	
10000	

(b) What is the largest natural number n such that $\sqrt{n+1} + \sqrt{n-1} < 100$?

2.6 More on Inequalities

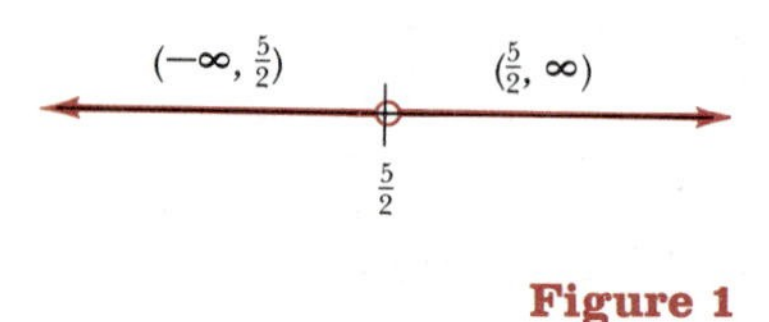

Figure 1

In this section we are going to solve inequalities involving polynomials and quotients of polynomials. Some observations about linear polynomials and inequalities will be helpful in introducing the key ideas. Consider, for example, the linear inequality $2x - 5 < 0$, along with its solution set $(-\infty, \frac{5}{2})$. Let us call a solution of the corresponding equality $2x - 5 = 0$ a *critical number* for the given inequality. So, in this case, the only critical number is $x = \frac{5}{2}$. As indicated in Figure 1, the critical number $\frac{5}{2}$ divides the number line into two intervals, namely $(-\infty, \frac{5}{2})$ and $(\frac{5}{2}, \infty)$. Notice that on each of the intervals determined by the critical number, the algebraic sign of $2x - 5$ is constant. (That is, for $x < \frac{5}{2}$ the value of $2x - 5$ is always negative; for $x > \frac{5}{2}$ the value of $2x - 5$ is always positive.) More generally, it can be shown that this same type of behavior regarding persistence of sign occurs with all polynomials, and indeed with quotients of polynomials as well. This important fact, along with the definition of a critical number, is presented in the box that follows.

> Let P and Q denote polynomials and consider the following four inequalities:
>
> $$\frac{P}{Q} < 0 \qquad \frac{P}{Q} \le 0 \qquad \frac{P}{Q} > 0 \qquad \frac{P}{Q} \ge 0$$
>
> The *critical numbers* for each of these inequalities are the real numbers for which $P = 0$ or $Q = 0$. It can be proved that the algebraic sign of P/Q is constant on each of the intervals determined by these critical numbers.

We'll show how this result is applied by solving the quadratic inequality

$$x^2 - 2x - 3 < 0$$

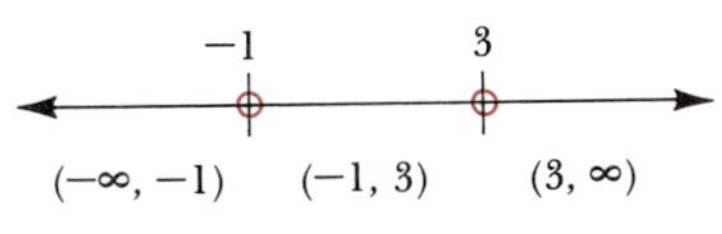

Figure 2

First we determine the critical numbers by solving the corresponding equation $x^2 - 2x - 3 = 0$. Since this can be rewritten as $(x + 1)(x - 3) = 0$, we see that the critical numbers are $x = -1$ and $x = 3$. Next we locate these numbers on a coordinate line. As Figure 2 indicates, this divides the coordinate line into three distinct intervals: $(-\infty, -1)$, $(-1, 3)$, and $(3, \infty)$. Now, according to the result stated in the box just prior to this example, no matter what x-value we choose in the interval $(-\infty, -1)$, the resulting sign of $x^2 - 2x - 3$ will always be the same. Thus, to see what that sign is, we choose any convenient *test number* from $(-\infty, -1)$, say $x = -2$, and substitute:

$$\text{If} \quad x = -2 \quad \text{then} \quad x^2 - 2x - 3 = (-2)^2 - 2(-2) - 3 = 5 > 0$$

(Whether you substitute in the original polynomial or in the factored form is merely a matter of preference or convenience.) From this we conclude that the values of $x^2 - 2x - 3$ are positive *throughout* the interval $(-\infty, -1)$, and consequently none of the numbers in this interval satisfy the given inequality. Next, what about the interval $(-1, 3)$? Choosing $x = 0$ for a test number, we have

$$x^2 - 2x - 3 = 0^2 - 2(0) - 3 = -3 < 0$$

Therefore the values of $x^2 - 2x - 3$ are negative throughout the interval $(-1, 3)$ and consequently all the numbers in this interval satisfy the inequality. Finally, for the interval $(3, \infty)$, we choose $x = 4$ as a test number. As you can check, the resulting value of the polynomial in this case is positive, and therefore none of the numbers in $(3, \infty)$ satisfy the given inequality. In summary then, the solution set for the inequality $x^2 - 2x - 3 < 0$ is $(-1, 3)$. Incidentally, it's worth noting that the work just carried out also provides us with three additional pieces of information:

The solution set for $x^2 - 2x - 3 \leq 0$ is $[-1, 3]$.

The solution set for $x^2 - 2x - 3 > 0$ is $(-\infty, -1) \cup (3, \infty)$.

The solution set for $x^2 - 2x - 3 \geq 0$ is $(-\infty, -1] \cup [3, \infty)$.

In the box we summarize the steps for solving polynomial inequalities:

Steps for Solving Polynomial Inequalities

1. If necessary, rewrite the inequality so that the polynomial is on the left-hand side and zero is on the right-hand side.
2. Find the critical numbers for the inequality and locate them on a number line.
3. List the intervals determined by the critical numbers.
4. From each interval choose a convenient test number. Then evaluate the polynomial with that test number to determine the sign of the polynomial throughout the interval.
5. Use the information obtained in the previous step to specify the required solution set. [Don't forget to take into account whether the original inequality involves strict inequalities ($<$ or $>$) or nonstrict inequalities ($\leq$ or $\geq$).]

Example 1 Solve $2x^2 - 5x \geq 3$.

Solution The inequality is equivalent to $2x^2 - 5x - 3 \geq 0$. The critical numbers are then found by solving the equation $2x^2 - 5x - 3 = 0$. Since this equation can be written $(2x + 1)(x - 3) = 0$, we see that the critical numbers are $x = -\frac{1}{2}$ and $x = 3$. As indicated in Figure 3, these numbers divide the number line into three distinct intervals. We need to choose one test number from each interval. From $(-\infty, -\frac{1}{2})$ we choose $x = -1$; from $(-\frac{1}{2}, 3)$ we choose $x = 0$; from $(3, \infty)$ we choose $x = 4$. Now, for each test number we check to see whether the value of the polynomial is positive or negative. Remember that this will tell us the sign of the polynomial *throughout* each interval. It's convenient to tabulate the work as follows.

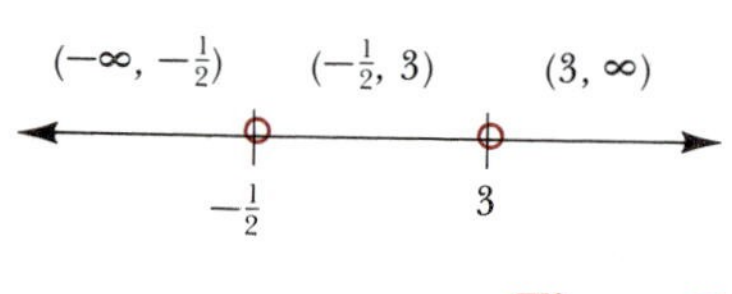

Figure 3

Interval	Test Number	Value of $2x^2 - 5x - 3$	Sign of $2x^2 - 5x - 3$
$\left(-\infty, -\frac{1}{2}\right)$	-1	$2(-1)^2 - 5(-1) - 3 = 4$	$+$
$\left(-\frac{1}{2}, 3\right)$	0	$2(0)^2 - 5(0) - 3 = -3$	$-$
$(3, \infty)$	4	$2(4)^2 - 5(4) - 3 = 9$	$+$

From these results we see that the values of the polynomial $2x^2 - 5x - 3$ are positive on both the intervals $(-\infty, -\frac{1}{2})$ and $(3, \infty)$. Furthermore, the value of the polynomial is zero when $x = -\frac{1}{2}$ and when $x = 3$. Thus, the required solution set is $(-\infty, -\frac{1}{2}] \cup [3, \infty)$. See Figure 4.

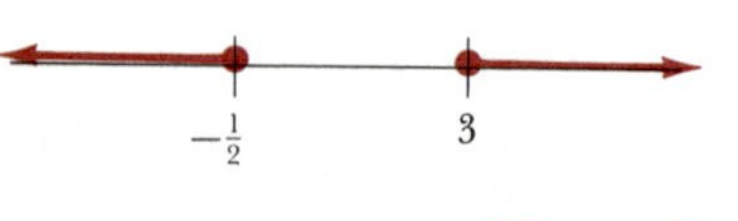

Figure 4

Sometimes it's simpler to use the factored form of a polynomial in determining whether the polynomial is positive or negative. Suppose, for instance, that we want to see whether $(x - 7)(x - 19)(x + 15)$ is positive or negative using the test value $x = 8$. With $x = 8$, we have

$$(x - 7)(x - 19)(x + 15) = \underbrace{(8 - 7)}_{\text{positive}}\ \underbrace{(8 - 19)}_{\text{negative}}\ \underbrace{(8 + 15)}_{\text{positive}}$$

From this we see without even carrying out the arithmetic that the product will be negative. This idea is used in the next example.

Example 2 Solve $x^4 - 14x^3 + 48x^2 < 0$.

Solution First, factor the left-hand side as follows:

$$x^4 - 14x^3 + 48x^2 < 0$$
$$x^2(x^2 - 14x + 48) < 0$$
$$x^2(x - 6)(x - 8) < 0$$

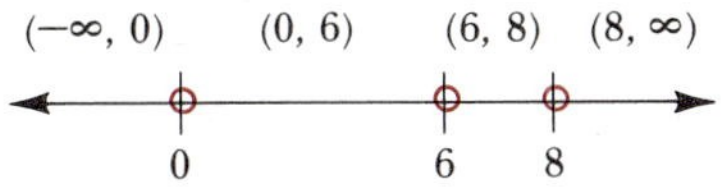

$(-\infty, 0)$ $(0, 6)$ $(6, 8)$ $(8, \infty)$

0 6 8

Figure 5

From this last line we see that the critical numbers are $x = 0$, 6, and 8. As indicated in Figure 5, these numbers divide the number line into four distinct intervals. Now we need to choose a test number from each interval and see whether the polynomial is positive or negative on that interval. This work is carried out in the following table. The table shows that the quantity $x^2(x - 6)(x - 8)$ is negative only for x-values in the interval (6, 8). Thus, the solution set for the given equality is the open interval (6, 8).

Interval	Test Number	Signs of Factors in $x^2(x - 6)(x - 8)$	Sign of Product
$(-\infty, 0)$	-1	$\underbrace{(-1)^2}_{+}\underbrace{(-1 - 6)}_{-}\underbrace{(-1 - 8)}_{-}$	+
(0, 6)	1	$\underbrace{(1)^2}_{+}\underbrace{(1 - 6)}_{-}\underbrace{(1 - 8)}_{-}$	+
(6, 8)	7	$\underbrace{(7)^2}_{+}\underbrace{(7 - 6)}_{+}\underbrace{(7 - 8)}_{-}$	−
$(8, \infty)$	9	$\underbrace{(9)^2}_{+}\underbrace{(9 - 6)}_{+}\underbrace{(9 - 8)}_{+}$	+

The technique used in Examples 1 and 2 can also be used to solve inequalities involving quotients of polynomials. For these cases recall that the definition of a critical number also includes the x-values for which the denominator is zero. For example, the critical numbers for the inequality $\dfrac{x + 3}{x - 4} \geq 0$ are -3 and 4.

Example 3 Solve $\dfrac{x + 3}{x - 4} \geq 0$.

Solution The critical numbers are -3 and 4. As indicated in Figure 6, these numbers divide the number line into three intervals. In the table that follows we've chosen a test number from each interval and determined the sign of the quotient $\dfrac{x + 3}{x - 4}$ on each interval.

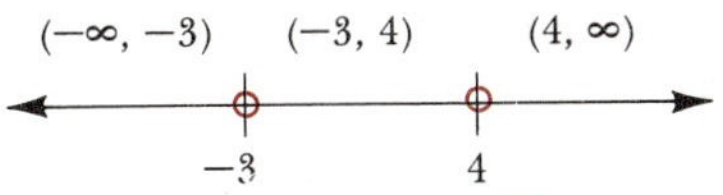

$(-\infty, -3)$ $(-3, 4)$ $(4, \infty)$

-3 4

Figure 6

Interval	Test Number	Value of $\frac{x+3}{x-4}$	Sign of $\frac{x+3}{x-4}$
$(-\infty, -3)$	-4	$\frac{-4+3}{-4-4} = \frac{1}{8}$	+
$(-3, 4)$	0	$\frac{0+3}{0-4} = -\frac{3}{4}$	−
$(4, \infty)$	5	$\frac{5+3}{5-4} = 8$	+

From these results we conclude that the solution set for $\frac{x+3}{x-4} \geq 0$ contains the two intervals $(-\infty, -3)$ and $(4, \infty)$. However, we still need to consider the two end points, -3 and 4. As you can easily check, the value $x = -3$ does satisfy the given inequality, but $x = 4$ does not. In summary then, the solution set is $(-\infty, -3] \cup (4, \infty)$.

Example 4 Solve $\frac{2x-1}{x-1} - \frac{7}{x+1} < 1$.

Solution Your first inclination here might be to multiply through by $(x-1)(x+1)$ to eliminate fractions. The problem with that is that we don't know whether the quantity $(x-1)(x+1)$ is positive or negative. Thus we begin instead by rewriting the inequality in an equivalent form, with zero on the right-hand side and a single fraction on the left-hand side:

$$\frac{2x-1}{x-1} - \frac{7}{x+1} - 1 < 0$$

$$\frac{2x-1}{x-1}\cdot\frac{x+1}{x+1} - \frac{7}{x+1}\cdot\frac{x-1}{x-1} - 1\cdot\frac{(x-1)(x+1)}{(x-1)(x+1)} < 0$$

$$\frac{2x^2 + x - 1 - 7(x-1) - (x^2-1)}{(x-1)(x+1)} < 0$$

$$\frac{x^2 - 6x + 7}{(x-1)(x+1)} < 0$$

The critical numbers are those x-values for which the denominator or the numerator is zero. By inspection, the denominator is zero when $x = 1$ or -1. On the other hand, if the numerator is zero we have $x^2 - 6x + 7 = 0$, and consequently (using the quadratic formula),

$$x = \frac{6 \pm \sqrt{36-28}}{2} = \frac{6 \pm \sqrt{8}}{2}$$

$$= \frac{6 \pm 2\sqrt{2}}{2} = 3 \pm \sqrt{2}$$

The four critical numbers are therefore ± 1 and $3 \pm \sqrt{2}$. As Figure 7 indicates, these numbers divide the number line into five distinct intervals.

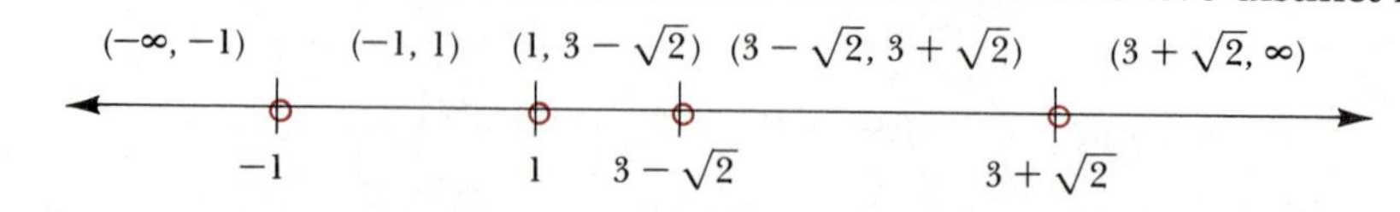

Figure 7

Now, just as in the previous example, we choose a test number from each interval and determine the sign of the quotient on that interval. (You should check for yourself each entry in the third column of the following table.)

Interval	Test Number	Value of $\frac{x^2 - 6x + 7}{(x - 1)(x + 1)}$	Sign of $\frac{x^2 - 6x + 7}{(x - 1)(x + 1)}$
$(-\infty, -1)$	-2	$\frac{23}{3}$	+
$(-1, 1)$	0	-7	−
$(1, 3 - \sqrt{2})$	$\frac{3}{2}$	$\frac{1}{5}$	+
$(3 - \sqrt{2}, 3 + \sqrt{2})$	3	$-\frac{1}{4}$	−
$(3 + \sqrt{2}, \infty)$	5	$\frac{1}{12}$	+

From these results we can see that the quotient $\frac{x^2 - 6x + 7}{(x - 1)(x + 1)}$ is negative on the two intervals $(-1, 1)$ and $(3 - \sqrt{2}, 3 + \sqrt{2})$. Now we need to check the end points of these two intervals. When $x = 3 \pm \sqrt{2}$ the quotient is zero, and so (in view of the original inequality) we exclude those two x-values from the solution set. Furthermore, the quotient is undefined when $x = \pm 1$, so we must also exclude those two values from the solution set. In summary, then, the solution set is $(-1, 1) \cup (3 - \sqrt{2}, 3 + \sqrt{2})$.

Exercise Set 2.6

Solve the inequalities in Exercises 1–54. Suggestion: *A calculator is useful for approximating the critical numbers in Exercises 19, 20, 33, 42, and 44.*

1. $x^2 + x - 6 < 0$
2. $x^2 + 4x - 32 < 0$
3. $x^2 - 11x + 18 > 0$
4. $2x^2 + 7x + 5 > 0$
5. $9x - x^2 \leq 20$
6. $3x^2 + x \leq 4$
7. $x^2 - 16 \geq 0$
8. $24 - x^2 \geq 0$
9. $16x^2 + 24x < -9$
10. $x^4 - 16 < 0$
11. $x^3 + 13x^2 + 42x > 0$
12. $2x^3 - 9x^2 + 4x \geq 0$
13. $225x \leq x^3$
14. $81x^2 \leq x^6$
15. $2x^2 + 1 \geq 0$
16. $1 + x^2 < 0$
17. $12x^3 + 17x^2 + 6x < 0$
18. $8x^4 < x^2 - 2x^3$
19. $x^2 + x - 1 > 0$
20. $2x^2 + 9x - 1 > 0$
21. $x^2 - 8x + 2 \leq 0$
22. $3x^2 - x + 5 \leq 0$
23. $(x - 1)(x + 3)(x + 4) \geq 0$
24. $x^4(x - 2)(x - 16) \geq 0$
25. $(x + 4)(x + 5)(x + 6) < 0$
26. $(x - \frac{1}{2})(x + \frac{1}{2})(x + \frac{3}{2}) < 0$
27. $(x - 2)^2(3x + 1)^3(3x - 1) > 0$
28. $(2x - 1)^3(2x - 3)^5(2x - 5) > 0$
29. $(x - 3)^2(x + 1)^4(2x + 1)^4(3x + 2) \leq 0$
30. $x^4 - 25x^2 + 144 \leq 0$
31. $x^4 - 9x^2 + 20 \geq 0$
32. $x(1 - x^2)^4 + (x + 3)(1 - x^2)^4 \geq 0$
33. $(x - 4)(2x^2 - 6x - 1) < 0$
34. $(x + 2)^3(x^2 - 4x - 2) < 0$
35. $x^3 + 2x^2 - x - 2 > 0$
36. $2x^4 + x^3 - 16x - 8 > 0$
37. $\frac{x - 1}{x + 1} \leq 0$
38. $\frac{x + 4}{2x - 5} \leq 0$
39. $\frac{2 - x}{3 - 2x} \geq 0$
40. $\frac{x^2 - 1}{x^2 + 8x + 15} \geq 0$
41. $\frac{x^2 - 8x - 9}{x} < 0$
42. $\frac{x^2 - 3x + 1}{1 - x} < 0$

43. $\dfrac{2x^3 + 5x^2 - 7x}{3x^2 + 7x + 4} > 0$

44. $\dfrac{x^2 - x - 1}{x^2 + x - 1} > 0$

45. $\dfrac{x + 2}{x + 5} \leq 1$

46. $\dfrac{2}{x - 4} - \dfrac{1}{x - 3} \geq \dfrac{2}{3}$

47. $\dfrac{1}{x - 2} - \dfrac{1}{x - 1} \geq \dfrac{1}{6}$

48. $\dfrac{2x}{x + 5} + \dfrac{x - 1}{x - 5} < \dfrac{1}{5}$

49. $\dfrac{1 + x}{1 - x} - \dfrac{1 - x}{1 + x} < -1$

50. $\dfrac{x + 1}{x + 2} > \dfrac{x - 3}{x + 4}$

51. $\dfrac{3 - 2x}{3 + 2x} > \dfrac{1}{x}$

52. $\dfrac{x}{x - 2} - \dfrac{3}{x + 1} \geq 2$

53. $1 + \dfrac{1}{x} \geq \dfrac{1}{1 + x}$

54. $x - \dfrac{10}{x - 1} \geq 4$

55. For which values of b will the equation $x^2 + bx + 1 = 0$ have real solutions?

56. The sum of the first n natural numbers is given by

$$1 + 2 + 3 + \cdots + n = \frac{n(n + 1)}{2}$$

For which values of n will the sum be less than 1225?

***57.** For which values of a is $x = 1$ a solution of the inequality $\dfrac{2a + x}{x - 2a} < 1$?

***58.** Solve $\dfrac{ax + b}{\sqrt{x}} > 2\sqrt{ab}$ $\quad (a > 0, b > 0)$

59. The two shorter sides in a right triangle have lengths x and $1 - x$, respectively ($x > 0$). For which values of x will the hypotenuse be less than $\dfrac{\sqrt{17}}{5}$?

60. A piece of wire 12 cm long is cut into two pieces. Denote the lengths of the two pieces by x and $12 - x$, respectively. Both pieces are then bent into squares. For which values of x will the combined areas of the squares exceed 5 cm^2?

61. Let V and S denote the volume and the total surface area, respectively, for a right circular cylinder of radius r and height 1. For which r-values will the ratio V/S be less than $\frac{1}{3}$?

***62.** Let V and S denote the volume and the total surface area, respectively, for a right circular cone of radius r and height 1. For which r-values will the ratio V/S be less than $\frac{4}{27}$?

63. **(a)** [C] Complete the table. (Round off the entries in the right-hand column to two decimal places.)

x	$x^2 + 1000$	$2x^2 + x$	$\dfrac{x^2 + 1000}{2x^2 + x}$
1			
2			
5			
10			
100			
200			
10,000			

(b) For which values of x will we have $\dfrac{x^2 + 1000}{2x^2 + x} < 0.5$?

(c) What is the smallest natural number n for which $\dfrac{n^2 + 1000}{2n^2 + n} < 0.5$?

[C] ***64.** Solve $x^3 + \dfrac{1}{x^3} \geq 3$.

[C] **65.** Solve $2.3x^2 - 0.2x - 5.3 \leq 0$. (Round off to one decimal place.)

66. **(a)** [C] Complete the table. (Don't round off; record the answers just as displayed on your calculator.)

x	$x^2 + 2x - 3$
1.9	
1.99	
1.999	
2.1	
2.01	
2.001	

(b) Find all positive numbers x for which $4.999 < x^2 + 2x - 3 < 5.001$. (Round off the x-values for the critical numbers to four decimal places.)

Chapter 2 Review Checklist

- □ Linear or first-degree equation (page 69)
- □ Solution or root of an equation (page 69)
- □ Equivalent equations (page 70)
- □ Quadratic equation (page 75)
- □ Zero-Product property (page 75)
- □ Square root method (page 76)
- □ Completing the square (pages 76–77)
- □ Quadratic formula (page 78)

- □ Discriminant (page 80)
- □ Equation of quadratic type (page 92)
- □ Extraneous root (page 95)
- □ Solution set of an inequality (page 97)
- □ ∞ (page 97)
- □ Properties of inequalities (page 98)
- □ $|u| < a$; $|u| > a$ (page 100)
- □ $\cup$ (Union of two sets) (page 101)
- □ Critical numbers of an inequality (page 105)
- □ Test number (page 105)

Chapter 2 Review Exercises

In Exercises 1–12, answer T if the statement is true without exception. Otherwise, answer F.

1. If $x < 3$, then $x + 7 < 10$.

2. If $x < y$, then $x^2 < y^2$.

3. If $x \geq -4$, then $x \leq 4$.

4. If $-x > y$, then $x < -y$.

5. If $x < 2$, then $\frac{1}{x} < \frac{1}{2}$.**6.** $x \leq x^2$

7. If $\sqrt{x + 1} = 3$, then $x + 1 = 9$.

8. If $0 < x < 1$, then $\frac{1}{x} > x$.

9. If $x < 2$ and $y < 3$, then $x < y$.

10. If $x < 2$ and $y > 3$, then $x < y$.

11. If $a - b \leq a^2 - b^2$, then $1 \leq a + b$.

12. If $0 < a - b \leq a^2 - b^2$, then $1 \leq a + b$.

Solve the equations in Exercises 13–72.

13. $5 - 9x = 2$

14. $\frac{x}{3} - \frac{3x}{5} = \frac{x}{6} - 13$

15. $(t - 4)(t + 3) = (t + 5)^2$

16. $\frac{1 - x}{1 + x} = 1$

17. $\frac{2t - 1}{t + 2} = 5$

18. $\frac{x - 2}{3} - \frac{4(x + 1)}{2} + \frac{1}{6} = 0$

19. $\frac{2}{x + 4} - \frac{1}{x - 4} = \frac{-7}{x^2 - 16}$

20. $\frac{1}{1 + \frac{1}{x + 1}} = \frac{5}{6}$

21. $\frac{2y - 5}{4y + 1} = \frac{y - 1}{2y + 5}$

22. $\frac{2x - 3}{x - 2} = \frac{1}{x - 2}$

23. $|y + 4| = 2$

24. $|3t - 6| = 0$

25. $|3x - 1| = 2$

26. $\sqrt{|x - 1|} = 3$

27. $t^2 - 2t - 99 = 0$

28. $x^2 + 24x + 144 = 0$

29. $12x^2 + 2x - 2 = 0$

30. $4y^2 - 21y = 18$

31. $\frac{1}{2}x^2 + x - 12 = 0$

32. $x^2 + \frac{13}{2}x + 10 = 0$

33. $\frac{1}{1 - x} + \frac{4}{2 - x} = \frac{11}{6}$

34. $\frac{2x + 1}{1 - 2x} + \frac{1 - 2x}{2x + 1} = 2$

35. $\frac{5}{5 - x} = \frac{-2}{11 - x}$

36. $\frac{x^2}{(x - 1)(x + 1)} = \frac{4}{x + 1} + \frac{4}{(x - 1)(x + 1)}$

37. $\frac{1}{x} - \frac{1}{x + 2} = \frac{1}{x + 3} - \frac{1}{x + 1}$

38. $\frac{x^2}{(x + 1)^2} = \frac{6}{x + 1} + \frac{1}{(x + 1)^2}$

39. $\frac{1}{3x - 7} - \frac{2}{5x - 5} - \frac{3}{3x + 1} = 0$

40. $x = \frac{5}{x + 1}$

41. $x^2 + 4x - 1 = 0$

42. $4x^2 + x - 2 = 0$

43. $t^2 + t - \frac{1}{2} = 0$

44. $\frac{1}{x - 1} = x\sqrt{2}$

45. $y^4 = 9$

46. $(y - 1)^4 = 81$

47. $x^4 - 7x^2 + 12 = 0$

48. $y^4 - 8y^2 + 11 = 0$

49. $x^5 - 2x^3 - 2x = 0$

50. $x^3 - 6x^2 + 7x = 0$

51. $1 + 14x^{-1} + 48x^{-2} = 0$

52. $x^{-2} - x^{-1} - 1 = 0$

53. $4x^{4/3} - 13x^{2/3} + 9 = 0$

C **54.** $512x^3 - 152x^{3/2} - 27 = 0$

55. $x^{1/2} - 13x^{1/4} + 36 = 0$

56. $y^{2/3} - 21y^{1/3} + 80 = 0$

57. $\sqrt{4 - 3x} = 5$

58. $\sqrt{1 + 2x} = x - 1$

59. $\sqrt{x} = \sqrt{x + 27} - 1$

60. $\sqrt{x} + 6 - x = 0$

61. $\sqrt{4x + 3} = \sqrt{11 - 8x} - 1$

62. $\sqrt{\sqrt{x - 4} + x} = 2$

63. $\sqrt{5 - 2x} - \sqrt{2 - x} - \sqrt{3 - x} = 0$

64. $2 - \sqrt{3\sqrt{2x - 1} + x} = 0$

65. $\sqrt{x + 48} - \sqrt{x} = 4$

66. $\sqrt{x + 6} - \sqrt{x} - \sqrt{2} = 0$

67. $3(4t - 1)^{1/2} - (2t)^{1/2} = (3 + 2t)^{1/2}$

68. $1 + \sqrt{2x^2 + 5x - 9} = \sqrt{2x^2 + 5x - 2}$

69. $\frac{2}{\sqrt{x^2 - 36}} + \frac{1}{\sqrt{x + 6}} - \frac{1}{\sqrt{x - 6}} = 0$

70. $\sqrt{x - \sqrt{1 - x}} + \sqrt{x} - 1 = 0$

71. $\sqrt{x + 7} - \sqrt{x + 2} = \sqrt{x - 1} - \sqrt{x - 2}$

72. $\dfrac{\sqrt{x} - 4}{\sqrt{x} - 18} = 3$

In Exercises 73–88, solve each equation for x *in terms of the other letters.*

73. $ax + b = cx + d \quad (a \neq c)$

74. $\dfrac{ax + b}{cx + d} = e \quad (a \neq ce)$

75. $ax = bx + a^2 - b^2 \quad (a \neq b)$

76. $(a^2 + b^2)x = a^3 + b^3 + abx$

77. $\dfrac{2}{x} = \dfrac{1}{a} + \dfrac{1}{b} \quad (a + b \neq 0)$

78. $\dfrac{x}{a} + \dfrac{a}{x} = 2$

79. $\frac{1}{4}x^2 + bx - 8b^2 = 0$

80. $\dfrac{1}{a + x} + \dfrac{1}{b + x} = \dfrac{a + b}{ab} \quad (a + b \neq 0)$

81. $4x^2y^2 - 4xy = -1 \quad (y \neq 0)$

82. $4x^4y^4 + 2x^2y^2 + \frac{1}{4} = 0 \quad (y \neq 0)$

83. $x + \dfrac{1}{a} - \dfrac{1}{b} = \dfrac{2}{a^2x} + \dfrac{2}{abx}$

84. $\dfrac{a}{x - b} + \dfrac{b}{x - a} = \dfrac{x - a}{b} + \dfrac{x - b}{a} \quad (a \neq -b)$

Suggestion: Before clearing fractions, carry out the indicated additions on each side of the equation.

85. $\dfrac{1}{x + a + b} = \dfrac{1}{x} + \dfrac{1}{a} + \dfrac{1}{b} \quad (a + b \neq 0)$

86. $\dfrac{1}{x} + \dfrac{1}{a - x} + \dfrac{1}{x + 3a} = 0 \quad (a \neq 0)$

87. $\dfrac{a^2 - b^2}{x} - \dfrac{2(a^2 + b^2)}{\sqrt{x}} = b^2 - a^2 \quad (a > b > 0)$

Hint: Let $t = \dfrac{1}{\sqrt{x}}$ and $t^2 = \dfrac{1}{x}$.

88. $x^2 + \dfrac{1}{x^2} = a^2 + \dfrac{1}{a^2}$

Solve the inequalities in Exercises 89–122.

89. $4 - 5x \leq -6$

90. $1 - (2 - x) > 2x + 5$

91. $4(1 + x) - 3(2x - 1) \geq 1$

92. $\dfrac{x}{2} - \dfrac{x - 1}{6} < 0$

93. $0 < 2x - 1 < 1$

94. $-1 < \dfrac{1 - 2(1 + x)}{3} < 1$

95. $3 < \dfrac{x - 1}{-2} < 4$

96. $\dfrac{3}{5} < \dfrac{3 - 2x}{-4} < \dfrac{4}{5}$

97. $|x| \geq 6$

98. $|x| \leq \frac{1}{2}$

99. $|t - 2| < 1$

100. $|x + 4| < \frac{1}{10}$

101. $|3 - 5x| < 2$

102. $|2x - 1| \geq 5$

103. $x^2 - 21x + 108 \leq 0$

104. $x^2 + 3x - 40 < 0$

105. $6x^2 + 13x > 8$

106. $x^2 \geq 15x$

107. $x^2 - 6x - 1 < 0$

108. $2x^2 + 2x - 5 \geq 0$

109. $(x + 12)(x - 1)(x - 8) < 0$

110. $(x - 4)^2(x + 8)^3 \geq 0$

111. $625 - x^4 \leq 0$

112. $x^4 - 34x^2 + 225 < 0$

113. $\dfrac{x + 12}{x - 5} > 0$

114. $\dfrac{(x - 7)^2}{(x + 2)^3} \geq 0$

115. $\dfrac{(x - 6)^2(x - 8)(x + 3)}{(x - 3)^2} \leq 0$

116. $\dfrac{x^2 - 10x + 9}{x^3 + 1} \leq 0$

117. $\dfrac{3x + 1}{x - 4} < 1$

118. $\dfrac{1 - 2x}{1 + 2x} \leq \dfrac{1}{2}$

119. $\dfrac{x}{x - 2} + \dfrac{1}{x - 1} \leq \dfrac{23}{12}$

120. $\dfrac{1}{x} + \dfrac{1}{x + 1} + \dfrac{1}{x + 2} \geq 0$

121. $x^2 + \dfrac{1}{x^2} > 3$ *Suggestion:* Use a calculator to evaluate the critical numbers.

122. $\sqrt{x} - \dfrac{5}{\sqrt{x}} \leq 4$

For Exercises 123–126, find the values of k *for which the roots of the equations are real numbers.*

123. $kx^2 - 6x + 5 = 0$

124. $x^2 + x + k^2 = 0$

125. $kx(x + 2) = -1$

126. $x^2 + (k + 1)x + 2k = 0$

127. How many pounds of an alloy that is 12% copper must be added to 1000 pounds of an alloy that is 9% copper to yield a new alloy that is 11.2% copper?

128. Kona coffee beans are sold for \$6 per pound and Colombian beans for \$5 per pound. How many pounds of each bean must be mixed to yield 280 pounds of a blend selling for \$5.60 per pound?

129. Find four consecutive odd integers whose sum is 2104.

130. The sum of the digits in a certain two-digit number is 11. If the order of the digits is reversed, the number is increased by 27. Find the original number.

131. Robin deposits \$6000, part at 10% and the remainder at 12%. If the net interest rate for the entire \$6000 is $11\frac{1}{3}$%, how much was deposited at each rate?

132. At 6 A.M. a train traveling at 50 mph leaves Los Angeles for San Francisco. Three hours later a train traveling 40 mph leaves San Francisco for Los Angeles. What time will the trains pass each other? Assume that the distance between the two cities is 450 miles.

133. Jennifer gives Lee a head start of d miles. If Jennifer runs at A mph and Lee runs at B mph (where $A > B$), how long will it take for Jennifer to catch Lee?

134. The width of a rectangle is one-sixth the length. If both dimensions are increased by 4 ft, the area increases by 72 ft^2. Find the area of the original rectangle.

135. The four corners of a square $ABCD$ have been cut off to form a regular octagon, as shown. If each side of the square is 1 cm long, how long is each side of the octagon?

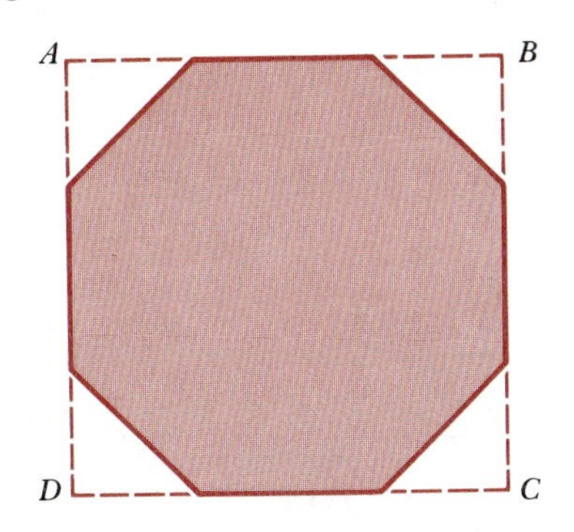

136. Determine p if the larger root of the equation $x^2 + px - 2 = 0$ is p.

137. A piece of wire x cm long is bent into a square. For which values of x will the area be (numerically) greater than the perimeter?

138. A piece of wire $6x$ cm long is bent into an equilateral triangle. For which values of x will the area be (numerically) less than the perimeter?

139. Find three consecutive positive integers such that the sum of their squares is 1454.

140. Find two consecutive even integers such that the sum of their reciprocals is $\frac{11}{60}$.

141. The sum of two numbers is 1 and the sum of their cubes is 91. What are the numbers?

142. In a 10-mile race, Joan covers the first 4 miles at a constant rate. Then she speeds up a bit and runs the last 6 miles at a rate that is $\frac{1}{2}$ mph faster. If the entire race had been run at this faster pace, Joan's overall time would have been 2 minutes better. What was Joan's speed for the first 4 miles, and what was her time for the race?

143. The length of a rectangular piece of tin exceeds the width by 8 cm. A 1-cm square is cut from each corner of the piece of tin, and then the resulting flaps are turned up to form a box with no top. What are the dimensions of the box if its volume is 48 cm^3?

144. The length and width of a rectangular flower garden are a and b, respectively. The garden is bordered on all four sides by a gravel path of uniform width. Find the width of the path, given that the area of the garden equals the area of the path.

145. Let α and β be the roots of the quadratic equation $ax^2 + bx + c = 0$. Verify each of the following statements.

(a) $\alpha + \beta = -b/a$ **(b)** $\alpha\beta = c/a$

(c) $\alpha^2 + \beta^2 = \dfrac{b^2 - 2ac}{a^2}$ *Suggestion:* Use (a) and (b) along with the fact that $\alpha^2 + \beta^2 = (\alpha + \beta)^2 - 2\alpha\beta$.

(d) $\dfrac{1}{\alpha^2} + \dfrac{1}{\beta^2} = \dfrac{b^2 - 2ac}{c^2}$ *Suggestion:* Add the fractions $\dfrac{1}{\alpha^2}$ and $\dfrac{1}{\beta^2}$.

If an object is thrown vertically upward from a height of h_0 *feet with an initial speed of* v_0 *ft/sec, then its height* h *(in feet) after* t *seconds is given by*

$$h = -16t^2 + v_0t + h_0$$

Make use of this formula in working Exercises 146–148.

146. A ball is thrown vertically upward from ground level with an initial speed of 64 ft/sec.

(a) At what time will the height of the ball be 15 ft? (Two answers.)

(b) For how long an interval of time will the height exceed 63 ft?

147. One ball is thrown vertically upward from a height of 50 ft with an initial speed of 40 ft/sec. At the same instant, another ball is thrown vertically upward from a height of 100 ft with an initial speed of 5 ft/sec. Which ball hits the ground first?

148. An object is projected vertically upward. Suppose that its height is H ft at t_1 seconds and again at t_2 seconds. Express the initial speed in terms of t_1 and t_2. *Answer:* $16(t_1 + t_2)$

149. A rectangle is inscribed in a semicircle of radius 1 cm, as shown. For which value of x is the area of the rectangle 1 cm^2? *Note:* x is defined in the figure.

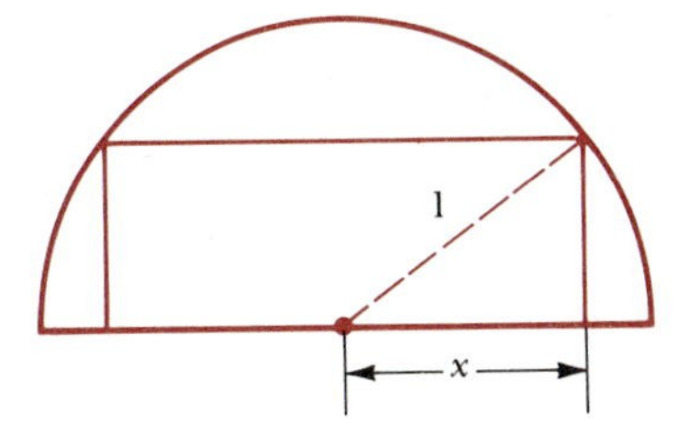

150. The height of an isosceles triangle is $a + b$ units ($a > b > 0$) and the area does not exceed $a^2 - b^2$ square units. What is the range of possible values for the base of the triangle?

Chapter 2 Test

Find all the real-number solutions of the equations in Problems 1–13.

1. $3(x - 4) + 2 = 1 - 2(1 - x)$

2. $\frac{x}{6} - \frac{4x}{5} = -38$

3. $\frac{x + a}{a - b} + \frac{x - a}{a + b} = \frac{2}{a^2 - b^2} \quad (a \neq 0)$

4. $\frac{1}{x - 4} + \frac{3}{x + 1} = \frac{5x}{x^2 - 3x - 4}$

5. $|3x - 8| = 1$

6. $8x^2 - 10x = 3$

7. $3x^2 + x - 1 = 0$

8. $\frac{2x}{1 + x} + 1 = \frac{1 + x}{x}$

9. $x^4 + 16x^2 - 225 = 0$

10. $x^2 - x = \frac{2}{x^2 - x} + 1$ *Hint:* Let $t = x^2 - x$.

11. $(x - 4)^{3/2} = 25$

12. $2 - \sqrt{4 + 3x} - \sqrt{3 + 2x} = 0$

13. $4x^{4/3} - 15x^{2/3} - 4 = 0$

14. A student in a chemistry laboratory has 85 cm^3 of a solution that is 80% alcohol (by volume). How much water should be added to yield a solution that is 68% alcohol?

15. Emily runs 2 mph faster than Andy in a 10-mile race. Find Emily's time for the race, given that she came in 25 minutes ahead of Andy.

16. James deposited a total of \$10,000 in two savings accounts, the first earning 12% (annual simple interest) and the second earning 8%. If the total annual interest is \$960, how much was deposited in each account?

17. Find a positive value for k such that the equation $3x^2 - kx + 5 = 0$ has exactly one real solution.

Determine the solution sets for the inequalities in Problems 18–20.

18. $|3 - 2x| < \frac{1}{10}$

19. $x^2 - 2x - 48 \leq 0$

20. $\frac{(x - 4)(x + 2)^2}{x - 1} \geq 0$

Chapter 3 Coordinates and Graphs

Introduction

In Chapter 1 we saw an important connection between algebra and geometry: The points on a line can be labeled using the real numbers. In Chapter 2 we exploited this connection in solving inequalities and in graphing their solution sets. Now, just as each point on a line can be labeled with a unique real number, so each point in a plane can be labeled with a unique *pair* of real numbers. The details of this are explained in Section 3.1, Cartesian Coordinates. These ideas are then used in the next three sections to graph equations. The basic theme running through the work on graphing equations is this: We start with a given equation relating two variables, say x and y. By choosing values for x, corresponding y-values are then determined. The resulting (x, y) number pairs (*ordered pairs*, they're called) are then interpreted as points in the plane. This provides another basic connection between algebra and geometry: With an equation in two variables we associate its geometric picture, its graph. Sometimes it is easier to work directly from a graph, other times from the equation. But in any event, having the latitude to think either algebraically or geometrically (or both) will clarify many problems later on.

3.1 Cartesian Coordinates

We begin by recalling the basic idea of a coordinate system in the plane. Draw two number lines, perpendicular to each other, as shown in Figure 1.

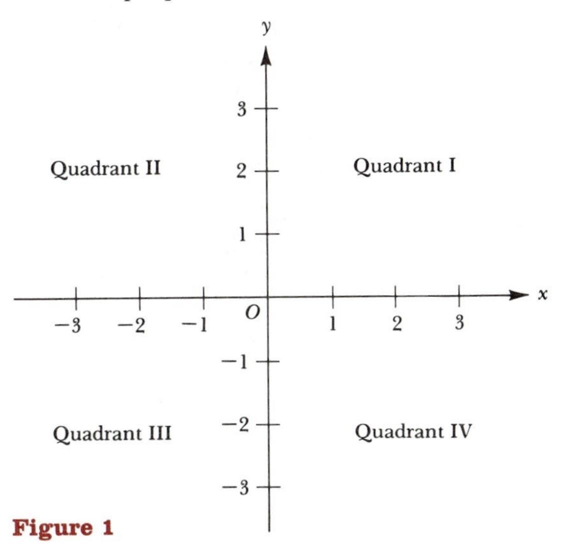

Figure 1

The point of intersection of the two number lines is called the *origin* and is denoted by the letter O. The horizontal number line is called the *x-axis*; the vertical number line is called the *y-axis*. Unless otherwise indicated, we assume that the same unit of length is used on both axes. Referring to Figure 1, notice that the axes divide the plane into four regions or *quadrants*, labeled with Roman numerals I through IV as shown.

Now look at the point P in Figure 2(a).

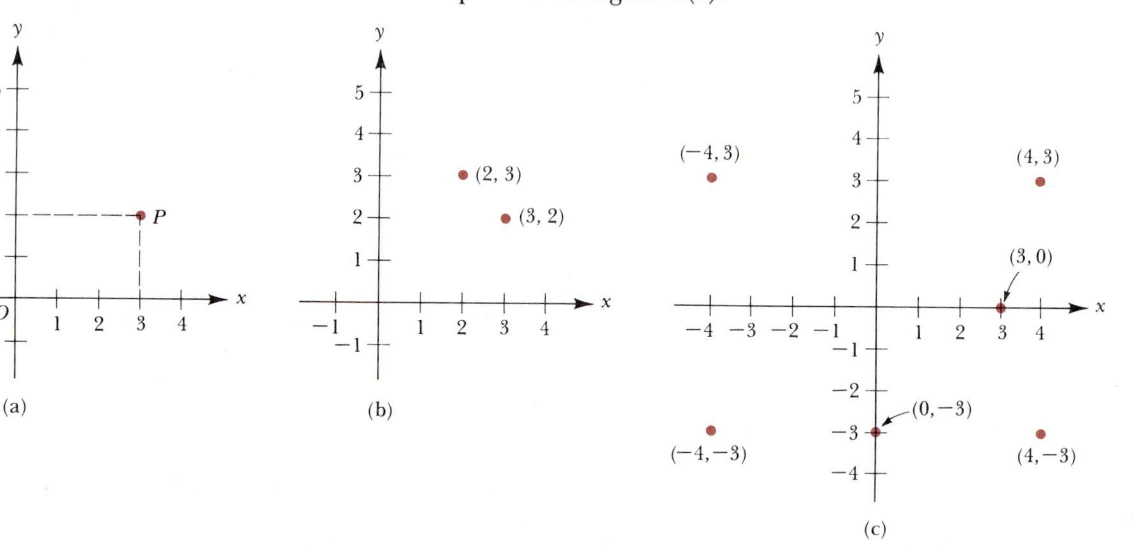

Figure 2

Starting from the origin O, one way to reach P is to move three units in the positive x-direction and then two units in the positive y-direction. That is, the location of P relative to the origin and the axes is "right 3, up 2." We say that the *coordinates* of P are $(3, 2)$. The first number within the parentheses conveys the information "right 3," and the second number conveys the information "up 2." We say that the *x-coordinate* of P is 3 and the *y-coordinate* of P is 2. Likewise, the coordinates of point Q in Figure 2(a) are $(-2, 4)$. With this coordinate notation in mind, observe in Figure 2(b) that $(3, 2)$ and $(2, 3)$ represent different points; that is, the order in which the two numbers appear within the parentheses affects the location of the point. Figure 2(c) displays various points with given coordinates; you should check for yourself that the coordinates are correct.

Some terminology and notation: The x-y coordinate system that we have described is often called a *rectangular* or *Cartesian* coordinate system. The word *rectangular* refers to the fact that the x- and y-axes are straight lines meeting at right angles. The term *Cartesian* is used in honor of René Descartes, the 17th-century French philosopher and mathematician who invented the coordinate system. The coordinates (x, y) of a point P are referred to as an *ordered pair*. Recall, for example, that $(3, 2)$ and $(2, 3)$ represent different points; that is, the order of the numbers matters. The x-coordinate of a point is sometimes referred to as the *abscissa* of the point; the y-coordinate is the *ordinate*. The notation $P(x, y)$ means that P is a point whose coordinates are (x, y). At times we abbreviate the phrase "the point whose coordinates are (x, y)" to simply "the point (x, y)."

Given two points $P(x_1, y_1)$ and $Q(x_2, y_2)$, there are occasions when we need to know the coordinates of the midpoint M of the line segment joining P and Q. See Figure 3(a). In the discussion that follows, and throughout this text, we shall use notation such as PQ to denote the line segment joining the points P and Q; the length of this line segment will be denoted by $\overline{PQ}$. There is a simple formula that gives us the coordinates of the midpoint M; we'll derive the formula using Figure 3(b). The dotted lines in that figure are drawn parallel to the y-axis. To understand our derivation, you'll need to recall (or, for now, just accept) the following theorem from geometry: If

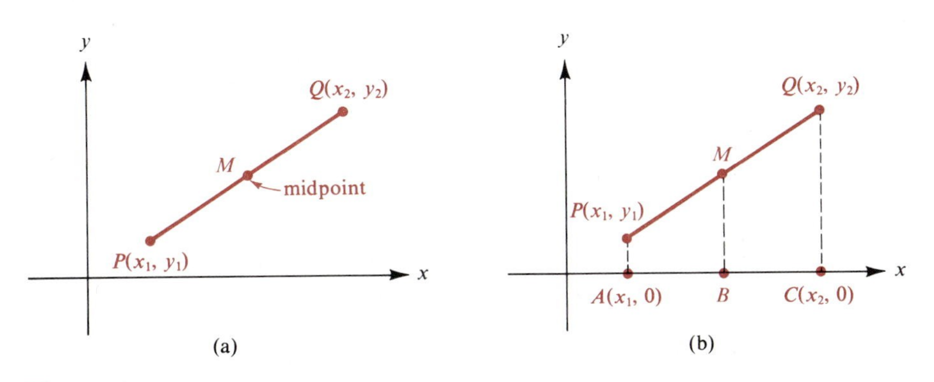

Figure 3

parallel lines intercept equal segments on one line, then they intercept equal segments on any line. Now, in Figure 3(b), we are assuming that M is the midpoint of PQ. Thus $\overline{PM} = \overline{MQ}$. From this last equation, it follows (by our theorem) that $\overline{AB} = \overline{BC}$, and therefore B is the midpoint of AC. At this point, we would like to know the x-coordinate of B, for if you look back at Figure 3(b), you'll see that the points B and M have the same x-coordinate. Since B is halfway between A and C on the x-axis, the x-coordinate of B must be exactly halfway between the numbers x_1 and x_2. Now the number halfway between x_1 and x_2 is found by taking the average of x_1 and x_2. Thus the x-coordinate of B (and therefore of M also) must be $\frac{x_1 + x_2}{2}$. Although Figure 3(b) shows P and Q in Quadrant I, it can be shown that the same result is obtained for all locations of P and Q. Furthermore, by following a similar line of reasoning, we also find that the y-coordinate of M is given by $\frac{y_1 + y_2}{2}$. For reference we summarize these facts as follows.

The Midpoint Formula

The midpoint M of the line segment joining the points $P(x_1, y_1)$ and $Q(x_2, y_2)$ is

$$\left(\frac{x_1 + x_2}{2}, \frac{y_1 + y_2}{2}\right)$$

Example 1 Determine the coordinates of the midpoint of the line segment joining the points $P(-3, 4)$ and $Q(2, -6)$.

Solution

$$x\text{-coordinate:} \quad \frac{x_1 + x_2}{2} = \frac{-3 + 2}{2} = \frac{-1}{2}$$

$$y\text{-coordinate:} \quad \frac{y_1 + y_2}{2} = \frac{4 + (-6)}{2} = -1$$

The required midpoint is therefore $(-1/2, -1)$. You should draw a sketch for yourself showing the line segment PQ and the midpoint $(-1/2, -1)$.

In the next example we use the Pythagorean Theorem (which was discussed in Section 1.5) to calculate the distance between two given points.

Example 2 Use the Pythagorean Theorem to calculate the distance d between the points $(2, 1)$ and $(6, 3)$.

Solution See Figure 4. Draw in the dotted lines shown, parallel to the axes, and apply the Pythagorean Theorem to the right triangle formed. The base of the triangle is four units long. You can see this by simply counting spaces, or by using absolute value, as discussed in Section 1.3: $|6 - 2| = 4$. The height of the triangle is found to be two units, either by counting spaces or by computing the absolute value: $|3 - 1| = 2$. Thus we have

$$d^2 = 4^2 + 2^2 = 20$$

$$d = \sqrt{20} = \sqrt{4}\sqrt{5}$$

$$d = 2\sqrt{5}$$

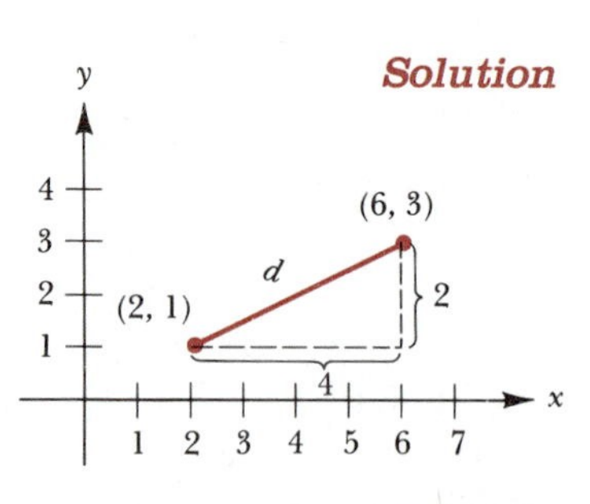

Figure 4

The method used in Example 2 above can be applied to derive a general formula for the distance d between any two points (x_1, y_1) and (x_2, y_2). See Figure 5.

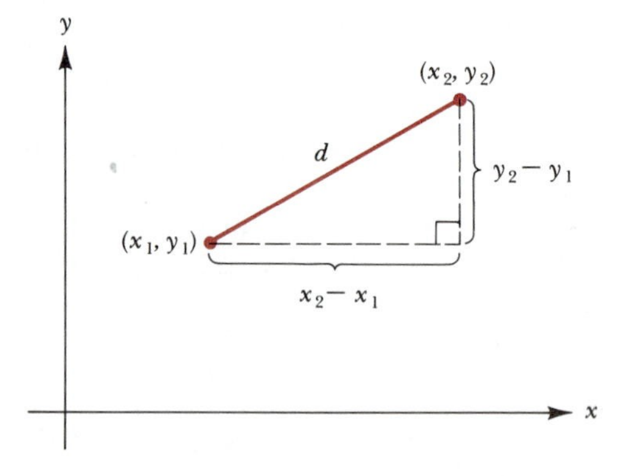

Figure 5

Just as before, we draw in the right triangle shown and apply the Pythagorean Theorem. We have

$$d^2 = (x_2 - x_1)^2 + (y_2 - y_1)^2$$

and therefore

$$d = \sqrt{(x_2 - x_1)^2 + (y_2 - y_1)^2}$$

This last equation is referred to as the *distance formula*. Although we've derived this formula using two points in the first quadrant, it's easy to check that the same result is obtained for any other locations of the two points.

The Distance Formula

The distance d between the points (x_1, y_1) and (x_2, y_2) is given by

$$d = \sqrt{(x_2 - x_1)^2 + (y_2 - y_1)^2}$$

Examples 3 and 4 demonstrate some simple calculations involving the distance formula. *Note:* In computing the distance between two given points, it does not matter which one you treat as (x_1, y_1) and which as (x_2, y_2). This is because quantities such as $(x_2 - x_1)$ and $(x_1 - x_2)$ are negatives of each other, and so their squares are equal.

Example 3 Calculate the distance between the points $(2, -6)$ and $(5, 3)$.

Solution Substituting $(2, -6)$ for (x_1, y_1) and $(5, 3)$ for (x_2, y_2) in the distance formula, we have

$$d = \sqrt{(5 - 2)^2 + [3 - (-6)]^2}$$
$$d = \sqrt{3^2 + 9^2} = \sqrt{9 + 81}$$
$$d = \sqrt{90} = \sqrt{9}\sqrt{10}$$
$$d = 3\sqrt{10}$$

You should check for yourself that the same answer is obtained using $(2, -6)$ as (x_2, y_2) and $(5, 3)$ as (x_1, y_1).

Example 4 Which point is closer to the origin, $(8, 3)$ or $(7, 5)$?

Solution The distance from $(8, 3)$ to the origin $(0, 0)$ is

$$\sqrt{(8 - 0)^2 + (3 - 0)^2} = \sqrt{64 + 9} = \sqrt{73}$$

The distance from $(7, 5)$ to the origin is

$$\sqrt{(7 - 0)^2 + (5 - 0)^2} = \sqrt{49 + 25} = \sqrt{74}$$

Thus $(8, 3)$ is closer to the origin since $\sqrt{73} < \sqrt{74}$.

As we've seen in this section, the basic idea behind coordinate geometry is to associate a unique ordered pair of numbers (x, y) with each point in the plane. This allows us to deal with geometric relationships in a purely numerical or algebraic manner. At times this can be a real advantage. For instance, by using the methods of coordinate geometry, proofs of theorems from elementary geometry can be constructed almost at will, without the need for arguments involving congruent triangles, angles, or the like. Example 5 illustrates this idea. *Note:* You needn't memorize the result in Example 5. Of primary importance for our purposes are the *methods* used to obtain that result.

Example 5 Use Figure 6 to prove the following theorem from geometry: The midpoint of the hypotenuse in a right triangle is equidistant from the vertices. (In Figure 6, $\triangle OBA$ is the right triangle and M is the midpoint of the hypotenuse.)

Solution With reference to Figure 6, we have to show that $\overline{OM} = \overline{MA} = \overline{MB}$. Now by hypothesis we already have that $\overline{MA} = \overline{MB}$, so we need only establish that $\overline{OM} = \overline{MA}$. The calculations are as follows.

$$\text{Coordinates of } M: \left(\frac{a}{2}, \frac{b}{2}\right)$$

$$\overline{OM} = \sqrt{\left(\frac{a}{2} - 0\right)^2 + \left(\frac{b}{2} - 0\right)^2} = \sqrt{\frac{a^2 + b^2}{4}} = \frac{\sqrt{a^2 + b^2}}{2}$$

$$\overline{MA} = \sqrt{\left(a - \frac{a}{2}\right)^2 + \left(0 - \frac{b}{2}\right)^2} = \sqrt{\frac{a^2}{4} + \frac{b^2}{4}} = \sqrt{\frac{a^2 + b^2}{4}} = \frac{\sqrt{a^2 + b^2}}{2}$$

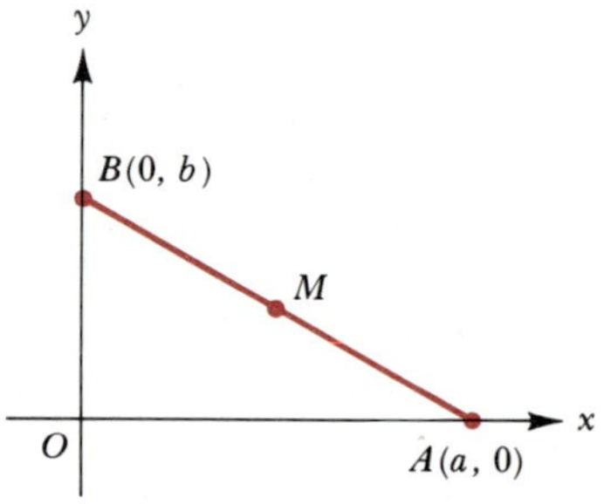

y, B(0, b), M, O, A(a, 0), x

Figure 6

From these calculations we conclude that $\overline{OM} = \overline{MA}$ (since both quantities are equal to $\frac{1}{2}\sqrt{a^2 + b^2}$). Thus $\overline{OM} = \overline{MA} = \overline{MB}$, as required. *Note:* We could have avoided working with fractions had we initially made a more clever choice for the coordinates of A and B. Since a and b were arbitrary, we could have used $(2a, 0)$ for A and $(0, 2b)$ for B.

Exercise Set 3.1

1. Plot the points $(5, 2)$, $(-4, 5)$, $(-4, 0)$, $(-1, -1)$, and $(5, -2)$.
2. Draw the square $ABCD$ whose vertices (corners) are $A(1, 0)$, $B(0, 1)$, $C(-1, 0)$, and $D(0, -1)$.
3. **(a)** Draw the right triangle PQR with vertices $P(1, 0)$, $Q(5, 0)$, and $R(5, 3)$.
   (b) Use the formula for the area of a triangle, $A = \frac{1}{2}bh$, to find the area of triangle PQR in part (a).

In Exercises 4–6, find the midpoint of the line segment joining the given points P *and* Q.

4. **(a)** $P(3, 2)$ and $Q(9, 8)$
   (b) $P(-4, 0)$ and $Q(5, -3)$
5. **(a)** $P(3, -6)$ and $Q(-1, -2)$
   (b) $P(12, 0)$ and $Q(12, 8)$
6. **(a)** $P(\frac{3}{5}, -\frac{2}{3})$ and $Q(0, 0)$
   (b) $P(1, \pi)$ and $Q(3, 3\pi)$
7. The coordinates of A and B are $(-1, 2)$ and $(5, -3)$, respectively. If B is the midpoint of line segment AC, what are the coordinates of C?
8. The coordinates of the points S and T are $S(4, 6)$ and $T(10, 2)$. If M is the midpoint of ST, find the midpoint of SM.
9. **(a)** Sketch the parallelogram with vertices $A(-7, -1)$, $B(4, 3)$, $C(7, 8)$, and $D(-4, 4)$.
   (b) Compute the midpoints of the diagonals AC and BD.
   (c) What conclusion can you draw from part (b)?

For Exercises 10–14, calculate the distance between the given points.

10. **(a)** $(0, 0)$ and $(-3, 4)$ **(b)** $(2, 1)$ and $(7, 13)$
11. **(a)** $(-1, -3)$ and $(-5, 4)$
    (b) $(6, -2)$ and $(-1, 1)$
12. **(a)** $(-5, 0)$ and $(5, 0)$ **(b)** $(0, -8)$ and $(0, 1)$
13. **(a)** $(-5, -3)$ and $(-9, -6)$
    (b) $(\frac{9}{2}, 3)$ and $(-2\frac{1}{2}, -1)$
14. $(1, \sqrt{3})$ and $(-1, -\sqrt{3})$
15. Use the distance formula to show that triangles with the given vertices are isosceles triangles.
    (a) $(0, 2)$, $(7, 4)$, $(2, -5)$
    (b) $(-1, -8)$, $(0, -1)$, $(-4, -4)$
    (c) $(-7, 4)$, $(-3, 10)$, $(1, 3)$
16. Which of the triangles with the given vertices are right triangles? *Hint:* Find the lengths of the sides and then use the *converse* of the Pythagorean Theorem. (The converse of the Pythagorean Theorem

states that if the lengths of the sides of a triangle satisfy a relationship of the form $a^2 + b^2 = c^2$, then the triangle is a right triangle. This fact is usually proved in elementary geometry courses.)

(a) $(7, -1), (-3, 5), (-12, -10)$
(b) $(4, 5), (-3, 9), (1, 3)$
(c) $(-8, -2), (1, -1), (10, 19)$

17. (a) Two of the three triangles specified in Exercise 16 are right triangles. Find their areas.
(b) Calculate the area of the remaining triangle in Exercise 16 by using the following formula for the area A of a triangle with vertices (x_1, y_1), (x_2, y_2), and (x_3, y_3).

$$A = \frac{1}{2}|x_1y_2 - x_2y_1 + x_2y_3 - x_3y_2 + x_3y_1 - x_1y_3|$$

The method used to derive this formula is indicated in Exercise 33.
(c) Use the formula given in part (b) to check your other answers in part (a).

18. Use the formula given in Exercise 17(b) to calculate the area of the triangle with vertices $(1, -4)$, $(5, 3)$, and $(13, 17)$. Conclusion?

19. Which point is farther from the origin?
(a) $(3, -2)$ or $(4, \frac{1}{2})$ **(b)** $(-6, 7)$ or $(9, 0)$

20. The coordinates of points A, B, and C are $A(-4, 6)$, $B(-1, 2)$, and $C(2, -2)$.
(a) Show that $\overline{AB} = \overline{BC}$, using the distance formula.
(b) Show that $\overline{AB} + \overline{BC} = \overline{AC}$, using the distance formula.
(c) What can you conclude from parts (a) and (b)?

21. The coordinates of points A, B, and C are $A(-5, -5)$, $B(-1, 1)$, and $C(5, 5)$.
(a) Show that $\overline{AB} = \overline{BC}$.
(b) Although a good sketch makes it seem obvious, prove in fact that A, B, and C are not collinear (i.e., do not lie on one straight line) by showing that $\overline{AB} + \overline{BC} \neq \overline{AC}$.

22. Find a value for t such that points $(0, 2)$ and $(12, t)$ are 13 units apart. *Hint:* By the distance formula, $13 = \sqrt{(12 - 0)^2 + (t - 2)^2}$. Now square both sides and solve for t.

23. (a) Find values for t such that the points $(-2, 3)$ and $(t, 1)$ are six units apart.
(b) Can you find a real number t such that the points $(-2, 3)$ and $(t, 1)$ are one unit apart? Why or why not?

24. The diagonals of a parallelogram bisect each other. Steps (a), (b), and (c) outline a proof of this theorem.
(a) (Refer to the figure at the top of the next column.) In the parallelogram $OABC$, check that the coordinates of B must be $(a + b, c)$.
(b) Use the midpoint formula to calculate the midpoints of diagonals OB and AC.

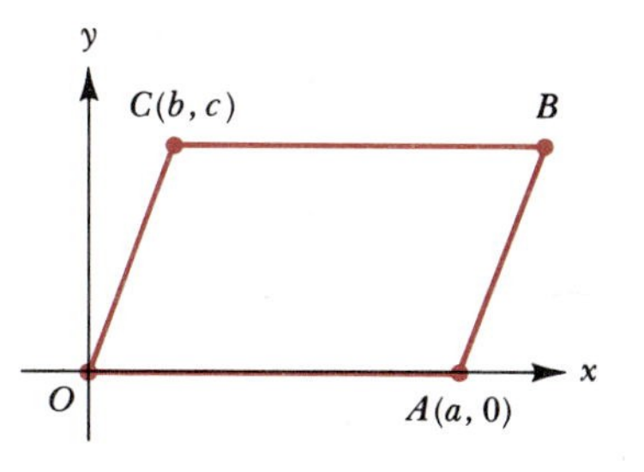

(c) Note that the two answers in part (b) are identical. This shows that the two diagonals do indeed bisect each other, as was to be proved.

25. Prove that, in a parallelogram, the sum of the squares of the lengths of the diagonals equals the sum of the squares of the lengths of the four sides. (Use the figure in Exercise 24.)

***26.** Use the figure in Exercise 24 to prove the following theorem: If the diagonals of a parallelogram are equal, then the parallelogram is a rectangle. *Hint:* Equate the two expressions for the lengths of the diagonals and conclude that $ab = 0$. The case in which $a = 0$ can be discarded.

***27.** The sum of the squares of the sides of any triangle equals four thirds of the sum of the squares of the medians. Prove this, taking the vertices to be $(0, 0)$, $(2, 0)$, and $(2a, 2b)$. *Language:* The expression *squares of the sides* means *squares of the lengths of the sides*. A *median* is a line drawn from a vertex of a triangle to the midpoint of the opposite side.

***28.** In $\triangle ABC$, M is the midpoint of side BC. Prove that

$$\overline{AB}^2 + \overline{AC}^2 = 2(\overline{BM}^2 + \overline{AM}^2)$$

Hint: Let the coordinates be $A(0, 0)$, $B(2, 0)$, and $C(2a, 2b)$.

29. If the point (x, y) is equidistant from the points $(-3, -3)$ and $(5, 5)$, show that x and y satisfy the equation $x + y = 2$. *Hint:* Use the distance formula.

30. The point (x, y) is four units away from $(0, 0)$. Show that x and y are related by the equation $x^2 + y^2 = 16$.

31. The point (x, y) is twice as far from $(0, 0)$ as it is from $(4, 0)$. Show that x and y are related by the equation $3x^2 + 3y^2 - 32x + 64 = 0$.

***32.** The point (x, y) is located in such a way that the sum of its distances from $(-1, 0)$ and from $(1, 0)$ is 4. Show that x and y are related by the equation $3x^2 + 4y^2 = 12$.

***33.** This problem indicates a method for calculating the area of a triangle when the coordinates of the three vertices are given.
(a) Calculate the area of $\triangle ABC$ in the figure on the next page. *Hint:* First calculate the area of the rectangle enclosing $\triangle ABC$ and then subtract the areas of the three right triangles.

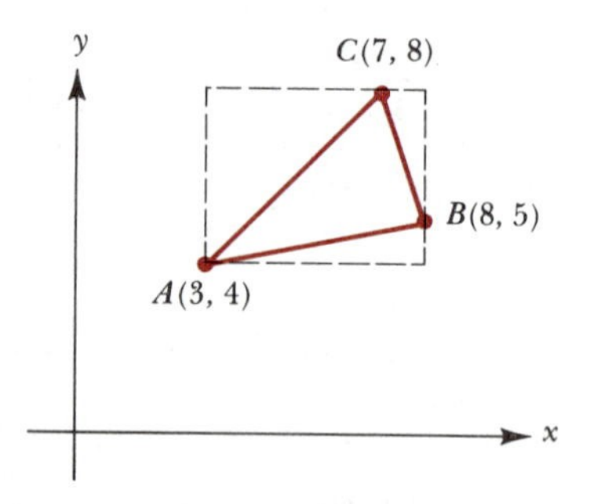

(b) Calculate the area of the triangle with vertices (1, 3), (4, 1), and (10, 4). *Hint:* Work with an enclosing rectangle and three right triangles, as in part (a).

(c) Using the same technique that you used in parts (a) and (b), show that the area of the triangle in the next column is given by

$$A = \frac{1}{2}(x_1y_2 - x_2y_1 + x_2y_3 - x_3y_2 + x_3y_1 - x_1y_3)$$

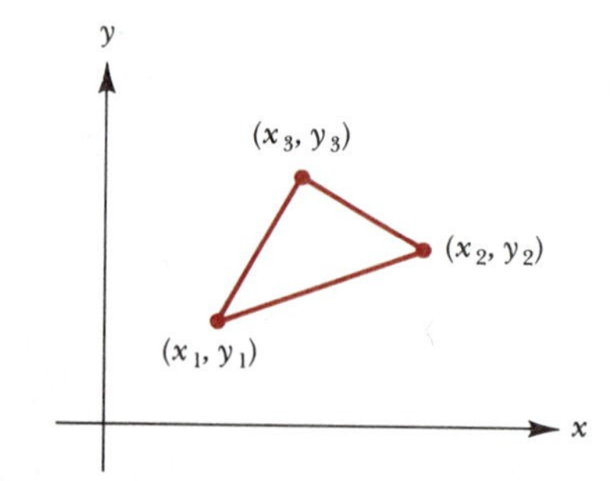

Remark: If we use absolute value signs instead of the parentheses, then the formula will hold regardless of the relative positions or quadrants of the three vertices. Thus, the area of a triangle with vertices (x_1, y_1), (x_2, y_2), (x_3, y_3) is given by

$$A = \frac{1}{2}|x_1y_2 - x_2y_1 + x_2y_3 - x_3y_2 + x_3y_1 - x_1y_3|$$

3.2 Graphs of Equations

Fermat and Descartes discovered that it is possible to combine the methods and concepts of geometry and algebra in a way that is useful in both branches of mathematics. Their basic idea was to associate algebraic equations in two unknowns with geometric curves in the plane. The geometric properties of a curve then correspond to the algebraic properties of its associated equation, and conversely. Using this correspondence, problems in geometry can be translated into problems solvable by algebraic manipulation, and insight into the nature of solutions of equations can be gained by consideration of the geometric interpretations.

Richard S. Hall in *About Mathematics*
(Englewood Cliffs, NJ: Prentice-Hall, 1973)

One of the most useful ways to obtain information about an equation relating two variables is through a graph. The key definition is as follows.

> **Definition: The Graph of an Equation**
>
> The graph of an equation in two variables is the set of all points whose coordinates satisfy the equation.

Suppose, for example, that we want to graph the equation $y = 3x - 2$. We begin by choosing (at will) various values for x and in each case computing the corresponding y-value from the equation $y = 3x - 2$. For example, if x is zero, then $y = 3(0) - 2 = -2$; y is -2. Table 1 summarizes the results of some of these calculations. The first line in Table 1 tells us that the point with coordinates $(0, -2)$ is on the graph of $y = 3x - 2$. Reading down the table, we see that some other points on the graph are $(1, 1)$, $(2, 4)$, $(3, 7)$, $(-1, -5)$, and $(-2, -8)$.

x	y
0	−2
1	1
2	4
3	7
−1	−5
−2	−8

Table 1
$y = 3x - 2$

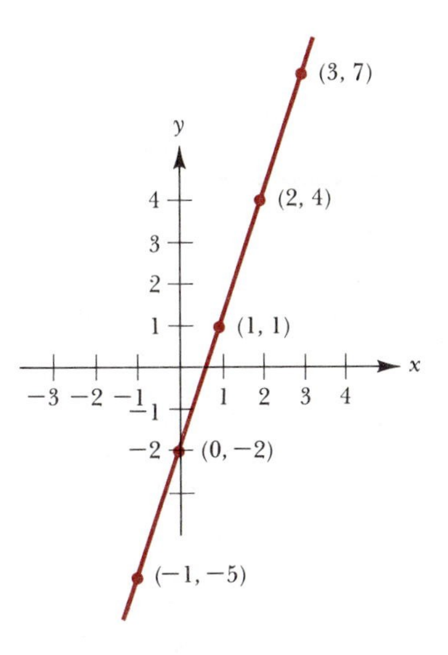

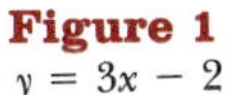
Figure 1
$y = 3x - 2$

We now plot (locate) the points we have determined and note, in this example, that they all appear to lie on a straight line. We draw the line indicated; this is the graph of $y = 3x - 2$. See Figure 1.

In graphing an equation, it's helpful to know where the curve or line crosses the x- or y-axis. By a *y-intercept* of a graph, we mean the y-coordinate of a point where the graph crosses the y-axis. For instance, returning to Figure 1, the y-intercept is -2. Notice that this y-intercept can be obtained algebraically just by setting x equal to zero in the given equation $y = 3x - 2$. In a similar fashion, an *x-intercept* of a graph is the x-coordinate of a point where the graph crosses the x-axis. In Figure 1 we can see that the x-intercept is a number between 0 and 1. To determine this x-intercept, set y equal to zero in the equation $y = 3x - 2$. That yields $0 = 3x - 2$, and consequently $x = \frac{2}{3}$. So the x-intercept of the graph in Figure 1 is $\frac{2}{3}$.

In elementary graphing there is always the question of how many points must be plotted before the essential features of a graph are clear. As you'll see throughout this text, there are a number of techniques and concepts that make it unnecessary to plot a large number of points. In this section we introduce the idea of symmetry as an aid in graphing. The graphs in Figures 2 through 4 present three types of symmetry.

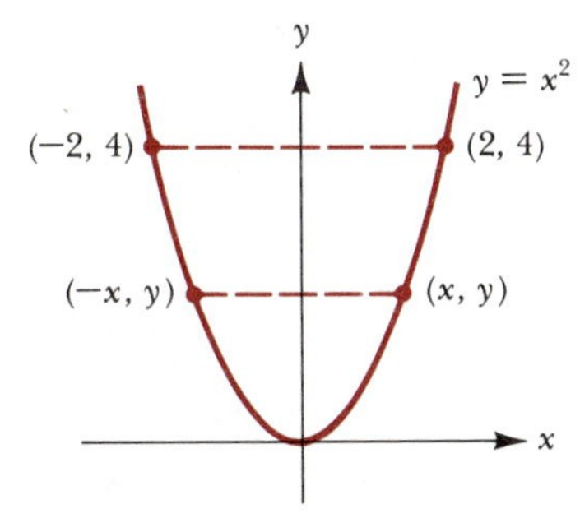

Figure 2
Symmetry with respect to the y-axis.

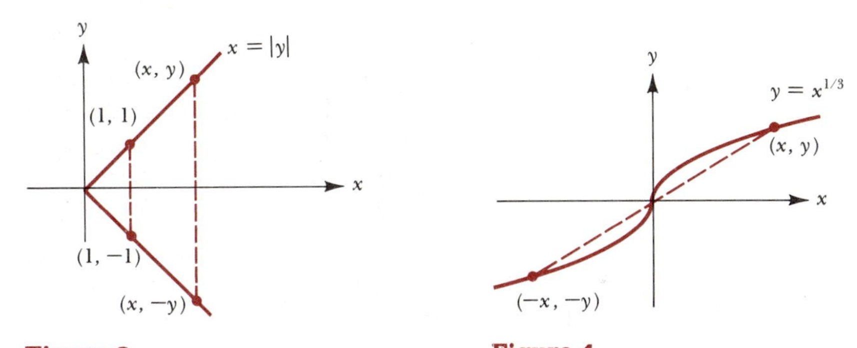

Figure 3
Symmetry with respect to the x-axis.

Figure 4
Symmetry with respect to the origin.

The graph of $y = x^2$ in Figure 2 is *symmetric with respect to the y-axis*. Geometrically this means that if you fold the paper along the y-axis, then the two portions of the curve will coincide. In terms of coordinates, this means that for each point (x, y) on the graph, the point $(-x, y)$ is also on the graph. Following these same ideas, we say that the graph of $x = |y|$ in Figure 3 is *symmetric with respect to the x-axis*; for each point (x, y) on the graph, the point $(x, -y)$ is also on the graph. Finally, the graph of $y = x^{1/3}$ in Figure 4 is *symmetric with respect to the origin*. In terms of coordinates this means that for each point (x, y) on the graph, the point $(-x, -y)$ is also on the graph. This type of symmetry derives its name from the fact that the origin is the midpoint of the line segment joining the points (x, y) and $(-x, -y)$. (Incidentally, you can relate symmetry about the origin to symmetry about the y- and x-axes as follows: If a curve is symmetric about the origin and you fold the paper along the y-axis and then along the x-axis, then the two portions of the curve will coincide.)

In the box that follows we list three rules for testing whether the graph of an equation possesses any of the types of symmetry just discussed. (The validity of each rule follows directly from the definitions of symmetry.)

Three Tests for Symmetry

1. The graph of an equation is symmetric with respect to the y-axis if replacing x with $-x$ yields an equivalent equation.
2. The graph of an equation is symmetric with respect to the x-axis if replacing y with $-y$ yields an equivalent equation.
3. The graph of an equation is symmetric with respect to the origin if replacing x and y with $-x$ and $-y$, respectively, yields an equivalent equation.

Example 1 (a) Test for symmetry with respect to the y-axis: $y = x^4 - 3x^2 + 1$.

(b) Test for symmetry with respect to the x-axis: $x = \dfrac{y^2}{y-1}$.

Solution (a) Replacing x with $-x$ in the original equation gives us

$$y = (-x)^4 - 3(-x)^2 + 1$$

or

$$y = x^4 - 3x^2 + 1$$

Since this last equation is the same as the original equation, we conclude that the graph is symmetric with respect to the y-axis.

(b) Replacing y with $-y$ in the original equation gives us

$$x = \frac{(-y)^2}{(-y) - 1}$$

or

$$x = \frac{y^2}{-y - 1}$$

Since this last equation is not equivalent to the original equation, we conclude that the graph is not symmetric with respect to the x-axis.

Example 2 Is the graph of $y = 2x^3 - x$ symmetric with respect to the origin?

Solution Replacing x and y with $-x$ and $-y$, respectively, gives us

$$-y = 2(-x)^3 - (-x)$$
$$-y = -2x^3 + x$$
$$y = 2x^3 - x \quad \text{multiplying through by } -1$$

This last equation is identical to the given equation. Therefore the graph is symmetric with respect to the origin.

In the next two examples we use the notions of symmetry and intercepts as guides for drawing the required graphs.

Example 3 Graph the equation $y = -x^2 + 4$.

Solution First we compute the x- and y-intercepts (if any).

$$x\text{-intercepts:}\quad 0 = -x^2 + 4 \qquad\qquad y\text{-intercept:}\quad y = -(0)^2 + 4$$
$$x^2 = 4 \qquad\qquad\qquad\qquad y = 4$$
$$x = \pm 2$$

Thus the x-intercepts are ± 2 and the y-intercept is 4. Using the tests for symmetry, we find that the graph is symmetric with respect to the y-axis, for replacing x with $-x$ in the given equation yields

$$y = -(-x)^2 + 4$$

or

$$y = -x^2 + 4$$

which is identical to the given equation. Thus we need only to be able to sketch the graph to the right of the y-axis; the portion to the left of the y-axis will be its mirror image (with the y-axis as the mirror). See Figure 5. (The curve in Figure 5 is a *parabola.* This type of curve will be studied in Chapter 5.)

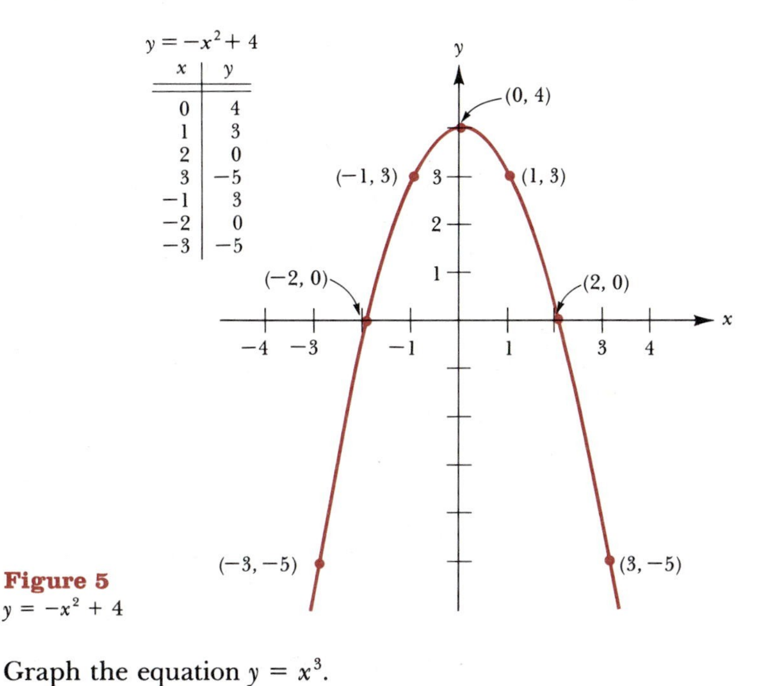

Figure 5
$y = -x^2 + 4$

Example 4 Graph the equation $y = x^3$.

Solution First notice that the curve must pass through the origin, because when $x = 0$ we have $y = 0^3 = 0$. Furthermore, the curve is symmetric with respect to the origin. (Verify this for yourself using the appropriate symmetry test.) Next let's set up a short table, using $x = 1$, 2, 3, and 4.

x	1	2	3	4
y	1	8	27	64

For each point (x, y) here, symmetry tells us that the point $(-x, -y)$ is also on the graph. Finally, before drawing the graph, we note how the y-values in the table become large rather quickly. In view of this, we use a different scale on the y-axis to accommodate the graph. See Figure 6 on the next page.

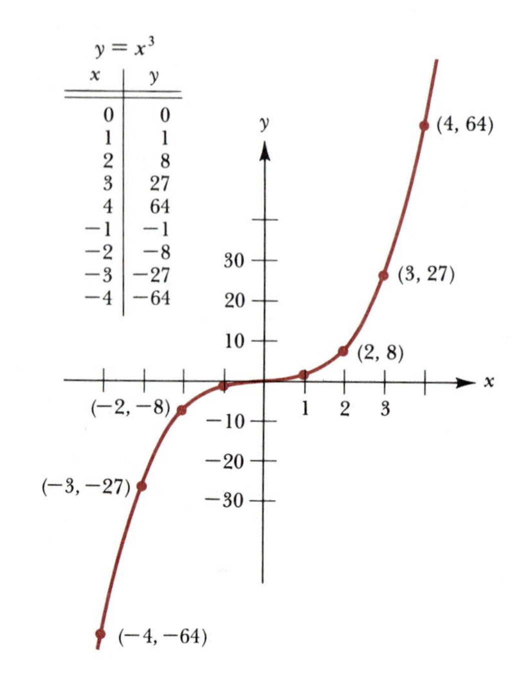

Figure 6
$y = x^3$

As we've been discussing symmetry in this section, it seems appropriate to conclude with a discussion of the circle since, in some sense, this is the most symmetric curve. We can use the distance formula from Section 3.1 to obtain the equation of a circle. Figure 7 shows a circle with center (h, k) and radius r.

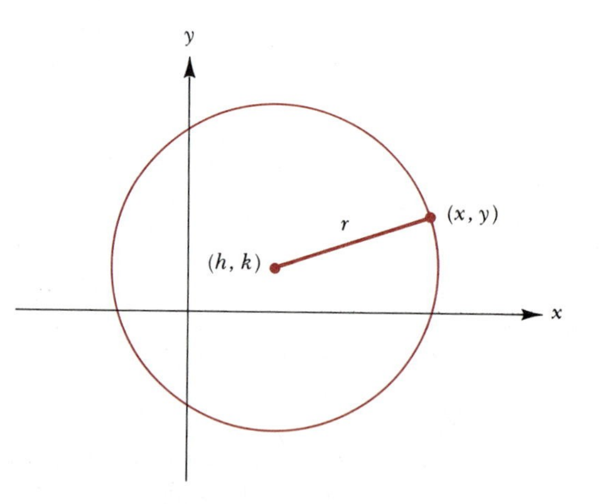

Figure 7

Now, by definition, a point (x, y) is on this circle if and only if the distance from (x, y) to (h, k) is r. Thus we have

$$\sqrt{(x-h)^2 + (y-k)^2} = r$$

or, after squaring both sides,

$$(x-h)^2 + (y-k)^2 = r^2$$

This last equation is called the *standard form* for the equation of a circle.

The Equation of a Circle in Standard Form

The equation of a circle with center (h, k) and radius r is

$$(x - h)^2 + (y - k)^2 = r^2$$

Example 5 Write the equation of the circle with center $(-2, 5)$ and radius 3.

Solution In the equation $(x - h)^2 + (y - k)^2 = r^2$, we substitute the given values $h = -2$, $k = 5$, $r = 3$. Thus

$$[x - (-2)]^2 + (y - 5)^2 = 3^2$$
$$(x + 2)^2 + (y - 5)^2 = 9$$

This is the standard form for the equation of the given circle. An alternative form of this answer is found by carrying out the indicated algebra. We have

$$(x + 2)^2 + (y - 5)^2 = 9$$

Therefore

$$x^2 + 4x + 4 + y^2 - 10y + 25 = 9$$

and consequently

$$x^2 + 4x + y^2 - 10y + 20 = 0$$

The disadvantage to this alternative form of the answer is that information regarding the center and radius is no longer so readily visible as it was when we wrote $(x + 2)^2 + (y - 5)^2 = 9$.

When the equation of a circle is not in standard form, the technique of completing the square (discussed in Section 2.2) can be used to convert the equation to standard form. Example 6 shows how this is done.

Example 6 Determine the center and radius of the circle

$$4x^2 - 24x + 4y^2 + 16y + 51 = 0$$

Solution First divide through by 4 so that the coefficients of x^2 and y^2 are both 1:

$$x^2 - 6x + y^2 + 4y + \frac{51}{4} = 0$$

Now, in preparation for completing the squares, we write

$$(x^2 - 6x + \underline{\quad\quad}) + (y^2 + 4y + \underline{\quad\quad}) = -\frac{51}{4}$$

To complete the square in x we need to add $(-\frac{6}{2})^2$, or 9. To complete the square in y we need to add $(\frac{4}{2})^2$, or 4. Thus we have

$$(x^2 - 6x + 9) + (y^2 + 4y + 4) = -\frac{51}{4} + 9 + 4$$

or $$(x-3)^2+(y+2)^2=-\frac{51}{4}+\frac{52}{4}=\frac{1}{4}$$

$$(x-3)^2+(y+2)^2=\left(\frac{1}{2}\right)^2$$

This is the equation in standard form. The center of the circle is $(3, -2)$; the radius is $\frac{1}{2}$.

Exercise Set 3.2

In Exercises 1–6, the graph of each equation is a straight line. Graph each equation after finding the x- *and* y-*intercepts.*

1. $3x + 4y = 12$
2. $3x - 4y = 12$
3. $y = 2x - 4$
4. $x = 2y - 4$
5. $x + y = 1$
6. $2x - 3y = 6$

For Exercises 7–16, determine any x- *or* y-*intercepts for the graph of each equation.* Note: *You're not asked to draw the graph.*

7. $y = x^2 - 5x - 6$
8. $y = x^2 - 4x - 12$
9. $y = x^2 + x - 1$
10. $y = 2x^2 + x + 1$
11. $x/2 + y/3 = 1$
12. $y = |x - 1|$
13. $3x - 5y = 10$
14. $xy = 1$
15. $y = x^3 - 8$
16. $y = (x - 8)^3 + 1$

In Exercises 17–26, test each equation for symmetry with respect to the x-*axis, the* y-*axis, and the origin.*

17. $3x^2 + y = 16$
18. $x^2y + xy^2 = 0$
19. $y = x^4 - 3$
20. $y = x^5 - 4x^3 + x$
21. $x^2 + y^2 = 16$
22. $y = |x - 2|$
23. $y = x^2 + x^3$
24. $x = y^2 - 4y^6$
25. $x + y = 1$
26. $|x| + |y| = 1$

For Exercises 27–46, graph each equation. Specify any x- *or* y-*intercepts or symmetry with respect to the* x-*axis, the* y-*axis, or the origin.*

27. $y = x^2$
28. $y = -x^3$
29. $y = 1/x$
30. $x = y^2 - 1$
31. $y = -x^2$
32. $y = 1/x^2$
33. $y = -1/x^2$
34. $y = |x|$
35. $y = \sqrt{x^2}$
36. $y = x + 1$
37. $y = x^2 - 2x + 1$
38. $x = y^3 - 1$
39. $y = 2^x$
40. $y = 2^{-x}$
41. $y = 2x^2 + x - 4$
42. **(a)** $y = x^2 - 4$ **(b)** $y = |x^2 - 4|$
43. **(a)** $y = 3x - 6$ **(b)** $y = |3x - 6|$
44. **(a)** $x + y = 2$ **(b)** $|x| + |y| = 2$
45. $y = \dfrac{x^2 - 9}{x - 3}$ *Hint:* When $x = 3$, y is undefined. So your graph must not contain any point with an x-coordinate of 3. On the other hand, for $x \neq 3$ you can simplify the expression $\dfrac{x^2 - 9}{x - 3}$ and thereby save some steps in making up the table.
46. $y = \dfrac{1}{1 + x^2}$
47. On the same set of axes, draw graphs of
    $y = 3x - 2 \quad y = 3x - 1 \quad y = 3x \quad y = 3x + 1$
    Label which is which.
48. On the same set of axes, draw graphs of
    $y = x \quad y = 2x \quad y = 3x \quad y = -x \quad y = -2x$
    Label which is which.
49. **(a)** Graph the equation $y = 3$. (In setting up your table, note that whatever value you assign to x, y is always 3.)
    (b) Graph $x = 3$. **(c)** Graph $x = -3$.
    (d) What is the graph of $y = 0$?
    (e) What is the graph of $x = 0$?
50. **(a)** On the same set of axes, draw the graphs of $y = x^2$, $y = x^2 + 1$, and $y = x^2 + 2$.
    (b) Based on your results in part (a), how would you say that the value of the constant K influences the graph of $y = x^2 + K$?
51. **(a)** On the same set of axes, draw the graphs of $y = |x|$, $y = |x| + 1$, and $y = |x| + 2$.
    (b) Based on your observations in part (a), draw the graphs of $y = |x| + 3$ and $y = |x| - 1$ without first setting up tables.

52. **(a)** On the same set of axes, draw the graphs of $y = x^2$, $y = (x - 1)^2$, and $y = (x + 1)^2$.
(b) Based on your observations in part (a), draw the graph of $y = (x + 2)^2$ without setting up a table.

53. **(a)** On the same set of axes, draw the graphs of $y = |x|$, $y = |x - 1|$, and $y = |x + 1|$.
(b) Based on your observations in part (a), draw the graphs of $y = |x - 2|$ and $y = |x + 2|$ without using tables.

54. **(a)** Graph the equation $v = 32t$, which relates the velocity v of a freely falling object and the length of time t it has been falling. In this formula, t is measured in seconds and v is in feet per second. (Use t for the horizontal axis. Assume that $t \geq 0$ so that your graph lies only in the first quadrant.)
(b) Indicate on the t-axis the time at which the velocity is 64 ft/sec.

55. **(a)** Graph the equation $s = 16t^2$, which gives the distance s (in feet) traveled by a freely falling object in t seconds. (The time $t = 0$ corresponds to the instant the object begins its fall; thus your graph should lie only in the first quadrant, where $t \geq 0$.)
(b) Indicate the portion of the s-axis that shows the distance covered during the first second, from $t = 0$ to $t = 1$.
(c) Indicate the portion of the s-axis that shows the distance covered during the next second, from $t = 1$ to $t = 2$.

56. Suppose that the formula $C = 5 + x$ gives the manufacturer's cost in dollars to produce x units of a certain commodity.
(a) Graph this equation.
(b) Indicate on the C-axis the cost corresponding to a production level of $x = 2$ units.
(c) Indicate on the x-axis the number of units corresponding to a cost of \$9.

57. Draw the following graphs on the same set of axes and in each case include only the portion of the graph between $x = 0$ and $x = 1$. Draw the graphs carefully enough to ensure that you can make accurate comparisons among them.
(a) $y = x$ **(b)** $y = x^2$
(c) $y = x^3$ **(d)** $y = x^4$
What is the pattern here? What would you say that $y = x^{100}$ must look like in this interval?

58. The figure shows the graph of $y = \sqrt{x}$. Use this graph to estimate the following quantities (to one decimal place).
(a) $\sqrt{2}$ **(b)** $\sqrt{3}$
(c) $\sqrt{6}$ *Hint:* $\sqrt{ab} = \sqrt{a}\sqrt{b}$.

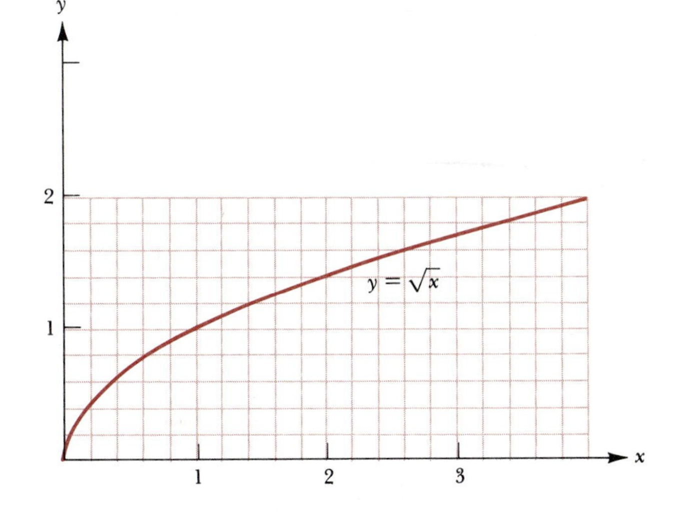

59. In a certain biology experiment, the number N of bacteria increases with time t as indicated in the figure.
(a) How many bacteria are initially present when $t = 0$?
(b) Approximately how long does it take for the original colony to double in size?
(c) For which value of t is the population N equal to about 5000?
(d) During which time interval does the population increase more rapidly, between $t = 0$ and $t = 1$ or between $t = 4$ and $t = 5$?

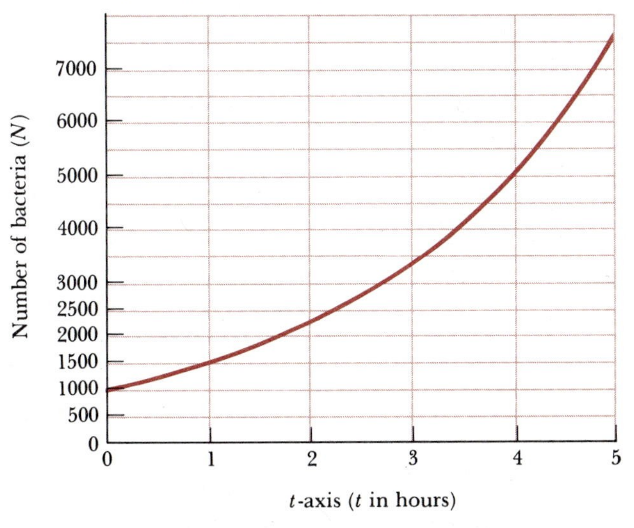

60. Specify the center and radius of each circle.
(a) $(x - 3)^2 + (y - 1)^2 = 25$
(b) $x^2 + (y + 1)^2 = 16$ **(c)** $x^2 + y^2 = 1$
(d) $(x + 2)^2 + (y - 1)^2 = \frac{1}{4}$ **(e)** $x^2 + y^2 = 2$
(f) $(x + \frac{3}{2})^2 + (y + \frac{2}{3})^2 = 20$
(g) $x^2 + y^2 = \sqrt{2}$

61. Write the equation of the circle specified in each case. Then sketch its graph.
(a) Center $(1, 3)$; radius 1.
(b) Center $(-1, 3)$; radius 2.
(c) Center $(0, 0)$; radius 3.

62. **(a)** Sketch the circle of radius 1 centered at the origin.
(b) Write the equation for this circle.
(c) Does the point $(\frac{3}{5}, \frac{4}{5})$ lie on this circle?
(d) Does the point $(-1/2, \sqrt{3}/2)$ lie on this circle?

63. The center of a circle is the point $(3, 2)$. If the point $(-2, -10)$ lies on this circle, find the equation of the circle.

For Exercises 64–69, determine the center, the radius, and the x- *and* y-*intercepts (if any) for each circle.*

64. $x^2 + y^2 - 10x + 2y + 17 = 0$

65. $x^2 + y^2 + 8x - 6y = -24.$

66. $4x^2 - 4x + 4y^2 - 63 = 0$

67. $9x^2 + 54x + 9y^2 - 6y + 64 = 0$

68. $2x^2 + 2y^2 + 4x + 4y + 3 = 0$

69. $x^2 + 3x + y^2 - 5y = \frac{1}{2}$

70. Use the formulas $A = \pi r^2$ and $C = 2\pi r$ to find the area and the circumference, respectively, of each circle.
(a) $x^2 + y^2 = 1$
(b) $x^2 + y^2 - 10x + 10y = -47$
(c) $4x^2 + 4y^2 - 4x + 8y - 4 = 0$
(d) $x^2 + y^2 - 8x - 20y = -116$
(e) $3x^2 + 3y^2 + 5x - 4y = 1$
(f) $x^2 - 2ax + y^2 - 2by = c^2 - a^2 - b^2$ (Assume that a, b, and c are constants and $c > 0$.)

71. **(a)** Find the equation of the circle tangent to the x-axis and with center $(3, 5)$.
(b) Find the equation of the circle tangent to the y-axis and with center $(3, 5)$.
(c) Find the equation of the circle with center $(3, 5)$ that passes through the origin.

72. In the same coordinate system, sketch the graphs of the circles $x^2 + y^2 - 10x - 6y = -33$ and $x^2 + y^2 - 10x - 6y = -18$. Then calculate the area of the region outside the smaller circle but within the larger circle.

73. Follow Exercise 72, but use the equations $x^2 + y^2 = 1$ and $x^2 + y^2 - 8x - 9 = 0$.

74. Given the two equations

$$x^2 + y^2 - 4y - 5 = 0$$

and

$$x^2 + y^2 - 14y + 33 = 0$$

form a third equation by subtracting the first equation from the second. Then, on the same set of axes, graph all three equations. What do you observe?

75. If the coordinates of P and Q are $P(-3, -4)$ and $Q(2, 8)$, find the equation of the circle having PQ as a diameter. Write the equation of the circle, in standard form.

***76.** Graph the equation $(x^2 + y^2 - 1)(x^2 + y^2 - 4) = 0$. *Hint:* Don't carry out the indicated multiplication.

***77.** Graph the equation $[(x - 1)^2 + y^2 - 1][(x - 3)^2 + y^2 - 1][(x - 5)^2 + y^2 - 1] = 0$.

***78.** Graph the equation $(y - x^2)(y + x^2)(x - y^2)(x + y^2) = 0$.

C **79.** Graph $y = \sqrt{x}$.

C **80.** Graph $y = \sqrt[3]{x}$.

C **81.** Determine the intercepts and symmetry, and graph the equation $y = x^{2/3}$.

C **82.** Find the intercepts and symmetry for $y = x^3 - 27x$. Then use a calculator (as necessary) to set up a table of x- and y-values, with x running from 0 to 6 in increments of $\frac{1}{2}$. Finally, graph the equation.

C **83.** Set up a table of x- and y-values for the equation $y = \dfrac{4x^2}{1 + x^2}$. Use x-values running from 0 to 5 in increments of $\frac{1}{2}$. Now use the results in your table, along with symmetry, to graph the equation.

C ***84.** Graph the equation $y = x + \dfrac{1}{\sqrt{x + 1}}$.

3.3 Slope

Figures 1(a) and (b) on the next page show two straight lines, both of which pass through a given point (2, 1). In what way do the lines differ from one another? Qualitatively, it's easy to see that the line in Figure 1(b) slants upward more sharply than the line in Figure 1(a). To make this idea of slant or direction of a line quantitative, we define a number called the *slope* of a line as follows.

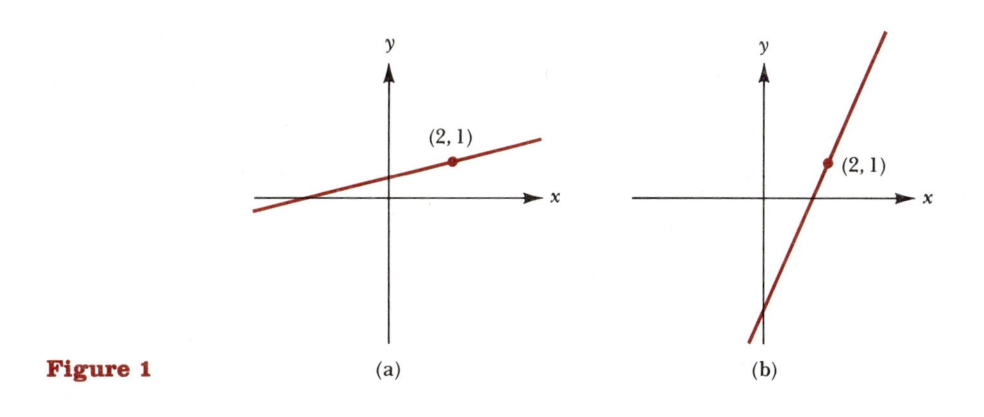

Figure 1

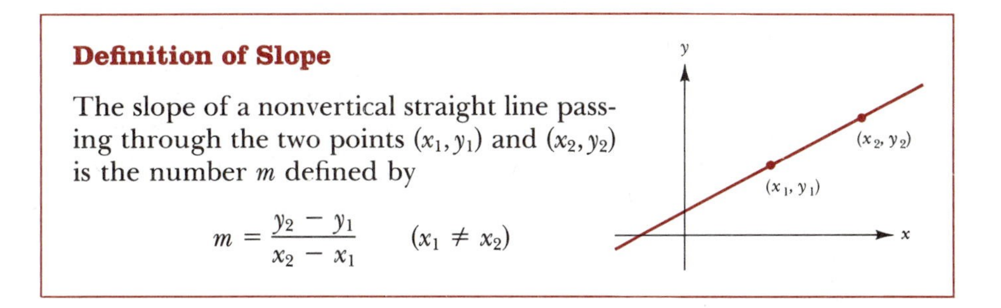

Definition of Slope

The slope of a nonvertical straight line passing through the two points (x_1, y_1) and (x_2, y_2) is the number m defined by

$$m = \frac{y_2 - y_1}{x_2 - x_1} \qquad (x_1 \neq x_2)$$

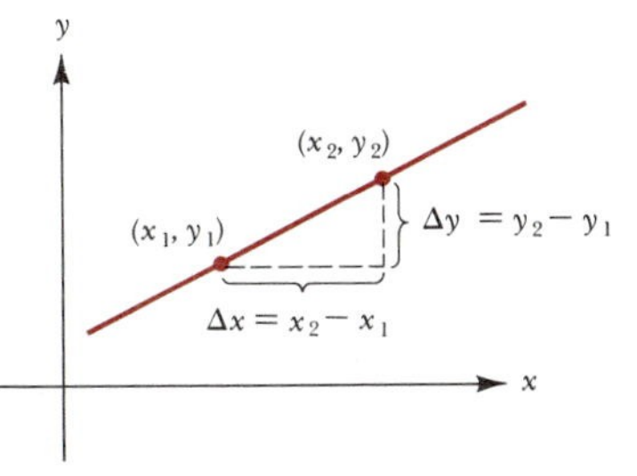

Figure 2

Note that the quantity $x_2 - x_1$ appearing in the definition of slope is the amount by which x changes as we move from (x_1, y_1) to (x_2, y_2) along the line. We denote this change in x by the symbol Δx (read *delta x*). Thus $\Delta x = x_2 - x_1$. See Figure 2. Similarly, the symbol Δy is defined to mean the change in y: $\Delta y = y_2 - y_1$. Using these ideas, we can rewrite our definition of slope as $m = \Delta y/\Delta x$.

Example 1 Compute the slope of the line in Figure 3.

Solution We use the formula $m = \dfrac{y_2 - y_1}{x_2 - x_1}$. Which point will serve as (x_1, y_1) and which as (x_2, y_2)? It will turn out not to matter how we label our points. Using $(-2, 1)$ as (x_1, y_1) and $(4, 2)$ as (x_2, y_2), we find $m = \dfrac{2 - 1}{4 - (-2)} = \dfrac{1}{6}$. If instead we use $(4, 2)$ as (x_1, y_1) and $(-2, 1)$ as (x_2, y_2), we find $m = \dfrac{1 - 2}{-2 - 4} = \dfrac{1}{6}$, the same result. This is not accidental, because in general $\dfrac{y_2 - y_1}{x_2 - x_1} = \dfrac{y_1 - y_2}{x_1 - x_2}$.

Reason: $\dfrac{y_2 - y_1}{x_2 - x_1} = \dfrac{-1(y_2 - y_1)}{-1(x_2 - x_1)} = \dfrac{-y_2 + y_1}{-x_2 + x_1} = \dfrac{y_1 - y_2}{x_1 - x_2}$

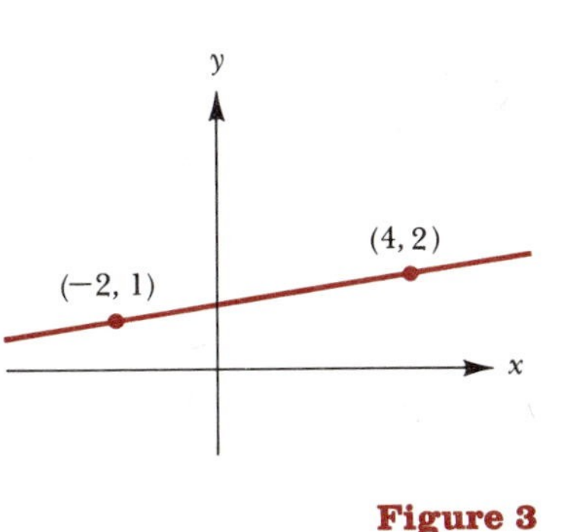

Figure 3

Let us also compute Δx and Δy for the two points given in Figure 3. If we begin at the point $(-2, 1)$ and move up the line to $(4, 2)$, then we have

$$\Delta x = 4 - (-2) = 6 \quad \text{and} \quad \Delta y = 2 - 1 = 1$$

Thus $m = \Delta y/\Delta x = 1/6$, as we obtained in Example 1. Note that if instead we start from the point $(4, 2)$ and move down the line to $(-2, 1)$, then

$$\Delta x = -2 - 4 = -6 \quad \text{and} \quad \Delta y = 1 - 2 = -1$$

Thus $m = \Delta y/\Delta x = -1/-6 = 1/6$, the same result as before, even though Δy and Δx each differ by a sign from their previous values. In summary, the individual values of Δx and Δy depend on which direction you move along the given line. However, the value obtained for the slope $\Delta y/\Delta x$ is the same in both cases.

Example 2 Calculate the slope of the line in Figure 4, first using the points (2, 1) and (3, 3) and then using (4, 5) and (7, 11).

Solution Using the points (2, 1) and (3, 3), we have $m = \dfrac{3-1}{3-2} = \dfrac{2}{1} = 2$. Using the points (4, 5) and (7, 11), we have $m = \dfrac{11-5}{7-4} = \dfrac{6}{3} = 2$. Note that the two results are the same.

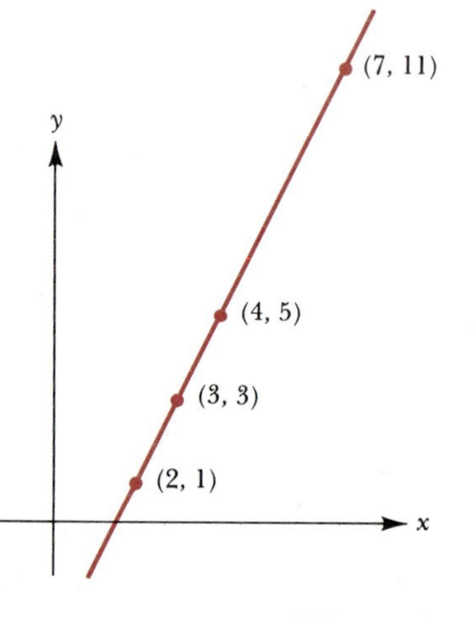

Figure 4

The point of Example 2 is to show you that the value obtained for the slope of a given line does not depend on which two points on the line are used to calculate the slope. To demonstrate this is true in general, consider Figure 5.

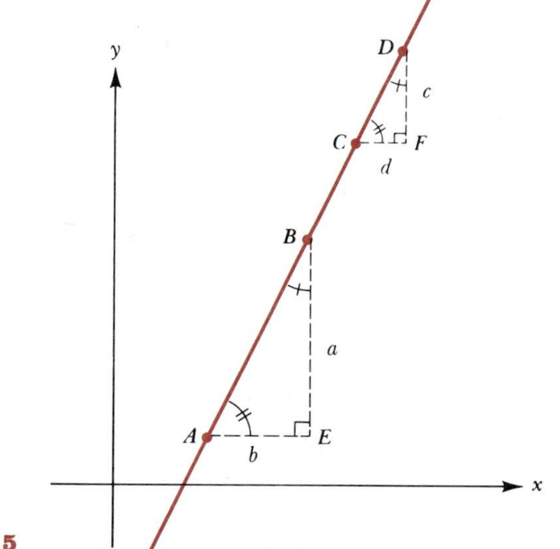

Figure 5

The two right triangles in the figure are similar (because the corresponding angles are equal). This implies that the corresponding sides of the two triangles are proportional, and so we have

$$\frac{a}{b} = \frac{c}{d}$$

Now just notice that the left-hand side of this equation represents the slope $\Delta y/\Delta x$ calculated using the points A and B, whereas the right-hand side represents the slope calculated using the points C and D. Thus the values we obtain for the slope are indeed equal. (See Exercise 28 at the end of this section for another perspective on this.)

Example 3 Compute and compare the slopes of the three lines shown in Figure 6.

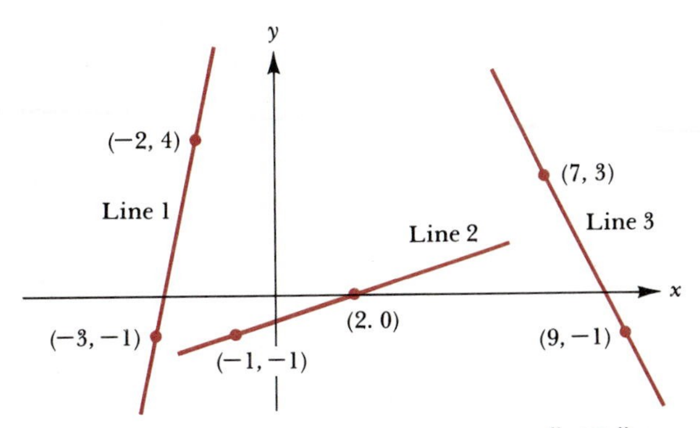

Figure 6

Solution The slopes are all calculated using the formula $m = \dfrac{y_2 - y_1}{x_2 - x_1}$. Writing m_1, m_2, m_3 for the respective slopes of lines 1, 2, and 3, we have

$$m_1 = \frac{4 - (-1)}{-2 - (-3)} = \frac{5}{1} = 5$$

$$m_2 = \frac{0 - (-1)}{2 - (-1)} = \frac{1}{3}$$

$$m_3 = \frac{3 - (-1)}{7 - 9} = \frac{4}{-2} = -2$$

Lines 1 and 2 both have positive slopes and slant upward to the right. Note that line 1 is steeper than line 2 and correspondingly has the larger slope. Line 3 has a negative slope and slants downward to the right.

The observations made in Example 3 are true in general. Lines with a positive slope slant upward to the right, the steeper line having the larger slope. Likewise, lines with a negative slope slant downward to the right. See Figure 7.

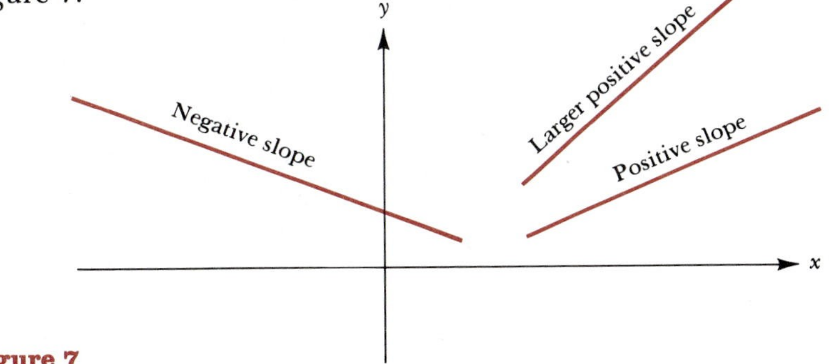

Figure 7

We have yet to mention slopes for horizontal or vertical lines. In Figure 8, line 1 is horizontal; it passes through the points (a, b) and (c, b). Note that the two y-coordinates must be the same in order for the line to be horizontal. Line 2 in Figure 8 is vertical; it passes through (d, e) and (d, f). Note that the two x-coordinates must be the same in order for the line to be vertical.

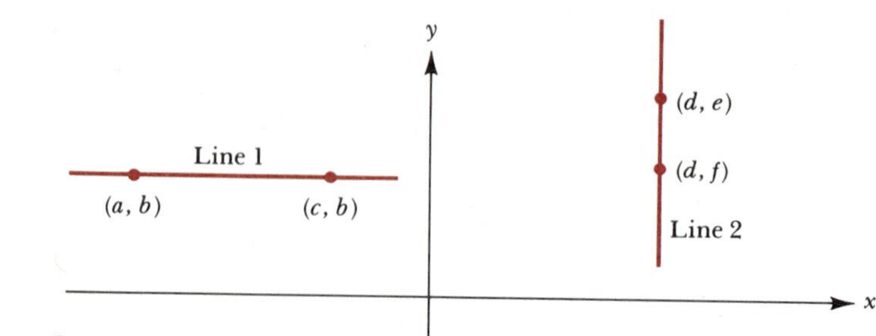

Figure 8

For the slope of line 1 we have $m = \dfrac{b - b}{c - a} = \dfrac{0}{c - a} = 0$ (provided $a \neq c$). Thus the slope of line 1, a horizontal line, is zero. For the vertical line in Figure 8, the calculation of slope begins with writing $\dfrac{e - f}{d - d}$. But then the denominator is zero, and since division by zero is undefined, we conclude that slope is undefined for vertical lines. Figure 9 shows for comparison some values of m for various lines.

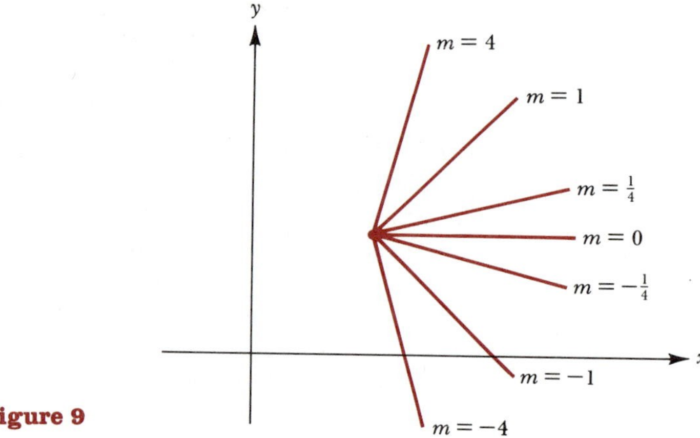

Figure 9

We conclude this section by looking at two ways in which slope is used in applications. As a first example, suppose that you are driving a car at a steady speed of 50 mph. So in 1 hr you travel 50 mi, in 2 hr you travel 100 mi, and so on. Letting d denote the distance in miles that you would cover in t hr at this speed, we can set up the following table.

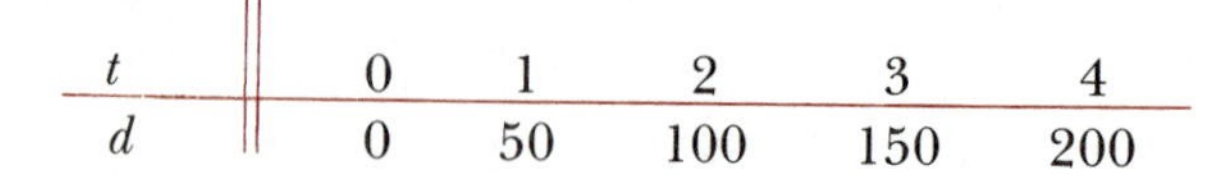

t	0	1	2	3	4
d	0	50	100	150	200

Now we draw a graph, using the data from this table, as shown in Figure 10. We can calculate the slope of the line in Figure 10 using the points (1,50) and (2,100). We have

$$m = \frac{100 - 50}{2 - 1} = \frac{50}{1} = 50$$

Figure 10

Observation: The value we have obtained here for the slope is the rate at which the car is traveling. Indeed, if we keep track of the units, our slope calculation will look like this:

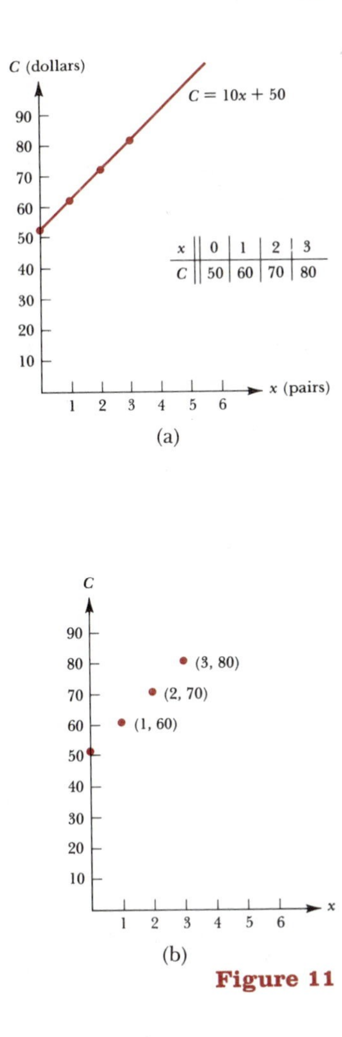

Figure 11

$$m = \frac{100 \text{ mi} - 50 \text{ mi}}{2 \text{ hr} - 1 \text{ hr}} = \frac{50 \text{ mi}}{1 \text{ hr}} = 50 \text{ mph}$$

Thus *slope is a rate of change.* In this example, slope is the *velocity* or rate of change of distance with respect to time.

As a second example of how slope arises in applications, let us suppose that a small manufacturer of handmade running shoes knows that her total cost C, in dollars, for producing x pairs of shoes each day is given by the formula

$$C = 10x + 50$$

Figure 11(a) displays a table and graph for this equation. Actually, since x represents the number of pairs of shoes, x can assume only whole number values. So technically the graph should be that given in Figure 11(b). However, it turns out to be useful in practice to make use of the graph in Figure 11(a), and we shall follow this convention.

Using the points $(0, 50)$ and $(1, 60)$, we calculate the slope of the line in Figure 11(a):

$$m = \frac{\$60 - \$50}{1 \text{ pair} - 0 \text{ pair}} = \frac{\$10}{1 \text{ pair}} = \$10 \text{ per pair}$$

Again the slope is a rate. In this case, the slope represents the rate of increase of cost; each additional pair of shoes produced costs the manufacturer \$10.

In the study of economics, an equation that gives the cost C for producing x units of a commodity is called a *cost equation* or *cost function.* When the graph of the cost equation is a straight line, we define the *marginal cost* as the additional cost to produce one more unit. Thus, in the example above, the marginal cost is \$10 per pair, and we see that the slope of the line in Figure 11(a) represents this marginal cost.

Exercise Set 3.3

1. In each case, compute the slope of the line passing through the two given points.
(a) $(-3, 2), (1, -6)$ **(b)** $(2, -5), (4, 1)$
(c) $(-2, 7), (1, 0)$ **(d)** $(4, 5), (5, 8)$
(e) $(-3, 0), (4, 9)$ **(f)** $(-1, 2), (3, 0)$
(g) $(\frac{1}{2}, -\frac{3}{5}), (\frac{3}{2}, \frac{3}{4})$ **(h)** $(\frac{17}{3}, -\frac{1}{2}), (-\frac{1}{2}, \frac{17}{3})$

2. In each case, compute the slope of the line passing through the two given points. Include a sketch.
(a) $(1, 1), (-1, -1)$ **(b)** $(-1, 1), (1, -1)$
(c) $(0, 5), (-8, 5)$ **(d)** $(a, b), (b, a)$

3. Compute the slope of the line in the figure to the right using the two points indicated:
(a) A and B **(b)** B and C
(c) A and C
The idea here is that no matter which pair of points on the line you choose, the slope is the same.

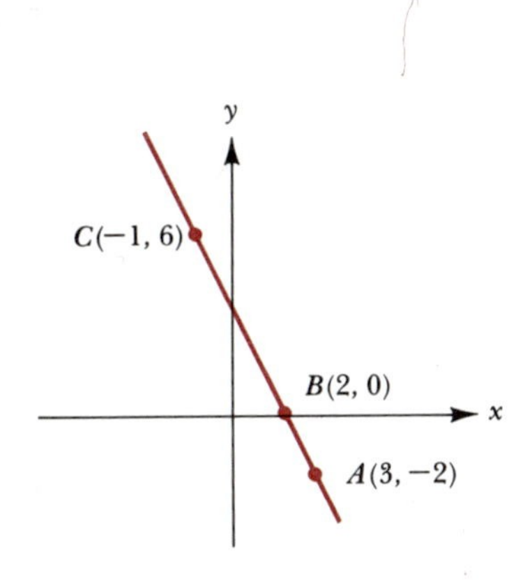

4. The slopes of four lines are indicated in the figure. List the numbers m_1, m_2, m_3, m_4 in order of increasing size.

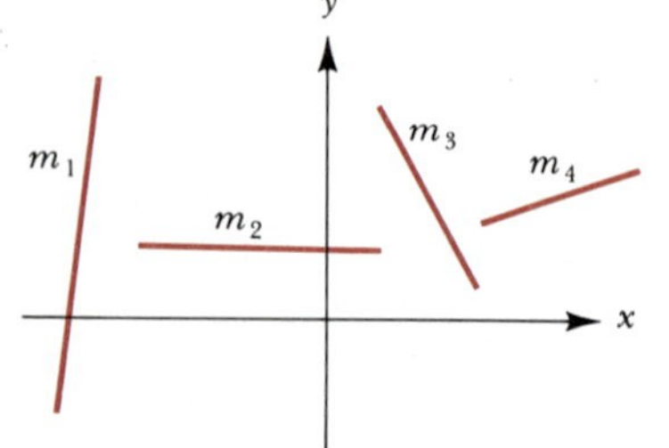

5. Determine the slopes of lines a through e. (Use the formula $m = \Delta y/\Delta x$.)

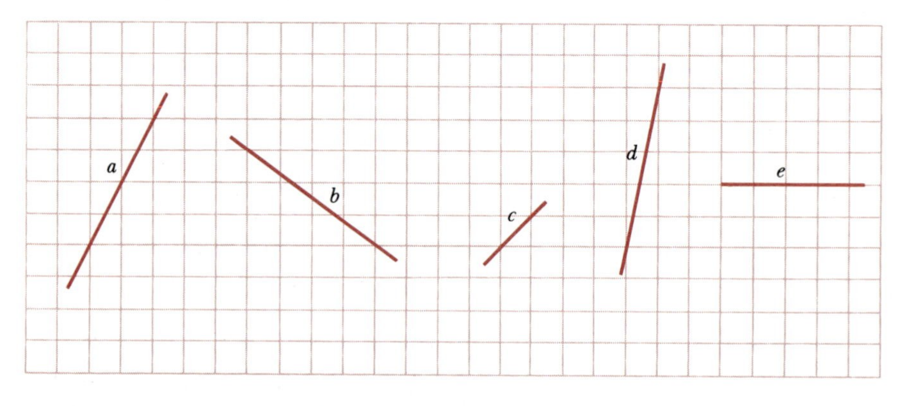

6. A straight line with a slope of -3 passes through the points $(-4, 5)$ and $(-3, y)$. Find y.

7. Compute the slopes of lines L_1, L_2, and L_3 in the figure below. (Assume the dotted lines are parallel to the axes.)

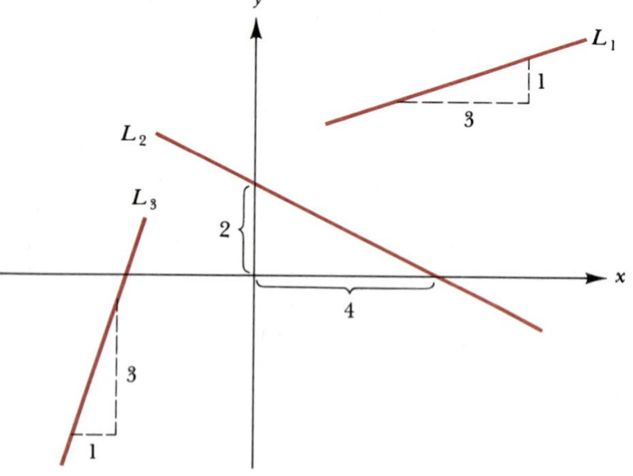

8. In each case, three points A, B, and C are specified. Find out if A, B, and C are collinear (lie on the same line) by checking to see whether the slope of AB equals the slope of BC.
   (a) $A(-8, -2)$, $B(2, \frac{1}{2})$, $C(11, -1)$
   (b) $A(4, -3)$, $B(-1, 0)$, $C(-4, 2)$
   (c) $A(0, -5)$, $B(3, 4)$, $C(-1, -8)$

9. If the area of the "triangle" formed by three points is zero, then the points must in fact be collinear. Use this observation, along with the formula provided at the end of Exercise Set 3.1, to rework Exercise 8(c).

10. **(a)** In the figure, find Δx, Δy, and $m = \Delta y/\Delta x$, assuming that you start at $(3, 1)$ and move along the line up to $(5, 3)$.
    (b) Same as part (a), but assuming that you start at $(5, 3)$ and move down the line to $(3, 1)$.

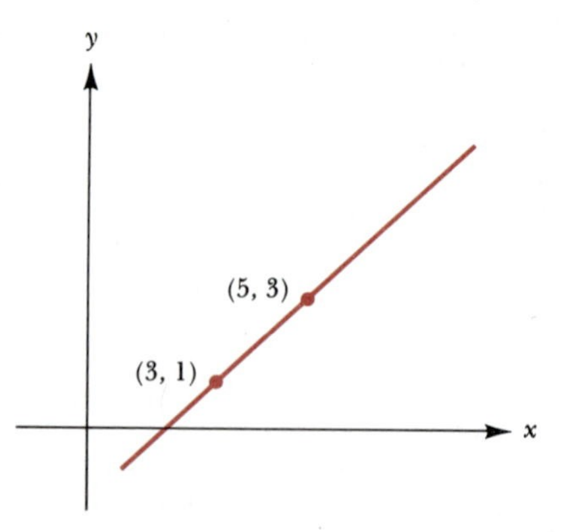

11. **(a)** In the figure at the top of page 137, find Δx, Δy, and $m = \Delta y/\Delta x$, assuming you start at $(6, 0)$ and move along the line to $(4, 3)$.
    (b) Same as part (a), but start at $(4, 3)$ and move down to $(6, 0)$.

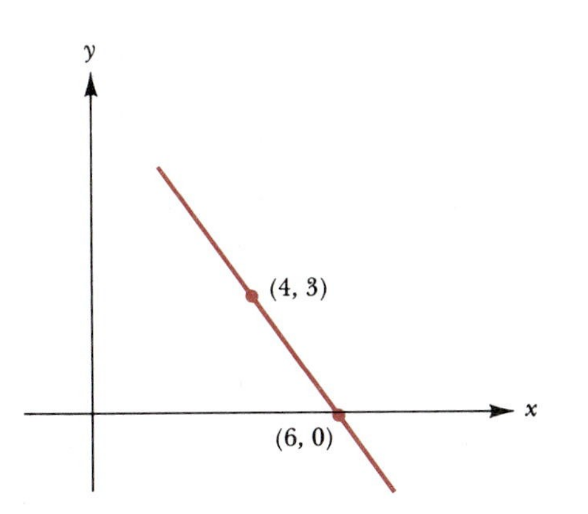

12. The graphs of the following equations are all straight lines. In each case, set up a table and draw the graph. Then pick any two points on the line and compute the slope.
(a) $y = 2x - 3$. **(b)** $y = 2x$
(c) $y = 2x + 1$ **(d)** $y = 3x - 4$
(e) $y = 3x$ **(f)** $y = 3x + 2$

13. Use the same instructions as in Exercise 12.
(a) $y = x$ **(b)** $y = -x$
(c) $y = 5x$ **(d)** $y = -5x$

14. Use the same instructions as in Exercise 12.
(a) $y = 4$ **(b)** $3x + 5y = 15$ **(c)** $y = 1 - x$

15. Show that the slope of the line passing through the two points $(3, 9)$ and $(3 + h, (3 + h)^2)$ is $6 + h$.

16. Show that the slope of the line passing through the two points (x, x^2) and $(x + h, (x + h)^2)$ is $2x + h$.

17. Graph the equation $d = 5t$, which relates the distance d, in feet, covered by an object moving for t sec at a speed of 5 ft/sec. Use t for the horizontal axis. Your graph should be a straight line. Pick two points on the line and compute the slope. Observation?

18. Graph the equation $d = 30t$, which relates the distance d, in miles, covered by an object moving for t hr at a rate of 30 mph. Your graph should be a straight line. Pick two points on the line and compute the slope. Observation?

19. In each case, a graph is given that relates the distance and time for a moving object. Determine the velocity in each case.

20. Suppose the cost C (in dollars) of producing x units of a certain kind of stereo receiver is given by $C = 400 + 50x$.
(a) Graph this equation.
(b) Using the coordinates of two points on this line, compute the slope.
(c) What is the marginal cost here?

21. Suppose the cost C (in dollars) of producing x record albums is given by $C = 0.5x + 500$. Graph this equation, determine the slope, and give the marginal cost.

***22.** Show that the slope of the line passing through the two points (x, x^3) and $(x + h, (x + h)^3)$ is $3x^2 + 3xh + h^2$.

***23.** **(a)** Show that the slope of the line passing through the points $(x, \sqrt{x})$ and $(x + h, \sqrt{x + h})$ is $\dfrac{\sqrt{x + h} - \sqrt{x}}{h}$.
(b) Show that the answer in part (a) can be written as $\dfrac{1}{\sqrt{x + h} + \sqrt{x}}$. *Suggestion:* Multiply the answer in part (a) by $\dfrac{\sqrt{x + h} + \sqrt{x}}{\sqrt{x + h} + \sqrt{x}}$.

24. Show that the slope of the line passing through the two points $\left(x, \dfrac{1}{x}\right)$ and $\left(x + h, \dfrac{1}{x + h}\right)$ is $\dfrac{-1}{x(x + h)}$.

***25.** Show that the slope of the line passing through the two points $(4, 2)$ and $(4 + h, \sqrt{4 + h})$ can be written as $\dfrac{1}{\sqrt{4 + h} + 2}$.

26. In a coordinate plane, assume that the x-coordinate gives the length of an object in inches, whereas the y-coordinate gives the length of the same object in centimeters. Find the slope of the line joining the points $(2, 5.08)$ and $(5, 12.70)$. What is the significance of this slope? *Hint:* What are the units?

27. Suppose that y represents the number of miles a certain car can travel on x gal of gas at a certain speed. Plot the two points in the x-y plane corresponding to the observations that the car travels 155 mi on 5 gal and 248 mi on 8 gal. Compute the slope of the line joining these two points. What is the significance of this slope? *Hint:* What are the units?

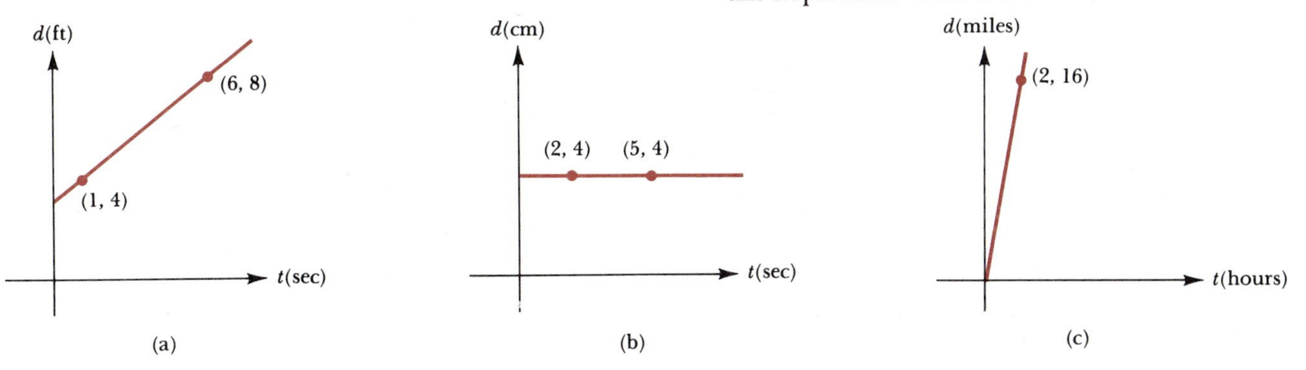

28. In the text, following Example 2, we showed that the slope of a given line does not depend on which two points on the line are used to calculate the slope. We summarize this fact by saying that slope is *well-defined*. However, if the original definition of slope had instead been given by $m = \dfrac{y_2 + y_1}{x_2 + x_1}$, then the concept would not be well-defined in the sense described. Verify this as follows: Use the expression $\dfrac{y_2 + y_1}{x_2 + x_1}$ to calculate the "slope" of the line in Figure 4 (page 132) using the points $(2, 1)$ and $(3, 3)$ and then using the points $(4, 5)$ and $(7, 11)$. Note that the answers are *not* the same.

29. Find a point P on the curve $y = x^2$ such that the slope of the line passing through P and $(2, 4)$ is 8. *Hint:* Denote the coordinates of P by (x, x^2).

***30.** Find a point P on the curve $y = \sqrt{x}$ such that the slope of the line through P and $(1, 1)$ is $\frac{1}{4}$.

31. Find a point P on the curve $y = 1/x$ such that the slope of the line through P and $(2, \frac{1}{2})$ is $-\frac{1}{16}$.

32. Find a point P on the circle $x^2 + y^2 = 20$ such that the slope of the radius drawn from $(0, 0)$ to P is 2. (Two answers.)

33. Find the slope of line L in the figure when the angle θ is as given.

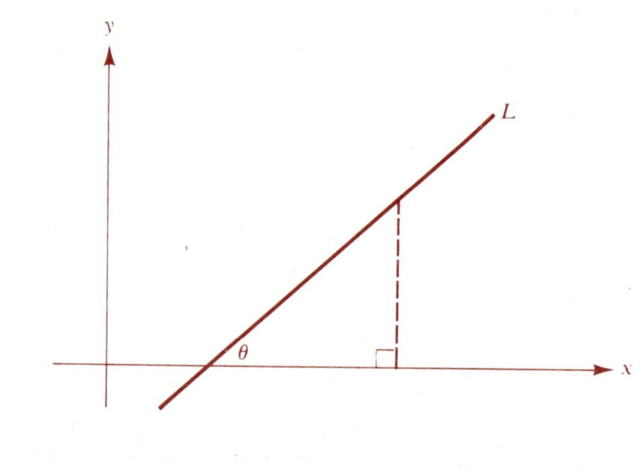

(a) $\theta = 30°$ *Hint:* In a 30°–60° right triangle, the side opposite the 30° angle is half the hypotenuse.
(b) $\theta = 60°$
(c) $\theta = 45°$

C **34.** Sketch the curve $y = x^2$ and indicate the point $T(3, 9)$ on the graph. In each case that follows you're given the x-coordinate of a point P on the curve. Find the corresponding y-coordinate, and then calculate the slope of the line through P and T. Display your results in a table.

(a) $x = 2.5$ **(b)** $x = 2.9$ **(c)** $x = 2.99$
(d) $x = 2.999$ **(e)** $x = 2.9999$

As P gets closer to T, what numerical value does the slope of PT seem to approach? What would you estimate to be the slope of the line that is tangent to the curve $y = x^2$ at the point T?

C **35.** Follow the instructions for Exercise 34 using the curve $y = x^3$ and the point $T(3, 27)$. Which curve has the steeper tangent when x is 3, $y = x^2$ or $y = x^3$?

C **36.** Follow the instructions for Exercise 34 using the curve $y = \sqrt{x}$, the point $T(1, 1)$, and the following sequence of x-values.

(a) $x = 1.1$ **(b)** $x = 1.01$ **(c)** $x = 1.001$
(d) $x = 1.0001$ **(e)** $x = 1.00001$

(Round off your answers to six decimal places.)

3.4 Equations of Lines

In this section we take a systematic look at equations of lines and their graphs. We begin by finding an equation for the line with slope m, passing through a given point (x_1, y_1). See Figure 1(a) on the next page. Let (x, y) be any other point on the line, as in Figure 1(b). Then the slope of the line is given by $m = \dfrac{y - y_1}{x - x_1}$. Thus $y - y_1 = m(x - x_1)$. Note that the given point (x_1, y_1) also satisfies this last equation, because in that case we have

$y_1 - y_1 = m(x_1 - x_1)$, or $0 = 0$. The equation $y - y_1 = m(x - x_1)$ is called the *point-slope formula*. We have shown that any point on the line satisfies this equation. (Conversely, it can be shown that if a point satisfies this equation, then the point does lie on the given line.)

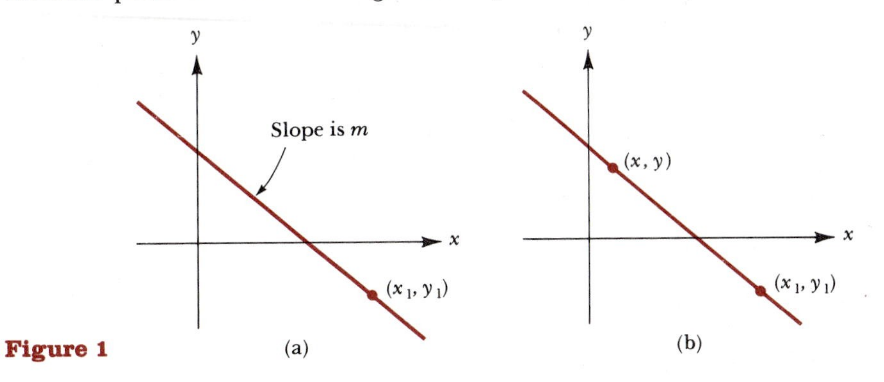

Figure 1

The Point-Slope Formula

The equation of a line with slope m passing through the point (x_1, y_1) is

$$y - y_1 = m(x - x_1)$$

Example 1 Write the equation of the line passing through $(-3, 1)$ with a slope of -2. Sketch a graph of the line.

Solution Since the slope and a point are given, we use the point-slope formula:

$$\begin{aligned} y - y_1 &= m(x - x_1) \\ y - 1 &= -2[x - (-3)] \\ y - 1 &= -2x - 6 \\ y &= -2x - 5 \end{aligned}$$

This is the required equation. One way to graph this line is to pick values for x and then compute the corresponding y-values, as done in Section 3.2. The table and graph are displayed in Figure 2. Notice that if we know ahead of time that the graph is a straight line, a table as brief as that in Figure 2 is sufficient.

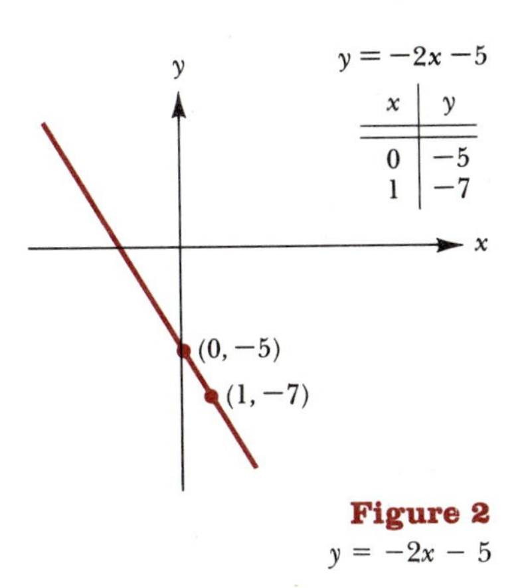

Figure 2
$y = -2x - 5$

There is another way to go about graphing the line $y = -2x - 5$ in Example 1. Since slope is (change in y)/(change in x), a slope of -2 (i.e., $-2/1$) can be interpreted as telling us that if we start at $(-3, 1)$ and let x increase by one unit, then y must decrease by two units to bring us back to the line. Following this path, in Figure 3, takes us from $(-3, 1)$ to $(-2, -1)$. We now draw the line through these two points, as shown in Figure 3.

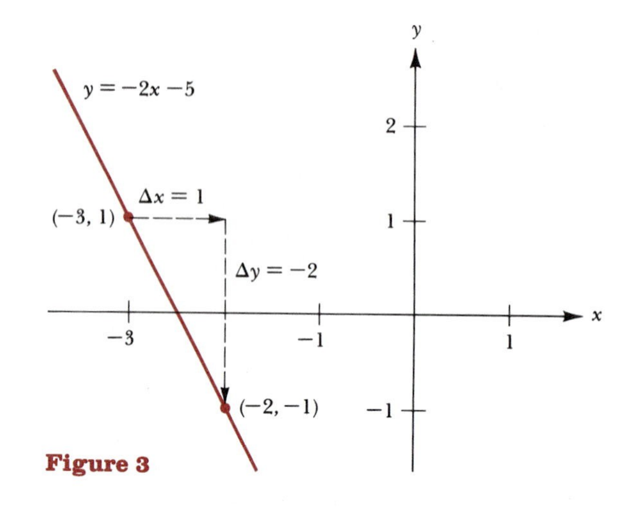

Figure 3

Example 2 Find the equation of the line passing through the points $(-2, -3)$ and $(2, 5)$.

Solution The slope of the line is

$$m = \frac{y_2 - y_1}{x_2 - x_1} = \frac{5 - (-3)}{2 - (-2)} = \frac{8}{4} = 2$$

Knowing the slope, we can apply the point-slope formula, making use of either $(-2, -3)$ or $(2, 5)$. Using the point $(2, 5)$ as (x_1, y_1), we have

$$\begin{aligned} y - y_1 &= m(x - x_1) \\ y - 5 &= 2(x - 2) \\ y - 5 &= 2x - 4 \\ y &= 2x + 1 \end{aligned}$$

Thus the required equation is $y = 2x + 1$. You should check for yourself that the same answer is obtained using the point $(-2, -3)$ instead of $(2, 5)$ in the last set of calculations.

Example 3 Find the equation of the horizontal line passing through the point $(4, -2)$. See Figure 4.

Solution As we saw in Section 3.3, the slope of a horizontal line is zero. So we have

$$\begin{aligned} y - y_1 &= m(x - x_1) \\ y - (-2) &= 0(x - 4) \\ y &= -2 \end{aligned}$$

Thus the equation of the horizontal line passing through $(4, -2)$ is $y = -2$.

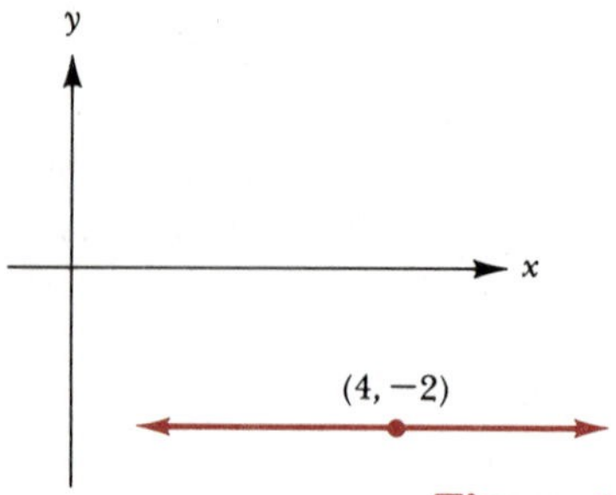

Figure 4

By using the point-slope formula exactly as we did in Example 3, we can show more generally that the equation of the horizontal line in Figure 5 passing through the point (a, b) is $y = b$. The calculations look as follows:

$$\begin{aligned} y - y_1 &= m(x - x_1) \\ y - b &= 0(x - a) \\ y &= b \end{aligned}$$

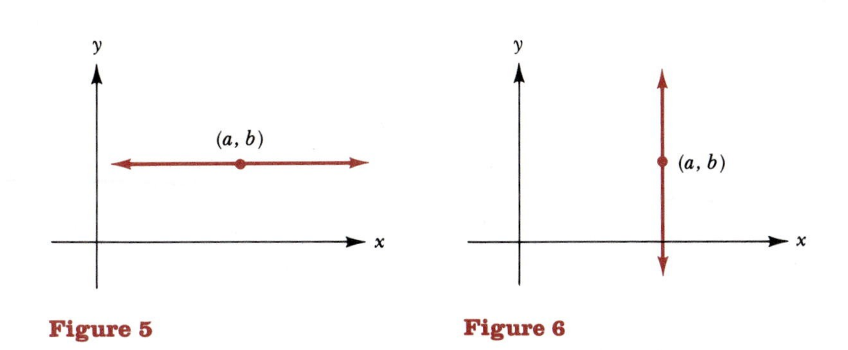

Figure 5 **Figure 6**

What about the equation of the vertical line in Figure 6 passing through the point (a, b)? Because slope is not defined for vertical lines, the point-slope formula is not applicable this time. However, looking at Figure 6, note that as you move along the vertical line, your x-coordinate is always a; it is only the y-coordinate that varies. The equation $x = a$ expresses exactly these two facts; it says that x must always be a and it places no restrictions on y.

In the box that follows we summarize our results concerning horizontal and vertical lines.

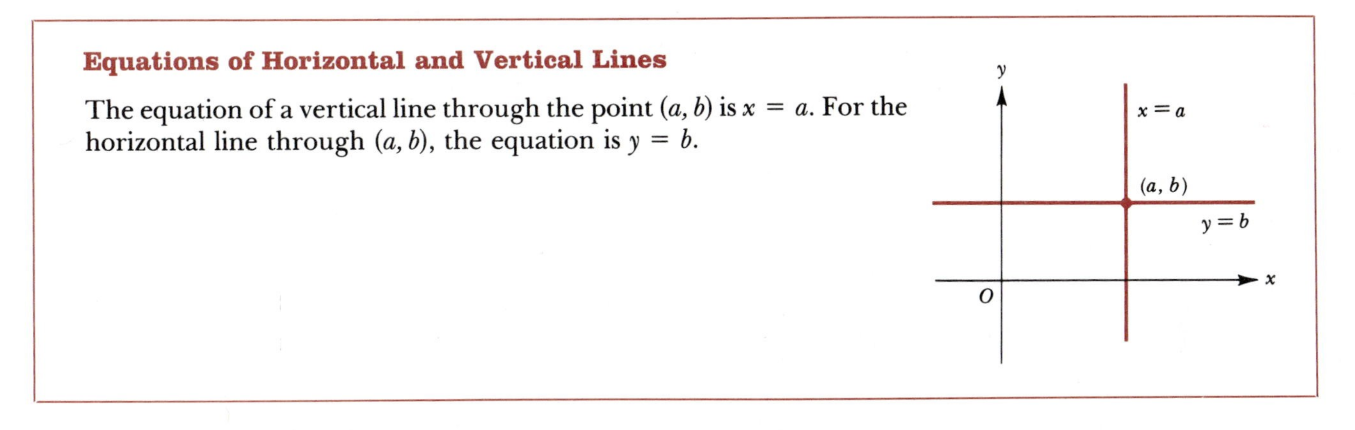

Equations of Horizontal and Vertical Lines

The equation of a vertical line through the point (a, b) is $x = a$. For the horizontal line through (a, b), the equation is $y = b$.

Example 4 What are the equations of the horizontal and vertical lines passing through the point $(-5, 7)$?

Solution The equation of the horizontal line through $(-5, 7)$ is $y = 7$. The equation of the vertical line is $x = -5$.

Next let us find an equation for a straight line with slope m and y-intercept b. See Figure 7. To say that the line has a y-intercept of b is the same as saying that the line passes through $(0, b)$. The point-slope formula is applicable now, using the slope m and the point $(0, b)$. We have

$$\begin{aligned} y - y_1 &= m(x - x_1) \\ y - b &= m(x - 0) \\ y - b &= mx \\ y &= mx + b \end{aligned}$$

Figure 7

This last equation is called the *slope-intercept formula*.

The Slope-Intercept Formula

The equation of a line with slope m and y-intercept b is

$$y = mx + b$$

Example 5 Write the equation of a line with slope of 4 and y-intercept of 1. Graph the line.

Solution Substituting $m = 4$ and $b = 1$ in the equation $y = mx + b$ yields

$$y = 4x + 1$$

This is the required equation. We could draw the graph by first setting up a simple table, but for purposes of emphasis and review we proceed as we did at the end of Example 1. Starting from the point (0, 1), we interpret the slope of 4 as saying that if x increases by 1, then y increases by 4. This takes us from (0, 1) to the point (1, 5) and the line can now be sketched as in Figure 8.

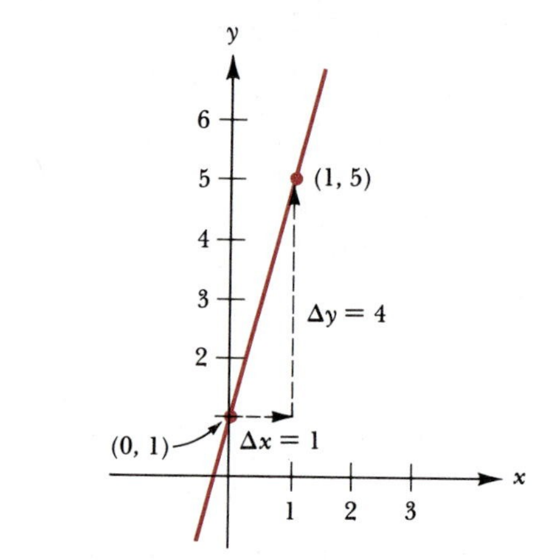

Figure 8
$y = 4x + 1$

Example 6 Find the slope and y-intercept of the line $3x - 5y = 15$.

Solution First solve for y so that the equation is in the form $y = mx + b$:

$$\begin{aligned} 3x - 5y &= 15 \\ -5y &= -3x + 15 \\ y &= \frac{3}{5}x - 3 \end{aligned}$$

The slope m and y-intercept b can now be read directly from the equation: $m = \frac{3}{5}$; $b = -3$.

There are two useful relationships regarding the slopes of parallel lines and the slopes of perpendicular lines. First, nonvertical parallel lines have the same slope. This should seem reasonable to you when you recall that slope is a number indicating the direction or slant of a line. The relationship concerning slopes of perpendicular lines is not so obvious. It is this: The slopes of two nonvertical perpendicular lines are negative reciprocals of each other. That is, if m_1 and m_2 denote the slopes of the two perpendicular lines, then $m_1 = -1/m_2$, or, equivalently, $m_1 m_2 = -1$. For example, if a line has a slope of $\frac{2}{3}$, then any line perpendicular to it must have a slope of $-\frac{3}{2}$. For reference we summarize these facts in the box that follows. (Proofs of these facts are outlined in detail in Exercises 41 and 42 at the end of this section.)

Parallel and Perpendicular Lines

Let m_1 and m_2 denote the slopes of two nonvertical lines. Then:

1. The lines are parallel if and only if $m_1 = m_2$;
2. The lines are perpendicular if and only if $m_1 = -1/m_2$.

Example 7 Are the two lines $3x - 6y - 8 = 0$ and $2y = x + 1$ parallel?

Solution By solving each equation for y, we can see what the slopes are:

$$\begin{aligned} 3x - 6y - 8 &= 0 \\ -6y &= -3x + 8 \\ y &= \frac{-3}{-6}x + \frac{8}{-6} \\ y &= \frac{1}{2}x - \frac{4}{3} \end{aligned} \qquad \begin{aligned} 2y &= x + 1 \\ y &= \frac{1}{2}x + \frac{1}{2} \end{aligned}$$

From this we see that both lines have the same slope, namely $m = \frac{1}{2}$. It follows therefore that the lines are parallel.

Example 8 Find the equation of the line parallel to $5x + 6y = 30$ and passing through the origin.

Solution First, find the slope of the line $5x + 6y = 30$:

$$\begin{aligned} 5x + 6y &= 30 \\ 6y &= -5x + 30 \\ y &= -\frac{5}{6}x + 5 \end{aligned}$$

The slope is the x-coefficient in the equation above: $m = -\frac{5}{6}$. A parallel line will have the same slope. Since the required line is to pass through $(0, 0)$, we have

$$\begin{aligned} y - y_1 &= m(x - x_1) \\ y - 0 &= -\frac{5}{6}(x - 0) \\ y &= -\frac{5}{6}x \end{aligned}$$

This is the equation of a line parallel to $5x + 6y = 30$ and passing through the origin, as required. See Figure 9.

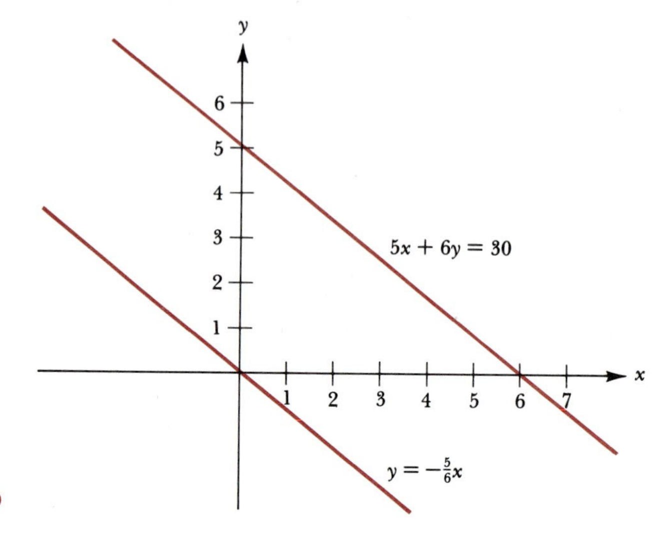

Figure 9

Example 9 **(a)** Sketch the circle $x^2 + y^2 = 25$.
(b) Verify that the point $(3, -4)$ lies on the circle.
(c) Find the equation of the line tangent to this circle at the point $(3, -4)$. Write the equation in the form $y = mx + b$.

Solution **(a)** and **(b)** From our work in Section 3.2 we know that the equation $x^2 + y^2 = 25$ represents a circle with center $(0, 0)$ and radius 5. To show that the point $(3, -4)$ lies on the circle, we need only check that the values $x = 3$, $y = -4$ satisfy the given equation. We have

$$\begin{aligned} x^2 + y^2 &= 25 \\ (3)^2 + (-4)^2 &= 25 \\ 9 + 16 &= 25 \qquad \text{True} \end{aligned}$$

This shows that the point $(3, -4)$ does lie on the circle. See Figure 10(a).

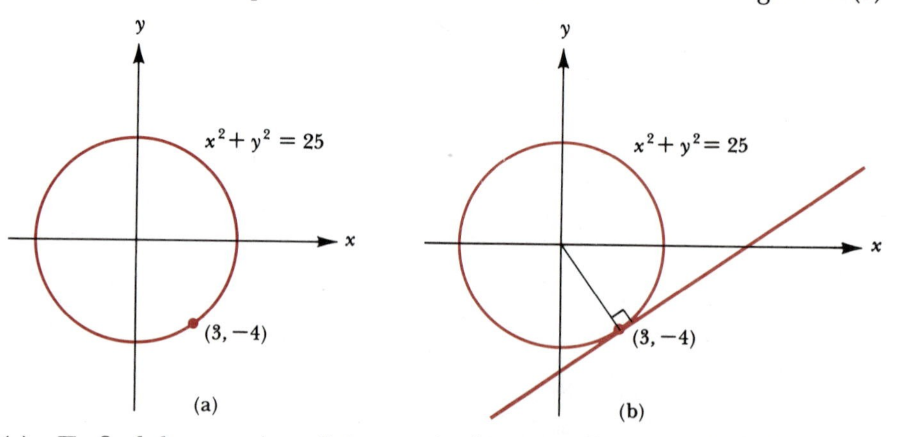

Figure 10

(c) To find the equation of the required tangent line, first draw the radius from $(0, 0)$ to $(3, -4)$ as in Figure 10(b). The slope of this radius is

$$\frac{-4 - 0}{3 - 0} = -\frac{4}{3}$$

From elementary geometry, it is known that the tangent line is perpendicular to this radius. Since the slope of the radius was found to be $-\frac{4}{3}$, it follows that the slope of the tangent line is $\frac{3}{4}$. The point-slope formula is now applicable:

$$y - y_1 = m(x - x_1)$$

$$y - (-4) = \frac{3}{4}(x - 3)$$

Check now for yourself that this can be simplified to $y = \frac{3}{4}x - \frac{25}{4}$, which is the required equation of the tangent line.

As a consequence of our work in this section, we can state that any equation of the form

$$Ax + By + C = 0$$

represents a straight line. (We are assuming that the constants A and B are not both simultaneously equal to zero.) Equations of the form $Ax + By + C = 0$ are called *linear equations.*

In the case when $A = 0$, we have $By + C = 0$, or $y = -C/B$, the graph of which is a horizontal line with y-intercept $-C/B$. In the case when $B = 0$, we have $Ax + C = 0$, so $x = -C/A$, the graph of which is a vertical line with x-intercept $-C/A$. Finally, if A and B are both nonzero, we have $By = -Ax - C$, or $y = -(A/B)x - C/B$, the graph of which we know is a straight line with slope $-A/B$ and y-intercept $-C/B$.

Exercise Set 3.4

In Exercises 1 and 2, find the equation for the line having the given slope and passing through the given point. Write your answers in the form $y = mx + b$.

1. **(a)** $m = -5$; through $(-2, 1)$
(b) $m = 4$; through $(4, -4)$
(c) $m = \frac{1}{3}$; through $(-6, -\frac{2}{3})$
(d) $m = -1$; through $(0, 1)$

2. **(a)** $m = 22$; through $(0, 0)$
(b) $m = -222$; through $(0, 0)$
(c) $m = \sqrt{2}$; through $(0, 0)$

3. Find an equation for the line passing through the two given points. Write your answer in the form $y = mx + b$.
(a) $(4, 8)$ and $(-3, -6)$ **(b)** $(-2, 0)$ and $(3, -10)$
(c) $(-3, -2)$ and $(4, -1)$
(d) $(7, 9)$ and $(-11, 9)$ **(e)** $(\frac{5}{4}, 2)$ and $(\frac{3}{4}, 3)$
(f) $(12, 13)$ and $(13, 12)$

4. **(a)** Write the equation of a vertical line passing through $(-3, 4)$.
(b) Write the equation of a horizontal line passing through $(-3, 4)$.
(c) What is the equation of the x-axis?
(d) What is the equation of the y-axis?

5. Find the equation of the line with the given slope and y-intercept.
(a) Slope of -4; y-intercept of 7.
(b) Slope of 2; y-intercept of $\frac{3}{2}$.
(c) Slope of $-\frac{4}{3}$; y-intercept of 14.
(d) Slope of 0; y-intercept of 14.
(e) Slope of 14, y-intercept of 0.

6. Find the slope and y-intercept of the given lines and draw the graphs.
(a) $y = -7x + 4$ **(b)** $3x - 4y = 12$
(c) $x - y - 1 = 0$ **(d)** $2x - 3 - 2y = 4y + 6$
(e) $x/2 + y/3 = 1$ **(f)** $y - \frac{5}{3} = 0$

7. Find an equation for the line described and write it in the form $y = mx + b$. Graph the line
(a) passing through $(-3, -1)$, with slope 4.
(b) passing through $(\frac{5}{2}, 0)$, with slope $\frac{1}{2}$.
(c) with x-intercept 6, y-intercept 5.
(d) with x-intercept -2, slope $\frac{3}{4}$.
(e) passing through $(1, 2)$ and $(2, 6)$.
(f) passing through $(-7, -2)$ and $(0, 0)$.
(g) passing through $(6, -3)$, with y-intercept 8.
(h) passing through $(0, -1)$ and with the same slope as the line $3x + 4y = 12$.

(i) passing through $(6, 2)$ and with the same x-intercept as the line $-2x + y = 1$.
(j) with x-intercept -6, y-intercept $\sqrt{2}$.

8. Find the equation of a line with the same slope as the line $3x + 5y = 15$ and that passes through the origin. Graph both lines.

9. Find the equation of a line with the same slope as the line $x + y = 0$ and that passes through $(-4, 0)$. Graph both lines.

10. Find the equation of a vertical line passing through $(4, -\frac{13}{2})$. Graph the line. What would the equation be if instead the line were horizontal?

11. Find a value for b in the equation $y = mx + b$ if the graph is to pass through the two points $(8, 14)$ and $(-2, 3)$.

12. Find the x- and y-intercepts and then graph the lines.
(a) $x/3 + y/4 = 1$ **(b)** $x/3 - y/4 = 1$
(c) $x/a + y/b = 1 \quad (a > 0, b > 0)$

13. Find the equations of the lines with the given intercepts. Write your answers in the form $y = mx + b$.
(a) x-intercept of 3; y-intercept of 7
(b) x-intercept of -1; y-intercept of -2
(c) x-intercept of $\frac{5}{4}$; y-intercept of $-\frac{1}{3}$
(d) x-intercept $= y$-intercept $= \pi$

14. A line with a slope of -5 passes through the point $(3, 6)$. Find the area of the triangle bounded by this line and the axes. *Hint:* First find the equation of the line; then use that to determine the x- and y-intercepts.

15. A line passes through the points $(-5, 1)$ and $(-3, 8)$. Find the area of the triangle in the second quadrant bounded by this line and the axes. (Your answer should be positive.)

16. The y-intercept of the line in the figure is 6. Find the slope of the line if the area of the shaded triangle is 72 square units.

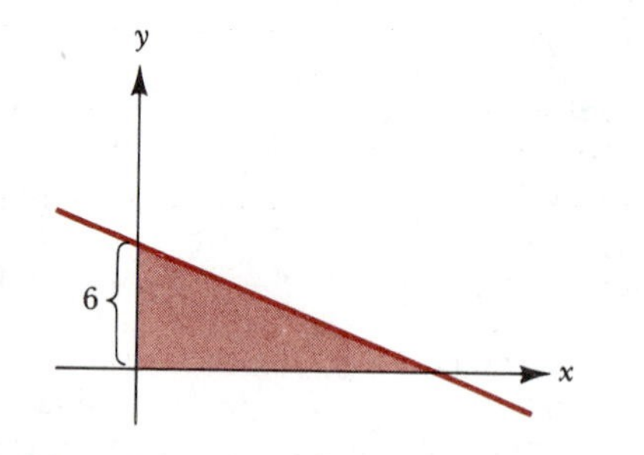

17. Suppose that the cost C in dollars of producing x electric fans is given by $C = 450 + 8x$.
(a) Find the cost to produce 10 fans.
(b) Find the cost to produce 11 fans.
(c) Use the results in (a) and (b) to find the marginal cost. (Then check that your answer is the slope of the line $C = 450 + 8x$.)

18. Suppose that the cost to a certain manufacturer of producing x units of a certain motorcycle is given by $C = 220x + 4000$, where C is in dollars.
(a) Find the marginal cost.
(b) Find the cost of producing 500 motorcycles.
(c) Use your answers in parts (a) and (b) to find the cost of producing 501 motorcycles.

19. The distance d in feet covered by a particular object in t sec is given by $d = 5t$.
(a) Find the velocity of the object.
(b) Find the distance covered in 15 sec.
(c) Use your answers in parts (a) and (b) to find the distance covered in 16 sec.

20. Two points labeled A and B move along the x-axis, their positions after t sec being given, respectively, by the equations

$$A: \quad x = 3t + 100$$
$$B: \quad x = 20t - 36$$

(a) Which point is traveling faster, A or B?
(b) Which point is farther to the right when $t = 0$?
(c) At what time t do A and B have the same x-coordinate?

21. Are the lines $y = x + 1$ and $y = 1 - x$ parallel, perpendicular, or neither?

22. State whether the two lines are parallel or perpendicular (or neither).
(a) $3x - 4y = 12$; $4x - 3y = 12$
(b) $y = 5x - 16$; $y = 5x + 2$
(c) $5x - 6y = 25$; $6x + 5y = 0$
(d) $y = -\frac{2}{3}x - 1$; $y = \frac{3}{2}x - 1$
(e) $-2x - 5y = 1$; $y - \frac{2}{5}x - 4 = 0$
(f) $x = 8y + 3$; $4y - \frac{1}{2}x = 32$
(g) $7x - y + 3 = 0$; $7x + y = 1$
(h) $x - 7y + 3 = 0$; $-7x + y = -1$

23. Find the equation of a line parallel to the line $2x - 5y = 10$ and passing through the point $(-1, 2)$. Graph both lines.

24. Follow Exercise 23, but use perpendicular instead of parallel.

25. Find the equation of a line passing through the origin and perpendicular to the line that passes through the two points $(5, 0)$ and $(0, -4)$. Graph both lines.

26. Follow Exercise 25, but use parallel instead of perpendicular.

27. Find the equation of a line parallel to the x-axis and with the same y-intercept as the line $-5x + 4y = 15$.

28. Find the equation of a line perpendicular to the x-axis and with the same x-intercept as the line $y = 6x + 11$.

29. Find the equation of the line that passes through the point $(-4, 1)$ and that is parallel to the line $y = 3x - 5$. Include a sketch.

30. Follow Exercise 29, but use perpendicular instead of parallel.

31. **(a)** Sketch the circle $x^2 + y^2 = 25$.
(b) Find the equation of the line tangent to this circle at the point $(-4, -3)$. Include a sketch.

32. **(a)** Tangents are drawn to the circle $x^2 + y^2 = 169$ at the points $(5, 12)$ and $(5, -12)$. Find the equations of the tangents.
(b) Find the coordinates of the point where these two tangents intersect.

33. **(a)** Sketch the circle with center $(0, 0)$ and radius 1. Write its equation.
(b) Verify that the point $(-1/2, \sqrt{3}/2)$ lies on the circle.
(c) Find the equation of the line tangent to this circle at the point $(-1/2, \sqrt{3}/2)$.

***34.** **(a)** Show that the equation of the line tangent to the circle $x^2 + y^2 = a^2$ at the point (x_1, y_1) on the circle is

$$x_1x + y_1y = a^2$$

(b) Use the result in part (a) to rework Exercises 31(b) and 33(c).

35. In each case, show that the line that passes through the two given points is perpendicular to the line $y = x$. Draw the graphs in each case.
(a) $(4, 1)$ and $(1, 4)$ **(b)** $(-3, 2)$ and $(2, -3)$
(c) $(7, 3)$ and $(3, 7)$
(d) (a, b) and (b, a) (Assume that $a \neq b$.)

***36.** Complete the details in the following steps to determine the slope m of the line in the figure.

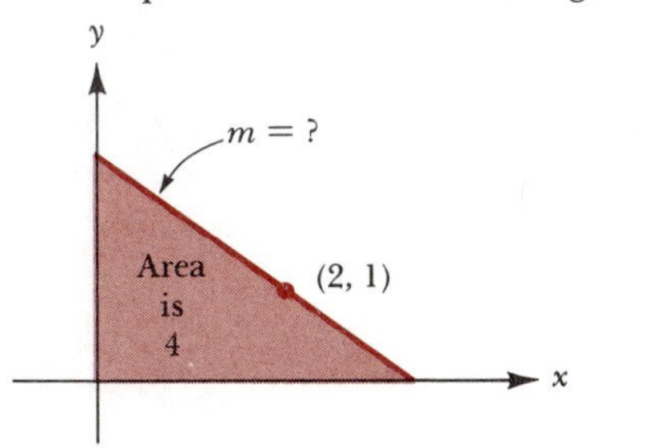

(a) Use the point-slope formula to show that the equation of the line is

$$y - 1 = m(x - 2)$$

(b) Show that the y-intercept for the line is $-2m + 1$. Show that the x-intercept is $\dfrac{2m - 1}{m}$.
(c) The area of the triangle, being one-half base times height, is $\dfrac{1}{2}\left(\dfrac{2m - 1}{m}\right)(-2m + 1)$.
(d) Set the expression for area in part (c) equal to 4, as given, and simplify the resulting equation to obtain $4m^2 + 4m + 1 = 0$.
(e) Solve the equation for m, either by factoring or by the quadratic formula. You should find that $m = -\frac{1}{2}$.

***37.** A line with a negative slope passes through the point $(3, 1)$. The area of the triangle bounded by this line and the axes is 6 square units. Find the possible slopes of the line. *Hint:* See Exercise 36.

***38.** Follow Exercise 37, but assume that the area is
(a) 8 square units. **(b)** 9 square units.
(c) 4 square units.

39. The accountant finds that Mr. Rich's taxable income is \$90,000. The income tax tables say that if your taxable income is over \$81,800 but not over \$108,300, then your tax is \$37,667 + 68% of the amount over \$81,800.
(a) Find a linear equation that gives the tax T on a taxable income of x dollars, assuming that x is in the given range.
(b) Use the equation found in part (a) to find Mr. Rich's tax.

***40.** By analyzing sales figures, the accountant for College Stereo Company knows that 280 units of a top-of-the-line turntable can be sold each month when the price is $P = \$195$ per unit. The figures also show that for each \$15 hike in the price 10 fewer units are sold monthly.
(a) Let x denote the number of units sold per month and P the price per unit. Find an equation that expresses x in terms of P, assuming that this relationship is linear. *Hint:* $\Delta x/\Delta P = -10/15 = -(2/3)$.
(b) Use the equation that you found in part (a) to determine how many units can be sold in a month when the price is \$270 per unit.
(c) What should the price be to sell 205 units per month?

41. This exercise outlines a proof of the fact that two nonvertical lines are parallel if and only if their slopes are equal. The proof relies on the following observation for the figure below: The lines $y = m_1x + b_1$ and $y = m_2x + b_2$ will be parallel if and only if the two vertical distances $\overline{AB}$ and $\overline{CD}$ are equal.

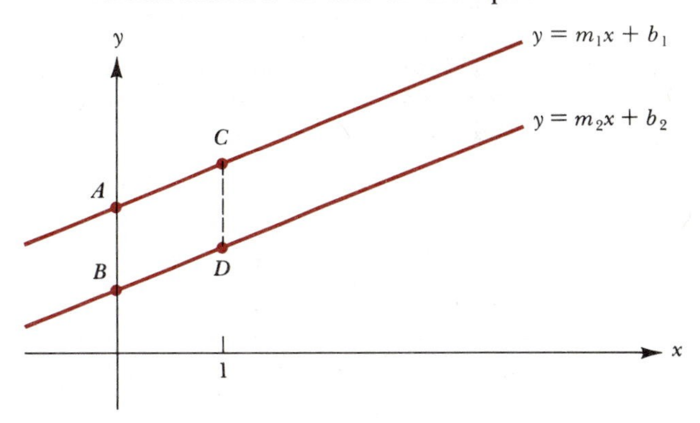

(a) Verify that the coordinates of A, B, C, and D are:

$A(0, b_1)$ $B(0, b_2)$ $C(1, m_1 + b_1)$ $D(1, m_2 + b_2)$

(b) Using part (a), check that $\overline{AB} = b_1 - b_2$ and $\overline{CD} = (m_1 + b_1) - (m_2 + b_2)$.

(c) Use part (b) to show that the equation $\overline{AB} = \overline{CD}$ is equivalent to $m_1 = m_2$.

42. This exercise outlines a proof of the fact that two nonvertical lines with slopes m_1 and m_2 are perpendicular if and only if $m_1 m_2 = -1$. In the accompanying figure, we've assumed that our two nonvertical lines $y = m_1 x$ and $y = m_2 x$ intersect at the origin. [If they did not intersect there, we could just as well work with lines parallel to these that do intersect at (0, 0), recalling that parallel lines have the same slope.] The proof relies on the following geometric fact:

$$OA \perp OB \quad \text{if and only if} \quad \overline{OA}^2 + \overline{OB}^2 = \overline{AB}^2$$

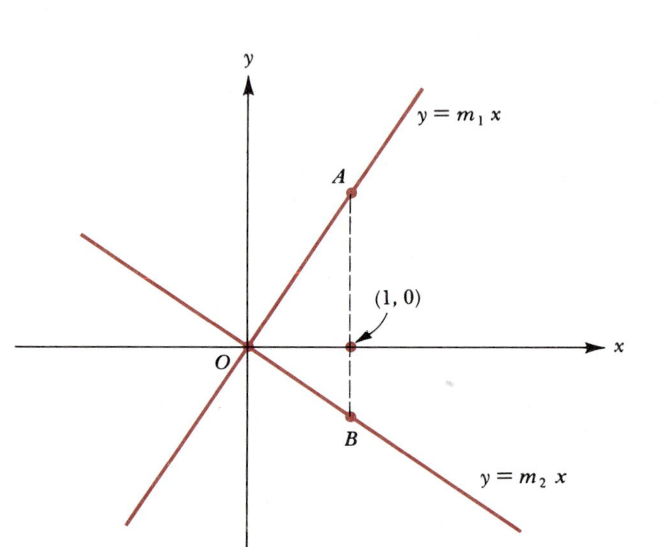

(a) Verify that the coordinates of A and B are $A(1, m_1)$ and $B(1, m_2)$.

(b) Show that

$$\overline{OA}^2 = 1 + m_1^2$$
$$\overline{OB}^2 = 1 + m_2^2$$
$$\overline{AB}^2 = m_1^2 - 2m_1 m_2 + m_2^2$$

(c) Use part (b) to show that the equation $\overline{OA}^2 + \overline{OB}^2 = \overline{AB}^2$ is equivalent to $m_1 m_2 = -1$.

***43.** In this exercise you'll derive a useful formula for the (perpendicular) distance d from the point (x_0, y_0) to the line $y = mx + b$. The formula is

$$d = \frac{|y_0 - mx_0 - b|}{\sqrt{1 + m^2}}$$

(a) Refer to the accompanying figure. Use similar triangles to show that $d / \overline{AB} = \dfrac{1}{\sqrt{1 + m^2}}$.

Therefore, $d = \dfrac{\overline{AB}}{\sqrt{1 + m^2}}$.

(b) Check that $\overline{AB} = \overline{AC} - \overline{BC} = y_0 - mx_0 - b$.

(c) Conclude from parts (a) and (b) that $d = \dfrac{y_0 - mx_0 - b}{\sqrt{1 + m^2}}$. For the general case (in which the point and line may not be situated as in our figure) we need to use the absolute value of the quantity in the numerator to assure that $\overline{AB}$ and d are nonnegative.

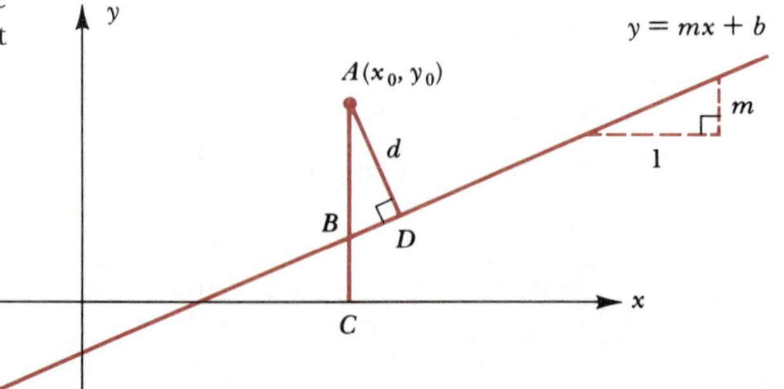

*44. Use the formula given in Exercise 43 to show that the distance from the point (x_0, y_0) to the line $Ax + By + C = 0$ is

$$d = \frac{|Ax_0 + By_0 + C|}{\sqrt{A^2 + B^2}}$$

45. Use the formula given in Exercise 43 to find the distance from the point $(1, 2)$ to the line $y = \frac{1}{2}x - 5$.

46. Use the formula given in Exercise 44 to find the distance from the point $(-1, -3)$ to the line $2x + 3y - 6 = 0$.

47. Find the equation of the circle with center $(2, 3)$ that is tangent to the line $x + y - 1 = 0$. *Hint:* Use the formula given in Exercise 44 to find the radius.

Chapter 3 Review Checklist

- □ Origin, x-axis, y-axis, quadrants (page 116)
- □ x-coordinate, y-coordinate, Cartesian coordinate system (page 116)
- □ Ordered pair (page 116)
- □ Midpoint formula (page 117)
- □ Distance formula (page 119)
- □ Graph of an equation (page 122)
- □ x-intercept and y-intercept (page 123)
- □ Symmetry with respect to the x-axis, the y-axis, or the origin (page 123)
- □ Three tests for symmetry (page 124)
- □ Equation of a circle in standard form (page 127)
- □ Slope:
  - definition (page 131)
  - as a rate of change (page 135)
- □ Cost equation and marginal cost (page 135)
- □ Point-slope formula (page 139)
- □ Equations of horizontal and vertical lines (page 141)
- □ Slope-intercept formula (page 142)
- □ Relationships between slopes of parallel lines and slopes of perpendicular lines (page 143)
- □ Linear equation (page 145)

Chapter 3 Review Exercises

In Exercises 1–18, find an equation for the line satisfying the given conditions. Write your answer in the form $y = mx + b$.

1. Passing through $(-4, 2)$ and $(-6, 6)$.
2. $m = -2$; y-intercept 5.
3. $m = \frac{1}{4}$; passing through $(-2, -3)$.
4. $m = \frac{1}{3}$; x-intercept -1.
5. x-intercept -4; y-intercept 8.
6. $m = -10$; y-intercept 0.
7. y-intercept -2; parallel to the x-axis.
8. Passing through $(0, 0)$ and parallel to $6x - 3y = 5$.
9. Passing through $(1, 2)$ and perpendicular to $x + y + 1 = 0$.
10. Passing through $(1, 1)$ and through the center of the circle $x^2 - 4x + y^2 - 8y + 16 = 0$.
11. Passing through the centers of the circles $x^2 + 4x + y^2 + 2y = 0$ and $x^2 - 4x + y^2 - 16y = 0$.
12. $m = 3$; the same x-intercept as the line $3x - 8y = 12$.
13. Passing through the origin and the midpoint of the line segment joining the points $(-2, -3)$ and $(6, -5)$.
14. Tangent to the circle $x^2 + y^2 = 20$ at the point $(-2, 4)$ on the circle.
15. Tangent to the circle $x^2 - 6x + y^2 + 8y = 0$ at the point $(0, 0)$.
16. Passing through $(2, 4)$; the y-intercept is twice the x-intercept.
17. Passing through $(2, -1)$; the sum of the x- and y-intercepts is 2.
18. Passing through $(4, 5)$; no x-intercept.
19. Find the distance between the points $(-1, 2)$ and $(4, -10)$.
20. (a) Find the perimeter of the triangle with vertices $A(3, 1)$, $B(7, 4)$, *and* $C(-2, 13)$.
    (b) Find the perimeter of the triangle formed by joining the midpoints of the sides of the triangle in part (a).
21. Which point is farther from the origin, $(15, 6)$ or $(16, 2)$?
22. Find the midpoint of the line segment joining
    (a) $(4, 9)$ and $(10, -2)$ (b) $(\frac{2}{3}, -1)$ and $(-\frac{3}{4}, -\frac{1}{3})$

In Exercises 23–32, test each equation for symmetry with respect to the x*-axis, the* y*-axis, and the origin.*

23. $y = x^4 - 2x^2$
24. $y = x^4 - 2x^2 + 1$
25. $y = x^3 + 5x$
26. $y = x^3 + 5x + 1$

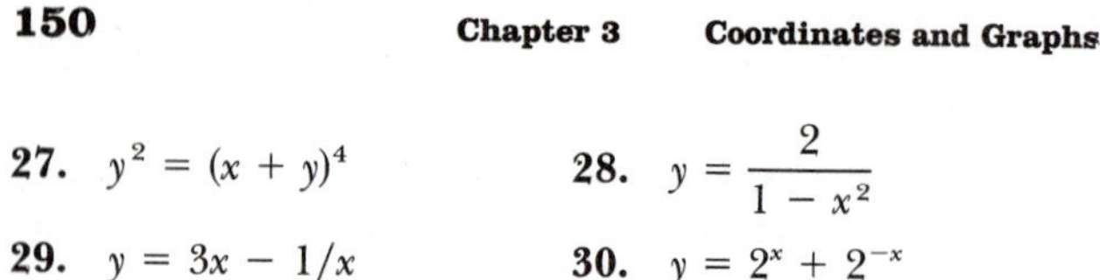

27. $y^2 = (x + y)^4$

28. $y = \dfrac{2}{1 - x^2}$

29. $y = 3x - 1/x$

30. $y = 2^x + 2^{-x}$

31. $y = 2^x - 2^{-x}$

32. $x^2 + y^2 + 2x = 2\sqrt{x^2 + y^2}$

For Exercises 33–42, tell whether each graph is symmetric with respect to the x*-axis, the* y*-axis, or the origin.*

33. **34.** **35.** **36.**

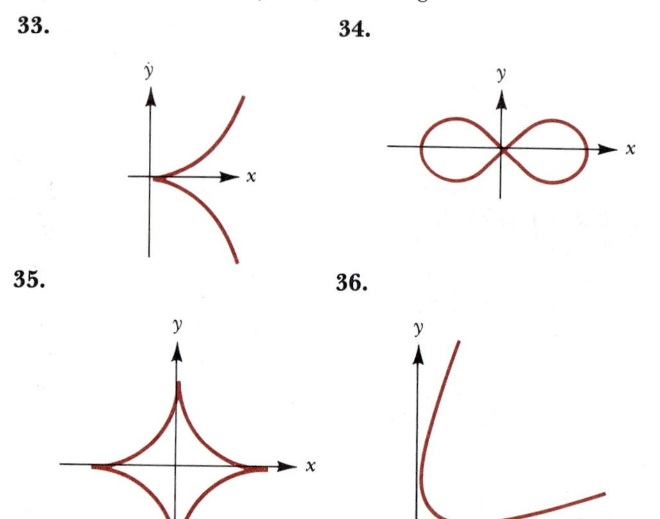

37. **38.** **39.** **40.**

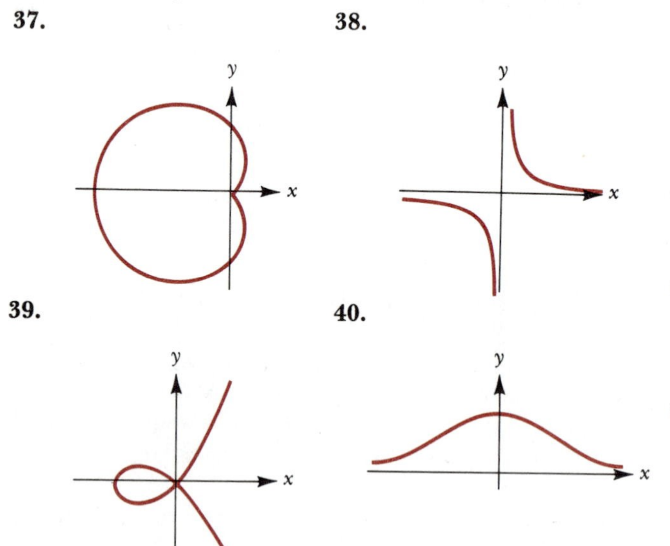

41. **42.**

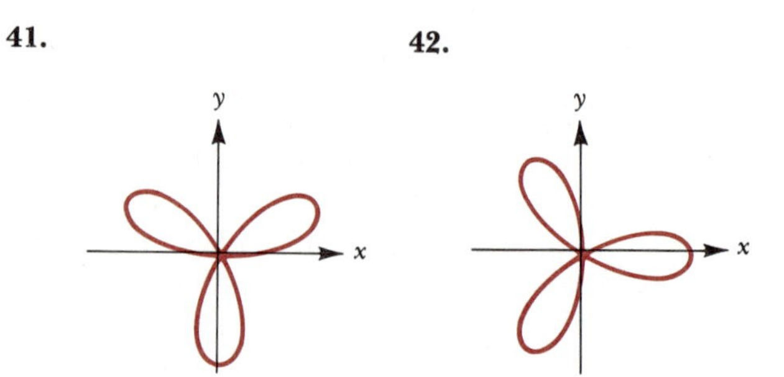

Graph the equations in Exercises 43–62.

43. $2x - 5y = 10$

44. $y = x^2 + 4x$

45. $x = 9 - y^2$

46. $y = -2x + 6$

47. $x^2 + y^2 = 1$

48. $(x + 2)^2 + y^2 = 1$

49. $y = |x| + 1$

50. $y = 1 - |x|$

51. $y = x^2 - 4x + 4$

52. $y = -x$

53. $y = |x - 2| + 2$

54. $y = |x - 2| - 2$

55. $3x + y = 0$

56. $y = \sqrt{x}$

57. $x^2 - 4x + y^2 + 6y = 0$

58. $x^2 + 2x + y^2 + 8y = 8$

59. $y = \dfrac{x^2 - 16}{x - 4}$

60. $y = 1 + 1/x$

61. $y = \dfrac{1}{x + 1}$

62. $(y - x)(y - x - 1)(y - x - 2) = 0$

63. Find a value for t such that the slope of the line passing through $(2, 1)$ and $(5, t)$ is 6.

64. Express in terms of x the slope of the line passing through (x, x^2) and $(-3, 9)$.

65. Express in terms of x the slope of the line passing through $(-2, -8)$ and (x, x^3).

66. The vertices of a parallelogram are $(0, 0)$, $(5, 2)$, $(8, 7)$, and $(3, 5)$. Find the midpoint of each diagonal. Observation?

67. The vertices of a right triangle are $A(0, 0)$, $B(0, 2b)$, and $C(2c, 0)$. Let M be the midpoint of the hypotenuse. Compute the three distances $\overline{MA}$, $\overline{MB}$, and $\overline{MC}$. What do you observe?

68. The vertices of parallelogram $ABCD$ are $A(-4, -1)$, $B(2, 1)$, $C(3, 3)$, and $D(-3, 1)$.

(a) Compute the sum of the squares of the two diagonals.

(b) Compute the sum of the squares of the four sides. What do you observe?

In Exercises 69–72 two points are given. In each case compute (a) the distance between the two points, (b) the slope of the line segment joining the two points, and (c) the midpoint of the line segment joining the two points.

69. $(2, 5)$ and $(5, -6)$

70. $(\frac{3}{2}, 1)$ and $(\frac{7}{2}, 9)$

71. $(\sqrt{3}/2, 1/2)$ and $(-\sqrt{3}/2, -1/2)$

72. $(\sqrt{3}, \sqrt{2})$ and $(\sqrt{2}, \sqrt{3})$

73. A line passes through the points $(1, 2)$ and $(4, 1)$. Find the area of the triangle bounded by this line and the axes.

74. Suppose that the cost C in dollars of producing x typewriters is $C = 2200 + 50x$.
- **(a)** Find the cost of producing 100 typewriters.
- **(b)** Find the cost of producing 101 typewriters.
- **(c)** Use your answers in parts (a) and (b) to compute the marginal cost.

75. Suppose that the cost C in dollars of producing x tires is $C = 8500 + 12x$.
- **(a)** What is the marginal cost?
- **(b)** Find the cost of producing 1000 tires.
- **(c)** Use your answers in parts (a) and (b) to find the cost of producing 1001 tires. Check your answer by letting $x = 1001$ in the given equation.

76. A point moves along the x-axis, and its x-coordinate after t seconds is $x = 4t + 10$. (Assume that x is in centimeters.)
- **(a)** What is the velocity?
- **(b)** What is the x-coordinate when $t = 2$ sec?
- **(c)** Use your answers in parts (a) and (b) to find the x-coordinate when $t = 3$ sec. Check your answer by letting $t = 3$ in the given equation.

77. Let $P_1(x_1, y_1)$ and $P_2(x_2, y_2)$ be two given points. Let Q be the point

$$\left(\tfrac{1}{3}x_1 + \tfrac{2}{3}x_2, \tfrac{1}{3}y_1 + \tfrac{2}{3}y_2\right)$$

- **(a)** Show that the points P_1, Q, and P_2 are collinear. *Hint:* Compute the slope of P_1Q and the slope of QP_2.
- **(b)** Show that $\overline{P_1Q} = \frac{2}{3}\overline{P_1P_2}$. (In other words, Q is on the line segment P_1P_2 and two thirds of the way from P_1 to P_2.)

78.
- **(a)** Let the vertices of $\triangle ABC$ be $A(-5, 3)$, $B(7, 7)$, and $C(3, 1)$. Find the point on each median that is two thirds of the way from the vertex to the midpoint of the opposite side. What do you observe? *Hint:* Use the result in Exercise 77.
- **(b)** Follow part (a) but take the vertices to be $A(0, 0)$, $B(2a, 0)$, and $C(2b, 2c)$. What do you observe? What does this prove?

79. Suppose that the circle $(x - h)^2 + (y - k)^2 = r^2$ has two x-intercepts, x_1 and x_2, and two y-intercepts, y_1 and y_2. Compute the quantity $x_1x_2 + y_1y_2$ (in terms of h, k, and r).

80. The point $(1, -2)$ is the midpoint of a chord of the circle $x^2 - 4x + y^2 + 2y = 15$. Find the length of the chord.

Chapter 3 Test

1. Let A and B be the points $(2, 1)$ and $(8, -3)$, respectively.
- **(a)** Find the distance from A to B.
- **(b)** Find the slope of the line through A and B.
- **(c)** Find the midpoint of the line segment AB.

2. Test for symmetry with respect to the x-axis, the y-axis, and the origin: $y = 4x^5 - x^3$.

3. Graph the equation $y = 1 - x^2$. Specify any x- or y-intercepts.

4. Write the equation of a circle with center $(-2, 3)$ and radius 3.

5. Determine the center and radius of the circle $9x^2 - 72x + 9y^2 + 18y + 149 = 0$.

6. Find the y-intercept and the slope of the line $4x - 3y = 12$. Graph the line.

7. Are the lines $y = 2x + 1$ and $y = 1 - 2x$ parallel, perpendicular, or neither?

8. Suppose that the cost C in dollars of producing x stereo receivers is $C = 1200 + 60x$.
- **(a)** What is the cost of producing 10 receivers?
- **(b)** What is the marginal cost?
- **(c)** Use your answers in parts (a) and (b) to calculate the cost of producing 11 receivers. Check your answer by substituting $x = 11$ in the given equation.

9. Find the slope of a line passing through the two points $(5, 25)$ and $(5 + h, (5 + h)^2)$. Express your answer in terms of h.

10. Find the equation of a line passing through $(-4, 3)$ and with a slope of $\frac{5}{2}$. Write your final answer in the form $y = mx + b$.

11. Find the equation of a line that passes through the origin and is parallel to $5x - 6y = 30$. Write your answer in the form $y = mx + b$.

12. Find the x-intercept of the line that is tangent to the circle $x^2 + y^2 = 25$ at the point $(-3, -4)$.

13.
- **(a)** Write the equation of a horizontal line passing through $(3, 7)$.
- **(b)** Write the equation of a vertical line with an x-intercept of $\frac{3}{4}$.

14. Which point is closer to $(-2, 1)$: $(9, 8)$ or $(7, 10)$?

15. Graph the equation $y = |4 - x^2|$. Specify symmetry and intercepts.

Chapter 4 Functions

From the beginning of modern mathematics in the 17th century the concept of function has been at the very center of mathematical thought.

Richard Courant and Fritz John in *Introduction to Calculus and Analysis* (New York: Wiley-Interscience, 1965)

The word "function" was introduced into mathematics by Leibniz, who used the term primarily to refer to certain kinds of mathematical formulas. It was later realized that Leibniz's idea of function was much too limited in scope, and the meaning of the word has since undergone many steps of generalization.

Tom M. Apostol in *Calculus*, second edition (New York: John Wiley & Sons, 1967)

Introduction

We begin the study of functions in this chapter. The first section deals with matters of definition and notation. Section 4.2 could aptly be subtitled "Translating English into Algebra." In that section we develop certain skills that you will need in solving many of the word problems that arise in subsequent sections of this text. In Sections 4.3 and 4.4 we discuss graphs of functions; several techniques are developed that allow us to graph functions with a minimum of calculation. In Sections 4.5 and 4.6 we find that two functions can be combined in various ways to produce new functions. One of these ways of combining two functions is known as *composition of functions*. As you'll see, this is the unifying theme between Sections 4.5 and 4.6. Finally, in Section 4.7 we study certain types of functions and formulas that occur so frequently in the sciences that a special terminology exists to describe them.

4.1 The Definition of a Function

There are numerous instances in mathematics and its applications in which one quantity corresponds to or depends on another according to some definite rule. Consider, for example, the equation $y = 3x - 2$. Each time that you select an x-value, a corresponding y-value is determined, in this case according to the rule *multiply by 3, then subtract 2*. In this sense, the equation $y = 3x - 2$ is an example of a *function*. It is a *rule* specifying a y-value corresponding to each x-value. It is useful to think of the x-values as inputs and the corresponding y-values as outputs. The function or rule then tells you what output results from a given input. This is indicated schematically in Figure 1. As another example, the area A of a circle depends on the radius r according to the rule or function $A = \pi r^2$. For each value of r a corresponding value for A is obtained by the rule *square r and multiply the result by* π. In this case the inputs are the values of r and the outputs are the corresponding values of A.

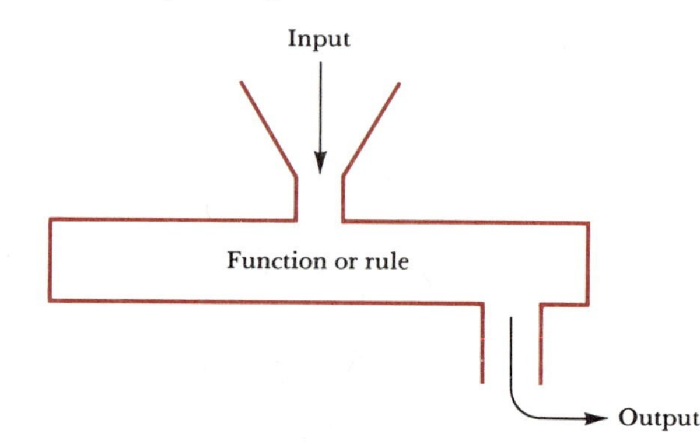

Figure 1

Mercury	0
Venus	0
Earth	1
Mars	2
Jupiter	16
Saturn	17
Uranus	5
Neptune	2
Pluto	1

Table 1

For most of the functions studied in this text (and in beginning calculus), the inputs and outputs are real numbers. And the function or rule is specified by means of an equation. This was the case with the two examples given above; however, this need not always be the case. Consider, for example, the correspondences set up in Table 1, which indicate the number of moons each planet in the solar system was known to have as of 1985. If we think of the inputs as the planets listed in the left-hand column of Table 1 and the outputs as the numbers in the right-hand column, then these correspondences constitute a function. The rule for this function may be stated *assign to each planet in the solar system the number of moons it was known to have in 1985.*

The following is one definition of the term *function*. As you will see, the definition is broad enough to encompass all the examples we have just looked at.

> **Definition of a Function**
>
> Let A and B be two nonempty sets. A function from A to B is a rule that assigns to each element in A exactly one element in B.

The set A in the definition we've just given is called the *domain* of the function. Think of the domain as the set of all inputs. For functions defined by equations (for example, $y = 3x - 2$) we'll agree to the following convention regarding the domain: Unless indicated otherwise, the domain is assumed to be the set of all real numbers that lead to unique real number outputs. Thus, the domain of the function defined by $y = 3x - 2$ is the set of all real numbers, whereas the domain of the function defined by

$y = \dfrac{1}{x-5}$ is the set of all real numbers except 5. (When $x = 5$, the expression $\dfrac{1}{x-5}$ is undefined, because the denominator is zero.)

Example 1 Find the domain of the function defined by each equation:

(a) $y = \sqrt{2x+6}$ **(b)** $s = \dfrac{1}{t^2 - 6t - 7}$

Solution **(a)** The quantity under the radical sign must be nonnegative. So we have

$$\begin{aligned} 2x + 6 &\geq 0 \\ 2x &\geq -6 \\ x &\geq -3 \end{aligned}$$

The domain is therefore the interval $[-3, \infty)$.

(b) Since division by zero is undefined, the domain of this function consists of all real numbers t except those for which the denominator is zero. Thus to find out which values of t to exclude, we solve the equation $t^2 - 6t - 7 = 0$. We have

$$t^2 - 6t - 7 = 0$$
$$(t-7)(t+1) = 0$$
$$t - 7 = 0 \quad \Big| \quad t + 1 = 0$$
$$t = 7 \quad \Big| \quad t = -1$$

It follows now that the domain of the function defined by $s = \dfrac{1}{t^2 - 6t - 7}$ is the set of all real numbers except $t = 7$ and $t = -1$.

The set of all outputs for a given function is called the *range* of the function. The next example demonstrates a method that's often useful in determining the range.

Example 2 Find the range of the function defined by $y = \dfrac{x+2}{x-3}$.

Solution The range of this function is the set of all outputs y. One way to see what restrictions the given equation imposes on y is to solve the equation for x as follows:

$$\begin{aligned} y(x-3) &= x + 2 && \text{multiplying by } (x-3) \\ yx - 3y &= x + 2 \\ yx - x &= 3y + 2 \\ x(y-1) &= 3y + 2 \\ x &= \frac{3y+2}{y-1} \end{aligned}$$

From this last equation we see that the value of y cannot be 1. (The denominator is zero when $y = 1$.) The range therefore consists of all real numbers except $y = 1$.

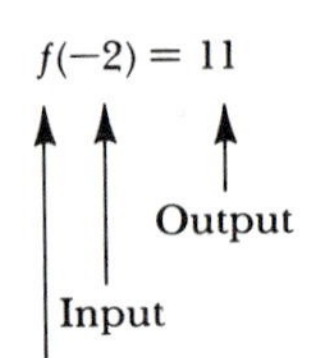

Figure 2

We often use single letters in order to name functions. If f is a function and x is an input for the function, then the resulting output is denoted by $f(x)$. This is read *f of x* or *the value of f at x*. As an example of this notation, suppose that f is the function defined by

$$f(x) = x^2 - 3x + 1 \tag{1}$$

Then $f(-2)$ denotes the output that results when the input is -2. To calculate this output, just replace x with -2 throughout equation (1). This yields

$$\begin{aligned} f(-2) &= (-2)^2 - 3(-2) + 1 \\ &= 4 + 6 + 1 = 11 \end{aligned}$$

That is, $f(-2) = 11$. Figure 2 summarizes this result and the notation.

Example 3 Let $f(x) = \dfrac{1}{x-1}$. Compute: **(a)** $f(0)$ **(b)** $f(t)$ **(c)** $f(x-1)$

Solution **(a)** $f(0) = \dfrac{1}{0-1} = -1$

(b) $f(t) = \dfrac{1}{t-1}$

(c) Replace x with the quantity $x - 1$ throughout the given equation. This yields

$$f(x-1) = \frac{1}{(x-1)-1} = \frac{1}{x-2}$$

Example 4 Let $g(x) = 1 - x^2$. Compute $g(x-1)$.

Solution In the equation $g(x) = 1 - x^2$, we substitute the *quantity* $(x - 1)$ in place of each occurrence of x. This gives us

$$\begin{aligned} g(x-1) &= 1 - (x-1)^2 \\ &= 1 - (x^2 - 2x + 1) \\ &= -x^2 + 2x \end{aligned}$$

Thus $g(x-1) = -x^2 + 2x$.

Example 5 Let $f(x) = x^2$. Show that $f(2+3) \neq f(2) + f(3)$.

Solution $f(2+3) = f(5) = (5)^2 = 25$; but $f(2) = (2)^2 = 4$ and $f(3) = (3)^2 = 9$. Since $25 \neq 4 + 9$, we have shown $f(2+3) \neq f(2) + f(3)$.

The next two examples using functional notation involve calculations with *difference quotients*. These are expressions of the form

$$\frac{f(x+h) - f(x)}{h}$$

or

$$\frac{f(x) - f(a)}{x - a}$$

For now we'll concentrate on the algebraic techniques used in calculating these quantities. (Later you'll see some applications, in Exercise Set 4.3, Exercises 43–45.)

Example 6 Let $f(x) = x^2 + 3x$. Compute $\dfrac{f(x) - f(2)}{x - 2}$.

Solution

$$\begin{aligned}\frac{f(x) - f(2)}{x - 2} &= \frac{(x^2 + 3x) - [2^2 + 3(2)]}{x - 2}\\ &= \frac{x^2 + 3x - 10}{x - 2}\\ &= \frac{(x - 2)(x + 5)}{x - 2} = x + 5\end{aligned}$$

The difference quotient is $x + 5$.

Example 7 Let $G(x) = 2/x$. Find $\dfrac{G(x + h) - G(x)}{h}$.

Solution

$$\frac{G(x + h) - G(x)}{h} = \frac{2/(x + h) - 2/x}{h}$$

An easy way to simplify this expression is to multiply it by $\dfrac{(x + h)x}{(x + h)x}$, which equals 1. This yields

$$\begin{aligned}\frac{G(x + h) - G(x)}{h} &= \frac{(x + h)x}{(x + h)x} \cdot \frac{2/(x + h) - 2/x}{h}\\ &= \frac{2x - 2(x + h)}{h(x + h)x}\\ &= \frac{2x - 2x - 2h}{h(x + h)x} = \frac{-2h}{h(x + h)x}\\ &= \frac{-2}{(x + h)x}\end{aligned}$$

The difference quotient is therefore $\dfrac{-2}{(x + h)x}$.

In Examples 2 through 7 in this section we've considered functions that were defined by means of equations. It is important to understand, however, that there are equations and rules that do not define functions. For example, consider the equation $y^2 = x$ and the input $x = 4$. Then we have $y^2 = 4$ and consequently $y = \pm 2$. So we have two outputs in this case, whereas the definition of a function requires that there be exactly one output. Example 8 provides some additional perspective on this.

Example 8 Let $A = \{b, g\}$ and $B = \{s, t, u, z\}$. Which of the four correspondences in Figure 3 represent functions from A to B? For those correspondences that do represent functions, specify the range in each case.

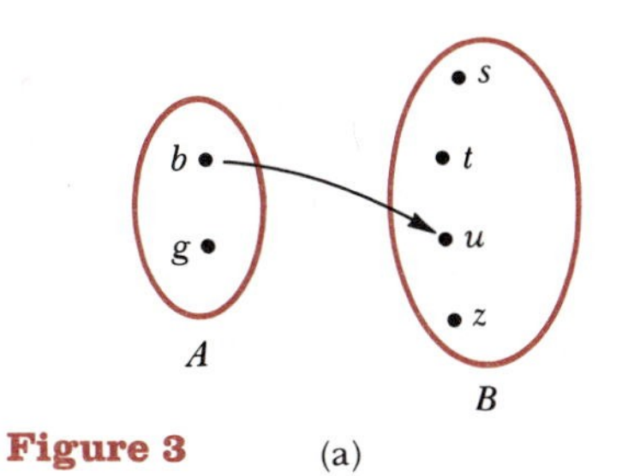

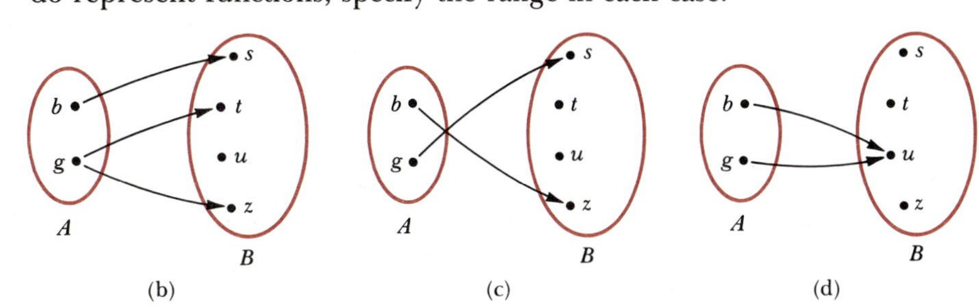

Figure 3 (a) (b) (c) (d)

Solution

Figure 2(a): Not a function. The definition requires that *each* element in A be assigned an element in B. The element g in this case has no assignment.

Figure 2(b): Not a function. The definition requires *exactly one* output for a given input. In this case there are two outputs for the input g.

Figures 2(c) and (d): Both of these rules qualify as functions from A to B. For each input there is exactly one output. [Regarding the function in Figure 2(d) in particular, notice that there is nothing in the definition of a function that prohibits two different inputs from producing the same output.] For the function in Figure 2(c), the outputs are s and z, and so the range is the set $\{s, z\}$. For the function in Figure 2(d), the only output is u, and consequently the range is the set $\{u\}$.

We conclude this section with some additional terminology that is often useful. When a function is defined by means of an equation, the letter representing elements from the domain is called the *independent variable.* For example, in the equation $y = 3x - 2$, x is the independent variable. The letter representing elements from the range is called the *dependent variable*; its value *depends* on x. This is also expressed by saying that y is a function of x.

Exercise Set 4.1

In Exercises 1–8, find the domains of the functions defined by the given equations.

1. **(a)** $y = -5x + 1$ **(b)** $y = \dfrac{x}{x-3}$
(c) $y = \sqrt{x}$ **(d)** $y = |x| + 1$

2. **(a)** $y = x^2 - 3x - 4$ **(b)** $y = \dfrac{-2}{x^2 - 3x - 4}$
(c) $y = \sqrt{x - 16}$ **(d)** $y = \dfrac{x}{x^2 + x + 1}$

3. **(a)** $y = 4x - 5$ **(b)** $s = 1/t^2$
(c) $s = \dfrac{1}{t^2 + 1}$ **(d)** $y = \dfrac{x-4}{x+4}$

4. **(a)** $y = \dfrac{4}{x^2 - 25}$ **(b)** $y = \sqrt{5 - 4x}$
(c) $y = x^3 - 14x^2 - 6x - 1$
(d) $y = x/|x|$

5. $y = \dfrac{1}{x^3 - 4x}$ **6.** $y = \dfrac{\sqrt{3x - 12}}{x^2 - 36}$

7. **(a)** $y = \sqrt{x^2 - 4x - 5}$ *Hint:* You need to solve the inequality $x^2 - 4x - 5 \geq 0$. The techniques for doing this are reviewed in Section 2.6.
(b) $y = \sqrt{\dfrac{x+2}{x-4}}$

8. $y = \dfrac{1}{x^3 + x^2 - 2x - 2}$

For Exercises 9–14, find the range of each function.

9. $y = \dfrac{x+3}{x-5}$ **10.** $y = \dfrac{3x - 2}{x + 4}$

11. $y = \dfrac{1-x}{1+x}$

12. $y = \dfrac{ax + b}{x}$ (a and b are real numbers and $b \neq 0$.)

13. $y = x^2 + 1$ **14.** $y = x^3$

15. Let $A = \{x, y, z\}$ and $B = \{1, 2, 3\}$. Which of the rules displayed in the figure represent functions from A to B?

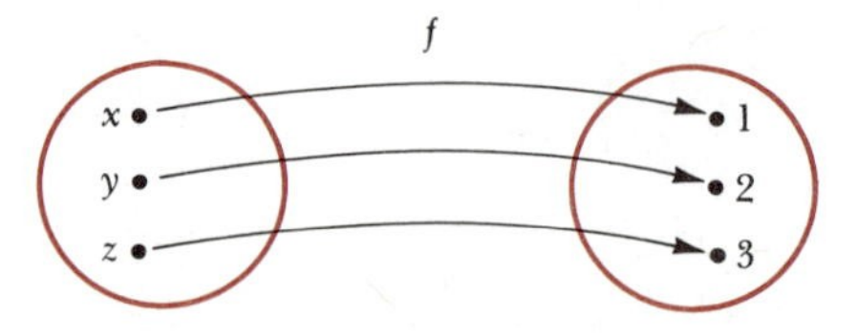

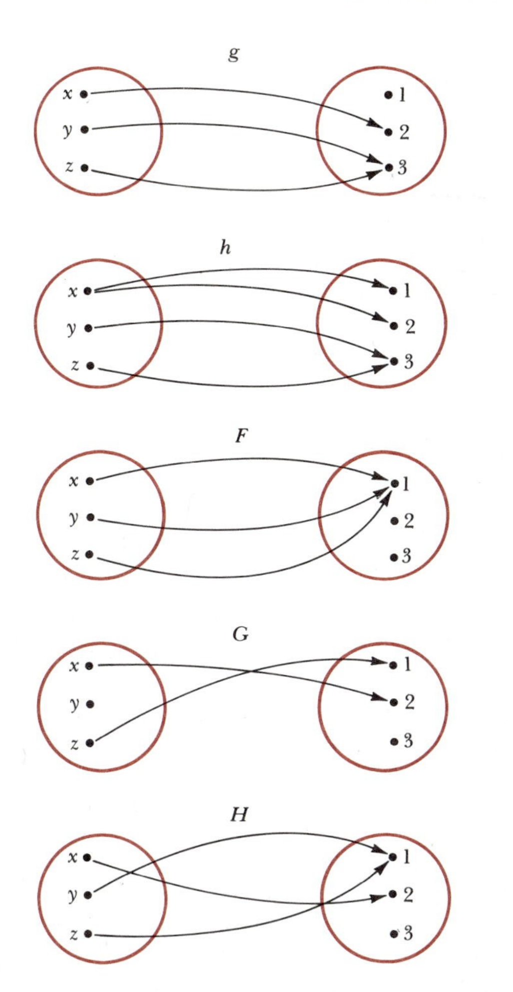

16. Let $D = \{a, b\}$ and $C = \{i, j, k\}$. Which of the rules displayed in the figure represent functions from D to C?

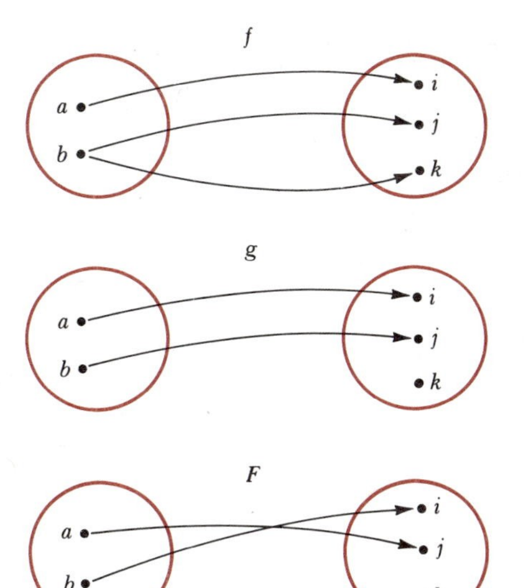

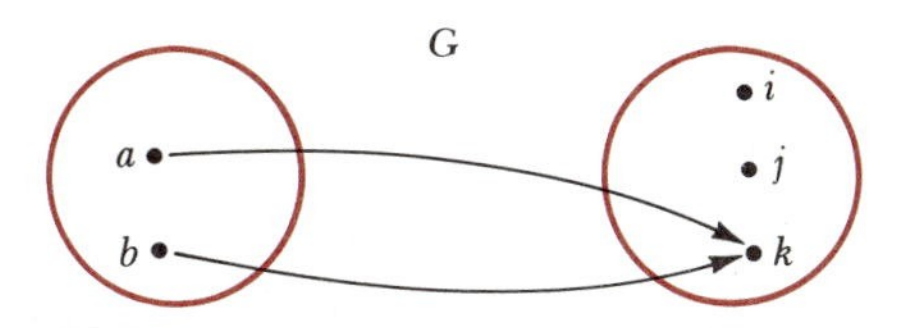

17. **(a)** Specify the range for each rule that represents a function in Exercise 15.

(b) Specify the range for each rule that represents a function in Exercise 16.

18. **(a)** Suppose that in Exercise 15 all the arrows were reversed. Which, if any, of the new rules would be functions from B to A?

(b) Suppose that in Exercise 16 all the arrows were reversed. Which, if any, of the new rules would be functions from C to D?

19. Each of the following rules defines a function whose domain is the set of all real numbers. Express each rule by means of an equation. *Example:* The rule *for each real number, compute its square* can be written as $y = x^2$.

(a) For each real number, subtract 3 and then square the result.

(b) For each real number, compute its square and then subtract 3 from the result.

(c) For each real number, multiply it by 3 and then square the result.

(d) For each real number, compute its square and then multiply the result by 3.

20. Each of the following rules defines a function whose domain is the set of all real numbers. Express each rule in words.

(a) $y = 2x^3 + 1$ **(b)** $y = 2(x + 1)^3$
(c) $y = (2x + 1)^3$ **(d)** $y = (2x)^3 + 1$

21. Let $f(x) = x^2 - 3x + 1$. Compute the following.

(a) $f(1)$ **(b)** $f(0)$ **(c)** $f(-1)$
(d) $f(\frac{3}{2})$ **(e)** $f(z)$ **(f)** $f(x + 1)$
(g) $f(a + 1)$ **(h)** $f(-x)$ **(i)** $|f(1)|$
(j) $f(\sqrt{3})$ **(k)** $f(1 + \sqrt{2})$ **(l)** $|1 - f(2)|$

22. Let $H(x) = 1 - x + x^2 - x^3$.

(a) Which number is larger, $H(0)$ or $H(1)$?

(b) Find $H(\frac{1}{2})$. Does $H(\frac{1}{2}) + H(\frac{1}{2}) = H(1)$?

23. Let $f(x) = 3x^2$. Find the following. *For checking:* Of the six answers, no two are the same.

(a) $f(2x)$ **(b)** $2f(x)$ **(c)** $f(x^2)$
(d) $[f(x)]^2$ **(e)** $f(x/2)$ **(f)** $f(x)/2$

24. Let $f(x) = 4 - 3x$. Find the following.

(a) $f(2)$ **(b)** $f(3)$ **(c)** $f(2) + f(3)$
(d) $f(2 + 3)$ **(e)** $f(2x)$ **(f)** $2f(x)$
(g) $f(x^2)$ **(h)** $f(1/x)$ **(i)** $f[f(x)]$
(j) $x^2f(x)$ **(k)** $1/f(x)$ **(l)** $f(-x)$
(m) $-f(x)$ **(n)** $-f(-x)$

25. Let $H(x) = 1 - 2x^2$. Find the following.
(a) $H(0)$ **(b)** $H(2)$
(c) $H(\sqrt{2})$ **(d)** $H(\frac{5}{6})$
(e) $H(1 - \sqrt{3})$ **(f)** $H(x^2)$
(g) $H(x + 1)$ **(h)** $H(x + h)$
(i) $H(x + h) - H(x)$ **(j)** $\dfrac{H(x + h) - H(x)}{h}$

26. **(a)** Let $f(x) = 2x + 1$; is $f(3 + 1) = f(3) + f(1)$?
(b) Let $f(x) = 2x$; is $f(3 + 1) = f(3) + f(1)$?
(c) Let $f(x) = x^3 - x^2$; is $f(3 + 1) = f(3) + f(1)$?
(d) Let $f(x) = \frac{1}{3}x$; is $f(3 + 1) = f(3) + f(1)$?
(e) Let $f(x) = \sqrt{x}$; is $f(3 + 1) = f(3) + f(1)$?
(f) Let $f(x) = 10^x$; is $f(3 + 1) = f(3) + f(1)$?

27. Let $R(x) = \dfrac{2x - 1}{x - 2}$. Find the following.
(a) The domain and range of R **(b)** $R(0)$
(c) $R(\frac{1}{2})$ **(d)** $R(-1)$
(e) $R(x^2)$ **(f)** $R(1/x)$
(g) $R(a)$ **(h)** $R(x - 1)$

28. Let $g(x) = 2$, for all x. Find the following.
(a) $g(0)$ **(b)** $g(5)$ **(c)** $g(x + h)$

29. Let $d(t) = -16t^2 + 96t$.
(a) Compute $d(1)$, $d(\frac{3}{2})$, $d(2)$, $d(t_0)$.
(b) For which values of t is $d(t) = 0$?
(c) For which values of t is $d(t) = 1$?

30. Let $A(x) = |x^2 - 1|$. Compute $A(2)$, $A(1)$, and $A(0)$.

31. Let $g(t) = |t - 4|$. Find $g(3)$. Find $g(x + 4)$.

32. Let $f(x) = x^2/|x|$.
(a) What is the domain of f?
(b) Find $f(2)$, $f(-2)$, $f(20)$, and $f(-20)$.
(c) What is the range of f?

33. Compute $\dfrac{f(x + h) - f(x)}{h}$ for the function f specified in each case.
(a) $f(x) = -3x + 6$ **(b)** $f(x) = x^2$
(c) $f(x) = 2x^2 - 3x + 1$ **(d)** $f(x) = x^3$
(e) $f(x) = 7$

34. Let $G(x) = 3x - 5$. Compute $\dfrac{G(x) - G(a)}{x - a}$.

35. Let $H(x) = 1 - x/4$. Compute $\dfrac{H(x) - H(1)}{x - 1}$.

36. Let $f(x) = x^2$. Compute $\dfrac{f(b) - f(a)}{b - a}$ using the values $a = 5$, $b = 3$.

37. In each case a pair of functions f and g are given. Find all real numbers x_0 for which $f(x_0) = g(x_0)$.
(a) $f(x) = 4x - 3$, $g(x) = 8 - x$
(b) $f(x) = x^2 - 4$, $g(x) = 4 - x^2$
(c) $f(x) = x^2$, $g(x) = x^3$
(d) $f(x) = 2x^2 - x$, $g(x) = 3$
(e) $f(x) = 1/x$, $g(x) = 5x$
(f) $f(x) = 3x^2 + 1$, $g(x) = -x$

38. Let $g(x) = x^2 - 2x + 1$. Compute the following.
(a) $\dfrac{g(1 + h) - g(1)}{h}$ **(b)** $\dfrac{g(x) - g(1)}{x - 1}$
(c) $\dfrac{g(1 + h) + g(1 - h)}{2h}$

39. Let $f(t) = t^2 + t$.
(a) Find $\dfrac{f(2 + h) - f(2)}{h}$. **(b)** Find $\dfrac{f(t) - f(2)}{t - 2}$.
(c) Find $\dfrac{f(t) - f(t_0)}{t - t_0}$.

40. Let $f(x) = \dfrac{x - a}{x + a}$.
(a) Find $f(a)$, $f(2a)$, and $f(3a)$. Is it true that $f(3a) = f(a) + f(2a)$?
(b) Show that $f(5a) = 2f(2a)$.

41. Let $M(x) = \dfrac{x - a}{x + a}$. Show that $M\left(\dfrac{1}{x}\right) = \dfrac{1 - ax}{1 + ax}$.

42. Let $g(t) = t^4 - 3t^2 + 1$. Find $g(\sqrt{t})$.

43. Let $\phi(y) = 2y - 3$. Show that $\phi(y^2) \neq [\phi(y)]^2$.

44. Let $k(x) = 5x^3 + 5/x^3 - x - 1/x$. Show that $k(x) = k(1/x)$.

45. If $p(x) = 2^x$, show that the following occurs:
(a) $2p(x) = p(x + 1)$ **(b)** $p(a + b) = p(a) \cdot p(b)$
(c) $p(x) \cdot p(-x) - 1 = 0$

***46.** Let $f(x) = 2x + 3$. Find values for a and b such that the equation $f(ax + b) = x$ is true for all values of x.

47. Let $F(x) = \dfrac{5x - 7}{2x + 1}$. Find $\dfrac{F(x + h) - F(x)}{h}$.

***48.** Let $f(t) = \dfrac{t - x}{t + y}$. Show that $f(x + y) + f(x - y) = \dfrac{-2y^2}{x^2 + 2xy}$.

49. Let $H(u) = \dfrac{2u - 1}{u + 3}$. Find $H(u/4)$.

50. Let $f(z) = \dfrac{3z - 4}{5z - 3}$. Find $f\left(\dfrac{3z - 4}{5z - 3}\right)$.

***51.** Let $F(x) = \dfrac{ax + b}{cx - a}$. Show that $F\left(\dfrac{ax + b}{cx - a}\right) = x$. (Assume that $a^2 + bc \neq 0$.)

52. If $f(x) = -2x^2 + 6x + k$ and $f(0) = -1$, find k.

53. If $g(x) = x^2 - 3xk - 4$ and $g(1) = -2$, find k.

54. Let $f(x) = x^2 - 5x - 6$.
(a) Find all values of x for which $f(x) = 0$.
(b) Find all values of x for which $f(x) = 1$.

55. If $f(x) = 1/x^2$, show that

$$\frac{f(x + h) - f(x)}{h} = \frac{-2x - h}{(x + h)^2x^2}$$

56. Let $C(x) = 7$, for all x. Find $\dfrac{C(x + h) - C(x)}{h}$.

57. Let $f(x) = \sqrt{x}$. Find the difference quotient $\dfrac{f(x + h) - f(x)}{h}$. Then, by multiplying by the quantity

$$\frac{\sqrt{x+h}+\sqrt{x}}{\sqrt{x+h}+\sqrt{x}}$$

show that the difference quotient can be written $\dfrac{1}{\sqrt{x+h}+\sqrt{x}}$.

58. Let $f(x) = ax^2 + bx + c$. Show that $\dfrac{f(x+h) - f(x)}{h} = 2ax + ah + b$.

***59.** Let $S(x) = \dfrac{3^x - 3^{-x}}{2}$ and $C(x) = \dfrac{3^x + 3^{-x}}{2}$. Show that the functions S and C possess the following properties.

(a) $S(0) = 0$; $C(0) = 1$; $S(1) = \frac{4}{3}$; $C(1) = \frac{5}{3}$.
(b) $[C(x)]^2 - [S(x)]^2 = 1$ for any number x.
(c) $S(-x) = -S(x)$ and $C(-x) = C(x)$ for any number x.
(d) $S(x + y) = S(x)C(y) + C(x)S(y)$ for any numbers x and y.
(e) $C(x + y) = C(x)C(y) + S(x)S(y)$ for any numbers x and y.
(f) $S(2x) = 2S(x)C(x)$ and $C(2x) = [C(x)]^2 + [S(x)]^2$ for all x.

60. By definition, a *fixed point* for the function f is a number x_0 such that $f(x_0) = x_0$. For instance, to find any fixed points for the function $f(x) = 3x - 2$, we write $3x_0 - 2 = x_0$. On solving this last equation we find that $x_0 = 1$. Thus 1 is a fixed point for f. Calculate the fixed points (if any) for the following functions.

(a) $f(x) = 6x + 10$ **(b)** $g(x) = x^2 - 2x - 4$
(c) $S(t) = t^2$ **(d)** $R(z) = \dfrac{z+1}{z-1}$
(e) $L(z) = \dfrac{z-1}{z+1}$ **(f)** $k(x) = \sqrt{x}$

61. Let $f(x) = \dfrac{3x - 4}{x - 3}$.

(a) Find $f[f(x)]$.
(b) Find $f[f(\frac{22}{7})]$. Try not to do it the hard way!

62. Consider the following two rules, F and G. F is the rule that assigns to each person his mother. G is the rule that assigns to each person his aunt. Explain why F is a function but G is not.

63. A prime number is a positive whole number with no factors other than itself and 1. For example, 2, 13, and 37 are primes, but 24 and 39 are not. By convention, 1 is not considered prime, so the list of the first few primes looks like this:

$$2, 3, 5, 7, 11, 13, 17, 19, 23, 29, \ldots$$

Let G be the rule that assigns to each number the nearest prime. For example, $G(8) = 7$, since 7 is the prime nearest 8. Explain why G is not a function. How could you alter the definition of G to make it a function?

64. Let f be the function that assigns to each natural number x the number of primes that are less than or equal to x. For example, $f(12) = 5$, because, as you can easily check, there are five primes that are less than or equal to 12. Similarly, $f(3) = 2$, because there are two primes that are less than or equal to 3. Find $f(8)$, $f(10)$, and $f(50)$.

***65.** If $P(x) = x^2 + x + 41$, find $P(1)$, $P(2)$, $P(3)$, and $P(4)$. Can you find a natural number x for which $P(x)$ is not prime?

66. A function needn't always be given by an algebraic formula. For example, let the function L be defined by the following rule: *$L(x)$ is the power to which 2 must be raised to yield x.* (For the moment we won't concern ourselves with the domain and range.) Then $L(8) = 3$, for example, since the power to which 2 must be raised to yield 8 is 3 ($8 = 2^3$). Find the following:

(a) $L(1)$ **(b)** $L(2)$ **(c)** $L(4)$ **(d)** $L(64)$
(e) $L(\frac{1}{2})$ **(f)** $L(\frac{1}{4})$ **(g)** $L(\frac{1}{64})$ **(h)** $L(\sqrt{2})$

The function L is called a *logarithm function.* The usual notation for $L(x)$ in this example is $\log_2 x$. Logarithm functions will be studied in Chapter 6.

***67.** Let $q(x) = ax^2 + bx + c$. Evaluate

$$q\left(\frac{-b + \sqrt{b^2 - 4ac}}{2a}\right)$$

C *For Exercises 68–70, let* f, g, *and* h *be the functions defined as follows:*

$$f(x) = x^2 \quad g(x) = 1/x \quad h(x) = \sqrt{x}$$

Make use of the $\boxed{x^2}$, $\boxed{1/x}$, *and* $\boxed{\sqrt{x}}$ *keys on your calculator to evaluate the given expressions. In cases in which the calculator yields only an approximation to an exact output, round off the final answer to three decimal places.*

68. **(a)** $f(0.791)$ **(b)** $f(\pi)$

69. **(a)** $\dfrac{g(5) - g(4.9)}{5 - 4.9}$ **(b)** $\dfrac{g(5) - g(4.99)}{5 - 4.99}$

70. **(a)** $\dfrac{h(5) - h(1)}{5 - 1}$ **(b)** $\dfrac{1}{h(5) + 1}$

(c) Why are the answers for parts (a) and (b) the same?

C 71. When \$1000 is deposited in a savings account at an annual rate of 12%, compounded quarterly, the amount in the account after t years is given by

$$A(t) = 1000\left(1 + \frac{0.12}{4}\right)^{4t}$$

(a) Compute $A(1) - A(0)$. This is the amount by which the account will grow in the first year.
(b) Compute $A(10) - A(9)$. This is the amount by which the account will grow in the tenth year.

C 72. Let $f(n) = (1 + 1/n)^n$.
(a) Complete the table. (Round off to three decimal places.)

n	1	2	5	10	15	20
$f(n)$						

(b) By trial and error, find the smallest natural number n such that $f(n) > 2.7$.

C 73. Let $g(n) = n^{1/n}$.
(a) Complete the table. (Round off to four decimal places.)

n	2	3	4	5	6	7	8
$g(n)$							

(b) By trial and error, find the smallest natural number n such that $g(n) < 1.2$.

4.2 Applied Functions: Setting Up Equations

I hope that I shall shock a few people in asserting that the most important single task of mathematical instruction in the secondary schools is to teach the setting up of equations to solve word problems.

George Polya (1887–1985)

One of the first steps in problem solving often involves defining a function. The function then serves to describe or summarize a given situation in a way that is both concise and (one hopes) revealing. In this section we want to practice setting up equations defining such functions. For many of the examples we look at, we'll rely on the following four-step procedure to set up the required equation. In practice you may eventually want to modify this procedure to fit your own style. The point is, however, that it's possible to approach these problems in a systematic manner. A word of advice: You're accustomed to working mathematics problems in which the answers are numbers. In this section the answers are functions (or, more precisely, equations defining functions); you'll need to get used to this.

Steps for Setting Up Equations

Step 1 After reading the problem carefully, draw a picture that conveys the given information.

Step 2 State in your own words, as specifically as you can, what the problem is asking for. (This usually requires rereading the problem.) Now, assuming that the problem asks you to find a particular quantity (or to find a formula for a particular quantity), assign a variable to denote that key quantity.

Step 3 Label any other quantities in your figure that appear relevant. Are there equations relating these quantities?

Step 4 Find an equation involving the key variable that you identified in Step 2. (Some people prefer to do this right after Step 2.) Now, as necessary, substitute in this equation using the auxiliary equations from Step 3 to obtain an equation involving only the required variables.

Example 1 A rectangle is inscribed in a circle of diameter 8 cm. Express the perimeter of the rectangle as a function of its width x.

Solution We'll follow the four-step procedure outlined above.

Step 1 See Figure 1.

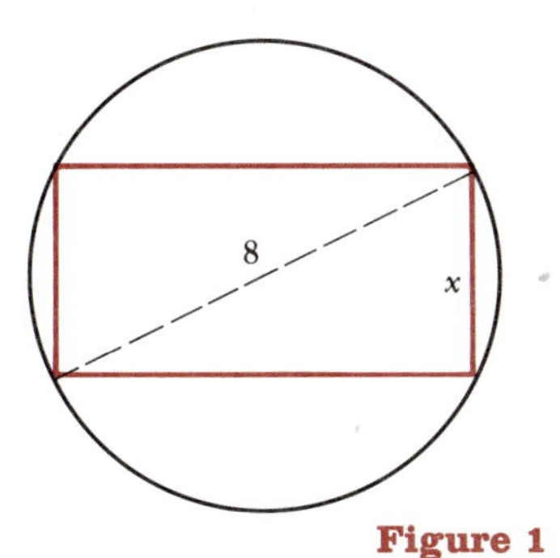

Figure 1

Step 2 The problem asks us to come up with a formula or function that gives the perimeter of the rectangle in terms of x, the width. Let P denote the perimeter.

Step 3 Let L denote the length of the rectangle. See Figure 2. Then by the Pythagorean Theorem we have

$$L^2 + x^2 = 8^2$$
$$L^2 = 64 - x^2$$
$$L = \sqrt{64 - x^2} \qquad (1)$$

Step 4 The perimeter of the rectangle is the sum of the lengths of the four sides. Thus

$$P = x + x + L + L$$
$$P = 2x + 2L \qquad (2)$$

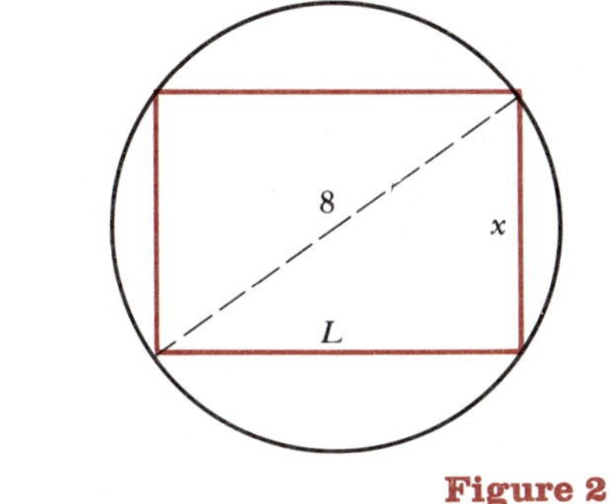
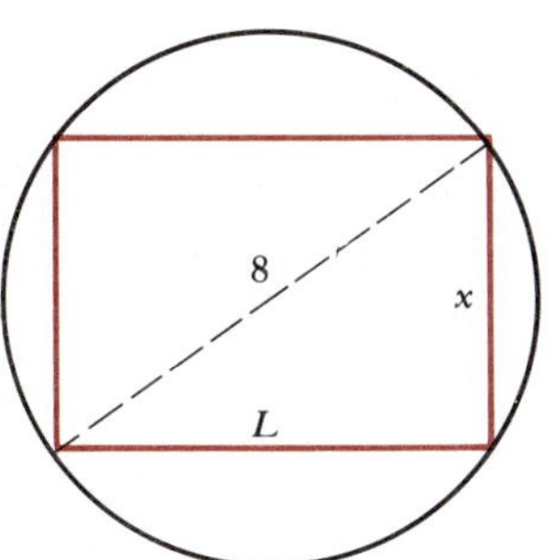

Figure 2

This expresses P in terms of x and L. However, the problem asks for P in terms of just x. Using equation (1) to substitute for L in equation (2), we have

$$P = 2x + 2\sqrt{64 - x^2} \qquad (3)$$

This is the required equation. It expresses the perimeter P as a function of the width x, meaning if you know x, you can calculate P. To emphasize this dependence of P on x, we can employ functional notation to rewrite equation (3) as

$$P(x) = 2x + 2\sqrt{64 - x^2} \qquad (4)$$

Before leaving this example, we need to specify the domain of the perimeter function in equation (4). An easy way to do this is to look back at Figure 1. Since x represents a width, we certainly want $x > 0$. Furthermore, Figure 1 tells us that $x < 8$, because in any right triangle, a leg is always shorter than the hypotenuse. Putting these observations together, we conclude that the domain of the perimeter function in equation (4) is the open interval $(0, 8)$. *Note:* Although it would make sense algebraically to use an input such as $x = -1$ in equation (4), it does not make sense in our geometric context, where x denotes the width.

Example 2 The perimeter of a rectangle is 100 cm. Express the area of the rectangle in terms of the width x.

Solution *Step 1* See Figure 3.

Step 2 We want to express the area of the rectangle in terms of x, the width. Let A stand for the area of the rectangle.

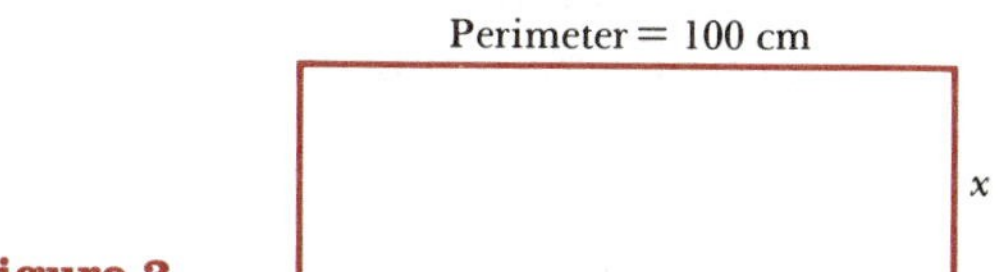

Figure 3

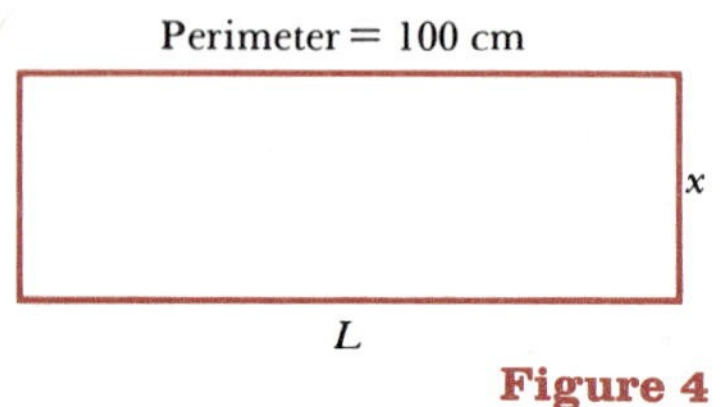

Figure 4

Step 3 Call the length of the rectangle L, as indicated in Figure 4. Then, since the perimeter is given as 100 cm, we have

$$2x + 2L = 100$$
$$x + L = 50$$
$$L = 50 - x \tag{5}$$

Step 4 Area of a rectangle equals width times length.

$$A = x \cdot L$$

$$A = x(50 - x)$$ We've substituted for L using equation (5).

$$A = 50x - x^2 \tag{6}$$

This is the required equation expressing the area of the rectangle in terms of the width x. To emphasize this dependence of A on x, we can use functional notation to rewrite equation (6) as

$$A(x) = 50x - x^2$$

The domain of this area function is the open interval $(0, 50)$. To see why this is so, first note that $x > 0$, because x denotes a width. Furthermore, in view of equation (5) we must have $x < 50$ (otherwise L, the length, would be zero or negative).

Example 3 $P(x, y)$ is a point on the curve $y = \sqrt{x}$. Express the distance from P to the point $(1, 0)$ as a function of x.

Solution *Step 1* See Figure 5.

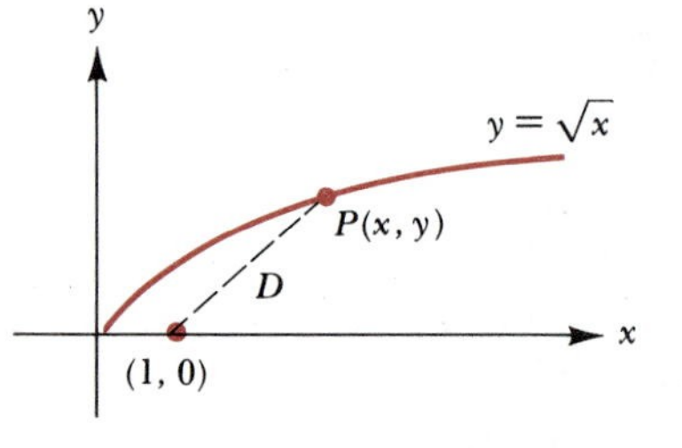

Figure 5

Step 2 We want to express the length of the dotted line in Figure 5 in terms of x. Call this length D.

Step 3 There are no other quantities in Figure 5 that need labeling. But don't forget we are given that

$$y = \sqrt{x} \tag{7}$$

Step 4 By the distance formula we have

$$D = \sqrt{(x_2 - x_1)^2 + (y_2 - y_1)^2}$$
$$D = \sqrt{(x - 1)^2 + (y - 0)^2}$$
$$D = \sqrt{x^2 - 2x + 1 + y^2} \tag{8}$$

Now use equation (7) to eliminate y in equation (8):

$$D = \sqrt{x^2 - 2x + 1 + (\sqrt{x})^2}$$
$$D = \sqrt{x^2 - 2x + 1 + x}$$
$$D = \sqrt{x^2 - x + 1} \tag{9}$$

Equation (9) expresses the distance as a function of x, as required. What about the domain of this distance function? Since the x-coordinate of a point on the curve $y = \sqrt{x}$ can be any nonnegative number, the domain of the distance function is $[0, \infty)$.

Example 4 A point $P(x, y)$ lies in the first quadrant on the parabola $y = 16 - x^2$, as indicated in Figure 6. Express the area of the triangular region in Figure 6 as a function of x.

Solution

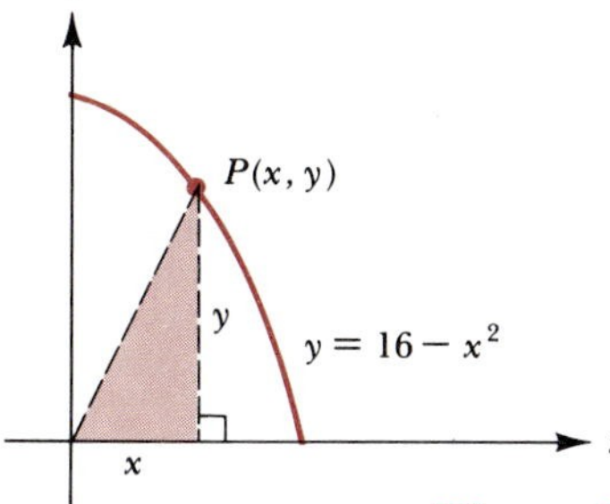

Figure 6

Step 1 Figure 6 is given.

Step 2 We want to express the area of the shaded triangle in the given figure in terms of x. Let A denote the area of this triangle.

Step 3 Since the coordinates of P are (x, y), the base of our triangle is x; the height of our triangle is y. Also, x and y are related by the given equation

$$y = 16 - x^2 \tag{10}$$

Step 4 Area of a triangle equals $\frac{1}{2}$(base)(height).

$$\begin{aligned} A &= \frac{1}{2}(x)(y) \\ &= \frac{1}{2}(x)\underbrace{(16 - x^2)} \qquad \text{We've substituted for } y \text{ using equation (10).} \\ &= 8x - \frac{1}{2}x^3 \end{aligned}$$

This last equation expresses the area of the triangle in Figure 6 as a function of x, as required. (Exercise 46 at the end of this section asks you to specify the domain of this area function.)

Example 5 A piece of wire x in. long is bent into the shape of a circle. Express the area of the circle in terms of x.

Solution

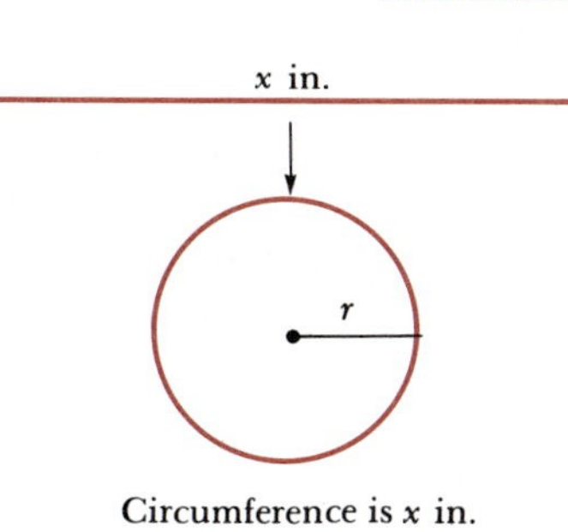

Circumference is x in.

Figure 7

Step 1 See Figure 7.

Step 2 We are supposed to express the area of the circle in terms of x, the circumference. Let A denote the area.

Step 3 We assume as known the formula for circumference in terms of radius:

$$C = 2\pi r$$

Since in our case the circumference is given to be x, our equation becomes

$$x = 2\pi r$$

or

$$r = \frac{x}{2\pi} \tag{11}$$

Step 4 We assume as known the formula for the area of a circle in terms of the radius:

$$A = \pi r^2 \tag{12}$$

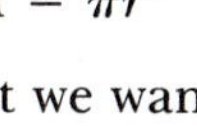

This expresses A in terms of r; but we want A in terms of x. So we replace r in equation (12) by the value given in equation (11). This yields

$$\begin{aligned} A &= \pi\left(\frac{x}{2\pi}\right)^2 \\ &= \frac{\pi x^2}{4\pi^2} = \frac{x^2}{4\pi} \end{aligned}$$

Thus we have

$$A = \frac{x^2}{4\pi}$$

This is the required equation. It expresses the area of the circle in terms of the circumference x. In other words, if we know the length of the piece of wire that is to be bent into a circle, we can calculate from that length what the area of the circle will be.

Example 6 Figure 8 displays a right circular cylinder, along with the formulas for the volume V and the total area S. Given that the volume is 10 cm³, express the total area S as a function of r, the radius of the base.

Solution *Step 1* We are given Figure 8.

Step 2 We're given a formula expressing the area S in terms of both r and h. We want to express S in terms of just r.

Step 3 We are given that $V = 10$ and also that $V = \pi r^2 h$. Thus

$$\pi r^2 h = 10$$

and consequently

$$h = \frac{10}{\pi r^2}$$

Step 4 Take the given formula for S, namely

$$S = 2\pi r^2 + 2\pi r h$$

and replace h using the value obtained in Step 3. You find

$$S = 2\pi r^2 + 2\pi r\left(\frac{10}{\pi r^2}\right)$$

$$S = 2\pi r^2 + \frac{20}{r}$$

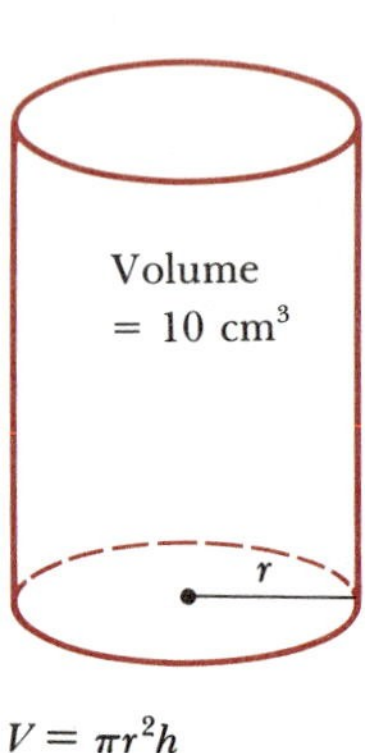

$V = \pi r^2 h$

$S = 2\pi r^2 + 2\pi r h$

Figure 8

This is the required equation. It expresses the surface area S in terms of the radius r. Since there are no restrictions on r other than it being positive, the domain of the area function here is $(0, \infty)$.

We have followed the same four-step procedure in Examples 1 through 6. Of course, no single method can cover all possible cases. As usual, some common sense or experience is often necessary. Also you should not feel compelled to follow this procedure at any cost. Keep this in mind as you study Examples 7, 8, and 9.

Example 7 Two numbers add to 8. Express the product P of these two numbers in terms of a single variable.

Solution If you call the two numbers x and $8 - x$, then their product P is given by

$$\begin{aligned} P &= x(8 - x) \\ &= 8x - x^2 \end{aligned}$$

That's it. This last equation expresses the product as a function of the variable x. Since there are no restrictions on x (other than it being a real number), the domain of this function is $(-\infty, \infty)$.

Example 8 Express the (length of the) radius of a circle as a function of the area of the circle.

Solution You already know a formula relating these two quantities:

$$A = \pi r^2$$

In essence, the problem at hand asks us to solve this equation for r instead of A. We have then

$$\pi r^2 = A$$

$$r^2 = \frac{A}{\pi}$$

$$r = \sqrt{\frac{A}{\pi}}$$

This is the required equation; it tells us how to calculate the radius of the circle if the area is known. *Question:* Are there any restrictions on A in this last equation?

Example 9 In economics, the revenue R generated by selling x units at a price of p dollars per unit is given by

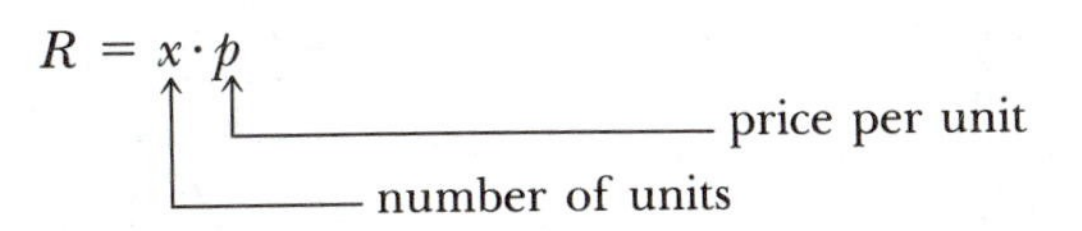

In Figure 9 we are given a hypothetical function relating the selling price of a certain item to the number of units sold. Such a function is called a *demand function*. Express the revenue in this case as a function of x.

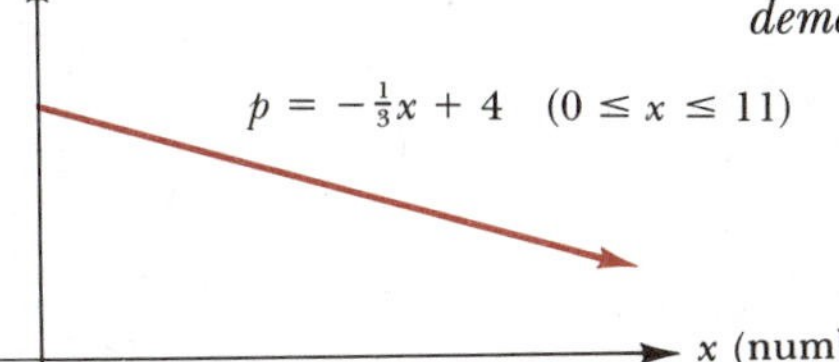

Figure 9

Solution More than anything else, this problem is an exercise in reading. After you've read the problem several times, you will find that it comes down to this:

given: $R = x \cdot p$ and $p = -\frac{1}{3}x + 4 \quad (0 \le x \le 11)$

find: an equation expressing R in terms of x

In view of this, we write

$$\begin{aligned} R &= x \cdot p \\ &= x\left(-\frac{1}{3}x + 4\right) \end{aligned}$$

Thus we have

$$R = -\frac{1}{3}x^2 + 4x \qquad (0 \le x \le 11)$$

This is the required function. It allows us to calculate the revenue when we know the number of units sold.

Exercise Set 4.2

1. A rectangle is inscribed in a circle of diameter 12 in.
   (a) Express the perimeter of the rectangle as a function of its width x. *Suggestion:* First reread Example 1 on pages 162–163.
   (b) Express the area of the rectangle as a function of its width x.
2. (a) The perimeter of a rectangle is 16 cm. Express the area of the rectangle in terms of the width x. *Suggestion:* Reread Example 2 on pages 163–164.
   (b) The area of a rectangle is 85 cm^2. Express the perimeter as a function of the width x.
3. $P(x, y)$ is a point on the curve $y = x^2 + 1$.
   (a) Express the distance from P to the origin as a function of x. *Suggestion:* First reread Example 3 on page 164.
   (b) In part (a) you expressed the length of a certain line segment as a function of x. Now express the slope of that line segment in terms of x.
4. A point $P(x, y)$ lies on the curve $y = \sqrt{x}$, as shown in the figure.
   (a) Express the area of the shaded triangle in terms of x. *Suggestion:* First reread Example 4 on pages 164–165.
   (b) Express the perimeter of the shaded triangle in terms of x.

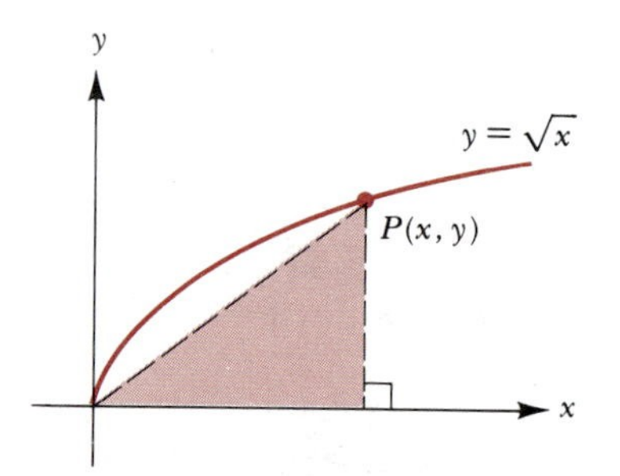

5. A piece of wire πy in. long is bent into a circle.
   (a) Express the area of the circle as a function of y. *Suggestion:* First reread Example 5 on page 165.
   (b) If the original piece of wire were bent into a square instead of a circle, how would you express the area in terms of y?
6. The volume of a certain right circular cylinder is 20 in^3. Express the total surface area of the cylinder as a function of r, the radius of the base. *Suggestion:* First reread Example 6, page 166.
7. Two numbers add to 16.
   (a) Express the product of the two numbers in terms of a single variable. *Suggestion:* First reread Example 7, page 166.
   (b) Express the sum of the squares of the two numbers in terms of a single variable.
   (c) Express the difference of the cubes of the two numbers in terms of a single variable. (Two answers)
   (d) What happens when you try to express the average of the two numbers in terms of one variable?
8. (a) Express the (length of the) radius of a circle as a function of the circumference. *Suggestion:* Reread Example 8, pages 166–167.
   (b) Express the (length of the) diameter of a circle as a function of the circumference.
9. Given the demand function $p = -\frac{1}{4}x + 8$, express the revenue as a function of x. (See Example 9 on page 167 for terminology and definitions.)
10. In Example 2 on pages 163–4, we considered a rectangle with a perimeter of 100 cm. We found that the area of such a rectangle is given by $A(x) = 50x - x^2$, where x is the width of the rectangle. Compute the numbers $A(1)$, $A(10)$, $A(20)$, $A(25)$, and $A(35)$. Which width x seems to yield the largest area $A(x)$?
11. In Example 1 on pages 162–163, we considered a rectangle inscribed in a circle of diameter 8 cm. We found that the perimeter of such a rectangle is given by

$$P(x) = 2x + 2\sqrt{64 - x^2}$$

   where x is the width of the rectangle.
   (a) Compute the perimeter for the cases when $x = 1, 2, 3, \ldots, 7$.
   (b) Use a calculator to list the following in order of increasing size:

$$P(1), P(2), P(3), \ldots, P(7)$$

12. In economics, the *demand function* for a given commodity tells us how the unit price p is related to the number of units sold x. Suppose we are given the demand function

$$p = 5 - \frac{x}{4} \quad (p \text{ is in dollars.})$$

   (a) Graph this demand function.
   (b) How many units can be sold when the unit price is \$3? Locate the point on the graph of the demand function that conveys this information.
   (c) In order to sell 12 items, how should the unit price be set? Locate the point on the graph of the demand function that conveys this information.

(d) Find the revenue function corresponding to the given demand function. (Use the formula $R = x \cdot p$ on page 167.) Graph the revenue function.

(e) Find the revenue when $x = 2$, when $x = 8$, and when $x = 14$.

(f) According to your graph in part (d), which x-value yields the greatest revenue? What is that revenue? What is the corresponding unit price?

13. Let $2s$ denote the length of the side of an equilateral triangle.

(a) Express the height of the triangle as a function of s.

(b) Express the area of the triangle as a function of s.

(c) Use the function you found in part (a) to determine the height of an equilateral triangle, each side of which is 8 cm long.

(d) Use the function you found in part (b) to determine the area of an equilateral triangle, each side of which has length 5 in.

14. If x denotes the length of a side of an equilateral triangle, express the area of the triangle as a function of x.

15. In a certain right circular cylinder, the height is twice the radius. Express the volume as a function of the radius.

16. In Exercise 15, express the radius as a function of the volume.

17. The volume of a right circular cylinder is 12π in^3.

(a) Express the height as a function of the radius.

(b) Express the total surface area as a function of the radius.

18. In a certain right circular cylinder, the total surface area is 14 in^2. Express the volume as a function of the radius.

19. The volume V and the surface area S of a sphere of radius r are given by the formulas

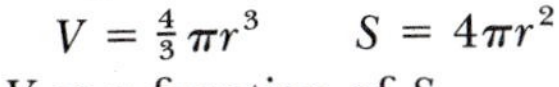

$$V = \tfrac{4}{3}\pi r^3 \qquad S = 4\pi r^2$$

Express V as a function of S.

20. The base of a rectangle lies on the x-axis, while the upper two vertices lie on the parabola $y = 10 - x^2$. Suppose that the coordinates of the upper right vertex of the rectangle are (x, y). Express the area of the rectangle as a function of x.

21. The hypotenuse of a right triangle is 20 cm. Express the area of the triangle as a function of the length x of one of the legs.

22. (a) (Refer to the figure at the top of the next column.) Express the area of the shaded triangle as a function of x.

(b) Express the perimeter of the shaded triangle as a function of x.

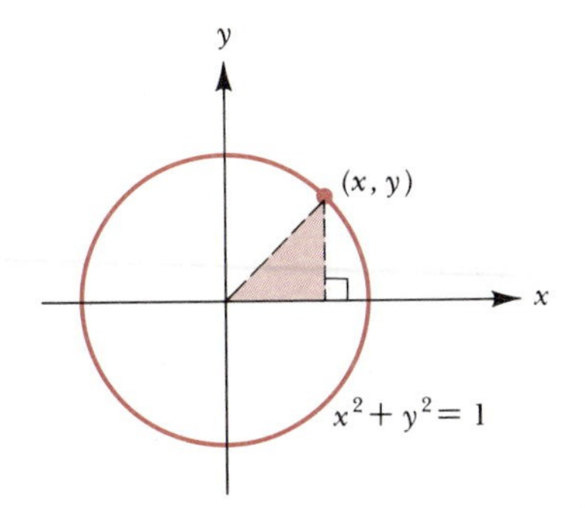

23. In the figure below, express the length of AB as a function of x. *Hint:* Similar triangles.

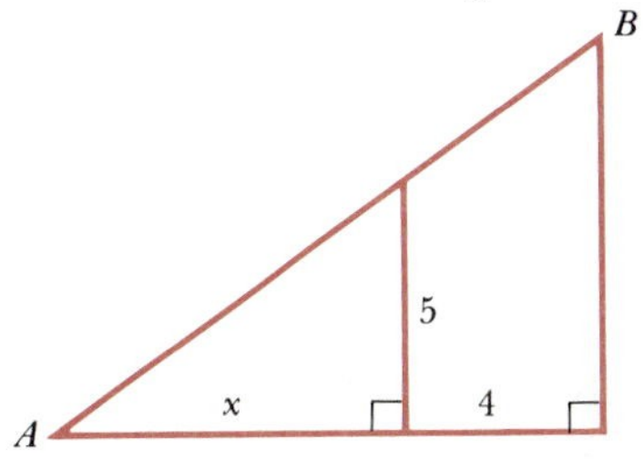

***24.** The figure below shows two concentric circles of radii r and R, respectively. Let A denote the area within the larger circle but outside the smaller one. Express A as a function of x (x is defined in the figure).

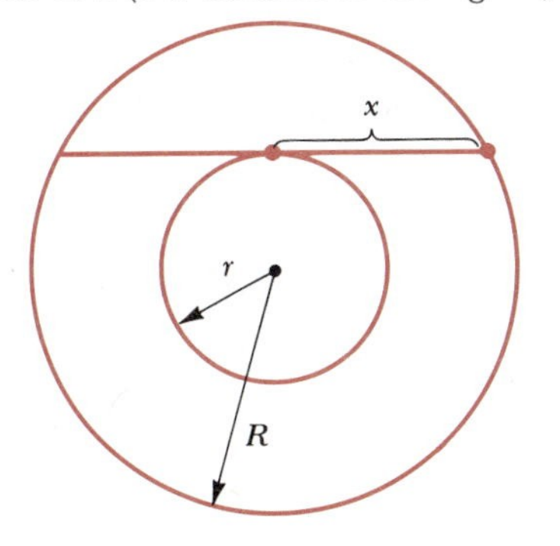

***25.** The figure shows the parabola $y = x^2$ and a line segment AP drawn from the point $A(0, -1)$ to the point $P(a, a^2)$ on the parabola. *(Continued on p. 170.)*

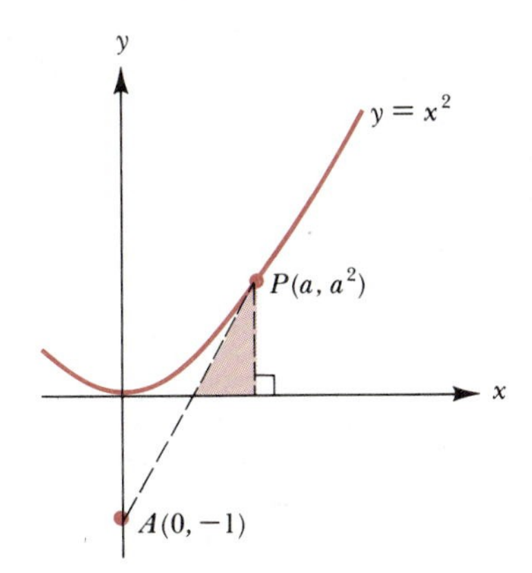

(a) Express the slope of AP in terms of a.
(b) Show that the area of the shaded triangle in the figure is given by

$$\text{area} = \frac{a^5}{2(a^2 + 1)}$$

For Exercises 26–29, refer to the figure below, which displays a right circular cone along with the formulas for the volume V and the lateral surface area S.

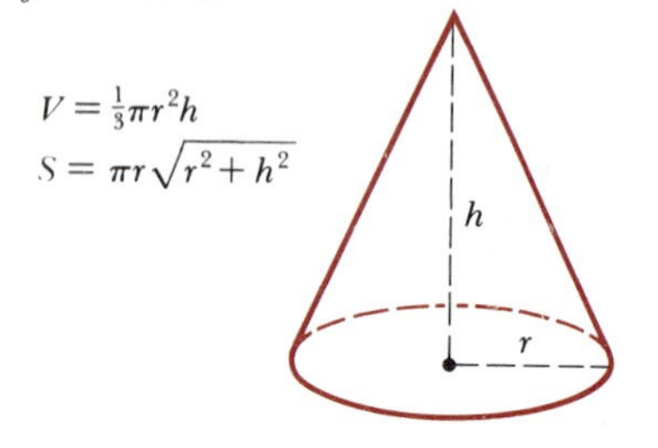

26. The volume of a right circular cone is 12π cm^3.
(a) Express the height as a function of the radius.
(b) Express the radius as a function of the height.

27. Suppose that, for a right circular cone, the height and radius are related by the equation $h = \sqrt{3}r$.
(a) Express the volume as a function of r.
(b) Express the lateral surface area as a function of r.

28. The volume of a right circular cone is 2 ft^3. Show that the lateral surface area as a function of r is given by

$$S = \frac{\sqrt{\pi^2 r^6 + 36}}{r}$$

29. In a certain right circular cone, the volume is numerically equal to the lateral surface area.
(a) Express the radius as a function of the height.
(b) Express the height as a function of the radius.

30. A line is drawn from the origin O to a point $P(x, y)$ in the first quadrant on the graph of $y = 1/x$. From point P, a line is drawn perpendicular to the x-axis, meeting the x-axis at B.
(a) Draw a figure of the situation described.
(b) Express the perimeter of triangle OPB as a function of x.
(c) Try to express the area of triangle OPB as a function of x. What happens?

***31.** A piece of wire 14 in. long is cut into two pieces. The first piece is bent into a circle, the second into a square. Express the combined total area of the circle and square as a function of x, where x denotes the length of the wire that is used for the circle.

***32.** A wire of length L is cut into two pieces. The first piece is bent into a square, the second into an equilateral triangle. Express the combined total area of the square and triangle as a function of a single variable. (L is to be considered a constant here, not another variable.)

33. An athletic field with a perimeter of $\frac{1}{4}$ mi consists of a rectangle with a semicircle at each end, as shown in the figure. Express the area of the field as a function of r, the radius of the semicircle.

34. A square of side x is inscribed in a circle. Express the area of the circle as a function of x.

***35.** An equilateral triangle of side x is inscribed in a circle. Express the area of the circle as a function of x.

36. Consider an object of mass m moving at a velocity v. The kinetic energy K is defined as $K = \frac{1}{2}mv^2$. The momentum p is defined as $p = mv$. Express the kinetic energy in terms of mass and momentum (but not velocity.)

***37.** An isosceles triangle is inscribed in a circle of radius R, where R is a constant. Express the area within the circle but outside the triangle as a function of h, where h denotes the height of the triangle.

38. Refer to the figure. An offshore oil rig is located at point A, 10 mi out to sea. An oil pipeline is to be constructed from A to a point C on the shore and then to an oil refinery at point D farther up the coast. If it costs \$800 per mi to lay the pipeline in the sea and \$200 per mi on land, express the cost of laying the pipeline in terms of x, where x is the distance from B to C.

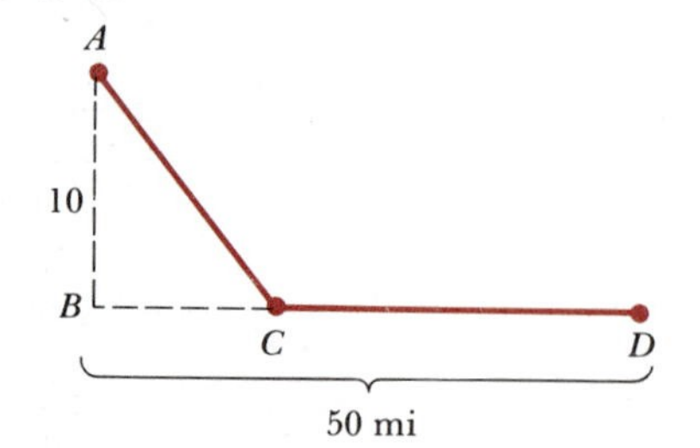

39. An open-top box is to be constructed from a 6 in. by 8 in. rectangular sheet of tin by cutting out equal squares at each corner and then folding up the flaps. (See the figure on the next page.) Express the volume of the box as a function of x, the length of the side of each square cut out.

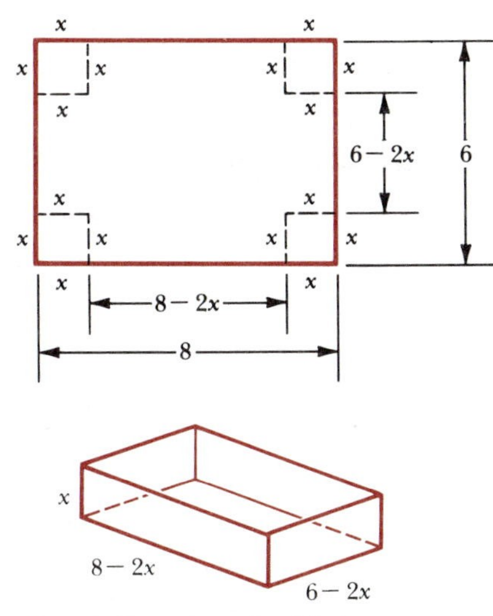

40. Follow Exercise 39, but assume the original piece of tin is a square, 12 in. on each side.

41. Five hundred feet of fencing are available to enclose a rectangular pasture alongside a river, the river serving as one side of the rectangle (so only three sides require fencing). Express the area of the rectangular pasture as a function of x. See the figure.

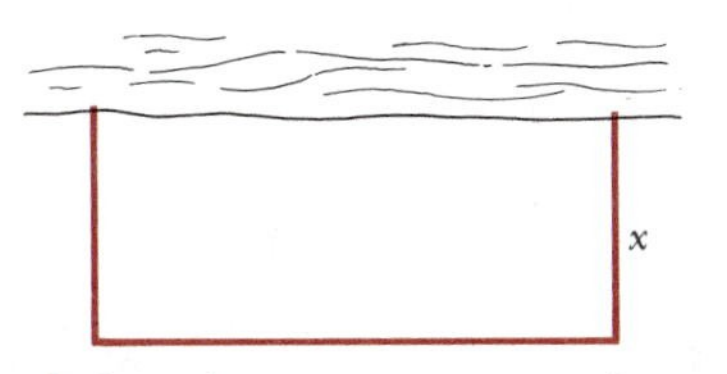

***42.** The figure below shows two concentric squares. Express the area of the shaded triangle in the figure as a function of x.

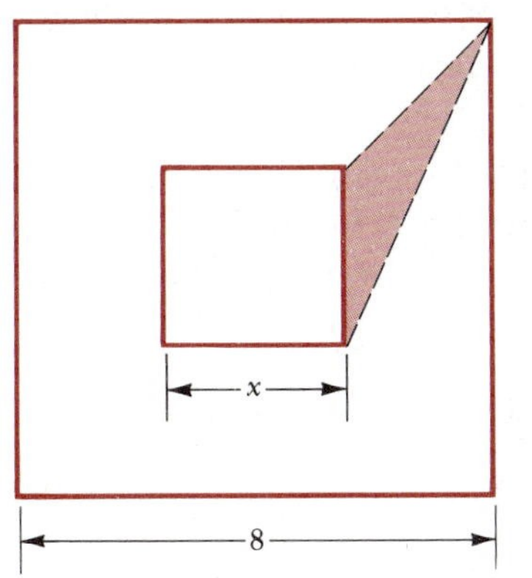

***43.** A Norman window is in the shape of a rectangle surmounted by a semicircle, as shown in the figure at the top of the next column. Assume that the perimeter of the window is 32 ft. Express the area of the window as a function of r, the radius of the semicircle.

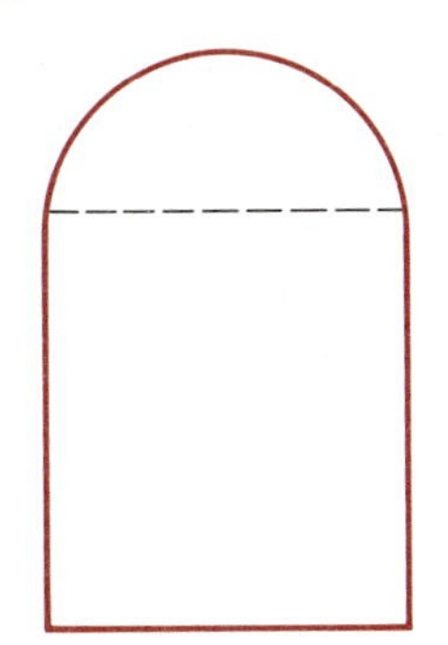

44. Refer to the figure below. Express the lengths of CB, CD, BD, and AB in terms of x. *Hint:* Recall the theorem from geometry stating that in a 30°-60°-90° right triangle, the side opposite the 30° angle is half the hypotenuse.

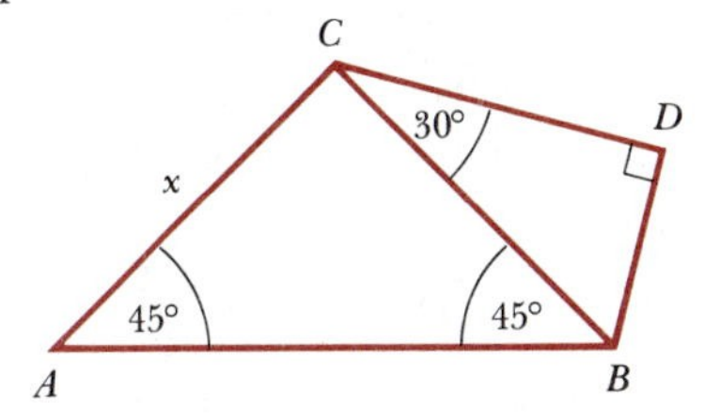

45. Refer to the figure below. Let s denote the ratio of y to z.
- **(a)** Express y as a function of s.
- **(b)** Express s as a function of y.
- **(c)** Express z as a function of s.
- **(d)** Express s as a function of z.

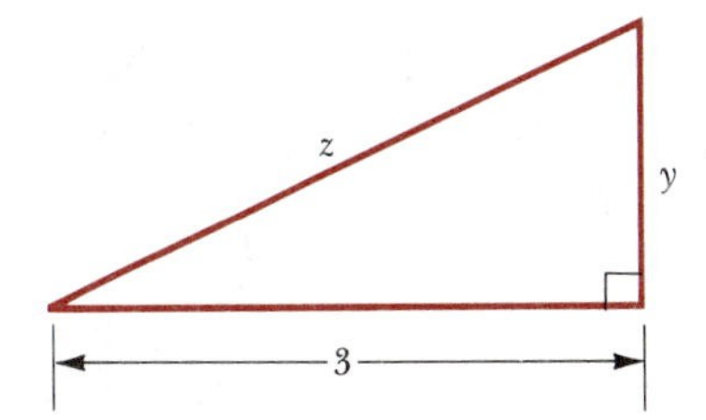

46. This exercise refers to Example 4, pages 164–165.
- **(a)** What is the x-intercept of the curve $y = 16 - x^2$ in Figure 6?
- **(b)** What is the domain of the area function in Example 4, assuming that the point P does not lie on the x- or y-axis?

***47.** A straight line with slope m ($m < 0$) passes through the point $(1, 2)$ and intersects the line $y = 4x$ at a point in the first quadrant. Let A denote the area of the triangle bounded by $y = 4x$, the x-axis, and the given line of slope m. Express A as a function of m.

C 48. After rereading Example 2 in the text, complete the following table. Which x-value in the table yields the largest area A? What is the corresponding value of L in that case?

x	5	10	20	24	24.8	24.9	25	25.1	25.2	45
$A(x)$										

C 49. After rereading Example 3 in the text, complete the following three tables. For each table, specify the x-value that yields the smallest distance D. (For Table 2, round off the values of D to two significant digits. For Table 3, round off the values of D to seven significant digits.)

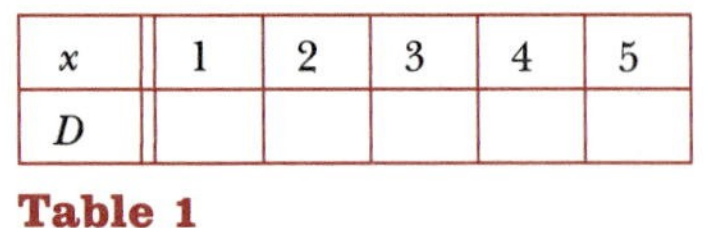

x	1	2	3	4	5
D					

Table 1

x	0.25	0.50	0.75
D			

Table 2

x	0.498	0.499	0.500	0.501	0.502
D					

Table 3

C 50. **(a)** After rereading Example 4 in the text, complete the following three tables. For each table, specify the x-value that yields the largest area A. [For Tables 2 and 3, round off the values of A to six significant digits. In part (b), you'll see why we are asking for that many digits.]

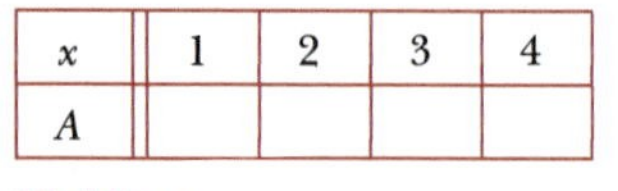

x	1	2	3	4
A				

Table 1

x	1.75	2.00	2.25	2.50	2.75
A					

Table 2

x	2.15	2.20	2.25	2.30	2.35
A					

Table 3

(b) Using calculus, it can be shown that the x-value yielding the largest area A is $x = 4\sqrt{3}/3$. Which x-value in the tables is closest to this x-value? To six significant digits, what is the area A when $x = 4\sqrt{3}/3$?

4.3 The Graph of a Function

In my own case, I got along fine without knowing the name of the distributive law until my sophomore year in college; meanwhile I had drawn lots of graphs.

Professor Donald E. Knuth (Professor Knuth is judged by many to be the world's preeminent computer scientist.)

When the domain and range of a function are sets of real numbers, the function can be graphed in the same way in which equations were graphed in Chapter 3. In graphing functions, the usual practice is to reserve the horizontal axis for the independent variable and the vertical axis for the dependent variable. The function or rule then tells you how you must pick your y-coordinate, once having selected an x-coordinate.

Definition of the Graph of a Function

The graph of a function f consists of those points (x, y) such that x is in the domain of f and $y = f(x)$. See Figure 1 on page 173.

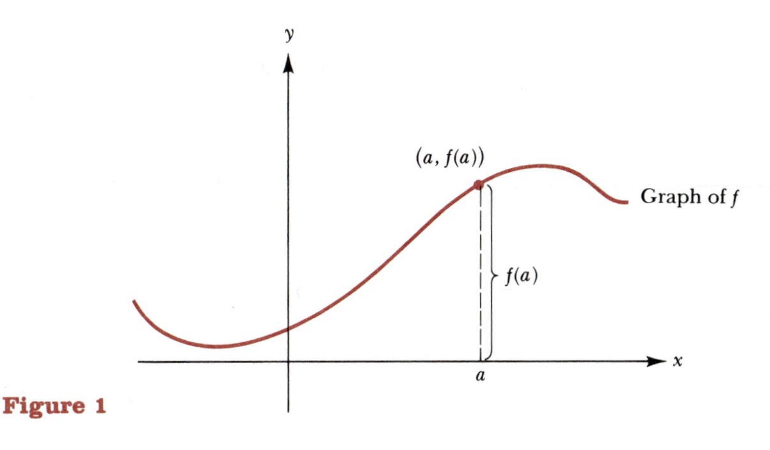

Figure 1

Example 1 Graph the function f defined by $f(x) = x^2 + 2$.

Solution The graph of this function is by definition the graph of the equation $y = x^2 + 2$. Whether we label the vertical axis y or $f(x)$ is immaterial. For the purposes of this example, we draw our graph by setting up a table and plotting points, as in Chapter 3. See Figure 2. However, later in this section and in the next section, we shall look at several techniques that allow us to sketch certain graphs without first setting up tables and without plotting many individual points.

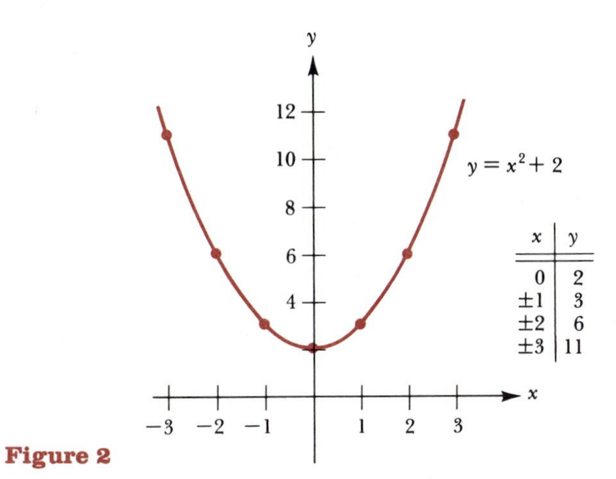

x	y
0	2
±1	3
±2	6
±3	11

Figure 2

Example 2 Specify the domain and the range of the function g whose graph appears in Figure 3(a) on page 174.

Solution The domain of g is just that portion of the x-axis (the inputs) utilized in graphing g. As Figure 3(b) indicates, this amounts to all real numbers from 1 to 5, inclusive: $1 \leq x \leq 5$. Recall from Section 1.3 that this set of numbers is denoted by $[1, 5]$. To find the range of g, you need to check which part of the y-axis is utilized in graphing g. As Figure 3(b) indicates, this is the set of all real numbers between 2 and 4, inclusive: $2 \leq y \leq 4$. Our shorthand for this interval of numbers is $[2, 4]$.

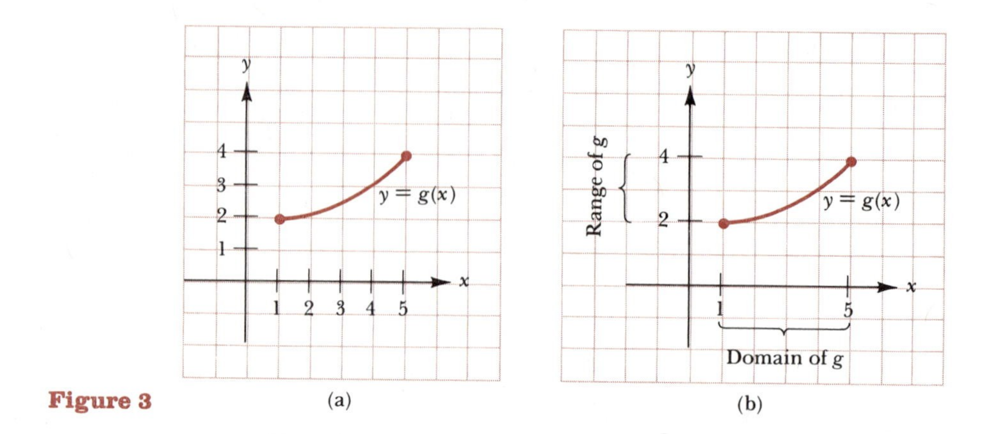

Figure 3

Example 3 A portion of the graph of a function h is shown in Figure 4. Find the following:

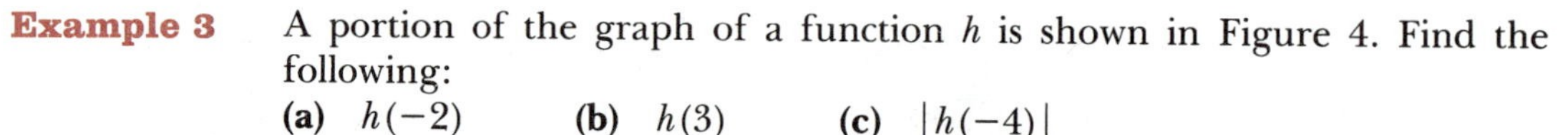

(a) $h(-2)$ **(b)** $h(3)$ **(c)** $|h(-4)|$

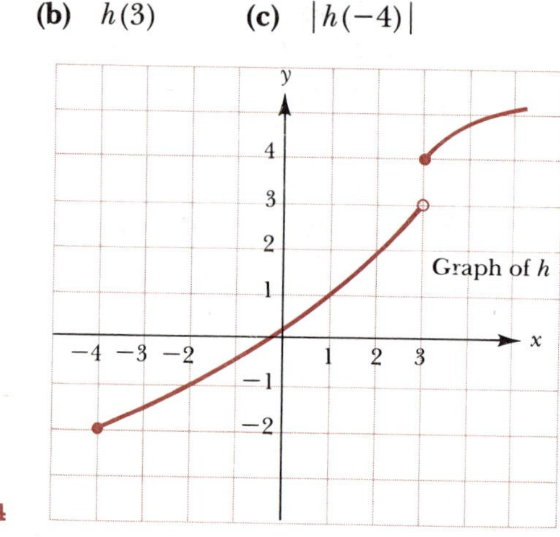

Figure 4

Solution **(a)** The functional notation $h(-2)$ stands for the y-coordinate of that point on the graph of h whose x-coordinate is -2. Since the point $(-2, -1)$ is on the graph of h, we conclude $h(-2) = -1$.

(b) $h(3) = 4$ because the point $(3, 4)$ lies on the graph of h. Note that h would not be considered a function if the point $(3, 3)$ were also part of the graph. (Why?)

(c) Since the point $(-4, -2)$ lies on the graph of h, we write $h(-4) = -2$. Thus $|h(-4)| = |-2| = 2$.

Many of the graphs that we looked at in Chapter 3 are graphs of functions. However, it's important to understand that there are graphs that do not represent functions. Consider, for example, the graph in Figure 5.

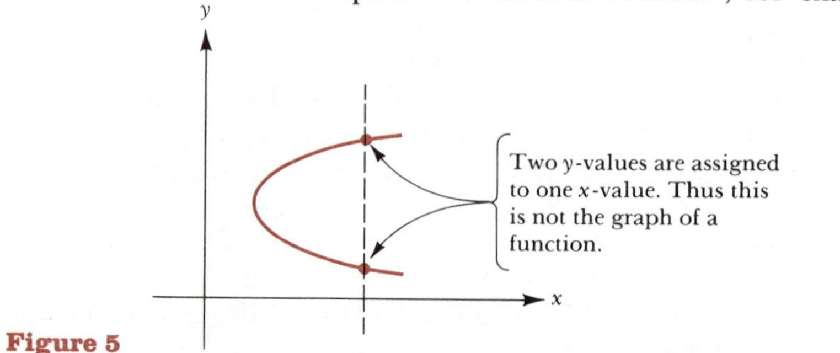

Figure 5

Figure 5 also shows a vertical line intersecting the graph in two distinct points. The specific coordinates of the two points are unimportant. What the vertical line helps us to see is that two different y-values have been assigned to the same x-value, and therefore the graph cannot be the graph of a function $y = f(x)$.

The preceding remarks can be summarized as follows.

> **Vertical Line Test**
>
> A graph in the x-y plane represents a function $y = f(x)$ provided that any vertical line intersects the graph in at most one point.

Example 4 The vertical line test implies that the graphs in Figures 6(a) and (c) represent functions, while those in Figures 6(b) and (d) do not.

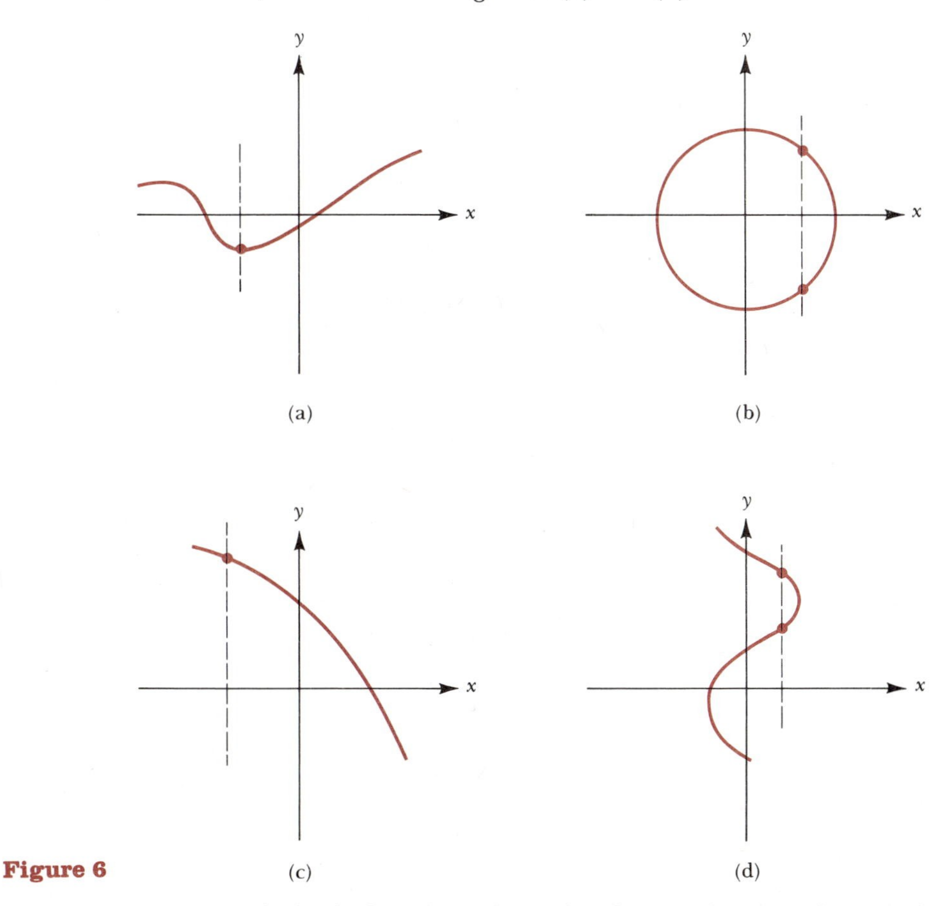

Figure 6

There are six basic functions that arise frequently enough to make it worth your while to memorize the basic shapes and features of their graphs. Figure 7 on the next page displays these graphs. Exercise 4 at the end of this section asks you to set up tables and verify for yourself that the graphs in Figure 7 are indeed correct. From now on (with the exception of Exercise 4), if you need to sketch a graph of one of these basic functions, you should do so from memory.

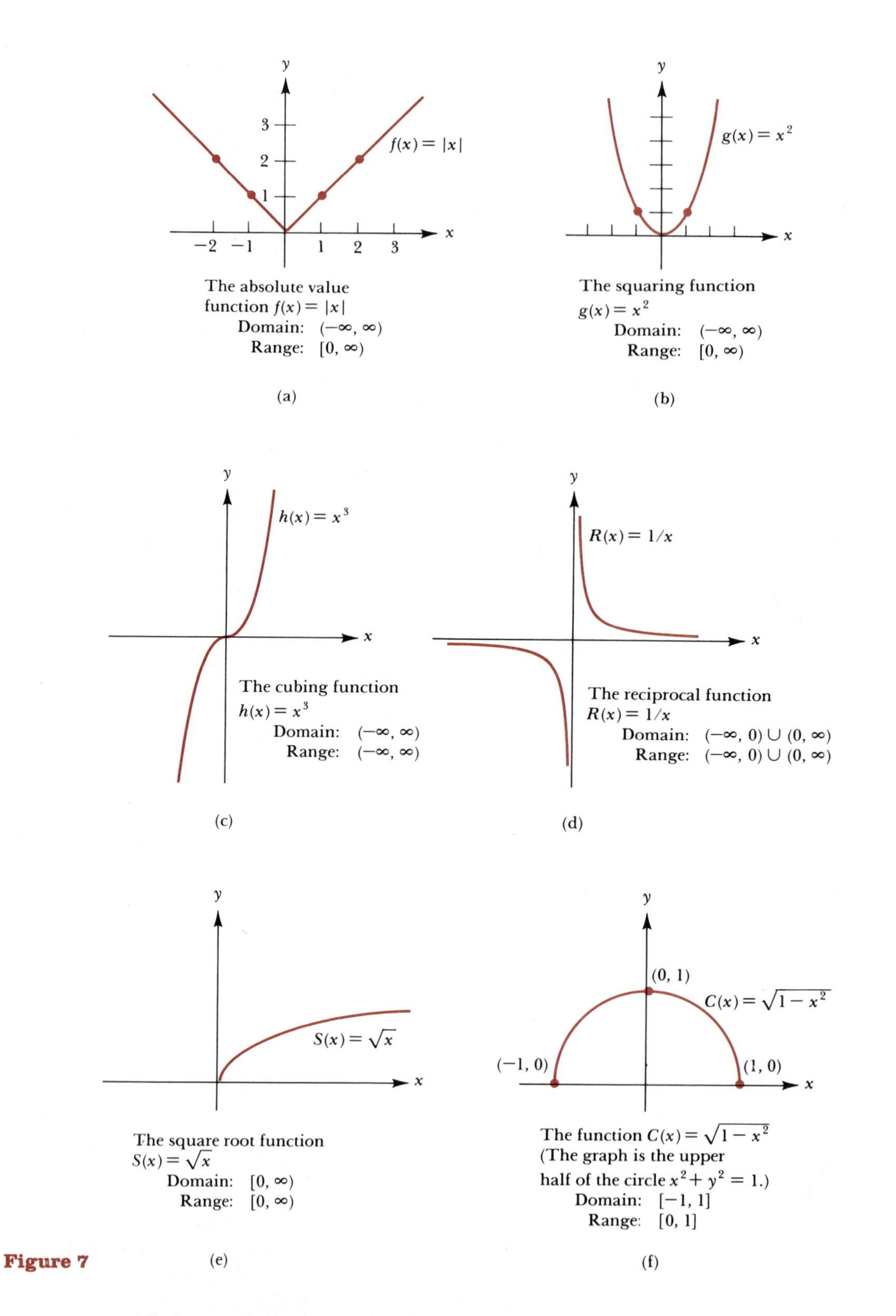

The absolute value function $f(x) = |x|$
Domain: $(-\infty, \infty)$
Range: $[0, \infty)$

(a)

The squaring function $g(x) = x^2$
Domain: $(-\infty, \infty)$
Range: $[0, \infty)$

(b)

The cubing function $h(x) = x^3$
Domain: $(-\infty, \infty)$
Range: $(-\infty, \infty)$

(c)

The reciprocal function $R(x) = 1/x$
Domain: $(-\infty, 0) \cup (0, \infty)$
Range: $(-\infty, 0) \cup (0, \infty)$

(d)

The square root function $S(x) = \sqrt{x}$
Domain: $[0, \infty)$
Range: $[0, \infty)$

(e)

The function $C(x) = \sqrt{1 - x^2}$ (The graph is the upper half of the circle $x^2 + y^2 = 1$.)
Domain: $[-1, 1]$
Range: $[0, 1]$

(f)

Figure 7

All the graphs in Figure 7 arise from functions defined by rather simple equations. In some instances, however, functions may be defined by combinations of equations. The next two examples display instances of this.

Example 5 A function g is defined by

$$g(x) = \begin{cases} x^2 & \text{if } x < 2 \\ \dfrac{1}{x} & \text{if } x \geq 2 \end{cases}$$

(a) Find $g(1)$, $g(2)$, and $g(3)$.
(b) Sketch the graph of g.

Solution **(a)** To find $g(1)$, do we substitute the value $x = 1$ in the expression x^2 or in the expression $1/x$? According to the instructions contained in the given definition of g, we should use the expression x^2 whenever the inputs are less than 2. Thus we have $g(1) = 1^2 = 1$; that is, $g(1) = 1$. On the other hand, the definition of g tells us to use the expression $1/x$ whenever the inputs are greater than or equal to 2. So in this case we have $g(2) = \frac{1}{2}$ and $g(3) = \frac{1}{3}$.

(b) For the graph of g, we look back at Figures 7(b) and (d) and we choose the appropriate portion of each. The result is displayed in Figure 8. The open circle in the figure is used to indicate that the point (2, 4) does not belong to the graph of g. The filled-in circle, on the other hand, is used to indicate that the point $(2, \frac{1}{2})$ does belong to the graph of g.

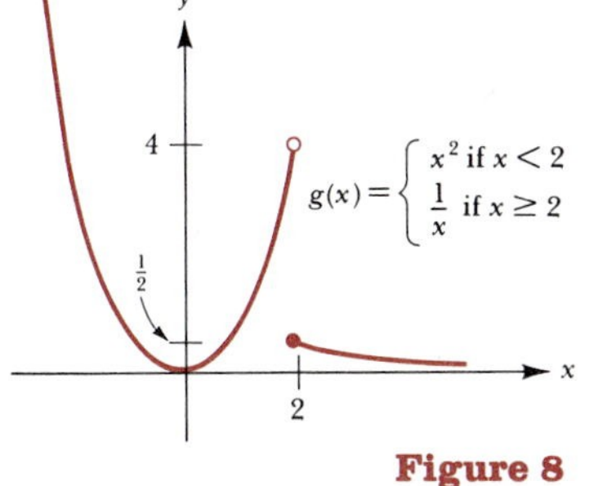

Figure 8

Example 6 Graph the function f defined by

$$f(x) = \begin{cases} x^2 - 1 & \text{if } x \leq 2 \\ 5 - x & \text{if } x > 2 \end{cases}$$

Solution In view of the definition of f, we'll set up two tables. For the first table, the inputs x will all be less than or equal to 2, and the outputs will be computed using the expression $x^2 - 1$. In the second table, the inputs x will be greater than 2, and the outputs will be computed from the expression $5 - x$. These tables and the graph derived from them are shown in Figure 9. We make

$y = x^2 - 1, x \leq 2$

x	−2	−1	0	1	2
y	3	0	−1	0	3

$y = 5 - x, x > 2$

x	3	4
y	2	1

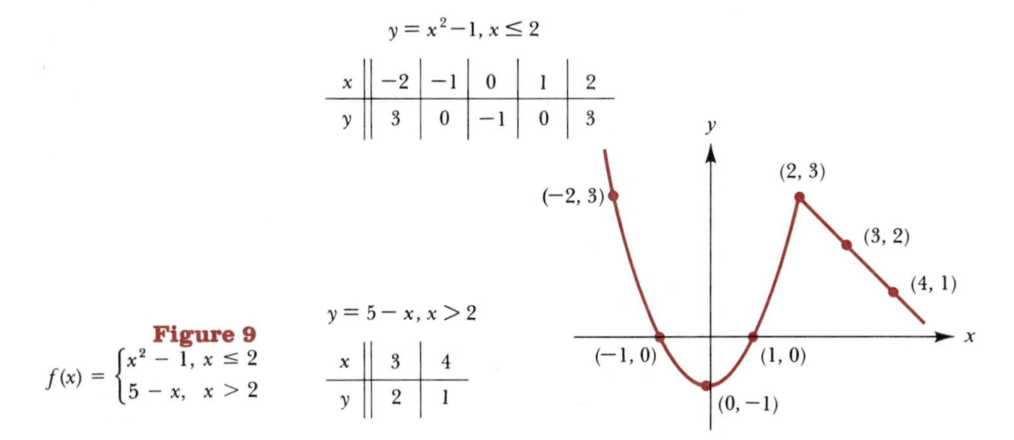

Figure 9

$$f(x) = \begin{cases} x^2 - 1, & x \leq 2 \\ 5 - x, & x > 2 \end{cases}$$

two observations here. First, notice that, in setting up a table for $y = 5 - x$ with $x > 2$, only two points are needed since the graph is a portion of a straight line. Second, notice that in this example, as opposed to the case in Example 5, the two portions of the graph together form a graph with no breaks or gaps in it. We say that the function f (whose graph appears in Figure 9) is *continuous* at $x = 2$ but that the function g (whose graph appears in Figure 8) is *discontinuous* at $x = 2$. A rigorous definition of continuity is properly a subject for calculus. However, even at the intuitive level at which we've presented the idea here, you'll find that the concept is useful in helping you to organize your thoughts about the graph of a function.

In Example 7 we are going to graph the *greatest integer function.* In order to define this function, we introduce the notation $[x]$ to denote the greatest (largest) integer that is less than or equal to x. For instance,

$[3] = 3$, since 3 is the greatest integer that is less than or equal to 3.

$[4\frac{1}{2}] = 4$, since 4 is the greatest integer that is less than or equal to $4\frac{1}{2}$.

$[-\frac{4}{3}] = -2$, since -2 is the largest integer that is less than or equal to $-\frac{4}{3}$.

The *greatest integer function* then is the function defined by the equation $y = [x]$. The domain of this function is the set of all real numbers.

Example 7 Graph the greatest integer function $y = [x]$.

Solution To help you to see what is happening, first consider the behavior of this function for values of x between 0 and 2. See Table 1.

Table 1

x	0	0.1	0.5	0.9	1.0	1.5	1.9	2.0
$y = [x]$	0	0	0	0	1	1	1	2

As indicated by Table 1, y is 0 throughout the interval $[0, 1)$. Then at $x = 1$, the value of y jumps up to 1. The value of y remains 1 throughout the interval $[1, 2)$. Then at $x = 2$, the value of y jumps up to 2. We can summarize this behavior as follows:

on the interval $[0, 1)$: $y = 0$

on the interval $[1, 2)$: $y = 1$

and, more generally,

on the interval $[p, p + 1)$: $y = p$ (for each integer p)

Now, using this information, we can sketch the graph of $y = [x]$ as shown in Figure 10. You should check the coordinates of a few points on the graph for yourself to be sure that you understand why the graph looks as it does. Although Figure 10 displays only that portion of the graph for which $-2 \leq x < 3$, it is understood that the pattern continues indefinitely in both directions. Finally, notice that the function is discontinuous whenever x is an integer.

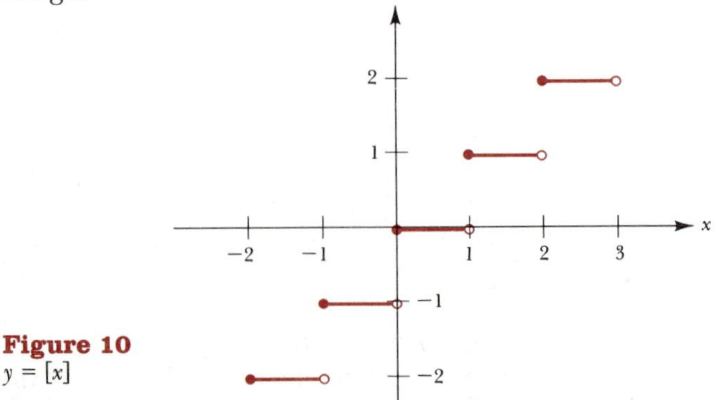

Figure 10
$y = [x]$

Exercise Set 4.3

1. The figure displays the graph of a function f.
   (a) Is $f(0)$ positive or negative?
   (b) Find $f(-2), f(1), f(2)$, and $f(3)$.
   (c) Which is larger, $f(2)$ or $f(4)$?
   (d) Find $f(4) - f(1)$.
   (e) Find $|f(4) - f(1)|$.
   (f) Write the domain and the range of f using the $[a, b]$ notation.

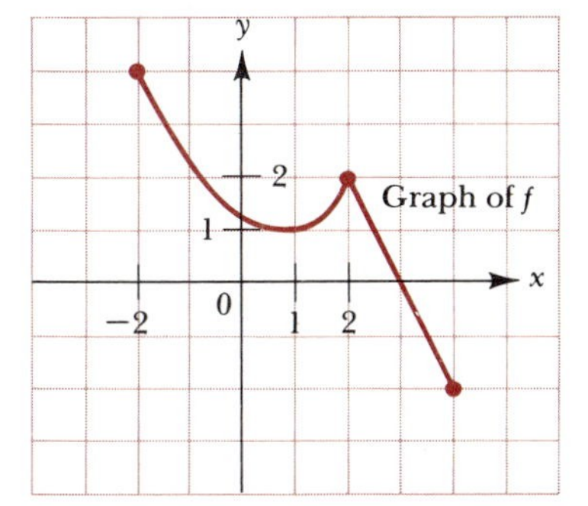

2. The figure below shows the graph of a function h.
   (a) Find $h(a)$, $h(b)$, $h(c)$, and $h(d)$.
   (b) Is $h(0)$ positive or negative?
   (c) For which values of x is $h(x) = 0$?
   (d) Which is larger, $h(b)$ or $h(0)$?
   (e) As x increases from c to d, do the corresponding values of $h(x)$ increase or decrease?
   (f) As x increases from a to b, do the corresponding values of $h(x)$ increase or decrease?

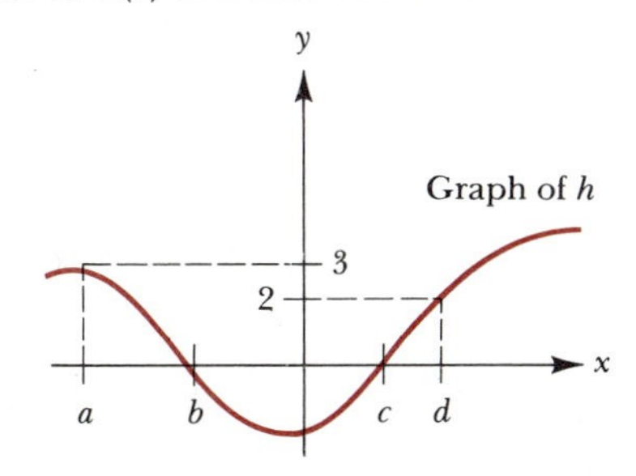

3. Let $f(x) = [x]$.
   (a) Compute $f(3), f(\pi), f(\frac{39}{10})$, and $f(-\frac{3}{2})$.
   (b) What is the range of the greatest integer function?

4. Set up tables and graph the given functions.
   (a) $f(x) = |x|$ (b) $g(x) = x^2$
   (c) $h(x) = x^3$ (d) $R(x) = 1/x$
   (e) $S(x) = \sqrt{x}$ (f) $C(x) = \sqrt{1 - x^2}$

5. Answer *true* or *false* for the following statements regarding the function f in the graph.
   (a) $f(1) = 1$ (b) $|f(1)| = 1$
   (c) $f(0) < 2$ (d) $f(1) + f(-1) = f(0)$
   (e) $|f(1) - f(-1)| = 5$

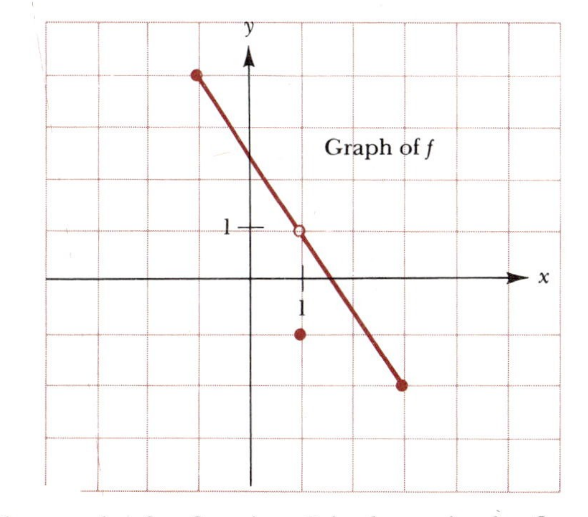

6. The graph of a function G is shown in the figure below. The domain of G is $[0, 9]$.
   (a) What is the range of g?
   (b) List the following numbers in order of increasing size: $G(0)$, $G(2)$, $G(3)$, $G(7)$, $G(9)$.
   (c) What are the coordinates of the highest point on the graph of G? The y-coordinate of this point is called the *maximum value* of G.
   (d) What are the coordinates of the lowest point on the graph of G? The y-coordinate of this point is called the *minimum value* of G.

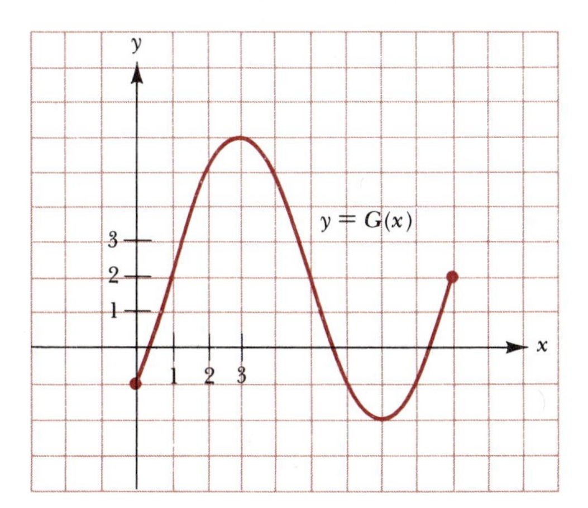

7. (First do Exercise 6.) Find the maximum and minimum values for each of the functions shown; assume the axes are marked off in one-unit intervals.

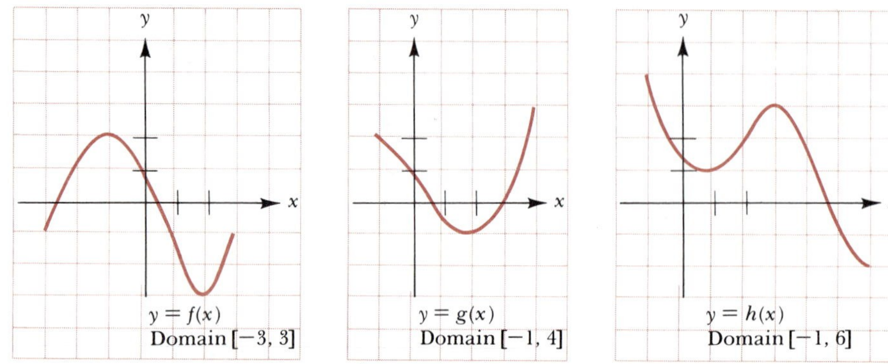

8. The graph of a function f is a straight line with an x-intercept of 4, as shown in the figure. Find $f(0)$ if the area of the triangle bounded by this line and the axes is 27 square units.

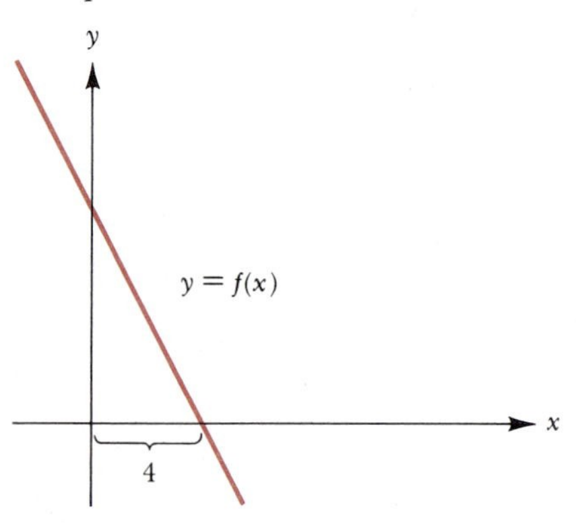

9. The graph of a function L is a straight line as shown in the figure. Find the x-intercept of this line given that $L(0) = 7$ and that the area of the triangle bounded by this line and the axes is 40 square units.

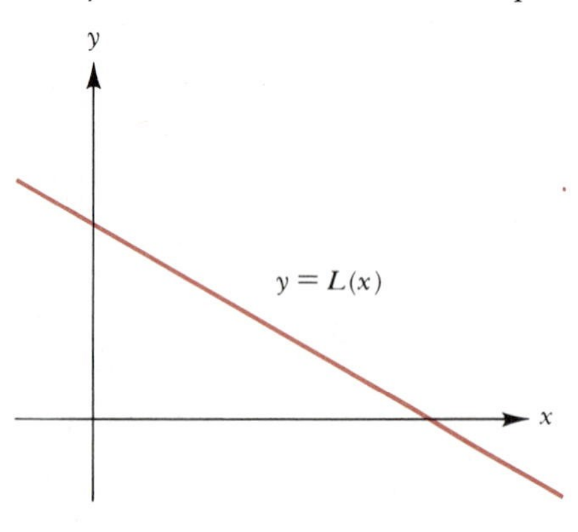

10. Use the vertical line test to determine which of the graphs in the next column and at the top of page 181 represent a function $y = f(x)$.

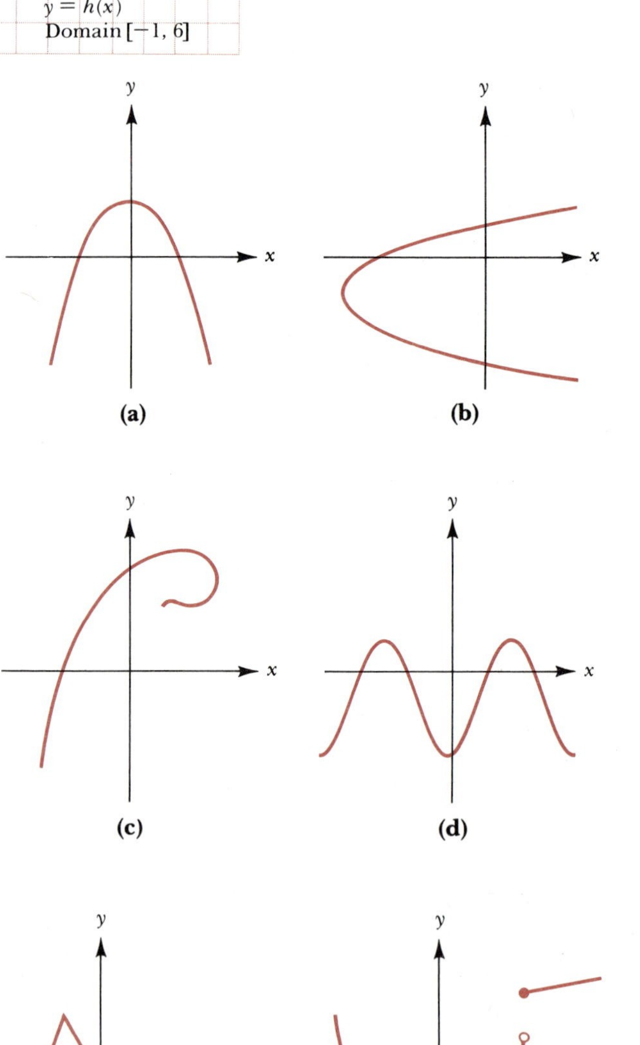

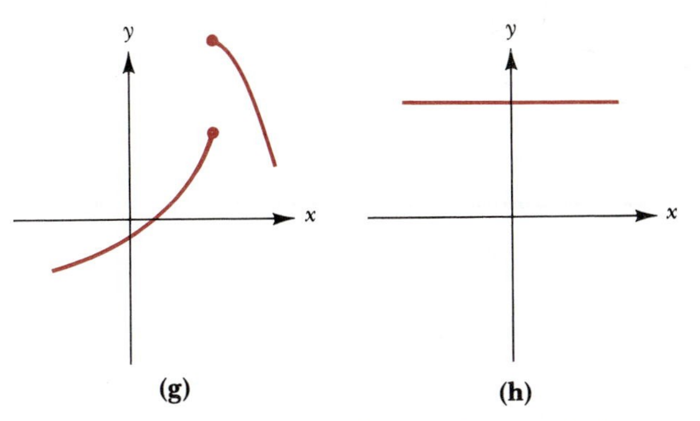

11. Let $f(x) = x^2$. Find the slope of the straight line that passes through the points $(3, f(3))$ and $(4, f(4))$. Include a sketch.

12. Let $G(x) = x^3$. Find the slope of the straight line that passes through the points $(0, G(0))$ and $(-2, G(-2))$. Include a sketch.

13. Let $T(x) = \sqrt{x}$. Which line has the larger slope, the line that passes through $(1, T(1))$ and $(4, T(4))$ or the line that passes through $(4, T(4))$ and $(9, T(9))$?

14. In each case, sketch a graph of the function on the given domain.
(a) $p(x) = x^2$ with domain $[-1, 2]$
(b) $q(x) = x^3$ with domain $[0, 1]$
(c) $r(x) = 1/x$ with domain $(0, \infty)$
(d) $s(x) = \sqrt{x}$ with domain $[1, 4]$

15. In each case, sketch a graph of the function on the given domain.
(a) $k(x) = \sqrt{1 - x^2}$ with domain $[0, 1]$
(b) $m(x) = \sqrt{1 - x^2}$ with domain $(0, 1)$
(c) $n(x) = |x|$ with domain consisting just of the numbers 0, 1, 2, 3, and 4
(d) $z(x) = x^2$ with domain consisting of the numbers $-2, -1, 0, 1$, and 4 as well as those numbers in the open interval $(2, 3)$

16. What are the domain and range of the function whose graph is the horizontal line $y = 3$?

In Exercises 17–30, graph the functions defined by the given rules.

17. $f(x) = \begin{cases} |x| & \text{if } x \le 0 \\ x^2 & \text{if } x > 0 \end{cases}$

18. **(a)** $g(x) = \begin{cases} |x| & \text{if } x \le 0 \\ x + 1 & \text{if } x > 0 \end{cases}$

(b) $F(x) = \begin{cases} |x| & \text{if } x < 0 \\ x + 1 & \text{if } x \ge 0 \end{cases}$

19. $A(x) = \begin{cases} x^3 & \text{if } -2 \le x \le -1 \\ x^2 & \text{if } x > -1 \end{cases}$

20. $B(x) = \begin{cases} \sqrt{1 - x^2} & \text{if } -1 \le x < 1 \\ 1/x & \text{if } x \ge 1 \end{cases}$

21. $C(x) = \begin{cases} x^3 & \text{if } x < 1 \\ \sqrt{x} & \text{if } x > 1 \end{cases}$

22. **(a)** $f(x) = \begin{cases} x^2/|x| & \text{if } x \ne 0 \\ 0 & \text{if } x = 0 \end{cases}$

(b) $F(x) = \begin{cases} x^2/|x| & \text{if } x \ne 0 \\ 1 & \text{if } x = 0 \end{cases}$

23. **(a)** $g(x) = \begin{cases} x/|x| & \text{if } x \ne 0 \\ 0 & \text{if } x = 0 \end{cases}$

(b) $G(x) = \begin{cases} x/|x| & \text{if } x \ne 0 \\ 1 & \text{if } x = 0 \end{cases}$

24. $U(x) = \begin{cases} 1 & \text{if } x \le -2 \\ -1 & \text{if } x > -2 \end{cases}$

25. $V(x) = \begin{cases} x & \text{if } x \le -2 \\ 1 & \text{if } x > -2 \end{cases}$

26. **(a)** $W(x) = \begin{cases} x & \text{if } x \le -2 \\ -2 & \text{if } x > -2 \end{cases}$

(b) $D(x) = \begin{cases} x & \text{if } x < -2 \\ -2 & \text{if } x > -2 \end{cases}$

27. $f(x) = \begin{cases} 1/x & \text{if } x < -1 \\ x & \text{if } -1 \le x \le 1 \\ 1/x & \text{if } x > 1 \end{cases}$

28. $g(x) = \begin{cases} 1/x & \text{if } x < -\frac{1}{2} \\ 1 & -\frac{1}{2} \le x \le 1 \\ x^3 & \text{if } x > 1 \end{cases}$

29. $y = \begin{cases} x & \text{if } 0 \le x < 1 \\ x - 1 & \text{if } 1 \le x < 2 \\ x - 2 & \text{if } 2 \le x < 3 \end{cases}$

30. $T(x) = \begin{cases} 2x + 2 & \text{if } -1 \le x < 0 \\ -2x + 2 & \text{if } 0 \le x < 1 \\ 2x - 2 & \text{if } 1 \le x < 2 \\ -2x + 6 & \text{if } 2 \le x \le 3 \end{cases}$

31. **(a)** Sketch the graph of $f(x) = \dfrac{x^2 - 25}{x + 5}$. What is the domain of f?
(b) Graph the function F defined by the rule

$$F(x) = \begin{cases} \dfrac{x^2 - 25}{x + 5} & \text{if } x \ne -5 \\ -10 & \text{if } x = -5 \end{cases}$$

32. In the figure, find the coordinates of the point T if the slope of the straight line passing through the two points P and T is as follows:
(a) 5 *Hint:* Let the coordinates of T be (x, x^2).
(b) 1000 **(c)** k

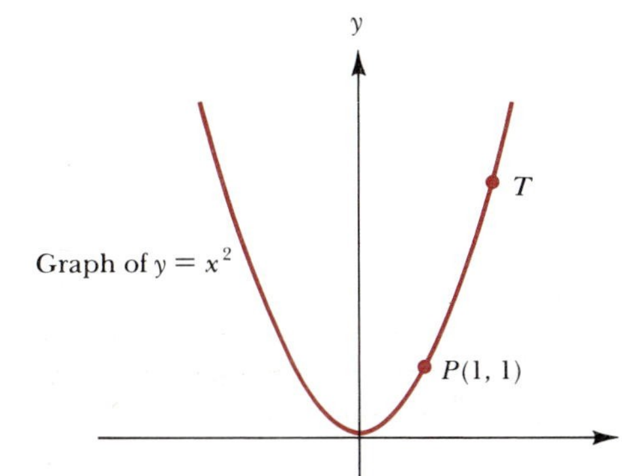

33. Let $f(x) = 1/x$. Find a number t such that the slope of the straight line passing through the two points (1, 1) and $(t, f(t))$ on the graph of f is $-\frac{1}{5}$. Include a sketch.

***34.** Find the coordinates of a point P on the graph of $f(x) = \sqrt{x}$ if the slope of the straight line passing through P and the point (1, 1) is $\frac{1}{7}$. Include a sketch.

35. (As background for this exercise, you should first look over Exercise 57 in Section 3.2.)

(a) Suppose that the thickness of the line drawn by a pencil is 0.01 cm and that the common unit of length marked off along both axes is 1 cm. Where is the graph of $y = x^2$ indistinguishable from the x-axis? That is, for which x-values will the corresponding y-values be less than 0.01?

(b) Same as part (a) but suppose you're using a felt-tip pen and the lines you draw are 0.1 cm wide.

(c) Same as part (a) but for the curve $y = x^4$ instead of $y = x^2$. (Make use of a table of square roots or a calculator.)

36. What are the domain and range of the greatest integer function $y = [x]$?

***37.** Graph the following functions.
(a) $p(x) = [2x]$ **(b)** $q(x) = [x/2]$

***38.** Graph each of the following functions.
(a) $r(x) = x - [x]$ **(b)** $R(x) = [x] - x$

***39.** Graph each of the following functions.
(a) $f(x) = \{[x]\}^2$ **(b)** $F(x) = [x^2]$

40. Let $f(x) = 2x + 1$.

(a) Graph the functions g and h that are defined as

$$g(x) = \frac{f(x) - f(-x)}{2} \qquad h(x) = \frac{f(x) + f(-x)}{2}$$

(b) Verify that the equation $f(x) = g(x) + h(x)$ holds for all values of x.

41. The graph of a function f is a straight line. As the figure shows, A and B are two points on the line and the x-coordinates of A and B are h units apart. The coordinates of A are $(x, f(x))$. What are the coordinates of B? Show that the slope of the line is $\dfrac{f(x + h) - f(x)}{h}$.

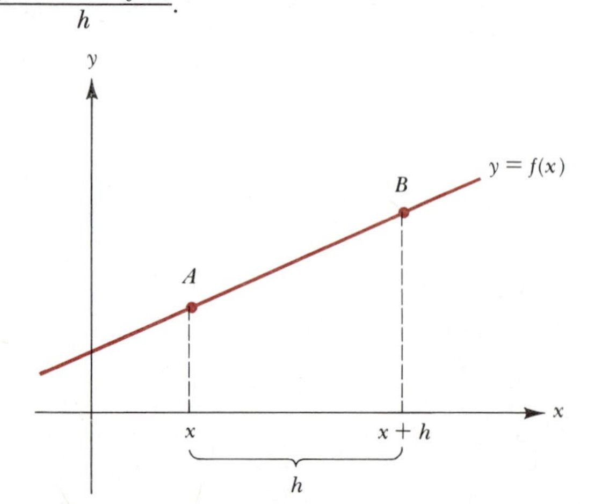

42. Let $F(x) = 2x + 3$. Using the values $x = 1$ and $h = \frac{1}{2}$, compute the value of $\dfrac{F(x + h) - F(x)}{h}$. Include a sketch of the type shown in Exercise 41.

43. The *average rate of change* of a function f between $x = a$ and $x = b$ is defined to be the quantity $\dfrac{f(b) - f(a)}{b - a}$.

(a) By means of a sketch, indicate why the average rate of change represents the slope of the line joining the two points on the graph of f whose x-coordinates are a and b, respectively.

(b) Which is larger, the average rate of change of $f(x) = x^2$ between 0 and 1 or between 10 and 11?

(c) Which function has the largest average rate of change between 3 and 4?

$$f(x) = x^2 \qquad g(x) = x^3 \qquad h(x) = \frac{1}{x}$$

(d) Find the average rate of change for the function $f(x) = x^2$ over the following intervals:
(i) $a = 1$ to $b = 1.1$
(ii) $a = 1$ to $b = 1.01$
(iii) $a = 1$ to $b = 1.001$
Looking at the results of your calculations, as b gets closer and closer to 1, what value does your average rate of change seem to be approaching? This "target value" is sometimes called the *instantaneous rate of change.*

44. Suppose that during the first 4 hr of a laboratory experiment the temperature of a certain substance is given by the formula $f(t) = t^3 - 6t^2 + 9t$, where t is measured in hours, with $t = 0$ corresponding to the time the experiment begins, and where $f(t)$ is the temperature (Fahrenheit) of the substance at time t. Calculate the average rate of change of temperature between the following times:
(a) $t = 0$ and $t = 1$
(b) $t = 1$ and $t = 2$
(c) $t = 0$ and $t = 3$
(d) $t = 0$ and $t = 4$
What are the units associated with your answers?

45. Assume that the distance covered by a falling object is given by the formula

$$f(t) = 16t^2$$

where t is measured in seconds, $f(t)$ is in feet, and $t = 0$ corresponds to the instant the object first begins to fall. Thus after 1 sec the object has fallen $16(1)^2$ or 16 ft; after 2 sec the object has fallen a total of $16(2)^2$ or 64 ft; and so on.

(a) Calculate $f(2) - f(1)$, the distance the object has fallen during the time interval from $t = 1$ to $t = 2$ (i.e., during the second second). Include a sketch of the graph of f. On the t-axis, indicate the time interval from $t = 1$ to $t = 2$; on the y- or $f(t)$-axis, indicate the interval corresponding to the distance $f(2) - f(1)$.

(b) Calculate $f(3) - f(2)$, the distance the object has fallen during the time interval from $t = 2$ to $t = 3$ (i.e., during the third second). Include a sketch as you did in part (a).

(c) The *average velocity* during the time interval from $t = a$ to $t = b$ is defined to be the quantity $\frac{f(b) - f(a)}{b - a}$. Note that this is just distance divided by elapsed time and that the units are feet per second. Calculate the average velocity from $t = 1$ to $t = 3$. On a graph of f, locate the two points $(1, f(1))$ and $(3, f(3))$. What is the slope of the line joining these two points?

(d) Calculate the average velocity over the given interval:

(i) $a = 1$ to $b = 1.1$
(ii) $a = 1$ to $b = 1.001$
(iii) $a = 1$ to $b = 1.00001$

Looking at the results of your calculations, as b comes closer and closer to 1, what value does your average velocity seem to be approaching? This "target value" is called the *instantaneous velocity* (or just *velocity*) at $t = 1$.

4.4 Techniques in Graphing

. . . geometrical figures are graphic formulas.

David Hilbert (1862–1943)

In this section we look at some techniques that allow us to sketch certain graphs without setting up tables. First, simply as a matter of motivation, we point out that setting up tables to do graphing has its limitations. For example, if you begin to set up a table to graph the function defined by

$$f(x) = x^4 - 6x^3 + 11x^2 - 6x$$

x	$f(x)$
0	0
1	0
2	0
3	0

Table 1
$f(x) = x^4 - 6x^3 + 11x^2 - 6x$

you obtain the results recorded in Table 1. At this point in our study of functions, the results in Table 1 probably raise more questions about the graph than they answer. So, for reasons such as this and others, we develop throughout the text various methods for graphing that allow us to bypass or at least supplement the use of tables. (Incidentally, you'll be able to graph the function f, defined above, after studying Section 10.4.)

Given the graph of a function f, it is useful to know how to sketch the graphs of the following closely related functions:

$y = f(x) + c$ (c denotes a constant throughout this discussion.)
$y = f(x + c)$
$y = -f(x)$
$y = f(-x)$

We shall approach this by looking at examples.

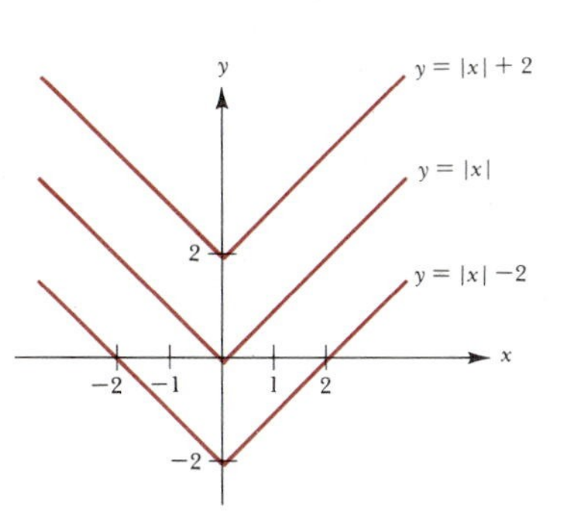

Figure 1

Figure 1 displays portions of the graphs of $y = |x|$, $y = |x| + 2$, and $y = |x| - 2$. All three graphs have identical shapes. We can view the graph of $y = |x| + 2$ as that obtained by moving the graph of $y = |x|$ up two units in the y-direction. Such a movement is called a *translation*. Similarly we can view the graph of $y = |x| - 2$ as that obtained by moving or translating the graph of $y = |x|$ down two units in the y-direction. We summarize these observations as follows.

> **How to Graph $y = f(x) + c$**
>
> If $c > 0$, move the graph of $y = f(x)$ a distance of c units in the positive y-direction.
>
> If $c < 0$, move the graph of $y = f(x)$ a distance of $|c|$ units in the negative y-direction.

Example 1 Graph $y = x^2 - 4$.

Solution Begin with the graph of $y = x^2$ and move it four units in the negative y-direction. See Figure 2.

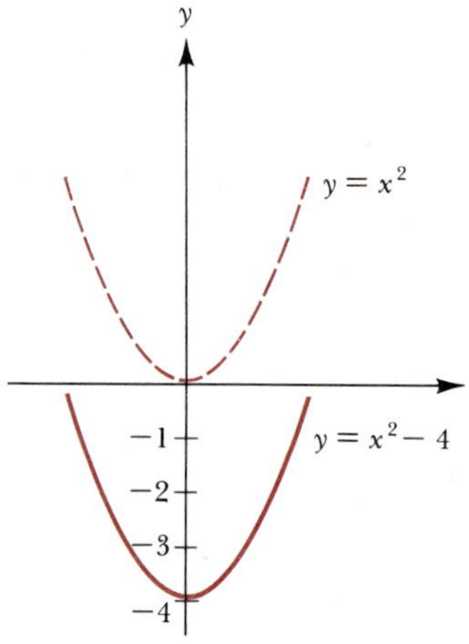

Figure 2

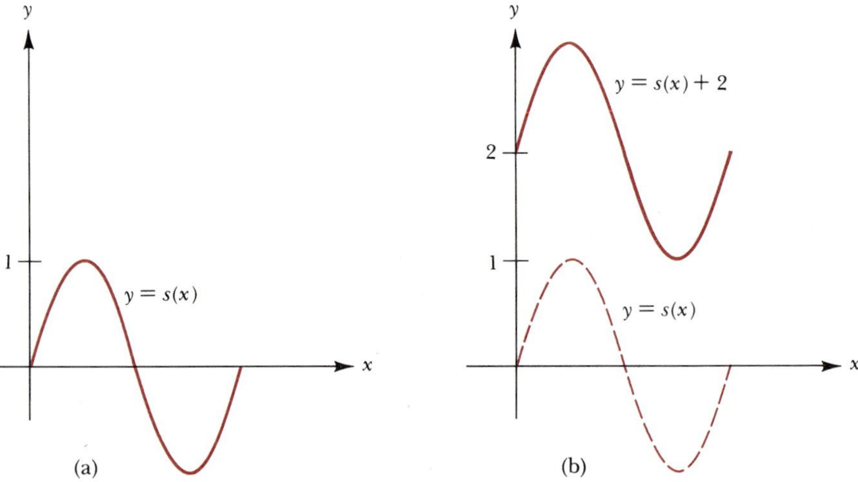

Figure 3

Example 2 Figure 3(a) shows the graph of a function s. Sketch the graph of $y = s(x) + 2$.

Solution To graph $y = s(x) + 2$, move the graph of s two units in the positive y-direction. See Figure 3(b).

Next we look at examples that indicate the relationship between the graphs of $y = f(x)$ and $y = f(x + c)$. Figure 4 displays portions of the graphs of $y = |x|$, $y = |x - 5|$, and $y = |x + 5|$. Again all three graphs have identical shapes, but this time the movement is in the x-direction. Furthermore, note that the graph of $y = |x - 5|$ is five units to the right, not left, of $y = |x|$. To help see why this is so, consider the output $y = 0$. To produce the output $y = 0$ using the function $y = |x|$, you clearly need an input of $x = 0$. But to produce that same output of $y = 0$ using $y = |x - 5|$, you need an input of $x = 5$, which is five units to the right of $x = 0$.

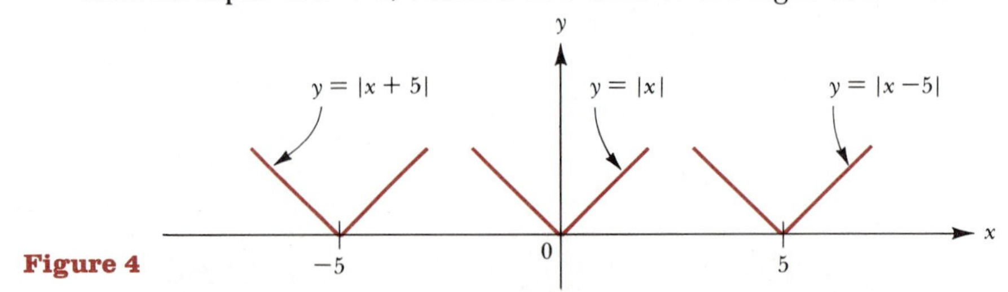

Figure 4

> **How to Graph $y = f(x + c)$**
>
> If $c > 0$, move the graph of $y = f(x)$ a distance of c units in the negative x-direction.
>
> If $c < 0$, move the graph of $y = f(x)$ a distance of $|c|$ units in the positive x-direction.

Example 3 Sketch the graph of $y = \dfrac{1}{x - 1}$.

Solution The graph of $y = 1/x$ is shown in Figure 7(d) on page 176. Moving this graph one unit to the right gives us the graph of $y = \dfrac{1}{x - 1}$ displayed in Figure 5.

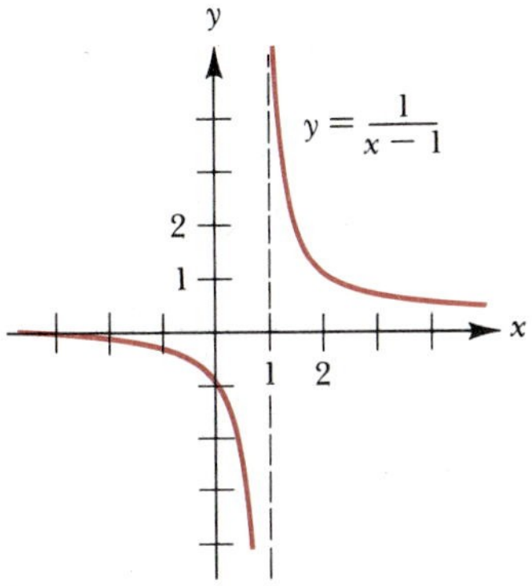

Figure 5

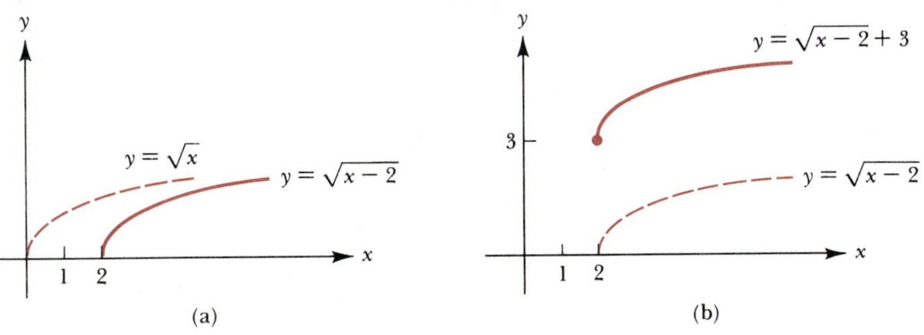

Figure 6

Example 4 Sketch the graphs of the following:

(a) $y = \sqrt{x - 2}$ **(b)** $y = \sqrt{x - 2} + 3$

Solution

(a) To graph $y = \sqrt{x - 2}$, move the graph of $y = \sqrt{x}$ two units to the right. See Figure 6(a).

(b) To graph $y = \sqrt{x - 2} + 3$, move the graph of $y = \sqrt{x - 2}$ up three units. See Figure 6(b).

Example 5 Graph $y = (x - 2)^2 + 1$.

Solution Begin with the graph of $y = x^2$ in Figure 7(a). Move the curve two units to the right to obtain the graph of $y = (x - 2)^2$; see Figure 7(b). Then move the curve in Figure 7(b) up one unit to obtain the graph of $y = (x - 2)^2 + 1$; see Figure 7(c).

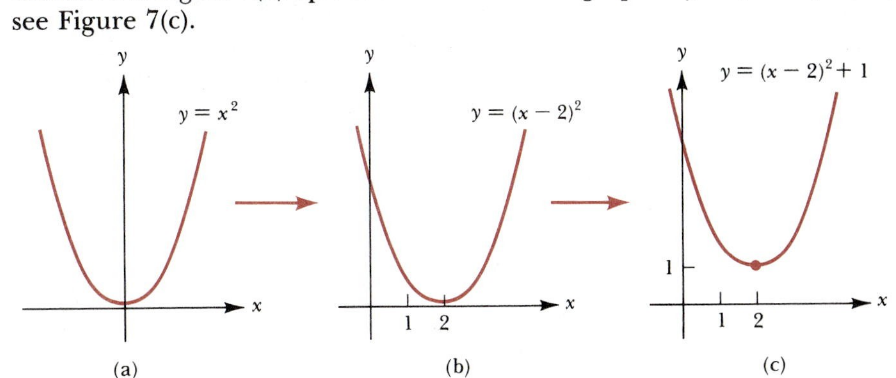

Figure 7

To compare the graphs of $y = f(x)$ and $y = f(-x)$, we return to the concept of symmetry introduced in Section 3.2. In Figure 8, we say that the points $(3, 4)$ and $(-3, 4)$ are *reflections* of each other in the y-axis. This is also described by saying that the points $(3, 4)$ and $(-3, 4)$ are mirror images of each other, the y-axis serving as the mirror or *axis of symmetry*. In general, the *reflection* of any point (x, y) in the y-axis is the point $(-x, y)$.

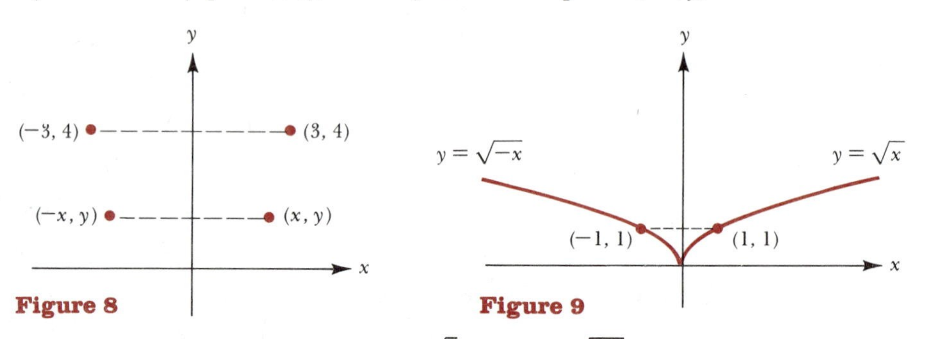

Figure 8 **Figure 9**

Now consider the graphs of $y = \sqrt{x}$ and $y = \sqrt{-x}$ in Figure 9. Notice that each point on the graph of $y = \sqrt{-x}$ is the reflection in the y-axis of a corresponding point on the graph of $y = \sqrt{x}$. For example, the point $(-1, 1)$ on $y = \sqrt{-x}$ is the reflection of $(1, 1)$ on $y = \sqrt{x}$. We say that the two graphs are reflections of one another in the y-axis. Another perspective on this is as follows. You know that x measures distances to the right and left of the y-axis. So if you replace x by $-x$ in an equation, you reverse the roles of left and right. Thus, where the graph of $y = \sqrt{x}$ begins at $(0, 0)$ and rises to the right, you expect the graph of $y = \sqrt{-x}$ to begin at $(0, 0)$ and rise to the left. We can summarize these observations as follows:

How to Graph $y = f(-x)$

Reflect (each point of) the graph of $y = f(x)$ in the y-axis to obtain the graph of $y = f(-x)$.

Example 6

Figure 10 shows the graph of a function f. Sketch a graph of $y = f(-x)$.

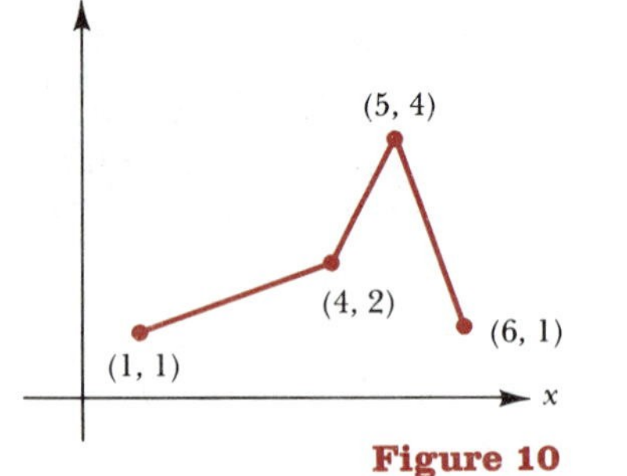

Figure 10

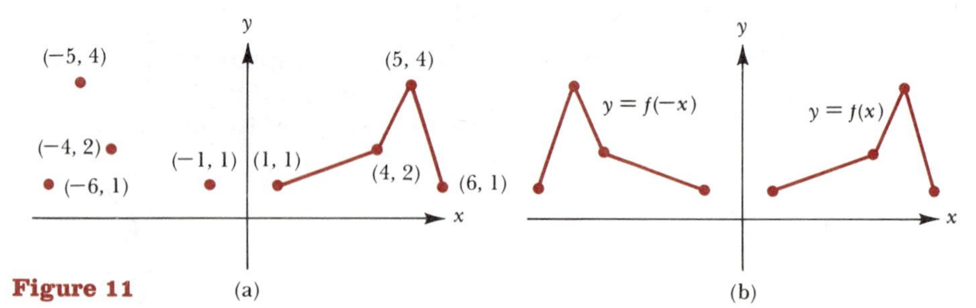

Figure 11 (a) (b)

Solution The graph of $y = f(-x)$ is obtained by reflecting the graph of $y = f(x)$ in the y-axis. It may be helpful to you to begin by reflecting the individual points whose coordinates are given in Figure 10. This gives the points indicated in Figure 11(a). The graph of $y = f(-x)$ is then drawn as in Figure 11(b).

Example 7 Figure 12 shows the graph of a function g. Sketch the following:
(a) $y = g(-x)$ **(b)** $y = g(-x) + 2$

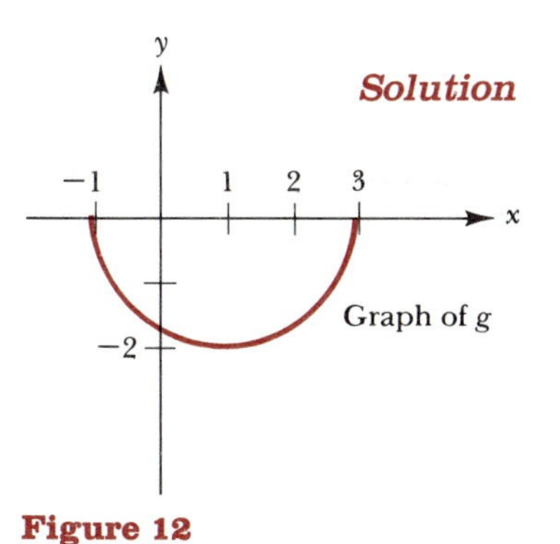

Figure 12

Solution

(a) Reflect the graph of g in the y-axis to obtain the graph of $y = g(-x)$. See Figure 13(a).

(b) To graph $y = g(-x) + 2$, move the graph obtained in part (a) up two units. See Figure 13(b).

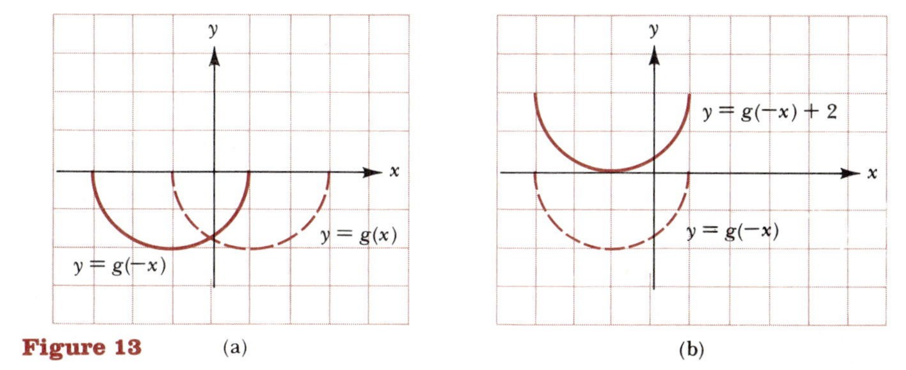

Figure 13 (a) (b)

In just the same way that the graphs of $y = \sqrt{x}$ and $y = \sqrt{-x}$ are reflections of each other in the y-axis, the graphs of $y = \sqrt{x}$ and $y = -\sqrt{x}$ are reflections of one another in the x-axis; see Figures 14(a) and (b). For each point (x, y) on the graph of $y = \sqrt{x}$ in Figure 14(b), there is a corresponding point $(x, -y)$ on the graph of $y = -\sqrt{x}$. For instance, (1, 1) is on the graph of $y = \sqrt{x}$ and $(1, -1)$ is on the graph of $y = -\sqrt{x}$. We can summarize these observations as follows:

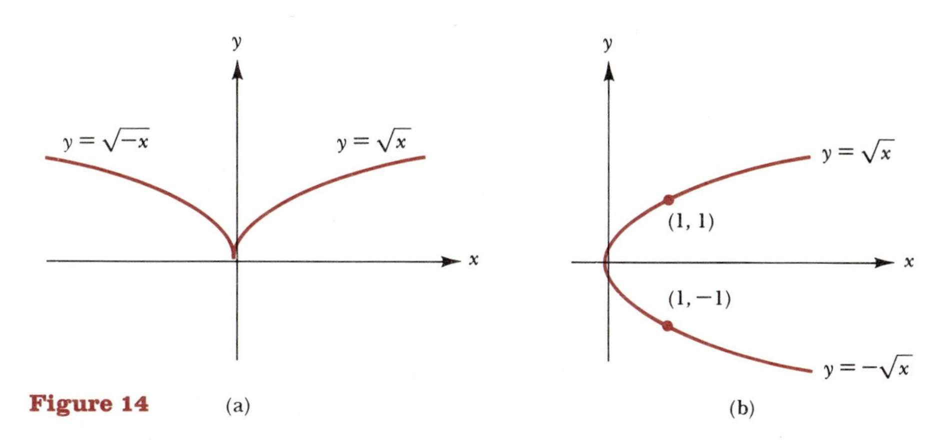

Figure 14 (a) (b)

> **How to Graph $y = -f(x)$**
>
> Reflect the graph of $y = f(x)$ in the x-axis.

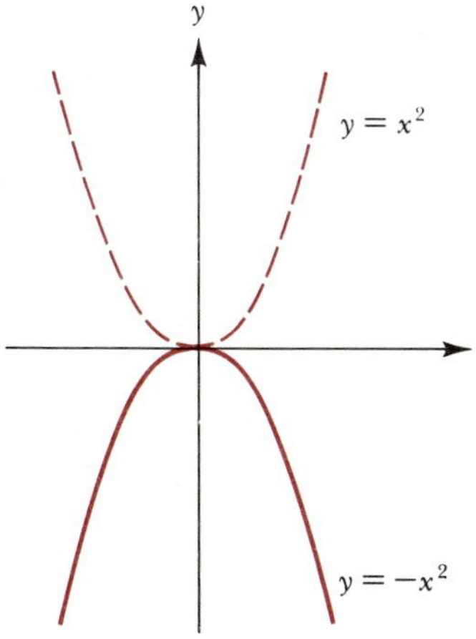

Figure 15

Example 8 Graph $y = -x^2$.

Solution See Figure 15. The dotted lines show the graph of $y = x^2$. By reflecting this graph in the x-axis, we obtain the graph of $y = -x^2$.

Example 9 Graph $y = -x^2 + 3$.

Solution First obtain the graph of $y = -x^2$ as in Example 8. Then move this graph up three units. See Figure 16.

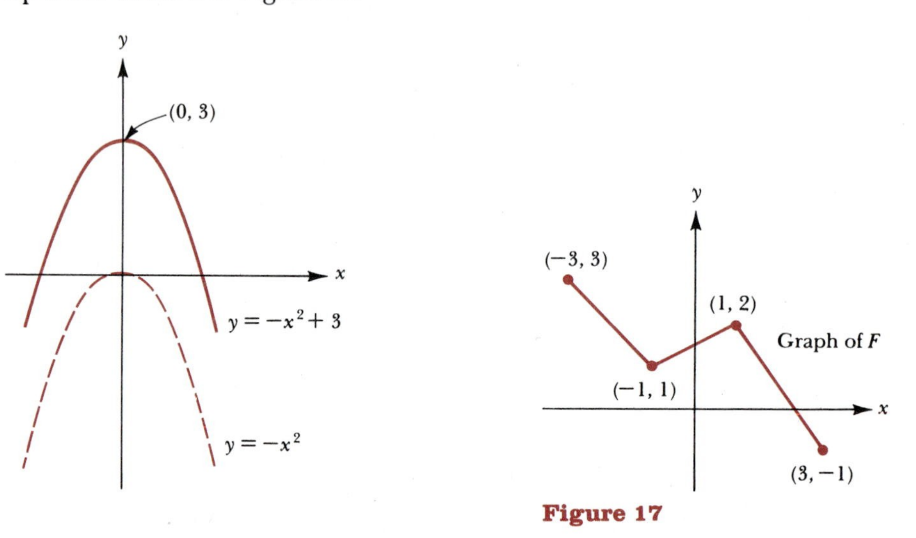

Figure 16

Figure 17

Example 10 The graph of a function F is shown in Figure 17. Sketch the graph of $y = -F(x)$.

Solution The graph of $y = -F(x)$ is obtained by reflecting the graph of $y = F(x)$ in the x-axis. Begin by reflecting the points whose coordinates are given. See Figure 18(a). After these points are obtained, the rest of the graph of $y = -F(x)$ can be drawn in as in Figure 18(b).

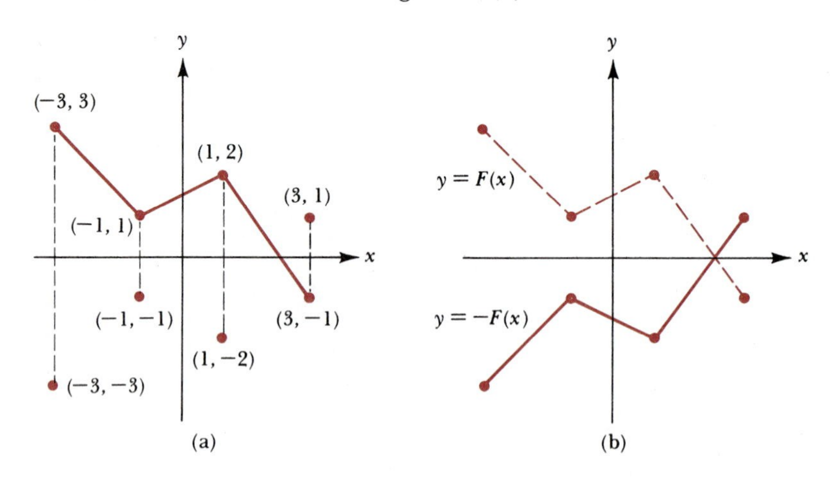

Figure 18

Example 11 Graph $y = -|x - 2| + 3$.

Solution Begin with the graph of $y = |x|$. Move the graph two units to the right to obtain the graph of $y = |x - 2|$. See Figure 19(a). Reflect this graph in the x-axis to obtain $y = -|x - 2|$. See Figure 19(b). And, finally, move this last graph up three units, as in Figure 19(c).

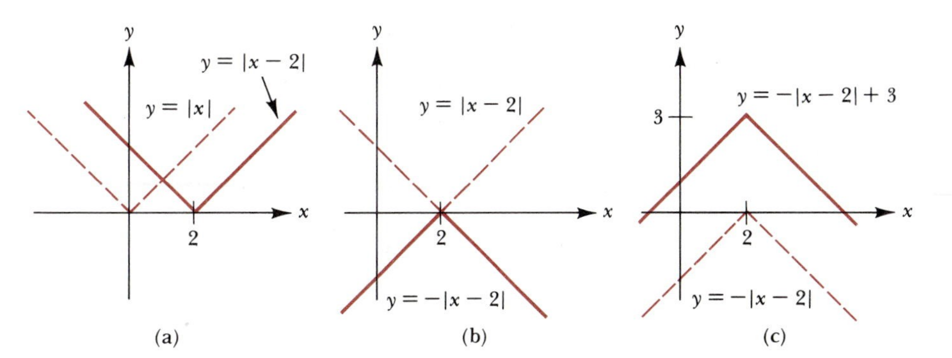

Figure 19

Example 12 Figure 20(a) shows the graph of $y = G(x)$. What equations describe the graphs in Figures 20(b), (c), and (d)?

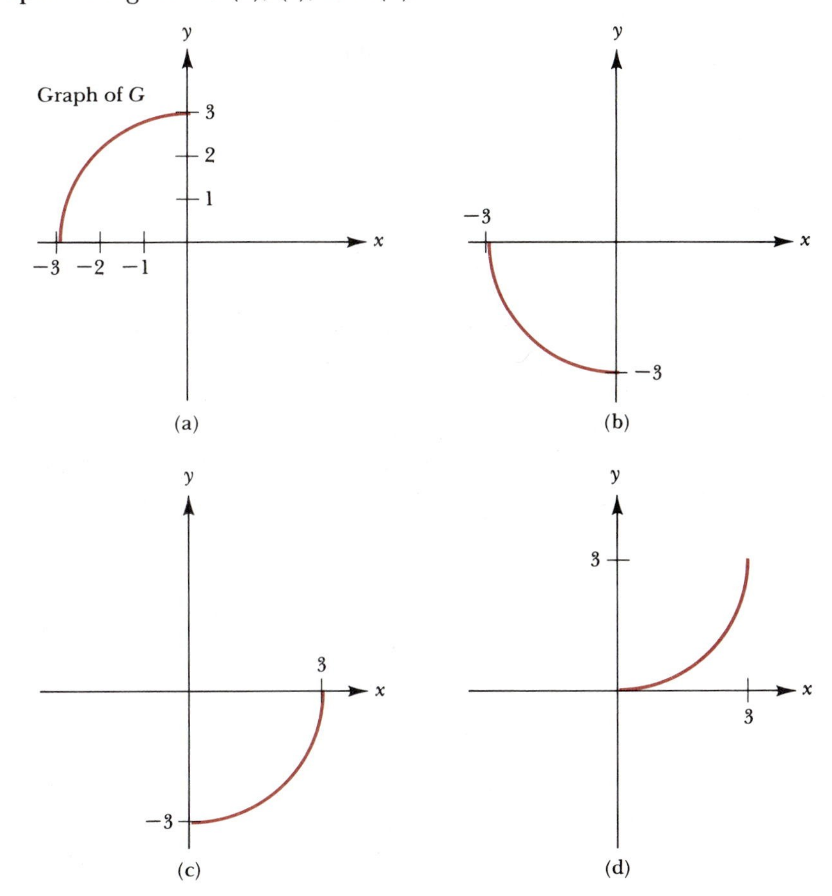

Figure 20

Solution

Graph in Given Figure	How Obtained	Equation
20(b)	Reflection in the x-axis of the graph of G	$y = -G(x)$
20(c)	Reflection in the y-axis of the graph of $y = -G(x)$	$y = -G(-x)$
20(d)	Translation of the graph of $y = -G(-x)$ by three units in the positive y-direction	$y = -G(-x) + 3$ or $y = 3 - G(-x)$

Exercise Set 4.4

For Exercises 1–20, sketch the graph of each function.

1. $y = x^2 - 3$ **2.** $y = x^2 + 3$

3. $y = (x + 4)^2$ **4.** $y = (x + 4)^2 - 3$

5. $y = (x - 4)^2$ **6.** $y = (x - 4)^2 + 1$

7. $y = -x^2$ **8.** $y = -x^2 - 3$

9. $y = -(x - 3)^2$ **10.** $y = -(x - 3)^2 - 3$

11. $y = \sqrt{x - 3}$ **12.** $y = \sqrt{x - 3} + 1$

13. $y = -\sqrt{x - 3}$ **14.** $y = \sqrt{x + 1}$

15. $y = -\sqrt{x + 1}$ **16.** $y = -\sqrt{x + 1} + 1$

17. $y = (x - 2)^3$ **18.** $y = (x - 2)^3 + 1$

19. $y = -x^3 + 4$ **20.** $y = -(x - 1)^3 + 4$

21. Let $f(x) = |x|$. Sketch the graphs of the following functions.
(a) $y = f(x - 5)$ (b) $y = -f(x - 5)$
(c) $y = 1 - f(x - 5)$

22. Let $F(x) = 1/x$. Sketch the graphs of the following functions.
(a) $y = F(x + 3)$ (b) $y = F(x) + 3$
(c) $y = -F(x + 3)$ (d) $y = F(-x) + 3$

23. Let $g(x) = \sqrt{1 - x^2}$. Graph the following functions.
(a) $y = g(x - 2)$ (b) $y = -g(x - 2)$
(c) $y = 1 - g(x - 2)$ (d) $y = g(-x)$

24. The figure shows the graph of the function $y = 10^x$. Sketch the graphs of the following functions.
(a) $y = 10^{-x}$ (b) $y = -10^x$
(c) $y = -10^{-x}$ (d) $y = 10^{x-1}$
(e) $y = 10^x + 1$ (f) $y = -10^{x-1} - 1$

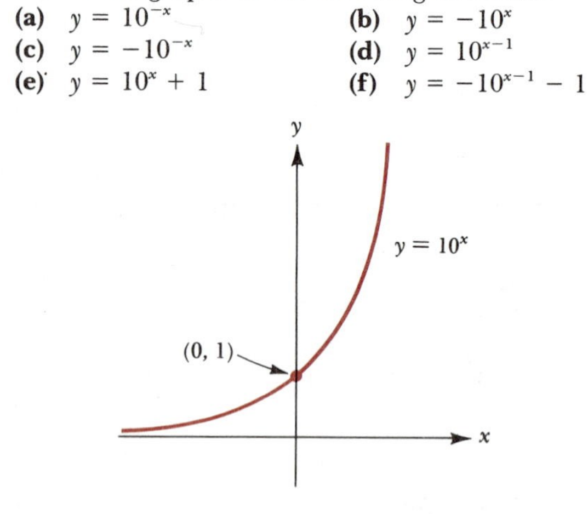

For Exercises 25 and 26, refer to the graph of the function f shown in the figure at the top of the next column.

25. Graph the following functions.
(a) $y = -f(x)$ (b) $y = f(-x)$
(c) $y = -f(-x)$

26. Graph the following functions.
(a) $y = f(x) - 2$ (b) $y = f(x - 2)$
(c) $y = f(x - 2) - 2$

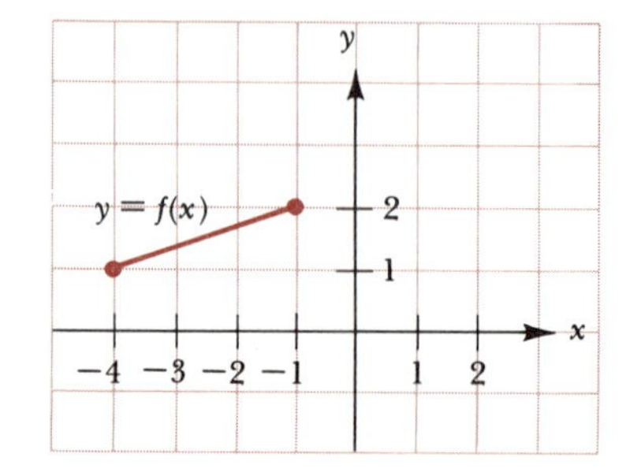

For Exercises 27 and 28, refer to the graph of the function g shown in the figure below.

27. Graph the following functions.
(a) $y = g(-x)$ (b) $y = -g(x)$
(c) $y = -g(-x)$

28. Graph the function $y = -g(x - 3) - 1$.

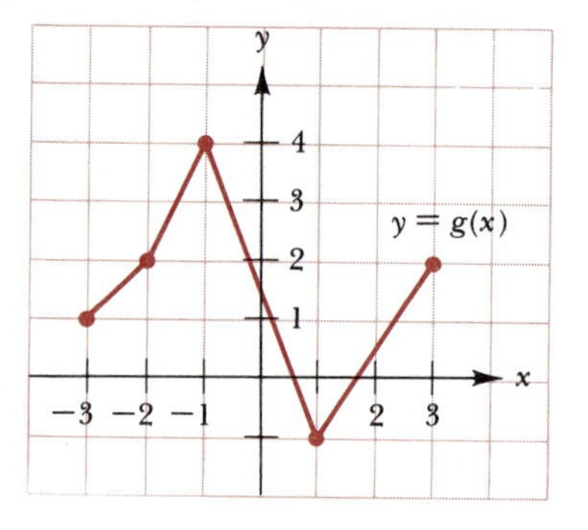

29. Show that $\frac{x}{x-1} = \frac{1}{x-1} + 1$, provided $x \neq 1$. Then use this fact to graph the function g defined by $g(x) = \frac{x}{x-1}$.

30. Verify that $-x^2 + 2x + 2 = -(x - 1)^2 + 3$. Then use this fact to graph the function h defined by $h(x) = -x^2 + 2x + 2$.

31. Show that $x^3 - 6x^2 + 12x - 9 = (x - 2)^3 - 1$. Then use this fact to sketch the graph of $y = x^3 - 6x^2 + 12x - 9$.

32. Show that $\frac{2 - x}{1 - \sqrt{x - 1}} = 1 + \sqrt{x - 1}$, provided $x \geq 1$ and $x \neq 2$. Then use this information to graph the function f defined by $f(x) = \frac{2 - x}{1 - \sqrt{x - 1}}$.

33. Let $f(x) = x^2$. In each of the following cases, sketch the graph of the given function and determine its x- and y-intercepts, if any.
(a) $y = f(x) + 4$ (b) $y = f(x) - 3$
(c) $y = f(x + 1)$ (d) $y = f(x - 4) + 3$
(e) $y = -f(x - 4) + 2$ (f) $y = f(x - 100)$

34. **(a)** A function f is said to be *even* if the equation $f(-x) = f(x)$ is satisfied by all values of x in the domain of f. Explain why the graph of an even function must be symmetric about the y-axis.

(b) Show that the following functions are even by computing $f(-x)$ and then noting that $f(x)$ and $f(-x)$ are equal.

(i) $f(x) = x^2$ **(ii)** $f(x) = 2x^4 - 6$

(iii) $f(x) = 3x^6 - 4/x^2 + 1$

35. **(a)** A function f is said to be *odd* if the equation $f(-x) = -f(x)$ is satisfied by all values of x in the domain of f. Show that if (x, y) is a point on the graph of an odd function f, then the point $(-x, -y)$ is also on the graph. (This implies that the graph of an odd function must be symmetric about the origin.)

(b) Show that the following are odd functions by computing $f(-x)$ as well as $-f(x)$ and then noting that the two expressions obtained are equal.

(i) $f(x) = x^3$ **(ii)** $f(x) = -2x^5 + 4x^3 - x$

(iii) $f(x) = \dfrac{|x|}{x + x^7}$

36. Are the following functions odd, even, or neither? (See Exercises 34 and 35 for the definitions.)

(a) $f(x) = \dfrac{1 - x^2}{2 + x^2}$ **(b)** $g(x) = \dfrac{x - x^3}{2x + x^3}$

(c) $h(x) = x^2 + x$ **(d)** $F(x) = (x^2 + x)^2$

(e) $G(x) = \begin{cases} 1 & \text{if } x > 0 \\ 0 & \text{if } x = 0 \\ -1 & \text{if } x < 0 \end{cases}$

Suggestion for part (e): Look at the graph.

4.5 Methods of Combining Functions

Given two numbers a and b, there are various ways in which they can be combined to produce a third number. For instance, we can form the sum $a + b$ or the difference $a - b$ or the product ab. Additionally, if $b \neq 0$, we can form the quotient a/b. Similarly, given two functions, these can be combined in various ways to produce a third function. Suppose, for example, that we start with the two functions $y = x^2$ and $y = x^3$. Then it would seem natural to define their sum, difference, product, and quotient as follows.

$$\begin{aligned} \text{sum:} \quad & y = x^2 + x^3 \\ \text{difference:} \quad & y = x^2 - x^3 \\ \text{product:} \quad & y = x^2x^3 \qquad \text{that is, } y = x^5 \\ \text{quotient:} \quad & y = \frac{x^2}{x^3} \qquad \text{that is, } y = \frac{1}{x} \end{aligned}$$

Indeed, this is just the idea behind the formal definitions that we now give.

Definition

Let f and g be two functions. Then the *sum* $f + g$, the *difference* $f - g$, the *product* fg, and the *quotient* f/g are defined by the following equations:

$$(f + g)(x) = f(x) + g(x) \tag{1}$$

$$(f - g)(x) = f(x) - g(x) \tag{2}$$

$$(fg)(x) = f(x) \cdot g(x) \tag{3}$$

$$(f/g)(x) = f(x)/g(x) \qquad \text{provided that } g(x) \neq 0 \tag{4}$$

For the functions defined by equations (1), (2), and (3), the domain in each case is the set of all inputs x belonging to both the domain of f and the domain of g. For the quotient function in equation (4), we impose the additional restriction that the domain exclude all inputs x for which $g(x) = 0$.

Example 1 Let $f(x) = 3x + 1$ and $g(x) = x - 1$. Compute the following: $(f + g)(x)$, $(f - g)(x)$, $(fg)(x)$, and $(f/g)(x)$.

Solution

$$(f + g)(x) = f(x) + g(x) = (3x + 1) + (x - 1) = 4x$$
$$(f - g)(x) = f(x) - g(x) = (3x + 1) - (x - 1) = 2x + 2$$
$$(fg)(x) = [f(x)][g(x)] = (3x + 1)(x - 1) = 3x^2 - 2x - 1$$
$$\left(\frac{f}{g}\right)(x) = \frac{f(x)}{g(x)} = \frac{3x + 1}{x - 1} \qquad \text{provided } x \neq 1$$

For the remainder of this section we are going to discuss a method for combining functions known as *composition of functions.* As you'll see, this method is based on the familiar algebraic process of substitution. We begin with an example. Suppose that f and g are two functions defined as follows.

$$f(x) = x^2 \qquad g(x) = 3x + 1$$

Choose any number in the domain of g, say $x = -2$ to be specific. We can compute $g(-2)$:

$$g(-2) = 3(-2) + 1 = -5$$

Now let's use the output -5 that g has given us as an *input* for f. We obtain $f(-5) = (-5)^2 = 25$, and consequently

$$f[g(-2)] = 25$$

So, beginning with the input -2, we've successively applied g and then f to obtain the output 25. Similarly, we could carry out this same procedure for any other numbers in the domain of g. Here is a summary of the procedure:

1. Start with an input x and calculate $g(x)$.
2. Use $g(x)$ as an input for f; that is, calculate $f[g(x)]$.

We use the notation $f \circ g$ to denote the function or rule that tells us to assign the output $f[g(x)]$ to the initial input x. In other words, $f \circ g$ denotes the rule consisting of two steps: First, apply g; then, apply f. We read the notation $f \circ g$ as *f circle g* or *f composed with g.*

When we write $g(x)$, we assume that x is in the domain of the function g. Likewise, for the notation $f[g(x)]$ to make sense, the outputs $g(x)$ must themselves be acceptable inputs for the function f. Our formal definition then for the composite function $f \circ g$ is as follows.

Composition of Functions. Definition of $f \circ g$

Given two functions f and g, the function $f \circ g$ is defined by

$$(f \circ g)(x) = f[g(x)]$$

The domain of $f \circ g$ consists of those inputs x (in the domain of g) for which $g(x)$ is in the domain of f.

Example 2 Let $f(x) = x^2$ and $g(x) = 3x + 1$. Compute $(f \circ g)(x)$ and also $(g \circ f)(x)$.

Solution

$$\begin{aligned}(f \circ g)(x) &= f[g(x)] && \text{definition of } f \circ g\\ &= f(3x + 1) && \text{definition of } g\\ &= (3x + 1)^2 && \text{definition of } f\\ &= 9x^2 + 6x + 1\end{aligned}$$

$$\begin{aligned}(g \circ f)(x) &= g[f(x)] && \text{definition of } g \circ f\\ &= g(x^2) && \text{definition of } f\\ &= 3(x^2) + 1 && \text{definition of } g\\ &= 3x^2 + 1\end{aligned}$$

Notice that the two results obtained in Example 2 are not the same. This shows that in general $f \circ g$ and $g \circ f$ represent different functions.

Example 3 Let f and g be defined as in Example 2: $f(x) = x^2$ and $g(x) = 3x + 1$. Compute $(f \circ g)(-2)$.

Solution We'll show two methods.

First Method Using the formula for $(f \circ g)(x)$ developed in Example 2, we have

$$(f \circ g)(x) = 9x^2 + 6x + 1$$

and therefore

$$\begin{aligned}(f \circ g)(-2) &= 9(-2)^2 + 6(-2) + 1\\ &= 36 - 12 + 1\\ &= 25\end{aligned}$$

Alternate Method Working directly from the definition of $f \circ g$, we have

$$\begin{aligned}(f \circ g)(-2) &= f[g(-2)]\\ &= f[3(-2) + 1]\\ &= f(-5)\\ &= (-5)^2 = 25\end{aligned}$$

In Examples 2 and 3, the domain of both f and g is the set of all real numbers. And, as you can easily check, the domain of both $f \circ g$ and $g \circ f$ is also the set of all real numbers. In Example 4, however, some care needs to be taken in describing the domain of the composite function.

Example 4 Let f and g be defined as follows.

$$f(x) = x^2 + 1 \qquad g(x) = \sqrt{x}$$

Compute $(f \circ g)(x)$. Find the domain of $f \circ g$ and sketch its graph.

Solution

$$\begin{aligned}(f \circ g)(x) &= f[g(x)]\\ &= f(\sqrt{x}) = (\sqrt{x})^2 + 1 = x + 1\end{aligned}$$

So we have $(f \circ g)(x) = x + 1$. Now what about the domain of $f \circ g$? Our first inclination may be to say (incorrectly!) that the domain is the set of all real numbers, since any real number can be used as an input in the expression $x + 1$. However, the definition of $f \circ g$ on page 192 tells us that the inputs for $f \circ g$ must first of all be acceptable inputs for g. In view of the definition of g then, we must require that x be nonnegative. On the other hand, for any nonnegative input x, the number $g(x)$ will be an acceptable input for f. (Why?) In summary then, the domain of $f \circ g$ is the interval $[0, \infty)$. The graph of $f \circ g$ is shown in Figure 1.

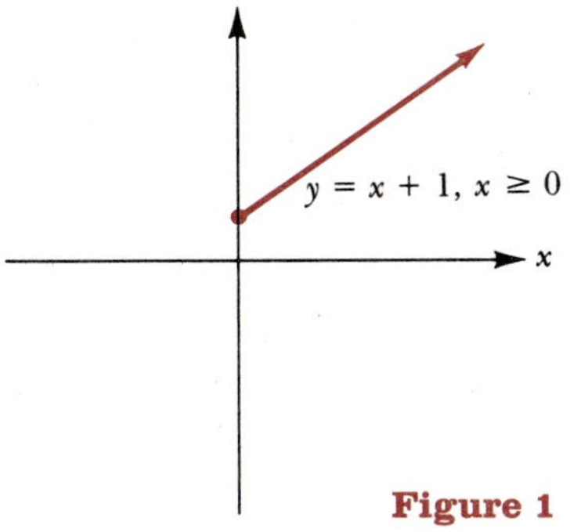

y
$y = x + 1, x \geq 0$
x

Figure 1
Graph of the function $f \circ g$ in Example 4.

One reason for studying composition of functions is that it provides a way in which a given function can be expressed in terms of simpler functions. Suppose, for example, that we wish to express the function C defined by

$$C(x) = (2x^3 - 5)^2$$

as a composition of simpler functions. That is, we want to come up with two functions f and g so that the equation

$$C(x) = (f \circ g)(x)$$

holds for every x in the domain of C.

Begin by thinking what you would do to compute $(2x^3 - 5)^2$ for a given value of x. First you would compute the quantity $2x^3 - 5$; then you would square the result. Therefore, recalling that the rule $f \circ g$ tells us to do g *first*, we let $g(x) = 2x^3 - 5$. And then since the next step is squaring, we let $f(x) = x^2$. Now let us see if these choices for f and g are correct; that is, let us calculate $(f \circ g)(x)$ and see if it really is the same as $C(x)$.

Using $f(x) = x^2$ and $g(x) = 2x^3 - 5$, we have

$$\begin{aligned}(f \circ g)(x) &= f[g(x)] \\ &= f(2x^3 - 5) \\ &= (2x^3 - 5)^2 \\ &= C(x)\end{aligned}$$

This shows that our choices for f and g were indeed correct and we have expressed C as a composition of two simpler functions.*

Before leaving this example, note that, in expressing C as $f \circ g$, we chose g to be the "inner" function, that is, the quantity inside parentheses: $g(x) = 2x^3 - 5$. This observation is used in Example 5.

Example 5 Let $s(x) = \sqrt{1 + x^4}$. Express the function s as a composition of two simpler functions f and g.

Solution Let g be the "inner" function; that is, let $g(x)$ be the quantity inside the radical:

$$g(x) = 1 + x^4$$

And let us take f to be the square root function:

$$f(x) = \sqrt{x}$$

Now we need to verify that these are appropriate choices for f and g; that is, we need to check that the equation $(f \circ g)(x) = s(x)$ is true for every x in the domain of s. We have

$$\begin{aligned}(f \circ g)(x) &= f[g(x)] \\ &= f(1 + x^4) \\ &= \sqrt{1 + x^4} \\ &= s(x)\end{aligned}$$

Thus $(f \circ g)(x) = s(x)$ as required.

* Other answers are possible too. For instance, if $F(x) = (x - 5)^2$ and $G(x) = 2x^3$, then (as you should verify for yourself) $C(x) = F[G(x)]$.

Exercise Set 4.5

For Exercises 1–10, compute the indicated expressions, given that the functions f, g, h, k, *and* m *are defined as follows:*

$f(x) = 2x - 1$ $\qquad k(x) = 2$

$g(x) = x^2 - 3x - 6$ $\qquad m(x) = x^2 - 9$

$h(x) = x^3$

1. **(a)** $(f+g)(x)$ **(b)** $(f-g)(x)$ **(c)** $(f-g)(0)$

2. **(a)** $(fh)(x)$ **(b)** $(h/f)(x)$ **(c)** $(f/h)(1)$

3. **(a)** $(m-f)(x)$ **(b)** $(f-m)(x)$

4. **(a)** $(fg)(x)$ **(b)** $(fg)(\frac{1}{2})$

5. **(a)** $(fk)(x)$ **(b)** $(kf)(x)$ **(c)** $(fk)(1) - (kf)(2)$

6. **(a)** $(g+m)(x)$ **(b)** $(g+m)(x) - (g-m)(x)$

7. **(a)** $(f/m)(x) - (m/f)(x)$ **(b)** $(f/m)(0) - (m/f)(0)$

8. **(a)** $[h \cdot (f+m)](x)$ *Note:* h and $(f+m)$ are two functions; the notation $h \cdot (f+m)$ denotes the product function.
(b) $(hf)(x) + (hm)(x)$

9. **(a)** $[m \cdot (k-h)](x)$ **(b)** $(mk)(x) - (mh)(x)$ **(c)** $(mk)(-1) - (mh)(-1)$

10. **(a)** $(g+g)(x)$ **(b)** $(g-g)(x)$ **(c)** $(kg)(x)$ **(d)** $(g+g)(-3) - (kg)(-3)$

11. Let $f(x) = 3x + 1$ and $g(x) = -2x - 5$. Compute the following.
(a) $(f \circ g)(x)$ **(b)** $(f \circ g)(10)$ **(c)** $(g \circ f)(x)$ **(d)** $(g \circ f)(10)$

12. Let $f(x) = 1 - 2x^2$ and $g(x) = x + 1$. Compute the following.
(a) $(f \circ g)(x)$ **(b)** $(f \circ g)(-1)$ **(c)** $(g \circ f)(x)$ **(d)** $(g \circ f)(-1)$ **(e)** $(f \circ f)(x)$ **(f)** $(g \circ g)(-1)$

13. Compute $(f \circ g)(x)$, $(f \circ g)(-2)$, $(g \circ f)(x)$, and $(g \circ f)(-2)$ for each pair of functions.
(a) $f(x) = 1 - x, \quad g(x) = 1 + x$
(b) $f(x) = x^2, \quad g(x) = 2x^2$
(c) $f(x) = x^2, \quad g(x) = x^3$
(d) $f(x) = x^2 - 3x - 4, \quad g(x) = 2 - 3x$
(e) $f(x) = x/3, \quad g(x) = 1 - x^4$
(f) $f(x) = 2^x, \quad g(x) = x^2 + 1$
(g) $f(x) = x, \quad g(x) = 3x^5 - 4x^2$
(h) $f(x) = 3x - 4, \quad g(x) = \dfrac{x+4}{3}$

14. Let $h(x) = 4x^2 - 5x + 1$, $k(x) = x$, $m(x) = 7$. Compute the following.
(a) $h[k(x)]$ **(b)** $k[h(x)]$ **(c)** $h[m(x)]$ **(d)** $m[h(x)]$ **(e)** $k[m(x)]$ **(f)** $m[k(x)]$

15. Let $F(x) = \dfrac{3x-4}{3x+3}$ and $G(x) = \dfrac{x+1}{x-1}$. Compute the following.
(a) $(F \circ G)(x)$ **(b)** $F[G(t)]$ **(c)** $(F \circ G)(2)$ **(d)** $(G \circ F)(x)$ **(e)** $G[F(y)]$ **(f)** $(G \circ F)(2)$

16. Let $f(x) = 1/x^2 + 1$ and $g(x) = 1/x - 1$.
(a) Compute $(f \circ g)(x)$.
(b) What is the domain of $f \circ g$?
(c) Graph the function $f \circ g$.

17. Let $M(x) = \dfrac{2x-1}{x-2}$.
(a) Compute $M(7)$ and then $M[M(7)]$.
(b) Compute $(M \circ M)(x)$.
(c) Compute $(M \circ M)(7)$ using the formula you obtained in part (b). Check that your answer agrees with that obtained in part (a).

18. Let $F(x) = (x+1)^5$, $f(x) = x^5$, $g(x) = x + 1$. Which is true for all x?

$$(f \circ g)(x) = F(x) \quad \text{or} \quad (g \circ f)(x) = F(x)$$

19. For this exercise, refer to the graphs of the functions f, g, and h to compute the required quantities. Assume that all the axes are marked off in one-unit lengths.
(a) $f[g(3)]$ **(b)** $g[f(3)]$ **(c)** $f[h(3)]$ **(d)** $(h \circ g)(2)$ **(e)** $h\{f[g(3)]\}$
(f) $(g \circ f \circ h \circ f)(2)$ This means first do f, then h, then f, then g.

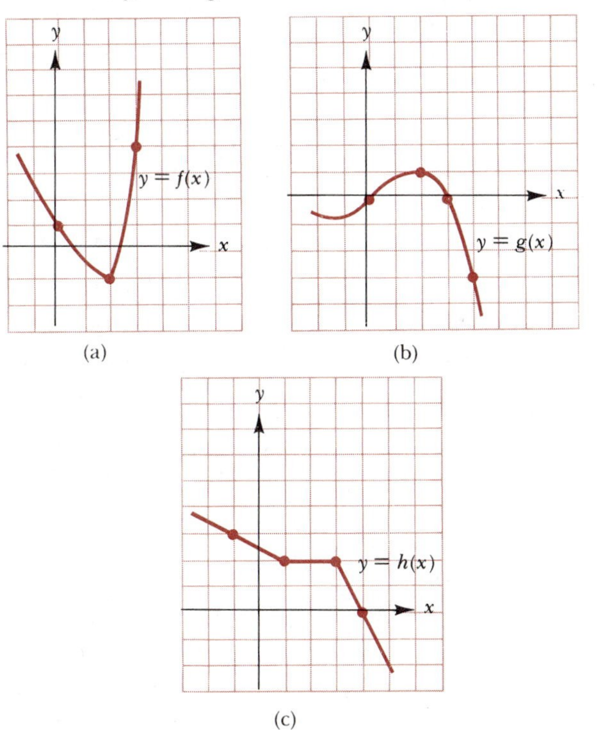

(a) (b) (c)

20. Let $f(x) = 2x - 1$. Find $f[f(x)]$. If $z = f[f(x)]$, find $f(z)$.

21. **(a)** Let $T(x) = 4x^3 - 3x^2 + 6x - 1$ and $I(x) = x$. Find $(T \circ I)(x)$ and $(I \circ T)(x)$.
(b) Let $G(x) = ax^2 + bx + c$ and $I(x) = x$. Find $(G \circ I)(x)$ and $(I \circ G)(x)$.
(c) What general conclusion do the results of parts (a) and (b) lead you to believe is true?

22. The figure shows the graphs of two functions F and G. Use the graphs to compute the following quantities.

(a) $(G \circ F)(1)$ **(b)** $(F \circ G)(1)$
(c) $(F \circ F)(1)$ **(d)** $(G \circ G)(-3)$
(e) $(G \circ F)(5)$ **(f)** $(F \circ F)(5)$
(g) $(F \circ G)(-1)$ **(h)** $(G \circ F)(-1)$
(i) $(G \circ F)(2)$
(j) What are the x- and y-intercepts of the graph of F? *Hint:* Point-slope formula.
(k) What are the x- and y-intercepts of the graph of G?

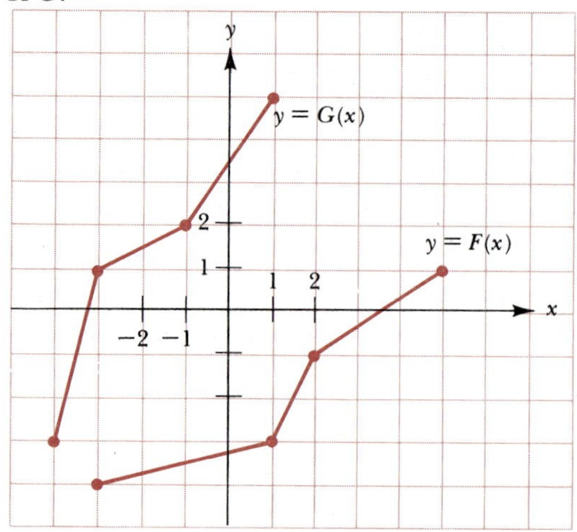

23. The domain of a function f consists of the numbers -1, 0, 1, 2, and 3. The table below shows which output f assigns to each input.

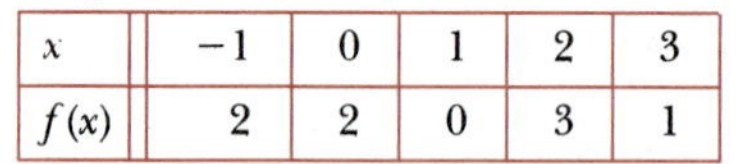

x	-1	0	1	2	3
$f(x)$	2	2	0	3	1

The domain of a function g consists of the numbers 0, 1, 2, 3, and 4. The table below shows which output g assigns to each input.

x	0	1	2	3	4
$g(x)$	3	2	0	4	-1

Use this information to complete the tables for $f \circ g$ and $g \circ f$ shown below. *Note:* Two of the entries will be undefined.

x	0	1	2	3	4
$(f \circ g)(x)$					

x	-1	0	1	2	3	4
$(g \circ f)(x)$						

24. The two tables below show certain pairs of inputs and outputs for the functions f and g.

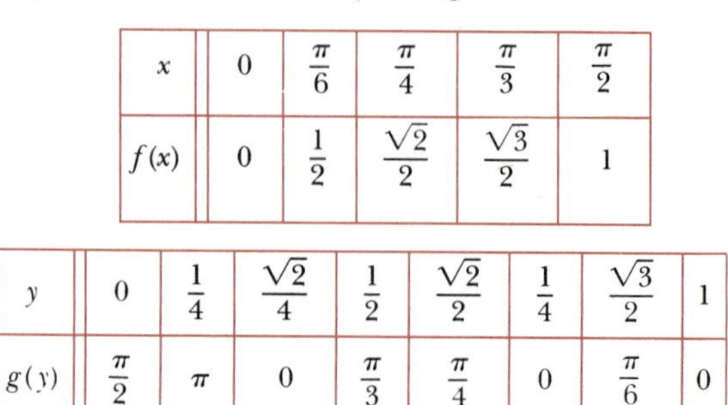

x	0	$\frac{\pi}{6}$	$\frac{\pi}{4}$	$\frac{\pi}{3}$	$\frac{\pi}{2}$
$f(x)$	0	$\frac{1}{2}$	$\frac{\sqrt{2}}{2}$	$\frac{\sqrt{3}}{2}$	1

y	0	$\frac{1}{4}$	$\frac{\sqrt{2}}{4}$	$\frac{1}{2}$	$\frac{\sqrt{2}}{2}$	$\frac{1}{4}$	$\frac{\sqrt{3}}{2}$	1
$g(y)$	$\frac{\pi}{2}$	π	0	$\frac{\pi}{3}$	$\frac{\pi}{4}$	0	$\frac{\pi}{6}$	0

Use this information to complete the following table of values for $(g \circ f)(x)$.

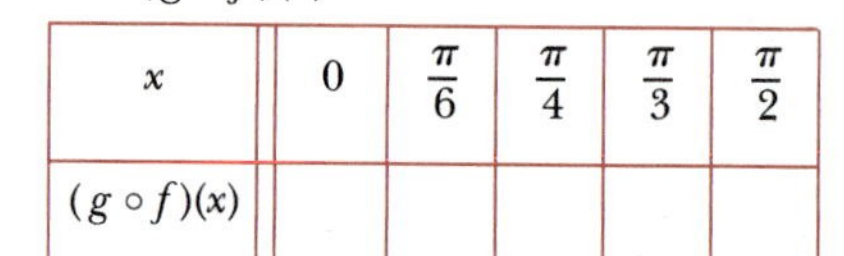

x	0	$\frac{\pi}{6}$	$\frac{\pi}{4}$	$\frac{\pi}{3}$	$\frac{\pi}{2}$
$(g \circ f)(x)$					

25. Let $f(x) = 2x + 1$ and $g(x) = 3x - 4$.
(a) Find $(f \circ g)(x)$ and graph the function $f \circ g$.
(b) Find $(g \circ f)(x)$ and graph the function $g \circ f$.

26. Let $F(x) = x^2$ and $G(x) = x - 4$.
(a) Find $F[G(x)]$ and graph the function $F \circ G$.
(b) Find $G[F(x)]$ and graph the function $G \circ F$.

27. Let $g(x) = \sqrt{x} - 3$ and $f(x) = x - 1$.
(a) Sketch a graph of g. Specify the domain and range.
(b) Sketch a graph of f. Specify the domain and range.
(c) Compute $(f \circ g)(x)$. Graph the function $f \circ g$ and specify its domain and range.
(d) Find a formula for $g[f(x)]$. Which values of x are acceptable inputs here? That is, what is the domain of $g \circ f$?
(e) Use the results of part (d) to sketch a graph of the function $g \circ f$.

28. The circumference of a circle with radius r is given by $C(r) = 2\pi r$. Suppose that the circle is shrinking in size and that the radius at time t is given by

$$r = f(t) = \frac{1}{t^2 + 1}$$

Assume that t is measured in minutes and that r or $f(t)$ is in feet. Find $(C \circ f)(t)$ and use this to find the circumference when $t = 3$ min.

29. A spherical weather balloon is being inflated in such a way that the radius is given by

$$r = g(t) = \tfrac{1}{2}t + 2$$

Assume that r is in meters and t is in seconds, with $t = 0$ corresponding to the time that inflation begins. If the volume of a sphere of radius r is given by

$$V(r) = \tfrac{4}{3}\pi r^3$$

compute $V[g(t)]$ and use this to find the time at which the volume of the balloon is 36π m^3.

30. Suppose that a manufacturer knows that his daily production cost to build x bicycles is given by the function C, where

$$C(x) = 100 + 90x - x^2 \qquad 0 \le x \le 40$$

That is, $C(x)$ represents the cost in dollars of building x bicycles. Furthermore, suppose that the number of bicycles that can be built in t hr is given by the function f, where

$$x = f(t) = 5t \qquad 0 \le t \le 8$$

(a) Compute $(C \circ f)(t)$.
(b) Compute the daily production cost on a day that the factory operates for $t = 3$ hr.
(c) If the factory runs for 6 hr instead of 3 hr, is the cost twice as much?

31. Suppose that, in a certain biology lab experiment, the number of bacteria is related to the temperature of the environment by the function

$$N(T) = -2T^2 + 240T - 5400 \qquad (40 \le T \le 90)$$

Here $N(T)$ represents the number of bacteria present when the temperature is T degrees Fahrenheit.

Also suppose that the temperature t hr after the experiment begins is given by

$$T(t) = 10t + 40 \qquad (0 \le t \le 5)$$

(a) Compute $N[T(t)]$.
(b) How many bacteria are present when $t = 0$? $t = 2$ hr? $t = 5$ hr?

32. Let $C(x) = (3x - 1)^4$. Express C as a composition of two simpler functions.

33. Let $C(x) = (1 + x^2)^3$. Find functions f and g so that $C(x) = (f \circ g)(x)$ is true for all values of x.

34. Express $G(x) = \dfrac{1}{1 + x^4}$ as a composition of two other functions, one of which is $f(x) = 1/x$.

35. Express each function as a composition of two functions.
(a) $F(x) = \sqrt[3]{3x + 4}$ **(b)** $G(x) = |2x - 3|$
(c) $H(x) = (ax + b)^5$ **(d)** $T(x) = 1/\sqrt{x}$

36. Let $a(x) = x^2$, $b(x) = |x|$, $c(x) = 3x - 1$. Express each of the following functions as a composition of two of the given functions.
(a) $f(x) = (3x - 1)^2$ **(b)** $g(x) = |3x - 1|$
(c) $h(x) = 3x^2 - 1$

37. Let $a(x) = 1/x$, $b(x) = \sqrt[3]{x}$, $c(x) = 2x + 1$, and $d(x) = x^2$. Express each of the following functions as a composition of two of the given functions.
(a) $f(x) = \sqrt[3]{2x + 1}$ **(b)** $g(x) = 1/x^2$
(c) $h(x) = 2x^2 + 1$ **(d)** $K(x) = 2\sqrt[3]{x} + 1$
(e) $l(x) = 2/x + 1$ **(f)** $m(x) = \dfrac{1}{2x + 1}$
(g) $n(x) = x^{2/3}$

38. Let $g(x) = 4x - 1$. Find $f(x)$ given that the equation $(g \circ f)(x) = x + 5$ is true for all values of x.

39. Let $g(x) = 2x + 1$. Find $f(x)$, given that $(g \circ f)(x) = 10x - 7$.

***40.** Let $f(x) = -2x + 1$ and $g(x) = ax + b$. Find values for a and b so that the equation

$$f[g(x)] = x$$

holds for all values of x.

41. Let $f(x) = \dfrac{3x - 4}{x - 3}$.
(a) Compute $(f \circ f)(x)$.
(b) Find $f[f(\frac{113}{355})]$. (Try not to do it the hard way.)

***42.** Let p and q be two numbers whose sum is 1. Define the function F by

$$F(x) = p - \frac{1}{x + q}$$

Show that $F\{F[F(x)]\} = x$.

43. Let $f(x) = x^2$ and $g(x) = 2x - 1$.
(a) Compute $\dfrac{f[g(x)] - f[g(a)]}{g(x) - g(a)}$.
(b) Compute $\dfrac{f[g(x)] - f[g(a)]}{x - a}$.

44. Suppose that $y = f(x)$ and $g(y) = x$. Show that $(g \circ f)(x) - (f \circ g)(y) = x - y$.

45. Given three functions f, g, and h, we define the function $f \circ g \circ h$ by the formula

$$(f \circ g \circ h)(x) = f\{g[h(x)]\}$$

For example, if $f(x) = x^2$, $g(x) = x + 1$, $h(x) = x/2$, we evaluate $(f \circ g \circ h)(x)$ as follows.

$$\begin{aligned}(f \circ g \circ h)(x) &= f\{g[h(x)]\}\\ &= f\left[g\left(\frac{x}{2}\right)\right]\\ &= f\left(\frac{x}{2} + 1\right)\\ &= \left(\frac{x}{2} + 1\right)^2 = \frac{x^2}{4} + x + 1\end{aligned}$$

Using $f(x) = x^2$, $g(x) = x + 1$, and $h(x) = x/2$, compute the following:
(a) $(g \circ h \circ f)(x)$ **(b)** $(h \circ f \circ g)(x)$
(c) $(g \circ f \circ h)(x)$ **(d)** $(f \circ h \circ g)(x)$
(e) $(h \circ g \circ f)(x)$
Hint for checking: No two answers are the same.

46. Let $F(x) = 2x - 1$, $G(x) = 4x$, $H(x) = 1 + x^2$. Compute the following:
(a) $(F \circ G \circ H)(x)$ **(b)** $(G \circ F \circ H)(x)$
(c) $(F \circ H \circ G)(x)$ **(d)** $(H \circ F \circ G)(x)$
(e) $(H \circ G \circ F)(x)$ **(f)** $(G \circ H \circ F)(x)$

47. Let $f(x) = x^2$, $g(x) = 1 - x$, $h(x) = 3x$. Express the following functions as compositions of f, g, and h.
(a) $p(x) = 1 - 9x^2$ **(b)** $q(x) = 3 - 3x^2$
(c) $r(x) = 1 - 6x + 9x^2$
(d) $s(x) = 3 - 6x + 3x^2$

48. Let $F(x) = x^2$, $G(x) = x + 1$, $H(x) = x/2$. Compute each of the following:
(a) $[F \circ (G - H)](x)$ **(b)** $[(F - G) \circ H](x)$
(c) $[(F \circ G) - H](x)$ **(d)** $[F - (G - H)](x)$
(e) $[(F - G) - H](x)$

49. Suppose that three functions f', g, and g' are defined as follows (f' is read *f prime*).

$$f'(x) = 3x^2 \qquad g(x) = 2x^2 + 5 \qquad g'(x) = 4x$$

(a) Find $f'[g(x)] \cdot g'(x)$. (The dot denotes multiplication, not composition.)
(b) Evaluate $f'[g(-1)] \cdot g'(-1)$.
(c) Find $f'[g(t)] \cdot g'(t)$.

50. Suppose that three functions F', G, and G' are defined by

$$F'(x) = \frac{1}{2\sqrt{x}} \qquad G(x) = x^2 + 2x + 2 \qquad G'(x) = 2x + 2$$

(a) Find $F'[G(x)] \cdot G'(x)$. (The dot denotes multiplication, not composition.)
(b) Find $F'[G(9)] \cdot G'(9)$.

4.6 Inverse Functions

We shall introduce the idea of inverse functions through an example. Consider the function f defined by

$$f(x) = 2x$$

$f(x) = 2x$

x	y
-2	-4
-1	-2
0	0
1	2
2	4

Figure 1

A short table of values for f and the graph of f are displayed in Figure 1.

Now we ask this question: What happens if you take the table for f and interchange the entries in the x and y columns to obtain a new table that looks like the following?

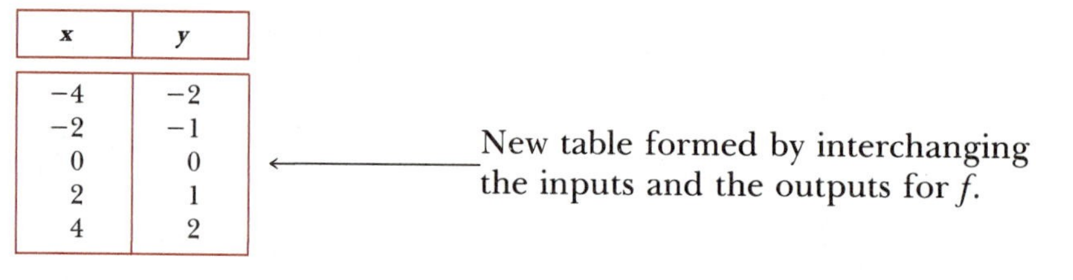

x	y
-4	-2
-2	-1
0	0
2	1
4	2

New table formed by interchanging the inputs and the outputs for f.

Several observations are in order. Our new table does (in this case) itself represent a function; for each input x there is exactly one output y. Let's call this new function g. Note that each output is half the corresponding input. So the function g can be described by the formula $g(x) = x/2$. In summary, we began with the "doubling function" $f(x) = 2x$; then, by interchanging the inputs with the outputs, we obtained the "halving function" $g(x) = x/2$.

The doubling function f and the halving function g are in a certain sense opposites, in that each reverses or "inverts" the effect of the other. That is, if you begin with x and then calculate $g[f(x)]$, you get x again; similarly, $f[g(x)]$ is also equal to x. (Verify these last two statements for yourself.)

The two functions f and g that we've been discussing are an example of a pair of inverse functions. The general definition for *inverse functions* is given in the box that follows.

Definition of Inverse Functions

Two functions f and g are said to be *inverses* of one another provided

$$f[g(x)] = x \qquad \text{for each } x \text{ in the domain of } g$$

and

$$g[f(x)] = x \qquad \text{for each } x \text{ in the domain of } f$$

Example 1 Verify that the functions f and g defined below are inverses.

$$f(x) = \frac{1}{3}x + 2 \qquad g(x) = 3x - 6$$

Solution In view of the definition, we must check that $f[g(x)] = x$ and $g[f(x)] = x$. We have

$$\begin{aligned} f[g(x)] &= f(3x - 6) \\ &= \frac{1}{3}(3x - 6) + 2 \\ &= (x - 2) + 2 \\ &= x \end{aligned}$$

Thus $f[g(x)] = x$. Now we still need to check that $g[f(x)] = x$. We have

$$\begin{aligned} g[f(x)] &= g\left(\frac{1}{3}x + 2\right) \\ &= 3\left(\frac{1}{3}x + 2\right) - 6 \\ &= (x + 6) - 6 \\ &= x \end{aligned}$$

Having shown that $f[g(x)] = x$ and $g[f(x)] = x$, we conclude that f and g are indeed inverse functions.

It's customary to use the notation f^{-1} (read *f inverse*) for the function that is the inverse of f. So in Example 1, for instance, we have

$$f(x) = \frac{1}{3}x + 2 \quad \text{and} \quad f^{-1}(x) = 3x - 6$$

Using this notation, the defining equations for inverse functions become

$$f[f^{-1}(x)] = x \qquad \text{for each } x \text{ in the domain of } f^{-1}$$

and

$$f^{-1}[f(x)] = x \qquad \text{for each } x \text{ in the domain of } f$$

Example 2 Suppose $f(x) = 4x^3 + 7$. Find $f[f^{-1}(5)]$. (Assume f^{-1} exists and that 5 is in its domain.)

Solution By definition $f[f^{-1}(x)] = x$. Substituting $x = 5$ on both sides of this identity directly gives us

$$f[f^{-1}(5)] = 5$$

which is the required solution. Note that we didn't need to make use of the formula $f(x) = 4x^3 + 7$, nor did we need to find a formula for $f^{-1}(x)$.

We'll postpone until the end of this section a discussion of which functions have inverses and which do not. For functions that do have inverses, however, there's a simple method that's often applicable for determining those inverses. This method is displayed in Examples 3 and 4.

Example 3 Let $f(x) = 3x - 4$. Find $f^{-1}(x)$.

Solution Begin by rewriting the given equation as

$$y = 3x - 4$$

We know that f^{-1} interchanges the inputs and outputs of f. So to determine f^{-1} we need only switch the x's and y's in the equation $y = 3x - 4$. That gives us $x = 3y - 4$. Now we solve for y as follows:

$$x = 3y - 4$$
$$x + 4 = 3y$$
$$\frac{x + 4}{3} = y$$

Thus the inverse function is $y = \frac{x + 4}{3}$. We can also write this as

$$f^{-1}(x) = \frac{x + 4}{3}$$

Actually, we should call this result just a candidate for f^{-1}, since certain technical matters remain to be discussed. However, if you compute $f[f^{-1}(x)]$ and $f^{-1}[f(x)]$ and find that they both do equal x, then the matter is settled, by definition. Exercise 13 at the end of this section asks you to carry out those calculations.

Here is a summary of our procedure for calculating $f^{-1}(x)$.

To Find $f^{-1}(x)$ for the Function $y = f(x)$:

1. Interchange x and y in the equation $y = f(x)$.
2. Solve the resulting equation for y.

Example 4 Let $f(x) = \frac{2x - 1}{3x + 5}$. Find $f^{-1}(x)$.

Solution *Step 1* Write the given function as $y = \frac{2x - 1}{3x + 5}$. Interchange x and y to get

$$x = \frac{2y - 1}{3y + 5}$$

Step 2 Solve the resulting equation for y.

$$x = \frac{2y - 1}{3y + 5}$$
$$x(3y + 5) = 2y - 1$$
$$3xy + 5x = 2y - 1$$
$$3xy - 2y = -5x - 1$$
$$y(3x - 2) = -5x - 1$$
$$y = \frac{-5x - 1}{3x - 2}$$

Thus the inverse function is given by

$$f^{-1}(x) = \frac{-5x - 1}{3x - 2}$$

There is a certain symmetry that always exists between the graphs of a function and its inverse. As background for this discussion we need the following definition of *symmetry about a line.*

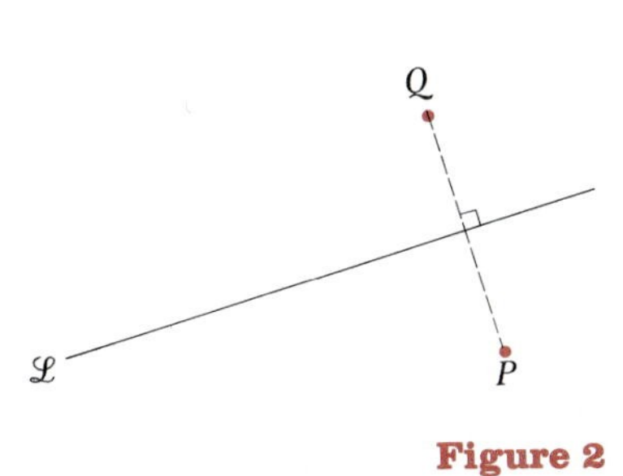

Figure 2

> **Definition: Symmetry about a Line**
>
> Refer to Figure 2. Two points P and Q are said to be *symmetric about the line* $\mathscr{L}$ provided
>
> $$PQ \text{ is perpendicular to } \mathscr{L}$$
>
> *and*
>
> $$\text{points } P \text{ and } Q \text{ are equidistant from } \mathscr{L}$$

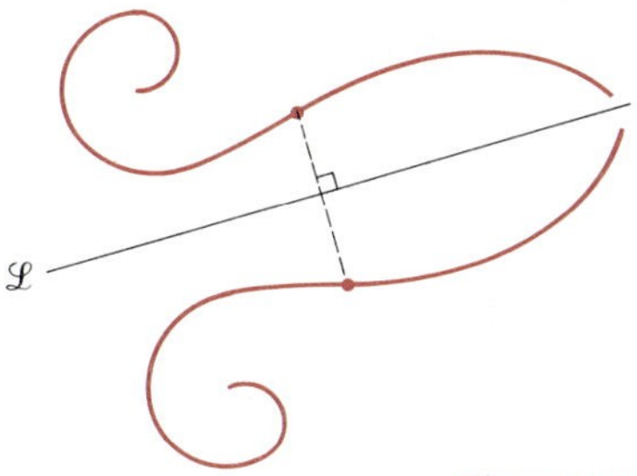

Figure 3

In other words, P and Q are symmetric about the line $\mathscr{L}$ if $\mathscr{L}$ is the perpendicular bisector of line segment PQ. We say that $\mathscr{L}$ is the *axis of symmetry* and that P and Q are mirror images or *reflections* of one another through $\mathscr{L}$. (Actually you've already encountered two instances of this type of symmetry in Section 3.2. In that section we discussed symmetry about the x- and y-axes.) Additionally, we say that two curves are symmetric about a line if each point on one curve is the reflection of a corresponding point on the other curve, and vice versa. See Figure 3.

Example 5 Use the definition of symmetry about a line to verify that the points $P(4, 1)$ and $Q(1, 4)$ are symmetric about the line $y = x$. (See Figure 4.)

Solution According to the definition, we first have to show that PQ is perpendicular to the line $y = x$. Now the slope of PQ is

$$m = \frac{4 - 1}{1 - 4} = -1$$

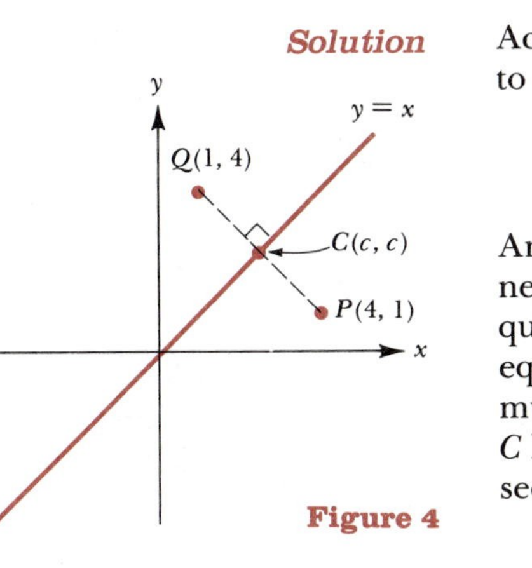

Figure 4

And on the other hand, the slope of $y = x$ is 1. Since these two slopes are negative reciprocals, we conclude that the lines are perpendicular, as required. Next, according to the definition, we must show that P and Q are equidistant from the line $y = x$. Looking at Figure 4 again, this means we must show that $\overline{PC} = \overline{QC}$. We shall use the distance formula to do this. Since C lies on the line $y = x$, we may write the coordinates of C as (c, c). As you'll see, it won't be necessary to find out what number c actually is. We have now

$$\begin{aligned} \overline{PC} &= \sqrt{(x_2 - x_1)^2 + (y_2 - y_1)^2} \\ &= \sqrt{(4 - c)^2 + (1 - c)^2} \end{aligned}$$

And similarly by the distance formula we find

$$\begin{aligned} \overline{QC} &= \sqrt{(x_2 - x_1)^2 + (y_2 - y_1)^2} \\ &= \sqrt{(1 - c)^2 + (4 - c)^2} \end{aligned}$$

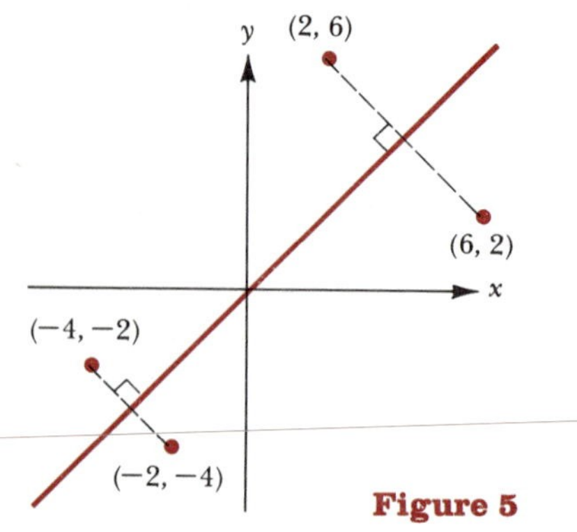

Figure 5

Looking at our expressions for $\overline{PC}$ and $\overline{QC}$, we conclude that $\overline{PC} = \overline{QC}$, as required. We have now verified that the points $P(4, 1)$ and $Q(1, 4)$ are symmetric about the line $y = x$.

In exactly the same way that we've shown (4, 1) and (1, 4) are symmetric about the line $y = x$, we can show in general that (a, b) and (b, a) are always symmetric about the line $y = x$. Figure 5 displays two examples of this, and Exercise 26 at the end of this section asks you to supply a proof for the general situation.

Now let us see how these ideas about symmetry relate to inverse functions. By way of example, consider the functions f and f^{-1} from Example 3:

$$f(x) = 3x - 4 \qquad f^{-1}(x) = \frac{1}{3}x + \frac{4}{3}$$

Figure 6 shows the graphs of f, f^{-1}, and the line $y = x$. *Observation:* The graphs of f and f^{-1} are symmetric about the line $y = x$. This kind of symmetry, in fact, always occurs for the graphs of any pair of inverse functions.

Why is it that the graphs of f and f^{-1} are always mirror images of each other about the line $y = x$? First, recall that the function f^{-1} switches the inputs and outputs of f. Thus (a, b) is on the graph of f if and only if (b, a) is on the graph of f^{-1}. But as we have stated, the points (a, b) and (b, a) are mirror images in the line $y = x$. It follows that the graphs of f and f^{-1} are mirror images of each other about the line $y = x$.

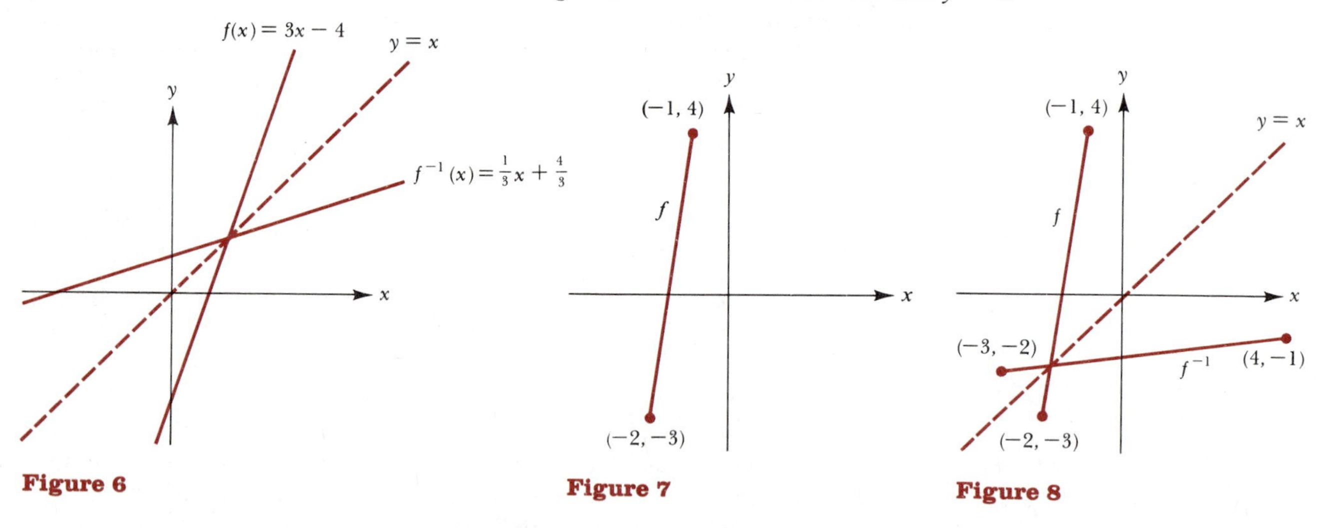

Figure 6 Figure 7 Figure 8

Example 6 The graph of a function f consists of the straight line segment joining the points $(-2, -3)$ and $(-1, 4)$, as shown in Figure 7. Sketch a graph of f^{-1}.

Solution The graph of f^{-1} is obtained by reflecting the graph of f in the line $y = x$. The reflection of $(-2, -3)$ is $(-3, -2)$ and the reflection of $(-1, 4)$ is $(4, -1)$. To graph f^{-1} then, we plot the reflected points $(-3, -2)$ and $(4, -1)$ and connect them with a straight line segment as shown in Figure 8. For reference, Figure 8 also shows the graphs of f and $y = x$.

Example 7 Let $g(x) = x^3$. Find $g^{-1}(x)$ and then, on the same set of axes, sketch the graphs of g, g^{-1}, and $y = x$.

Solution Begin with $y = x^3$. Now switch x and y and solve for y. We have first

$$x = y^3$$

To solve this equation for y, take the cube root of both sides to obtain

$$\sqrt[3]{x} = y$$

So the inverse function is $g^{-1}(x) = \sqrt[3]{x}$. We could graph the function g^{-1} by plotting points. But the easier way is to reflect the graph of $g(x) = x^3$ in the line $y = x$. See Figure 9.

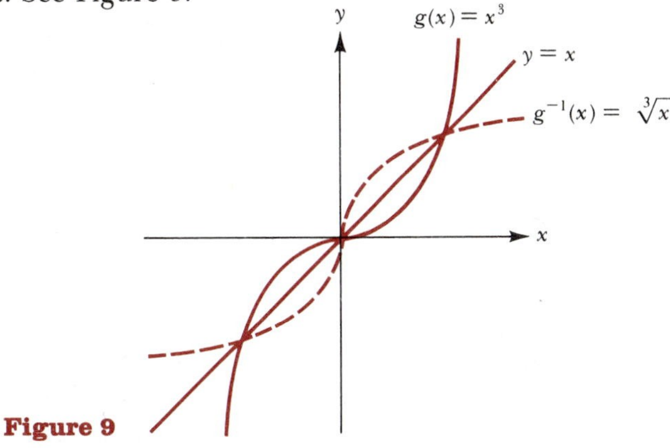

Figure 9

x	y
1	1
2	4
3	9
−1	1
−2	4
−3	9

⟶ reverse inputs and outputs

x	y
1	1
4	2
9	3
1	−1
4	−2
9	−3

Table 1
(a) $F(x) = x^2$ **(b) The table formed by reversing the inputs and outputs for $F(x) = x^2$**

Earlier in this section we mentioned that not every function has an inverse function. For instance, Tables 1(a) and (b) indicate what happens if we begin with the function defined by $F(x) = x^2$ and interchange the inputs and outputs. The resulting table [Table 1(b)] does not define a function. For instance, the input 4 in Table 1(b) produces two distinct outputs, 2 and -2, whereas the definition of a function requires exactly one output for a given input.

Why didn't this difficulty arise when we looked for the inverse of $f(x) = 2x$ at the beginning of this section? The answer is that there is an essential difference between the functions defined by $f(x) = 2x$ and $F(x) = x^2$. With $f(x) = 2x$, distinct inputs never yield the same output. That is, if $x_1 \neq x_2$, then $f(x_1) \neq f(x_2)$. This condition guarantees that interchanging the inputs and outputs of f will yield a function. With $F(x) = x^2$, however, distinct inputs can yield the same output. For instance,

$$2 \neq -2$$

but

$$F(2) = F(-2) \qquad \text{because } F(2) = 4 = F(-2)$$

It's useful to have a name for those functions in which distinct inputs always yield distinct outputs. We call these functions *one-to-one functions.* So in view of the discussion in the previous paragraph, the function defined by $f(x) = 2x$ is one-to-one, but $F(x) = x^2$ is not. Using graphs, there is an easy way to tell which functions are one-to-one.

Horizontal Line Test

A function f is one-to-one if and only if each horizontal line intersects the graph of $y = f(x)$ in at most one point.

Example 8 Which of the following functions are one-to-one?
(a) $F(x) = 2x$ **(b)** $H(x) = x^2$ **(c)** $h(x) = (\sqrt{x})^4$

Solution Sketch the graphs and apply the horizontal line test, as in Figure 10. The horizontal line test tells us that F and h are one-to-one, but H is not.

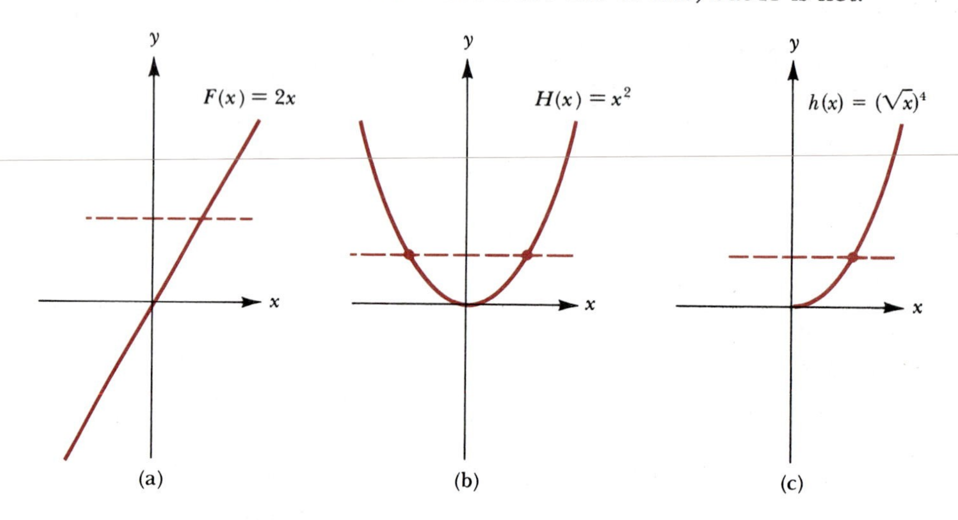

Figure 10

The following theorem tells us that the functions with inverses are precisely the one-to-one functions. (The proof of the theorem isn't difficult, but we'll omit giving it here.)

> **Theorem**
>
> A function f has an inverse function if and only if f is one-to-one.

Example 9 Which of the three functions in Example 8 have inverse functions?

Solution According to the theorem stated just prior to this example, the functions with inverses are those that are one-to-one. In Example 8 we found that F and h were one-to-one, but that H was not. So F and h each has an inverse, but H does not.

Exercise Set 4.6

1. Verify that the given pairs of functions are inverse functions.
   (a) $f(x) = 3x;\quad g(x) = x/3$
   (b) $f(x) = 4x - 1;\quad g(x) = \dfrac{x+1}{4}$
   (c) $g(x) = \sqrt{x};\quad h(x) = x^2$ Assume the domain of both g and h is $[0, \infty)$.
2. Which pairs of functions are inverses?
   (a) $f(x) = -3x + 2;\quad g(x) = \dfrac{2-x}{3}$
   (b) $F(x) = 2x + 1;\quad G(x) = \frac{1}{2}x - 1$
   (c) $G(x) = x^3;\quad H(x) = 1 - x^3$
   (d) $f(t) = t^3;\quad g(t) = \sqrt[3]{t}$
3. Let $f(x) = x^3 + 2x + 1$ and assume that the domain of f^{-1} is $(-\infty, \infty)$.
   (a) Find $f[f^{-1}(4)]$. **(b)** Find $f^{-1}[f(-1)]$.
   (c) Find $(f \circ f^{-1})(\sqrt{2})$. **(d)** Find $f[f^{-1}(t + 1)]$.
   (e) Find $f(0)$ and then use your answer to find $f^{-1}(1)$.
   (f) Find $f(-1)$ and then use your answer to find $f^{-1}(-2)$.

4. Let $f(x) = 2x + 1$.
(a) Find $f^{-1}(x)$.
(b) Calculate $f^{-1}(5)$ and $1/f(5)$. Are your answers the same?

5. Let $f(x) = 3x - 1$.
(a) Compute $f^{-1}(x)$.
(b) Verify that $f[f^{-1}(x)] = x$ and that $f^{-1}[f(x)] = x$.
(c) On the same set of axes, sketch the graphs of f, f^{-1}, and the line $y = x$. Note that the graphs of f and f^{-1} are symmetric about the line $y = x$.

6. Follow Exercise 5, but use $f(x) = \dfrac{x - 6}{3}$.

7. Follow Exercise 5, but use $f(x) = \sqrt{x - 1}$. [The domain of f^{-1} will be $[0, \infty)$.]

8. Follow Exercise 5, but use $f(x) = 1/x$.

9. Let $f(x) = \dfrac{x + 2}{x - 3}$.
(a) Find the domain and range of the function f.
(b) Find $f^{-1}(x)$.
(c) Find the domain and range of the function f^{-1}. What do you observe?

10. Let $f(x) = \dfrac{2x - 3}{x + 4}$. Find $f^{-1}(x)$. Find the domain and range for f and for f^{-1}. What do you observe?

11. Let $f(x) = 2x^3 + 1$. Find $f^{-1}(x)$.

12. In our preliminary discussion at the beginning of this section, we considered the functions $f(x) = 2x$ and $g(x) = x/2$. Verify that $f[g(x)] = x$ and that $g[f(x)] = x$. On the same set of axes, sketch the graphs of f, g, and $y = x$.

13. This exercise refers to the comments made at the end of Example 3 on page 200. Compute $f[f^{-1}(x)]$ and $f^{-1}[f(x)]$ using the functions $f(x) = 3x - 4$ and $f^{-1}(x) = \dfrac{x + 4}{3}$. The point here is actually to carry out the calculations and to see in each case that the answer is indeed x. Then, on the same set of axes, sketch the graphs of f, f^{-1}, and the line $y = x$.

14. Let $f(x) = (x - 3)^2$ and take the domain to be $[3, \infty)$.
(a) Find $f^{-1}(x)$. Give the domain of f^{-1}.
(b) On the same set of axes, sketch the graphs of f and f^{-1}.

15. Let $f(x) = (x - 3)^3 - 1$.
(a) Compute $f^{-1}(x)$.
(b) On the same set of axes, sketch the graphs of f, f^{-1}, and the line $y = x$. *Hint:* Sketch f using the ideas of Section 4.4. Then reflect f to get f^{-1}.

16. The figure at the top of the next column shows the graph of a function f. (The axes are marked off in one-unit intervals.) Graph the following functions.
(a) $y = f^{-1}(x)$ **(b)** $y = f(x - 2)$
(c) $y = f(x) - 2$ **(d)** $y = f(-x)$
(e) $y = -f(x)$ **(f)** $y = f^{-1}(x - 2)$
(g) $y = f^{-1}(x) - 2$ **(h)** $y = f^{-1}(x - 2) - 2$
(i) $y = f^{-1}(-x)$ **(j)** $y = -f^{-1}(x)$

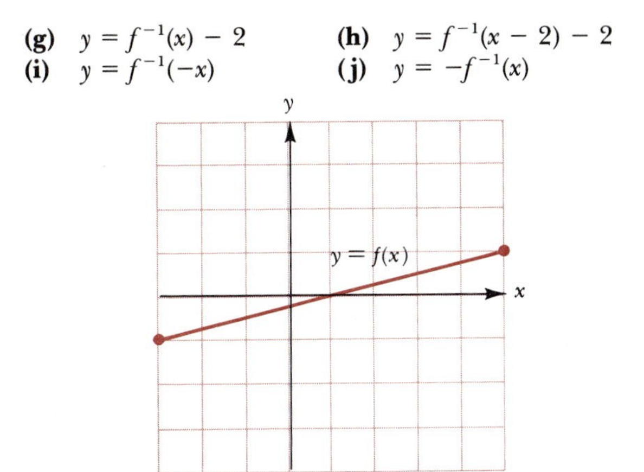

17. The figure below shows the graph of a function g. (The axes are marked off in one-unit intervals.) Graph the following functions.
(a) $y = g^{-1}(x)$ **(b)** $y = g^{-1}(x) - 1$
(c) $y = g^{-1}(x - 1)$ **(d)** $y = g^{-1}(-x)$
(e) $y = -g^{-1}(x)$ **(f)** $y = -g^{-1}(-x)$

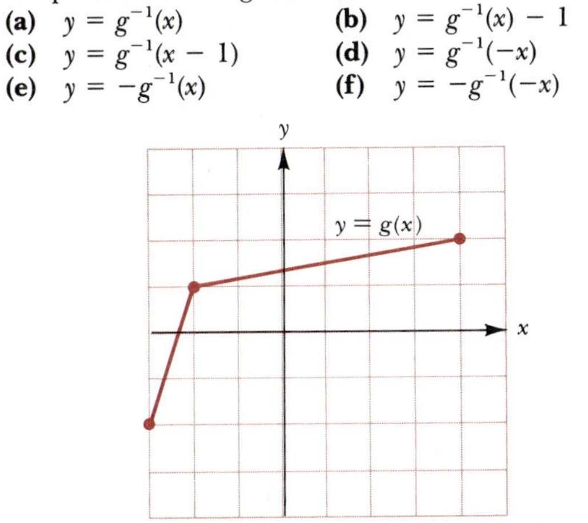

18. The figure below shows the graph of a function $y = h(x)$.
(a) Sketch a graph that is the reflection of h in the y-axis. What is the equation of the graph you obtain?
(b) On the same set of axes, sketch graphs of h, h^{-1}, and the line $y = x$.
(c) Let k be the function whose graph is the reflection of h in the x-axis. Sketch a graph of the function k^{-1}.

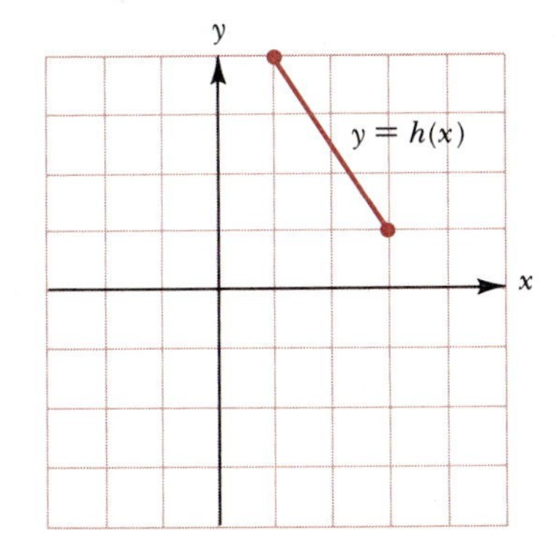

19. Let $f(x) = \dfrac{3x - 2}{5x - 3}$. Show that $f[f(x)] = x$. This says that f is its own inverse.

20. Let $f(x) = \dfrac{ax + b}{cx - a}$. Show that f is its own inverse. (Assume $a^2 + bc \neq 0$.)

21. Let $f(x) = \sqrt{x}$.

(a) Find $f^{-1}(x)$. What is the domain of f^{-1}? [The domain is not $(-\infty, \infty)$.]

(b) In each case, determine whether the given point lies on the graph of f or f^{-1}.

(i) $(4, 2)$ **(ii)** $(2, 4)$

(iii) $(5, \sqrt{5})$ **(iv)** $(\sqrt{5}, 5)$

(v) $(a, f(a))$ (Assume $a \geq 0$.)

(vi) $(f(a), a)$

(vii) $(b, f^{-1}(b))$ (Assume $b \geq 0$.)

(viii) $(f^{-1}(b), b)$

***22.** Let $F(x) = x^3 + 7x - 5$. Find $F^{-1}(-5)$. [Assume that the domain of F^{-1} is $(-\infty, \infty)$.]

23. Let $M(x) = \dfrac{ax + b}{cx + d}$.

(a) Find $M^{-1}(x)$. **(b)** Find $M[M(x)]$.

(c) Find $M[M(0)]$. **(d)** Find $M^{-1}[M(x)]$.

(e) Find $M^{-1}[M(0)]$.

24. Reread the definition of symmetry about a line given on page 201. Then use the method shown in Example 5 to show that the points $(7, -1)$ and $(-1, 7)$ are symmetric about the line $y = x$.

25. Use the method shown in Example 5 to show that $(5, 2)$ and $(2, 5)$ are symmetric about $y = x$.

26. Use the method of Example 5 to show that the two points (a, b) and (b, a) are symmetric about the line $y = x$.

In Exercises 27–38, use the horizontal line test to determine which functions are one-to-one (and therefore have inverses).

27. $y = x^2 + 1$ **28.** $y = \sqrt{x}$

29. $f(x) = 1/x$ **30.** $g(x) = |x|$

31. $y = x^3$ **32.** $y = 1 - x^3$

33. $y = \sqrt{1 - x^2}$ **34.** $y = 3x$

35. $g(x) = 5$ (for all x) **36.** $y = mx + b$ $(m \neq 0)$

37. $f(x) = \begin{cases} x^2 & \text{if } -1 \leq x \leq 0 \\ x^2 + 1 & \text{if } x > 0 \end{cases}$

38. $g(x) = \begin{cases} x^2 & \text{if } -1 \leq x < 0 \\ x^2 + 1 & \text{if } x \geq 0 \end{cases}$

***39.** Determine values for m and b such that the points $(8, 2)$ and $(4, 8)$ are symmetric about the line $y = mx + b$.

***40.** Let $f(x) = 2x^3 - 5$. Define a function F by $F(x) = f(x - 1)$. Find $F[f^{-1}(x) + 1]$.

***41.** Let f be a function whose domain is $(-\infty, \infty)$ and suppose that f^{-1} exists. Let F be the function defined by $F(x) = f(x - 1)$. Show that $F^{-1}(x) = f^{-1}(x) + 1$.

***42.** Reflect the point $(2, 1)$ in the line $y = 3x$. What are the coordinates of the point you obtain?

43. Points P and Q are both reflected in the line $y = x$ to obtain points P' and Q', respectively. Does the distance from P to Q equal the distance from P' to Q'?

44. Pick any three points on the line $y = \frac{1}{2}x + 3$. Reflect each point in the line $y = x$. Show that the three reflected points all lie on one line. What is the equation of that line?

4.7 Variation

Certain formulas or relationships between variables occur so frequently in the applications of mathematics that there is a special terminology used to describe them. For example, if $y = 9x$, we say that *y varies directly as x* or that *y is directly proportional to x*. The number 9 in this case is called the *constant of proportionality*. As another example of this special terminology, suppose that the variables A and B are related by the equation $A = 3/B$. Then we say that *A varies inversely as B* or that *A is inversely proportional to B*. The constant of proportionality in this case is the number 3. Before going on to list other forms of this terminology, let's look at examples of what we have just described.

Example 1 Suppose that F varies directly as a. If $F = 60$ when $a = 5$, find F when $a = 6$.

Solution Since F varies directly as a, we write

$$F = ka$$

To find k, we substitute the given pair of values $F = 60$ and $a = 5$ in this equation. This yields

$$60 = k(5)$$
$$k = 12$$

Now that we know $k = 12$, the equation $F = ka$ becomes

$$F = 12a$$

From this equation we can calculate F whenever a is given. In particular then, if $a = 6$, we obtain $F = 12(6)$. Thus $F = 72$, as required. The formula $F = 12a$ is called the *law of variation* for this problem.

Example 2 At a constant speed, the distance traveled varies directly as the time. Suppose a runner covers 1 mile in 7 minutes. At this rate, what distance would be covered in 30 minutes?

Solution Let d and t denote distance and time, respectively. The statement that distance varies directly as time is written algebraically as

$$d = kt$$

As in Example 1, we first use the given data to calculate k. Since we are given that $d = 1$ when $t = 7$, we obtain

$$1 = k(7)$$
$$k = \frac{1}{7}$$

Now that we know $k = \frac{1}{7}$, the equation $d = kt$ becomes

$$d = \frac{1}{7}t$$

Using this law of variation, we can calculate d whenever t is given. In particular then, if $t = 30$, we obtain $d = \frac{1}{7}(30)$. Thus in 30 minutes, the runner covers $\frac{30}{7} \approx 4.3$ miles.

Example 3 Given that y varies inversely as x and that y is 6 when x is -2, find y when x is 4.

Solution The statement that y varies inversely as x translates algebraically as

$$y = \frac{k}{x}$$

To first find k, we substitute the given pair of values $x = -2$ and $y = 6$ in this equation. This yields

$$6 = \frac{k}{-2}$$
$$k = -12$$

Now that we know $k = -12$, the equation $y = k/x$ becomes

$$y = \frac{-12}{x}$$

Using this law of variation, we can calculate y whenever x is given. In particular then, if $x = 4$, we obtain

$$y = \frac{-12}{4} = -3$$

Thus when x is 4, y is -3; this is the required solution.

The list that follows shows some common formulas that can be expressed in the language of variation. In this list, k denotes a constant in each case, the so-called *constant of proportionality.*

Formula or Law of Variation	Terminology
1. $y = kx^n$	y varies directly as x^n or y is directly proportional to x^n.
2. $y = \frac{k}{x^n}$	y varies inversely as x^n or y is inversely proportional to x^n.
3. $y = kxz$	y varies jointly as x and z or y varies directly as the product of x and z.
4. $y = \frac{kx}{z}$	y varies directly as x and inversely as z.

The rest of this section and the exercises that follow deal with this terminology. Notice that we have already discussed formulas 1 and 2 for the case $n = 1$.

Example 4 The volume of a sphere varies directly as the cube of the radius. When the radius is 2 cm, the volume is $32\pi/3$ cm^3. Find the constant of proportionality k. Write the law of variation.

Solution Let V denote the volume and r the radius. The statement that the volume varies directly as the cube of the radius translates algebraically as

$$V = kr^3$$

To find the constant of proportionality k, we substitute the given values for r and V in this equation to obtain

$$\frac{32\pi}{3} = k \cdot 2^3$$

$$\frac{32\pi}{3} = 8k$$

$$\frac{32\pi}{3 \cdot 8} = k \quad \text{or} \quad k = \frac{4\pi}{3}$$

Now that we have found the constant k, we can write the law of variation simply by substituting this value of k in the equation $V = kr^3$. Thus the law of variation is $V = (4\pi/3)r^3$.

Example 5 B varies inversely as the square of C. What happens to the value of B if C is tripled?

Solution The statement that B varies inversely as the square of C is written algebraically as

$$B = \frac{k}{C^2}$$

To see what happens when C is tripled, replace C by the quantity $(3C)$ in our equation.

$$B = \frac{k}{(3C)^2} = \frac{k}{9C^2}$$

Now compare the original expression for B with the new expression.

original expression for B: $\dfrac{k}{C^2}$

new expression for B after C is tripled: $\dfrac{1}{9} \cdot \dfrac{k}{C^2}$

From this we can see that if C is tripled, B is one ninth of its original value. This is the required solution.

Example 5 differs from those preceding it in this section in that we did not need to evaluate the constant of proportionality k specifically. Indeed, the problem did not supply enough information to calculate a value for k. Example 6 is similar in this respect.

Example 6 The force of attraction between two objects varies directly as the product of their masses and inversely as the square of the distance between them. (This is Newton's Law of Gravitation.) What happens to the force if both masses are tripled and the distance is doubled?

Solution Begin by assigning letters to denote the relevant quantities and then writing the law of variation. Let F, m, M, and d denote, respectively, the force, the masses, and the distance. Then the given statement translates algebraically as

$$F = \frac{kmM}{d^2}$$

We want to find out what happens to F when both masses are tripled and the distance is doubled. To do this, we make the following substitutions in our law of variation:

replace m by $(3m)$
replace M by $(3M)$
replace d by $(2d)$

We then have

$$F = \frac{k(3m)(3M)}{(2d)^2}$$

$$= \frac{9kmM}{4d^2}$$

Now compare the original expression for F with the expression obtained after tripling the masses and doubling the distance.

$$\text{original expression for } F: \quad \frac{kmM}{d^2}$$

$$\text{new expression:} \quad \frac{9}{4}\left(\frac{kmM}{d^2}\right)$$

From this we can see that the new force is nine fourths of the original force. In other words, the new force is $2\frac{1}{4}$ times as strong as the original. This is the required solution.

Example 7 Suppose that A varies jointly as X and Y. Also, suppose that X varies inversely as the square of Y. Show that A varies inversely as Y.

Solution Translating the two given statements into equations gives us

$$A = kXY \tag{1}$$

and

$$X = \frac{c}{Y^2} \tag{2}$$

Both k and c are constants here. We do need the two different letters since there is no reason to suppose that $k = c$. Now we are asked to show that A varies inversely as Y. To do this, just substitute for X in equation (1) using the value given in equation (2). We have

$$\begin{aligned} A &= kXY \\ &= k\left(\frac{c}{Y^2}\right)Y \\ &= \frac{kc}{Y} \end{aligned}$$

This last equation says that A equals a constant divided by Y; that is, A varies inversely as Y. If we want, we can denote kc by the single letter K. Then we have $A = K/Y$, which again says that A varies inversely as Y.

Exercise Set 4.7

In Exercises 1–5, translate the given statement into an equation.

1. **(a)** y varies directly as x.
   (b) A varies inversely as B.
2. **(a)** y varies directly as the cube of z.
   (b) s varies inversely as t^2.
3. **(a)** x varies jointly as u and v^2.
   (b) z varies jointly as A^2 and B^3.
4. **(a)** y varies directly as u and inversely as v.
   (b) y varies directly as the square root of x and inversely as the product of w and z.
5. **(a)** F is inversely proportional to r^2.
   (b) V^2 is directly proportional to the sum of U^2 and T^2.
6. If y varies directly as x, and y is 20 when $x = 4$, find y when $x = -6$.
7. A varies inversely as B, and $A = -1$ when $B = 2$.
   (a) Find the constant of proportionality.
   (b) Write the law of variation.
   (c) Find A when $B = \frac{5}{4}$.
8. T varies directly as V^2. Also $T = 8$ when $V = 4$.
   (a) Find the constant of proportionality and write the law of variation.
   (b) Find T when $V = 10$.

(c) Find T when $V = \sqrt{10}$.

9. x varies jointly as y and z. Also $x = 9$ when $y = 2$ and $z = -3$. Find x when both y and z are 4.

10. F varies directly as the product of q and Q and inversely as the square of d. When $q = 2$, $Q = 4$, and $d = \frac{1}{2}$, the value of F is 8.

(a) Find the constant of proportionality and write the law of variation.

(b) Find F when $q = 3$, $Q = 6$, and $d = 2$.

11. A varies jointly as B and C. If B is tripled and C is doubled, what happens to A?

12. S varies inversely as the square of T. What happens to S if T is cut in half?

13. x varies jointly as B and C and inversely as the square root of A. What happens to x if A, B, and C are all quadrupled?

14. Suppose that $xy = \frac{3}{4}$. Does y vary directly or inversely with x? What is the constant of proportionality in this case?

15. The surface area of any sphere varies directly as the square of the radius. When the radius is 2 cm, the surface area is 16π cm^2.

(a) Find the constant of proportionality and write the law of variation.

(b) Find the surface area when the radius is $\sqrt{3}$ cm.

16. For motion at a constant speed, the distance varies directly as the time. Suppose a car covers 10 mi in 15 min. At this rate, what distance would be covered in 50 min?

17. The force of attraction between two objects varies jointly as their masses and inversely as the square of the distance between them. What happens to the force if one mass is tripled, the other mass is quadrupled, and the distance is cut in half?

18. The magnitude of the force between two electrical charges Q_1 and Q_2 varies directly as the product of Q_1 and Q_2 and inversely as the square of the distance between the charges. (This is Coulomb's Law, named after the French physicist Charles Coulomb, who verified the law experimentally in 1785.) What happens to the magnitude of the force if Q_1 is doubled, Q_2 is quadrupled, and the distance is tripled?

19. Neglecting air resistance, the distance any object falls starting from rest is directly proportional to the square of the time. (This law was discovered experimentally by Galileo, sometime before the year 1590. In his first experiments, Galileo used his own pulse to measure time.) Given that an object starting from rest will fall 490 m in 10 sec, how far will an object fall in 5 sec, starting from rest?

20. The volume of a right circular cone varies jointly as the height and the square of the radius. When the radius and height are both 3 cm, the volume is 9π cm^3. Find the volume of a cone having a height of 4 cm and a radius of $\frac{3}{2}$ cm.

21. (Boyle's Law) In 1662, Robert Boyle observed that, for a sample of gas at a constant temperature, the volume is inversely proportional to the pressure.

(a) Suppose that, at a certain fixed temperature, the pressure of a 2-liter sample of gas is 1.025 atm. (atm is the abbreviation for atmospheres, a unit of pressure. 1 atm = 14.7 lb/in^2, the air pressure at sea level.) Calculate the constant of proportionality and write the law of variation.

(b) Suppose that, with the temperature still fixed, the pressure on our sample of gas is reduced to 1 atm. Calculate the new volume of the gas.

22. (Charles' Law) In 1787, Jacques Charles stated that, for a sample of gas held at a constant pressure, the volume is directly proportional to the Kelvin temperature. Suppose that initially you have 300 ml of a gas at 300 K. Then, holding the pressure constant, you heat the gas to 310 K. What is the new volume?

23. (a) For a satellite in a circular orbit about the earth, the kinetic energy varies directly as the mass of the satellite and inversely as the radius of the orbit. What happens to the kinetic energy if we cut the radius in half and triple the mass?

(b) The velocity of an earth satellite in a circular orbit varies inversely as the square root of the radius of the orbit. What happens to the velocity if we cut the radius in half?

24. (Hooke's Law) In 1676, the English physicist Robert Hooke found that the distance a spring is stretched varies directly as the force pulling on the spring. Suppose that, for a certain spring, a pull of 5 lb results in a stretch of 3 in.

(a) Find the constant of proportionality and write the law of variation.

(b) How far will a pull of 12 lb stretch the spring?

25. The weight of an object above the earth's surface varies inversely as the square of its distance from the center of the earth. How much would an astronaut weigh 500 mi above the surface of the earth, assuming her weight at the surface is 140 lb? (Assume that the radius of the earth is 4000 mi.)

26. One of the ways in which the human body adjusts blood flow is through changes in the radii of small arteries called *arterioles.* According to Poiseuille's Law [named after the French physiologist J. L. Poiseuille (1779–1869)], the blood flow in a given arteriole (measured, for example, in cm^3/sec) varies directly as the fourth power of the radius. By what fraction is the blood flow through an arteriole multiplied if the radius is decreased by 20%?

***27.** Suppose that A varies directly as B. Show that $A^2 + B^2$ varies directly as $A^2 - B^2$ $(k \neq \pm 1)$.

***28.** Suppose that x varies directly as z and that y varies directly as z.

(a) Does $x + y$ vary directly as z?

(b) Does xy vary directly as z?

(c) Does $\sqrt{xy}$ vary directly as z? (Assume x, y, and z are positive.)

***29.** Suppose that A varies directly as B. Show that AB varies directly as $A^2 + B^2$.

30. The graphs of six functions are displayed below. Which graph best represents the given statement?

(a) y varies directly as x, where the constant of proportionality is negative.

(b) y varies inversely as x.

(c) y varies directly as x^2, with a positive constant of proportionality.

(d) y varies directly as x, with a positive constant of proportionality.

(e) y varies directly as x^2, with a negative constant of proportionality.

33. At highway speeds, the wind resistance against a car varies directly as the square of the speed.

(a) If the speed is increased from 40 mph to 60 mph, by what factor is the wind resistance multiplied?

(b) At what speed is the wind resistance twice as much as that at 50 mph?

34. The intensity of illumination due to a small light source varies inversely as the square of the distance from the light. Suppose a book is 6 ft from a lamp. At what distance from the lamp will the illumination be twice as great?

35. The volume of a coin varies jointly as the thickness and the square of the radius. Suppose that two silver dollars are melted down (illegal!) and cast into one coin of the same thickness as a silver dollar. Show that the diameter of the new coin is $\sqrt{2}$ times the diameter of a silver dollar.

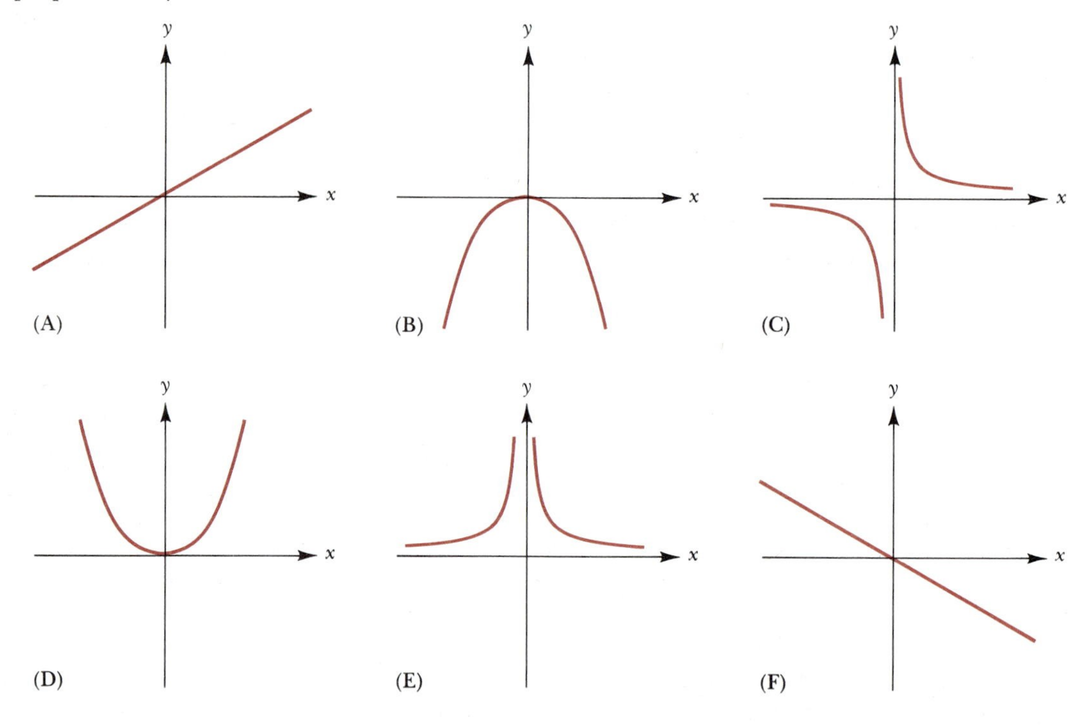

31. Three solid lead spheres of radii a, b, and c, respectively, are melted down and recast into one large sphere of radius R. Show that $R = \sqrt[3]{a^3 + b^3 + c^3}$. Use the fact that the volume of a sphere varies directly as the cube of the radius.

***32.** The *period* of a pendulum is the time required for one complete swing, forward and back. If the period of a pendulum varies directly as the square root of the length of the pendulum, by how much should a pendulum of length L be shortened in order to cut the period in half? (*Answer:* $\frac{3}{4}L$)

36. Show that the area of the shaded triangle in the figure varies jointly as m and the square of x.

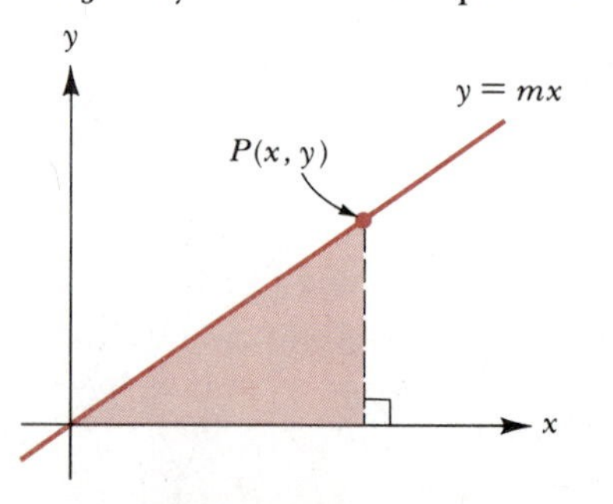

37. Show that the area of the shaded triangle in the figure varies directly as a^2.

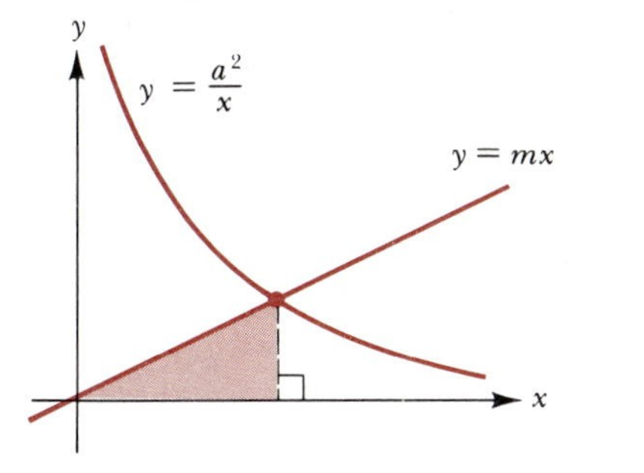

***38.** For this exercise, refer to the figure.

(a) Show that the area of triangle OAB varies inversely as b.

(b) Show that the area of triangle I varies directly as a and inversely as the product of b and $a - b$.

(c) Show that the area of triangle II varies inversely as $a - b$.

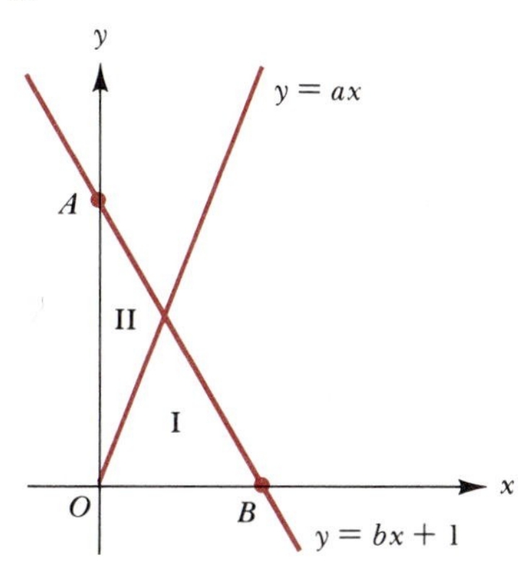

***39.** Suppose that $x + y$ varies directly as $x - y$. Also suppose that the constant of proportionality is not ± 1.

(a) Show that y varies directly as x.

(b) Show that $x^2 + y^2$ varies jointly as x and y.

(c) Show that $x^3 + y^3$ varies jointly as x, y, and $x + y$.

***40.** Suppose that $A^3 + 1/B^3$ varies directly as $A^3 - 1/B^3$. Also suppose that the constant of proportionality is not ± 1. Show that B varies inversely as A.

***41.** The weight of an object on or above the earth's surface varies inversely as the square of the object's distance from the center of the earth. At what height x above the earth's surface will an object weigh half as much as it does on the surface? Assume the radius of the earth is 4000 mi. [*Answer:* $4000(\sqrt{2} - 1) \approx 1656.9$ mi]

C **42.** In the year 1619, Johannes Kepler announced his so-called Third Law of Planetary Motion. This law states that the square of the time it takes for a planet to complete one revolution about the sun varies directly as the cube of the planet's average distance from the sun. A table giving the average distance of each planet from the sun is shown below. Use a calculator to find the time it takes for each planet to complete one revolution about the sun. For earth, assume the time is 365.2564 days.

Planet	Average Distance from Sun Measured in Astronomical Units
Mercury	0.387099
Venus	0.723332
Earth	1.000000
Mars	1.523691
Jupiter	5.204377
Saturn	9.577971
Uranus	19.26020
Neptune	30.09421
Pluto	39.82984

C **43.** Most astronomers accept the theory that the universe is expanding and that the galaxies are receding from one another at velocities that vary directly as their distances. Given that a galaxy in Virgo, which is 78,000,000 light years from us, is receding at 745 mi/sec, estimate the distance from us of the Hydra galaxy, which is receding from us at 37,881 mi/sec.

Chapter 4 Review Checklist

- □ Function (page 154)
- □ Domain and range (pages 154–155)
- □ $f(x)$ notation (page 156)
- □ Difference quotients (pages 156–157)
- □ Independent variable and dependent variable (page 158)
- □ Graph of a function (page 172)
- □ Vertical line test (page 175)
- □ Greatest integer function $y = [x]$ (page 178)
- □ Graph of $y = f(x) + c$ (page 184)
- □ Graph of $y = f(x + c)$ (page 185)
- □ Graph of $y = f(-x)$ (page 186)
- □ Graph of $y = -f(x)$ (page 187)
- □ Sum, difference, product, and quotient functions (page 191)
- □ Composition of two functions (page 192)
- □ Inverse functions (page 198)
- □ Symmetry about a line (page 201)
- □ One-to-one function (page 203)
- □ Horizontal line test (page 203)
- □ Direct variation, inverse variation, and joint variation (page 208)

Chapter 4 Review Exercises

In Exercises 1–23, sketch the graphs and specify any x- *or* y-*intercepts.*

1. $y = 4 - x^2$

2. $y = -(x - 1)^2 + 2$

3. $y = \frac{1}{x} + 1$

4. $f(x) = \frac{1}{x + 1}$

5. $y = \frac{1}{x + 1} + 1$

6. $y = |x + 3|$

7. $g(x) = -\sqrt{x - 4}$

8. $h(x) = \sqrt{1 - x^2}$

9. $f(x) = \sqrt{1 - x^2} + 1$

10. $y = 1 - (x + 1)^3$

11. $y = 4 - \sqrt{-x}$

12. $y = (\sqrt{x})^2$

13. $f \circ g$ where $g(x) = x + 3$ and $f(x) = x^2$

14. $f \circ g$ where $g(x) = \sqrt{x - 1}$ and $f(x) = -x^2$

15. $f(x) = \begin{cases} \sqrt{1 - x^2} & \text{if } -1 \le x \le 0 \\ \sqrt{x} + 1 & \text{if } x > 0 \end{cases}$

16. $F(x) = \begin{cases} -\sqrt{1 - x^2} & \text{if } -1 \le x < 0 \\ \sqrt{x} & \text{if } x \ge 0 \end{cases}$

17. $y = \begin{cases} |x - 1| & \text{if } 0 \le x \le 2 \\ |x - 3| & \text{if } 2 < x \le 4 \end{cases}$

18. $y = \begin{cases} 1/x & \text{if } 0 < x \le 1 \\ 1/(x - 1) & \text{if } 1 < x \le 2 \end{cases}$

19. $(y + |x| - 1)(y - |x| + 1) = 0$

20. f^{-1} where $f(x) = \frac{x + 1}{2}$

21. g^{-1} where $g(x) = \sqrt[3]{x + 2}$

22. $f \circ f^{-1}$ where $f(x) = \sqrt{x - 2}$

23. $f^{-1} \circ f$ where $f(x) = \sqrt{x - 2}$

For Exercises 24–27, assume that the graph of the function f *is the straight line segment joining the points $(-4, -1)$ and $(0, 1)$.*

24. Find the slope of the line segment described by $y = -f(x)$.

25. Find the slope of the line segment described by $y = f(-x)$.

26. Find the midpoint of the line segment described by $y = -f(x + 2) + 5$.

27. Find the midpoint of the line segment described by $y = f^{-1}(x) + 1$.

Find the domains of the given functions in Exercises 28–37.

28. $y = \frac{1}{x^2 - 9}$

29. $y = x^3 - x^2$

30. $y = \sqrt{8 - 2x}$

31. $y = \frac{x}{6x^2 + 7x - 3}$

32. $y = \sqrt{|2 - 5x|}$

33. $y = \frac{25 - x^2}{\sqrt{x^2 + 1}}$

34. $y = \sqrt{x^2 - 2x - 3}$

35. $y = \sqrt{5 - x^2}$

36. $y = x + 1/x$

37. $y = \frac{1}{x - [x]}$ ($[x]$ denotes the greatest integer not exceeding x, as defined in Section 4.3.)

In Exercises 38–43, determine the range of each of the given functions.

38. $y = \frac{x + 4}{3x - 1}$

39. $y = \frac{2x - 3}{x - 2}$

40. $f \circ g$ where $f(x) = 1/x$ and $g(x) = 3x + 4$

41. $g \circ f$ where $f(x) = \frac{x + 2}{x - 1}$ and $g(x) = \frac{x + 1}{x + 4}$

42. f^{-1} where $f(x) = \frac{x}{3x - 6}$

43. f^{-1} where $f(x) = \frac{5 - x}{1 + x}$

In Exercises 44–51, express the functions as a composition of two or more of the following functions:

$$f(x) = \frac{1}{x} \quad g(x) = x - 1 \quad F(x) = |x| \quad G(x) = \sqrt{x}$$

44. $a(x) = \frac{1}{x - 1}$

45. $b(x) = \frac{1}{x} - 1$

46. $c(x) = \sqrt{x - 1}$

47. $d(x) = \sqrt{x} - 1$

48. $A(x) = \frac{1}{\sqrt{x}} - 1$

49. $B(x) = |x - 2|$

50. $C(x) = \sqrt[4]{x} - 1$

51. $D(x) = \frac{1}{\sqrt{x - 3}}$

For Exercises 52–85, compute each of the indicated quantities using the functions f, g, *and* F *defined as follows:*

$$f(x) = x^2 - x \quad g(x) = 1 - 2x \quad F(x) = \frac{x - 3}{x + 4}$$

52. $f(-3)$

53. $f(1 + \sqrt{2})$

54. $F(\frac{3}{4})$

55. $f(t)$

56. $f(-t)$

57. $g(2x)$

58. $f(x - 2)$

59. $g(x + h)$

60. $g(2) - g(0)$

61. $f(x) - g(x)$

62. $|f(1) - f(3)|$

63. $f(x + h) - f(x)$

64. $f(x^2)$

65. $f(x)/x \quad (x \ne 0)$

66. $[f(x)][g(x)]$

67. $f[f(x)]$

68. $f[g(x)]$

69. $g[f(3)]$

70. $(g \circ f)(x)$

71. $(g \circ f)(x) - (f \circ g)(x)$

72. $(F \circ g)(x)$

73. $\frac{g(x + h) - g(x)}{h}$

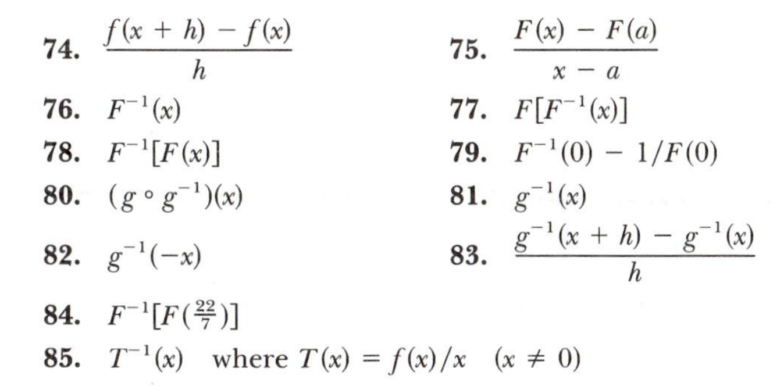

74. $\dfrac{f(x+h)-f(x)}{h}$

75. $\dfrac{F(x)-F(a)}{x-a}$

76. $F^{-1}(x)$

77. $F[F^{-1}(x)]$

78. $F^{-1}[F(x)]$

79. $F^{-1}(0) - 1/F(0)$

80. $(g \circ g^{-1})(x)$

81. $g^{-1}(x)$

82. $g^{-1}(-x)$

83. $\dfrac{g^{-1}(x+h)-g^{-1}(x)}{h}$

84. $F^{-1}[F(\frac{22}{7})]$

85. $T^{-1}(x)$ where $T(x) = f(x)/x \quad (x \neq 0)$

In Exercises 86–97, refer to the graph of the function f in the figure. (The axes are marked off in one-unit intervals.)

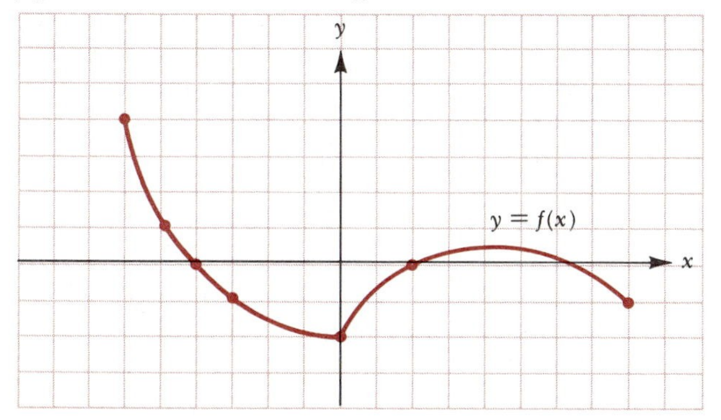

86. Is $f(0)$ positive or negative?

87. Specify the domain and range of f.

88. Find $f(-3)$.

89. Which is larger, $f(-\frac{5}{2})$ or $f(-\frac{1}{2})$?

90. Compute $f(0) - f(8)$.

91. Compute $|f(0) - f(8)|$.

92. As x increases from -4 to 0, do the corresponding y-values increase or decrease?

93. For which x-values is it true that $1 \leq f(x) \leq 4$? (Your answer should be a certain closed interval.)

94. What is the largest value of $f(x)$ when $|x| \leq 2$?

95. Is f a one-to-one function?

96. Does f possess an inverse function?

97. Compute $f[f(-4)]$.

For Exercises 98–109, refer to the graphs of the functions f and g in the figure at the top of the next column. Assume that the domain of each function is [0, 10].

98. For which x-value is $f(x) = g(x)$?

99. For which x-values is it true that $g(x) \leq f(x)$?

100. **(a)** For which x-value is $f(x) = 0$?
(b) For which x-value is $g(x) = 0$?

101. Compute $f(0) + g(0)$.

102. Compute each of the following.
(a) $(f + g)(8)$ **(b)** $(f - g)(8)$
(c) $(fg)(8)$ **(d)** $(f/g)(8)$

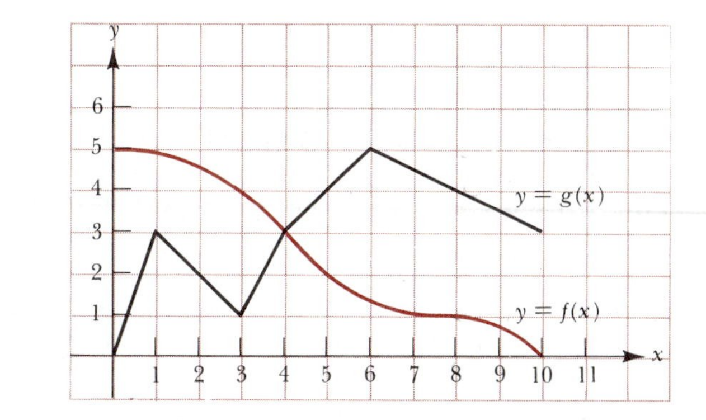

103. Compute each of the following.
(a) $g[f(5)]$ **(b)** $f[g(5)]$
(c) $(g \circ f)(5)$ **(d)** $(f \circ g)(5)$

104. Which is larger, $(f \circ f)(10)$ or $(g \circ g)(10)$?

105. Compute $g[f(10)] - f[g(10)]$.

106. For which x-values is it true that $f(x) \geq 3$?

107. For which x-values is it true that $|f(x) - 3| \leq 1$?

108. What is the largest number in the range of g?

109. Specify the coordinates of the highest point on the graph of each of the following equations.
(a) $y = g(-x)$ **(b)** $y = -g(x)$
(c) $y = g(x - 1)$ **(d)** $y = f(-x)$
(e) $y = -f(x)$ **(f)** $y = -f(-x)$

110. A varies jointly as x and y^2. What happens to the value of A when x is tripled and y is doubled?

111. B varies directly as x and inversely as y^3. When $x = 6$ and $y = 2$, the value of B is 9. Find B when $x = 3$ and $y = 4$.

112. The intensity of illumination from a small light source varies inversely as the square of the distance from the light. Suppose that a book is 10 ft from a lamp. At what distance from the lamp will the illumination be five times as great?

113. The velocity of an earth satellite in a circular orbit varies inversely as the square root of the radius of the orbit. What happens to the velocity if the radius is decreased by 10 percent?

114. The length of a side in an equilateral triangle varies directly as the square root of the area of the triangle. When the area is $16\sqrt{3}$ cm^2, the length of a side is 8 cm. Find the length of a side when the area is 20 cm^2.

115. If $f^{-1}(3) = 4$, what is $f(4)$?

116. Let $f(x) = 3x + 1$ and $g(x) = 2x + b$. Find b, given that the equation $f[g(x)] = g[f(x)]$ holds for all values of x.

117. Let $F(x) = x^2$ and $G(x) = 1/x$. Assume that the domain of both F and G is the set of all nonzero real numbers.
(a) Compute $F[G(x)]$ and $G[F(x)]$.
(b) Are F and G inverse functions?

118. Let $F(x) = \dfrac{2^x - 2^{-x}}{2}$ and $G(x) = \dfrac{2^x + 2^{-x}}{2}$. Show that $F(2x) = 2F(x)G(x)$.

119. A circle of radius r is inscribed in a square. Express the area of the square as a function of r.

120. Let $P(x, y)$ be a point on the semicircle $y = \sqrt{1 - x^2}$.
(a) Express the distance from P to the point $(-1, 0)$ as a function of x.
(b) Use the formula that you obtained in part (a) to determine the coordinates of a point Q on the semicircle that is one unit from P.

121. The sum of two numbers is 8.
(a) Express the sum of the squares of the two numbers in terms of a single variable.
(b) Express the sum of the cubes of the two numbers in terms of a single variable.

122. The perimeter of a rectangle is 42 cm. Express the area as a function of the width x.

123. Suppose that we are given a demand function $p = 6 - x/2$. (See Exercise 12 in Exercise Set 4.2 for the terminology. Assume that p is in dollars.)
(a) How many units can be sold when the unit price is \$2? What is the corresponding revenue R? (Use the formula $R = xp$ given in Example 9 in Section 4.2.)
(b) How many units can be sold when the unit price is \$4? What is the corresponding revenue?

124. (Continuation of Exercise 123.) Graph the revenue function $R = xp = x(6 - x/2)$. According to your graph, which x-value yields the largest revenue? What is that revenue? What is the corresponding unit price?

125. A line with slope m $(m < 0)$ passes through the point $(3, 4)$. Express the area of the triangular region in the first quadrant bounded by this line and the axes as a function of m.

126. A piece of wire of length L (a constant) is cut in two. Call the lengths of the two pieces x and $L - x$, respectively. Each piece of wire is then bent into a circle. Express the sum of the areas of the two circles as a function of x.

127. Let $f(x) = -\sqrt{x - 3} + 2$. Compute $f^{-1}(x)$. On the same set of axes graph f, f^{-1}, and the line $y = x$.

128. Let F be the function that assigns to each person their shoe size. Let G be the function that assigns to each person their set of fingerprints. Which function has an inverse function?

Chapter 4 Test

1. Find the domain of the function defined by $y = \sqrt{12 - 3x}$.

2. Find the range of the function defined by $y = \dfrac{4 - x}{3x + 1}$.

3. Let $f(x) = 3x^2 - 2x$.
(a) Find $f(-1)$. **(b)** Find $f(1 - \sqrt{2})$.

4. Let $g(x) = 2x^2 + 1$. Compute (and simplify, as usual) $\dfrac{g(x + h) - g(x)}{h}$.

5. Let $F(x) = 1/x$. Compute (and simplify, as usual) $\dfrac{F(x) - F(a)}{x - a}$.

6. The figure shows a portion of the graph of a function f. (Assume the axes are marked off in one-unit intervals.)

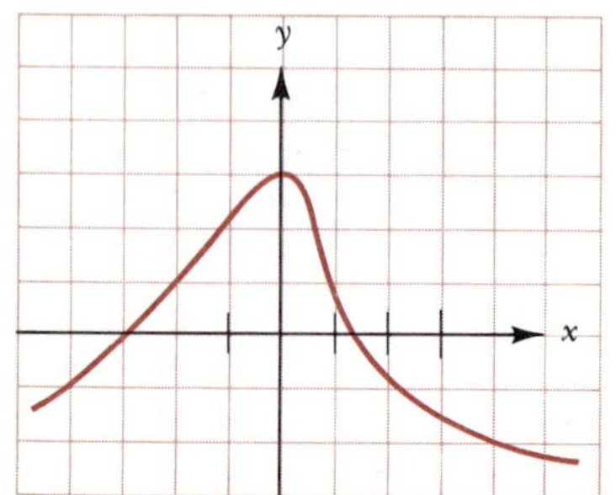

(a) Which is larger, $f(-2)$ or $f(2)$?
(b) Compute $f(-3)$.

7. Graph the function G defined as follows:

$$G(x) = \begin{cases} x^3 & \text{if } x < 1 \\ \sqrt{x} & \text{if } x \geq 1 \end{cases}$$

8. Graph the function $y = [x]$ on the interval $[-3, 0]$.

9. Graph the function $y = |x - 3| + 1$.

10. Graph the function $y = -\sqrt{x + 1}$.

11. The figure shows the graph of a function $y = f(x)$. Sketch the graph of $y = f(-x)$.

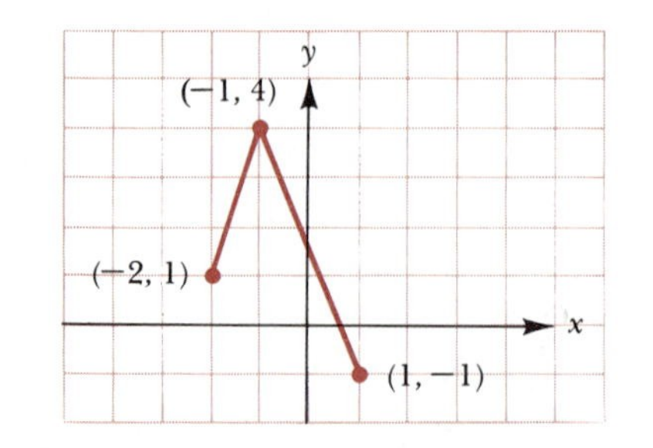

12. Let $f(x) = 2x^2 - 5x + 1$ and $g(x) = 1 + 5x$.
(a) Find $(f + g)(x)$. **(b)** Compute $(f/g)(1)$.

13. Let $f(x) = 3x^2$ and $g(x) = 2x - 3$.
(a) Compute $(f \circ g)(x)$. **(b)** Compute $f[g(0)]$.

14. Let $F(x) = 1 - 2x$. Find $F^{-1}(x)$. Then, on the same set of axes, sketch the graphs of F, F^{-1}, and the line $y = x$.

15. The figure below displays the graph of a function f. Sketch the graph of f^{-1}.

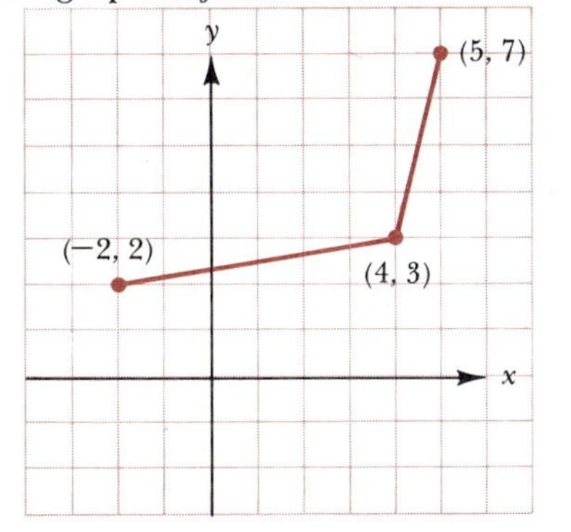

16. Suppose that F varies directly as a. If $F = 120$ when $a = 10$, find F when $a = 15$.

17. The gravitational force of attraction between two objects varies jointly as their masses and inversely as the square of the distance between them. What happens to the force if both masses are doubled and the distance is cut in half?

18. Let $P(x, y)$ be a point [other than $(-1, -1)$] on the graph of $f(x) = x^3$. Express the slope of the line passing through the points P and $(-1, -1)$ as a function of x. Simplify your answer as far as possible.

19. A rectangle is inscribed in a circle. The circumference of the circle is 12 cm. Express the perimeter of the rectangle as a function of its width w.

20. Suppose that T varies inversely as U^2. If $T = 5$ when $U = 1$, find T when $U = 5$.

Chapter 5 Polynomial Functions, Rational Functions, and Conic Sections

Introduction In the previous two chapters we studied some rather general rules for graphing functions and equations. Now we focus our attention on the graphs of several specific types of functions and equations. Section 5.1 begins with a brief consideration of linear functions; the remainder of the section then covers quadratic functions and their graphs. In the next two sections a number of techniques are presented for graphing polynomial functions and rational functions. Finally, in the last two sections of this chapter, we discuss the graphs of the conic sections.

5.1 Linear and Quadratic Functions

We are going to define the terms *linear function* and *quadratic function.* As you will soon see, many of the methods for dealing with these functions have already been developed in previous chapters. By a *linear function,* we mean a function defined by an equation of the form

$$f(x) = Ax + B$$

where A and B are constants. In this chapter the constants A and B will always be real numbers. From our work in Chapter 3 we know that the graph of $y = Ax + B$ is a straight line. So, aside from terminology and the use of functional notation, there is nothing really new here.

Example 1 f is a linear function. If $f(1) = 0$ and $f(2) = 3$, find an equation defining f.

Solution From the statement of the problem we know that the graph of f is a straight line passing through the points $(1, 0)$ and $(2, 3)$. Thus the slope of the line is

$$m = \frac{y_2 - y_1}{x_2 - x_1} = \frac{3 - 0}{2 - 1} = 3$$

Now we can use the point-slope formula to find the required equation. We have

$$\begin{aligned} y - y_1 &= m(x - x_1) \\ y - 0 &= 3(x - 1) \\ y &= 3x - 3 \end{aligned}$$

This is the equation defining f. If we wish, we can rewrite it using functional notation: $f(x) = 3x - 3$.

An important application of linear functions occurs in business and economics with *linear* or *straight line depreciation*. In this situation we assume that the value V of an asset (such as a machine or an apartment building) decreases linearly over time t; that is, $V = mt + b$, where the slope m is negative.

Example 2 A factory owner buys a new machine for \$8000. After 10 years the machine has a salvage value of \$500.

(a) Assuming linear depreciation (as indicated in Figure 1), find a formula for the value V of the machine after t years, where $0 \le t \le 10$.

(b) Use the formula derived in part (a) to find the value of the machine after 5 years.

Solution

(a) We need to determine m and b in the equation $V = mt + b$. From Figure 1 we see that the V-intercept of the line segment is $b = 8000$. Furthermore, since the line segment passes through the two points (10, 500) and (0, 8000), the slope m is $\dfrac{8000 - 500}{0 - 10}$, or -750. So we have $m = -750$ and $b = 8000$, and the required equation is

$$V = -750t + 8000 \qquad (0 \le t \le 10)$$

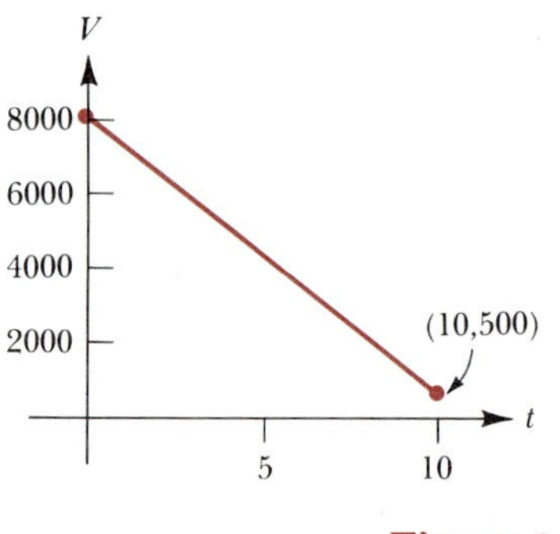

Figure 1

(b) Substituting $t = 5$ in the equation $V = -750t + 8000$ yields

$$\begin{aligned} V &= -750(5) + 8000 \\ &= -3750 + 8000 = 4250 \end{aligned}$$

Thus the value of the machine after 5 years is \$4250.

Next we turn to *quadratic functions*. These are functions defined by equations of the form

$$f(x) = ax^2 + bx + c \qquad (a \neq 0)$$

where a, b, and c are constants and (as already indicated to the right of the equation) a is not zero. In this chapter the constants a, b, and c will always be real numbers. What we are going to see is that the graph of any quadratic function is a curve called a *parabola*, which is similar or identical in shape to the graph of $y = x^2$. Figure 2 displays the graphs of two typical parabolas.

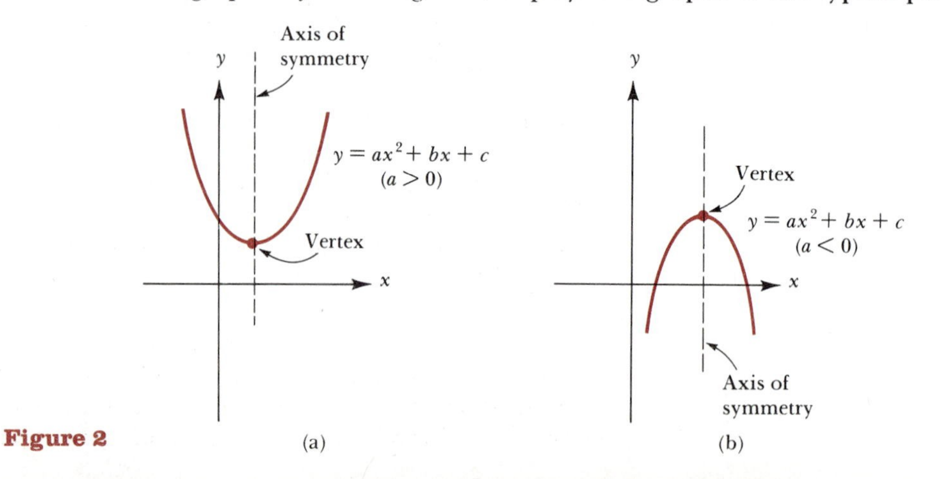

Figure 2

Subsequent examples will indicate that the parabola $y = ax^2 + bx + c$ opens upward when $a > 0$ and downward when $a < 0$. As Figure 2 indicates, the low or high point on the parabola is called the *vertex*. The *axis of symmetry* of the parabola $y = ax^2 + bx + c$ is the vertical line passing through the vertex.

At the beginning of this section we mentioned that many of the methods for dealing with quadratic functions were covered in previous chapters. In particular, the following two topics are prerequisites for understanding the rest of the examples in this section:

1. Completing the square. For a review, see Section 2.2.
2. Translations and reflections. For a review, see Section 4.4.

Example 3 Graph the function $y = x^2 - 2x + 3$.

Solution The idea here is to use the technique of completing the square; this will enable us to obtain the required graph simply by shifting the basic $y = x^2$ graph. Begin by writing the given equation as

$$y = x^2 - 2x \qquad + 3$$

To complete the square for the x-terms we want to add 1. (Check this.) Of course, to keep the equation in balance we have to account for this. This is accomplished by writing

$$y = (x^2 - 2x + 1) + 3 - 1 \qquad \text{We've added zero to the right side.}$$

or

$$y = (x - 1)^2 + 2$$

Now, as we know from Section 4.4, the graph of this last equation is obtained by moving the parabola $y = x^2$ one unit in the positive x-direction and two units in the positive y-direction. See Figure 3. *Note:* As a guide before sketching the graph yourself, you'll want to know the y-intercept. To find the y-intercept, substitute $x = 0$ in the given equation to obtain $y = 3$. Then, given the vertex (1, 2) and the point (0, 3), a reasonably accurate graph can be quickly sketched. [Actually, once you find that (0, 3) is on the graph, you also know that the reflection of this point about the axis of symmetry is on the graph. This is why the point (2, 3) is shown in Figure 3.]

Figure 3

As Figure 3 indicates, the lowest point on the graph of the function $y = x^2 - 2x + 3$ is the point (1, 2). We say that 2 is the *minimum value* of this function, and we say that this minimum occurs at $x = 1$. More generally, given a function f, we say that f has a *minimum* at $x = x_0$ if $f(x_0) \leq f(x)$ for all x in the domain of f. The number $f(x_0)$ in this case is called a *minimum value* of the function. We can also talk about a maximum value for a function. Figure 4, for example, shows the graph of $f(x) = -x^2 + 4$. The *maximum value* of f in this case is 4, and this maximum occurs when $x = 0$. More generally, given a function f, we say that f has a *maximum* at $x = x_0$ if $f(x_0) \geq f(x)$ for all x in the domain of f. The number $f(x_0)$ in this case is called a *maximum value* of the function.

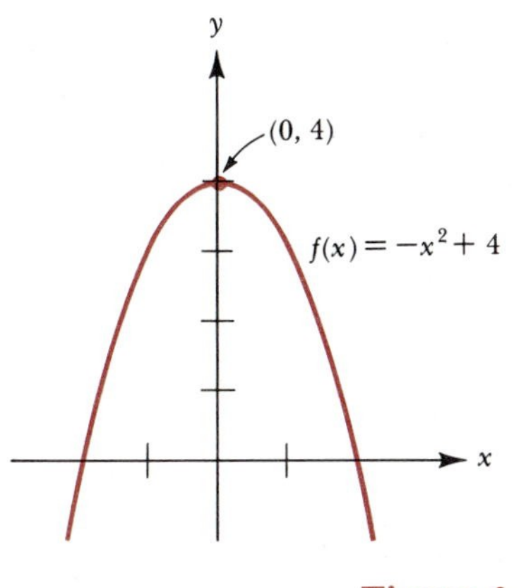

Figure 4

Now we want to compare the graphs of $y = x^2$, $y = 2x^2$, and $y = \frac{1}{2}x^2$. The last two equations were not discussed in Chapter 4, so for the moment you can think about graphing them by first setting up tables. Figure 5 displays the graphs. All three graphs are parabolas that open upward; but notice

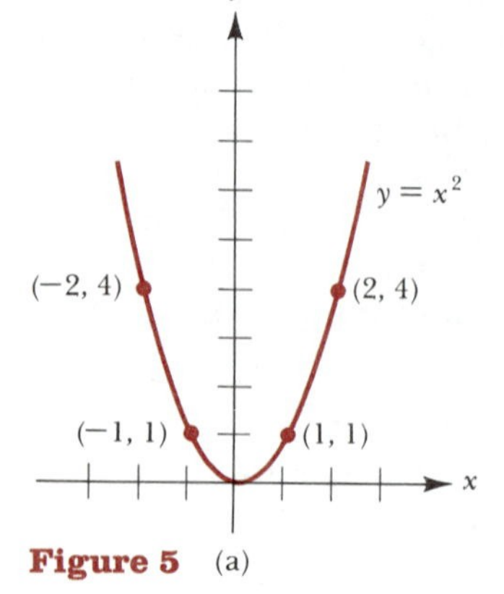

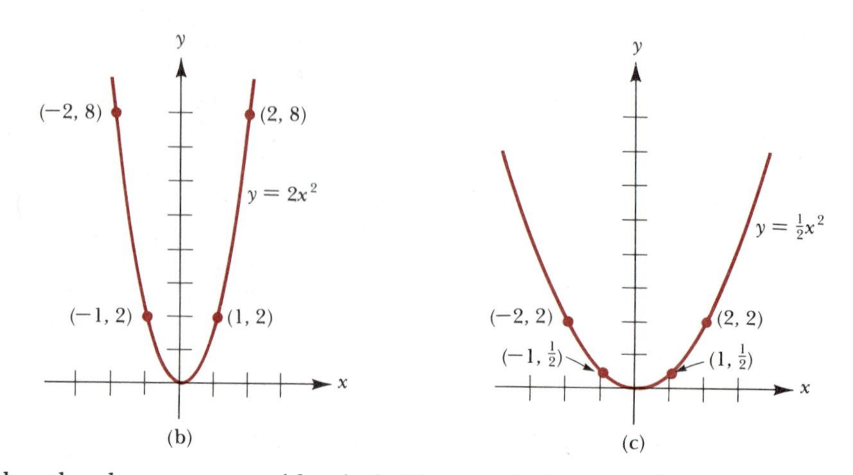

Figure 5 (a)

that the shapes are not identical. The parabola $y = 2x^2$ is narrower than $y = x^2$, while $y = \frac{1}{2}x^2$ is wider than $y = x^2$. These same observations about shape would also apply to $y = -x^2$, $y = -2x^2$, and $y = -\frac{1}{2}x^2$. In these cases, though, the parabolas open downward rather than upward. The observations that we have now made can be generalized as indicated in the box that follows.

1. The graph of $y = ax^2$ is a parabola with vertex at the origin. It is similar in shape to $y = x^2$.
2. $y = ax^2$ opens upward if $a > 0$; downward if $a < 0$
3. $y = ax^2$ is narrower than $y = x^2$ if $|a| > 1$; wider than $y = x^2$ if $|a| < 1$

Example 4 Graph the function $f(x) = -2x^2 + 12x - 16$ and specify the following: vertex, axis of symmetry, maximum or minimum value of f, and x- and y-intercepts.

Solution The idea is to complete the square, as in Example 3. We have

$$
\begin{aligned}
y &= -2x^2 + 12x \qquad - 16 \\
&= -2(x^2 - 6x \qquad) - 16 \\
&= -2(x^2 - 6x + 9) - 16 + 18 \leftarrow \\
&= -2(x - 3)^2 + 2
\end{aligned}
$$

Since the $+9$ within parentheses is multiplied by -2, we needed to add 18 to keep the equation in balance.

The graph of this last equation is obtained simply by shifting the graph of $y = -2x^2$ "right 3, up 2" so that the vertex is (3, 2). As a guide for actually sketching the graph, we could obtain the y-intercept. You should check for yourself that it is -16. As an alternative to this, however, let us calculate the x-intercepts. To do this, we replace y by 0 in the original equation $y = -2x^2 + 12x - 16$ to obtain

$$-2x^2 + 12x - 16 = 0$$
$$x^2 - 6x + 8 = 0$$
$$(x - 4)(x - 2) = 0$$
$$x - 4 = 0 \quad \Big| \quad x - 2 = 0$$
$$x = 4 \quad \Big| \quad x = 2$$

Thus the x-intercepts are $x = 4$ and $x = 2$. Knowing these intercepts and the vertex, we can sketch the graph as in Figure 6. You should check for yourself that the information accompanying Figure 6 is correct.

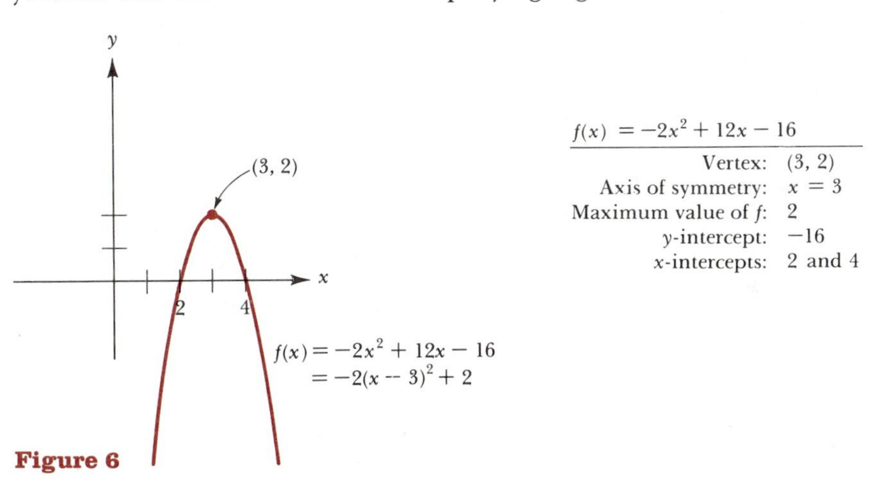

Figure 6

There is a simple formula that gives the x-coordinate for the vertex of any parabola $y = ax^2 + bx + c$:

> **Vertex Formula**
>
> The x-coordinate of the vertex of the parabola $y = ax^2 + bx + c$ is given by
>
> $$x = \frac{-b}{2a}$$

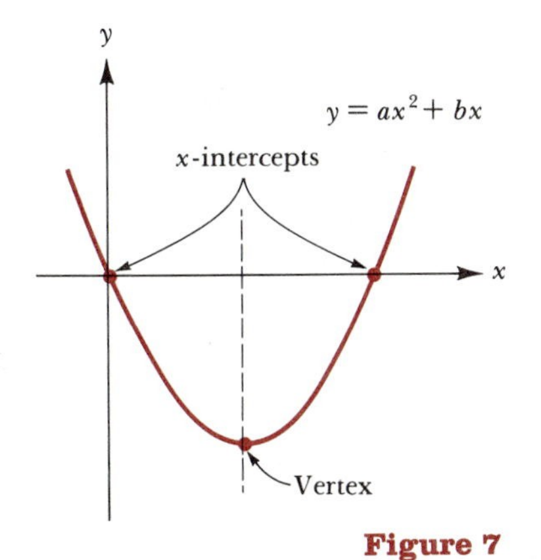

Figure 7

One way to derive this formula is to begin with the equation $y = ax^2 + bx + c$ and complete the square, exactly as done in Example 4. Exercise 36 at the end of this section asks you to carry out the details. Here is a different way to show that the x-coordinate of the vertex is $-b/2a$. The parabola $y = ax^2 + bx + c$ will have the same x-coordinate for its vertex as does $y = ax^2 + bx$. This is because the two graphs are just vertical translates of one another by a distance $|c|$. But because of symmetry it's easy to find the x-coordinate for the vertex of $y = ax^2 + bx$. As Figure 7 indicates, the x-coordinate of the vertex is midway between the two x-intercepts. To find these intercepts we have

$$ax^2 + bx = 0$$
$$x(ax + b) = 0$$
$$x = 0 \quad \Big| \quad ax + b = 0$$
$$ax = -b$$
$$x = -\frac{b}{a}$$

Now we want the x-value halfway between $x = 0$ and $x = -b/a$: namely $\frac{1}{2}(-b/a)$. Thus x is $-b/2a$, which is what we wanted to show.

In each of the following examples we use the vertex formula to answer questions about the maximum or minimum value of a quadratic function.

Example 5 Among all possible inputs for the function

$$g(x) = 3x^2 - 2x - 6$$

which yields the smallest output?

Solution Think in terms of a graph. Since the graph of this function is a parabola opening upward, the input we want is just the x-coordinate of the vertex. Thus

$$x = \frac{-b}{2a} = \frac{-(-2)}{2(3)} = \frac{1}{3}$$

So among all possible inputs, $x = \frac{1}{3}$ will produce the smallest output. Another way to say this is that the minimum value of g occurs when $x = \frac{1}{3}$.

Example 6 Two numbers add to 9. What is the largest possible value for their product?

Solution Call the two numbers x and $9 - x$. Then their product P is given by

$$P = x(9 - x)$$
$$= 9x - x^2$$

The graph of this quadratic function is the parabola in Figure 8. Note the accompanying calculations for the vertex. As Figure 8 and the accompanying calculations show, the largest value of the product P is 20.25. This is the required solution. (Note from the graph that there is no smallest value of P.)

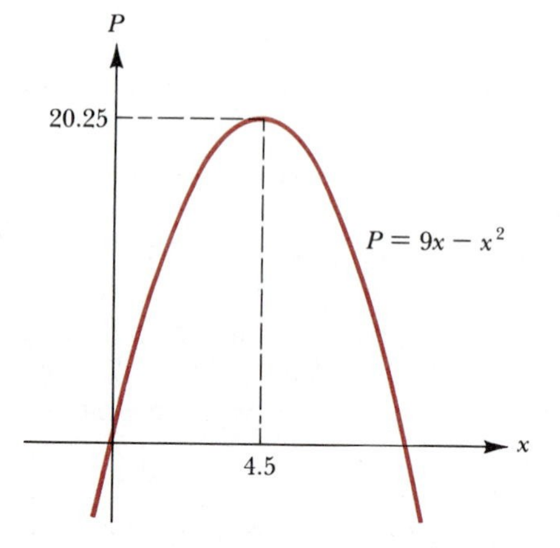

Vertex Calculations

x-coordinate: $\quad x = \frac{-b}{2a} = \frac{-9}{2(-1)} = \frac{9}{2}$

$x = 4.5$

P-coordinate: $\quad P = 9x - x^2$

$$= 9\left(\frac{9}{2}\right) - \left(\frac{9}{2}\right)^2$$
$$= \frac{81}{2} - \frac{81}{4}$$
$$= \frac{162}{4} - \frac{81}{4}$$
$$= \frac{81}{4} = 20.25$$

Figure 8

Example 7 Suppose that you have 600 m of fencing available with which to build two adjacent rectangular corrals. The two corrals are to share a common fence on one side, as shown in Figure 9. Find the dimensions x and y so that the total enclosed area is as large as possible.

Solution You have 600 m of fencing to be set up as shown in Figure 9. The question is how you should choose x and y so that the total area is a maximum. Since the total length of fencing is 600 m, we can relate x and y by writing

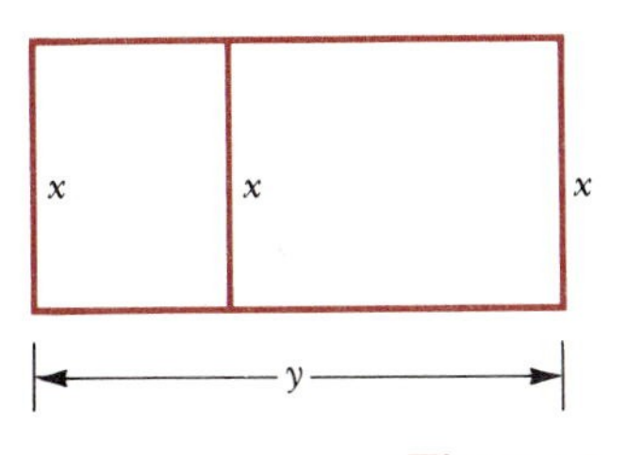

Figure 9

$$\begin{aligned} x + x + x + y + y &= 600 \\ 3x + 2y &= 600 \\ 2y &= 600 - 3x \\ y &= 300 - \frac{3}{2}x \qquad (1) \end{aligned}$$

Letting A denote the total area, we then have

$$\begin{aligned} A &= xy \\ &= x\left(300 - \frac{3}{2}x\right) \\ &= 300x - \frac{3}{2}x^2 \end{aligned}$$

We've used equation (1) to substitute for y.

This last equation expresses the area A as a function of x. Note that the graph of this function is a parabola opening downward, so it does make sense to talk about a maximum. We find the x-coordinate of the vertex as usual:

$$\begin{aligned} x &= \frac{-b}{2a} \\ &= \frac{-300}{2(-\frac{3}{2})} = \frac{300}{3} = 100 \end{aligned}$$

Now that we have the x-value that maximizes the area A, we can calculate y using equation (1). We have

$$\begin{aligned} y &= 300 - \frac{3}{2}x \\ &= 300 - \frac{3}{2}(100) = 300 - 3(50) \\ &= 150 \end{aligned}$$

Thus, by choosing x to be 100 m and y to be 150 m, the total area in Figure 9 will be as large as possible. Incidentally, note that the exact location of the fence dividing the two corrals does not influence our work or final answer.

For the three examples just completed, keep in mind that we were able to find the maximum or minimum easily only because the functions were quadratics. By way of contrast, you cannot expect to find the minimum of $y = x^4 - 8x$ by using our vertex formula, for this is not a quadratic function. In general, the techniques of calculus are required to find maxima or minima for functions other than quadratics. There are some instances, however, in which our present method can be adapted to functions that are closely related to quadratics. The next example shows an instance of this. (You'll see other examples in the exercises.)

Example 8 Which point on the curve $y = \sqrt{x}$ is closest to the point $(1, 0)$?

Solution See Figure 10, in which we let D denote the distance from a point (x, y) on the curve to the point $(1, 0)$. We are asked to find out exactly which point (x, y) will make the distance D as small as possible. Using the distance formula, we have

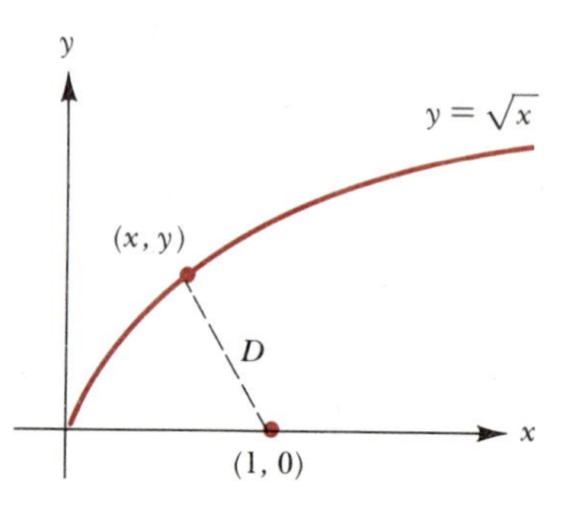

Figure 10

$$D = \sqrt{(x-1)^2 + (y-0)^2}$$
$$= \sqrt{x^2 - 2x + 1 + y^2} \qquad (2)$$

This expresses D in terms of both x and y. To express D in terms of x alone, use the given equation $y = \sqrt{x}$ to substitute for y in equation (2). This yields

$$D = \sqrt{x^2 - 2x + 1 + (\sqrt{x})^2}$$
$$= \sqrt{x^2 - x + 1}$$

Now the same x-value that makes D as small as it can be will also cause the quantity $D^2 = x^2 - x + 1$ to be its smallest. That is, we need only find the x-value that minimizes the quantity $x^2 - x + 1$. We have therefore

$$x = \frac{-b}{2a} = \frac{-(-1)}{2(1)} = \frac{1}{2}$$

Now that we have x, we want to calculate y. (The problem did not ask for D; we needn't calculate it.) Since $y = \sqrt{x}$ and $x = \frac{1}{2}$, we have

$$y = \sqrt{\frac{1}{2}} = \frac{\sqrt{1}}{\sqrt{2}} = \frac{1}{\sqrt{2}}$$
$$= \frac{1}{\sqrt{2}} \cdot \frac{\sqrt{2}}{\sqrt{2}} = \frac{\sqrt{2}}{2}$$

Thus, among all points (x, y) on the curve $y = \sqrt{x}$, the point closest to $(1, 0)$ is

$$\left(\frac{1}{2}, \frac{\sqrt{2}}{2}\right)$$

Question: Why would it not have made sense to ask instead for the point farthest from $(1, 0)$?

Exercise Set 5.1

In Exercises 1–8, find the linear functions satisfying the given conditions.

1. $f(-1) = 0$ and $f(5) = 4$
2. $f(3) = 2$ and $f(-3) = -4$
3. $g(0) = 0$ and $g(1) = \sqrt{2}$
4. The graph passes through the points $(2, 4)$ and $(3, 9)$.
5. $f(\frac{1}{2}) = -3$ and the graph of f is a line parallel to the line $x - y = 1$.
6. $g(2) = 1$ and the graph of g is perpendicular to the line $6x - 3y = 2$.
7. The graph of f is a horizontal line that passes through the larger y-intercept of the circle $x^2 - 2x + y^2 - 3 = 0$.

8. The x- and y-intercepts of the graph of g are 1 and 4, respectively.

9. Let $f(x) = 3x - 4$ and $g(x) = 1 - 2x$. Find out if the function $f \circ g$ is linear.

10. Explain why there is no linear function whose graph passes through all three of the points $(-3, 2)$, $(1, 1)$, and $(5, 2)$.

11. A factory owner buys a new machine for \$20,000. After 8 years the machine has a salvage value of \$1000. Find a formula for the value of the machine after t years, where $0 \le t \le 8$.

12. A manufacturer buys a new machine costing \$120,000. It is estimated that the machine has a useful lifetime of 10 years and a salvage value of \$4000 at the end of that time.

(a) Find a formula for the value of the machine after t years, where $0 \le t \le 10$.

(b) Find the value of the machine after 8 years.

13. A factory owner installs a new machine costing \$60,000. Its expected lifetime is 5 years, and at the end of that time the machine has no salvage value.

(a) Find a formula for the value of the machine after t years, where $0 \le t \le 5$.

(b) Complete the following depreciation schedule.

End of Year	Yearly Depreciation	Accumulated Depreciation	Value V
0	0	0	60,000
1			
2			
3			
4			
5		60,000	0

14. Let x denote the temperature on the Celsius scale and let y denote the corresponding temperature on the Fahrenheit scale.

(a) Find a linear function relating x and y; use the facts that 32°F corresponds to 0°C and 212°F corresponds to 100°C. Write your answer in the form $y = Ax + B$.

(b) What Celsius temperature corresponds to 98.6°F?

(c) Find a number z for which z°F = z°C.

For Exercises 15–28, graph the quadratic functions. In each case, specify the vertex, axis of symmetry, maximum or minimum value, and the x- and y-intercepts.

15. $y = (x + 2)^2$ **16.** $y = -(x + 2)^2$

17. $y = 2(x + 2)^2$ **18.** $y = 2(x + 2)^2 + 4$

19. $y = -2(x + 2)^2 + 4$ **20.** $y = x^2 + 6x - 1$

21. $f(x) = x^2 - 4x$ **22.** $F(x) = x^2 - 3x + 4$

23. $g(x) = 1 - x^2$ **24.** $y = 2x^2 + \sqrt{2}x$

25. $y = x^2 - 2x - 3$ **26.** $y = 2x^2 + 3x - 2$

27. $y = -x^2 + 6x + 2$ **28.** $y = -3x^2 + 12x$

29. For the given function, find the maximum or minimum value (whichever is appropriate). In each case, state whether the value is a maximum or minimum.

(a) $y = x^2 - 8x + 3$

(b) $f(x) = -2x^2 - 3x + 1$

(c) $g(x) = -12x^2 + 1000$

30. Among all possible inputs for the given function, which yields the smallest or largest output (whichever is appropriate)? In each case, state whether the output is largest or smallest.

(a) $y = 2x^2 - 4x + 11$ **(b)** $f(x) = 8x^2 + x - 5$

(c) $g(x) = -8x^2 + x - 5$

(d) $s = -16t^2 + 196t + 80$

(e) $y = x^2 - 10$ **(f)** $y = -x^2 + \frac{3}{2}x + 1$

31. How far from the origin is the vertex of the parabola $y = x^2 - 6x + 13$?

32. Find the x-coordinate of the vertex of the parabola $y = (x - a)(x - b)$, where a and b are constants.

33. Let $f(x) = 2x - 3$, $g(x) = x^2 + 4x - 1$, and $h(x) = 1 - 2x^2$. Classify each of the following functions as linear, quadratic, or neither.

(a) f **(b)** g **(c)** h

(d) $f \circ g$ **(e)** $g \circ f$ **(f)** $g \circ h$

In Exercises 34–35, find quadratic functions satisfying the given conditions.

34. (a) The graph passes through the origin and the vertex is $(2, 2)$.

(b) The graph is obtained by translating $y = x^2$ four units in the negative x-direction and three units in the positive y-direction.

35. (a) The vertex is $(3, -1)$ and one x-intercept is $x = 1$.

(b) The axis of symmetry is the line $x = 1$. The y-intercept is 1. There is only one x-intercept.

36. By completing the square, show that the coordinates of the vertex of the parabola $y = ax^2 + bx + c$ are

$$\left(\frac{-b}{2a}, \frac{4ac - b^2}{4a}\right)$$

Hint: Follow the steps in Example 4 on page 222.

37. Find the equation of the circle that passes through the origin and whose center is located at the vertex of the parabola $y = 2x^2 + 12x + 14$.

38. For which value of c will the minimum value of the function $f(x) = x^2 + 2x + c$ be $\sqrt{2}$?

39. Two numbers add to 5. What is the largest possible value of their product?

40. Find two numbers adding to 20 such that the sum of their squares is as small as possible.

41. The difference of two numbers is 1. What is the smallest possible value for the sum of their squares?

42. For each quadratic function specified below, state whether it would make sense to look for a highest or a lowest point on the graph. Then determine the coordinates of that point.
(a) $y = 2x^2 - 8x + 1$ **(b)** $y = -3x^2 - 4x - 9$
(c) $h = -16t^2 + 256t$ **(d)** $f(x) = 1 - (x + 1)^2$
(e) $g(t) = t^2 + 1$
(f) $f(x) = 1000x^2 - x + 100$

43. Among all rectangles having a perimeter of 25 m, find the dimensions of the one with the largest area.

44. What is the largest possible area for a rectangle whose perimeter is 80 cm?

45. What is the largest possible area for a right triangle in which the sum of the lengths of the two shorter sides is 100 in.?

46. The perimeter of a rectangle is 12 m. Find the dimensions for which the diagonal is as short as possible.

47. Two numbers add to 6.
(a) Let T denote the sum of the squares of the two numbers. What is the smallest possible value for T?
(b) Let S denote the sum of the first number and the square of the second. What is the smallest possible value for S?
(c) Let U denote the sum of the first number and twice the square of the second number. What is the smallest possible value for U?
(d) Let V denote the sum of the first number and the square of twice the second. What is the smallest possible value for V?

48. Suppose that the height of an object shot straight up is given by $h = 512t - 16t^2$. (h is in feet and t is in seconds.) Find the maximum height and the time at which the object hits the ground.

49. A baseball is thrown straight up, and its height as a function of time is given by the formula $h = -16t^2 + 32t$. (h is in feet; t is in seconds.)
(a) Find the height of the ball when $t = 1$ and when $t = \frac{3}{2}$.
(b) Find the maximum height of the ball and the time at which that height is attained.
(c) At what times is the height 7 ft?

50. Find the point on the curve $y = \sqrt{x}$ that is nearest to the point (3, 0).

51. Which point on the curve $y = \sqrt{x - 2} + 1$ is closest to the point (4, 1)? What is this minimum distance?

52. Find the coordinates of the point on the line $y = 3x + 1$ closest to (4, 0).

53. **(a)** What number exceeds its square by the greatest amount?
(b) What number exceeds twice its square by the greatest amount?

54. Suppose that you have 1800 m of fencing available with which to build three adjacent rectangular corrals as shown in the figure. Find the dimensions so that the total enclosed area is as large as possible.

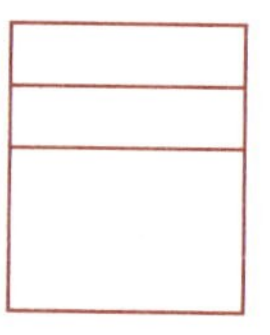

55. Five hundred feet of fencing is available for a rectangular pasture alongside a river, the river serving as one side of the rectangle (so only three sides require fencing). Find the dimensions yielding the greatest area.

56. Let $A = 3x^2 + 4x - 5$ and $B = x^2 - 4x - 1$. Find the minimum value of $A - B$.

57. Let $R = 0.4x^2 + 10x + 5$ and $C = 0.5x^2 + 2x + 101$. For which value of x is $R - C$ a maximum?

58. Suppose that the revenue generated by selling x units of a certain commodity is given by $R = -\frac{1}{5}x^2 + 200x$. Assume that R is in dollars. What is the maximum revenue possible in this situation?

59. Suppose that the function $p = -\frac{1}{4}x + 30$ relates the selling price p of an item to the quantity sold x. Assume p is in dollars. For which value of x will the corresponding revenue be a maximum? What is this maximum revenue and what is the unit price in this case?

***60.** A piece of wire 200 cm long is to be cut into two pieces of lengths x and $200 - x$. The first piece is to be bent into a circle and the second piece into a square. For which value of x is the combined area of the circle and square as small as possible?

***61.** A 30-in. piece of string is to be cut into two pieces. The first piece will be formed into the shape of an equilateral triangle and the second piece into a square. Find the length of the first piece if the combined area of the triangle and square is to be as small as possible.

62. **(a)** Same as Exercise 61, except both pieces are to be formed into squares.
(b) Could you have guessed the answer to part (a)?

63. The action of sunlight on automobile exhaust produces air pollutants known as *photochemical oxidants.* In a study of cross-country runners in Los Angeles, it was shown that running performances can be adversely affected when the oxidant level reaches 0.03 parts per million. Let us suppose that on a given day the oxidant level L is approximated by the formula

$$L = 0.059t^2 - 0.354t + 0.557 \qquad (0 \le t \le 7)$$

Here, t is measured in hours, with $t = 0$ corresponding to 12 noon, and L is in parts per million. At what time is the oxidant level L a minimum? At this time, is the oxidant level high enough to affect a runner's performance?

64. If $x + y = 1$, find the largest possible value of the quantity $x^2 - 2y^2$.

65. **(a)** Find the smallest possible value of the quantity $x^2 + y^2$ under the restriction that $2x + 3y = 6$.
(b) Find the radius of the circle whose center is at the origin and that is tangent to the line $2x + 3y = 6$. How does this answer relate to your answer in part (a)?

66. Through a type of chemical reaction known as *autocatalysis,* the human body produces the enzyme trypsin from the enzyme trypsinogen. (Trypsin then breaks down proteins into amino acids, which the body needs for growth.) Let r denote the rate of this chemical reaction in which trypsin is formed from trypsinogen. It has been shown experimentally that $r = kx(a - x)$,

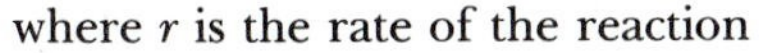

where r is the rate of the reaction
k is a positive constant
a is the initial amount of trypsinogen
x is the amount of trypsin produced (so x increases as the reaction proceeds)

Show that the reaction rate r is a maximum when $x = a/2$. In other words, the speed of the reaction is the greatest when the amount of trypsin formed is half of the original amount of trypsinogen.

67. **(a)** Let $x + y = 15$. Find the minimum value of the quantity $x^2 + y^2$.
(b) Let C be a constant and $x + y = C$. Show that the minimum value of $x^2 + y^2$ is $C^2/2$. Then use this result to check your answer in part (a).

68. Suppose that A, B, and C are positive constants and that $x + y = C$. Show that the minimum value of $Ax^2 + By^2$ occurs when $x = \dfrac{BC}{A + B}$ and $y = \dfrac{AC}{A + B}$.

***69.** The figure shows a square inscribed within a unit square. For which value of x is the area of the inscribed square a minimum? What is the minimum area? *Hint:* Denote the lengths of the two segments comprising the base of the unit square by t and $1 - t$, respectively. Now use the Pythagorean Theorem and congruent triangles to express x in terms of t.

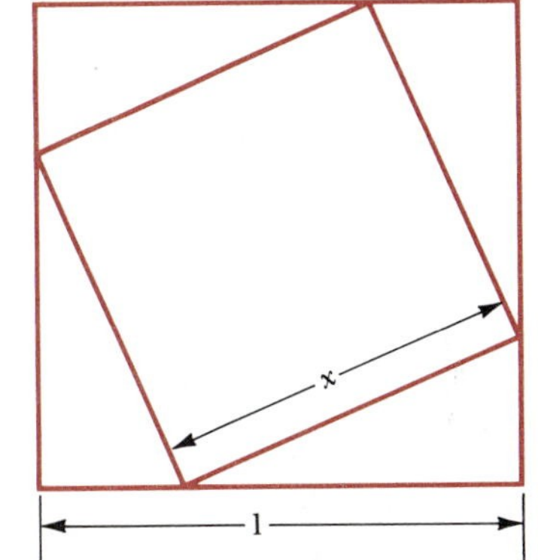

70. **(a)** Find the coordinates of the point on the line $y = mx + b$ that is closest to the origin.

$$\left[\textit{Answer:} \quad \left(\frac{-mb}{1 + m^2}, \frac{b}{1 + m^2}\right)\right]$$

(b) Show that the perpendicular distance from the origin to the line $y = mx + b$ is $\dfrac{|b|}{\sqrt{1 + m^2}}$.
Suggestion: Use the result in part (a).
(c) Use part (b) to show that the perpendicular distance from the origin to the line $Ax + By + C = 0$ is $\dfrac{|C|}{\sqrt{A^2 + B^2}}$.

71. The point P lies in the first quadrant on the graph of the line $y = 7 - 3x$. From the point P, perpendiculars are drawn to the x-axis and y-axis, respectively. What is the largest possible area for the rectangle thus formed?

72. Show that the largest possible area for the shaded rectangle shown in the figure below is $-b^2/4m$. Then use this to check your answer in Exercise 71.

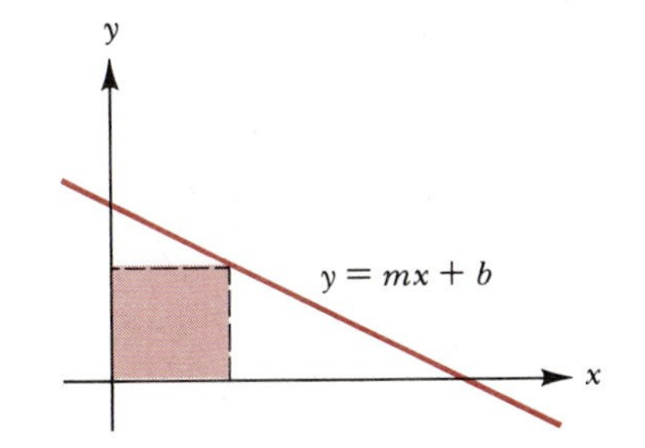

***73.** Show that the maximum possible area for a rectangle inscribed in a circle of radius R is $2R^2$. *Hint:* Maximize the square of the area.

***74.** A Norman window is in the shape of a rectangle surmounted by a semicircle, as shown in the figure. Assume that the perimeter of the window is P, a constant. Show that the area of the window is a maximum when both x and r are equal to $\dfrac{P}{\pi + 4}$. Show that this maximum area is $\dfrac{P^2}{2(\pi + 4)}$.

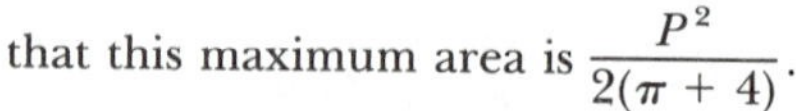

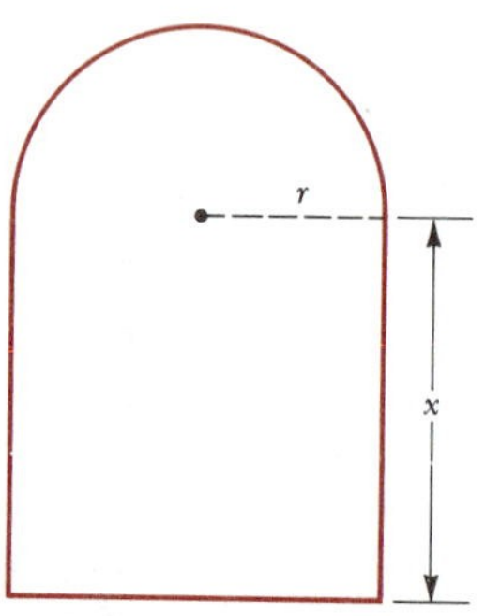

***75.** An athletic field with a perimeter of $\frac{1}{4}$ mi consists of a rectangle with a semicircle at each end, as shown in the figure below. Find the dimensions x and r that yield the greatest possible area for the rectangular region.

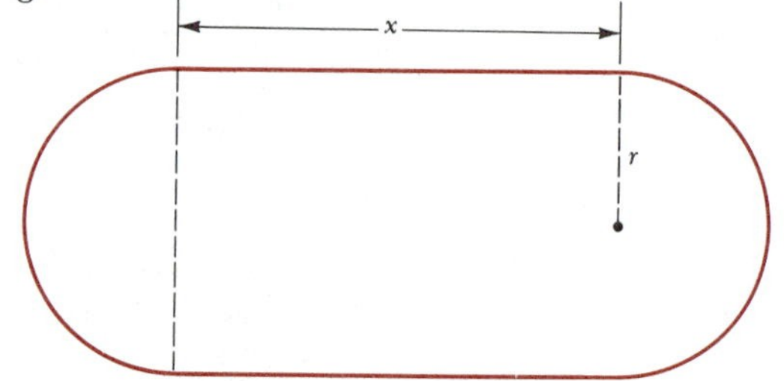

76. A rancher, who wishes to fence off a rectangular area, finds that the fencing in the east-west direction will require extra reinforcement due to strong prevailing winds. Because of this, the cost of fencing in the east-west direction will be \$12 per (linear) yard, as opposed to a cost of \$8 per yard for fencing in the north-south direction. Find the dimensions of the largest possible rectangular area that can be fenced for \$4800.

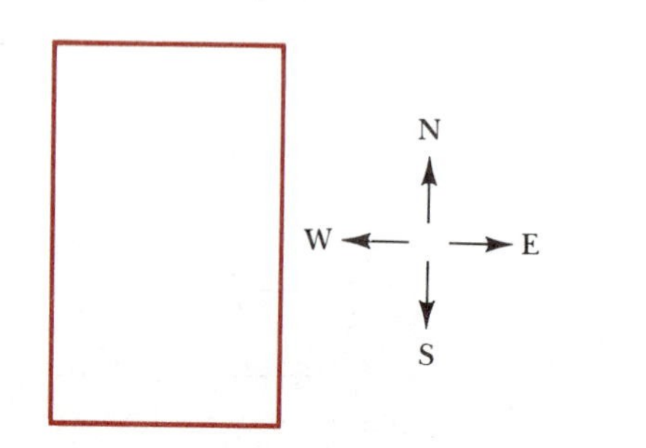

77. Let $f(x) = (x - a)^2 + (x - b)^2 + (x - c)^2$, where a, b, and c are constants. Show that $f(x)$ will be a minimum when x is the average of a, b, and c.

78. Let $y = a_1(x - x_1)^2 + a_2(x - x_2)^2$, where a_1, a_2, x_1, and x_2 are all constants. Further, suppose that a_1 and a_2 are both positive. Show that the minimum of this function occurs when

$$x = \frac{a_1x_1 + a_2x_2}{a_1 + a_2}$$

79. Among all rectangles with a given perimeter P, find the dimensions of the one with the shortest diagonal.

***80.** By analyzing sales figures, the economist for a stereo manufacturer knows that 150 units of a top-of-the-line turntable can be sold each month when the price is set at $p = \$200$ per unit. The figures also show that for each \$10 hike in price, 5 fewer units are sold each month.

(a) Let x denote the number of units sold per month and let p denote the price per unit. Find a linear function relating p and x. *Hint:* $\Delta p/\Delta x = 10/-5 = -2$.

(b) The revenue R is given by $R = xp$. What is the maximum revenue? At what level should the price be set to achieve this maximum revenue?

81. Let $f(x) = x^2 + px + q$, and suppose that the minimum value of this function is 0. Show that $q = p^2/4$.

82. For which numbers t will the value of $16t^2 - 4t^4$ be as large as possible? *Hint:* Let $t^2 = x$ and $t^4 = x^2$. Then the resulting expression $16x - 4x^2$ is quadratic in x.

83. Find the minimum value of the function $f(t) = t^4 - 8t^2$. (See the hint in Exercise 82.)

84. Among all possible inputs for the function $f(t) = -t^4 + 6t^2 - 6$, which ones yield the largest output? (See the hint in Exercise 82.)

85. In this exercise we find the minimum value of the function $f(x) = x^2 + 1/x^2$.

(a) Verify that $x^2 + 1/x^2 = (x - 1/x)^2 + 2$.

(b) Find the minimum value of $(x - 1/x)^2 + 2$. *Hint:* Let $t = x - 1/x$.

***86.** Let $f(x) = x^2 + bx + 1$. Find a positive value for b such that the distance from the origin to the vertex of the parabola is as small as possible.

87. Let $g(x) = x^2 + 2(a + b)x + 2(a^2 + b^2)$, where a and b are constants and $a \neq b$.

(a) Show that the minimum value of g is $(a - b)^2$.

(b) Use the result in part (a) to explain why the graph of g has no x-intercepts.

C 88. Determine the linear function whose graph passes through the points $(1.27, -0.83)$ and $(2.06, 5.42)$. Round off the decimals in your answer to two places.

C 89. Determine a value for c such that the graph of $g(x) = x^2 + x + c$ passes through the point $(4.51, -3.02)$. Round off your answer to two decimal places.

C 90. The minimum value of the function $f(x) = x^2 - bx$ is -0.168. What are the possible values for b? Round off your answer(s) to three decimal places.

C 91. A factory owner installs a new machine costing \$26,450. After 7 years the machine has no salvage value; in fact, it will cost the owner \$1900 to have the machine disassembled and removed from its second-floor location. Assuming linear depreciation, find a formula for the value of the machine after t years, where $0 \le t \le 7$.

5.2 Polynomial Functions

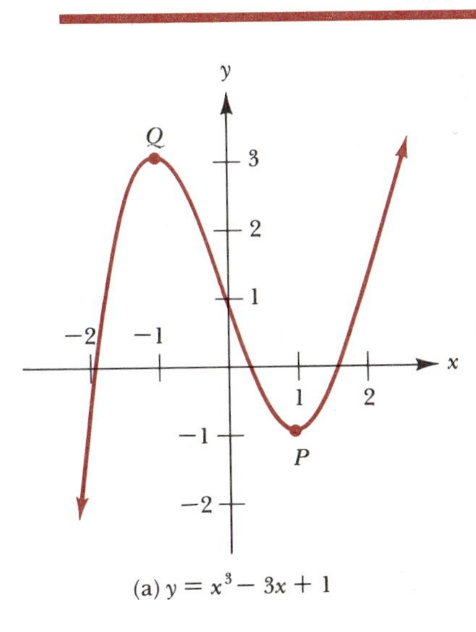

(a) $y = x^3 - 3x + 1$

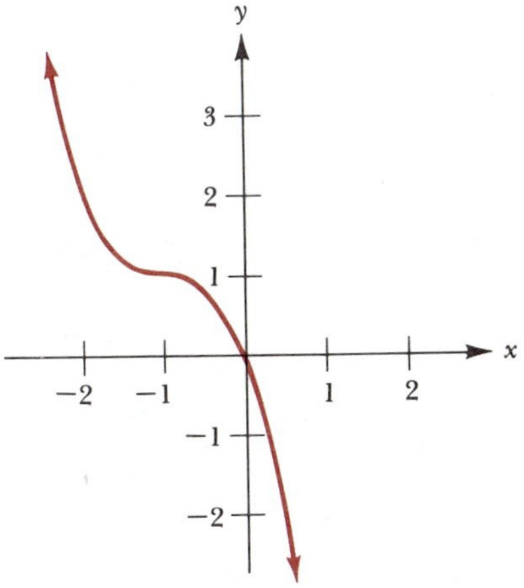

(b) $y = -(x + 1)^3 + 1$

Figure 1

We can rephrase the definitions of linear and quadratic functions using the terminology for polynomials given in Section 1.7.

f is a linear function if $f(x)$ is a polynomial of degree 1:

$$f(x) = a_1x + a_0 \tag{1}$$

f is a quadratic function if $f(x)$ is a polynomial of degree 2:

$$f(x) = a_2x^2 + a_1x + a_0 \tag{2}$$

In view of equation (1), linear functions are sometimes called *polynomial functions of degree 1*. Similarly, in view of equation (2), quadratic functions are *polynomial functions of degree 2*. More generally, by a *polynomial function of degree n*, we mean a function defined by an equation of the form

$$f(x) = a_nx^n + a_{n-1}x^{n-1} + \cdots + a_1x + a_0 \tag{3}$$

where n is a nonnegative integer and $a_n \neq 0$. Throughout the remainder of this chapter, the coefficients a_k will always be real numbers.

In principle at least, we could obtain the graph of any polynomial function by setting up a table and plotting a sufficient number of points. Indeed, this is just the way a computer equipped with a curve plotter operates. However, in order to *understand* why the graphs look as they do, we want to discuss some additional methods for graphing polynomial functions.

There are three facts that we shall need. By way of example, consider the graphs of the polynomial functions in Figure 1. First, notice that both graphs are unbroken smooth curves, with no "corners." As is shown in calculus, this is the case for the graph of every polynomial function. By way of contrast, the graphs in Figures 2 and 3 cannot represent polynomial functions. The graph in Figure 2 has a break in it; the graph in Figure 3 has a corner, or *cusp*.

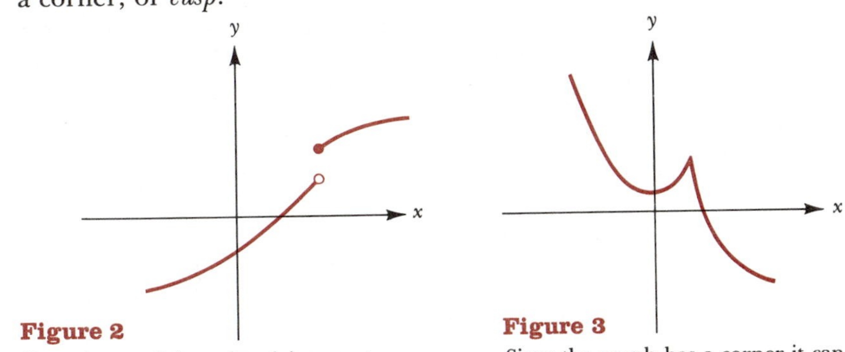

Figure 2
Since the graph has a break it cannot represent a polynomial function.

Figure 3
Since the graph has a corner it cannot represent a polynomial function.

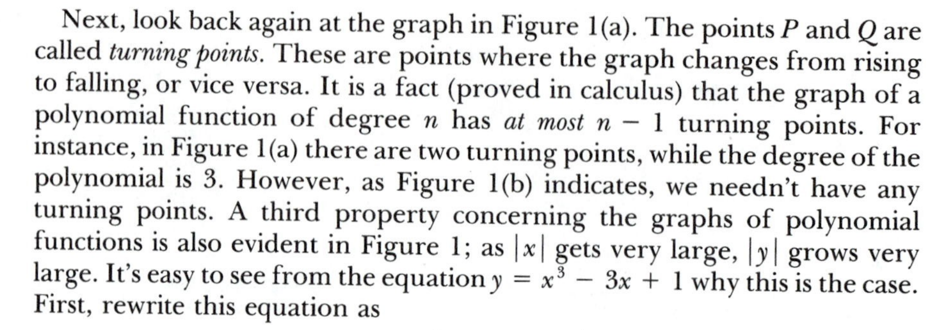

Next, look back again at the graph in Figure 1(a). The points P and Q are called *turning points.* These are points where the graph changes from rising to falling, or vice versa. It is a fact (proved in calculus) that the graph of a polynomial function of degree n has *at most* $n - 1$ turning points. For instance, in Figure 1(a) there are two turning points, while the degree of the polynomial is 3. However, as Figure 1(b) indicates, we needn't have any turning points. A third property concerning the graphs of polynomial functions is also evident in Figure 1; as $|x|$ gets very large, $|y|$ grows very large. It's easy to see from the equation $y = x^3 - 3x + 1$ why this is the case. First, rewrite this equation as

$$y = x^3\left(1 - \frac{3}{x^2} + \frac{1}{x^3}\right)$$

Now when $|x|$ is very large, the quantity in parentheses is close to 1 (because the terms $3/x^2$ and $1/x^3$ are then close to zero). So in this case we have the approximation $y \approx x^3$, which clearly shows that when $|x|$ is very large, $|y|$ is larger still. In view of this property, neither of the graphs in Figure 4 can represent polynomial functions.

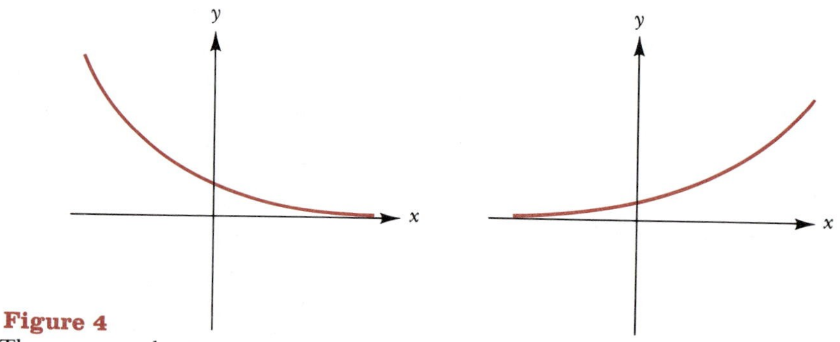

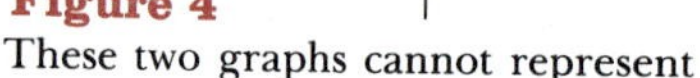

Figure 4
These two graphs cannot represent polynomial functions.

The simplest polynomial functions to graph are those of the form $y = x^n$. From our work in Chapters 3 and 4, we already know about the graphs of $y = x$, $y = x^2$, and $y = x^3$. For reference, these are shown in Figure 5.

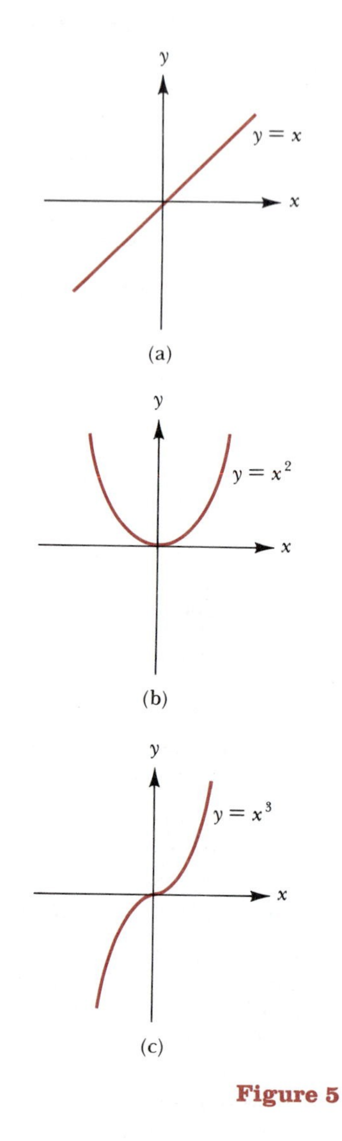

Figure 5

The graph of $y = x^n$, when n is greater than 3, resembles the graph of $y = x^2$ or $y = x^3$, depending upon whether n is even or odd. Consider, for instance, the graph of $y = x^4$, shown in Figure 6 along with the graph of $y = x^2$. Just as with $y = x^2$, the graph of $y = x^4$ is a symmetric, U-shaped curve passing through the three points (0, 0), (1, 1), and (−1, 1). However, in the interval $-1 < x < 1$, the graph of $y = x^4$ is flatter than that of $y = x^2$. Similarly, the graph of $y = x^6$ in this interval would be flatter still. The data in Table 1 show why this is true. Figure 7 displays the graphs of $y = x^2$, $y = x^4$, and $y = x^6$ on the interval $0 \leq x \leq 1$.

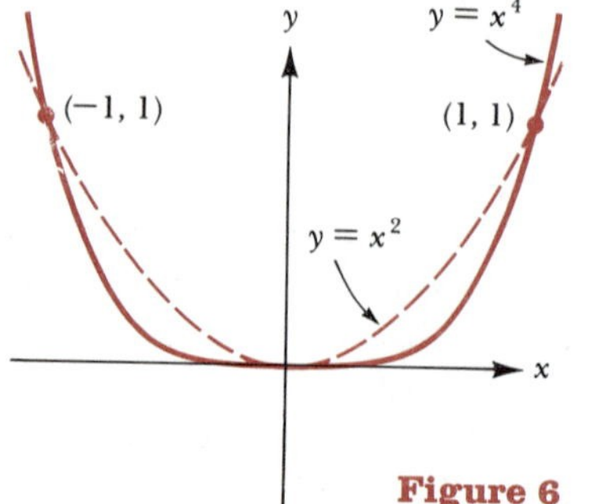

Figure 6

x	0.2	0.4	0.6	0.8	1.0
x^2	0.04	0.16	0.36	0.64	1.0
x^4	0.0016	0.0256	0.1296	0.4096	1.0
x^6	0.000064	0.004096	0.046656	0.262144	1.0

Table 1

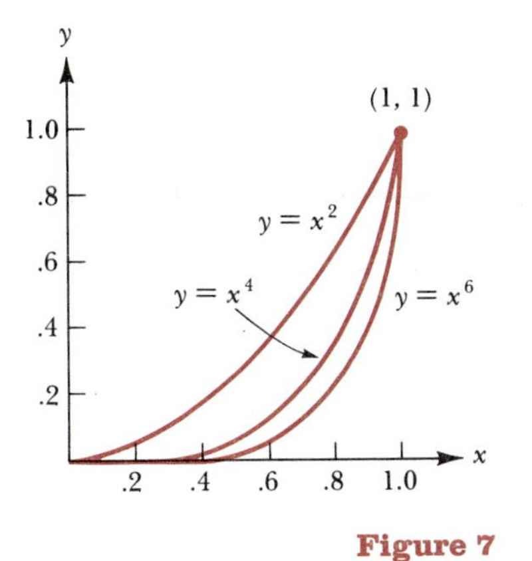

Figure 7

Incidentally, Figure 7 indicates one of the practical difficulties you may encounter in trying to draw an accurate graph of $y = x^n$. Suppose, for instance, that you want to graph $y = x^6$, and the thickness of the lines you draw is 0.01 cm. Also, to be definite, suppose that you use the same scale on both axes, taking the common unit to be 1 cm. Then in the first quadrant, your graph of $y = x^6$ will be indistinguishable from the x-axis when $x^6 < 0.01$, or $x < \sqrt[6]{0.01} \approx 0.46$ cm (using a calculator). This explains why sections of the graphs in Figure 7 appear horizontal.

For $|x| > 1$, the graph of $y = x^4$ rises more rapidly than that of $y = x^2$. Similarly, the graph of $y = x^6$ rises still more rapidly. This is shown in Figures 8(a) and (b). (Note the different scales used on the y-axes in the two figures.)

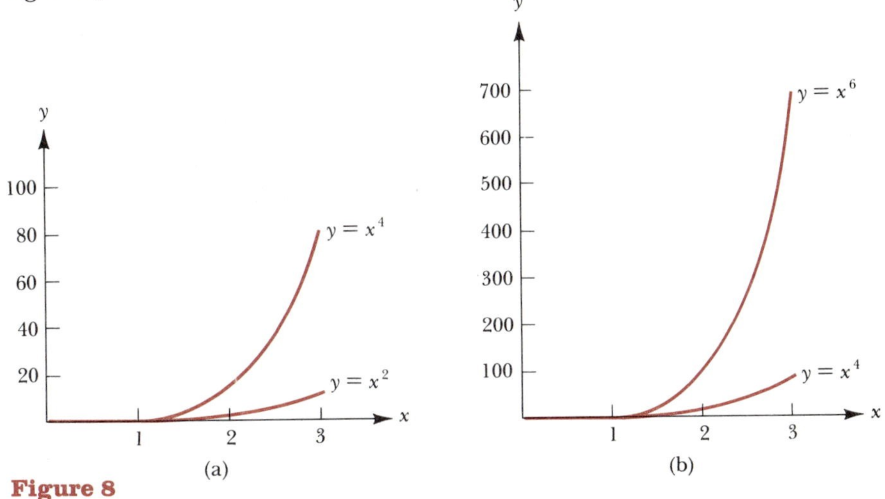

Figure 8

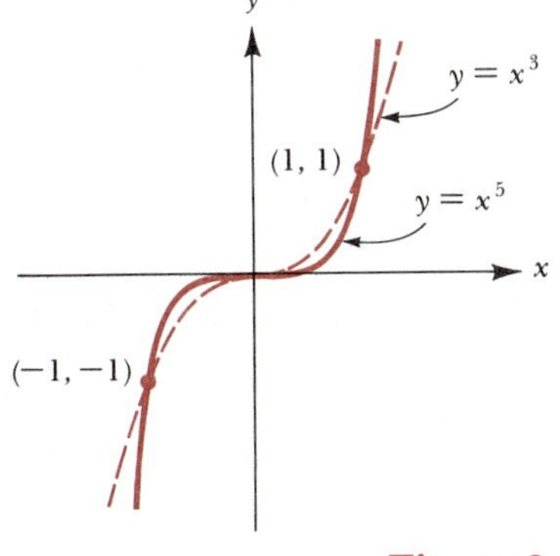

Figure 9

As mentioned earlier, when n is odd, the graph of $y = x^n$ resembles that of $y = x^3$. In Figure 9, for instance, we compare the graphs of $y = x^3$ and $y = x^5$. Notice that both curves pass through (0, 0), (1, 1), and $(-1, -1)$. For reasons similar to those explained above, the graph of $y = x^5$ is flatter than that of $y = x^3$ in the interval $-1 < x < 1$; the graph of $y = x^7$ would be flatter still. For $|x| > 1$, the graph of $y = x^5$ is steeper than that of $y = x^3$; $y = x^7$ would be steeper still.

In Section 5.1 (on page 222) we noted the effect of the constant a on the graph of $y = ax^2$. Those same comments apply to the graph of $y = ax^n$. For instance, the graph of $y = \frac{1}{2}x^4$ is wider than that of $y = x^4$, while the graph of $y = -\frac{1}{2}x^4$ is obtained by reflecting $y = \frac{1}{2}x^4$ in the x-axis.

Example 1 Sketch the graph of $y = (x + 2)^5$ and specify the y-intercept.

Solution The graph of $y = (x + 2)^5$ is obtained by moving the graph of $y = x^5$ two units to the left. As a guide to drawing the curve, we recall that $y = x^5$ passes through the points (1, 1) and $(-1, -1)$. Thus $y = (x + 2)^5$ must pass through $(-1, 1)$ and $(-3, -1)$. See Figure 10 on page 234. Although the curve rises and falls very sharply, it is important to realize that it is never really vertical. For instance, the curve eventually crosses the y-axis. To find the y-intercept, we set $x = 0$ to obtain $y = 2^5 = 32$. Thus the y-intercept is 32.

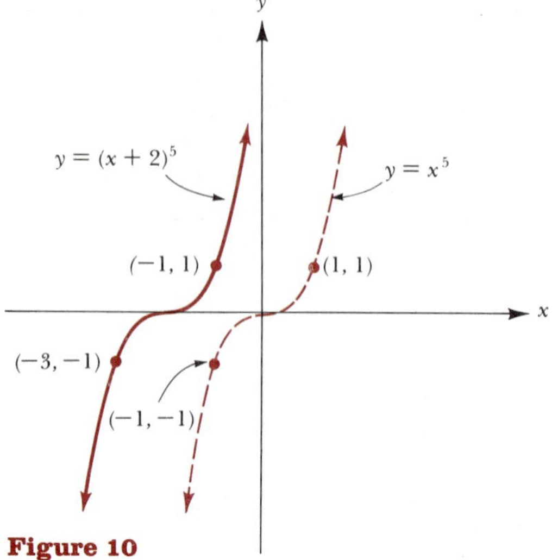

Figure 10

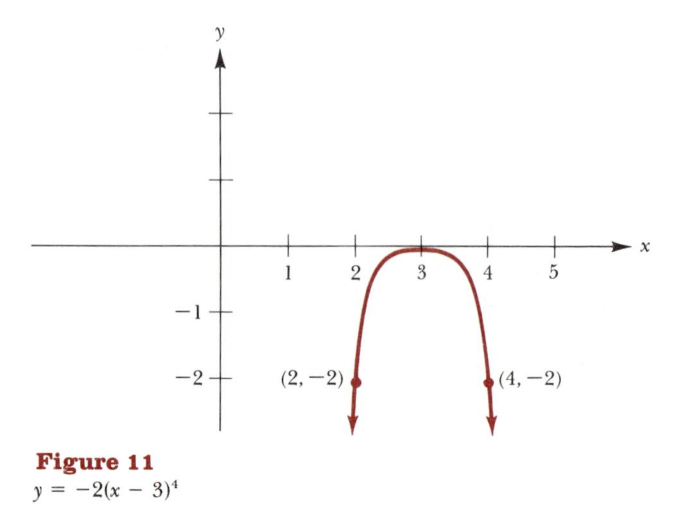

Figure 11
$y = -2(x - 3)^4$

Example 2 Graph the function $y = -2(x - 3)^4$.

Solution Begin with the graph of $y = x^4$. By translating this graph three units to the right we obtain the graph of $y = (x - 3)^4$. Next, by reflecting the graph of $y = (x - 3)^4$ in the x-axis, we obtain the graph of $y = -(x - 3)^4$. Finally, the graph of $y = -2(x - 3)^4$ will resemble that of $y = -(x - 3)^4$, but it will be narrower. As a guide as to just how narrow to draw the graph, we compute the y-values corresponding to $x = 4$ and $x = 2$. As you can readily check, the calculations show that the points $(4, -2)$ and $(2, -2)$ are on the graph. Figure 11 shows the required graph.

In general, the most efficient ways to graph a given polynomial function rely on the methods of calculus. However, if the polynomial can be written as a product of linear factors with real coefficients, then a fairly accurate graph can be obtained without the use of calculus. By way of example, let us graph the function

$$f(x) = (x + 2)(x - 1)(x - 3)$$

By inspection we see that $f(x) = 0$ when $x = -2$, $x = 1$, and $x = 3$. These are the x-intercepts of the graph. Next, we want to find out what the graph looks like in the immediate vicinity of each x-intercept. First, suppose that x is very close to -2. We have

$$f(x) = (x + 2)\underbrace{(x - 1)}\underbrace{(x - 3)}$$

If x is close to -2, this is close to -3. If x is close to -2, this is close to -5.

Thus if x is very close to -2, we have

$$f(x) \approx (x + 2)(-3)(-5)$$
$$f(x) \approx 15(x + 2)$$
$$f(x) \approx 15x + 30$$

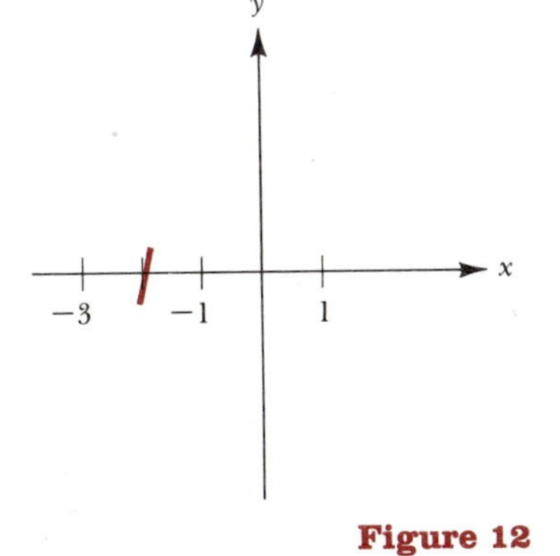

Figure 12
Near $x = -2$ the graph of $f(x) = (x+2)(x-1)(x-3)$ is nearly vertical; it resembles the line $y = 15x + 30$.

(Notice the technique used to obtain this approximation: We retained the factor corresponding to the intercept -2 and we estimated the remaining factors.) We conclude from this that when x is very close to -2, the graph of f resembles the line $y = 15x + 30$. In particular, the graph is quite steep and rising to the right. See Figure 12.

Next, we carry out similar analyses to see how the graph of f looks near its other x-intercepts.

x near 1

$f(x) = (x+2)(x-1)(x-3)$

$f(x) \approx (1+2)(x-1)(1-3)$

$f(x) \approx (3)(x-1)(-2)$

$f(x) \approx -6x + 6$

x near 3

$f(x) = (x+2)(x-1)(x-3)$

$f(x) \approx (3+2)(3-1)(x-3)$

$f(x) \approx (5)(2)(x-3)$

$f(x) \approx 10x - 30$

Thus for x very close to 1, the graph of f resembles the line $y = -6x + 6$, while for x very close to 3, the graph resembles the line $y = 10x - 30$. Figure 13 summarizes our information about the behavior of the graph of f near its x-intercepts. With this as a guide, we can then draw a rough sketch of the graph of f, as shown in Figure 14.

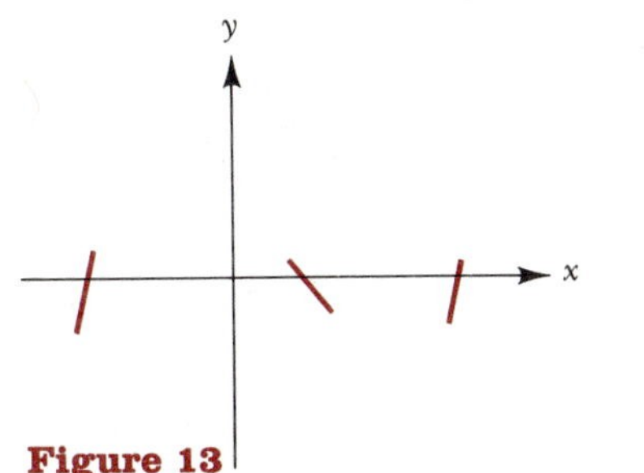

Figure 13
The graph of f near its x-intercepts.

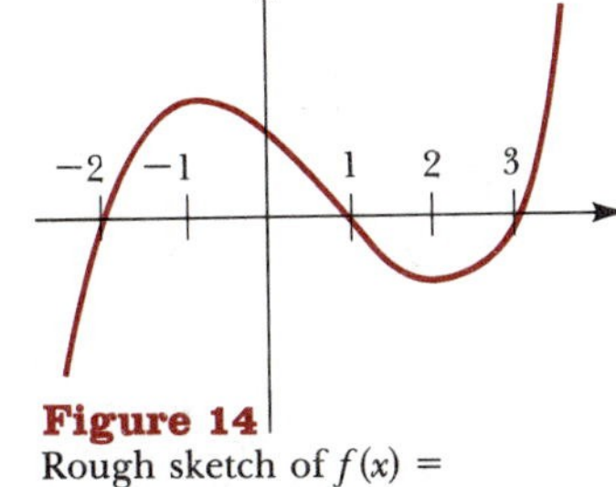

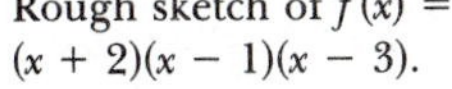

Figure 14
Rough sketch of $f(x) = (x+2)(x-1)(x-3)$.

Notice that in order to draw a smooth curve satisfying the conditions of Figure 13, you need at least two turning points: one between $x = -2$ and $x = 1$, and one between $x = 1$ and $x = 3$. On the other hand, since the degree of $f(x)$ is 3, there can be no more than two turning points. This is why Figure 14 shows exactly two turning points. While the precise location of the turning points is a matter for calculus, we can nevertheless improve upon the sketch in Figure 14 by computing $f(x)$ for several specific values of x. Some reasonable choices here are $x = -1$, $x = 0$, and $x = 2$. As you can check, the resulting calculations show that the points $(-1, 8)$, $(0, 6)$, and $(2, -4)$ are on the graph. We can now sketch the graph of $f(x) = (x+2)(x-1)(x-3)$ as shown in Figure 15.

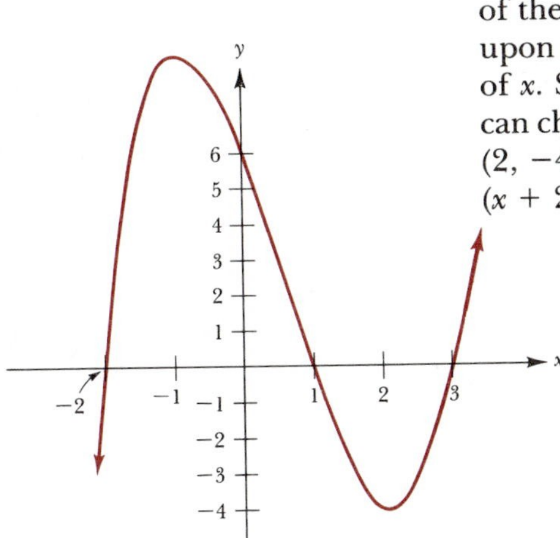

Figure 15
$f(x) = (x+2)(x-1)(x-3)$

The next example further illustrates our technique for graphing those polynomial functions that can be expressed as a product of linear factors with real coefficients.

Example 3 Sketch the graph of $y = x^3(x + 1)(x - 2)$.

Solution From the given equation we can see that the x-intercepts are $x = 0, x = -1$, and $x = 2$. Now, just as in the previous example, we want to see how the graph looks in the immediate vicinity of the x-intercepts.

x near 0

$y \approx x^3(0 + 1)(0 - 2)$

$y \approx -2x^3$

x near -1

$y \approx (-1)^3(x + 1)(-1 - 2)$

$y \approx 3(x + 1)$

$y \approx 3x + 3$

x near 2

$y \approx 2^3(2 + 1)(x - 2)$

$y \approx 24(x - 2)$

$y \approx 24x - 48$

Figure 16
The graph of $y = x^3(x + 1)(x - 2)$ near its x-intercepts.

These results are displayed in Figure 16. Then with this as a guide, we can draw a rough sketch of the required graph, as shown in Figure 17. To obtain a slightly more accurate sketch, we compute the y-values corresponding to $x = -\frac{1}{2}$ and $x = 1$. As you can check, we find that the curve passes through the points $(-\frac{1}{2}, \frac{5}{32})$ and $(1, -2)$. This additional information is used to draw the graph shown in Figure 18.

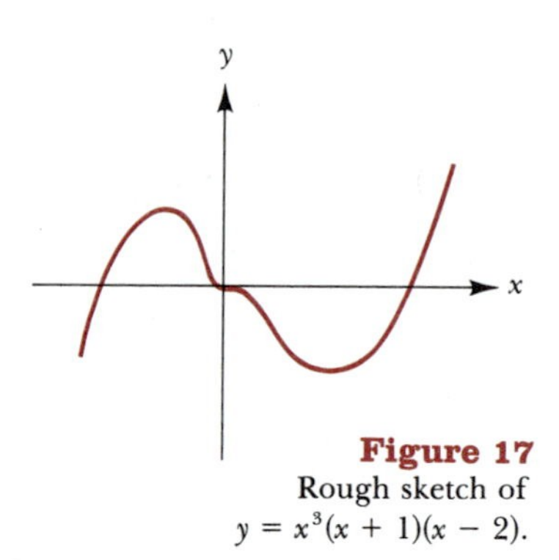

Figure 17
Rough sketch of $y = x^3(x + 1)(x - 2)$.

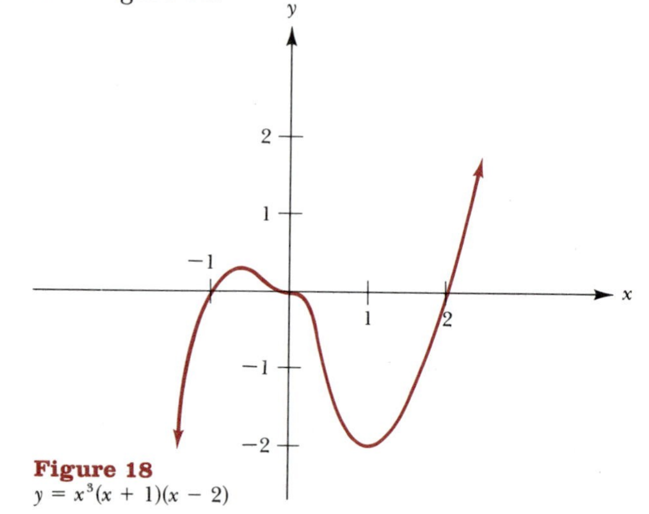

Figure 18
$y = x^3(x + 1)(x - 2)$

Exercise Set 5.2

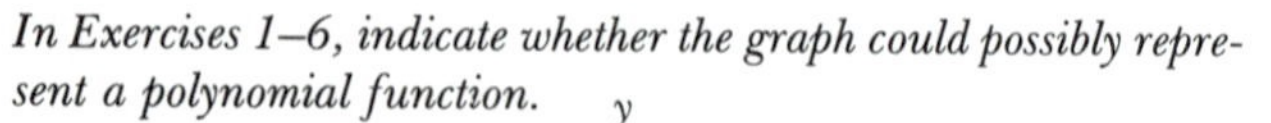

In Exercises 1–6, indicate whether the graph could possibly represent a polynomial function.

1.

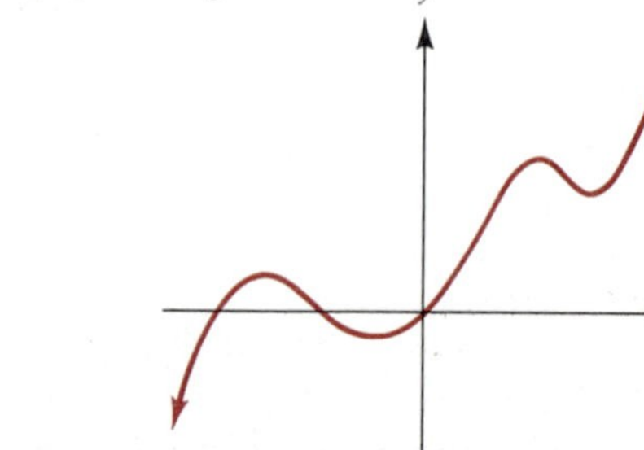

2.

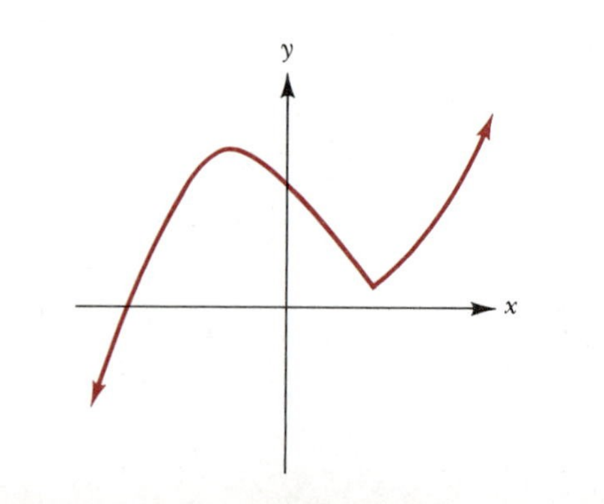

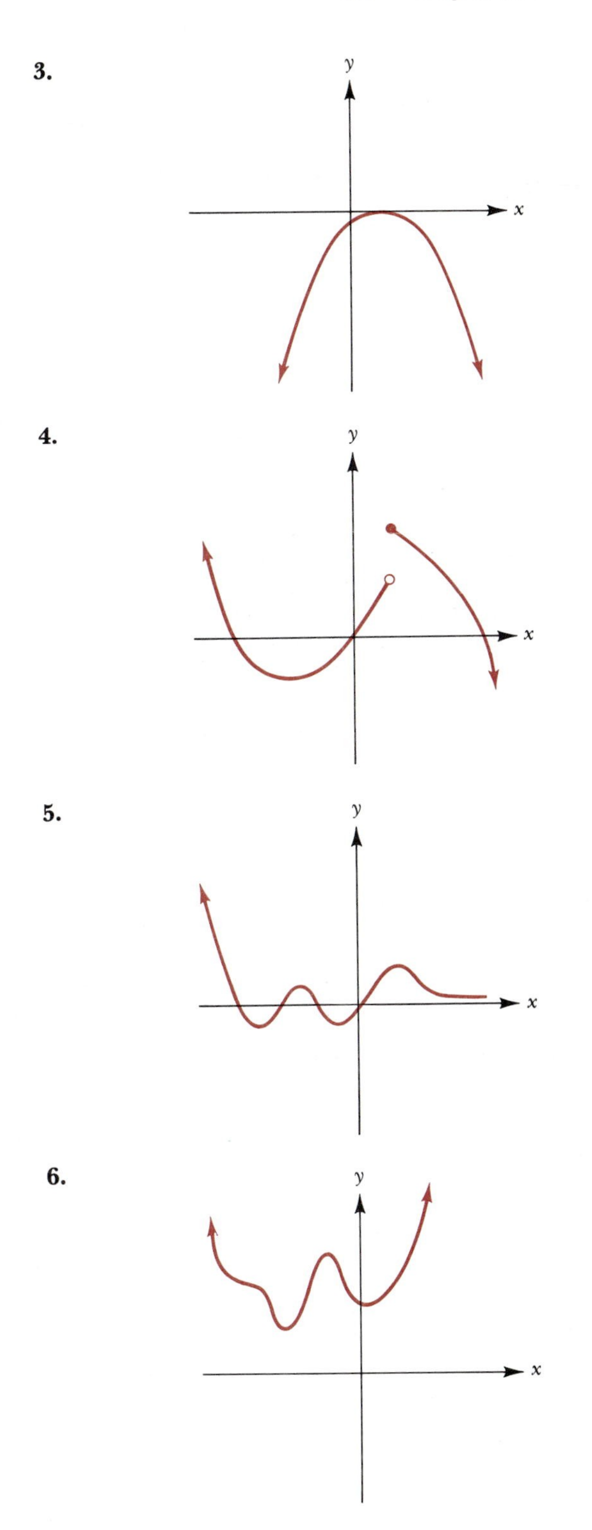

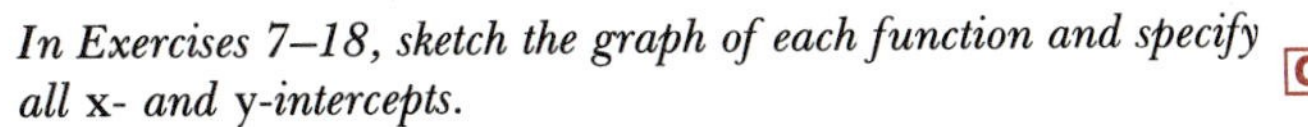

In Exercises 7–18, sketch the graph of each function and specify all x- *and* y-*intercepts.*

7. $y = (x - 2)^2 + 1$

8. $y = -3x^4$

9. $y = -(x - 1)^4$

10. $y = -(x + 2)^3$

11. $y = (x - 4)^3 - 2$

12. $y = -(x - 4)^3 - 2$

13. $y = -2(x + 5)^4$

14. $y = -2x^4 + 5$

15. $y = \frac{1}{2}(x + 1)^5$

16. $y = \frac{1}{2}x^5 + 1$

17. $y = -(x - 1)^3 - 1$

18. $y = x^8$

In Exercises 19–32, first sketch the graph in the immediate vicinity of each of the x-*intercepts. Then sketch the graph.*

19. $y = (x - 2)(x - 1)(x + 1)$

20. $y = (x - 3)(x + 2)(x + 1)$

21. $y = 2x(x + 1)(x + 3)$

22. $y = -x^2(x + 2)$

23. $y = x^3(x + 2)$

24. $y = (x - 1)(x - 4)^2$

25. $y = 2(x - 1)(x - 4)^3$

26. $y = (x - 1)^2(x - 4)^2$

27. $y = (x + 1)^2(x - 1)(x - 3)$

28. $y = x^2(x - 4)(x + 2)$

29. $y = -x^3(x - 4)(x + 2)$

30. $y = 4(x - 2)^2(x + 2)^3$

31. $y = -4x(x - 2)^2(x + 2)^3$

32. $y = -3x^3(x + 1)^4$

For Exercises 33–41, six functions are defined, as follows:

$$f(x) = x \quad g(x) = x^2 \quad h(x) = x^3$$
$$F(x) = x^4 \quad G(x) = x^5 \quad H(x) = x^6$$

Refer also to the following figure.

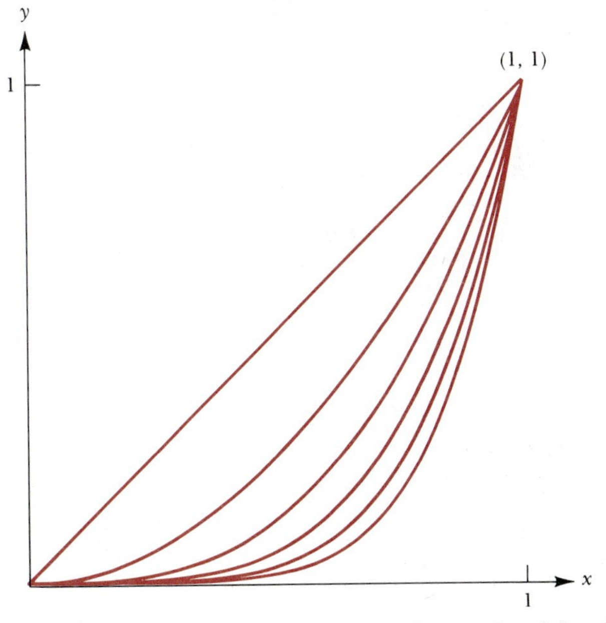

33. The six graphs in the figure are the graphs of the six given functions on the interval [0, 1], but the graphs are not labeled. Which is which?

C **34.** For which x values in [0, 1] will the graph of g lie strictly below the horizontal line $y = 0.1$? Use a calculator to evaluate your answer. Round off the result to two significant figures.

C **35.** Follow Exercise 34, using the function H instead of g.

36. Find a number t in $[0, 1]$ such that the vertical distance between $f(t)$ and $g(t)$ is $\frac{1}{4}$.

***37.** Is there a number t in $[0, 1]$ such that the vertical distance between $g(t)$ and $F(t)$ is 0.26?

C **38.** Find all numbers t in $[0, 1]$ such that $F(t) = G(t) + H(t)$.

C **39.** Using the graphs, explain why there are no numbers t in the open interval $(0, 1)$ for which $H(t) = F(t) + G(t)$.

***40.** Let A and B denote the points $(0, 0)$ and $(1, 1)$, respectively. Find a point P on the graph of $y = g(x)$ $(0 \leq x \leq 1)$ such that the triangle ABP is isosceles.

C ***41.** For which numbers t in $[0, 1]$ will the vertical distance between $h(t)$ and $H(t)$ exceed $\frac{1}{8}$?

5.3 Graphs of Rational Functions

We shall not attempt to explain the numerous situations in life sciences where the study of such graphs is important nor the chemical reactions which give rise to the rational functions r(x). . . . Let it suffice to mention that, to the experimental biochemist, theoretical results concerning the shapes of rational functions are of considerable interest.

W. G. Bardsley and R. M. W. Wood in the article "Critical Points and Sigmoidicity of Positive Rational Functions" in *The American Mathematical Monthly,* vol. 92 (1985), pp. 37–42.

After the polynomial functions, the next simplest functions are the *rational functions.* These are functions defined by equations of the form

$$y = \frac{f(x)}{g(x)}$$

where $f(x)$ and $g(x)$ are polynomials. In general, when we write $y = f(x)/g(x)$, we will assume that $f(x)$ and $g(x)$ contain no common factors (other than constants). The domain of the rational function defined by $y = f(x)/g(x)$ consists of all real numbers for which $g(x)$ is not zero.

Example 1 Specify the domain of the rational function defined by

$$y = \frac{3x - 2}{x^2 - 1}$$

Also, find the x-intercepts (if any) for the graph of this function.

Solution By factoring the denominator, we can rewrite the given equation as

$$y = \frac{3x - 2}{(x - 1)(x + 1)}$$

Since the denominator is zero when $x = 1$ and when $x = -1$, it follows that the domain of the given function consists of all real numbers except 1 and -1. To determine the x-intercepts of the graph, we set $y = 0$ (as usual) in the given equation to obtain

$$\frac{3x - 2}{x^2 - 1} = 0 \qquad (x \neq 1;\ x \neq -1)$$

$$3x - 2 = 0$$

We multiplied both sides of the equation by $x^2 - 1$; the quantity $x^2 - 1$ is nonzero, because $x \neq \pm 1$.

$$x = \frac{2}{3}$$

Thus the only x-intercept is $x = \frac{2}{3}$.

For reference, in the box that follows we summarize the ideas used in Example 1 for determining the domain and the x-intercepts of a rational function.

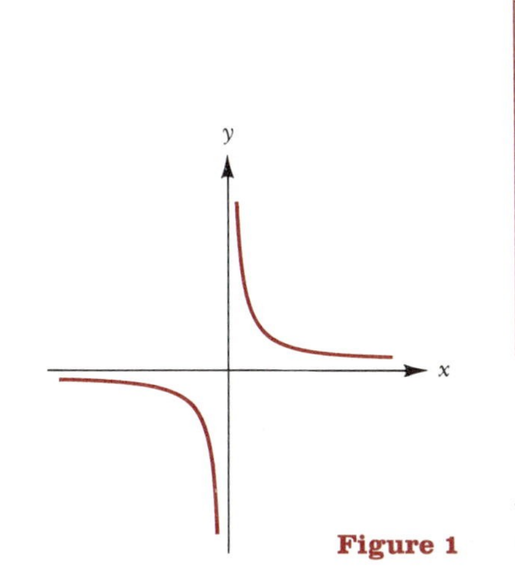

Figure 1

> Let R be a rational function defined by
>
> $$R(x) = \frac{f(x)}{g(x)}$$
>
> where $f(x)$ and $g(x)$ are polynomials with no common factors (other than constants). Then the domain of R consists of all real numbers for which $g(x) \neq 0$; the x-intercepts (if any) are the real solutions of the equation $f(x) = 0$.

You already know about the graph of one rational function, for in Chapter 4 we graphed $y = 1/x$. For reference, the graph is shown again in Figure 1.

The graph in Figure 1 differs from the graph of every polynomial function in two important aspects. First, the graph has a break in it; it is composed of two distinct pieces or *branches*. Recall that in Section 5.2 we learned that the graph of a polynomial function never has a break in it. In general, the graph of a rational function has one more branch than the number of distinct real zeros of the denominator. For instance, referring to the function in Example 1, we can expect the graph of $y = \dfrac{3x - 2}{x^2 - 1}$ to have three branches. The second way in which the graph in Figure 1 differs from that of a polynomial function has to do with the asymptotes. We say that a line is an *asymptote* for a curve if the distance between the line and the curve approaches zero as we move out farther and farther along the line. Thus the x-axis is a horizontal asymptote for the graph in Figure 1, while the y-axis is a vertical asymptote. It can be shown that the graph of a polynomial function never has an asymptote. Figure 2 displays further examples of curves with asymptotes.

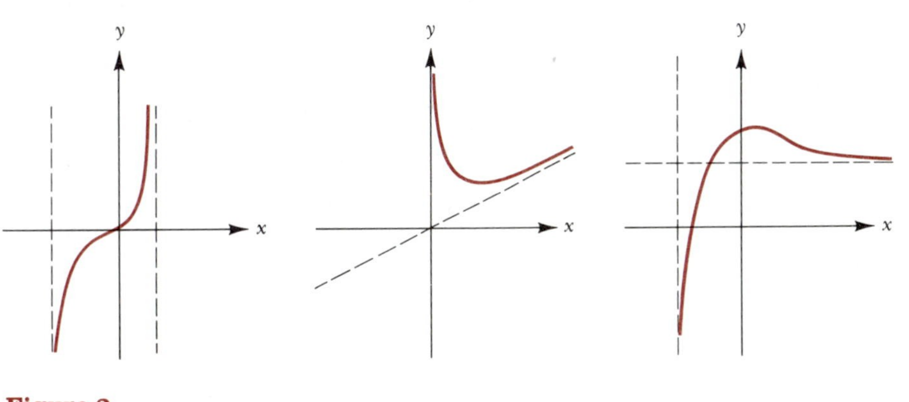

Figure 2
Curves with asymptotes.

If k is a positive constant, the graph of $y = k/x$ resembles that of $y = 1/x$. Figure 3 shows the graphs of $y = 1/x$ and $y = 4/x$. Once we know about the graph of $y = k/x$, we can graph any rational function of the form

$$y = \frac{ax + b}{cx + d}$$

The next three examples show how this is done. These examples make use of the techniques of translation and reflection developed in Chapter 4.

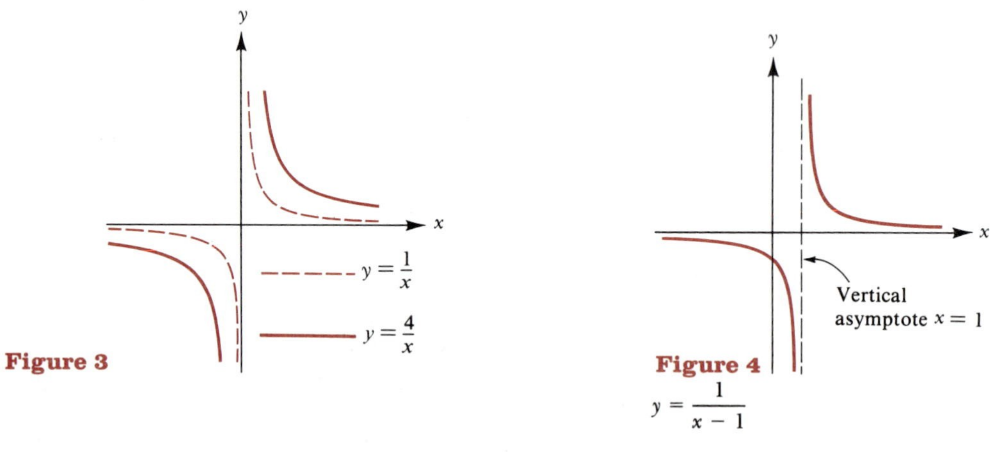

Figure 3

Figure 4 $y = \dfrac{1}{x - 1}$

Example 2 Graph $y = \dfrac{1}{x - 1}$.

Solution By translating the graph of $y = 1/x$ one unit to the right we obtain the graph of $y = \dfrac{1}{x - 1}$ shown in Figure 4. Notice that the horizontal asymptote is unchanged by the translation. However, the vertical asymptote moves one unit to the right.

Example 3 Graph:

(a) $y = \dfrac{2}{x - 1}$ **(b)** $y = \dfrac{-2}{x - 1}$

Solution The graph of $y = \dfrac{2}{x - 1}$ has the same basic shape and location as the graph of $y = \dfrac{1}{x - 1}$ shown in Figure 4. As a further guide to sketching $y = \dfrac{2}{x - 1}$, we can pick several convenient x-values on either side of the asymptote $x = 1$ and then compute the corresponding y-values. After doing this, we obtain the graph shown in Figure 5(a). By reflecting this graph in the x-axis, we obtain the graph of $y = \dfrac{-2}{x - 1}$, which is shown in Figure 5(b).

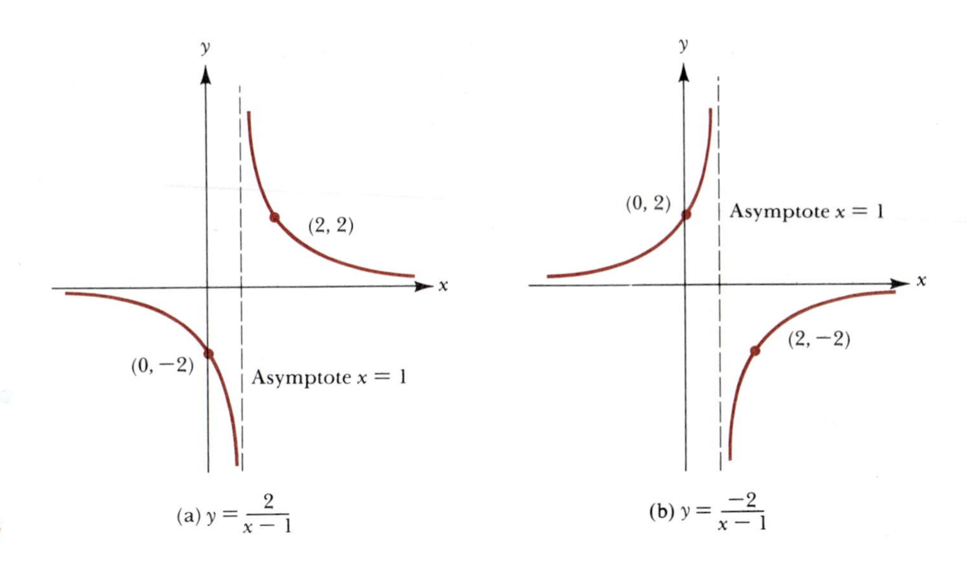

Figure 5

Example 4 Graph $y = \dfrac{4x - 2}{x - 1}$.

Solution Using long division, we find that

$$\frac{4x - 2}{x - 1} = 4 + \frac{2}{x - 1}$$

We conclude from this that the required graph can be obtained by moving the graph of $y = \dfrac{2}{x - 1}$ up four units in the y-direction. See Figure 6. Notice that the vertical asymptote is still $x = 1$, but the horizontal asymptote is now $y = 4$.

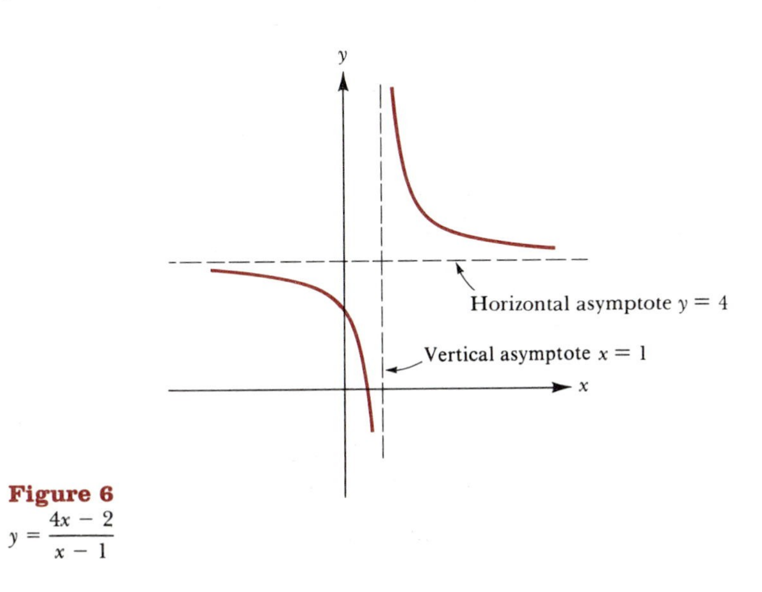

Figure 6
$y = \dfrac{4x - 2}{x - 1}$

(a)

(b) **Figure 7**

Next we look at rational functions of the form $y = 1/x^n$. First consider $y = 1/x^2$. As with $y = 1/x$, the domain consists of all real numbers except $x = 0$. For $x \neq 0$, $1/x^2$ is always positive. This means that the graph will always lie above the x-axis. As $|x|$ becomes very large, the quantity $1/x^2$ approaches zero; this is true whether x itself is positive or negative. So when $|x|$ is very large, we expect the graph of $y = 1/x^2$ to look as shown in Figure 7(a). On the other hand, when x is a very small fraction close to zero, either negative or positive, the quantity $1/x^2$ is very large. For instance, if $x = \frac{1}{10}$, we find that $y = 100$, and if $x = \frac{1}{100}$ we find that $y = 10{,}000$. Thus as x approaches zero, from the right or from the left, the graph of $y = 1/x^2$ must look as shown in Figure 7(b). Now by plotting several points and taking figures 7(a) and 7(b) into account, we obtain the graph of $y = 1/x^2$ shown in Figure 8. Also, by following a similar line of reasoning, we find that the graph of $y = 1/x^3$ looks as shown in Figure 9.

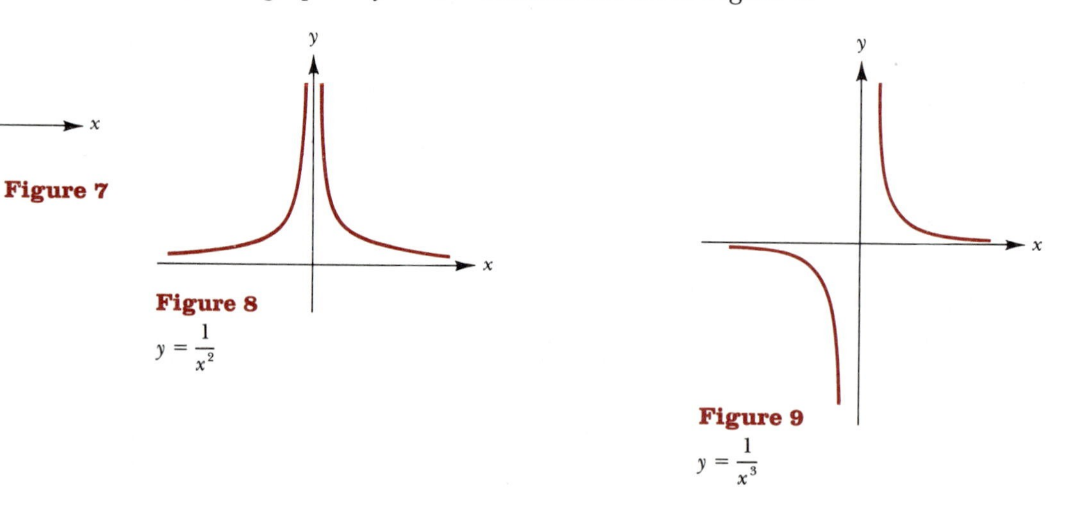

Figure 8
$y = \dfrac{1}{x^2}$

Figure 9
$y = \dfrac{1}{x^3}$

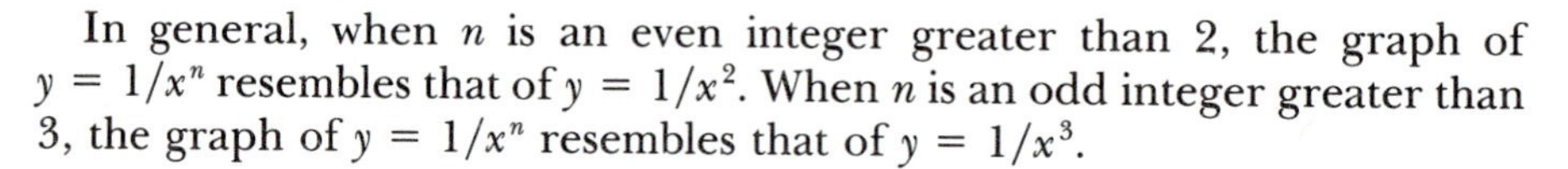
In general, when n is an even integer greater than 2, the graph of $y = 1/x^n$ resembles that of $y = 1/x^2$. When n is an odd integer greater than 3, the graph of $y = 1/x^n$ resembles that of $y = 1/x^3$.

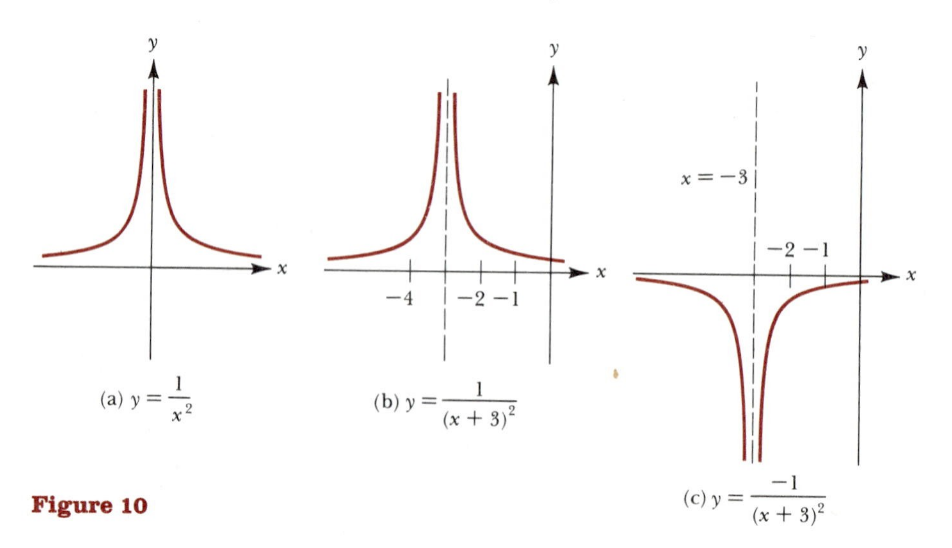

(a) $y = \dfrac{1}{x^2}$

(b) $y = \dfrac{1}{(x+3)^2}$

(c) $y = \dfrac{-1}{(x+3)^2}$

Figure 10

Example 5 Graph $y = \frac{-1}{(x+3)^2}$.

Solution (Refer to Figure 10.) Begin with the graph of $y = 1/x^2$. By moving the graph three units to the left, we obtain the graph of $y = \frac{1}{(x+3)^2}$. Then, by reflecting the graph of $y = \frac{1}{(x+3)^2}$ in the x-axis, we get the graph of $y = \frac{-1}{(x+3)^2}$. See Figure 10.

Example 6 Graph $y = \frac{4}{(x-2)^3}$.

Solution Moving the graph of $y = 1/x^3$ two units to the right gives us the graph of $y = \frac{1}{(x-2)^3}$. The graph of $y = \frac{4}{(x-2)^3}$ will then have the same basic shape and location as this. As a further guide to sketching the required graph, we can pick several convenient x-values near the asymptote $x = 2$ and compute the corresponding y-values. For instance, using $x = 0$, $x = 1$, and $x = 3$, we find that the points $(0, -\frac{1}{2})$, $(1, -4)$, and $(3, 4)$ are on the graph. Using this information, the graph can be sketched as in Figure 11.

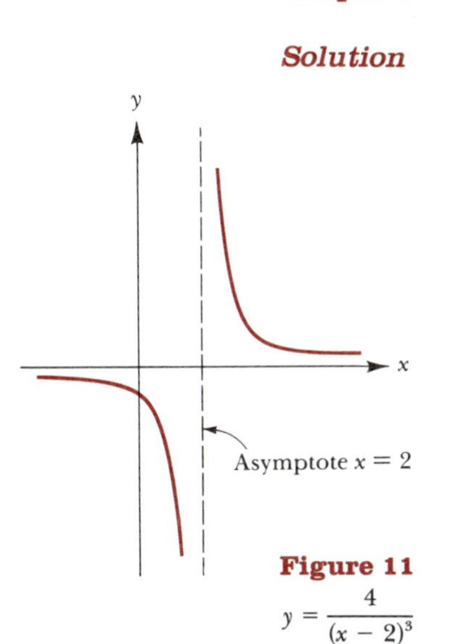

Figure 11 $y = \frac{4}{(x-2)^3}$

In the previous section, you learned a technique for graphing those polynomial functions that could be expressed as a product of linear factors. That technique can also be used to graph rational functions in which both the numerator and denominator can be expressed as products of linear factors. In addition to investigating the behavior of the function near its x-intercepts, now we will also want to see how the function looks near its asymptotes. As the previous examples indicate, the vertical asymptotes of $y = f(x)/g(x)$ are known once we find the values of x for which $g(x) = 0$. For instance, the vertical asymptotes of $y = \frac{3x^2}{x^2 - 1}$ are the lines $x = 1$ and $x = -1$. To find the horizontal asymptotes, we divide the numerator and denominator by the highest power of x that appears. (You'll see the reason for doing this in a moment.) In the case of $y = \frac{3x^2}{x^2 - 1}$ we divide numerator and denominator by x^2 to obtain

$$y = \frac{3}{1 - 1/x^2}$$

Now when $|x|$ grows very large, $1/x^2$ approaches zero and we have therefore

$$y \approx \frac{3}{1 - 0} \qquad \text{when } |x| \text{ is very large}$$

This tells us that the line $y = 3$ is a horizontal asymptote for the graph of $y = \frac{3x^2}{x^2 - 1}$. You'll see another example of this type of calculation in Example 7.

Example 7 Graph the function $y = \frac{(x-3)(x+2)}{(x+1)(x-2)}$.

Solution By inspection we see that the x-intercepts are $x = 3$ and $x = -2$, while the vertical asymptotes are the lines $x = -1$ and $x = 2$. To find the horizontal asymptote we write

$$y = \frac{x^2 - x - 6}{x^2 - x - 2}$$

$$= \frac{1 - \frac{1}{x} - \frac{6}{x^2}}{1 - \frac{1}{x} - \frac{2}{x^2}} \approx \frac{1 - 0 - 0}{1 - 0 - 0} \qquad \text{when } |x| \text{ is large}$$

The horizontal asymptote is therefore the line $y = 1$.

Now we want to see how the graph looks in the immediate vicinity of the x-intercepts and the vertical asymptotes. To do this we'll use the approximation technique explained in the previous section in connection with polynomial functions. Let's start with the x-intercept $x = 3$. We have

$$x \text{ near } 3: \quad y = \frac{(x - 3)(x + 2)}{(x + 1)(x - 2)} \approx \frac{(x - 3)(3 + 2)}{(3 + 1)(3 - 2)} = \frac{5}{4}(x - 3)$$

So, in the immediate vicinity of the x-intercept $x = 3$, the required graph will closely resemble the graph of the line $y = \frac{5}{4}x - \frac{15}{4}$. The remaining calculations for approximating the graph near the other x-intercept and near the two vertical asymptotes are carried out in a similar manner. As Exercise 33 asks you to verify, the results are as follows. (For completeness we'll repeat the result just obtained.)

$$x \text{ near } 3: \qquad y \approx \frac{5}{4}x - \frac{15}{4}$$

$$x \text{ near } -2: \qquad y \approx -\frac{5}{4}x - \frac{5}{2}$$

$$x \text{ near } -1: \qquad y \approx \frac{\frac{4}{3}}{x + 1}$$

$$x \text{ near } 2: \qquad y \approx \frac{-\frac{4}{3}}{x - 2}$$

We can summarize these results as follows. As the graph passes through the points $(3, 0)$ and $(-2, 0)$, it resembles the lines $y = \frac{5}{4}x - \frac{15}{4}$ and $y = -\frac{5}{4}x - \frac{5}{2}$, respectively. Near the vertical asymptote $x = -1$, the graph has the same basic shape as $y = \dfrac{1}{x + 1}$. Near the vertical asymptote $x = 2$, the graph has the same basic shape as $y = \dfrac{-1}{x - 2}$. See Figure 12. Next we need to see how the graph approaches the horizontal asymptote $y = 1$. When $|x|$ is large and $x > 0$, we find that y is less than 1. (Check this for yourself using $x = 10$ or $x = 100$.) This means that to the right of the origin, the graph approaches the asymptote $y = 1$ from below. Similarly, when $|x|$ is large and $x < 0$, we find that y is again less than 1. This means that to the left of the origin, the graph approaches the asymptote $y = 1$ from below. In Figure 13 we summarize what we have discovered up to this point about the graph. We shall use Figure 13 as a guide for sketching the required graph. But before doing that, it would be helpful to know where

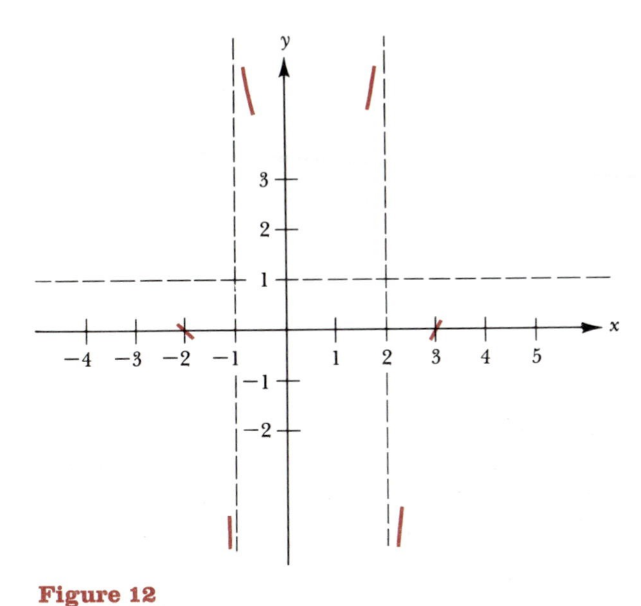

Figure 12

The graph of $y = \dfrac{(x-3)(x+2)}{(x+1)(x-2)}$ near its x-intercepts and vertical asymptotes.

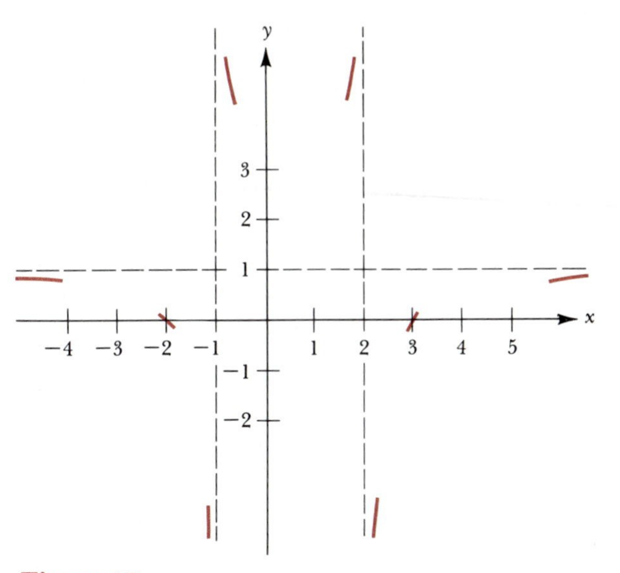

Figure 13

The graph of $y = \dfrac{(x-3)(x+2)}{(x+1)(x-2)}$ near its x-intercepts and horizontal and vertical asymptotes.

the graph crosses the y-axis. To find the y-intercept, we set $x = 0$ in the original equation to obtain

$$y = \frac{(0-3)(0+2)}{(0+1)(0-2)} = 3$$

The graph can now be sketched as in Figure 14. *Note:* A method for finding the precise coordinates of the lowest point on the middle branch of the graph is studied in calculus. Those coordinates can also be found, however, by applying techniques already developed in previous chapters. See Exercise 35.

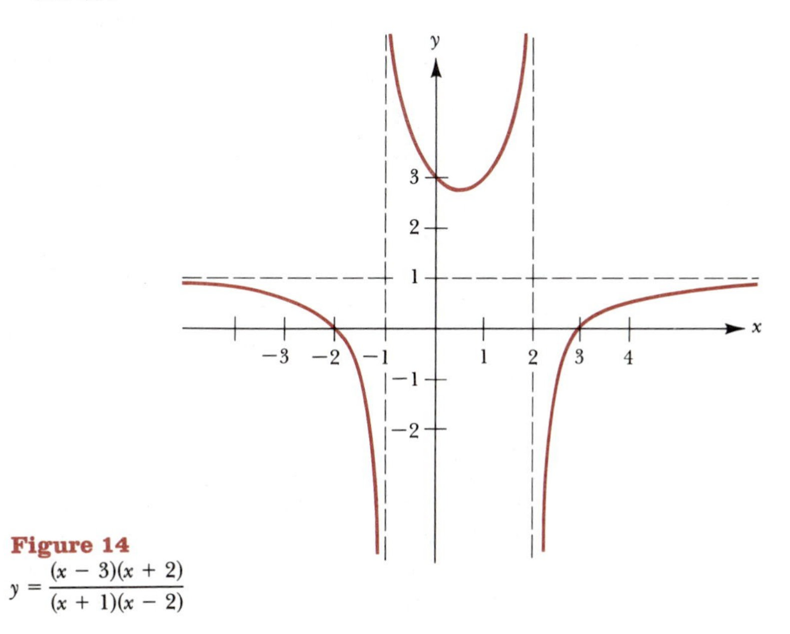

Figure 14

$y = \dfrac{(x-3)(x+2)}{(x+1)(x-2)}$

Exercise Set 5.3

In Exercises 1–6, find the domain and the x-intercepts for each rational function.

1. $y = \dfrac{3x + 15}{4x - 12}$
2. $y = \dfrac{(x + 6)(x + 4)}{(x - 1)^2}$
3. $y = \dfrac{x^2 - 8x - 9}{x^2 - x - 6}$
4. $y = \dfrac{2x^2 + x - 5}{x^2 + 1}$
5. $y = \dfrac{(x^2 - 4)(x^3 - 1)}{x^6}$
6. $y = \dfrac{x^5 - 2x^4 - 9x + 18}{8x^3 + 2x^2 - 3x}$

In Exercises 7–31, sketch the graph of each rational function. Specify the intercepts and the asymptotes.

7. $y = \dfrac{1}{x + 4}$
8. $y = \dfrac{-1}{x + 4}$
9. $y = \dfrac{3}{x + 2}$
10. $y = \dfrac{-3}{x + 2}$
11. $y = \dfrac{-2}{x - 3}$
12. $y = \dfrac{x - 1}{x + 1}$
13. $y = \dfrac{x - 3}{x - 1}$
14. $y = \dfrac{2x}{x + 3}$
15. $y = \dfrac{4x - 2}{2x + 1}$
16. $y = \dfrac{3x + 2}{x - 3}$
17. $y = \dfrac{1}{(x - 2)^2}$
18. $y = \dfrac{-1}{(x - 2)^2}$
19. $y = \dfrac{3}{(x + 1)^2}$
20. $y = \dfrac{-3}{(x + 1)^2}$
21. $y = \dfrac{1}{(x + 2)^3}$
22. $y = \dfrac{-1}{(x + 2)^3}$
23. $y = \dfrac{-4}{(x + 5)^3}$
24. $y = \dfrac{x}{(x + 1)(x - 3)}$
25. $y = \dfrac{-x}{(x + 2)(x - 2)}$
26. $y = \dfrac{2x}{(x + 1)^2}$
27. $y = \dfrac{x}{(x - 1)(x + 3)}$
28. $y = \dfrac{x^2 + 1}{x^2 - 1}$
29. $y = \dfrac{(x - 4)(x + 2)}{(x - 1)(x - 3)}$
30. $y = \dfrac{(x - 1)(x - 3)}{(x + 1)^2}$
31. $y = \dfrac{(x - 1)(x - 3)}{(x + 1)^3}$

***32.** Let $y = \dfrac{(x - 2)(x - 3)}{x - c}$. If the range of this function is the set of all real numbers, show that $2 \le c \le 3$.

33. (This exercise refers to Example 7.) Let $y = \dfrac{(x - 3)(x + 2)}{(x + 1)(x - 2)}$. Verify each of the following approximations.

(a) When x is close to -2, $y \approx -\frac{5}{4}x - \frac{5}{2}$.

(b) When x is close to -1, $y \approx \dfrac{\frac{4}{3}}{x + 1}$.

(c) When x is close to 2, $y \approx \dfrac{-\frac{4}{3}}{x - 2}$.

34. In the examples for this section, and in the exercises preceding this one, the two polynomials making up each rational function contain no common factors (other than constants). In this exercise you are asked to graph two rational functions in which the numerator and denominator do contain common factors. Be careful to pay attention to the domain in each case here.

(a) Graph $y = \dfrac{x - 1}{(x - 1)(x - 2)}$.

(b) Graph $y = \dfrac{(x - 1)(x - 2)(x - 3)}{(x - 1)(x - 2)(x - 3)(x - 4)}$.

***35.** This exercise shows you how to find the coordinates of the lowest point on the middle branch of

$$y = \frac{(x - 3)(x + 2)}{(x + 1)(x - 2)}$$

(We graphed this function in Example 7.) The basic idea involves finding the range of this function.

(a) Verify that the original equation is equivalent to $y = \dfrac{x^2 - x - 6}{x^2 - x - 2}$.

(b) By clearing of fractions and then combining like terms, show that the equation in part (a) can be written $(y - 1)x^2 - (y - 1)x + (6 - 2y) = 0$.

(c) Solve the equation in part (b) for x in terms of y by using the quadratic formula.

(d) Working from the result in part (c), show that the range of the original function is the set $(-\infty, 1) \cup [\frac{25}{9}, \infty)$. This result, along with the graph of the function that we obtained in Example 7, implies that the y-coordinate of the lowest point on the middle branch is $\frac{25}{9}$. Now find the corresponding x-coordinate.

5.4 The Conic Sections (Part 1: The Parabola and the Ellipse)

The Greeks knew the properties of the curves given by cutting a cone with a plane—the ellipse, the parabola and hyperbola. Kepler discovered by analysis of astronomical observations, and Newton proved mathematically . . . that the planets move in ellipses. The geometry of Ancient Greece thus became the cornerstone of modern astronomy.

John Lighton Synge (b. 1897)

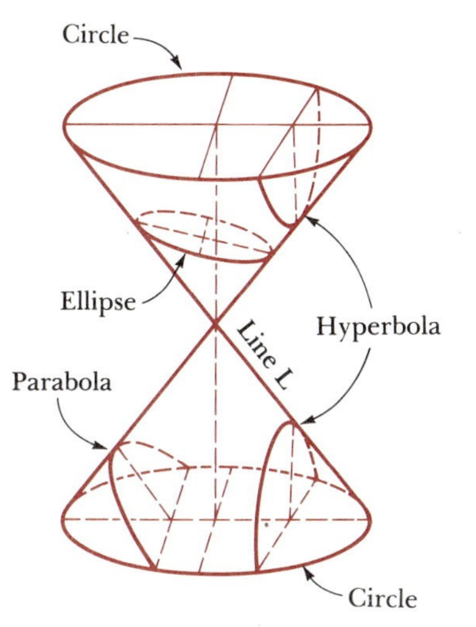

Figure 1

The *conic sections* are the curves formed when a plane intersects the surface of a right circular cone. As Figure 1 indicates, these curves are the *circle,* the *parabola,* the *ellipse,* and the *hyperbola.* For the parabola in Figure 1, the cutting plane is parallel to line L; for the hyperbola, the cutting plane is parallel to the axis of the cone. In this section and the next, we'll discuss the equations describing these curves in an x-y plane. We won't, however, derive these equations from the geometry of Figure 1.*

In Section 5.1 we saw that the graph of any quadratic function $y = ax^2 + bx + c$ is a parabola. We learned that by completing the square, this equation can always be rewritten in the general form

$$y = A(x - h)^2 + k \tag{1}$$

In this form, we know that the vertex of the parabola is (h, k) and the axis of symmetry is the vertical line $x = h$. In order to emphasize the similarities between the equation of the parabola and other equations that will appear in this section, we are going to rewrite equation (1) as

$$y - k = A(x - h)^2 \tag{2}$$

Figures 2(a) and (b) serve to summarize our remarks up to this point.

If we reverse the roles of x and y in the equation of a parabola $y = ax^2 + bx + c$, we obtain $x = ay^2 + by + c$. The graph of this equation is also a parabola, but now the axis of symmetry will be horizontal rather than vertical. By completing the square in this last equation, we obtain an equation of the form

$$x - h = A(y - k)^2 \tag{3}$$

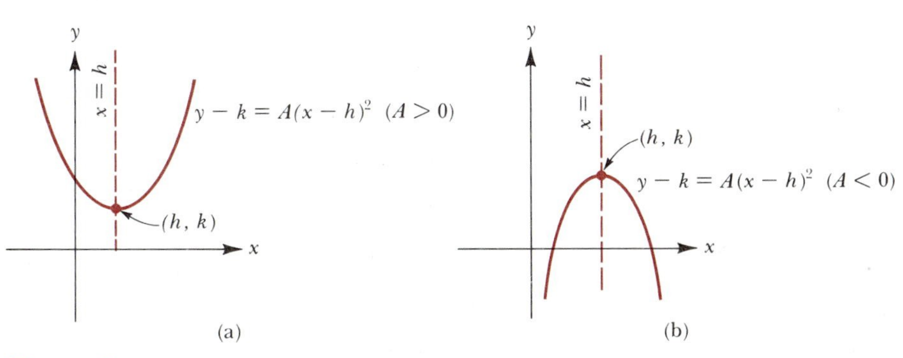

Figure 2

* For a proof that the equations given in this section and the next do describe the curves in Figure 1, see *Calculus and Analytic Geometry,* 3rd ed., by Al Shenk (Glenview, IL: Scott, Foresman & Company, 1984). The derivations given there do not use calculus.

As indicated in Figure 3, the vertex of this parabola is (h, k). Incidentally, as you can see by applying the vertical line test in Figure 3, the graphs there do not represent functions.

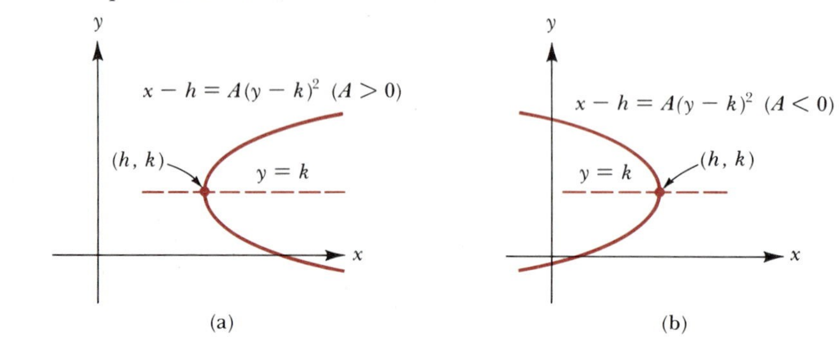

Figure 3

Example 1 Graph the parabola $x = 2y^2 - 4y + 5$. Specify the vertex and the axis of symmetry.

Solution By completing the square we'll be able to rewrite the equation in the form of equation (3). We have

$$
\begin{aligned}
x - 5 &= 2y^2 - 4y \\
x - 5 &= 2(y^2 - 2y \quad\ ) \\
x - 5 + 2 &= 2(y^2 - 2y + 1) \qquad \text{We added 2 to both sides.} \\
x - 3 &= 2(y - 1)^2
\end{aligned}
$$

By comparing this last equation with equation (3), we see that the vertex of this parabola is $(3, 1)$ and that the axis is the horizontal line $y = 1$. Also, since the coefficient of y^2 is 2, which is positive, the parabola opens to the right. Before sketching the parabola, notice in the original equation that if $y = 0$, then $x = 5$. So the x-intercept of the curve is 5. See Figure 4.

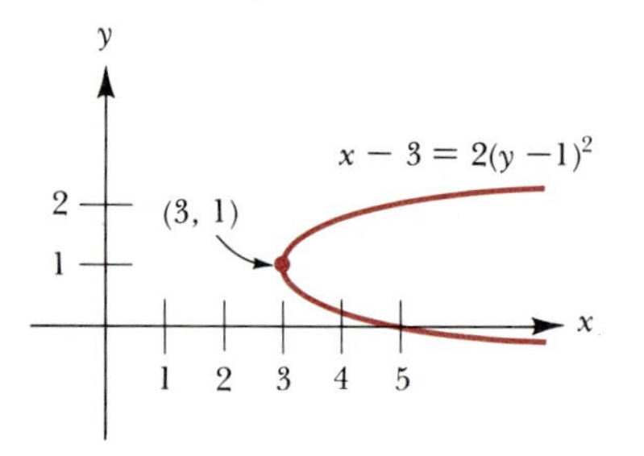

Figure 4

Before concluding our discussion of the parabola, we point out that there are numerous applications of this curve in the sciences. Parabolic reflectors, for instance, are used in communications systems, in telescope mirrors, and in automobile headlights. In the case of the telescope mirror, for example, incoming light rays that are parallel to the axis of the parabola are all reflected through a single point called the *focus* of the parabola, as indicated in Figure 5.

In Section 3.2 we saw that the equation of a circle with radius r and center (h, k) can be written

$$(x - h)^2 + (y - k)^2 = r^2$$

If we divide both sides of this equation by r^2, we obtain the equivalent equation

$$\frac{(x - h)^2}{r^2} + \frac{(y - k)^2}{r^2} = 1 \tag{4}$$

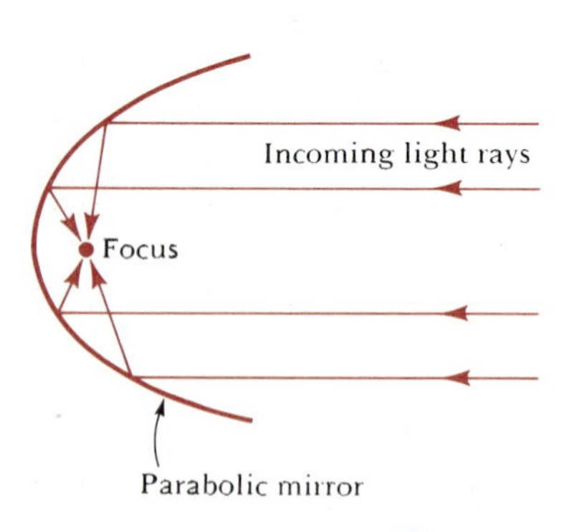

Figure 5

Equation (4) is actually a particular case of the following, more general type of equation:

$$\frac{(x - h)^2}{a^2} + \frac{(y - k)^2}{b^2} = 1 \qquad (a > 0, b > 0) \tag{5}$$

The graph of equation (5) is an *ellipse*. To study this curve, let's first look at the case in which both h and k are zero. That is, we are considering the equation

$$\frac{x^2}{a^2} + \frac{y^2}{b^2} = 1$$

As you can check by using the symmetry tests described in Section 3.2, the graph of this last equation must be symmetric about both coordinate axes and also about the origin. The intercepts of the graph are easy to find. For the x-intercept, we just set $y = 0$ to obtain

$$\frac{x^2}{a^2} = 1$$
$$x^2 = a^2$$
$$x = \pm a$$

So the x-intercepts are a and $-a$. In a similar fashion now, you should be able to check for yourself that the y-intercepts are b and $-b$. Figure 6 shows the graph of this ellipse. (Calculator exercises at the end of this section will help convince you that the general shapes of the curves in Figure 6 are correct.)

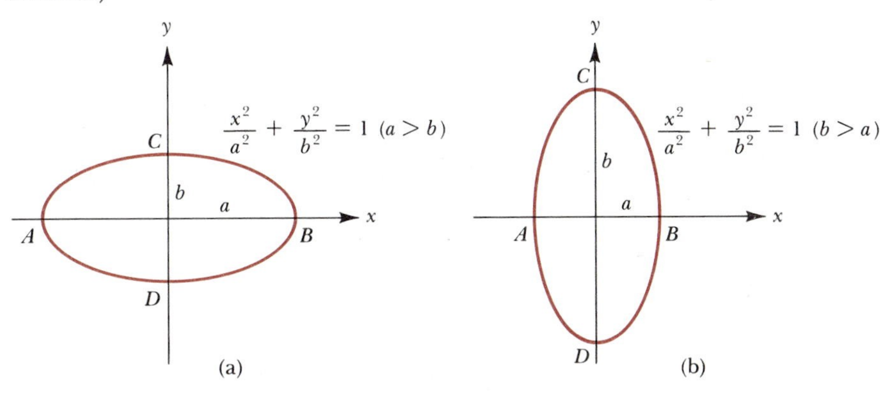

Figure 6

The line segments AB and CD in Figure 6(a) are called the *axes* of the ellipse. The longer of the two segments is the *major axis;* the shorter is the *minor axis*. For the ellipse in Figure 6(a) (where $a > b$), AB is the major axis and CD is the minor axis. In Figure 6(b) (where $b > a$), AB is the minor axis and CD is the major axis. The point where the two axes of an ellipse meet is called the *center* of the ellipse. So, for the ellipses in Figure 6, the centers coincide with the origin in each case.

In general, if A, B, and C are positive numbers, then the graph of the equation

$$Ax^2 + By^2 = C$$

will be an ellipse. By studying the next example you'll see why this is so.

Example 2 Graph the ellipse $9x^2 + 16y^2 = 144$.

Solution If we divide both sides of the given equation by 144, we obtain

$$\frac{9x^2}{144} + \frac{16y^2}{144} = 1$$
$$\frac{x^2}{16} + \frac{y^2}{9} = 1$$
$$\frac{x^2}{4^2} + \frac{y^2}{3^2} = 1$$

Since the form of this last equation is $x^2/a^2 + y^2/b^2 = 1$, we know that the graph is an ellipse. See Figure 7. The lengths of the major and the minor axes of this ellipse are 8 and 6, respectively.

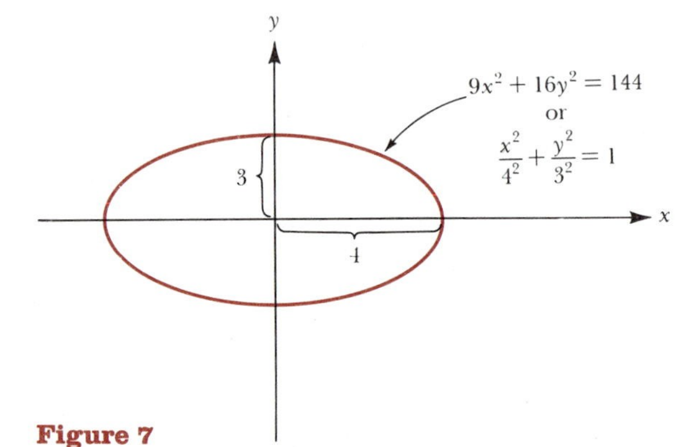

Figure 7

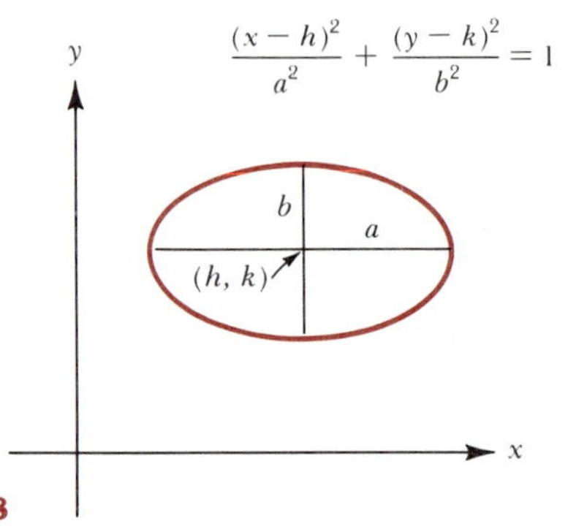

Figure 8

Just as the equation $(x - h)^2 + (y - k)^2 = r^2$ represents a circle with center (h, k), so the equation

$$\frac{(x-h)^2}{a^2} + \frac{(y-k)^2}{b^2} = 1 \tag{6}$$

represents an ellipse with center (h, k). As indicated in Figure 8, the axes of the ellipse are parallel to the coordinate axes.

We refer to equation (6) as the *standard form* for the equation of an ellipse. In the next example we use the completing the square technique to convert an equation to standard form. Just as with the circle, when the equation is in standard form the graph is easily obtained.

Example 3 The graph of the equation $5x^2 - 30x + 2y^2 - 16y + 57 = 0$ is an ellipse. Convert this equation to standard form by completing the squares, and then graph the ellipse.

Solution

$$5(x^2 - 6x \qquad) + 2(y^2 - 8y \qquad) = -57 \qquad \text{factoring}$$
$$5(x^2 - 6x + 9) + 2(y^2 - 8y + 16) = -57 + 45 + 32 \qquad \text{completing the squares}$$
$$5(x-3)^2 + 2(y-4)^2 = 20$$
$$\frac{(x-3)^2}{4} + \frac{(y-4)^2}{10} = 1 \qquad \text{dividing by 20}$$
$$\frac{(x-3)^2}{2^2} + \frac{(y-4)^2}{(\sqrt{10})^2} = 1$$

This last equation is the required standard form for the ellipse. (Notice that it was not necessary to know in advance that the original equation actually represented an ellipse; this fact emerged after we completed the square.)

The center of the ellipse is (3, 4). Since $\sqrt{10} > 2$, the major axis is parallel to the y-axis and the minor axis is parallel to the x-axis. To find the end points of the minor axis, we start at the center, (3, 4), and then locate the points two units to the right and two units to the left. Moving two units to the right of (3, 4) takes us to (5, 4); moving two units to the left of (3, 4) takes us to (1, 4). In a similar manner, by moving up and then down from (3,4) by $\sqrt{10}$ units, we find that the end points of the major axis are $(3, 4 + \sqrt{10})$ and $(3, 4 - \sqrt{10})$. See Figure 9.

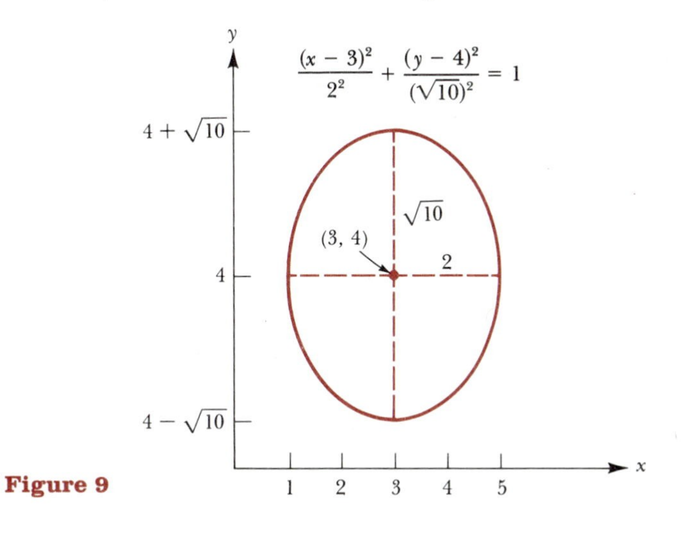

Figure 9

Exercise Set 5.4

In Exercises 1–18, graph the parabolas. In each case specify the axis and the vertex.

1. $x^2 = 4y$
2. $x^2 = 16y$
3. $y^2 = -8x$
4. $y^2 = 12x$
5. $x^2 = -20y$
6. $x^2 - y = 0$
7. $y^2 + 28x = 0$
8. $4y^2 + x = 0$
9. $x^2 = 6y$
10. $y^2 = -10x$
11. $y^2 - 6y - 4x + 17 = 0$
12. $y^2 + 2y + 8x + 17 = 0$
13. $x^2 - 8x - y + 18 = 0$
14. $x^2 + 6y + 18 = 0$
15. $y^2 + 2y - x + 1 = 0$
16. $2y^2 - x + 1 = 0$
17. $2x^2 - 12x - y + 18 = 0$
18. $y + \sqrt{2} = (x - 2\sqrt{2})^2$

For Exercises 19–40, graph the ellipses. In each case specify the center and the lengths of the major and minor axes.

19. $4x^2 + 9y^2 = 36$
20. $4x^2 + 25y^2 = 100$
21. $x^2 + 16y^2 = 16$
22. $9x^2 + 25y^2 = 225$
23. $x^2 + 2y^2 = 2$
24. $2x^2 + 3y^2 = 3$
25. $16x^2 + 9y^2 = 144$
26. $25x^2 + y^2 = 25$
27. $15x^2 + 3y^2 = 5$
28. $9x^2 + y^2 = 4$
29. $\frac{(x-5)^2}{5^2} + \frac{(y+1)^2}{3^2} = 1$
30. $\frac{(x-1)^2}{2^2} + \frac{(y+4)^2}{3^2} = 1$
31. $\frac{(x-1)^2}{1^2} + \frac{(y-2)^2}{2^2} = 1$
32. $\frac{x^2}{4^2} + \frac{(y-3)^2}{2^2} = 1$
33. $\frac{(x+3)^2}{3^2} + \frac{y^2}{1^2} = 1$
34. $\frac{(x-2)^2}{2^2} + \frac{(y-2)^2}{2^2} = 1$
35. $3x^2 + 4y^2 - 6x + 16y + 7 = 0$
36. $16x^2 + 64x + 9y^2 - 54y + 1 = 0$
37. $5x^2 + 3y^2 - 40x - 36y + 188 = 0$
38. $x^2 + 16y^2 - 160y + 384 = 0$
39. $16x^2 + 25y^2 - 64x - 100y + 564 = 0$

40. $4x^2 + 4y^2 - 32x + 32y + 127 = 0$

41. The area A of an ellipse $x^2/a^2 + y^2/b^2 = 1$ is given by the formula $A = \pi ab$. Use this formula to compute the area of the ellipse $5x^2 + 6y^2 = 60$.

42. Show that the coordinates of the vertex of the parabola $Ax^2 + Cx + Dy + E = 0$ are given by

$$x = -\frac{C}{2A} \quad \text{and} \quad y = \frac{C^2 - 4AE}{4AD}$$

C 43. Solve the equation $x^2/3^2 + y^2/2^2 = 1$ to obtain $y = \frac{\pm\sqrt{36 - 4x^2}}{3}$. Then complete the following table and use the results to graph the given equation. (Use the fact that the graph must be symmetric about the y-axis.)

x	0	0.5	1.0	1.5	2.0	2.5	3
y	±2						0

C 44. Solve the equation $x^2/1^2 + y^2/4^2 = 1$ for y to obtain $y = \pm 4\sqrt{1 - x^2}$. Then complete the following table and use the results to graph the given equation. (After plotting the points obtained from your table, use the fact that the graph must be symmetric about the y-axis.)

x	0	0.1	0.2	0.3	0.4	0.5	0.6	0.7	0.8	0.9	1.0
y	±4										0

C 45. As you know, there is a simple expression for the circumference of a circle of radius a: $2\pi a$. However, there is no similar type of elementary expression for the circumference of an ellipse. (The circumference of an ellipse can be computed to as many decimal places as required using the methods of calculus.) Nevertheless, there are some interesting elementary formulas that allow us to approximate the circumference of an ellipse quite closely. Three such formulas follow, along with the names of their discoverers and the dates of discovery. Each formula yields an approximate value for the circumference of the ellipse $x^2/a^2 + y^2/b^2 = 1$.

$$C_1 = \pi[a + b + \tfrac{1}{2}(\sqrt{a} - \sqrt{b})^2]$$
Giuseppe Peano, 1887

$$C_2 = \pi[3(a + b) - \sqrt{(a + 3b)(3a + b)}]$$
Srinivasa Ramanujan, 1914

$$C_3 = \frac{\pi}{2}(a + b + \sqrt{2(a^2 + b^2)})$$
R. A. Johnson, 1930

Use these formulas to complete the following table of approximations for the circumference of the ellipse $x^2/5^2 + y^2/3^2 = 1$. Round off the values of C_1, C_2, and C_3 to six decimal places. In order to complete the right-hand column of the table, you need two facts. First, the actual circumference of the ellipse, rounded off to six decimal places, is 25.526999. Second, percentage error in an approximation is given by

$$\frac{|(\text{true value}) - (\text{approximation})|}{\text{true value}} \times 100.$$

Round off the percentage errors to two significant digits. Which of the three approximations is the best?

	Approximation Obtained	Percentage Error
C_1		
C_2		
C_3		

5.5 The Conic Sections (Part 2: The Hyperbola)

We have seen in the previous section that the graph of the equation

$$\frac{x^2}{a^2} + \frac{y^2}{b^2} = 1$$

is an ellipse. Now we consider the equation

$$\frac{x^2}{a^2} - \frac{y^2}{b^2} = 1 \qquad (a > 0, b > 0) \tag{1}$$

The graph of this equation is a *hyperbola.* According to the symmetry tests in Section 3.2, the graph will be symmetric about both coordinate axes and about the origin. To find the x-intercepts, we set $y = 0$ in equation (1) to obtain

$$\frac{x^2}{a^2} = 1$$
$$x^2 = a^2$$
$$x = \pm a$$

Thus the hyperbola crosses the x-axis at the points $(-a, 0)$ and $(a, 0)$. On the other hand, the curve does not cross the y-axis, for if we set $x = 0$ in equation (1) we obtain

$$-\frac{y^2}{b^2} = 1$$

For any nonzero real number y, the left-hand side of this last equation will be negative, whereas the right-hand side is certainly positive. Consequently the equation has no real solutions, and the hyperbola has no y-intercepts.

The two lines $y = (b/a)x$ and $y = -(b/a)x$ bear a special relationship to the hyperbola; they are both asymptotes. We can see why this is so as follows. First solve equation (1) for y:

$$\frac{x^2}{a^2} - \frac{y^2}{b^2} = 1$$
$$b^2x^2 - a^2y^2 = a^2b^2 \qquad \text{multiplying by } a^2b^2$$
$$-a^2y^2 = a^2b^2 - b^2x^2$$
$$y^2 = \frac{b^2x^2 - a^2b^2}{a^2} = \frac{b^2(x^2 - a^2)}{a^2}$$
$$y = \pm\frac{b}{a}\sqrt{x^2 - a^2} \tag{2}$$

x	$\sqrt{x^2 - 5^2}$
100	99.875
1000	999.987
10000	9999.999

Table 1

Now as x grows arbitrarily large, the value of the quantity $\sqrt{x^2 - a^2}$ becomes closer and closer to x itself. Table 1 provides some empirical evidence for this statement in the case when $a = 5$. (A formal proof of the statement properly belongs to calculus.) In summary then, we have the approximation $\sqrt{x^2 - a^2} \approx x$ as x grows arbitrarily large. So, in view of equation (2), we have

$$y = \pm\frac{b}{a}\sqrt{x^2 - a^2} \approx \pm\frac{b}{a}x \qquad \text{as } x \text{ grows arbitrarily large}$$

In other words, the two lines $y = \pm(b/a)x$ are asymptotes for the hyperbola. (For the record, we again point out that the argument here is intuitive; calculus would be required to make it rigorous.)

A simple way to sketch the two asymptotes and then graph the hyperbola is as follows. First draw the rectangle with vertices (a, b), $(-a, b)$, $(-a, -b)$, and $(a, -b)$, as indicated in Figure 1(a) on the next page. The slopes of the diagonals in this rectangle are b/a and $-b/a$. Thus by extending these diagonals as in Figure 1(b), we obtain the two asymptotes $y = \pm(b/a)x$. Now, since the x-intercepts of the hyperbola are a and $-a$, we can sketch the curve as shown in Figure 1(c).

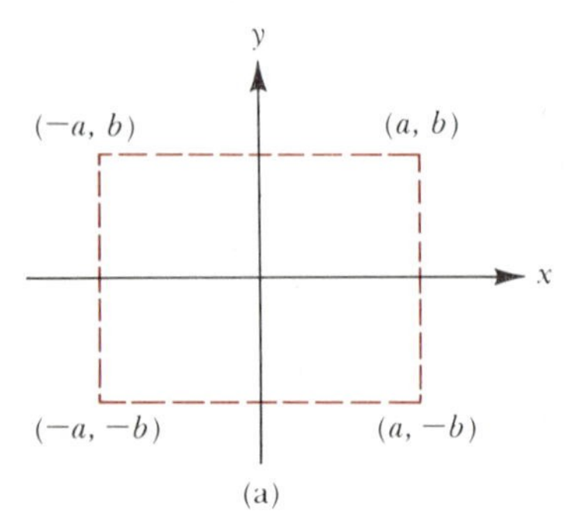

For the hyperbola in Figure 1(c), the two points $(\pm a, 0)$, where the curve meets the x-axis, are referred to as *vertices*. The midpoint of the line segment joining the two vertices is called the *center* of the hyperbola. (Equivalently, we can define the center as the point of intersection of the two asymptotes.) For the hyperbola in Figure 1(c), the center coincides with the origin. The line segment joining the vertices of a hyperbola is the *transverse axis* of the hyperbola.

In general, if A, B, and C are positive numbers, then the graph of an equation of the form

$$Ax^2 - By^2 = C$$

will be a hyperbola of the type shown in Figure 1(c). Example 1 shows why this is so.

Figure 1

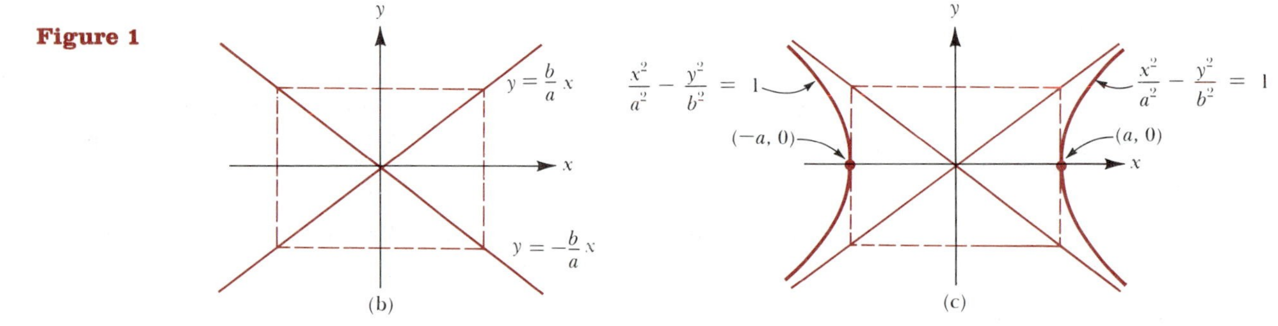

Example 1 Graph the hyperbola $25x^2 - 16y^2 = 400$.

Solution If we divide both sides of the given equation by 400, we obtain

$$\frac{25x^2}{400} - \frac{16y^2}{400} = 1$$

$$\frac{x^2}{16} - \frac{y^2}{25} = 1$$

$$\frac{x^2}{4^2} - \frac{y^2}{5^2} = 1$$

Since this last equation has the form $x^2/a^2 - y^2/b^2 = 1$, the graph is a hyperbola of the type shown in Figure 1(c). Using the technique indicated in Figure 1, we can now draw the required graph. See Figure 2.

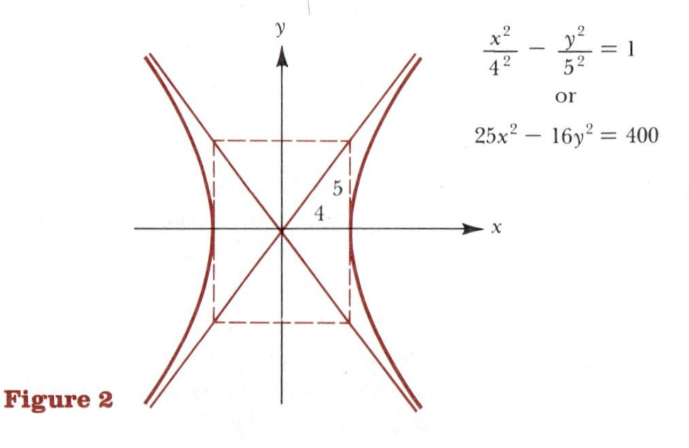

Figure 2

The graph of the equation

$$\frac{y^2}{b^2} - \frac{x^2}{a^2} = 1 \qquad (a > 0, b > 0) \tag{3}$$

is also a hyperbola. As you can check, the graph in this case has two y-intercepts, b and $-b$, but no x-intercepts. As Figure 3 indicates, the vertices of this hyperbola are $(0, b)$ and $(0, -b)$, and the asymptotes are the two lines $y = \pm(b/a)x$.

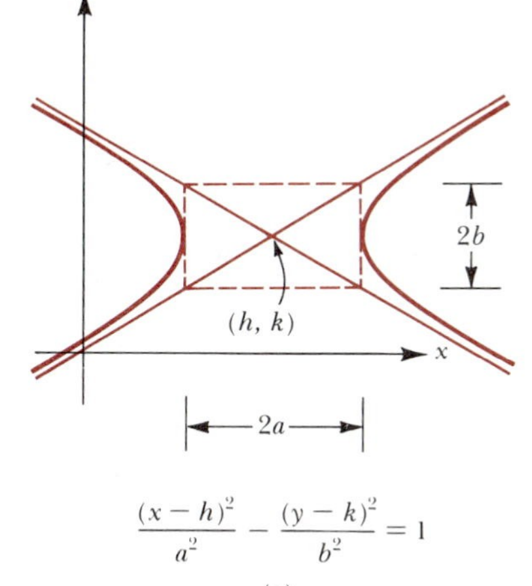

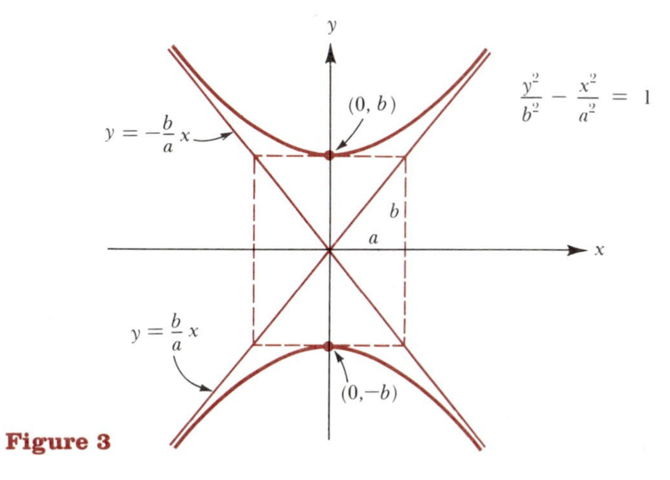

Figure 3

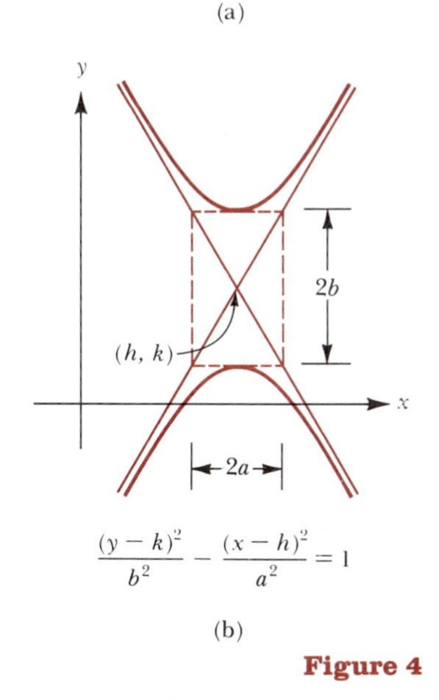

Figure 4

Just as the equation

$$\frac{(x-h)^2}{a^2} + \frac{(y-k)^2}{b^2} = 1$$

represents an ellipse with center (h, k), so the equations

$$\frac{(x-h)^2}{a^2} - \frac{(y-k)^2}{b^2} = 1 \tag{4}$$

and

$$\frac{(y-k)^2}{b^2} - \frac{(x-h)^2}{a^2} = 1 \tag{5}$$

represent hyperbolas whose centers both are (h, k). For the hyperbola defined by equation (4), the transverse axis is horizontal, as indicated in Figure 4(a) in the margin. For the hyperbola defined by equation (5), the transverse axis is vertical, as shown in Figure 4(b). We refer to equations (4) and (5) as the *standard forms* for the equation of a hyperbola with center (h, k).

Example 2 Use the technique of completing the square to show that the graph of the following equation is a hyperbola.

$$16x^2 + 32x - y^2 + 4y + 28 = 0$$

Graph the hyperbola and specify the center, the vertices, the length of the transverse axis, and the equations of the asymptotes.

Solution

$$16(x^2 + 2x \qquad) - (y^2 - 4y \qquad) = -28 \qquad \text{factoring}$$

$$16(x^2 + 2x + 1) - (y^2 - 4y + 4) = -28 + 16 - 4 \qquad \text{completing the squares}$$

$$16(x + 1)^2 - (y - 2)^2 = -16$$

$$\frac{(y - 2)^2}{16} - \frac{(x + 1)^2}{1} = 1 \qquad \text{dividing by } -16 \text{ and rearranging}$$

$$\frac{(y - 2)^2}{4^2} - \frac{(x + 1)^2}{1^2} = 1$$

This last equation is the standard form for the equation of a hyperbola whose transverse axis is vertical. By comparing this equation with equation (5), we see that

$$h = -1 \qquad k = 2 \qquad a = 1 \qquad b = 4$$

So the center of the hyperbola is $(-1, 2)$. Now, using the values $a = 1$ and $b = 4$, we can construct our preliminary rectangle and the resulting asymptotes as shown in Figure 5(a). The hyperbola itself can then be sketched as indicated in Figure 5(b). You should check for yourself that the information accompanying Figure 5(b) is correct.

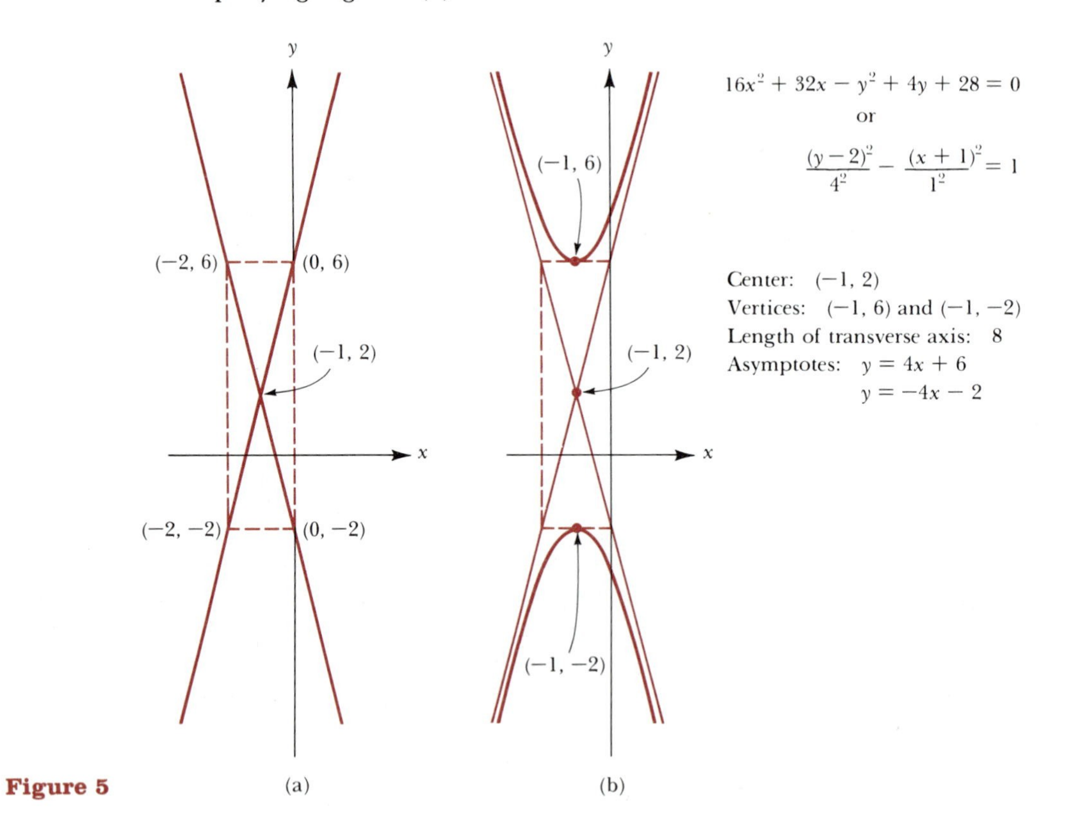

Figure 5

If you reread the example just completed, you'll see that it was not necessary to know in advance that the given equation represented a hyperbola. Rather, this fact emerged naturally after we completed the square.

Indeed, completing the square is a useful technique for identifying the graph of any equation of the form

$$Ax^2 + By^2 + Cx + Dy + E = 0$$

Example 3 Identify the graph of the equation

$$4x^2 - 32x - y^2 + 2y + 63 = 0$$

Solution As before, we complete the squares:

$$4(x^2 - 8x) - (y^2 - 2y) = -63$$
$$4(x^2 - 8x + 16) - (y^2 - 2y + 1) = -63 + 4(16) - 1$$
$$4(x - 4)^2 - (y - 1)^2 = 0$$

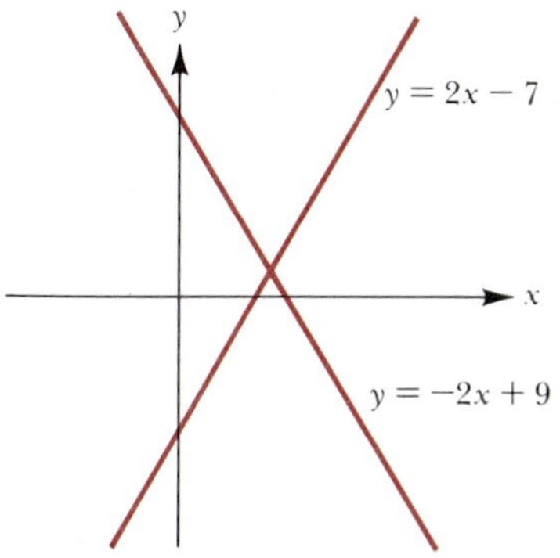

Since the right-hand side of this last equation is 0 rather than 1, dividing both sides by 4 will not bring the equation into one of the standard forms. Indeed, if we factor the left-hand side of the equation as a difference of two squares, we obtain

$$[2(x - 4) - (y - 1)][2(x - 4) + (y - 1)] = 0$$

or

$$(2x - y - 7)(2x + y - 9) = 0$$

$2x - y - 7 = 0$	$2x + y - 9 = 0$
$-y = -2x + 7$	$y = -2x + 9$
$y = 2x - 7$	

y
x
$y = 2x - 7$
$y = -2x + 9$

$4x^2 - 22x - y^2 + 2y + 63 = 0$
or
$(2x - y - 7)(2x + y - 9) = 0$

Figure 6

Thus the given equation is equivalent to the two equations $y = 2x - 7$ and $y = -2x + 9$. These two lines, taken together, constitute the graph. See Figure 6.

The two lines that we graphed in Figure 6 are actually the asymptotes for the hyperbola $4(x - 4)^2 - (y - 1)^2 = 1$. (Verify this for yourself.) For that reason, the graph in Figure 6 is referred to as a *degenerate hyperbola*. There are other cases similar to this that can arise in graphing equations of the form $Ax^2 + By^2 + Cx + Dy + E = 0$. For instance, as you can check for yourself by completing the squares, the graph of the equation

$$x^2 - 2x + 4y^2 - 16y + 17 = 0$$

consists of ths single point (1, 2). We refer to the graph in this case as a *degenerate ellipse*. Similarly, as you can check by completing the squares, the equation $x^2 - 2x + 4y^2 - 16y + 18 = 0$ has no graph; there are no points whose coordinates satisfy the equation.

We conclude this section by summarizing (in Figures 7–9) the graphs we have considered in this section and in the previous section. Except for the degenerate cases, the graph of any equation of the form

$$Ax^2 + By^2 + Cx + Dy + E = 0$$

where A and B are not both zero, will always be one of these conic sections. (We are omitting the circle from our summary because it can be considered as a special type of ellipse.)

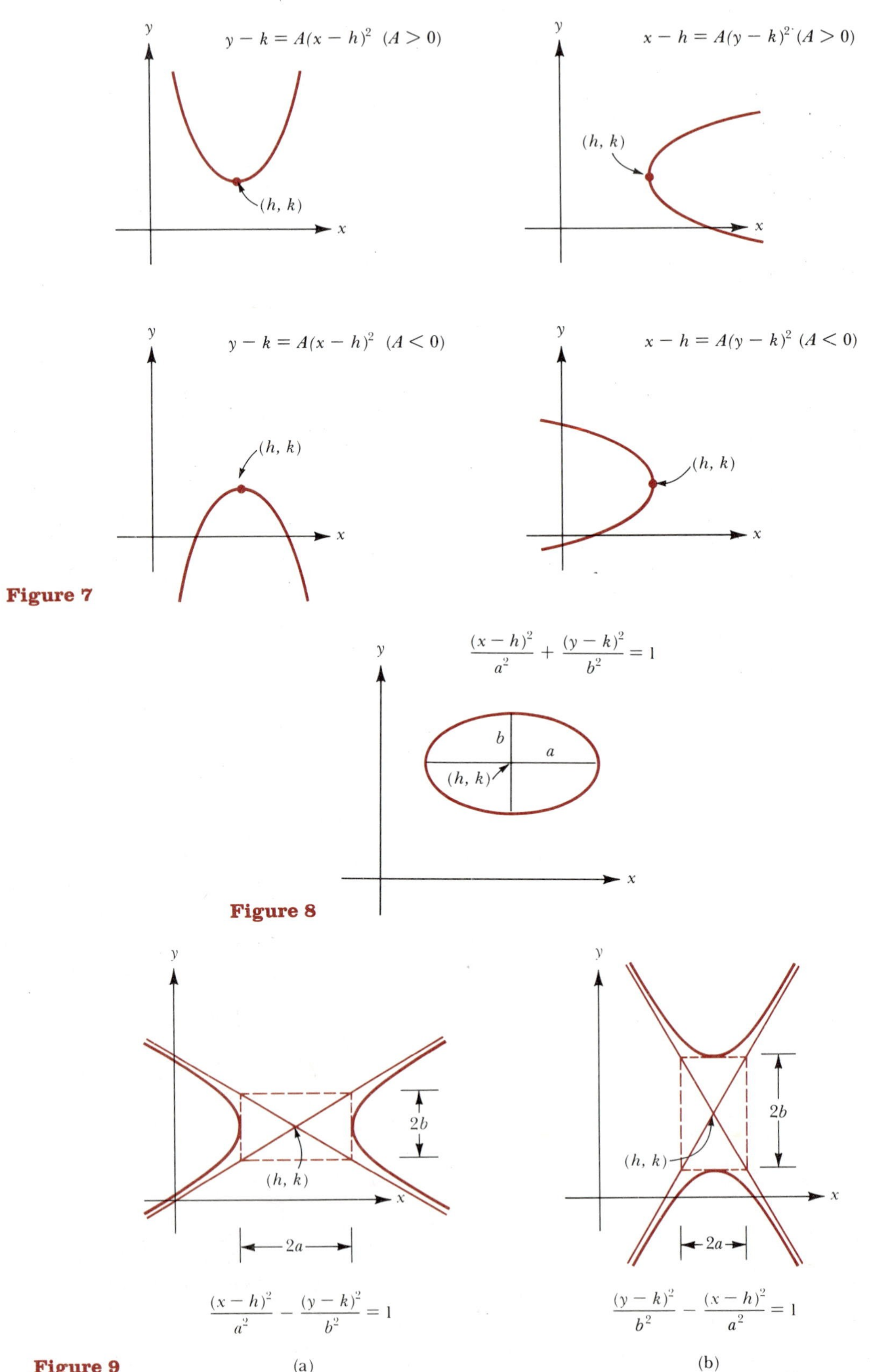

Figure 7

Figure 8

Figure 9 (a) (b)

Exercise Set 5.5

For Exercises 1–24, graph the hyperbolas. In each case in which the hyperbola is nondegenerate, specify the center, the vertices, the length of the transverse axis, and the equations of the asymptotes.

1. $x^2 - 4y^2 = 4$
2. $y^2 - x^2 = 1$
3. $y^2 - 4x^2 = 4$
4. $25x^2 - 9y^2 = 225$
5. $16x^2 - 25y^2 = 400$
6. $9x^2 - y^2 = 36$
7. $2y^2 - 3x^2 = 1$
8. $x^2 - y^2 = 9$
9. $4y^2 - 25x^2 = 100$
10. $x^2 - 3y^2 = 3$
11. $\dfrac{(x-5)^2}{5^2} - \dfrac{(y+1)^2}{3^2} = 1$
12. $\dfrac{(x-5)^2}{3^2} - \dfrac{(y+1)^2}{5^2} = 1$
13. $\dfrac{(y-2)^2}{2^2} - \dfrac{(x-1)^2}{1^2} = 1$
14. $\dfrac{(y-3)^2}{2^2} - \dfrac{x^2}{1^2} = 1$
15. $\dfrac{(x+3)^2}{4^2} - \dfrac{(y-4)^2}{4^2} = 1$
16. $\dfrac{(x+1)^2}{5^2} - \dfrac{(y+2)^2}{3^2} = 1$
17. $x^2 - y^2 + 2y - 5 = 0$
18. $16x^2 - 32x - 9y^2 + 90y - 353 = 0$
19. $x^2 - y^2 - 4x + 2y - 6 = 0$
20. $x^2 - 8x - y^2 + 8y - 25 = 0$
21. $y^2 - 25x^2 + 8y - 9 = 0$
22. $9y^2 - 18y - 4x^2 - 16x - 43 = 0$
23. $x^2 + 7x - y^2 - y + 12 = 0$
24. $9x^2 + 9x - 16y^2 + 4y + 2 = 0$
25. If the equation $Ax^2 + By^2 + Cx + Dy + E = 0$ represents an ellipse or a hyperbola, show that the center is the point $(-C/2A, -D/2B)$.
26. Let $P(x, y)$ be a point in the first quadrant on the hyperbola $x^2/2^2 - y^2/1^2 = 1$. Let Q be the point in the first quadrant with the same x-coordinate as P and lying on an asymptote to the hyperbola. Show that $\overline{PQ} = \frac{1}{2}(x - \sqrt{x^2 - 4})$.

C 27. The distance $\overline{PQ}$ in Exercise 26 represents the vertical distance between the hyperbola and the asymptote. Complete the following table to see numerical evidence that this separation distance approaches zero as x gets larger and larger. (Round off each entry to one significant digit.)

x	10	50	100	500	1000	10000
$\overline{PQ}$						

Chapter 5 Review Checklist

- □ Linear function (page 219)
- □ Quadratic function (page 220)
- □ Parabola (page 220)
- □ Vertex and axis of symmetry of a parabola (page 221)
- □ Maximum value and minimum value of a function (page 221)
- □ Vertex formula for a parabola (page 223)
- □ Polynomial function (page 231)
- □ Turning point (page 232)
- □ Rational function (page 238)
- □ Asymptote (page 239)
- □ Conic sections (page 247)
- □ Ellipse (page 249)
- □ Center and axes of an ellipse (page 249)
- □ Hyperbola (page 253)
- □ Vertices, center, and transverse axis of a hyperbola (page 254)

Chapter 5 Review Exercises

In Exercises 1–6, find equations for the linear functions satisfying the given conditions. Write each answer in the form $f(x) = mx + b$.

1. $f(3) = 5$ and $f(-2) = 0$.
2. $f(1) = 5$ and the graph of f passes through the origin.
3. $f(4) = -1$ and the graph of f is parallel to the line $3x - 8y = 16$.
4. The graph passes through (6, 1) and the x-intercept is twice the y-intercept.
5. $f(-3) = 5$ and the graph of the inverse function passes through (2, 1).

6. The graph of f passes through the vertices of the two parabolas $y = x^2 + 4x + 1$ and $x = 2y^2 - 8y - 1$.

For Exercises 7–12, graph the quadratic functions. In each case specify the vertex and the x- *and* y-*intercepts.*

7. $y = x^2 + 2x - 3$

8. $f(x) = x^2 - 2x - 15$

9. $y = -x^2 + 2\sqrt{3}\,x + 3$

10. $f(x) = 2x^2 - 2x + 1$

11. $y = -3x^2 + 12x$

12. $f(x) = -4x^2 + 16x$

13. Find the distance between the vertices of the two parabolas $y = x^2 - 4x + 6$ and $y = -x^2 - 4x - 5$.

14. Find the value of a given that the minimum value of the function $f(x) = ax^2 + 3x - 4$ is 5.

15. The sum of two numbers is $\sqrt{3}$. Find the largest possible value for their product.

16. The sum of two numbers is $\frac{2}{3}$. What is the smallest possible value for the sum of their squares?

17. Suppose that an object is thrown vertically upward (from ground level) with an initial velocity of v_0 ft/sec. Then it can be shown that the height h (in feet) after t seconds is given by the formula

$$h = v_0 t - 16t^2$$

(a) At what time does the object reach its maximum height? What is that maximum height?
(b) At what time does the object strike the ground?

18. Let $f(x) = 4x^2 - x + 1$ and $g(x) = \dfrac{x - 3}{2}$.

(a) For which input will the value of the function $f \circ g$ be a minimum?
(b) For which input will the value of $g \circ f$ be a minimum?

19. (a) Let P be a point on the parabola $y = x^2$. Express the distance from P to the point $(0, 2)$ in terms of x. *Hint:* The coordinates of P can be written (x, x^2).
(b) Which point in the second quadrant on the parabola $y = x^2$ is closest to the point $(0, 2)$? *Hint:* You want to minimize the expression under the radical in your answer for part (a). That expression is not a quadratic, but if you make the substitutions $x^2 = t$ and $x^4 = t^2$, you'll obtain a quadratic that you can minimize using the methods of Section 5.1.

20. Find the coordinates of the point on the line $y = 2x - 1$ closest to $(-5, 0)$.

21. Find all values of b such that the minimum distance from the point $(2, 0)$ to the line $y = \frac{4}{3}x + b$ is 5.

22. What number exceeds one-half its square by the greatest amount?

23. Suppose that $x + y = \sqrt{2}$. Find the minimum value of the quantity $x^2 + y^2$.

24. For which numbers t will the value of $9t^2 - t^4$ be as large as possible?

25. Find the maximum area possible for a right triangle with a hypotenuse of 15 cm. *Hint:* Let x denote the length of one leg. Show that the area is $A = \frac{1}{2}x\sqrt{225 - x^2}$. Now work with A^2.

26. For which points (x, y) on the curve $y = 1 - x^2$ is the sum $x + y$ a maximum?

27. Let $f(x) = x^2 - (a^2 + 2a)x + 2a^3$, where $0 < a < 2$. For which value of a will the distance between the x-intercepts of f be a maximum?

28. Suppose that the revenue R (in dollars) generated by selling x units of a certain product is given by $R = 300x - \frac{1}{4}x^2$ $(0 \le x \le 1200)$. How many units should be sold to maximize the revenue? What is that maximum revenue?

29. Suppose that the function $p = 160 - \frac{1}{5}x$ relates the selling price p of an item to the quantity sold x. Assume that p is in dollars. For which value of x will the revenue R $(= xp)$ be a maximum? What is the selling price p in this case?

30. A piece of wire 16 cm long is to be cut into two pieces. Let x denote the length of the first piece and $16 - x$ the length of the second. The first piece is to be formed into a rectangle in which the length is twice the width. The second piece of wire will also be formed into a rectangle, but with the length three times the width. For which value of x will the total area of the two rectangles be a minimum?

In Exercises 31–46, graph each function and specify the x- *and* y-*intercepts.*

31. $y = (x + 4)(x - 2)$

32. $y = (x + 4)(x - 2)^2$

33. $y = -(x + 5)^3$

34. $y = -x(x + 1)$

35. $y = -x^2(x + 1)$

36. $y = -x^3(x + 1)$

37. $y = x(x - 2)(x + 2)$

38. $y = (x - 3)(x + 1)(x + 5)$

39. $y = \dfrac{3x + 1}{x}$

40. $y = \dfrac{1 - 2x}{x}$

41. $y = \dfrac{-1}{(x - 1)^2}$

42. $y = \dfrac{x + 1}{x + 2}$

43. $y = \dfrac{x - 2}{x - 3}$

44. $y = \dfrac{x}{(x - 2)(x + 4)}$

45. $y = \dfrac{x^2 - 2x + 1}{x^2 - 4x + 4}$

46. $y = \dfrac{x(x - 2)}{(x - 4)(x + 4)}$

47. Let $f(x) = x^2 + 2bx + 1$.

(a) If $b = 1$, find the distance from the vertex of the parabola to the origin.
(b) If $b = 2$, find the distance from the vertex of the parabola to the origin.
(c) For which real numbers b will the distance from the vertex to the origin be as small as possible?

48. Find the range of the function $f(x) = -2x^2 + 12x - 5$. *Hint:* Look at the graph.

49. The range of the function $y = x^2 - 2x + k$ is the interval $[5, \infty)$. Find the value of k.

50. Find a value for b such that the range of the function $f(x) = x^2 + bx + b$ is the interval $[-15, \infty)$.

51. Find the range of the function $y = \dfrac{(x-1)(x-3)}{x-4}$.
Hints: Solve the equation for x in terms of y using the quadratic formula. If you're careful with the algebra, you will find that the expression under the resulting radical sign is $y^2 - 8y + 4$. The range of the given function can then be found by solving the inequality $y^2 - 8y + 4 \geq 0$.

In Exercises 52–60, use the technique of completing the square to rewrite each equation in standard form. Then sketch the graph in each case.

52. $x^2 - 12x - y + 37 = 0$

53. $4x^2 + 25y^2 - 50y - 75 = 0$

54. $x^2 - y^2 - 2x - 2y + 1 = 0$

55. $x^2 + y^2 - 6x - 8y = 0$

56. $4x^2 + y^2 - 32x - 2y + 49 = 0$

57. $2y^2 - x + 12y + 17 = 0$

58. $9y^2 - 16x^2 + 18y - 128x - 391 = 0$

59. $y^2 - 2x + 2y + 3 = 0$

60. $9x^2 + 4y^2 - 36x - 8y + 40 = 0$

***61.** In the figure, triangle OAB is equilateral and AB is parallel to the x-axis. Find the length of a side and the area of triangle OAB.

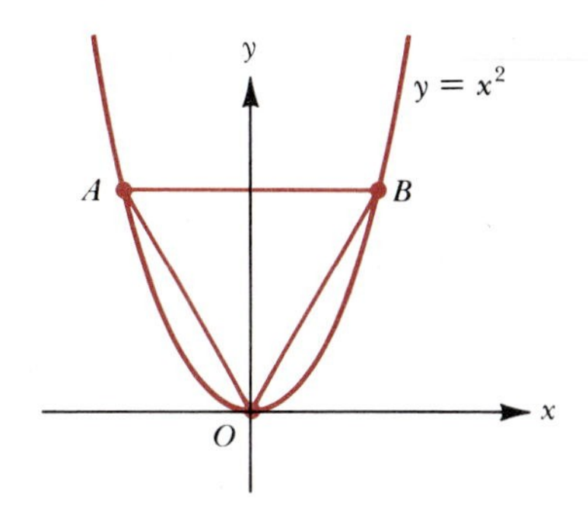

62. P is a point in the first quadrant on the graph of the ellipse $x^2/5^2 + y^2/3^2 = 1$. Furthermore, the x- and y-coordinates of P are equal. Find the area of the right triangle in the figure.

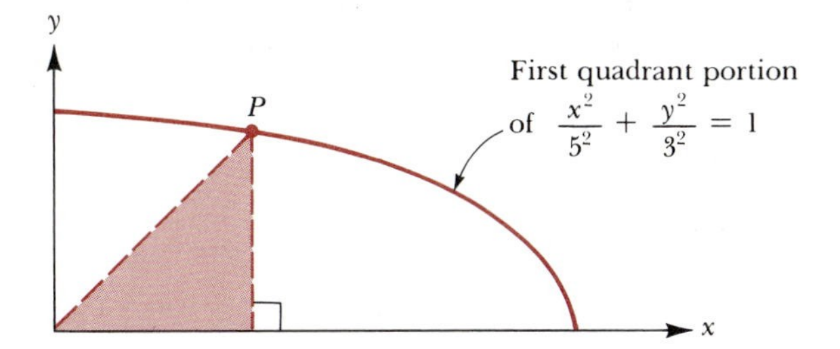

Chapter 5 Test

1. G is a linear function such that $G(1) = -2$ and $G(-2) = -11$. Find $G(0)$.

2. Let $f(x) = 3x^2 + 6x - 10$. For which input x is the value of the function a minimum? What is that minimum value?

3. Suppose the function $p = -\frac{1}{8}x + 100$ $(0 \leq x \leq 12)$ relates the selling price p of an item to the quantity sold x. Assume that p is in dollars. What is the maximum revenue possible in this situation? (Make use of the formula from the text: Revenue equals number of units sold times price per unit.)

4. Graph the function $y = \dfrac{-1}{(x+1)^3}$.

5. Graph the function $y = (x-4)(x-1)(x+1)$.

6. Graph the function $f(x) = x^2 + 4x - 5$. Specify the vertex, the x- and y-intercepts, and the axis of symmetry.

7. A factory owner buys a new machine for \$1000. After 5 years the machine has a salvage value of \$100. Assuming linear depreciation, find a formula for the value V of the machine after t years, where $0 \leq t \leq 5$.

8. Graph the function $y = 2(x-3)^4$. Does the graph cross the y-axis? If so, where?

9. Graph the function $y = \dfrac{3x+5}{x+2}$. Specify all intercepts and asymptotes.

10. What is the largest area possible for a right triangle in which the sum of the lengths of the two shorter sides is 12 cm?

11. Graph the function $\frac{x}{(x+2)(x-4)}$.

12. Graph the equation $x - 4 = -2(y - 1)^2$.

For Problems 13–15, graph each equation. If the graph is an ellipse, specify the center and the lengths of the major and minor axes. If the graph is a parabola, specify the vertex and the axis of symmetry. If the graph is a hyperbola, specify the center, the vertices, and the equations of the asymptotes.

13. $4x^2 + 40x + y^2 + 84 = 0$

14. $x^2 - 8x - 4y^2 + 16y - 4 = 0$

15. $x = 4y^2 + 12y$

Chapter 6 Exponential and Logarithmic Functions

Introduction

In this chapter we study exponential functions and their inverses. Section 6.1 begins with an example involving perhaps the simplest exponential function, $y = 2^x$. (Notice that this is not a polynomial function; don't confuse it with $y = x^2$.) As you'll see, the function $y = 2^x$ grows or increases extremely rapidly. This is one of the key features of many exponential functions. In Section 6.1 we also define exponential functions in general, and we look at a number of specific graphs. Then in Sections 6.2 and 6.3 we study the inverses of these functions. These inverse functions are called logarithmic functions. Section 6.4 is devoted to some of the applications of exponential and logarithmic functions. The applications we consider are quite diverse. They include interest rates in banking, population growth, and nuclear energy.

6.1 Exponential Functions

We begin with an example. Suppose that your mathematics instructor, in an effort to improve classroom attendance, offers to pay you each day for attending class! Say you are to receive 2¢ on the first day you attend class, 4¢ the second day, 8¢ the third day, and so on, as indicated in Table 1. How much money would you supposedly receive for attending class on the 30th day?

x (day number)	y (amount earned that day)
1	2¢ $(= 2^1)$
2	4¢ $(= 2^2)$
3	8¢ $(= 2^3)$
4	16¢ $(= 2^4)$
5	32¢ $(= 2^5)$
⋮	⋮
x	2^x

Table 1

As you can see by looking at Table 1, the amount y earned on day x is given by the rule or *exponential function*

$$y = 2^x$$

Thus on the 30th day (when $x = 30$), you receive

$$y = 2^{30} \text{ cents}$$

If you use a calculator, you will find this amount to be well over 10 million dollars. The point here is simply this. Although we begin with a small amount, $y = 2¢$, repeated doubling quickly leads to a very large amount. Put in other terms, the exponential function grows very rapidly.

Before leaving this example, we point out a simple method for quickly estimating numbers such as 2^{30} (or any power of two) in terms of the more familiar powers of ten. Begin by observing that

$$2^{10} \approx 10^3 \qquad \text{(a useful coincidence, worth remembering)}$$

Now just cube both sides to obtain

$$(2^{10})^3 \approx (10^3)^3$$

or

$$2^{30} \approx 10^9$$

Thus 2^{30} is about one billion. To convert this number of cents to dollars, you divide by 100 or 10^2 to obtain

$$\frac{10^9}{10^2} = 10^7 \text{ dollars}$$

which is 10 million dollars, as mentioned before.

Example 1 Estimate each number in terms of integral powers of 10.
(a) 2^{40} **(b)** 2^{43}

Solution **(a)** Take the basic approximation $2^{10} \approx 10^3$ and raise both sides to the fourth power. This yields

$$(2^{10})^4 \approx (10^3)^4$$

or

$$2^{40} \approx 10^{12} \qquad \text{as required}$$

(b) To estimate 2^{43}, we have

$$2^{43} = 2^3 \times 2^{40}$$

and therefore

$$2^{43} \approx 8 \times 10^{12} \qquad \text{as required}$$

Example 2 Estimate the power to which 10 must be raised to yield 2.

Solution We begin with our approximation

$$10^3 \approx 2^{10}$$

Raising both sides to the power $\frac{1}{10}$ yields

$$(10^3)^{1/10} \approx (2^{10})^{1/10}$$

and consequently

$$10^{3/10} \approx 2$$

Thus the power to which 10 must be raised to yield 2 is approximately $\frac{3}{10}$.

In Section 1.6 we defined the expression b^x where x is a rational number. We also stated that if x is irrational, then b^x can be defined in such a way that the usual properties of exponents continue to apply. Although a rigorous definition of irrational exponents requires concepts from calculus, we can nevertheless indicate the basic idea by means of an example. (We need to do this before we give the general definition for exponential functions.)

How shall we assign a meaning to $2^{\sqrt{2}}$, for example? The basic idea is to successively evaluate the expression 2^x using rational numbers x that are closer and closer to $\sqrt{2}$. The following table displays the results of some calculations along these lines.

x	1.4	1.41	1.414	1.4142	1.41421	1.414213
2^x	2.6 . . .	2.65 . . .	2.664 . . .	2.6651 . . .	2.66514 . . .	2.665143 . . .

Table 1 **Values of 2^x for rational numbers x approaching $\sqrt{2}$. ($\sqrt{2}$ = 1.41421356 . . .)**

The data in the table suggest that as x approaches $\sqrt{2}$, the corresponding values of 2^x approach a unique real number, call it t, whose decimal expansion begins as 2.665. Furthermore, by continuing this process we could obtain (in theory, at least) as many places in the decimal expansion of t as we wished. The value of the expression $2^{\sqrt{2}}$ is then defined to be this number t. The following theorem (stated here without proof) serves to summarize this discussion and also to pave the way for the definition we will give for exponential functions.

Function	Exponential Function?	
	Yes	*No*
$y = 10^x$	✓	
$y = 2^x$	✓	
$y = x^2$		✓
$y = (\frac{1}{2})^x$	✓	
$y = \pi^x$	✓	
$y = 2^{x^2}$		✓

Table 2

Theorem

Let b denote an arbitrary positive real number. Then:

1. For each real number x, b^x is a unique real number.
2. When x is irrational, we can approximate b^x as closely as we wish by evaluating b^r, where r is a rational number sufficiently close to x.
3. The properties of rational exponents continue to hold for irrational exponents.

For the remainder of this section, b denotes an arbitrary positive constant other than 1. We now define the *exponential function with base b* by the equation

$$y = b^x$$

Table 2 provides some examples of this terminology.

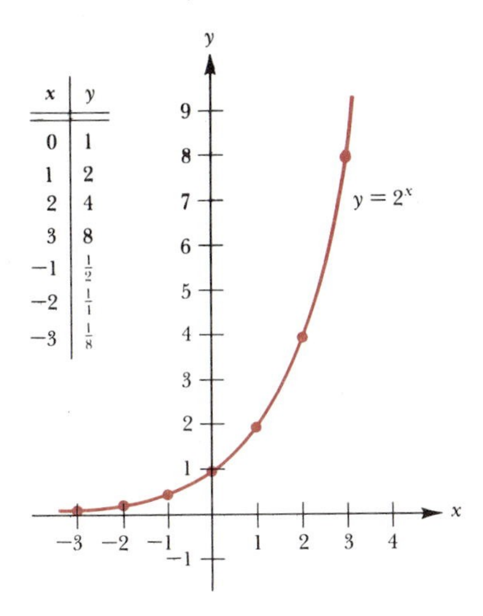

Figure 1

To help with our analysis of exponential functions, let's set up a table and use it to graph the exponential function $y = 2^x$. This is done in Figure 1. (In drawing a smooth and unbroken curve we are actually relying on the theorem that we just stated.) The key features of the exponential function $y = 2^x$ and its graph are these:

1. The domain of $y = 2^x$ is the set of all real numbers. The range is the set of all positive real numbers.
2. The y-intercept of the graph is 1. There is no x-intercept.
3. For $x > 0$, the function increases or grows very rapidly. For $x < 0$, the graph approaches the x-axis; the x-axis is a horizontal asymptote for the graph. (Recall from Section 5.3 that a line is an asymptote for a curve if the separation distance between the curve and the line approaches zero as we move out farther and farther along the line.)

You should memorize the basic shape and features of the graph of $y = 2^x$ so that you can sketch it as needed without first setting up a table. The next example shows why this is useful.

Example 3 Graph the following functions. In each case specify the domain, range, intercept(s), and asymptote.
(a) $y = -2^x$ **(b)** $y = 2^{-x}$ **(c)** $y = (\frac{1}{2})^x$ **(d)** $y = 2^{-x} - 2$

Solution **(a)** Recall that -2^x means $-(2^x)$, not $(-2)^x$. The graph of $y = -2^x$ is obtained by reflecting the graph of $y = 2^x$ in the x-axis. See Figure 2(a).

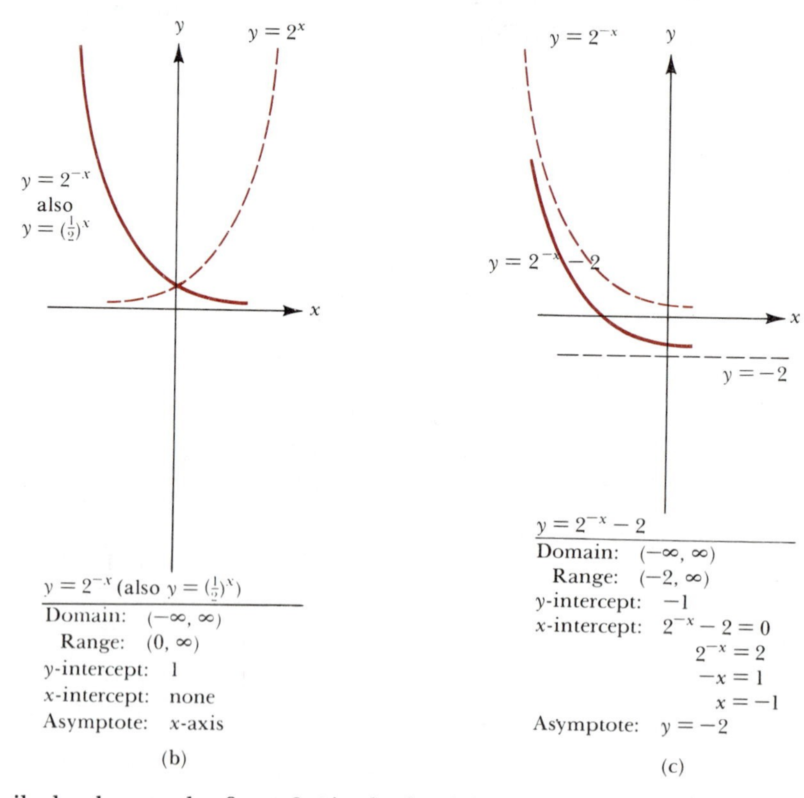

Figure 2

(b) Similarly, the graph of $y = 2^{-x}$ is obtained from the graph of $y = 2^x$ by reflection in the y-axis. See Figure 2(b).
(c) Next, regarding $y = (\frac{1}{2})^x$, observe that

$$\left(\frac{1}{2}\right)^x = \frac{1^x}{2^x} = \frac{1}{2^x} = 2^{-x}$$

That is, we have

$$\left(\frac{1}{2}\right)^x = 2^{-x}$$

In other words, $y = (\frac{1}{2})^x$ is really the same function as $y = 2^{-x}$, and this we already graphed in Figure 2(b).

(d) Finally, to graph $y = 2^{-x} - 2$, take the graph of $y = 2^{-x}$ in Figure 2(b) and move it two units in the negative y-direction. Note that the asymptote and y-intercept will also move down two units. See Figure 2(c). [Note the calculations accompanying Figure 2(c) for the x-intercept.]

In the next example we apply our knowledge about the graph of $y = 2^x$ to solve an equation. In particular, we use the fact that the graph of $y = 2^x$ always lies above the x-axis; for no value of x is 2^x ever zero.

Example 4 Solve the equation $x^2 2^x - 2^x = 0$.

Solution First, factor the left-hand side of the equation; the common term is 2^x. This gives us

$$2^x(x^2 - 1) = 0$$

Since 2^x is never zero, we may now divide both sides of this last equation by 2^x to obtain

$$x^2 - 1 = 0$$
$$(x - 1)(x + 1) = 0$$

and consequently

$$x = 1 \quad \text{or} \quad x = -1$$

Thus the required solutions are $x = 1$, $x = -1$. (You should check for yourself that both of these numbers satisfy the original equation.)

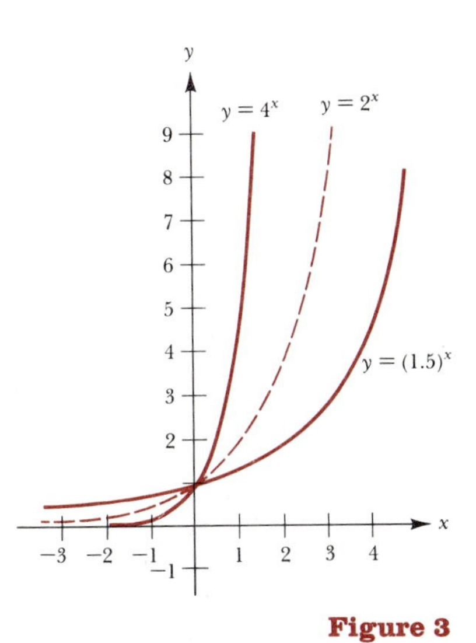

Figure 3

Now what about exponential functions with bases other than 2? As Figure 3 indicates, there is nothing essentially new here. As you would expect, the graph of $y = 4^x$ rises more rapidly than $y = 2^x$ when x is positive. For negative x-values, the graph of $y = 4^x$ is below that of $y = 2^x$. You can see why this must be by taking $x = -1$, for example, and comparing the values of 4^x and 2^x. If $x = -1$, then

$$2^x = 2^{-1} = \frac{1}{2} \quad \text{but} \quad 4^x = 4^{-1} = \frac{1}{4}$$

Therefore $4^x < 2^x$, when $x = -1$. Notice also in Figure 3 that all three graphs have the same y-intercept of 1. This follows from the fact that $b^0 = 1$ for any positive number b.

The exponential functions in Figure 3 all have bases larger than 1. To see examples in which the bases are less than 1, you need only reflect those graphs in Figure 3 in the y-axis. For instance, the reflection of $y = 4^x$ in the y-axis gives us the graph of $y = (\frac{1}{4})^x$. [We discussed the idea behind this in Example 3; see Figure 2(b) for instance.]

In the box that follows, we summarize what we've learned up to this point regarding the exponential function $y = b^x$.

Property Summary

The Exponential Function $y = b^x$

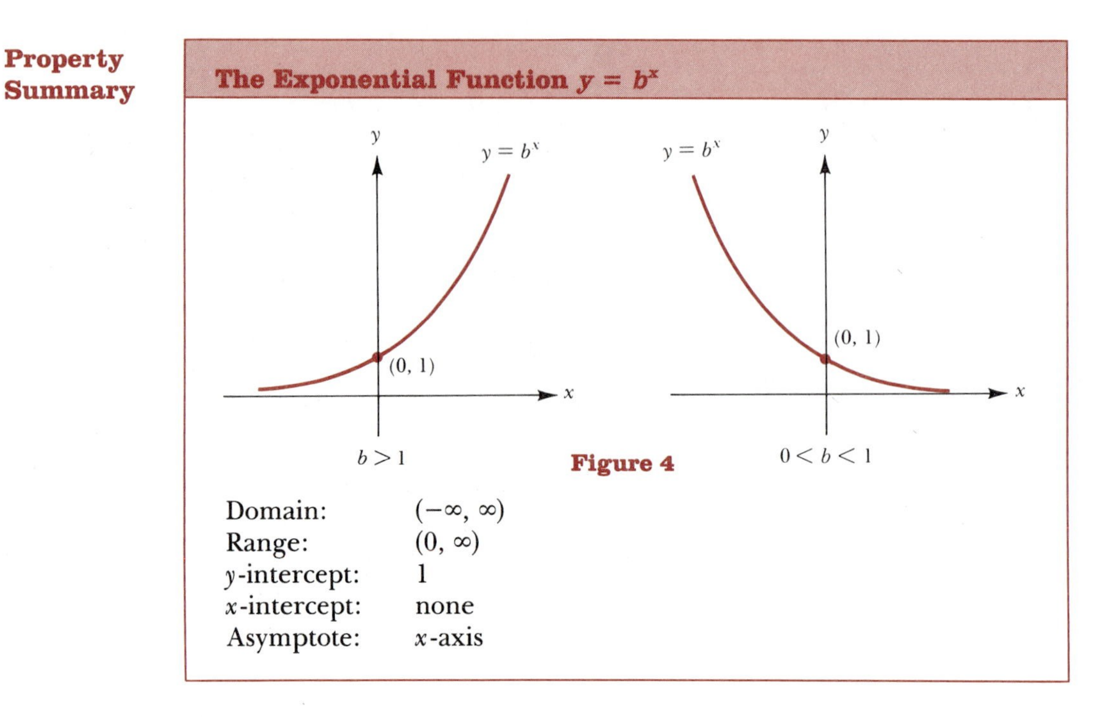

Figure 4

Domain:	$(-\infty, \infty)$
Range:	$(0, \infty)$
y-intercept:	1
x-intercept:	none
Asymptote:	x-axis

Example 5 Graph the function $y = -3^{-x} + 1$.

Solution The required graph is obtained by reflecting and translating the graph of $y = 3^x$ as shown in Figure 5.

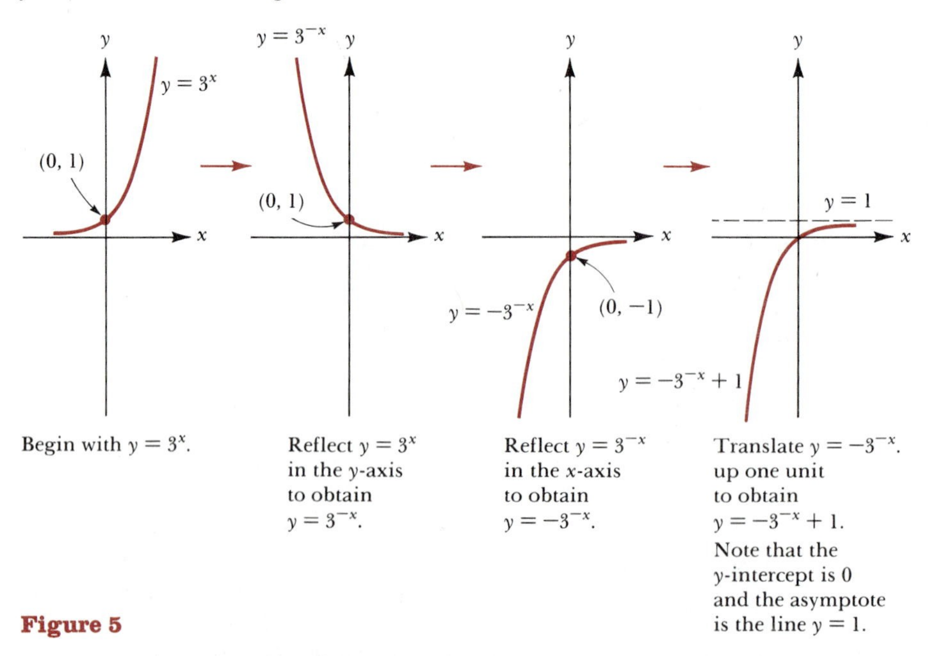

Figure 5

From the standpoint of calculus and scientific applications, there is one base for exponential functions that is by far the most useful. This base is a certain irrational number that lies between 2 and 3 and that is denoted by

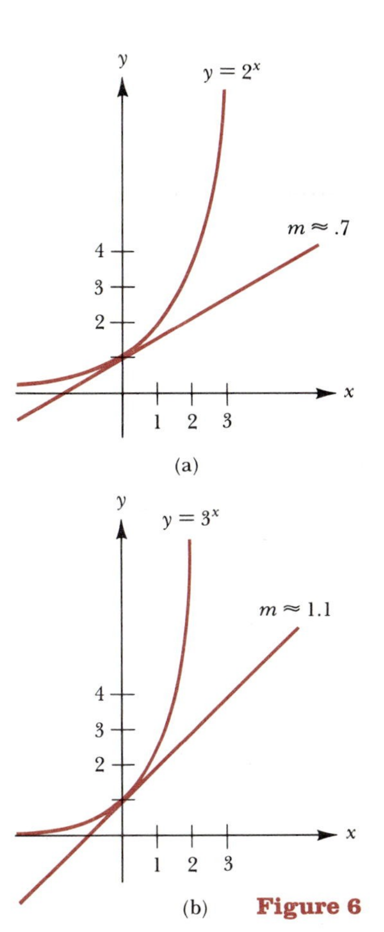

Figure 6

the letter e.* For purposes of approximation, you'll need to know that

$$e \approx 2.7$$

At the precalculus level, it's hard to escape the feeling that $y = 2^x$ or $y = 10^x$ is by far more simple and more natural than $y = e^x$. So as we work with e and e^x now and again in this chapter, you will need to take it on faith that, in the long run, the number e and the function $y = e^x$ make life simpler, not more complex.

Here is one way to define the number e. (For another approach see Exercise 36 at the end of this section.) You know that the graph of each exponential function $y = b^x$ passes through the point (0, 1). Figure 6(a) shows the exponential function $y = 2^x$ along with a straight line that is tangent to the curve at the point (0, 1). By carefully measuring rise and run, it can be shown that the slope of this tangent line is about 0.7. Figure 6(b) shows a similar situation with the curve $y = 3^x$. There the slope of the tangent line through (0, 1) is approximately 1.1. Now since the slope of the tangent to $y = 2^x$ is a bit less than 1, while that for $y = 3^x$ is a bit more than 1, it seems reasonable to suppose that there is a number between 2 and 3, call it e, with the property that the slope of the tangent to $y = e^x$ through (0, 1) is exactly 1. See Figure 7.

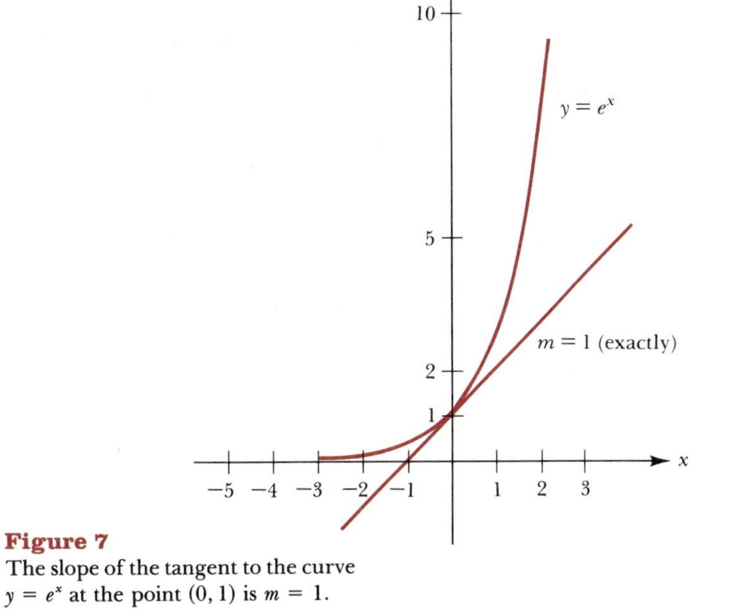

Figure 7
The slope of the tangent to the curve $y = e^x$ at the point (0, 1) is $m = 1$.

We now turn to some applications involving the number e. In Section 6.4 we'll look at these and other applications in greater detail. (Thus your immediate goals in studying the remainder of this section are these: Begin to get used to seeing formulas involving e^x; master the algebra that you will be seeing in Examples 6 and 7.)

*The Swiss mathematician Leonhard Euler introduced the letter e to denote this number in the year 1741. To six decimal places, the value of e is 2.718281 . . . .

Under ideal conditions involving unlimited food and space, a colony of bacteria increases according to the *growth law*

$$N = N_0 e^{kt}$$

In this formula, N is the population at time t and k is a positive constant related to the growth rate of the population. N_0 is also a constant; it represents the size of the population at time $t = 0$. Sometimes to emphasize the fact that N depends on (is a function of) t, we employ functional notation to rewrite the growth law as

$$N(t) = N_0 e^{kt}$$

Example 6 Suppose that at the start of an experiment in a biology laboratory, there are 1200 bacteria in a colony. Four hours later, the population is found to be 3600. How many bacteria are there three hours after the experiment began?

Solution Begin with the growth law $N = N_0 e^{kt}$. Our strategy will be to evaluate the quantity e^k; then we can determine the required value of N. (*Note:* In Section 6.2 we'll see how to find k itself, rather than e^k.) We are given that $N_0 = 1200$. Thus we can write

$$N = 1200e^{kt} \tag{1}$$

We are also given that $N = 3600$ when $t = 4$. Substituting these values in equation (1) gives us

$$3600 = 1200e^{k(4)}$$

Now divide both sides of this last equation by 1200 to obtain

$$3 = e^{4k} \tag{2}$$

To isolate e^k, raise both sides of the equation to the power $\frac{1}{4}$.

$$3^{1/4} = (e^{4k})^{1/4}$$

Therefore

$$3^{1/4} = e^k \tag{3}$$

Now we can use this value of e^k to determine the size of the colony after 3 hours, as required. Take equation (1) and replace t by 3. We have

$$\begin{aligned} N &= 1200e^{k(3)} \\ &= 1200(e^k)^3 \\ &= 1200(3^{1/4})^3 \qquad \text{We used equation (3) to substitute for } e^k. \\ &= 1200(3^{3/4}) \end{aligned}$$

Using a calculator, we find that $1200(3^{3/4}) \approx 2735$. Rather than claim that there are precisely 2735 bacteria after 3 hours, we follow common sense and round off our answer to the nearest hundred. Thus we say that after 3 hours there are about 2700 bacteria in the colony.

In the example just concluded, we used the function $N = N_0 e^{kt}$ to describe population growth. It is a remarkable fact, shown in calculus, that the same general equation also governs radioactive decay, but in that case the constant k is negative, not positive. Thus we will assume here the following *decay law* for radioactive substances:

$$N = N_0 e^{kt}$$

where N_0 is the original amount present at time $t = 0$, N is the amount present at time t, and k is a negative constant having to do with the rate of decay of the substance. In discussing radioactive decay, it is convenient to introduce the term half-life.

Definition of Half-life

The *half-life* of a radioactive substance is the time required for half of a given sample to disintegrate. The half-life is an intrinsic property of the substance; it does not depend on the given sample size.

Example

Iodine-131 is a radioactive substance with a half-life of 8 days. Suppose that 2 gm are initially present. Then:

at $t = 0$, 2 gm are present;
at $t = 8$ days, 1 gm is left;
at $t = 16$ days, $\frac{1}{2}$ gm is left;
at $t = 24$ days, $\frac{1}{4}$ gm is left;
at $t = 32$ days, $\frac{1}{8}$ gm is left.

Example 7 Hospitals utilize the radioactive substance iodine-131 in the diagnosis of the thyroid gland. If a hospital acquires 2 gm of iodine-131, how much of this sample will remain after 30 days? As stated earlier, the half-life of iodine-131 is 8 days.

Solution First, let's estimate the answer to get a feeling for the situation. We noted just prior to this example that, after 24 days, $\frac{1}{4}$ gm will remain, while after 32 days, $\frac{1}{8}$ gm will remain. Since 30 is between 24 and 32, it follows that, after 30 days, the amount remaining will be something between one quarter and one eighth of a gram.

Our actual calculations for the answer begin with the decay law

$$N = N_0 e^{kt} \tag{4}$$

We are asked to find N when $t = 30$. As in Example 6, we shall first determine the quantity e^k. To do this, we use the half-life information. This says that when $t = 8$, the value of N is $\frac{1}{2}N_0$. Using these values in equation (4), we find

$$\frac{1}{2}N_0 = N_0 e^{k(8)}$$

$$\frac{1}{2} = e^{8k}$$

$$\left(\frac{1}{2}\right)^{1/8} = (e^{8k})^{1/8}$$

$$\left(\frac{1}{2}\right)^{1/8} = e^k \tag{5}$$

Returning to equation (4) now, we substitute $N_0 = 2$ and $t = 30$ to obtain

$$\begin{aligned} N &= 2e^{k(30)} \\ &= 2(e^k)^{30} \\ &= 2\left[\left(\frac{1}{2}\right)^{1/8}\right]^{30} \end{aligned}$$

We used equation (5) to substitute for e^k.

Using the properties for exponents, this can be simplified to

$$N = 2^{-11/4}$$

(Exercise 38 at the end of this section asks you to verify this.) Using a calculator now, we obtain

$$N = 0.148\ldots$$

Thus, after 30 days approximately 0.15 gm of the iodine-131 remains. As you can check, this figure is indeed between one quarter and one eighth, as we first estimated.

Exercise Set 6.1

1. Estimate each of the following quantities in terms of integral powers of 10, as was done in Example 1.
   (a) 2^{20} **(b)** 2^{50} **(c)** 2^{21} **(d)** 2^{59}
2. On the same set of axes, graph the functions $y = 2^x$ and $y = 3^x$.
3. On the same set of axes, graph the functions $y = 2^x$ and $y = e^x$.
4. On the same set of axes, graph the functions $y = e^x$ and $y = 3^x$.
5. Complete the table. After completing the table, determine which seems to be larger "in the long run," x^2 or 2^x.

x	x^2	2^x	Which Is Larger, 2^x or x^2?
0			
1			
2			
3			
4			
5			
6			
10			

In Exercises 6–15, graph each function and specify the domain, range, intercept(s), and asymptote.

6. **(a)** $y = e^x$ **(b)** $y = e^{-x}$ **(c)** $y = -e^x$ **(d)** $y = -e^{-x}$
7. **(a)** $y = -2^x + 1$ **(b)** $y = -2^x + 2$
8. **(a)** $y = 3^{-x} + 1$ **(b)** $y = 3^{-x} - 3$
9. **(a)** $y = 1 - e^x$ **(b)** $y = e - e^x$
10. $y = 2^{x-1}$
11. **(a)** $y = 3^{x+1}$ **(b)** $y = 3^{x+1} + 1$ **(c)** $y = -3^{x+1}$
12. $y = e^{x-1}$
13. **(a)** $y = e^{x+1}$ **(b)** $y = -e^{x+1}$
14. **(a)** $y = 4^x$ **(b)** $y = (\frac{1}{4})^x$
15. **(a)** $y = (\frac{1}{2})^x$ **(b)** $y = (\frac{1}{2})^{-x}$
16. Solve for x.
    (a) $(3x)2^x + 2^x = 0$ **(b)** $4x^2 2^x - 9(2^x) = 0$
17. Solve for x.
    (a) $x^2e^x - 16e^x = 0$
    (b) $3e^x - 5xe^x + 2x^2e^x = 0$
18. Solve for t.
    (a) $\dfrac{(t+1)10^t}{t+3} = 0$
    (b) $t^2e^t + te^t = e^t$
19. Solve for x. $(2x)10^{-x} - 10^{1-x} = 0$.
20. At the start of an experiment there are 2000 bacteria in a colony. In 2 hours the population has tripled. Assume that the growth law $N = N_0e^{kt}$ applies.
    (a) Determine e^k. *Answer:* $\sqrt{3}$
    (b) Determine the population 10 hours after the start of the experiment. *Answer:* 486,000
21. At the start of an experiment there are 2×10^4 bacteria in a colony. After 8 hours the population is 3×10^4. Assume that the growth law $N = N_0e^{kt}$ applies.

(a) Determine the population 1 hour after it was 3×10^4.
(b) Determine the population 24 hours after the start of the experiment.

22. A colony of bacteria grows according to the law $N = N_0e^{kt}$. It is known that the value of e^k *is* $\sqrt[5]{2}$. If the initial population is 3200, what will the population be 5 hours later?

23. The half-life of a certain radioactive substance is 5 days. Initially there are 8 gm present. How many grams are present after
(a) 5 days? (b) 10 days? (c) 15 days?
(d) 20 days? (e) 50 days?

24. The half-life of iodine-131 is 8 days. How much of a 1-gm sample will remain after 7 days?
Answer: $(\frac{1}{2})^{7/8} \approx 0.55$ gm

25. The half-life of strontium-90 is 28 years. How much of a 10-gm sample will remain after
(a) 1 year? (b) 10 years?

26. Plutonium-239 is a product of nuclear reactors. The half-life of plutonium-239 is about 24,000 years. What percentage of a given sample will remain after 1000 years?

27. Krypton-91 has a half-life of 10 seconds. What percentage of a given sample will remain after 5 minutes?

28. Let $f(x) = 2^x$. Show that

$$\frac{f(x+h) - f(x)}{h} = 2^x\left(\frac{2^h - 1}{h}\right)$$

29. If $E(x) = e^x$, show that

$$\frac{E(x+h) - E(x)}{h} = e^x\left(\frac{e^h - 1}{h}\right)$$

***30.** Let $S(x) = \dfrac{e^x - e^{-x}}{2}$ and $C(x) = \dfrac{e^x + e^{-x}}{2}$.

Prove the following identities.
(a) $[C(x)]^2 - [S(x)]^2 = 1$
(b) $S(-x) = -S(x)$
(c) $C(-x) = C(x)$
(d) $S(x+y) = S(x)C(y) + C(x)S(y)$
(e) $C(x+y) = C(x)C(y) + S(x)S(y)$
(f) $S(2x) = 2S(x)C(x)$
(g) $C(2x) = [C(x)]^2 + [S(x)]^2$
(h) $[S(x)]^2 = \frac{1}{2}[C(2x) - 1]$
(i) $[C(x)]^2 = \frac{1}{2}[C(2x) + 1]$
(j) $[C(x) + S(x)]^2 = C(2x) + S(2x)$

C **31.** Set up a table and graph the function

$$S(x) = \frac{e^x - e^{-x}}{2}$$

C **32.** Set up a table and graph the function

$$C(x) = \frac{e^x + e^{-x}}{2}$$

33. Look back at the graph of $y = e^x$ in Figure 7.
(a) What is the equation of the tangent line shown there?
(b) Provided that x is close to zero, e^x may be approximated by the quantity $x + 1$. Use part (a) to explain why.
C (c) Complete the table. What do you observe?

x	$x+1$	e^x
1		
0.5		
0.1		
0.01		
0.001		
0.0001		

34. Set up a table to graph each of the following functions. Indicate asymptotes and intercepts, if any.
(a) $y = 2^{-x^2}$ (b) $y = 2^{|x|}$

35. Let $f(x) = 2^x$ and let g denote the function that is the inverse of f.
(a) On the same set of axes, sketch the graphs of f, g, and the line $y = x$.
(b) On the basis of your graph in part (a), specify the domain, range, intercept, and asymptote for the function g.

C **36.** Complete the following table. It can be shown that if the table is continued indefinitely, the numbers in the right-hand column approach the value e. (In fact this is one way of defining e in calculus.)

n	$\left(1 + \frac{1}{n}\right)$	$\left(1 + \frac{1}{n}\right)^n$
1		
10		
100		
1000		
10000		
100000		

C **37.** Assume the growth law $N = N_0e^{kt}$, with which we described bacterial growth, also approximates the earth's human population.
(a) Use the following data to estimate the population for the year 1980. In 1850 (call this $t = 0$), the population was estimated to be 1 billion. In 1930, it was 2 billion.
(b) Statistics from the United Nations say that the actual 1980 population was about 4.2 billion. How does your estimate from part (a) compare with this? Does the formula $N = N_0e^{kt}$ predict too high or too low a figure?

38. Carry out the simplification referred to in Example 7. That is, show

$$2[(\tfrac{1}{2})^{1/8}]^{30} = 2^{-11/4}$$

***39.** Let $\phi(t) = 1 + a^t$. Show that

$$\frac{1}{\phi(t)} + \frac{1}{\phi(-t)} = 1$$

40. Let $f(x) = e^x$ and $g(x) = x - 1$. Graph the functions F and G defined by

$$F(x) = (f \circ g)(x)$$
$$G(x) = (g \circ f)(x)$$

Specify any intercepts or asymptotes.

41. Let $f(x) = e^x$. Let L denote the function that is the inverse of f.

(a) On the same set of axes, sketch the graphs of f and L.

(b) Specify the domain, range, intercept, and asymptote for the function L and its graph.

(c) Graph each of the following. Specify the intercept and asymptote in each case.

(i) $y = -L(x)$
(ii) $y = L(-x)$
(iii) $y = L(x - 1)$

42. This exercise serves as a preview of the work on logarithms in the next section. Follow steps (a) through (f) to complete the table. (Notice that one entry in the table is already filled in. Reread Example 2 in the text to see how that entry was obtained.)

(a) Fill in the entries corresponding to $x = 1$ and $x = 10$.

(b) 4 and 8 are powers of 2. Use these facts along with the approximation $10^{0.3} \approx 2$ to find the entries in the table corresponding to $x = 4$ and $x = 8$.

(c) Find the entry corresponding to $x = 5$.
Hint: $5 = \frac{10}{2} \approx \frac{10}{10^{0.3}}$.

(d) Find the entry corresponding to $x = 7$.
Hint: $7^2 \approx 50 = (5)(10)$. Now make use of your answer in part (c).

(e) Find the entry corresponding to $x = 3$.
Hint: $3^4 \approx 80 = 8 \times 10$.

(f) Find the entries corresponding to $x = 6$ and $x = 9$. *Hint:* $6 = 3 \times 2$ and $9 = 3^2$.

Remark: This table of powers of 10 is called a table of logarithms to the base 10. We say, for example, that the logarithm of 2 to the base 10 is (about) 0.3. We write this symbolically as $\log_{10} 2 \approx 0.3$.

x	Power to Which 10 Must Be Raised to Yield x
1	
2	≈0.3
3	
4	
5	
6	
7	
8	
9	
10	

6.2 Logarithmic Functions

In the previous section we studied exponential functions. Now we're going to consider functions that are inverses of exponential functions. These inverse functions are called *logarithmic functions.*

Having said this, let's back up for a moment to review briefly some of the basic ideas behind inverse functions (as discussed in Section 4.6). Start with a given function F, say $F(x) = 3x$ for example, that is one-to-one. (That is, for each output there is but one input.) Then, by interchanging the inputs and outputs, we obtain a new function, the so-called inverse function. In the case of $F(x) = 3x$, it's easy to find an equation defining the inverse function. Just interchange x and y in the equation $y = 3x$ to obtain $x = 3y$. Solving for y in this last equation then gives us $y = \frac{1}{3}x$, which defines the inverse function. Using functional notation, we can summarize the situation by writing $F(x) = 3x$ and $F^{-1}(x) = \frac{1}{3}x$.

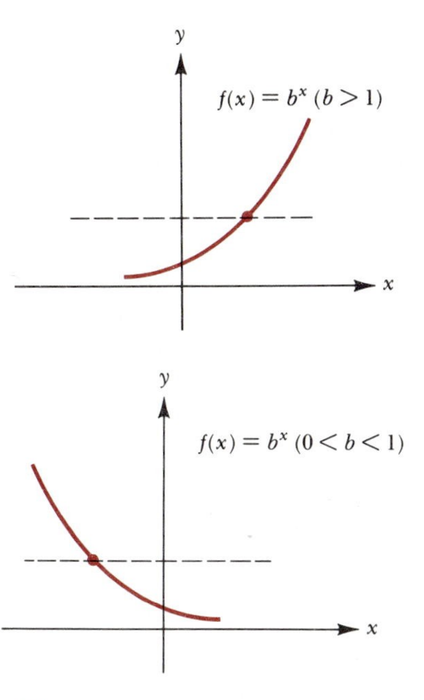

Figure 1
The exponential function $y = b^x$ is one-to-one.

In the preceding paragraph we saw that a particular linear function had an inverse, and we found a formula for that inverse. Now let's repeat that same reasoning beginning with an exponential function. First, the exponential function f defined by $f(x) = b^x$ is one-to-one. We can see this by applying the horizontal line test, as indicated in Figure 1. Next, since $f(x) = b^x$ is one-to-one, it has an inverse function. Let's study this inverse.

We begin by writing the exponential function $f(x) = b^x$ in the form

$$y = b^x \tag{1}$$

Then, to obtain an equation for f^{-1}, we interchange x and y in equation (1). This gives us

$$x = b^y \tag{2}$$

The crucial step now is to express equation (2) in words:

$$y \text{ is the power to which } b \text{ must be raised to yield } x \tag{3}$$

Equation (3) defines the function that is the inverse of $y = b^x$. Now we introduce a notation that will allow us to write this statement in a more compact form.

Definition of $\log_b x$	**Examples**
We define the expression $\log_b x$ to mean "the power to which b must be raised to yield x." ($\log_b x$ is read "log base b of x" or "the logarithm of x to the base b.")	**(a)** $\log_2 8 = 3$, since 3 is the power to which 2 must be raised to yield 8. **(b)** $\log_{10} \frac{1}{10} = -1$, since -1 is the power to which 10 must be raised to yield $\frac{1}{10}$. **(c)** $\log_5 1 = 0$, since 0 is the power to which 5 must be raised to yield 1.

Using this notation, equation (3) becomes

$$y = \log_b x$$

Since equations (2) and (3) are equivalent, we have the following important relationship.

> $y = \log_b x$ is equivalent to $x = b^y$.

We say that the equation $y = \log_b x$ is in *logarithmic form* and that the equivalent equation $x = b^y$ is in *exponential form.* Table 1 displays some examples.

Table 1

Exponential Form of Equation	Logarithmic Form of Equation
$8 = 2^3$	$\log_2 8 = 3$
$\frac{1}{9} = 3^{-2}$	$\log_3 \frac{1}{9} = -2$
$1 = e^0$	$\log_e 1 = 0$
$a = b^c$	$\log_b a = c$

Let us now summarize our discussion up to this point.

1. According to the horizontal line test, the function $f(x) = b^x$ is one-to-one and therefore it possesses an inverse. This inverse function is written as
$$f^{-1}(x) = \log_b x$$
2. $\log_b a = c$ means that $a = b^c$.

To graph the function $y = \log_b x$, we recall from Chapter 4 that the graph of a function and its inverse are reflections of one another through the line $y = x$. Thus, to graph $y = \log_b x$, we need only reflect the curve $y = b^x$ through the line $y = x$. This is shown in Figure 2. *Note:* For the rest of this chapter we assume that the base b is greater than 1 when we use the expression $\log_b x$.

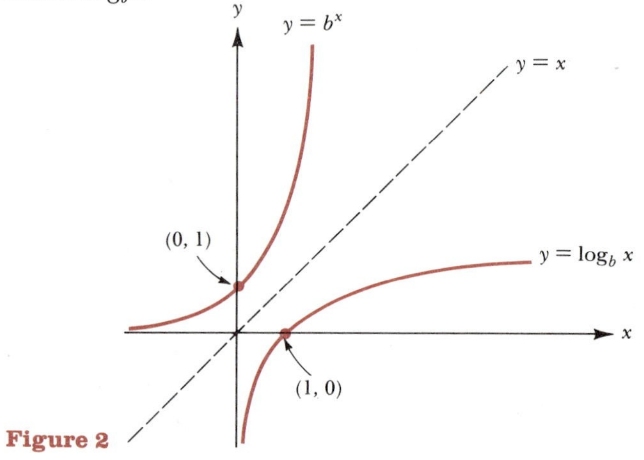

Figure 2

With the aid of Figure 2, we can make the following observations about the function $y = \log_b x$.

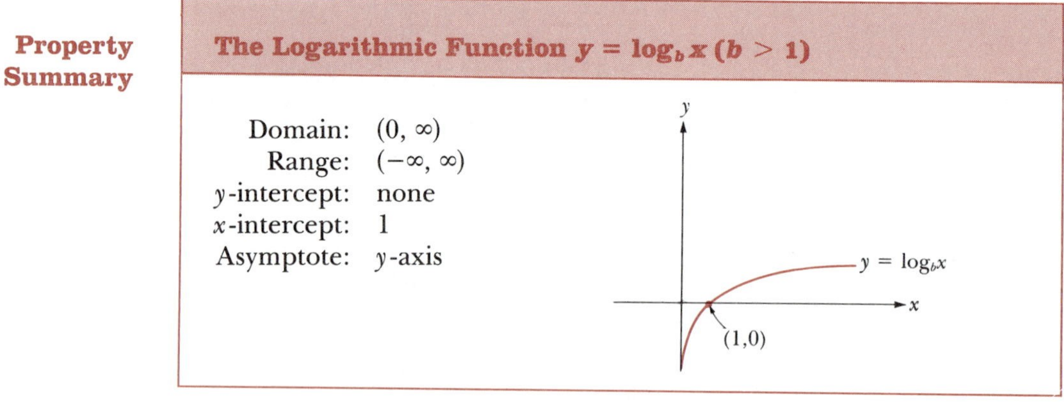

Property Summary

The Logarithmic Function $y = \log_b x$ ($b > 1$)

Domain: $(0, \infty)$
Range: $(-\infty, \infty)$
y-intercept: none
x-intercept: 1
Asymptote: y-axis

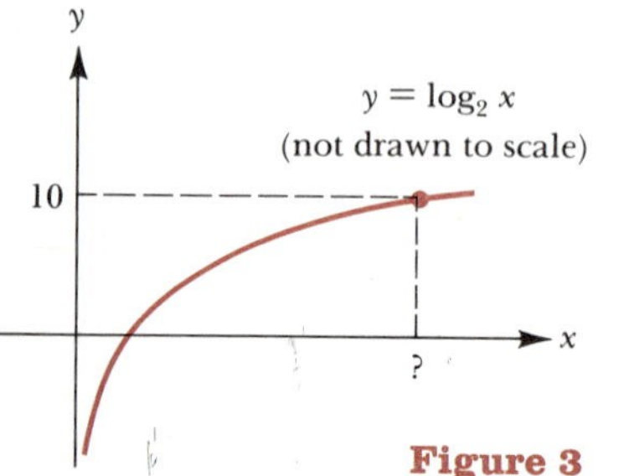

Figure 3

One aspect of the function $y = \log_b x$ may not be immediately apparent to you from Figure 2. The function grows or increases *very* slowly. Consider $y = \log_2 x$, for example. Let us ask how large x must be before the curve reaches the height $y = 10$. See Figure 3.

To answer this question, we substitute $y = 10$ in the equation $y = \log_2 x$. This gives us
$$10 = \log_2 x$$
Writing this equation in exponential form yields

$$x = 2^{10}$$

or

$$x = 1024$$

In other words, you must go out beyond 1000 on the x-axis before the curve $y = \log_2 x$ reaches a height of 10 units. Exercise 47 at the end of this section asks you to show that the graph of $y = \log_2 x$ doesn't reach a height of 100 until x is greater than 10^{30}. (Numbers as large as 10^{30} rarely occur in any of the sciences. For instance, the distance in inches to the Andromeda galaxy is less than 10^{24}.) The point we are emphasizing here is this: The graph of $y = \log_2 x$ (or $\log_b x$) is always rising, but very slowly.

We conclude this section with a number of examples involving logarithms and logarithmic functions. In one way or another, each example makes use of the key fact that the equation $\log_b a = c$ is equivalent to $b^c = a$.

Example 1 Which quantity is the larger: $\log_3 10$ or $\log_7 40$?

Solution First estimate the quantity $\log_3 10$. $\text{Log}_3\, 10$ represents the power to which 3 must be raised to yield 10. Since $3^2 = 9$ (less than 10), but $3^3 = 27$ (more than 10), we conclude that the quantity $\log_3 10$ lies between 2 and 3. In a similar way we can estimate the quantity $\log_7 40$. $\text{Log}_7\, 40$ represents the power to which 7 must be raised to yield 40. Since $7^1 = 7$ (less than 40), while $7^2 = 49$ (more than 40), we conclude that the quantity $\log_7 40$ lies between 1 and 2. It now follows from the two estimates we've made that $\log_3 10$ is larger than $\log_7 40$.

In the next example we shall need to make use of the following property of the real numbers: If b denotes a positive constant other than 1, then

$$b^x = b^y \quad \text{if and only if} \quad x = y$$

Example 2 Evaluate $\log_4 32$.

Solution Let $y = \log_4 32$. Writing this equation in its equivalent exponential form gives us

$$\begin{aligned} 4^y &= 32 \\ (2^2)^y &= 2^5 \\ 2^{2y} &= 2^5 \\ 2y &= 5 \\ y &= \frac{5}{2} \quad \text{as required} \end{aligned}$$

Example 3 Solve $\log_{10}(x^2 + 3x + 12) = 1$ for x.

Solution Writing the given equation in exponential form yields

$$\begin{aligned} x^2 + 3x + 12 &= 10^1 \\ x^2 + 3x + 2 &= 0 \\ (x + 2)(x + 1) &= 0 \end{aligned}$$

Consequently we have

$$x + 2 = 0 \quad \text{or} \quad x + 1 = 0$$

and therefore

$$x = -2 \quad \text{or} \quad x = -1$$

Check: With $x = -2$ the given equation reads $\log_{10}(4 - 6 + 12) = 1$, or $\log_{10} 10 = 1$. Since this last equation is equivalent to $10^1 = 10$, we conclude that the value $x = -2$ does satisfy the given equation. In a similar manner, we find that $x = -1$ also satisfies the equation. In summary then, the equation has two solutions, $x = -2$ and $x = -1$.

Example 4 Graph the following.
(a) $y = \log_{10} x$ **(b)** $y = -\log_{10} x$

Solution **(a)** The function $y = \log_{10} x$ is the inverse function for the exponential function $y = 10^x$. Thus we obtain the graph of $y = \log_{10} x$ by reflecting the graph of $y = 10^x$ in the line $y = x$. See Figure 4(a).
(b) To graph $y = -\log_{10} x$, we reflect $y = \log_{10} x$ in the x-axis. See 4(b).

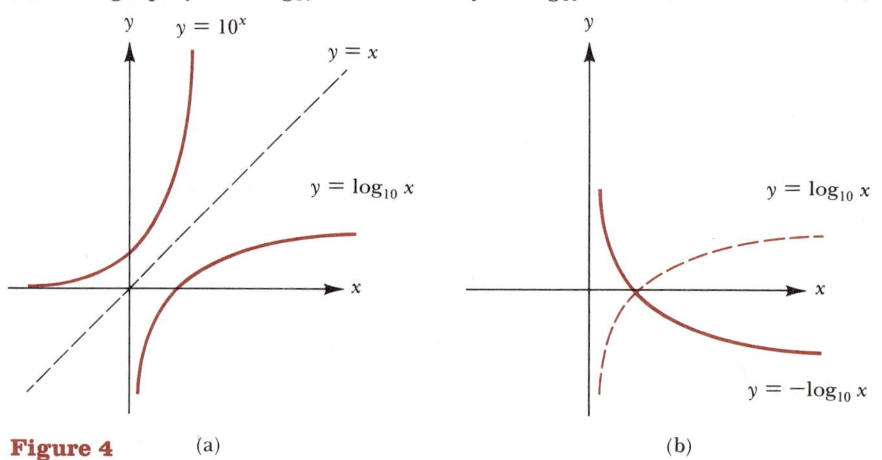

Figure 4 (a) (b)

Example 5 Find the domain of the function defined by $y = \log_2(12 - 4x)$.

Solution As you can see by looking back at Figure 2 on page 276, the inputs for the logarithmic function must be positive. So, in the case at hand, we require that the quantity $12 - 4x$ be positive. Consequently we have

$$\begin{aligned} 12 - 4x &> 0 \\ -4x &> -12 \\ x &< 3 \end{aligned}$$

Therefore the domain of the function defined by $y = \log_2(12 - 4x)$ is the interval $(-\infty, 3)$.

The next example concerns the exponential function $y = e^x$ and its inverse function $y = \log_e x$. Many books, as well as calculators, abbreviate the expression $\log_e x$ by $\ln x$, read "natural log of x." For reference, we repeat this as follows.

Definition	$\ln x$ means $\log_e x$.

Example 6 Graph the following.
(a) $y = \ln x$ **(b)** $y = \ln(x - 1)$

Solution
(a) The function $y = \ln x$ ($= \log_e x$) is the inverse of $y = e^x$. Thus its graph is obtained by reflecting $y = e^x$ in the line $y = x$, as in Figure 5(a).
(b) To graph $y = \ln(x - 1)$, take the graph of $y = \ln x$ and move it one unit in the positive x-direction. See Figure 5(b).

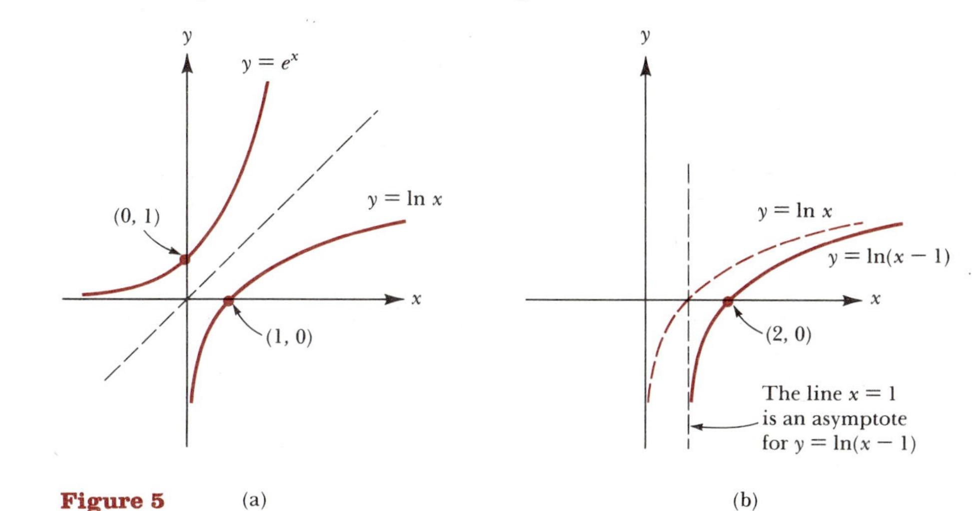

Figure 5 (a) (b)

Example 7 Simplify each of the following.
(a) $\ln e$ **(b)** $\ln 1$

Solution
(a) $\ln e$ denotes the power to which e must be raised to yield e. Clearly this power is 1. Thus $\ln e = 1$.
(b) Similarly, $\ln 1$ denotes the power to which e must be raised to yield 1. Since $e^0 = 1$, zero is the required power. Thus $\ln 1 = 0$.

Example 8 Solve the following equations.
(a) $10^{2x} = 200$ **(b)** $e^{3t-1} = 2$

Solution
(a) Write the equation $10^{2x} = 200$ in logarithmic form to obtain

$$2x = \log_{10} 200$$
$$x = \frac{\log_{10} 200}{2}$$

Without a calculator or tables, we leave the answer in this form. On the other hand, using a calculator we obtain

$$x \approx 1.15$$

(b) To solve $e^{3t-1} = 2$, write the equation in logarithmic form:

$$3t - 1 = \log_e 2$$

That is

$$3t - 1 = \ln 2$$
$$3t = 1 + \ln 2$$
$$t = \frac{1 + \ln 2}{3}$$

Without a calculator or tables, this is the final answer. On the other hand, using a calculator we obtain

$$t \approx 0.56$$

In the previous section, we used the equation $N = N_0 e^{kt}$ to describe radioactive decay as well as population growth. If you look back at the examples and problems there, you'll see that every question was of the same type: Given t, find N. In other words, we were always asked "how much" or "how many," but never "when." Now that we have defined logarithms, we can solve a much wider range of problems, as the next example indicates.

Example 9 Suppose that the half-life of a certain radioactive substance is 4 days.

(a) Compute the so-called *decay constant* k in the formula $N = N_0 e^{kt}$.

(b) If you begin with a 2-gm sample, how long will it be until only 0.01 gm remains?

Solution **(a)** The half-life is 4 days. This means that when t is 4, N is $\frac{1}{2}N_0$. Using these values for t and N in the formula $N = N_0 e^{kt}$, we obtain

$$\frac{1}{2}N_0 = N_0 e^{k4}$$

and therefore

$$\frac{1}{2} = e^{4k}$$

We solve for k by writing this last equation in logarithmic form:

$$4k = \ln \frac{1}{2}$$

$$k = \frac{\ln \frac{1}{2}}{4} = \frac{\ln 0.5}{4}$$

If a numerical value for k is required, we can use tables or a calculator to obtain

$$k \approx -0.17$$

(b) We are given that $N_0 = 2$ and we want to find the time t at which N is 0.01 gm. Substituting the values $N_0 = 2$ and $N = 0.01$ into the decay law $N = N_0 e^{kt}$ yields

$$0.01 = 2e^{kt} \qquad \text{where } k \text{ has the value determined in (a)}$$

or, dividing by 2,

$$0.005 = e^{kt}$$

Converting this last equation into its equivalent logarithmic form gives us

$$kt = \ln(0.005)$$

Therefore

$$t = \frac{\ln(0.005)}{k}$$

Finally, upon replacing k by the expression $\frac{\ln 0.5}{4}$, we obtain

$$t = \frac{\ln(0.005)}{(\ln 0.5)/4}$$

$$= \frac{4\ln(0.005)}{\ln 0.5}$$

$$\approx 30.58 \qquad \text{(using a calculator)}$$

Thus after about $30\frac{1}{2}$ days, only 0.01 gm remains of the original 2-gm sample.

Exercise Set 6.2

Exercises 1–6 are review exercises dealing with inverse functions.

1. Which of the following functions are one-to-one and therefore have inverses?
(a) $y = x^2 + 1$ **(b)** $y = 3x$ **(c)** $y = (x + 1)^3$

2. Do either of the following functions have an inverse?
(a) $f(x) = \begin{cases} x^2 & \text{if } -1 \le x \le 0 \\ x^2 + 1 & \text{if } x > 0 \end{cases}$
(b) $g(x) = \begin{cases} x^2 & \text{if } -1 \le x < 0 \\ x^2 + 1 & \text{if } x \ge 0 \end{cases}$

3. Let $f(x) = \frac{2x - 1}{3x + 4}$.
(a) Find $f^{-1}(x)$. **(b)** Find $1/f(x)$.
(c) Find $f^{-1}(0)$. **(d)** Find $1/f(0)$.

4. Let $f(x) = x^3 + 2x + 1$. Evaluate $f[f^{-1}(5)]$. [Assume that the domain of f^{-1} is $(-\infty, \infty)$.]

5. The graph of $y = f(x)$ is a straight line segment joining the two points $(3, -2)$ and $(-1, 5)$. What are the corresponding end points for the graph of $y = f^{-1}(x - 1)$?

6. Which (if either) of the following two conditions tells us that a function is one-to-one?
(a) For each input there is exactly one output.
(b) For each output there is exactly one input.

7. Write each of the following equations in logarithmic form.
(a) $9 = 3^2$ **(b)** $1000 = 10^3$ **(c)** $7^3 = 343$
(d) $\sqrt{2} = 2^{1/2}$

8. Write each equation in logarithmic form.
(a) $\frac{1}{125} = 5^{-3}$ **(b)** $e^0 = 1$ **(c)** $5^x = 6$
(d) $e^{3t} = 8$

9. Write each equation in exponential form.
(a) $\log_2 32 = 5$ **(b)** $\log_{10} 1 = 0$
(c) $\log_e \sqrt{e} = \frac{1}{2}$ **(d)** $\log_3 \frac{1}{81} = -4$
(e) $\log_t u = v$

10. Complete the table.

x	1	10	10^2	10^3	10^4	10^{-1}	10^{-2}	10^{-3}
$\log_{10} x$								

11. Which quantity is the larger: $\log_5 30$ or $\log_8 60$?

12. Which quantity is the larger:
(a) $\log_{10} 90$ or $\log_e e^5$? **(b)** $\log_2 3$ or $\log_3 2$?

13. Evaluate each expression.
(a) $\log_9 27$ **(b)** $\log_4 \frac{1}{32}$ **(c)** $\log_5(5\sqrt{5})$

14. Evaluate each expression.
(a) $\log_{25} \frac{1}{625}$ **(b)** $\log_{16} \frac{1}{64}$ **(c)** $\log_{10} 10$
(d) $\log_2 8\sqrt{2}$

15. Solve each equation for x.
(a) $\log_x 256 = 8$ **(b)** $\log_5 x = -1$

16. Solve each equation for x.
(a) $\log_{10}(x^2 + 36) = 2$
(b) $\log_2(x^2 - 8x + 1) = 0$
(c) $\log_{10}(x^2 - 5x + 14) = 1$

17. Find the domain of each function.
(a) $y = \log_4 5x$
(b) $y = \log_{10}(3 - 4x)$
(c) $y = \ln(x^2)$
(d) $y = (\ln x)^2$
(e) $y = \ln(x^2 - 25)$

18. Find the domain of each function.
(a) $y = \ln(2 - x - x^2)$
(b) $y = \log_{10} \frac{2x + 3}{x - 5}$

***19.** Find the domain and the range of the function defined by $y = \frac{1}{1 - \ln x}$.

In Exercises 20–24, graph the given function and specify the domain, range, intercept(s), and asymptote.

20. $y = \log_2(x - 3)$

21. **(a)** $y = \log_{10}(x + 1)$
(b) $y = -\log_{10}(x + 1)$

22. $y = 1 + \log_2 x$

23. **(a)** $y = \ln x$
(b) $y = \ln(-x)$
(c) $y = -1 + \ln(-x)$

24. $y = 1 - \log_2(x - 1)$

25. Simplify each expression.
(a) $\ln e^4$ **(b)** $\ln(1/e)$ **(c)** $\ln \sqrt{e}$

26. **(a)** Solve the equation $10^x = 25$, leaving your answer in terms of logs.
(b) Use a calculator or tables to evaluate your answer in part (a).

27. **(a)** Solve $10^{(x^2)} = 40$.
(b) Solve $(10^x)^2 = 40$.
(c) Without using tables or a calculator, determine which of the two answers is the larger.

28. **(a)** Solve the equation $e^{2t+3} = 10$, expressing your answer in terms of logarithms.
(b) Use a calculator or tables to evaluate your answer in part (a).

29. There are no real numbers that satisfy the equation $2^{5x} = -3$. Why?

30. Determine if there are any real numbers that satisfy the equation

$$e^{1-t} = -100$$

31. Find all positive solutions to the equation

$$e^{4t^2-1} = 6$$

C **32.** The half-life of radium-226 is 1620 years.
(a) Find the decay constant k.
(b) How much of a 0.1-gm sample remains after 100 years?

C **33.** The half-life of uranium-238 is 4.5×10^9 years.
(a) Find the decay constant k.
(b) What percentage of a given amount N_0 remains after 1000 years?

C **34.** Compute the half-life of carbon-14 if the value of the decay constant is $k = -0.00012$. (Assume that the units of t are years.)

C **35.** The half-life of a certain radioactive substance is 1 year.
(a) Find the decay constant k.
(b) Find the time required for 90% of a given 4-gm sample to decay away.

C **36.** The initial population of a bacteria culture is 1.5×10^5. Three hours later the population is found to be 2×10^5. Assume the growth law $N = N_0e^{kt}$ applies.
(a) Find the growth constant k.
(b) How long will it take for the initial population to double?

C **37.** Initially a bacteria culture has a population of 2×10^7. Two hours later the population is 3×10^8.
(a) How long does it take for the culture to double its original size?
(b) When does the population reach 1 billion?

38. Which function is the inverse of $y = e^x$? On the same set of axes, graph $y = e^x$ and its inverse.

***39.** Let $f(x) = e^{x+1}$. Find $f^{-1}(x)$ and sketch its graph. Specify any intercepts or asymptotes.

***40.** Let $g(t) = \ln(t - 1)$. Find $g^{-1}(t)$ and draw its graph. Specify any intercepts or asymptotes.

41. Sketch the region bounded by the curves $y = e^x$, $y = e^{-x}$, the x-axis, and the vertical lines $x = \pm 1$. Why must the area of this region be less than 2 square units?

42. Let $N = N_0e^{kt}$ and suppose that when $t = t_1$, $N = N_1$. Show that

$$k = \frac{\ln(N_1/N_0)}{t_1}$$

C **43.** **(a)** If the half-life of a certain radioactive substance is T, show that

$$k = \frac{\ln \frac{1}{2}}{T}$$

(b) The half-life of radon gas is 3.8 days. Find the decay constant k.
(c) How long will it take for 99% of a given sample of radon gas to disintegrate?

***44.** Solve $2^{2x} - 2^{x+1} - 15 = 0$ for x. *Hint:* First show the equation is equivalent to $(2^x)^2 - 2(2^x) - 15 = 0$.

45. Solve $e^{2x} - 5e^x - 6 = 0$ for x.

46. Solve $e^x - 3e^{-x} = 2$ for x. *Hint:* Multiply both sides by e^x.

47. Estimate a value for x such that $\log_2 x = 100$. Use the approximation $10^3 \approx 2^{10}$ to express your answer as a power of 10. *Answer:* 10^{30}

48. **(a)** How large must x be before the graph of $y = \ln x$ reaches a height of $y = 100$?
(b) How large must x be before the graph of $y = e^x$ reaches a height of **(i)** $y = 100$? **(ii)** $y = 10^6$?

***49.** Suppose that $A = A_0e^{k_1t}$ and $B = B_0e^{k_2t}$. Find the value of t for which $A = B$.

Answer: $t = \dfrac{\ln(A_0/B_0)}{k_2 - k_1}$

How would you interpret this question and answer in terms of radioactive decay?

50. Chemists define pH by the formula

$$\text{pH} = -\log_{10}[\text{H}^+]$$

where $[H^+]$ is the hydrogen ion concentration, measured in moles/liter. For example, if $[H^+] = 10^{-5}$ then pH = 5. Solutions with a pH of 7 are called *neutral*; below 7 is *acidic* and above 7 *basic*.

(a) $[H^+]$ for some fruit juices is 3×10^{-4}. Determine the pH and classify as acid or base. (Use a calculator.)

(b) $[H^+]$ for sulfuric acid is 1. Find the pH.

(c) An unknown substance has a hydrogen ion concentration of 3.5×10^{-9}. Classify it as acid or base. (Use a calculator.)

***51.** **(a)** Find the domain of the function f defined by

$$f(x) = \ln(\ln x)$$

(b) Find $f^{-1}(x)$ for the function f in part (a).

6.3 Properties of Logarithms

There are four basic properties of logarithms that we shall find useful. Our procedure in this section will be to state these properties, discuss their proofs, and then look at examples.

Property Summary

Properties of Logarithms

1. $\log_b PQ = \log_b P + \log_b Q$
   (The log of a product is the sum of the logs.)
2. $\log_b (P/Q) = \log_b P - \log_b Q$
   (The log of a quotient is the log of the numerator minus the log of the denominator.)
3. $\log_b P^n = n \log_b P$
4. $b^{\log_b P} = P$

Note: P and Q are assumed to be positive in all four properties.

First let's prove property 1. We begin by letting x denote the expression $\log_b P$. That is, let $x = \log_b P$. The equivalent exponential form of this equation is

$$P = b^x \tag{1}$$

Similarly, let $y = \log_b Q$. The exponential form of this equation is

$$Q = b^y \tag{2}$$

Now if we multiply equation (1) by equation (2), we have

$$PQ = b^x b^y$$

and therefore

$$PQ = b^{x+y} \tag{3}$$

Next we write equation (3) in its equivalent logarithmic form. This yields

$$\log_b PQ = x + y$$

But in view of the definitions of x and y, this last equation is equivalent to

$$\log_b PQ = \log_b P + \log_b Q$$

That completes the proof of property 1.

The proof of property 2 is quite similar to that just given for property 1. Exercise 52 at the end of this section asks you to carry out the proof.

We turn now to the proof of property 3. We begin by letting $x = \log_b P$. In exponential form, this last equation becomes

$$b^x = P \tag{4}$$

Now we raise both sides of equation (4) to the power n. This yields

$$(b^x)^n = P^n$$

and therefore

$$b^{nx} = P^n$$

The logarithmic form of this last equation is

$$\log_b P^n = nx$$

or (in view of the definition of x)

$$\log_b P^n = n \log_b P$$

The proof of property 3 is now complete since this last equation is what we wanted to show.

Property 4 is actually just a restatement of the meaning of the expression $\log_b P$. To derive this property, let

$$x = \log_b P$$

Therefore

$$b^x = P$$

Now, in this last equation, just replace x with $\log_b P$. The result is

$$b^{\log_b P} = P$$

as required.

By way of review, we point out that the demonstration just given for property 4 can be recast in terms of functional notation and inverse functions. To do this we define two functions f and g as follows:

$$f(P) = b^P \quad \text{and} \quad g(P) = \log_b P$$

Then, as explained in Section 6.2, f and g are inverses. We therefore have

$$f[g(P)] = P \qquad \text{by definition of inverse functions}$$

and consequently

$$f(\log_b P) = P \qquad \text{by definition of } g$$

Now in view of the definition of f, this last equation can be written

$$b^{\log_b P} = P \qquad \text{as we wished to show}$$

Now let's look at some examples of how these properties are used. To begin with, we display a simple numerical application of each property in the box that follows.

Property Summary

Property	Example
$\log_b P + \log_b Q = \log_b PQ$	Simplify: $\log_{10} 50 + \log_{10} 2$ Solution: $\log_{10} 50 + \log_{10} 2 = \log_{10}(50 \cdot 2)$ $= \log_{10} 100$ $= 2$
$\log_b P - \log_b Q = \log_b \frac{P}{Q}$	Simplify: $\log_8 56 - \log_8 7$ Solution: $\log_8 56 - \log_8 7 = \log_8 \frac{56}{7}$ $= \log_8 8$ $= 1$
$\log_b P^n = n \log_b P$	Simplify: $\log_2 \sqrt[5]{16}$ Solution: $\log_2 \sqrt[5]{16} = \log_2(16^{1/5})$ $= \frac{1}{5} \log_2 16$ $= \frac{1}{5} \cdot 4$ $= \frac{4}{5}$
$b^{\log_b P} = P$	Simplify: $3^{\log_3 7}$ Solution: $3^{\log_3 7} = 7$

These examples show how to simplify or shorten certain expressions involving logarithms. The next example is also of this type.

Example 1 Express as a single logarithm with a coefficient of 1:

$$\tfrac{1}{2} \log_b x - \log_b(1 + x^2)$$

Solution

$$\frac{1}{2} \log_b x - \log_b(1 + x^2) = \log_b x^{1/2} - \log_b(1 + x^2) \qquad \text{using property 3}$$

$$= \log_b \frac{x^{1/2}}{1 + x^2} \qquad \text{using property 2}$$

This last expression is the required answer.

Property 1 says that the logarithm of a product of two factors is equal to the sum of the logarithms of the two factors. This can be generalized to any number of factors. For instance, with three factors we have

$$\log_b(ABC) = \log_b[A(BC)]$$

$$= \log_b A + \log_b BC \qquad \text{using property 1}$$

$$= \log_b A + \log_b B + \log_b C$$

The next example makes use of this idea.

Example 2 Express as a single logarithm with a coefficient of 1:

$$\ln(x^2 - 9) + 2\ln\frac{1}{x+3} + 4\ln x \qquad (x > 3)$$

Solution

$$\begin{aligned}\ln(x^2 - 9) + 2\ln\frac{1}{x+3} + 4\ln x &= \ln(x^2 - 9) + \ln\left[\left(\frac{1}{x+3}\right)^2\right] + \ln x^4\\ &= \ln(x^2 - 9) + \ln\frac{1}{(x+3)^2} + \ln x^4\\ &= \ln\left[(x^2 - 9)\cdot\frac{1}{(x+3)^2}\cdot x^4\right]\\ &= \ln\frac{(x^2 - 9)x^4}{(x+3)^2}\end{aligned}$$

This last expression can be simplified still further by writing $x^2 - 9$ as $(x - 3)(x + 3)$. Then the factor $x + 3$ can be divided out of the numerator and denominator of the fraction. The result is

$$\begin{aligned}\ln(x^2 - 9) + 2\ln\frac{1}{x+3} + 4\ln x &= \ln\frac{(x-3)x^4}{x+3}\\ &= \ln\frac{x^5 - 3x^4}{x+3} \qquad \text{as required}\end{aligned}$$

In the examples considered so far, we've used the properties of logarithms to shorten given expressions. There are times, however, when we use the properties to expand an expression. The next two examples show such cases.

Example 3 Write each of the following using sums and differences of simpler logarithmic expressions. Express the answer in such a way that no logarithms of products, quotients, or powers appear.

(a) $\log_{10}\sqrt[3]{\dfrac{2x}{3x^2+1}}$ **(b)** $\ln\dfrac{x^2\sqrt{2x-1}}{(2x+1)^{3/2}}$

Solution

(a)
$$\begin{aligned}\log_{10}\sqrt[3]{\frac{2x}{3x^2+1}} &= \log_{10}\left[\left(\frac{2x}{3x^2+1}\right)^{1/3}\right]\\ &= \frac{1}{3}\log_{10}\frac{2x}{3x^2+1}\\ &= \frac{1}{3}[\log_{10}2x - \log_{10}(3x^2+1)]\\ &= \frac{1}{3}[\log_{10}2 + \log_{10}x - \log_{10}(3x^2+1)] \qquad \text{as required}\end{aligned}$$

(b)
$$\begin{aligned}\ln\frac{x^2\sqrt{2x-1}}{(2x+1)^{3/2}} &= \ln\frac{x^2(2x-1)^{1/2}}{(2x+1)^{3/2}}\\ &= \ln x^2 + \ln(2x-1)^{1/2} - \ln(2x+1)^{3/2}\\ &= 2\ln x + \frac{1}{2}\ln(2x-1) - \frac{3}{2}\ln(2x+1) \qquad \text{as required}\end{aligned}$$

Example 4 Given that $\log_{10} A = a$, $\log_{10} B = b$, and $\log_{10} C = c$, express $\log_{10} \frac{A^3}{B^4\sqrt{C}}$ in terms of a, b, and c.

Solution

$$\begin{aligned}
\log_{10} \frac{A^3}{B^4\sqrt{C}} &= \log_{10} \frac{A^3}{B^4C^{1/2}} \\
&= \log_{10} A^3 - \log_{10} B^4C^{1/2} \\
&= \log_{10} A^3 - (\log_{10} B^4 + \log_{10} C^{1/2}) \\
&= 3 \log_{10} A - \left(4 \log_{10} B + \frac{1}{2} \log_{10} C\right) \\
&= 3 \log_{10} A - 4 \log_{10} B - \frac{1}{2} \log_{10} C \\
&= 3a - 4b - \frac{1}{2} c \qquad \text{as required}
\end{aligned}$$

Our four properties for logarithms are also useful in solving equations. The next two examples show instances of this.

Example 5 Use logarithms to the base e to solve $10 = 3e^{1-2x}$ for x.

Solution Take the natural (base e) logarithm of both sides of the given equation to obtain

$$\begin{aligned}
\ln 10 &= \ln(3e^{1-2x}) \\
\ln 10 &= \ln 3 + \ln e^{1-2x} && \text{property 1} \\
\ln 10 &= \ln 3 + (1 - 2x) && \text{definition of ln} \\
2x &= \ln 3 - \ln 10 + 1 \\
x &= \frac{\ln 3 - \ln 10 + 1}{2}
\end{aligned}$$

This is the required solution. If we wish, we can use property 2 to rewrite this solution as

$$x = \frac{\ln \frac{3}{10} + 1}{2}$$

(There is yet another way to write this answer. Exercise 53 at the end of this section asks you to show that $x = \frac{1}{2} + \ln\sqrt{0.3}$.)

Example 6 Solve $\log_3 x + \log_3(x + 2) = 1$ for x.

Solution By using property 1, we can write the given equation as

$$\log_3[x(x + 2)] = 1$$

or

$$\log_3(x^2 + 2x) = 1$$

Writing this last equation in exponential form yields

$$\begin{aligned}
x^2 + 2x &= 3^1 \\
x^2 + 2x - 3 &= 0 \\
(x + 3)(x - 1) &= 0
\end{aligned}$$

Thus we have

$$x + 3 = 0 \quad \text{or} \quad x - 1 = 0$$

and consequently

$$x = -3 \quad \text{or} \quad x = 1$$

Now let us check these values in the original equation to see if they are indeed solutions.

If $x = -3$, the equation becomes	*If $x = 1$, the equation becomes*
$\underbrace{\log_3(-3)} + \underbrace{\log_3(-1)} \stackrel{?}{=} 1$	$\log_3 1 + \log_3 3 \stackrel{?}{=} 1$
Neither expression is defined since the domain of the logarithm function does not contain negative numbers.	$0 + 1 \stackrel{?}{=} 1$ True

Thus the value $x = 1$ is a solution of the original equation, but $x = -3$ is not. How is it possible that an extraneous solution ($x = -3$) was generated along with the correct solution $x = 1$? The difficulty arose in the second line of our solution when we used the property $\log_b PQ = \log_b P + \log_b Q$. This property is valid only when P and Q are both positive.

Occasionally it is necessary to convert logarithms in one base to logarithms in another base. After the next example, we'll state a formula for this. However, as the next example indicates, it's easy to work this type of problem from the basics, without relying on a formula.

Example 7 Express the quantity $\log_2 5$ in terms of base ten logarithms.

Solution Let $z = \log_2 5$. The exponential form of this equation is

$$2^z = 5$$

Now take the base ten logarithm of each side of this last equation to obtain

$$\begin{aligned} \log_{10} 2^z &= \log_{10} 5 \\ z \log_{10} 2 &= \log_{10} 5 \qquad \text{property 3} \\ z &= \frac{\log_{10} 5}{\log_{10} 2} \end{aligned}$$

In view of the definition of z, this last equation can be written

$$\log_2 5 = \frac{\log_{10} 5}{\log_{10} 2}$$

This is the required answer.

The method used in Example 7 can be used to convert between any two bases. Exercise 52(b) at the end of this section asks you to follow the method using letters rather than numbers to arrive at a general formula, which we list here for reference.

Change of Base Formula

$$\log_a x = \frac{\log_b x}{\log_b a}$$

Up to now, all of the examples in this section have dealt with applications of the four properties of logarithms. However, you also need to understand what the properties *don't* say. For instance, property 2 does not apply to an expression such as $\frac{\log_{10} 5}{\log_{10} 2}$. (Property 2 would apply if the expression were $\log_{10} \frac{5}{2}$.)

Example 8 How does the statement $(\log_b P)^n = n \log_b P$ differ from property 3? Give an example showing that this equation is not an identity.

Solution In property 3, the exponent n on the left-hand side of the equation applies only to P, not to the quantity $\log_b P$. To provide the required example, we choose (almost any) convenient values for b, P, and n for which both sides of the equation can be easily evaluated. Using $b = 10$, $P = 10$, and $n = 2$, the equation reads

$$(\log_{10} 10)^2 \stackrel{?}{=} 2 \log_{10} 10$$
$$1^2 \stackrel{?}{=} (2)(1)$$
$$1 \stackrel{?}{=} 2 \qquad \text{No!}$$

Thus the given statement is not always true. To repeat, the point here is that you must learn not only what the properties say, but also what they don't say.

Exercise Set 6.3

In Exercises 1–10, simplify the given expressions by using the definition and properties for logarithms.

1. $\log_{10} 70 - \log_{10} 7$
2. $\log_{10} 40 + \log_{10} \frac{5}{2}$
3. $\log_7 \sqrt{7}$
4. $\log_9 25 - \log_9 75$
5. $\log_3 108 + \log_3 \frac{3}{4}$
6. $\ln e^3 - \ln e$
7. $-\frac{1}{2} + \ln \sqrt{e}$
8. $e^{\ln 3} + e^{\ln 2} - e^{\ln e}$
9. $2^{\log_2 5} - 3 \log_5 \sqrt[3]{5}$
10. $\log_b b^b$

In Exercises 11–19, write the given expression as a single logarithm with a coefficient of 1.

11. $\log_{10} 30 + \log_{10} 2$
12. $2 \log_{10} x - 3 \log_{10} y$
13. $\log_5 6 + \log_5 \frac{1}{3} + \log_5 10$
14. $p \log_b A - q \log_b B + r \log_b C$
15. (a) $\ln 3 - 2 \ln 4 + \ln 32$
    (b) $\ln 3 - 2(\ln 4 + \ln 32)$
16. (a) $\log_{10}(x^2 - 16) - 3 \log_{10}(x + 4) + 2 \log_{10} x$
    (b) $\log_{10}(x^2 - 16) - 3[\log_{10}(x + 4) + 2 \log_{10} x]$
17. $\log_b 4 + 3[\log_b(1 + x) - \frac{1}{2} \log_b(1 - x)]$
18. $\ln(x^3 - 1) - \ln(x^2 + x + 1)$
19. $4 \log_{10} 3 - 6 \log_{10}(x^2 + 1) + \frac{1}{2}[\log_{10}(x + 1) - 2 \log_{10} 3]$

In Exercises 20–26, write each quantity using sums and differences of simpler logarithmic expressions. Express each answer so that logarithms of products, quotients, and powers do not appear.

20. (a) $\log_{10}\sqrt{(x + 1)(x + 2)}$
    (b) $\ln \sqrt{\frac{(x + 1)(x + 2)}{(x - 1)(x - 2)}}$
21. (a) $\log_{10} \frac{x^2}{1 + x^2}$ (b) $\ln \frac{x^2}{\sqrt{1 + x^2}}$
22. (a) $\log_b \frac{\sqrt{1 - x^2}}{x}$ (b) $\ln \frac{x\sqrt[3]{4x + 1}}{\sqrt{2x - 1}}$
23. (a) $\log_{10} \sqrt{9 - x^2}$ (b) $\ln \frac{\sqrt{4 - x^2}}{(x - 1)(x + 1)^{3/2}}$
24. (a) $\log_b \sqrt[3]{\frac{x + 3}{x}}$ (b) $\ln \frac{1}{\sqrt{x^2 + x + 1}}$
25. (a) $\log_b \sqrt{\frac{x}{b}}$
    (b) $2 \ln \sqrt{(1 + x^2)(1 + x^4)(1 + x^6)}$
26. (a) $\log_b \sqrt[3]{\frac{(x - 1)^2(x - 2)}{(x + 2)^2(x + 1)}}$ (b) $\ln \left(\frac{e - 1}{e + 1}\right)^{3/2}$
27. Suppose that $\log_{10} A = a$, $\log_{10} B = b$, and $\log_{10} C = c$. Express each of the following in terms of a, b, and c.
    (a) $\log_{10} AB^2C^3$ (b) $\log_{10} \sqrt{10ABC}$

(c) $\dfrac{\log_{10} 10A}{\sqrt{BC}}$ (d) $\dfrac{\log_{10} 100A^2}{B^4 \sqrt[3]{C}}$ (e) $\dfrac{\log_{10} (AB)^5}{C}$

28. If $\ln x = t$ and $\ln y = u$, write each of the following in terms of t and u.
(a) $\ln xy - \ln x^2$ (b) $\ln \sqrt{xy} + \ln (x/e)$
(c) $\ln (e^2 x^3 \sqrt{y})$

In Exercises 29–35, solve the given equations. Express your answers in terms of base e *logarithms.*

29. $5 = 2e^{2x-1}$ **30.** $100 = 3e^x$ **31.** $3e^{1+t} = 2$
32. $4e^{1-2t} = 7$ **33.** $2^x = 9$ **34.** $5^{3x-1} = 27$
35. $10 \cdot 2^x = 5^x$

For Exercises 36–45, solve each equation for x*; check your answers to remove any extraneous roots.*

36. $\log_6 x + \log_6(x + 1) = 1$
37. $\log_9(x + 1) = \frac{1}{2} + \log_9 x$
38. $\log_2(x + 4) = 2 - \log_2(x + 1)$
39. $\log_{10}(2x + 4) + \log_{10}(x - 2) = 1$
40. $\ln x + \ln(x + 1) = \ln 12$
41. $\log_{10}(x + 3) - \log_{10}(x - 2) = 2$
42. $\ln(x + 1) = 2 + \ln(x - 1)$
43. $\log_b(x + 1) = 2 \log_b(x - 1)$
44. $\log_2(2x^2 + 4) = 5$
45. $\log_{10}(x - 6) + \log_{10}(x + 3) = 1$
46. Solve for x in terms of a:

$$\log_2(x + a) - \log_2(x - a) = 1$$

47. Solve for x in terms of y.
(a) $\log_{10} x - y = \log_{10}(3x - 1)$
(b) $\log_{10}(x - y) = \log_{10}(3x - 1)$
48. Solve for x in terms of b: $\log_b(1 - 3x) = 3 + \log_b x$.
49. (a) Express each of the following quantities in terms of base ten logarithms.
(i) $\log_2 5$ (ii) $\log_5 10$ (iii) $\ln 3$
(iv) $\log_b 2$ (v) $\log_2 b$
(b) Express each of the following in terms of natural logarithms.
(i) $\log_{10} 6$ (ii) $\log_2 10$ (iii) $\log_{10}(\log_{10} x)$
50. Give specific examples showing the following rules are *false*.
(a) $\log(x + y) = \log x + \log y$
(b) $(\log x)/(\log y) = \log x - \log y$
(c) $(\log x)(\log y) = \log x + \log y$
(d) $(\log x)^k = k \log x$
51. True or false:
(a) $\log_{10} A + \log_{10} B - \frac{1}{2} \log_{10} C = \log_{10} AB/\sqrt{C}$
(b) $\log_e \sqrt{e} = \frac{1}{2}$
(c) $\ln \sqrt{e} = \frac{1}{2}$
(d) $\ln x^3 = \ln 3x$
(e) $\ln x^3 = 3 \ln x$
(f) $\ln 2x^3 = 3 \ln 2x$
(g) $\log_a c = b$ means $a^b = c$.
(h) $\log_5 24$ is between 5^1 and 5^2.
(i) $\log_5 24$ is between 1 and 2.
(j) $\log_5 24$ is closer to 1 than 2.
(k) The domain of $g(x) = \ln x$ is the set of all real numbers.
(l) The range of $g(x) = \ln x$ is the set of all real numbers.
(m) The function $g(x) = \ln x$ is one-to-one.
52. (a) Prove that $\log_b (P/Q) = \log_b P - \log_b Q$.
Hint: Study the proof of property 1 in the text.
(b) Prove the change of base formula

$$\log_a x = \frac{\log_b x}{\log_b a}$$

Hint: Use the method of Example 7 in the text.
53. Show that $\dfrac{\ln 3 - \ln 10 + 1}{2} = \dfrac{1}{2} + \ln \sqrt{0.3}$.
***54.** Solve $\log_b(2x + 1) = 2 + \log_b x$ for x (in terms of b). What restrictions does your solution impose on b?
***55.** Show that $\log_b \dfrac{\sqrt{3} + \sqrt{2}}{\sqrt{3} - \sqrt{2}} = 2 \log_b (\sqrt{3} + \sqrt{2})$.
***56.** (a) Show that $\log_b (P/Q) + \log_b (Q/P) = 0$.
(b) Simplify $\log_a x + \log_{1/a} x$.
57. Simplify $b^{3 \log_b x}$.
58. Let $\log_{10} 2 = a$ and let $\log_{10} 3 = b$. Express each of the following in terms of a and (or) b.
(a) $\log_{10} 4$ (b) $\log_{10} 8$ (c) $\log_{10} 5$
(d) $\log_{10} 6$ (e) $\log_{10} \frac{5}{64}$ (f) $\log_{10} 108$
(g) $\log_{10} \sqrt[3]{12}$ (h) $\log_{10} 0.0027$
59. Simplify $\log_2 \sqrt[5]{4\sqrt{2}}$.
60. Solve for x in terms of α and β: $\alpha \ln x + \ln \beta = 0$.
61. Solve for x in terms of α and β: $3 \ln x = \alpha + 3 \ln \beta$.
Answer: $x = \beta e^{\alpha/3}$
***62.** Is there a constant k such that the equation $e^x = 2^{kx}$ holds for all values of x?
***63.** Prove that $(\log_a x)/(\log_{ab} x) = 1 + \log_a b$.
***64.** Prove that $\log_b a = 1/(\log_a b)$.
***65.** Simplify $(\log_2 3)(\log_3 4)(\log_4 5)$.
***66.** Simplify a^x when $x = [\log_b(\log_b a)]/(\log_b a)$.

C *Use a calculator for Exercises 67–81.*

67. Check property 1 using the values $b = 10$, $P = \pi$, and $Q = \sqrt{2}$.
68. Let $P = 3$ and $Q = 4$. Show that $\ln(P + Q) \neq \ln P + \ln Q$.
69. Check property 2 using the values $b = 10$, $P = 2$, and $Q = 3$.

70. Let $P = 10$ and $Q = 20$. Show that $\ln(PQ) \neq (\ln P)(\ln Q)$.

71. Check property 2 using natural logarithms and the values $P = 19$, $Q = 86$.

72. Show that $(\log_{10} 19)/(\log_{10} 86) \neq \log_{10} 19 - \log_{10} 86$.

73. Show that $(\ln 19)/(\ln 86) \neq \ln 19 - \ln 86$.

74. Check property 3 using the values $b = 10$, $P = \pi$, and $n = 7$.

75. Using the values given for b, P, and n in Exercise 74, show that $\log_b P^n \neq (\log_b P)^n$.

76. Verify property 4 using the values $b = 10$ and $P = 1776$.

77. Verify that $\ln 2 + \ln 3 + \ln 4 = \ln(24)$.

78. Verify that $\log_{10} A + \log_{10} B + \log_{10} C = \log_{10}(ABC)$ using the values $A = 11$, $B = 12$, and $C = 13$.

79. Let $f(x) = e^x$ and $g(x) = \ln x$. Compute $f[g(2345.6)]$.

80. Let $f(x) = 10^x$ and $g(x) = \log_{10} x$. Compute $g[f(0.123456)]$.

*81. Let $y = \ln[\ln(\ln x)]$. First, complete the table using a calculator. After doing that, disregard the evidence in your table and prove (without a calculator, of course) that the range of the given function is actually the set of all real numbers.

x	100	1000	10^6	10^{20}	10^{50}	10^{99}
y						

6.4 Applications

S = Pe^{rt}. This result is remarkable both because of its simplicity and the occurrence of e. (Who would expect that number to pop up in finance theory?)

Philip Gillett in *Calculus and Analytic Geometry,* 2nd ed. (Lexington, MA: D. C. Heath & Company, 1984)

We begin this section by considering the way in which money accumulates in a savings account. Eventually this will lead us back to the number e and the growth law $N = N_0e^{kt}$, both of which arose in different contexts in Section 6.1.

The following idea from arithmetic is a prerequisite for our discussion. To increase a given quantity by 15%, for example, multiply the quantity by the factor 1.15. For instance, suppose you want to increase \$100 by 15%. The calculations can be written as

$$\begin{aligned} \$100 + 15\% \text{ of } \$100 &= 100 + 0.15(100) \\ &= 100(1 + 0.15) \quad \text{factoring} \\ &= 100(1.15) \quad \text{as stated above} \end{aligned}$$

Similarly, to increase a quantity by 30%, you would multiply by the factor 1.30, and so on. The next example displays some calculations involving percentage increase. The results may surprise you unless you're already familiar with this topic.

Example 1 An amount of \$100 is increased by 15% and then the new amount is increased by 15%. Is this the same as an overall increase of 30%?

Solution To increase \$100 by 15%, we multiply by 1.15 to obtain \$100(1.15). Now to increase this amount by 15%, we multiply it by 1.15 to obtain

$$\begin{aligned} [(\$100)(1.15)](1.15) &= \$100(1.15)^2 \\ &= \$132.25 \end{aligned}$$

On the other hand, if we increase the original \$100 by 30%, we obtain

$$\$100(1.30) = \$130$$

Comparing, we see that the result of two successive 15% increases is greater than the result of a single 30% increase.

Now let us look at another example and use it to introduce some terminology. Suppose that you place \$1000 in a savings account at 10% interest, compounded annually. This means that at the end of each year, the bank contributes to your account 10% of the amount then in it. Interest compounded in this manner is called *compound interest.* The original deposit of \$1000 is called the *principal* P. The interest rate, expressed as a decimal, is denoted by r. Thus $r = 0.10$ in this example. The variable A is used to denote the *amount* in the account at any given time. The calculations displayed in Table 1 show how the account grows.

Table 1

Time Period	Algebra	Arithmetic
After 1 year	$A = P(1 + r)$	$A = 1000(1.10)$ $= \$1100$
After 2 years	$A = [P(1 + r)](1 + r)$ $A = P(1 + r)^2$	$A = 1000(1.10)^2$ $= \$1210$
After 3 years	$A = [P(1 + r)^2](1 + r)$ $A = P(1 + r)^3$	$A = 1000(1.10)^3$ $= \$1331$

If you look over the algebra in Table 1, you can see what the general formula ought to be for the amount after t years.

$$A = P(1 + r)^t$$

where P is the principal, r is the interest rate, compounded annually, t is the number of years, and A is the amount after t years.

Example 2 Suppose that \$2000 is invested at $7\frac{1}{2}$% interest, compounded annually. How many years will it take for the money to double?

Solution In the formula $A = P(1 + r)^t$, we use the given values $P = \$2000$ and $r = 0.075$. We want to find how long it will take for the money to double; that is, we want to find t when $A = \$4000$. Making these substitutions in the formula, we obtain

$$4000 = 2000(1 + 0.075)^t$$

and therefore

$$2 = (1.075)^t \quad \text{dividing by 2000}$$

We can solve this exponential equation by taking the logarithm of both sides. We shall use base e logarithms. (Base 10 would be just as convenient here.) This yields

$$\ln 2 = \ln(1.075)^t$$

and consequently

$$\ln 2 = t \ln(1.075) \quad \text{(Why?)}$$

Now to isolate t, we divide both sides of this last equation by ln(1.075). This yields

$$t = \frac{\ln 2}{\ln(1.075)}$$

or (using a calculator)

$$t \approx 9.6 \text{ years}$$

$A = 2000(1.075)^t$	
t *(years)*	A *(dollars)*
9	3834.48
10	4122.06

Table 2

Now, assuming that the bank computes the compound interest only at the end of the year, we must round off the preliminary answer of 9.6 years and say that when $t = 10$ years, the initial \$2000 will have more than doubled. Table 2 adds some perspective to this. The table shows that after 9 years, something less than \$4000 is in the account, whereas after 10 years, the amount exceeds \$4000.

In Example 2 the interest was compounded annually. In practice though, the interest is usually computed more often. For instance, a bank may advertise a rate of 10% per year, compounded semiannually. This means that after one half year, the interest is compounded at 5%, and then after another half year, the interest is again compounded at 5%. Now if you review Example 1, you'll see that one compounding at the rate r is not the same as two compoundings, each at the rate $r/2$. The formula $A = P(1 + r)^t$ can be generalized to cover such cases where interest is compounded more than once each year.

> Suppose that a principal of P dollars is invested at an annual rate r, compounded n times per year. Then the amount A after t years is given by
>
> $$A = P\left(1 + \frac{r}{n}\right)^{nt}$$

Example 3 \$1000 is placed in a savings account at 10% per annum. How much is in the account at the end of one year if the interest is

(a) compounded once each year ($n = 1$)?
(b) compounded quarterly ($n = 4$)?

Solution We use the formula $A = P(1 + r/n)^{nt}$.

(a) For $n = 1$ we obtain

$$\begin{aligned} A &= 1000\left(1 + \frac{0.10}{1}\right)^{(1)(1)} \\ &= 1000(1.1) \\ &= \$1100 \end{aligned}$$

(b) For $n = 4$ we obtain

$$\begin{aligned} A &= 1000\left(1 + \frac{0.10}{4}\right)^{4(1)} \\ &= 1000(1.025)^4 \\ &= \$1103.81 \text{(using a calculator)} \end{aligned}$$

Notice here that compounding the interest quarterly rather than annually yields the greater amount. This is in agreement with our observations in Example 1.

The results in Example 3 will serve to illustrate some additional terminology used by financial institutions. In that example, the interest for the year, under quarterly compounding, was

$$\$1103.81 - \$1000 = \$103.81$$

Now \$103.81 is 10.381% of \$1000. We say in this case that the *effective rate* of interest is 10.381%. The given rate of 10% per annum, compounded once a year, is called the *nominal rate*. The next example further illustrates these ideas.

Example 4 A bank offers an interest rate of 12% per annum for certain accounts. (This is the nominal rate.) Compute the effective rate if interest is compounded monthly.

Solution Let P denote the principal earning 12% ($r = 0.12$), compounded monthly ($n = 12$). Then with $t = 1$ our formula yields

$$\begin{aligned} A &= P\left(1 + \frac{0.12}{12}\right)^{12(1)} \\ &= P(1.01)^{12} \end{aligned}$$

Now, using a calculator to approximate the quantity $(1.01)^{12}$, we obtain

$$A \approx P(1.12683)$$

This shows that the effective interest rate is about 12.68%.

There are two rather natural questions to ask when you first encounter compound interest calculations.

Question 1. For a fixed period of time, say one year, does more and more frequent compounding of interest continue to yield greater and greater amounts?

Question 2. Is there a limit on how much money can accumulate in a year when interest is compounded more and more frequently?

The answer to both of these questions is "yes." If you look back over Example 1, you'll see evidence for the yes answer to Question 1. For more evidence, and for Question 2, let's do some calculations. To keep things as simple as possible, suppose a principal of \$1.00 is invested for 1 year at the nominal rate of 100% per annum. (More realistic figures could be used here, but the algebra becomes more cluttered.) With this data, our formula becomes

$$A = 1\left(1 + \frac{1}{n}\right)^{n(1)}$$

that is,

$$A = \left(1 + \frac{1}{n}\right)^{n}$$

Table 3 shows the results of compounding the interest more and more frequently.

Number of Compoundings (n)	Amount $\left(1 + \frac{1}{n}\right)^n$
$n = 1$ (annually)	$\left(1 + \frac{1}{1}\right)^1 = 2$
$n = 2$ (semiannually)	$\left(1 + \frac{1}{2}\right)^2 = 2.25$
$n = 4$ (quarterly)	$\left(1 + \frac{1}{4}\right)^4 \approx 2.44$
$n = 12$ (monthly)	$\left(1 + \frac{1}{12}\right)^{12} \approx 2.61$
$n = 365$ (daily)	$\left(1 + \frac{1}{365}\right)^{365} \approx 2.7146$
$n = 8760$ (hourly)	$\left(1 + \frac{1}{8760}\right)^{8760} \approx 2.7181$
$n = 525{,}600$ (each minute)	$\left(1 + \frac{1}{525{,}600}\right)^{525{,}600} \approx 2.71827$
$n = 31{,}536{,}000$ (each second)	$\left(1 + \frac{1}{31{,}536{,}000}\right)^{31{,}536{,}000} \approx 2.71828$

Table 3

Table 3 shows, on the one hand, that the amount does increase with the number of compoundings. But on the other hand, assuming the bank rounds off to the nearest penny, Table 3 shows that there is no difference between compounding hourly, or each minute, or each second. In each case, the amount, when rounded off, is $2.72.

The data in Table 3 suggest that the quantity $(1 + 1/n)^n$ gets closer and closer to the number e as n becomes larger and larger. In symbols,

$$\left(1 + \frac{1}{n}\right)^n \approx e \qquad \text{when } n \text{ is large}$$

Indeed, in many calculus books, the number e is defined as the "limiting value" or "limit" of the quantity $(1 + 1/n)^n$ as n grows ever larger. Admittedly, we have not defined here the meaning of "limiting value" or "limit." That is a topic for calculus. Nevertheless, Table 3 should give you a reasonable, if intuitive, appreciation of the idea.

Some banks advertise interest compounded not monthly, daily, nor even hourly, but *continuously,* that is, at each instant. The formula that can be derived for the amount earned under continuous compounding of interest is as follows.

$$A = Pe^{rt}$$

where P is the principal, r is the annual interest rate (the nominal rate), t is the number of years, and A is the amount after t years.

Example 5 A sum of $100 is placed in a savings account at 5% per annum, compounded continuously. Assuming no subsequent withdrawals or deposits, when will the balance reach $150?

Solution Substitute the values $A = 150$, $P = 100$, and $r = 0.05$ in the formula $A = Pe^{rt}$ to obtain

$$150 = 100e^{0.05t}$$

and therefore

$$1.5 = e^{0.05t}$$

To solve this last equation for t, rewrite it in its equivalent exponential form:

$$0.05t = \ln 1.5$$

$$t = \frac{\ln 1.5}{0.05} \text{ years}$$

Using a calculator, we find that $t \approx 8.1$ years. In other words, it will take slightly more than 8 years and 1 month for the balance to reach \$150.

In the next example we compare the nominal rate with the effective rate under continuous compounding of interest.

Example 6 **(a)** Given a nominal rate of 8% per annum, compounded continuously, compute the effective interest rate.
(b) Given an effective rate of 8% per annum, compute the nominal rate.

Solution **(a)** With the values $r = 0.08$ and $t = 1$, the formula $A = Pe^{rt}$ yields

$$A = Pe^{0.08(1)}$$

$$A \approx P(1.08329) \qquad \text{(using a calculator)}$$

This shows that the effective interest rate is approximately 8.33% per year, as required.

(b) We now wish to compute the nominal rate r, given an effective rate of 8% per year. An effective rate of 8% means that the initial principal P grows to $P(1.08)$ by the end of the year. Thus, in the formula $A = Pe^{rt}$, we make the substitutions $A = P(1.08)$ and $t = 1$. This yields

$$P(1.08) = Pe^{r(1)}$$

Upon dividing both sides of this last equation by P, we have

$$1.08 = e^{r}$$

To solve this equation for r, just rewrite it in its equivalent logarithmic form:

$$r = \ln(1.08)$$

$$r \approx 0.07696 \qquad \text{(using a calculator)}$$

Thus a nominal rate of about 7.70% per annum yields an effective rate of 8%. This is the required answer for part (b). Table 4 summarizes these results.

Nominal Rate (per annum)	Effective Rate (per annum)
8%	8.33%
7.70%	8%

Table 4
Comparison of nominal and effective rates in Example 6

Now we come to one of the remarkable and characteristic features of growth governed by the formula $A = Pe^{rt}$. By the *doubling time* we mean, as the name implies, the amount of time required for a given principal to double. The surprising fact here is that the doubling time does not depend on the principal P. To see why this is so, we begin with the formula

$$A = Pe^{rt}$$

We are interested in the time t at which $A = 2P$. Replacing A by $2P$ in the formula yields

$$\begin{aligned} 2P &= Pe^{rt} \\ 2 &= e^{rt} \\ rt &= \ln 2 \\ t &= \frac{\ln 2}{r} \end{aligned}$$

Denoting the doubling time by T_2, we have shown

$$\text{Doubling time} = T_2 = \frac{\ln 2}{r}.$$

As you can see, the formula for the doubling time T_2 does not involve P, but only r. Thus, for instance, at a given rate under continuous compounding, $2 and $2000 would both take the same amount of time to double. (This idea takes some getting used to.)

Example 7 Compute the doubling time T_2 when a sum is invested at an interest rate of 4% per annum, compounded continuously.

Solution

$$T_2 = \frac{\ln 2}{r} = \frac{\ln 2}{0.04} \approx 17.3 \text{ years} \qquad \text{(using a calculator)}$$

There is a convenient approximation that will allow you to easily estimate doubling times. Using a calculator, you'll see that

$$\ln 2 \approx 0.7$$

Using this approximation, we have the following rule of thumb for estimating doubling times.

$$T_2 \approx \frac{0.7}{r}$$

Let's use this rule to rework Example 7. With $r = 0.04$, we obtain

$$T_2 \approx \frac{0.7}{0.04} = \frac{70}{4} = 17.5 \text{ years}$$

Notice that this estimation is quite close to the actual doubling time obtained in Example 7.

The idea of a doubling time turns out to be useful in graphing the function $A = Pe^{rt}$. In this discussion, we assume that P and r are constants, so that the amount A is a function of the time t. Suppose, for example, that a principal of \$1000 is invested at 10% per annum, compounded continuously. Then the function we wish to graph is

$$A = 1000e^{0.1t}$$

Now the doubling time in this situation is

$$T_2 \approx \frac{0.7}{r} = \frac{0.7}{0.1} = 7 \text{ years}$$

Table 5 shows the results of doubling a principal of \$1000 every 7 years.

Table 5

t (years)	0	7	14	21	28
A (dollars)	1000	2000	4000	8000	16000

In view of the data in Table 5, we'll mark off units on the t-axis in multiples of 7; on the A-axis we'll use multiples of 2000. Figure 1 shows the result of plotting the points in Table 5 and then joining them with a smooth curve. Notice that the domain in this context is $[0, \infty)$.

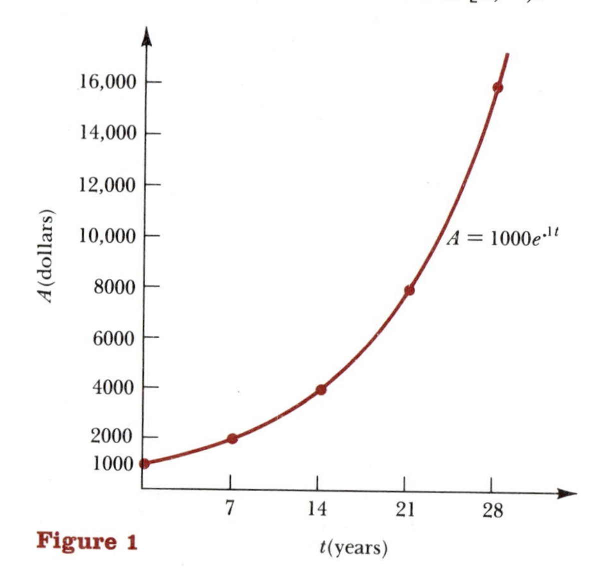

Figure 1

You may have already noticed that aside from the arbitrary choice of letters, there is no difference between the functions $A = Pe^{rt}$ and $N = N_0e^{kt}$. Both formulas have the same form. Thus this single function serves as a model for phenomena as diverse as continuous compounding of interest, population growth, and radioactive decay.

In newspapers and in everyday speech, the term "exponential growth" is used rather loosely to describe any situation involving rapid growth. In the sciences, however, the term *exponential growth* refers specifically to growth governed by functions of the form $N = N_0e^{kt}$, with $k > 0$. So, for example, since the function $A = Pe^{rt}$ has this form, we say that money grows exponentially under continuous compounding of interest. Similarly, in the sciences, *exponential decay* refers specifically to decay governed by functions of the form $N = N_0e^{kt}$, with $k < 0$.

In the next example we predict the population of the world in the year 2000, assuming that the population grows exponentially. In this context, it can be shown that the constant k in the formula $N = N_0e^{kt}$ represents the *relative growth rate** of the population N. Algebraically, we'll work with k just as we did in the corresponding calculations with compound interest, where k represented an interest rate.

Example 8 Statistics from the United Nations show that for the last several decades the relative growth rate for the world's population has been 1.9% per year. Assuming that this rate remains at 1.9%, use the formula $N = N_0e^{kt}$ to predict the world population in the year 2000, given that the population in 1975 was 4.09 billion.

Solution Let $t = 0$ correspond to the year 1975. Then the year 2000 corresponds to the value $t = 25$. Our given data therefore is

$$k = 0.019$$
$$t = 25$$
$$N_0 = 4.09 \qquad \text{(in units of 1 billion)}$$

Using these values in the formula $N = N_0e^{kt}$, we obtain

$$N = 4.09e^{(0.019)25}$$
$$N \approx 6.6 \qquad \text{using a calculator}$$

Thus our prediction for the world's population in the year 2000 is 6.6 billion.

In previous sections we used the function $N = N_0e^{kt}$, with $k < 0$, to describe radioactive decay. Let's return to that idea now. Just as we used the idea of a doubling time to graph an exponential growth function, we can use the half-life concept, introduced in Section 6.1, to graph an exponential decay function. Consider, for example, the radioactive element iodine-131, which has a half-life of about 8 days. Table 6 shows what fraction of an initial amount remains at 8-day intervals. Using the data in this table, we can draw the graph for the decay function $N = N_0e^{kt}$ for iodine-131. See Figure 2 on the next page. Note that we were able to construct this graph without specifically evaluating the decay constant k.

t (days)	0	8	16	24	32
N (amount)	N_0	$\frac{1}{2}N_0$	$\frac{1}{4}N_0$	$\frac{1}{8}N_0$	$\frac{1}{16}N_0$

Table 6

*The relative growth rate is $(\Delta N/N)/(\Delta t)$; it is the fractional change in N per unit time. This is distinct from (but sometimes confused with) the actual growth rate $\Delta N/\Delta t$.

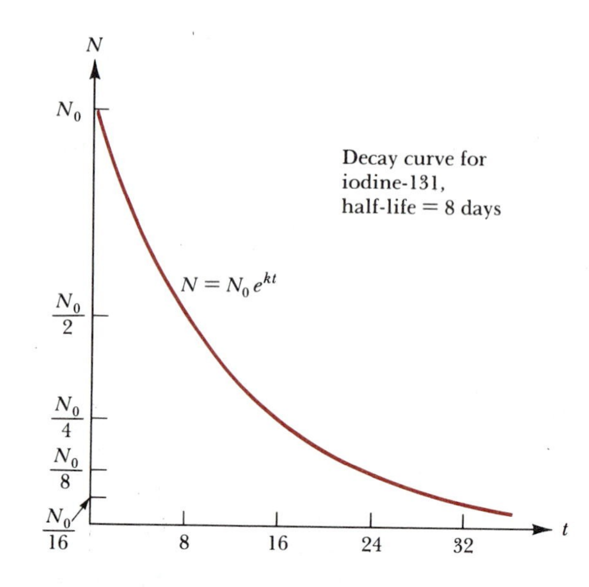

Figure 2

The next example touches on the subject of nuclear energy. The point of the example is not to present an argument for or against the use of nuclear power. Rather, the point is to show you that with an understanding of exponential decay, you will be better equipped to read about and to evaluate the issues on your own.

Example 9 An article on nuclear energy appeared in the January 1976 issue of *Scientific American* magazine. The author of the article was Hans Bethe, a Nobel prize-winning physicist. At one point in the article, Professor Bethe is discussing the disposal (through burial) of radioactive waste material from a nuclear reactor. The particular waste product under discussion is plutonium-239.

> . . . Plutonium-239 has a half-life of nearly 25,000 years, and 10 half-lives are required to cut the radioactivity by a factor of 1000. Thus the buried wastes must be kept out of the biosphere for 250,000 years.

(a) Supply the detailed calculations behind the statement that 10 half-lives are required before the radioactivity is reduced by a factor of 1000.
(b) Show how the figure of 10 half-lives can be obtained by estimation, as opposed to detailed calculation.

Solution **(a)** Let N_0 denote the initial amount of plutonium-239 at time $t = 0$. Then the amount N present at time t is given by $N = N_0e^{kt}$. We wish to determine t when $N = \frac{1}{1000}N_0$. First, we determine the value of k by using the half-life information. To say that the half-life is 25,000 years is to say that $N = \frac{1}{2}N_0$ when $t = 25{,}000$. Substituting these values into the formula $N = N_0e^{kt}$ yields

$$\frac{1}{2}N_0 = N_0e^{k(25{,}000)}$$

$$\frac{1}{2} = e^{25{,}000k} \quad \text{dividing by } N_0$$

By writing this last equation in its equivalent logarithmic form we obtain

$$25{,}000k = \ln\frac{1}{2}$$

and therefore

$$k = \frac{\ln \frac{1}{2}}{25{,}000}$$

Now we can find t when $N = \frac{1}{1000} N_0$, as required. Substituting $\frac{1}{1000} N_0$ for N in the decay law gives us

$$\frac{1}{1000} N_0 = N_0 e^{kt} \qquad \text{where } k \text{ is } \frac{\ln \frac{1}{2}}{25{,}000}$$

$$\frac{1}{1000} = e^{kt}$$

$$kt = \ln \frac{1}{1000}$$

$$t = \frac{\ln \frac{1}{1000}}{k} = \frac{\ln \frac{1}{1000}}{(\ln \frac{1}{2})/25{,}000}$$

$$t = \frac{(\ln \frac{1}{1000})(25{,}000)}{\ln \frac{1}{2}}$$

Using a calculator here, we obtain the value 249,144.6 years; however, given the time scale involved, it would be ludicrous to announce the answer in this form. Instead, we round off to the nearest thousand years and we say that after 249,000 years, the radioactivity will have decreased by a factor of 1000. Notice now that this result confirms Professor Bethe's ballpark estimate of 10 half-lives or 250,000 years.

(b) After one half-life: $N = \frac{N_0}{2}$

After two half-lives: $N = \frac{1}{2}\left(\frac{N_0}{2}\right) = \frac{N_0}{2^2}$

After three half-lives: $N = \frac{1}{2}\left(\frac{N_0}{2^2}\right) = \frac{N_0}{2^3}$

Following this pattern, we see that after 10 half-lives we should have

$$N = \frac{N_0}{2^{10}}$$

But as we noted in the first section of this chapter, 2^{10} is approximately 1000. Therefore we have

$$N \approx \frac{N_0}{1000} \qquad \text{after 10 half-lives}$$

This is in agreement with Professor Bethe's statement.

Exercise Set 6.4

1. You invest \$800 at 6% interest, compounded annually. How much is in the account after 4 years, assuming you make no subsequent withdrawals or deposits?

2. A sum of \$1000 is invested at an interest rate of $5\frac{1}{2}\%$, compounded annually. How many years will it take before the sum exceeds \$2500? (First find out when the amount equals \$2500, then round off as in Example 2.)

3. At what interest rate, compounded annually, will a sum of $4000 grow to $6000 in 5 years?

4. A bank pays 7% interest, compounded annually. What principal will grow to $10,000 in 10 years?

5. You place $500 in a savings account at 5%, compounded annually. After 4 years you withdraw all of your money and take it to a different bank, which advertises a rate of 6%, compounded annually. What is the balance in this new account after 4 more years? (As usual, assume that no subsequent withdrawals or deposits are made.)

6. A sum of $3000 is placed in a savings account at 6% per annum. How much is in the account after 1 year if the interest is
   (a) compounded annually?
   (b) compounded semiannually?
   (c) compounded daily?

7. A sum of $1000 is placed in a savings account at 7% per annum. How much is in the account after 20 years if the interest is
   (a) compounded annually?
   (b) compounded quarterly?

8. Your friend invests $2000 at $5\frac{1}{4}$% per annum, compounded semiannually. You invest an equal amount at the same yearly rate, but compounded daily. How much larger is your account than your friend's after 8 years?

9. You invest $100 at 6% per annum, compounded quarterly. How long will it take for your balance to exceed $120? (Round off your answer to the nearest next quarter.)

10. A bank offers an interest rate of 7% per annum, compounded daily. What is the effective rate?

11. What principal should you deposit at $5\frac{1}{2}$% per annum, compounded semiannually, so as to have $6000 after 10 years?

12. You place a sum of $800 in a savings account at 6% per annum, compounded continuously. Assuming you make no subsequent withdrawals or deposits, how much is in the account after 1 year? Also, when will the balance reach $1000?

13. A bank offers an interest rate of $6\frac{1}{2}$% per annum, compounded continuously. What principal will grow to $5000 in 10 years under these conditions?

14. Given a nominal rate of 6% per annum, compute the effective rate under continuous compounding of interest.

15. Suppose that under continuous compounding of interest, the effective rate is 6% per annum. Compute the nominal rate.

16. You have two savings accounts, each with an initial principal of $1000. The nominal rate on both accounts is $5\frac{1}{4}$% per annum. In the first account, interest is compounded semiannually. In the second, interest is compounded continuously. How much more is in the second account after 12 years?

17. You want to invest $10,000 for 5 years and you have a choice between two accounts. The first pays 6% per annum, compounded annually. The second pays 5% per annum, compounded continuously. Which is the better investment for you under these conditions?

18. Suppose that a certain principal is invested at 6% per annum, compounded continuously.
   (a) Use the rule $T_2 \approx 0.7/r$ to estimate the doubling time.
   (b) Compute the doubling time using the formula $T_2 = (\ln 2)/r$.
   (c) Do your answers in (a) and (b) differ by more than two months?

19. $1500 is invested at 5% per annum, compounded continuously.
   (a) Estimate the doubling time.
   (b) Compute the actual doubling time.
   (c) Let d_1 and d_2 denote the actual and the estimated doubling times, respectively. Define d by
   $$d = |d_1 - d_2|$$
   What percent is d of the actual doubling time?

20. $5000 is invested at 10% per annum, compounded continuously.
   (a) Estimate the doubling time.
   (b) Estimate the time required for the $5000 to grow to $40,000.

21. After carrying out the calculations in this problem, you'll see one of the reasons why the government imposes inheritance taxes and why there are laws that prohibit savings accounts from being passed from generation to generation, indefinitely. Suppose that a family invests $1000 at 8% per annum, compounded continuously. If this account were to remain intact, being passed from generation to generation for 300 years, how much would be in the account at the end of those 300 years?

22. A principal of $500 is invested at 7% per annum, compounded continuously.
   (a) Estimate the doubling time.
   (b) Sketch a graph, similar to the one in Figure 1, showing how the amount increases with time.

23. A principal of $7000 is invested at 5% per annum, compounded continuously.
   (a) Estimate the doubling time.
   (b) Sketch a graph showing how the amount increases with time.

24. In one savings account, a principal of $1000 is deposited at 5% per annum. In a second account, a principal of $500 is deposited at 10% per annum. Both accounts compound interest continuously.
   (a) Estimate the doubling time for each account.

(b) On the same set of axes, sketch graphs showing the amount of money in each account over time. Give the (approximate) coordinates of the point where the two curves meet. In financial terms, what is the significance of this point? (In working this problem, assume that the initial deposits in each account were made at the same time.)

(c) During what period of time does the first account have the larger balance?

The statistics presented in Exercises 25–27 are taken from the Global 2000 Report to the President. *The following statement by then President Jimmy Carter provides some background on the report.*

I am directing the Council on Environmental Quality and the Department of State, working in cooperation with . . . other appropriate agencies, to make a one-year study of the probable changes in the world's population, natural resources, and environment through the end of the century.

President Jimmy Carter
May 23, 1977

25. Complete the following table on world population projections, assuming that the populations grow exponentially and that the indicated relative growth rates are valid through the year 2000.

Region	1975 Population (billions)	Percent of Population in 1975	Relative Growth Rate (percent per year)	Year 2000 Population (billions)	Percent of World Population in 2000
World	4.090	100	1.8	?	?
More Developed Regions	1.131	?	0.6	?	?
Less Developed Regions	2.959	?	2.1	?	?

26. Complete the following table. Assume that the growth is exponential and that the given relative growth rates are in effect through the year 2000.

Country	1975 Population (billions)	Relative Growth Rate (percent per year)	Year 2000 Population (billions)	Percent Increase in Population
United States	0.214	0.6	?	?
People's Republic of China	0.935	1.4	?	?
Mexico	0.060	3.1	?	?

27. *The Global 2000 Report* actually contains three sets of world population estimates and projections, as shown in the table below. Complete the table.

Projection	1975 Population (billions)	Relative Growth Rate (percent per year)	Year 2000 Population (billions)	Percent Increase in Population
Low	4.043	1.5	?	?
Medium	4.090	1.8	?	?
High	4.134	2.0	?	?

28. According to the United States Bureau of the Census, the population of the United States grew most rapidly during the period 1800–1810, and least rapidly during the period 1930–1940. Use the following data to compute the relative growth rate (percent per year) for each of these two periods. In 1800, the population was 5,308,483; in 1810, the population was 7,239,881. In 1930, the population was 123,202,624; in 1940 it was 132,164,569.

29. According to the United States Census Bureau, the population of the United States in the year 1850 was 23,191,876; in 1900 the population was 62,947,714.

(a) Assuming that the population grew exponentially during this period, compute the growth constant k.

(b) Assuming continued growth at the same rate, predict the 1950 population.

(c) The actual population for 1950, according to the Census Bureau, was 150,697,361. How does this compare to the prediction in part (b)? Was the actual growth over that period (1900–1950) faster or slower than exponential growth with a fixed growth constant k?

In Exercises 30–32, make use of the data regarding populations from the United States Census contained in the following table.

State	1930	1940	1950
California	5,677,251	6,907,387	10,586,223
New York	12,588,066	13,479,142	14,830,192
North Dakota	680,845	641,935	619,636

30. **(a)** Assume that the population of California grew exponentially over the period 1930–1940. Compute the growth constant k and express it as a percent (per year).
(b) Assume that the population of California grew exponentially over the period 1940–1950. Compute the growth constant k and express it as a percent.
(c) What would the 1980 California population have been if the relative growth rate obtained in part (b) had remained in effect throughout the period 1950–1980? *Hint:* Let $t = 0$ correspond to 1950. Then you want to find N when $t = 30$.
(d) How does your prediction in part (c) compare to the actual 1980 population of 23,668,562?

31. Repeat Exercise 30 for New York. [For part (d): the 1980 population of New York was 17,557,288.]

32. Repeat Exercise 30 for North Dakota. [For part (d): the 1980 population of North Dakota was 652,695.]

33. **(a)** The half-life of radium-226 is 1620 years. Draw a graph of the decay function for radium-226, similar to that shown in Figure 2.
(b) The half-life of radium A is 3 minutes. Draw a graph of the decay function for radium A.

34. **(a)** The half-life of thorium-232 is 1.4×10^{10} years. Draw a graph of the decay function.
(b) The half-life of thorium A is 0.16 seconds. Draw a graph of the decay function.

35. The half-life of plutonium-241 is 13 years.
(a) Compute the decay constant k.
(b) What fraction of an initial sample N_0 would remain after 10 years? after 100 years?

36. The half-life of thorium-229 is 7340 years.
(a) Compute the time required for a given sample to be reduced by a factor of 1000. Show detailed calculations as in Example 9(a).
(b) Express your answer in part (a) in terms of half-lives.
(c) As in Example 9(b), estimate the time required for a given sample of thorium-229 to be reduced by a factor of 1000. Compare your answer with that obtained in part (b).

37. Strontium-90, with a half-life of 28 years, is one of the waste products from nuclear fission reactors. One of the reasons great care is taken in the storage and disposal of this substance stems from the fact that strontium-90 is, in some chemical respects, similar to ordinary calcium. Thus strontium-90 in the biosphere, entering the food chain via plants or animals, would eventually be absorbed into our bones. (In fact, everyone already has a measurable amount of strontium-90 in their bones due to fallout from atmospheric nuclear tests.)
(a) Compute the decay constant k for strontium-90.
(b) Compute the time required if a given quantity of strontium-90 is to be stored until the radioactivity is reduced by a factor of 1000.
(c) Using half-lives, estimate the time required for a given sample to be reduced by a factor of 1000. Compare your answer with that in (b).

38. **(a)** Suppose that country X violates the ban against above-ground nuclear testing, and as a result an island is contaminated with debris containing the radioactive substance iodine-131. A team of scientists from the United Nations wants to visit the island to look for clues in determining which country was involved. However, the level of radioactivity from the iodine-131 is estimated to be 30,000 times the safe level. Approximately how long must the team wait before it is safe to visit the island? The half-life of iodine-131 is 8 days.
(b) Rework part (a), assuming instead that the radioactive substance is strontium-90 rather than iodine-131. The half-life of strontium-90 is 28 years. Assume as before that the initial level of radioactivity is 30,000 times the safe level. (The purpose of this problem is to underscore the difference between a half-life of 8 days and one of 28 years.)

39. In 1969, the United States National Academy of Sciences issued a report entitled *Resources and Man.* One conclusion in the report is that a world population of 10 billion "is close to (if not above) the maximum that an intensively managed world might hope to support with some degree of comfort and individual choice." (The figure "10 billion" is sometimes referred to as the "carrying capacity" of the earth.)
(a) When the report was issued in 1969, the world population was about 3.6 billion, with a relative growth rate of 2% per year. Assuming continued exponential growth at this rate, estimate the year in which the earth's carrying capacity of 10 billion might be reached.
(b) The data in the table at the top of the next page, from the *Global 2000 Report,* show the world population estimates for 1975. Use the "low" figures to estimate the year in which the earth's carrying capacity of 10 billion might be reached.
(c) Use the "high" figures to estimate the year in which the earth's carrying capacity of 10 billion might be reached.

Projection	Population (billions)	Relative Growth Rate (percent per year)
Low	4.043	1.5
High	4.134	2.0

40. *Depletion of Nonrenewable Energy Resources* Suppose that the world population grows exponentially. Then, as a first approximation, it is reasonable to assume that the use of energy resources such as petroleum and coal also grows exponentially. Under these conditions, the following formula can be derived (using calculus):

$$A = \frac{A_0}{k}\left(e^{kT} - 1\right)$$

where A is the amount of the resource consumed from time $t = 0$ to $t = T$, A_0 is the amount of the resource consumed during the year $t = 0$, and k is the relative growth rate of annual consumption.

(a) Show that the result of solving the given formula for T is

$$T = \frac{\ln[(Ak/A_0) + 1)]}{k}$$

This formula gives the "life expectancy" T for a given resource. In the formula, A_0 and k are as defined above and A represents the total amount of the resource available.

(b) In the *Global 2000 Report to the President,* it was estimated that the 1976 worldwide consumption of oil was 21.7 billion barrels and that the total remaining oil resources ultimately available were 1661 billion barrels. Compute the "life expectancy" for oil

(i) if the relative growth rate k is 1% per year.

(ii) if $k = 2\%$ per year.

(iii) if $k = 3\%$ per year. (This is, in fact, the predicted rate for the next decade.)

(c) In the *Global 2000 Report,* it was estimated that the 1976 worldwide consumption of natural gas was approximately 50 tcf (trillions of cubic feet), while the total remaining gas ultimately available was 8493 tcf. Compute the "life expectancy" for natural gas if $k = 2\%$ per year.

41. In this exercise, the term "nonrenewable energy resources" refers collectively to petroleum, natural gas, coal, shale oil, and uranium. In the *Global 2000 Report,* it was estimated that in 1976, the total worldwide consumption of these nonrenewable resources amounted to 250 quadrillion Btu. It was also estimated that in 1976, the total nonrenewable resources ultimately available were 161,241 quadrillion Btu. Use the formula for T in Exercise 40 to compute the "life expectancy" for these nonrenewable energy resources

(a) if $k = 1\%$ per year.

(b) if $k = 2\%$ per year.

42. The data in the table, from the *Global 2000 Report,* show the world reserves for selected minerals. The term "world reserves" refers to known resources available with current technology. Use the formula for T derived in Exercise 40 to estimate the "depletion date" for each mineral under these conditions. (Depletion date = T + 1976.)

Mineral	1976 World Reserves	1976 Consumption	Relative Growth Rate (percent per year)
Fluorine (million short tons)	37	2.1	4.58
Silver (million troy ounces)	6,100	305	2.33
Tin (thousand metric tons)	10,000	241	2.05
Copper (million short tons)	503	8.0	2.94
Phosphate rock (million metric tons)	25,732	107	5.17
Aluminum in bauxite (million short tons)	5,610	18	4.29

43. In 1956, the scientists Gutenberg and Richter developed a formula that can be used to estimate the amount of energy E released in an earthquake. The formula is

$$\log_{10} E = 11.4 + (1.5)M$$

where E is the energy in units of ergs and M is the Richter magnitude.

(a) Solve the formula for E.

(b) The table shows the Richter magnitudes for two large earthquakes that occurred in 1976. If E_1 denotes the energy of the quake in Italy and E_2 denotes the energy of the quake in China, compute the ratio E_2/E_1 to compare the energies of the two earthquakes.

Date	Location	M (magnitude)
July 27, 1976	China	7.6
May 6, 1976	Italy	6.5

44. The age of some rocks can be estimated by measuring the ratio of the amounts of certain chemical elements within the rock. One method known as the *rubidium-strontium method* will be discussed here. This method has been used in dating the moon rocks brought back on the Apollo missions.

Rubidium-87 is a radioactive substance with a half-life of 4.7×10^{10} years. Rubidium-87 decays into strontium-87, which is stable (nonradioactive). We are going to derive the following formula for the age of a rock:

$$T = \frac{\ln(N_s/N_r + 1)}{-k}$$

where T is the age of the rock, k is the decay constant for rubidium-87, N_s is the number of atoms of strontium-87 now present in the rock, and N_r is the number of atoms of rubidium-87 now present in the rock.

(a) Assume that initially, when the rock was formed, there were N_0 atoms of rubidium-87 and none of strontium-87. Then as time goes by, some of the rubidium atoms decay into strontium atoms, but the total number of atoms must still be N_0. Thus after T years we have

$$N_0 = N_r + N_s$$

or, equivalently,

$$N_s = N_0 - N_r \qquad (1)$$

However, according to the law of exponential decay for the rubidium-87, we must have $N_r = N_0e^{kT}$. Solve this equation for N_0 and then use the result to eliminate N_0 from equation (1). Show that the result can be written

$$N_s = N_re^{-kT} - N_r \qquad (2)$$

(b) Solve equation (2) for T to obtain the formula given at the beginning of this exercise.

45. (Continuation of Exercise 44.) The half-life of rubidium-87 is 4.7×10^{10} years. Compute the decay constant k.

46. (Continuation of Exercise 44.) Analysis of lunar rock samples taken on the Apollo 11 mission showed the strontium-rubidium ratio to be

$$\frac{N_s}{N_r} = 0.0588$$

Estimate the age of these lunar rocks.

47. (Continuation of Exercise 44.) Analysis of the so-called genesis rock sample taken on the Apollo 15 mission revealed a strontium-rubidium ratio of 0.0636. Estimate the age of this rock.

48. *Radiocarbon Dating* Because rubidium-87 decays so slowly, the technique of rubidium-strontium dating discussed in the previous exercises is generally considered effective only for objects older than 10 million years. By way of contrast, archeologists and geologists rely on the method of *radiocarbon dating* in assigning ages ranging from 500 to 50,000 years.

There are two types of carbon that occur naturally in our environment. The first is carbon-12, which is nonradioactive; the second is carbon-14, which has a half-life of 5730 years. All living plant and animal tissue contains both types of carbon, always in the same ratio. (The ratio is one part carbon-14 to 10^{12} parts carbon-12.) As long as the plant or animal is living, this ratio is maintained. When the organism dies, however, no new carbon-14 is absorbed and the amount of carbon-14 begins to decrease exponentially. Since the amount of carbon-14 decreases exponentially, it follows that the level of radioactivity also must decrease exponentially. The formula describing this situation is

$$N = N_0e^{kT}$$

where T is the age of the sample, N is the present level of radioactivity (in units of disintegrations per hour per gram of carbon), and N_0 is the level of radioactivity T years ago, when the organism was alive. Given that the half-life of carbon-14 is 5730 years and that $N_0 = 920$ disintegrations per hour per gram, show that the age T of a sample is given by

$$T = \frac{5730 \ln (N/920)}{\ln \frac{1}{2}}$$

In Exercises 49–52, use the formula derived in Exercise 48 to estimate the age of each sample. Note: *There are some technical complications that arise in interpreting such results. Studies have shown that the ratio of carbon-12 to carbon-14 in the air, and therefore in living matter, has not in fact been constant over time. For instance, air pollution from factory smokestacks tends to increase the level of carbon-12. In the other direction, nuclear bomb testing increases the level of carbon-14.*

49. Prehistoric cave paintings were discovered in the Lascaux cave in France. Charcoal from the site was analyzed and the level of radioactivity was found to be $N = 141$ disintegrations per hour per gram. Estimate the age of the paintings.

50. Before radiocarbon dating was used, historians estimated that the tomb of Vizier Hemaka, in Egypt, was constructed about 4900 years ago. After radiocarbon dating became available, wood samples from the tomb were analyzed and it was determined that the radioactivity level was 510 disintegrations per hour per gram. Estimate the age of the tomb on the basis of this reading and compare your answer to the figure mentioned above.

51. Analyses of the oldest campsites of ancient man in the Western Hemisphere reveal a carbon-14 radioactivity level of $N = 226$ disintegrations per hour per gram. Show that this implies an age of 11,500 years, to the nearest 500 years. (This would correspond to the last Ice Age. At that time, the sea level was significantly lower than today. It is believed that at the time of the last Ice Age there was land across the Bering Strait, and that man first entered the Western Hemisphere by this route.)

52. The Dead Sea Scrolls are a collection of ancient manuscripts discovered in caves along the west bank of the Dead Sea. (The discovery occurred by accident when an Arab herdsman of the Taamireh tribe was searching for a stray goat.) When the linen wrappings on the scrolls were analyzed, the carbon-14 radioactivity level was found to be 723 disintegrations per hour per gram. Estimate the age of the scrolls using this information. Historical evidence suggests that some of the scrolls date back somewhere between 150 B.C. and A.D. 40. How do these dates compare with the estimate using radiocarbon dating?

Chapter 6 Review Checklist

- □ Exponential function with base b (pages 265–266)
- □ e (pages 268–269)
- □ $N = N_0e^{kt}$, $k > 0$ (pages 270, 298)
- □ $N = N_0e^{kt}$, $k < 0$ (pages 271, 299)
- □ Half-life (page 271)
- □ $\log_b x$ (page 275)
- □ Logarithmic function (page 276)
- □ $\ln x$ (page 278)
- □ Properties of logarithms (pages 283, 285)
- □ Change of base formula (page 288)
- □ Compound interest, principal, and amount (page 292)
- □ Formulas involving compound interest:
  - $A = P(1 + r)^t$ (page 298)
  - $A = P(1 + r/n)^{nt}$ (page 293)
  - $A = Pe^{rt}$ (page 295)
- □ Doubling time (page 297)
- □ Exponential growth and decay (pages 298–299)
- □ Relative growth rate (page 299)

Chapter 6 Review Exercises

In Exercises 1–12, graph each function and specify any asymptotes or intercepts.

1. $y = e^x$ **2.** $y = -e^{-x}$

3. $y = \ln x$ **4.** $y = \ln(x + 2)$

5. $y = 2^{x+1} + 1$ **6.** $y = \log_{10}(-x)$

7. $y = (1/e)^x$ **8.** $y = (\frac{1}{2})^{-x}$

9. $y = e^{x+1} + 1$ **10.** $y = -\log_2(x + 1)$

11. $y = \ln(e^x)$ **12.** $y = e^{\ln x}$

For Exercises 13–29, solve each equation for x. *(When logarithms appear in your answer, leave the answer in that form, rather than use a calculator.)*

13. $\log_4 x + \log_4(x - 3) = 1$

14. $\log_3 x + \log_3(2x + 5) = 1$

15. $\ln x + \ln(x + 2) = \ln 15$ **16.** $\log_6 \dfrac{x + 4}{x - 1} = 1$

17. $\log_2 x + \log_2(3x + 10) - 3 = 0$

18. $2 \ln x - 1 = 0$ **19.** $3 \log_9 x = \frac{1}{2}$

20. $e^{2x} = 6$ **21.** $e^{1-5x} = \sqrt{e}$

22. $2^x = 100$

23. $\log_{10} x - 2 = \log_{10}(x - 2)$

24. $\log_{10}(x^2 - x - 10) = 1$

25. $\ln(x + 2) = \ln x + \ln 2$

26. $\ln(2x) = \ln 2 + \ln x$ **27.** $\ln (x^4) = 4 \ln x$

28. $(\ln x)^4 = 4 \ln x$ **29.** $\log_{10} x = \ln x$

30. Solve for x: $(\ln x)/(\ln 3) = \ln x - \ln 3$. Use a calculator to evaluate your result; round off to the nearest integer. *Answer:* 206765

In Exercises 31–46, simplify each expression (without using a calculator).

31. $\log_{10} \sqrt{10}$ **32.** $\log_7 1$

33. $\ln \sqrt[5]{e}$ **34.** $\log_3 54 - \log_3 2$

35. $\log_{10} \pi - \log_{10} 10\pi$ **36.** $\log_2 2$

37. 10^t where $t = \log_{10} 16$

38. $e^{\ln 5}$ **39.** $\ln(e^4)$

40. $\log_{10}(10^{\sqrt{2}})$ **41.** $\log_{12} 2 + \log_{12} 18 + \log_{12} 4$

42. $(\log_{10} 8)/(\log_{10} 2)$ **43.** $(\ln 100)/(\ln 10)$

44. $\log_5 2 + \log_{1/5} 2$ **45.** $\log_2 \sqrt[7]{16\sqrt[3]{2\sqrt{2}}}$

46. $\log_{1/8}(\frac{1}{16}) + \log_5(0.02)$

In Exercises 47–50, express the given quantities in terms of a, b, and c, *where* a = $\log_{10}$ A, b = $\log_{10}$ B, and c = $\log_{10}$ C.

47. $\log_{10} A^2 B^3 \sqrt{C}$ **48.** $\log_{10} \sqrt[3]{AC/B}$

49. $16 \log_{10} \sqrt{A}\,\sqrt[4]{B}$ **50.** $6 \log_{10} [B^{1/3}/(A\sqrt{C})]$

For Exercises 51–56, find consecutive integers n *and* n + 1 *such that the given expression lies between* n *and* n + 1. *Do not use a calculator.*

51. $\log_{10} 209$ **52.** $\ln 2$ **53.** $\log_6 100$

54. $\log_{10}(\frac{1}{12})$ **55.** $\log_{10}(0.003)$ **56.** $\log_3 244$

57. **(a)** In which quadrant do the graphs of the two curves $y = \ln(x + 2)$ and $y = \ln(-x) - 1$ intersect?
(b) Find the x-coordinate of the intersection point.

58. The curve $y = ae^{bt}$ passes through the point (3, 4) and has a y-intercept of 2. Find a and b.

59. A certain radioactive substance has a half-life of T years. Find the decay constant (in terms of T).

60. Find the half-life of a certain radioactive substance if it takes T years for one third of a given sample to disintegrate. (Your answer will be in terms of T.)

61. A radioactive substance has a half-life of M minutes. What percentage of a given sample will remain after $4M$ minutes?

62. At the start of an experiment there are a bacteria in a colony. After b hours there are c bacteria. Determine the population d hours after the start of the experiment.

63. The half-life of a radioactive substance is d days. If you begin with a sample weighing b grams, how long will it be until c grams remain?

64. Find the half-life of a radioactive substance if it takes D days for P percent of a given sample to disintegrate.

In Exercises 65–70, write the given expression as a single logarithm with a coefficient of 1.

65. $\log_{10} 8 + \log_{10} 3 - \log_{10} 12$

66. $4 \log_{10} x - 2 \log_{10} y$

67. $\ln 5 - 3 \ln 2 + \ln 16$

68. $\ln(x^4 - 1) - \ln(x^2 - 1)$

69. $a \ln x + b \ln y$

70. $\ln(x^3 + 8) - \ln(x + 2) - 2 \ln(x^2 - 2x + 4)$

For Exercises 71–78, write each quantity using sums and differences of simpler logarithmic expressions. Express each answer so that logarithms of products, quotients, and powers do not appear.

71. $\ln \sqrt{(x - 3)(x + 4)}$

72. $\log_{10} \dfrac{x^2 - 4}{x + 3}$

73. $\log_{10} \dfrac{x^3}{\sqrt{1 + x}}$

74. $\ln \dfrac{x^2\sqrt{2x + 1}}{2x - 1}$

75. $\log_{10} \sqrt[3]{x/100}$

76. $\log_{10} \sqrt{\dfrac{x^2 + 8}{(x^2 + 9)(x + 1)}}$

77. $\ln\left(\dfrac{1 + 2e}{1 - 2e}\right)^3$

78. $\log_{10} \sqrt[3]{x\sqrt{1 + y^2}}$

79. Suppose that A dollars are invested at R%, compounded annually. How many years will it take for the money to double?

80. \$2800 is placed in a savings account at 9% per annum. How much is in the account after two years if the interest is compounded quarterly?

81. A bank offers an interest rate of 9.5% per annum, compounded monthly. Compute the effective interest rate.

82. A sum of D dollars is placed in an account at R% per annum, compounded continuously. When will the balance reach E dollars?

83. Compute the doubling time for a sum of D dollars invested at R% per annum, compounded continuously.

84. Your friend invests D dollars at R% per annum, compounded semiannually. You invest an equal amount at the same yearly rate, but compounded daily. How much larger is your account than your friend's after T years?

85. You invest D dollars at R% per annum, compounded quarterly. How long will it take for your balance to reach nD dollars? (Assume that $n > 1$.)

In Exercises 86–90, find the domain of each function.

86. **(a)** $y = \ln x$
(b) $y = e^x$

87. **(a)** $y = \log_{10} \sqrt{x}$
(b) $y = \sqrt{\log_{10} x}$

88. **(a)** $f \circ g$ where $f(x) = \ln x$ and $g(x) = e^x$
(b) $g \circ f$ where f and g are as in part (a)

89. **(a)** $y = \log_{10}|x^2 - 2x - 15|$
(b) $y = \log_{10}(x^2 - 2x - 15)$

90. $y = (2 + \ln x)/(2 - \ln x)$

91. Find the range of the function defined by $y = \dfrac{e^x + 1}{e^x - 1}$.

For Exercises 92–102, use the graph shown and the properties of logarithms to estimate each quantity to the nearest one tenth.

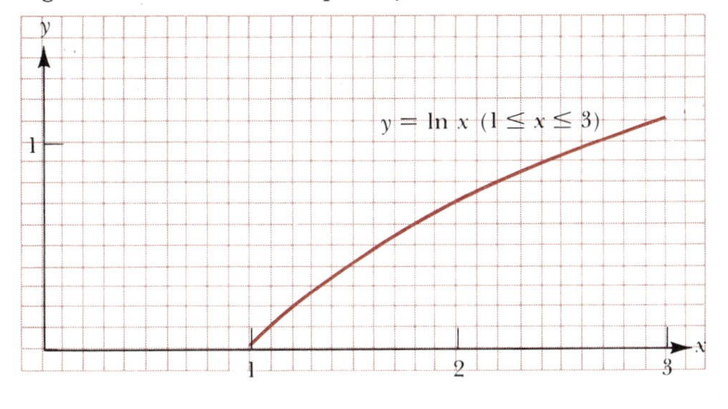

92. ln 2 **93.** ln(0.5) **94.** ln 3

95. $\ln(\frac{1}{9})$ **96.** ln 6 **97.** ln 72

98. ln(40.5) *Hint:* $40.5 = \frac{81}{2}$

99. e *Hint:* If $x = e$, then $\ln x = 1$.

100. $e^{1.1}$ *Hint:* If $x = e^{1.1}$, then $\ln x = 1.1$.

101. $\log_2 3$ **102.** $\log_3 6$

Chapter 6 Test

1. Which is larger, $\log_5 126$ or $\log_{10} 999$?
2. Graph the function $y = 3^{-x} - 3$. Specify the domain, range, intercepts, and asymptote.
3. Suppose that the population of a colony of bacteria increases exponentially. If the population at the start of an experiment is 8000, and 4 hours later it is 10,000, how long (from the start of the experiment) will it take for the population to reach 12,000? (Express your answer in terms of base e logarithms.)
4. Express $\log_{10} 2$ in terms of base e logarithms.
5. Let f be the function defined by

$$f(x) = \begin{cases} 2^{-x} & \text{if } x < 0 \\ x^2 & \text{if } x \geq 0 \end{cases}$$

Sketch the graph of f and then use the horizontal line test to determine whether f is one-to-one.

6. Estimate 2^{60} in terms of an integral power of 10.
7. Solve for x: $\ln(x + 1) - 1 = \ln(x - 1)$.
8. Suppose that \$5000 is invested at 8% interest, compounded annually. How many years will it take for the money to double? Make use of the approximations $\ln 2 \approx 0.7$ and $\ln 1.08 \approx 0.07$ to obtain a numerical answer.
9. On the same set of axes sketch the graphs of $y = e^x$ and $y = \ln x$. Specify the domain and range for each function.
10. Solve for x: $xe^x - 2e^x = 0$.
11. Simplify $\log_9 \frac{1}{27}$.
12. Given that $\ln A = a$, $\ln B = b$, and $\ln C = c$, express $\ln [(A^2\sqrt{B})/C^3]$ in terms of a, b, and c.
13. The half-life of plutonium-241 is 13 years. What is the decay constant? Use the approximation $\ln \frac{1}{2} \approx -0.7$ to obtain a numerical answer.
14. Express as a single logarithm with a coefficient of 1: $3 \log_{10} x - \log_{10}(1 - x)$.
15. Solve the following equation for x, leaving your answer in terms of base e logarithms: $5e^{2-x} = 12$.
16. Let $f(x) = e^{x+1}$. Find a formula for $f^{-1}(x)$ and specify the domain of f^{-1}.
17. Suppose that in 1980 the population of a certain country was 2 million and increasing with a relative growth rate of 2% per year. Estimate the year in which the population will reach 3 million. (Use the approximation $\ln 1.5 \approx 0.40$ in order to obtain a numerical answer.)
18. Simplify $\ln e + \ln \sqrt{e} + \ln 1 + \ln e^{\ln 10}$
19. A principal of \$1000 is deposited at 10% per annum, compounded continuously. Estimate the doubling time and then sketch a graph showing how the amount increases with time.
20. Simplify $\ln(\log_8 56 - \log_8 7)$.

Chapter 7 Trigonometric Functions of Angles

Introduction

There are two general approaches to the study of trigonometry at the precalculus level. The first approach centers around the study of triangles. Indeed, the word "trigonometry" is derived from two Greek words, *trigonon,* meaning triangle, and *metria,* meaning measurement. This is the approach that we shall follow in this chapter. The second approach, which we take up in Chapter 8, is in a sense a generalization of the first. Before getting down to specifics, we offer the following overview of Chapters 7 and 8, in terms of functions.

	Inputs for the Functions	Outputs	Emphasis
Chapter 7	Angles	Certain real numbers	The trigonometric functions are used to study triangles.
Chapter 8	Certain real numbers	Certain real numbers	The objects of study are the trigonometric functions themselves.

7.1 Trigonometric Functions of Acute Angles

Most of the functions that we have considered up to this point were defined by means of equations that involved only the basic operations of algebra. Examples of such functions are

$$f(x) = x^2 \qquad g(x) = mx + b \qquad h(x) = 2^x$$

In this chapter and the next, we will study the so-called trigonometric functions. The definitions of these functions involve some geometry, as well as algebra.

Let θ denote an acute angle in a right triangle, as indicated in Figure 1. Then we define the six trigonometric functions of the acute angle θ in the box that follows. Notice that sin θ and csc θ are reciprocals of one another. Similarly, cos θ and sec θ are reciprocals, as are tan θ and cot θ.

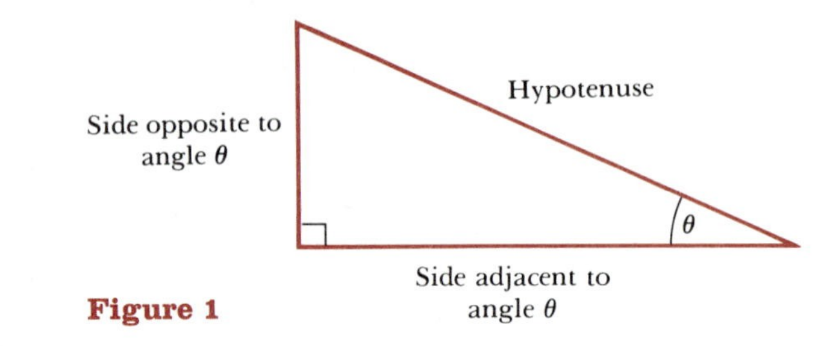

Figure 1

Name of Function	Abbreviation	Definition
sine	sin	$\sin\theta = \dfrac{\text{length of side opposite angle } \theta}{\text{length of hypotenuse}}$
cosine	cos	$\cos\theta = \dfrac{\text{length of side adjacent angle } \theta}{\text{length of hypotenuse}}$
tangent	tan	$\tan\theta = \dfrac{\text{length of side opposite angle } \theta}{\text{length of side adjacent angle } \theta}$
cosecant	csc	$\csc\theta = \dfrac{\text{length of hypotenuse}}{\text{length of side opposite angle } \theta}$
secant	sec	$\sec\theta = \dfrac{\text{length of hypotenuse}}{\text{length of side adjacent angle } \theta}$
cotangent	cot	$\cot\theta = \dfrac{\text{length of side adjacent angle } \theta}{\text{length of side opposite angle } \theta}$

Example 1 Let θ be the acute angle indicated in Figure 2. Determine the six quantities sin θ, cos θ, tan θ, csc θ, sec θ, and cot θ.

Solution Use the definitions.

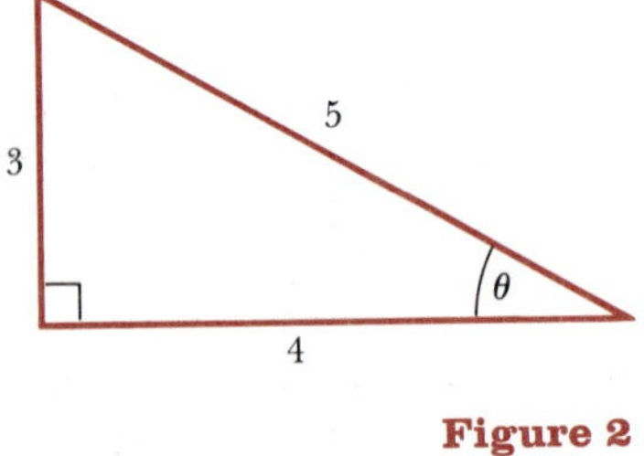

Figure 2

$$\sin\theta = \frac{\text{opposite}}{\text{hypotenuse}} = \frac{3}{5} \qquad \csc\theta = \frac{\text{hypotenuse}}{\text{opposite}} = \frac{5}{3}$$

$$\cos\theta = \frac{\text{adjacent}}{\text{hypotenuse}} = \frac{4}{5} \qquad \sec\theta = \frac{\text{hypotenuse}}{\text{adjacent}} = \frac{5}{4}$$

$$\tan\theta = \frac{\text{opposite}}{\text{adjacent}} = \frac{3}{4} \qquad \cot\theta = \frac{\text{adjacent}}{\text{opposite}} = \frac{4}{3}$$

Note the pairs of answers that are reciprocals. (This helps in memorizing the definitions.)

Example 2 Let β be the acute angle indicated in Figure 3. Find sin β and cos β.

Solution In view of the definitions, we need to know the length of the hypotenuse in Figure 3. If we call this length x, then by the Pythagorean Theorem we have

$$x^2 = 3^2 + 1^2 = 10$$
$$x = \sqrt{10}$$

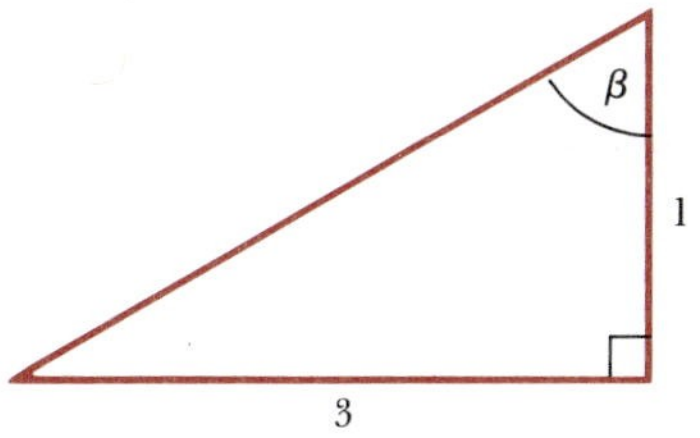

Figure 3

Therefore

$$\sin \beta = \frac{\text{opposite}}{\text{hypotenuse}} = \frac{3}{\sqrt{10}} \qquad \left(= \frac{3\sqrt{10}}{10} \right)$$

and

$$\cos \beta = \frac{\text{adjacent}}{\text{hypotenuse}} = \frac{1}{\sqrt{10}} \qquad \left(= \frac{\sqrt{10}}{10} \right)$$

Example 3 Figure 4 shows two right triangles. The first right triangle has sides 5, 12, and 13. The second right triangle is similar to the first (the angles are the same), but each side is 10 times longer than the corresponding side in the first triangle. Calculate and compare the values of sin θ, cos θ, and tan θ for both triangles.

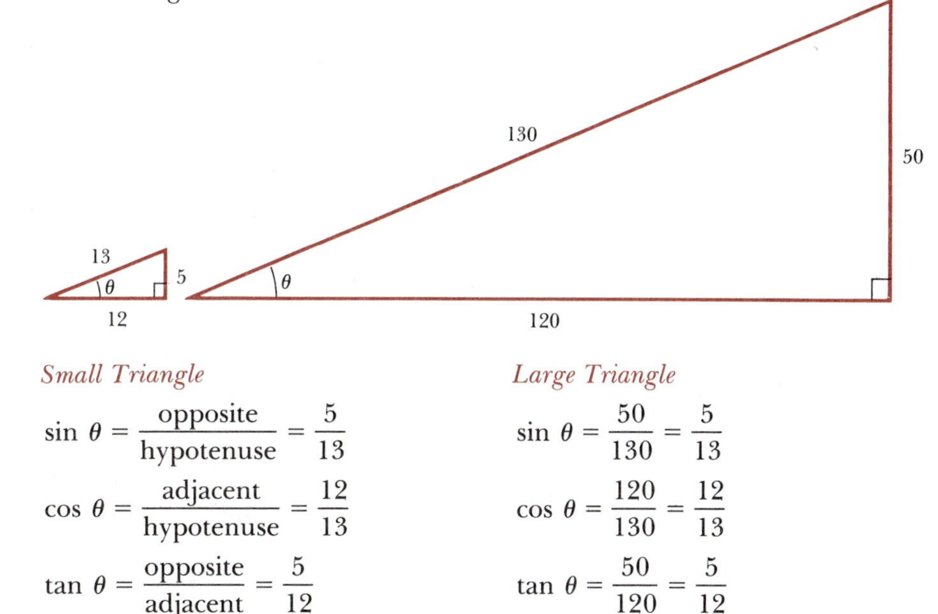

Figure 4

Solution

Small Triangle

$$\sin \theta = \frac{\text{opposite}}{\text{hypotenuse}} = \frac{5}{13}$$
$$\cos \theta = \frac{\text{adjacent}}{\text{hypotenuse}} = \frac{12}{13}$$
$$\tan \theta = \frac{\text{opposite}}{\text{adjacent}} = \frac{5}{12}$$

Large Triangle

$$\sin \theta = \frac{50}{130} = \frac{5}{13}$$
$$\cos \theta = \frac{120}{130} = \frac{12}{13}$$
$$\tan \theta = \frac{50}{120} = \frac{5}{12}$$

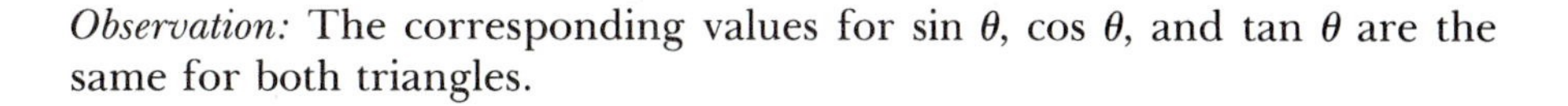

Observation: The corresponding values for sin θ, cos θ, and tan θ are the same for both triangles.

The point of Example 3 is this. For a given acute angle θ, the values of the trigonometric functions depend on the *ratios* of the lengths of the sides, but not on the size of the particular right triangle in which θ resides.

There are certain angles θ for which we can evaluate the trigonometric functions without the use of calculators or tables. Two such angles are 30° and 60°. To do this, we rely on the following theorem from geometry. (For a proof of this theorem, see Exercise 46 at the end of this section.)

Theorem

In a 30°-60° right triangle, the side opposite the 30° angle is half the hypotenuse.

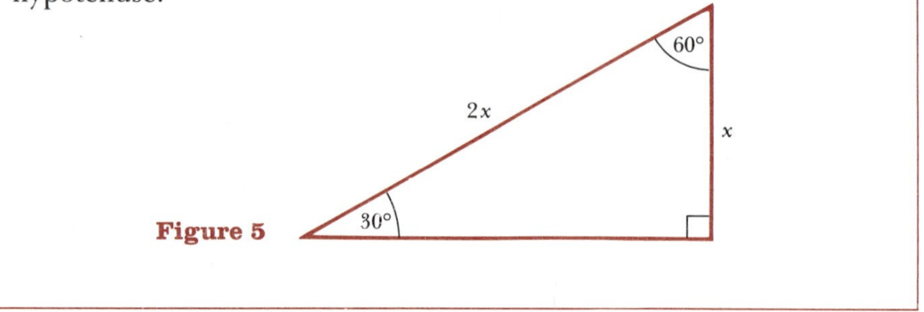

Figure 5

Let y denote the length of the unmarked side of the triangle in Figure 5. Then we have (according to the Pythagorean Theorem)

$$y^2 + x^2 = (2x)^2$$
$$y^2 = 4x^2 - x^2$$
$$y^2 = 3x^2$$
$$y = \sqrt{3}\,x$$

Our 30°-60° right triangle can now be labeled as in Figure 6. From this figure, the required trigonometric values can be written down by inspection. We will list only the values of the sine, cosine, and tangent, since the remaining three values are the reciprocals of these.

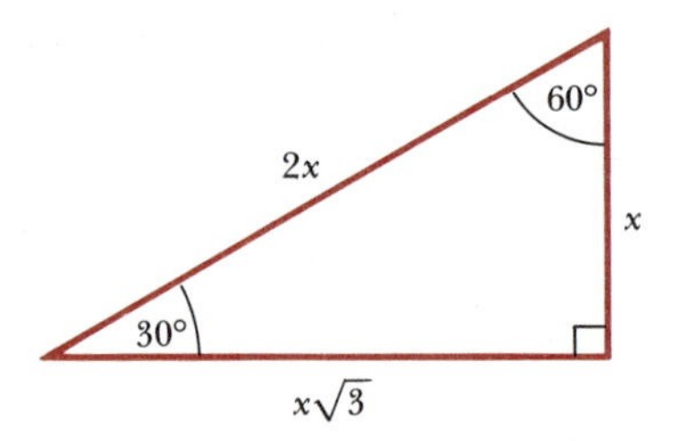

Figure 6

$$\sin 30° = \frac{x}{2x} = \frac{1}{2} \qquad \sin 60° = \frac{x\sqrt{3}}{2x} = \frac{\sqrt{3}}{2}$$

$$\cos 30° = \frac{x\sqrt{3}}{2x} = \frac{\sqrt{3}}{2} \qquad \cos 60° = \frac{x}{2x} = \frac{1}{2}$$

$$\tan 30° = \frac{x}{x\sqrt{3}} = \frac{1}{\sqrt{3}} = \frac{\sqrt{3}}{3} \qquad \tan 60° = \frac{x\sqrt{3}}{x} = \sqrt{3}$$

There are two observations to be made here. First, the final answers do not involve x. Once again this shows that for a given angle, it is the ratios of the sides that determine the trigonometric values, not the particular triangle. Second, notice that

$$\sin 30° = \cos 60°$$

and

$$\sin 60° = \cos 30°$$

It is easy to see that this is no coincidence. The terms "opposite" and "adjacent" are relative ones. As a glance at Figure 6 shows, the side opposite to one acute angle is automatically adjacent to the other acute angle.

Another angle for which we can compute the trigonometric values without the aid of a calculator or tables is 45°. Figure 7(a) and the calculations that accompany it show that if one acute angle in a right triangle is 45°, then the other acute angle is also 45°. It then follows from a theorem of geometry that the triangle is isosceles (that is, it has two equal legs), as indicated in

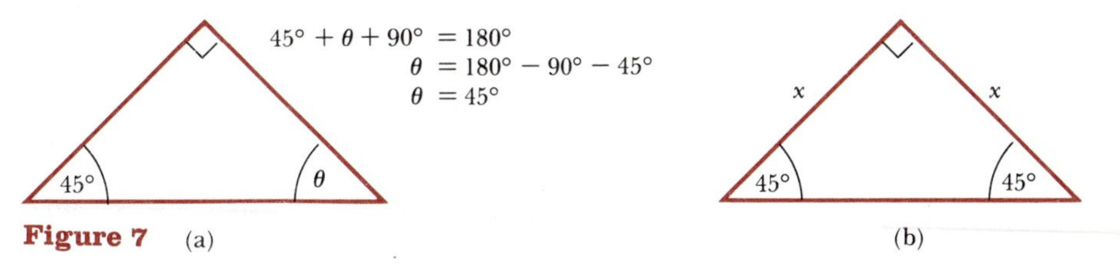

$$45° + \theta + 90° = 180°$$
$$\theta = 180° - 90° - 45°$$
$$\theta = 45°$$

Figure 7 (a) (b)

Figure 7(b). If we use y to denote the length of the unmarked side in Figure 7(b), then we have

$$y^2 = x^2 + x^2$$
$$y^2 = 2x^2$$
$$y = \sqrt{2}\,x$$

Our 45°-45° right triangle can then be labeled as in Figure 8. From this figure, the required trigonometric values can be obtained by inspection. As before, we list only the values for sine, cosine, and tangent, since the remaining three values are just the reciprocals of these.

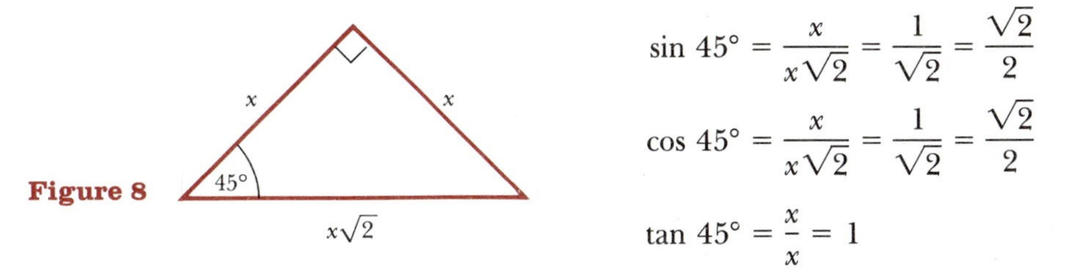

Figure 8

$$\sin 45° = \frac{x}{x\sqrt{2}} = \frac{1}{\sqrt{2}} = \frac{\sqrt{2}}{2}$$

$$\cos 45° = \frac{x}{x\sqrt{2}} = \frac{1}{\sqrt{2}} = \frac{\sqrt{2}}{2}$$

$$\tan 45° = \frac{x}{x} = 1$$

Table 1 summarizes the results we've now obtained for angles of 30, 45, and 60 degrees. These results should be memorized.

Table 1

θ	$\sin\theta$	$\cos\theta$	$\tan\theta$
30°	$\frac{1}{2}$	$\frac{\sqrt{3}}{2}$	$\frac{\sqrt{3}}{3}$
45°	$\frac{\sqrt{2}}{2}$	$\frac{\sqrt{2}}{2}$	1
60°	$\frac{\sqrt{3}}{2}$	$\frac{1}{2}$	$\sqrt{3}$

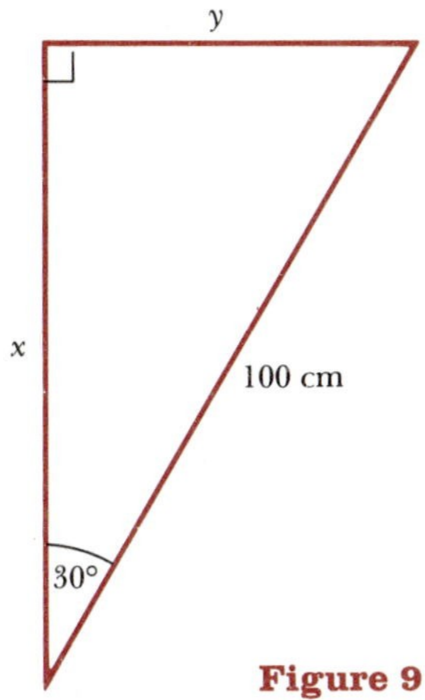

Figure 9

We now turn to some first applications of the trigonometric functions.

Example 4 Find x in Figure 9.

Solution Relative to the given 30° angle, x is the adjacent side. The length of the hypotenuse is 100 cm. Since the adjacent side and the hypotenuse are involved, we employ the cosine function here. We have

$$\cos 30° = \frac{\text{adjacent}}{\text{hypotenuse}} = \frac{x}{100}$$

That is,

$$\cos 30° = \frac{x}{100}$$

and consequently

$$x = 100 \cos 30°$$

We can substitute for cos 30° using Table 1. This yields

$$x = 100\left(\frac{\sqrt{3}}{2}\right)$$
$$= 50\sqrt{3} \text{ cm}$$

This is the result we were looking for.

We used the cosine function in Example 4 because the adjacent side and the hypotenuse were involved. We could instead use the secant. In that case, with reference again to Figure 9, the calculations look like this:

$$\sec 30° = \frac{\text{hypotenuse}}{\text{adjacent}} = \frac{100}{x}$$
$$\frac{2}{\sqrt{3}} = \frac{100}{x}$$
$$2x = 100\sqrt{3}$$
$$x = \frac{100\sqrt{3}}{2} = 50\sqrt{3} \text{ cm} \qquad \text{as obtained previously}$$

Example 5 Find y in Figure 9.

Solution As you can see from Figure 9, the side of length y is opposite the 30° angle. Furthermore, we are given the length of the hypotenuse. Since the opposite side and the hypotenuse are involved, we employ the sine function. This yields

$$\sin 30° = \frac{\text{opposite}}{\text{hypotenuse}} = \frac{y}{100}$$

That is,

$$\sin 30° = \frac{y}{100}$$

and therefore

$$y = 100 \sin 30°$$

We now substitute for sin 30° using Table 1.

$$y = (100)\left(\frac{1}{2}\right)$$
$$= 50 \text{ cm}$$

This is the required answer. Actually, we could have obtained this particular result much faster by recalling that in the 30°-60° right triangle, the side opposite the 30° angle, namely y, is half the hypotenuse. That is, $y = \frac{100}{2} = 50$ cm, as obtained previously.

Example 6 A ladder, which is leaning against the side of a building, forms an angle of 50° with the ground. If the foot of the ladder is 12 ft from the base of the building, how far up the side of the building does the ladder reach? See Figure 10.

Solution In Figure 10 we have used y to denote the required distance. Notice y is opposite the 50° angle, while the given side is adjacent to that angle. Since the opposite and adjacent sides are involved, we employ the tangent function. (The cotangent function could also be used.) We have

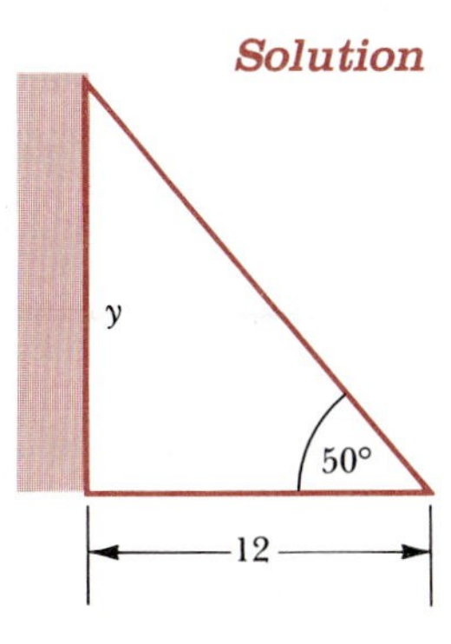

Figure 10

$$\tan 50° = \frac{y}{12}$$

and therefore

$$y = 12 \tan 50°$$

Without the use of a calculator or tables, this is our final answer. On the other hand, using a calculator we find that tan 50° is approximately 1.19. Thus we have

$$y \approx (12)(1.19) = 14 \text{ ft} \qquad \text{to the nearest foot}$$

Example 7 Show that the area of the triangle in Figure 11 is given by $A = \frac{1}{2}ab \sin \theta$.

Solution Draw an altitude as shown in Figure 12. Call the length of this altitude h. Then we have

Figure 11

$$\sin \theta = \frac{h}{a}$$

and therefore

$$h = a \sin \theta$$

This value for h can now be used in the usual formula for area, $A = \frac{1}{2}bh$. This yields

$$A = \frac{1}{2}b(a \sin \theta)$$

$$= \frac{1}{2}ab \sin \theta \qquad \text{as required}$$

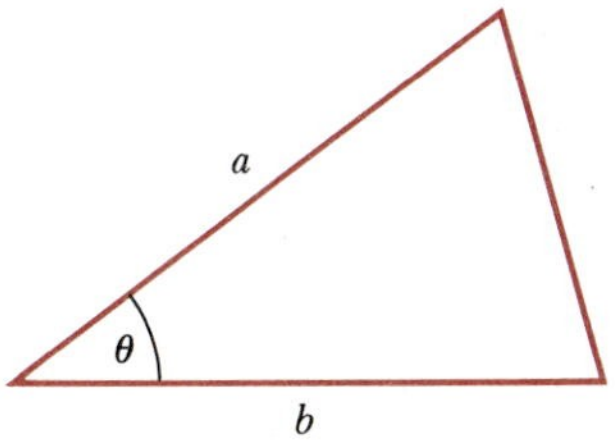

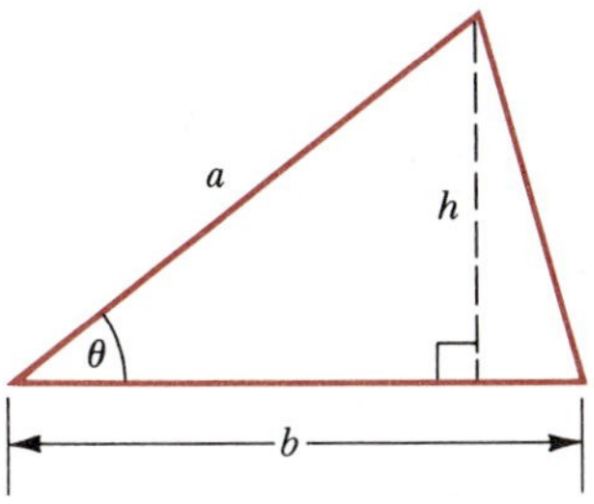

Figure 12

The formula that we just derived in Example 7 is worth remembering. In words, the formula states that the area of a triangle is equal to half the product of the lengths of two of the sides times the sine of the included acute angle. In Section 7.3 we'll see that the formula remains valid even when the included angle is not an acute angle.

Example 8 Find the area of the triangle in Figure 13.

Solution In the formula $A = \frac{1}{2}ab \sin \theta$, we let $a = 3$ cm, $b = 12$ cm, and $\theta = 60°$. This gives us

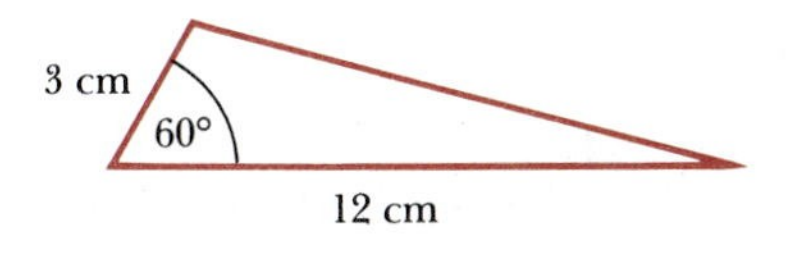

Figure 13

$$A = \frac{1}{2}(3)(12) \sin 60°$$

$$= (18)\frac{\sqrt{3}}{2}$$

$$= 9\sqrt{3} \text{ cm}^2$$

This is the required area.

Example 9 Figure 14 on the following page shows a regular pentagon inscribed in a circle of radius 2 in. ("Regular" means that all of the sides are equal; all of the angles are equal.) Find the area of the pentagon.

Solution The idea here is to first find the area of $\triangle BOA$ using the area formula from Example 7. Then, since the pentagon is composed of five such identical triangles, the area of the pentagon will be five times the area of $\triangle BOA$. We will assume as known the result from geometry that states that, in a regular n-sided figure, the central angle is $360°/n$. In our case we therefore have

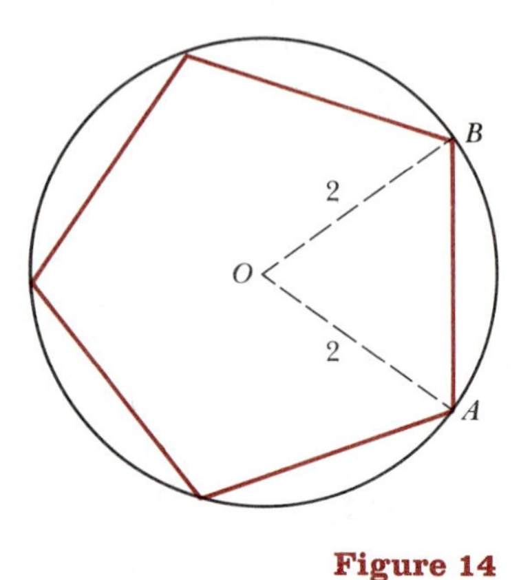

Figure 14

$$\angle BOA = \frac{360°}{5} = 72°$$

The area of $\triangle BOA$ can now be found. We have

$$\begin{aligned} \text{area} &= \frac{1}{2}ab \sin \theta \\ &= \frac{1}{2}(2)(2) \sin 72° \\ &= 2 \sin 72° \text{ in}^2 \end{aligned}$$

The area of the pentagon is five times this, or $10 \sin 72° \text{ in}^2$. (Using a calculator, this is about 9.51 in^2.)

Here is some terminology that will be used in the next two examples. Suppose that a surveyor sights an object at a point above the horizontal, as indicated in Figure 15(a). Then the angle between the line of sight and the horizontal is called the *angle of elevation.* The *angle of depression* is defined similarly, as shown in Figure 15(b).

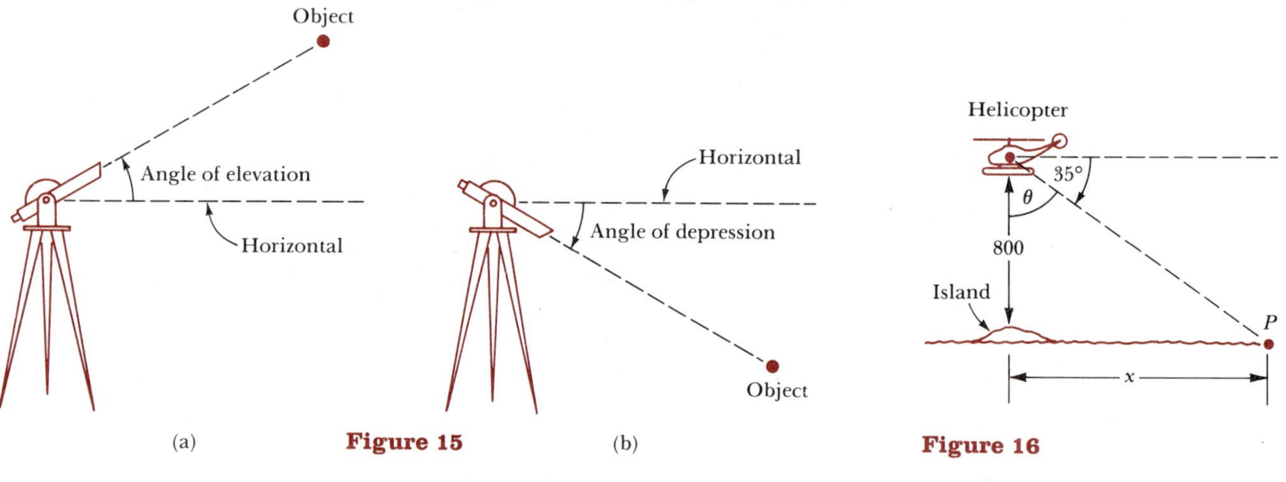

Figure 15

Figure 16

Example 10 A helicopter hovers 800 ft directly above a small island that is off the California coast. From the helicopter, the pilot takes a sighting to a point P directly ashore on the mainland, at the water's edge. If the angle of depression is 35°, how far off the coast is the island? See Figure 16.

Solution Let x denote the required distance from the island to the mainland. Then, as you can see from Figure 16, we have $\theta + 35° = 90°$, from which it follows that $\theta = 55°$. Now we can write

$$\tan 55° = \frac{x}{800}$$

or

$$x = 800 \tan 55^\circ$$

$$\approx 1150 \text{ ft} \qquad \text{(using a calculator and rounding off to the nearest 50 ft)}$$

Example 11 A surveyor stands 30 yd from the base of a building. On top of the building is a vertical radio antenna. Let α denote the angle of elevation when the surveyor sights to the top of the building. Let β denote the angle of elevation when the surveyor sights to the top of the antenna. Express the length of the antenna in terms of the angles α and β.

Solution First draw a picture conveying the relevant information. See Figure 17. Now let h denote the length of the antenna. We are supposed to express h in terms of α and β. Let x denote the height of the building. If we look at the larger of the two right triangles in Figure 17, we have

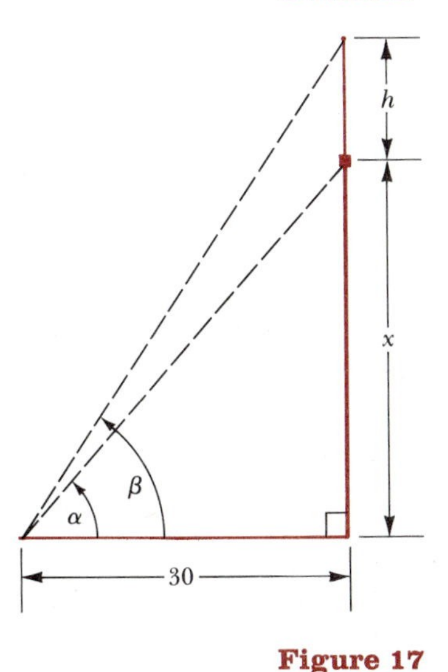

Figure 17

$$\tan \beta = \frac{x + h}{30}$$

$$x + h = 30 \tan \beta$$

$$h = 30 \tan \beta - x \tag{1}$$

Equation (1) expresses h in terms of β and x. However, we are required to express h in terms of α and β. Toward this end, consider the smaller of the two right triangles in Figure 17. In that triangle we have

$$\tan \alpha = \frac{x}{30}$$

$$x = 30 \tan \alpha$$

This result can be substituted for x in equation (1). Doing so gives us

$$h = 30 \tan \beta - 30 \tan \alpha$$

$$= 30 (\tan \beta - \tan \alpha)$$

This expresses the length of the tower (in yards) in terms of α and β, as required.

Example 12 The arc shown in Figure 18 is a portion of the circle $x^2 + y^2 = 1$. Express the following quantities in terms of θ.

(a) $\overline{OA}$ **(b)** $\overline{AB}$ **(c)** Area of $\triangle OAB$ **(d)** $\overline{OC}$

Solution **(a)** In right triangle OAB:

$$\cos \theta = \frac{\overline{OA}}{\overline{OB}} = \frac{\overline{OA}}{1}$$

$$\overline{OA} = \cos \theta$$

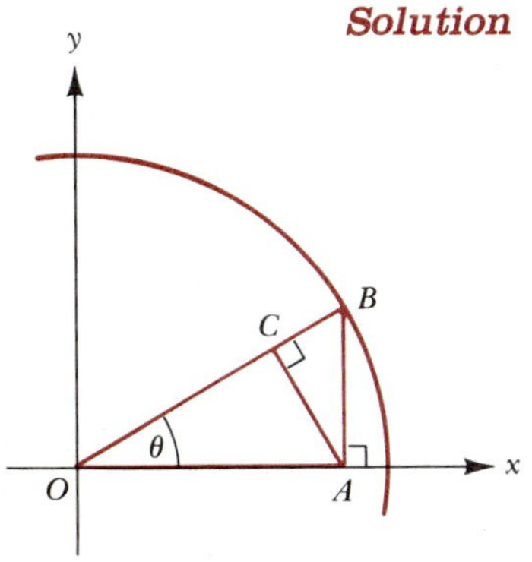

(b) In right triangle OAB:

$$\sin \theta = \frac{\overline{AB}}{\overline{OB}} = \frac{\overline{AB}}{1}$$

$$\overline{AB} = \sin \theta$$

(c) $$\text{Area } \triangle OAB = \tfrac{1}{2}(\overline{OA})(\overline{OB}) \sin \theta$$

$$= \tfrac{1}{2}(\cos \theta)(\sin \theta)$$

Figure 18

It is customary to omit the parentheses here and write

$$\text{area } \triangle OAB = \frac{1}{2} \cos \theta \sin \theta$$

(d) In right triangle *OAC:*

$$\cos \theta = \frac{\overline{OC}}{\overline{OA}}$$

$$\overline{OC} = \overline{OA} \cos \theta$$

$$= (\cos \theta)(\cos \theta) = (\cos \theta)^2$$

In trigonometry, it is customary to write $\cos^2 \theta$ in place of $(\cos \theta)^2$. Thus we have finally

$$\overline{OC} = \cos^2 \theta$$

Example 13 Determine angle θ in Figure 19.

Solution $\tan \theta = \frac{3}{4} = 0.75$. From here, you have three choices.

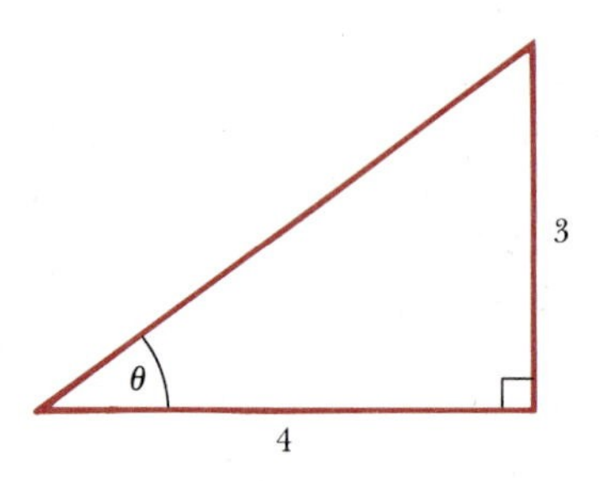

Figure 19

Choice 1 Take the easy way out and say "θ is the acute angle whose tangent is $\frac{3}{4}$." True, this doesn't tell you what θ really is in degrees; nevertheless, this form of the answer will often suffice in calculus.

Choice 2 Using a table of the trigonometric functions, hunt until you find an angle whose tangent is as close to 0.75 as possible.

Choice 3 Use a calculator to do all the work in choice 2. For instance, with a Texas Instruments calculator, you push the buttons

0.75 [INV] [TAN]

The result in this case is

$$\theta \approx 36.8°$$

Note: When performing the calculations in Example 13, check to see that your calculator is set in the "degree mode," rather than the "radian mode." (Radian measure for angles is discussed in the next chapter, as are the inverse trigonometric functions.)

Exercise Set 7.1

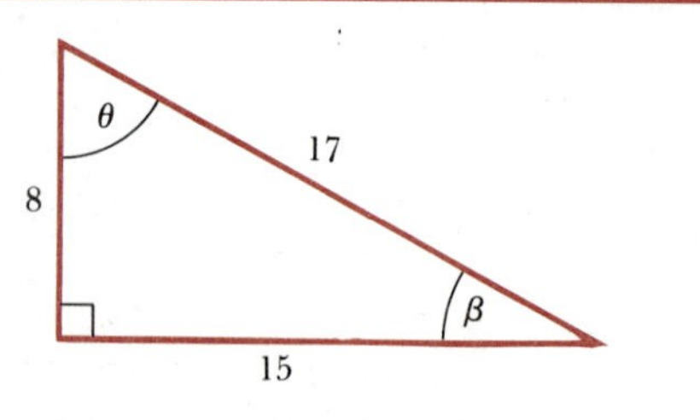

1. **(a)** Check that $8^2 + 15^2 = 17^2$. This shows that the triangle in the figure is indeed a right triangle.
(b) Find $\sin \theta$ and $\cos \beta$.
(c) Find $\cos \theta$ and $\sin \beta$.
(d) Find $\tan \theta$, $\csc \theta$, $\sec \theta$, and $\cot \theta$.
(e) Find $\tan \beta$, $\csc \beta$, $\sec \beta$, and $\cot \beta$.

2. (a) Find x.
(b) Find the values of the six trigonometric functions of θ.
(c) Find the values of the six trigonometric functions of β.

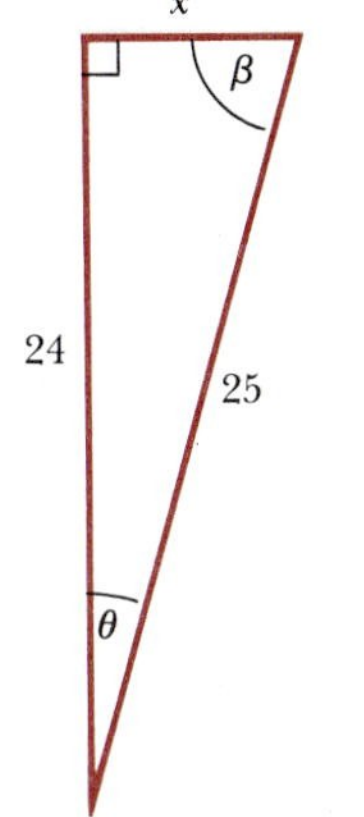

3. The two shorter sides of a right triangle are 2 cm and 5 cm, respectively. Let θ denote the angle opposite the 2-cm side.
(a) Find $\sin\theta$ and $\cos\theta$.
(b) Evaluate $\sin^2\theta + \cos^2\theta$.
Note: $\sin^2\theta$ means $(\sin\theta)^2$ and $\cos^2\theta$ means $(\cos\theta)^2$.
(c) Find $\tan\theta$. Also find $(\sin\theta)/(\cos\theta)$.
(d) Find $\csc\theta$, $\sec\theta$, and $\cot\theta$.

4. (a) Find x given that $\sin\theta = \frac{7}{8}$.
(b) Find the perimeter of the triangle.

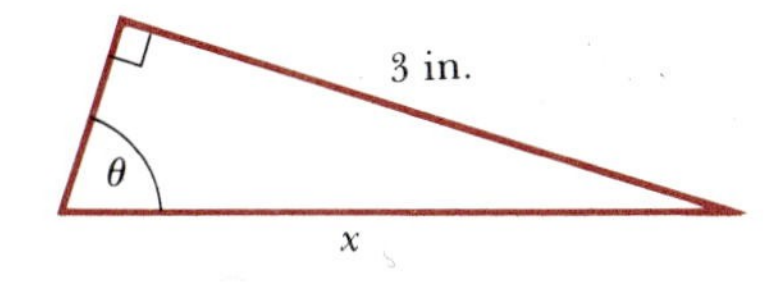

5. Find the other two sides of the given triangle.

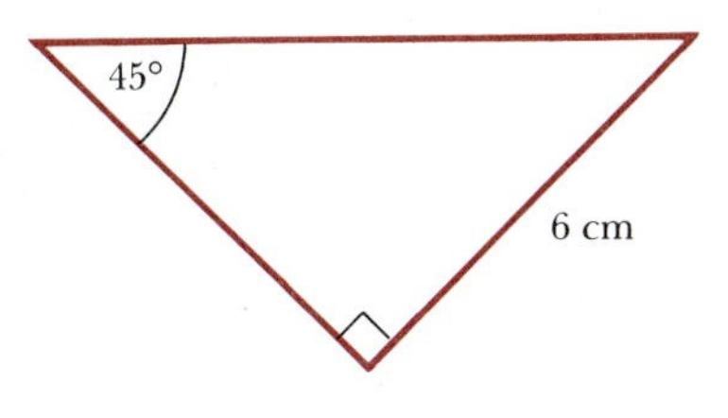

In Exercises 6–10, refer to the following figure. (However, each problem is independent of the others.)

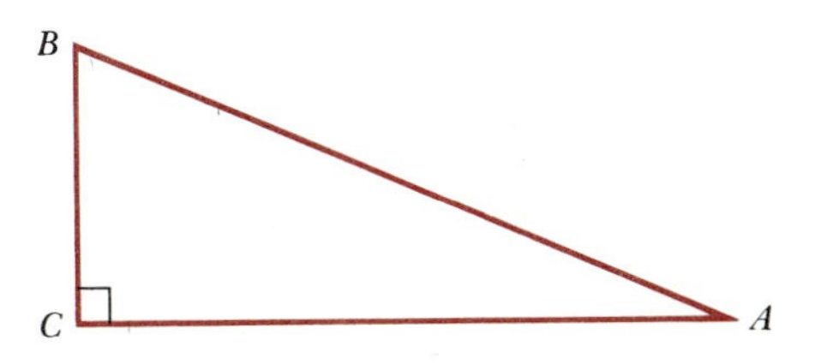

6. If $\overline{BC} = 10$ and $\tan A = 2$, find $\overline{AC}$.
7. If $\overline{AB} = 1$ and $\sec B = \frac{3}{2}$, find $\overline{BC}$.
8. Find $\overline{BC}$ if $\overline{AC} = 5$ and $\cos A = \frac{5}{6}$.
9. Find $\csc A$ given that $\overline{AC} = 20$ and $\overline{AB} = 29$.
10. If $\angle B = 60°$ and $\overline{AB} = 18$, find $\overline{BC}$ and $\overline{AC}$.

The purpose of Exercises 11 and 12, which follow, is twofold. First, doing the problems will help you memorize the values in Table 1. Second, the exercises serve as an algebra review.

11. Verify the following equations.
(a) $\cos 60° = \cos^2 30° - \sin^2 30°$
(b) $\cos 60° = 1 - 2\sin^2 30°$
(c) $\sin^2 30° + \sin^2 45° + \sin^2 60° = \frac{3}{2}$
(d) $\sin 30° \cos 60° + \cos 30° \sin 60° = 1$
(e) $2\sin 30° \cos 30° = \sin 60°$
(f) $2\sin 45° \cos 45° = 1$

12. Verify the following equations.
(a) $\sin 30° = \sqrt{\dfrac{1 - \cos 60°}{2}}$
(b) $\cos 30° = \sqrt{\dfrac{1 + \cos 60°}{2}}$
(c) $\tan 30° = \dfrac{\sin 60°}{1 + \cos 60°}$
(d) $\tan 30° = \dfrac{1 - \cos 60°}{\sin 60°}$
(e) $1 + \tan^2 45° = \sec^2 45°$
(f) $1 + \cot^2 60° = \csc^2 60°$
(g) $\sin^2 15° + \cos^2 15° = 1$, given that $\sin 15° = \frac{1}{4}(\sqrt{6} - \sqrt{2})$ and $\cos 15° = \frac{1}{4}(\sqrt{6} + \sqrt{2})$.
(h) $\sin^2 18° + \cos^2 18° = 1$, given that $\sin 18° = \frac{1}{4}(\sqrt{5} - 1)$ and $\cos 18° = \frac{1}{4}\sqrt{10 + 2\sqrt{5}}$.

13. Estimate the height of the Washington Monument, given that, at a point level with and 1000 ft away from the base, the angle of elevation to the top of the monument is approximately 29°.

14. A ladder 30 ft long leans against a building. The ladder forms an angle of 65° with the ground.
(a) How high up the side of the building does the ladder reach?
(b) How far from the base of the building is the foot of the ladder?

C 15. Refer to the figure on the next page. At certain times, the planets Earth and Mercury line up in such a way that $\angle EMS$ is a right angle, as shown in the figure. At such times, $\angle SEM$ is found to be 21.16°. Use this information to estimate the distance $\overline{MS}$ of Mercury from the Sun. Assume that the distance from the Earth to the Sun is 93 million miles. (Round off your answer to the nearest million miles. Because Mercury's orbit is not exactly circular, the actual distance of Mercury from the Sun varies from about 28 million miles to 43 million miles.)

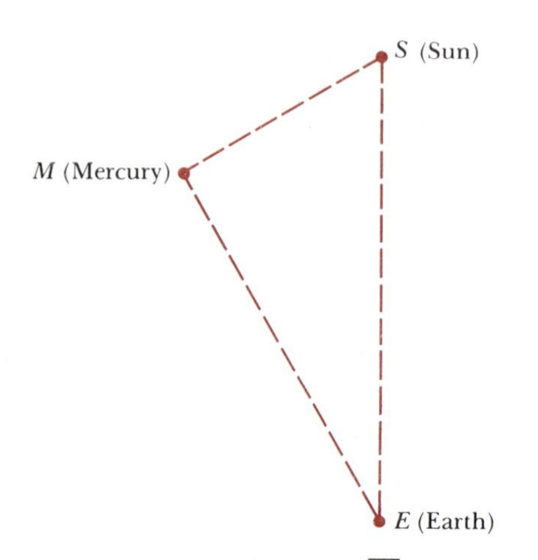

16. **(a)** Determine the distance $\overline{AB}$ across the lake shown in the figure, using the following data. $\overline{AC} = 400$ m, $\angle C = 90°$, and $\angle CAB = 40°$.

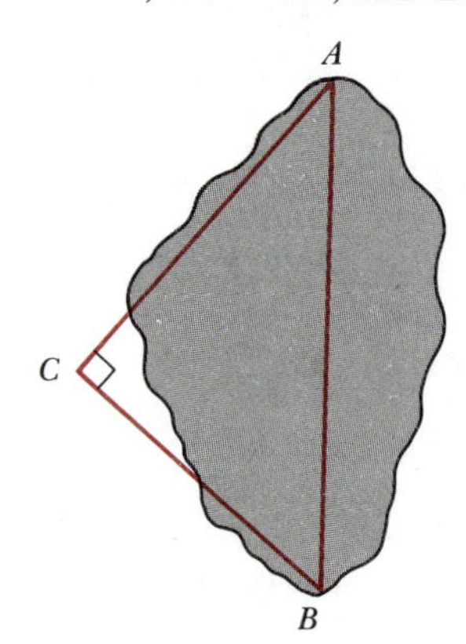

(b) Determine the distance $\overline{AB}$ across the lake shown in the figure, using the following data. $\overline{BC} = 640$ ft, $\angle B = 90°$, and $\angle C = 34°$.

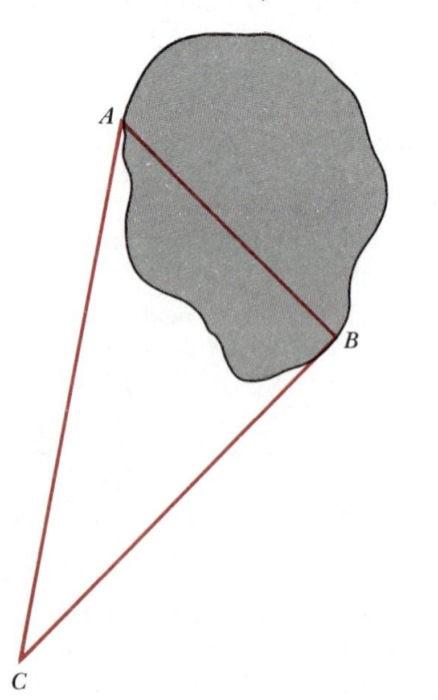

17. Find the area of each triangle.

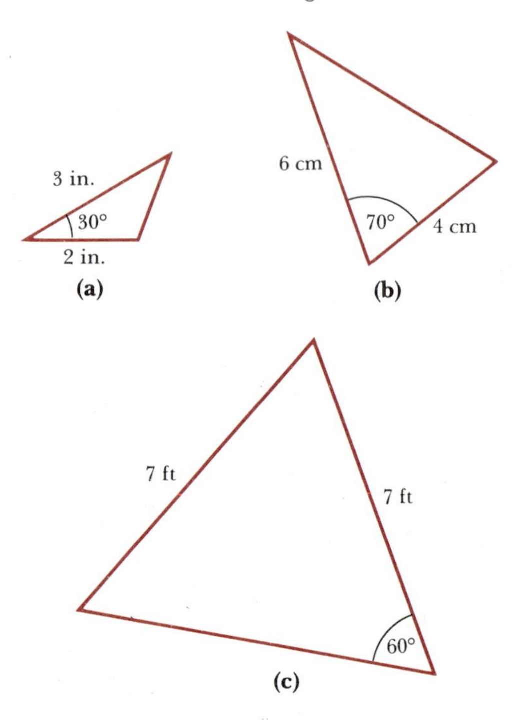

18. A regular decagon (10 sides) is inscribed in a circle of radius 20 cm. Find the area of the decagon.

19. In a triangle OAB, $\overline{OA} = \overline{OB} = 6$ in. and $\angle AOB = 72°$. Find $\overline{AB}$. *Hint:* Draw a perpendicular from O to AB.

20. Find the length of a side of a regular seven-sided polygon inscribed in a circle of radius 9 cm. (See the hint in the previous problem.)

21. The figure shows two ships at points P and Q, which are in the same vertical plane with an airplane at point R. When the height of the airplane is 3500 ft, the angle of depression to P is 48° and to Q is 25°. Find the distance between the two ships.

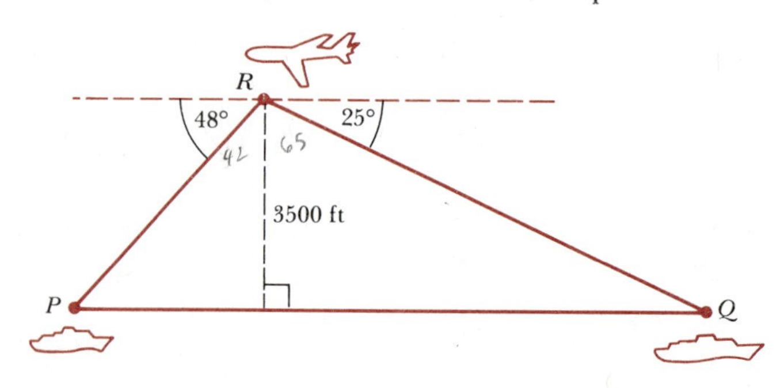

22. In the figure at the top of the following page, $\overline{AC} = 18$ units, $\angle BAC = 45°$, and $\angle DAC = 60°$. Find the length of BD.

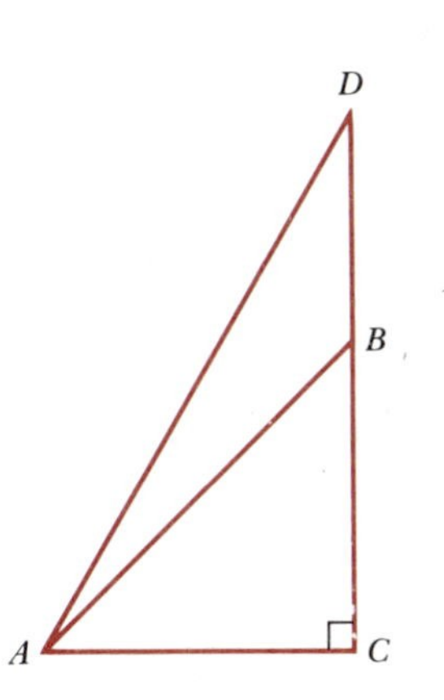

23. The radius of the circle in the figure below is 1 unit. Express the lengths of OA, AB, and DC in terms of the angle α.

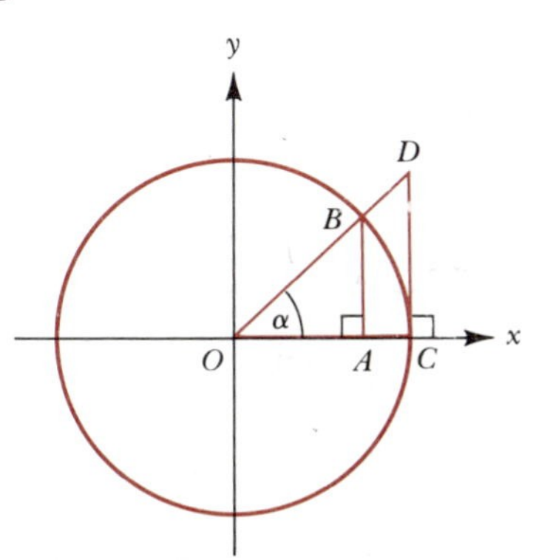

C 24. An observer in a lighthouse is 66 ft above the surface of the water. The observer sees a ship and finds the angle of depression to be 0.7°. Estimate the distance of the ship from the base of the lighthouse.

25. From a point on ground level, you measure the angle of elevation to the top of a mountain to be 38°. Then you walk 200 miles further away from the mountain and find the angle of elevation now is 20°. Find the height of the mountain.

26. (First look over Example 13.) Use a calculator or tables to estimate the angle θ in each triangle.

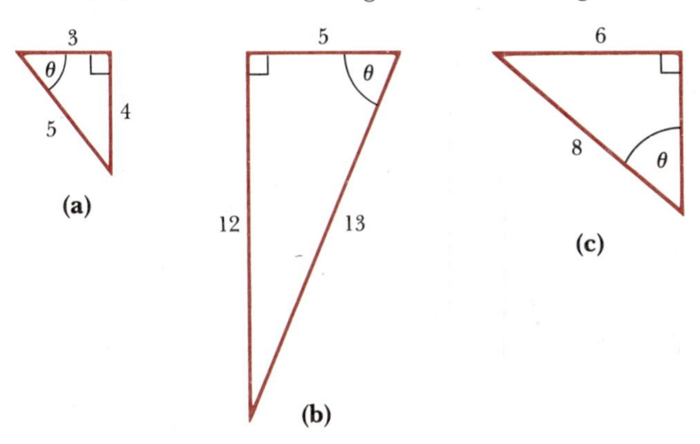

27. First, use the Pythagorean Theorem to find the lengths of OA, OB, OC, OD, and OE shown in the figure. Then use a calculator or tables to estimate the angles α, β, γ, δ, and ϵ. (Actually, you don't need any aid for two of the angles.)

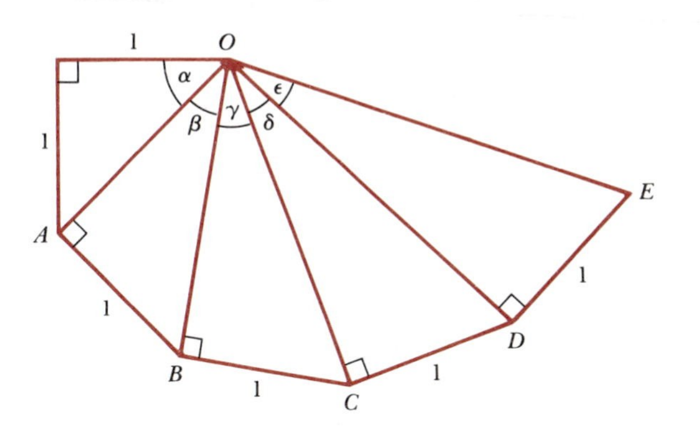

28. **(a)** Suppose that θ is an acute angle and $\cos\theta = x$. Express the other five trigonometric functions in terms of x. *Hint:* Draw a right triangle and call one of the acute angles θ. Let x denote the length of the side adjacent to θ. What must the length of the hypotenuse be if we are to have $\cos\theta = x$? Now use the Pythagorean Theorem to find the length of the third side of the triangle in terms of x.

(b) Suppose $\cos\theta = \frac{2}{7}$, where θ is an acute angle. Use your results in part (a) to find $\sin\theta$ and $\tan\theta$.

(c) Suppose $\cos\theta = \frac{1}{4}$, where θ is an acute angle. Use your results in part (a) to determine $\csc\theta$ and $\cot\theta$.

29. Suppose β is an acute angle and $\sec\beta = \sqrt{1 + a^2}$. Express the other five trigonometric functions in terms of a.

30. Express the length of each line segment shown in the figure in terms of x, the length of AD. When radicals appear in an answer, leave the answer in that form, rather than using decimal approximations.

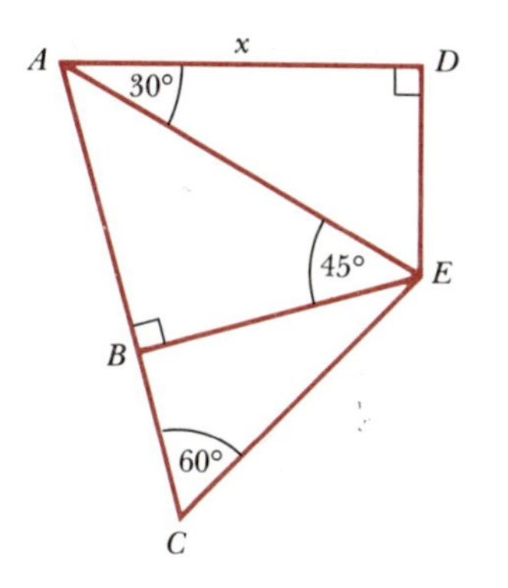

31. Express $\overline{AC}$ as a function of θ.

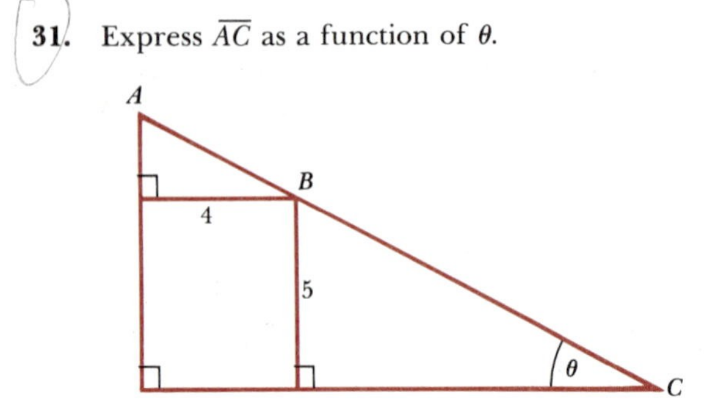

32. $\overline{AB}$ = 8 in. Express x as a function of θ.

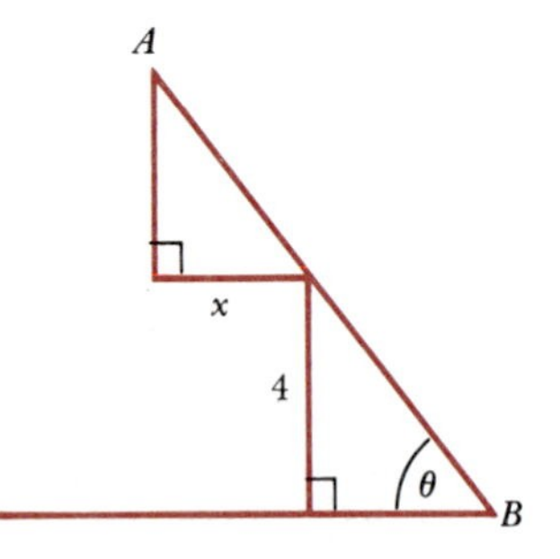

***33.** In the figure, line segment BA is tangent to the circle at A. Also, CF is tangent to the circle at F. Express the length of each of the following segments in terms of θ.

(a) DE (*Answer:* $\sin\theta$) **(b)** OE
(c) CF **(d)** OC **(e)** AB **(f)** OB

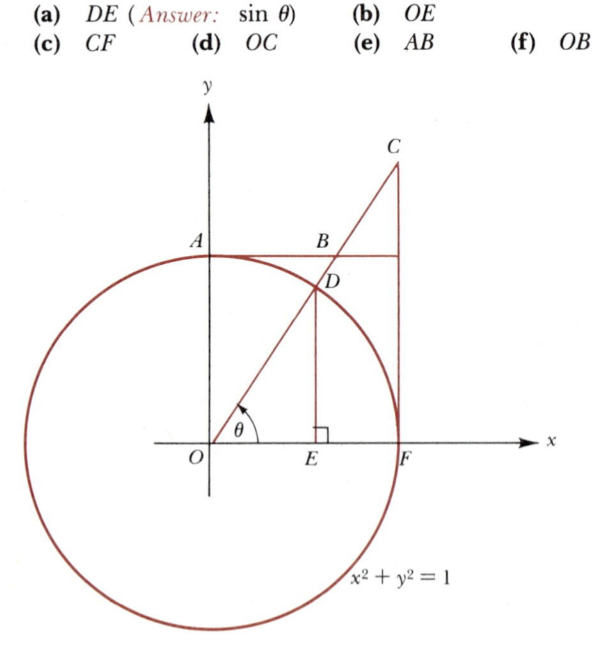

34. Let $\theta = 60°$. Evaluate the following.

(a) $\sin\theta$ **(b)** $\sin\dfrac{\theta}{2}$ **(c)** $\dfrac{\sin\theta}{2}$
(d) $2\sin\theta$ **(e)** $\sin^2\theta$

Check: Of the five answers, no two are the same.

35. Use the figure and the Pythagorean Theorem to prove that the following equations are true for any acute angle θ.

(a) $\sin^2\theta + \cos^2\theta = 1$ **(b)** $\tan\theta = \dfrac{\sin\theta}{\cos\theta}$
(c) $\tan^2\theta + 1 = \sec^2\theta$
(d) $\sin\theta = \cos(90° - \theta)$
(e) $\cos\theta = \sin(90° - \theta)$

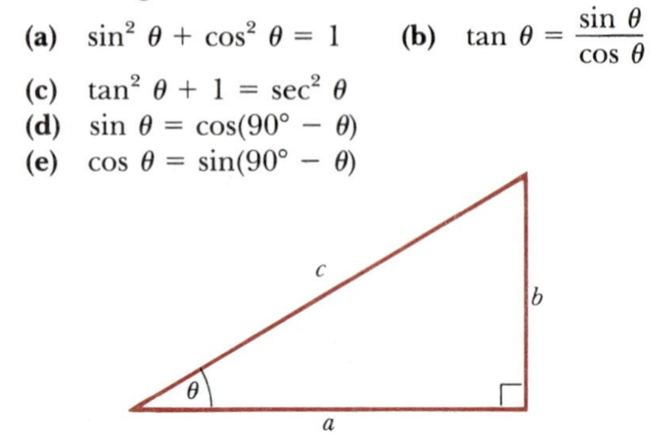

***36.** The figure below shows a semicircle with radius 1 unit.

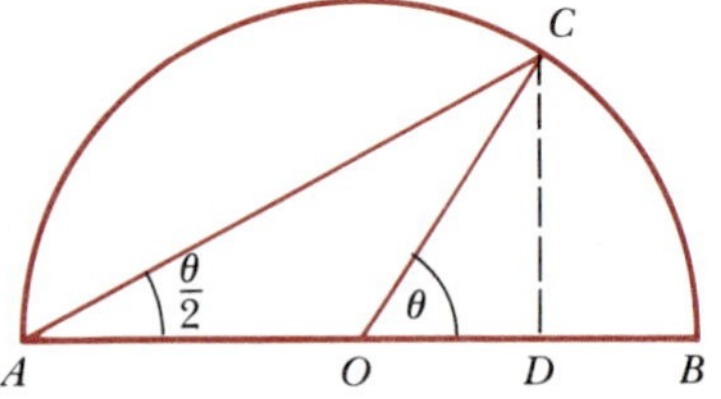

(a) Use the figure to derive the formula

$$\tan\frac{\theta}{2} = \frac{\sin\theta}{1 + \cos\theta}$$

Hints: Show $\overline{CD} = \sin\theta$ and $\overline{OD} = \cos\theta$. Then look at right triangle ADC to find $\tan(\theta/2)$.

(b) Use the formula developed in part (a) to show

(i) $\tan 15° = \dfrac{1}{2 + \sqrt{3}}$

(ii) $\tan 22.5° = \sqrt{2} - 1$

(c) In the figure, show that $\overline{AC} = \sqrt{2 + 2\cos\theta}$. (You will need to make use of the result given in part (a) of Exercise 35.)

(d) Show that $\sin\dfrac{\theta}{2} = \dfrac{\sin\theta}{\sqrt{2 + 2\cos\theta}}$.

(e) Show that $\sin 15° = \dfrac{1}{2\sqrt{2 + \sqrt{3}}}$.

(f) Show that $\dfrac{1}{2\sqrt{2 + \sqrt{3}}} = \dfrac{\sqrt{6} - \sqrt{2}}{4}$.

(g) Show that $\cos 15° = \dfrac{\sqrt{6} + \sqrt{2}}{4}$, using parts (e) and (f) along with the formula in part (a) of Exercise 35.

***37.** In the figure at the top of the next page, the smaller circle is tangent to the larger circle. Line PQ is a common tangent and line PR passes through the centers of both circles. If the radius of the smaller circle is a and that of the larger b, show that

$$\sin\theta = \frac{b - a}{a + b}$$

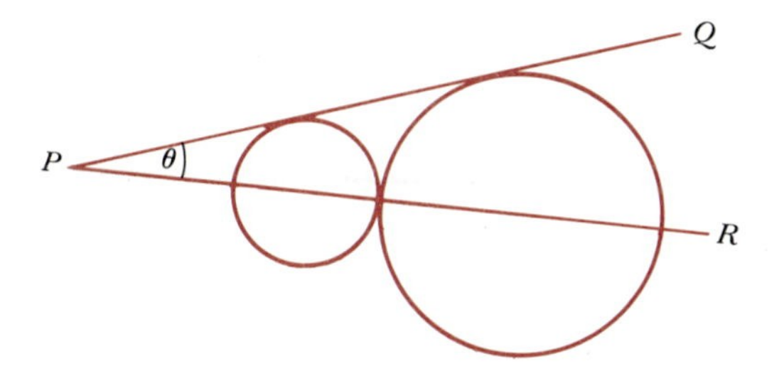

Then, using this result and the formula in part (a) of Exercise 35, show that

$$\cos \theta = \frac{2\sqrt{ab}}{a + b}$$

***38.** Suppose that β is an acute angle and

$$\sin \beta = \frac{m^2 - n^2}{m^2 + n^2} \quad (m > n > 0)$$

Show that

$$\cos \beta = \frac{2mn}{m^2 + n^2} \quad \text{and} \quad \tan \beta = \frac{m^2 - n^2}{2mn}$$

Hint: See Exercise 35.

39. **(a)** Show that the area of a regular n-gon inscribed in a circle of radius 1 unit is given by

$$A_n = \frac{n}{2} \sin \frac{360°}{n}$$

C **(b)** Use a calculator and the formula in part (a) to complete the following table.

n	5	10	50	100	1000	5000	10,000
A_n							

(c) Explain why the successive values of A_n in your table get closer and closer to π.

***40.** A vertical tower of height h stands on level ground. From a point P at ground level and due south of the tower, the angle of elevation to the top of the tower is θ. From a point Q at ground level and due west of the tower, the angle of elevation of the top of the tower is β. If d is the distance between P and Q, show that

$$h = \frac{d}{\sqrt{\cot^2 \theta + \cot^2 \beta}}$$

C ***41.** There is a simple construction with a straightedge and compass for inscribing a regular hexagon in a circle. You probably know the method from high school geometry. It turns out that there is no corresponding method, using only a straightedge and compass, for inscribing a regular heptagon (seven sides) in a circle. About 2000 years ago, Heron of Alexandria used the length of the apothem of a regular hexagon to approximate the length of the side of a regular heptagon. (The *apothem* is a line drawn from the center of a regular polygon perpendicular to a side.) Is this a good approximation? (Assume the radius of the circle is 1 unit in both cases and use a calculator where necessary.)

42. At point P on the earth's surface, the moon is observed to be directly overhead, while at the same time at T, the moon is just visible. See the figure.

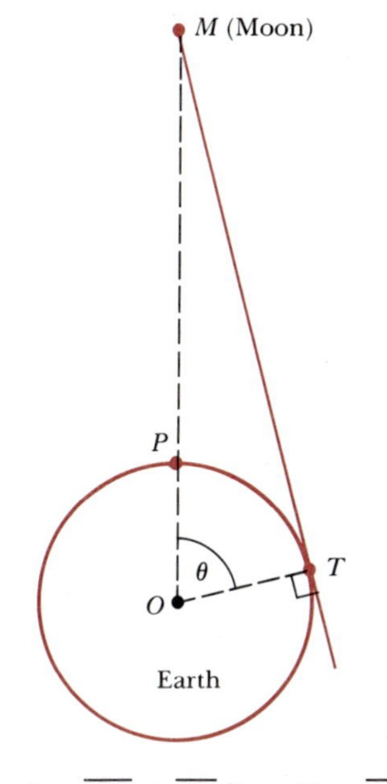

(a) Show that $\overline{MP} = \overline{OT}/(\cos \theta) - \overline{OP}$.

C **(b)** Use a calculator and the following data to estimate the distance $\overline{MP}$ from the earth to the moon. $\theta = 89.05°$, and $\overline{OT} = \overline{OP} = 4000$ mi. (Round off your answer to the nearest thousand miles. Because the moon's orbit is not really circular, the actual distance varies from about 216,400 mi to 247,000 mi.)

43. Refer to the figure at the top of the following page. Let r denote the radius of the moon.

(a) Show that $r = \left(\frac{\sin \theta}{1 - \sin \theta}\right) \overline{PS}$.

C **(b)** Use a calculator and the following data to estimate the radius r of the moon. $\overline{PS} = 238{,}857$ miles and $\theta = 0.257°$.

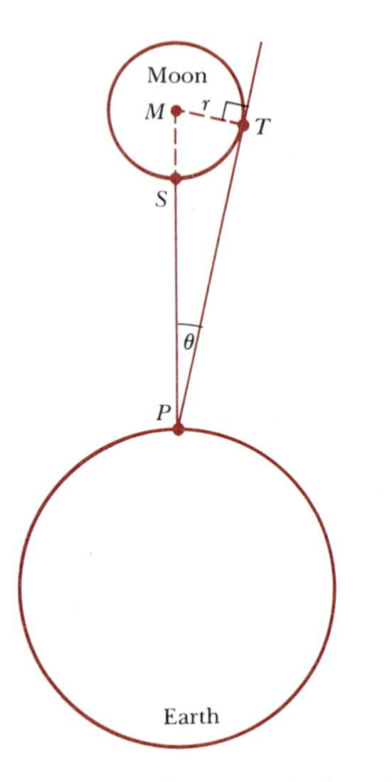

44. This exercise shows how to calculate sin 18° and cos 18° using the figure below.

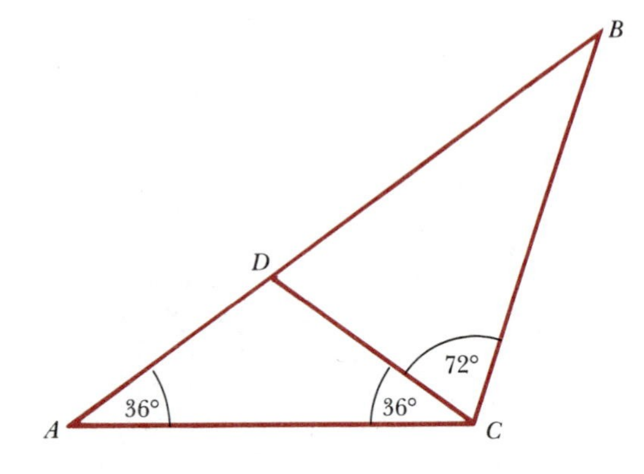

(a) Find $\angle B$, $\angle BDC$, and $\angle ADC$.
(b) Why does $\overline{AC} = \overline{BC} = \overline{BD}$?
For the rest of this problem, assume $\overline{AD} = 1$.
(c) Why does $\overline{CD} = 1$?
(d) Let x denote the common lengths of $\overline{AC}$, $\overline{BC}$, and $\overline{BD}$. Use similar triangles to deduce that $\dfrac{x}{1+x} = \dfrac{1}{x}$. Then show $x = \dfrac{1+\sqrt{5}}{2}$.
(e) In $\triangle ACD$, draw an altitude from D to AC, meeting AC at E. Use right triangle DEC to conclude that $\cos 36° = \dfrac{1+\sqrt{5}}{4}$.
(f) In $\triangle BDC$, draw an altitude from B to DC, meeting DC at F. Use right triangle BFC to conclude that $\sin 18° = \dfrac{1}{1+\sqrt{5}}$. Also show $\dfrac{1}{1+\sqrt{5}} = \dfrac{\sqrt{5}-1}{4}$.
(g) Verify that $\cos 36° - \sin 18° = \frac{1}{2}$.
(h) Show that $\sin 36° = \frac{1}{4}\sqrt{10 - 2\sqrt{5}}$.
Hint: Use the formula in Exercise 35(a).
(i) Show that $\cos 18° = \frac{1}{4}\sqrt{10 + 2\sqrt{5}}$.
(j) Verify that $\sin 36° = 2 \sin 18° \cos 18°$.

C **45.** Use a calculator to check the results given in Exercise 44, parts (g), (h), and (j).

46. The figure below shows a 30°-60° right triangle $\triangle ABC$. Prove that $\overline{AC} = 2\overline{AB}$. *Suggestion:* Construct $\triangle DBC$ as shown, congruent to $\triangle ABC$. Then note that $\triangle ADC$ is equilateral.

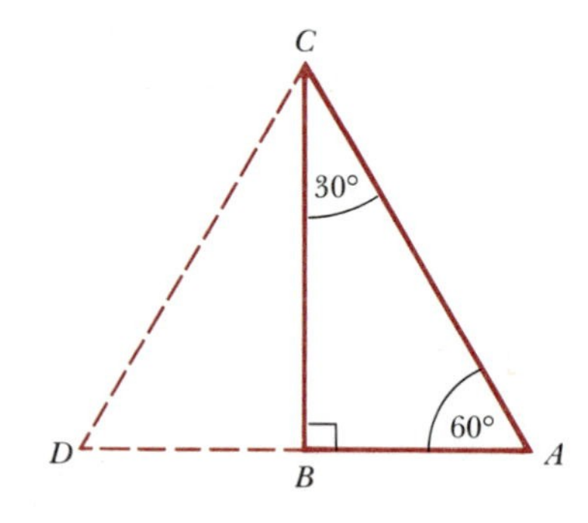

7.2 Algebra and the Trigonometric Functions

In this section, we are going to concentrate on the algebra involved in manipulating certain trigonometric expressions.

Some Notational Conventions

1. The quantity $(\sin\theta)^n$ is usually written as $\sin^n\theta$. For example, $(\sin\theta)^2$ is written as $\sin^2\theta$. The same convention also applies to the other five trigonometric functions.*
2. Parentheses are often omitted in multiplication. For example, the product $(\sin\theta)(\cos\theta)$ is usually written as $\sin\theta\cos\theta$. Similarly, $2(\sin\theta)$ is written $2\sin\theta$.
3. An expression such as $\sin\theta$ really means $\sin(\theta)$, where sin or sine is the name of the function and θ is an input. It is for historical rather than mathematical reasons that the parentheses are suppressed. An exception to this, however, occurs in expressions such as $\sin(A + B)$; here the parentheses are necessary.

Example 1 Multiply $(1 - \sin\theta)^2$.

Solution Do this the same way you would expand $(1 - s)^2$. Since

$$(1 - s)^2 = 1 - 2s + s^2$$

it follows that

$$(1 - \sin\theta)^2 = 1 - 2\sin\theta + \sin^2\theta$$

Example 2 Add $\sin\theta + \dfrac{1}{\cos\theta}$.

Solution The least common denominator is the quantity $\cos\theta$. We have

$$\begin{aligned}\sin\theta + \frac{1}{\cos\theta} &= \frac{\sin\theta}{1} + \frac{1}{\cos\theta}\\ &= \frac{\sin\theta}{1}\cdot\frac{\cos\theta}{\cos\theta} + \frac{1}{\cos\theta}\\ &= \frac{\sin\theta\cos\theta}{\cos\theta} + \frac{1}{\cos\theta}\\ &= \frac{\sin\theta\cos\theta + 1}{\cos\theta}\end{aligned}$$

This is the required result.

Example 3 Show that the statement $\cos A + \cos B = \cos(A + B)$ is not true in general.

Solution Back up for a minute. How would you convince a beginning algebra student that the statement $(x + y)^2 = x^2 + y^2$ is not true in general? One way would be to simply pick specific values for x and y and then show the equation fails in that case. For instance, using $x = 1$ and $y = 1$ would give

$$(1 + 1)^2 \stackrel{?}{=} 1^2 + 1^2$$

$$4 \stackrel{?}{=} 2 \qquad \text{No!}$$

*A single exception to this convention is the case $n = -1$. The meaning of expressions such as $\sin^{-1} x$ will be explained in Section 7 of Chapter 8.

We can do the same thing in the problem at hand. Let $A = 30°$ and $B = 30°$. Then we have

$$\cos A + \cos B \stackrel{?}{=} \cos(A + B)$$
$$\cos 30° + \cos 30° \stackrel{?}{=} \cos(30° + 30°)$$
$$\frac{\sqrt{3}}{2} + \frac{\sqrt{3}}{2} \stackrel{?}{=} \cos 60°$$
$$\sqrt{3} \stackrel{?}{=} \frac{1}{2} \qquad \text{No!}$$

We conclude that the statement $\cos A + \cos B = \cos(A + B)$ is not true in general. [To see what $\cos(A + B)$ does equal, see Exercise 41 at the end of this section.]

Example 4 Factor $\tan^2 A + 5 \tan A + 6$.

Preliminary solution To help you focus on the algebra that is actually involved, replace each occurrence of the quantity $\tan A$ by the letter T. Then

$$T^2 + 5T + 6 = (T + 3)(T + 2)$$

Actual solution

$$\tan^2 A + 5 \tan A + 6 = (\tan A + 3)(\tan A + 2)$$

Note: After you are accustomed to working with trigonometric expressions, you can eliminate the preliminary step that we have used above.

Example 5 Simplify:

$$\frac{\dfrac{\sin\theta + 1}{\sin\theta} + 1}{\dfrac{\sin\theta - 1}{\sin\theta} - 1}$$

Preliminary solution

$$\frac{\dfrac{S+1}{S} + 1}{\dfrac{S-1}{S} - 1} = \frac{S\left(\dfrac{S+1}{S} + 1\right)}{S\left(\dfrac{S-1}{S} - 1\right)} = \frac{S + 1 + S}{S - 1 - S} = \frac{2S + 1}{-1} = -2S - 1$$

Actual solution

$$\frac{\dfrac{\sin\theta + 1}{\sin\theta} + 1}{\dfrac{\sin\theta - 1}{\sin\theta} - 1} = \frac{\sin\theta\left(\dfrac{\sin\theta + 1}{\sin\theta} + 1\right)}{\sin\theta\left(\dfrac{\sin\theta - 1}{\sin\theta} - 1\right)} = \frac{\sin\theta + 1 + \sin\theta}{\sin\theta - 1 - \sin\theta} = \frac{2\sin\theta + 1}{-1} = -2\sin\theta - 1 \text{ as required}$$

There are numerous identities involving the trigonometric functions. For now we'll state and prove three of these identities. In the next chapter we take up this subject in greater detail. Recall that an *identity* is an equation that is satisfied by all relevant values of the variables concerned. Two examples of identities are $x^2 - y^2 = (x - y)(x + y)$ and $x^3 = x^4/x$. The first of these is true no matter what real numbers are used for x and y. The second of these is true for all real numbers except $x = 0$.

Theorem

Let θ be an acute angle. Then

1. $\sin^2\theta + \cos^2\theta = 1$
2. $\dfrac{\sin\theta}{\cos\theta} = \tan\theta$
3. $\sin(90° - \theta) = \cos\theta$ and $\cos(90° - \theta) = \sin\theta$

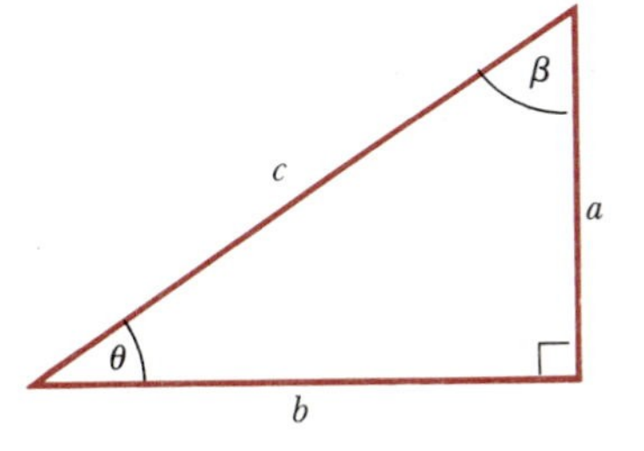

Figure 1

All three of these identities can be proved by referring to Figure 1.

Proof that $\sin^2\theta + \cos^2\theta = 1$ Looking at Figure 1, we have $\sin\theta = a/c$ and $\cos\theta = b/c$. Thus

$$\begin{aligned}\sin^2\theta + \cos^2\theta &= \left(\frac{a}{c}\right)^2 + \left(\frac{b}{c}\right)^2\\ &= \frac{a^2}{c^2} + \frac{b^2}{c^2}\\ &= \frac{a^2+b^2}{c^2}\\ &= \frac{c^2}{c^2} \qquad \text{since } a^2+b^2=c^2, \text{ by the Pythagorean Theorem}\\ &= 1 \qquad \text{as required}\end{aligned}$$

Proof that $(\sin\theta)/(\cos\theta) = \tan\theta$ Again, with reference to Figure 1, we have $\sin\theta = a/c$, $\cos\theta = b/c$, and $\tan\theta = a/b$. Therefore

$$\begin{aligned}\frac{\sin\theta}{\cos\theta} &= \frac{a/c}{b/c}\\ &= \frac{a}{c}\times\frac{c}{b}\\ &= \frac{a}{b}\\ &= \tan\theta \qquad \text{as required}\end{aligned}$$

Proof that $\sin(90° - \theta) = \cos\theta$ First of all, since the sum of the angles in any triangle is 180°, we have

$$\begin{aligned}\theta + \beta + 90° &= 180°\\ \beta &= 180° - 90° - \theta\\ \beta &= 90° - \theta\end{aligned}$$

Then $\sin(90° - \theta) = \sin\beta = b/c$. But also $\cos\theta = b/c$. Thus

$$\sin(90° - \theta) = \cos\theta$$

since both expressions equal b/c. This is what we wanted to prove.

The proof that $\cos(90° - \theta) = \sin\theta$ is entirely similar and we shall omit it. We can conveniently summarize these last two results by recalling the notion of complementary angles. Two acute angles are said to be *complementary* provided their sum is 90°. Thus the two angles θ and $90° - \theta$ are complementary. In view of this, we can restate the last two results as follows.

> If two angles are complementary, then the sine of (either) one equals the cosine of the other.

Incidentally, this result gives us an insight into the origin of the term "cosine." "Cosine" is a shortened form of the phrase "complement's sine."

Example 6 Multiply $(\sin\theta + \cos\theta)^2$.

Solution

$$(\sin\theta + \cos\theta)^2 = \underbrace{\sin^2\theta}_{} + 2\sin\theta\cos\theta + \underbrace{\cos^2\theta}_{}$$

The sum of these two terms is 1.

$$= 1 + 2\sin\theta\cos\theta$$

Example 7 Suppose that B is an acute angle and $\cos B = \frac{2}{5}$. Find $\sin B$ and $\tan B$.

Solution We substitute the value given for $\cos B$ in the identity $\sin^2 B + \cos^2 B = 1$.

$$\sin^2 B + \left(\frac{2}{5}\right)^2 = 1$$

$$\sin^2 B = 1 - \frac{4}{25} = \frac{21}{25}$$

$$\sin B = \sqrt{\frac{21}{25}}$$

$$\sin B = \frac{\sqrt{21}}{5} \quad \text{as required}$$

Notice that we've chosen the positive square root here. This is because the values of the trigonometric functions of an acute angle are by definition positive. *Caution:* In the next section we'll extend the definitions of the trigonometric functions to include angles that are not acute. In those cases, some of the trigonometric values will not be positive; we'll then need to pay more attention to the matter of signs.

Next, to calculate $\tan B$ we have

$$\tan B = \frac{\sin B}{\cos B}$$

$$= \frac{\frac{\sqrt{21}}{5}}{\frac{2}{5}} = \frac{\sqrt{21}}{5} \times \frac{5}{2} = \frac{\sqrt{21}}{2}$$

The required value of tan B is therefore $\frac{\sqrt{21}}{2}$.

Example 8 Combine and simplify $\frac{\sin A}{\cos A} + \frac{\cos A}{\sin A}$.

Solution The common denominator is cos A sin A. We have

$$\begin{aligned}\frac{\sin A}{\cos A} + \frac{\cos A}{\sin A} &= \frac{\sin A}{\cos A} \cdot \frac{\sin A}{\sin A} + \frac{\cos A}{\sin A} \cdot \frac{\cos A}{\cos A} \\ &= \frac{\sin^2 A}{\cos A \sin A} + \frac{\cos^2 A}{\sin A \cos A} \\ &= \frac{\sin^2 A + \cos^2 A}{\cos A \sin A} \\ &= \frac{1}{\cos A \sin A}\end{aligned}$$

This is the required result. If we wish, we can write this answer in an alternative form that doesn't involve fractions:

$$\begin{aligned}\frac{1}{\cos A \sin A} &= \frac{1}{\cos A} \cdot \frac{1}{\sin A} \\ &= \sec A \csc A\end{aligned}$$

The identities that were stated in the theorem on page 329 should be memorized. This is because they are used so frequently in trigonometry, as well as in calculus and other areas where trigonometry is applied. On the other hand, the identities that follow in Examples 9 through 14 should not be memorized; they are too specialized. In these examples you should concentrate on the proofs themselves, noting where the basic definitions and the results from our theorem are used.

Example 9 Prove that the equation csc A tan A cos A = 1 is an identity.

Solution Begin with the left-hand side and express each factor in terms of sines or cosines.

$$\begin{aligned}\csc A \tan A \cos A &= \frac{1}{\cancel{\sin A}} \cdot \frac{\cancel{\sin A}}{\cancel{\cos A}} \cdot \cancel{\cos A} \\ &= 1 \qquad \text{as required}\end{aligned}$$

Example 10 Prove $(\cot \theta)/(\csc \theta) = \sin(90° - \theta)$.

Solution Again we begin with the left-hand side and express everything in terms of sin θ and cos θ.

$$\begin{aligned}\frac{\cot \theta}{\csc \theta} &= \frac{\frac{\cos \theta}{\sin \theta}}{\frac{1}{\sin \theta}} \qquad \text{Since } \tan \theta = \frac{\sin \theta}{\cos \theta}, \text{ it follows that } \cot \theta = \frac{\cos \theta}{\sin \theta}. \\ &= \frac{\cos \theta}{\cancel{\sin \theta}} \cdot \frac{\cancel{\sin \theta}}{1}\end{aligned}$$

$$= \cos\theta$$
$$= \sin(90° - \theta) \qquad \text{as required}$$

Example 11 Prove $\cos^2 B - \sin^2 B = \dfrac{1 - \tan^2 B}{1 + \tan^2 B}$.

Solution We begin with the right-hand side this time; it is the more complicated expression. As in the previous examples, we express everything in terms of sines and cosines.

$$\frac{1 - \tan^2 B}{1 + \tan^2 B} = \frac{1 - \dfrac{\sin^2 B}{\cos^2 B}}{1 + \dfrac{\sin^2 B}{\cos^2 B}}$$

$$= \frac{\cos^2 B\left(1 - \dfrac{\sin^2 B}{\cos^2 B}\right)}{\cos^2 B\left(1 + \dfrac{\sin^2 B}{\cos^2 B}\right)}$$

$$= \frac{\cos^2 B - \sin^2 B}{\cos^2 B + \sin^2 B}$$

$$= \frac{\cos^2 B - \sin^2 B}{1}$$

$$= \cos^2 B - \sin^2 B \qquad \text{as required}$$

Example 12 Prove $\dfrac{\sec A + 1}{\sin A + \tan A} = \csc A$.

Solution Again, the technique is to begin with the more complicated side and to express everything in terms of $\sin A$ and $\cos A$.

$$\frac{\sec A + 1}{\sin A + \tan A} = \frac{\dfrac{1}{\cos A} + 1}{\sin A + \dfrac{\sin A}{\cos A}}$$

$$= \frac{\cos A\left(\dfrac{1}{\cos A} + 1\right)}{\cos A\left(\sin A + \dfrac{\sin A}{\cos A}\right)}$$

$$= \frac{1 + \cos A}{\cos A \sin A + \sin A}$$

$$= \frac{\overset{1}{\cancel{1 + \cos A}}}{\sin A\,\underset{1}{\cancel{(\cos A + 1)}}}$$

$$= \frac{1}{\sin A}$$

$$= \csc A \qquad \text{as required}$$

Example 13 Prove $\frac{\cos\theta}{1-\sin\theta} = \frac{1+\sin\theta}{\cos\theta}$.

Solution The suggestions given in the previous examples are not applicable here. Everything is already in terms of sines and cosines. Furthermore, neither side appears more complicated than the other. A technique that does work here is to begin with the left-hand side and multiply numerator *and* denominator by the same quantity, namely $1 + \sin\theta$. Doing so gives us

$$\begin{aligned}\frac{\cos\theta}{1-\sin\theta} &= \frac{\cos\theta}{1-\sin\theta}\cdot\frac{1+\sin\theta}{1+\sin\theta}\\ &= \frac{\cos\theta(1+\sin\theta)}{1-\sin^2\theta}\\ &= \frac{\cos\theta(1+\sin\theta)}{\cos^2\theta}\\ &= \frac{1+\sin\theta}{\cos\theta} \quad \text{as required}\end{aligned}$$

The general strategy for each of the proofs in Examples 9 through 13 was the same. In each case we worked with one side of the given equation, and we transformed it (into equivalent expressions) until it was identical to the other side of the equation. This is not the only strategy that can be used. For instance, an alternate way to establish the identity in Example 13 is as follows. The given equation is equivalent to

$$\frac{\cos\theta}{1-\sin\theta} - \frac{1+\sin\theta}{\cos\theta} = 0 \qquad (1)$$

Now we show that equation (1) is an identity. To do this, we combine the two fractions on the left-hand side, using the common denominator $\cos\theta(1-\sin\theta)$. We have

$$\begin{aligned}\frac{\cos\theta}{1-\sin\theta} - \frac{1+\sin\theta}{\cos\theta} &= \frac{\cos^2\theta - (1+\sin\theta)(1-\sin\theta)}{\cos\theta(1-\sin\theta)}\\ &= \frac{\cos^2\theta - (1-\sin^2\theta)}{\cos\theta(1-\sin\theta)}\\ &= \frac{\cos^2\theta + \sin^2\theta - 1}{\cos\theta(1-\sin\theta)} = \frac{1-1}{\cos\theta(1-\sin\theta)}\\ &= 0\end{aligned}$$

This shows that equation (1) is an identity, and thus the equation given in Example 13 is an identity.

Another strategy that can be used in establishing trigonometric identities is to work independently with each side of the given equation until a common expression is obtained. This is the strategy used in Example 14.

Example 14 Prove $\frac{\cos A \tan A}{\tan(90^\circ - A)} = \sec A - \cos A$.

Solution

Left-hand side	*Right-hand side*
$\dfrac{\cos A \tan A}{\tan(90^\circ - A)} = \dfrac{(\cancel{\cos A})\left(\dfrac{\sin A}{\cancel{\cos A}}\right)}{\dfrac{\sin(90^\circ - A)}{\cos(90^\circ - A)}}$	$\sec A - \cos A = \dfrac{1}{\cos A} - \cos A$
$= \dfrac{\sin A}{\left(\dfrac{\cos A}{\sin A}\right)}$	$= \dfrac{1 - \cos^2 A}{\cos A}$
$= \dfrac{\sin^2 A}{\cos A}$	$= \dfrac{\sin^2 A}{\cos A}$

We've now established the required identity by showing that both sides are equal to the same quantity [namely, $(\sin^2 A)/(\cos A)$].

Exercise Set 7.2

1. Perform the indicated operations and simplify where possible.
(a) $(1 - \cos\theta)^2$ **(b)** $(\sin\theta + \cos\theta)^2$
(c) $\cos\theta + \dfrac{1}{\sin\theta}$
(d) $(\sin A + \cos A)(\csc A + \sec A)$
(e) $(\sin^2 B + 1)(\cos^2 B + 1)$

2. Show that each statement is *not* true in general.
(a) $\sin(A + B) = \sin A + \sin B$
(b) $\cos 2\theta = 2\cos\theta$ (Use $\theta = 30^\circ$.)
(c) $\sin\dfrac{\theta}{2} = \dfrac{\sin\theta}{2}$ (Use $\theta = 60^\circ$.)

3. Factor each expression.
(a) $\tan^2\theta - 5\tan\theta - 6$
(b) $\sin^2 B - \cos^2 B$
(c) $\cos^2 A + 2\cos A + 1$
(d) $3\sin\theta\cos^2\theta + 6\sin\theta$
(e) $\csc^2\alpha - 2\csc\alpha - 3$
(f) $6\sin^2\theta + 7\sin\theta + 2$

4. Simplify each expression.
(a) $\dfrac{\dfrac{\cos\theta + 1}{\cos\theta} + 1}{\dfrac{\cos\theta - 1}{\cos\theta} - 1}$ **(b)** $\dfrac{\dfrac{\cot\beta + 1}{\cot\beta} + 1}{\dfrac{\cot\beta - 1}{\cot\beta} - 1}$

5. Simplify $\dfrac{1 - \tan\theta}{\dfrac{\sin\theta}{\cos\theta} - 1}$.

6. Simplify each expression.
(a) $\dfrac{2 + \dfrac{1}{\cos\theta}}{\dfrac{1}{\cos^2\theta}}$ *Hint:* Multiply numerator and denominator by $\cos^2\theta$.
(b) $\dfrac{\sin^2 A - \cos^2 A}{\sin A - \cos A}$ *Hint:* Factor the numerator.
(c) $\dfrac{\sin^2 A - \cos^2 A}{\cos A - \sin A}$

7. Simplify each expression.
(a) $\sin^2\theta\cos\theta\csc^3\theta\sec\theta$
(b) $\sin\theta\csc\theta\tan\theta$
(c) $\cot B\sin^2 B\cot B$

8. Simplify each expression.
(a) $\dfrac{3\sin\theta + 6}{\sin^2\theta - 4}$
(b) $\dfrac{\cos^2 A + \cos A - 12}{\cos A - 3}$
(c) $\dfrac{x - y}{y - x}$
(d) $\dfrac{\sin\theta - \cos\theta}{\cos\theta - \sin\theta}$

9. Verify each statement using the figure.
(a) $\sin^2\theta + \cos^2\theta = 1$
(b) $\sin^2\theta + \cos^2\beta \neq 1$
(c) $\tan\theta = \dfrac{\sin\theta}{\cos\theta}$
(d) $\sin\theta = \cos\beta$ and $\cos\theta = \sin\beta$

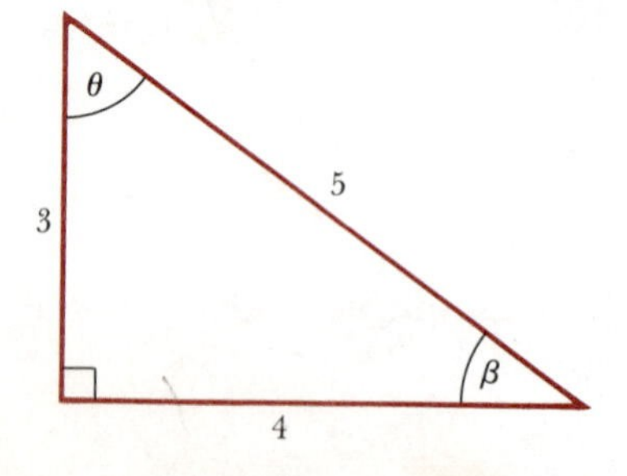

10. Verify each statement.
(a) $\sin^2 30° + \cos^2 30° = 1$
(b) $\tan 60° = \dfrac{\sin 60°}{\cos 60°}$

11. Suppose that θ is an acute angle and $\cos \theta = \frac{5}{13}$. Use the method of Example 7 to determine $\sin \theta$ and $\tan \theta$.

12. In Example 7 we were given that $\cos B = \frac{2}{5}$ and we found $\sin B$ using the identity $\sin^2 B + \cos^2 B = 1$. A different method for determining $\sin B$ is as follows. Since $\cos B = \dfrac{2}{5} = \dfrac{\text{adjacent}}{\text{hypotenuse}}$, you can draw the diagram below. Now use the Pythagorean Theorem to determine the third side of this triangle. After this, the value of $\sin B$ (or of any other trigonometric function of B) can be read off from the figure. Check that the result you obtain in this manner for $\sin B$ agrees with that obtained in Example 7.

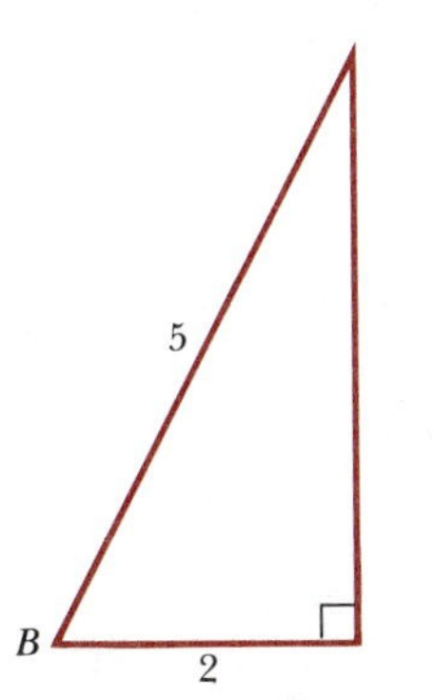

13. Assume that θ is an acute angle and $\cos \theta = \frac{4}{5}$. Evaluate the other five trigonometric functions of θ. (Use either the method of Example 7 or the method of Exercise 12.)

14. Assume that A is an acute angle and $\sec A = 3$. Evaluate the other five trigonometric functions of A. *Hint:* $\sec A = \dfrac{\text{hypotenuse}}{\text{adjacent}} = \dfrac{3}{1}$. Draw a sketch of this.

15. **(a)** Given that $\sin 18° = \frac{1}{4}(\sqrt{5} - 1)$, find $\cos 18°$, using the method of Example 7.
(b) Find $\sin 72°$ and $\cos 72°$.

16. **(a)** Given that $\sin 15° = \frac{1}{4}(\sqrt{6} - \sqrt{2})$ and $\cos 15° = \frac{1}{4}(\sqrt{6} + \sqrt{2})$, find $\tan 15°$.
(b) Show that the expression you've obtained for $\tan 15°$ is in fact equal to $2 - \sqrt{3}$. *Hint:* Rationalize your answer in part (a) by multiplying by

$$\frac{\sqrt{6} - \sqrt{2}}{\sqrt{6} - \sqrt{2}}$$

which equals 1.

17. Prove the identity $\sin \theta \cos \theta \sec \theta \csc \theta = 1$.

18. Prove the identity $\tan^2 A + 1 = \sec^2 A$.

19. Prove the identity $(\sin \theta \sec \theta)/(\tan \theta) = 1$.

In Exercises 20–35, prove that the equations are identities.

20. $\tan \beta \sin \beta = \sec \beta - \cos \beta$

21. $\dfrac{1 - 5 \sin x}{\cos x} = \sec x - 5 \tan x$

22. $\dfrac{1}{\sin \theta} - \sin \theta = \cot \theta \cos \theta$

23. $\cos A(\sec A - \cos A) = \sin^2 A$

24. $\dfrac{\sin \theta}{\csc \theta} + \dfrac{\cos \theta}{\sec \theta} = 1$

25. $(1 - \sin \theta)(\sec \theta + \tan \theta) = \cos \theta$

26. $(\cos \theta - \sin \theta)^2 + 2 \sin \theta \cos \theta = 1$

27. $(\sec \alpha - \tan \alpha)^2 = \dfrac{1 - \sin \alpha}{1 + \sin \alpha}$

28. $\dfrac{\sin B}{1 + \cos B} + \dfrac{1 + \cos B}{\sin B} = 2 \csc B$

29. $\sin A + \cos A = \dfrac{\sin A}{1 - \cot A} - \dfrac{\cos A}{\tan A - 1}$

30. $(1 - \cos C)(1 + \sec C) = \tan C \sin C$

31. $\csc^2 \theta + \sec^2 \theta = \csc^2 \theta \sec^2 \theta$

32. $1 - \sin(90° - \theta) \cos \theta = \sin^2 \theta$

33. $\dfrac{2 \sin^3 \beta}{1 - \cos \beta} = 2 \sin \beta + 2 \sin \beta \cos \beta$
Hint: Write $\sin^3 \beta$ as $(\sin \beta)(\sin^2 \beta)$.

34. $\dfrac{\cos B \tan B}{\tan(90° - B)} = \sin^2 B \sec B$

35. $\dfrac{\sin^3 \theta + \cos^3 \theta}{\sin \theta + \cos \theta} = 1 - \sin \theta \cos \theta$

36. Prove the identity $\dfrac{\sin \theta}{1 - \cos \theta} = \dfrac{1 + \cos \theta}{\sin \theta}$ in two different ways.
(a) Adapt the method of Example 13.
(b) Begin with the left-hand side and multiply numerator and denominator by $\sin \theta$.

37. Prove the identity $\dfrac{\sec \theta - \csc \theta}{\sec \theta + \csc \theta} = \dfrac{\tan \theta - 1}{\tan \theta + 1}$.

***38.** **(a)** Factor the expression $\cos^3 \theta - \sin^3 \theta$.
(b) Prove the identity $\dfrac{\cos \phi \cot \phi - \sin \phi \tan \phi}{\csc \phi - \sec \phi} = 1 + \sin \phi \cos \phi$.

39. Prove the identity

$$(r \sin \theta \cos \phi)^2 + (r \sin \theta \sin \phi)^2 + (r \cos \theta)^2 = r^2.$$

40. *Formula for* $\sin(\alpha + \beta)$ In the figure, $AD \perp BC$ and $AD = 1$.

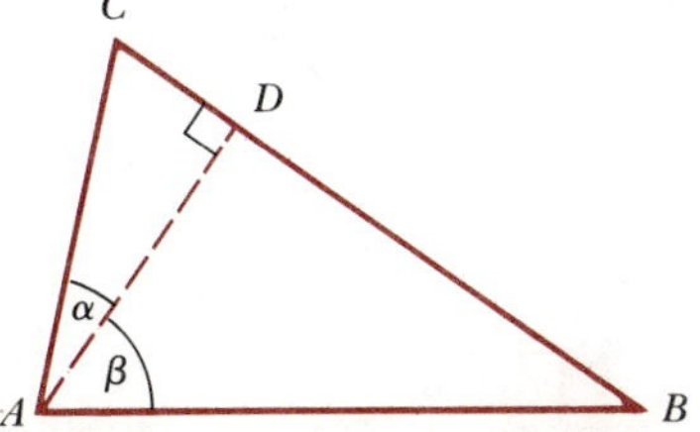

(a) Show that $\overline{AC} = \sec\alpha$ and $\overline{AB} = \sec\beta$.

(b) Show that

$$\text{area } \triangle ADC = \tfrac{1}{2}\sec\alpha\sin\alpha$$

$$\text{area } \triangle ADB = \tfrac{1}{2}\sec\beta\sin\beta$$

$$\text{area } \triangle ABC = \tfrac{1}{2}\sec\alpha\sec\beta\sin(\alpha+\beta)$$

Hint: Use the formula given in Example 7 of Section 1 in this chapter.

(c) The sum of the areas of the two smaller triangles in part (b) equals the area of $\triangle ABC$. Use this fact and the expressions given in part (b) to show that

$$\sin(\alpha+\beta) = \sin\alpha\cos\beta + \cos\alpha\sin\beta$$

(d) Find sin 75°. *Hint:* 75° = 30° + 45°.

(e) Show that sin 75° ≠ sin 30° + sin 45°.

(f) Find cos 15°.

41. *Formula for* $\cos(\alpha+\beta)$ The figure shown is constructed as follows. Begin with right triangle ABC in which $AC = 1$ and $\alpha + \beta = \angle A$. Extend line AP and draw CE perpendicular to it. Draw ED perpendicular to BC. Finally, from E, draw a perpendicular to AB extended, meeting this extension in F.

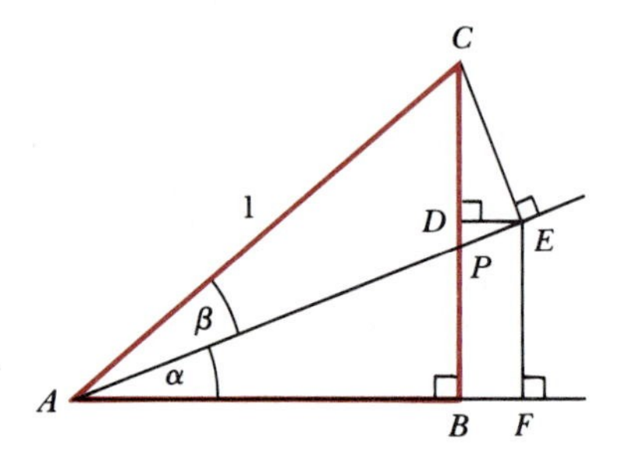

(a) Show $\overline{CE} = \sin\beta$. (Look at triangle AEC.)

(b) Show $\angle DCE = \alpha$.

(c) Show $\overline{DE} = \sin\alpha\sin\beta$. (Look at triangle CDE.)

(d) Show $\overline{AE} = \cos\beta$.

(e) Show $\overline{AF} = \cos\alpha\cos\beta$.

(f) Show $\cos(\alpha+\beta) = \overline{AF} - \overline{DE}$.
Hint: $\overline{DE} = \overline{BF}$.

(g) On the basis of steps (f), (e), and (c), conclude that

$$\cos(\alpha+\beta) = \cos\alpha\cos\beta - \sin\alpha\sin\beta$$

(h) Find cos 75°. *Hint:* 75° = 30° + 45°.

(i) Show that cos 75° ≠ cos 30° + cos 45°.

42. Let $f(\theta) = \tan\theta + \cot\theta$ $(0° < \theta < 90°)$.

C (a) Complete the tables, rounding off the values of $f(\theta)$ to three decimal places.

θ	10°	20°	30°	40°	50°	60°	70°	80°
$f(\theta)$								

θ	40°	42°	44°	45°	46°	48°	50°
$f(\theta)$							

(b) What is the smallest value that you obtained for $f(\theta)$ in part (a)?

(c) Use the inequality $\sqrt{ab} \le (a+b)/2$ [given in Exercise Set 2.5, Exercise 64(b)] to prove that $\tan\theta + \cot\theta \ge 2$ $(0° < \theta < 90°)$. *Hint:* In the inequality from Exercise Set 2.5, make the substitutions $a = \tan\theta$ and $b = \cot\theta$.

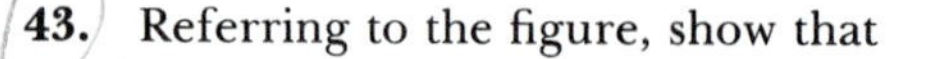

43. Referring to the figure, show that

$$x = \cos\theta$$

$$y = \cos^2\theta$$

$$z = \cos^3\theta$$

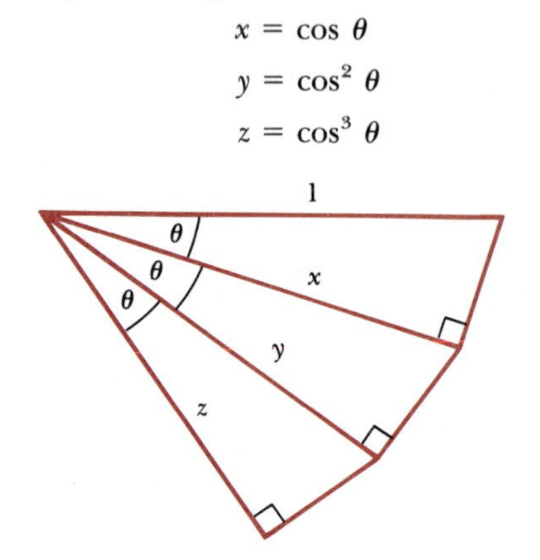

***44.** Prove the identity $\dfrac{\tan\theta + \sec\theta - 1}{\tan\theta - \sec\theta + 1} = \dfrac{1+\sin\theta}{\cos\theta}$.

45. Suppose that

$$A\sin\theta + \cos\theta = 1$$

and

$$B\sin\theta - \cos\theta = 1$$

Show that $AB = 1$. *Hint:* Solve the first equation for A, the second for B, and then compute AB.

46. (a) Make the substitution $u = a\sin\theta$ in the expression $\sqrt{a^2 - u^2}$ and show that the expression becomes $a\cos\theta$. (Assume $0 < u < a$.)

(b) Make the substitution $u = a\tan\theta$ in the expression $\sqrt{a^2 + u^2}$ and show that the result is $a\sec\theta$. (Assume $u > 0$ and $a > 0$.)

(c) Make the substitution $x = 2\sin\theta$ in the expression $\dfrac{x^2}{(4 - x^2)^{3/2}}$ and show that the result is $\tfrac{1}{2}\tan^2\theta\sec\theta$.

47. Prove the identity

$$\tan\theta(1 - \cot^2\theta) + \cot\theta(1 - \tan^2\theta) = 0$$

48. In the figure, show that the equation of line L is $x\cos\theta + y\sin\theta = d$.

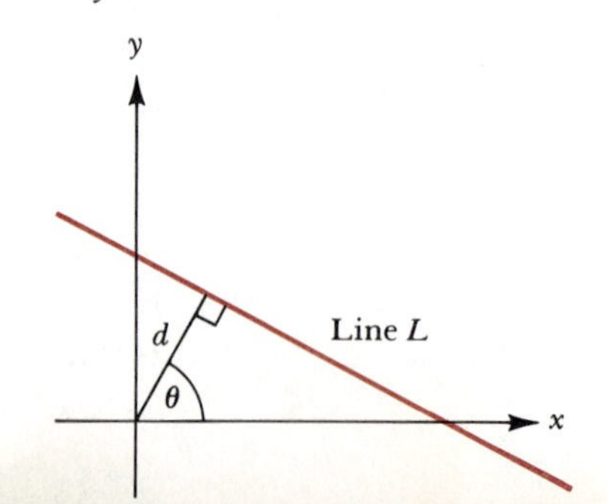

7.3 Trigonometric Functions of General Angles

There is nothing strange in the circle being the origin of any and every marvel.
Aristotle

Recall that the definitions of the trigonometric functions given in Section 7.1 apply only to acute angles. That is, as defined on page 312, the domains of trigonometric functions consist of all angles θ such that $0° < \theta < 90°$. In this section, we want to expand our definitions of the trigonometric functions to include angles of any size. To do this, we will need to be more explicit than we have been about angles and their size or measure.

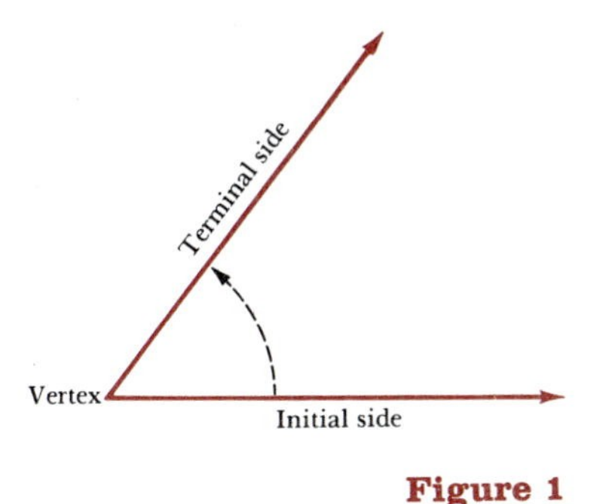

Figure 1

An angle is formed by two rays, or half-lines, issuing from a common point called the *vertex*. For our purposes, it is useful to think of the two rays as originally coincident. Then, while one ray is held fixed, the other is rotated to create the given angle. As Figure 1 indicates, the fixed ray is called the *initial side* of the angle, while the rotated ray is called the *terminal side*. By convention, we take the measure of an angle to be positive if the rotation is counterclockwise, and negative if the rotation is clockwise. Figure 2 shows some examples of this.

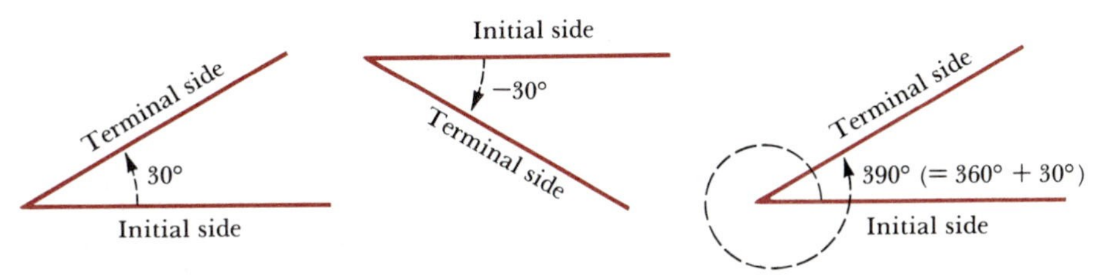

Figure 2

We say that an angle is in *standard position* provided:

(a) The vertex of the angle coincides with the origin in a rectangular coordinate system.

(b) The initial side of the angle lies along the positive horizontal axis.

Figure 3 displays examples of angles in standard position.

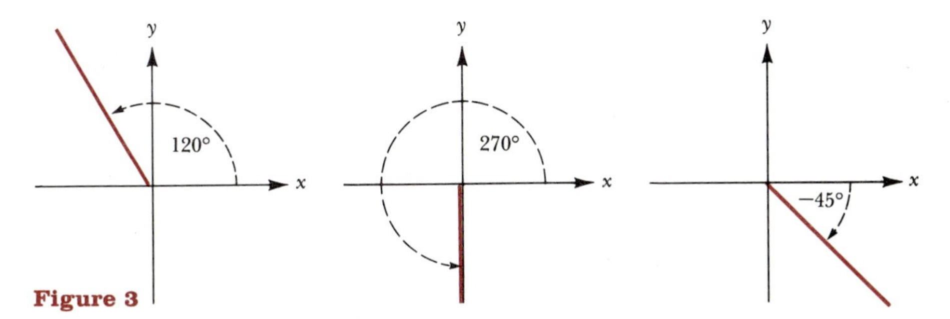

Figure 3

Now, to extend our definitions of the trigonometric functions to accommodate any angle θ, we begin by placing the angle θ in standard position; see Figure 4(a) on page 338. Next we draw the *unit circle* $x^2 + y^2 = 1$, as shown in Figure 4(b). (Recall from Chapter 3 that the equation $x^2 + y^2 = 1$ represents a circle of radius 1, with center at the origin.) As Figure 4(b) indicates, $P(x, y)$ denotes the point where the terminal side of angle θ intersects the unit circle. Then with this notation, we state the following definitions.

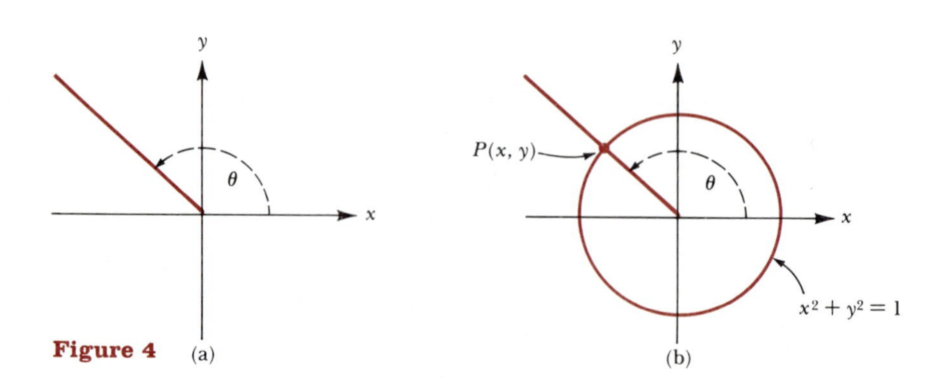

Figure 4

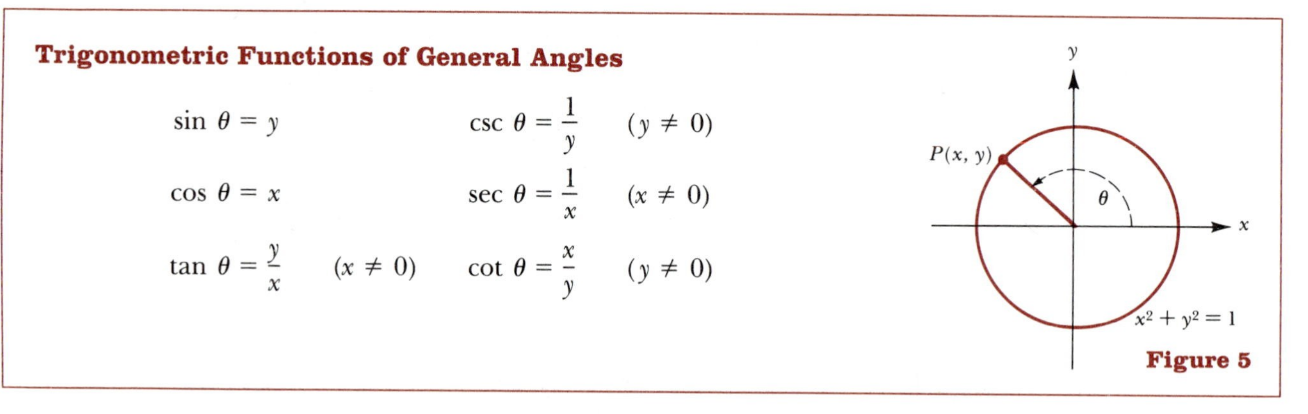

Trigonometric Functions of General Angles

$$\sin\theta = y \qquad \csc\theta = \frac{1}{y} \quad (y \neq 0)$$

$$\cos\theta = x \qquad \sec\theta = \frac{1}{x} \quad (x \neq 0)$$

$$\tan\theta = \frac{y}{x} \quad (x \neq 0) \qquad \cot\theta = \frac{x}{y} \quad (y \neq 0)$$

Figure 5

The above definitions of sine and cosine can be summarized this way:

Sin θ is the second coordinate of that point where the terminal side of angle θ intersects the unit circle; cos θ is the first coordinate of that point.

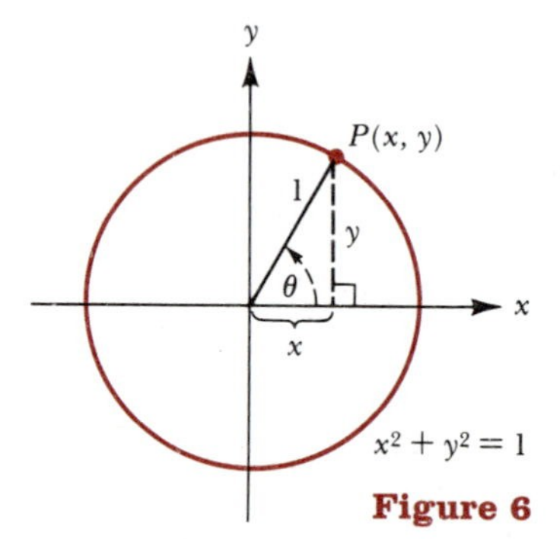

Figure 6

For the cases in which the angle θ is acute, these definitions are really equivalent to the original right triangle definitions on page 312. Consider, for instance, the acute angle θ in Figure 6. According to our "new" definition of sine, we have

$$\sin\theta = y$$

On the other hand, applying the original right triangle definition in Figure 6 yields

$$\sin\theta = \frac{\text{opposite}}{\text{hypotenuse}} = \frac{y}{1} = y$$

Thus, in both cases we obtain the same result. Using Figure 6, you should check for yourself that the same agreement also occurs for the other trigonometric functions.

In fact, these definitions agree in other ways too with our previous work. Notice the definitions imply that $\tan\theta = (\sin\theta)/(\cos\theta)$. Additionally, from the definitions we see that $\sin\theta$ and $\csc\theta$ are reciprocals, as are $\cos\theta$ and $\sec\theta$, as well as $\tan\theta$ and $\cot\theta$.

Example 1 Compute sin 90°, cos 90°, tan 90°, csc 90°, sec 90°, and cot 90°.

Solution Place the angle $\theta = 90°$ in standard position. Then, as Figure 7 indicates, the terminal side of the angle meets the unit circle at the point (0, 1). Now apply the definitions.

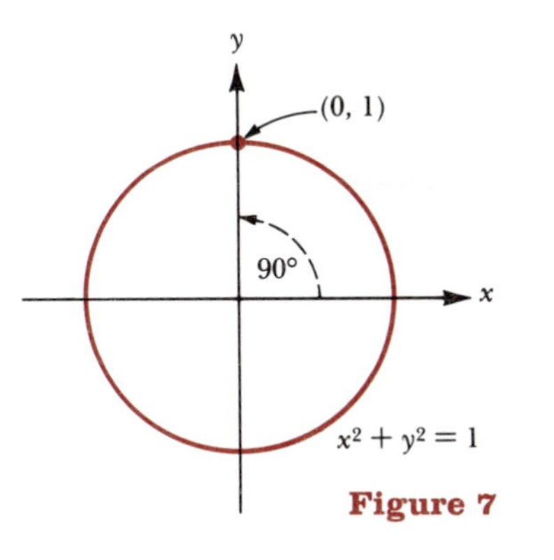

Figure 7

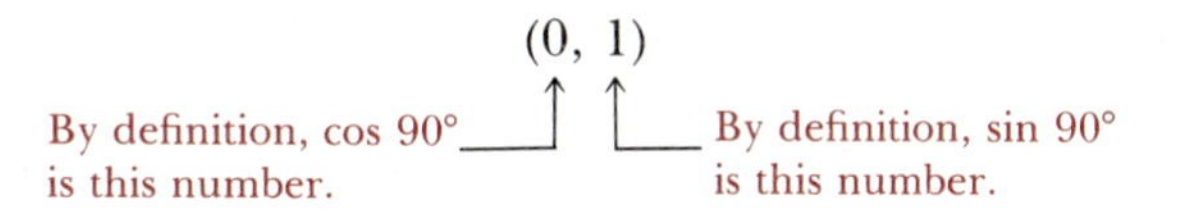

Thus,

$$\sin 90° = 1 \quad \text{and} \quad \cos 90° = 0$$

For the remaining trigonometric functions of 90°, we have

$$\tan 90° = \frac{y}{x} = \frac{1}{0} \qquad \text{which is undefined}$$

$$\csc 90° = \frac{1}{y} = \frac{1}{1} = 1$$

$$\sec 90° = \frac{1}{x} = \frac{1}{0} \qquad \text{which is undefined}$$

$$\cot 90° = \frac{x}{y} = \frac{0}{1} = 0$$

These are the required results.

Example 2 Evaluate the trigonometric functions of $-180°$.

Solution The results can be read off from Figure 8.

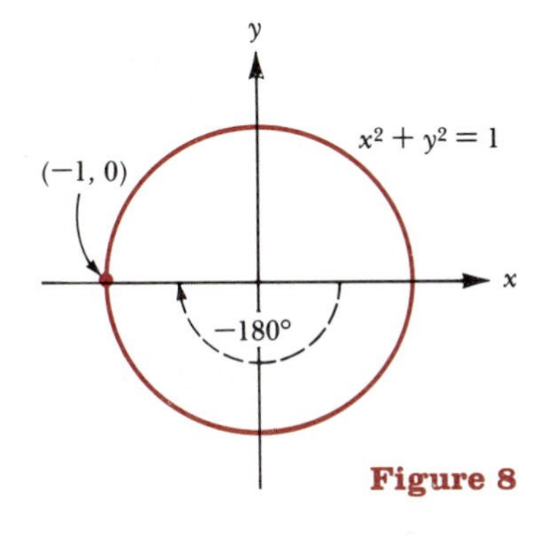

Figure 8

$$\sin(-180°) = y = 0$$

$$\cos(-180°) = x = -1$$

$$\tan(-180°) = \frac{y}{x} = \frac{0}{-1} = 0$$

$$\csc(-180°) = \frac{1}{y} = \frac{1}{0} \qquad \text{which is undefined}$$

$$\sec(-180°) = \frac{1}{x} = \frac{1}{-1} = -1$$

$$\cot(-180°) = \frac{x}{y} = \frac{-1}{0} \qquad \text{which is undefined}$$

In the two examples just concluded, we evaluated the trigonometric functions for $\theta = 90°$ and for $\theta = -180°$. In the same manner, the trigonometric functions can be evaluated just as easily for any angle that is an integral multiple of 90°. Table 1 shows the results of such calculations. Exercise 3 at the end of this section asks you to make these calculations for yourself.

θ	$\sin\theta$	$\cos\theta$	$\tan\theta$	$\csc\theta$	$\sec\theta$	$\cot\theta$
0°	0	1	0	undefined	1	undefined
90°	1	0	undefined	1	undefined	0
180°	0	−1	0	undefined	−1	undefined
270°	−1	0	undefined	−1	undefined	0
360°	0	1	0	undefined	1	undefined

Table 1

In order to evaluate the trigonometric functions for angles that are not multiples of 90°, we introduce the notion of a reference angle. Given an angle θ that is not a multiple of 90°, the *reference angle* associated with θ is the acute angle formed by the x-axis and the terminal side of θ. Figure 9 displays examples of angles and their respective reference angles.

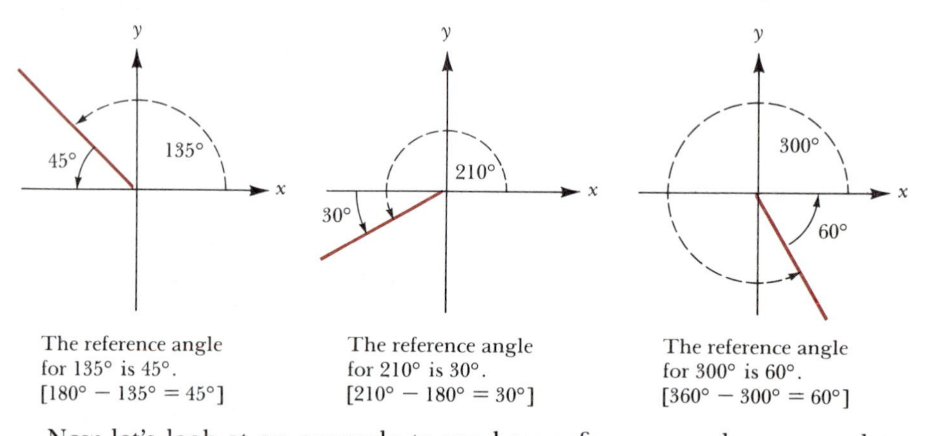

Figure 9

Now let's look at an example to see how reference angles are used to evaluate the trigonometric functions. Suppose we want to evaluate cos 150°. In Figure 10(a) we've placed the angle $\theta = 150°$ in standard position. As you can also see in Figure 10(a), the reference angle for 150° is 30°. By definition, the value of cos 150° is the x-coordinate of the point P in Figure 10(a). To determine this x-coordinate, we'll use Figure 10(b), in which the line segment PQ is parallel to the x-axis. There are two facts that follow from the symmetry of the situation in Figure 10(b). (Exercise 33 outlines a detailed verification of these two facts.)

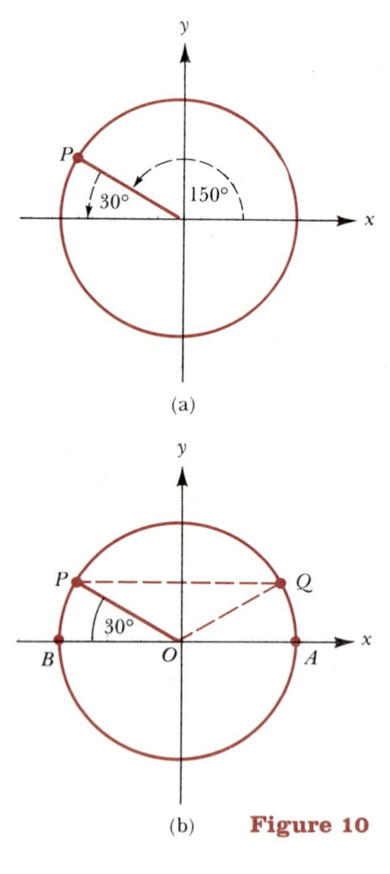

Figure 10

First fact The x-coordinate of P is just the negative of the x-coordinate of Q.

Second fact The reference angle POB is equal to angle QOA, and therefore angle QOA is 30°.

Now we apply these two facts. Since angle QOA is 30°, the x-coordinate of Q is by definition cos 30°, or $\sqrt{3}/2$. The x-coordinate of P is then the negative of this. That is, the x-coordinate of P is $-\sqrt{3}/2$. It follows now, again by definition, that the value of cos 150° is $-\sqrt{3}/2$.

The same method that we have just used to evaluate cos 150° can be used to evaluate any of the trigonometric functions when the angles are not multiples of 90°. The following three steps summarize this method.

Step 1 Determine the reference angle associated with the given angle.

Step 2 Evaluate the given trigonometric function using the reference angle for the input.

Step 3 Affix the appropriate sign to the number found in Step 2.

The next three examples illustrate this procedure.

Example 3 Evaluate the following.

(a) sin 135° **(b)** cos 135° **(c)** tan 135°

Solution As Figure 11 indicates, the reference angle associated with 135° is 45°.

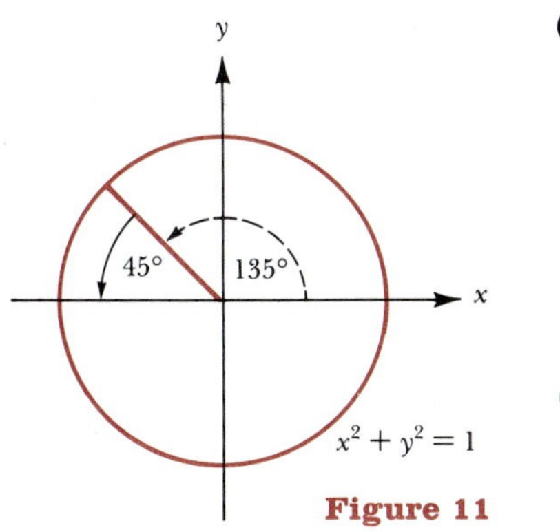

Figure 11

(a) *Step 1* The reference angle is 45°.

Step 2 $\sin 45° = \sqrt{2}/2$

Step 3 Sin θ is the y-coordinate. In the second quadrant y-coordinates are positive. Thus sin 135° is positive, since the terminal side of $\theta = 135°$ lies in the second quadrant. We have therefore

$$\sin 135° = \frac{\sqrt{2}}{2}$$

(b) *Step 1* The reference angle is 45°.

Step 2 $\cos 45° = \sqrt{2}/2$

Step 3 Cos θ is the x-coordinate. In the second quadrant x-coordinates are negative. Thus cos 135° is negative, since the terminal side of $\theta = 135°$ lies in the second quadrant. We have therefore

$$\cos 135° = -\frac{\sqrt{2}}{2}$$

(c) *Step 1* The reference angle is 45°.

Step 2 $\tan 45° = 1$

Step 3 By definition, $\tan \theta = y/x$. Now the terminal side of $\theta = 135°$ lies in the second quadrant, in which y is positive and x is negative. Thus tan 135° is negative. We have therefore

$$\tan 135° = -1$$

Example 4 Evaluate cos 240°.

Solution As Figure 12 shows, the reference angle for 240° is 60°.

Step 1 The reference angle for 240° is 60°.

Step 2 $\cos 60° = \frac{1}{2}$

Step 3 Cos θ is the x-coordinate. In the third quadrant x-coordinates are negative. Thus cos 240° is negative, since the terminal side of $\theta = 240°$ lies in the third quadrant. It follows now that

$$\cos 240° = -\frac{1}{2}$$

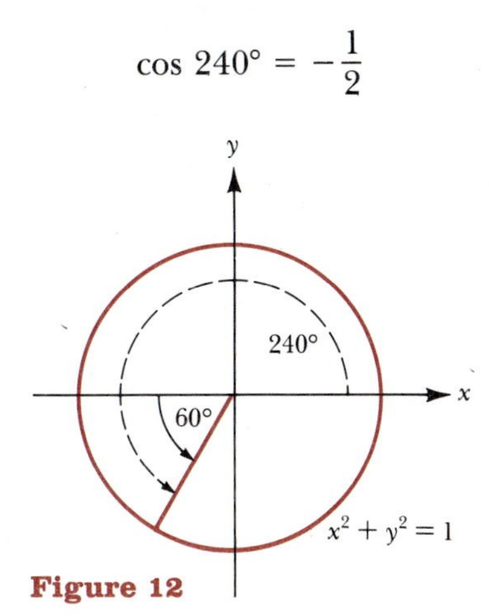

Figure 12

Example 5 Evaluate $\cot 330°$.

Solution As Figure 13 shows, the reference angle for 330° is 30°.

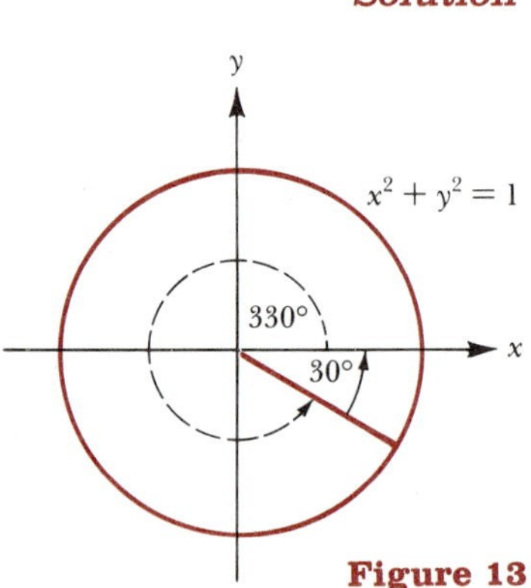

Figure 13

Step 1 The reference angle for 330° is 30°.

Step 2 $\cot 30° = \sqrt{3}$

Step 3 By definition, $\cot \theta = x/y$. Now the terminal side of $\theta = 330°$ lies in the fourth quadrant, in which x is positive and y is negative. Thus $\cot 330°$ is negative. It follows now that

$$\cot 330° = -\sqrt{3}$$

In the previous section, we showed that the identity $\sin^2 \theta + \cos^2 \theta = 1$ is valid for all values of θ in the range $0° < \theta < 90°$. Now we're in a position to show that this identity holds for all angles, not just acute angles. Also in the previous section we showed that the sine of an angle is equal to the cosine of its complement. Here we are going to prove an analogous (but not identical) result concerning supplementary angles. Recall that two angles are *supplementary* provided their sum is 180°. Thus θ and $180° - \theta$ form a pair of supplementary angles. Our theorem is as follows.

> **Theorem**
>
> **1.** For any angle θ, we have
>
> $$\sin^2 \theta + \cos^2 \theta = 1$$
>
> **2.** If $0° < \theta < 180°$, then
>
> $$\cos(180° - \theta) = -\cos \theta$$
>
> and
>
> $$\sin(180° - \theta) = \sin \theta$$

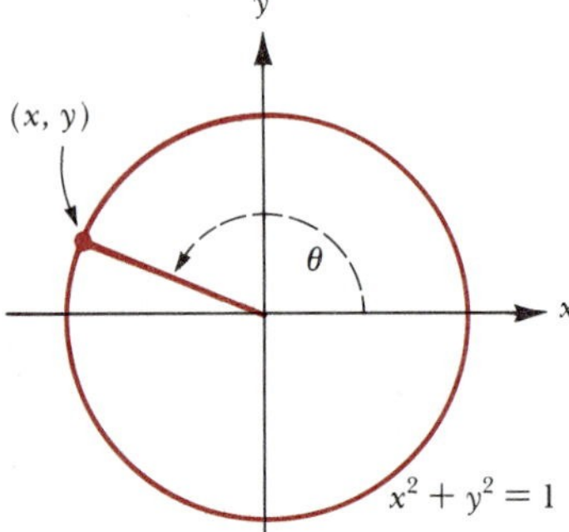

Figure 14

To prove that $\sin^2 \theta + \cos^2 \theta = 1$, we need only use the unit circle definitions of sine and cosine. See Figure 14. Since the point (x, y) lies on the unit circle, we have

$$x^2 + y^2 = 1$$

Replacing x by $\cos \theta$ and y by $\sin \theta$ in this equation then yields

$$(\cos \theta)^2 + (\sin \theta)^2 = 1$$

That is,

$$\sin^2 \theta + \cos^2 \theta = 1 \qquad \text{as required}$$

We will prove the second part of our theorem for the case in which θ is an acute angle. (See Exercises 34 and 35 at the end of this section for the cases in which θ is an angle between 90° and 180°.) Now, referring to Figure 15, notice that the reference angle for $(180° - \theta)$ is θ itself. Thus we have

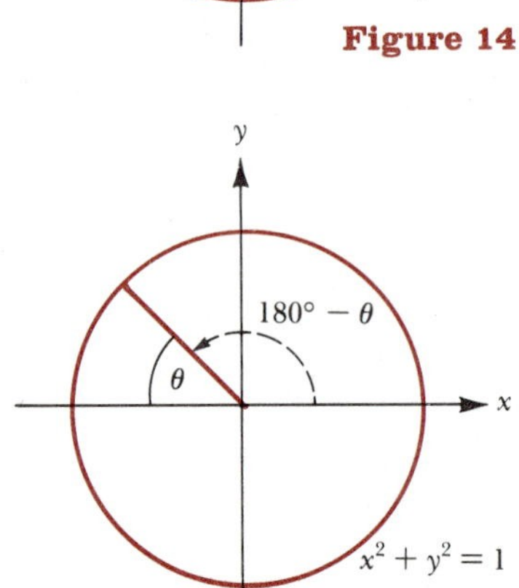

Figure 15

$$\cos(180° - \theta) = -\cos \theta$$

(reference angle for $(180° - \theta)$ — points to θ; negative sign since the terminal side of $(180° - \theta)$ lies in the second quadrant — points to the minus sign) and

$$\sin(180^\circ - \theta) = \underbrace{\sin \theta}$$

↓ reference angle for $(180^\circ - \theta)$

↑ positive since the terminal side of $(180^\circ - \theta)$ lies in the second quadrant

These observations complete the proof of the second part of our theorem.

Example 6 Given that $\sin \theta = \frac{2}{3}$ and $90^\circ < \theta < 180^\circ$, find $\cos \theta$.

Solution Make the substitution $\sin \theta = \frac{2}{3}$ in the identity $\sin^2 \theta + \cos^2 \theta = 1$. This yields

$$\left(\frac{2}{3}\right)^2 + \cos^2 \theta = 1$$

$$\cos^2 \theta = 1 - \frac{4}{9} = \frac{5}{9}$$

$$\cos \theta = \pm\sqrt{\frac{5}{9}} = \frac{\pm\sqrt{5}}{3}$$

To decide whether to choose the positive or negative value, note that the given inequality, $90^\circ < \theta < 180^\circ$, tells us that the terminal side of θ lies in the second quadrant. Since in the second quadrant x-coordinates are negative, we choose the negative value here. Thus

$$\cos \theta = \frac{-\sqrt{5}}{3} \qquad \text{as required}$$

Example 7 Show that the area A of the triangle in Figure 16(a) is given by

$$A = \frac{1}{2}ab \sin \theta$$

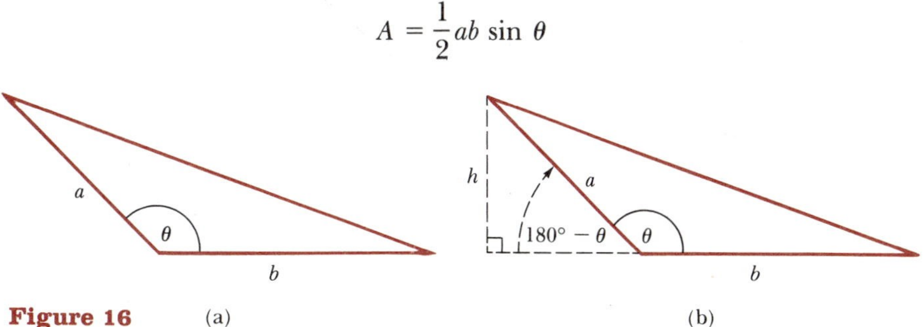

Figure 16 (a) (b)

Note: In Section 7.1 we proved this formula for the case in which θ is an acute angle.

Solution The area A of any triangle is given by

$$A = \frac{1}{2}(\text{base})(\text{height})$$

Thus, referring to Figure 16(b) we have

$$A = \frac{1}{2}bh \tag{1}$$

Also from Figure 16(b) we have

$$\begin{aligned} \sin(180° - \theta) &= \frac{\text{opposite}}{\text{hypotenuse}} = \frac{h}{a} \\ h &= a \sin(180° - \theta) \\ h &= a \sin \theta \qquad (2) \end{aligned}$$

Now we can use equation (2) to substitute for h in equation (1). This yields

$$\begin{aligned} A &= \frac{1}{2}b(a \sin \theta) \\ &= \frac{1}{2}ab \sin \theta \qquad \text{as required} \end{aligned}$$

In words: The area of a triangle equals half the product of two sides times the sine of the included angle. We will use this result in the next section to prove the so-called *law of sines.*

Exercise Set 7.3

1. Sketch the following angles in standard position. In each case, specify the reference angle.
(a) 110° **(b)** 240° **(c)** 182°
(d) 300° **(e)** 1000° **(f)** −15°

2. Evaluate the following. If the value is not defined, say so.
(a) cos 0° **(b)** sin 450°
(c) sin 270° **(d)** sin(−270°)
(e) tan 180° **(f)** sec(−90°)
(g) csc(−90°) **(h)** cot 720°

3. Verify each of the entries in Table 1.

4. In the figure, the x-coordinate of P is $\frac{3}{4}$.
(a) What is cos θ?
(b) Find sin θ and tan θ.

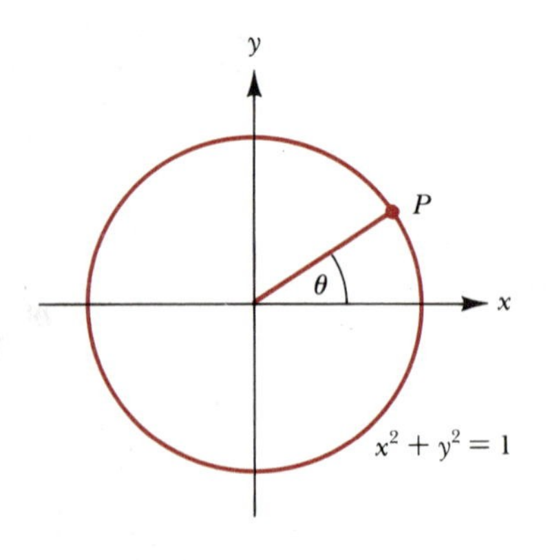

5. In the figure, the y-coordinate of P is $\frac{2}{3}$. Find cot θ.

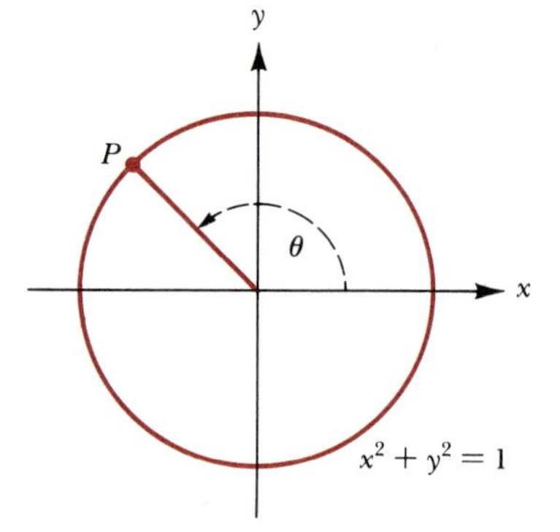

6. Which is larger, cos 325° or sin 325°? (No calculations are necessary; draw a sketch.)

7. Use the method of Example 3 to evaluate the following.
(a) sin 315° **(b)** cos 300° **(c)** tan 330°
(d) cos 210° **(e)** sec 210° **(f)** csc 225°
(g) tan 135° **(h)** cot 120° **(i)** cot 480°

8. Evaluate the six trigonometric functions of $\theta = -30°$.

9. Evaluate the six trigonometric functions of $\theta = -45°$.

10. List the following quantities in order of increasing size.
(a) sin 90° **(b)** sin 180°
(c) cos 60° **(d)** tan 315°

11. Complete the following tables.

(a)

θ	0°	90°	180°	270°	360°	450°	540°	630°	720°
$\sin\theta$									
$\cos\theta$									

(b)

θ	30°	60°	90°	120°	150°	180°	210°	240°	270°	300°	330°	360°
$\sin\theta$												

12. Complete the following tables.

(a)

θ	$\sin\theta$	$\cos\theta$	$\tan\theta$
0°			
30°			
45°			
60°			
90°			
120°			
135°			
150°			
180°			

(b)

θ	$\sin\theta$	$\cos\theta$	$\tan\theta$
180°			
210°			
225°			
240°			
270°			
300°			
315°			
330°			
360°			

13. **(a)** Complete the following chart, using the words "positive" or "negative" as appropriate.

	Terminal Side of Angle θ Lies in			
	Quadrant I	Quadrant II	Quadrant III	Quadrant IV
$\sin\theta$	positive	positive		
$\cos\theta$				
$\tan\theta$				

(b) The mnemonic (memory device) ASTC (*a*ll *s*tudents *t*ake *c*alculus) is sometimes used to recall the signs of the trigonometric values in each quadrant.

A . . . . . all are positive in the first quadrant.

S . . . . . sine is positive in the second quadrant.

T . . . . . tangent is positive in the third quadrant.

C . . . . . cosine is positive in the fourth quadrant.

Check the validity of this mnemonic against your chart in part (a).

14. Find $\cos\theta$ given that $\sin\theta = \frac{1}{5}$ and $90° < \theta < 180°$.

15. Find $\sin\theta$ given that $\cos\theta = -\frac{3}{5}$ and $180° < \theta < 270°$.

16. Find $\tan\theta$ given that $\sin\theta = -\frac{12}{13}$ and $270° < \theta < 360°$.

17. Find $\sec\beta$ given that $\sin\beta = \frac{1}{4}$ and $90° < \beta < 180°$.

18. List three angles for which the sine of each is equal to 1.

19. List three angles for which the cosine of each is equal to $-\frac{1}{2}$.

20. Find the area of each triangle. Draw a sketch in each case.

(a) Two of the sides are 5 and 7 cm, respectively, and the angle between those sides is 120°.

(b) Two of the sides are 3 and 6 in., respectively, and the included angle is 150°.

***21.** An equilateral triangle is inscribed in a circle of radius 1 unit. Find the area of the triangle.

22. Use the figure at the top of the next page to estimate the following quantities.

(a) $\sin 10°$; $\cos 10°$ **(b)** $\sin 20°$; $\cos 20°$
(c) $\sin 30°$; $\cos 30°$ **(d)** $\sin 40°$; $\cos 40°$
(e) $\sin 50°$; $\cos 50° \longrightarrow$ *Hint:* $\sin\theta = \cos(90 - \theta)$.
(f) $\sin 70°$; $\cos 70°$ **(g)** $\sin 80°$; $\cos 80°$
(h) $\sin 100°$; $\cos 100°$ **(i)** $\sin 130°$; $\cos 130°$

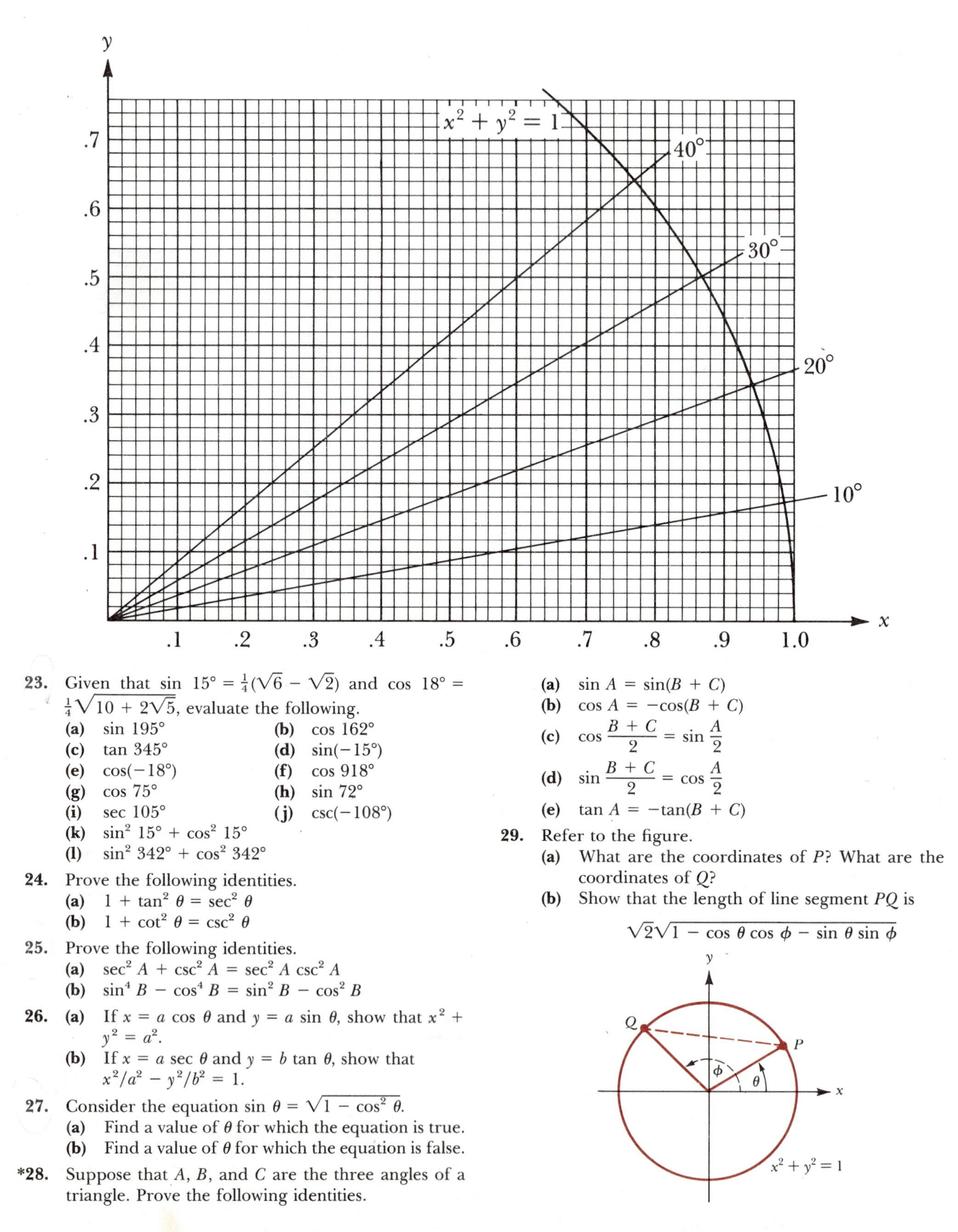

23. Given that $\sin 15° = \frac{1}{4}(\sqrt{6} - \sqrt{2})$ and $\cos 18° = \frac{1}{4}\sqrt{10 + 2\sqrt{5}}$, evaluate the following.

(a) $\sin 195°$ (b) $\cos 162°$
(c) $\tan 345°$ (d) $\sin(-15°)$
(e) $\cos(-18°)$ (f) $\cos 918°$
(g) $\cos 75°$ (h) $\sin 72°$
(i) $\sec 105°$ (j) $\csc(-108°)$
(k) $\sin^2 15° + \cos^2 15°$
(l) $\sin^2 342° + \cos^2 342°$

24. Prove the following identities.

(a) $1 + \tan^2 \theta = \sec^2 \theta$
(b) $1 + \cot^2 \theta = \csc^2 \theta$

25. Prove the following identities.

(a) $\sec^2 A + \csc^2 A = \sec^2 A \csc^2 A$
(b) $\sin^4 B - \cos^4 B = \sin^2 B - \cos^2 B$

26. (a) If $x = a \cos \theta$ and $y = a \sin \theta$, show that $x^2 + y^2 = a^2$.
(b) If $x = a \sec \theta$ and $y = b \tan \theta$, show that $x^2/a^2 - y^2/b^2 = 1$.

27. Consider the equation $\sin \theta = \sqrt{1 - \cos^2 \theta}$.

(a) Find a value of θ for which the equation is true.
(b) Find a value of θ for which the equation is false.

***28.** Suppose that A, B, and C are the three angles of a triangle. Prove the following identities.

(a) $\sin A = \sin(B + C)$
(b) $\cos A = -\cos(B + C)$
(c) $\cos \dfrac{B + C}{2} = \sin \dfrac{A}{2}$
(d) $\sin \dfrac{B + C}{2} = \cos \dfrac{A}{2}$
(e) $\tan A = -\tan(B + C)$

29. Refer to the figure.

(a) What are the coordinates of P? What are the coordinates of Q?
(b) Show that the length of line segment PQ is

$$\sqrt{2}\sqrt{1 - \cos \theta \cos \phi - \sin \theta \sin \phi}$$

***30.** Use the figure to prove the following theorem: The area of a quadrilateral is equal to half the product of the diagonals times the sine of the included angle. *Hint:* Find the areas of each of the four triangles that make up the quadrilateral. Show that the sum of those areas is $\frac{1}{2}\sin\theta[qs + rq + rp + ps]$. Then factor the quantity within the brackets.

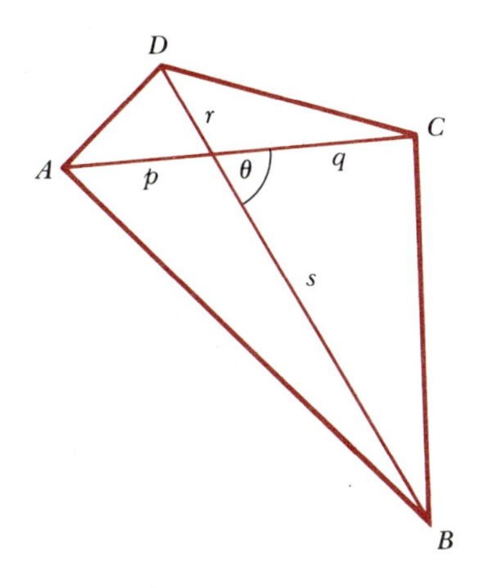

31. Prove the identity
$2\log_{10}\sin\theta = \log_{10}(1 - \cos\theta) + \log_{10}(1 + \cos\theta)$, $0° < \theta < 180°$. Why is the restriction $0° < \theta < 180°$ necessary?

32. Use the identity $\sin^2\theta + \cos^2\theta = 1$ to explain why for any angle θ we have

$$|\sin\theta| \le 1 \quad \text{and} \quad |\cos\theta| \le 1$$

33. In this exercise we are going to verify two statements that were made in the text concerning Figure 10(b). In particular, with reference to that figure, we'll show that angle QOA is 30°, and also that the x-coordinate of P is the negative of the x-coordinate of Q.

(a) Since PQ is parallel to BA, the two angles BOP and QPO are equal. Thus angle QPO is 30°. Now explain why angle PQO is 30°. *Hint:* The base angles of an isosceles triangle are equal.

(b) Since angle PQO is 30°, it follows that angle QOA is 30°. Why?

(c) Let C denote the point where the line segment PQ meets the y-axis. Show that triangle OPC is congruent to triangle OQC.

(d) Use the result in part (c) to explain why the x-coordinate of P is the negative of the x-coordinate of Q.

34. In this exercise we prove that the formulas

$$\cos(180° - \theta) = -\cos\theta$$

and

$$\sin(180° - \theta) = \sin\theta$$

are valid when $90° < \theta < 180°$. (Recall that the proof in the text assumes θ is an acute angle.)

The statements that follow refer to the figure. Supply the reason or reasons for each statement.

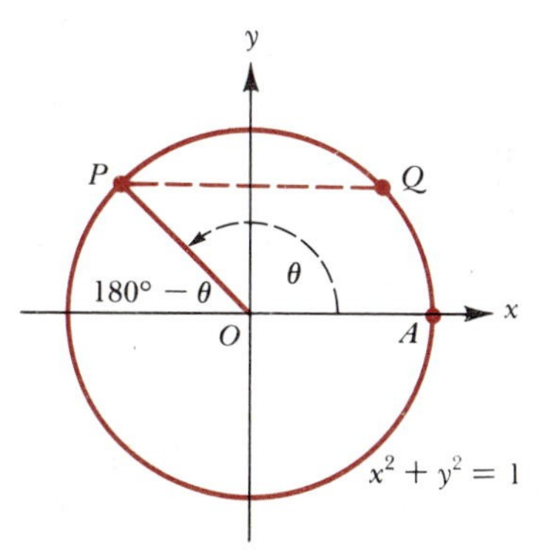

(a) The coordinates of P are $(\cos\theta, \sin\theta)$.

(b) Draw PQ parallel to the x-axis, as shown in the figure. Then the coordinates of Q are $(-\cos\theta, \sin\theta)$.

(c) $\angle QOA = 180° - \theta$

(d) The coordinates of Q are $(\cos(180° - \theta), \sin(180° - \theta))$.

(e) $\cos(180° - \theta) = -\cos\theta$ and $\sin(180° - \theta) = \sin\theta$.

35. Check that the two formulas in part (e) of Exercise 34 are valid in the cases $\theta = 0°$, $\theta = 90°$, and $\theta = 180°$.

7.4 The Law of Sines and the Law of Cosines

The law of sines and the law of cosines are, in essence, formulas relating the sides and angles in any triangle. These formulas can be used to determine an unknown side or angle using given information about the triangle. As you will see, which formula to apply in a particular case depends upon what data are initially given. To help you keep track of this, we'll use the following notation in our examples.

Notation	*Explanation*
SAA	One side and two angles are given.
SSA	Two sides and an angle opposite one of those sides is given.
SAS	Two sides and the included angle are given.
SSS	Three sides of the triangle are given.

Also in this section, we shall often follow the convention of denoting the angles of a triangle by A, B, and C and the lengths of the corresponding opposite sides by a, b, and c. With this notation, we're ready to state the law of sines.

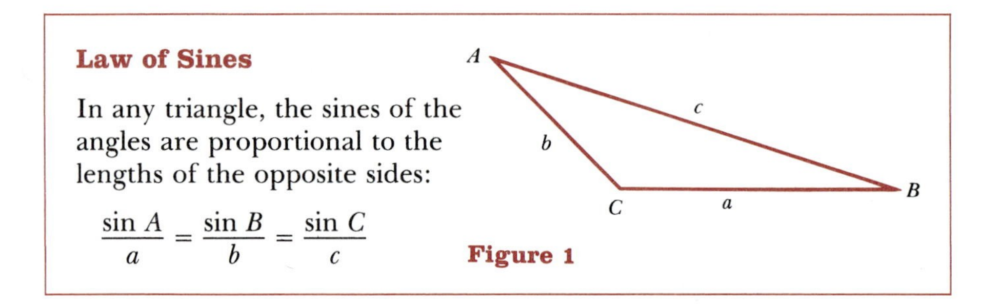

Law of Sines

In any triangle, the sines of the angles are proportional to the lengths of the opposite sides:

$$\frac{\sin A}{a} = \frac{\sin B}{b} = \frac{\sin C}{c}$$

Figure 1

The proof of the law of sines is easy. We use the result that was stated at the end of the previous section: The area of any triangle is equal to half the product of two sides times the sine of the included angle. Thus, with reference to Figure 1, we have

$$\frac{1}{2}bc \sin A = \frac{1}{2}ac \sin B = \frac{1}{2}ab \sin C$$

since each of these three expressions equals the area of triangle ABC. Now we just multiply through by the quantity $2/abc$. As you can check, this yields

$$\frac{\sin A}{a} = \frac{\sin B}{b} = \frac{\sin C}{c} \qquad \text{as we wished to show}$$

Example 1 (SAA) Find the length x in Figure 2.

Solution We can determine x directly by applying the law of sines. We have

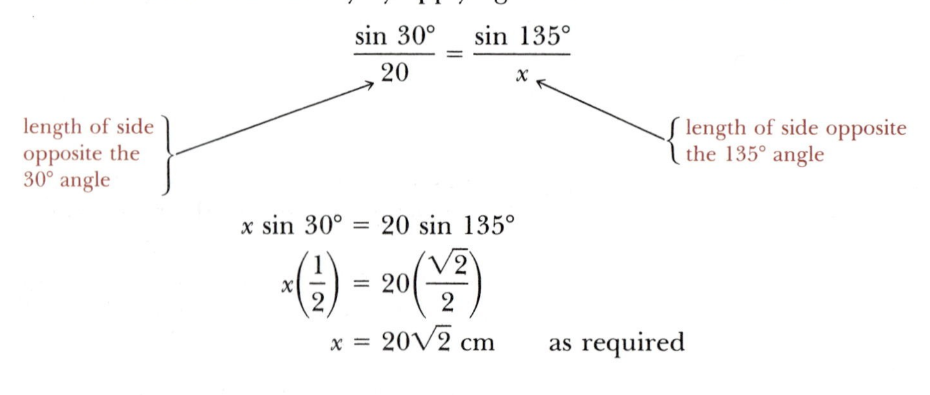

$$\frac{\sin 30^\circ}{20} = \frac{\sin 135^\circ}{x}$$

$$x \sin 30^\circ = 20 \sin 135^\circ$$

$$x\left(\frac{1}{2}\right) = 20\left(\frac{\sqrt{2}}{2}\right)$$

$$x = 20\sqrt{2} \text{ cm} \qquad \text{as required}$$

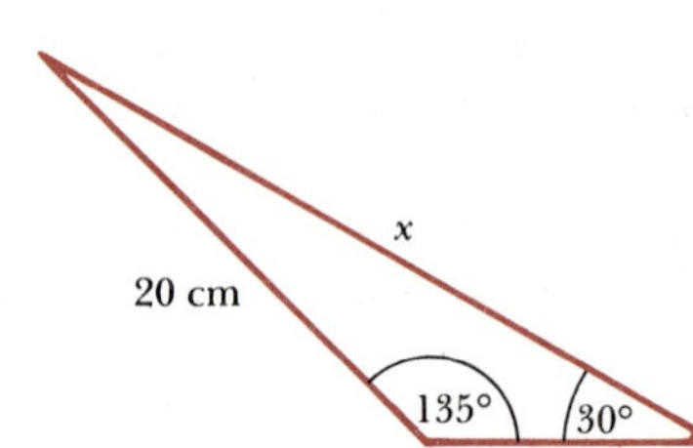

Figure 2

Example 2 (SAA) Find the length y in Figure 3.

Solution To determine y using the law of sines, it is necessary to know the opposite angle, denoted by θ, in Figure 3. Since the sum of the angles in any triangle is 180°, we have

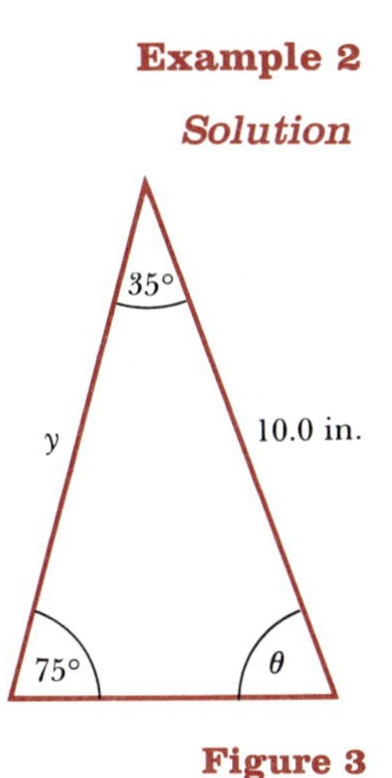

Figure 3

$$\theta + 75° + 35° = 180°$$

and consequently

$$\theta = 180° - 75° - 35° = 70°$$

Then, by the law of sines, we obtain

$$\frac{\sin 75°}{10.0} = \frac{\sin 70°}{y}$$

$$y \sin 75° = (10.0) \sin 70°$$

$$y = \frac{(10.0)(\sin 70°)}{\sin 75°} \text{ in.}$$

Without the use of a calculator or tables, this is the final form of the answer. On the other hand, a calculator gives us

$$y \approx \frac{(10.0)(0.939)}{0.965}$$

$$\approx 9.7 \text{ in.}$$

Example 3 (SSA) In $\triangle ABC$, $\angle C = 45°$, $b = 4\sqrt{2}$ ft, and $c = 8$ ft. Determine the remaining sides and angles.

Solution First draw a preliminary sketch conveying the given data. See Figure 4. (The sketch must be considered tentative. At the outset, we don't know whether the other angles are acute or even whether the given data are compatible.) To find angle B, we have (according to the law of sines)

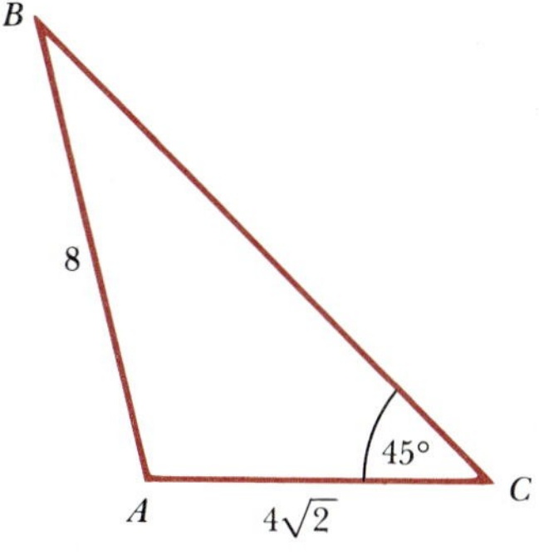

Figure 4

$$\frac{\sin B}{4\sqrt{2}} = \frac{\sin 45°}{8}$$

and therefore

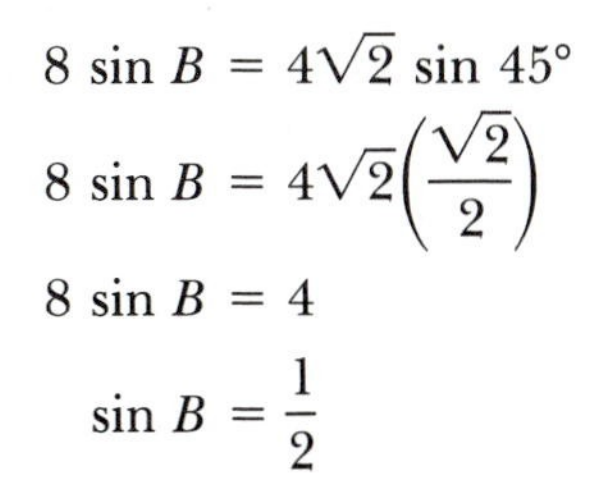

$$8 \sin B = 4\sqrt{2} \sin 45°$$

$$8 \sin B = 4\sqrt{2}\left(\frac{\sqrt{2}}{2}\right)$$

$$8 \sin B = 4$$

$$\sin B = \frac{1}{2}$$

From our work in Section 7.1, we know that one possibility for B is 30°, since $\sin 30° = \frac{1}{2}$.* However, there is another possibility. Since the reference angle for 150° is 30°, we know that sin 150° is also equal to $\frac{1}{2}$. Which angle do we want? For the problem at hand, this is easy to answer. Since angle C was given as 45°, angle B cannot equal 150°, for the sum of 45° and 150° exceeds 180°. We conclude that

* If the value were not so familiar, say $\sin B = \frac{2}{3}$, we could use a calculator or tables, as in Example 13 on page 320.

$$\angle B = 30°$$

Next, since $\angle B = 30°$ and $\angle C = 45°$, we have

$$\begin{aligned}\angle A &= 180° - 30° - 45° \\ &= 105°\end{aligned}$$

Finally, since angle A is opposite to side BC, the length of BC can be determined using the law of sines:

$$\begin{aligned}\frac{\sin C}{c} &= \frac{\sin A}{a} \\ \frac{\sin 45°}{8} &= \frac{\sin 105°}{a} \\ a \sin 45° &= 8 \sin 105° \\ a &= \frac{8 \sin 105°}{\sin 45°} = \frac{8 \sin 105°}{(\sqrt{2}/2)} \\ &= \frac{16}{\sqrt{2}} \sin 105° = \frac{16\sqrt{2}}{2} \sin 105° \\ &= 8\sqrt{2} \sin 105° \text{ ft}\end{aligned}$$

Using a calculator, this expression for a can be evaluated directly. The result is approximately $a = 11$ ft, to the nearest foot. On the other hand, using tables, it would first be necessary to use the identity $\sin(180° - \theta) = \sin \theta$ to write

$$\sin 105° = \sin(180° - 105°) = \sin 75°$$

This is because trigonometric tables generally do not include values of angles beyond 90°.

In the example just concluded, two possibilities arose for angle B, 30° and 150°. However, it turned out that the value 150° was incompatible with the given information in the problem. In Exercise 9 at the end of this section, you will see a case in which both of two possibilities are compatible with the given data. This results in two distinct solutions to the problem. In contrast to this, Exercise 8 shows a case in which there is no triangle fulfilling the given conditions. For these reasons, the case SSA is sometimes referred to as the *ambiguous case.*

Now we turn to the law of cosines.

Law of Cosines

In any triangle, the square of any side equals the sum of the squares of the other two sides minus twice the product of those other two sides times the cosine of their included angle.

Figure 5

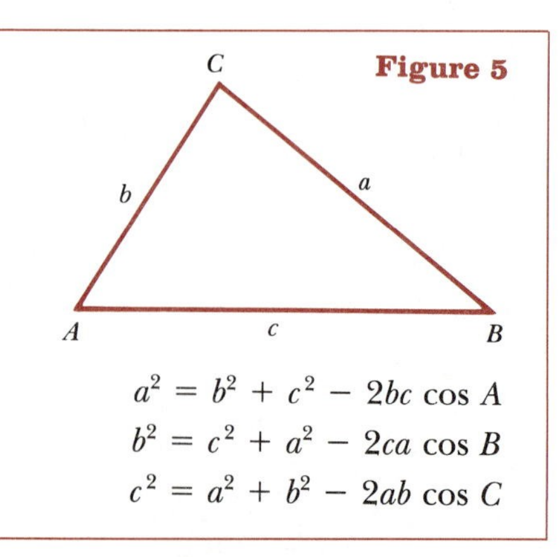

$$\begin{aligned}a^2 &= b^2 + c^2 - 2bc \cos A \\ b^2 &= c^2 + a^2 - 2ca \cos B \\ c^2 &= a^2 + b^2 - 2ab \cos C\end{aligned}$$

Before looking at a proof of this law, we make some preliminary comments. First, it is important to understand that the three equations in the box all follow the same pattern. For example, look at the first equation.

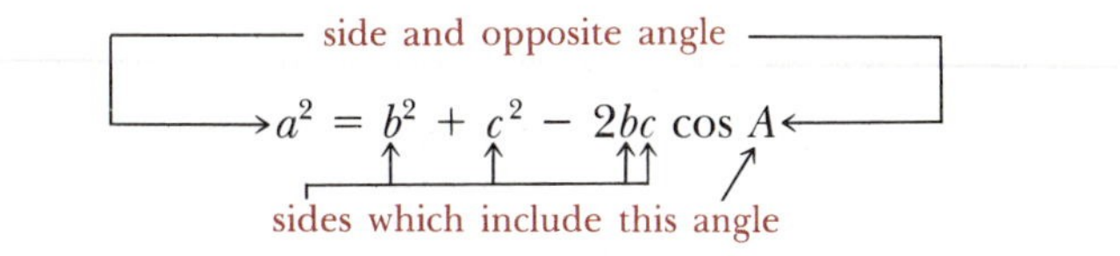

Now check for yourself to see that the other two equations follow this pattern also. It is the pattern that is important here; after all, not every triangle will be labeled $\triangle ABC$.

The second preliminary observation is that the law of cosines is a generalization of the Pythagorean Theorem. In fact, look what happens to the equation

$$a^2 = b^2 + c^2 - 2bc \cos A$$

when angle A is a right angle:

$$a^2 = b^2 + c^2 - 2bc \underbrace{\cos 90^\circ}_{0}$$

$$a^2 = b^2 + c^2 \qquad \text{which is the Pythagorean Theorem}$$

Now let us prove the law of cosines:

$$a^2 = b^2 + c^2 - 2bc \cos A$$

(The other equations would be proved in the same way. Indeed, just relabeling the figure we use would suffice.) The proof that we give uses coordinate geometry in a very nice way to complement the trigonometry. We begin by placing angle A in standard position. Then, as Figure 6 and the accompanying calculations show, the coordinates of C are

$$(b \cos A, b \sin A)$$

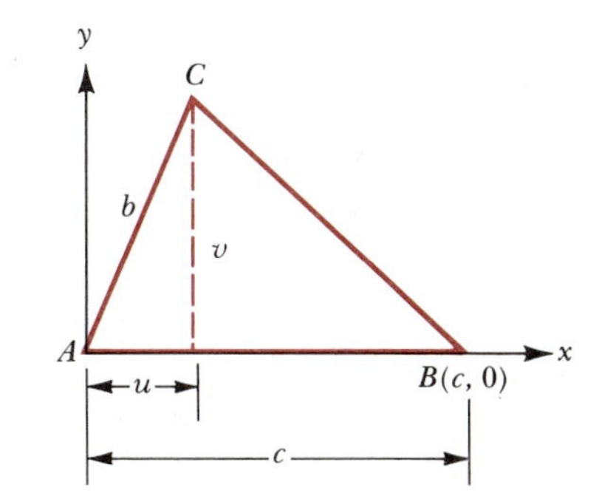

Figure 6

If u and v denote the lengths indicated in Figure 6, the coordinates of C are (u, v). We have

$$\cos A = \frac{\text{adjacent}}{\text{hypotenuse}} = \frac{u}{b}$$

Therefore

$$u = b \cos A$$

Similarly, we have

$$\sin A = \frac{\text{opposite}}{\text{hypotenuse}} = \frac{v}{b}$$

Therefore

$$v = b \sin A$$

Thus the coordinates of C are

$$(b \cos A, b \sin A)$$

(Exercise 24 at the end of this section asks you to check that these still represent the coordinates of C when angle A is not acute.) Now we use the distance formula

$$d = \sqrt{(x_2 - x_1)^2 + (y_2 - y_1)^2}$$

to compute the required distance between the points $C(b \cos A, b \sin A)$ and $B(c, 0)$. We have

$$\begin{aligned} a &= \sqrt{(b \cos A - c)^2 + (b \sin A - 0)^2} \\ &= \sqrt{b^2 \cos^2 A - 2bc \cos A + c^2 + b^2 \sin^2 A} \\ &= \sqrt{b^2 \underbrace{(\cos^2 A + \sin^2 A)}_{1} - 2bc \cos A + c^2} \\ &= \sqrt{b^2 + c^2 - 2bc \cos A} \end{aligned}$$

Squaring both sides of this last equation gives us the law of cosines, as we wished to prove.

Example 4 (SAS) Compute the length x in Figure 7.

Solution The law of cosines is directly applicable. We have

$$\begin{aligned} x^2 &= 7^2 + 8^2 - 2(7)(8) \cos 120° \\ x^2 &= 49 + 64 - 112\left(-\frac{1}{2}\right) \\ x^2 &= 113 + 56 = 169 \\ x &= \sqrt{169} \\ x &= 13 \text{ cm} \qquad \text{as required} \end{aligned}$$

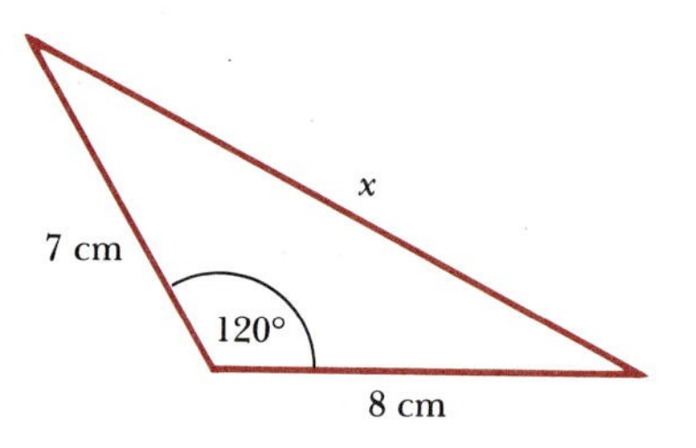

Figure 7

If the equation $a^2 = b^2 + c^2 - 2bc \cos A$ is solved for $\cos A$, the result is

$$\cos A = \frac{b^2 + c^2 - a^2}{2bc}$$

This expresses the cosine of an angle in a triangle in terms of the lengths of the sides. In a similar fashion, we obtain the corresponding formulas

$$\cos B = \frac{c^2 + a^2 - b^2}{2ca}$$

and

$$\cos C = \frac{a^2 + b^2 - c^2}{2ab}$$

These alternative forms for the law of cosines are used in the next example.

Example 5 (SSS) In $\triangle ABC$, $a = 3$ units, $b = 5$ units, and $c = 7$ units. Find the angles.

Solution Figure 8 summarizes the given data. We now have

$$\begin{aligned} \cos A &= \frac{b^2 + c^2 - a^2}{2bc} \\ &= \frac{5^2 + 7^2 - 3^2}{2(5)(7)} = \frac{65}{70} = \frac{13}{14} \end{aligned}$$

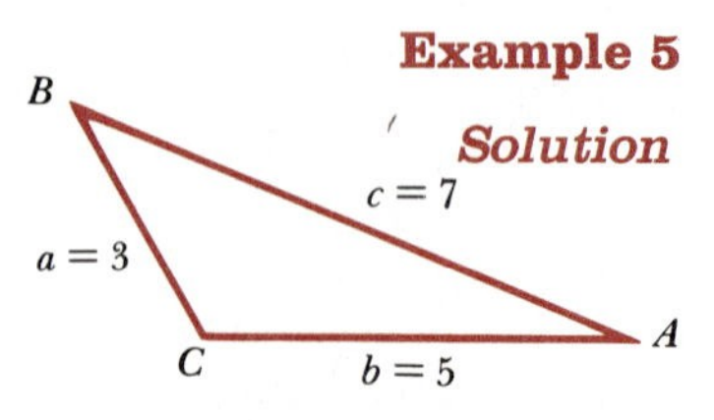

Figure 8

Using a calculator (set in the degree mode), we obtain $\angle A \approx 21.8°$. For example, the sequence of keystrokes on a Texas Instruments calculator is as follows:

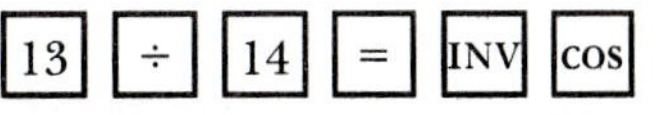

In a similar manner, we have

$$\cos B = \frac{c^2 + a^2 - b^2}{2ca}$$

$$= \frac{7^2 + 3^2 - 25}{2(7)(3)} = \frac{33}{42} = \frac{11}{14}$$

Thus $\cos B = \frac{11}{14}$ and, using a calculator, we find $\angle B \approx 38.2°$.

Finally, working in the same manner to compute $\angle C$, we have

$$\cos C = \frac{a^2 + b^2 - c^2}{2ab}$$

$$= \frac{3^2 + 5^2 - 7^2}{2(3)(5)} = \frac{-15}{30} = -\frac{1}{2}$$

Thus $\cos C = -\frac{1}{2}$. A calculator is not needed in this case. On the basis of our work in the previous section, we conclude that $\angle C = 120°$. The required angles are thus

$$\angle A \approx 21.8°$$
$$\angle B \approx 38.2°$$
$$\angle C = 120°$$

Exercise Set 7.4

1. Use the law of sines to find x in each case. Leave your answers in terms of radicals rather than decimals.

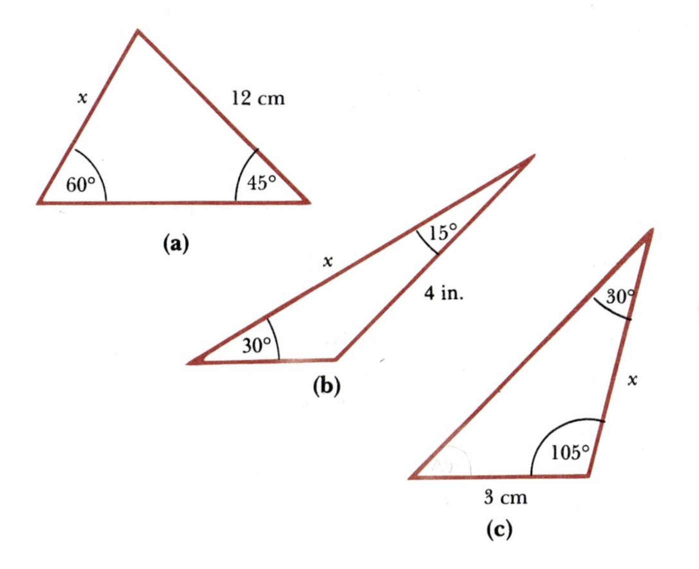

2. Find the lengths of the unknown sides in each case.

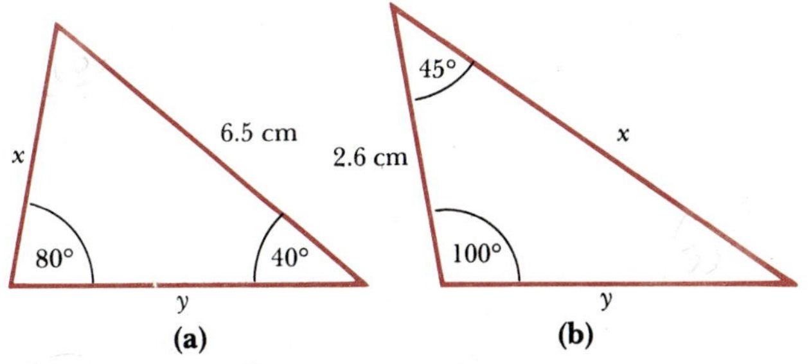

3. Express the lengths b and c in terms of x.

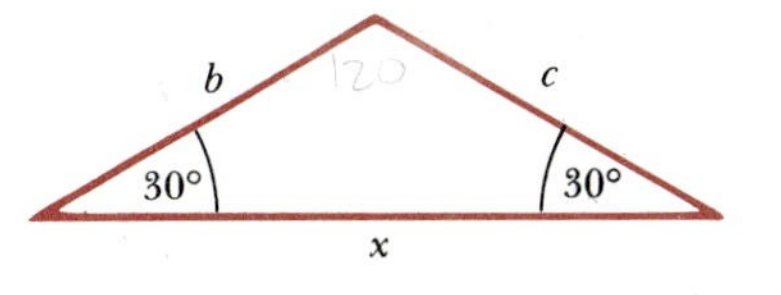

4. Find the lengths a, b, c, and d. Leave your answers in terms of the trigonometric functions (rather than decimals) as necessary.

Partial answer: $a = \dfrac{2 \sin 110°}{\sin 20°}$

Do you see why this is equivalent to

$$a = \frac{2 \sin 70°}{\sin 20°}?$$

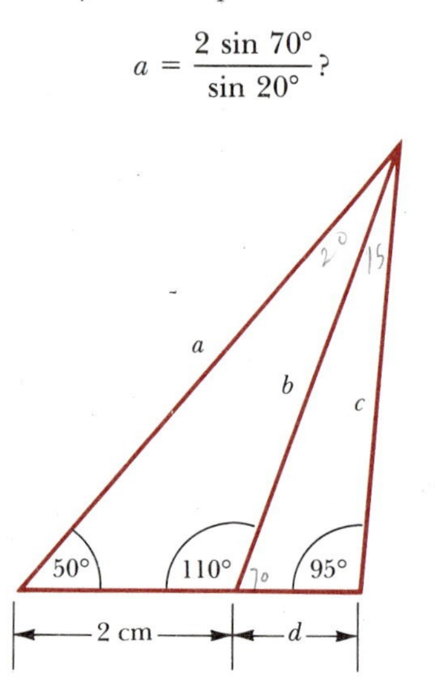

5. Find the area of $\triangle ABC$ if $c = 7$ in., $\angle A = 40°$, and $\angle B = 85°$.

6. (a) In $\triangle ABC$, $\sin B = \sqrt{2}/2$. What are the possible values for $\angle B$?
   (b) In $\triangle ABC$, if $\cos A = -\sqrt{3}/2$, what are the possible values for $\angle A$?
   C (c) In $\triangle CDE$, $\sin D = \frac{1}{4}$. Using a calculator, determine the possible values for angle D.

C 7. In each case, determine the remaining sides and angles of $\triangle ABC$ using the given data.
   (a) $a = 29.4$, $b = 30.1$, $\angle B = 66°$.
   (b) $a = 52$, $c = 42$, $\angle A = 125°$.

8. Show that there is no triangle for which $a = 23$, $b = 50$, and $\angle A = 30°$. *Hint:* Try computing $\sin B$ using the law of sines.

9. Let $b = 1$, $a = \sqrt{2}$, and $\angle B = 30°$.
   (a) Use the law of sines to show $\sin A = \sqrt{2}/2$.
   (b) Conclude that $A = 45°$ or $A = 135°$.
   (c) Assuming that $A = 45°$, determine the remaining parts of $\triangle ABC$.
   (d) Assuming that $A = 135°$, determine the remaining parts of the triangle.

C 10. Let $a = 30$, $b = 36$, and $\angle A = 20°$. Find the remaining parts of $\triangle ABC$. (There are two distinct triangles possible.)

11. Use the law of cosines to determine x in each case. [For part (b), use a calculator or tables.]

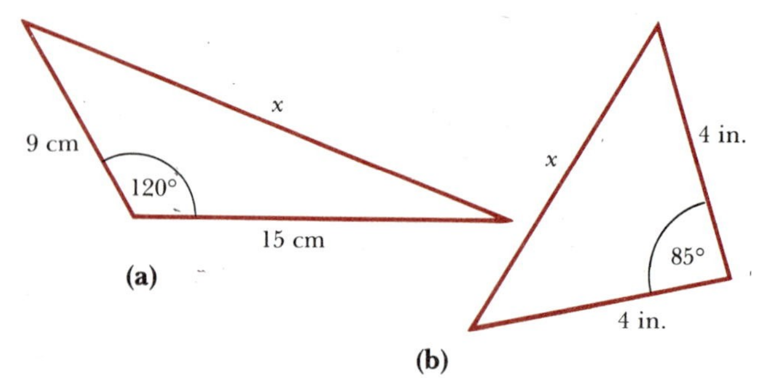

12. In applying the law of cosines to the figure shown, a student incorrectly writes

$$x^2 = 5^2 + 11^2 - 2(5)(11) \cos 95°.$$

Why is this incorrect? What is the correct equation?

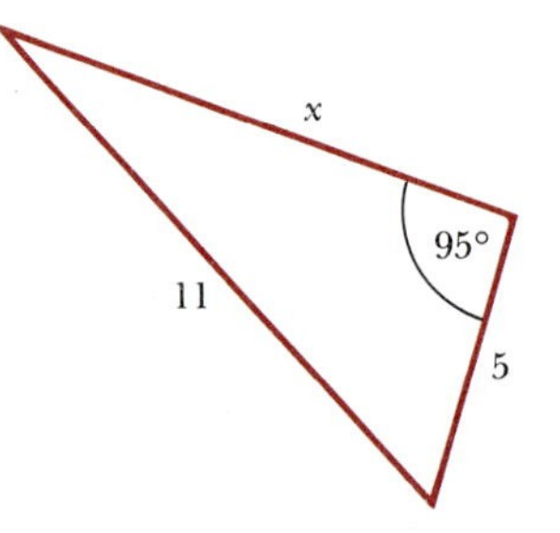

C 13. A regular pentagon is inscribed in a circle of unit radius. Find the perimeter of the pentagon. (First find the length of a side using the law of cosines.)

14. In $\triangle ABC$, $a = 6$, $b = 7$, and $c = 10$. Find the cosine of each angle.

C 15. In $\triangle ABC$, $a = 2$, $b = 3$, and $c = 4$. Find each angle.

C 16. In $\triangle ABC$, $\angle A = 40°$, $b = 6.1$ cm, and $c = 3.2$ cm.
   (a) Find a using the law of cosines.
   (b) Find $\angle C$ using the law of sines.
   (c) Find $\angle B$.

C 17. In parallelogram $ABCD$, $\overline{AB} = 6$ in., $\overline{AD} = 4$ in., and $\angle A = 40°$. Find the length of each diagonal.

18. $ABCD$ is a square in which each side has unit length. P is a point on diagonal AC such that $\overline{AP} = 1$. Find the distance from P to D. (Express your answer using radicals.)

C 19. P and Q are two points on opposite sides of a river. From P to another point R on the same side is 300 ft. Angles PRQ and RPQ are measured and found to be 20° and 120°, respectively. Compute the distance from P to Q, across the river. (Round off your answer to the nearest foot.)

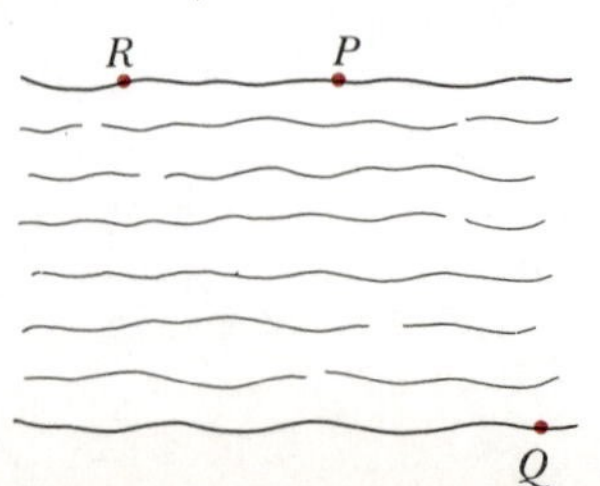

***20.** **(a)** Let m and n be positive numbers with $m > n$. Furthermore, suppose that in $\triangle ABC$ the lengths a, b, and c are given by

$$a = 2mn + n^2 \qquad b = m^2 - n^2 \qquad c = m^2 + n^2 + mn$$

Show then that $\cos C = -\frac{1}{2}$ and conclude $\angle C = 120°$.

(b) Give an example of a triangle in which the lengths of the sides are whole numbers and in which one of the angles is 120°. (Specify the three sides; you needn't find the other angles.)

21. If the lengths of two adjacent sides of a parallelogram are a and b, respectively, and the acute angle formed by these two sides is θ, show that the product of the lengths of the two diagonals is given by the expression

$$\sqrt{(a^2 + b^2)^2 - 4a^2b^2 \cos^2 \theta}$$

***C22.** Compute the length x in the figure. *Suggestion:* Draw the line segment joining the vertices of the 80° and 50° angles.

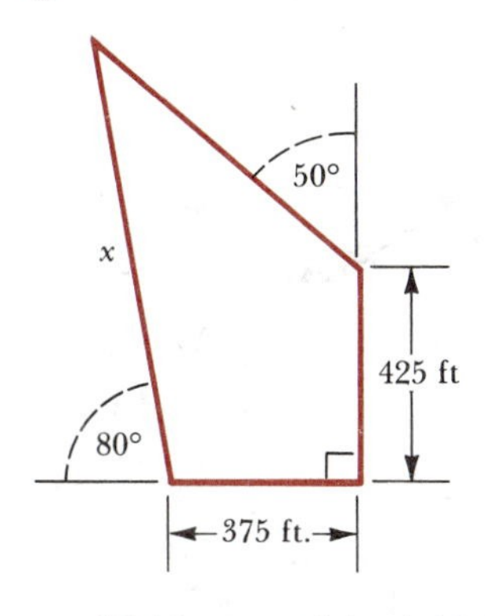

***23.** In the figure, AD bisects $\angle A$ in $\triangle ABC$. Show that

$$\frac{n}{m} = \frac{c}{b}$$

In words, this says that the bisector of an angle in a triangle divides the opposite side into segments that are proportional to the adjacent sides. *Hint:* Apply the law of sines in $\triangle ABD$ and in $\triangle ACD$.

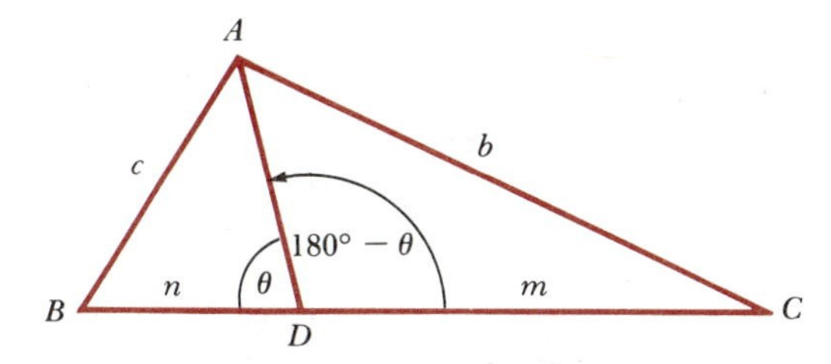

24. Let the positive numbers u and v denote the lengths indicated in the figure so that the coordinates of C are $(-u, v)$. Show that $u = -b \cos A$ and $v = b \sin A$. Conclude from this that the coordinates of C are

$$(b \cos A,\ b \sin A)$$

Hint: Use the right triangle definitions for cosine and sine along with the formulas for $\cos(180° - \theta)$ and $\sin(180° - \theta)$.

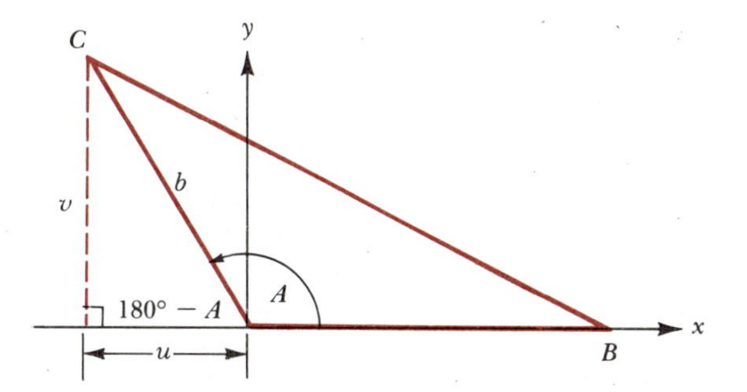

***25.** (Heron's Formula) Approximately 2000 years ago, Heron of Alexandria derived a formula for the area of a triangle in terms of the lengths of the sides. A more modern derivation of Heron's formula using the law of cosines is indicated in the steps that follow.

(a) Solve the equation $c^2 = a^2 + b^2 - 2ab \cos C$ for $\cos C$ to obtain

$$\cos C = \frac{a^2 + b^2 - c^2}{2ab}$$

(b) Show that $1 - \cos C = \dfrac{(c + a - b)(c - a + b)}{2ab}$.

Suggestion: Work with $1 - \cos C$. You can factor the expression $c^2 - (a - b)^2$ using the difference of squares technique.

(c) Show that $1 + \cos C = \dfrac{(a + b + c)(a + b - c)}{2ab}$.

(d) Show that $\sin^2 C = \dfrac{(a + b + c)(-a + b + c)(a - b + c)(a + b - c)}{4a^2b^2}$.

(e) Using the formula

$$\text{area } \triangle ABC = \tfrac{1}{2}ab \sin C$$

show that area $\triangle ABC$

$$= \tfrac{1}{4}\sqrt{(a + b + c)(-a + b + c)(a - b + c)(a + b - c)}$$

(f) The formula in part (e) expresses the area in terms of a, b, and c. We can simplify this formula somewhat by introducing the following notation.

Let s denote one-half the perimeter of $\triangle ABC$. That is, let

$$s = \tfrac{1}{2}(a + b + c)$$

Using this notation, check that

(i) $a + b + c = 2s$

(ii) $-a + b + c = 2(s - a)$

(iii) $a - b + c = 2(s - b)$
(iv) $a + b - c = 2(s - c)$
Then show that with this notation we obtain

$$\text{area } \triangle ABC = \sqrt{s(s-a)(s-b)(s-c)}$$

This is *Heron's formula.* [For historical background and a purely geometric proof, see *An Introduction to the History of Mathematics,* 5th edition, by Howard Eves (Saunders College Publishing, 1983), pp. 133 and 147.]

***26.** The figure shows a quadrilateral, with sides a, b, c, and d, inscribed in a circle. If λ denotes the length of the diagonal indicated in the figure, prove that

$$\lambda^2 = \frac{(ab + cd)(ac + bd)}{bc + ad}$$

This result is known as *Brahmagupta's Theorem.* It is named after its discoverer, a 7th-century Hindu mathematician. *Hints:* Assume as given the theorem from geometry stating that when a quadrilateral is inscribed in a circle, the opposite angles are supplementary. Apply the law of cosines in both of the triangles in the figure to obtain expressions for λ^2. Then eliminate $\cos\theta$ from one equation.

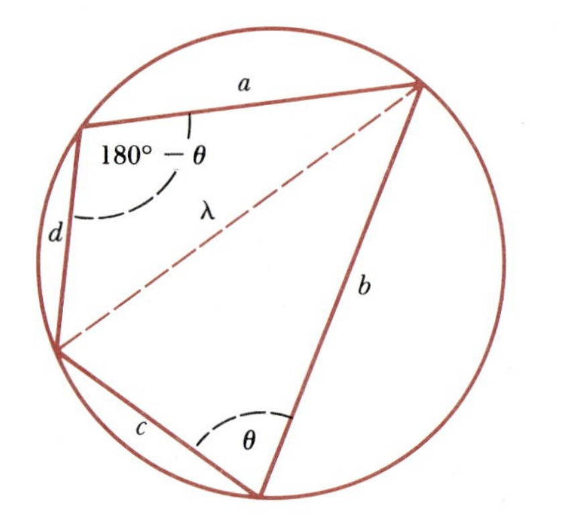

27. (Alternate derivation for the law of sines) The figure at the top of the next column shows $\triangle ABC$ inscribed in a circle of radius r.

(a) Show that $\angle AOD = \angle C$. *Hint:* Use the following theorem from geometry: The measure of an angle inscribed in a circle is half the measure of the corresponding central angle. (In the figure, angle C is the inscribed angle and angle AOB is the central angle.)

(b) Show $r = c/(2 \sin C)$.

(c) Using the reasoning in parts (a) and (b), conclude also that

$$r = \frac{a}{2 \sin A} \quad \text{and} \quad r = \frac{b}{2 \sin B}$$

Then show

$$\frac{\sin A}{a} = \frac{\sin B}{b} = \frac{\sin C}{c}$$

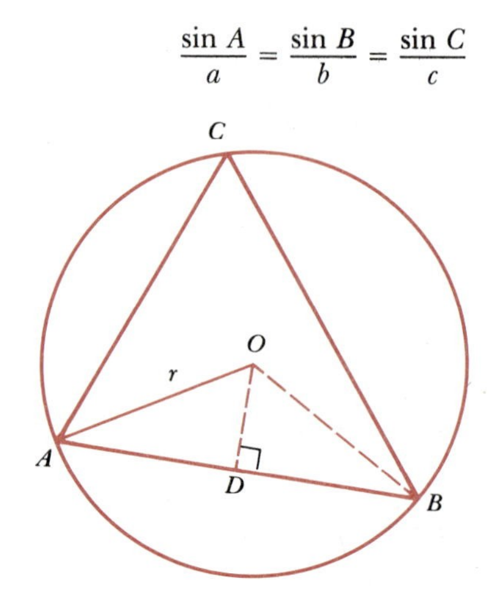

(d) In the figure we used, the center of the circle falls within $\triangle ABC$. Draw a figure in which the center lies outside of $\triangle ABC$ and prove that the equation $r = c/(2 \sin C)$ is true in this case too.

28. The equations

$$\begin{cases} a = b\cos C + c\cos B \\ b = c\cos A + a\cos C \\ c = a\cos B + b\cos A \end{cases}$$

are sometimes referred to as the *projection laws* for a triangle. Use the law of cosines to give a proof of the projection laws as follows. To obtain the formula

$$a = b\cos C + c\cos B$$

add the two equations

$$b^2 = a^2 + c^2 - 2ac\cos B$$

and

$$c^2 = a^2 + b^2 - 2ab\cos C$$

In the resulting equation, combine like terms and solve for a. The other projection laws are obtained in a similar manner.

***29.** **(a)** In $\triangle ABC$, prove that

$$\sin C = \sin A\cos B + \cos A\sin B$$

Hint: In the equation $c = a\cos B + b\cos A$ (obtained in Exercise 28), multiply the term c by $(\sin C)/c$; multiply the term $a\cos B$ by $(\sin A)/a$; multiply the term $b\cos A$ by $(\sin B)/b$.

(b) Use the formula in part (a) and the figure at the top of the next page to show that

$$\sin 15^\circ = \frac{\sqrt{6} - \sqrt{2}}{4}$$

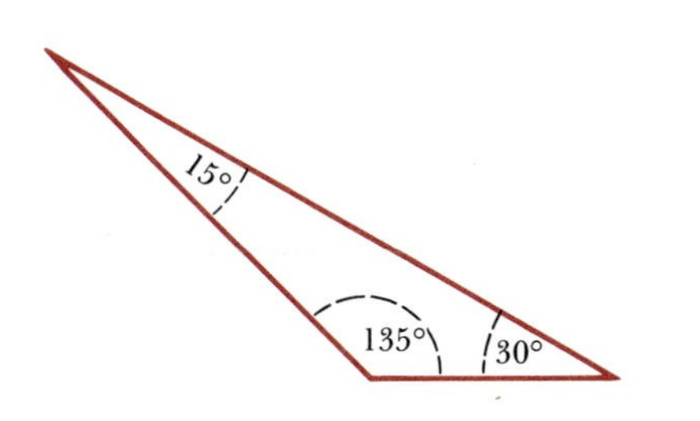

(c) Express $\sin 75°$ in terms of radicals. *Hint:* $75° + 60° + 45° = 180°$.

(d) Express $\sin 105°$ in terms of radicals.

Chapter 7 Review Checklist

- □ Trigonometric functions:
  Right triangle definitions (page 312)
  Unit circle definitions (page 338)
- □ $\sin^n \theta$ (and other notational conventions) (page 327)
- □ $\sin(90° - \theta) = \cos \theta$; $\cos(90° - \theta) = \sin \theta$ (page 329)
- □ $\tan \theta = \dfrac{\sin \theta}{\cos \theta}$ (page 329)
- □ $\sin^2 \theta + \cos^2 \theta = 1$ (page 329)
- □ $\cos(180° - \theta) = -\cos \theta$; $\sin(180° - \theta) = \sin \theta$ (page 342)
- □ Area of a triangle: $A = \frac{1}{2}ab \sin \theta$ (page 343)
- □ Law of sines: $\dfrac{\sin A}{a} = \dfrac{\sin B}{b} = \dfrac{\sin C}{c}$ (page 348)
- □ Law of cosines: $a^2 = b^2 + c^2 - 2bc \cos A$ (page 350)

Chapter 7 Review Exercises

In Exercises 1–18, the lengths of the three sides of a triangle are denoted by a, b, *and* c*; the angles opposite these sides are* A, B, *and* C, *respectively. In each exercise, use the given information to find the required quantities.*

	Given	*Find*
1.	$B = 90°, A = 30°, b = 1$	a and c
2.	$B = 90°, A = 60°, a = 1$	c and b
3.	$B = 90°, \sin A = \frac{2}{5}, a = 7$	b
4.	$B = 90°, \sec C = 4, c = 2^{1/2}$	b
5.	$B = 90°, \cos A = \frac{3}{8}$	$\sin A$ and $\cot A$
6.	$B = 90°, b = 1, \tan C = 5^{1/2}$	a
7.	$b = 4, c = 5, A = 150°$	a and area of triangle ABC
8.	$A = 120°, b = 8$, area of $\triangle ABC = 12\sqrt{3}$	c
9.	$a = 6, b = 3, c = 5$	$\cos A$ and $\sin A$
10.	$a = 1, c = 2^{1/2}, B = 60°$	b
11.	$c = 4, a = 2, B = 90°$	$\sin^2 A + \cos^2 B$
12.	$B = 90°, 2a = b$	A
13.	$B = 45°, C = 30°, b = 3\sqrt{2}$	c
14.	$A = 30°, B = 120°, b = 16$	a
15.	$a = 3, c = 12, C = 135°$	$\sin A$, $\cos A$, and B
16.	$a = 7, b = 8, \sin C = \frac{1}{4}$	area of triangle ABC
17.	$b = 4, a = 5, \cos C = \frac{1}{8}$	c
18.	$a = b = 5, \sin(C/2) = \frac{9}{10}$	C

For Exercises 19–34, evaluate each expression. (Don't use a calculator.)

19. $\sin 135°$
20. $\cos(-60°)$
21. $\tan 240°$
22. $\sin 450°$
23. $\csc 210°$
24. $\sec 225°$
25. $\sin 270°$
26. $\cot 330°$
27. $\cos 315°$
28. $\cos 180°$
29. $\cos 1800°$
30. $\sec 120°$
31. $\csc 240°$
32. $\cot 45°$
33. $\sec 780°$
34. $\sin^2 33° + \sin^2 57°$

35. A student who was asked to simplify the expression $\sin^2 33° + \sin^2 57°$ wrote $\sin^2 33° + \sin^2 57° = \sin^2(33° + 57°) = \sin^2 90° = 1^2 = 1$.
(a) Where is the error?
(b) What is the correct answer?

36. If $\theta = 45°$, find each of the following.
(a) $\cos \theta$ **(b)** $\cos^3 \theta$ **(c)** $\cos 2\theta$
(d) $\cos 3\theta$ **(e)** $\cos(2\theta/3)$ **(f)** $(\cos 3\theta)/3$
(g) $\cos(-\theta)$ **(h)** $\cos^3(5\theta)$

In Exercises 37–46, use the given information to find the required quantities.

	Given	Find
37.	$\cos \theta = \frac{3}{5}$, $0° < \theta < 90°$	$\sin \theta$ and $\tan \theta$
38.	$\sin \theta = -\frac{12}{13}$, $180° < \theta < 270°$	$\cos \theta$
39.	$\sec \theta = \frac{25}{7}$, $270° < \theta < 360°$	$\tan \theta$
40.	$\cot \theta = \frac{4}{3}$, $0° < \theta < 90°$	$\sin \theta$
41.	$\csc \theta = \frac{13}{12}$, $0° < \theta < 90°$	$\cot \theta$
42.	$\sin \theta = \frac{1}{5}$, $0° < \theta < 90°$	$\cos(90° - \theta)$
43.	$\cos \theta = \frac{3}{4}$, $0° < \theta < 90°$	$\tan(90° - \theta)$
44.	$\sin \theta = \sqrt{5}/6$, $0° < \theta < 90°$	$\sin(180° - \theta)$
45.	$\cos \theta = \frac{7}{9}$, $0° < \theta < 90°$	$\cos(180° - \theta)$
46.	$\cos \theta = -\frac{1}{4}$, $90° < \theta < 180°$	$\cos(180° - \theta)$

47. Simplify: $\dfrac{\sin^4 \theta - \cos^4 \theta}{\sin^2 \theta - \cos^2 \theta} \div \dfrac{1 + \sin \theta \cos \theta}{\sin^3 \theta - \cos^3 \theta}$

48. Simplify: $\dfrac{1 + \sin \theta + \sin^2 \theta}{1 - \sin^3 \theta}$

For Exercises 49–58, convert each expression into one involving only sines and cosines and then simplify. (Leave your answers in terms of sines and/or cosines.)

49. $\dfrac{\sin A + \cos A}{\sec A + \csc A}$
50. $\dfrac{\csc A \sec A}{\sec^2 A + \csc^2 A}$
51. $\dfrac{\sin A \sec A}{\tan A + \cot A}$
52. $\cos A + \tan A \sin A$
53. $\dfrac{\cos A}{1 - \tan A} + \dfrac{\sin A}{1 - \cot A}$
54. $\dfrac{1}{\sec A - 1} \div \dfrac{1}{\sec A + 1}$
55. $(\sec A + \csc A)^{-1}[(\sec A)^{-1} + (\csc A)^{-1}]$
56. $\dfrac{\dfrac{\tan^2 A - 1}{\tan^3 A + \tan A}}{\dfrac{\tan A + 1}{\tan^2 A + 1}}$
57. $\dfrac{\dfrac{\sin A + \cos A}{\sin A - \cos A} - \dfrac{\sin A - \cos A}{\sin A + \cos A}}{\dfrac{\sin A + \cos A}{\sin A - \cos A} + \dfrac{\sin A - \cos A}{\sin A + \cos A}}$
58. $\dfrac{\sin A \tan A - \cos A \cot A}{\sec A - \csc A}$

In Exercises 59–68, evaluate each expression in terms of a, *where* a = *cos 20°.*

59. $\sin 20°$
60. $\tan 20°$
61. $\cos 70°$
62. $\sin 70°$
63. $\cos 160°$
64. $\cos 340°$
65. $\cos(-160°)$
66. $\cos 200°$
67. $\sin 200°$
68. $\cot 200°$

For Exercises 69–72, express your answers in terms of the trigonometric functions (rather than using a calculator).

69. The lengths of two adjacent sides of a parallelogram are 2 cm and 8 cm, respectively. If the angle formed by these two sides is 30°, find the length of the shorter diagonal.

70. A 100-foot vertical antenna is on the roof of a building. From a point on the ground, the angles of elevation of the top and the bottom of the antenna are 51° and 37°, respectively. Find the height of the building.

71. In an isosceles triangle, the two base angles are each 35°, and the length of the base is 120 cm. Find the area of the triangle.

72. Find the perimeter and the area of a regular pentagon inscribed in a circle of radius 9 cm.

73. In triangle ABC, let h denote the length of the altitude from A to BC. Show that $h = \dfrac{a}{\cot B + \cot C}$.

74. The length of each side of an equilateral triangle is $2a$. Show that the radius of the inscribed circle is $a/\sqrt{3}$ and the radius of the circumscribed circle is $2a/\sqrt{3}$.

75. From a helicopter h ft above the sea, the angle of depression to the pilot's horizon is θ. Show that $\cot\theta = \dfrac{R}{\sqrt{2Rh + h^2}}$, where R is the radius of the earth.

76. In triangle ABC, angle C is a right triangle. Show that $\dfrac{\sin A - \sin B}{\sin A + \sin B} = \dfrac{a - b}{a + b}$.

77. In this exercise we prove that the equation given in Exercise 76 holds for any triangle, not just a right triangle.

(a) First, verify that if $\frac{a}{b} = \frac{c}{d}$ then $\dfrac{a-b}{a+b} = \dfrac{c-d}{c+d}$.

(b) From the law of sines $a/b = \sin A/\sin B$. Now apply the result in part (a).

78. In triangle ABC, $\overline{BC} = t(4 + t)$, $\overline{AC} = 4 - t^2$, and $\overline{AB} = t^2 + 2t + 4$. Compute $\cos C$. What is angle C? (Assume that $0 < t < 2$.)

For Exercises 79–106, show that each equation is an identity.

79. $\cos A \cot A = \csc A - \sin A$

80. $\sec A - 1 = \sec A(1 - \cos A)$

81. $\dfrac{\cot A - 1}{\cot A + 1} = \dfrac{\cos A - \sin A}{\cos A + \sin A}$

82. $\dfrac{\sin A}{\csc A - \cot A} = 1 + \cos A$

83. $\cos^2\theta - \sin^2\theta = 2\cos^2\theta - 1$

84. $\cos^2\theta - \sin^2\theta = 1 - 2\sin^2\theta$

85. $\sin A \tan A = \dfrac{1 - \cos^2 A}{\cos A}$

86. $\dfrac{\cot A - 1}{\cot A + 1} = \dfrac{1 - \tan A}{1 + \tan A}$

87. $\cot^2 A + \csc^2 A = -\cot^4 A + \csc^4 A$

88. $\sin^3 A + \cos^3 A = (\sin A + \cos A)(1 - \sin A \cos A)$

89. $\dfrac{\cot^2 A - \tan^2 A}{(\cot A + \tan A)^2} = 2\cos^2 A - 1$

90. $\dfrac{\sin A - \cos A}{\sin A} + \dfrac{\cos A - \sin A}{\cos A} = 2 - \sec A \csc A$

91. $\dfrac{\cos A - \sin A}{\cos A + \sin A} = \dfrac{1 - \tan A}{1 + \tan A}$

92. $\dfrac{\cos A - \sin A}{\sec A \cot A - \csc A \tan A} = \cos A \sin A$

93. $\dfrac{1 - \sin A}{\cos A} - \dfrac{1}{\sec A + \tan A} = 0$

94. $\dfrac{\cos A}{1 - \tan A} + \dfrac{\sin A}{1 - \cot A} = \cos A + \sin A$

95. $\tan A \tan B = \dfrac{\tan A + \tan B}{\cot A + \cot B}$

96. $\dfrac{\sin A}{1 + \cos A} + \dfrac{1 + \cos A}{\sin A} = 2\csc A$

97. $\tan A - \dfrac{\sec A \sin^3 A}{1 + \cos A} = \sin A$

98. $\dfrac{1}{1 - \cos A} + \dfrac{1}{1 + \cos A} = 2 + 2\cot^2 A$

99. $\dfrac{1}{\csc A - \cot A} - \dfrac{1}{\csc A + \cot A} = 2\cot A$

100. $\dfrac{\sin^3 A}{\cos A - \cos^3 A} = \tan A$

101. $\dfrac{1}{1 + \sin^2 A} + \dfrac{1}{1 + \cos^2 A} + \dfrac{1}{1 + \sec^2 A} + \dfrac{1}{1 + \csc^2 A} = 2$

102. $\dfrac{\sec A - \tan A}{\sec A + \tan A} = \left(\dfrac{\cos A}{1 + \sin A}\right)^2$

103. $\dfrac{\sec A - \csc A}{\sec A + \csc A} = \dfrac{\sin A - \cos A}{\sin A + \cos A}$

104. $\sin^6 A + \cos^6 A = 1 - 3\sin^2 A \cos^2 A$

105. $\sin A \cos A = \dfrac{\tan A}{1 + \tan^2 A}$

106. $\dfrac{2\tan A}{1 - \tan^2 A} + \dfrac{1}{\cos^2 A - \sin^2 A} = \dfrac{\cos A + \sin A}{\cos A - \sin A}$

107. If θ is an acute angle in a right triangle and $\cos\theta = \dfrac{a - b}{a + b}$, find $\sin\theta$. (Assume that $a > b > 0$.)

108. In triangle ABC, $B = 60°$, $\overline{AC} = \frac{3}{2}$, and $\overline{BC} = \cos A$. Find angle A.

109. For any triangle ABC, show that $a^2 + b^2 + c^2 = 2(ab \cos C + bc \cos A + ca \cos B)$. *Suggestion:* Use the three formulas accompanying Figure 5 in Section 7.4.

110. The radius of the circle in the figure is 1 unit. PT is tangent to the circle at P, and PN is perpendicular to OT. Express each of the following quantities in terms of θ.

(a) $\overline{PN}$ **(b)** $\overline{ON}$ **(c)** $\overline{PT}$
(d) $\overline{OT}$ **(e)** $\overline{NA}$ **(f)** $\overline{NT}$

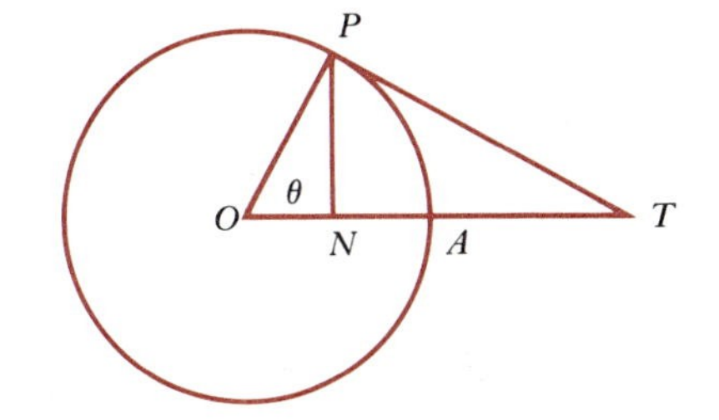

111. Refer to the figure accompanying Exercise 110.

(a) Show that the ratio of the area of triangle ONP to the area of triangle NPT is $\cot^2\theta$.

(b) Show that the ratio of the area of triangle NPT to the area of triangle OPT is $\sin^2 \theta$.

C *For Exercises 112–120, refer to the figure. In each case determine the indicated quantity. Use a calculator, and round off your result to two decimal places.*

112. $\overline{BE}$

113. $\overline{BC}$

114. Area of triangle BCE

115. $\overline{BD}$

116. Area of triangle ABD

117. $\overline{DE}$

118. $\overline{CD}$

119. $\overline{AD}$

120. $\overline{AC}$

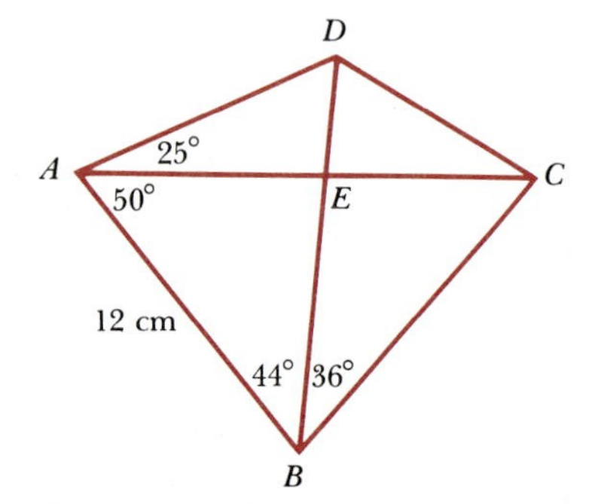

C **121.** (A curious approximation) In an article appearing in *The American Mathematical Monthly* in 1946, J. S. Frame noted that the expression $\dfrac{11a - 3a^3}{7 + a^2}$ could be used to approximate the values of the sine function. Complete the following table to see evidence of this. (Round off your results to three decimal places.)

a	$a \cdot 90°$	$\sin(a \cdot 90°)$	$\dfrac{11a - 3a^3}{7 + a^2}$
0.0			
0.1			
0.2			
0.3			
0.4			
0.5			
0.6			
0.7			
0.8			
0.9			
1.0			

122. (Refer to the figure.) Show that

$$y = x[\tan(\alpha + \beta) - \tan \beta]$$

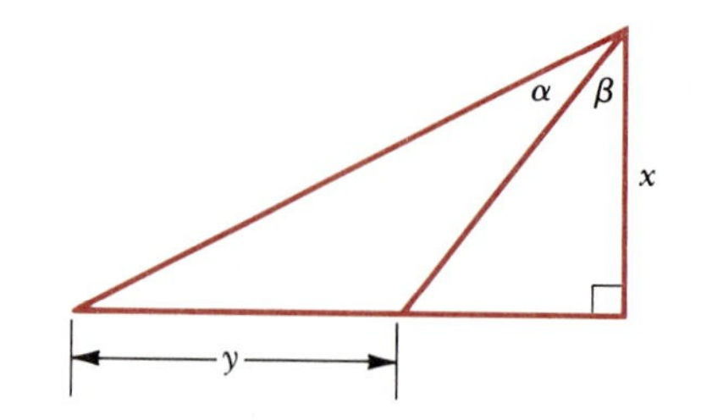

123. (Refer to the figure.) Show that

$$y = \frac{x}{\cot \alpha - \cot \beta}$$

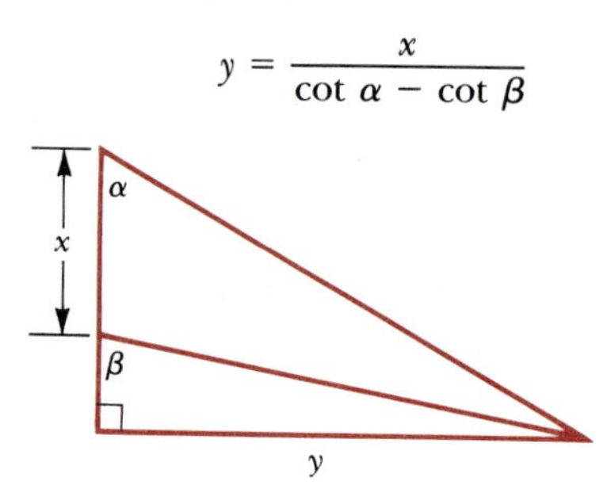

124. (Refer to the figure.) Show that $\cot \theta = a/b + \cot \alpha$.

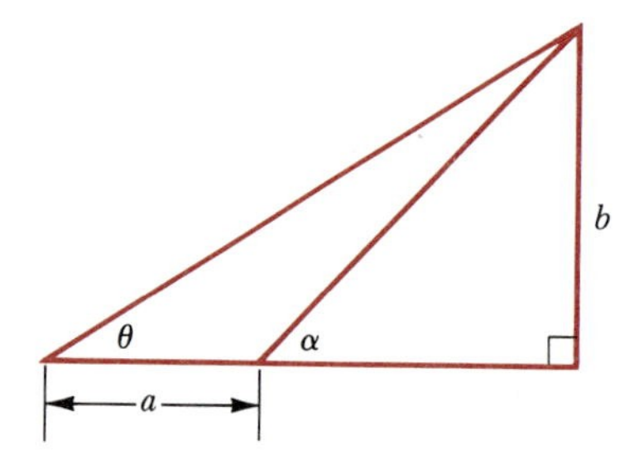

Chapter 7 Test

1. Specify a value for each of the following expressions: $\cos 30°$, $\tan 60°$, and $\sin^2 7° + \cos^2 7°$.

2. Evaluate each of the following.
(a) $\sin(-270°)$ **(b)** $\cos 540°$ **(c)** $\cot 450°$

3. Factor $2\cos^2\theta + 11\cos\theta + 12$.

4. Find the area of a triangle in which the lengths of two sides are 5 cm and 6 cm, respectively, and the angle included between these sides is 135°.

5. In $\triangle ABC$, $A = 120°$, $b = 5$ cm, and $c = 3$ cm. Find a.

6. The sides of a triangle are 2 cm, 3 cm, and 4 cm, respectively. Determine the cosine of the angle opposite the longest side. On the basis of your answer, explain whether or not the angle opposite the longest side is an acute angle.

7. Evaluate each of the following.
   (a) cos 330° **(b)** tan 135°

8. If $\sin\theta = -\frac{1}{3}$, and $270° < \theta < 360°$, determine $\cos\theta$ and $\cot\theta$.

9. Compute $\tan\theta$ in the figure shown.

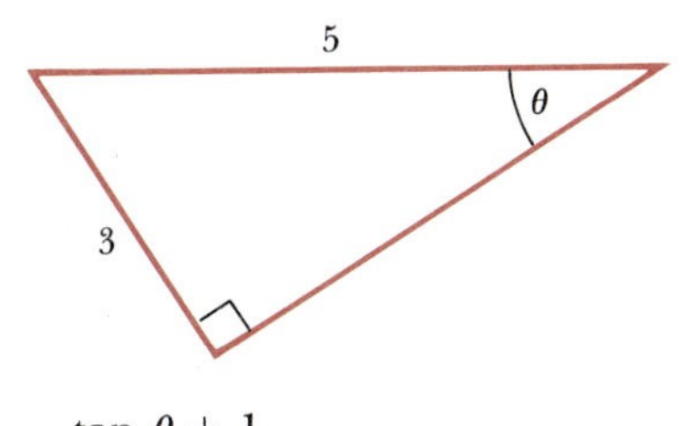

10. Simplify

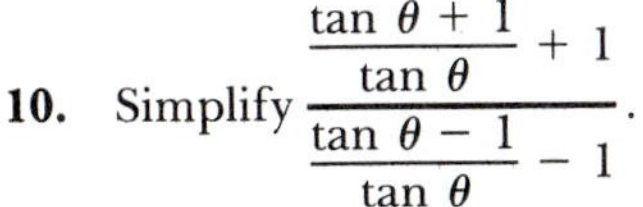

$$\frac{\dfrac{\tan\theta + 1}{\tan\theta} + 1}{\dfrac{\tan\theta - 1}{\tan\theta} - 1}.$$

11. Two of the angles in a triangle are 30° and 45°. If the side opposite the 45° angle is $20\sqrt{2}$ cm, find the side opposite the 30° angle.

12. A 10-ft ladder, which is leaning against the side of a building, makes an angle of 60° with the ground, as shown. How far up the building does the ladder reach?

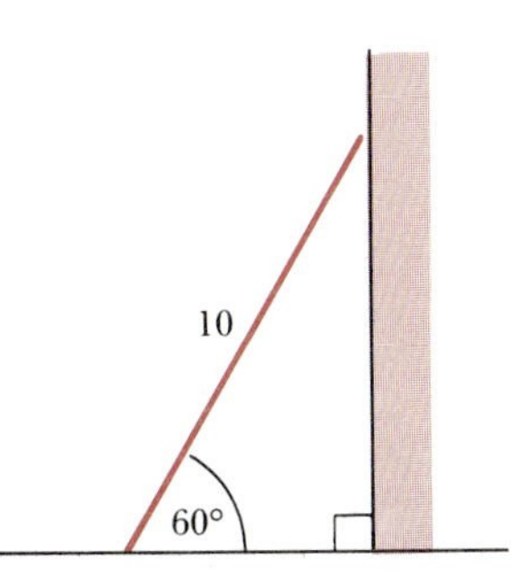

13. Prove that the following equation is an identity.

$$\frac{\cos\theta + 1}{\csc\theta + \cot\theta} = \sin\theta$$

14. The arc in the figure is a portion of the unit circle. Express $\overline{AB}$ and $\overline{BP}$ in terms of θ.

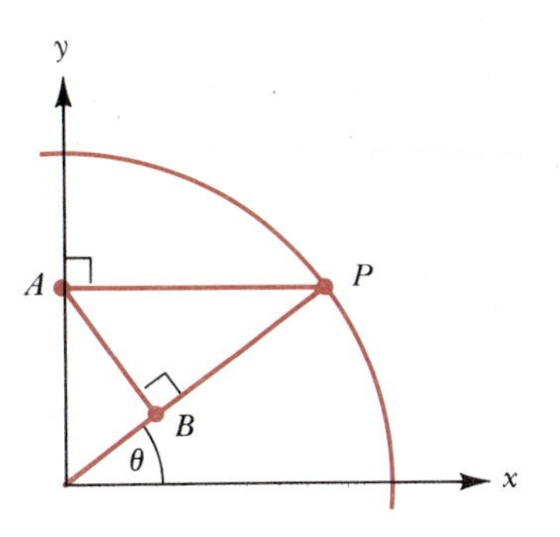

15. Prove that the following equation is an identity.

$$\frac{\sin\theta}{1 + \cos(180° - \theta)} = \frac{1 + \cos\theta}{\sin(180° - \theta)}$$

16. Simplify $\dfrac{(\cos\theta + \sin\theta)^2}{1 + 2\sin\theta\cos\theta}$.

17. If $\cos 15° = \frac{1}{4}(\sqrt{6} + \sqrt{2})$, find sin 15°.

18. Each side of the square $STUV$ is 8 cm long. P is a point on diagonal SU such that $\overline{SP} = 2$ cm. Find the distance from P to V.

19. In the figure, $\angle DAB = 25°$, $\angle CAB = 55°$, and $\overline{AB} = 50$ cm. Find $\overline{CD}$. (Leave your answer in terms of the trigonometric functions rather than decimals.)

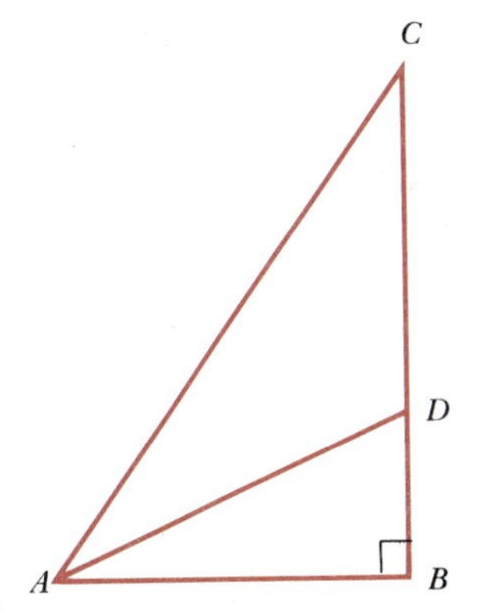

20. A regular nine-sided polygon is inscribed in a circle of radius 2 m. Find the area of the polygon. (Leave your answer in terms of the trigonometric functions rather than decimals.)

Chapter 8 Trigonometric Functions of Real Numbers

Introduction

In the previous chapter we measured angles using the familiar units of degrees. However, there is another unit for measuring angles, called the *radian,* that has proven to be more useful in calculus and in many of the sciences. In Section 8.1 we define radian measure. Then, in Section 8.2, we use radian measure to restate the definitions of the trigonometric functions in such a way that the domains are sets of real numbers. We want to do this for two reasons. First, for analytical work, the domains of the trigonometric functions are sets of real numbers. Second, if the domains are sets of real numbers, then the graphing techniques developed in Chapter 4 can be used to analyze these functions. With this in mind, we discuss the graphs of the trigonometric functions in Section 8.5. In the two sections prior to this, Sections 8.3 and 8.4, we develop the basic trigonometric identities. Then in Section 8.6 we apply some of these identities in solving trigonometric equations. Finally, in Section 8.7 we consider the inverse trigonometric functions.

8.1 Radian Measure

. . . I wrote to him [i.e., to Alexander J. Ellis, in 1874], and he declared at once for the form "radian," on the ground that it could be viewed as a contraction for "radial angle. . . ."

Thomas Muir, in a letter appearing in the April 7, 1910 issue of *Nature*

I shall be very pleased to send Dr. Muir a copy of my father's examination questions of June, 1873, containing the word radian. . . . *It thus appears that* radian *was thought of independently by Dr. Muir and my father, and, what is really more important than the exact form of the name, they both independently thought of the necessity of giving a name to the unit-angle.*

James Thomson, in a letter appearing in the June 16, 1910 issue of *Nature*

For the portion of trigonometry dealing with triangles, the units of degrees are quite suitable for measuring angles. For the more analytical portions of trigonometry, however, radian measure is used. The radian measure of an angle is defined as follows.

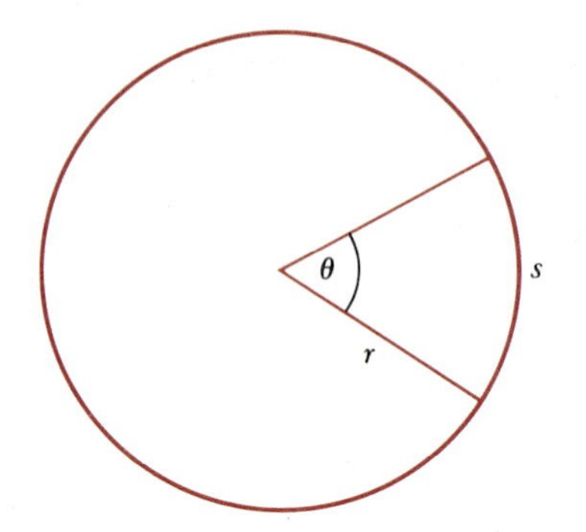

Figure 1

Definition of the Radian Measure of an Angle

Place the vertex of the angle θ at the center of a circle of radius r. Let s denote the length of the arc intercepted by θ, as indicated in Figure 1. Then the *radian measure* of θ is the ratio of the arc length s to the radius r. In symbols,

$$\theta = \frac{s}{r}$$

At first it may appear to you that the radian measure of θ depends upon the radius of the particular circle that we use. However, as you'll see in the next paragraph, this is not the case.

To gain some experience in working with the definition of radian measure, let us calculate the radian measure of the right angle in Figure 2. We begin with the formula

$$\theta = \frac{s}{r} \tag{1}$$

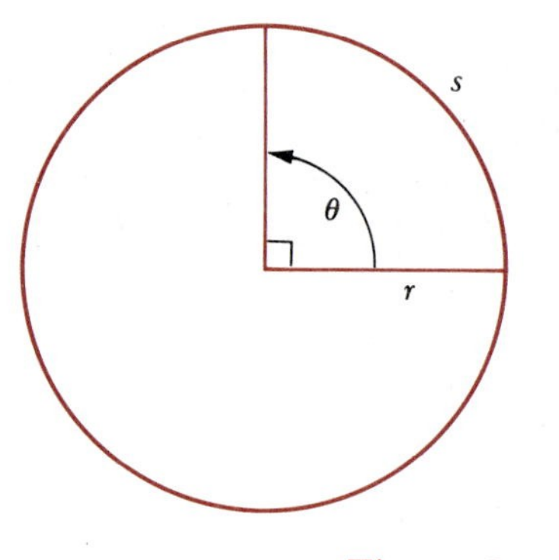

Figure 2

Now, since θ is a right angle, the arc length s, in this case, is one quarter of the entire circumference. Thus

$$s = \frac{1}{4}(2\pi r)$$

$$s = \frac{\pi r}{2} \tag{2}$$

Using equation (2) to substitute for s in equation (1) then gives us

$$\theta = \frac{\left(\frac{\pi r}{2}\right)}{r}$$

$$= \frac{\pi r}{2} \times \frac{1}{r}$$

$$= \frac{\pi}{2}$$

In other words

$$90° = \frac{\pi}{2} \text{ radians}$$

Notice that the radius r does not appear in our answer.

For practical reasons, we would like to be able to convert rapidly between degree and radian measure. First of all, multiplying both sides of the equation $90° = \pi/2$ radians by 2 yields

$$180° = \pi \text{ radians} \tag{3}$$

Equation (3) is useful and should be memorized. For instance, dividing both sides of equation (3) by 6 yields

$$\frac{180°}{6} = \frac{\pi}{6}$$

That is,

$$30° = \frac{\pi}{6} \text{ radians}$$

Similarly, dividing both sides of equation (3) by 4 and 3, respectively, yields

$$45° = \frac{\pi}{4} \text{ radians} \qquad \text{and} \qquad 60° = \frac{\pi}{3} \text{ radians}$$

Again, multiplying both sides of equation (3) by 2 gives us

$$360° = 2\pi \text{ radians}$$

Figure 3 summarizes some of these results.

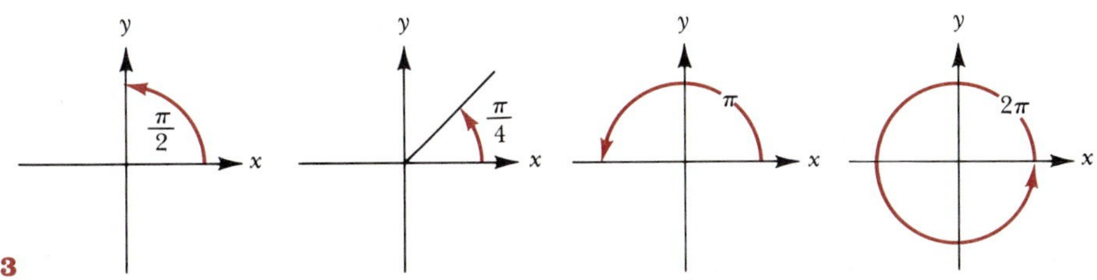

Figure 3

Example 1 **(a)** Express 1° in radian measure.
(b) Express 1 radian in terms of degrees.

Solution **(a)** Dividing both sides of the equation 180° = π radians by 180 yields

$$1° = \frac{\pi}{180} \text{ radian} \qquad \text{as required}$$

(b) Dividing both sides of the equation 180° = π radians by π yields

$$\frac{180°}{\pi} = 1 \text{ radian}$$

In other words, one radian is approximately 180°/3.14 or 57.3°.

On the basis of the results in Example 1, we have the following rules for converting between radians and degrees.

> To convert from degrees to radians, multiply by $\frac{\pi}{180°}$. To convert from radians to degrees, multiply by $\frac{180°}{\pi}$.

Example 2 Convert 150° to radians.

Solution

$$150°\left(\frac{\pi}{180°}\right) = \frac{5\pi}{6} \qquad \text{reducing the fraction}$$

Thus

$$150° = \frac{5\pi}{6} \text{ radians} \qquad \text{as required}$$

Example 3 Convert $11\pi/6$ radians to degrees.

Solution

$$\frac{11\pi}{6}\left(\frac{180°}{\pi}\right) = \frac{(11)(180°)}{6}$$
$$= 11(30°)$$
$$= 330°$$

Thus

$$\frac{11\pi}{6} \text{ radians} = 330° \qquad \text{as required}$$

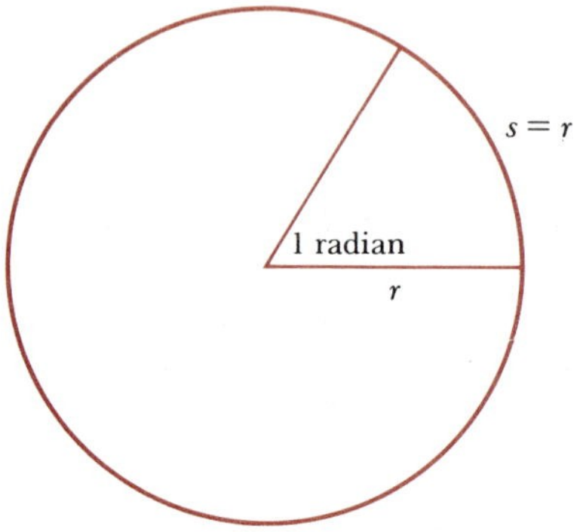

Figure 4
An angle of one radian intercepts an arc whose length is the same as the length of the radius.

We saw in Example 1(b) that one radian is approximately 57°. It is also useful to be able to visualize an angle of one radian without thinking in terms of degree measure. This is done as follows. In the equation $\theta = s/r$, we let $\theta = 1$. This yields $1 = s/r$, and consequently $r = s$. In other words, in a circle, one radian is the central angle that intercepts an arc equal in length to the radius of the circle. See Figure 4.

From now on, when we specify the measure of an angle, we shall assume that the units are radians, unless the degree symbol is explicitly used. (This convention is used in calculus also.) For instance, in this context, the equation $\theta = 2$ means that θ is two radians. Similarly, the expression $\sin \pi/6$ refers to the sine of $\pi/6$ radians.

Example 4 Evaluate the following.

(a) $\sin \frac{\pi}{6}$ **(b)** $\cos 2\pi$

Solution **(a)**

$$\frac{\pi}{6} \text{ radians} = 30°$$

Therefore

$$\sin \frac{\pi}{6} = \sin 30°$$
$$= \frac{1}{2} \qquad \text{because } \sin 30° = \frac{1}{2}$$

(b)

$$2\pi \text{ radians} = 360°$$

Therefore

$$\cos 2\pi = \cos 360°$$
$$= 1 \qquad \text{because } \cos 360° = 1$$

Let us now return to the defining equation for radians, $\theta = s/r$, and rewrite it as

$$s = r\theta \tag{4}$$

This useful equation expresses arc length on a circle in terms of the radius r and the central angle θ subtended by the arc. In applying this equation, we must be certain that θ is expressed in radians.

Example 5 Find the arc length s in Figure 5.

Solution We have seen previously that $30° = \pi/6$ radians. Thus

$$\begin{aligned} s &= r\theta \\ &= (10)\left(\frac{\pi}{6}\right) \\ &= \frac{5\pi}{3}\text{ cm} \end{aligned}$$

Note that the units for s and for r are the same.

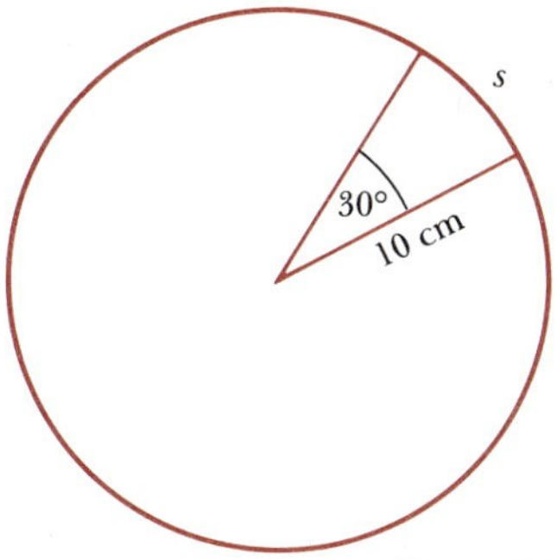

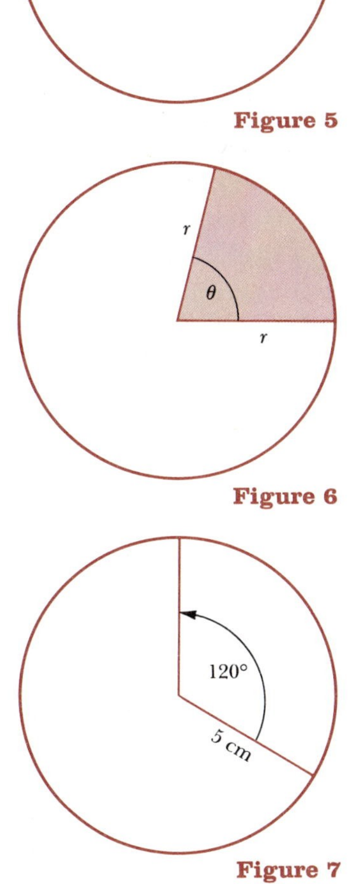

Figure 5

Figure 6

Figure 7

One of the advantages of using radian measure is that many formulas then take on particularly simple forms. The formula $s = r\theta$ is one example of this. As another example, let us derive a formula for the area of a sector of a circle. (In Figure 6, the shaded region is the sector.) From geometry, we know that the area A of the sector is directly proportional to the measure of its central angle θ. In symbols,

$$A = k\theta \tag{5}$$

To determine the constant k, we use the fact that when θ is 2π, we have a complete circle. Thus, when θ is 2π, the area must be πr^2. Substituting the values $\theta = 2\pi$ and $A = \pi r^2$ in equation (5) yields

$$\pi r^2 = k(2\pi)$$

and therefore

$$\frac{1}{2}r^2 = k$$

Now using this value for k in equation (5) gives us the following result:

$$A = \frac{1}{2}r^2\theta$$

This is the formula for the area of a sector of a circle in terms of the central angle θ and the radius r. The formula is valid only when θ is expressed in radians.

Example 6 Compute the area of the sector in Figure 7.

Solution First convert 120° to radians:

$$120°\left(\frac{\pi}{180°}\right) = \frac{2\pi}{3}$$

Thus

$$120° = \frac{2\pi}{3}\text{ radians}$$

We then have

$$\begin{aligned} A &= \frac{1}{2}r^2\theta \\ &= \frac{1}{2}(5^2)\frac{2\pi}{3} \\ &= \frac{25\pi}{3}\text{ cm}^2 \qquad \text{as required} \end{aligned}$$

Example 7 Express the area of the shaded region in Figure 8 as a function of θ.

Solution Let S denote the area of the shaded region in Figure 8. Then we have

$$S = (\text{area of sector } OAB) - (\text{area } \triangle OAB)$$

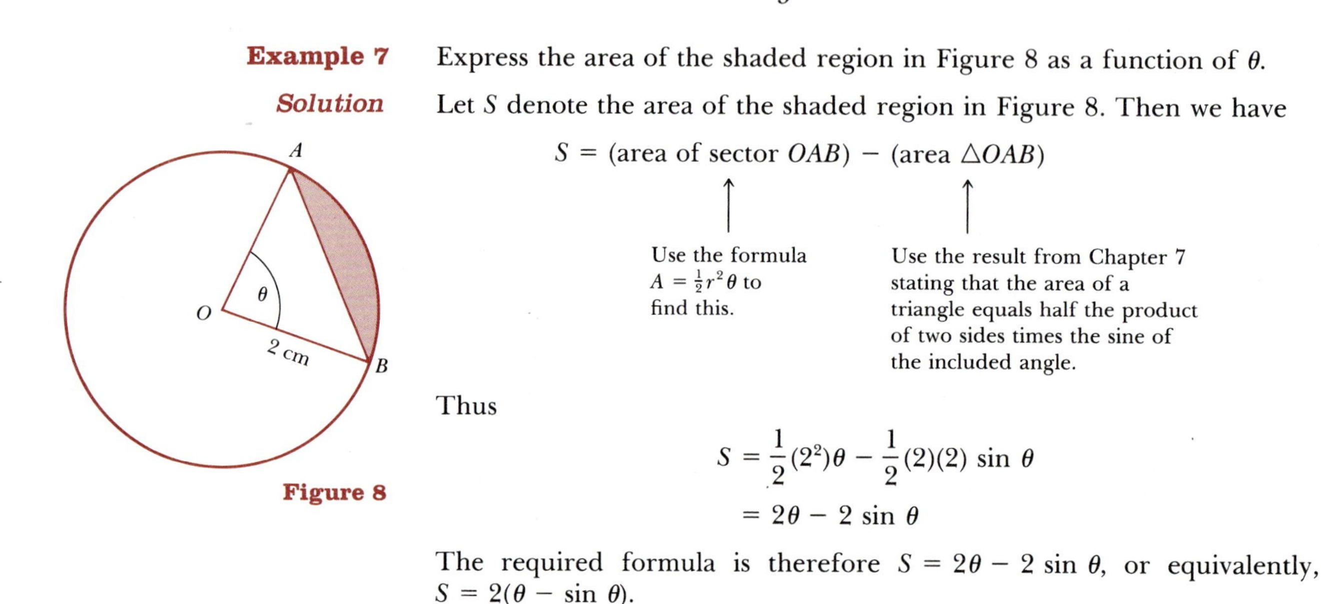

Figure 8

Use the formula $A = \frac{1}{2}r^2\theta$ to find this.

Use the result from Chapter 7 stating that the area of a triangle equals half the product of two sides times the sine of the included angle.

Thus

$$\begin{aligned} S &= \frac{1}{2}(2^2)\theta - \frac{1}{2}(2)(2)\sin\theta \\ &= 2\theta - 2\sin\theta \end{aligned}$$

The required formula is therefore $S = 2\theta - 2\sin\theta$, or equivalently, $S = 2(\theta - \sin\theta)$.

Exercise Set 8.1

1. Convert to radian measure.
(a) 60° **(b)** 225° **(c)** 36°
(d) 450° **(e)** 0°

2. Convert to degree measure.
(a) $\frac{\pi}{12}$ **(b)** $\frac{3\pi}{2}$ **(c)** 6π
(d) $\frac{\pi}{10}$ **(e)** $\frac{\pi}{2}$

3. Convert to radian measure.
(a) 35° **(b)** 22.5° **(c)** 2° **(d)** 100°

4. Convert to degree measure.
(a) $\frac{11\pi}{6}$ **(b)** $\frac{3\pi}{7}$ **(c)** 2π
(d) 2 **(e)** $\frac{1}{\pi}$

5. Suppose that $\theta = \frac{3}{2}$. Without using a calculator or tables, decide whether θ is larger or smaller than a right angle.

6. Complete the following table.

θ	0	$\frac{\pi}{2}$	π	$\frac{3\pi}{2}$	2π
$\sin\theta$					

7. Complete the following table.

θ	0	$\frac{\pi}{2}$	π	$\frac{3\pi}{2}$	2π
$\cos\theta$					

8. Complete the following table.

θ	0	$\frac{\pi}{6}$	$\frac{\pi}{4}$	$\frac{\pi}{3}$	$\frac{\pi}{2}$
$\tan\theta$					

9. Find the arc length s in each case.

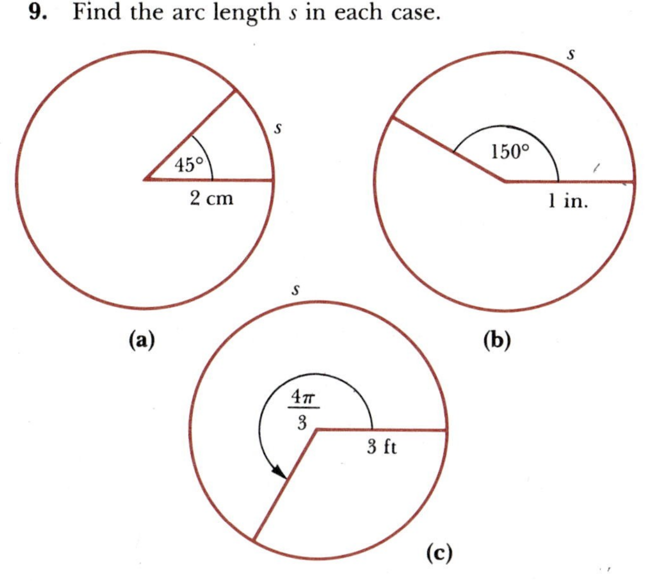

10. In a circle of radius 6 cm, an arc of length 1 cm subtends an angle θ at the center of the circle. What is the radian measure of θ?

11. In a circle of radius 4 in., an arc of length 4 in. subtends an angle θ at the center of the circle. What is the radian measure of θ?

12. Find the area of each sector.

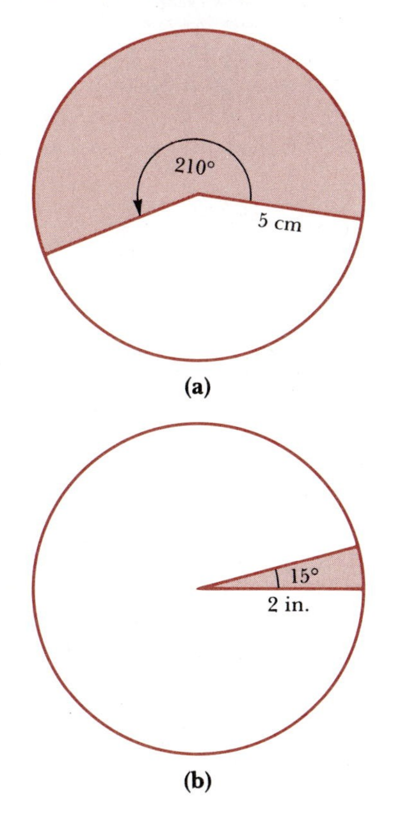

13. In a circle of radius 1 cm, the area of a certain sector is $\pi/5$ cm^2. Find the central angle of this sector.

14. Find the area of the shaded region in the figure.

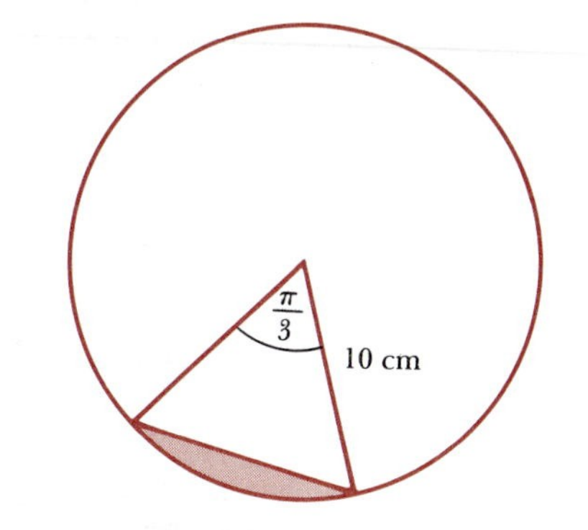

15. Express the area of the shaded region in the figure as a function of θ. *Hint:* The region is composed of a sector and a triangle.

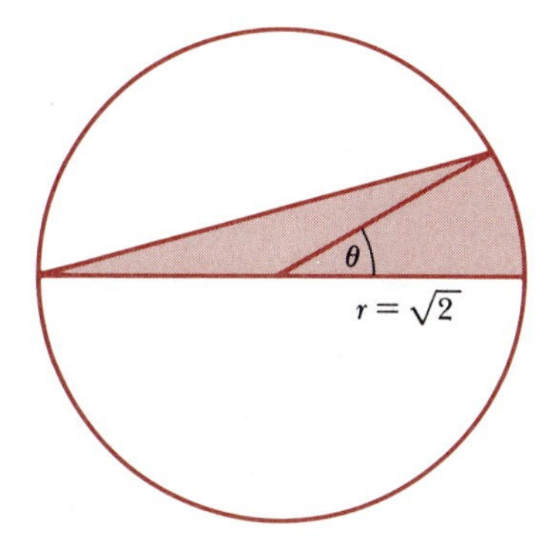

16. Suppose that we have two sticks and a piece of string, each of length 1 ft, fastened at the ends to form an equilateral triangle. See Figure (a). If side BC is bent out to form an arc of a circle with center A, then the angle at A will decrease from 60° to something less. See Figure (b). What is the measure of this new angle at A? (Assume the string does not stretch; its length remains 1 ft.)

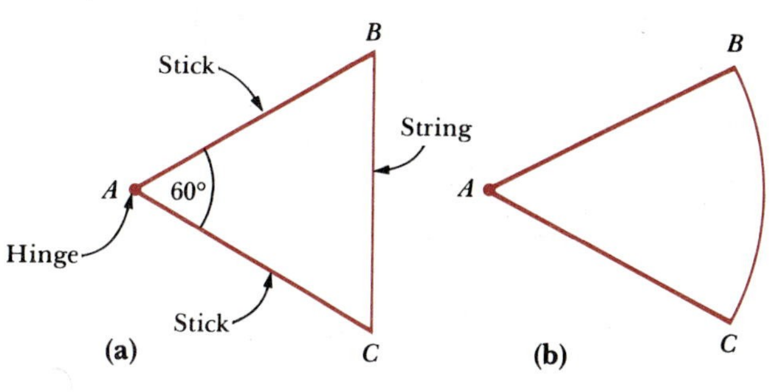

17. **(a)** When a clock reads 4:00, what is the radian measure of the (smaller) angle between the hour hand and minute hand?

(b) When a clock reads 5:30, what is the radian measure of the (smaller) angle between the hour hand and minute hand?

18. A wheel in a motor makes 1000 revolutions per minute.

(a) Through how many radians does the wheel turn in 1 min?

(b) Through how many radians does the wheel turn in 2 hr?

19. A wheel 3 ft in diameter makes 45 revolutions. Find the distance traveled by a point on the rim.

20. Suppose that the radius of a tire on a truck is 2 ft. In traveling 1000 mi, how many revolutions does the tire make?

***21.** The figure shows a semicircle of radius 1 unit and two adjacent sectors, AOC and COB.

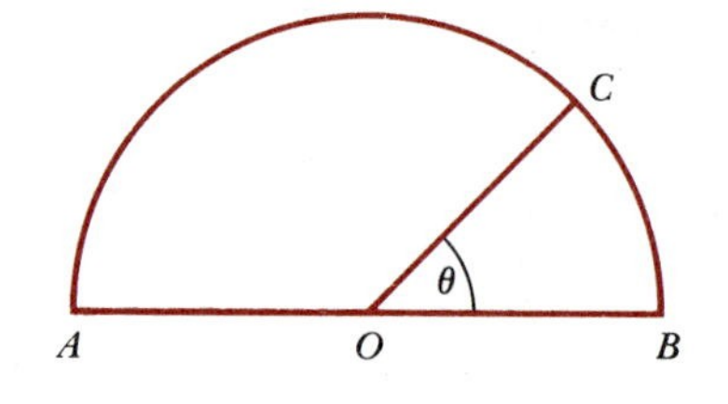

(a) Show that the product P of the areas of the two sectors is given by

$$P = \frac{\pi\theta}{4} - \frac{\theta^2}{4}$$

Is this a quadratic function?

(b) For what value of θ is P a maximum?

***22.** The following problem is taken from *A Treatise on Plane Trigonometry,* 7th edition, by E. W. Hobson (Cambridge University Press, 1928).

Two circles, the sum of whose radii is a, are placed in the same plane, with their centers at a distance $2a$, and an endless string, quite stretched, partly surrounds the circles and crosses itself between them. Show that the length of the string is $\left(\frac{4\pi}{3} + 2\sqrt{3}\right)a$.

8.2 Trigonometric Functions of Real Numbers

The definitions of the trigonometric functions in Section 7.3 are based on the unit circle. Let us look at radian measure in the context of the unit circle. By definition, the radian measure of θ is given by $\theta = s/r$; so when $r = 1$, we obtain

$$\theta = \frac{s}{1}$$

or

$$\theta = s$$

Thus, in the unit circle, the arc length *is* the radian measure of the angle θ. See Figure 1.

Figure 1
In the unit circle, the length of the intercepted arc is the radian measure of the central angle.

In view of this observation, we can restate the definitions of the trigonometric functions in such a way that the domains are sets of real numbers, rather than angles. As explained in the introduction to this chapter, there are two reasons for doing this. First, for analytical work, the domains of the trigonometric functions are sets of real numbers. Second, if the domains are sets of real numbers, then the techniques for graphing that we developed in Chapter 4 can be used to analyze these functions.

In the definitions that follow, you may think of θ either as the measure of an arc length or as the radian measure of an angle. But in both cases, and this is the point, θ denotes a real number. The conventions regarding the measurement of arc length on the unit circle are the same as those we used previously for angles. We measure from the point (1, 0) and we assume that the positive direction is counterclockwise. The notation $P(x, y)$ in the definition is used to denote the point on the unit circle whose arc length from (1, 0) is θ. Or, equivalently, $P(x, y)$ denotes the point where the terminal side of the angle with radian measure θ intersects the unit circle. (See Figure 2.)

Trigonometric Functions of Real Numbers

$\sin \theta = y$	$\csc \theta = \dfrac{1}{y} \; (y \neq 0)$
$\cos \theta = x$	$\sec \theta = \dfrac{1}{x} \; (x \neq 0)$
$\tan \theta = \dfrac{y}{x} \; (x \neq 0)$	$\cot \theta = \dfrac{x}{y} \; (y \neq 0)$

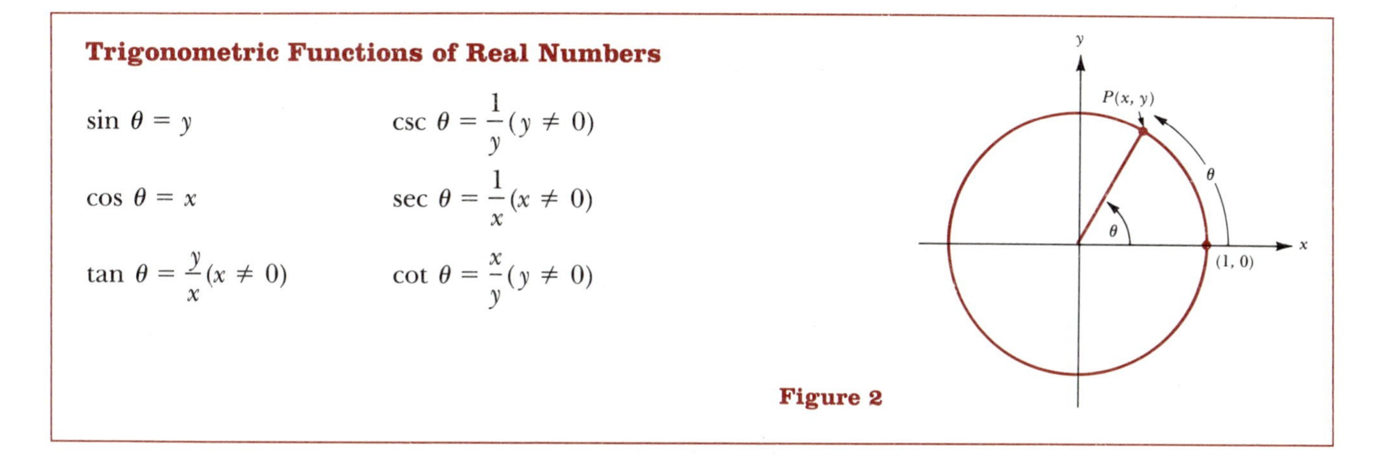

Figure 2

Let us immediately note that there is nothing new here as far as evaluating the trigonometric functions. For instance, $\sin \pi/6$ is still equal to $\frac{1}{2}$. (If this seems unfamiliar, review Example 4(a) on page 366.) Also, whether we are working in radians or degrees, our three-step procedure for evaluating the trigonometric functions is applicable. For instance, in Example 4 on page 341, we computed $\cos 240°$ ($\cos 4\pi/3$). Here is how that same procedure looks when radian measure is used. You should carefully compare each of the corresponding steps.

Step 1 The reference angle* for $4\pi/3$ is $\pi/3$. ($4\pi/3 - \pi = \pi/3$.)

Step 2 $\cos \pi/3 = \frac{1}{2}$

Step 3 Cos θ is the x-coordinate. In the third quadrant, x-coordinates are negative. Thus $\cos 4\pi/3$ is negative since the terminal side of $\theta = 4\pi/3$ lies in the third quadrant. (Equivalently, the arc of length $4\pi/3$ terminates in

*Some texts refer to this as the reference number, instead of reference angle, to emphasize the fact that $4\pi/3$ is a real number.

the third quadrant.) Thus we have

$$\cos \frac{4\pi}{3} = -\frac{1}{2}$$

There are many identities involving the trigonometric functions. (In fact, that is one reason why these functions are so useful in calculus.) We shall be concerned with these identities, in one form or another, throughout the rest of this chapter. Many of the identities should seem familiar to you from the text and exercises in Chapter 7. However, you should keep in mind that now the identities will be true for all real numbers θ for which the expressions are defined, rather than for a limited range of angles.

First of all, there are five trigonometric identities that are immediate consequences of the definitions. These are as follows.

$$\csc \theta = \frac{1}{\sin \theta} \qquad \tan \theta = \frac{\sin \theta}{\cos \theta}$$

$$\sec \theta = \frac{1}{\cos \theta} \qquad \cot \theta = \frac{\cos \theta}{\sin \theta}$$

$$\cot \theta = \frac{1}{\tan \theta}$$

The next identities that we consider are the three Pythagorean identities.

The Pythagorean Identities

$$\sin^2 \theta + \cos^2 \theta = 1$$
$$\tan^2 \theta + 1 = \sec^2 \theta$$
$$\cot^2 \theta + 1 = \csc^2 \theta$$

To prove the identity $\sin^2 \theta + \cos^2 \theta = 1$, we refer to Figure 2 (on page 371.) Since (x, y) is a point on the unit circle, we have

$$x^2 + y^2 = 1$$

Replacing x by $\cos \theta$ and y by $\sin \theta$ then gives us

$$\cos^2 \theta + \sin^2 \theta = 1$$

which is essentially what we wished to show. Incidentally, you should also become familiar with the equivalent forms of this identity:

$$\cos^2 \theta = 1 - \sin^2 \theta \quad \text{and} \quad \sin^2 \theta = 1 - \cos^2 \theta$$

To prove the second of the Pythagorean identities, begin with $\sin^2 \theta + \cos^2 \theta = 1$ and divide both sides by the quantity $\cos^2 \theta$ to obtain

$$\frac{\sin^2 \theta}{\cos^2 \theta} + \frac{\cos^2 \theta}{\cos^2 \theta} = \frac{1}{\cos^2 \theta} \qquad (\text{assuming } \cos \theta \neq 0)$$

and consequently

$$\tan^2 \theta + 1 = \sec^2 \theta \qquad \text{as required}$$

The proof of the third Pythagorean identity is similar to this and we omit giving it here.

Example 1 If $\sin\theta = \frac{2}{3}$ and $\pi/2 < \theta < \pi$, compute $\cos\theta$ and $\tan\theta$.

Solution

$$\begin{aligned}\cos^2\theta &= 1 - \sin^2\theta && \text{by the first Pythagorean identity}\\ &= 1 - \left(\frac{2}{3}\right)^2 && \text{substitution}\\ &= 1 - \frac{4}{9} = \frac{5}{9}\end{aligned}$$

Consequently

$$\cos\theta = \frac{\sqrt{5}}{3} \quad \text{or} \quad \cos\theta = \frac{-\sqrt{5}}{3}$$

Now since $\pi/2 < \theta < \pi$, it follows that $\cos\theta$ is negative. (Why?) Thus

$$\cos\theta = \frac{-\sqrt{5}}{3} \qquad \text{as required}$$

To compute $\tan\theta$, we use the identity $\tan\theta = \sin\theta/\cos\theta$ to obtain

$$\begin{aligned}\tan\theta &= \frac{\frac{2}{3}}{-\sqrt{5}/3} = \frac{2}{\cancel{3}} \times \frac{\cancel{3}}{-\sqrt{5}}\\ &= -\frac{2}{\sqrt{5}}\end{aligned}$$

If required, we can rationalize the denominator (by multiplying by $\sqrt{5}/\sqrt{5}$) to obtain

$$\tan\theta = -\frac{2\sqrt{5}}{5}$$

Example 2 If $\sec\theta = -\frac{5}{3}$ and $\pi < \theta < 3\pi/2$, compute $\cos\theta$ and $\tan\theta$.

Solution Cos θ is the reciprocal of $\sec\theta$. Thus $\cos\theta = -\frac{3}{5}$. We can compute $\tan\theta$ using the second Pythagorean identity as follows.

$$\begin{aligned}\tan^2\theta &= \sec^2\theta - 1\\ &= \left(-\frac{5}{3}\right)^2 - 1\\ &= \frac{25}{9} - \frac{9}{9} = \frac{16}{9}\end{aligned}$$

Therefore

$$\tan\theta = \frac{4}{3} \quad \text{or} \quad \tan\theta = -\frac{4}{3}$$

Since θ is between π and $3\pi/2$, $\tan\theta$ is positive. Thus

$$\tan\theta = \frac{4}{3} \qquad \text{as required}$$

In calculus, certain radical expressions can often be simplified through an appropriate trigonometric substitution. The next example shows how this works.

Example 3 In the expression $\dfrac{u}{\sqrt{u^2-1}}$, make the substitution $u = \sec\theta$ and show that the resulting expression is equal to $\csc\theta$. (Assume that $0 < \theta < \pi/2$.)

Solution Replacing u by $\sec\theta$ in the given expression yields

$$\begin{aligned}
\frac{u}{\sqrt{u^2-1}} &= \frac{\sec\theta}{\sqrt{\sec^2\theta-1}}\\
&= \frac{\sec\theta}{\sqrt{\tan^2\theta}} \qquad \text{by the second Pythagorean identity}\\
&= \frac{\sec\theta}{\tan\theta}\\
&= \frac{1/\cos\theta}{\sin\theta/\cos\theta} = \frac{1}{\cos\theta}\times\frac{\cos\theta}{\sin\theta}\\
&= \frac{1}{\sin\theta}\\
&= \csc\theta \qquad \text{as required}
\end{aligned}$$

Question: Where was the condition $0 < \theta < \pi/2$ used?

The next three identities indicate the effects of replacing θ by $-\theta$ in the expressions $\sin\theta$, $\cos\theta$, and $\tan\theta$.

$$\begin{aligned}
\cos(-\theta) &= \cos\theta\\
\sin(-\theta) &= -\sin\theta\\
\tan(-\theta) &= -\tan\theta
\end{aligned}$$

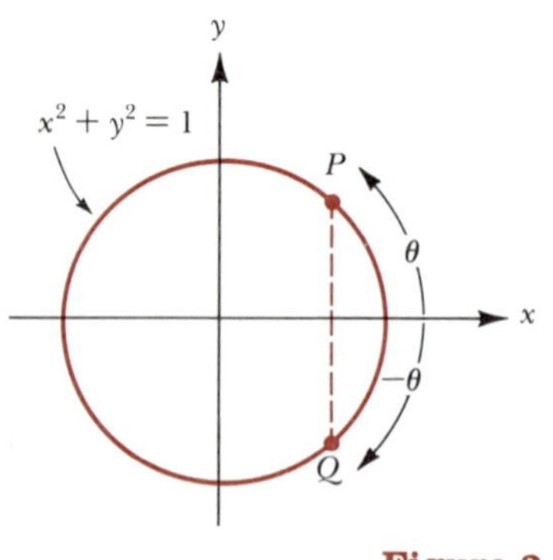

Figure 3

To see why the first two of these identities are true, consider Figure 3. (Although Figure 3 shows an arc of length θ terminating in the first quadrant, the same kind of argument we use will work when θ terminates in any quadrant.) Now by definition, the coordinates of P and Q are as follows.

$$P: \quad (\cos\theta, \sin\theta)$$
$$Q: \quad (\cos(-\theta), \sin(-\theta))$$

However, as you can see by looking at Figure 3, the x-coordinates of P and Q are the same, while the y-coordinates are negatives of one another. Thus

$$\cos(-\theta) = \cos\theta$$

and

$$\sin(-\theta) = -\sin\theta \qquad \text{as we wished to show}$$

Now we can establish the third identity involving $\tan(-\theta)$ as follows.

$$\begin{aligned}\tan(-\theta) &= \frac{\sin(-\theta)}{\cos(-\theta)}\\ &= \frac{-\sin\theta}{\cos\theta} = -\frac{\sin\theta}{\cos\theta}\\ &= -\tan\theta \qquad \text{as required}\end{aligned}$$

The last set of identities that we are going to discuss in this section are simply consequences of the fact that the circumference of the unit circle is 2π*. Thus if you begin at any point P on the unit circle and travel a distance of 2π units along the circle, you return to the same point P. In other words, arc lengths of θ and $\theta + 2\pi$ [measured from (1, 0), as usual] yield the same terminal point P on the unit circle. Since the trigonometric functions are defined in terms of the coordinates of that point P, we conclude that

$$\sin(\theta + 2\pi) = \sin\theta$$
$$\cos(\theta + 2\pi) = \cos\theta$$

These two identities are true for all real numbers θ. The same kind of identities also hold for the other trigonometric functions in their respective domains:

$$\tan(\theta + 2\pi) = \tan\theta \qquad \csc(\theta + 2\pi) = \csc\theta$$
$$\cot(\theta + 2\pi) = \cot\theta \qquad \sec(\theta + 2\pi) = \sec\theta$$

Example 4 Evaluate $\sin 5\pi/2$.

Solution First, simply as a matter of arithmetic, observe that $5\pi/2 = \pi/2 + 2\pi$. Thus

$$\begin{aligned}\sin\frac{5\pi}{2} &= \sin\left(\frac{\pi}{2} + 2\pi\right)\\ &= \sin\frac{\pi}{2}\\ &= 1\end{aligned}$$

The preceding set of identities can be generalized as follows. If we start at a point P on the unit circle and make two complete revolutions, the arc length we travel is

$$2\pi + 2\pi = 4\pi$$

Similarly, the arc length covered in three complete revolutions is

$$2\pi + 2\pi + 2\pi = 6\pi$$

and the arc length for k complete revolutions is $2k\pi$. (When k is positive, the revolution is counterclockwise; when k is negative, the revolution is clock-

*This is because $C = 2\pi r$. Thus when $r = 1$ we obtain $C = 2\pi$, as stated.

wise.) It follows that arc lengths θ and $\theta + 2\pi k$ yield the same terminal point P. Thus, for any integer k we have

$$\sin(\theta + 2\pi k) = \sin\theta$$
$$\cos(\theta + 2\pi k) = \cos\theta$$

Example 5 Evaluate $\cos(-17\pi)$.

Solution

$$\begin{aligned}\cos(-17\pi) &= \cos(\pi - 18\pi)\\ &= \cos\pi\\ &= -1\end{aligned}$$

Exercise Set 8.2

1. Complete the following table.

θ	$\sin\theta$	$\cos\theta$	$\tan\theta$	$\csc\theta$	$\sec\theta$	$\cot\theta$
0						
$\frac{\pi}{6}$						
$\frac{\pi}{4}$						
$\frac{\pi}{3}$						
$\frac{\pi}{2}$						
$\frac{2\pi}{3}$						
$\frac{3\pi}{4}$						
$\frac{5\pi}{6}$						
π						

2. Complete the following table.

θ	$\sin\theta$	$\cos\theta$	$\tan\theta$	$\csc\theta$	$\sec\theta$	$\cot\theta$
π						
$\frac{7\pi}{6}$						
$\frac{5\pi}{4}$						
$\frac{4\pi}{3}$						
$\frac{3\pi}{2}$						
$\frac{5\pi}{3}$						
$\frac{7\pi}{4}$						
$\frac{11\pi}{6}$						

3. If $\sin\theta = -\frac{3}{5}$ and $\pi < \theta < 3\pi/2$, compute $\cos\theta$ and $\tan\theta$.
4. If $\cos\theta = \frac{5}{13}$ and $3\pi/2 < \theta < 2\pi$, compute $\sin\theta$ and $\cot\theta$.
5. If $\sin t = \sqrt{3}/4$ and $\pi/2 < t < \pi$, compute $\tan t$.
6. If $\sec\theta = -\sqrt{13}/2$ and $\sin\theta > 0$, compute $\tan\theta$.
7. If $\tan\alpha = \frac{12}{5}$ and $\cos\alpha > 0$, compute $\sec\alpha$, $\cos\alpha$, and $\sin\alpha$.
8. If $\cot\theta = -1/\sqrt{3}$ and $\cos\theta < 0$, compute $\csc\theta$ and $\sin\theta$.

9. In the expression $\sqrt{9 - x^2}$, make the substitution $x = 3 \sin \theta\,(0 < \theta < \pi/2)$ and show that the result is $3 \cos \theta$.

10. Make the substitution $u = 2 \cos \theta$ in the expression $\dfrac{1}{\sqrt{4 - u^2}}$ and simplify the result. (Assume $0 < \theta < \pi$.)

11. In the expression $\dfrac{1}{(u^2 - 25)^{3/2}}$, make the substitution $u = 5 \sec \theta\,(0 < \theta < \pi/2)$ and show that the result is $(\cot^3 \theta)/125$.

12. In the expression $\dfrac{1}{(x^2 + 5)^2}$, replace x by $\sqrt{5} \tan \theta$ and show that the result is $(\cos^4 \theta)/25$.

13. In the expression $\dfrac{1}{\sqrt{u^2 + 7}}$, let $u = \sqrt{7} \tan \theta$ $(0 < \theta < \pi/2)$ and simplify the result.

14. **(a)** If $\sin \theta = 0.136$, find $\sin(-\theta)$.
(b) If $\cos \alpha = \sqrt{3}/4$, find $\cos(-\alpha)$.

15. If $\cos \theta = -\frac{1}{3}$ and $\pi/2 < \theta < \pi$, compute the following.
(a) $\sin(-\theta) + \cos(-\theta)$
(b) $\sin^2(-\theta) + \cos^2(-\theta)$

16. If $\sin(-\theta) = \frac{3}{5}$ and $\pi < \theta < 3\pi/2$, compute the following.
(a) $\sin \theta$ **(b)** $\cos(-\theta)$
(c) $\cos \theta$ **(d)** $\tan \theta + \tan(-\theta)$

17. Evaluate the following.
(a) $\cos\left(\dfrac{\pi}{4} + 2\pi\right)$ **(b)** $\sin\left(\dfrac{\pi}{3} + 2\pi\right)$
(c) $\sin\left(\dfrac{\pi}{2} - 6\pi\right)$

18. Evaluate the following.
(a) $\sin\left(\dfrac{17\pi}{4}\right)$ **(b)** $\sin\left(-\dfrac{17\pi}{4}\right)$
(c) $\cos(11\pi)$ **(d)** $\cos\left(\dfrac{53\pi}{4}\right)$
(e) $\tan\left(\dfrac{-7\pi}{4}\right)$ **(f)** $\cot\left(\dfrac{7\pi}{4}\right)$
(g) $\sec\left(\dfrac{11\pi}{6} + 2\pi\right)$ **(h)** $\csc\left(2\pi - \dfrac{\pi}{3}\right)$

19. In the figure the x-coordinate of P is $-\frac{8}{17}$.

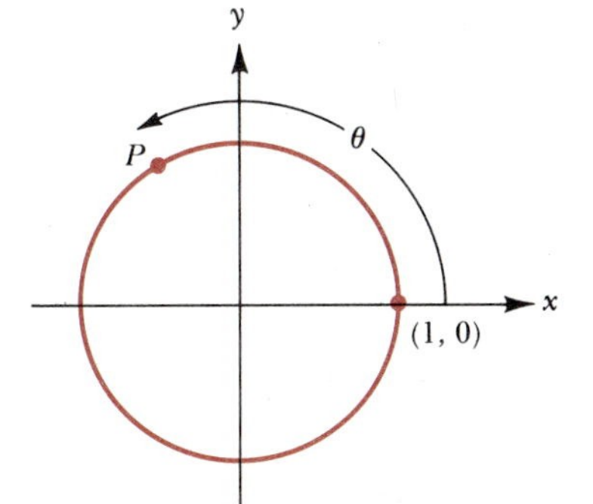

(a) Evaluate the six trigonometric functions of θ.
(b) Evaluate the six trigonometric functions of $-\theta$.

20. Prove the following trigonometric identities.
(a) $\csc \theta = \sin \theta + \cot \theta \cos \theta$
(b) $\sin^2 \theta - \cos^2 \theta = \dfrac{1 - \cot^2 \theta}{1 + \cot^2 \theta}$
(c) $\dfrac{1}{1 + \sec \theta} + \dfrac{1}{1 - \sec \theta} = -2 \cot^2 \theta$
(d) $\dfrac{1 + \tan \theta}{1 - \tan \theta} = \dfrac{\sec^2 \theta + 2 \tan \theta}{2 - \sec^2 \theta}$
(e) $\cot \theta + \tan \theta + 1 = \dfrac{\cot \theta}{1 - \tan \theta} + \dfrac{\tan \theta}{1 - \cot \theta}$
(f) $\dfrac{\sec s + \cot s \csc s}{\cos s} = \csc^2 s \sec^2 s$
(g) $\tan \theta(1 - \cot^2 \theta) + \cot \theta(1 - \tan^2 \theta) = 0$
(h) $(\cos \alpha \cos \beta - \sin \alpha \sin \beta) \times$ $(\cos \alpha \cos \beta + \sin \alpha \sin \beta) = \cos^2 \alpha - \sin^2 \beta$

21. Suppose that $\tan \theta = \dfrac{b^2 - a^2}{2ab}$, where a and b are positive and $\pi/2 < \theta < \pi$. Show that $\cos \theta = \dfrac{-2ab}{a^2 + b^2}$.

22. In the equation $x^4 + 6x^2y^2 + y^4 = 32$, make the substitutions

$$x = X \cos \frac{\pi}{4} - Y \sin \frac{\pi}{4}$$

$$y = X \sin \frac{\pi}{4} + Y \cos \frac{\pi}{4}$$

and show that the resulting equation simplifies to

$$X^4 + Y^4 = 16$$

23. Suppose $\tan \theta = 2$ and $0 < \theta < \pi/2$.
(a) Compute $\sin \theta$ and $\cos \theta$.
(b) Using the values obtained in part (a), make the substitutions

$$x = X \cos \theta - Y \sin \theta$$

$$y = X \sin \theta + Y \cos \theta$$

in the expression $7x^2 - 8xy + y^2$ and simplify the result.

24. Another reason for using radian measure is that the trigonometric functions can be closely approximated by very simple polynomial functions. To see an example of this, complete the table below. The values for $\cos \theta$ in the table were obtained using a calculator.

θ	$1 - \dfrac{\theta^2}{2}$	$\cos \theta$
0.1		0.995004...
0.2		0.980066...
0.3		0.955336...
1.0		0.540302...

Note: The approximating polynomials in Exercises 24 and 25 are known as *Taylor polynomials,* after the English mathematician Brook Taylor (1685–1731). The theory of Taylor polynomials is developed in calculus.

25. Complete the following table, as in Exercise 24.

θ	$\theta - \frac{\theta^3}{6}$	$\sin\theta$
0.1		0.099833...
0.2		0.198669...
0.3		0.295520...
1.0		0.841470...

26. Let $a = \sin^2\theta + \csc^2\theta$, $b = \cos^2\theta + \sec^2\theta$, and $c = \tan^2\theta + \cot^2\theta$. Show that $a + b - c = 3$.

27. All of the trigonometric functions can be expressed in terms of any one of them. For instance, to express the other five functions in terms of $\sin\theta$, we have

$$\cos\theta = \pm\sqrt{1 - \sin^2\theta}$$

Then

$$\tan\theta = \frac{\sin\theta}{\cos\theta} = \frac{\sin\theta}{\pm\sqrt{1 - \sin^2\theta}}$$

The remaining three functions, csc, sec, and cot, are then found by taking reciprocals. By following a similar procedure, express the remaining five trigonometric functions in terms of $\sec\theta$.

***28.** Let $f(\theta) = \sin\theta\cos\theta$ $(0 \le \theta \le \pi/2)$.

C **(a)** Set your calculator in the radian mode and complete the following table.

θ	0	$\frac{\pi}{10}$	$\frac{\pi}{5}$	$\frac{\pi}{4}$	$\frac{3\pi}{10}$	$\frac{2\pi}{5}$	$\frac{\pi}{2}$
$f(\theta)$							

(b) What is the largest value of $f(\theta)$ in your table in part (a)?

(c) Show that $\sin\theta\cos\theta \le \frac{1}{2}$ for all real numbers θ in the interval $0 \le \theta \le \pi/2$. *Hint:* Use the inequality $\sqrt{ab} \le \frac{a+b}{2}$ [given in Exercise Set 2.5, Exercise 64 (b)], with $a = \sin\theta$ and $b = \cos\theta$.

(d) Does the inequality $\sin\theta\cos\theta \le \frac{1}{2}$ hold for all real numbers θ?

29. C **(a)** Use a calculator to complete the following table. (Set your calculator in the radian mode.)

θ	$\sin\theta$	Which is larger, θ or $\sin\theta$?
0.1		
0.2		
0.3		
0.4		
0.5		

(b) From the figure explain why

$$\overline{PQ} < \overline{PR} < \theta$$

Then use this to show that if $0 < \theta < \pi/2$, then

$$\sin\theta < \theta$$

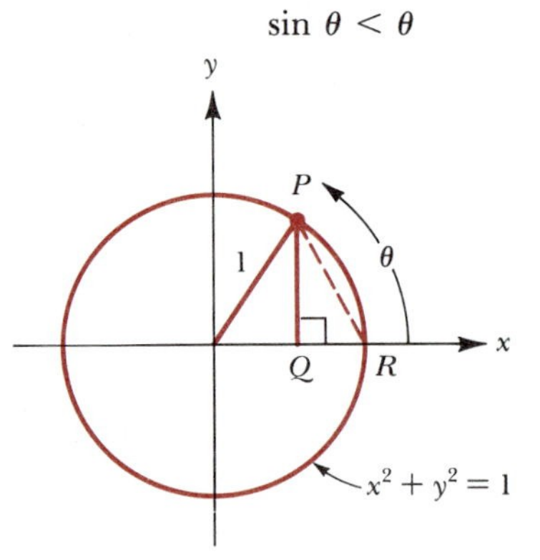

30. **(a)** List four distinct real numbers θ for which $\sin\theta = 0$.

(b) What is the domain of the function $f(\theta) = 1/\sin\theta$?

(c) What is the domain of the function $g(\theta) = 1/\sin\pi\theta$?

8.3 The Addition Formulas

For any real numbers r, s, and t, it is always true that $r(s + t) = rs + rt$. This is the so-called *distributive law* for real numbers. For functions, however, there is no corresponding version of the distributive law. By way of example, consider the cosine function. Then it is not true in general that $\cos(s + t) = \cos s + \cos t$. For instance with $s = \pi/6$ and $t = \pi/3$, we have

$$\cos\left(\frac{\pi}{6} + \frac{\pi}{3}\right) \stackrel{?}{=} \cos\frac{\pi}{6} + \cos\frac{\pi}{3}$$

$$\cos\frac{\pi}{2} \stackrel{?}{=} \cos\frac{\pi}{6} + \cos\frac{\pi}{3}$$

$$0 \stackrel{?}{=} \frac{\sqrt{3}}{2} + \frac{1}{2} \qquad \text{No!}$$

In this section we shall see just what $\cos(s + t)$ does equal. The correct formula for $\cos(s + t)$ is one of a group of four important trigonometric identities called the *addition formulas*.

The Addition Formulas

$$\sin(s + t) = \sin s \cos t + \cos s \sin t$$
$$\sin(s - t) = \sin s \cos t - \cos s \sin t$$
$$\cos(s + t) = \cos s \cos t - \sin s \sin t$$
$$\cos(s - t) = \cos s \cos t + \sin s \sin t$$

Our strategy for deriving these formulas will be as follows. First we'll prove the fourth formula—this takes some effort. The other formulas are then relatively easy consequences of the fourth formula.

To prove the formula $\cos(s - t) = \cos s \cos t + \sin s \sin t$, we use Figure 1. The idea behind the proof is simply this. There are two distinct ways to calculate the quantity $\overline{PQ}^2$ in Figure 1: we can calculate $\overline{PQ}^2$ using the law of cosines in $\triangle QOP$; or we can calculate $\overline{PQ}^2$ using the distance formula $d = \sqrt{(x_2 - x_1)^2 + (y_2 - y_1)^2}$. Of course the results of these two calculations must agree because they'll both represent $\overline{PQ}^2$. It is by equating the two results that we'll arrive at the required formula for $\cos(s - t)$.

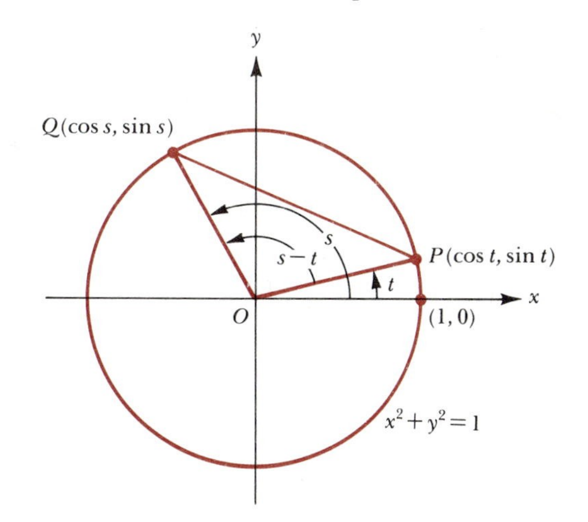

Figure 1

Applying the law of cosines in $\triangle QOP$, we have

$$\begin{aligned}\overline{PQ}^2 &= \overline{OP}^2 + \overline{OQ}^2 - 2\overline{OP}\cdot\overline{OQ}\cos(s-t)\\ &= 1^2 + 1^2 - 2(1)(1)\cos(s-t)\\ &= 2 - 2\cos(s-t) \qquad (1)\end{aligned}$$

On the other hand, using the distance formula we find

$$\overline{PQ} = \sqrt{(\cos s - \cos t)^2 + (\sin s - \sin t)^2}$$

Therefore

$$\begin{aligned}\overline{PQ}^2 &= (\cos s - \cos t)^2 + (\sin s - \sin t)^2\\ &= \cos^2 s - 2\cos s\cos t + \cos^2 t + \sin^2 s - 2\sin s\sin t + \sin^2 t\\ &= 2 - 2\cos s\cos t - 2\sin s\sin t \qquad (2)\end{aligned}$$

(Annotations: $\cos^2 t$ and $\sin^2 t$ — These two terms add to 1. $\cos^2 s$ and $\sin^2 s$ — These two terms add to 1.)

Now we equate the two expressions for $\overline{PQ}^2$ obtained in equations (1) and (2). This yields

$$\begin{aligned}2 - 2\cos(s-t) &= 2 - 2\cos s\cos t - 2\sin s\sin t\\ -2\cos(s-t) &= -2\cos s\cos t - 2\sin s\sin t\\ \cos(s-t) &= \cos s\cos t + \sin s\sin t\end{aligned}$$

This completes the proof of the fourth addition formula. Before deriving the other three addition formulas, let us look at some applications of this one.

Example 1 Simplify $\cos(\theta - \pi)$.

Solution Use the formula for $\cos(s - t)$, with s and t replaced by θ and π, respectively. This yields

$$\cos(s - t) = \cos s\cos t + \sin s\sin t$$

$$\updownarrow \quad \updownarrow$$

$$\begin{aligned}\cos(\theta - \pi) &= \cos\theta\cos\pi + \sin\theta\sin\pi\\ &= (\cos\theta)(-1) + (\sin\theta)(0)\\ &= -\cos\theta\end{aligned}$$

Thus the required simplification is $\cos(\theta - \pi) = -\cos\theta$.

In the example just completed we found that $\cos(\theta - \pi) = -\cos\theta$. This type of identity is often referred to as a *reduction formula*. The next example develops two useful reduction formulas that we shall need to use later in this section.

Example 2 Prove the following identities.

(a) $\cos\left(\frac{\pi}{2} - \alpha\right) = \sin\alpha$ **(b)** $\sin\left(\frac{\pi}{2} - \beta\right) = \cos\beta$

Solution **(a)**

$$\cos\left(\frac{\pi}{2} - \alpha\right) = \cos\frac{\pi}{2}\cos\alpha + \sin\frac{\pi}{2}\sin\alpha$$
$$= (0)\cos\alpha + (1)\sin\alpha$$
$$= \sin\alpha$$

This proves the identity in (a). Incidentally, using degree instead of radian measure, this identity states that

$$\cos(90° - \alpha) = \sin\alpha$$

as we saw earlier, in Chapter 7.

(b) Since the identity $\cos(\pi/2 - \alpha) = \sin\alpha$ holds for all values of α, we may replace α by the quantity $(\pi/2 - \beta)$ to obtain

$$\cos\left[\frac{\pi}{2} - \left(\frac{\pi}{2} - \beta\right)\right] = \sin\left(\frac{\pi}{2} - \beta\right)$$
$$\cos\left(\frac{\pi}{2} - \frac{\pi}{2} + \beta\right) = \sin\left(\frac{\pi}{2} - \beta\right)$$
$$\cos\beta = \sin\left(\frac{\pi}{2} - \beta\right)$$

This proves the identity (b), as required.

The two identities in Example 2 are worth memorizing, so for convenience we are going to restate them. For simplicity, we replace both α and β by the single letter θ.

$$\cos\left(\frac{\pi}{2} - \theta\right) = \sin\theta$$
$$\sin\left(\frac{\pi}{2} - \theta\right) = \cos\theta$$

We developed the fourth addition formula under the assumption that the angles were measured in radians. However, the formula is still valid when the angles are measured in degrees. In the next example we apply the formula in just such a case.

Example 3 Use the formula for $\cos(s - t)$ to determine the exact value of $\cos 15°$.

Solution First observe that $15° = 45° - 30°$. Then we have

$$\cos 15° = \cos(45° - 30°)$$
$$= \cos 45° \cos 30° + \sin 45° \sin 30°$$
$$= \left(\frac{\sqrt{2}}{2}\right)\left(\frac{\sqrt{3}}{2}\right) + \left(\frac{\sqrt{2}}{2}\right)\left(\frac{1}{2}\right)$$
$$= \frac{\sqrt{6}}{4} + \frac{\sqrt{2}}{4}$$
$$= \frac{\sqrt{6} + \sqrt{2}}{4}$$

We used the formula for $\cos(s - t)$ with $s = 45°$ and $t = 30°$.

Thus the exact value of $\cos 15°$ is $\dfrac{\sqrt{6}+\sqrt{2}}{4}$.

Now let us return to the main line of development in this section. Using the fourth addition formula, we can easily derive the third formula as follows. In the formula

$$\cos(s-t) = \cos s \cos t + \sin s \sin t$$

we replace t by the quantity $(-t)$. This is permissible since the formula holds for all real numbers. We obtain

$$\cos[s-(-t)] = \cos s \cos(-t) + \sin s \sin(-t)$$

On the right-hand side of this equation, we can use the identities developed for $\cos(-t)$ and $\sin(-t)$ in the previous section. Doing this yields

$$\cos(s+t) = \cos s(\cos t) + \sin s(-\sin t)$$

which is equivalent to

$$\cos(s+t) = \cos s \cos t - \sin s \sin t$$

This is the third addition formula, as we wished to prove.

Next we derive the formula for $\sin(s+t)$. We have

$$\begin{aligned}
\sin(s+t) &= \cos\left[\frac{\pi}{2}-(s+t)\right] \\
&= \cos\left[\left(\frac{\pi}{2}-s\right)-t\right] \\
&= \cos\left(\frac{\pi}{2}-s\right)\cos t + \sin\left(\frac{\pi}{2}-s\right)\sin t \qquad \text{(Why?)} \\
&= \sin s \cos t + \cos s \sin t \qquad \text{(Why?)}
\end{aligned}$$

In the identity $\sin\theta = \cos\left(\frac{\pi}{2}-\theta\right)$, we replaced θ by the quantity $(s+t)$.

This proves the first addition formula.

Finally, we can use the first addition formula to prove the second as follows.

$$\begin{aligned}
\sin(s-t) &= \sin[s+(-t)] \\
&= \sin s \cos(-t) + \cos s \sin(-t) \\
&= \sin s(\cos t) + \cos s(-\sin t) \\
&= \sin s \cos t - \cos s \sin t \qquad \text{as required}
\end{aligned}$$

This completes the proofs of the four addition formulas.

Example 4 If $\sin s = \frac{3}{5}$, $0 < s < \pi/2$, and $\sin t = -\sqrt{3}/4$, $\pi < t < 3\pi/2$, compute $\sin(s-t)$.

Solution

$$\sin(s-t) = \underbrace{\sin s}_{\text{given as } \frac{3}{5}} \cos t - \cos s \underbrace{\sin t}_{\text{given as } \frac{-\sqrt{3}}{4}} \tag{3}$$

In view of the above, we need only to find $\cos t$ and $\cos s$. These can be determined using the Pythagorean identity $\cos^2\theta = 1 - \sin^2\theta$. We have

$$\cos^2 t = 1 - \sin^2 t$$
$$= 1 - \left(\frac{-\sqrt{3}}{4}\right)^2$$
$$= 1 - \frac{3}{16} = \frac{13}{16}$$

Therefore

$$\cos t = \frac{\sqrt{13}}{4} \quad \text{or} \quad \cos t = \frac{-\sqrt{13}}{4}$$

We choose the negative value here for cosine, since it is given that $\pi < t < 3\pi/2$. Thus

$$\cos t = \frac{-\sqrt{13}}{4}$$

Similarly, to find $\cos s$ we have

$$\cos^2 s = 1 - \sin^2 s$$
$$= 1 - \left(\frac{3}{5}\right)^2$$
$$= 1 - \frac{9}{25} = \frac{16}{25}$$

Therefore

$$\cos s = \frac{4}{5}$$ $\cos s$ is positive since $0 < s < \pi/2$.

Finally, we substitute the values we've obtained for $\cos t$ and $\cos s$, along with the given data, back into equation (3). This yields

$$\sin(s - t) = \left(\frac{3}{5}\right)\left(\frac{-\sqrt{13}}{4}\right) - \left(\frac{4}{5}\right)\left(\frac{-\sqrt{3}}{4}\right)$$
$$= \frac{-3\sqrt{13}}{20} + \frac{4\sqrt{3}}{20}$$
$$= \frac{-3\sqrt{13} + 4\sqrt{3}}{20} \quad \text{as required}$$

Exercise Set 8.3

1. Simplify each term (as was done in Example 1).
   (a) $\sin\left(\theta - \frac{3\pi}{2}\right)$ **(b)** $\cos(\theta + \pi)$
   (c) $\cos\left(\frac{3\pi}{2} + \theta\right)$ **(d)** $\sin(\theta - \pi)$
2. **(a)** Expand $\sin(\theta + 2\pi)$ using the appropriate addition formula and check to see that your answer agrees with the formula for $\sin(\theta + 2\pi)$ given on page 375.
   (b) Same as part (a), but using $\cos(\theta + 2\pi)$.
3. Use the formula for $\cos(s + t)$ to compute the exact value of $\cos 75°$.
4. Use the formula for $\sin(s - t)$ to compute the exact value of $\sin(\pi/12)$.
5. Use the formula for $\sin(s + t)$ to find $\sin(7\pi/12)$.
6. If $\sin s = \frac{4}{5}$, $0 < s < \pi/2$, and $\sin t = -\frac{1}{4}$, $\pi < t < 3\pi/2$, compute $\sin(s - t)$.
7. If $\sin \alpha = \frac{12}{13}$, $\pi/2 < \alpha < \pi$, and $\cos \beta = \frac{3}{5}$, $3\pi/2 < \beta < 2\pi$, compute the following.

(a) $\sin(\alpha + \beta)$ (b) $\sin(\alpha - \beta)$
(c) $\cos(\alpha + \beta)$ (d) $\cos(\alpha - \beta)$
(e) $\tan(\alpha + \beta)$ (f) $\tan(\alpha - \beta)$

8. Suppose that $\sin \theta = \frac{1}{5}$ and $0 < \theta < \pi/2$.
(a) Compute $\cos \theta$.
(b) Compute $\sin 2\theta$. *Hint:* $\sin 2\theta = \sin(\theta + \theta)$.

9. Suppose that $\cos \theta = \frac{12}{13}$ and $3\pi/2 < \theta < 2\pi$.
(a) Compute $\sin \theta$.
(b) Compute $\cos 2\theta$. *Hint:* $\cos 2\theta = \cos(\theta + \theta)$.

10. Simplify $\sin(\pi/4 + s) - \sin(\pi/4 - s)$.

11. Given $\tan \theta = -\frac{2}{3}$, $\pi/2 < \theta < \pi$, and $\csc \beta = 2$, $0 < \beta < \pi/2$, find $\sin(\theta + \beta)$ and $\cos(\beta - \theta)$.

12. Given $\sec s = \frac{5}{4}$, $\sin s < 0$, $\cot t = -1$, $\pi/2 < t < \pi$, find $\sin(s - t)$ and $\cos(s + t)$.

13. In the figure, $\overline{AB} = \overline{CD} = 1$ and $\overline{BC} = 2$. Find $\alpha + \beta$ and express your answer using radian measure. *Hint:* Find $\cos(\alpha + \beta)$.

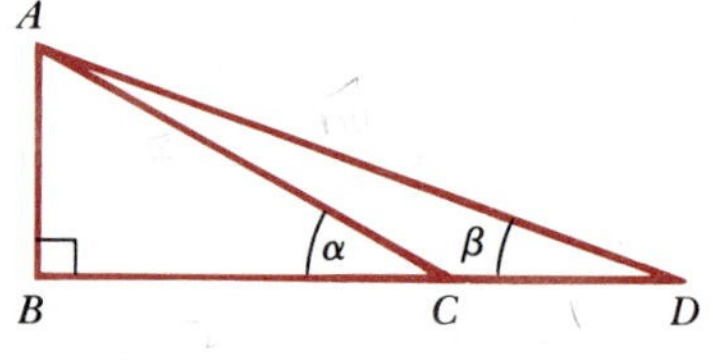

14. Prove the following trigonometric identities.
(a) $\cos(A - B) - \cos(A + B) = 2 \sin A \sin B$
(b) $\cos(A - B) + \cos(A + B) = 2 \cos A \cos B$
(c) $\sin(A - B) + \sin(A + B) = 2 \sin A \cos B$
(d) $\sin(A - B) - \sin(A + B) = -2 \cos A \sin B$

15. Simplify $\cos(\theta + \pi/3) + \cos(\theta - \pi/3)$.

16. Simplify the following.
(a) $\sin 35° \cos 25° + \cos 35° \sin 25°$
(b) $\cos \frac{\pi}{5} \cos \frac{3\pi}{10} - \sin \frac{\pi}{5} \sin \frac{3\pi}{10}$

17. Prove that $\cos(\alpha + \beta) \cos(\alpha - \beta) = \cos^2 \alpha - \sin^2 \beta$.

18. Simplify $\cos^2 \theta + \cos^2\left(\frac{2\pi}{3} + \theta\right) + \cos^2\left(\frac{2\pi}{3} - \theta\right)$.

19. Let A, B, and C be the angles of a triangle.
(a) Show that $\sin(A + B) = \sin C$.
(b) Show that $\cos(A + B) = -\cos C$.
(c) Show that $\tan(A + B) = -\tan C$.

20. Let f be the function defined by $f(\theta) = \sin \theta$. Prove that

$$\frac{f(\theta + h) - f(\theta)}{h} = \sin \theta\left(\frac{\cos h - 1}{h}\right) + \cos \theta\left(\frac{\sin h}{h}\right)$$

21. Prove the following identities.
(a) $\sin 2\theta = 2 \sin \theta \cos \theta$
Hint: $\sin 2\theta = \sin(\theta + \theta)$.
(b) $\cos 2\theta = \cos^2 \theta - \sin^2 \theta$
(c) $\sin 3\theta = 3 \sin \theta \cos^2 \theta - \sin^3 \theta$
Hint: $\sin 3\theta = \sin(2\theta + \theta)$.
(d) $\cos 3\theta = \cos^3 \theta - 3 \cos \theta \sin^2 \theta$

***22.** Prove the addition formulas for tangent.
(a) $\tan(\alpha + \beta) = \dfrac{\tan \alpha + \tan \beta}{1 - \tan \alpha \tan \beta}$
Hint: After completing the obvious first steps, divide both numerator and denominator by the quantity $\cos \alpha \cos \beta$.
(b) $\tan(\alpha - \beta) = \dfrac{\tan \alpha - \tan \beta}{1 + \tan \alpha \tan \beta}$

***23.** Suppose that both α and β are between 0 and $\pi/2$. If $\tan \alpha = \dfrac{a}{a + 1}$ and $\tan \beta = \dfrac{1}{2a + 1}$, prove that $\alpha + \beta = \pi/4$. *Hint:* Use the formula for $\tan(\alpha + \beta)$ given in the previous exercise.

***24.** Prove that

$$\frac{\sin(\alpha - \beta)}{\cos \alpha \cos \beta} + \frac{\sin(\beta - \gamma)}{\cos \beta \cos \gamma} + \frac{\sin(\gamma - \alpha)}{\cos \gamma \cos \alpha} = 0$$

***25.** Suppose that A, B, and C are the angles of a triangle. Show that

$$\cos^2 A + \cos^2 B + \cos^2 C + 2 \cos A \cos B \cos C = 1$$

***26.** Prove the following identities.
(a) $\sin(A + B + C) = \sin A \cos B \cos C + \sin B \cos C \cos A + \sin C \cos A \cos B - \sin A \sin B \sin C$
(b) $\cos(A + B + C) = \cos A \cos B \cos C - \cos A \sin B \sin C - \cos B \sin C \sin A - \cos C \sin A \sin B$

27. Suppose that $a^2 + b^2 = 1$ and $c^2 + d^2 = 1$. Prove that $|ac + bd| \leq 1$. *Hint:* Let $a = \cos \theta$, $b = \sin \theta$, $c = \cos \phi$, and $d = \sin \phi$.

***28.** If $\triangle ABC$ is not a right triangle, prove that

$$\tan A + \tan B + \tan C = \tan A \tan B \tan C$$

Suggestion: Use the formulas in Exercise 22.

***29.** If $\triangle ABC$ is not a right triangle, and $\cos A = \cos B \cos C$, show that $\tan B \tan C = 2$.

8.4 Further Identities

We are going to prove the following identities. All of them are consequences of the addition formulas for sine and cosine, which we established in the previous section.

Addition Formulas for Tangent

(a) $\tan(s + t) = \dfrac{\tan s + \tan t}{1 - \tan s \tan t}$

(b) $\tan(s - t) = \dfrac{\tan s - \tan t}{1 + \tan s \tan t}$

Double-Angle Formulas

(a) $\sin 2\theta = 2 \sin \theta \cos \theta$

(b) $\cos 2\theta = \cos^2 \theta - \sin^2 \theta$

(c) $\tan 2\theta = \dfrac{2 \tan \theta}{1 - \tan^2 \theta}$

Half-Angle Formulas

(a) $\sin \dfrac{s}{2} = \pm\sqrt{\dfrac{1 - \cos s}{2}}$

(b) $\cos \dfrac{s}{2} = \pm\sqrt{\dfrac{1 + \cos s}{2}}$

(c) $\tan \dfrac{s}{2} = \dfrac{\sin s}{1 + \cos s}$

The proof of the formula for $\tan(s + t)$ is as follows.

$$\begin{aligned} \tan(s + t) &= \frac{\sin(s + t)}{\cos(s + t)} \\ &= \frac{\sin s \cos t + \cos s \sin t}{\cos s \cos t - \sin s \sin t} \qquad (1) \end{aligned}$$

Now we divide both numerator and denominator on the right-hand side of equation (1) by the quantity $\cos s \cos t$. This yields

$$\begin{aligned} \tan(s + t) &= \frac{\dfrac{\sin s \cancel{\cos t}}{\cos s \cancel{\cos t}} + \dfrac{\cancel{\cos s} \sin t}{\cancel{\cos s} \cos t}}{\dfrac{\cos s \cancel{\cos t}}{\cos s \cancel{\cos t}} - \dfrac{\sin s \sin t}{\cos s \cos t}} \\ &= \frac{\tan s + \tan t}{1 - \tan s \tan t} \end{aligned}$$

This proves the formula for $\tan(s + t)$. The formula for $\tan(s - t)$ can be deduced from this with the aid of the identity $\tan(-t) = -\tan t$, which was derived in the previous section. We have

$$\begin{aligned} \tan(s - t) &= \tan[s + (-t)] \\ &= \frac{\tan s + \tan(-t)}{1 - \tan s \tan(-t)} \\ &= \frac{\tan s + (-\tan t)}{1 - \tan s(-\tan t)} \\ &= \frac{\tan s - \tan t}{1 + \tan s \tan t} \qquad \text{as required} \end{aligned}$$

Example 1 Compute $\tan 7\pi/12$, using the fact that $7\pi/12 = \pi/4 + \pi/3$.

Solution

$$\tan\frac{7\pi}{12} = \tan\left(\frac{\pi}{4} + \frac{\pi}{3}\right) = \frac{\tan \pi/4 + \tan \pi/3}{1 - \tan \pi/4 \tan \pi/3}$$

We used the formula for $\tan(s + t)$ with $s = \pi/4$ and $t = \pi/3$.

$$= \frac{1 + \sqrt{3}}{1 - (1)(\sqrt{3})}$$

Thus $\tan\frac{7\pi}{12} = \frac{1 + \sqrt{3}}{1 - \sqrt{3}}$. We can write this answer in a more compact form by rationalizing the denominator as follows:

$$\frac{1 + \sqrt{3}}{1 - \sqrt{3}} = \frac{1 + \sqrt{3}}{1 - \sqrt{3}} \cdot \frac{1 + \sqrt{3}}{1 + \sqrt{3}}$$

$$= \frac{1 + \sqrt{3} + \sqrt{3} + 3}{1 - \sqrt{3} + \sqrt{3} - 3} = \frac{4 + 2\sqrt{3}}{-2}$$

$$= -2 - \sqrt{3}$$

Now let us prove the double-angle formulas. We have

$$\begin{aligned} \sin 2\theta &= \sin(\theta + \theta) \\ &= \sin\theta\cos\theta + \cos\theta\sin\theta \\ &= 2\sin\theta\cos\theta \end{aligned}$$

This establishes the formula for $\sin 2\theta$. The formulas for $\cos 2\theta$ and $\tan 2\theta$ are proved in the same manner:

$$\begin{aligned} \cos 2\theta &= \cos(\theta + \theta) \\ &= \cos\theta\cos\theta - \sin\theta\sin\theta \\ &= \cos^2\theta - \sin^2\theta \quad \text{as required} \end{aligned}$$

$$\begin{aligned} \tan 2\theta &= \tan(\theta + \theta) \\ &= \frac{\tan\theta + \tan\theta}{1 - \tan\theta\tan\theta} \\ &= \frac{2\tan\theta}{1 - \tan^2\theta} \quad \text{as required} \end{aligned}$$

Example 2 If $\sin\theta = \frac{4}{5}$ and $\pi/2 < \theta < \pi$, find $\cos\theta$, $\sin 2\theta$, and $\cos 2\theta$.

Solution We have

$$\begin{aligned} \cos^2\theta &= 1 - \sin^2\theta \\ &= 1 - \left(\frac{4}{5}\right)^2 = 1 - \frac{16}{25} \\ &= \frac{9}{25} \end{aligned}$$

Consequently

$$\cos\theta = \frac{3}{5} \quad \text{or} \quad \cos\theta = -\frac{3}{5}$$

We want the negative value here since $\pi/2 < \theta < \pi$. Thus

$$\cos\theta = -\frac{3}{5}$$

Now that we know the values of $\cos\theta$ and $\sin\theta$, the double-angle formulas can be used to determine $\sin 2\theta$ and $\cos 2\theta$. We have

$$\sin 2\theta = 2 \sin\theta \cos\theta = 2\left(\frac{4}{5}\right)\left(-\frac{3}{5}\right) = -\frac{24}{25}$$

Also

$$\cos 2\theta = \cos^2\theta - \sin^2\theta = \left(-\frac{3}{5}\right)^2 - \left(\frac{4}{5}\right)^2 = \frac{9}{25} - \frac{16}{25} = -\frac{7}{25}$$

The required values are therefore $\cos\theta = -\frac{3}{5}$, $\sin 2\theta = -\frac{24}{25}$, and $\cos 2\theta = -\frac{7}{25}$.

Example 3 Prove the following identities.

(a) $\cos 3\theta = \cos^3\theta - 3\sin^2\theta\cos\theta$ **(b)** $\cos 3\theta = 4\cos^3\theta - 3\cos\theta$

Solution **(a)**

$$\begin{aligned}\cos 3\theta &= \cos(2\theta + \theta)\\ &= \underbrace{\cos 2\theta}_{(\cos^2\theta - \sin^2\theta)} \cos\theta - \underbrace{\sin 2\theta}_{(2\sin\theta\cos\theta)} \sin\theta\\ &= (\cos^2\theta - \sin^2\theta)\cos\theta - (2\sin\theta\cos\theta)\sin\theta\\ &= \cos^3\theta - \sin^2\theta\cos\theta - 2\sin^2\theta\cos\theta\end{aligned}$$

Collecting like terms now gives us

$$\cos 3\theta = \cos^3\theta - 3\sin^2\theta\cos\theta \quad \text{as required}$$

(b) Using the identity established in part (a), we replace $\sin^2\theta$ by the quantity $1 - \cos^2\theta$. This yields

$$\begin{aligned}\cos 3\theta &= \cos^3\theta - 3(1 - \cos^2\theta)\cos\theta\\ &= \cos^3\theta - 3\cos\theta + 3\cos^3\theta\\ &= 4\cos^3\theta - 3\cos\theta \quad \text{as required}\end{aligned}$$

There are several alternate ways of writing the formula for $\cos 2\theta$ that are often useful.

Equivalent Forms of the Formula $\cos 2\theta = \cos^2\theta - \sin^2\theta$

(a) $\cos 2\theta = 2\cos^2\theta - 1$ **(b)** $\cos 2\theta = 1 - 2\sin^2\theta$

(c) $\cos^2\theta = \dfrac{1 + \cos 2\theta}{2}$ **(d)** $\sin^2\theta = \dfrac{1 - \cos 2\theta}{2}$

One way to prove identity (a) in the above list is as follows.

$$\begin{aligned}\cos 2\theta &= \cos^2\theta - \sin^2\theta\\ &= \cos^2\theta - (1 - \cos^2\theta)\\ &= \cos^2\theta - 1 + \cos^2\theta\\ &= 2\cos^2\theta - 1 \quad \text{as required}\end{aligned}$$

If in this last equation we add 1 to both sides, and then divide by 2, the result is identity (c). (Verify this.) The proofs for (b) and (d) are similar and are left to the exercises.

The last three formulas that we prove in this section are the half-angle formulas:

$$\cos\frac{s}{2} = \pm\sqrt{\frac{1+\cos s}{2}}$$

$$\sin\frac{s}{2} = \pm\sqrt{\frac{1-\cos s}{2}}$$

$$\tan\frac{s}{2} = \frac{\sin s}{1+\cos s}$$

To derive the formula for $\cos(s/2)$, we begin with one of the alternate forms of the cosine double-angle formula:

$$\cos^2\theta = \frac{1+\cos 2\theta}{2}$$

or

$$\cos\theta = \pm\sqrt{\frac{1+\cos 2\theta}{2}}$$

Since this identity holds for all values of θ, we may replace θ by $s/2$ to obtain

$$\cos\frac{s}{2} = \pm\sqrt{\frac{1+\cos 2\left(\frac{s}{2}\right)}{2}}$$

$$= \pm\sqrt{\frac{1+\cos s}{2}}$$

This is the required formula for $\cos(s/2)$. The derivation of the formula for $\sin(s/2)$ follows exactly the same procedure, except that you begin with the identity $\sin^2\theta = \dfrac{1-\cos 2\theta}{2}$. (Exercise 25 on page 391 asks you to complete the details.) In both formulas, the sign before the radical is determined by considering in which quadrant the arc (or angle) $s/2$ terminates.

Example 4 Evaluate $\cos 105°$ using the half-angle formula.

Solution

$$\cos 105° = \cos\frac{210°}{2} = \pm\sqrt{\frac{1+\cos 210°}{2}}$$

We used the formula for $\cos(s/2)$ with $s = 210°$.

$$= \pm\sqrt{\frac{1+(-\sqrt{3}/2)}{2}}$$

$$= \pm\sqrt{\frac{1-(\sqrt{3}/2)}{2}\cdot\frac{2}{2}} = \pm\sqrt{\frac{2-\sqrt{3}}{4}}$$

$$= \frac{\pm\sqrt{2-\sqrt{3}}}{2}$$

We want to choose the negative value here since the terminal side of 105° lies in the second quadrant. Thus we finally obtain

$$\cos 105° = \frac{-\sqrt{2-\sqrt{3}}}{2}$$

Our last task for this section is to establish the formula for tan $(s/2)$. To do this we first prove the identity $\tan\theta = \dfrac{\sin 2\theta}{1+\cos 2\theta}$.

Proof that $\tan\theta = \dfrac{\sin 2\theta}{1+\cos 2\theta}$

$$\frac{\sin 2\theta}{1+\cos 2\theta} = \frac{2\sin\theta\cos\theta}{2\cos^2\theta}$$

In the denominator we used the identity $\cos 2\theta = 2\cos^2\theta - 1$ from page 387.

$$= \frac{\sin\theta}{\cos\theta}$$

$$= \tan\theta \quad \text{as we wished to prove}$$

If we now replace θ by $s/2$ in the identity $\tan\theta = \dfrac{\sin 2\theta}{1+\cos 2\theta}$, the result is

$$\tan\frac{s}{2} = \frac{\sin s}{1+\cos s}$$

This is the half-angle formula for tangent.

We conclude this section with a summary of the principal trigonometric identities developed in this and the previous section. For completeness, the list also includes two sets of trigonometric identities that we did not discuss in the text, but that are occasionally useful. These are the so-called *product-to-sum formulas* and the *sum-to-product formulas.* Proofs and applications of these formulas are included in the exercises.

Property Summary

Principal Trigonometric Identities

I. (Consequences of the Definitions)

(a) $\csc\theta = \dfrac{1}{\sin\theta}$ **(b)** $\sec\theta = \dfrac{1}{\cos\theta}$

(c) $\cot\theta = \dfrac{1}{\tan\theta}$ **(d)** $\tan\theta = \dfrac{\sin\theta}{\cos\theta}$

(e) $\cot\theta = \dfrac{\cos\theta}{\sin\theta}$

II. Pythagorean Identities

(a) $\sin^2\theta + \cos^2\theta = 1$ **(b)** $\tan^2\theta + 1 = \sec^2\theta$

(c) $\cot^2\theta + 1 = \csc^2\theta$

III. Opposite-Angle Formulas

(a) $\sin(-\theta) = -\sin\theta$ **(b)** $\cos(-\theta) = \cos\theta$

(c) $\tan(-\theta) = -\tan\theta$

IV. Reduction Formulas

(a) $\sin(\theta + 2\pi k) = \sin\theta$ **(b)** $\cos(\theta + 2\pi k) = \cos\theta$

(c) $\sin\left(\dfrac{\pi}{2} - \theta\right) = \cos\theta$ **(d)** $\cos\left(\dfrac{\pi}{2} - \theta\right) = \sin\theta$

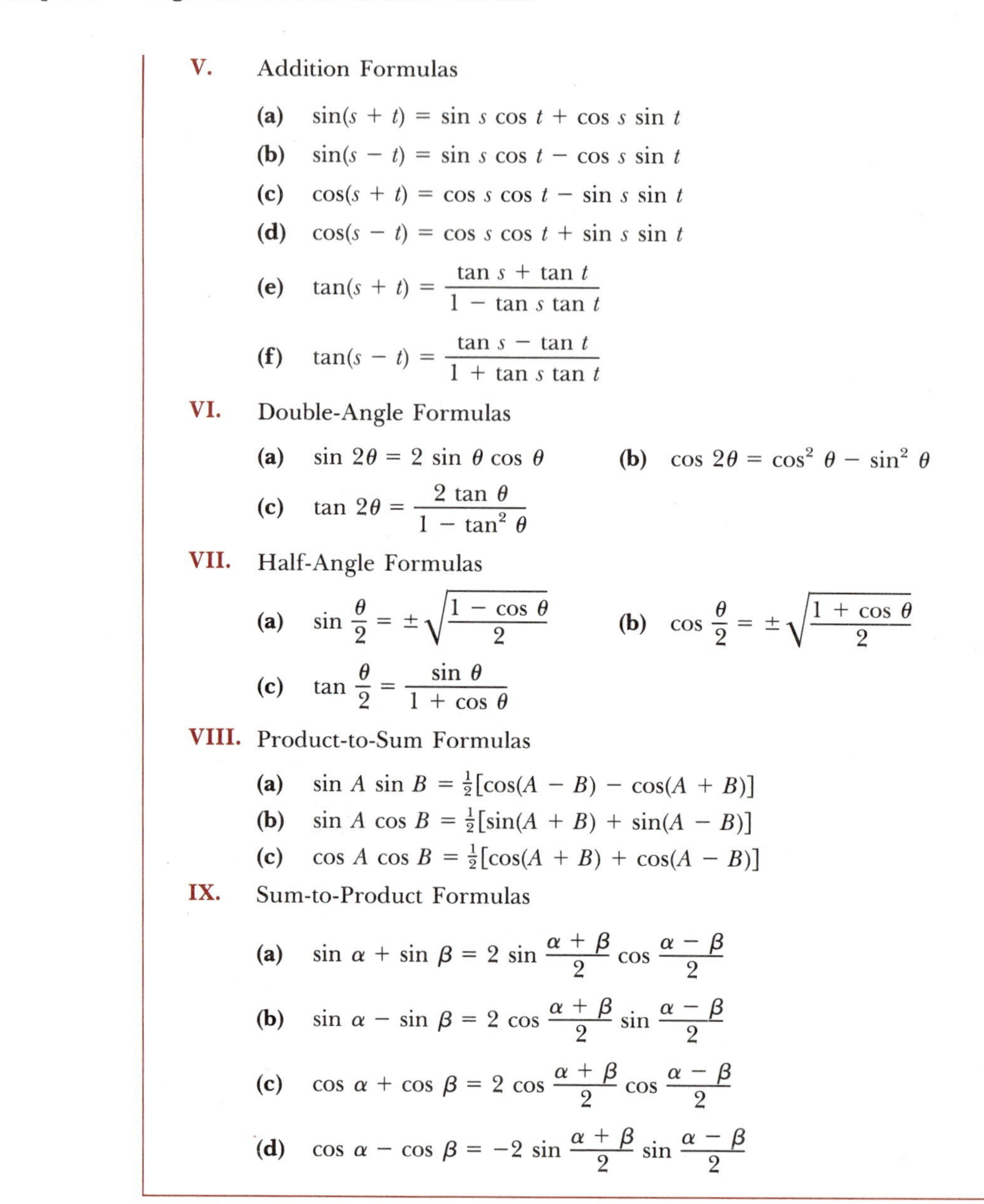

V. Addition Formulas

(a) $\sin(s + t) = \sin s \cos t + \cos s \sin t$

(b) $\sin(s - t) = \sin s \cos t - \cos s \sin t$

(c) $\cos(s + t) = \cos s \cos t - \sin s \sin t$

(d) $\cos(s - t) = \cos s \cos t + \sin s \sin t$

(e) $\tan(s + t) = \dfrac{\tan s + \tan t}{1 - \tan s \tan t}$

(f) $\tan(s - t) = \dfrac{\tan s - \tan t}{1 + \tan s \tan t}$

VI. Double-Angle Formulas

(a) $\sin 2\theta = 2 \sin \theta \cos \theta$ (b) $\cos 2\theta = \cos^2 \theta - \sin^2 \theta$

(c) $\tan 2\theta = \dfrac{2 \tan \theta}{1 - \tan^2 \theta}$

VII. Half-Angle Formulas

(a) $\sin \dfrac{\theta}{2} = \pm\sqrt{\dfrac{1 - \cos \theta}{2}}$ (b) $\cos \dfrac{\theta}{2} = \pm\sqrt{\dfrac{1 + \cos \theta}{2}}$

(c) $\tan \dfrac{\theta}{2} = \dfrac{\sin \theta}{1 + \cos \theta}$

VIII. Product-to-Sum Formulas

(a) $\sin A \sin B = \frac{1}{2}[\cos(A - B) - \cos(A + B)]$

(b) $\sin A \cos B = \frac{1}{2}[\sin(A + B) + \sin(A - B)]$

(c) $\cos A \cos B = \frac{1}{2}[\cos(A + B) + \cos(A - B)]$

IX. Sum-to-Product Formulas

(a) $\sin \alpha + \sin \beta = 2 \sin \dfrac{\alpha + \beta}{2} \cos \dfrac{\alpha - \beta}{2}$

(b) $\sin \alpha - \sin \beta = 2 \cos \dfrac{\alpha + \beta}{2} \sin \dfrac{\alpha - \beta}{2}$

(c) $\cos \alpha + \cos \beta = 2 \cos \dfrac{\alpha + \beta}{2} \cos \dfrac{\alpha - \beta}{2}$

(d) $\cos \alpha - \cos \beta = -2 \sin \dfrac{\alpha + \beta}{2} \sin \dfrac{\alpha - \beta}{2}$

Exercise Set 8.4

In Exercises 1–7, refer to the two triangles and compute the quantities indicated.

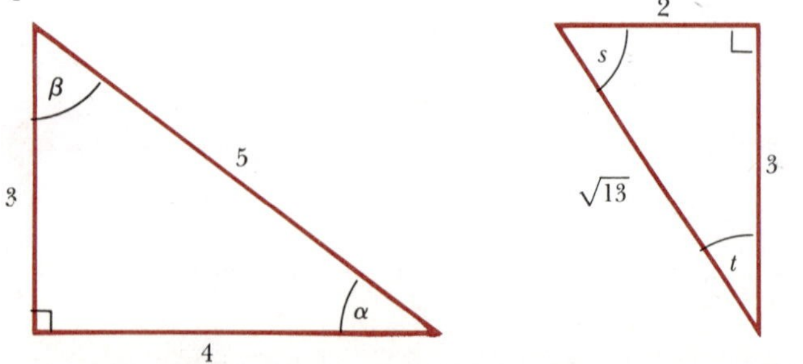

1. (a) $\tan(\beta + s)$ (b) $\tan(\beta - s)$
   (c) $\tan(\alpha - s)$ (d) $\tan(\alpha - \beta)$
2. (a) $\sin 2\beta$ and $\cos 2\beta$ (b) $\sin 2\alpha$ and $\cos 2\alpha$
   (c) $\cos(2\alpha + \beta)$ (d) $\sin(\alpha - 2\beta)$
   (e) $\sin 4\alpha$
3. (a) $\tan 2\alpha$ (b) $\tan 2\beta$
   (c) $\tan 2s$ (d) $\tan 2t$
4. (a) $\tan(2\beta + \alpha)$ (b) $\tan(2\alpha - \beta)$
5. (a) $\sin \dfrac{\alpha}{2}$ and $\cos \dfrac{\alpha}{2}$ (b) $\sin \dfrac{s}{2}$ and $\cos \dfrac{s}{2}$

(c) $\sin\left(\alpha + \frac{s}{2}\right)$ (d) $\cos\left(\frac{\beta}{2} - t\right)$

6. (a) $\tan \frac{s}{2}$ (b) $\tan\left(\frac{s}{2} + t\right)$

7. (a) $\tan \frac{\alpha}{2}$ (b) $\tan\left(\frac{\beta}{2} + \frac{\pi}{4}\right)$

8. Referring to the triangle below, prove the following statements.

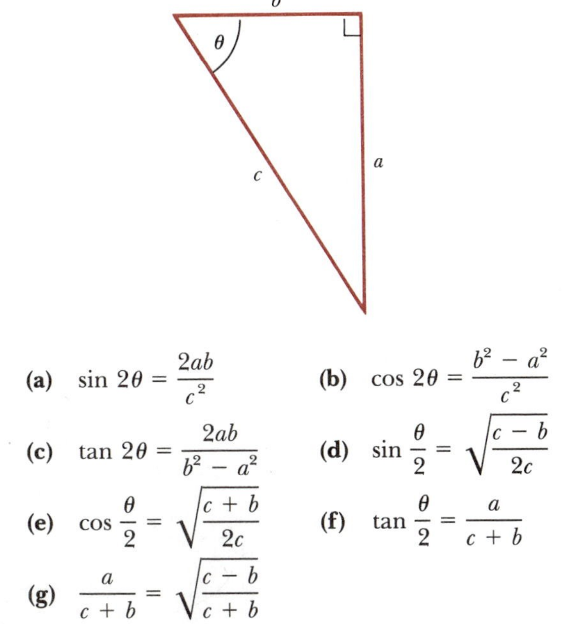

(a) $\sin 2\theta = \dfrac{2ab}{c^2}$ (b) $\cos 2\theta = \dfrac{b^2 - a^2}{c^2}$

(c) $\tan 2\theta = \dfrac{2ab}{b^2 - a^2}$ (d) $\sin \dfrac{\theta}{2} = \sqrt{\dfrac{c - b}{2c}}$

(e) $\cos \dfrac{\theta}{2} = \sqrt{\dfrac{c + b}{2c}}$ (f) $\tan \dfrac{\theta}{2} = \dfrac{a}{c + b}$

(g) $\dfrac{a}{c + b} = \sqrt{\dfrac{c - b}{c + b}}$

9. Show that $\tan (\pi/8) = \sqrt{2} - 1$.

10. Compute $\cos 15°$ first using the formula for $\cos(A - B)$, then using the formula for $\cos (\theta/2)$. Can you show that your two answers are the same?

11. Referring to the triangle below, show that

$$\tan \frac{\theta}{2} = \frac{b}{a}$$

and

$$\tan \frac{\phi}{2} = \frac{a - b}{a + b}$$

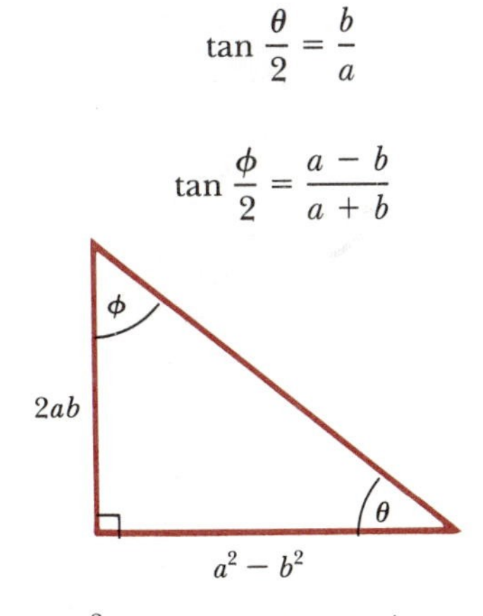

12. If $\sin \theta = -\frac{2}{3}$ and $\pi < \theta < 3\pi/2$, compute $\sin 2\theta$ and $\sin (\theta/2)$.

13. If $\sec \theta = z$ and $0 < \theta < \pi/2$, express the following in terms of z.

(a) $\sin \frac{\theta}{2}$ (b) $\sin 2\theta$ (c) $\cos 2\theta$

14. Use an appropriate half-angle formula to evaluate each of the following.

(a) $\sin \frac{\pi}{12}$ (b) $\cos 22.5°$

(c) $\sin 75°$ (d) $\tan \frac{5\pi}{12}$

In Exercises 15–35, prove that the given equations are identities.

15. $\cos 2s = \dfrac{1 - \tan^2 s}{1 + \tan^2 s}$

16. $1 + \cos 2t = \cot t \sin 2t$

17. $2 \sin \dfrac{\theta}{2} \cos \dfrac{\theta}{2} = \sin \theta$

18. $\cos \theta = 2 \cos^2 \dfrac{\theta}{2} - 1$

19. $\dfrac{\sin 2\theta}{\sin \theta} - \dfrac{\cos 2\theta}{\cos \theta} = \sec \theta$

20. $\sin^4 \theta = \dfrac{3 - 4 \cos 2\theta + \cos 4\theta}{8}$

21. $\sin 3\theta = 3 \sin \theta \cos^2 \theta - \sin^3 \theta$

22. $\sin 3\theta = 3 \sin \theta - 4 \sin^3 \theta$

23. $\cos 2\theta = 1 - 2 \sin^2 \theta$

24. $\sin^2 \theta = \dfrac{1 - \cos 2\theta}{2}$

25. $\sin \dfrac{\theta}{2} = \pm\sqrt{\dfrac{1 - \cos \theta}{2}}$. *Hint:* Use the identity in Exercise 24.

26. $\sin 2\theta = \dfrac{2 \tan \theta}{1 + \tan^2 \theta}$

27. $2 \csc 2\theta = \dfrac{\csc^2 \theta}{\cot \theta}$

28. $\sin 2\theta = 2 \sin^3 \theta \cos \theta + 2 \sin \theta \cos^3 \theta$

29. $\cot \theta = \dfrac{1 + \cos 2\theta}{\sin 2\theta}$

30. $\dfrac{1 + \tan (\theta/2)}{1 - \tan (\theta/2)} = \tan \theta + \sec \theta$

31. $\tan \theta + \cot \theta = 2 \csc 2\theta$

32. $2 \sin^2(45° - \theta) = 1 - \sin 2\theta$

33. $(\sin \theta - \cos \theta)^2 = 1 - \sin 2\theta$

34. $1 + \tan \theta \tan 2\theta = \tan 2\theta \cot \theta - 1$

35. $\tan\left(\dfrac{\pi}{4} + \theta\right) - \tan\left(\dfrac{\pi}{4} - \theta\right) = 2 \tan 2\theta$

36. Let $\triangle ABC$ be a right triangle, with $\angle C = 90°$. Prove the following statements.

(a) $\sin 2A = \sin 2B$ (b) $\cos 2A + \cos 2B = 0$

(c) $\sin 3A = \dfrac{3ab^2 - a^3}{c^3}$

37. Prove that $\tan \dfrac{\theta}{2} = \dfrac{1 - \cos \theta}{\sin \theta}$.

38. If $\cos(\alpha + \beta) = 0$, show that $\sin(\alpha + 2\beta) = \sin \alpha$.

***39.** **(a)** Prove that $\tan\theta\,\tan(60° - \theta)\,\tan(60° + \theta) = \tan 3\theta$.
(b) Prove that $\tan 20°\,\tan 40°\,\tan 80° = \sqrt{3}$.

***40.** Is it possible to find a triangle whose dimensions are as indicated in the figure? *Hint:* Apply the law of sines, followed by the formula for $\sin 2\theta$.

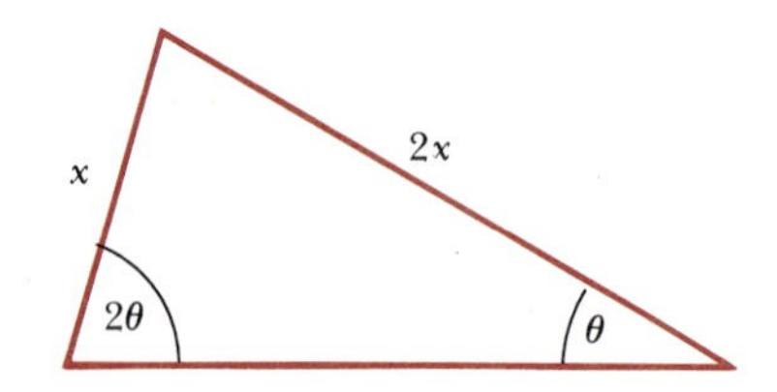

***41.** (Calculation of sin 18°, cos 18° and sin 3°)
(a) Prove that $\cos 3\theta = 4\cos^3\theta - 3\cos\theta$.
(b) Supply a reason for each statement.
(i) $\sin 36° = \cos 54°$
(ii) $2\sin 18°\cos 18° = 4\cos^3 18° - 3\cos 18°$
(iii) $2\sin 18° = 4\cos^2 18° - 3$
(c) In equation (iii) above, replace $\cos^2 18°$ by $1 - \sin^2 18°$ and then solve the resulting equation for $\sin 18°$. Thus show that

$$\sin 18° = \tfrac{1}{4}(\sqrt{5} - 1)$$

(d) Show that

$$\cos 18° = \tfrac{1}{4}\sqrt{10 + 2\sqrt{5}}$$

(e) Show that $\sin 3° = \frac{1}{16}[(\sqrt{5} - 1)(\sqrt{6} + \sqrt{2}) - 2(\sqrt{3} - 1)\sqrt{5 + \sqrt{5}}]$

Hint: $3° = 18° - 15°$.

42. If $\sin\theta = \dfrac{a^2 - b^2}{a^2 + b^2}$, show that $\tan\dfrac{\theta}{2} = \dfrac{a - b}{a + b}$. (Assume θ is an acute angle and a and b are positive.)

43. Let $z = \tan\theta$. Show that $\cos 2\theta = \dfrac{1 - z^2}{1 + z^2}$ and $\sin 2\theta = \dfrac{2z}{1 + z^2}$.

44. Prove the following three formulas, known as the product-to-sum formulas.
(a) $\sin A \sin B = \frac{1}{2}[\cos(A - B) - \cos(A + B)]$
(b) $\sin A \cos B = \frac{1}{2}[\sin(A + B) + \sin(A - B)]$
(c) $\cos A \cos B = \frac{1}{2}[\cos(A + B) + \cos(A - B)]$

45. Express each product as a sum or difference of sines or cosines.
(a) $4\sin 6\theta\cos 2\theta$ **(b)** $2\sin A\sin 3A$
(c) $\cos 4\theta\cos 2\theta$

46. **(a)** Show that $\sin\dfrac{\pi}{4}\cos\dfrac{\pi}{12} = \dfrac{1}{4}(\sqrt{3} + 1)$.
(b) Evaluate $2(\sin 82.5°)(\cos 37.5°)$ without using a calculator.

47. Prove the following identities.
(a) $\sin(A + B)\sin(A - B) = \sin^2 A - \sin^2 B$
(b) $\cos(A + B)\cos(A - B) = \cos^2 A - \sin^2 B$

48. In the formula $\sin A\cos B = \frac{1}{2}[\sin(A + B) + \sin(A - B)]$, make the substitutions $A + B = \alpha$ and $A - B = \beta$ and deduce that

$$\sin\alpha + \sin\beta = 2\sin\frac{\alpha + \beta}{2}\cos\frac{\alpha - \beta}{2}$$

This is one of the so-called sum-to-product formulas mentioned in the text.

49. Prove the following sum-to-product identities. *Hint:* See the previous exercise.
(a) $\sin\alpha - \sin\beta = 2\cos\dfrac{\alpha + \beta}{2}\sin\dfrac{\alpha - \beta}{2}$
(b) $\cos\alpha + \cos\beta = 2\cos\dfrac{\alpha + \beta}{2}\cos\dfrac{\alpha - \beta}{2}$
(c) $\cos\alpha - \cos\beta = -2\sin\dfrac{\alpha + \beta}{2}\sin\dfrac{\alpha - \beta}{2}$

50. **(a)** Show that $\cos\dfrac{2\pi}{9} + \cos\dfrac{\pi}{9} = \sqrt{3}\cos\dfrac{\pi}{18}$.
(b) Show that $\sin 105° + \sin 15° = \dfrac{\sqrt{6}}{2}$.

51. Prove that $\dfrac{\sin\theta + \sin 3\theta}{\cos\theta + \cos 3\theta} = \tan 2\theta$. *Hint:* Use the sum-to-product formulas.

52. Prove that $\dfrac{\sin 7\theta - \sin 5\theta}{\cos 7\theta + \cos 5\theta} = \tan\theta$.

53. Prove that $\tan(x + y) = \dfrac{\sin 2x + \sin 2y}{\cos 2x + \cos 2y}$.

54. Angles A and B are acute angles in right triangle ABC.
C (a) Complete the following table. Which triangle yields the largest value for the quantity $\cos A + \cos B$?

a	b	c	cos A + cos B
3	4	5	
5	12	13	
7	24	25	
20	21	29	
8	15	17	
1	$\sqrt{3}$	2	
1	1	$\sqrt{2}$	
$\sqrt{2}$	$\sqrt{3}$	$\sqrt{5}$	
696	697	985	

(b) Using the sum-to-product formula, prove that $\cos A + \cos B = \sqrt{2}\cos\dfrac{A - B}{2}$.
(c) Use the result in part (b) to deduce that $\cos A + \cos B \le \sqrt{2}$, with equality occurring when $A = B = \pi/4$.

***55.** Let f denote the length of the bisector of angle C in $\triangle ABC$, as shown in the figure on the next page.

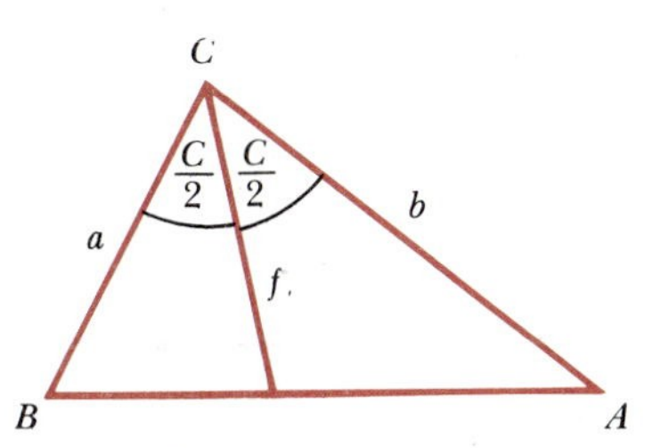

(a) Explain why

$$\frac{1}{2}af\sin\frac{C}{2}+\frac{1}{2}bf\sin\frac{C}{2}=\frac{1}{2}ab\sin C$$

Hint: Use areas.

(b) Show that $f=\dfrac{2ab\cos(C/2)}{a+b}$.

(c) By the cosine law, $\cos C=\dfrac{a^2+b^2-c^2}{2ab}$. Use this to show that

$$\cos\frac{C}{2}=\frac{1}{2}\sqrt{\frac{(a+b-c)(a+b+c)}{ab}}$$

(d) Show that the length of the angle bisector in terms of the sides is given by

$$f=\frac{\sqrt{ab}}{a+b}\sqrt{(a+b-c)(a+b+c)}$$

***56.** In $\triangle XYZ$ (shown in the figure), SX bisects $\angle ZXY$ and TY bisects $\angle ZYX$. In this exercise you are going to prove the following theorem, known as the Steiner-Lehmus Theorem: If the lengths of the angle bisectors SX and TY are equal, then $\triangle XYZ$ is isosceles (with $XZ = YZ$).

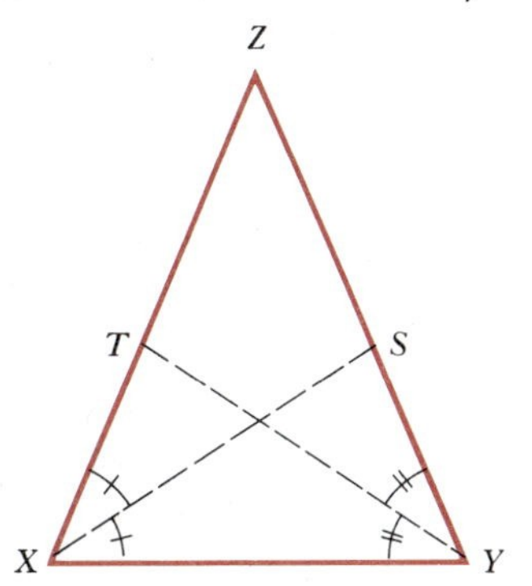

(a) Let x, y, and z denote the lengths of the sides YZ, XZ, and XY, respectively. Use the formula in Exercise 55(d) to show that the equation $\overline{TY}=\overline{SX}$ is equivalent to

$$\frac{\sqrt{xz}}{x+z}\sqrt{(x+z-y)(x+z+y)}=\frac{\sqrt{yz}}{y+z}\sqrt{(y+z-x)(y+z+x)} \quad (1)$$

(b) What common factors do you see on both sides of equation (1)? Divide both sides of equation (1) by those common factors. You should obtain

$$\frac{\sqrt{x}}{x+z}\sqrt{x+z-y}=\frac{\sqrt{y}}{y+z}\sqrt{y+z-x} \quad (2)$$

(c) Clear equation (2) of fractions, and then square both sides. After combining like terms and then grouping, the equation can be written

$$(3x^2yz-3xy^2z)+(x^3y-xy^3)+(x^2z^2-y^2z^2)+(xz^3-yz^3)=0 \quad (3)$$

(d) Show that equation (3) can be written

$$(x-y)[3xyz+xy(x+y)+x^2(x+y)+z^3]=0$$

Now notice that the quantity in brackets in this last equation must be positive. (Why?) Consequently $x-y=0$, and so $x=y$, as required. *Remark:* This theorem has a fascinating history, beginning in 1840 when C. L. Lehmus first proposed the theorem to the great Swiss geometer Jacob Steiner (1796–1863). For background (and for much shorter proofs!) see either of the following references: *Scientific American*, vol. 204 (1961), pp. 166–168; *American Mathematical Monthly*, vol. 70 (1963), pp. 79–80.

8.5 Graphs of the Trigonometric Functions

We are going to study the graphs of the trigonometric functions, with emphasis on the graphs of the sine and cosine functions and their variations. As preparation for this discussion, we want to first understand what is meant by the term *periodic function*. By way of example, all of the functions in Figure 1 (shown on the next page) are periodic. That is, their graphs repeat themselves at regular intervals.

In Figure 1(a) the graph of the function f repeats itself every six units. We say that the period of f is 6. Similarly, the period of g in Figure 1(b) is 2, while the period of h in Figure 1(c) is 2π. Notice that in each case, the period represents the minimum number of units along the horizontal axis that we

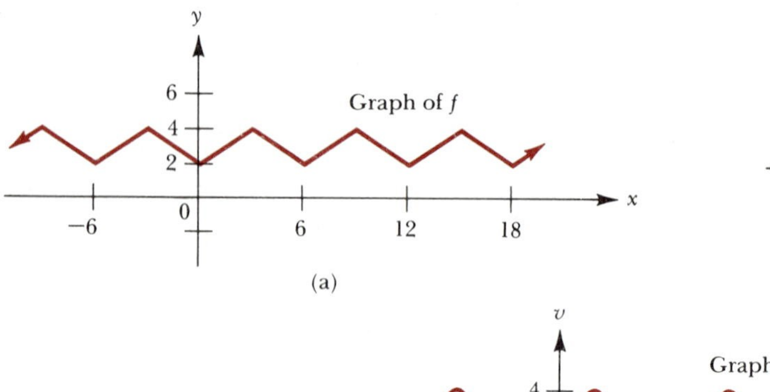

(a)

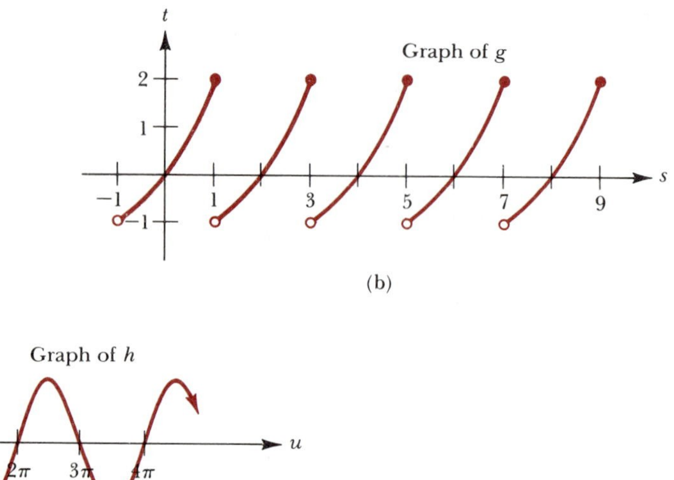

(b)

(c)

Figure 1

must travel before the graph begins to repeat itself. Because of this, we can state the definition of a periodic function as follows.

> **Definition of a Periodic Function and its Period**
>
> A function f is said to be *periodic* provided there is a number $p > 0$ such that
>
> $$f(x + p) = f(x)$$
>
> for all x in the domain of f. The smallest such number p is called the *period* of f.

We also want to define the term *amplitude*, as it applies to periodic functions. For a function such as h in Figure 1(c), in which the graph is centered about the horizontal axis, the amplitude is simply the maximum height of the graph above that horizontal axis. Thus the amplitude of h is 4. More generally, we define the amplitude of any periodic function as follows.

> **Definition of Amplitude**
>
> Let f be a periodic function and let m and M denote, respectively, the smallest and largest values of the function. Then, by definition, the *amplitude* of f is the number
>
> $$\frac{M - m}{2}$$

For the function h in Figure 1(c), this definition tells us that the amplitude is $\dfrac{4 - (-4)}{2} = 4$, which agrees with our previous value. Now check for yourself that the amplitudes of the functions f and g in Figure 1(a) and (b) are, respectively, 1 and $\frac{3}{2}$.

Let us now graph the sine function $f(\theta) = \sin\theta$. We are assuming that θ is in radians, so that the domain of the sine function is the set of all real numbers. Even before making any calculations, we can gain a strong intuitive insight into how the graph must look by carrying out the following experiment. After drawing the unit circle $x^2 + y^2 = 1$, place your fingertip at the point (1, 0) and then move your finger counterclockwise around the circle. As you do this, keep track of what happens to the y-coordinate of your fingertip. (The y-coordinate is $\sin\theta$.) Figure 2 indicates the general results.

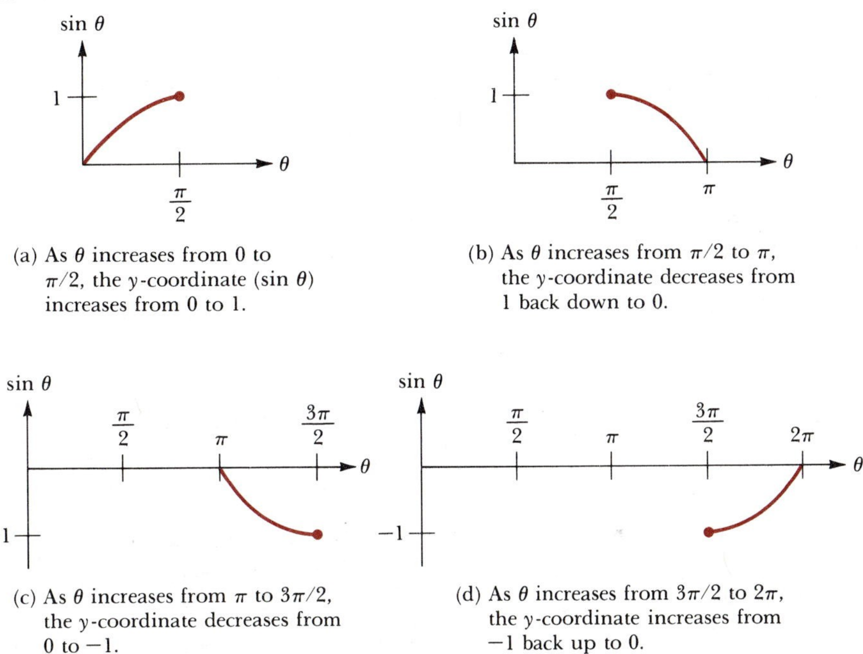

(a) As θ increases from 0 to $\pi/2$, the y-coordinate ($\sin\theta$) increases from 0 to 1.

(b) As θ increases from $\pi/2$ to π, the y-coordinate decreases from 1 back down to 0.

(c) As θ increases from π to $3\pi/2$, the y-coordinate decreases from 0 to -1.

(d) As θ increases from $3\pi/2$ to 2π, the y-coordinate increases from -1 back up to 0.

Figure 2

In summary then, as θ increases from 0 to 2π, the graph of $y = \sin\theta$ must appear as in Figure 3. Furthermore, since at $\theta = 2\pi$ we've returned to our starting point (1, 0), additional counterclockwise trips around the unit circle will then just result in repetitions of the pattern established in Figure 3. In other words, the period of $y = \sin\theta$ is 2π.

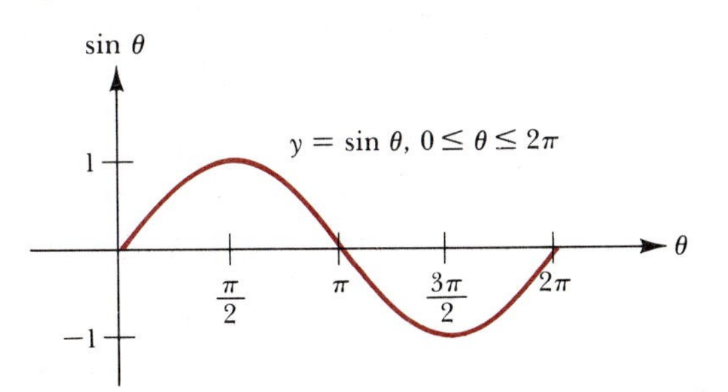

Figure 3

Now let's set up a table and use it to sketch the graph of $y = \sin\theta$ more accurately. Since (as we've just observed) the sine function is periodic, with period 2π, our table need contain only values of θ between 0 and 2π. This will establish the basic pattern for the graph. See Table 1 on page 396.

θ	0	$\frac{\pi}{6}$	$\frac{\pi}{3}$	$\frac{\pi}{2}$	$\frac{2\pi}{3}$	$\frac{5\pi}{6}$	π	$\frac{7\pi}{6}$	$\frac{4\pi}{3}$	$\frac{3\pi}{2}$	$\frac{5\pi}{3}$	$\frac{11\pi}{6}$	2π
$\sin\theta$	0	$\frac{1}{2}$	$\frac{\sqrt{3}}{2}$	1	$\frac{\sqrt{3}}{2}$	$\frac{1}{2}$	0	$-\frac{1}{2}$	$-\frac{\sqrt{3}}{2}$	-1	$-\frac{\sqrt{3}}{2}$	$-\frac{1}{2}$	0

Table 1

In plotting the points obtained from this table, we have used the approximation $\sqrt{3}/2 \approx 0.87$. Rather than approximating π, however, we mark off units on the horizontal axis in terms of π. The graph obtained is shown in

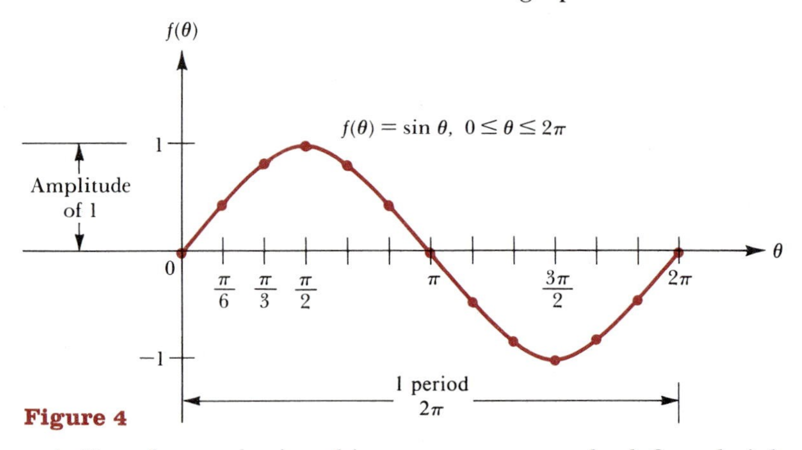

Figure 4

Figure 4. Now, by continuing this same pattern to the left and right, we obtain the complete graph of $f(\theta) = \sin\theta$, as indicated in Figure 5.

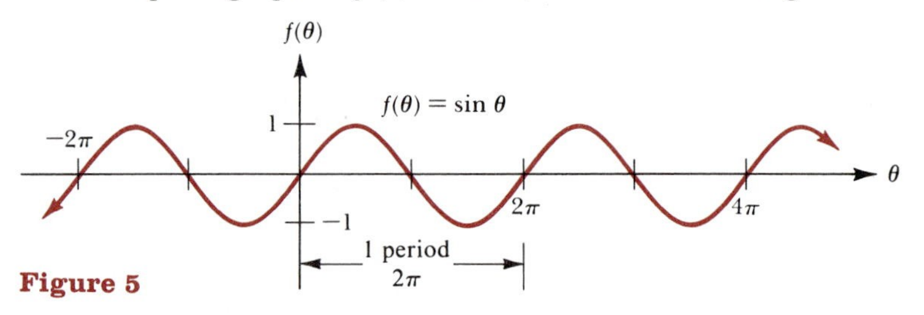

Figure 5

Before going on to analyze the sine function or to study the graphs of the other trigonometric functions, we are going to make a slight change in the notation we have been using. To conform with common usage, we will use x instead of θ on the horizontal axis; we will use y for the vertical axis. The sine function is then written simply as $y = \sin x$, where the real number x denotes the radian measure of an angle, or equivalently, the length of the corresponding arc. For reference, we redraw Figures 4 and 5 using this familiar x-y notation. See Figure 6.

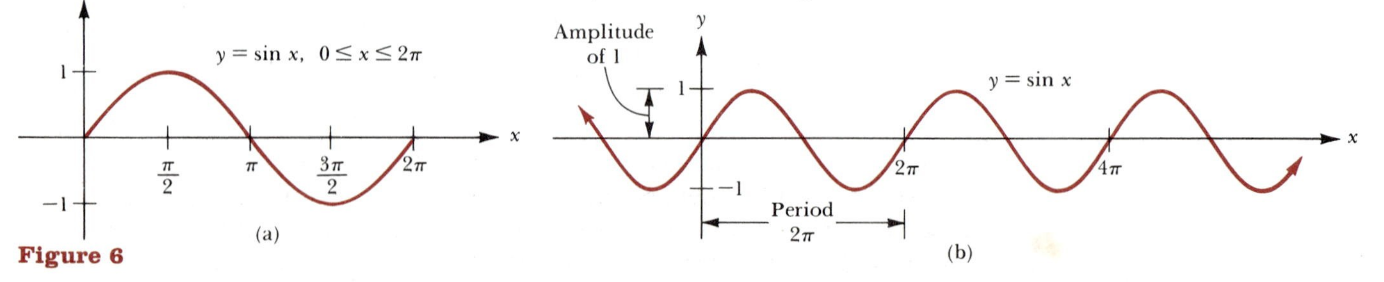

Figure 6

You should memorize the graph of the sine curve in Figure 6(b) so that you can sketch it without first setting up a table. Of course, once you know the shape of the basic sine wave in Figure 6(a), you automatically know the graph of the full sine curve in Figure 6(b).

We can use the graphs in Figure 6 to help us list some of the key properties of the sine function.

1. The domain of the sine function is the set of all real numbers. The range of the sine function consists of all real numbers y such that $-1 \leq y \leq 1$. Another way to express this last fact is

$$-1 \leq \sin x \leq 1, \qquad \text{for all } x$$

2. The sine function is periodic, with period 2π. The amplitude is 1.
3. (Refer to Figure 7.) The basic sine wave crosses the x-axis at the beginning, midpoint, and end of the period. The curve reaches its highest point after one quarter of the period and its lowest point after three quarters of the period.

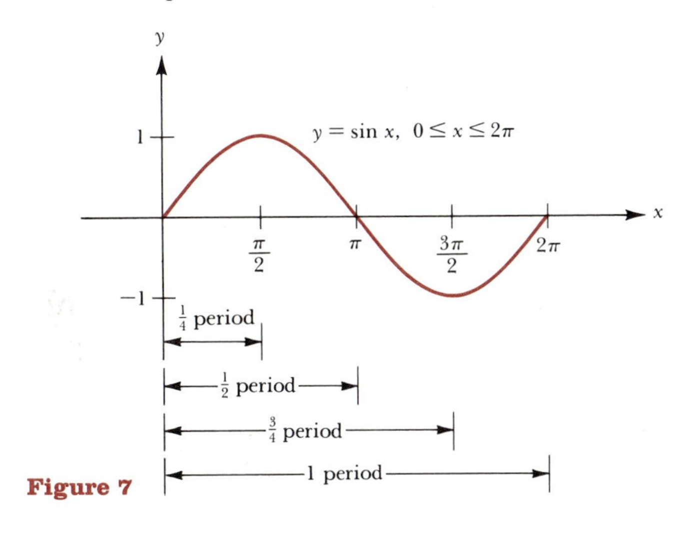

Figure 7

The graph of the cosine function is readily obtained from that of the sine function through the use of the identity

$$\cos x = \sin\left(x + \frac{\pi}{2}\right) \tag{1}$$

First, to verify that this is indeed an identity, we have

$$\begin{aligned}\sin\left(x + \frac{\pi}{2}\right) &= \sin x \cos \frac{\pi}{2} + \cos x \sin \frac{\pi}{2}\\ &= (\sin x)(0) + (\cos x)(1)\\ &= \cos x \qquad \text{as required}\end{aligned}$$

From equation (1), it follows that the graph of $y = \cos x$ is obtained by translating (shifting) the sine curve $\pi/2$ units to the left. The result is shown in Figure 8(a) on the following page. Figure 8(b) displays one complete cycle of the cosine curve, from $x = 0$ to $x = 2\pi$.

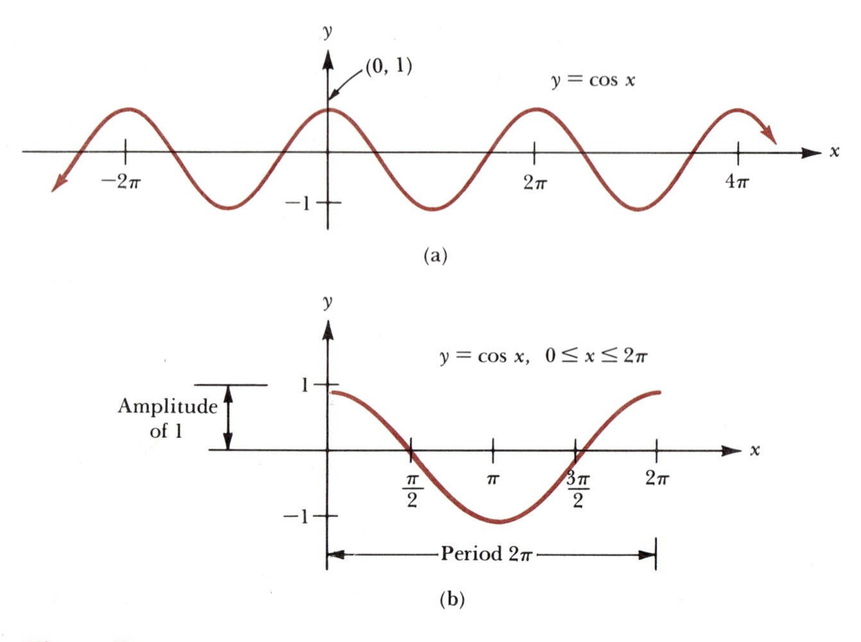

Figure 8

As with the sine function, you should memorize the graph and basic features of the cosine function. The key features of the cosine are as follows.

1. The domain of the cosine function is the set of all real numbers, while the range consists of all real numbers y such that $-1 \leq y \leq 1$. Another way to express this last fact is

$$-1 \leq \cos x \leq 1 \qquad \text{for all } x$$

2. The cosine function is periodic, with period 2π. The amplitude of the cosine function is 1.
3. (Refer to Figure 9.) At the beginning and at the end of the period, the basic cosine wave reaches its highest point. At the midpoint of the period, the curve reaches its lowest point. The x-intercepts occur one quarter and three quarters of the way through the period.

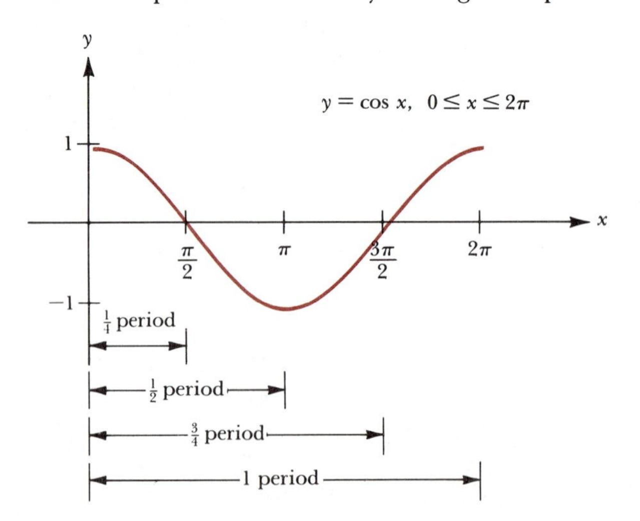

Figure 9

Now that we have looked at the graphs of $y = \sin x$ and $y = \cos x$, we consider graphs of functions of the form

$$y = A \sin (Bx + C) \quad \text{and} \quad y = A \cos (Bx + C)$$

where A, B, and C are constants and neither A nor B is zero. These functions are used throughout the sciences to analyze a variety of periodic phenomena ranging from the vibrations of an electron in an atom to the oscillations in the size of a particular animal population as it interacts with its environment.

First consider $y = A \sin x$. To obtain the graph of $y = A \sin x$ from that of $y = \sin x$, we multiply each y-coordinate on $y = \sin x$ by A. In particular, this changes the amplitude from 1 to $|A|$, but it will not affect the period, which remains 2π. Here are two examples.

Example 1 Graph the function $y = 2 \sin x$ over one period.

Solution The amplitude is 2 and the period is 2π. For purposes of comparison, Figure 10 displays the graphs of $y = 2 \sin x$ and $y = \sin x$.

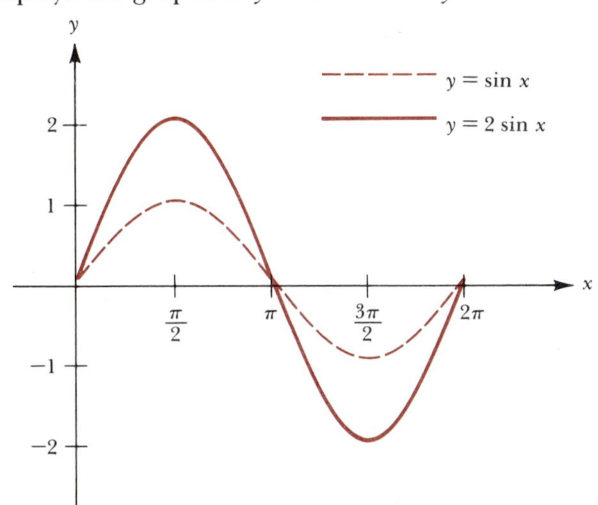

Figure 10

Example 2 Graph the function $y = -2 \sin x$ over one period.

Solution In Chapter 4, we saw that the graph of $y = -f(x)$ is obtained from that of $y = f(x)$ by reflection in the x-axis. Thus we need only take the graph of $y = 2 \sin x$ from the previous example and reflect it in the x-axis. See Figure 11. Note that both functions have an amplitude of 2 and a period of 2π.

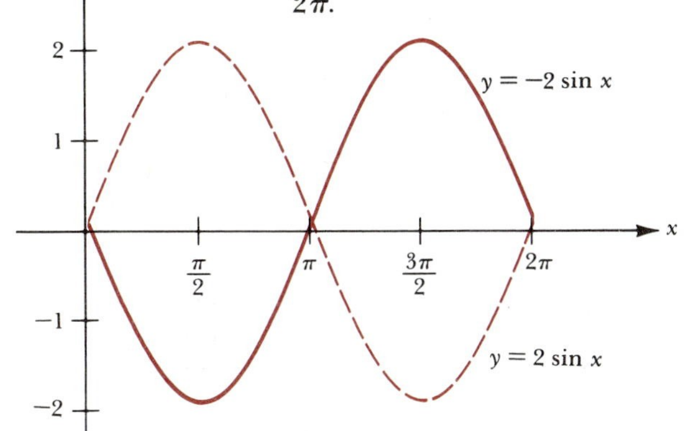

Figure 11

Next let us look at functions of the form $y = \sin Bx$, with $B > 0$. To find the period of this function, we reason as follows. We know that $y = \sin x$ begins its basic pattern when $x = 0$ and completes that pattern when $x = 2\pi$. Thus $y = \sin Bx$ will begin its basic pattern when $Bx = 0$ and it will complete that pattern when $Bx = 2\pi$. From the equation $Bx = 0$ we conclude $x = 0$, and from the equation $Bx = 2\pi$ we conclude $x = 2\pi/B$. Thus the function $y = \sin Bx$ begins its basic pattern at $x = 0$ and ends at $x = 2\pi/B$. In other words, the period of $y = \sin Bx$ is $2\pi/B$. Now combining this with our earlier observations about amplitude yields the graph shown in Figure 12.

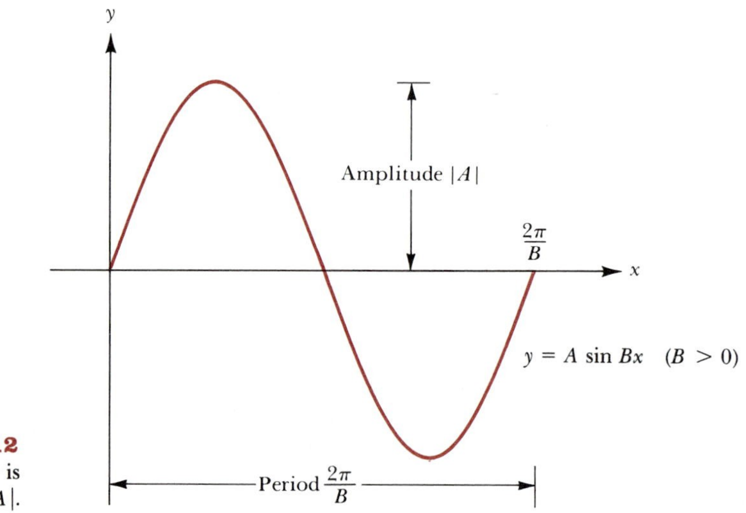

Figure 12
The period of $y = A \sin Bx$ is $2\pi/B$; the amplitude is $|A|$.

Before turning to some examples, we point out that the same analysis we have used for $y = A \sin Bx$ applies equally well to $y = A \cos Bx$. See Figure 13.

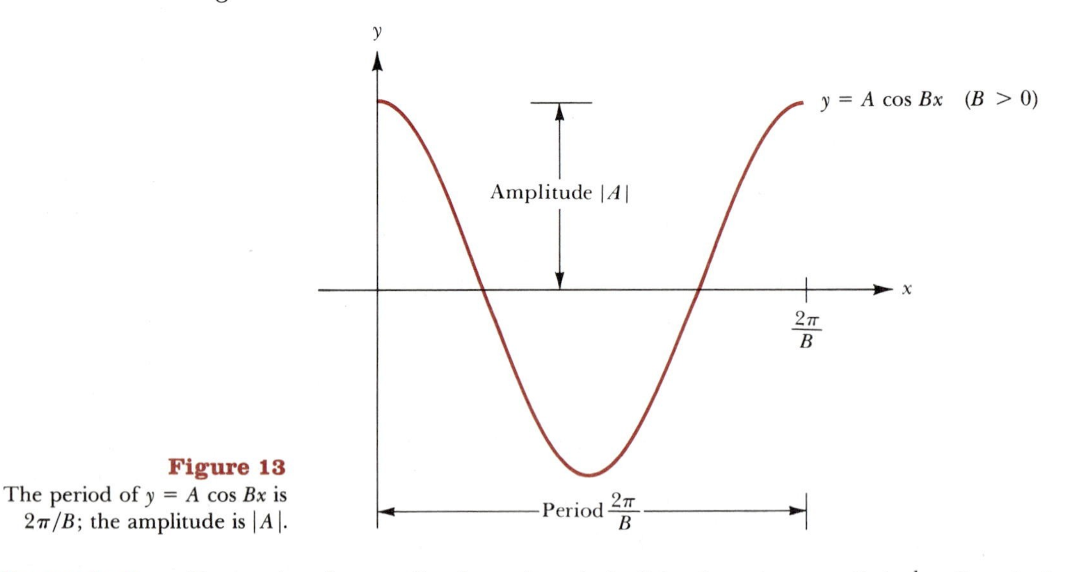

Figure 13
The period of $y = A \cos Bx$ is $2\pi/B$; the amplitude is $|A|$.

Example 3 Determine the amplitude and period of the function $y = 3 \sin \frac{1}{2}x$. Graph the function over one period and indicate the coordinates of the highest and lowest points on the graph.

Solution Comparing the equations

$$y = A \sin Bx$$

and

$$y = 3 \sin \tfrac{1}{2}x$$

we see that $A = 3$ and $B = \frac{1}{2}$. Thus we have

$$\text{period} = \frac{2\pi}{B} = \frac{2\pi}{\left(\frac{1}{2}\right)} = 4\pi$$

Also the amplitude is $|A| = |3| = 3$. The required graph is thus a sine wave with period 4π and amplitude 3. This is displayed in Figure 14.

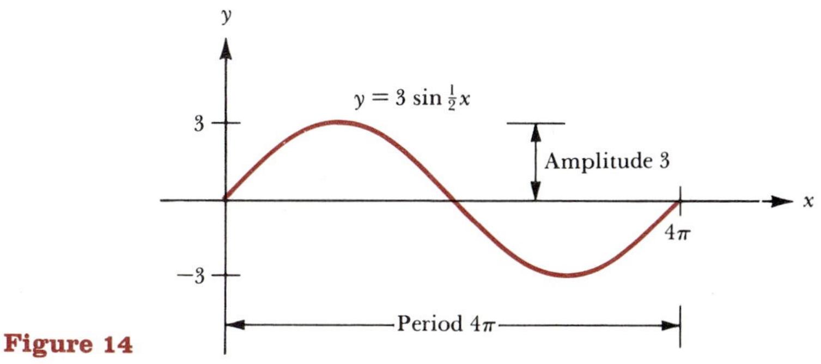

Figure 14

The highest point on the sine wave occurs when x is one quarter of the way through the period, that is

$$x = \frac{1}{4}(4\pi) = \pi$$

Since the largest y-value is (in this case) the amplitude, we conclude that the coordinates of the highest point on the graph in Figure 14 are $(\pi, 3)$. In a similar fashion, the lowest point occurs when x is three quarters of the way through the period, that is

$$x = \frac{3}{4}(4\pi) = 3\pi$$

Thus the lowest point on the graph is $(3\pi, -3)$.

Example 4 Determine the amplitude and period of $y = \cos \pi x$. Sketch the graph of this function over one period and specify the x-intercepts.

Solution For the function $y = \cos \pi x$ we have $A = 1$ and $B = \pi$. Thus

$$\text{amplitude} = |A| = 1$$

and

$$\text{period} = \frac{2\pi}{B} = \frac{2\pi}{\pi} = 2$$

Figure 15 on the following page shows the graph of a cosine wave with these specifications. Since the x-intercepts of the cosine wave occur one quarter and three quarters of the way through the period, we compute these intercepts by writing

$$\frac{1}{4}(2) = \frac{1}{2} \quad \text{and} \quad \frac{3}{4}(2) = \frac{3}{2}$$

These x-intercepts are shown in Figure 15.

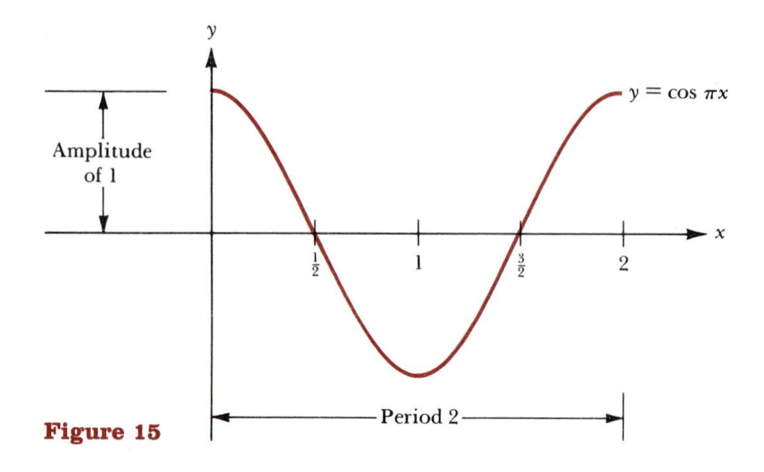

Figure 15

We have at our disposal now all of the tools needed to graph functions of the form

$$y = A \sin (Bx + C) \quad \text{or} \quad y = A \cos (Bx + C) \qquad \text{with } B > 0$$

First, as a matter of algebra, observe that

$$A \sin (Bx + C) = A\left[\sin B\left(x + \frac{C}{B}\right)\right]$$

The conclusion we draw from this is that the graph of $y = A \sin (Bx + C)$ is obtained by taking the graph of $y = A \sin Bx$ and translating it horizontally a distance of $|C/B|$ units. If C/B is positive, the movement is to the left; if C/B is negative, the movement is to the right. The quantity $-C/B$ is called the *phase shift.* As Figure 16(a) indicates, the phase shift gives you the x-coordinate at which to begin drawing $y = A \sin (Bx + C)$ over one period. Note that the amplitude and period are the same as they were for $y = A \sin Bx$.

The remarks we've made concerning $y = A \sin (Bx + C)$ apply equally well to $y = A \cos (Bx + C)$. See Figure 16(b).

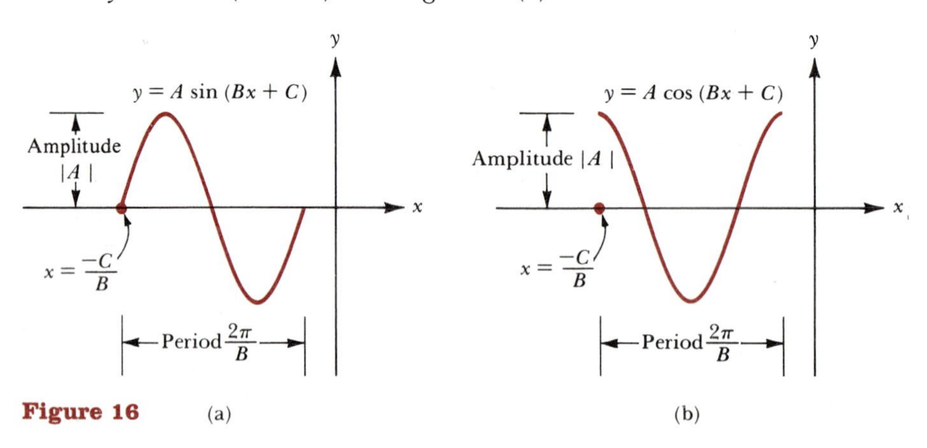

Figure 16 (a) (b)

Example 5 Determine the amplitude, period, and phase shift for the function $y = \frac{1}{2} \sin\left(2x - \frac{2\pi}{3}\right)$. Sketch the graph over one period, indicating the x-intercepts and the coordinates of the highest and lowest points on the graph over this period.

Solution

$$\text{Amplitude:}\quad |A| = \frac{1}{2}$$

$$\text{Period:}\quad \frac{2\pi}{B} = \frac{2\pi}{2} = \pi$$

$$\text{Phase shift:}\quad \frac{-C}{B} = \frac{2\pi/3}{2} = \frac{\pi}{3}$$

From this data, we conclude that this sine wave begins at $x = \pi/3$ and ends the period at

$$x = \frac{\pi}{3} + \pi = \frac{4\pi}{3}$$

Thus two of the x-intercepts are $\pi/3$ and $4\pi/3$. Since the third x-intercept is midway between these two, we may compute it by taking the average:

$$\frac{\frac{\pi}{3} + \frac{4\pi}{3}}{2} = \frac{5\pi/3}{2} = \frac{5\pi}{6}$$

The three x-intercepts are therefore $\pi/3$, $5\pi/6$, and $4\pi/3$. Next, the x-coordinate of the highest point on the graph is midway between $\pi/3$ and $5\pi/6$. Computing the average of these two numbers, we have

$$\frac{\frac{\pi}{3} + \frac{5\pi}{6}}{2} = \frac{\frac{2\pi}{6} + \frac{5\pi}{6}}{2} = \frac{7\pi/6}{2} = \frac{7\pi}{12}$$

Thus, the coordinates of the highest point on the graph are $(7\pi/12, \frac{1}{2})$. Finally, in a similar manner, the x-coordinate of the lowest point on the graph can be obtained by averaging the numbers $5\pi/6$ and $4\pi/3$. As you can check, this average is $13\pi/12$. It follows that the lowest point on the graph is $(13\pi/12, -\frac{1}{2})$. Figure 17 displays the required graph.

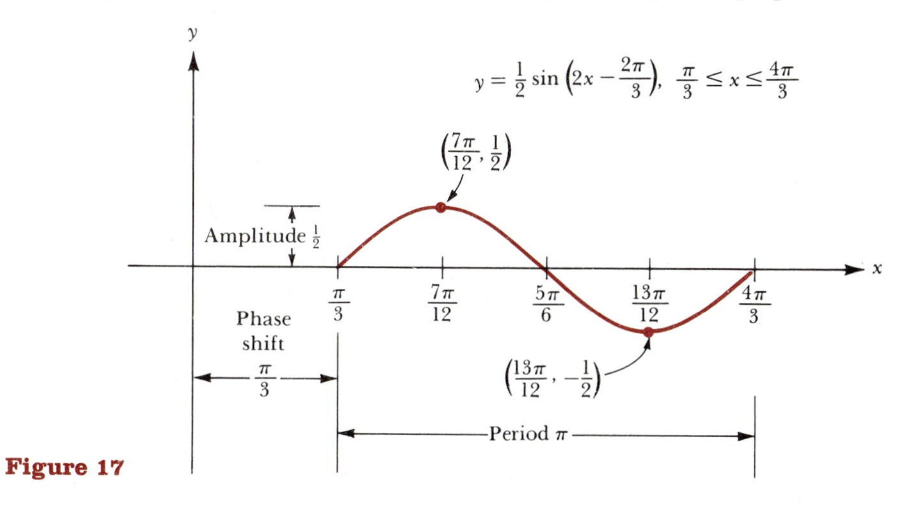

Figure 17

We conclude this section by displaying in graphs of the functions $y = \tan x$, $y = \cot x$, $y = \csc x$, and $y = \sec x$ in Figure 18 on the following page. Exercises 34 through 37 at the end of this section will help you to see why these graphs appear as they do. Of these four graphs, the graph of $y = \tan x$ is used the most frequently in applications; in fact, the other three graphs are needed only rarely. Notice that the period of the tangent function is π (rather than 2π).

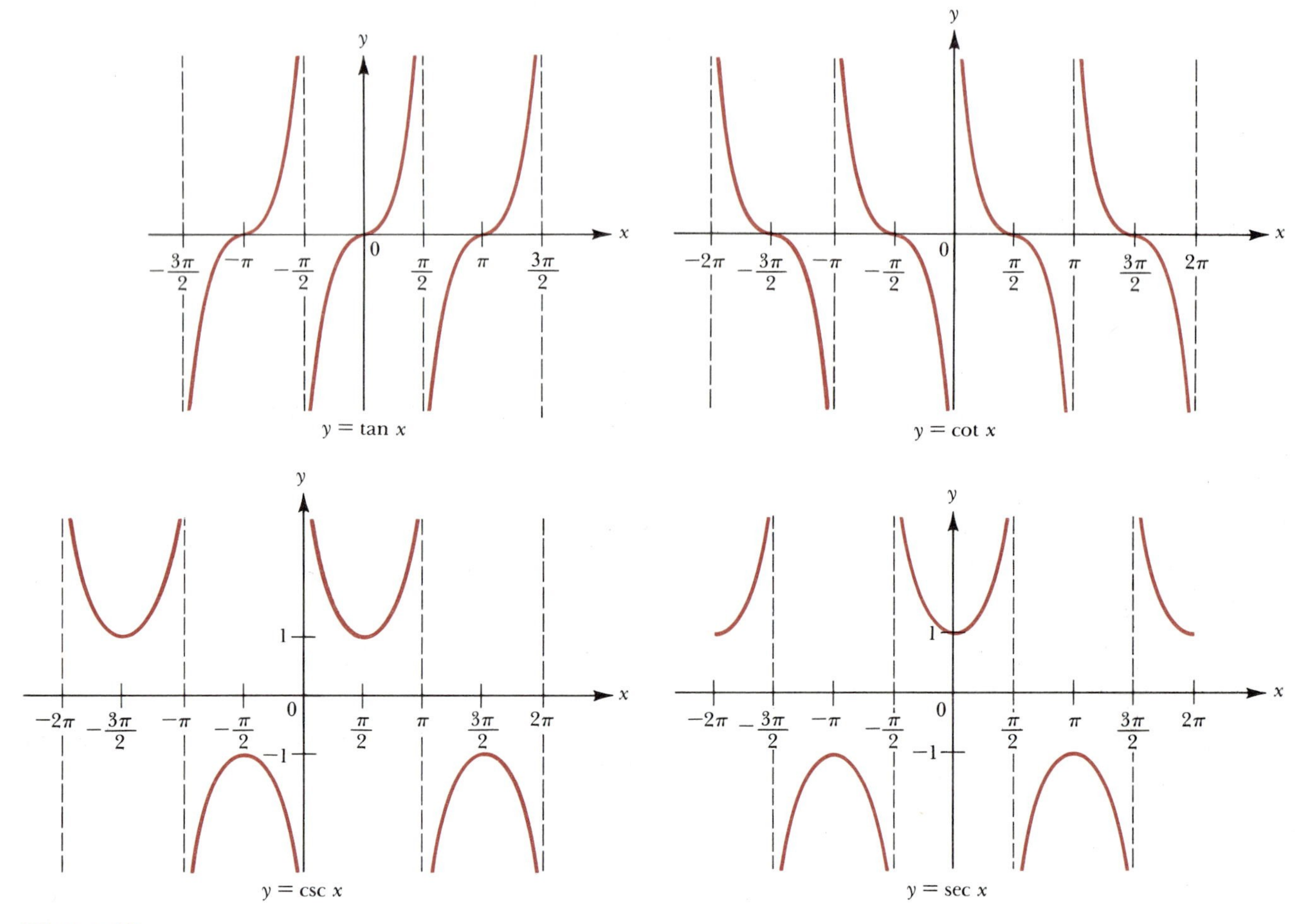

Figure 18

Exercise Set 8.5

1. Specify the period and amplitude for each function whose graph appears below and on page 405.

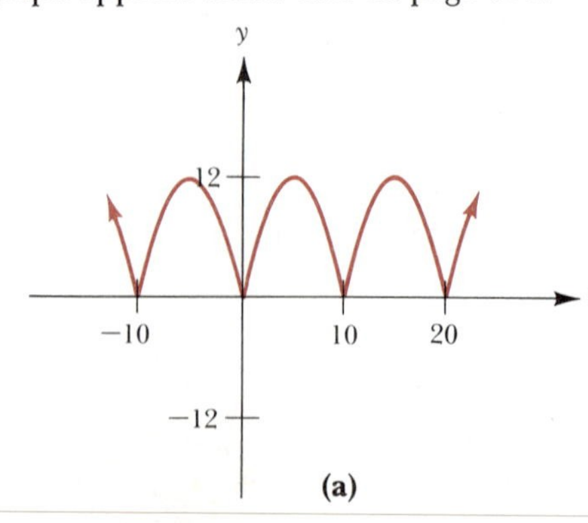
(a)

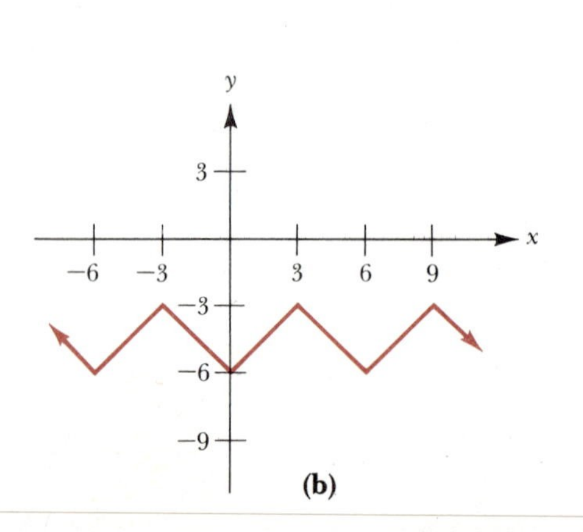
(b)

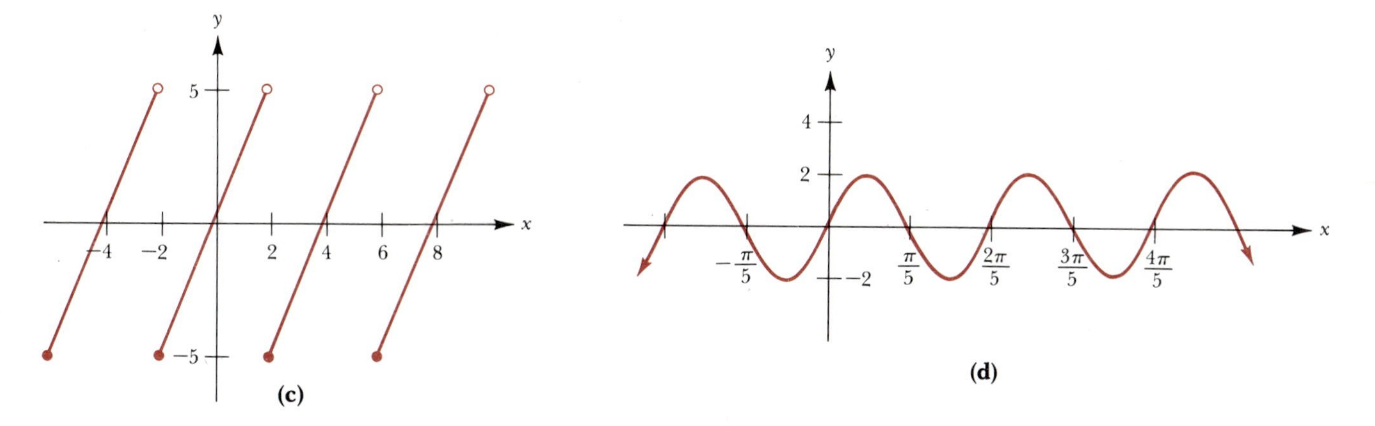

In Exercises 2–9, graph the function over one period. In each case, specify the amplitude, period, x-intercepts, and highest and lowest points on the graph.

2. **(a)** $y = 2 \sin x$
(b) $y = \sin 2x$
(c) $y = 2 \sin 2x$

3. **(a)** $y = 3 \cos x$
(b) $y = \cos 3x$
(c) $y = -\cos 3x$

4. **(a)** $y = \sin \frac{x}{2}$
(b) $y = 3 \sin \frac{x}{2}$

5. **(a)** $y = \cos \pi x$
(b) $y = -\cos \pi x$
(c) $y = 1 - \cos \pi x$

6. **(a)** $y = 4 \sin \frac{\pi x}{2}$
(b) $y = 4 \cos \frac{\pi x}{2}$

7. **(a)** $y = -\cos 2x$
(b) $y = 2 - \cos 2x$

8. **(a)** $y = 4 \sin \frac{x}{4}$
(b) $y = 4 \sin 4x$

9. **(a)** $y = 4 + 4 \sin 4x$
(b) $y = 4 - 4 \sin 4x$

In Exercises 10–19, determine the amplitude, period, and phase shift for the given function. Graph the function over one period. Indicate the x-intercepts and the coordinates of the highest and lowest points on the graph.

10. $y = \sin(2x - \pi)$

11. $y = 2 \cos\left(x - \frac{\pi}{2}\right)$

12. $y = 3 \sin\left(\frac{1}{2}x + \frac{\pi}{6}\right)$

13. $y = -2 \sin(\pi x + \pi)$

14. $y = 4 \cos\left(3x - \frac{\pi}{4}\right)$

15. $y = \cos(x + 1)$

16. $y = \frac{1}{2}\sin\left(\frac{\pi x}{2} - \pi^2\right)$

17. $y = \cos\left(2x - \frac{\pi}{3}\right) + 1$

18. $y = 1 - \cos\left(2x - \frac{\pi}{3}\right)$

19. $y = 3 \cos\left(\frac{2x}{3} + \frac{\pi}{6}\right)$

In Exercises 20–29, graph the functions over the indicated intervals.

20. $y = -\tan x; \quad -\pi \le x \le \pi$

21. $y = \tan\left(x - \frac{\pi}{2}\right); \quad -\pi < x < \pi$

22. $y = \tan \pi x; \quad 0 \le x \le 2$

23. $y = \tan\left(\pi x + \frac{\pi}{2}\right); \ 0 < x < 1$

24. $y = -\cot x; \quad -\frac{\pi}{2} \le x \le \frac{\pi}{2}$

25. $y = \cot 2\pi x; \quad 0 < x < 1$

26. $y = -\csc x; \quad -\pi < x < 0$

27. $y = -\csc\left(x - \frac{\pi}{2}\right); \quad 0 \le x < \frac{\pi}{2}$

28. $y = |\sec x|; \quad -\frac{3\pi}{2} < x < \frac{3\pi}{2}$

29. $y = -\sec \frac{\pi x}{2}; \quad 0 \le x < 1$

30. **(a)** Graph the function $y = \sin^2 x$ over one period. *Hint:* Make use of the formula for $\sin^2 \theta$ given on page 387.
(b) Graph the function $y = \cos^2 x$ over one period. *Hint:* Make use of the formula $\cos^2 x = 1 - \sin^2 x$.

31. Graph the function $y = \sin x \cos x$ over one period. *Hint:* $\sin 2\theta = 2 \sin \theta \cos \theta$.

32. The figure shows a simple pendulum consisting of a string with a weight attached at one end, while the other end is suspended from a fixed point. If the pendulum is displaced 0.1 radian from the equilibrium (vertical) position, and then released, it can be shown that the size of the angle θ at time t is very closely approximated by $\theta = 0.1 \cos \sqrt{g/L}\,t$.

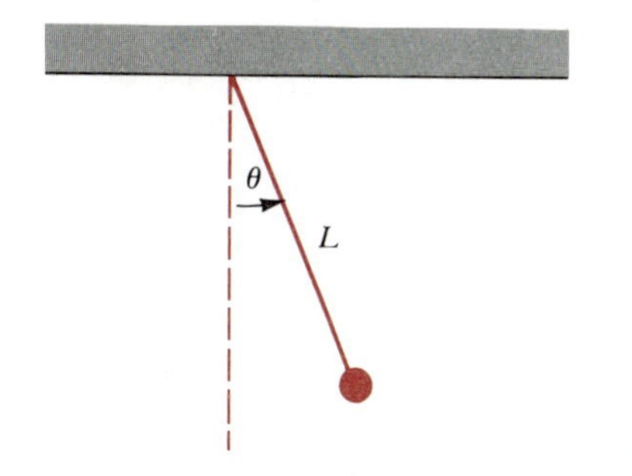

(continued)

In this formula, t is measured in seconds, L is the length of the pendulum, and g is a constant (the acceleration due to gravity). Show that under these conditions, the time T required for one complete oscillation of the pendulum is given by

$$T = 2\pi\sqrt{\frac{L}{g}}$$

(Two conclusions that follow from this formula are these. We may adjust a pendulum clock by altering the length L of the pendulum. On the other hand, altering the amplitude slightly will have no effect on the period, since the value of T does not depend upon the amplitude.)

***33.** A weight is attached to a spring hung from the ceiling, as shown in the figure. The coordinate system has been chosen so that the equilibrium position of the weight corresponds to $s = 0$. Now suppose that the weight is pulled down to the position where $s = 3$, and then it is released, so that it oscillates up and down in a periodic fashion. Assume that the position of the weight at time t is given by the function

$$s = 3\cos\frac{\pi t}{3}$$

where s is in feet and t is in seconds.

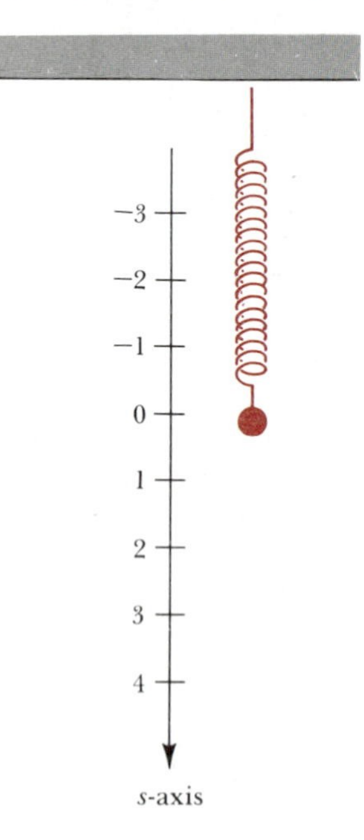

(a) Find the amplitude and period of this function and sketch its graph over the interval $0 \le t \le 12$.

(b) Use your graph to determine at what times during this interval the weight is furthest from the origin.

(c) When during this interval is the weight passing through the origin? (Specify the values of t.)

(d) Now assume that the velocity v of the weight, under these conditions, is given by the function

$$v = -\pi\sin\frac{\pi t}{3}$$

where t is in seconds and v is in units of ft/sec. (The sign of the velocity indicates the direction of the motion. For instance, a velocity of -2 ft/sec would indicate motion toward the ceiling, in this example.) Graph this velocity function over the interval $0 \le t \le 12$. Specify the amplitude and period.

(e) Use your graph to find the times during this interval when the velocity is 0. At these times, what is the position (s-coordinate) of the weight?

(f) At what times during this interval is the weight moving downward?

(g) Use your graph to find the times when the velocity is greatest and least. What is the numerical value of the velocity in those cases? (Use $\pi \approx 3.14$.) Where is the weight located at those times when the velocity is greatest or least?

(h) On the same set of axes, graph the velocity function and the position function over the interval $0 \le t \le 12$.

34. (a) According to Figure 18 on page 404, what is the period of $y = \tan x$?

(b) Confirm your answer in part (a) by proving the identity

$$\tan(x + \pi) = \tan x$$

(c) List several real numbers x for which $\cos x = 0$. Why is $\tan x$ undefined in these cases? What is the domain of the function $y = \tan x$?

(d) Complete the following table. *Suggestion:* Make use of the result $\tan(-x) = -\tan x$. Then plot the seven points thus obtained.

x	$-\frac{\pi}{3}$	$-\frac{\pi}{4}$	$-\frac{\pi}{6}$	0	$\frac{\pi}{6}$	$\frac{\pi}{4}$	$\frac{\pi}{3}$
$\tan x$							

(e) The following table of values was obtained using a calculator. After taking these results into account, use the table in part (d) to draw the graph of $y = \tan x$ over the interval $-\pi/2 < x < \pi/2$.

x	89°	89.9°	89.99°
$\tan x$	≈ 57	≈ 573	≈ 5730

(f) Check that the portion of the graph that you drew in part (e) agrees with what is shown in Figure 18.

35. (a) Prove the identity $\cot x = -\tan(x - \pi/2)$.

(b) The identity in part (a) implies that the graph of $y = \cot x$ is obtained by first reflecting the graph of $y = \tan x$ in the x-axis and then translating horizontally $\pi/2$ to the right. Use this observation to graph $y = \cot x$ and compare your result with Figure 18.

36. (a) You know that the graph of $y = \sin x$ is contained between the lines $y = 1$ and $y = -1$. What does this tell you about the graph of $y = \csc x$?

(b) Show that $\csc(x + 2\pi) = \csc x$.

(c) Complete the following table and use it to sketch $y = \csc x$ on the interval $0 < x < \pi$.

x	$\frac{\pi}{24}$	$\frac{\pi}{12}$	$\frac{\pi}{6}$	$\frac{\pi}{3}$	$\frac{\pi}{2}$	$\frac{2\pi}{3}$	$\frac{5\pi}{6}$	$\frac{11\pi}{12}$	$\frac{23\pi}{24}$
$\csc x$	≈7.66	≈3.86	≈0.00	≈0.00	≈0.00	≈0.00	≈0.00	≈3.86	≈7.66

(d) Prove the identity $\csc(x + \pi) = -\csc x$. Then use this result to graph $y = \csc x$ in the interval $\pi < x < 2\pi$. Now compare this graph and that obtained in part (c) with Figure 18.

37. (a) Prove the identity $\sec x = \csc(x + \pi/2)$.

(b) Use the result in part (a) to graph the function $y = \sec x$. Compare your graph with Figure 18.

***38.** (a) Show that $3 \sin x + 4 \cos x = 5 \sin(x + \delta)$, where δ is the acute angle shown below.

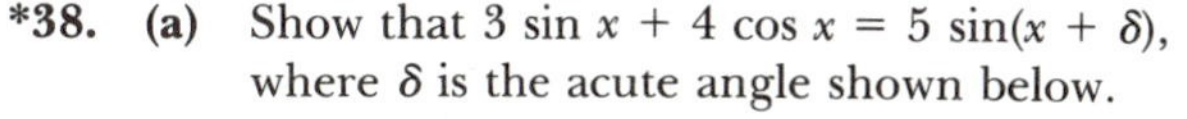

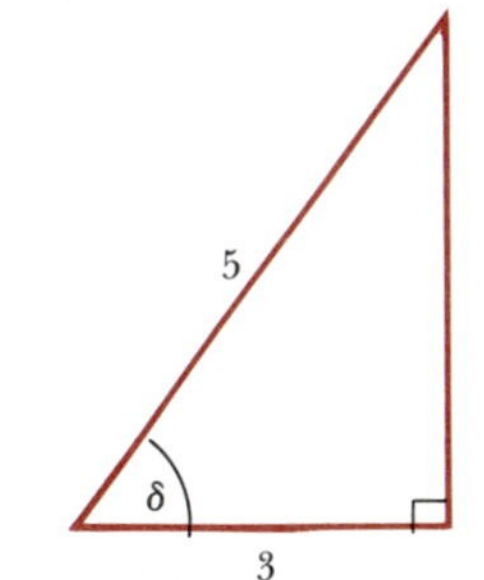

(b) Graph the function $y = 3 \sin x + 4 \cos x$ by using the identity in part (a). *Hint:* See Example 13 on page 320 for methods of determining δ. If you use a calculator, be sure it is set in the radian mode.

***39.** (a) Determine the amplitude and period of the function

$$y = \sin 2x + \sqrt{3} \cos 2x$$

Hint: Write $\sin 2x + \sqrt{3} \cos 2x = A \sin(2x + \delta)$ and solve for A and δ.

(b) What is the smallest possible value of the expression $\sin 2x + \sqrt{3} \cos 2x$?

8.6 Trigonometric Equations

In this section we consider some techniques for solving equations involving the trigonometric functions. As usual, by a *solution* of an equation, we mean a value of the variable for which the equation becomes a true statement.

Example 1 Consider the trigonometric equation $\sin x + \cos x = 1$. Is $x = \pi/4$ a solution? Is $x = \pi/2$ a solution?

Solution To see if the value $x = \pi/4$ satisfies the given equation, we write

$$\sin \frac{\pi}{4} + \cos \frac{\pi}{4} \stackrel{?}{=} 1$$

$$\frac{\sqrt{2}}{2} + \frac{\sqrt{2}}{2} \stackrel{?}{=} 1$$

$$\sqrt{2} \stackrel{?}{=} 1 \qquad \text{No!}$$

Thus $x = \pi/4$ is not a solution. In a similar fashion, we can check to see if $x = \pi/2$ is a solution:

$$\sin\frac{\pi}{2} + \cos\frac{\pi}{2} \stackrel{?}{=} 1$$

$$1 + 0 \stackrel{?}{=} 1 \qquad \text{Yes!}$$

Thus $x = \pi/2$ is a solution.

The example that we have just concluded serves to point out the difference between trigonometric identities, which we've studied previously, and the trigonometric equations that we study in this section. Trigonometric identities are true for all values of the variable in the given domain. Trigonometric equations, on the other hand, are only true for certain values of the unknown. Indeed, for some trigonometric equations, there may be no values of the unknown that satisfy the equation. For instance, the equation $\sin\theta = 2$ has no solution since, as you know, the value of $\sin\theta$ never exceeds 1. In general there is no single technique that can be used to solve every trigonometric equation. In the examples that follow, we illustrate some of the more common approaches to solving trigonometric equations.

Example 2 Solve the equation $\sin\theta = \frac{1}{2}$.

Solution First of all, you know that $\sin\pi/6 = \frac{1}{2}$. Thus one solution is certainly $\theta = \pi/6$. To find another solution, we note that $\sin\theta$ is positive in the second quadrant, as well as in the first. In the second quadrant, the angle with a reference angle of $\pi/6$ is $5\pi/6$ (150°). Thus, $\sin 5\pi/6 = \frac{1}{2}$ and we conclude that $\theta = 5\pi/6$ is also a solution to the given equation. Since $\sin\theta$ is negative in the third and fourth quadrants, we needn't look there for solutions. Nevertheless, there are other solutions, because the sine function is periodic, with period 2π. Therefore, since $\theta = \pi/6$ is a solution, we know that

$$\theta = \frac{\pi}{6} + 2\pi$$

is another solution. So is $\theta = \pi/6 + 4\pi$ a solution, and so on. Similarly, adding multiples of 2π to $\theta = 5\pi/6$ produces solutions also. We summarize as follows. The solutions of the equation $\sin\theta = \frac{1}{2}$ are

$$\frac{\pi}{6} + 2\pi k \quad \text{and} \quad \frac{5\pi}{6} + 2\pi k$$

where k is any integer. If we wish, we may express these solutions in degrees by writing them as

$$30° + 360°k \quad \text{and} \quad 150° + 360°k$$

The next example shows how factoring can be used to solve a trigonometric equation.

Example 3 Solve the equation $\cos^2 x + \cos x - 2 = 0$.

Solution By factoring the expression on the left-hand side of the equation, we obtain

$$(\cos x + 2)(\cos x - 1) = 0$$

Therefore

$$\cos x = -2 \quad \text{or} \quad \cos x = 1$$

We discard the result $\cos x = -2$ since the value of $\cos x$ is never less than -1. From the equation $\cos x = 1$ we conclude that $x = 0$ is one solution. After $x = 0$, the next time we have $\cos x = 1$ is when $x = 2\pi$. (You can see this by looking at the graph of $y = \cos x$, or by considering the unit circle definition of the cosine.) In general then, the solutions of the equation $\cos^2 x + \cos x - 2 = 0$ are given by $x = 2\pi k$, where k is an integer.

In some equations, more than one trigonometric function is present. A common approach here is to express the various functions in terms of a single one. The next example demonstrates this technique.

Example 4 Find all solutions of the equation $3\tan^2 x - \sec^2 x - 5 = 0$.

Solution Use the Pythagorean identity

$$\tan^2 x + 1 = \sec^2 x$$

to substitute for $\sec^2 x$ in the given equation. This gives

$$\begin{aligned} 3\tan^2 x - (\tan^2 x + 1) - 5 &= 0 \\ 3\tan^2 x - \tan^2 x - 1 - 5 &= 0 \\ 2\tan^2 x - 6 &= 0 \\ 2\tan^2 x &= 6 \\ \tan^2 x &= 3 \\ \tan x &= \pm\sqrt{3} \end{aligned}$$

Since the period of the tangent function is π, we need only to find values of x between 0 and π satisfying $\tan x = \pm\sqrt{3}$. The other solutions will then be obtained by adding multiples of π to these. Now in the first quadrant, the tangent is positive and we know $\tan x = \sqrt{3}$ when $x = \pi/3$. In the second quadrant, $\tan x$ is negative and we know $\tan 2\pi/3 = -\sqrt{3}$, since the reference angle for $2\pi/3$ is $\pi/3$. Thus the solutions between 0 and π are $x = \pi/3$ and $x = 2\pi/3$. It follows that all of the solutions to the equation $3\tan^2 x - \sec^2 x - 5 = 0$ are given by

$$x = \frac{\pi}{3} + \pi k \quad \text{or} \quad x = \frac{2\pi}{3} + \pi k$$

where k is an integer.

The technique (used in Example 4) of expressing the various functions in terms of a single function is most useful when it does not involve introducing a radical expression. For instance, consider the equation $\sin s + \cos s = 1$. Although we could begin by replacing $\cos s$ by the expression $\pm\sqrt{1 - \sin^2 s}$, it turns out to be easier in this situation to begin by squaring both sides of the given equation. This is done in the next example.

Example 5 Find all solutions of the equation $\sin s + \cos s = 1$ satisfying $0° \leq s < 360°$.

Solution Squaring both sides of the equation yields

$$\begin{aligned} (\sin s + \cos s)^2 &= 1^2 \\ \sin^2 s + 2\sin s\cos s + \cos^2 s &= 1 \end{aligned}$$

These add to 1.

Consequently we have

$$2 \sin s \cos s = 0$$

From this last equation we conclude that $\sin s = 0$ or $\cos s = 0$. When $\sin s = 0$, we know that $s = 0°$ or $s = 180°$. And when $\cos s = 0$, we know that $s = 90°$ or $s = 270°$. Now we must go back and check which (if any) of these values is a solution to the *original* equation. This must be done whenever we square both sides in the process of solving an equation.

$s = 0°$:	$\sin 0° + \cos 0° \stackrel{?}{=} 1$	
	$0 + 1 \stackrel{?}{=} 1$	True
$s = 90°$:	$\sin 90° + \cos 90° \stackrel{?}{=} 1$	
	$1 + 0 \stackrel{?}{=} 1$	True
$s = 180°$:	$\sin 180° + \cos 180° \stackrel{?}{=} 1$	
	$0 + (-1) \stackrel{?}{=} 1$	False
$s = 270°$:	$\sin 270° + \cos 270° \stackrel{?}{=} 1$	
	$-1 + 0 \stackrel{?}{=} 1$	False

We conclude now that the only solutions of the equation $\sin s + \cos s = 1$ on the interval $0° \leq s < 360°$ are $s = 0°$ and $s = 90°$.

In the example that follows, we consider an equation that involves a multiple of the unknown angle. (Methods for solving other types of equations involving multiple angles are indicated in the exercises.)

Example 6 Solve the equation $\sin 3x = 1$ on the interval $0 \leq x \leq 2\pi$.

Solution You know that $\sin \pi/2 = 1$. Thus, one solution can be found by writing

$$3x = \frac{\pi}{2}$$

from which we conclude

$$x = \frac{\pi}{6}$$

We can look for other solutions in the required interval by writing more generally

$$3x = \frac{\pi}{2} + 2\pi k$$

from which it follows that

$$x = \frac{\pi}{6} + \frac{2\pi k}{3}$$

Thus, when $k = 1$ we obtain

$$x = \frac{\pi}{6} + \frac{2\pi}{3} = \frac{\pi}{6} + \frac{4\pi}{6} = \frac{5\pi}{6}$$

When $k = 2$ we obtain

$$x = \frac{\pi}{6} + \frac{2\pi(2)}{3} = \frac{\pi}{6} + \frac{8\pi}{6}$$

$$= \frac{9\pi}{6} = \frac{3\pi}{2}$$

When $k = 3$ we obtain

$$x = \frac{\pi}{6} + \frac{2\pi(3)}{3}$$

$$= \frac{\pi}{6} + 2\pi \qquad \text{which is greater than } 2\pi$$

We conclude now that the solutions of $\sin 3x = 1$ on the interval $0 \le x \le 2\pi$ are

$$\frac{\pi}{6}, \quad \frac{5\pi}{6}, \quad \text{and} \quad \frac{3\pi}{2}$$

In Example 6, notice that we did not need to make use of a formula for $\sin 3\theta$, even though the expression $\sin 3\theta$ did appear in the given equation. In the next example, however, we do make use of the identity $\sin 2\theta = 2 \sin \theta \cos \theta$.

Example 7 Solve the equation $\sin x \cos x = 1$.

Solution You could begin by squaring both sides. (Exercise 45 at the end of this section asks you to use this approach.) However, with the double-angle formula for sine in mind, we can proceed instead as follows. We multiply both sides of the given equation by 2. This yields

$$2 \sin x \cos x = 2$$

and consequently

$$\sin 2x = 2 \qquad \text{We used the double-angle formula.}$$

This last equation has no solution, since the value of the sine function never exceeds 1. Thus the equation $\sin x \cos x = 1$ has no solution.

For the last example in this section, we look at a case in which tables or a calculator are required. As background for this, you should go back and review Example 13 on page 320.

Example 8 Find all angles θ between 0° and 360° satisfying the equation $\sin \theta = 2 \cos \theta$.

Solution We first want to rewrite the given equation using a single function. The easiest way to do this is to divide both sides by $\cos \theta$. (Nothing is lost here in assuming that $\cos \theta \neq 0$. If $\cos \theta$ were 0, then θ would be 90° or 270°; but neither of those angles is a solution of the given equation.) Dividing through by $\cos \theta$, we obtain

$$\frac{\sin \theta}{\cos \theta} = \frac{2 \cos \theta}{\cos \theta} \quad \text{or} \quad \tan \theta = 2$$

From experience, we know that none of the angles with which we are familiar (the multiples of 30° and 45°) have a value of 2 for their tangent. Thus a calculator or a table is required at this point. To find an angle whose tangent is 2 using a Texas Instruments calculator (set in the degree mode), the keystrokes are as follows.

Upon doing this, we find that

$$\theta \approx 63.4°$$

(Were we to retain all of the decimals displayed by the calculator, that would still be only an approximate solution.) To find another value of θ, we note that the tangent is also positive in the third quadrant. Thus a second (approximate) solution is given by

$$\theta = 180° + 63.4° = 243.4°$$

We conclude that the solutions between 0° and 360° to the equation $\sin \theta = 2 \cos \theta$ are approximately

$$63.4° \quad \text{and} \quad 243.4°$$

Note that the calculator only provided one value for θ; we still needed to work out the second value ourselves using the reference angle concept.

Exercise Set 8.6

1. Is $\theta = \pi/2$ a solution of the equation $2 \cos^2 \theta - 3 \cos \theta = 0$?
2. Is $x = 15°$ a solution of $(\sqrt{3}/3) \cos 2x + \sin 2x = 1$?
3. Is $x = 3\pi/4$ a solution of $\tan^2 x - 3 \tan x + 2 = 0$?
4. Is $t = 2\pi/3$ a solution of $2 \sin t + 2 \cos t = \sqrt{3} - 1$?
5. Is $\theta = 20°$ a solution of $\cos^4 3\theta - \cos^3 3\theta - 3 \cos^2 3\theta + \cos 3\theta - \sin^2 3\theta = 0$?

In Exercises 6–17, determine all solutions of the given equations. Express your answers using radian measure.

6. $\sin \theta = \sqrt{3}/2$
7. $\cos \theta = -1$
8. $\tan x = 0$
9. $2 \sin^2 x - 3 \sin x + 1 = 0$
10. $2 \cos^2 \theta + \cos \theta = 0$
11. $\sin^2 x - \sin x - 6 = 0$
12. $\cos^2 t \sin t - \sin t = 0$
13. $\cos \theta + 2 \sec \theta = -3$
14. $2 \cos^2 x - \sin x - 1 = 0$
15. $2 \cot^2 x + \csc^2 x - 2 = 0$
16. $\sqrt{3} \sin t - \sqrt{1 + \sin^2 t} = 0$
17. $\sec \alpha + \tan \alpha = \sqrt{3}$

In Exercises 18–25, determine all solutions in the interval $0° \le \theta < 360°$.

18. $\cos 3\theta = 1$
19. $\tan 2\theta = -1$
20. $\sin 3\theta = \dfrac{-\sqrt{2}}{2}$
21. $\sin \dfrac{\theta}{2} = \dfrac{1}{2}$
22. $\sin \theta = \cos \dfrac{\theta}{2}$

    Hint: $\sin \theta = 2 \sin \dfrac{\theta}{2} \cos \dfrac{\theta}{2}$.
23. $2 \sin^2 \theta - \cos 2\theta = 0$
24. $\sin 2\theta = \sqrt{3} \cos 2\theta$

    Hint: Divide by $\cos 2\theta$.
25. $\sin 2\theta = -2 \cos \theta$

C *In Exercises 26–31, use a calculator where necessary to find all solutions in the interval $0° \le \theta < 360°$.*

26. $\sin \theta = \frac{1}{4}$
27. $3 \cos \theta = 5$
28. $2 \tan \theta = -4$
29. $3 \sin^2 \theta - 2 \sin \theta - 1 = 0$
30. $\cos^2 x - \cos x - 1 = 0$
31. $2 \sin x = 3$

*32. Find all solutions of the equation $\tan 3x - \tan x = 0$ in the interval $0 \le x < 2\pi$. *Hint:* Write $\tan 3x = \tan(2x + x)$ and use the addition formula for tangent.

C 33. Find all solutions of the equation $2 \sin x = 1 - \cos x$ in the interval $0° \le x < 360°$.

34. Find all solutions of the equation $\cos (x/2) = 1 + \cos x$ in the interval $0 \le x < 2\pi$.

35. Find all solutions of the equation $\sin 3x \cos x + \cos 3x \sin x = \sqrt{3}/2$ in the interval $0 < x < 2\pi$.

36. Find all real numbers θ for which $\sec 4\theta + 2 \sin 4\theta = 0$.

Each of the equations in Exercises 37–42 can be solved with one of the sum-to-product identities listed on page 390. Find all solutions of the equations in Exercises 37–42 on the interval $0 \le x \le 2\pi$.

37. $\sin 5x = \sin 3x$

38. $\cos 3x = \cos 2x$

39. $\sin 3x = \cos 2x$

Hint: $\cos 2x = \sin\left(\frac{\pi}{2} - 2x\right)$.

40. $\sin 5x - \sin 3x + \sin x = 0$
Hint: First subtract $\sin x$ from both sides.

41. $\sin\left(x + \frac{\pi}{18}\right) = \cos\left(x - \frac{2\pi}{9}\right)$

***42.** $\cos 3x + \cos 2x + \cos x = 0$

C **43.** Find a solution of the equation $4 \sin \theta - 3 \cos \theta = 2$ in the interval $0° < \theta < 90°$. *Hint:* Add $3 \cos \theta$ to both sides, and then square.

***44.** Find all solutions of the equation $\sin^3 \theta \cos \theta - \sin \theta \cos^3 \theta = -\frac{1}{4}$ in the interval $0 < \theta < \pi$. *Hint:* Factor the left-hand side, and then use the double-angle formulas.

45. Consider the equation $\sin x \cos x = 1$.

(a) Square both sides, and then replace $\cos^2 x$ by $1 - \sin^2 x$. Show that the resulting equation can be written as $\sin^4 x - \sin^2 x + 1 = 0$.

(b) Show that the equation $\sin^4 x - \sin^2 x + 1 = 0$ has no solutions. Conclude from this that the original equation has no solutions.

8.7 The Inverse Trigonometric Functions

According to the horizontal line test (page 203), the function $y = \sin x$ is not one-to-one. See Figure 1.

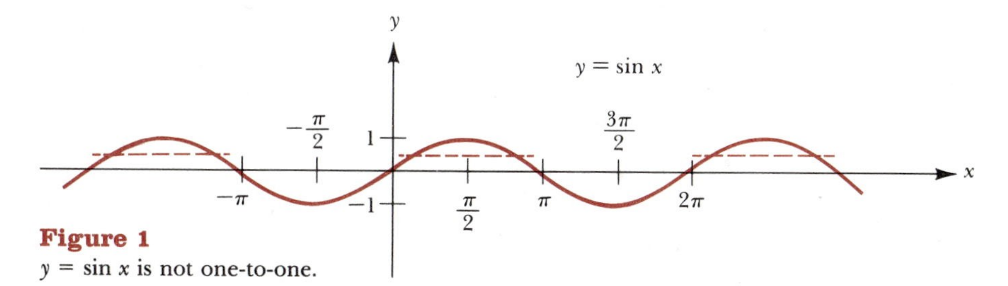

Figure 1
$y = \sin x$ is not one-to-one.

Since the function is not one-to-one, there is no inverse function. However, let us consider now the *restricted sine function*

$$y = \sin x \qquad -\frac{\pi}{2} \le x \le \frac{\pi}{2}$$

As Figure 2 indicates, this restricted sine function is one-to-one, according to the horizontal line test.

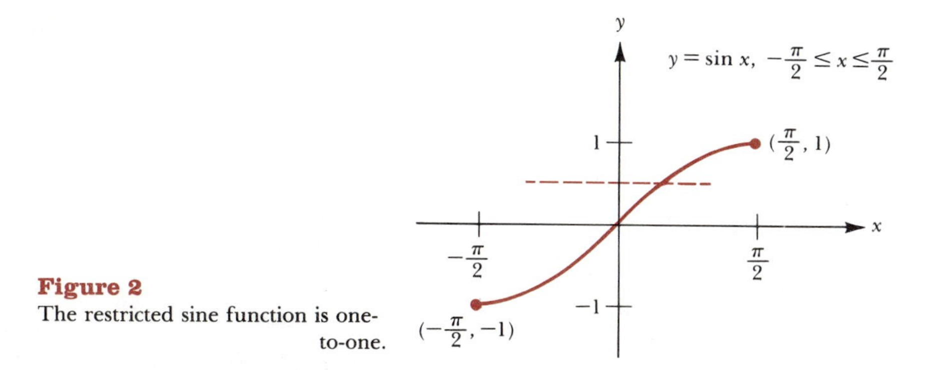

Figure 2
The restricted sine function is one-to-one.

Since the restricted sine function is one-to-one, the inverse function does exist. There are two notations commonly used to denote this inverse:

$$y = \sin^{-1} x \quad \text{and} \quad y = \arcsin x$$

Initially at least, we will use the notation $y = \sin^{-1} x$. The graph of $y = \sin^{-1} x$ is easily obtained using the fact that the graph of a function and its inverse are reflections of one another through the line $y = x$. Figure 3(a) shows the graph of the restricted sine function and its inverse $y = \sin^{-1} x$. Figure 3(b) shows the graph of $y = \sin^{-1} x$ alone. From Figure 3(b), we see that the domain of $y = \sin^{-1} x$ is the interval $[-1, 1]$, while the range is $[-\pi/2, \pi/2]$.

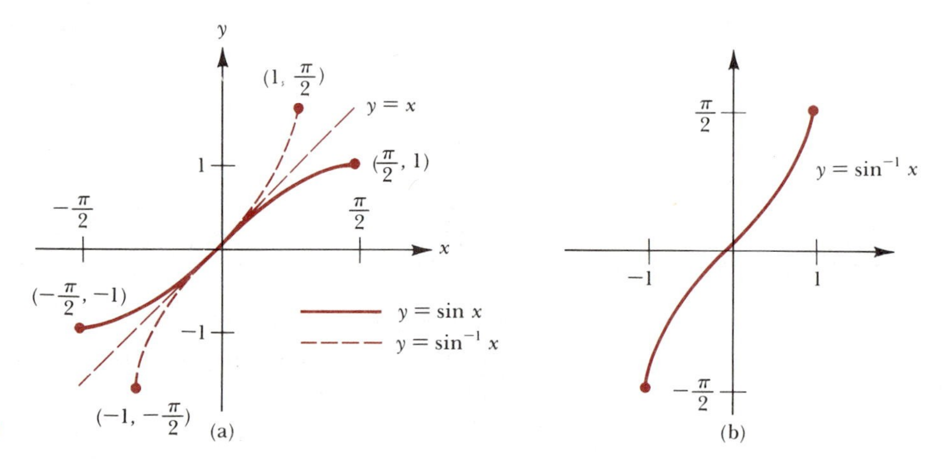

Figure 3

Values of $\sin^{-1} x$ are computed according to the following rule.

> $\sin^{-1} x$ is that number in the interval $[-\pi/2, \pi/2]$ whose sine is x.

To see why this is the case, we begin with the restricted sine function

$$y = \sin x \qquad -\frac{\pi}{2} \leq x \leq \frac{\pi}{2}$$

Then, as explained in Chapter 4, the inverse function is obtained by interchanging x and y (the inputs and the outputs). So for the inverse function we have

$$x = \sin y \qquad -\frac{\pi}{2} \leq y \leq \frac{\pi}{2} \tag{1}$$

Equation (1) tells us that y is the number in the interval $[-\pi/2, \pi/2]$ whose sine is x. This is what we wished to show.

Example 1 Evaluate $\sin^{-1} \frac{1}{2}$.

Solution $\sin^{-1} \frac{1}{2}$ is that number in the interval $[-\pi/2, \pi/2]$ whose sine is $\frac{1}{2}$. Since $\sin \pi/6 = \frac{1}{2}$, we conclude that $\sin^{-1} \frac{1}{2} = \pi/6$.

Using the arcsin notation, the result in Example 1 can be written as

$$\arcsin \frac{1}{2} = \frac{\pi}{6}$$

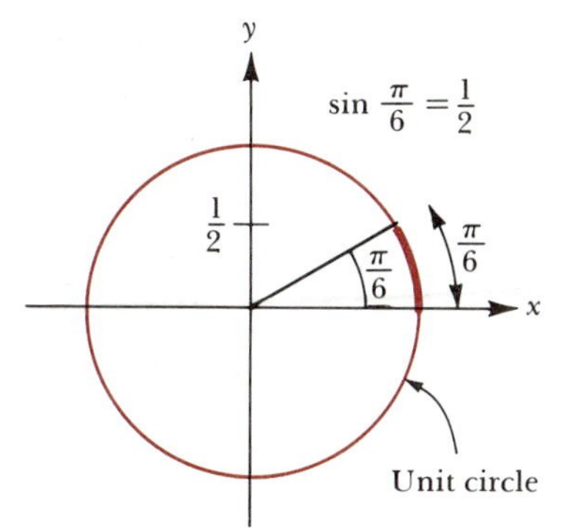

Figure 4

Now, as Figure 4 indicates, the arc length with a sine of $\frac{1}{2}$ is $\pi/6$. This is the idea behind the arcsin notation.

Example 2 Evaluate arcsin $\frac{3}{4}$.

Solution

arcsin $\frac{3}{4}$ is that number (or arc length, or angle in radians) in the interval $[-\pi/2, \pi/2]$ whose sine is $\frac{3}{4}$. Since we are not familiar with an angle with a sine of $\frac{3}{4}$, we use a calculator. For instance, using a Texas Instruments calculator, we set the calculator to the radian mode and then use the keystrokes

0.75 [INV] [SIN]

The result in this case is

$$\arcsin\frac{3}{4}\approx 0.85$$

Example 3 Show that the following two expressions are not equal.

(a) $\sin^{-1} 0$ **(b)** $\dfrac{1}{\sin 0}$

Solution $\sin^{-1} 0$ is that number in the interval $[-\pi/2, \pi/2]$ whose sine is 0. Since $\sin 0 = 0$, we conclude that

$$\sin^{-1} 0 = 0$$

On the other hand, since $\sin 0 = 0$, the expression $1/\sin 0$ is not even defined. Thus the two given expressions certainly are not equal.

If f and f^{-1} are any pair of inverse functions, then by definition we have that

$$f[f^{-1}(x)] = x \qquad \text{for every } x \text{ in the domain of } f^{-1}$$

and

$$f^{-1}[f(x)] = x \qquad \text{for every } x \text{ in the domain of } f$$

Applying these facts to the restricted sine function $y = \sin x$ and its inverse $y = \sin^{-1} x$, we obtain the following two identities.

$$\sin(\sin^{-1} x) = x \qquad -1 \le x \le 1$$
$$\sin^{-1}(\sin x) = x \qquad -\frac{\pi}{2} \le x \le \frac{\pi}{2}$$

The example that follows indicates that the domain restrictions accompanying these two identities cannot be disregarded.

Example 4 Compute each of the following that is defined.

(a) $\sin^{-1}\left(\sin\frac{\pi}{4}\right)$ **(b)** $\sin^{-1}(\sin \pi)$

(c) $\sin(\sin^{-1} 2)$ **(d)** $\sin\left[\sin^{-1}\left(-\frac{1}{\sqrt{5}}\right)\right]$

Solution

(a) Since $\pi/4$ lies in the domain of the restricted sine function, the identity $\sin^{-1}(\sin x) = x$ is applicable here. Thus

$$\sin^{-1}\left(\sin\frac{\pi}{4}\right) = \frac{\pi}{4}$$

Check: $\sin \pi/4 = \sqrt{2}/2$. Therefore $\sin^{-1}(\sin \pi/4) = \sin^{-1}(\sqrt{2}/2) = \pi/4$.

(b) The number π is not in the domain of the restricted sine function, so the identity $\sin^{-1}(\sin x) = x$ does not apply in this case. However, since $\sin \pi = 0$, we have

$$\sin^{-1}(\sin \pi) = \sin^{-1} 0 = 0$$

Thus $\sin^{-1}(\sin \pi)$ is equal to 0, not π.

(c) The number 2 is not in the domain of the inverse sine function. Thus the expression $\sin(\sin^{-1} 2)$ is undefined.

(d) The identity $\sin(\sin^{-1} x) = x$ is applicable here. (Why?) Thus

$$\sin\left[\sin^{-1}\left(-\frac{1}{\sqrt{5}}\right)\right] = -\frac{1}{\sqrt{5}}$$

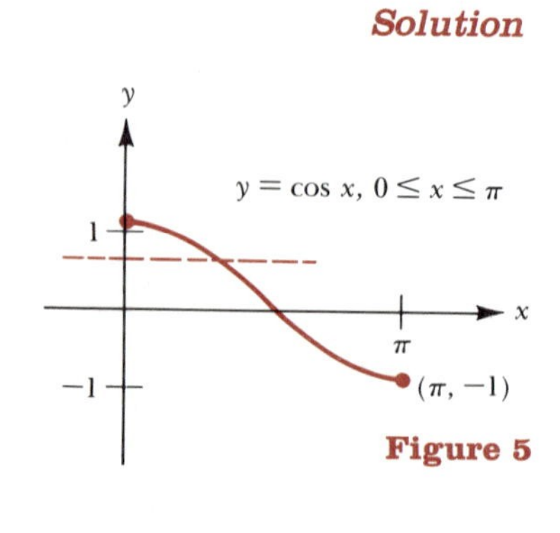

Figure 5

To define the inverse cosine function, we begin by defining the *restricted cosine function:*

$$y = \cos x \qquad 0 \le x \le \pi$$

As indicated by the horizontal line test in Figure 5, this function is one-to-one. Since the restricted cosine function is one-to-one, the inverse function exists. We denote this inverse by

$$y = \cos^{-1} x \quad \text{or} \quad y = \arccos x$$

The graph of $y = \cos^{-1} x$ is obtained by reflecting the restricted cosine function in the line $y = x$. Figure 6(a) displays the graph of the restricted cosine function along with $y = \cos^{-1} x$. Figure 6(b) shows the graph of $y = \cos^{-1} x$ alone. From Figure 6(b), we see that the domain of $y = \cos^{-1} x$ is the interval $[-1, 1]$, while the range is $[0, \pi]$. By reasoning in the same way we did for the inverse sine, we obtain the following results for the inverse cosine.

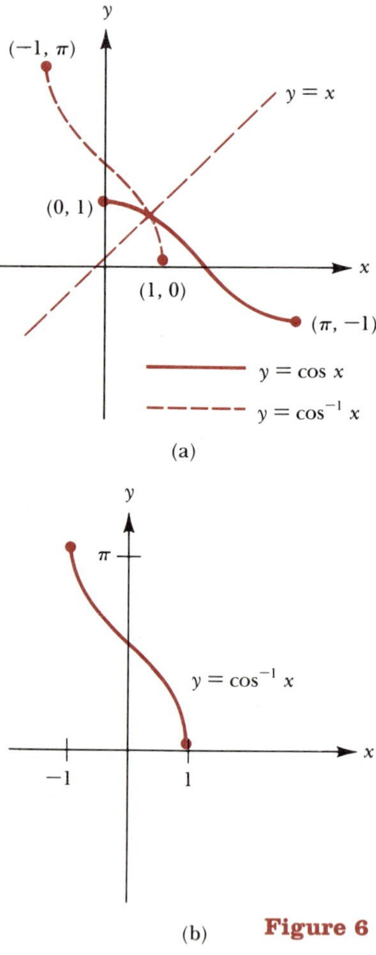

Figure 6

$\cos^{-1} x$ is that number in the interval $[0, \pi]$ whose cosine is x.

$$\cos(\cos^{-1} x) = x \qquad -1 \le x \le 1$$
$$\cos^{-1}(\cos x) = x \qquad 0 \le x \le \pi$$

Example 5 Compute $\cos^{-1}(0)$.

Solution $\cos^{-1}(0)$ is that number in the interval $[0, \pi]$ whose cosine is 0. Since $\cos \pi/2 = 0$, we have

$$\cos^{-1}(0) = \frac{\pi}{2}$$

Example 6 Compute $\arccos(\cos \frac{2}{5})$.

Solution Since the number $\frac{2}{5}$ is in the domain of the restricted cosine function, the identity $\arccos(\cos x) = x$ is applicable. Thus

$$\arccos\left(\cos \frac{2}{5}\right) = \frac{2}{5}$$

Example 7 Show that $\sin(\cos^{-1} x) = \sqrt{1 - x^2}$ for $-1 \leq x \leq 1$.

Solution We use the identity

$$\sin y = \sqrt{1 - \cos^2 y} \qquad \text{which is valid for } 0 \leq y \leq \pi$$

Substituting $\cos^{-1} x$ for y in this identity, we obtain

$$\sin(\cos^{-1} x) = \sqrt{1 - [\cos(\cos^{-1} x)]^2}$$

or

$$\sin(\cos^{-1} x) = \sqrt{1 - x^2} \qquad \text{as required}$$

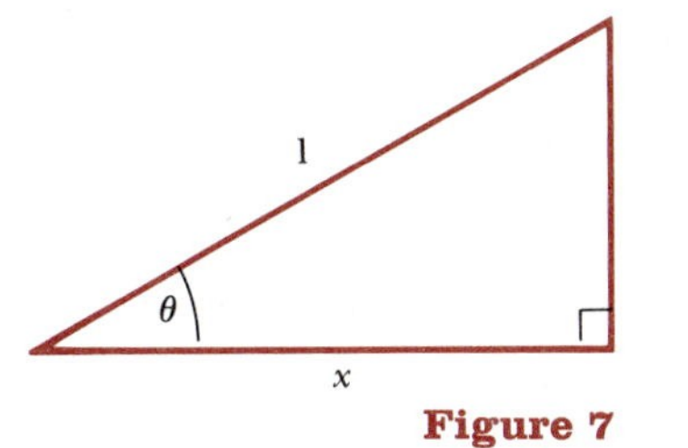

Figure 7

Before leaving Example 7, we point out an alternate method of solution that is useful when the restriction on x is $0 < x < 1$. In this case we let $\theta = \cos^{-1} x$. Then θ is the radian measure of the acute angle whose cosine is x. We can sketch a portrait of θ as shown in Figure 7. The sides of the triangle in Figure 7 have been labeled in such a way that $\cos \theta = x$. Then by the Pythagorean Theorem we find that the third side of the triangle is $\sqrt{1 - x^2}$. We therefore have

$$\sin \theta = \frac{\text{opposite}}{\text{hypotenuse}} = \frac{\sqrt{1 - x^2}}{1} = \sqrt{1 - x^2} \qquad \text{as required}$$

Now let us turn to the definition of the inverse tangent function. We begin by defining the *restricted tangent function* as

$$y = \tan x \qquad -\frac{\pi}{2} < x < \frac{\pi}{2}$$

Figure 8 shows the graph of this function. As you can check by applying the horizontal line test, the restricted tangent function is one-to-one. Since the

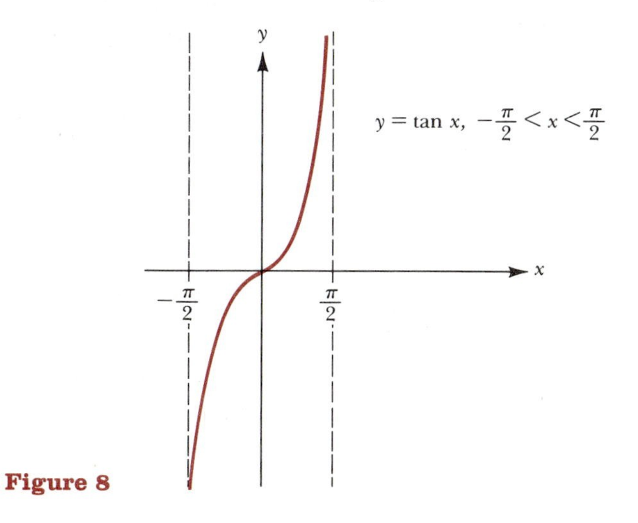

Figure 8

restricted tangent function is one-to-one, the inverse function exists. We denote the inverse tangent function by

$$y = \tan^{-1} x \quad \text{or} \quad y = \arctan x$$

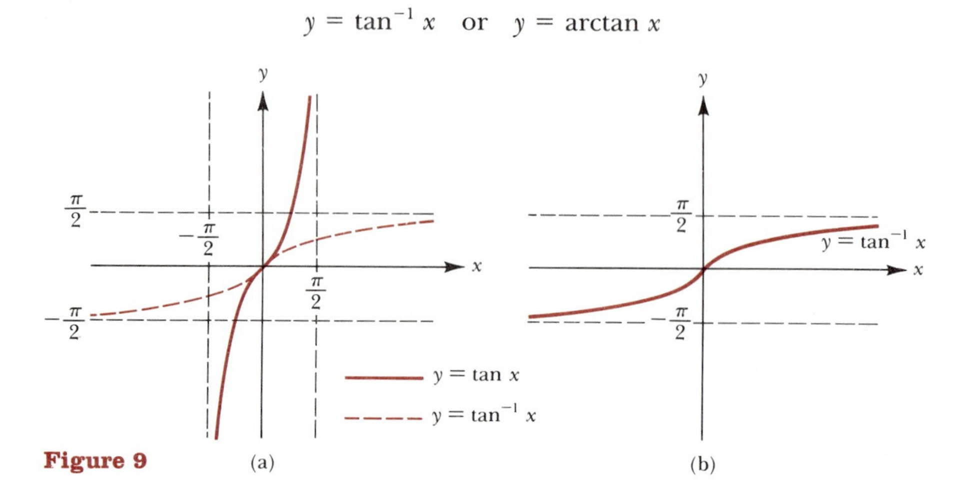

Figure 9

The graph of $y = \tan^{-1} x$ is obtained by reflecting the restricted tangent function in the line $y = x$. Figure 9(a) shows the graphs of the restricted tangent function and its inverse, while Figure 9(b) shows the graph of $y = \tan^{-1} x$ alone. From Figure 9(b), we see that the domain of $y = \tan^{-1} x$ is the set of all real numbers, while the range is the open interval $(-\pi/2, \pi/2)$. Applying the same type of reasoning used for the inverse sine, we obtain the following results for the inverse tangent.

> $\tan^{-1} x$ is that number in the interval $(-\pi/2, \pi/2)$ whose tangent is x.
>
> $$\tan(\tan^{-1} x) = x \qquad \text{for all } x$$
>
> $$\tan^{-1}(\tan x) = x \qquad -\frac{\pi}{2} < x < \frac{\pi}{2}$$

Example 8 Compute the following.

(a) $\tan^{-1}(-1)$ **(b)** $\tan(\tan^{-1} \sqrt{5})$

Solution **(a)** $\tan^{-1}(-1)$ is that number in the interval $(-\pi/2, \pi/2)$ whose tangent is -1. Since $\tan(-\pi/4) = -1$, we have

$$\tan^{-1}(-1) = -\frac{\pi}{4}$$

(b) The identity $\tan(\tan^{-1} x) = x$ holds for all real numbers x. We have therefore

$$\tan(\tan^{-1} \sqrt{5}) = \sqrt{5}$$

When suitable restrictions are placed on the domains of the cosecant, secant, and cotangent functions, corresponding inverse functions can be defined. However, since these three inverses are rarely, if ever, encountered in calculus, we omit a discussion of them here. Instead, we look at two additional examples involving the inverse sine and the inverse tangent.

Example 9 Evaluate $\cos(\sin^{-1} \frac{3}{5})$.

Solution Let $\theta = \sin^{-1} \frac{3}{5}$. Then θ is the measure of the acute angle whose sine is $\frac{3}{5}$, as indicated in Figure 10. Using the Pythagorean Theorem, we find that the length of the third side of the triangle is 4. Thus

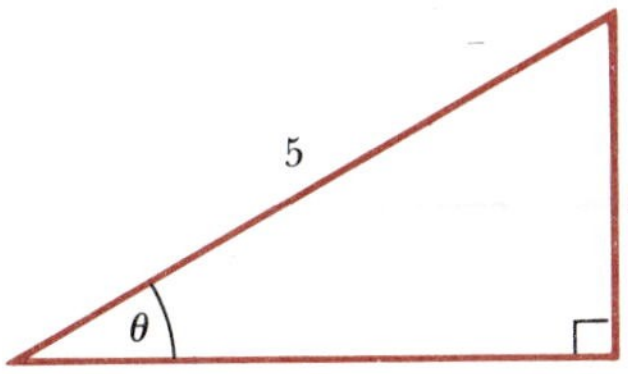

$$\cos \theta = \frac{\text{adjacent}}{\text{hypotenuse}} = \frac{4}{5}$$

Figure 10

In view of the definition of θ, this last equation states that

$$\cos\left(\sin^{-1} \frac{3}{5}\right) = \frac{4}{5}$$

This is the required answer.

Example 10 Show that $\pi = 4(\tan^{-1} \frac{1}{2} + \tan^{-1} \frac{1}{5} + \tan^{-1} \frac{1}{8})$.

Solution The given equation is equivalent to

$$\underbrace{\frac{\pi}{4} - \tan^{-1} \frac{1}{2}}_{A} = \underbrace{\tan^{-1} \frac{1}{5} + \tan^{-1} \frac{1}{8}}_{B}$$

Then, using the letters A and B as indicated, we wish to show that $A = B$. We will accomplish this if we can show that $\tan A = \tan B$, since both A and B lie between 0 and $\pi/2$. (Why?) Using the formula $\tan(x - y) = \dfrac{\tan x - \tan y}{1 + \tan x \tan y}$, we have

$$\begin{aligned}
\tan A &= \tan\left(\frac{\pi}{4} - \tan^{-1} \frac{1}{2}\right) \\
&= \frac{\tan \frac{\pi}{4} - \tan(\tan^{-1} \frac{1}{2})}{1 + \tan \frac{\pi}{4} \tan(\tan^{-1} \frac{1}{2})} \\
&= \frac{1 - \frac{1}{2}}{1 + 1(\frac{1}{2})} = \frac{\frac{1}{2}}{\frac{3}{2}} \\
&= \frac{1}{3}
\end{aligned}$$

Similarly, using the formula $\tan(x + y) = \dfrac{\tan x + \tan y}{1 - \tan x \tan y}$, we have

$$\begin{aligned}
\tan B &= \tan\left(\tan^{-1} \frac{1}{5} + \tan^{-1} \frac{1}{8}\right) \\
&= \frac{\tan(\tan^{-1} \frac{1}{5}) + \tan(\tan^{-1} \frac{1}{8})}{1 - [\tan(\tan^{-1} \frac{1}{5})][\tan(\tan^{-1} \frac{1}{8})]} \\
&= \frac{\frac{1}{5} + \frac{1}{8}}{1 - (\frac{1}{5})(\frac{1}{8})} \\
&= \frac{1}{3} \qquad \text{(Check the arithmetic.)}
\end{aligned}$$

We have now shown that $\tan A = \tan B$, since both quantities are equal to $\frac{1}{3}$. As pointed out above, it now follows in this case that $A = B$, as we wished to show.

Exercise Set 8.7

In Exercises 1–20, evaluate each of the quantities that is defined, without using a calculator or tables. If a quantity is undefined, state this.

1. $\sin^{-1}\frac{\sqrt{3}}{2}$ **2.** $\cos^{-1}(-1)$

3. $\tan^{-1}\sqrt{3}$ **4.** $\arccos\left(-\frac{\sqrt{2}}{2}\right)$

5. $\arctan\left(-\frac{1}{\sqrt{3}}\right)$ **6.** $\arcsin(-1)$

7. $\tan^{-1} 1$ **8.** $\sin^{-1} 0$

9. $\cos^{-1} 2\pi$ **10.** $\arctan 0$

11. $\sin\left(\sin^{-1}\frac{1}{4}\right)$ **12.** $\cos\left(\cos^{-1}\frac{4}{3}\right)$

13. $\cos\left(\cos^{-1}\frac{3}{4}\right)$ **14.** $\tan(\arctan 3\pi)$

15. $\arctan\left[\tan\left(-\frac{\pi}{7}\right)\right]$ **16.** $\sin(\arcsin 2)$

17. $\arcsin\left(\sin\frac{\pi}{2}\right)$ **18.** $\arccos\left(\cos\frac{\pi}{8}\right)$

19. $\arccos(\cos 2\pi)$ **20.** $\sin^{-1}\left(\sin\frac{3\pi}{2}\right)$

In Exercises 21–30, evaluate the given quantities without using a calculator or tables.

21. $\tan\left(\sin^{-1}\frac{4}{5}\right)$ **22.** $\cos\left(\arcsin\frac{2}{7}\right)$

23. $\sin(\tan^{-1} 1)$ **24.** $\sin[\tan^{-1}(-1)]$

25. $\tan\left(\arccos\frac{5}{13}\right)$ **26.** $\cos\left(\sin^{-1}\frac{2}{3}\right)$

27. $\cos(\arctan\sqrt{3})$ **28.** $\sin\left(\cos^{-1}\frac{1}{3}\right)$

29. $\sin\left[\arccos\left(-\frac{1}{3}\right)\right]$ **30.** $\tan\left(\arcsin\frac{20}{29}\right)$

C **31.** Use a calculator to evaluate each of the following quantities. Express your answers to two decimal places; don't round off.
(a) $\sin^{-1}\frac{3}{4}$ (b) $\cos^{-1}\frac{2}{3}$
(c) $\tan^{-1}\pi$ (d) $\tan^{-1}(\tan^{-1}\pi)$

In Exercises 32–34, evaluate the given expressions without using a calculator or tables.

32. $\csc\left(\sin^{-1}\frac{1}{2} - \cos^{-1}\frac{1}{2}\right)$

33. $\sec\left[\cos^{-1}\frac{\sqrt{2}}{2} + \sin^{-1}(-1)\right]$

34. $\cot\left[\cos^{-1}\left(-\frac{1}{2}\right) + \cos^{-1}(0) + \tan^{-1}\frac{1}{\sqrt{3}}\right]$

35. Show that $\cos(\sin^{-1} x) = \sqrt{1 - x^2}$ for $-1 \le x \le 1$. *Suggestion:* Use the method of Example 7 in the text.

36. Evaluate $\sin(2\tan^{-1} 4)$.
Hint: $\sin 2\theta = 2 \sin\theta \cos\theta$.

37. Evaluate $\cos\left(2\sin^{-1}\frac{5}{13}\right)$.

38. Evaluate $\sin\left(\arccos\frac{3}{5} - \arctan\frac{7}{13}\right)$. *Suggestion:* Use the formula for $\sin(x - y)$.

39. Show that $\sin\left(\sin^{-1}\frac{1}{3} + \sin^{-1}\frac{1}{4}\right) = \frac{\sqrt{15} + 2\sqrt{2}}{12}$.

40. Evaluate $\cos(\tan^{-1} 2 + \tan^{-1} 3)$.

***41.** (a) Explain why $0 < \tan^{-1}\frac{1}{3} + \tan^{-1}\frac{1}{2} < \pi/2$.
(b) Show that $\pi/4 = \tan^{-1}\frac{1}{3} + \tan^{-1}\frac{1}{2}$.

***42.** Show that $\tan^{-1}\frac{4}{3} - \tan^{-1}\frac{1}{7} = \pi/4$.

43. Show that $\sin^{-1} x + \cos^{-1} x = \pi/2$. *Suggestions:* Let $\theta = \sin^{-1} x$ and $\beta = \cos^{-1} x$. First explain why $-\pi/2 \le \theta + \beta \le 3\pi/2$. Then show $\sin(\theta + \beta) = 1$.

***44.** Show that $\sin^{-1} x = \tan^{-1}\frac{x}{\sqrt{1 - x^2}}$ for $-1 < x < 1$.

***45.** Show that $\arctan x + \arctan y = \arctan\frac{x + y}{1 - xy}$, when x and y are positive and $xy < 1$.

***46.** Show that $\tan^{-1}\frac{1}{7} + 2\tan^{-1}\frac{1}{3} = \pi/4$.

***47.** Evaluate $\arcsin\frac{4}{5} + \arctan\frac{3}{4}$.

***48.** Show that $\sin(2\sin^{-1} x) = 2x\sqrt{1 - x^2}$ for $-1 \le x \le 1$.

***49.** Show that $\tan^{-1} 1 + \tan^{-1} 2 + \tan^{-1} 3 = \pi$.

***50.** Suppose that $\arccos A + \arccos B + \arccos C = \pi$. Show that $A^2 + B^2 + C^2 + 2ABC = 1$.

***51.** (a) Show that $2\tan^{-1} x = \tan^{-1}\frac{2x}{1 - x^2}$ for $0 \le x < 1$.
(b) Show that $4\tan^{-1}\frac{1}{5} = \tan^{-1}\frac{120}{119}$.
(c) Show that $4\tan^{-1}\frac{1}{5} - \tan^{-1}\frac{1}{239} = \pi/4$. (In 1706, John Machin made use of this formula in calculating the value of π to 100 decimal places.)

Chapter 8 Review Checklist

- □ Definition of radian measure (page 364)
- □ Arc length formula (page 366)

Chapter 8 Review Exercises

In Exercises 1–8, convert the degree measures to radians.

1. $360°$ **2.** $36°$ **3.** $1°$ **4.** $330°$

5. $7.5°$ **6.** $\pi°$ **7.** $\left(\dfrac{1}{\pi}\right)^{\circ}$ **8.** $\left(\dfrac{\pi}{180}\right)^{\circ}$

For Exercises 9–16, convert the radian measures to degrees.

9. $\dfrac{3\pi}{4}$ **10.** $\dfrac{\pi}{15}$ **11.** 5π **12.** $\dfrac{\pi}{3}$

13. $\dfrac{5\pi}{6}$ **14.** 1 **15.** 2 **16.** $\dfrac{180}{\pi}$

In Exercises 17–24, use the arc length formula $s = r\theta$ and the sector area formula $A = \frac{1}{2}r^2\theta$ to find the required quantities.

	Given	Find
17.	$\theta = \frac{\pi}{8}$, $r = 16$ cm	s and A
18.	$\theta = 120°$, $r = 12$ cm	s and A
19.	$r = s = 1$ cm	A
20.	$r = s = \sqrt{3}$ cm	θ
21.	$s = 4$ cm, $\theta = 36°$	r
22.	$\theta = \frac{\pi}{10}$, $A = 200\pi$ cm^2	r
23.	$s = 12$ cm, $\theta = r + 1$	r and θ
24.	$\theta = 50°$, $A = 20\pi$ cm^2	s

For Exercises 25–36, evaluate each expression without using a calculator or tables.

25. $\cos \pi$ **26.** $\sin\left(-\dfrac{3\pi}{2}\right)$ **27.** $\csc \dfrac{2\pi}{3}$

28. $\tan \dfrac{\pi}{3}$ **29.** $\cot \dfrac{11\pi}{6}$ **30.** $\cos 0$

31. $\sin \dfrac{\pi}{6}$ **32.** $\sec \dfrac{3\pi}{4}$ **33.** $\cot \dfrac{5\pi}{4}$

34. $\tan\left(\dfrac{-7\pi}{4}\right)$ **35.** $\csc\left(-\dfrac{5\pi}{6}\right)$ **36.** $\sin^2 \dfrac{\pi}{7} + \cos^2 \dfrac{\pi}{7}$

37. Two angles in a triangle are 40° and 70°, respectively. What is the radian measure of the third angle?

38. The radian measures of two angles in a triangle are $\frac{1}{6}$ and $\frac{5}{12}$, respectively. What is the radian measure of the third angle?

C *Exercises 39–48 are calculator exercises. (Set your calculator to the radian mode.) In Exercises 39–44, where numerical answers are required, round off your results to three decimal places.*

39. Evaluate $\sin 1$. **40.** Evaluate $\cos 2$.

41. Evaluate $\sin(3\pi/2)$. **42.** Evaluate $\sin(0.78)$.

43. Evaluate $\sin(\sin 0.0123)$.

44. Evaluate $\sin[\sin(\sin 0.0123)]$.

45. Verify that $\sin^2 1986 + \cos^2 1986 = 1$.

46. Verify that $\sin 14 = 2 \sin 7 \cos 7$.

47. Verify that $\cos(0.5) = \cos^2(0.25) - \sin^2(0.25)$.

48. Verify that $\cos(0.3) = [\frac{1}{2}(1 + \cos 0.6)]^{1/2}$.

49. In the expression $\sqrt{25 - x^2}$, make the substitution $x = 5 \sin \theta$ $(0 < \theta < \pi/2)$ and simplify the result.

50. In the expression $(49 + x^2)^{1/2}$, make the substitution $x = 7 \tan \theta$ $(0 < \theta < \pi/2)$ and simplify the result.

51. In the expression $(x^2 - 100)^{1/2}$, make the substitution $x = 10 \sec \theta$ $(0 < \theta < \pi/2)$ and simplify the result.

52. In the expression $(x^2 - 4)^{-3/2}$, make the substitution $x = 2 \sec \theta$ $(0 < \theta < \pi/2)$ and simplify the result.

53. In the expression $(x^2 + 5)^{-1/2}$, make the substitution $x = \sqrt{5} \tan \theta$ $(0 < \theta < \pi/2)$ and simplify the result.

54. If $\sin \theta = -\frac{5}{13}$ and $\pi < \theta < 3\pi/2$, compute $\cos \theta$.

For Exercises 55–57, 59, and 60, assume that $0 < \theta < 2\pi$.

55. If $\sin \theta = \frac{3}{5}$ and $\tan \theta$ is positive, compute $\sin(\theta/2)$.

56. If $\cos\theta = \frac{8}{17}$ and $\sin\theta$ is negative, compute $\tan\theta$.

57. If $\tan\theta = \frac{7}{24}$ and $\cos\theta$ is positive, compute $\cos 2\theta$.

58. If $\sec\theta = -\frac{25}{7}$ and $\pi < \theta < 3\pi/2$, compute $\cot\theta$.

59. If $\csc\theta = \frac{29}{20}$ and $\cos\theta$ is negative, compute $\cos(\theta/2)$.

60. If $\cos\theta = \frac{3}{5}$ and $\sin\theta$ is positive, compute
(a) $\sin 3\theta$ **(b)** $\sin(\theta/4)$.

61. If $\tan 2\theta = 7/24$ and $\pi < 2\theta < 3\pi/2$, compute $\sin\theta$.

62. Express the area of the shaded region in the figure in terms of r and θ.

63. In the figure, $ABCD$ is a square, each side of which is 1 cm. The two arcs are each portions of circles with radii of 1 cm and with centers A and C, respectively. Find the area of the shaded region. *Hint:* draw BD and use the result in Exercise 62.

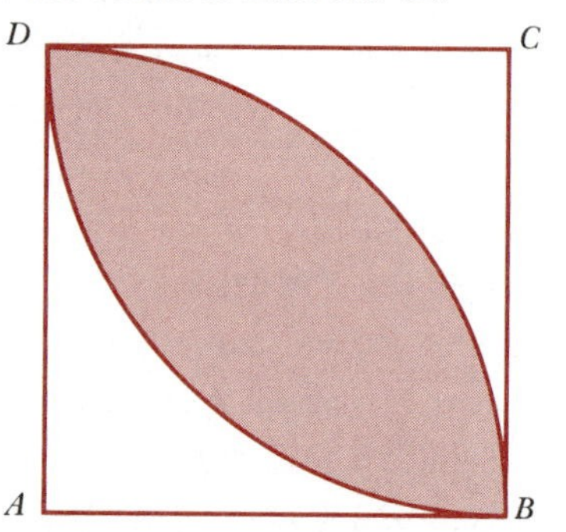

64. In the figure the three circles are mutually tangent, and each circle has a radius of 2 cm. Compute the area of the shaded region.

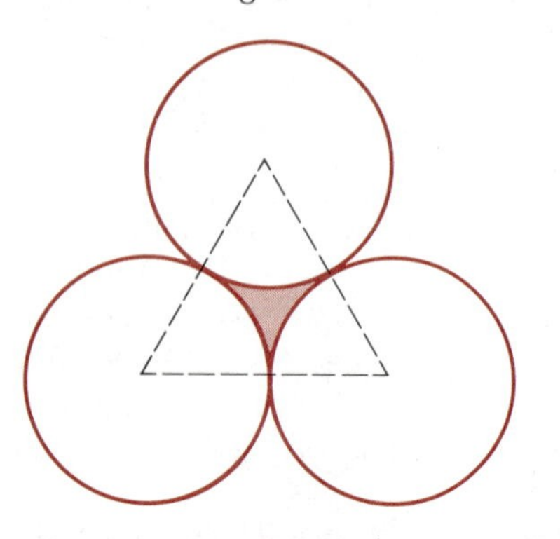

C 65. The following identity is known as *Euler's Formula:*

$$\cos(36^\circ + A) + \cos(36^\circ - A) = \cos A + \sin(18^\circ + A) + \sin(18^\circ - A)$$

Use a calculator to evaluate each side of the formula for the following values of A. In each case, round off your final results to three decimal places.
(a) 1° **(b)** 4° **(c)** 6°

66. In Euler's formula (given in Exercise 65), let $A = 0^\circ$. Then use one of the double-angle formulas for cosine to transform the equation you've obtained into an equation that is quadratic in the quantity $\sin 18^\circ$. Solve this equation to obtain the exact value of $\sin 18^\circ$ in radical form.

In Exercises 67–73, prove that the given equations are identities.

67. $(\sin x + \cos x)^2 + (\sin x - \cos x)^2 = 2$

68. $(\sin x + \cos x)^3 - (\sin x - \cos x)^3 = 6\cos x - 4\cos^3 x$

69. $(\cos x + \sin x - 1)(\cos x + \sin x + 1) = 2\sin x\cos x$

70. $\dfrac{1}{1-\sin x} + \dfrac{1}{1+\sin x} + \dfrac{\sin x}{1+\cos x} + \dfrac{1+\cos x}{\sin x} = 2(\csc x + \sec^2 x)$

71. $(9 - 4\sin x - \cos^2 x)(9 + 4\sin x - \cos^2 x) = \sin^4 x + 64$

72. $(\sin x - \cos x)(1 + \tan x + \cot x) = \sin^2 x\cos^2 x(\sec^3 x - \csc^3 x)$

73. $(a\sin x - b\cos x)^2 + (a\cos x + b\sin x)^2 = a^2 + b^2$

74. If A and B are acute angles such that $\tan A = \frac{12}{5}$ and $\tan B = \frac{3}{4}$, compute $\sin(A + B)$.

75. Using the data in Exercise 74, compute $\cos(A + B)$.

For Exercises 76–85, use the addition formulas to simplify the expressions.

76. $\sin\left(x + \dfrac{3\pi}{2}\right)$ **77.** $\sin(\pi - x)$ **78.** $\cos(\pi - x)$

79. $\sin 10^\circ\cos 80^\circ + \cos 10^\circ\sin 80^\circ$

80. $\cos 175^\circ\cos 25^\circ + \sin 175^\circ\sin 25^\circ$

81. $\cos\dfrac{2\pi}{5}\cos\dfrac{\pi}{10} - \sin\dfrac{2\pi}{5}\sin\dfrac{\pi}{10}$

82. $\cos\left(x - \dfrac{2\pi}{3}\right) - \cos\left(x + \dfrac{2\pi}{3}\right)$

83. $\sin\left(x + \dfrac{\pi}{6}\right) - \sin\left(x - \dfrac{\pi}{6}\right)$

84. $\tan(x + 45^\circ)\tan(x - 45^\circ)$

85. $\tan x\tan y + \tan y\tan z + \tan z\tan x$, where $y = x + 60^\circ$ and $z = x - 60^\circ$

86. **(a)** Show that $\tan(\pi/12) = 2 - \sqrt{3}$. *Hint:* $\pi/12 = \pi/4 - \pi/6$.
(b) Show that $\cot(\pi/12) = 2 + \sqrt{3}$.

87. If x, y, and z are acute angles with sines $\frac{3}{5}$, $\frac{5}{13}$, and $\frac{8}{17}$, respectively, compute $\sin(x + y + z)$.

88. In triangle ABC, suppose that $\sin^2 A + \sin^2 B = \sin^2 C$. Show that angle C is a right angle.

89. Suppose that $\sin x = \sqrt{5}/5$ and $\sin y = \sqrt{10}/10$, where $0 < x < \pi/2$ and $0 < y < \pi/2$. Compute $\tan(x + y)$.

For Exercises 90–94, a function of the form $y = A \sin Bx$ or $y = A \cos Bx$ is graphed for one period. Specify the equation in each case.

90. **91.** **92.** **93.**

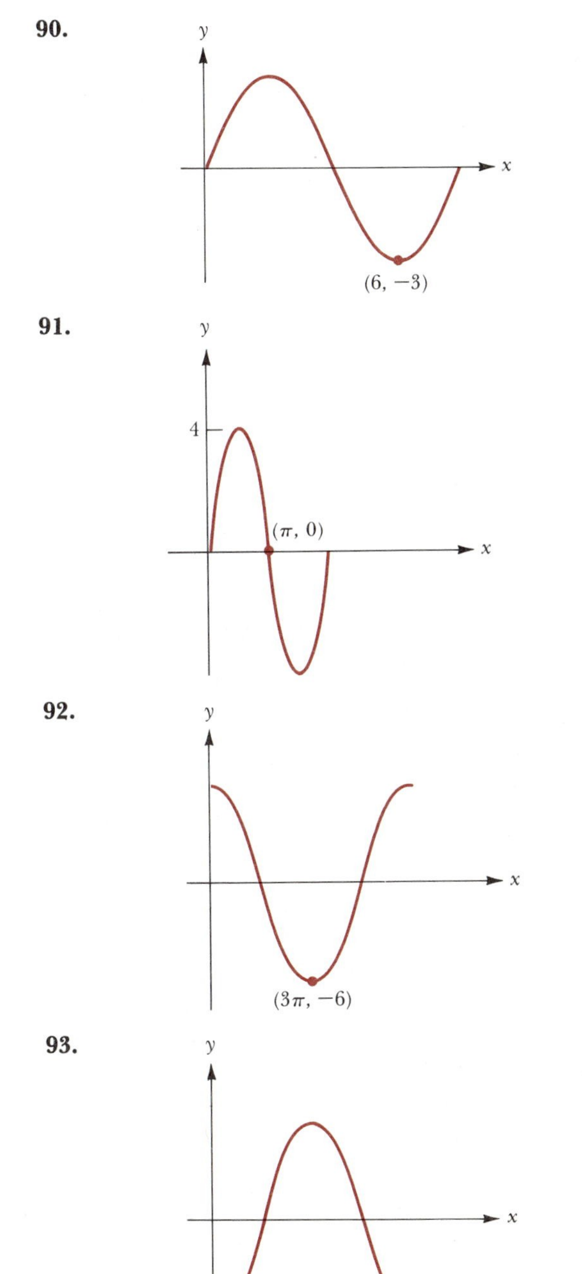

94.

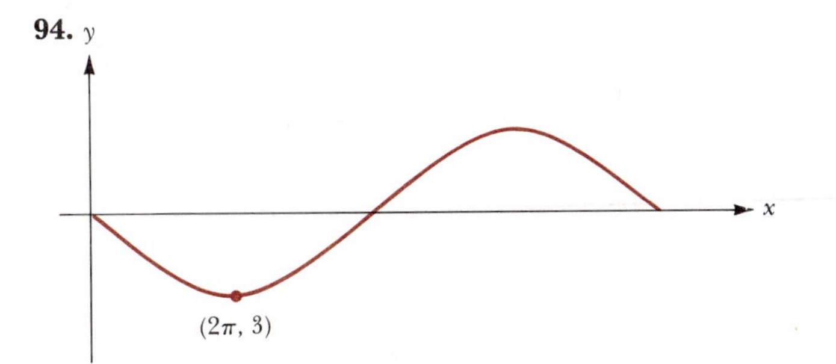

95. Find the amplitude of the function $y = A \sin\left(\frac{3\pi x}{2} - \pi\right)$ given that the graph passes through the point $(1, \sqrt{2})$.

96. Find the amplitude of the function $y = -3 \sin(2x - 3) \cos(2x - 3)$. *Hint:* Make use of the formula for $\sin 2\theta$.

In Exercises 97–102, sketch the graph of each function for one complete cycle. In each case, specify the x-intercepts and the coordinates of the highest and lowest points on the graph.

97. $y = 2 \sin\left(\frac{\pi x}{2} - \frac{\pi}{4}\right)$

98. $y = -\sin(x - 1)$

99. $y = 3 \cos\left(\frac{\pi x}{3} - \frac{\pi}{3}\right)$

100. $y = -2 \cos(x + \pi)$

101. $y = -\cos 2x \sin 2x$

102. $y = 1 - 2 \sin^2 x$

For Exercises 103–108, sketch the graphs over the indicated intervals.

103. $y = \tan x \quad (0 \le x < \pi/2)$

104. $y = -\cot x \quad (0 < x < 2\pi)$

105. $y = |\csc x| \quad (0 < x < 2\pi)$

106 $y = \sec x \quad (0 \le x \le \pi)$

107. $y = \tan(x - 2) \quad \left(-\frac{\pi}{2} + 2, \frac{\pi}{2} + 2\right)$

108. $y = \dfrac{\sin 2x}{1 + \cos 2x} \quad \left(0 \le x < \frac{\pi}{2}\right)$

In Exercises 109–134, prove that the equations are identities.

109. $\cot(x + y) = \dfrac{\cot x \cot y - 1}{\cot x + \cot y}$

110. $\cos 2x = \dfrac{1 - \tan^2 x}{1 + \tan^2 x}$

111. $\sin 2x = \dfrac{2 \tan x}{1 + \tan^2 x}$

112. $\sin^2 x - \sin^2 y = \sin(x + y) \sin(x - y)$

113. $\tan^2 x - \tan^2 y = \dfrac{\sin(x + y) \sin(x - y)}{\cos^2 x \cos^2 y}$

114. $2 \csc 2x = \sec x \csc x$

115. $\sin x\left(\tan \frac{x}{2} + \cot \frac{x}{2}\right) = 2$

116. $\tan\left(x + \frac{\pi}{4}\right) = \frac{1 + \tan x}{1 - \tan x}$

117. $\tan\left(\frac{\pi}{4} + x\right) - \tan\left(\frac{\pi}{4} - x\right) = 2 \tan 2x$

118. $\frac{\cot x - 1}{\cot x + 1} = \frac{1 - \sin 2x}{\cos 2x}$

119. $2 \sin\left(\frac{\pi}{4} - \frac{x}{2}\right) \cos\left(\frac{\pi}{4} - \frac{x}{2}\right) = \cos x$

120. $\frac{\tan(x + y) - \tan y}{1 + \tan(x + y) \tan y} = \tan x$

121. $\tan 2x + \sec 2x = \frac{\cos x + \sin x}{\cos x - \sin x}$

122. $\cos^4 x - \sin^4 x = \cos 2x$

123. $2 \sin x + \sin 2x = \frac{2 \sin^3 x}{1 - \cos x}$

124. $1 + \tan x \tan(x/2) = \sec x$

125. $\tan \frac{x}{2} = \frac{1 - \cos x + \sin x}{1 + \cos x + \sin x}$

126. $\frac{\sin 3x}{\sin x} - \frac{\cos 3x}{\cos x} = 2$

127. $\sin(x + y) \cos y - \cos(x + y) \sin y = \sin x$

128. $\frac{\sin x + \sin 2x}{\cos x - \cos 2x} = \cot \frac{x}{2}$

129. $\frac{1 - \tan^2(x/2)}{1 + \tan^2(x/2)} = \cos x$

130. $4 \sin \frac{x}{4} \cos \frac{x}{4} \cos \frac{x}{2} = \sin x$

131. $\sin 4x = 4 \sin x \cos x - 8 \sin^3 x \cos x$

132. $\cos 4x = 8 \cos^4 x - 8 \cos^2 x + 1$

133. $\sin 5x = 16 \sin^5 x - 20 \sin^3 x + 5 \sin x$

134. $\cos 5x = 16 \cos^5 x - 20 \cos^3 x + 5 \cos x$

For Exercises 135–148, establish the identities by applying the sum-to-product formulas.

135. $\sin 80° - \sin 20° = \cos 50°$

136. $\sin 65° + \sin 25° = \sqrt{2} \cos 20°$

137. $\frac{\cos x - \cos 3x}{\sin x + \sin 3x} = \tan x$

138. $\frac{\sin 3° + \sin 33°}{\cos 3° + \cos 33°} = \tan 18°$

139. $\sin \frac{5\pi}{12} + \sin \frac{\pi}{12} = \frac{\sqrt{6}}{2}$

140. $\cos 10° - \sin 10° = \sqrt{2} \sin 35°$

141. $\frac{\cos 3y + \cos(2x - 3y)}{\sin 3y + \sin(2x - 3y)} = \cot x$

142. $\frac{\sin 10° - \sin 50°}{\cos 50° - \cos 10°} = \sqrt{3}$

143. $\frac{\sin 40° - \sin 20°}{\cos 20° - \cos 40°} = \frac{\sin 10° - \sin 50°}{\cos 50° - \cos 10°}$

144. $\cos x + \cos 3x + \cos 5x = \cos 3x(4 \cos^2 x - 1)$

145. $\cos 2x + \cos 4x + \cos 6x + \cos 8x$
$= 4 \cos 5x \cos 2x \cos x$

146. $\frac{\cos x + \sin x}{\cos x - \sin x} - \frac{\cos x - \sin x}{\cos x + \sin x} = 2 \tan 2x$

147. $\frac{\sin x + \sin 3x + \sin 5x}{\cos x + \cos 3x + \cos 5x} = \tan 3x$

148. $\cos \frac{\pi}{18} + \cos \frac{11\pi}{18} + \cos \frac{13\pi}{18} = 0$

149. Suppose that in triangle ABC we have $\sin A + \sin B = \cos A + \cos B$. Show that angle C is a right angle. *Suggestion:* Begin with the sum-to-product formulas.

In Exercises 150–170, find all solutions of each equation in the range $0 \le x < 2\pi$.

150. $\tan^2 x - 3 = 0$

151. $\cot^2 x - \cot x = 0$

152. $1 + \sin x = \cos x$

153. $2 \sin 3x - \sqrt{3} = 0$

154. $\sin x - \cos 2x + 1 = 0$

155. $\sin x + \sin 2x = 0$

156. $3 \csc x - 4 \sin x = 0$

157. $2 \sin^2 x + \sin x - 1 = 0$

158. $2 \sin^4 x - 3 \sin^2 x + 1 = 0$

159. $\sec^2 x - \sec x - 2 = 0$

160. $4 \cos^3 x - 3 \cos x = 0$

161. $\sin 2x - 2 \cos x = 0$

162. $\cos 4x + \cos 2x - \cos x = 0$

163. $\cos 3x + 8 \cos^3 x = 0$

164. $\sin 2x - \sin 4x + \cos 3x = 0$

165. $\sin\left(x + \frac{\pi}{15}\right) + \sin\left(x - \frac{2\pi}{45}\right) = \sin \frac{\pi}{9}$

166. $\sin^4 x + \cos^4 x = \frac{5}{8}$

167. $4 \sin^2 2x + \cos 2x - 2 \cos^2 x - 2 = 0$

168. $\cos x + \sin \frac{x}{2} = 0$

169. $\sin x - \cos x = 1$

170. $\cot x + \csc x + \sec x = \tan x$ *Suggestion:* Using sines and cosines, the given equation becomes $\cos^2 x - \sin^2 x + \cos x + \sin x = 0$. The left-hand side of this last equation can be factored.

171. If A and B both are solutions of the equation $a \cos x + b \sin x = c$, show that $\tan \frac{A + B}{2} = \frac{b}{a}$.
Hint: The given information yields two equations. After subtracting one of those equations from the other and rearranging, you will have $\frac{\cos A - \cos B}{\sin A - \sin B} = -\frac{b}{a}$. Now use the sum-to-product formulas.

172. If $a \sin x + b \cos x = a \csc x + b \sec x$, show that $\tan x = -\sqrt[3]{a/b}$.

173. Evaluate $\cos \tan^{-1} \sin \tan^{-1}(\sqrt{2}/2)$.

For Exercises 174–200, evaluate each expression (without using a calculator or tables.)

174. $\cos^{-1}\left(-\frac{\sqrt{2}}{2}\right)$

175. $\arctan\left(\frac{\sqrt{3}}{3}\right)$

176. $\sin^{-1} 0$ **177.** $\arcsin \frac{1}{2}$

178. $\arctan \sqrt{3}$

179. $\cos^{-1} \frac{1}{2}$

180. $\tan^{-1}(-1)$

181. $\cos^{-1}(-\frac{1}{2})$

182. $\arccos 0$

183. $\tan^{-1} 0$

184. $\sin(\sin^{-1} 1)$

185. $\cos[\cos^{-1}(\frac{2}{7})]$

186. $\sin[\arccos(-\frac{1}{2})]$

187. $\sin[\tan^{-1}(-1)]$

188. $\cot\left(\cos^{-1} \frac{1}{2}\right)$

189. $\sec\left(\cos^{-1} \frac{\sqrt{2}}{3}\right)$

190. $\sin\left[\frac{\pi}{2} + \sin^{-1}\left(\frac{1}{5}\right)\right]$

191. $\cos\left(\frac{\pi}{2} - \cos^{-1} \frac{1}{4}\right)$

192. $\sin\left(\frac{3\pi}{2} + \arccos \frac{3}{5}\right)$

193. $\tan\left[\frac{\pi}{4} + \sin^{-1}\left(\frac{5}{13}\right)\right]$

194. $\sin[2 \sin^{-1}(\frac{4}{5})]$

195. $\tan(2 \tan^{-1} 2)$

196. $\sin^{-1}\left(\sin \frac{\pi}{7}\right)$

197. $\cos[\frac{1}{2} \cos^{-1}(\frac{4}{5})]$

198. $\sin\left(\sin^{-1} \frac{\pi}{\pi + 1}\right)$

199. $\sin(\arctan \frac{1}{2} + \arctan \frac{1}{3})$

200. $\tan[\frac{1}{2}(\arcsin \frac{1}{3} + \arccos \frac{1}{2})]$

In Exercises 201–210, show that each equation is an identity.

201. $\tan(\tan^{-1} x + \tan^{-1} y) = \frac{x + y}{1 - xy}$

202. $\tan^{-1} \frac{x}{\sqrt{1 - x^2}} = \sin^{-1} x$

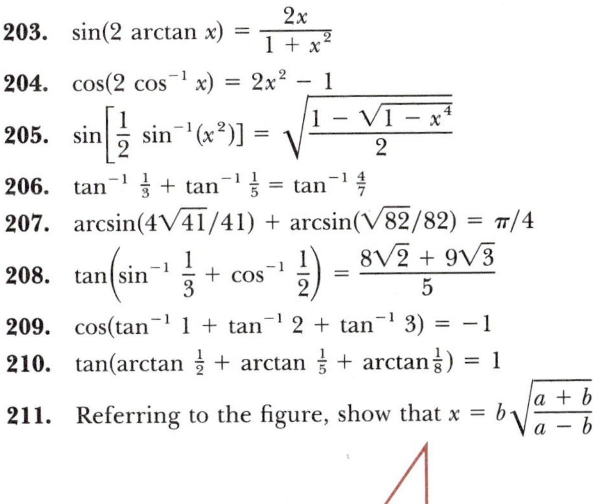

203. $\sin(2 \arctan x) = \frac{2x}{1 + x^2}$

204. $\cos(2 \cos^{-1} x) = 2x^2 - 1$

205. $\sin\left[\frac{1}{2} \sin^{-1}(x^2)\right] = \sqrt{\frac{1 - \sqrt{1 - x^4}}{2}}$

206. $\tan^{-1} \frac{1}{3} + \tan^{-1} \frac{1}{5} = \tan^{-1} \frac{4}{7}$

207. $\arcsin(4\sqrt{41}/41) + \arcsin(\sqrt{82}/82) = \pi/4$

208. $\tan\left(\sin^{-1} \frac{1}{3} + \cos^{-1} \frac{1}{2}\right) = \frac{8\sqrt{2} + 9\sqrt{3}}{5}$

209. $\cos(\tan^{-1} 1 + \tan^{-1} 2 + \tan^{-1} 3) = -1$

210. $\tan(\arctan \frac{1}{2} + \arctan \frac{1}{5} + \arctan \frac{1}{8}) = 1$

211. Referring to the figure, show that $x = b\sqrt{\frac{a + b}{a - b}}$

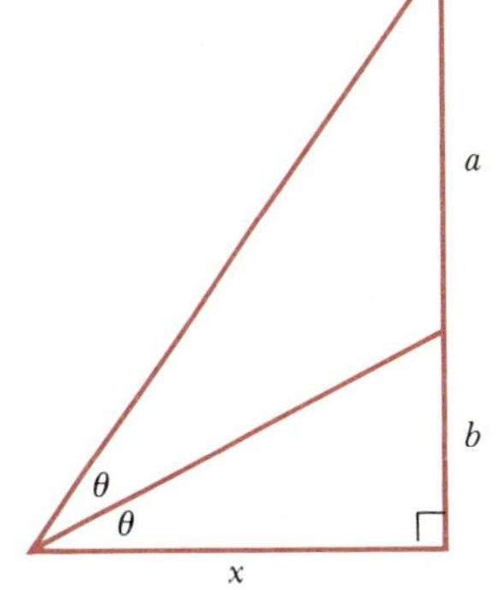

212. C **(a)** Use a calculator to compute the quantity $\cos 20° \cos 40° \cos 60° \cos 80°$. Give your answer to as many decimal places as is shown on your calculator.

(b) Now use a product-to-sum formula to *prove* that the display on your calculator is the exact value of the given expression, not an approximation.

Chapter 8 Test

1. Evaluate the following.
   (a) $\sin \frac{5\pi}{3}$ **(b)** $\cot \frac{11\pi}{6}$
2. Graph the function $y = 3 \cos 3\pi x$ over the interval $-\frac{1}{3} \le x \le \frac{1}{3}$. Specify the x-intercepts and the coordinates of the highest point on the graph.
3. Use an appropriate addition formula to simplify the expression $\sin(\theta + 3\pi/2)$.
4. Referring to the figure, compute the following.
   (a) $\cos 2\beta$ **(b)** $\tan \frac{\alpha}{2}$

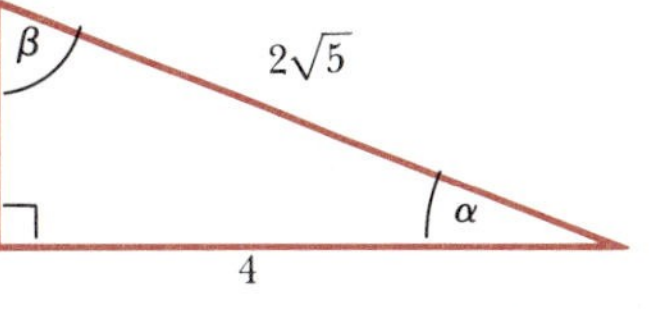

5. B and C are two points on a circle of radius $\sqrt{5}$ cm. The center of the circle is A, and angle BAC is 75°.
   (a) Find the length of the (shorter) arc of the circle joining B to C.
   (b) Find the area of the (smaller) sector determined by angle BAC.

6. In the expression $\dfrac{1}{\sqrt{4-t^2}}$, make the substitution $t = 2\cos x$ and simplify the result. Assume that $0 < x < \pi$.

7. Is an angle of $\dfrac{\pi^2}{180}$ radians larger or smaller than an angle of 3°? Give reasons for your answer. (A calculator is not needed.)

8. Find all solutions of the equation

$$2\sin^2 x + 7\sin x + 3 = 0$$

on the interval $0 \le x \le 2\pi$.

9. If $\cos\alpha = 2/\sqrt{5}$, $3\pi/2 < \alpha < 2\pi$, and $\sin\beta = \frac{4}{5}$, $\pi/2 < \beta < \pi$, compute $\sin(\beta - \alpha)$.

10. Prove that the following equation is an identity: $\cos 4\theta = 8\cos^4\theta - 8\cos^2\theta + 1$.

11. If $\sec t = -\frac{5}{3}$ and $\pi < t < 3\pi/2$, compute $\cot t$.

12. Find all solutions of the equation $\sin(x + 30°) = \sqrt{3}\sin x$ on the interval $0° < x < 90°$.

13. Graph the function $y = -\sin(2x - \pi)$ for one complete period. Specify the amplitude, the period, and the phase shift.

14. If $\csc\theta = -3$ and $\pi < \theta < 3\pi/2$, compute $\sin(\theta/2)$.

15. Simplify the expression

$$\sin 15° \cos 75° + \cos 15° \sin 75°$$

16. In the figure, P is the center of the circle. The radius of the circle is $\sqrt{2}$ units. If the radian measure of angle BPA is θ, express the area of the shaded region in terms of θ. Simplify your answer as much as possible.

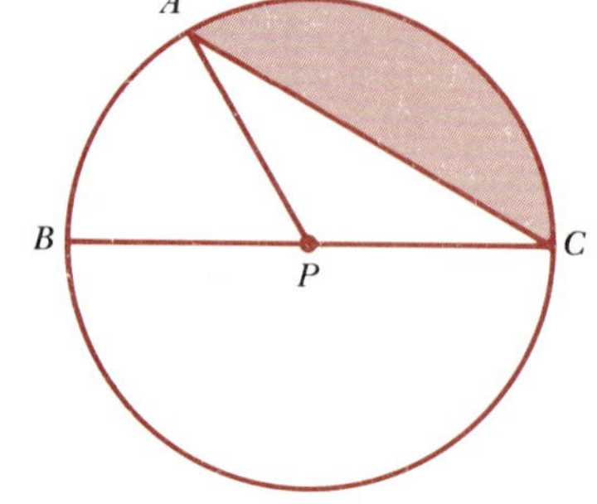

17. On the same set of axes, sketch the graphs of the restricted sine function and the function $y = \sin^{-1} x$. Specify the domain and the range for each function.

18. Compute each of the following.

(a) $\sin^{-1}\left(\sin\dfrac{\pi}{10}\right)$ **(b)** $\sin^{-1}(\sin 2\pi)$

19. Compute $\cos(\arcsin\frac{3}{4})$.

20. Prove that the following equation is an identity:

$$\tan\left(\frac{\pi}{4} + \frac{\theta}{2}\right) = \frac{1 + \cos\theta + \sin\theta}{1 + \cos\theta - \sin\theta}$$

Chapter 9 Systems of Equations

Introduction

In this chapter we consider systems of equations. Roughly speaking, a system of equations is just a collection of equations with a common set of unknowns. In solving such systems, we try to find values for the unknowns that simultaneously satisfy each of the equations in the system. In Section 9.1 we review two techniques that are often presented in intermediate algebra for solving systems involving two linear equations in two unknowns. An important technique for solving larger systems of equations is developed in Section 9.2; this technique is known as *Gaussian elimination.* After that, in Section 9.3, we introduce matrices as a tool for reducing the amount of bookkeeping involved in Gaussian elimination. Another technique for solving systems of linear equations is explained in Section 9.4; this involves determinants and Cramer's Rule. Section 9.5 contains a brief discussion of nonlinear systems of equations. In the last section of this chapter, Section 9.6, we consider systems of inequalities. As an application of this and of the earlier work in the chapter, we conclude Section 9.6 with an introduction to linear programming.

9.1 Systems of Two Linear Equations in Two Unknowns

Both in theory and in applications, it's often necessary to solve two equations in two unknowns. You may have been introduced to this topic of simultaneous equations in a previous course in algebra; however, to put matters on a firm foundation, we begin here with the basic definitions. By a *linear equation in two variables,* we mean an equation of the form

$$ax + by = c$$

where the constants a and b are not both zero. The two variables needn't always be denoted by the letters x and y, of course; it is the *form* of the equation that matters. Table 1 displays some examples.

Table 1

Equation in Two Variables	Is It Linear?	
	Yes	No
$3x - 8y = 12$	✓	
$-s + 4t = 0$	✓	
$2x - 3y^2 = 1$		✓
$y = 4 - 2x$	✓	
$\frac{4}{u} + \frac{5}{v} = 3$		✓

An ordered pair of numbers (x_0, y_0) is said to be a *solution* of the linear equation $ax + by = c$ provided that we obtain a true statement when we replace x and y in the equation by x_0 and y_0, respectively. For example, the ordered pair (3, 2) is a solution of the equation $x - y = 1$ since $3 - 2 = 1$. On the other hand, (2, 3) is not a solution of $x - y = 1$ since $2 - 3 \neq 1$.

Now consider a pair or *system* of two linear equations in two unknowns:

$$\begin{cases} ax + by = c \\ dx + ey = f \end{cases}$$

If we can find an ordered pair that is a solution to both equations, then we say that this ordered pair is a solution of the system. Sometimes, to emphasize the fact that a solution must satisfy both equations, we refer to the system as a pair of *simultaneous equations*. A system that has at least one solution is said to be *consistent*. If there are no solutions, the system is *inconsistent*.

Example 1 Consider the system $\begin{cases} x + y = 2 \\ 2x - 3y = 9 \end{cases}$

(a) Is (1, 1) a solution of the system?
(b) Is (3, −1) a solution of the system?

Solution **(a)** Although (1, 1) is a solution of the first equation, it is not a solution of the system because it does not also satisfy the second equation. (Check this for yourself.)

(b) (3, −1) satisfies the first equation:

$$3 + (-1) = 2$$

(3, −1) satisfies the second equation:

$$2(3) - 3(-1) \stackrel{?}{=} 9$$
$$6 + 3 \stackrel{?}{=} 9 \qquad \text{True}$$

Since (3, −1) satisfies both equations, it is a solution of the system.

We can gain an important perspective on systems of linear equations by looking at Example 1 in graphical terms. Table 2 shows how each of the statements in that example can be rephrased using geometric ideas with which you are already familiar.

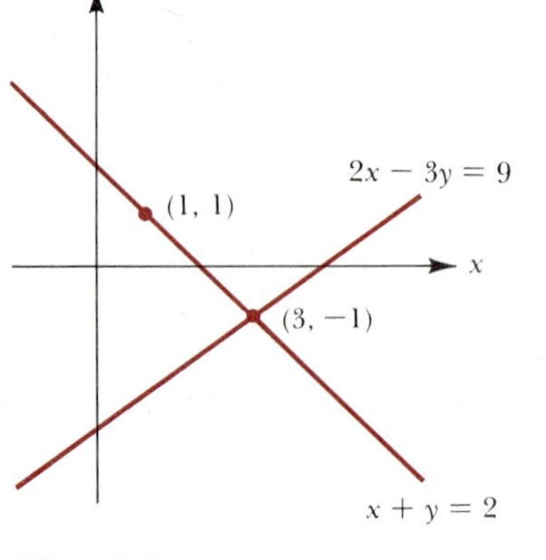

Figure 1

Table 2

Algebraic Idea	Corresponding Geometric Idea
1. The ordered pair (1, 1) is a solution of the equation $x + y = 2$.	**1.** The point (1, 1) lies on the line $x + y = 2$. See Figure 1.
2. The ordered pair (1, 1) is not a solution of the equation $2x - 3y = 9$.	**2.** The point (1, 1) does not lie on the line $2x - 3y = 9$. See Figure 1.
3. The ordered pair (3, −1) is a solution of the system $\begin{cases} x + y = 2 \\ 2x - 3y = 9 \end{cases}$	**3.** The point (3, −1) lies on both of the lines $x + y = 2$ and $2x - 3y = 9$. See Figure 1.

In Example 1 we verified that $(3, -1)$ was a solution of the system $\begin{cases} x + y = 2 \\ 2x - 3y = 9 \end{cases}$. Are there other solutions of this particular system? No; Figure 1 shows us that there are no other solutions since $(3, -1)$ is clearly the only point common to both lines. In a moment we'll look at some systematic methods for solving pairs of linear equations in two unknowns. But in fact, even before we have these methods at our disposal, we can say something about the solutions of these systems.

Property Summary

Given a system of two linear equations in two unknowns, exactly one of the following cases must occur.

Case 1 The graphs of the two linear equations intersect in exactly one point. Thus there is exactly one solution to the system. See Figure 2.

Case 2 The graphs of the two linear equations are parallel lines. Therefore the lines do not intersect, and the system has no solution. See Figure 3.

Case 3 The two equations actually represent the same line. Thus there are infinitely many points of intersection and correspondingly infinitely many solutions. See Figure 4.

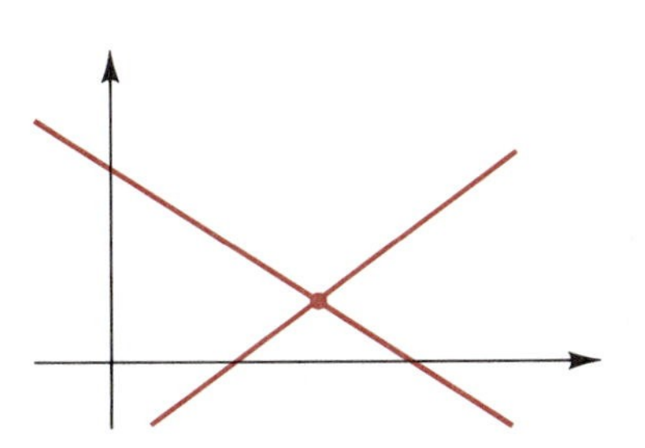

Figure 2

A consistent system with exactly one solution.

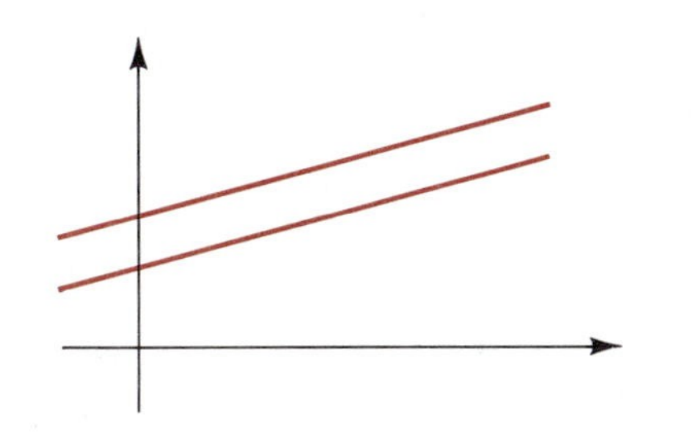

Figure 3

An inconsistent system has no solution.

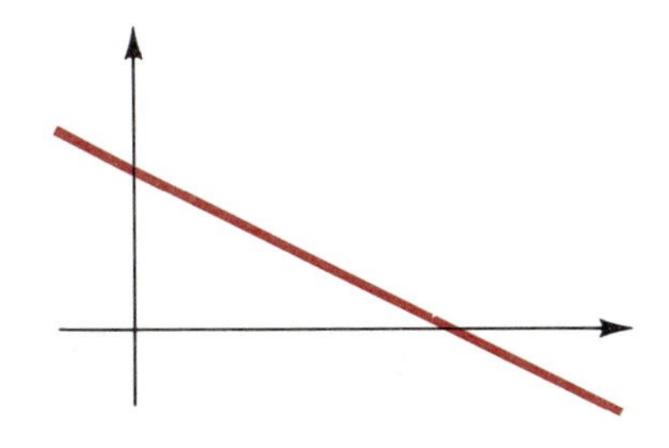

Figure 4

A consistent system with infinitely many solutions.

We are going to develop two methods in this section for solving systems of two linear equations in two unknowns. These methods are the *substitution method* and the *addition-subtraction method*. We'll begin by demonstrating the substitution method. Consider the system

$$\begin{cases} 3x + 2y = 17 & (1) \\ 4x - 5y = -8 & (2) \end{cases}$$

With the substitution method, we begin by choosing one of the two equations and then using it to express one of the variables in terms of the other. In the case at hand, neither equation appears particularly simpler than the other, so let's just start with the first equation and solve for x in terms of y. We have

$$3x = 17 - 2y$$

$$x = \frac{1}{3}(17 - 2y) \qquad (3)$$

Now we use equation (3) to substitute for x in the equation that we have not used yet, namely equation (2). This yields

$$\begin{aligned} 4\left[\frac{1}{3}(17 - 2y)\right] - 5y &= -8 \\ 4(17 - 2y) - 15y &= -24 \qquad \text{multiplying by 3} \\ -23y &= -92 \\ y &= 4 \end{aligned}$$

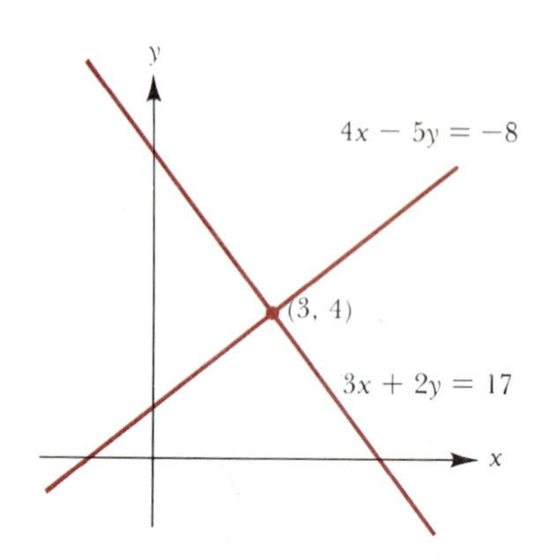

Figure 5

The value $y = 4$ that we have just obtained can be used in equation (3) to find x. Replacing y with 4 in equation (3) yields

$$x = \frac{1}{3}[17 - 2(4)] = \frac{1}{3}(9) = 3$$

We have now found that $x = 3$ and $y = 4$. As you can easily check, this pair of values indeed satisfies both of the original equations. We write our solution as the ordered pair (3, 4). Figure 5 provides a graphical summary of the situation. The figure shows that the system is consistent and that (3, 4) is the only solution.

Generally speaking, it is not necessary to graph the equations in a given system in order to decide whether the system is consistent. Rather, this information will emerge as you attempt to follow an algebraic method of solution. Examples 2 and 3 will illustrate this.

Example 2 Solve the system

$$\begin{cases} \frac{3}{2}x - 3y = -9 \\ x - 2y = 4 \end{cases}$$

Solution We use the substitution method. Since it is easy to solve the second equation for x, we begin there.

$$\begin{aligned} x - 2y &= 4 \\ x &= 4 + 2y \end{aligned}$$

Now we substitute this result in the first equation of our system to obtain

$$\begin{aligned} \frac{3}{2}(4 + 2y) - 3y &= -9 \\ \frac{3}{2}\cdot 4 + \frac{3}{2}\cdot 2y - 3y &= -9 \\ 6 + 3y - 3y &= -9 \\ 6 &= -9 \qquad \text{False} \end{aligned}$$

Since the substitution process leads us to this obviously false statement, we conclude that the given system has no solution; that is, the system is inconsistent. *Question:* What can you say about the graphs of the two given equations?

Example 3 Solve the system

$$\begin{cases} 3x + 4y = 12 \\ 2y = 6 - \frac{3}{2}x \end{cases}$$

Solution We use the method of substitution. Since it is easy to solve the second equation for y, we begin there.

$$2y = 6 - \frac{3}{2}x$$

$$y = 3 - \frac{3}{4}x$$

Now use this result to substitute for y in the first equation of the original system. The result is

$$3x + 4\left(3 - \frac{3}{4}x\right) = 12$$

$$3x + 12 - 3x = 12$$

$$3x - 3x = 12 - 12$$

$$0 = 0 \qquad \text{Always true}$$

This last identity imposes no restrictions on x. Graphically speaking, this says that our two lines intersect for every value of x. In other words, the two lines coincide. We could have foreseen this initially had we solved both equations for y. As you can verify, the result in both cases is

$$y = -\frac{3}{4}x + 3$$

Every point on this line yields a solution to our system of equations. In summary then, our system is consistent and the solutions to the system have the form $(x, -\frac{3}{4}x + 3)$, where x can be any real number. For instance, when $x = 0$, we obtain the solution $(0, 3)$. When $x = 1$, we obtain the solution $(1, -\frac{3}{4}(1) + 3)$, which simplifies to $(1, \frac{9}{4})$. The point here is that for *each* value of x, we obtain a solution; thus there are infinitely many solutions.

Now let us turn to the addition-subtraction method of solving systems of equations. By way of example, consider the system

$$\begin{cases} 2x + 3y = 5 \\ 4x - 3y = -1 \end{cases}$$

Notice that if we add these two equations, the result will be an equation involving only the unknown x. Carrying this out gives us

$$(2x + 3y) + (4x - 3y) = 5 + (-1)$$

$$6x = 4$$

$$x = \frac{4}{6} = \frac{2}{3}$$

There are several ways now in which the corresponding value of y may be obtained. As you can easily check, substituting the value $x = \frac{2}{3}$ in either the first equation or the second equation leads us to the result $y = \frac{11}{9}$.

Another way to find y is as follows. Multiply both sides of the first equation by -2. (You'll see why in a moment.) We display the work this way:

$$2x + 3y = 5 \xrightarrow{\text{multiply by } -2} -4x - 6y = -10$$

$$4x - 3y = -1 \xrightarrow{\text{no change}} 4x - 3y = -1$$

Adding the last two equations then gives us

$$(-4x - 6y) + (4x - 3y) = -10 + (-1)$$
$$-9y = -11$$
$$y = \frac{11}{9}$$

The required solution therefore is $(\frac{2}{3}, \frac{11}{9})$.

In the previous example, we were able to find x directly by adding the two equations. As the next example shows, it may be necessary to first multiply both sides of each equation by an appropriate constant.

Example 4 Solve the system
$$\begin{cases} 5x - 3y = 4 \\ 2x + 4y = 1 \end{cases}$$

Solution To eliminate x, we could multiply the second equation by $\frac{5}{2}$ and then subtract the resulting equation from the first equation. However, to avoid working with fractions, we proceed as follows.

$$5x - 3y = 4 \xrightarrow{\text{multiply by 2}} 10x - 6y = 8$$
$$2x + 4y = 1 \xrightarrow{\text{multiply by 5}} 10x + 20y = 5$$

Subtracting the last two equations then yields

$$(10x - 6y) - (10x + 20y) = 8 - 5$$
$$-26y = 3$$
$$y = -\frac{3}{26}$$

Now to find x, we return to the original system and work similarly.

$$5x - 3y = 4 \xrightarrow{\text{multiply by 4}} 20x - 12y = 16$$
$$2x + 4y = 1 \xrightarrow{\text{multiply by 3}} 6x + 12y = 3$$

Upon adding the last two equations we obtain

$$26x = 19$$
$$x = \frac{19}{26}$$

The solution of the given system of equations is therefore $(\frac{19}{26}, -\frac{3}{26})$.

We conclude this section with some examples of problems that can be solved by using simultaneous equations.

Example 5 Determine the constants b and c so that the parabola $y = x^2 + bx + c$ passes through the points $(-3, 1)$ and $(1, -2)$.

Solution Since the point $(-3, 1)$ lies on the curve $y = x^2 + bx + c$, the coordinates must satisfy the equation. Thus we have

$$1 = (-3)^2 + b(-3) + c$$
$$-8 = -3b + c \qquad (4)$$

This gives us one equation in two unknowns. We need another equation involving b and c. Since the point $(1, -2)$ lies on the graph of

$$y = x^2 + bx + c$$

we must have

$$-2 = 1^2 + b(1) + c$$

$$-3 = b + c \qquad (5)$$

Rewriting equations (4) and (5), we have the system

$$\begin{cases} -3b + c = -8 & (6) \\ \quad b + c = -3 & (7) \end{cases}$$

Subtracting equation (7) from (6) then yields

$$-4b = -5$$

$$b = \frac{5}{4}$$

One way to obtain the corresponding value of c now is just to take equation (7) and replace b by $\frac{5}{4}$. This yields

$$\frac{5}{4} + c = -3$$

$$c = -3 - \frac{5}{4} = -\frac{12}{4} - \frac{5}{4}$$

$$c = -\frac{17}{4}$$

The required values of b and c are therefore

$$b = \frac{5}{4} \qquad c = -\frac{17}{4}$$

(Exercise 54 at the end of this section asks you to check that the parabola $y = x^2 + \frac{5}{4}x - \frac{17}{4}$ indeed passes through the given points.)

Example 6 Find the area of the triangular region bounded by the lines $y = 3x$, $y = -\frac{1}{2}x + 7$, and the x-axis. See Figure 6.

Solution To compute the area of the triangle we will need to know the base and the height. To determine the base we compute the x-intercept of the line $y = -\frac{1}{2}x + 7$:

$$0 = -\frac{1}{2}x + 7$$

$$\frac{1}{2}x = 7$$

$$x = 14$$

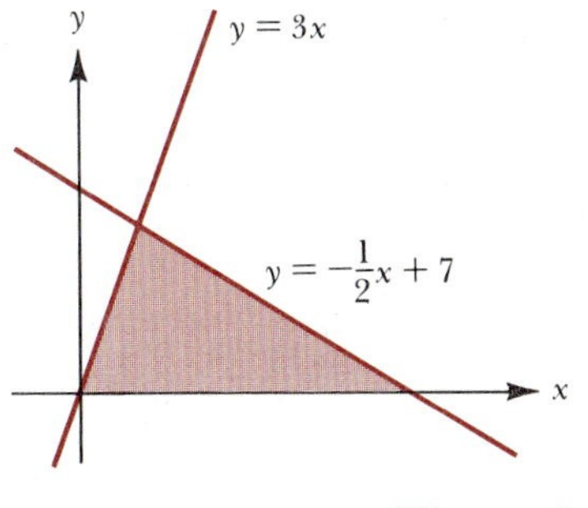

Figure 6

Now the height of the triangle is the y-coordinate of the point where the lines $y = 3x$ and $y = -\frac{1}{2}x + 7$ intersect. (Look at Figure 6.) The substitution method can be used to solve this pair of simultaneous equations for y. From the equation $y = 3x$ we obtain $x = \frac{1}{3}y$. Then substituting this in the equation $y = -\frac{1}{2}x + 7$ yields

$$y = -\frac{1}{2}\left(\frac{1}{3}y\right) + 7$$

$$y = -\frac{1}{6}y + 7$$

$$6y = -y + 42$$

$$7y = 42$$

$$y = 6$$

Now that we know the base is 14 and the height is 6, we can compute the required area.

$$\begin{aligned} A &= \frac{1}{2}bh \\ &= \frac{1}{2}(14)(6) \\ &= 42 \text{ square units} \end{aligned}$$

In the next example we solve a mixture problem using a system of two equations in two unknowns. (In Section 2.3 we solved this same problem using one equation and one unknown.)

Example 7 Suppose that a chemistry student can obtain two acid solutions from the stockroom. The first solution is 20% acid and the second solution is 45% acid. (The percents are by volume.) How many cm^3 of each solution should the student mix together to obtain 100 cm^3 of a 30% acid solution?

Solution Begin by assigning letters to denote the required quantities.

Let x denote the number of cm^3 of the 20% solution to be used.

Let y denote the number of cm^3 of the 45% solution to be used.

Then the data can be summarized as follows.

Type of Solution	Number of cm^3	Percent of Acid	Total Acid (cm^3)
First solution (20% acid)	x	20	$(0.20)x$
Second solution (45% acid)	y	45	$(0.45)y$
Mixture	$x + y$	30	$(0.30)(x + y)$

Since the final mixture is to total 100 cm^3, we have the equation

$$x + y = 100 \tag{8}$$

This gives one equation in two unknowns. However, we need a second equation. Looking at the data in the right-hand column of the table, we may write

$$\underbrace{0.20x}_{\substack{\text{Amount of acid} \\ \text{in } x \text{ cm}^3 \text{ of the} \\ 20\% \text{ solution.}}} + \underbrace{0.45y}_{\substack{\text{Amount of acid} \\ \text{in } y \text{ cm}^3 \text{ of the} \\ 45\% \text{ solution.}}} = \underbrace{(0.30)(x + y)}_{\substack{\text{Amount of acid in} \\ \text{the final mixture.}}}$$

Thus

$$\begin{aligned} 0.20x + 0.45y &= 0.30(x + y) \\ 20x + 45y &= 30(x + y) \\ 4x + 9y &= 6(x + y) \\ 4x + 9y &= 6x + 6y \\ -2x + 3y &= 0 \end{aligned} \tag{9}$$

Equations (8) and (9) may be solved using either the substitution method or the addition method. As Exercise 55 at the end of this section asks you to show, the results are

$$x = 60 \text{ cm}^3 \quad \text{and} \quad y = 40 \text{ cm}^3$$

These are the required values.

In the next example neither of the given equations is linear. By using appropriate substitutions, however, we can reduce the problem (in this particular case) to one involving linear equations.

Example 8 Solve the system

$$\begin{cases} \dfrac{2}{u} - \dfrac{5}{v} = -2 \\ \dfrac{3}{u} + \dfrac{4}{v} = 5 \end{cases}$$

Note that neither of these equations is linear.

Solution Rewrite the system this way:

$$\begin{cases} 2\left(\dfrac{1}{u}\right) - 5\left(\dfrac{1}{v}\right) = -2 \\ 3\left(\dfrac{1}{u}\right) + 4\left(\dfrac{1}{v}\right) = 5 \end{cases}$$

Now make the substitutions $1/u = x$ and $1/v = y$. The system becomes

$$\begin{cases} 2x - 5y = -2 \\ 3x + 4y = 5 \end{cases}$$

This last system can be solved using either of the methods of this section. As Exercise 56 at the end of this section asks you to show, the solutions are

$$x = \frac{17}{23} \quad \text{and} \quad y = \frac{16}{23}$$

Since by definition x and y are the reciprocals of u and v, respectively, we then have

$$u = \frac{23}{17} \quad \text{and} \quad v = \frac{23}{16}$$

These are the required values for u and v.

Exercise Set 9.1

1. Which of the following are linear equations in two variables?
   (a) $3x + 3y = 10$ (b) $2x + 4xy + 3y = 1$
   (c) $u - v = 1$ (d) $x = 2y + 6$

2. Which of the following are linear equations in two variables?
   (a) $y = x$ (b) $y = x^2$
   (c) $\frac{4}{x} - \frac{3}{y} = -1$ (d) $2w + 8z = -4w + 3$

3. Is (5, 1) a solution of the following system?
$$\begin{cases} 2x - 8y = 2 \\ 3x + 7y = 22 \end{cases}$$

4. Is (14, −2) a solution of the following system?
$$\begin{cases} x + y = 12 \\ x - y = 4 \end{cases}$$

5. Is (0, −4) a solution of the following system?
$$\begin{aligned} \tfrac{1}{6}x + \tfrac{1}{2}y &= -2 \\ \tfrac{2}{3}x + \tfrac{3}{4}y &= 2 \end{aligned}$$

6. Is (12, −8) a solution of the system in Exercise 5?

7. Is (3, −2) a solution of the following system?
$$\begin{aligned} \tfrac{2}{7}x - \tfrac{1}{5}y &= \tfrac{44}{35} \\ \tfrac{1}{3}x - \tfrac{5}{4}y &= \tfrac{7}{2} \end{aligned}$$

In Exercises 8–18, use the substitution method to find all solutions of each system.

8. $\begin{cases} 4x - y = 7 \\ -2x + 3y = 9 \end{cases}$

9. $\begin{cases} 3x - 2y = -19 \\ x + 4y = -4 \end{cases}$

10. $\begin{cases} 6x - 2y = -3 \\ 5x + 3y = 4 \end{cases}$

11. $\begin{cases} 4x + 2y = 3 \\ 10x + 4y = 1 \end{cases}$

12. $\begin{cases} \frac{3}{2}x - 5y = 1 \\ x + \frac{3}{4}y = -1 \end{cases}$

13. $\begin{cases} 13x - 8y = -3 \\ -7x + 2y = 0 \end{cases}$

14. $\begin{cases} 4x + 6y = 3 \\ -6x - 9y = -\frac{9}{2} \end{cases}$

15. $\begin{cases} -\frac{2}{5}x + \frac{1}{4}y = 3 \\ \frac{1}{4}x - \frac{2}{5}y = -3 \end{cases}$

16. $\begin{cases} 0.02x - 0.03y = 1.06 \\ 0.75x + 0.50y = -0.01 \end{cases}$

17. $\begin{cases} \sqrt{2}x - \sqrt{3}y = \sqrt{3} \\ \sqrt{3}x - \sqrt{8}y = \sqrt{2} \end{cases}$

18. $\begin{cases} 7x - 3y = -12 \\ \frac{14}{3}x - 2y = 2 \end{cases}$

In Exercises 19–28, use the addition-subtraction method to find all solutions of each system of equations.

19. $\begin{cases} 5x + 6y = 4 \\ 2x - 3y = -3 \end{cases}$

20. $\begin{cases} -8x + y = -2 \\ 4x - 3y = 1 \end{cases}$

21. $\begin{cases} 4x + 13y = -5 \\ 2x - 54y = -1 \end{cases}$

22. $\begin{cases} 16x - 3y = 100 \\ 16x + 10y = 10 \end{cases}$

23. $\begin{cases} \frac{1}{4}x - \frac{1}{3}y = 4 \\ \frac{2}{7}x - \frac{1}{7}y = \frac{1}{10} \end{cases}$

    Suggestion: First clear both equations of fractions.

24. $\begin{cases} 2.1x - 3.5y = 1.2 \\ 1.4x + 2.6y = 1.1 \end{cases}$

25. $\begin{cases} 8x + 16y = 5 \\ 2x + 5y = \frac{5}{4} \end{cases}$

26. $\begin{cases} 8x + 16y = 5 \\ 2x + 4y = 1 \end{cases}$

27. $\begin{cases} \sqrt{6}x - \sqrt{3}y = 3\sqrt{2} - \sqrt{3} \\ \sqrt{2}x - \sqrt{5}y = \sqrt{6} + \sqrt{5} \end{cases}$

28. $\begin{cases} 125x - 40y = 45 \\ \frac{1}{10}x + \frac{1}{10}y = \frac{3}{10} \end{cases}$

29. Find b and c given that the parabola $y = x^2 + bx + c$ passes through (0, 4) and (2, 14).

30. Determine the constants a and b given that the parabola $y = ax^2 + bx + 1$ passes through (−1, 11) and (3, 1).

31. Determine the constants A and B given that the line $Ax + By = 2$ passes through the points (−4, 5) and (7, −9).

32. Find the area of the triangular region in the first quadrant bounded by the lines $y = 5x$, $y = -3x + 6$, and the x-axis.

33. Find the area of the triangular region in the first quadrant bounded by the x-axis and the lines $y = 2x - 5$ and $y = -\frac{1}{2}x + 3$.

34. Find the area of the triangular region in the first quadrant bounded by the y-axis and the lines $y = 2x + 2$ and $y = -\frac{3}{2}x + 9$.

35. A student in a chemistry laboratory has available to her two acid solutions. The first solution is 10% acid and the second is 35% acid. (The percents are by volume.) How many cm^3 of each should she mix together to obtain 200 cm^3 of a 25% acid solution?

36. One salt solution is 15% salt and another is 20% salt. How many cm^3 of each solution must be mixed to obtain 50 cm^3 of a 16% salt solution?

37. A shopkeeper has two types of coffee beans on hand. One type sells for \$5.20 a pound, the other for \$5.80 a pound. How many pounds of each type must be mixed to produce 16 lb of a blend that sells for \$5.50 a pound?

38. A certain alloy contains 10% tin and 30% copper. (The percents are by weight.) How many pounds of tin and how many pounds of copper must be melted with 1000 lb of the given alloy to yield a new alloy containing 20% tin and 35% copper?

39. Find x and y in terms of a and b:

$$\begin{cases} \dfrac{x}{a} + \dfrac{y}{b} = 1 \\ \dfrac{x}{b} + \dfrac{y}{a} = 1 \end{cases}$$

Does your solution impose any conditions upon a and b?

40. Solve the following system for x and y in terms of a and b.

$$\begin{cases} ax + by = \dfrac{1}{a} \\ b^2x + a^2y = 1 \end{cases} \qquad a \neq b$$

41. Solve the following system for x and y in terms of a and b.

$$\begin{cases} ax + a^2y = 1 \\ bx + b^2y = 1 \end{cases} \qquad a \neq b$$

Does your solution impose any additional conditions upon a and b?

42. Solve the following system for s and t.

$$\begin{cases} \dfrac{3}{s} - \dfrac{4}{t} = 2 \\ \dfrac{5}{s} + \dfrac{1}{t} = -3 \end{cases}$$

43. Solve the following system.

$$\begin{cases} \dfrac{5}{x} - \dfrac{5}{y} = \dfrac{1}{3} \\ \dfrac{2}{x} - \dfrac{3}{y} = 4 \end{cases}$$

44. Find all solutions of the following system.

$$\begin{cases} -\dfrac{4}{x} + \dfrac{3}{y} = 12 \\ -\dfrac{8}{x} - \dfrac{9}{y} = -1 \end{cases}$$

In Exercises 45–49, find all solutions of the given systems. (Use either of the methods of this section.)

45. $$\begin{cases} \dfrac{2w-1}{3} + \dfrac{z+2}{4} = 4 \\ \dfrac{w+3}{2} - \dfrac{w-z}{3} = 3 \end{cases}$$

46. $$\begin{cases} 2x = 7 + 3y \\ x = \dfrac{5-y}{3} \end{cases}$$

47. $$\begin{cases} \dfrac{1}{2u} - \dfrac{1}{2v} = -10 \\ \dfrac{2}{u} + \dfrac{3}{v} = 5 \end{cases}$$

48. $$\begin{cases} 0.5x - 0.8y = 0.3 \\ 0.4x - 0.1y = 0.9 \end{cases}$$

49. $\dfrac{x-y}{2} = \dfrac{x+y}{3} = 1$

50. The sum of two numbers is 64. Twice the larger plus five times the smaller is 20. Find the two numbers. (Let x denote the larger and let y denote the smaller.)

51. In a two-digit number, the sum of the digits is 14. Twice the tens' digit exceeds the units' digit by one. Find the number.

52. The sum of the digits in a two-digit number is 14. Furthermore, the number itself is two greater than 11 times the tens' digit. Find the number.

53. The perimeter of a rectangle is 34 in. If the length is 2 in. more than twice the width, find the length and the width.

54. Verify that the parabola $y = x^2 + \frac{5}{4}x - \frac{17}{4}$ indeed passes through $(-3, 1)$ and $(1, -2)$.

55. Consider the system $\begin{cases} x + y = 100 \\ -2x + 3y = 0. \end{cases}$

(a) Solve this system using the method of substitution. *Answer:* $(60, 40)$

(b) Solve the system using the addition-subtraction method.

56. Consider the system $\begin{cases} 2x - 5y = -2 \\ 3x + 4y = 5. \end{cases}$

(a) Solve the system using the substitution method.

(b) Solve the system using the addition-subtraction method. *Answer:* $(\frac{17}{23}, \frac{16}{23})$

(c) Verify that your solution indeed satisfies both equations of the system.

57. Solve the following system for x and y.

$$\begin{cases} by = x + ab \\ cy = x + ac \end{cases}$$

(Assume that a, b, and c are constants and that $b \neq c$.)

***58.** Given that the lines $7x + 5y = 4$, $x + ky = 3$, and $5x + y + k = 0$ are concurrent (pass through a common point), what are the possible values for k?

Answer: 2 and $\frac{-29}{7}$

59. Solve the following system for x and y in terms of a and b.

$$\begin{cases} \dfrac{a}{bx} + \dfrac{b}{ay} = a + b \\ \dfrac{b}{x} + \dfrac{a}{y} = a^2 + b^2 \end{cases} \qquad a \neq b$$

***60.** Solve the following system for x and y in terms of a and b.

$$\begin{cases} \dfrac{x+y-1}{x-y+1} = a \\ \dfrac{y-x+1}{x-y+1} = ab \end{cases} \qquad ab \neq -1$$

Answer: $\left(\dfrac{a+1}{ab+1}, \dfrac{a(b+1)}{ab+1}\right)$

***61.** Solve the following system for x and y in terms of a and b. Assume a and b are nonzero and $a \neq \pm b$.

$$\begin{cases} (a+b)x + (a^2+b^2)y = a^3+b^3 \\ (a-b)x + (a^2-b^2)y = a^3-b^3 \end{cases}$$

***62.** The vertices of the triangle ABC are $A(2a, 0)$, $B(2b, 0)$, and $C(0, 2)$.

(a) Show that the equations of the sides are $x + by = 2b$, $x + ay = 2a$, and $y = 0$.

(b) Show that the equations of the medians are $x + (2a - b)y = 2a$, $x + (2b - a)y = 2b$, and $2x + (a + b)y = 2(a + b)$.

(c) Show that all three medians intersect at the point $\left(\frac{2(a+b)}{3}, \frac{2}{3}\right)$.

(d) Show that the equations of the altitudes are $bx - y = 2ab$, $ax - y = 2ab$, and $x = 0$.

(e) Show that all three altitudes intersect at the point $(0, -2ab)$.

(f) Show that the equations of the perpendicular bisectors of the sides are $x = a + b$, $bx - y = b^2 - 1$, and $ax - y = a^2 - 1$.

(g) Show that all three perpendicular bisectors intersect at the point $(a + b, ab + 1)$.

(h) Show that the three points found in parts (c), (e), and (g) all lie on a line. This is the so-called *Euler line* of the triangle.

63. Solve for x and y in terms of a, b, c, d, e, and f:

$$\begin{cases} ax + by = c \\ dx + ey = f \end{cases}$$

(Assume that $ae - bd \neq 0$.)

9.2 Gaussian Elimination

A method of solution is perfect if we can foresee from the start, and even prove, that following that method we shall obtain our aim.

Gottfried Wilhelm von Leibniz (1646–1716)

In the previous section we solved systems of linear equations in two unknowns. In this section we introduce the technique known as *Gaussian elimination** for solving systems of linear equations in which there are more than two unknowns.

As a first example, consider the following system of three linear equations in the three unknowns x, y, and z.

$$\begin{cases} 3x + 2y - z = -3 \\ \quad\;\; 5y - 2z = 2 \\ \qquad\quad\;\; 5z = 20 \end{cases}$$

This system is easy to solve. Dividing both sides of the third equation by 5 yields $z = 4$. Then substituting $z = 4$ back into the second equation gives us

$$\begin{aligned} 5y - 2(4) &= 2 \\ 5y &= 10 \\ y &= 2 \end{aligned}$$

Finally, substituting the values $z = 4$ and $y = 2$ back into the first equation yields

$$\begin{aligned} 3x + 2(2) - 4 &= -3 \\ 3x &= -3 \\ x &= -1 \end{aligned}$$

*The technique is named in honor of Carl Friedrich Gauss (1777–1855). However, the technique itself was in existence before Gauss' time. Indeed, the essentials of the method appear in a Chinese text, "Nine Chapters on the Mathematical Art" (*Chiu Chang Suan Shu*) written approximately two thousand years ago.

We have now found that $x = -1$, $y = 2$, and $z = 4$. If you go back and check, you will find that these values indeed satisfy all three equations in the given system. Furthermore, the algebra we've just carried out shows that these are the only possible values for x, y, and z satisfying all three equations. We summarize by saying that the *ordered triple* $(-1, 2, 4)$ is the solution of the given system.

The system we have just considered is easy to solve because of the particular form in which it is written. This form is called *echelon form*. Because the equations were in echelon form, we were able to solve the third equation for z directly. Then we substituted back into the second equation to obtain y. Lastly, we substituted back into the first equation to obtain x. The definition of echelon form for a system with three unknowns is given below. Although this definition of echelon form refers to systems with three unknowns, the same type of definition can be given for systems with any number of unknowns. Table 1 displays examples of systems that are in echelon form.

Echelon Form (Three Variables)

A system of linear equations in x, y, and z is said to be in echelon form provided x appears in no equation after the first and y appears in no equation after the second. (It is possible that y may not even appear in the second equation.)

Table 1
Examples of Systems in Echelon Form

2 Unknowns: x, y	3 Unknowns: x, y, z	4 Unknowns: x, y, z, t
$\begin{cases} 3x + 5y = 7 \\ 8y = 5 \end{cases}$	$\begin{cases} 4x - 3y + 2z = -5 \\ 7y + z = 9 \\ -4z = 3 \end{cases}$	$\begin{cases} x - y + z - 4t = 1 \\ 3y - 2z + t = -1 \\ 3z - 5t = 4 \\ 6t = 7 \end{cases}$
	$\begin{cases} 15x - 2y + z = 1 \\ 3z = -8 \end{cases}$	$\begin{cases} 2x + y + 2z - t = -3 \\ 4z + 3t = 1 \\ 5t = 6 \end{cases}$
		$\begin{cases} 8x + 3y - z + t = 2 \\ 2y + z - 4t = 1 \end{cases}$

When we solved linear systems of equations with two unknowns in the previous section, we observed that there were three possibilities: a unique solution; infinitely many solutions; no solution. The next three examples indicate that the situation is similar when dealing with larger systems. Notice as you read that the systems in Examples 1 and 2 are presented in echelon form, but the system in Example 3 is not in echelon form.

Example 1 Find all solutions of the system

$$\begin{cases} x + y + 2z = 2 \\ 3y - 4z = -5 \\ 6z = 3 \end{cases}$$

Solution Dividing the third equation by 6 yields $z = \frac{1}{2}$. Substituting this value for z back into the second equation then yields

$$3y - 4\left(\frac{1}{2}\right) = -5$$
$$3y = -3$$
$$y = -1$$

Now substituting the values $z = \frac{1}{2}$ and $y = -1$ back in the first equation, we obtain

$$x + (-1) + 2\left(\frac{1}{2}\right) = 2$$
$$x = 2$$

As you can easily check, the values $x = 2$, $y = -1$, and $z = \frac{1}{2}$ indeed satisfy all three equations. Furthermore, the algebra we've just carried out shows that these are the only possible values for x, y, and z satisfying all three equations. We summarize by saying that the unique solution to our system is the ordered triple $(2, -1, \frac{1}{2})$.

Example 2 Find all solutions of the system

$$\begin{cases} -2x + y + 3z = 6 \\ 2z = 10 \end{cases}$$

Solution Solving the second equation for z yields $z = 5$. Then replacing z by 5 in the first equation gives us

$$-2x + y + 3(5) = 6$$
$$-2x + y = -9$$
$$y = 2x - 9$$

At this point we've made use of both equations in the given system. There is no third equation to provide additional restrictions on x, y, or z. Now we know from Section 9.1 that the equation $y = 2x - 9$ has infinitely many solutions, all of the form

$$(x, 2x - 9) \qquad \text{where } x \text{ is a real number}$$

It follows then that there are infinitely many solutions to the given system and that they may be written as

$$(x, 2x - 9, 5) \qquad \text{where } x \text{ is a real number}$$

For instance, choosing in succession $x = 0$, $x = 1$, and $x = 2$ yields the solutions $(0, -9, 5)$, $(1, -7, 5)$, and $(2, -5, 5)$. (We remark in passing that any linear system in echelon form in which the number of unknowns exceeds the number of equations will always have infinitely many solutions.)

Example 3 Find all solutions of the system

$$\begin{cases} 4x - 7y + 3z = 1 \\ 3x + y - 2z = 4 \\ 4x - 7y + 3z = 6 \end{cases}$$

(Note that the system is not in echelon form.)

Solution Look at the left-hand sides of the first and the third equations. They are identical. Thus if there were values for x, y, and z that satisfied both equations, it would follow that $1 = 6$, which is clearly impossible. We conclude that the given system has no solutions.

As Examples 1 and 2 demonstrate, systems in echelon form can be readily solved. In view of this, it would be useful to have a technique for converting a given system into an equivalent system in echelon form. (The expression "equivalent system" here means a system with exactly the same set of solutions as the original system.) *Gaussian elimination* is one such technique. We'll demonstrate this technique in Examples 4, 5, and 6. In using Gaussian elimination, we rely on the three so-called *elementary operations* that are listed in the box that follows. These are operations that can be performed on an equation in a system without altering the set of solutions of the system.

The Elementary Operations

1. Multiply both sides of an equation by a nonzero constant.
2. Interchange the order in which two equations of a system are listed.
3. To one equation add a multiple of another equation in the system.

Example 4 Find all solutions of the system

$$\begin{cases} x + 2y + z = 3 \\ 2x + y + z = 16 \\ x + y + 2z = 9 \end{cases}$$

Solution First we want to eliminate x from the second and third equations. To eliminate x from the second equation, add to it -2 times the first equation. The result is the equivalent system

$$\begin{cases} x + 2y + z = 3 \\ -3y - z = 10 \\ x + y + 2z = 9 \end{cases}$$

To eliminate x from the third equation, add to it -1 times the first equation. The result is the equivalent system

$$\begin{cases} x + 2y + z = 3 \\ -3y - z = 10 \\ -y + z = 6 \end{cases}$$

Now to bring the system into echelon form, we need to eliminate y from the third equation. We could (but won't) do this by adding to the third equation $-\frac{1}{3}$ times the second equation. However, to avoid working with fractions as long as possible, we proceed instead to interchange the second and third equations to obtain the equivalent system

$$\begin{cases} x + 2y + z = 3 \\ -y + z = 6 \\ -3y - z = 10 \end{cases}$$

Now add to the last equation -3 times the second to obtain the equivalent system

$$\begin{cases} x + 2y + z = 3 \\ -y + z = 6 \\ -4z = -8 \end{cases}$$

The system is now in echelon form. From the third equation we obtain $z = 2$. Substituting this value back into the second equation then gives us

$$-y + 2 = 6$$
$$y = -4$$

Finally, substituting $z = 2$ and $y = -4$ into the first equation yields $x = 9$. (Check this for yourself.) The required solution is therefore $(9, -4, 2)$.

In Example 5, which follows, it will be convenient to shorten certain expressions used in the explanations. For instance, an expression such as "to the first equation add -1 times the second equation" can be shortened to "subtract the second equation from the first."

Example 5 Solve the system
$$\begin{cases} 4x - 3y + 2z = 40 \\ 5x + 9y - 7z = 47 \\ 9x + 8y - 3z = 97 \end{cases}$$

Solution

Subtract the second equation from the first.
$$\longrightarrow \begin{cases} -x - 12y + 9z = -7 \\ 5x + 9y - 7z = 47 \\ 9x + 8y - 3z = 97 \end{cases}$$

Add 5 times the first equation to the second. Add 9 times the first to the third.
$$\longrightarrow \begin{cases} -x - 12y + 9z = -7 \\ -51y + 38z = 12 \\ -100y + 78z = 34 \end{cases} \longrightarrow$$

Divide the third equation by 2.
$$\longrightarrow \begin{cases} -x - 12y + 9z = -7 \\ -51y + 38z = 12 \\ -50y + 39z = 17 \end{cases}$$

Subtract the third equation from the second. (The motivation for this step is the desire to work with smaller but integral coefficients.)
$$\longrightarrow \begin{cases} -x - 12y + 9z = -7 \\ -y - z = -5 \\ -50y + 39z = 17 \end{cases} \longrightarrow$$

Subtract 50 times the second equation from the third.
$$\longrightarrow \begin{cases} -x - 12y + 9z = -7 \\ -y - z = -5 \\ 89z = 267 \end{cases}$$

The system is now in echelon form. Solving the third equation we obtain $z = 3$. Substituting this value back into the second yields $y = 2$. (Check this for yourself.) Finally, substituting $z = 3$ and $y = 2$ back into the first equation yields $x = 10$. (Again, check this for yourself.) The solution to the system is therefore $(10, 2, 3)$.

Example 6 Solve the system

$$\begin{cases} x + 2y + 4z = 0 \\ x + 3y + 9z = 0 \end{cases}$$

Solution This system is similar to the one in Example 2 in that there are fewer equations than there are unknowns. By subtracting the first equation from the second, we readily obtain an equivalent system in echelon form:

$$\begin{cases} x + 2y + 4z = 0 & (1) \\ y + 5z = 0 & (2) \end{cases}$$

Although the system is now in echelon form, notice that equation (2) does not determine y or z uniquely; that is, there are infinitely many number pairs (y, z) satisfying equation (2). We can solve equation (2) for y in terms of z; the result is $y = -5z$. Now we replace y with $-5z$ in equation (1) to obtain

$$x + 2(-5z) + 4z = 0$$

or

$$x = 6z$$

At this point, we've used both of the equations in the system to express x and y in terms of z. Furthermore, there is no third equation in the system to provide additional restrictions on x, y, or z. We conclude therefore that the given system has infinitely many solutions. These solutions have the form

$$(6z, -5z, z) \qquad \text{where } z \text{ can be any real number}$$

In Examples 4 through 6, we have used Gaussian elimination to solve the given systems. We've emphasized this technique because it is systematic and because it readily lends itself to computer solutions. It should be pointed out, however, that there are other techniques you can use. In Section 9.4 for instance, we'll use a technique known as *Cramer's Rule* to solve certain systems. In the next example also, we solve a system without first reducing it to echelon form.

Example 7 Find all solutions of the system

$$\begin{cases} x + y + z = 26 \\ x - y \quad\;\; = 4 \\ x \quad\;\; - z = 6 \end{cases}$$

Solution Adding all three equations yields

$$3x = 36$$
$$x = 12$$

Substituting $x = 12$ back into the third equation yields

$$12 - z = 6$$
$$z = 6$$

Similarly, substituting $x = 12$ back into the second equation yields

$$12 - y = 4$$
$$y = 8$$

Thus the solution of the given system is (12, 8, 6).

In the next two examples we introduce the subject of *partial fractions.* The basic goal here is to write a given fractional expression as a sum or difference of two or more *simpler* fractions. For instance, in Example 8 we will be given the fraction $\dfrac{1}{(x-1)(x+1)}$ and we will be asked to find constants A and B so that

$$\frac{1}{(x-1)(x+1)} = \frac{A}{x-1} + \frac{B}{x+1}$$

When A and B are determined, we refer to the right-hand side of the above equation as the *partial fraction decomposition* of the given fraction. Detailed techniques for working with partial fractions are developed as the need arises in calculus courses. The two examples and the exercises that follow will provide you with ample background for that later work.

Example 8 Determine constants A and B so that the following equation is an identity. (That is, the equation is to hold for all values of x for which the denominators are not zero.)

$$\frac{1}{(x-1)(x+1)} = \frac{A}{x-1} + \frac{B}{x+1} \tag{3}$$

Solution First, to clear equation (3) of fractions, multiply both sides of the equation by the quantity $(x-1)(x+1)$. This yields

$$\begin{aligned}(x-1)(x+1)\cdot\frac{1}{(x-1)(x+1)}& \\ &= (x-1)(x+1)\cdot\frac{A}{x-1} + (x-1)(x+1)\cdot\frac{B}{x+1}\\ 1 &= (x+1)A + (x-1)B\\ 1 &= Ax + A + Bx - B\\ 1 &= (A+B)x + A - B \qquad (4)\end{aligned}$$

Since this last equation is to be an identity, we now reason as follows.

$$\begin{bmatrix}\text{Coefficient of } x \text{ on the right-}\\ \text{hand side of equation (4)}\end{bmatrix} = \begin{bmatrix}\text{Coefficient of } x \text{ on the left-}\\ \text{hand side of equation (4)}\end{bmatrix}$$

$$\text{therefore} \quad A + B = 0 \tag{5}$$

Similarly, we have

$$\begin{bmatrix}\text{Constant term on the right-}\\ \text{hand side of equation (4)}\end{bmatrix} = \begin{bmatrix}\text{Constant term on the left-}\\ \text{hand side of equation (4)}\end{bmatrix}$$

$$\text{therefore} \quad A - B = 1 \tag{6}$$

Rewriting equations (5) and (6) gives us a system of two linear equations in the unknowns A and B:

$$\begin{cases} A + B = 0 \\ A - B = 1 \end{cases}$$

Adding these two equations yields

$$\begin{aligned} 2A &= 1 \\ A &= \frac{1}{2} \end{aligned}$$

Subtracting the two equations gives us

$$\begin{aligned} B - (-B) &= 0 - 1 \\ 2B &= -1 \\ B &= -\frac{1}{2} \end{aligned}$$

We've now found that $A = \frac{1}{2}$, and $B = -\frac{1}{2}$, as required. The partial fraction decomposition of $\dfrac{1}{(x-1)(x+1)}$ is therefore

$$\begin{aligned}\frac{1}{(x-1)(x+1)} &= \frac{\frac{1}{2}}{x-1} + \frac{-\frac{1}{2}}{x+1}\\ &= \frac{1}{2(x-1)} - \frac{1}{2(x+1)}\end{aligned}$$

You should check for yourself that the result of combining the two fractions on the right-hand side of this last equation is indeed $\dfrac{1}{(x-1)(x+1)}$.

In Example 8 we were able to find the partial fraction decomposition by solving a system of two linear equations. There is a shortcut that is sometimes useful in such problems. As before, multiply both sides of equation (3) by $(x - 1)(x + 1)$ to obtain

$$1 = (x+1)A + (x-1)B \qquad (7)$$

Now equation (7) is to hold for all values of x; in particular it must hold when x is -1. Replacing x by -1 in the equation then yields

$$1 = -2B$$

$$B = -\frac{1}{2} \qquad \text{as obtained previously}$$

The value of A can be obtained in a similar way. Replace x by 1 in equation (7). As you can check, this readily yields $A = \frac{1}{2}$, which again agrees with the result obtained previously.

Example 9 Determine the constants A, B, and C so that the following equation is an identity.

$$\frac{2x+1}{(x+2)(x-3)^2} = \frac{A}{x+2} + \frac{B}{(x-3)} + \frac{C}{(x-3)^2}$$

Solution First, to clear the given equation of fractions, multiply both sides by the quantity $(x + 2)(x - 3)^2$. After dividing out the common factors, we have

$$\begin{aligned}2x+1 &= A(x-3)^2 + B(x+2)(x-3) + C(x+2)\\ 2x+1 &= A(x^2-6x+9) + B(x^2-x-6) + C(x+2)\\ 2x+1 &= Ax^2 - 6Ax + 9A + Bx^2 - Bx - 6B + Cx + 2C\\ 2x+1 &= (A+B)x^2 + (-6A-B+C)x + 9A - 6B + 2C\end{aligned}$$

Equating the coefficients of x^2 on both sides of this identity yields

$$A + B = 0$$

Similarly, equating the x-coefficients on both sides yields

$$-6A - B + C = 2$$

Finally, equating the constant terms on both sides gives us

$$9A - 6B + 2C = 1$$

We now have a system of three linear equations in the three unknowns A, B, and C:

$$\begin{cases} A + B = 0 \\ -6A - B + C = 2 \\ 9A - 6B + 2C = 1 \end{cases}$$

Exercise 12 at the end of this section asks you to verify that the required values here are $A = -\frac{3}{25}$, $B = \frac{3}{25}$, and $C = \frac{7}{5}$. The partial fraction decomposition is therefore

$$\frac{2x + 1}{(x + 2)(x - 3)^2} = \frac{-3}{25(x + 2)} + \frac{3}{25(x - 3)} + \frac{7}{5(x - 3)^2}$$

Here is an alternate method for working Example 9 that relies on the shortcut mentioned earlier. Again, first multiply the given equation by $(x + 2)(x - 3)^2$ to obtain

$$2x + 1 = A(x - 3)^2 + B(x + 2)(x - 3) + C(x + 2) \tag{8}$$

Now just replace x by 3 in this equation. This yields

$$7 = 5C$$

$$C = \frac{7}{5} \quad \text{as obtained previously}$$

The value of A can be obtained in a similar manner. Replace x by -2 in equation (8). As you can check, this readily yields $A = -\frac{3}{25}$, which again agrees with the value obtained previously. We've now computed A and C using the short method. But as you'll see if you try, the value of B cannot be obtained in the same way. One way to find B here is as follows. Substitute the values found for A and C in equation (8) to obtain

$$2x + 1 = -\frac{3}{25}(x - 3)^2 + B(x + 2)(x - 3) + \frac{7}{5}(x + 2)$$

This equation is to hold for all values of x; in particular it must hold for the convenient value $x = 0$. Substituting $x = 0$ in this last equation yields

$$1 = -\frac{3}{25}(-3)^2 + B(2)(-3) + \frac{7}{5}(2)$$

$$25 = -27 - 150B + 70 \quad \text{multiplying by 25}$$

$$150B = 18 \quad \text{collecting like terms}$$

$$B = \frac{18}{150} = \frac{3}{25} \quad \text{dividing by 150 and simplifying}$$

We've now found that $B = \frac{3}{25}$, in agreement with the value obtained previously.

Exercise Set 9.2

In Exercises 1–10, the systems of linear equations are in echelon form. Find all solutions of each system.

1. $\begin{cases} 2x + y + z = -9 \\ 3y - 2z = -4 \\ 8z = -8 \end{cases}$

2. $\begin{cases} -3x + 7y + 2z = -19 \\ y + z = 1 \\ -2z = -2 \end{cases}$

3. $\begin{cases} 8x + 5y + 3z = 1 \\ 3y + 4z = 2 \\ 5z = 3 \end{cases}$

4. $\begin{cases} 2x + 7z = -4 \\ 5y - 3z = 6 \\ 6z = 18 \end{cases}$

5. $\begin{cases} -4x + 5y = 0 \\ 3y + 2z = 1 \\ 3z = -1 \end{cases}$

6. $\begin{cases} 3x - 2y + z = 4 \\ 3z = 9 \end{cases}$

7. $\begin{cases} -x + 8y + 3z = 0 \\ 2z = 0 \end{cases}$

8. $\begin{cases} -x + y + z + w = 9 \\ 2y - z - w = 9 \\ 3z + 2w = 1 \\ 11w = 22 \end{cases}$

9. $\begin{cases} 2x + 3y + z + w = -6 \\ y + 3z - 4w = 23 \\ 6z - 5w = 31 \\ -2w = 10 \end{cases}$

10. $\begin{cases} 7x - y - z + w = 3 \\ 2y - 3z - 4w = -2 \\ 3w = 6 \end{cases}$

In Exercises 11–30, find all solutions of each system.

11. $\begin{cases} x + y + z = 12 \\ 2x - y - z = -1 \\ 3x + 2y + z = 22 \end{cases}$

12. $\begin{cases} A + B = 0 \\ -6A - B + C = 2 \\ 9A - 6B + 2C = 1 \end{cases}$ *Answer:* $(\frac{-3}{25}, \frac{3}{25}, \frac{7}{5})$

13. $\begin{cases} 2x - 3y + 2z = 4 \\ 4x + 2y + 3z = 7 \\ 5x + 4y + 2z = 7 \end{cases}$

14. $\begin{cases} x + 2z = 5 \\ y - 30z = -16 \\ x - 2y + 4z = 8 \end{cases}$

15. $\begin{cases} 3x + 3y - 2z = 13 \\ 6x + 2y - 5z = 13 \\ 7x + 5y - 3z = 26 \end{cases}$

16. $\begin{cases} 2x + 5y - 3z = 4 \\ 4x - 3y + 2z = 9 \\ 5x + 6y - 2z = 18 \end{cases}$

17. $\begin{cases} x + y + z = 1 \\ -2x + y + z = -2 \\ x - 2y + z = -2 \end{cases}$

18. $\begin{cases} 3x + y + 2z = 0 \\ x + y + z = 0 \\ x + y - 3z = -8 \end{cases}$

19. $\begin{cases} 2x - y + z = -1 \\ x + 3y - 2z = 2 \\ 3x - 2y + 3z = 1 \end{cases}$

20. $\begin{cases} -2x + 2y - z = 0 \\ 4y + z = 0 \\ x + y + z = 0 \end{cases}$

21. $\begin{cases} 2x - y + z = 4 \\ x + 3y + 2z = -1 \\ 7x + 5z = 11 \end{cases}$

22. $\begin{cases} 3x + y - z = 10 \\ 8x - y - 6z = -3 \\ 5x - 2y - 5z = 1 \end{cases}$

23. $\begin{cases} x + y + z + w = 4 \\ x - 2y - z - w = 3 \\ 2x - y + z - w = 2 \\ x - y + 2z - 2w = -7 \end{cases}$

24. $\begin{cases} x + y - 3z + 2w = 0 \\ -2x - 2y + 6z + w = -5 \\ -x + 3y + 3z + 3w = -5 \\ 2x + y - 3z - w = 4 \end{cases}$

25. $\begin{cases} 2x + 3y + 2z = 5 \\ x + 4y - 3z = 1 \end{cases}$

26. $\begin{cases} 4x - y - 3z = 2 \\ 6x + 5y - z = 0 \end{cases}$

27. $\begin{cases} x - 2y - 2z + 2w = -10 \\ 3x + 4y - z - 3w = 11 \\ -4x - 3y - 3z + 8w = -21 \end{cases}$

28. $\begin{cases} 2x + y + z + w = 1 \\ x + 3y - 3z - 3w = 0 \\ -3x - 4y + 2z + 2w = -1 \end{cases}$

29. $\begin{cases} 4x - 2y + 3z = -2 \\ 6y - 4z = 6 \end{cases}$

30. $\begin{cases} 6x - 2y - 5z + w = 3 \\ 5x - y + z + 5w = 4 \\ -4y - 7z - 9w = -5 \end{cases}$

31. Find all solutions of the system

$$\begin{cases} x + y + z = 1 \\ ax + y + z = a \\ x + ay + z = a \end{cases}$$

Assume that the constant a is not 1.

32. Solve the following system for x, y, and z in terms of the constants a, b, and c.

$$\begin{cases} x + y = a \\ x + z = b \\ y + z = c \end{cases}$$

33. Find all solutions of the system

$$\begin{cases} x + y + z = a \\ x + y - z = b \\ x - y + z = c \end{cases}$$

(As usual, a, b, and c denote constants.)

In Exercises 34–44, determine the constants (denoted by the capital letters) so that each equation is an identity.

34. $\dfrac{1}{(x-2)(x+2)} = \dfrac{A}{x-2} + \dfrac{B}{x+2}$

35. $\dfrac{1}{(x-2)(x+2)^2} = \dfrac{A}{x-2} + \dfrac{B}{x+2} + \dfrac{C}{(x+2)^2}$

36. $\dfrac{3}{(x-4)(x+4)} = \dfrac{A}{x-4} + \dfrac{B}{x+4}$

37. $\dfrac{3}{(x-4)(x+4)^2} = \dfrac{A}{x-4} + \dfrac{B}{x+4} + \dfrac{C}{(x+4)^2}$

38. $\dfrac{1}{(x+1)(x^2-x+1)} = \dfrac{A}{x+1} + \dfrac{Bx+C}{x^2-x+1}$

39. $\dfrac{1}{(x+1)^2(x^2-x+1)}$

$$= \frac{A}{x+1} + \frac{B}{(x+1)^2} + \frac{Cx+D}{x^2-x+1}$$

*40. $\dfrac{1}{(x+1)(x^2-x+1)^2}$

$$= \frac{A}{x+1} + \frac{Bx+C}{x^2-x+1} + \frac{Dx+E}{(x^2-x+1)^2}$$

41. $\dfrac{4}{x(1-x)} = \dfrac{A}{x} + \dfrac{B}{1-x}$

42. $\dfrac{4}{x^2(1-x)} = \dfrac{A}{x} + \dfrac{B}{x^2} + \dfrac{C}{1-x}$

43. $\dfrac{1}{(x^2+x+1)(x^2-x+1)}$

$$= \frac{Ax+B}{x^2+x+1} + \frac{Cx+D}{x^2-x+1}$$

44. $\dfrac{1}{x^3 - 1} = \dfrac{A}{x - 1} + \dfrac{Bx + C}{x^2 + x + 1}$

45. The figure displays three circles that are mutually tangent. The line segments joining the centers have lengths a, b, and c, as shown. Let r_1, r_2, and r_3 denote the radii of the circles, as indicated in the figure.

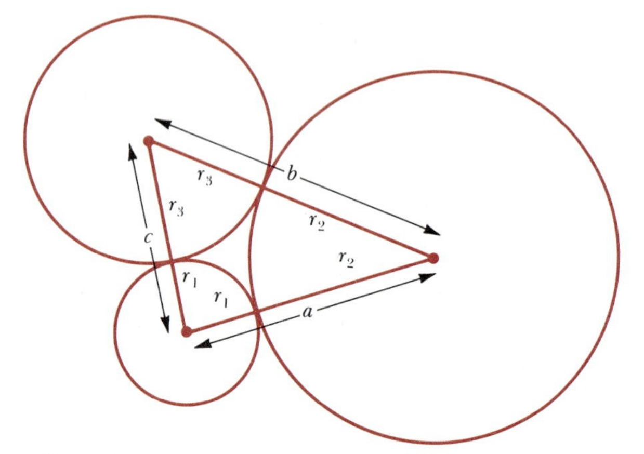

Show that

$$r_1 = \frac{a + c - b}{2}$$

$$r_2 = \frac{a + b - c}{2}$$

$$r_3 = \frac{b + c - a}{2}$$

***46.** Consider the following system. Assume $a \neq 0$.

$$\begin{cases} a^2x - ay + z = a^2 \\ ax - y + z = a^2 \\ 3a^2x - 2ay - z = 0 \end{cases}$$

(a) Under the additional assumption that $a \neq 1$, show that the solution of this system is $(1, a, a^2)$.

(b) If $a = 1$, show that the solutions of the system are $(3z - 2, 4z - 3, z)$, where z can be any real number.

47. Determine constants a, b, and c so that the parabola $y = ax^2 + bx + c$ passes through the points $(-1, -2)$, $(1, -10)$, and $(2, -17)$.

48. Suppose that the height of an object as a function of time is given by $f(t) = at^2 + bt + c$, where t is time in seconds, $f(t)$ is the height in feet at time t, and a, b, and c are certain constants. If after 1, 2, and 3 seconds the corresponding heights are 184 ft, 136 ft, and 56 ft, find the time at which the object is at ground level (height = 0).

49. Find the equation of a circle that passes through the origin and the points $(8, -4)$ and $(7, -1)$. Specify the center and radius.

50. Find the equation of the circle passing through the points $(-1, -5)$, $(1, 2)$, and $(-2, 0)$. Write your answer in the form $Ax^2 + Ay^2 + Bx + Cy + D = 0$.

51. For what values of a does the following system have a solution other than $(0, 0, 0)$?

$$\begin{cases} x + y + z = 0 \\ (1 + a)y + z = 0 \\ (1 - a)x + 2y - z = 0 \end{cases}$$

52. Find all solutions of the system

$$\begin{cases} x + ay + a^2z = -a^3 \\ x + by + b^2z = -b^3 \\ x + cy + c^2z = -c^3 \end{cases}$$

(Assume that a, b, and c are all distinct.)

53. Find all solutions of the system

$$cy + bz = az + cx = bx + ay = abc$$

(Assume that $abc \neq 0$.)

54. Find all solutions of the system

$$\begin{cases} x + y = a \\ x + z = b \\ y + z = c \end{cases}$$

55. Solve the following system for x, y, z, and w in terms of a, b, c, and d.

$$\begin{cases} x + y + z = a \\ x + y + w = b \\ x + z + w = c \\ y + z + w = d \end{cases}$$

56. Solve the following system for x, y, z, and w in terms of a, b, c, and d.

$$\begin{cases} x + y + z + w = 2a \\ x + y - z - w = 2b \\ x - y + z - w = 2c \\ x - y - z + w = 2d \end{cases}$$

57. Consider the system

$$\begin{cases} \lambda x + y + z = a \\ x + \lambda y + z = b \\ x + y + \lambda z = c \end{cases}$$

(a) Assuming that the value of the constant λ is neither 1 nor -2, find all solutions of the system.

(b) If $\lambda = -2$, how must a, b, and c be related for the system to have solutions? Find these solutions.

(c) If $\lambda = 1$, how must a, b, and c be related for the system to have solutions? Find these solutions.

In Exercises 58–62, the lowercase letters a, b, p, *and* q *denote given constants. In each case determine the values of* A *and* B *so that the equation is an identity.*

58. $\dfrac{px + q}{(x - a)(x - b)} = \dfrac{A}{x - a} + \dfrac{B}{x - b} \quad (a \neq b)$

59. $\dfrac{1}{(x - a)(x + a)} = \dfrac{A}{x - a} + \dfrac{B}{x + a} \quad (a \neq 0)$

60. $\dfrac{1}{(x-a)(x-b)} = \dfrac{A}{x-a} + \dfrac{B}{x-b} \quad (a \neq b)$

61. $\dfrac{px+q}{(x-a)(x+a)} = \dfrac{A}{x-a} + \dfrac{B}{x+a} \quad (a \neq 0)$

62. $\dfrac{px+q}{(x-a)^2} = \dfrac{A}{x-a} + \dfrac{B}{(x-a)^2}$

The following exercise appears in Algebra for Colleges and Schools, *by H. S. Hall and S. R. Knight, revised by F. L. Sevenoak (New York: The Macmillan Company, 1906).*

***63.** A, B, and C are three towns forming a triangle. A man has to walk from one to the next, ride thence to the next, and drive thence to his starting point. He can walk, ride, and drive a mile in a, b, and c minutes, respectively. If he starts from B he takes $a + c - b$ hours, if he starts from C he takes $b + a - c$ hours, and if he starts from A he takes $c + b - a$ hours. Find the length of the circuit.

9.3 Matrices

Matrices, which are the objects of study in this section, are used extensively in the sciences. They are essential tools in fields as diverse as economics and physics. At the most basic level, matrices are closely related to systems of linear equations; this is the approach with which we begin here.

A *matrix* (plural: matrices) is simply a rectangular array of numbers, enclosed in parentheses or brackets. Here are three examples.

$$\begin{pmatrix} 2 & 3 \\ -5 & 4 \end{pmatrix} \qquad \begin{pmatrix} -6 & 0 & 1 & \frac{1}{4} \\ \frac{2}{3} & 1 & 5 & 8 \end{pmatrix} \qquad \begin{pmatrix} \pi & 0 & 0 \\ 0 & 1 & .9 \\ -1 & -2 & 3 \\ -4 & 8 & 6 \end{pmatrix}$$

The particular numbers constituting a matrix are called its *entries* or *elements*. For instance, the entries in the matrix $\begin{pmatrix} 2 & 3 \\ -5 & 4 \end{pmatrix}$ are the four numbers 2, 3, -5, and 4. In this section the entries will always be real numbers. However, it's also possible to consider matrices with complex numbers for the entries.

It is convenient to agree on a standard system for labeling the rows and the columns of a matrix. The rows are numbered from top to bottom and the columns from left to right, as indicated in the following example.

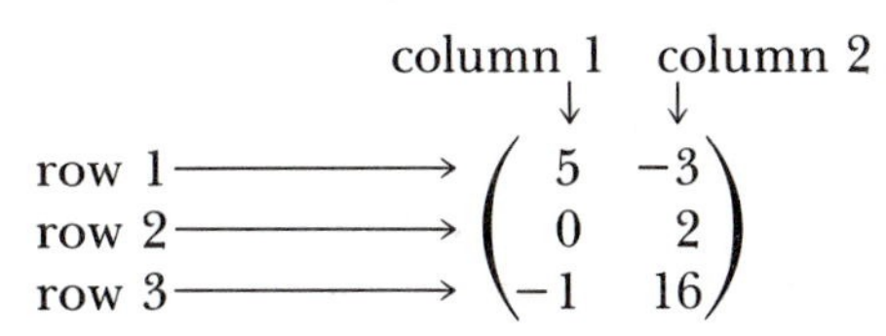

The *size* or *dimension* of a matrix is expressed by specifying the number of rows and the number of columns, in that order. For instance, we would say that the matrix

$$\begin{pmatrix} 5 & -3 \\ 0 & 2 \\ -1 & 16 \end{pmatrix}$$

is a 3×2 (read: 3 by 2) matrix, not 2×3. The following example will help fix in your mind the terminology that we have introduced.

Example 1 Consider the matrix

$$\begin{pmatrix} 1 & 3 & 5 \\ 7 & 9 & 11 \end{pmatrix}$$

(a) List the entries.
(b) What is the size of the matrix?
(c) Which element is in the second row, third column?

Solution

(a) The entries are 1, 3, 5, 7, 9, and 11.
(b) Since there are two rows and three columns, this is a 2×3 matrix.
(c) To locate the element in the second row, third column, draw lines through the second row and the third column and see where they intersect:

$$\begin{pmatrix} 1 & 3 & 5 \\ 7 & 9 & 11 \end{pmatrix}$$

Thus the entry in row 2, column 3 is 11.

There is a natural way to use matrices to describe and to solve systems of linear equations. Consider for example the following system of linear equations in *standard form* (with the x, y, and z terms lined up and the constant terms on the right-hand side of each equation).

$$\begin{cases} x + 2y - 3z = 4 \\ 3x \phantom{{}+2y} + z = 5 \\ -x - 3y + 4z = 0 \end{cases}$$

The *coefficient matrix* of this system is the matrix

$$\begin{pmatrix} 1 & 2 & -3 \\ 3 & 0 & 1 \\ -1 & -3 & 4 \end{pmatrix}$$

As the name implies, the coefficient matrix of the system is just the matrix whose entries are the coefficients of x, y, and z, written in the same relative positions as they appear in the system. Notice the zero appearing in the second row, second column of the matrix. It is there because the coefficient of y in the second equation is in fact zero. The *augmented matrix* of the system of equations under consideration is

$$\begin{pmatrix} 1 & 2 & -3 & 4 \\ 3 & 0 & 1 & 5 \\ -1 & -3 & 4 & 0 \end{pmatrix}$$

As you can see, the augmented matrix is formed by *augmenting* the coefficient matrix with the column of constant terms taken from the right-hand side of the given system of equations. Simply as a visual aid in relating the augmented matrix to the original system of equations, we will write the augmented matrix this way:

$$\left(\begin{array}{ccc|c} 1 & 2 & -3 & 4 \\ 3 & 0 & 1 & 5 \\ -1 & -3 & 4 & 0 \end{array}\right)$$

Example 2 Write the coefficient matrix and the augmented matrix for the system

$$\begin{cases} 8x - 2y + z = 1 \\ 3x - 4z + y = 2 \\ 12y - 3z - 6 = 0 \end{cases}$$

Solution First write the system in standard form, with the x, y, and z terms lined up, and the constant terms on the right. This yields

$$\begin{cases} 8x - 2y + z = 1 \\ 3x + y - 4z = 2 \\ 12y - 3z = 6 \end{cases}$$

The coefficient matrix is then

$$\begin{pmatrix} 8 & -2 & 1 \\ 3 & 1 & -4 \\ 0 & 12 & -3 \end{pmatrix}$$

while the augmented matrix is

$$\left(\begin{array}{ccc|c} 8 & -2 & 1 & 1 \\ 3 & 1 & -4 & 2 \\ 0 & 12 & -3 & 6 \end{array}\right)$$

As you know from the previous section, there can be a good deal of bookkeeping involved in using Gaussian elimination to solve systems of equations. We can lessen the load somewhat by working with the augmented matrix rather than with the initial system of equations. First though, we need to express the three elementary operations described in the previous section in the language of matrices. This is done in Table 1. We call the three corresponding matrix operations *elementary row operations.*

Table 1

Elementary Operations for a System of Linear Equations	Corresponding Elementary Row Operation for a Matrix
1. Multiply both sides of an equation by a nonzero constant.	1′. Multiply each entry in a given row by a nonzero constant.
2. Interchange two equations.	2′. Interchange two rows.
3. To one equation, add a multiple of another equation.	3′. To one row, add a multiple of another row.

Table 2, which follows, displays an example of each elementary row operation. The table also introduces a convenient shorthand for describing row operations.

Table 2

Examples of the Elementary Row Operations	Comments
$\begin{pmatrix} 1 & 2 & 3 \\ 4 & 5 & 6 \\ 7 & 8 & 9 \end{pmatrix} \xrightarrow{10R_1} \begin{pmatrix} 10 & 20 & 30 \\ 4 & 5 & 6 \\ 7 & 8 & 9 \end{pmatrix}$	$10R_1$ means multiply row 1 by 10.
$\begin{pmatrix} 1 & 2 & 3 \\ 4 & 5 & 6 \\ 7 & 8 & 9 \end{pmatrix} \xrightarrow[R_2 \text{ and } R_3]{\text{interchange}} \begin{pmatrix} 1 & 2 & 3 \\ 7 & 8 & 9 \\ 4 & 5 & 6 \end{pmatrix}$	R_2 means row 2. R_3 means row 3.
$\begin{pmatrix} 1 & 2 & 3 \\ 4 & 5 & 6 \\ 7 & 8 & 9 \end{pmatrix} \xrightarrow{R_2 - 4R_1} \begin{pmatrix} 1 & 2 & 3 \\ 0 & -3 & -6 \\ 7 & 8 & 9 \end{pmatrix}$	From each entry in row 2, subtract four times the corresponding entry in row 1.

We are now ready to employ matrices as a tool in solving systems of linear equations. In the example that follows, we'll use the same system of equations used in Example 5 of Section 9.2. *Suggestion:* After reading the next example, carefully compare each step with the corresponding one taken in Example 5 of Section 9.2.

Example 3 Solve the system

$$\begin{cases} 4x - 3y + 2z = 40 \\ 5x + 9y - 7z = 47 \\ 9x + 8y - 3z = 97 \end{cases}$$

Solution

$$\left(\begin{array}{rrr|r} 4 & -3 & 2 & 40 \\ 5 & 9 & -7 & 47 \\ 9 & 8 & -3 & 97 \end{array}\right) \xrightarrow{R_1 - R_2} \left(\begin{array}{rrr|r} -1 & -12 & 9 & -7 \\ 5 & 9 & -7 & 47 \\ 9 & 8 & -3 & 97 \end{array}\right)$$

$$\xrightarrow[R_3 + 9R_1]{R_2 + 5R_1} \left(\begin{array}{rrr|r} -1 & -12 & 9 & -7 \\ 0 & -51 & 38 & 12 \\ 0 & -100 & 78 & 34 \end{array}\right) \xrightarrow{\frac{1}{2}R_3} \left(\begin{array}{rrr|r} -1 & -12 & 9 & -7 \\ 0 & -51 & 38 & 12 \\ 0 & -50 & 39 & 17 \end{array}\right)$$

$$\xrightarrow{R_2 - R_3} \left(\begin{array}{rrr|r} -1 & -12 & 9 & -7 \\ 0 & -1 & -1 & -5 \\ 0 & -50 & 39 & 17 \end{array}\right) \xrightarrow{R_3 - 50R_2} \left(\begin{array}{rrr|r} -1 & -12 & 9 & -7 \\ 0 & -1 & -1 & -5 \\ 0 & 0 & 89 & 267 \end{array}\right)$$

This last augmented matrix represents a system of equations in echelon form:

$$\begin{cases} -x - 12y + 9z = -7 \\ \quad\;\; -y - \;\; z = -5 \\ \qquad\qquad 89z = 267 \end{cases}$$

As you should now check for yourself, this yields the values $z = 3$, then $y = 2$, then $x = 10$. The solution of the original system is therefore $(10, 2, 3)$.

For the remainder of this section we are going to study matrices, without reference to systems of equations. As mentioned at the beginning of this section, matrices serve as essential tools in many of the sciences. For example, a knowledge of matrices and their properties is needed for work in computer graphics. To begin with, we need to say what it means for two matrices to be equal.

Definition of Equality for Matrices

Two matrices are equal provided that they have the same size (same number of rows, same number of columns) and that the corresponding entries are equal.

Examples

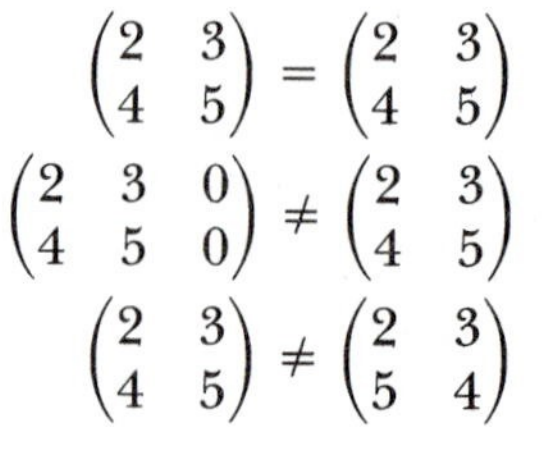

$$\begin{pmatrix} 2 & 3 \\ 4 & 5 \end{pmatrix} = \begin{pmatrix} 2 & 3 \\ 4 & 5 \end{pmatrix}$$

$$\begin{pmatrix} 2 & 3 & 0 \\ 4 & 5 & 0 \end{pmatrix} \neq \begin{pmatrix} 2 & 3 \\ 4 & 5 \end{pmatrix}$$

$$\begin{pmatrix} 2 & 3 \\ 4 & 5 \end{pmatrix} \neq \begin{pmatrix} 2 & 3 \\ 5 & 4 \end{pmatrix}$$

Now we define matrix addition and subtraction. These operations are only defined between matrices of the same size.

Definition of Matrix Addition

To add (or subtract) two matrices of the same size, add (or subtract) the corresponding entries.

Examples

$$\begin{pmatrix} 2 & 3 \\ -1 & 4 \end{pmatrix} + \begin{pmatrix} 6 & 1 \\ 0 & -4 \end{pmatrix} = \begin{pmatrix} 8 & 4 \\ -1 & 0 \end{pmatrix}$$

$$\begin{pmatrix} 2 & 3 \\ -1 & 4 \\ 9 & 10 \end{pmatrix} - \begin{pmatrix} 6 & 1 \\ 0 & -4 \\ 3 & 2 \end{pmatrix} = \begin{pmatrix} -4 & 2 \\ -1 & 8 \\ 6 & 8 \end{pmatrix}$$

Many of the properties that we listed in Chapter 1 for the real numbers also apply to matrices. For instance, matrix addition is commutative:

$$A + B = B + A$$ whenever A and B are matrices of the same size

Matrix addition is also associative:

$$A + (B + C) = (A + B) + C$$ whenever A, B, and C are matrices of the same size

The next example shows particular instances of this.

Example 4 Let

$$A = \begin{pmatrix} 1 & 2 \\ 3 & 4 \end{pmatrix} \quad B = \begin{pmatrix} 0 & -5 \\ 8 & -1 \end{pmatrix} \quad C = \begin{pmatrix} 6 & 7 \\ 8 & 9 \end{pmatrix}$$

(a) Show that $A + C = C + A$.
(b) Show that $A + (B + C) = (A + B) + C$.

Solution

(a)
$$A + C = \begin{pmatrix} 1 & 2 \\ 3 & 4 \end{pmatrix} + \begin{pmatrix} 6 & 7 \\ 8 & 9 \end{pmatrix} = \begin{pmatrix} 7 & 9 \\ 11 & 13 \end{pmatrix}$$
$$C + A = \begin{pmatrix} 6 & 7 \\ 8 & 9 \end{pmatrix} + \begin{pmatrix} 1 & 2 \\ 3 & 4 \end{pmatrix} = \begin{pmatrix} 7 & 9 \\ 11 & 13 \end{pmatrix}$$

This shows that $A + C = C + A$, since both $A + C$ and $C + A$ represent the matrix $\begin{pmatrix} 7 & 9 \\ 11 & 13 \end{pmatrix}$.

(b) First compute $A + (B + C)$.

$$\begin{aligned} A + (B + C) &= \begin{pmatrix} 1 & 2 \\ 3 & 4 \end{pmatrix} + \left[\begin{pmatrix} 0 & -5 \\ 8 & -1 \end{pmatrix} + \begin{pmatrix} 6 & 7 \\ 8 & 9 \end{pmatrix}\right] \\ &= \begin{pmatrix} 1 & 2 \\ 3 & 4 \end{pmatrix} + \begin{pmatrix} 6 & 2 \\ 16 & 8 \end{pmatrix} \\ &= \begin{pmatrix} 7 & 4 \\ 19 & 12 \end{pmatrix} \end{aligned}$$

Next compute $(A + B) + C$.

$$\begin{aligned}(A + B) + C &= \left[\begin{pmatrix}1 & 2\\ 3 & 4\end{pmatrix} + \begin{pmatrix}0 & -5\\ 8 & -1\end{pmatrix}\right] + \begin{pmatrix}6 & 7\\ 8 & 9\end{pmatrix}\\ &= \begin{pmatrix}1 & -3\\ 11 & 3\end{pmatrix} + \begin{pmatrix}6 & 7\\ 8 & 9\end{pmatrix}\\ &= \begin{pmatrix}7 & 4\\ 19 & 12\end{pmatrix}\end{aligned}$$

We conclude from these calculations that $A + (B + C) = (A + B) + C$ since both sides of that equation represent the matrix $\begin{pmatrix}7 & 4\\ 19 & 12\end{pmatrix}$.

Next we define an operation on matrices called *scalar multiplication*. First of all, the word *scalar* here just means *real number*. So we are talking about multiplying a matrix by a real number. (In more advanced work, nonreal complex scalars are also considered.)

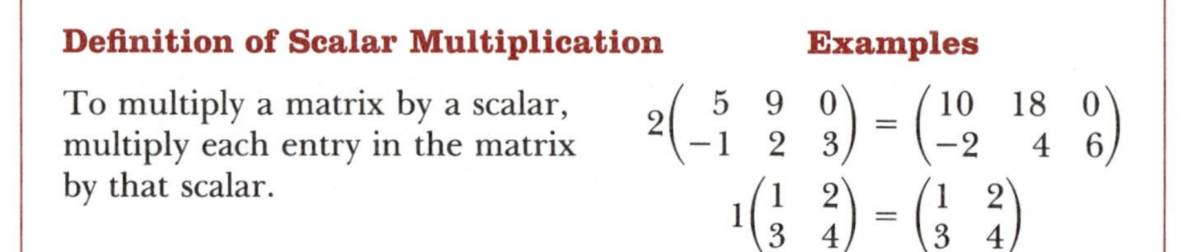

Definition of Scalar Multiplication

To multiply a matrix by a scalar, multiply each entry in the matrix by that scalar.

Examples

$$2\begin{pmatrix}5 & 9 & 0\\ -1 & 2 & 3\end{pmatrix} = \begin{pmatrix}10 & 18 & 0\\ -2 & 4 & 6\end{pmatrix}$$

$$1\begin{pmatrix}1 & 2\\ 3 & 4\end{pmatrix} = \begin{pmatrix}1 & 2\\ 3 & 4\end{pmatrix}$$

There are two simple but useful properties of scalar multiplication that are worth noting at this point. We'll omit the proofs of these two properties; however, the next example does ask us to verify the properties for some particular cases.

Property Summary

Properties of Scalar Multiplication

1. $c(kM) = (ck)M$ for all scalars c and k and any matrix M.
2. $c(M + N) = cM + cN$ where c is any scalar and M and N are any matrices of the same size.

Example 5 Let $c = 2$, $k = 3$, $M = \begin{pmatrix}1 & 2\\ 3 & 4\end{pmatrix}$, and $N = \begin{pmatrix}5 & 6\\ 7 & 8\end{pmatrix}$.

(a) Show that $c(kM) = (ck)M$.
(b) Show that $c(M + N) = cM + cN$.

Solution

(a) $c(kM) = 2\left[3\begin{pmatrix}1 & 2\\ 3 & 4\end{pmatrix}\right] = 2\begin{pmatrix}3 & 6\\ 9 & 12\end{pmatrix} = \begin{pmatrix}6 & 12\\ 18 & 24\end{pmatrix}$

$(ck)M = (2 \cdot 3)\begin{pmatrix}1 & 2\\ 3 & 4\end{pmatrix} = 6\begin{pmatrix}1 & 2\\ 3 & 4\end{pmatrix} = \begin{pmatrix}6 & 12\\ 18 & 24\end{pmatrix}$

Thus $c(kM) = (ck)M$, since in both cases the result is $\begin{pmatrix}6 & 12\\ 18 & 24\end{pmatrix}$.

(b) $c(M + N) = 2\left[\begin{pmatrix} 1 & 2 \\ 3 & 4 \end{pmatrix} + \begin{pmatrix} 5 & 6 \\ 7 & 8 \end{pmatrix}\right]$

$$= 2\begin{pmatrix} 6 & 8 \\ 10 & 12 \end{pmatrix}$$

$$= \begin{pmatrix} 12 & 16 \\ 20 & 24 \end{pmatrix}$$

$$cM + cN = 2\begin{pmatrix} 1 & 2 \\ 3 & 4 \end{pmatrix} + 2\begin{pmatrix} 5 & 6 \\ 7 & 8 \end{pmatrix}$$

$$= \begin{pmatrix} 2 & 4 \\ 6 & 8 \end{pmatrix} + \begin{pmatrix} 10 & 12 \\ 14 & 16 \end{pmatrix}$$

$$= \begin{pmatrix} 12 & 16 \\ 20 & 24 \end{pmatrix}$$

Thus $c(M + N) = cM + cN$, since in both cases the result is $\begin{pmatrix} 12 & 16 \\ 20 & 24 \end{pmatrix}$.

A matrix with zeros for all of its entries plays the same role in matrix addition as does the number zero in ordinary addition. For instance, in the case of 2×2 matrices we certainly have

$$\begin{pmatrix} a & b \\ c & d \end{pmatrix} + \begin{pmatrix} 0 & 0 \\ 0 & 0 \end{pmatrix} = \begin{pmatrix} a & b \\ c & d \end{pmatrix}$$

and

$$\begin{pmatrix} 0 & 0 \\ 0 & 0 \end{pmatrix} + \begin{pmatrix} a & b \\ c & d \end{pmatrix} = \begin{pmatrix} a & b \\ c & d \end{pmatrix} \qquad \text{for all real numbers } a, b, c, \text{ and } d$$

The matrix $\begin{pmatrix} 0 & 0 \\ 0 & 0 \end{pmatrix}$ is called the *additive identity* for addition of 2×2 matrices. Similarly, any matrix with all zero entries is the additive identity for matrices of that size. At times it is convenient to denote an additive identity matrix by a boldface zero: **0**. With this notation we can write

$$A + \mathbf{0} = \mathbf{0} + A = A \qquad \text{for any matrix } A$$

In writing the above matrix equation, it is understood that the size of the matrix **0** is the same as the size of A. With this notation we also have

$$A - A = \mathbf{0} \qquad \text{for any matrix } A$$

These observations are used in the next example to solve a matrix equation.

Example 6 Let

$$A = \begin{pmatrix} -1 & 2 & 8 \\ 0 & 4 & 3 \end{pmatrix} \qquad B = \begin{pmatrix} 6 & 0 & 3 \\ 2 & 1 & -6 \end{pmatrix}$$

Solve the following matrix equation for X. (To solve for X means to find the entries in the matrix X.)

$$3X + A = 4B$$

Solution We will show two different methods.

First Method Subtracting A from both sides of the given equation yields

$$3X + A - A = 4B - A$$
$$3X + \mathbf{0} = 4B - A$$
$$3X = 4B - A$$

Now multiply both sides by the scalar $\frac{1}{3}$ to obtain

$$X = \frac{1}{3}(4B - A)$$

Substituting for B and A then gives us

$$\begin{aligned} X &= \frac{1}{3}\left[4\begin{pmatrix} 6 & 0 & 3 \\ 2 & 1 & -6 \end{pmatrix} - \begin{pmatrix} -1 & 2 & 8 \\ 0 & 4 & 3 \end{pmatrix}\right] \\ &= \frac{1}{3}\left[\begin{pmatrix} 24 & 0 & 12 \\ 8 & 4 & -24 \end{pmatrix} - \begin{pmatrix} -1 & 2 & 8 \\ 0 & 4 & 3 \end{pmatrix}\right] \\ &= \frac{1}{3}\begin{pmatrix} 25 & -2 & 4 \\ 8 & 0 & -27 \end{pmatrix} \\ &= \begin{pmatrix} \frac{25}{3} & -\frac{2}{3} & \frac{4}{3} \\ \frac{8}{3} & 0 & -9 \end{pmatrix} \quad \text{as required} \end{aligned}$$

Alternate Method The unknown matrix X must have two rows and three columns as is the case with A and B. (Why?) Let $X = \begin{pmatrix} u & v & w \\ x & y & z \end{pmatrix}$. Then we have

$$3X + A = 4B$$

$$3\begin{pmatrix} u & v & w \\ x & y & z \end{pmatrix} + \begin{pmatrix} -1 & 2 & 8 \\ 0 & 4 & 3 \end{pmatrix} = 4\begin{pmatrix} 6 & 0 & 3 \\ 2 & 1 & -6 \end{pmatrix}$$

$$\begin{pmatrix} 3u & 3v & 3w \\ 3x & 3y & 3z \end{pmatrix} + \begin{pmatrix} -1 & 2 & 8 \\ 0 & 4 & 3 \end{pmatrix} = \begin{pmatrix} 24 & 0 & 12 \\ 8 & 4 & -24 \end{pmatrix}$$

$$\begin{pmatrix} 3u - 1 & 3v + 2 & 3w + 8 \\ 3x & 3y + 4 & 3z + 3 \end{pmatrix} = \begin{pmatrix} 24 & 0 & 12 \\ 8 & 4 & -24 \end{pmatrix}$$

For this last matrix equation to hold, all of the corresponding entries must be equal. This gives us six equations to solve, one for each letter. For instance, equating the entries in the first row, first column of each matrix yields

$$\begin{aligned} 3u - 1 &= 24 \\ 3u &= 25 \\ u &= \frac{25}{3} \end{aligned}$$

As you can check, the results of equating the other corresponding entries are

$$v = -\frac{2}{3}, \qquad w = \frac{4}{3}, \qquad x = \frac{8}{3}, \qquad y = 0, \quad \text{and} \quad z = -9$$

It follows that

$$X = \begin{pmatrix} \frac{25}{3} & -\frac{2}{3} & \frac{4}{3} \\ \frac{8}{3} & 0 & -9 \end{pmatrix}$$

which agrees with the answer obtained using the first method.

Our last topic in this section is matrix multiplication. We will begin with the simplest case and work up to the more general. By convention, a matrix with only one row is called a *row vector.* Thus, examples of row vectors are

$$(2 \quad 13), \qquad (-1 \quad 4 \quad 3), \quad \text{and} \quad (0 \quad 0 \quad 0 \quad 1)$$

Similarly, a matrix with only one column is called a *column vector.* Examples of column vectors are

$$\begin{pmatrix} 2 \\ 13 \end{pmatrix}, \quad \begin{pmatrix} -1 \\ 4 \\ 3 \end{pmatrix}, \quad \text{and} \quad \begin{pmatrix} 0 \\ 0 \\ 0 \\ 1 \end{pmatrix}$$

The following definition tells us how to multiply a row vector and a column vector when they have the same number of entries.

Inner Product of a Row Vector and a Column Vector

Let A be a row vector, B a column vector, and assume that the number of columns in A is the same as the number of rows in B. Then the *inner product* $A \cdot B$ is defined to be the number obtained by first multiplying the corresponding entries and then adding the results.

Examples

$$(1 \quad 2 \quad 3) \cdot \begin{pmatrix} 4 \\ 5 \\ 6 \end{pmatrix} = 1 \cdot 4 + 2 \cdot 5 + 3 \cdot 6 = 32$$

$(1 \quad 2) \cdot \begin{pmatrix} 4 \\ 5 \\ 6 \end{pmatrix}$ is not defined

$(1 \quad 2 \quad 3) \cdot \begin{pmatrix} 4 \\ 5 \end{pmatrix}$ is not defined

An important observation here is that the end result of taking the inner product is always just a number. The definition of matrix product that we now give depends upon this observation.

The Product of Two Matrices

Let A and B be two matrices and assume that the number of columns in A is the same as the number of rows in B. Then the product matrix AB is computed according to the following rule.

The entry in the ith row and the jth column of AB is the inner product of the ith row of A with the jth column of B.

The matrix AB will have as many rows as A and as many columns as B.

As an example of matrix multiplication, we will compute the product AB, where $A = \begin{pmatrix} 1 & 2 \\ 3 & 4 \end{pmatrix}$ and $B = \begin{pmatrix} 5 & 6 & 0 \\ 7 & 8 & 1 \end{pmatrix}$. In other words we will compute

$\begin{pmatrix} 1 & 2 \\ 3 & 4 \end{pmatrix}\begin{pmatrix} 5 & 6 & 0 \\ 7 & 8 & 1 \end{pmatrix}$. Before carrying out the calculations, however, we should check on two points.

1. Is the product defined? That is, does the number of columns in A equal the number of rows in B? In this case, yes; the common number is two.
2. What is the size of the product? According to the definition, the product AB will have as many rows as does A and as many columns as B. Thus the size of AB will be 2×3.

Schematically then, the situation looks like this:

$$\begin{pmatrix} 1 & 2 \\ 3 & 4 \end{pmatrix}\begin{pmatrix} 5 & 6 & 0 \\ 7 & 8 & 1 \end{pmatrix} = \begin{pmatrix} ? & ? & ? \\ ? & ? & ? \end{pmatrix}$$

We have six positions to fill. Here are the computations.

Position	How to Compute	Computation
row 1, column 1	inner product of row 1 and column 1 $\begin{pmatrix} 1 & 2 \\ 3 & 4 \end{pmatrix}\begin{pmatrix} 5 & 6 & 0 \\ 7 & 8 & 1 \end{pmatrix}$	$1 \cdot 5 + 2 \cdot 7 = 19$
row 1, column 2	inner product of row 1 and column 2 $\begin{pmatrix} 1 & 2 \\ 3 & 4 \end{pmatrix}\begin{pmatrix} 5 & 6 & 0 \\ 7 & 8 & 1 \end{pmatrix}$	$1 \cdot 6 + 2 \cdot 8 = 22$
row 1, column 3	inner product of row 1 and column 3 $\begin{pmatrix} 1 & 2 \\ 3 & 4 \end{pmatrix}\begin{pmatrix} 5 & 6 & 0 \\ 7 & 8 & 1 \end{pmatrix}$	$1 \cdot 0 + 2 \cdot 1 = 2$
row 2, column 1	inner product of row 2 and column 1 $\begin{pmatrix} 1 & 2 \\ 3 & 4 \end{pmatrix}\begin{pmatrix} 5 & 6 & 0 \\ 7 & 8 & 1 \end{pmatrix}$	$3 \cdot 5 + 4 \cdot 7 = 43$
row 2, column 2	inner product of row 2 and column 2 $\begin{pmatrix} 1 & 2 \\ 3 & 4 \end{pmatrix}\begin{pmatrix} 5 & 6 & 0 \\ 7 & 8 & 1 \end{pmatrix}$	$3 \cdot 6 + 4 \cdot 8 = 50$
row 2, column 3	inner product of row 2 and column 3 $\begin{pmatrix} 1 & 2 \\ 3 & 4 \end{pmatrix}\begin{pmatrix} 5 & 6 & 0 \\ 7 & 8 & 1 \end{pmatrix}$	$3 \cdot 0 + 4 \cdot 1 = 4$

We've now completed all of the computations for the entries in the product AB. The result is

$$AB = \begin{pmatrix} 1 & 2 \\ 3 & 4 \end{pmatrix}\begin{pmatrix} 5 & 6 & 0 \\ 7 & 8 & 1 \end{pmatrix} = \begin{pmatrix} 19 & 22 & 2 \\ 43 & 50 & 4 \end{pmatrix}$$

Example 7 Let

$$A = \begin{pmatrix} 1 & 2 \\ 3 & 4 \end{pmatrix} \qquad B = \begin{pmatrix} 5 & 6 \\ 7 & 8 \end{pmatrix}$$

By computing AB and then BA, show that $AB \neq BA$. This shows that, in general, matrix multiplication is not commutative.

Solution

$$AB = \begin{pmatrix} 1 & 2 \\ 3 & 4 \end{pmatrix}\begin{pmatrix} 5 & 6 \\ 7 & 8 \end{pmatrix} = \begin{pmatrix} 1\cdot5+2\cdot7 & 1\cdot6+2\cdot8 \\ 3\cdot5+4\cdot7 & 3\cdot6+4\cdot8 \end{pmatrix} = \begin{pmatrix} 19 & 22 \\ 43 & 50 \end{pmatrix}$$

$$BA = \begin{pmatrix} 5 & 6 \\ 7 & 8 \end{pmatrix}\begin{pmatrix} 1 & 2 \\ 3 & 4 \end{pmatrix} = \begin{pmatrix} 5\cdot1+6\cdot3 & 5\cdot2+6\cdot4 \\ 7\cdot1+8\cdot3 & 7\cdot2+8\cdot4 \end{pmatrix} = \begin{pmatrix} 23 & 34 \\ 31 & 46 \end{pmatrix}$$

Comparing the two matrices AB and BA, we conclude that $AB \neq BA$.

Exercise Set 9.3

In Exercises 1–3, specify the size of each matrix.

1. **(a)** $\begin{pmatrix} -4 & 0 & 5 \\ 2 & 8 & -1 \end{pmatrix}$ **(b)** $\begin{pmatrix} 7 & 1 \\ 4 & -3 \\ 0 & 0 \end{pmatrix}$

2. **(a)** $\begin{pmatrix} 1 & 0 \\ 0 & -1 \end{pmatrix}$ **(b)** $\begin{pmatrix} 1 \\ 6 \\ 8 \\ 1 \end{pmatrix}$

3. $\begin{pmatrix} 1 & a & b & c \\ a & 1 & 0 & a \\ b & 0 & 1 & b \\ c & a & b & 1 \\ 0 & 0 & 0 & 1 \end{pmatrix}$

In Exercises 4–7, write the coefficient matrix and the augmented matrix for each system.

4. $\begin{cases} 2x + 3y + 4z = 10 \\ 5x + 6y + 7z = 9 \\ 8x + 9y + 10z = 8 \end{cases}$

5. $\begin{cases} 5x - y + z = 0 \\ 4y + 2z = 1 \\ 3x + y + z = -1 \end{cases}$

6. $\begin{cases} x + z + w = -1 \\ x + y + 2w = 0 \\ y + z + w = 1 \\ 2x - y - z = 2 \end{cases}$

7. $\begin{cases} 8x - 8y = 5 \\ x - y + z = 1 \end{cases}$

In Exercises 8–22, use matrices to solve the systems of equations.

8. $\begin{cases} x + y + z = -4 \\ 2x - 3y + z = -1 \\ 4x + 2y - 3z = 33 \end{cases}$

9. $\begin{cases} x - y + 2z = 7 \\ 3x + 2y - z = -10 \\ -x + 3y + z = -2 \end{cases}$

10. $\begin{cases} 2x - 3y + 4z = 14 \\ 3x - 2y + 2z = 12 \\ 4x + 5y - 5z = 16 \end{cases}$

11. $\begin{cases} x + z = -2 \\ -3x + 2y = 17 \\ x - y - z = -9 \end{cases}$

12. $\begin{cases} 5x + y + 10z = 23 \\ 4x + 2y - 10z = 76 \\ 3x - 4y = 18 \end{cases}$

13. $\begin{cases} \frac{1}{2}x - \frac{1}{3}y + \frac{1}{6}z = 10 \\ \frac{1}{3}x + \frac{1}{4}y - \frac{1}{2}z = -10 \\ \frac{1}{6}x + \frac{1}{6}y - \frac{1}{6}z = -4 \end{cases}$

14. $\begin{cases} 2x + 3y - 4z = 7 \\ x - y + z = -\frac{3}{2} \\ 6x - 5y - 2z = -7 \end{cases}$

15. $\begin{cases} 3x - 2y + 6z = 0 \\ x + 3y + 20z = 15 \\ 10x - 11y - 10z = -9 \end{cases}$

16. $\begin{cases} 3A - 3B + C = 4 \\ 6A + 9B - 3C = -7 \\ A - 2B - 2C = -3 \end{cases}$

17. $\begin{cases} 4x - 3y + 3z = 2 \\ 5x + y - 4z = 1 \\ 9x - 2y - z = 3 \end{cases}$

18. $\begin{cases} 6x + y - z = -1 \\ -3x + 2y + 2z = 2 \\ 5y + 3z = 1 \end{cases}$

19. $\begin{cases} x - y + z + w = 6 \\ x + y - z + w = 4 \\ x + y + z - w = -2 \\ -x + y + z + w = 0 \end{cases}$

20. $\begin{cases} x + 2y - z - 2w = 5 \\ 2x + y + 2z + w = -7 \\ -2x - y - 3z - 2w = 10 \\ z + w = -3 \end{cases}$

21. $\begin{cases} 15A + 14B + 26C = 1 \\ 18A + 17B + 32C = -1 \\ 21A + 20B + 38C = 0 \end{cases}$

22. $\begin{cases} A + B + C + D + E = -1 \\ 3A - 2B - 2C + 3D + 2E = 13 \\ 3C + 4D - 4E = 7 \\ 5A - 4B + E = 30 \\ C - 2E = 3 \end{cases}$

In Exercises 23–50, the matrices A, B, C, D, E, F, and G are defined as follows.

$$A = \begin{pmatrix} 2 & 3 \\ -1 & 4 \end{pmatrix} \quad B = \begin{pmatrix} 1 & -1 \\ 3 & 0 \end{pmatrix} \quad C = \begin{pmatrix} 1 & 0 \\ 0 & 1 \end{pmatrix}$$

$$D = \begin{pmatrix} -1 & 2 & 3 \\ 4 & 0 & 5 \end{pmatrix} \quad E = \begin{pmatrix} 2 & 1 \\ 8 & -1 \\ 6 & 5 \end{pmatrix} \quad F = \begin{pmatrix} 5 & -1 \\ -4 & 0 \\ 2 & 3 \end{pmatrix}$$

$$G = \begin{pmatrix} 0 & 0 \\ 0 & 0 \\ 0 & 0 \end{pmatrix}$$

In each exercise, carry out the indicated matrix operations if they are defined. If an operation is not defined, state this.

23. $A + B$ **24.** $A - B$ **25.** $2A + 2B$
26. $2(A + B)$ **27.** AB **28.** BA
29. AC **30.** CA **31.** $3D + E$
32. $E + F$ **33.** $2F - 3G$ **34.** DE
35. ED **36.** DF **37.** FD
38. $A + D$ **39.** $G + A$ **40.** DG
41. GD **42.** $(A + B) + C$ **43.** $A + (B + C)$
44. CD **45.** DC **46.** $5E - 3F$
47. A^2 $(= AA)$ **48.** A^2A **49.** AA^2
50. C^2

In Exercises 51–54, solve the matrix equations given that

$$A = \begin{pmatrix} 1 & 0 & 3 \\ -2 & 4 & 5 \end{pmatrix} \quad B = \begin{pmatrix} 0 & 1 & -1 \\ 5 & 3 & -2 \end{pmatrix}$$

$$C = \begin{pmatrix} 9 & 4 & 0 \\ 2 & 7 & 3 \end{pmatrix}$$

51. $2X - A = 3B$ **52.** $3A + X = C - B$
53. $2X - A = A - B + C$ **54.** $3(X - C) = A + 3B$

55. Let

$$A = \begin{pmatrix} -1 & 3 & 4 \\ 3 & 2 & -3 \\ 9 & 1 & 6 \end{pmatrix} \quad B = \begin{pmatrix} 7 & 0 & 1 \\ 0 & 0 & 3 \\ -1 & 2 & 4 \end{pmatrix}$$

$$C = \begin{pmatrix} 4 & 6 & 1 \\ 2 & 1 & 3 \\ -1 & -1 & 2 \end{pmatrix}$$

(a) Compute $A(B + C)$. **(b)** Compute $AB + AC$.
(c) Compute $(AB)C$. **(d)** Compute $A(BC)$.

56. Let

$$A = \begin{pmatrix} 1 & 2 \\ 3 & 4 \end{pmatrix} \quad B = \begin{pmatrix} 5 & 6 \\ 7 & 8 \end{pmatrix}$$

Let A^2 and B^2 denote the matrix products AA and BB, respectively. Compute each of the following.
(a) $(A + B)(A + B)$ **(b)** $A^2 + 2AB + B^2$
(c) $A^2 + AB + BA + B^2$

57. Let

$$A = \begin{pmatrix} 3 & 5 \\ 7 & 9 \end{pmatrix} \quad B = \begin{pmatrix} 2 & 4 \\ 6 & 8 \end{pmatrix}$$

Compute each of the following.
(a) $A^2 - B^2$ **(b)** $(A - B)(A + B)$
(c) $(A + B)(A - B)$ **(d)** $A^2 + AB - BA - B^2$

58. Let

$$A = \begin{pmatrix} 1 & 0 \\ 0 & 1 \end{pmatrix} \quad B = \begin{pmatrix} 1 & 0 \\ 0 & -1 \end{pmatrix}$$

$$C = \begin{pmatrix} -1 & 0 \\ 0 & 1 \end{pmatrix} \quad D = \begin{pmatrix} -1 & 0 \\ 0 & -1 \end{pmatrix}$$

Complete the multiplication table.

	A	B	C	D
A				
B			D	
C				
D				

Hint: D is the proper entry in the second row, third column because (as you can check) $BC = D$.

59. For this exercise let us agree to write the coordinates (x, y) of a point in the plane as the 2×1 matrix $\begin{pmatrix} x \\ y \end{pmatrix}$.

(a) Let $A = \begin{pmatrix} 1 & 0 \\ 0 & -1 \end{pmatrix}$ and $Z = \begin{pmatrix} x \\ y \end{pmatrix}$. Compute the matrix AZ. After computing AZ, observe that it represents the point obtained by reflecting $\begin{pmatrix} x \\ y \end{pmatrix}$ in the x-axis.

(b) Let $B = \begin{pmatrix} -1 & 0 \\ 0 & 1 \end{pmatrix}$ and $Z = \begin{pmatrix} x \\ y \end{pmatrix}$. Compute the matrix BZ. After computing BZ, observe that it represents the point obtained by reflecting $\begin{pmatrix} x \\ y \end{pmatrix}$ in the y-axis.

(c) Let A, B, and Z represent the matrices defined in parts (a) and (b). Compute the matrix $(AB)Z$, and then interpret it in terms of reflections in the axes.

60. The *trace* of a 2×2 matrix $\begin{pmatrix} a & b \\ c & d \end{pmatrix}$ is defined by

$$\text{tr}\begin{pmatrix} a & b \\ c & d \end{pmatrix} = a + d$$

(a) If $A = \begin{pmatrix} 1 & 2 \\ 3 & 4 \end{pmatrix}$ and $B = \begin{pmatrix} 5 & 6 \\ 7 & 8 \end{pmatrix}$, verify that $\text{tr}(A + B) = \text{tr } A + \text{tr } B$.

(b) If $A = \begin{pmatrix} a & b \\ c & d \end{pmatrix}$ and $B = \begin{pmatrix} e & f \\ g & h \end{pmatrix}$, show that $\text{tr}(A + B) = \text{tr } A + \text{tr } B$.

61. Let $A = \begin{pmatrix} a & b \\ c & d \end{pmatrix}$. The transpose of A is the matrix denoted by A^T and defined by

$$A^T = \begin{pmatrix} a & c \\ b & d \end{pmatrix}$$

In other words, A^T is obtained by switching the columns and the rows of A. Show that the following equations hold for all 2×2 matrices A and B.

(a) $(A + B)^T = A^T + B^T$ **(b)** $(A^T)^T = A$

(c) $(AB)^T = B^T A^T$

62. Find an example of two 2×2 matrices A and B for which $AB = \mathbf{0}$, but neither A nor B is $\mathbf{0}$.

63. **(a)** If $A = \begin{pmatrix} a & b \\ c & d \end{pmatrix}$ and $I = \begin{pmatrix} 1 & 0 \\ 0 & 1 \end{pmatrix}$ show that

$$AI = IA = A$$

(The matrix I is called the identity matrix of order 2.)

(b) Let $A = \begin{pmatrix} a & b \\ c & d \end{pmatrix}$ and assume that $ad - bc \neq 0$. Define the matrix B as follows.

$$B = \frac{1}{ad - bc}\begin{pmatrix} d & -b \\ -c & a \end{pmatrix}$$

Verify that $AB = BA = I$, where I is the matrix defined in part (a).

(c) Solve the matrix equation

$$AX = C$$

when $A = \begin{pmatrix} 3 & -2 \\ -5 & 4 \end{pmatrix}$ and $C = \begin{pmatrix} 1 & 6 \\ 2 & 8 \end{pmatrix}$.

Hint: Calculate the matrix B defined in part (b). What happens when both sides of the given equation are multiplied by B?

9.4 Determinants and Cramer's Rule

A matrix in which there are as many rows as there are columns is called a *square matrix*. Examples of square matrices are

$$\begin{pmatrix} 1 & 2 \\ 3 & 4 \end{pmatrix} \quad \begin{pmatrix} 1 & 2 & 3 \\ 4 & 5 & 6 \\ 7 & 8 & 9 \end{pmatrix} \quad \begin{pmatrix} 3 & 7 & 8 & 9 \\ 5 & 6 & 4 & 3 \\ -9 & 9 & 0 & 1 \\ 1 & 3 & -2 & 1 \end{pmatrix}$$

For each square matrix A, we will assign a number called the *determinant of* A. The determinant of a matrix A may be denoted by det A or $|A|$. Determinants are also denoted simply by replacing the parentheses of matrix notation with vertical lines. Thus, the determinants of the three matrices specified above are, respectively,

$$\begin{vmatrix} 1 & 2 \\ 3 & 4 \end{vmatrix} \quad \begin{vmatrix} 1 & 2 & 3 \\ 4 & 5 & 6 \\ 7 & 8 & 9 \end{vmatrix} \quad \begin{vmatrix} 3 & 7 & 8 & 9 \\ 5 & 6 & 4 & 3 \\ -9 & 9 & 0 & 1 \\ 1 & 3 & -2 & 1 \end{vmatrix}$$

A determinant with n rows and n columns is said to be an *nth-order determinant*. Therefore the determinants we've just written are, respectively, second-, third-, and fourth-order determinants. As with matrices, we speak of the numbers in a determinant as its *entries*. We also number the rows and the columns of a determinant as we do with matrices. However, unlike matrices, each determinant has a definite value. The value of a second-order determinant is defined as follows.

$$\begin{vmatrix} a & b \\ c & d \end{vmatrix} = ad - bc$$

Table 1 provides some examples of how this definition is used to evaluate or *expand* a second-order determinant.

Determinant	Value of Determinant
$\begin{vmatrix} 3 & 7 \\ 5 & 10 \end{vmatrix}$	$\begin{vmatrix} 3 & 7 \\ 5 & 10 \end{vmatrix} = 3(10) - 7(5) = 30 - 35 = -5$
$\begin{vmatrix} 3 & -7 \\ 5 & 10 \end{vmatrix}$	$\begin{vmatrix} 3 & -7 \\ 5 & 10 \end{vmatrix} = 3(10) - (-7)(5) = 30 + 35 = 65$
$\begin{vmatrix} a & a^3 \\ 1 & a^2 \end{vmatrix}$	$\begin{vmatrix} a & a^3 \\ 1 & a^2 \end{vmatrix} = a(a^2) - a^3(1) = a^3 - a^3 = 0$

Table 1

In general, the value of an nth-order determinant ($n > 2$) is defined in terms of certain ($n - 1$)st-order determinants. For instance, the value of a third-order determinant is defined in terms of second-order determinants. We'll use the following example to introduce the necessary terminology here.

$$\begin{vmatrix} 8 & 3 & 5 \\ 2 & 4 & 6 \\ 9 & 1 & 7 \end{vmatrix}$$

Pick a given entry, say 8, and imagine crossing out all entries occupying the same row or the same column as does 8.

$$\begin{vmatrix} \not{8} & \not{3} & \not{5} \\ \not{2} & 4 & 6 \\ \not{9} & 1 & 7 \end{vmatrix}$$

Then you are left with the second-order determinant $\begin{vmatrix} 4 & 6 \\ 1 & 7 \end{vmatrix}$. This second-order determinant is called the *minor* of the entry 8. Similarly, to find the minor of the entry 6 in the original determinant, imagine crossing out all entries that occupy the same row or the same column as does 6.

$$\begin{vmatrix} 8 & 3 & \not{5} \\ \not{2} & \not{4} & \not{6} \\ 9 & 1 & \not{7} \end{vmatrix}$$

You are left with the second-order determinant $\begin{vmatrix} 8 & 3 \\ 9 & 1 \end{vmatrix}$, which by definition is the minor of the entry 6. In the same manner, *the minor of any element is obtained by crossing out the entries occupying the same row or column as the given element.*

Closely related to the minor of an entry in a determinant is the *cofactor* of that entry. The *cofactor* of an entry is defined to be the minor multiplied by $+1$ or -1 according to the scheme displayed in Figure 1.

$$\begin{vmatrix} + & - & + \\ - & + & - \\ + & - & + \end{vmatrix}$$

Figure 1

After looking at an example, we'll give a more formal rule for computing cofactors that will not rely on a figure and that will also apply to larger determinants.

Example 1 Consider the determinant

$$\begin{vmatrix} 1 & 2 & 3 \\ 4 & 5 & 6 \\ 7 & 8 & 9 \end{vmatrix}$$

Compute the minor and the cofactor of the entry 4.

Solution By definition, we have

$$\begin{aligned} (\text{minor of } 4) &= \begin{vmatrix} 2 & 3 \\ 8 & 9 \end{vmatrix} \\ &= 18 - 24 = -6 \end{aligned}$$

Thus the minor of the entry 4 is -6. To compute the cofactor of 4, we first notice that 4 is located in the second row and first column of the given determinant. Upon checking the corresponding position in the display for Figure 1, we see a minus sign. Therefore, we have by definition

$$\begin{aligned} (\text{cofactor of } 4) &= (-1)(\text{minor of } 4) \\ &= (-1)(-6) = 6 \end{aligned}$$

The cofactor of 4 is therefore 6.

The following rule tells us how cofactors may be computed without relying on Figure 1. For reference we also restate the definition of a minor. (You should verify for yourself that this rule yields results that are consistent with Figure 1.)

Minors and Cofactors

The *minor* of an entry b in a determinant is the determinant formed by suppressing the entries in the row and in the column in which b appears.

Suppose that the entry b is in the ith row and the jth column. Then the cofactor of b is given by the expression

$$(-1)^{i+j}(\text{minor of } b)$$

We are prepared now to state the definition that tells us how to evaluate a third-order determinant. Actually, as you'll see later, the definition is quite general and may be applied to determinants of any size.

Definition for the Value of a Determinant

Multiply each entry in the first row of the determinant by its cofactor; then add the results. The value of the determinant is defined to be this sum.

To see how this definition is used, we will evaluate the determinant

$$\begin{vmatrix} 1 & 2 & 3 \\ 4 & 5 & 6 \\ 7 & 8 & 9 \end{vmatrix}$$

The definition tells us to multiply each entry in the first row by its cofactor and then add the results. Carrying out this procedure, we have

$$\begin{aligned}\begin{vmatrix} 1 & 2 & 3 \\ 4 & 5 & 6 \\ 7 & 8 & 9 \end{vmatrix} &= 1\begin{vmatrix} 5 & 6 \\ 8 & 9 \end{vmatrix} - 2\begin{vmatrix} 4 & 6 \\ 7 & 9 \end{vmatrix} + 3\begin{vmatrix} 4 & 5 \\ 7 & 8 \end{vmatrix} \\ &= 1(45 - 48) - 2(36 - 42) + 3(32 - 35) \\ &= -3 - 2(-6) + 3(-3) \\ &= -3 + 12 - 9 \\ &= 0\end{aligned}$$

So the value of this particular determinant is zero. The procedure we've used here is referred to as "expanding the determinant along its first row." The following theorem (stated here without proof) tells us that the value of a determinant can be obtained by expanding along any row or along any column; the results are the same in all cases.

Theorem

Select any row or any column in a determinant and multiply each element in that row or column by its cofactor. Then add the results. The number obtained will be the value of the determinant. (In other words, the number obtained will be the same as that obtained by expanding the determinant along its first row.)

According to this theorem, we could have evaluated the determinant

$$\begin{vmatrix} 1 & 2 & 3 \\ 4 & 5 & 6 \\ 7 & 8 & 9 \end{vmatrix}$$

by expanding it along any row or any column. Let's expand it along the second column and check to see that the result agrees with the value obtained earlier. Expanding along the second column, we have

$$\begin{aligned}\begin{vmatrix} 1 & 2 & 3 \\ 4 & 5 & 6 \\ 7 & 8 & 9 \end{vmatrix} &= -2\begin{vmatrix} 4 & 6 \\ 7 & 9 \end{vmatrix} + 5\begin{vmatrix} 1 & 3 \\ 7 & 9 \end{vmatrix} - 8\begin{vmatrix} 1 & 3 \\ 4 & 6 \end{vmatrix} \\ &= -2(36 - 42) + 5(9 - 21) - 8(6 - 12) \\ &= -2(-6) + 5(-12) - 8(-6) \\ &= 12 - 60 + 48 \\ &= 0 \qquad \text{as obtained previously}\end{aligned}$$

We would have obtained the same result had we chosen to begin with any other row or column. (Exercise 13 at the end of this section asks you to verify this.)

Our principal use for determinants in this section will be in connection with Cramer's Rule for solving systems of equations. However, as the next example demonstrates, determinants are useful in other ways too.

Example 2 Show that the equation of the straight line passing through the two distinct points (x_1, y_1) and (x_2, y_2) may be written as

$$\begin{vmatrix} x & y & 1 \\ x_1 & y_1 & 1 \\ x_2 & y_2 & 1 \end{vmatrix} = 0$$

Solution Expanding the determinant along its first row shows that the given equation is equivalent to

$$x(y_1 - y_2) - y(x_1 - x_2) + 1(x_1y_2 - y_1x_2) = 0$$

or

$$\underbrace{(y_1 - y_2)}_{A}x + \underbrace{(x_2 - x_1)}_{B}y + \underbrace{(x_1y_2 - y_1x_2)}_{C} = 0 \qquad (1)$$

Letting the constants A, B, and C be defined as indicated, we see that equation (1) has the form $Ax + By + C = 0$. Now this equation represents a straight line provided that A and B are not both zero. That A and B are not both zero follows from their definitions and the fact that the points (x_1, y_1) and (x_2, y_2) are distinct. Furthermore, the line described by equation (1) passes through both of the given points because both sets of coordinates satisfy the equation. (Verify this last statement for yourself.) We have now shown that the given equation represents a straight line passing through the given points. This completes the demonstration.

There are a number of rules that make determinants easier to evaluate. Here are three of the most useful. (Suggestions for proving these and other rules may be found in the exercises at the end of this section.)

Three Rules for Manipulating Determinants	**Examples**
1. If each entry in a given row is multiplied by the constant k, then the value of the determinant is multiplied by k. This is true for columns also.	$10\begin{vmatrix} 1 & 3 & 4 \\ 1 & 2 & 3 \\ 4 & 5 & 6 \end{vmatrix} = \begin{vmatrix} 10 & 30 & 40 \\ 1 & 2 & 3 \\ 4 & 5 & 6 \end{vmatrix}$ $k\begin{vmatrix} a & b & c \\ d & e & f \\ g & h & i \end{vmatrix} = \begin{vmatrix} a & kb & c \\ d & ke & f \\ g & kh & i \end{vmatrix}$
2. If a multiple of one row is added to another row, the value of the determinant is not changed. This applies to columns also.	$\begin{vmatrix} a & b & c \\ d & e & f \\ g & h & i \end{vmatrix} = \begin{vmatrix} a + kb & b & c \\ d + ke & e & f \\ g + kh & h & i \end{vmatrix}$
3. If two rows are interchanged, or two columns are interchanged, then the value of the determinant is multiplied by -1.	$\begin{vmatrix} 1 & 2 & 3 \\ 4 & 5 & 6 \\ a & b & c \end{vmatrix} = -\begin{vmatrix} 4 & 5 & 6 \\ 1 & 2 & 3 \\ a & b & c \end{vmatrix}$

Example 3 Evaluate the determinant

$$\begin{vmatrix} 15 & 14 & 26 \\ 18 & 17 & 32 \\ 21 & 20 & 42 \end{vmatrix}$$

Solution

$$\begin{vmatrix} 15 & 14 & 26 \\ 18 & 17 & 32 \\ 21 & 20 & 42 \end{vmatrix} = (3 \times 2)\begin{vmatrix} 5 & 14 & 13 \\ 6 & 17 & 16 \\ 7 & 20 & 21 \end{vmatrix}$$

We used Rule 1 to factor a 3 from the first column and a 2 from the third column.

$$= 6\begin{vmatrix} 5 & 1 & 13 \\ 6 & 1 & 16 \\ 7 & -1 & 21 \end{vmatrix}$$

We used Rule 2 to subtract the third column from the second column.

$$= 6\begin{vmatrix} 12 & 0 & 34 \\ 13 & 0 & 37 \\ 7 & -1 & 21 \end{vmatrix}$$

We used Rule 2 to add the third row to the first and second rows.

$$= 6\left[-\left(-1\begin{vmatrix} 12 & 34 \\ 13 & 37 \end{vmatrix}\right)\right]$$

We expanded the determinant along the second column.

$$= 6\begin{vmatrix} 12 & 34 \\ 1 & 3 \end{vmatrix}$$

We used Rule 2 to subtract the first row from the second row.

$$= 6(36 - 34) = 12$$

The value of the given determinant is therefore 12. Notice the general strategy employed here. We used Rules 1 and 2 until one column contained two zeros. At that point it was a simple matter to expand the determinant along that column.

Example 4 Show that

$$\begin{vmatrix} 1 & 1 & 1 \\ a & b & c \\ a^2 & b^2 & c^2 \end{vmatrix} = (b - a)(c - a)(c - b)$$

Solution

$$\begin{vmatrix} 1 & 1 & 1 \\ a & b & c \\ a^2 & b^2 & c^2 \end{vmatrix} = \begin{vmatrix} 1 & 0 & 0 \\ a & b-a & c-a \\ a^2 & b^2-a^2 & c^2-a^2 \end{vmatrix}$$

We used Rule 2 to subtract the first column from the second and the third columns.

$$= \begin{vmatrix} 1 & 0 & 0 \\ a & b-a & c-a \\ a^2 & (b-a)(b+a) & (c-a)(c+a) \end{vmatrix}$$

$$= (b - a)(c - a)\begin{vmatrix} 1 & 0 & 0 \\ a & 1 & 1 \\ a^2 & b+a & c+a \end{vmatrix}$$

We used Rule 1 to factor $(b - a)$ from the second column and $(c - a)$ from the third column.

$$= (b - a)(c - a)[(c + a) - (b + a)]$$
$$= (b - a)(c - a)(c + a - b - a)$$

We expanded the determinant along the first row.

$$= (b - a)(c - a)(c - b) \quad \text{as required}$$

The definition on page 463 gives us a procedure for evaluating any third-order determinant. That same procedure can also be used to define fourth-order (or larger) determinants. Consider for example the fourth-order determinant A given by

$$A = \begin{vmatrix} 25 & 40 & 5 & 10 \\ 9 & 0 & 3 & 6 \\ -2 & 3 & 11 & -17 \\ -3 & 4 & 7 & 2 \end{vmatrix}$$

By definition then, we could evaluate this determinant by selecting the first row, multiplying each entry by its cofactor, and then adding the results. This yields

$$A = 25\begin{vmatrix} 0 & 3 & 6 \\ 3 & 11 & -17 \\ 4 & 7 & 2 \end{vmatrix} - 40\begin{vmatrix} 9 & 3 & 6 \\ -2 & 11 & -17 \\ -3 & 7 & 2 \end{vmatrix} + 5\begin{vmatrix} 9 & 0 & 6 \\ -2 & 3 & -17 \\ -3 & 4 & 2 \end{vmatrix} - 10\begin{vmatrix} 9 & 0 & 3 \\ -2 & 3 & 11 \\ -3 & 4 & 7 \end{vmatrix}$$

The problem now is reduced to evaluating four third-order determinants. Had we instead expanded down the second column, the ensuing work would be somewhat less because of the zero in that column. Nevertheless, there would still be three third-order determinants to evaluate.

Because of the amount of computation involved, we in fact rarely evaluate fourth-order determinants directly from the definition. Instead we use the three rules (on page 465) to first simplify matters. In the example that follows, we show how this is done. (We will accept the fact that the three rules are applicable in evaluating determinants of any size.)

Example 5 Evaluate the determinant A given by

$$A = \begin{vmatrix} 25 & 40 & 5 & 10 \\ 9 & 0 & 3 & 6 \\ -2 & 3 & 11 & -17 \\ -3 & 4 & 7 & 2 \end{vmatrix}$$

Solution

$$A = 5 \times 3\begin{vmatrix} 5 & 8 & 1 & 2 \\ 3 & 0 & 1 & 2 \\ -2 & 3 & 11 & -17 \\ -3 & 4 & 7 & 2 \end{vmatrix} = 15\begin{vmatrix} 2 & 8 & 0 & 0 \\ 3 & 0 & 1 & 2 \\ -35 & 3 & 0 & -39 \\ -24 & 4 & 0 & -12 \end{vmatrix}$$

Factor 5 from the first row and 3 from the second row.

Subtract the second row from the first. Subtract 11 times the second row from the third. Subtract seven times the second row from the fourth.

$$= 15 \times 2 \times 4\begin{vmatrix} 1 & 4 & 0 & 0 \\ 3 & 0 & 1 & 2 \\ -35 & 3 & 0 & -39 \\ -6 & 1 & 0 & -3 \end{vmatrix} = 120\left[-1\begin{vmatrix} 1 & 4 & 0 \\ -35 & 3 & -39 \\ -6 & 1 & -3 \end{vmatrix}\right]$$

Factor 2 from the first row. Factor 4 from the fourth row.

Expand the determinant along the third column.

$$= -120(-3)\begin{vmatrix} 1 & 4 & 0 \\ -35 & 3 & 13 \\ -6 & 1 & 1 \end{vmatrix} = 360\begin{vmatrix} 1 & 0 & 0 \\ -35 & 143 & 13 \\ -6 & 25 & 1 \end{vmatrix}$$

Factor (-3) from the third column.

Subtract four times the first column from the second.

$$= 360\begin{vmatrix} 143 & 13 \\ 25 & 1 \end{vmatrix} = 360\begin{vmatrix} -7 & 7 \\ 25 & 1 \end{vmatrix}$$

Expand the determinant along the first row.

Subtract 6 times the second row from the first row.

$$= 360(-7 - 175) = -65520$$

Do the arithmetic!

The value of the given fourth-order determinant is therefore -65520.

Determinants can be used to solve certain systems of linear equations in which there are as many unknowns as there are equations. In the box that follows we state *Cramer's Rule** for solving a system of three linear equations in three unknowns. A more general, but entirely similar, version of Cramer's Rule holds for n equations in n unknowns.

Cramer's Rule

Consider the system

$$\begin{cases} a_1x + b_1y + c_1z = d_1 \\ a_2x + b_2y + c_2z = d_2 \\ a_3x + b_3y + c_3z = d_3 \end{cases}$$

Let the four determinants D, D_x, D_y, and D_z be defined as follows.

$$D = \begin{vmatrix} a_1 & b_1 & c_1 \\ a_2 & b_2 & c_2 \\ a_3 & b_3 & c_3 \end{vmatrix} \qquad D_x = \begin{vmatrix} d_1 & b_1 & c_1 \\ d_2 & b_2 & c_2 \\ d_3 & b_3 & c_3 \end{vmatrix}$$

$$D_y = \begin{vmatrix} a_1 & d_1 & c_1 \\ a_2 & d_2 & c_2 \\ a_3 & d_3 & c_3 \end{vmatrix} \qquad D_z = \begin{vmatrix} a_1 & b_1 & d_1 \\ a_2 & b_2 & d_2 \\ a_3 & b_3 & d_3 \end{vmatrix}$$

Then if $D \neq 0$, the unique values of x, y, and z satisfying the system are given by

$$x = \frac{D_x}{D} \qquad y = \frac{D_y}{D} \qquad z = \frac{D_z}{D}$$

Before discussing a proof of Cramer's Rule, let us look at an example showing how the rule is applied.

*The rule is named after one of its discoverers, the Swiss mathematician Gabriel Cramer (1704–1752).

Example 6 Use Cramer's Rule to find all solutions of the system of equations

$$\begin{cases} 2x + 2y - 3z = -20 \\ x - 4y + z = 6 \\ 4x - y + 2z = -1 \end{cases}$$

Solution The determinants D, D_x, D_y, and D_z are as follows.

$$D = \begin{vmatrix} 2 & 2 & -3 \\ 1 & -4 & 1 \\ 4 & -1 & 2 \end{vmatrix} \qquad D_x = \begin{vmatrix} -20 & 2 & -3 \\ 6 & -4 & 1 \\ -1 & -1 & 2 \end{vmatrix}$$

$$D_y = \begin{vmatrix} 2 & -20 & -3 \\ 1 & 6 & 1 \\ 4 & -1 & 2 \end{vmatrix} \qquad D_z = \begin{vmatrix} 2 & 2 & -20 \\ 1 & -4 & 6 \\ 4 & -1 & -1 \end{vmatrix}$$

We will show the calculations for evaluating D.

$$D = \begin{vmatrix} 2 & 2 & -3 \\ 1 & -4 & 1 \\ 4 & -1 & 2 \end{vmatrix} = \begin{vmatrix} 0 & 10 & -5 \\ 1 & -4 & 1 \\ 0 & 15 & -2 \end{vmatrix}$$

Subtract twice the second row from the first. Subtract four times the second row from the third.

Since we now have two zeros in the first column, it is an easy matter to expand D along that column to obtain

$$D = -1\begin{vmatrix} 10 & -5 \\ 15 & -2 \end{vmatrix}$$

$$= -1[-20 - (-75)] = -1(55) = -55$$

The value of D is therefore -55. (Since this value is nonzero, Cramer's Rule does apply.) As Exercise 34 at the end of this section asks you to check, the values of the other three determinants are

$$D_x = 144 \qquad D_y = 61 \qquad D_z = -230$$

By Cramer's Rule then, the unique values of x, y, and z that satisfy the system are

$$x = \frac{D_x}{D} = \frac{144}{-55} = -\frac{144}{55}$$

$$y = \frac{D_y}{D} = \frac{61}{-55} = -\frac{61}{55}$$

$$z = \frac{D_z}{D} = \frac{-230}{-55} = \frac{46}{11}$$

One way in which Cramer's Rule can be proven involves using Gaussian elimination to solve the system

$$\begin{cases} a_1x + b_1y + c_1z = d_1 \\ a_2x + b_2y + c_2z = d_2 \\ a_3x + b_3y + c_3z = d_3 \end{cases} \tag{2}$$

A much shorter and simpler proof, however, has been found by D. E. Whitford and M. S. Klamkin.* This is the proof we give here. The proof makes effective use of the rules employed in this section for manipulating determinants.

Consider the system of equations (2) and assume that $D \neq 0$. We shall show that if x, y, and z satisfy the system, then in fact $x = D_x/D$ with similar equations giving y and z. (Exercise 67 at the end of this section then shows how to check that these values indeed satisfy the given system.) We have

$$D_x = \begin{vmatrix} d_1 & b_1 & c_1 \\ d_2 & b_2 & c_2 \\ d_3 & b_3 & c_3 \end{vmatrix}$$ by definition

$$= \begin{vmatrix} (a_1x + b_1y + c_1z) & b_1 & c_1 \\ (a_2x + b_2y + c_2z) & b_2 & c_2 \\ (a_3x + b_3y + c_3z) & b_3 & c_3 \end{vmatrix}$$ We used the equations in (2) to substitute for d_1, d_2, and d_3.

$$= \begin{vmatrix} a_1x & b_1 & c_1 \\ a_2x & b_2 & c_2 \\ a_3x & b_3 & c_3 \end{vmatrix}$$ From the first column we subtracted y times the second column as well as z times the third column.

$$= x\begin{vmatrix} a_1 & b_1 & c_1 \\ a_2 & b_2 & c_2 \\ a_3 & b_3 & c_3 \end{vmatrix}$$ We factored x out of the first column.

$$= xD$$ by definition

We now have $D_x = xD$, which is equivalent to $x = \dfrac{D_x}{D}$, as required. The formulas for y and z are obtained similarly.

Exercise Set 9.4

In Exercises 1–6, evaluate the determinants.

1. **(a)** $\begin{vmatrix} 2 & -17 \\ 1 & 6 \end{vmatrix}$ **(b)** $\begin{vmatrix} 1 & 6 \\ 2 & -17 \end{vmatrix}$

2. **(a)** $\begin{vmatrix} 1 & 0 \\ 0 & 1 \end{vmatrix}$ **(b)** $\begin{vmatrix} 0 & 1 \\ 0 & 1 \end{vmatrix}$

3. **(a)** $\begin{vmatrix} 5 & 7 \\ 500 & 700 \end{vmatrix}$ **(b)** $\begin{vmatrix} 5 & 500 \\ 7 & 700 \end{vmatrix}$

4. **(a)** $\begin{vmatrix} -8 & -3 \\ 4 & -5 \end{vmatrix}$ **(b)** $\begin{vmatrix} -3 & -8 \\ -5 & 4 \end{vmatrix}$

5. $\begin{vmatrix} \sqrt{2} - 1 & \sqrt{2} \\ \sqrt{2} & \sqrt{2} + 1 \end{vmatrix}$

6. $\begin{vmatrix} \sqrt{3} + \sqrt{2} & 1 + \sqrt{5} \\ 1 - \sqrt{5} & \sqrt{3} - \sqrt{2} \end{vmatrix}$

In Exercises 7–12, refer to the following determinant.

$$\begin{vmatrix} -6 & 3 & 8 \\ 5 & -4 & 1 \\ 10 & 9 & -10 \end{vmatrix}$$

7. Evaluate the minor of the entry 3.

8. Evaluate the cofactor of the entry 3.

9. Evaluate the minor of -10.

10. Evaluate the cofactor of -10.

11. **(a)** Multiply each entry in the first row by its minor and find the sum of the results.
(b) Multiply each entry in the first row by its cofactor and find the sum of the results.
(c) Does the answer in part (a) or part (b) give you the value of the determinant?

12. **(a)** Multiply each entry in the first column by its cofactor and find the sum of the results.
(b) Same as part (a) but use the second column.
(c) Same as part (a) but use the third column.

13. Let $A = \begin{vmatrix} 1 & 2 & 3 \\ 4 & 5 & 6 \\ 7 & 8 & 9 \end{vmatrix}$. Evaluate A by expanding it along the indicated row or column.

*The proof was published in the *American Mathematical Monthly,* Vol. 60 (1953), pp. 186–187.

(a) second row **(b)** third row
(c) first column **(d)** third column

In Exercises 14–21, evaluate the determinants.

14. $\begin{vmatrix} 1 & 2 & -1 \\ 2 & -1 & 1 \\ 4 & 0 & 2 \end{vmatrix}$ **15.** $\begin{vmatrix} 5 & 10 & 15 \\ 1 & 2 & 3 \\ -9 & 11 & 7 \end{vmatrix}$

16. $\begin{vmatrix} 8 & 4 & 2 \\ 3 & 9 & 3 \\ -2 & 8 & 6 \end{vmatrix}$ **17.** $\begin{vmatrix} 1 & 2 & -3 \\ 4 & 5 & -9 \\ 0 & 0 & 1 \end{vmatrix}$

18. $\begin{vmatrix} 3 & 0 & 0 \\ 0 & 19 & 0 \\ 0 & 0 & 10 \end{vmatrix}$ **19.** $\begin{vmatrix} -6 & -8 & 18 \\ 25 & 12 & 15 \\ -9 & 4 & 13 \end{vmatrix}$

20. $\begin{vmatrix} 23 & 0 & 47 \\ -37 & 0 & 18 \\ 14 & 0 & 25 \end{vmatrix}$ **21.** $\begin{vmatrix} 16 & 0 & -64 \\ -8 & 15 & -12 \\ 30 & -20 & 10 \end{vmatrix}$

22. Use the method of Example 4 in the text to show that

$$\begin{vmatrix} 1 & 1 & 1 \\ a & b & c \\ a^3 & b^3 & c^3 \end{vmatrix} = (b-a)(c-a)(c^2-b^2+ac-ab)$$
$$= (b-a)(c-a)(c-b)(a+b+c)$$

23. Use the method of Example 4 in the text to express the following determinant as a product.

$$\begin{vmatrix} 1 & x & x^2 \\ 1 & y & y^2 \\ 1 & z & z^2 \end{vmatrix}$$

24. Show that

$$\begin{vmatrix} 1 & 1 & 1 \\ a^2 & b^2 & c^2 \\ a^3 & b^3 & c^3 \end{vmatrix}$$
$$= (b-a)(c-a)(bc^2-b^2c+ac^2-ab^2)$$
$$= (b-a)(c-a)(c-b)(bc+ac+ab)$$

25. Simplify the following.

$$\begin{vmatrix} 1 & 1 & 1 \\ 1 & 1+x & 1 \\ 1 & 1 & 1+y \end{vmatrix}$$

26. Use the method of Example 4 in the text to express the following determinant as a product of three factors.

$$\begin{vmatrix} 1 & 1 & 1 \\ a & b & c \\ bc & ca & ab \end{vmatrix}$$

Example 2 in the text shows that the equation of a nonvertical line passing through (x_1, y_1) *and* (x_2, y_2) *is given by*

$$\begin{vmatrix} x & y & 1 \\ x_1 & y_1 & 1 \\ x_2 & y_2 & 1 \end{vmatrix} = 0$$

Use this result in Exercises 27–29 to find the equation of the line passing through the given points. Write your answer in the form $y = mx + b$.

27. $(7, -6)$ and $(10, 3)$ **28.** $(-8, 5)$ and $(-6, 1)$
29. $(0, 0)$ and $(\sqrt{3}, 1)$

In Exercises 30–33, evaluate the determinants.

30. $\begin{vmatrix} 1 & -1 & 0 & 2 \\ 0 & 1 & -1 & 0 \\ 2 & 1 & 0 & -1 \\ -2 & 2 & 1 & 1 \end{vmatrix}$ **31.** $\begin{vmatrix} 3 & -2 & 3 & 4 \\ 1 & 4 & -3 & 2 \\ 6 & 3 & -6 & -3 \\ -1 & 0 & 1 & 5 \end{vmatrix}$

32. $\begin{vmatrix} 2 & 0 & 0 & 0 \\ 0 & 3 & 0 & 0 \\ 0 & 0 & 4 & 0 \\ 0 & 0 & 0 & 5 \end{vmatrix}$ **33.** $\begin{vmatrix} 7 & -8 & 1 & 2 \\ 21 & 4 & 3 & -1 \\ -35 & 8 & 3 & -2 \\ 14 & 16 & 0 & 1 \end{vmatrix}$

34. Verify the following statements.

(a) $\begin{vmatrix} -20 & 2 & -3 \\ 6 & -4 & 1 \\ -1 & -1 & 2 \end{vmatrix} = 144$

(b) $\begin{vmatrix} 2 & -20 & -3 \\ 1 & 6 & 1 \\ 4 & -1 & 2 \end{vmatrix} = 61$

(c) $\begin{vmatrix} 2 & 2 & -20 \\ 1 & -4 & 6 \\ 4 & -1 & -1 \end{vmatrix} = -230$

35. Consider the following system.

$$\begin{cases} x + y + 3z = 0 \\ x + 2y + 5z = 0 \\ x - 4y - 8z = 0 \end{cases}$$

(a) Without doing any calculations, find one obvious solution of this system.
(b) Calculate the determinant D.
(c) List all solutions of this system.

In Exercises 36–45, use Cramer's Rule to solve those systems for which $D \neq 0$. *In cases where* $D = 0$, *use Gaussian elimination or matrix methods.*

36. $\begin{cases} 3x + 4y - z = 5 \\ x - 3y + 2z = 2 \\ 5x - 6z = -7 \end{cases}$ **37.** $\begin{cases} 3A - B - 4C = 3 \\ A + 2B - 3C = 9 \\ 2A - B + 2C = -8 \end{cases}$

38. $\begin{cases} 3x + 2y - z = -6 \\ 2x - 3y - 4z = -11 \\ x + y + z = 5 \end{cases}$ **39.** $\begin{cases} 5x - 3y - z = 16 \\ 2x + y - 3z = 5 \\ 3x - 2y + 2z = 5 \end{cases}$

40. $\begin{cases} 2x + 5y + 2z = 0 \\ 3x - y - 4z = 0 \\ x + 2y - 3z = 0 \end{cases}$ **41.** $\begin{cases} 4u + 3v - 2w = 14 \\ u + 2v - 3w = 6 \\ 2u - v + 4w = 2 \end{cases}$

42. $\begin{cases} 12x - 11z = 13 \\ 6x + 6y - 4z = 26 \\ 6x + 2y - 5z = 13 \end{cases}$ **43.** $\begin{cases} 3x + 4y + 2z = 1 \\ 4x + 6y + 2z = 7 \\ 2x + 3y + z = 11 \end{cases}$

44. $\begin{cases} x + y + z + w = -7 \\ x - y + z - w = -11 \\ 2x - 2y - 3z - 3w = 26 \\ 3x + 2y + z - w = -9 \end{cases}$

45. $\begin{cases} 2A - B - 3C + 2D = -2 \\ A - 2B + C - 3D = 4 \\ 3A - 4B + 2C - 4D = 12 \\ 2A + 3B - C - 2D = -4 \end{cases}$

46. Find all values of x for which

$$\begin{vmatrix} x-4 & 0 & 0 \\ 0 & x+4 & 0 \\ 0 & 0 & x+1 \end{vmatrix} = 0$$

47. Find all values of x for which

$$\begin{vmatrix} 1 & x & x^2 \\ 1 & 1 & 1 \\ 4 & 5 & 0 \end{vmatrix} = 0$$

48. Evaluate the following.

(a) $\begin{vmatrix} 0 & 1 & 1 \\ 1 & 0 & 1 \\ 1 & 1 & 0 \end{vmatrix}$ **(b)** $\begin{vmatrix} 0 & 1 & 1 & 1 \\ 1 & 0 & 1 & 1 \\ 1 & 1 & 0 & 1 \\ 1 & 1 & 1 & 0 \end{vmatrix}$

49. By expanding the determinant $\begin{vmatrix} a & b & c \\ a & b & c \\ d & e & f \end{vmatrix}$ down the first column, show that its value is zero.

50. By expanding the determinant $\begin{vmatrix} ka & kb & kc \\ d & e & f \\ g & h & i \end{vmatrix}$ along its first row, show that it is equal to

$$k\begin{vmatrix} a & b & c \\ d & e & f \\ g & h & i \end{vmatrix}$$

51. Show that

$$\begin{vmatrix} a_1 + A_1 & b_1 & c_1 \\ a_2 + A_2 & b_2 & c_2 \\ a_3 + A_3 & b_3 & c_3 \end{vmatrix} = \begin{vmatrix} a_1 & b_1 & c_1 \\ a_2 & b_2 & c_2 \\ a_3 & b_3 & c_3 \end{vmatrix} + \begin{vmatrix} A_1 & b_1 & c_1 \\ A_2 & b_2 & c_2 \\ A_3 & b_3 & c_3 \end{vmatrix}$$

52. Show that

$$\begin{vmatrix} a_1 & b_1 & c_1 \\ a_2 & b_2 & c_2 \\ a_3 & b_3 & c_3 \end{vmatrix} = \begin{vmatrix} a_1 + kb_1 & b_1 & c_1 \\ a_2 + kb_2 & b_2 & c_2 \\ a_3 + kb_3 & b_3 & c_3 \end{vmatrix}$$

53. By expanding each determinant along a row or column, show that

$$\begin{vmatrix} a_1 & b_1 & c_1 \\ a_2 & b_2 & c_2 \\ a_3 & b_3 & c_3 \end{vmatrix} = -\begin{vmatrix} a_2 & b_2 & c_2 \\ a_1 & b_1 & c_1 \\ a_3 & b_3 & c_3 \end{vmatrix}$$

54. Solve for x in terms of a, b, and c.

$$\begin{vmatrix} a & a & x \\ c & c & c \\ b & x & b \end{vmatrix} = 0 \qquad (c \neq 0)$$

***55.** Show that

$$\begin{vmatrix} 1+a & 1 & 1 & 1 \\ 1 & 1+b & 1 & 1 \\ 1 & 1 & 1+c & 1 \\ 1 & 1 & 1 & 1+d \end{vmatrix} = abcd\left(\frac{1}{a} + \frac{1}{b} + \frac{1}{c} + \frac{1}{d} + 1\right)$$

56. Show that

$$\begin{vmatrix} 1 & a & a^2 \\ a^2 & 1 & a \\ a & a^2 & 1 \end{vmatrix} = (a^3 - 1)^2$$

57. Consider the determinant

$$D = \begin{vmatrix} a_1 & b_1 & c_1 \\ a_2 & b_2 & c_2 \\ a_3 & b_3 & c_3 \end{vmatrix}$$

Let A_1 denote the cofactor of a_1, let B_1 denote the cofactor of b_1, and so on. Prove that

$$\begin{vmatrix} B_2 & C_2 \\ B_3 & C_3 \end{vmatrix} = a_1 D$$

58. Show that

$$\begin{vmatrix} 1 & bc & b+c \\ 1 & ca & c+a \\ 1 & ab & a+b \end{vmatrix} = (b-c)(c-a)(a-b)$$

59. Find the value of each determinant

(a) $\begin{vmatrix} 1 & 0 & 0 \\ x & 1 & 0 \\ x & y & 1 \end{vmatrix}$ **(b)** $\begin{vmatrix} 1 & 0 & 0 & 0 \\ x & 1 & 0 & 0 \\ x & y & 1 & 0 \\ x & y & z & 1 \end{vmatrix}$

60. Evaluate the determinant

$$\begin{vmatrix} a & b & c \\ a & a+b & a+b+c \\ a & 2a+b & 3a+2b+c \end{vmatrix}$$

61. Show that

$$\begin{vmatrix} 1 & a & a & a \\ 1 & b & a & a \\ 1 & a & b & a \\ 1 & a & a & b \end{vmatrix} = (b-a)^3$$

62. Show that

$$\begin{vmatrix} a & 1 & 1 & 1 \\ 1 & a & 1 & 1 \\ 1 & 1 & a & 1 \\ 1 & 1 & 1 & a \end{vmatrix} = (a-1)^3(a+3)$$

63. Solve the system for x, y, and z. (Assume that the values of a, b, and c are all distinct and all nonzero.)

$$\begin{cases} ax + by + cz = k \\ a^2x + b^2y + c^2z = k^2 \\ a^3x + b^3y + c^3z = k^3 \end{cases}$$

64. Show that

$$\begin{vmatrix} x^2 & y^2 & 2xy \\ y^2 & 2xy & x^2 \\ 2xy & x^2 & y^2 \end{vmatrix} = -(x^3 + y^3)^2$$

***65.** Show that

$$\begin{vmatrix} x+a & a & a & a \\ b & x+b & b & b \\ c & c & x+c & c \\ d & d & d & x+d \end{vmatrix} = x^3(a + b + c + d + x)$$

Hint: To the first row add each of the three other rows. For convenience, let $t = a + b + c + d + x$.

***66.** Show that the area of the shaded triangle in figure (a) is

$$\frac{1}{2}\begin{vmatrix} a & b \\ c & d \end{vmatrix}$$

Hint: Figure (b) indicates how the required area may be found using a rectangle and three right triangles.

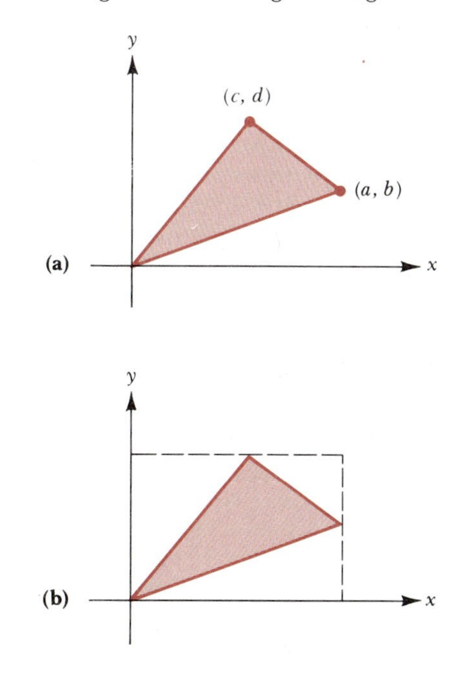

67. This exercise completes some aspects of the derivation of Cramer's Rule in the text. Using the same notation, and assuming $D \neq 0$, we need to show that the values $x = D_x/D$, $y = D_y/D$, and $z = D_z/D$ satisfy the equations in (2). We shall show that these values satisfy the first equation in (2), the verification for the other equations being entirely similar.

(a) Check that substituting the values $x = D_x/D$, $y = D_y/D$, and $z = D_z/D$ in the first equation of (2) yields an equation equivalent to

$$a_1D_x + b_1D_y + c_1D_z - d_1D = 0$$

(b) Show that the equation in (a) may be written as

$$a_1\begin{vmatrix} b_1 & c_1 & d_1 \\ b_2 & c_2 & d_2 \\ b_3 & c_3 & d_3 \end{vmatrix} - b_1\begin{vmatrix} a_1 & c_1 & d_1 \\ a_2 & c_2 & d_2 \\ a_3 & c_3 & d_3 \end{vmatrix} + c_1\begin{vmatrix} a_1 & b_1 & d_1 \\ a_2 & b_2 & d_2 \\ a_3 & b_3 & d_3 \end{vmatrix} - d_1\begin{vmatrix} a_1 & b_1 & c_1 \\ a_2 & b_2 & c_2 \\ a_3 & b_3 & c_3 \end{vmatrix} = 0$$

(c) Show that the equation in (b) may be written as

$$\begin{vmatrix} a_1 & b_1 & c_1 & d_1 \\ a_1 & b_1 & c_1 & d_1 \\ a_2 & b_2 & c_2 & d_2 \\ a_3 & b_3 & c_3 & d_3 \end{vmatrix} = 0$$

(d) Now explain why the equation in (c) indeed holds.

68. Assuming that $p + q + r = 0$, solve the following equation for x.

$$\begin{vmatrix} p-x & r & q \\ r & q-x & p \\ q & p & r-x \end{vmatrix} = 0$$

Hint: After adding certain rows or columns, the quantity $p + q + r$ will appear.

***69.** The figure shows a parabola that passes through the origin and through the two points (x_1, y_1) and (x_2, y_2). Show that the x-coordinate of the vertex is given by

$$x = \frac{\begin{vmatrix} x_1^2 & y_1 \\ x_2^2 & y_2 \end{vmatrix}}{2\begin{vmatrix} x_1 & y_1 \\ x_2 & y_2 \end{vmatrix}}$$

y

(x_1, y_1)

(x_2, y_2)

x

9.5 Nonlinear Systems of Equations

In the previous sections of this chapter, we looked at several techniques for solving systems of linear equations. In particular, we saw that if solutions exist they can always be found by Gaussian elimination. In the present section we consider *nonlinear systems* of equations; that is, systems in which at least one of the equations is not linear. There is no single technique that serves to solve all nonlinear systems. However, simple substitution will often suffice. The work in this section focuses on examples showing some of the more common approaches. In all the examples and the exercises, we will be concerned exclusively with solutions (x, y) in which both x and y are real numbers. In Example 1, the system consists of one linear and one quadratic equation. Such a system can always be solved by substitution.

Example 1 Find all solutions (x, y) of the following system, where x and y are real numbers.

$$\begin{cases} 2x + y = 1 \\ y = 4 - x^2 \end{cases}$$

Solution Use the second equation to substitute for y in the first equation. This will yield an equation with only one unknown:

$$\begin{aligned} 2x + (4 - x^2) &= 1 \\ -x^2 + 2x + 3 &= 0 \\ x^2 - 2x - 3 &= 0 \\ (x - 3)(x + 1) &= 0 \end{aligned}$$

$$x - 3 = 0 \quad\Big|\quad x + 1 = 0$$
$$x = 3 \quad\Big|\quad x = -1$$

The values $x = 3$ and $x = -1$ can now be substituted back into either of the original equations. Substituting $x = 3$ in the equation $y = 4 - x^2$ yields $y = -5$. Similarly, substituting $x = -1$ in the equation $y = 4 - x^2$ gives us $y = 3$. We've now determined the two ordered pairs $(3, -5)$ and $(-1, 3)$. As you can easily check, both of these are solutions of the given system. Figure 1 displays the graphical interpretation of this result. The line $2x + y = 1$ intersects the parabola $y = 4 - x^2$ at the points $(-1, 3)$ and $(3, -5)$.

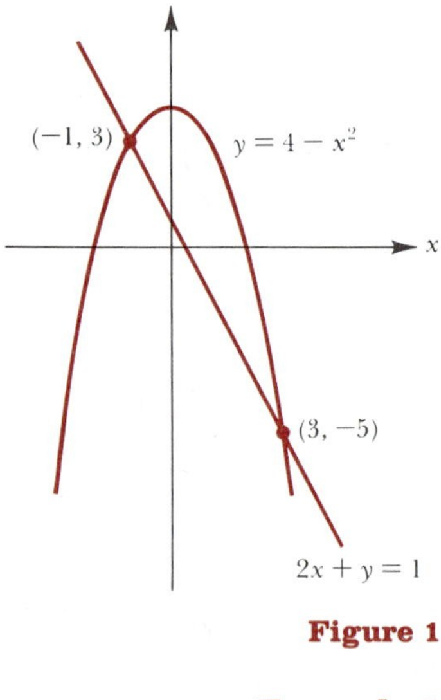

Figure 1

Example 2 Where do the graphs of the parabola $y = x^2$ and the circle $x^2 + y^2 = 1$ intersect? See Figure 2.

Solution The system that we wish to solve is

$$\begin{cases} y = x^2 & (1) \\ x^2 + y^2 = 1 & (2) \end{cases}$$

In view of equation (1), we may replace the x^2 term of equation (2) by y. Doing this yields

$$y + y^2 = 1$$

or

$$y^2 + y - 1 = 0$$

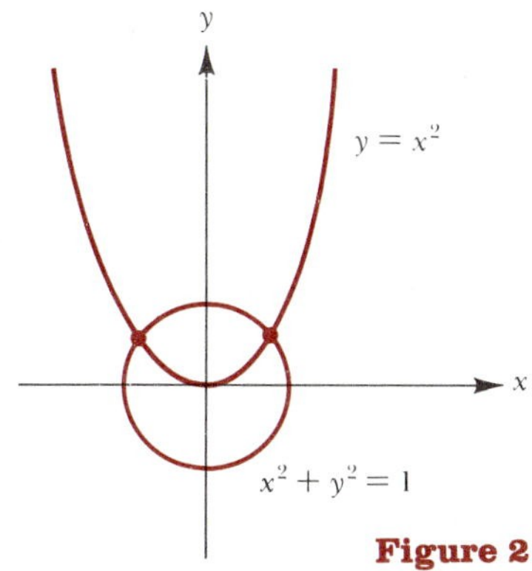

Figure 2

This last equation can be solved by using the quadratic formula with $a = 1$, $b = 1$, and $c = -1$. As you can check, the results are

$$y = \frac{-1 + \sqrt{5}}{2} \quad \text{and} \quad y = \frac{-1 - \sqrt{5}}{2}$$

However, from Figure 2 it is clear that the y-coordinate at each intersection point is positive. Therefore we discard the negative number $y = \dfrac{-1 - \sqrt{5}}{2}$ from further consideration in this context. Substituting the positive number $y = \dfrac{-1 + \sqrt{5}}{2}$ back in the equation $y = x^2$ gives us

$$\frac{-1 + \sqrt{5}}{2} = x^2$$

Therefore

$$\pm\sqrt{\frac{-1 + \sqrt{5}}{2}} = x$$

By choosing the positive square root here, we obtain the x-coordinate for the intersection point in the first quadrant. The point is therefore

$$\left(\sqrt{\frac{-1 + \sqrt{5}}{2}}, \frac{-1 + \sqrt{5}}{2}\right) \approx (0.79, 0.62) \qquad \text{using a calculator}$$

Similarly, the negative square root yields the x-coordinate for the intersection point in the second quadrant. That point then is

$$\left(-\sqrt{\frac{-1 + \sqrt{5}}{2}}, \frac{-1 + \sqrt{5}}{2}\right) \approx (-0.79, 0.62) \qquad \text{using a calculator}$$

We have now found the two intersection points, as required.

Example 3 Find all solutions (x, y) of the following system, where x and y are real numbers.

$$\begin{cases} xy = 1 & (3) \\ y = 3x + 1 & (4) \end{cases}$$

Solution Since these equations are easy to graph, we do so, because that will tell us something about the required solutions. As Figure 3 indicates, there are two intersection points, one in the first quadrant, the other in the third quadrant. One way to begin now would be to solve equation (3) for one unknown in terms of the other. However, to avoid introducing fractions at the outset, let us use equation (4) to substitute for y in equation (3). This yields

$$x(3x + 1) = 1$$

or

$$3x^2 + x - 1 = 0$$

This last equation can be solved by using the quadratic formula. As you should verify, the solutions are

$$x = \frac{-1 + \sqrt{13}}{6} \qquad x = \frac{-1 - \sqrt{13}}{6}$$

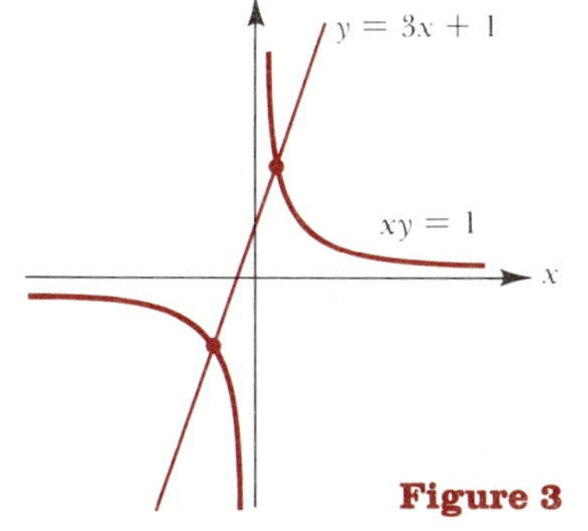

Figure 3

The corresponding y-values may now be obtained by substituting for x in either of the given equations. We will substitute in the equation $y = 3x + 1$. (Exercise 23 at the end of this section asks you to substitute in the other equation, $y = 1/x$, and then to show that the very different-looking answers obtained in that way are in fact equal to those found here.) Substituting $x = \frac{-1 + \sqrt{13}}{6}$ in the equation $y = 3x + 1$ gives us

$$\begin{aligned} y &= 3\left(\frac{-1 + \sqrt{13}}{6}\right) + 1 \\ &= \frac{-1 + \sqrt{13}}{2} + 1 \\ &= \frac{-1 + \sqrt{13}}{2} + \frac{2}{2} = \frac{-1 + \sqrt{13} + 2}{2} \\ &= \frac{1 + \sqrt{13}}{2} \end{aligned}$$

Thus one of the intersection points is

$$\left(\frac{-1 + \sqrt{13}}{6}, \frac{1 + \sqrt{13}}{2}\right)$$

Notice that this must be the first-quadrant point of intersection since both coordinates are positive. The intersection point in the third quadrant is obtained in exactly the same manner. As you can check, substituting the value $x = \frac{-1 - \sqrt{13}}{6}$ in the equation $y = 3x + 1$ yields $y = \frac{1 - \sqrt{13}}{2}$. Thus the other intersection point is

$$\left(\frac{-1 - \sqrt{13}}{6}, \frac{1 - \sqrt{13}}{2}\right)$$

As Figure 3 indicates, there are no other solutions.

Example 4 Find all real numbers x and y satisfying the system of equations

$$\begin{cases} y = \sqrt{x} \\ (x + 2)^2 + y^2 = 1 \end{cases}$$

Solution Use the first equation to substitute for y in the second equation. This yields

$$(x + 2)^2 + (\sqrt{x})^2 = 1$$

or

$$x^2 + 4x + 4 + x = 1$$

or

$$x^2 + 5x + 3 = 0$$

Using the quadratic formula to solve this last equation, we have

$$\begin{aligned} x &= \frac{-5 \pm \sqrt{5^2 - 4(1)(3)}}{2(1)} \\ &= \frac{-5 \pm \sqrt{13}}{2} \end{aligned}$$

The two values of x are thus

$$\frac{-5+\sqrt{13}}{2} \quad \text{and} \quad \frac{-5-\sqrt{13}}{2}$$

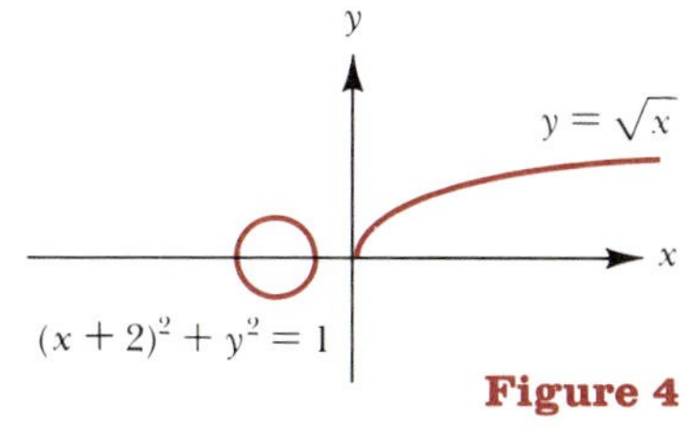

Figure 4

However, notice that both of these quantities are negative. (The second is obviously negative; without using a calculator can you explain why the first is negative?) Thus neither of these quantities is an appropriate x-input in the equation $y = \sqrt{x}$ since we are looking for y-values that are real numbers. We conclude from this that there are no pairs of real numbers x and y satisfying the given system. In geometric terms, this means that the two graphs do not intersect. See Figure 4.

In the next example we look at a system that can be reduced to a linear system through appropriate substitutions. (Actually we've already seen one instance of this technique in Example 9 of Section 9.1.)

Example 5 Solve the system

$$\begin{cases} \dfrac{2}{x^2} - \dfrac{3}{y^2} = -6 \\ \dfrac{3}{x^2} + \dfrac{4}{y^2} = 59 \end{cases}$$

Solution Let $u = \dfrac{1}{x^2}$ and $v = \dfrac{1}{y^2}$ so that the system becomes

$$\begin{cases} 2u - 3v = -6 \\ 3u + 4v = 59 \end{cases}$$

We've learned several methods for solving this linear system. For instance, using the addition-subtraction method, we have

$$2u - 3v = -6 \xrightarrow{\text{multiply by 4}} 8u - 12v = -24$$

$$3u + 4v = 59 \xrightarrow{\text{multiply by 3}} 9u + 12v = 177$$

Adding the last two equations now, we have

$$17u = 153$$

$$u = \frac{153}{17} = 9$$

To determine v, we can substitute the value just obtained for u in the equation $2u - 3v = -6$. As you can check, this yields $v = 8$. In view of the definitions of u and v, we now have

$$\frac{1}{x^2} = 9 \qquad\qquad \frac{1}{y^2} = 8$$

$$x^2 = \frac{1}{9} \qquad\qquad y^2 = \frac{1}{8}$$

$$x = \pm\frac{1}{3} \qquad\qquad y = \pm\frac{1}{\sqrt{8}} = \pm\frac{1}{\sqrt{(4)(2)}} = \pm\frac{1}{2\sqrt{2}}$$

$$y = \pm\frac{1}{2\sqrt{2}} \cdot \frac{\sqrt{2}}{\sqrt{2}} = \pm\frac{\sqrt{2}}{4}$$

This gives us four possible solutions for the original system:

$$\left(\frac{1}{3}, \frac{\sqrt{2}}{4}\right) \quad \left(\frac{1}{3}, -\frac{\sqrt{2}}{4}\right) \quad \left(-\frac{1}{3}, \frac{\sqrt{2}}{4}\right) \quad \left(-\frac{1}{3}, -\frac{\sqrt{2}}{4}\right)$$

As you can check, all four of these pairs satisfy the given system.

Example 6 Determine all solutions (x, y) of the following system, where x and y are real numbers.

$$\begin{cases} y = 3^x & (5) \\ y = 3^{2x} - 2 & (6) \end{cases}$$

Solution We'll use the substitution method. First we rewrite equation (6) as

$$y = (3^x)^2 - 2 \qquad (7)$$

Now, in view of equation (5), we can replace 3^x with y in equation (7) to obtain

$$\begin{aligned} y &= y^2 - 2 \\ 0 &= y^2 - y - 2 \\ 0 &= (y + 1)(y - 2) \end{aligned}$$

From this last equation we see that $y = -1$ or $y = 2$. With $y = -1$, equation (5) becomes $-1 = 3^x$, contrary to the fact that 3^x is positive for all real numbers x. Thus we discard the case with $y = -1$. On the other hand, if $y = 2$, equation (5) becomes

$$2 = 3^x$$

We can solve this exponential equation by taking the logarithm of both sides. Using base e logarithms, we have

$$\begin{aligned} \ln 2 &= \ln 3^x \\ \ln 2 &= x \ln 3 \end{aligned}$$

and consequently

$$x = \frac{\ln 2}{\ln 3} \qquad \textit{Caution: } \frac{\ln 2}{\ln 3} \neq \ln 2 - \ln 3.$$

We've now found that $x = (\ln 2)/(\ln 3)$ and $y = 2$. As Exercise 44 asks you to verify, this pair of values indeed satisfies the given system.

Exercise Set 9.5

In Exercises 1–22, find all solutions (x, y) of the systems, where x and y are real numbers.

1. $\begin{cases} y = 3x \\ y = x^2 \end{cases}$

2. $\begin{cases} y = x + 3 \\ y = 9 - x^2 \end{cases}$

3. $\begin{cases} x^2 + y^2 = 25 \\ 24y = x^2 \end{cases}$

4. $\begin{cases} 3x + 4y = 12 \\ x^2 - y + 1 = 0 \end{cases}$

5. $\begin{cases} xy = 1 \\ y = -x^2 \end{cases}$

6. $\begin{cases} x + 2y = 0 \\ xy = -2 \end{cases}$

7. $\begin{cases} 2x^2 + y^2 = 17 \\ x^2 + 2y^2 = 22 \end{cases}$

8. $\begin{cases} x - 2y = 1 \\ y^2 - x^2 = 3 \end{cases}$

9. $\begin{cases} y = 1 - x^2 \\ y = x^2 - 1 \end{cases}$

10. $\begin{cases} xy = 4 \\ y = 4x \end{cases}$

11. $\begin{cases} xy = 4 \\ y = 4x + 1 \end{cases}$

12. $\begin{cases} \dfrac{2}{x^2} + \dfrac{5}{y^2} = 3 \\ \dfrac{3}{x^2} - \dfrac{2}{y^2} = 1 \end{cases}$

13. $\begin{cases} \dfrac{1}{x^2} - \dfrac{3}{y^2} = 14 \\ \dfrac{2}{x^2} + \dfrac{1}{y^2} = 35 \end{cases}$

14. $\begin{cases} y = -\sqrt{x} \\ (x-3)^2 + y^2 = 4 \end{cases}$

15. $\begin{cases} y = -\sqrt{x-1} \\ (x-3)^2 + y^2 = 4 \end{cases}$

16. $\begin{cases} y = -\sqrt{x-6} \\ (x-3)^2 + y^2 = 4 \end{cases}$

17. $\begin{cases} y = 2^x \\ y = 2^{2x} - 12 \end{cases}$

18. $\begin{cases} y = e^{4x} \\ y = e^{2x} + 6 \end{cases}$

19. $\begin{cases} 2(\log_{10} x)^2 - (\log_{10} y)^2 = -1 \\ 4(\log_{10} x)^2 - 3(\log_{10} y)^2 = -11 \end{cases}$

20. $\begin{cases} y = \log_2(x+1) \\ y = 5 - \log_2(x-3) \end{cases}$

21. $\begin{cases} 2^x \cdot 3^y = 4 \\ x + y = 5 \end{cases}$

***22.** $\begin{cases} a^{2x} + a^{2y} = 10 \\ a^{x+y} = 4 \end{cases} \quad (a > 0)$

Hint: Use the substitutions $a^x = t$ and $a^y = u$.

23. Let $x = \dfrac{-1 + \sqrt{13}}{6}$. Using this x-value, show that the equations $y = 3x + 1$ and $y = 1/x$ yield the same y-value.

24. A sketch shows that the line $y = 100x$ intersects the parabola $y = x^2$ at the origin. Are there any other intersection points? If so, find them. If not, explain why not.

25. Solve the following system for x and y. Assume that neither a nor b is zero.

$$ax + by = 2$$
$$abxy = 1$$

***26.** Solve the following system for x and y. Assume that a and b are positive.

$$\begin{cases} \dfrac{1}{x^2} + \dfrac{1}{xy} = \dfrac{1}{a^2} \\ \dfrac{1}{y^2} + \dfrac{1}{xy} = \dfrac{1}{b^2} \end{cases}$$

27. Find all solutions of the system

$$\begin{cases} x^3 + y^3 = 3473 \\ x + y = 23 \end{cases}$$

28. Solve the following system for x, y, and z. Assume that p, q, and r are positive constants.

$$\begin{cases} yz = p^2 \\ zx = q^2 \\ xy = r^2 \end{cases}$$

29. If the diagonal of a rectangle has length d and the perimeter is $2p$, express the lengths of the sides in terms of d and p.

30. Solve the following system for x and y.

$$\begin{cases} xy + pq = 2px \\ x^2y^2 + p^2q^2 = 2q^2y^2 \end{cases} \quad (pq \neq 0)$$

Hint: Square the first equation, then subtract the second. This results in an equation that can be written as $(2px + qy)(px - qy) = 0$.

***31.** Solve the following system for u and v.

$$\begin{cases} u^2 - v^2 = 9 \\ \sqrt{u+v} + \sqrt{u-v} = 4 \end{cases}$$

Hint: Square the second equation.

***32.** Solve for x and y in terms of p and q. Assume that all of the constants and variables are positive.

$$q^{\ln x} = p^{\ln y}$$
$$(px)^{\ln a} = (qy)^{\ln b}$$

33. Solve for x, y, and z in terms of p, q and r. Assume that p, q, and r are nonzero.

$$x(x + y + z) = p^2$$
$$y(x + y + z) = q^2$$
$$z(x + y + z) = r^2$$

Hint: Denote $x + y + z$ by w; then add the three equations.

34. **(a)** Find the points where the line $y = -2x - 2$ intersects the parabola $y = \frac{1}{2}x^2$.
(b) On the same set of axes, sketch the line $y = -2x - 2$ and the parabola $y = \frac{1}{2}x^2$. Be certain that your sketch is consistent with the results obtained in part (a).

35. The area of a right triangle is 180 cm^2 and the hypotenuse is 41 cm. Find the lengths of the two legs.

36. The sum of two numbers is 8, while their product is -128. What are the two numbers?

37. The perimeter of a rectangle is 46 cm and the area is 60 cm^2. Find the length and the width.

***38.** Find all right triangles for which the perimeter is 24 units and the area is 24 square units.

39. Solve the following system for x and y using the substitution method.

$$\begin{cases} x^2 + y^2 = 5 & (1) \\ xy = 2 & (2) \end{cases}$$

40. The substitution method in Exercise 39 leads to a quadratic equation. Here is an alternative approach to solving that system; this approach leads to linear equations. Multiply equation (2) by 2 and add the resulting equation to equation (1). Now take square roots to conclude that $x + y = \pm 3$. Next multiply equation (2) by 2 and subtract the resulting equation from equation (1). Take square roots to conclude that

$x - y = \pm 1$. You now have the following four linear systems, each of which can be solved (with almost no work) by the addition-subtraction method.

$$\begin{cases} x + y = 3 \\ x - y = 1 \end{cases} \quad \begin{cases} x + y = 3 \\ x - y = -1 \end{cases} \quad \begin{cases} x + y = -3 \\ x - y = 1 \end{cases} \quad \begin{cases} x + y = -3 \\ x - y = -1 \end{cases}$$

Solve these systems and compare your results with those obtained in Exercise 39.

41. Solve the following system using the method explained in Exercise 40.

$$\begin{cases} x^2 + y^2 = 7 \\ xy = 3 \end{cases}$$

42. Solve the following system using the substitution method. (Begin by solving the first equation for y.)

$$\begin{cases} 3xy - 4x^2 = 2 \\ -5x^2 + 3y^2 = 7 \end{cases}$$

43. Here is an alternative approach for solving the system in Exercise 42. Let $y = mx$, where m is a constant to be determined. Replace y with mx in both equations of the system to obtain the following pair of equations:

$$x^2(3m - 4) = 2 \qquad (1)$$

$$x^2(-5 + 3m^2) = 7 \qquad (2)$$

Now divide equation (2) by (1). The resulting equation can be written (after clearing of fractions and simplifying) $2m^2 - 7m + 6 = 0$. Solve this last equation by factoring. The values of m can then be used in equation (1) to determine values for x. In each case the corresponding y-values are determined by the equation $y = mx$.

44. Verify that the pair of values $x = (\ln 2)/(\ln 3)$, $y = 2$ satisfies the system in Example 6 on page 478.

***45.** Solve the following system for x and y.

$$\begin{cases} x^4 = y^6 \\ \ln \dfrac{x}{y} = \dfrac{\ln x}{\ln y} \end{cases}$$

9.6 Systems of Inequalities. Linear Programming

In the first section of this chapter we solved systems of equations in two unknowns. We begin the present section with a discussion of systems of inequalities in two unknowns. The technique we develop will then be applied to solving certain types of optimization problems known as *linear programming problems.* In these problems we want to maximize or minimize a given quantity or resource, subject to a set of given constraints. Consider, for instance, the manufacturer who must ship a certain number of items from various warehouses to various retail outlets. The problem here is to complete the required deliveries while minimizing shipping costs. As another example, suppose an advertising agency wishes to mount a radio and television campaign for a particular product or political candidate. Within certain specified constraints (the budget, for one), how shall the time between radio and television be divided to reach the largest number of persons?

Let a, b, and c denote real numbers and assume that a and b are not both zero. All of the following are called *linear inequalities:*

$$ax + by + c < 0 \quad ax + by + c > 0 \quad ax + by + c \le 0 \quad ax + by + c \ge 0$$

The first two inequalities here are called *strict* inequalities. An ordered pair of numbers (x_0, y_0) is said to be a *solution* of a given inequality (linear or not) provided that we obtain a true statement upon substituting x_0 and y_0 for x and y, respectively. For instance $(-1, 1)$ is a solution of the inequality $2x + 3y - 6 < 0$ since substitution yields

$$2(-1) + 3(1) - 6 < 0$$

$$-5 < 0 \qquad \text{True}$$

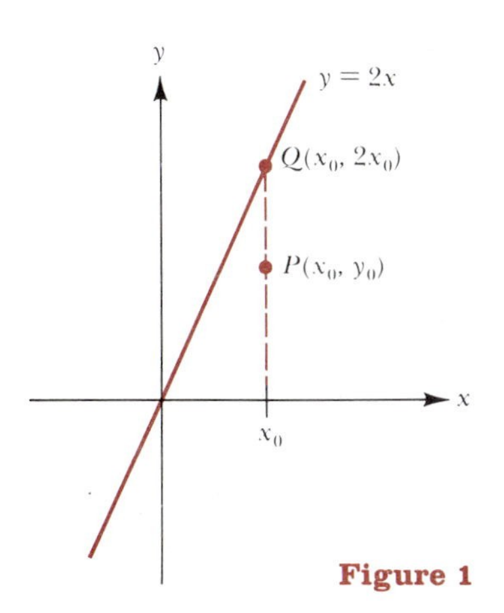

Figure 1

As with a linear equation in two unknowns, a linear inequality has infinitely many solutions. For this reason, we often represent the solutions graphically. When we do this, we say we are graphing the inequality.

As an example, let us graph the inequality $y < 2x$. First observe that the coordinates of the points on the line $y = 2x$ do not, by definition, satisfy this strict inequality. So it remains to consider points above the line and points below the line. We shall show that the required graph in this case consists of all points that lie below the line. Take any point $P(x_0, y_0)$ not on the line $y = 2x$. Let $Q(x_0, 2x_0)$ be the point on $y = 2x$ with the same first coordinate as P. Then, as indicated in Figure 1, the point P lies below the line if and only if the y-coordinate of P is less than the y-coordinate of Q. In other words, (x_0, y_0) lies below the line if and only if

$$y_0 < 2x_0$$

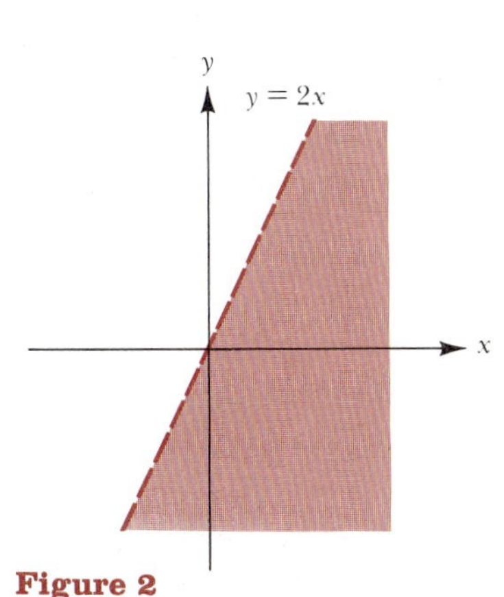

Figure 2
The graph of $y < 2x$.

This last statement is equivalent to saying that (x_0, y_0) satisfies the inequality $y < 2x$. This shows that the graph of $y < 2x$ consists of all points below the line $y = 2x$. See Figure 2. The dashed line is used in Figure 2 to indicate that the points on $y = 2x$ are not part of the required graph. Had the original inequality instead been $y \leq 2x$, then we would use a solid rather than dashed line to indicate that the line is included in the graph. Again, if the original inequality had been $y > 2x$, the graph would have been the region above the line.

Just as the graph of $y < 2x$ is the region below the line $y = 2x$, it is so in general that the graph of $y < f(x)$ is the region below the graph of the function f. For example, Figures 3 and 4 display the graphs of $y < x^2$ and $y \geq x^2$.

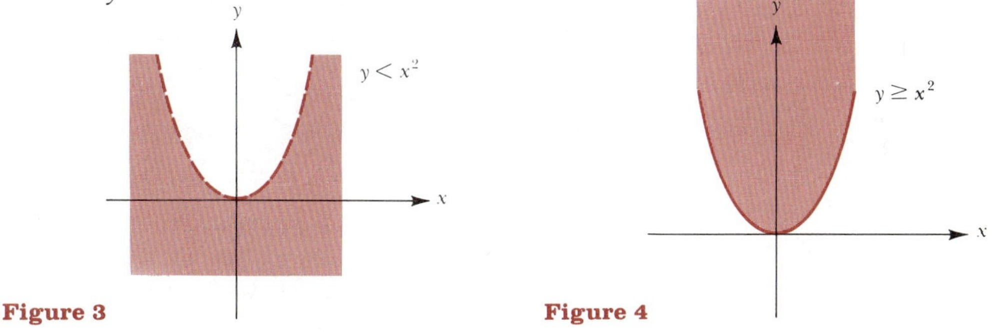

Figure 3 **Figure 4**

Example 1 serves to summarize the technique developed up to this point for graphing an inequality. After this example we will point out a useful alternative method.

Example 1 Graph the inequality $4x - 3y \leq 12$.

Solution The graph will include the line $4x - 3y = 12$ and either the region above or the region below the line. To decide which region, solve the inequality for y:

$$4x - 3y \leq 12$$

$$-3y \leq -4x + 12$$

$$y \geq \frac{4}{3}x - 4$$

The sense of an inequality is reversed when we multiply or divide by a negative number.

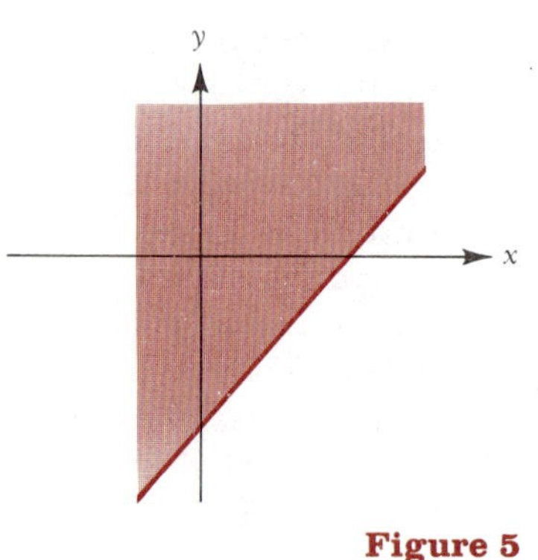

Figure 5
The graph of $4x - 3y \leq 12$.

This last inequality tells us that we want the region above the line, as well as the line itself. Figure 5 displays the required graph.

Another method that can be used in Example 1 to determine which side of the line $4x - 3y = 12$ is appropriate involves choosing a *test point.* We pick any convenient point that is not on the line $4x - 3y = 12$. Then we test to see if this point satisfies the given inequality. Let us pick the point (0, 0). Substituting these coordinates in the inequality $4x - 3y \leq 12$ yields

$$4(0) - 3(0) \stackrel{?}{\leq} 12$$
$$0 \stackrel{?}{\leq} 12 \qquad \text{True}$$

We conclude from this that the required side of the line is the side on which the point (0, 0) resides. In agreement with Figure 5, we see this is the region above the line.

Next we discuss systems of inequalities in two unknowns. As with systems of equations, a *solution* of a system of inequalities is an ordered pair (x_0, y_0) that satisfies all of the inequalities in the system. As a first example, let us graph the points that satisfy the following nonlinear system:

$$\begin{cases} y - x^2 \geq 0 \\ x^2 + y^2 < 1 \end{cases}$$

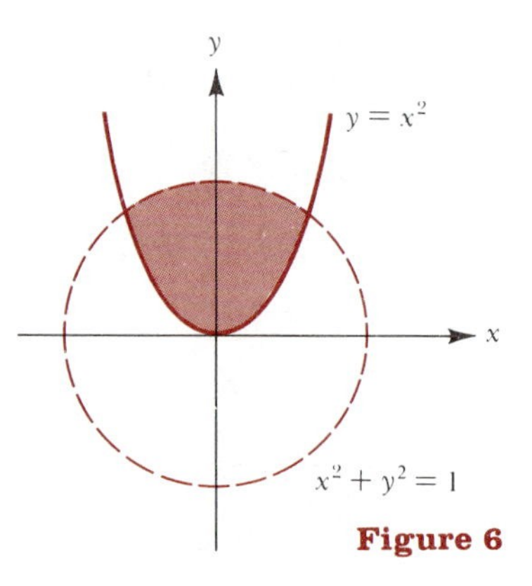

Figure 6

By writing the first inequality as $y \geq x^2$, we see that it describes the set of points on or above the parabola $y = x^2$. For the second inequality, we must decide whether it describes the points within or outside the circle $x^2 + y^2 = 1$. One way to do this is to choose (0, 0) as a test point. Substituting the values $x = 0$ and $y = 0$ in the inequality $x^2 + y^2 < 1$ yields the true statement $0^2 + 0^2 < 1$. Then since (0, 0) lies within the circle and satisfies the inequality, we conclude that $x^2 + y^2 < 1$ describes the set of all points within the circle. Now we put our information together. We wish to graph the points that lie on or above the parabola $y = x^2$ but within the circle $x^2 + y^2 = 1$. This is the shaded region shown in Figure 6.

In the next example we graph a system of linear inequalities. As you will see, it is this type of system that we deal with in subsequent linear programming problems.

Example 2 Graph the system

$$\begin{cases} -x + 3y \leq 12 \\ x + y \leq 8 \\ x \geq 0 \\ y \geq 0 \end{cases}$$

Solution Solving the first inequality for y, we have

$$3y \leq x + 12$$
$$y \leq \frac{1}{3}x + 4$$

Thus the first inequality is satisfied by the points on or below the line $-x + 3y = 12$. Similarly, by solving the second inequality for y, we see that it describes the set of points on or below the line $x + y = 8$. The third inequality, $x \geq 0$, describes the points on or to the right of the y-axis. Similarly, the fourth inequality, $y \geq 0$, describes the points on or above the

x-axis. In Figure 7(a), we summarize these statements. The arrows there indicate which side of the lines we wish to consider. Finally, Figure 7(b) shows the graph of the given system. The coordinates of each point in the shaded region satisfy all four of the given inequalities. (Exercise 25 at the end of this section asks you to show that the coordinates listed in the figure are correct.)

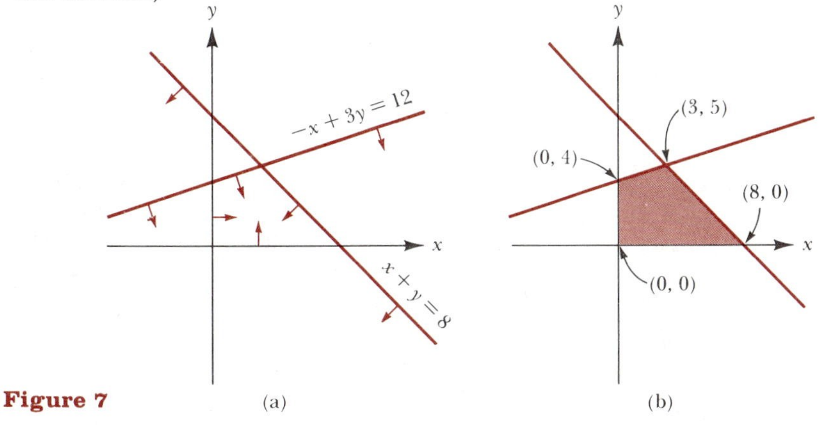

Figure 7

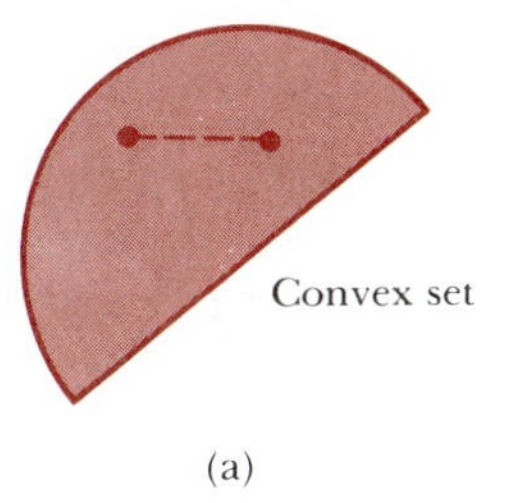

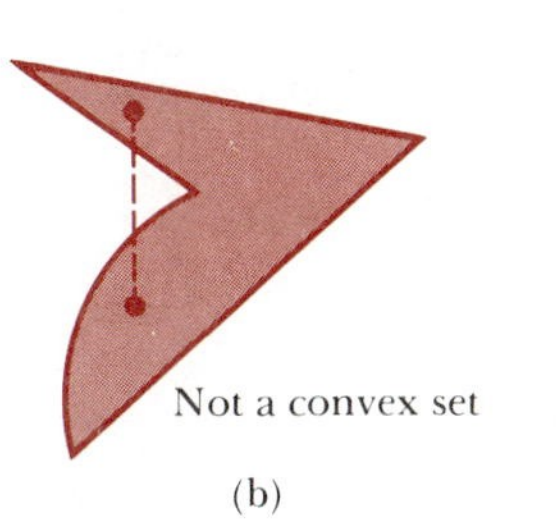

Figure 8

We can use Figure 7(b) to introduce some terminology that is useful in describing sets of points in the plane. The *vertices* of a region are the corners or points where the adjacent bounding sides meet. Thus the vertices of the shaded region in Figure 7(b) are the four points (0, 0), (8, 0), (3, 5), and (0, 4). The shaded region in Figure 7(b) is *convex.* This means that given any two points in that region, the straight line segment joining these two points lies wholly within the region. Figure 8(a) displays another example of a convex set, while Figure 8(b) shows a set that is not convex. The shaded region in Figure 8(b) is also an example of a *bounded region.* By this we mean that the region can be wholly contained within some (sufficiently large) circle. Perhaps the simplest example of a region that is not bounded is the entire x-y plane itself. Figure 9 shows another example of an unbounded region.

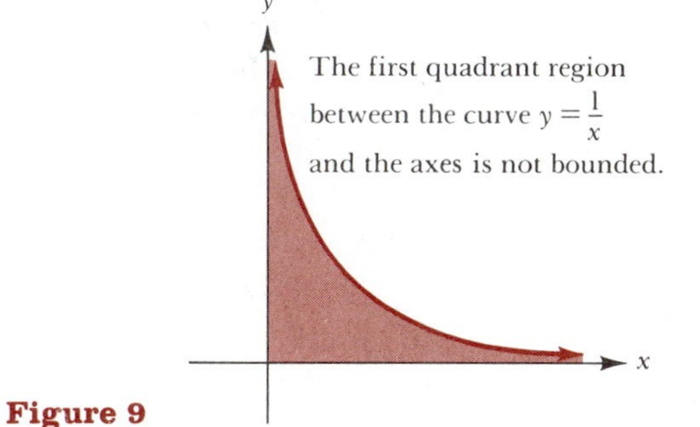

Figure 9

We are prepared now to consider a typical problem in linear programming.

Problem: Let $C = x + 3y$. Find the values of x and y that yield the largest possible value of C, given that x and y satisfy the following set of inequalities.

$$\begin{cases} -x + 3y \leq 12 \\ x + y \leq 8 \\ x \geq 0 \\ y \geq 0 \end{cases}$$

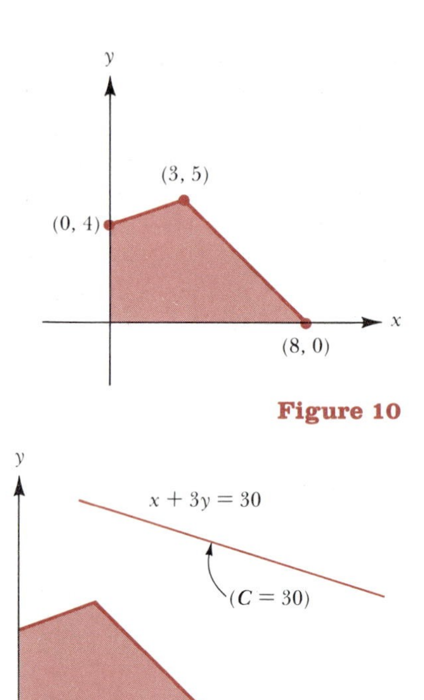

Figure 10

Figure 11

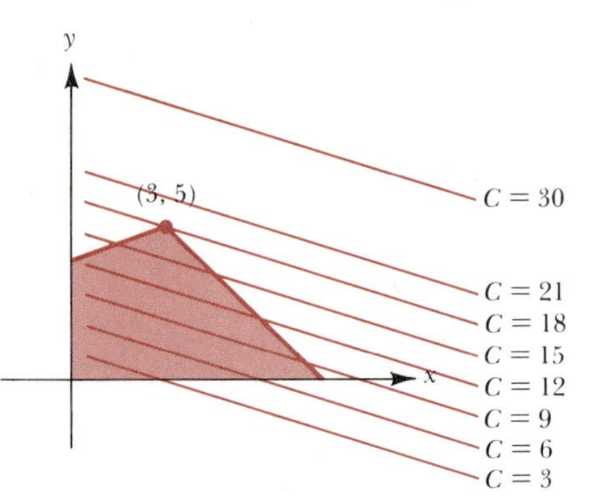

Figure 12

The function $C = x + 3y$ in this problem is known as the *objective function.* The given inequalities are called the *constraints.* In analyzing this problem we will take advantage of the fact that we've already graphed this system of inequalities in Example 2. For convenience, we repeat the graph here (Figure 10). Since the coordinates of each point in the shaded region of Figure 10 satisfy the constraints of the present problem, we refer to the region as the set of *feasible solutions.* The problem then is to find that point among the feasible solutions that yields the largest possible value for C. A feasible solution (x_0, y_0) that yields the largest value for C is called an *optimal solution.* Some linear programming problems ask for a smallest rather than a largest value of the objective function. A feasible solution yielding the smallest value for C is also called an *optimal solution.*

We will see subsequently that there is a general method for solving this type of linear programming problem. But as a start, let us experiment. How large can the value of C be, subject to the given constraints? For instance, can C equal 30? With $C = 30$, we have

$$x + 3y = 30$$

Figure 11 shows the graph of this line. Since the line has no points in common with the set of feasible solutions, we conclude that C cannot equal 30 under the conditions of the problem.

Still experimenting, in Figure 12 we have graphed the lines $x + 3y = C$ for various values of C. Observe that the lines are all parallel. To see why this is so, solve the equation $x + 3y = C$ for y. This yields

$$y = -\frac{1}{3}x + \frac{1}{3}C$$

As you can see, the slope here is $m = -\frac{1}{3}$, independent of C. Since the lines all have the same slope, they are parallel. On the other hand, the equation $y = -\frac{1}{3}x + \frac{1}{3}C$ shows how the value of C does influence the graph. The y-intercept of the line is $C/3$. So as C increases, the line moves *up*, but the slope is always $-\frac{1}{3}$.

Now as you can see in Figure 12, with $C = 18$ the line $x + 3y = 18$ intersects the set of feasible solutions at the vertex (3, 5). However, it is also evident from the figure and the remarks in the previous paragraph that when C is greater than 18, the line $x + 3y = C$ will not intersect the set of feasible solutions. We conclude therefore that the vertex (3, 5) is the optimal solution. That is, subject to the given constraints, the value of C is largest when $x = 3$ and $y = 5$.

In the example just concluded, we found that the optimal solution to the linear programming problem occurred at a vertex. The following theorem (stated here without proof) tells us that the optimal solution, when it exists, always occurs at a vertex.

Linear Programming Theorem

Consider the objective function $C = ax + by + c$, where x and y are subject to constraints specified by nonstrict linear inequalities. (Nonstrict inequalities involve the symbols $\leq$ or $\geq$ rather than $<$ or $>$.) Then if C assumes a maximum (or a minimum) value, this occurs at a vertex. Furthermore, if the set of feasible solutions is both convex and bounded, then a maximum and minimum always exist.

In view of this theorem, we can solve linear programming problems by first finding the vertices of the feasibility set and then computing and comparing the values of the objective function at those vertices. This is done in Example 3, which follows. Technically speaking, the theorem we've stated only guarantees a maximum and a minimum when the feasibility set is convex and bounded. However, this is not as great a restriction as it might at first seem. As it turns out, the graph of any set of linear inequalities forms a convex set. (For a look at cases in which the feasibility set is not bounded, see Exercises 44 and 45 at the end of this section.)

Example 3 Let $C = 3x - 2y$. Find the maximum and the minimum value of C subject to the following set of constraints.

$$\begin{cases} 6y - x \geq 5 \\ 4x + 3y \leq 34 \\ 2y - x \leq 8 \\ 4x - y \geq 3 \end{cases}$$

What are the optimal solutions in each case?

Solution In Figure 13, we have graphed the given system of inequalities using the method discussed in Example 2. Now we need to determine the coordinates of the four vertices. To find the coordinates of E, we solve the system

$$\begin{cases} 6y - x = 5 & (1) \\ 4x - y = 3 & (2) \end{cases}$$

Solving equation (1) for x, we have

$$6y - 5 = x \qquad (3)$$

Now we use equation (3) to substitute for x in equation (2). This yields

$$\begin{aligned} 4(6y - 5) - y &= 3 \\ 24y - 20 - y &= 3 \\ 23y &= 23 \\ y &= 1 \end{aligned}$$

Finally, substituting $y = 1$ in equation (3) gives us

$$\begin{aligned} 6(1) - 5 &= x \\ x &= 1 \end{aligned}$$

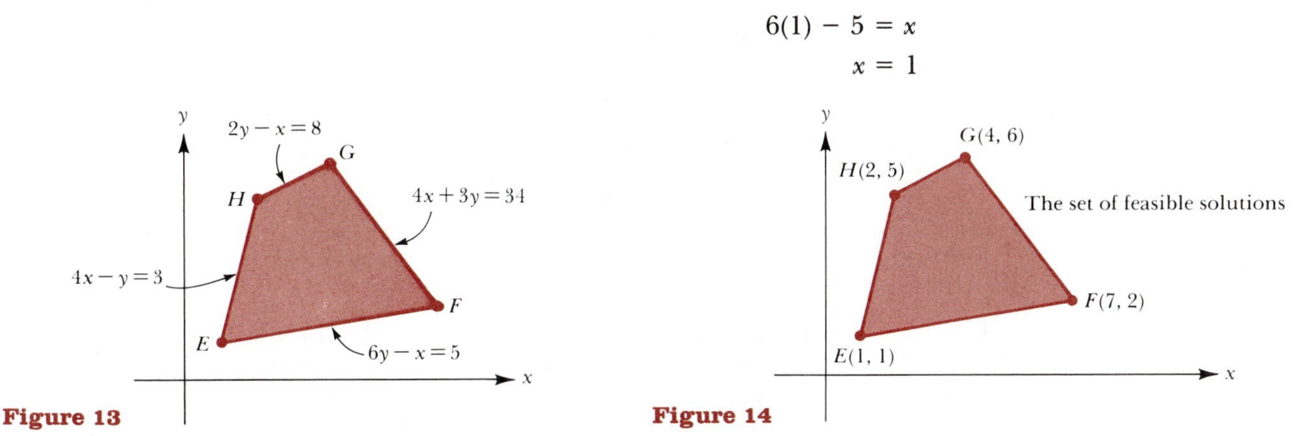

Figure 13 **Figure 14**

We have now found that the coordinates of the vertex E are (1, 1). The coordinates of F, G, and H can be found in a similar fashion. The results are displayed in Figure 14. Notice that the set of feasible solutions forms a convex and bounded set. Thus the objective function attains a maximum and a minimum value at certain vertices. We compute the value of C at each vertex and tabulate the results as follows.

Vertex	$C = 3x - 2y$
$E(1, 1)$	$3(1) - 2(1) = 1$
$F(7, 2)$	$3(7) - 2(2) = 17$
$G(4, 6)$	$3(4) - 2(6) = 0$
$H(2, 5)$	$3(2) - 2(5) = -4$

The largest value of C in this table is 17. This is the required maximum. The corresponding optimal solution is (7, 2). Similarly, the smallest value of C is -4. This is the required minimum. The corresponding optimal solution is (2, 5).

The next example deals with a type of linear programming problem known as the *transportation problem.* In this type of problem we are required to minimize the cost of shipping goods from warehouses to retail outlets.

Example 4 A motorcycle manufacturer must fill orders from two dealers. The first dealer, D_1, has ordered 20 motorcycles, while the second dealer, D_2, has ordered 30 motorcycles. The manufacturer has the motorcycles stored in two warehouses, W_1 and W_2. There are 40 motorcycles in W_1 and 15 in W_2. The shipping costs, per motorcycle, are as follows: \$15 from W_1 to D_1; \$13 from W_1 to D_2; \$14 from W_2 to D_1; \$16 from W_2 to D_2. Under these conditions, find the number of motorcycles to be shipped from each warehouse to each dealer if the total shipping cost is to be a minimum. What is this minimum cost?

Solution The first step is to summarize the data. We do this as follows.

Shipping Cost per Motorcycle	
W_1 to D_1:	\$15
W_1 to D_2:	\$13
W_2 to D_1:	\$14
W_2 to D_2:	\$16

Supply and Demand	
W_1 (40 motorcycles in stock)	W_2 (15 motorcycles in stock)
D_1 (wants 20 motorcycles)	D_2 (wants 30 motorcycles)

The problem asks how many motorcycles should be shipped from each warehouse to each dealer.

1. Let x denote the number of motorcycles to be shipped from W_1 to D_1.
2. Then $20 - x$ denotes the number of motorcycles to be shipped from W_2 to D_1. (Why?)
3. Let y denote the number of motorcycles to be shipped from W_1 to D_2.
4. Then $30 - y$ denotes the number of motorcycles to be shipped from W_2 to D_2.

If C denotes the total shipping cost, we then have

$$C = 15x + 13y + 14(20 - x) + 16(30 - y)$$

(shipping cost per unit: 15, 13, 14, 16; number of units: x, y, $20 - x$, $30 - y$)

As you can check, the above equation simplifies to

$$C = x - 3y + 760$$

This is the objective function that we must minimize.

Now let us examine the constraints of the problem. The number of items shipped from the warehouse certainly must be a nonnegative number in each case. Thus we must have

$$x \geq 0$$
$$y \geq 0$$
$$20 - x \geq 0$$
$$30 - y \geq 0$$

Furthermore, since at most 40 motorcycles can be shipped from W_1, we must have

$$x + y \leq 40$$

Similarly, since at most 15 motorcycles can be shipped from W_2, we have

$$(20 - x) + (30 - y) \leq 15$$

or

$$50 - x - y \leq 15$$
$$-x - y \leq -35$$
$$x + y \geq 35$$

In summary then, we have the following set of constraints:

$$\begin{cases} x \geq 0 \\ y \geq 0 \\ 20 - x \geq 0 \\ 30 - y \geq 0 \\ x + y \leq 40 \\ x + y \geq 35 \end{cases}$$

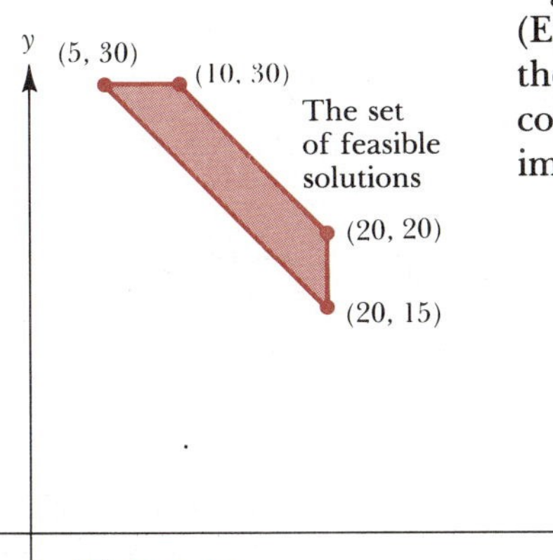

Figure 15

Figure 15 shows the graph of the region described by this set of constraints. (Exercise 33 at the end of this section asks you to verify that the graph and the coordinates of the vertices are as indicated.) Since the region is both convex and bounded, we know that the objective function assumes its minimum value at one of the vertices. We compute the value of C at each vertex.

Vertex	$C = x - 3y + 760$
(20, 15)	$C = 20 - 3(15) + 760 = 735$
(20, 20)	$C = 20 - 3(20) + 760 = 720$
(10, 30)	$C = 10 - 3(30) + 760 = 680$
(5, 30)	$C = 5 - 3(30) + 760 = 675$

As the table shows, the minimum shipping cost is $675. The corresponding vertex is (5, 30). Therefore, the shipping cost will be minimized if the deliveries are made as follows.

$x = 5$: ship 5 motorcycles from W_1 to D_1
$20 - x = 15$: ship 15 motorcycles from W_2 to D_1
$y = 30$: ship 30 motorcycles from W_1 to D_2
$30 - y = 0$: ship no motorcycles from W_2 to D_2

The linear programming problems that we have considered involve only two unknowns, x and y. Because there were only two unknowns, we were able to utilize graphical methods in solving the problems. As you might guess, linear programming problems actually arising in business and industry may involve dozens or even thousands of unknowns. For such cases, there is a purely algebraic method of solution called the *Simplex method,* which makes use of matrices. Furthermore, this method can be programmed for computer solution. Several commercial firms now market computer packages for solving linear programming problems.*

Exercise Set 9.6

In Exercises 1–3, decide whether or not the ordered pair is a solution of the given inequality.

1. $4x - 6y + 3 \geq 0$
   (a) $(1, 2)$ (b) $(0, \frac{1}{2})$
2. $5x + 2y < 1$
   (a) $(-1, 3)$ (b) $(0, 0)$
3. $y < x^2 - 2x + 1$
   (a) $(0, 0)$ (b) $(1, 0)$

In Exercises 4–17, graph the given inequalities.

4. $2x - 3y < 6$
5. $2x - 3y \geq 6$
6. $2x + 3y \leq 6$
7. $x - y < 0$
8. $y \leq \frac{1}{2}x - 1$
9. $x \geq 1$
10. $y < 0$
11. $x > 0$
12. $y \leq \sqrt{x}$
13. $y > x^3 + 1$
14. $y \leq |x - 2|$
15. $x^2 + y^2 \geq 25$
16. $y \leq e^x$
17. $y < \ln x$

In Exercises 18–23, graph the systems of inequalities.

18. $\begin{cases} y \leq x^2 \\ x^2 + y^2 \leq 1 \end{cases}$
19. $\begin{cases} y < x \\ x^2 + y^2 < 1 \end{cases}$
20. $\begin{cases} y \geq 1 \\ y \leq |x| \end{cases}$
21. $\begin{cases} x \geq 0 \\ y \geq 0 \\ y < \sqrt{x} \\ x \leq 4 \end{cases}$
22. $\begin{cases} x \geq 0 \\ y \geq 0 \\ y \leq \pi \\ y \leq \sin x \end{cases}$
23. $\begin{cases} y < 2x \\ y > \frac{1}{2}x \end{cases}$

In Exercises 24–34, graph the systems of linear inequalities. In each case specify the vertices. Is the region convex? Is the region bounded?

24. $\begin{cases} y \leq x + 5 \\ y \leq -2x + 14 \\ x \geq 0 \\ y \geq 0 \end{cases}$
25. $\begin{cases} -x + 3y \leq 12 \\ x + y \leq 8 \\ x \geq 0 \\ y \geq 0 \end{cases}$
26. $\begin{cases} y \geq 2x \\ y \geq -x + 6 \end{cases}$
27. $\begin{cases} 0 \leq 2x - y + 3 \\ x + 3y \leq 23 \\ 5x + y \leq 45 \end{cases}$
28. $\begin{cases} 0 \leq 2x - y + 3 \\ x + 3y \leq 23 \\ 5x + y \leq 45 \\ x \geq 0 \\ y \geq 0 \end{cases}$
29. $\begin{cases} 5x + 6y < 30 \\ y > 0 \end{cases}$

*For an excellent discussion of the Simplex method as well as methods of computer solution, see *An Introduction to Management Science,* Fourth Edition, by David R. Anderson et al. (St. Paul: West Publishing Company, 1985).

30. $\begin{cases} 5x + 6y < 30 \\ x > 0 \end{cases}$

31. $\begin{cases} 5x + 6y < 30 \\ x > 0 \\ y > 0 \end{cases}$

32. $\begin{cases} 2x + 3y \geq 6 \\ 2x + 3y \leq 12 \end{cases}$

33. $\begin{cases} x \geq 0 \\ y \geq 0 \\ 20 - x \geq 0 \\ 30 - y \geq 0 \\ x + y \leq 40 \\ x + y \geq 35 \end{cases}$

34. $\begin{cases} x \geq 0 \\ y \geq 0 \\ 3x - y + 1 \geq 0 \\ 0 \leq x - y + 3 \\ y \leq 5 \\ x \leq \dfrac{17 - y}{3} \\ x \leq \dfrac{y + 8}{2} \end{cases}$

In Exercises 35–43, find the maximum and the minimum values of the objective function subject to the given constraints. Specify the corresponding optimal solutions in each case.

35. Objective function: $C = 3y - x$

Constraints: $\begin{cases} y \leq x + 1 \\ y \leq -2x + 10 \\ x \geq 0 \\ y \geq 0 \end{cases}$

36. Objective function: $C = 2x - y + 8$

Constraints: $\begin{cases} 4x + 3y \geq 12 \\ 6x + 5y \leq 30 \\ x \geq 0 \\ y \geq 0 \end{cases}$

37. Objective function: $C = 2x + y$

Constraints: $\begin{cases} x \geq -1 \\ x \leq 1 \\ y \geq -2 \\ 2y - x \leq 3 \end{cases}$

38. Objective function: $C = 2x - y$

Constraints: Use the same constraints given in Exercise 37.

39. Objective function: $C = 10y + 9x - 1$

Constraints: $\begin{cases} x + y \geq 3 \\ y - x + 1 \geq 0 \\ y \leq -x + 12 \\ y \leq 6 \\ x \geq 1 \end{cases}$

40. Objective function: $C = 10y + 11x - 1$

Constraints: Use the same constraints given in Exercise 39.

41. Objective function: $C = 10y + 3x + 100$

Constraints: $\begin{cases} 0 \leq 4x - y + 4 \\ x + 5y \leq 41 \\ x + 3y \leq 27 \\ 2x + y \leq 24 \\ x \geq 0 \\ y \geq 0 \end{cases}$

42. Objective function: $C = 20y + 7x + 100$

Constraints: Use the same ones given in Exercise 41.

43. **(a)** Objective function: $C = 19x + 100y$

Constraints: Use the same constraints given in Exercise 41.

(b) Objective function: $C = 21x + 100y$

Constraints: Use the same constraints given in Exercise 41.

(This exercise shows that a slight change in the objective function may yield a relatively large change in the optimal solution.)

44. Consider the objective function $C = 4y - x$ subject to the constraints

$$\begin{cases} y \geq -\frac{3}{2}x + 4 \\ 2y \geq x \\ y \geq 0 \\ x \geq 0 \end{cases}$$

(a) Given that there is an optimal solution yielding a minimum value of C, find that solution and the corresponding value of C.
(b) Find a feasible solution for which $C = 100$.
(c) Find a feasible solution for which $C = 1000$.
(d) Find a feasible solution for which $C = 10^6$.

45. Consider the objective function $C = x + y$ subject to the constraints

$$\begin{cases} 3y + x \leq 18 \\ x \geq 0 \\ x \leq 9 \end{cases}$$

(a) Given that there is an optimal solution yielding a maximum value of C, find that solution and the corresponding value of C.
(b) Find a feasible solution for which $C = -100$.
(c) Find a feasible solution for which $C = -1000$.
(d) Find a feasible solution for which $C = -10^6$.

46. An automobile manufacturer must fill orders from two dealers. The first dealer, D_1, has ordered 40 cars, while the second dealer, D_2, has ordered 25 cars. The manufacturer has the cars stored in two locations, W_1 and W_2. There are 30 cars in W_1 and 50 cars in W_2. The shipping costs per car are as follows: \$180 from W_1 to D_1; \$150 from W_1 to D_2; \$160 from W_2 to D_1; \$170 from W_2 to D_2. Under these conditions, how many cars should be shipped from each storage location to each dealer so as to minimize the total shipping cost? What is this minimum cost?

47. A television manufacturer must fill orders from two retailers. The first retailer, R_1, has ordered 55 television sets, while the second retailer, R_2, has ordered 75 sets. The manufacturer has the television sets stored in two warehouses, W_1 and W_2. There are 100 sets in W_1 and 120 sets in W_2. The shipping costs, per television set, are as follows: \$8 from W_1 to R_1; \$12 from W_1 to R_2; \$13 from W_2 to R_1; \$7 from W_2 to R_2. Under these conditions, find the number of television sets to be shipped from each warehouse to each retailer if the total shipping cost is to be a minimum. What is this minimum cost?

48. A factory produces two products, *A* and *B*. The production time (in hours) for each unit is as follows.

Product	Assembly	Inspection and Packing
A	0.40	0.20
B	0.50	0.40

Each day at most 500 man-hours of assembly time and 300 man-hours of inspection-packing time are available. The profit realized on each unit of *A* is \$0.60; on each unit of *B* it is \$0.80. Find the maximum daily profit under these conditions. *Hints:* Let x denote the number of units of *A* produced each day and let y denote the number of units of *B* produced each day. After graphing the constraints you'll find that the coordinates of one vertex are not whole numbers. Round these off to the nearest whole number. (Why?)

49. Due to increased energy costs, the owner of the factory described in Exercise 48 decides to limit production. In particular, the owner decides that no more than 800 units of *A* and 500 units of *B* should be produced daily. Under these additional constraints, how many units of *A* and of *B* should be produced daily to maximize profits? What is the maximum profit?

50. A farmer can allocate at most 600 acres for growing cherry tomatoes and regular tomatoes. The labor requirements and the profits are as follows.

	Cherry Tomatoes	Regular Tomatoes
Labor (hours per acre)	3	2
Profit (dollars per acre)	50	36

What is the farmer's maximum profit if 1350 hr of labor are available? How many acres should be devoted to each crop to achieve the maximum profit?

51. Consider the situation described in Exercise 50. Now additionally suppose that due to market considerations, the farmer decides to devote at most 550 acres to regular tomatoes and at most 100 acres to cherry tomatoes. What is the farmer's maximum profit in this case?

52. Suppose that you inherit \$12,000 but there are stipulations attached. Some (or all) of the money must be invested in two stocks, *A* and *B*. *A* is a low-risk stock and *B* is medium-risk. The stipulations are that at most \$6,000 be invested in *A* and at least \$2,000 be invested in *B*. Furthermore, the amount invested in *B* is not to exceed twice the amount invested in *A*. The expected returns on the stocks are \$0.06 per dollar for *A* and \$0.08 per dollar for *B*. Under these conditions, how much should you invest in each stock so as to maximize your expected returns?

53. You have \$10,000 to invest in three stocks, *A*, *B*, and *C*. At least \$1,000 must be invested in each stock. The combined investment in *A* and *B* must be at least \$5,000, while the investment in *B* cannot exceed five times that in *A*. The expected returns on the stocks are as follows: \$0.04 per dollar for *A*; \$0.05 per dollar for *B*; \$0.06 per dollar for *C*. Under these conditions, how much should you invest in each stock so as to maximize the expected returns? *Hint:* Let x and y denote the dollar amounts invested in *A* and *B*, respectively. Then $10,000 - x - y$ denotes the number of dollars invested in *C*.

54. In Exercise 53, suppose instead that the expected returns are \$0.05 per dollar on *A* and \$0.04 per dollar on *B*. The return on *C* remains \$0.06 per dollar. Now how much should be invested in each stock so as to maximize the expected returns?

55. In Exercise 53, suppose instead that the expected return on *B* is \$0.07 per dollar. The expected returns on *A* and *C* remain the same. Now how much should be invested to maximize the expected returns?

56. A veterinarian wishes to prepare a supply of dog food by mixing two commercially available foods, *A* and *B*. The mixture is to contain at least 50 units of carbohydrates, at least 36 units of protein, and at least 40 units of fat. The specifications for the two commercially available products are listed below.

Product	Units of Carbohydrates (per pound)	Units of Protein (per pound)	Units of Fat (per pound)
A	3	2	2
B	5	4	8

The veterinarian can purchase food *A* for \$0.44 per pound and *B* for \$0.80 per pound. How many pounds of each food should be mixed to meet the dietary requirements at the least cost?

57. A factory assembles two types of motors, small and large. The cost of material is \$15 for a small motor and \$30 for a large motor. Two hours of labor are required to assemble a small motor and 6 hours are needed for a large motor. The cost of labor is \$8 per hour. The owner of the factory can allocate at most \$1500 per day for material and \$2000 per day for labor. Find the largest daily profit that the owner can make if the small motors are sold for \$40 each and the large motors for \$104 each. Assume that every motor produced is sold. *Hint:* Profit equals revenue minus cost.

58. Suppose that the factory owner in Exercise 57 finds it necessary to limit production to a total of at most 65 motors per day. Find the largest daily profit under this additional constraint.

Chapter 9 Review Checklist

- □ Linear equation in two variables (page 427)
- □ Solution of a linear equation (page 428)
- □ Consistent system; inconsistent system (page 428)
- □ Echelon form (page 439)
- □ The elementary operations (page 441)
- □ Gaussian elimination—examples (Examples 4–6, pages 441–443)
- □ Partial fraction decomposition (page 444)
- □ Matrix (page 449)
- □ Coefficient matrix (page 450)
- □ Augmented matrix (page 450)
- □ Matrix equality (page 452)
- □ Matrix addition and subtraction (page 453)
- □ Scalar multiplication (page 454)
- □ Matrix multiplication (page 457)
- □ Second-order determinant (page 461)
- □ Minor and cofactor (pages 462, 463)
- □ Determinant (pages 463–464)
- □ Cramer's rule (page 468)
- □ Linear inequality in two variables (page 480)
- □ Vertices (of a region in the plane) (page 483)
- □ Convex region (page 483)
- □ Bounded region (page 483)
- □ Objective function; constraints (page 484)
- □ Feasible solution; optimal solution (page 484)
- □ Linear Programming Theorem (page 484)

Chapter 9 Review Exercises

In Exercises 1–38, solve each system of equations. If there are no solutions in a particular case, state this. In cases where there are literal (rather than numerical) coefficients, specify any restrictions that your solutions impose upon those coefficients.

1. $\begin{cases} x + y = -2 \\ x - y = 8 \end{cases}$

2. $\begin{cases} x - y = 1 \\ x + y = 5 \end{cases}$

3. $\begin{cases} 2x + y = 2 \\ x + 2y = 7 \end{cases}$

4. $\begin{cases} 3x + 2y = 6 \\ 5x + 4y = 4 \end{cases}$

5. $\begin{cases} 7x + 2y = 9 \\ 4x + 5y = 63 \end{cases}$

6. $\begin{cases} \dfrac{x}{2} + \dfrac{y}{3} = 9 \\ \dfrac{x}{5} - \dfrac{y}{2} = -4 \end{cases}$

7. $\begin{cases} 2x - \dfrac{y}{2} = -8 \\ \dfrac{x}{3} + \dfrac{y}{8} = -1 \end{cases}$

8. $\begin{cases} 3x - 14y - 1 = 0 \\ -6x + 28y - 3 = 0 \end{cases}$

9. $\begin{cases} 3x + 5y - 1 = 0 \\ 9x - 10y - 8 = 0 \end{cases}$

10. $\begin{cases} 9x + 15y - 1 = 0 \\ 6x + 10y + 1 = 0 \end{cases}$

11. $\begin{cases} \frac{2}{3}x = -\frac{1}{2}y - 12 \\ \dfrac{x}{2} = y + 2 \end{cases}$

12. $\begin{cases} 0.1x + 0.2y = -5 \\ -0.2x - 0.5y = 13 \end{cases}$

13. $\begin{cases} \dfrac{1}{x} + \dfrac{1}{y} = -1 \\ \dfrac{2}{x} + \dfrac{5}{y} = -14 \end{cases}$

14. $\begin{cases} \dfrac{2}{x} + \dfrac{15}{y} = -9 \\ \dfrac{1}{x} + \dfrac{10}{y} = -2 \end{cases}$

15. $\begin{cases} ax + (1 - a)y = 1 \\ (1 - a)x + y = 0 \end{cases}$

16. $\begin{cases} ax - by - 1 = 0 \\ (a - 1)x + by + 2 = 0 \end{cases}$

17. $\begin{cases} 2x - y = 3a^2 - 1 \\ 2y + x = 2 - a^2 \end{cases}$

18. $\begin{cases} 3ax + 2by = 3a^2 - ab + 2b^2 \\ 3bx + 2ay = 2a^2 + 5ab - 3b^2 \end{cases}$

19. $\begin{cases} 5x - y = 4a^2 - 6b^2 \\ 2x + 3y = 5a^2 + b^2 \end{cases}$

20. $\begin{cases} \dfrac{2b}{x} - \dfrac{3}{y} = 7ab \\ \dfrac{4a}{x} + \dfrac{5a}{by} = 3a^2 \end{cases}$

21. $\begin{cases} px - qy = q^2 \\ qx + py = p^2 \end{cases}$

22. $\begin{cases} x - y = \dfrac{a - b}{a + b} \\ x + y = 1 \end{cases}$

23. $\begin{cases} \dfrac{4a}{x} - \dfrac{3b}{y} = a - 7b \\ \dfrac{3a^2}{x} - \dfrac{2b^2}{y} = (3a + b)(a - 2b) \end{cases}$

24. $\begin{cases} 6b^2x^{-1/2} - 5a^2y^{-1/2} - a^2b^2 = 0 \\ 2bx^{-1/2} + a^2y^{-1/2} - 3a^2b = 0 \end{cases}$

Hint: Let $u = x^{-1/2}$ and $v = y^{-1/2}$.

25. $\begin{cases} x + y + z = 9 \\ x - y - z = -5 \\ 2x + y - 2z = -1 \end{cases}$

26. $\begin{cases} x - 4y + 2z = 9 \\ 2x + y + z = 3 \\ 3x - 2y - 3z = -18 \end{cases}$

27. $\begin{cases} 4x - 4y + z = 4 \\ 2x + 3y + 3z = -8 \\ x + y + z = -3 \end{cases}$

28. $\begin{cases} x - 8y + z = 1 \\ 5x + 16y + 3z = 3 \\ 4x - 4y - 4z = -4 \end{cases}$

29. $\begin{cases} -2x + y + z = 1 \\ x - 2y + z = -2 \\ x + y - 2z = 4 \end{cases}$

30. $\begin{cases} -x + y + z = 1 \\ x - y + z = -1 \\ x + y - z = 1 \end{cases}$

31. $\begin{cases} 4x + 2y - 3z = 15 \\ 2x + y + 3z = 3 \end{cases}$

32. $\begin{cases} 3x + 2y + 17z = 1 \\ x + 2y + 3z = 3 \end{cases}$

33. $\begin{cases} x + 2y - 3z = -2 \\ 2x - y + z = 1 \\ 3x - 4y + 5z = 1 \end{cases}$

34. $\begin{cases} 9x + y + z = 0 \\ -3x + y - z = 0 \\ 3x - 5y + 3z = 0 \end{cases}$

35. $\begin{cases} x + y + z = a + b \\ 2x - y + 2z = -a + 5b \\ x - 2y + z = -2a + 4b \end{cases}$

36. $\begin{cases} ax + by - 2az = 4ab + 2b^2 \\ x + y + z = 4a + 2b \\ bx + ay + 4az = 5a^2 + b^2 \end{cases}$

37. $\begin{cases} x + y + z + w = 8 \\ 3x + 3y - z - w = 20 \\ 4x - y - z + 2w = 18 \\ 2x + 5y + 5z - 5w = 8 \end{cases}$

38. $\begin{cases} x - 2y + 3z + w = 1 \\ x + y + z + w = 5 \\ 2x + 3y + 2z - w = 3 \\ 3x + y - z + 2w = 4 \end{cases}$

For Exercises 39–50, determine the constants (denoted by capital letters) so that each equation is an identity.

39. $\dfrac{1}{(x - 10)(x + 10)} = \dfrac{A}{x - 10} + \dfrac{B}{x + 10}$

40. $\dfrac{x}{(x - 5)(x + 5)} = \dfrac{A}{x - 5} + \dfrac{B}{x + 5}$

41. $\dfrac{2x}{(x + 1)^2} = \dfrac{A}{x + 1} + \dfrac{B}{(x + 1)^2}$

42. $\dfrac{x + 2}{(x - 1)^2} = \dfrac{A}{x - 1} + \dfrac{B}{(x - 1)^2}$

43. $\dfrac{5}{x(x - 4)} = \dfrac{A}{x} + \dfrac{B}{x - 4}$

44. $\dfrac{6x}{(x + 2)(x + 3)} = \dfrac{A}{x + 2} + \dfrac{B}{x + 3}$

45. $\dfrac{1}{(x - 1)(x + 3)^2} = \dfrac{A}{x - 1} + \dfrac{B}{x + 3} + \dfrac{C}{(x + 3)^2}$

46. $\dfrac{x^2}{(x + 2)(x - 2)^2} = \dfrac{A}{x + 2} + \dfrac{B}{x - 2} + \dfrac{C}{(x - 2)^2}$

47. $\dfrac{4x^2 + 2x + 15}{(x - 1)(x^2 + x + 5)} = \dfrac{A}{x - 1} + \dfrac{Bx + C}{x^2 + x + 5}$

48. $\dfrac{1}{x(x^2 + 16)} = \dfrac{A}{x} + \dfrac{Bx + C}{x^2 + 16}$

49. $\dfrac{1}{x^3 + 64} = \dfrac{A}{x + 4} + \dfrac{Bx + C}{x^2 - 4x + 16}$

50. $\dfrac{1}{x^3 + 3x^2 + 3x + 1} = \dfrac{A}{x + 1} + \dfrac{B}{(x + 1)^2} + \dfrac{C}{(x + 1)^3}$

In Exercises 51–54, the lowercase letters a *and* b *denote nonzero constants, with* $a \neq b$. *In each case, determine the values of* A *and* B *(and* C*, if applicable) so that the equation is an identity.*

51. $\dfrac{x}{(x - a)^3} = \dfrac{A}{x - a} + \dfrac{B}{(x - a)^2} + \dfrac{C}{(x - a)^3}$

52. $\dfrac{ax(1 - a)}{(x - a)(x + a^2)} = \dfrac{A}{x - a} + \dfrac{B}{x + a^2}$

53. $\dfrac{(a - b)(a + b - x)}{(x - a)(x - b)} = \dfrac{A}{x - a} + \dfrac{B}{x - b}$

54. $\dfrac{4a^2x + 6a^3}{(x + a)(x + 2a)(x + 3a)} = \dfrac{A}{x + a} + \dfrac{B}{x + 2a} + \dfrac{C}{x + 3a}$

For Exercises 55–62, evaluate each of the determinants.

55. $\begin{vmatrix} 1 & 5 \\ -6 & 4 \end{vmatrix}$

56. $\begin{vmatrix} \frac{1}{6} & 1 \\ 2 & 12 \end{vmatrix}$

57. $\begin{vmatrix} 4 & 0 & 3 \\ -2 & 1 & 5 \\ 0 & 2 & -1 \end{vmatrix}$

58. $\begin{vmatrix} 2 & 6 & 4 \\ 6 & 18 & 24 \\ 15 & 5 & -10 \end{vmatrix}$

59. $\begin{vmatrix} 1 & 5 & 7 \\ 1 & 5 & 7 \\ 17 & 19 & 21 \end{vmatrix}$

60. $\begin{vmatrix} 0 & 2 & 4 & 0 \\ 4 & 0 & 6 & 2 \\ 0 & 0 & 1 & 1 \\ 14 & 7 & 1 & 0 \end{vmatrix}$

61. $\begin{vmatrix} 1 & 0 & 0 & 0 \\ 0 & 2 & 0 & 0 \\ 0 & 0 & 3 & 0 \\ 0 & 0 & 0 & 4 \end{vmatrix}$

62. $\begin{vmatrix} 1 & a & b & c \\ 0 & 2 & d & e \\ 0 & 0 & 3 & f \\ 0 & 0 & 0 & 4 \end{vmatrix}$

63. Show that $\begin{vmatrix} a & b & c \\ b & c & a \\ c & a & b \end{vmatrix} = 3abc - a^3 - b^3 - c^3$.

64. Show that $\begin{vmatrix} 1 & 1 & 1 \\ 1 & 1 + x & 1 \\ 1 & 1 & 1 + x^2 \end{vmatrix} = x^3$.

65. Show that $\begin{vmatrix} a^2+x & b & c & d \\ -b & 1 & 0 & 0 \\ -c & 0 & 1 & 0 \\ -d & 0 & 0 & 1 \end{vmatrix}$

$= a^2 + b^2 + c^2 + d^2 + x.$

66. In a two-digit number, the sum of the digits is 11. Four times the units' digit exceeds the tens' digit by 4. Find the number.

67. Find two numbers whose sum and difference are 52 and 10, respectively.

68. Determine constants a and b so that the parabola $y = ax^2 + bx - 1$ passes through the points $(-2, 5)$ and $(2, 9)$.

69. The vertices of triangle ABC are $A(-2, 0)$, $B(4, 0)$, and $C(0, 6)$. Let A_1, B_1, and C_1 denote the midpoints of sides BC, AC, and AB, respectively.

(a) Find the point where the line segments AA_1 and BB_1 intersect. *Answer:* $(\frac{2}{3}, 2)$

(b) Follow part (a) using BB_1 and CC_1.

(c) Follow part (a) using AA_1 and CC_1.

(d) Let P denote the point $(\frac{2}{3}, 2)$ that you found in part (a). Compute each of the following ratios: $\overline{AP}/\overline{PA_1}$, $\overline{BP}/\overline{PB_1}$, $\overline{CP}/\overline{PC_1}$. What do you observe?

70. An *altitude* of a triangle is a line segment drawn from a vertex perpendicular to the opposite side. Suppose that the vertices of triangle ABC are as given in Exercise 69. Find the intersection point for each pair of altitudes. What do you observe about the three answers?

71. [This exercise appears in *Plane and Solid Analytic Geometry* by W. F. Osgood and W. G. Graustein (New York: Macmillan, 1920).] Let P be any point (a, a) of the line $x - y = 0$, other than the origin. Through P draw two lines, of arbitrary slopes m_1 and m_2, intersecting the x-axis in A_1 and A_2, and the y-axis in B_1 and B_2, respectively. Prove that the lines A_1B_2 and A_2B_1 will, in general, meet on the line $x + y = 0$.

72. Determine constants h, k, and r so that the circle $(x - h)^2 + (y - k)^2 = r^2$ passes through the three points $(0, 0)$, $(0, 1)$, and $(1, 0)$.

73. The vertices of a triangle are the points of intersection of the lines $y = x - 1$, $y = -x - 2$, and $y = 2x + 3$. Find the equation of the circle passing through these three intersection points.

74. The vertices of a triangle are $A(-4, 0)$, $B(2, 0)$, and $C(0, 6)$. Let M_1, M_2, and M_3 be the midpoints of AB, BC, and AC, respectively. Let H_1, H_2, and H_3 be the feet of the altitudes on sides AB, BC, and AC, respectively.

(a) Find the equation of the circle passing through M_1, M_2, and M_3. *Answer:* $3x^2 + 3y^2 + 3x - 11y = 0$

(b) Find the equation of the circle passing through H_1, H_2, and H_3. *Answer:* $3x^2 + 3y^2 + 3x - 11y = 0$

(c) Find the point P at which the three altitudes intersect. *Answer:* $(0, \frac{4}{3})$

(d) Let N_1, N_2, and N_3 be the midpoints of AP, BP, and CP, respectively. Show that the circle obtained in parts (a) and (b) passes through N_1, N_2, and N_3. *Note:* This circle is called the *nine-point circle* of triangle ABC.

(e) The *circumcircle* of triangle ABC is the circle passing through the three points A, B, and C. Show that the coordinates of the center Q of the circumcircle are $Q(-1, \frac{7}{3})$. *Hint:* Use the result from geometry that the perpendicular bisectors of the sides of the triangle intersect at the center of the circumcircle.

(f) For triangle ABC, show that the radius of the nine-point circle is one-half the radius of the circumcircle.

(g) Show that the midpoint of line segment QP is the center of the nine-point circle.

(h) Find the point G where the three medians of triangle ABC intersect.

(i) Show that G lies on line segment QP and that $\overline{PG} = 2\overline{GQ}$.

In Exercises 75–106, the matrices A, B, C, D, E, *and* F *are defined as follows.*

$$A = \begin{pmatrix} 3 & -2 \\ 1 & 5 \end{pmatrix} \quad B = \begin{pmatrix} 2 & 1 \\ 1 & 8 \end{pmatrix} \quad C = \begin{pmatrix} -1 & 0 \\ 0 & -1 \end{pmatrix}$$

$$D = \begin{pmatrix} -4 & 0 & 6 \\ 1 & 3 & 2 \end{pmatrix} \quad E = \begin{pmatrix} 3 & -1 \\ 4 & 1 \\ -5 & 9 \end{pmatrix} \quad F = \begin{pmatrix} 2 & 6 \\ 5 & 3 \\ 5 & 8 \end{pmatrix}$$

In each exercise, carry out the indicated matrix operations if they are defined. If an operation is not defined, state this.

75. $2A + 2B$

76. $2(A + B)$

77. $4B$

78. $B + 4$

79. AB

80. BA

81. $AB - BA$

82. $B + E$

83. $B + C$

84. $A(B + C)$

85. $AB + AC$

86. $(B + C)A$

87. $BA + CA$

88. $D + E$

89. DE

90. $(EE)D$

91. $E(ED)$

92. $E + F$

93. EF

94. $3E - 2F$

95. $(A + B) + C$

96. $A + (B + C)$

97. $(AB)C$

98. $A(BC)$

99. $2A + (3B - C)$

100. $(2A + 3B) - C$

101. $-2(F - E)$

102. $-2F + 2E$

103. $E - E$

104. $D - D$

105. $AA - BB$

106. $(A - B)(A + B)$

107. For a square matrix A, the notation A^2 means AA. Similarly, A^3 means AAA. If $A = \begin{pmatrix} 1 & 1 \\ 0 & 1 \end{pmatrix}$, verify that $A^2 = \begin{pmatrix} 1 & 2 \\ 0 & 1 \end{pmatrix}$ and $A^3 = \begin{pmatrix} 1 & 3 \\ 0 & 1 \end{pmatrix}$.

108. Let A be the matrix $\begin{pmatrix} 0 & 0 & 0 \\ a & 0 & 0 \\ b & c & 0 \end{pmatrix}$. Compute A^2 and A^3.

For Exercises 109–112, solve each matrix equation given that A, B, *and* C *are defined as follows.*

$$A = \begin{pmatrix} 2 & -1 & 1 \\ 0 & 3 & 2 \end{pmatrix} \quad B = \begin{pmatrix} 4 & 1 & 1 \\ 5 & 2 & 5 \end{pmatrix} \quad C = \begin{pmatrix} 9 & 2 & 8 \\ -1 & -6 & 0 \end{pmatrix}$$

109. $X - A = B$

110. $3X + A = B$

111. $2X - C = A - 2B$

112. $4(X - C) = 2(A - X) + B$

In Exercises 113–120, compute D, D_x, D_y, D_z *(and* D_w *where appropriate) for each system of equations. Use Cramer's Rule to solve the systems in which* $D \neq 0$. *If* $D = 0$, *solve the system using Gaussian elimination or matrix methods.*

113. $\begin{cases} 2x - y + z = 1 \\ 3x + 2y + 2z = 0 \\ x - 5y - 3z = -2 \end{cases}$

114. $\begin{cases} x + 2y - z = -1 \\ 2x - 3y + 3z = 3 \\ 2x + 3y + z = 1 \end{cases}$

115. $\begin{cases} x + 2y + 3z = -1 \\ 4x + 5y + 6z = 2 \\ 7x + 8y + 9z = -3 \end{cases}$

116. $\begin{cases} 3x + 2y - 2z = 0 \\ 2x + 3y - z = 0 \\ 8x + 7y - 5z = 0 \end{cases}$

117. $\begin{cases} 3x + 2y - 2z = 1 \\ 2x + 3y - z = -2 \\ 8x + 7y - 5z = 0 \end{cases}$

118. $\begin{cases} x + y + z + w = 5 \\ x - y - z + w = 3 \\ 2x + 3y + 3z + 2w = 21 \\ 4z - 3w = -7 \end{cases}$

119. $\begin{cases} 2x - y + z + 3w = 15 \\ x + 2y + 2w = 12 \\ 3y + 3z + 4w = 12 \\ -4x + y - 4z = -11 \end{cases}$

120. $\begin{cases} x + y + z = (a + b)^2 \\ \dfrac{bx}{a} + \dfrac{ay}{b} - z = 0 \\ x + y - z = (a - b)^2 \end{cases}$

For Exercises 121–134, find all solutions (x, y) *for each system, where* x *and* y *are real numbers.*

121. $\begin{cases} y = 6x \\ y = x^2 \end{cases}$

122. $\begin{cases} y = 4x \\ y = x^3 \end{cases}$

123. $\begin{cases} y = 9 - x^2 \\ y = x^2 - 9 \end{cases}$

124. $\begin{cases} x^3 - y = 0 \\ xy - 16 = 0 \end{cases}$

125. $\begin{cases} x^2 - y^2 = 9 \\ x^2 + y^2 = 16 \end{cases}$

126. $\begin{cases} 2x + 3y = 6 \\ y = \sqrt{x + 1} \end{cases}$

127. $\begin{cases} x^2 + y^2 = 1 \\ y = \sqrt{x} \end{cases}$

128. $\begin{cases} \dfrac{x}{11} + \dfrac{y}{12} = 2 \\ xy/132 = 1 \end{cases}$

129. $\begin{cases} x^2 + y^2 = 1 \\ y = 2x^2 \end{cases}$

130. $\begin{cases} x^2 - 3xy + y^2 = -11 \\ 2x^2 + xy - y^2 = 8 \end{cases}$

131. $\begin{cases} x^2 + 2xy + 3y^2 = 68 \\ 3x^2 - xy + y^2 = 18 \end{cases}$

132. $\begin{cases} \dfrac{x^2}{a^2} + \dfrac{y^2}{b^2} = 1 \\ \dfrac{x^2}{b^2} + \dfrac{y^2}{a^2} = 1 \quad (a > b > 0) \end{cases}$

133. $\begin{cases} 2(x - 3)^2 - (y + 1)^2 = -1 \\ -3(x - 3)^2 + 2(y + 1)^2 = 6 \end{cases}$

Hint: Let $u = x - 3$ and $v = y + 1$.

134. $\begin{cases} x^4 = y - 1 \\ y - 3x^2 + 1 = 0 \end{cases}$

Exercises 135–140 appear (in German) in an algebra text by Leonhard Euler, first published in 1770. The English versions given here are taken from the translated version Elements of Algebra, *Fifth Edition, by Leonhard Euler (London: Longman, Orme, and Company, 1840). (This in turn has been reprinted by Springer-Verlag, New York, 1984.)*

135. Required two numbers, whose sum may be s, and their proportion as a to b.

Answer: $\dfrac{as}{a + b}$ and $\dfrac{bs}{a + b}$

136. The sum $2a$, and the sum of the squares $2b$, of two numbers being given; to find the numbers.

Answer: $a - \sqrt{b - a^2}$ and $a + \sqrt{b - a^2}$

137. To find three numbers, so that (the sum of) one-half of the first, one-third of the second, and one-quarter of the third, shall be equal to 62; one-third of the first, one-quarter of the second, and one-fifth of the third, equal to 47; and one-quarter of the first, one-fifth of the second, and one-sixth of the third, equal to 38.

Answer: 24, 60, 120

138. Required two numbers, whose product may be 105, and whose squares (when added) may together make 274.

139. Required two numbers, whose product may be m, and the sum of the squares n ($n \geq 2m$).

140. Required two numbers such that their sum, their product, and the difference of their squares may all be equal.

In Exercises 141–146, graph each system of inequalities and specify whether the region is convex or bounded.

141. $\begin{cases} x^2 + y^2 \geq 1 \\ y - 4x \leq 0 \\ y - x \geq 0 \\ x \geq 0 \\ y \geq 0 \end{cases}$

142. $\begin{cases} x^2 + y^2 \leq 1 \\ x \geq 0 \\ y \geq 0 \end{cases}$

143. $\begin{cases} y - \sqrt{x} \leq 0 \\ y \geq 0 \\ x \geq 1 \\ x - 4 \leq 0 \end{cases}$

144. $\begin{cases} y - |x| \leq 0 \\ x + 1 \geq 0 \\ x - 1 \leq 0 \\ y + 1 \geq 0 \end{cases}$

145. $\begin{cases} y \leq 1/x \\ y \geq 0 \\ x \geq 1 \end{cases}$

146. $\begin{cases} y - 100x \leq 0 \\ y - x^2 \geq 0 \\ x \geq 0 \\ y \geq 0 \end{cases}$

For Exercises 147–150, find the maximum and minimum values of the objective functions subject to the given constraints. Specify the corresponding optimal solutions in each case.

147. Objective function: $C = 3x + 2y$

Constraints: $\begin{cases} 2y + x - 8 \leq 0 \\ 2x + y - 10 \leq 0 \\ x \geq 0 \\ y \geq 0 \end{cases}$

148. Objective function: $C = 4y + 3x + 12$

Constraints: $\begin{cases} x + 3y - 15 \leq 0 \\ x + y - 7 \leq 0 \\ 3x + y - 15 \leq 0 \\ x \geq 0 \\ y \geq 0 \end{cases}$

149. Objective function: $C = 20y + 15x + 49$

Constraints: $\begin{cases} 2x + 3y - 48 \leq 0 \\ x + 3y - 42 \leq 0 \\ x + y - 19 \leq 0 \\ 8x + y - 96 \leq 0 \\ x \geq 0 \\ y \geq 0 \end{cases}$

150. Objective function: $C = 5y - 11x$

Constraints: $\begin{cases} 2y - x - 7 \leq 0 \\ y - 2x - 2 \leq 0 \\ y \leq 5 \\ x + y - 9 \leq 0 \\ x \geq 0 \\ y \geq 0 \end{cases}$

151. The backers of a candidate for mayor wish to run a series of television commercials the day before the election. Through television they wish to reach an overall audience of at least 64,000 voters. Furthermore, the backers require that the number of unmarried people in the overall voter audience be at least 17,000. For daytime television, it is estimated that each showing of a commercial reaches 2000 voters who have not seen the commercial previously. For evening television, it is estimated that each showing of a commercial reaches 8000 voters who have not seen a commercial previously. Similarly, the number of unmarried voters reached by each commercial, whether daytime or evening, is estimated to be 1000. If the cost of running one commercial is \$2000 on daytime television and \$6000 on evening television, what is the least cost of running the television commercials? With this minimum budget, how many commercials will be shown on daytime television and how many will be shown evenings?

Chapter 9 Test

1. Determine all solutions of the system

$$\begin{cases} 3x + 4y = 12 \\ y = x^2 + 2x + 3 \end{cases}$$

2. Find all solutions of the system

$$\begin{cases} x - 2y = 13 \\ 3x + 5y = -16 \end{cases}$$

3. (a) Find all solutions of the following system using Gaussian elimination.

$$\begin{cases} x + 4y - z = 0 \\ 3x + y + z = -1 \\ 4x - 4y + 5z = -7 \end{cases}$$

(b) Compute D, D_x, D_y, and D_z for the system in part (a). [Then check your answer in part (a) using Cramer's Rule.]

4. Suppose that the matrices A and B are defined as follows:

$$A = \begin{pmatrix} 1 & -3 \\ 2 & -1 \end{pmatrix} \qquad B = \begin{pmatrix} 0 & 4 \\ 1 & 3 \end{pmatrix}$$

(a) Compute $2A - B$ **(b)** Compute BA.

5. Determine the area of the triangular region in the first quadrant that is bounded by the x-axis and the lines $y = 2x$ and $y = -x + 6$.

6. Find all solutions of the system

$$\begin{cases} \dfrac{1}{2x} + \dfrac{1}{3y} = 10 \\ \dfrac{5}{x} - \dfrac{4}{y} = -4 \end{cases}$$

7. Specify the coefficient matrix for the system

$$\begin{cases} x + y - z = -1 \\ 2x - y + 2z = 11 \\ x - 2y + z = 10 \end{cases}$$

Also specify the augmented matrix for this system.

8. Use matrix methods to find all solutions of the system displayed in the previous problem.

9. Find the equation of a line that passes through the point of intersection of the lines $x + y = 11$ and $3x + 2y = 7$, and that is perpendicular to the line $2x - 4y = 7$.

10. Determine the constants A, B, and C such that the following equation is an identity.

$$\frac{x-2}{(x+1)(x-1)^2} = \frac{A}{x+1} + \frac{B}{x-1} + \frac{C}{(x-1)^2}$$

11. Consider the determinant

$$\begin{vmatrix} 2 & 3 & -1 \\ 0 & 1 & 4 \\ 5 & -2 & 6 \end{vmatrix}$$

(a) What is the minor of the entry in the third row, second column?

(b) What is the cofactor of the entry in the third row, second column?

12. Evaluate the determinant

$$\begin{vmatrix} 4 & -5 & 0 \\ -8 & 10 & 7 \\ 16 & 20 & 14 \end{vmatrix}$$

13. Find all solutions of the system

$$\begin{cases} x^2 + y^2 = 15 \\ xy = 5 \end{cases}$$

14. Find the solutions of the system

$$\begin{cases} A + 2B + 3C = 1 \\ 2A - B - C = 2 \end{cases}$$

15. Let

$$A = \begin{pmatrix} 2 & 0 & -1 \\ 1 & 3 & 0 \end{pmatrix} \qquad B = \begin{pmatrix} 1 & -4 & 6 \\ 2 & 1 & 1 \end{pmatrix}$$

Solve the matrix equation $2X + B = 3A$.

16. Graph the inequality $5x - 6y \geq 30$.

17. Determine the constants P and Q given that the parabola $y = Px^2 + Qx - 5$ passes through the two points $(-2, -1)$ and $(-1, -2)$.

18. Graph the following system of inequalities, and specify the vertices.

$$\begin{cases} x \geq 0 \\ y \geq 0 \\ 2y - x \leq 14 \\ x + 3y \leq 36 \\ 9x + y \leq 99 \end{cases}$$

19. Determine the maximum value of the function $C = 3x + y + 1$, subject to the set of constraints displayed in the previous problem.

20. Determine the constant k given that the three lines $kx + 3y = -4$, $x - 2y = -3$, and $y = x$ are concurrent (pass through a common point).

Chapter 10 Roots of Polynomial Equations

Introduction

In this chapter we continue the work begun in Chapter 2 on solving polynomial equations. We begin in Section 10.1 with a second look at the long division process for polynomials. In this section you'll see how synthetic division is used to abbreviate the long division process. Section 10.2 presents two theorems about polynomials: the Remainder Theorem and the Factor Theorem. It is in the application of the former theorem that you'll begin to appreciate the utility of synthetic division. The Remainder Theorem and the Factor Theorem are then applied repeatedly in the next section (Section 10.3) to develop some fundamental results regarding polynomial equations and their solutions. You can view much of the material in this section as a kind of generalization of what you already know about quadratic equations. The last two sections of this chapter (Sections 10.4 and 10.5) present additional results that are useful in actually solving polynomial equations, or in determining the nature of their solutions.

10.1 More on Division of Polynomials

In Section 1.7 we studied the long division process for polynomials. There is a theorem, commonly referred to as the *Division Algorithm*, that summarizes rather nicely the key results of that long division process. We state the theorem here without proof.

The Division Algorithm

Let $p(x)$ and $d(x)$ be polynomials and assume that $d(x)$ is not the zero polynomial. Then there are unique polynomials $q(x)$ and $R(x)$ such that

$$p(x) = d(x) \cdot q(x) + R(x)$$

and where either $R(x)$ is the zero polynomial or the degree of $R(x)$ is less than the degree of $d(x)$.

The polynomials $p(x)$, $d(x)$, $q(x)$, and $R(x)$ are referred to, respectively, as the *dividend, divisor, quotient,* and *remainder*. In the case where $R(x) = 0$, we say that $d(x)$ and $q(x)$ are *factors* of $p(x)$.

Example 1 Let $p(x) = x^3 + 2x^2 - 4$ and $d(x) = x - 3$. Use the long division process to find the polynomials $q(x)$ and $R(x)$ such that

$$p(x) = d(x) \cdot q(x) + R(x)$$

and where either $R(x) = 0$ or the degree of $R(x)$ is less than the degree of $d(x)$.

Solution After inserting the term $0x$ in the dividend $p(x)$, we use long division to divide $p(x)$ by $d(x)$.

$$\begin{array}{r|l} & x^2 + 5x + 15 \\ \hline x - 3 & x^3 + 2x^2 + 0x - 4 \\ & \underline{x^3 - 3x^2} \\ & \quad 5x^2 + 0x \\ & \quad \underline{5x^2 - 15x} \\ & \qquad 15x - 4 \\ & \qquad \underline{15x - 45} \\ & \qquad\qquad 41 \end{array}$$

We now have that

$$\underbrace{x^3 + 2x^2 - 4}_{p(x)} = \underbrace{(x - 3)}_{d(x)}\underbrace{(x^2 + 5x + 15)}_{q(x)} \underbrace{+ 41}_{R(x)}$$

Thus $q(x) = x^2 + 5x + 15$ and $R(x) = 41$, as required. Notice that the degree of $R(x)$ is less than the degree of $d(x)$.

The long division procedure for polynomials can be streamlined when the divisor is of the form $x - a$. This shortened version, known as *synthetic division*, will be useful to us in subsequent sections when we are solving polynomial equations.

We'll explain the idea behind synthetic division using the long division that we just carried out in Example 1:

$$\begin{array}{r|l} & x^2 + 5x + 15 \\ \hline x - 3 & x^3 + 2x^2 + 0x - 4 \\ & \underline{x^3 - 3x^2} \\ & \quad 5x^2 + 0x \\ & \quad \underline{5x^2 - 15x} \\ & \qquad 15x - 4 \\ & \qquad \underline{15x - 45} \\ & \qquad\qquad 41 \end{array}$$

The basic idea behind synthetic division is that it is the coefficients of the various polynomials in the long division process that carry all of the necessary information. In our example, for instance, the quotient and remainder can be abbreviated by writing down the sequence of four numbers

$$1 \quad 5 \quad 15 \quad 41$$

By studying the long division process, you'll find that these numbers are obtained through the following four steps.

Step 1 Write down the first coefficient of the dividend. This will be the first coefficient of the quotient.

Step 2 Multiply the 1 obtained in the previous step by the -3 (in the divisor). Then subtract this from the second coefficient of the dividend.

$$-3 \times 1 = -3$$
$$2 - (-3) = 5 \qquad \textit{Result } \boxed{5}$$

Step 3 Multiply the 5 obtained in the previous step by the -3 (in the divisor). Then subtract this from the third coefficient of the dividend.

$$-3 \times 5 = -15$$
$$0 - (-15) = 15 \qquad \textit{Result } \boxed{15}$$

Step 4 Multiply the 15 obtained in the previous step by the -3 (in the divisor). Then subtract this from the fourth coefficient of the dividend.

$$-3 \times 15 = -45$$
$$-4 - (-45) = 41 \qquad \textit{Result } \boxed{41}$$

A convenient format for setting up this process is as follows:

$$\begin{array}{r|rrrr} -3 & 1 & 2 & 0 & -4 \\ & & & & \\ \hline & & & & \end{array}$$

Now using this format, let us again go through the four steps we have just described.

Step 1 Bring down the 1.

$$\begin{array}{r|rrrr} -3 & 1 & 2 & 0 & -4 \\ & & & & \\ \hline & 1 & & & \end{array}$$

Step 2

$$-3 \times 1 = -3$$
$$2 - (-3) = 5$$

$$\begin{array}{r|rrrr} -3 & 1 & 2 & 0 & -4 \\ & & -3 & & \\ \hline & 1 & 5 & & \end{array}$$

Step 3

$$-3 \times 5 = -15$$
$$0 - (-15) = 15$$

$$\begin{array}{r|rrrr} -3 & 1 & 2 & 0 & -4 \\ & & -3 & -15 & \\ \hline & 1 & 5 & 15 & \end{array}$$

Step 4

$$-3 \times 15 = -45$$
$$-4 - (-45) = 41$$

$$\begin{array}{r|rrrr} -3 & 1 & 2 & 0 & -4 \\ & & -3 & -15 & -45 \\ \hline & 1 & 5 & 15 & 41 \end{array}$$

Although we've now obtained the required sequence of numbers 1 5 15 41, there is one further simplification that can be made. In Steps 2 through 4, we could have added instead of subtracted if we'd used 3 instead of -3 in the initial format. (You'll see the motivation for this in the next section when we discuss the Remainder Theorem.) With this change, let us now summarize the technique of synthetic division using the example with which we've been working. The method is applicable for any polynomial division in which the form of the divisor is $x - a$.

To Divide $x^3 + 2x^2 - 4$ by $x - 3$ Using Synthetic Division

		Comments
Format	$\begin{array}{r\|rrrr} 3 & 1 & 2 & 0 & -4 \\ \hline & & & & \end{array}$	Since the divisor is $x - 3$, the format begins with 3. The coefficients from the dividend are written in the order corresponding to decreasing powers of x. A zero coefficient is inserted as place holder.
Procedure	$\begin{array}{r\|rrrr} 3 & 1 & 2 & 0 & -4 \\ & & 3 & 15 & 45 \\ \hline & 1 & 5 & 15 & 41 \end{array}$	Step 1: Bring down the 1. Step 2: $3 \times 1 = 3$; $2 + 3 = 5$ Step 3: $3 \times 5 = 15$; $0 + 15 = 15$. Step 4: $3 \times 15 = 45$;$-4 + 45 = 41$.
Answer	Quotient: $x^2 + 5x + 15$ Remainder: 41	The degree of the first term in the quotient is one less than the degree of the first term of the dividend.

As we've just seen, when the divisor is $x - 3$, we write $+3$ in the synthetic division format. In general, if the divisor is $x - a$, we write a. The next example shows what to do when the form of the divisor is $x + a$.

Example 2 Use synthetic division to divide $x^4 - 2x^3 + 5x^2 - 4x + 3$ by $x + 1$.

Solution We need to first write the divisor $x + 1$ in the form $x - a$. We have

$$x + 1 = x - (-1)$$

In other words, a is -1 and this is the value we use to set up the synthetic division format. The format then is

$$\begin{array}{r|rrrrr} -1 & 1 & -2 & 5 & -4 & 3 \\ \hline & & & & & \end{array}$$

Now we carry out the synthetic division procedure.

$$\begin{array}{r|rrrrr} -1 & 1 & -2 & 5 & -4 & 3 \\ & & -1 & 3 & -8 & 12 \\ \hline & 1 & -3 & 8 & -12 & 15 \end{array}$$

The quotient therefore is $x^3 - 3x^2 + 8x - 12$ and the remainder is 15. We can summarize this result by writing

$$x^4 - 2x^3 + 5x^2 - 4x + 3 = (x + 1)(x^3 - 3x^2 + 8x - 12) + 15$$

(Notice that the degree of the remainder is less than the degree of the divisor, in agreement with the Division Algorithm.)

Exercise Set 10.1

In Exercises 1–22, use synthetic division to find the quotient and the remainder.

1. $\dfrac{x^2 - 6x - 2}{x - 5}$

2. $\dfrac{3x^2 + 4x - 1}{x - 1}$

3. $\dfrac{4x^2 - x - 5}{x + 1}$

4. $\dfrac{x^2 - 1}{x + 2}$

5. $\dfrac{6x^3 - 5x^2 + 2x + 1}{x - 4}$

6. $\dfrac{x^4 - 4x^3 + 6x^2 - 4x + 1}{x - 1}$

7. $\dfrac{x^3 - 1}{x - 2}$

8. $\dfrac{x^3 - 8}{x - 2}$

9. $\dfrac{x^5 - 1}{x + 2}$

10. $\dfrac{x^3 - 8x^2 - 1}{x + 3}$

11. $\dfrac{x^4 - 6x^3 + 2}{x + 4}$

12. $\dfrac{3x^3 - 2x^2 + x + 1}{x - \frac{1}{2}}$

13. $\dfrac{x^3 - 4x^2 - 3x + 6}{x - 10}$

14. $\dfrac{1 + 3x + 3x^2 + x^3}{x + 1}$

15. $\dfrac{x^3 - x^2}{x + 5}$

16. $\dfrac{5x^4 - 4x^3 + 3x^2 - 2x + 1}{x + \frac{1}{2}}$

17. $\dfrac{14 - 27x - 27x^2 + 54x^3}{x - \frac{2}{3}}$

18. $\dfrac{14 - 27x - 27x^2 + 54x^3}{x + \frac{2}{3}}$

19. $\dfrac{x^4 + 3x^2 + 12}{x - 3}$

20. **(a)** $\dfrac{x^2 - a^2}{x - a}$ **(b)** $\dfrac{x^3 - a^3}{x - a}$ **(c)** $\dfrac{x^4 - a^4}{x - a}$ **(d)** $\dfrac{x^5 - a^5}{x - a}$

21. $\dfrac{ax^2 + bx + c}{x - r}$

22. $\dfrac{ax^3 + bx^2 + cx + d}{x - r}$

In Exercises 23 and 24, use synthetic division to determine the quotient and remainder.

23. $\dfrac{3x^2 - 8x + 1}{3x - 4}$

    Hint: $\dfrac{3x^2 - 8x + 1}{3x - 4} = \dfrac{1}{3}\left(\dfrac{3x^2 - 8x + 1}{x - \frac{4}{3}}\right)$

24. $\dfrac{4x^3 + 6x^2 - 6x - 5}{2x - 3}$

25. When $x^3 + kx + 1$ is divided by $x + 1$, the remainder is -4. Find k.

26. **(a)** Show that when $x^3 + kx + 6$ is divided by $x + 3$, the remainder is $-21 - 3k$.
    (b) Determine a value of k such that $x + 3$ will be a factor of $x^3 + kx + 6$.

10.2 Roots of Polynomial Equations. The Remainder and Factor Theorems

The techniques for solving polynomial equations of degree 1 and degree 2 were discussed in Chapter 2. Now we want to extend those ideas. Our focus in this section and in the remainder of this chapter is on solving polynomial equations, that is, equations of the form

$$f(x) = a_nx^n + a_{n-1}x^{n-1} + \cdots + a_1x + a_0 = 0 \tag{1}$$

Here (as in Chapter 2), by a *root* or *solution* of equation (1) we mean a number r that when substituted for x leads to a true statement. Thus, r is a root of equation (1) provided that $f(r) = 0$. We also refer to the number r in this case as a *zero* of the function f.

Example 1 **(a)** Is -3 a zero of the function f defined by $f(x) = x^4 + x^2 - 6$?
(b) Is $\sqrt{2}$ a root of the equation $x^4 + x^2 - 6 = 0$?

Solution **(a)** By definition, -3 will be a zero of f if $f(-3) = 0$. Let us compute. We have

$$f(-3) = (-3)^4 + (-3)^2 - 6 = 81 + 9 - 6 = 84$$

or

$$f(-3) = 84 \ (\neq 0)$$

Thus -3 is not a zero of the function f.

(b) To check if $\sqrt{2}$ is a root of the given equation, we have

$$(\sqrt{2})^4 + (\sqrt{2})^2 - 6 \stackrel{?}{=} 0$$
$$4 + 2 - 6 \stackrel{?}{=} 0$$
$$0 \stackrel{?}{=} 0 \qquad \text{True}$$

Thus $\sqrt{2}$ is a root of the equation $x^4 + x^2 - 6 = 0$.

There are cases in which a root of an equation is what we call a *repeated root.* Consider for instance the equation $x(x - 1)(x - 1) = 0$. We have

$$x(x - 1)(x - 1) = 0$$
$$x = 0 \quad \Big| \quad x - 1 = 0 \quad \Big| \quad x - 1 = 0$$
$$x = 1 \qquad\qquad x = 1$$

The roots of the equation are therefore 0, 1, and 1. The repeated root here is $x = 1$. We say in this case that 1 is a *double root* or, equivalently, that 1 is a *root of multiplicity two.* More generally, if a root is repeated k times we call it a *root of multiplicity k.*

Example 2 What is the multiplicity of each root of the equation

$$(x - 4)^2(x - 5)^3 = 0$$

Solution Writing out the equation, we have

$$(x - 4)(x - 4)(x - 5)(x - 5)(x - 5) = 0$$

Now by setting each factor equal to zero, we obtain the roots 4, 4, 5, 5, and 5. From this we see that the root 4 has multiplicity two, while the root 5 has multiplicity three. Notice that it is not really necessary to write out all the factors as we did here; the exponents of the factors in the original equation give you the required multiplicities.

There are two simple but important theorems that will form the basis for much of our subsequent work with polynomials. These are the *Remainder Theorem* and the *Factor Theorem.* We begin with a statement of the Remainder Theorem.

The Remainder Theorem

When the polynomial $f(x)$ is divided by $x - r$, the remainder is $f(r)$.

Before turning to a proof of the Remainder Theorem, let us see what the theorem is saying in two particular cases. First, suppose that we divide the polynomial $f(x) = 2x^2 - 3x + 4$ by $x - 1$. Then according to the Remainder Theorem, the remainder in this case should be the number $f(1)$. Let's check.

$$\begin{array}{r|rrr} 1 & 2 & -3 & 4 \\ & & 2 & -1 \\ \hline & 2 & -1 & \boxed{3} \end{array} \qquad \begin{aligned} f(x) &= 2x^2 - 3x + 4 \\ f(1) &= 2(1)^2 - 3(1) + 4 \\ f(1) &= \boxed{3} \end{aligned}$$

As the calculations show, the remainder is indeed equal to $f(1)$. As a second example, let us divide the polynomial $g(x) = ax^2 + bx + c$ by $x - r$. According to the Remainder Theorem, the remainder in this case should be $g(r)$. Again, let us check.

$$\begin{array}{r|lll} r & a & b & c \\ & & ar & ar^2 + br \\ \hline & a & ar + b & \boxed{ar^2 + br + c} \end{array} \qquad \begin{aligned} g(x) &= ax^2 + bx + c \\ g(r) &= \boxed{ar^2 + br + c} \end{aligned}$$

The calculations show the remainder is equal to $g(r)$, as we wished to check.

In the example just concluded, we verified that the Remainder Theorem holds for any quadratic polynomial $g(x) = ax^2 + bx + c$. A general proof of the Remainder Theorem can easily be given along those same lines. The only drawback there is that it becomes slightly cumbersome to carry out the synthetic division process when the dividend is

$$a_nx^n + a_{n-1}x^{n-1} + \cdots + a_1x + a_0$$

For this reason, mathematicians often prefer to base the proof of the Remainder Theorem upon the Division Algorithm given in the previous section. This is the path we shall follow here.

To prove the Remainder Theorem, we must show that when the polynomial $f(x)$ is divided by $x - r$, the remainder is $f(r)$. Now according to the Division Algorithm, we may write

$$f(x) = (x - r) \cdot q(x) + R(x) \tag{2}$$

In this identity, either $R(x)$ is the zero polynomial or the degree of $R(x)$ is less than that of $x - r$. But since the degree of $x - r$ is 1, we must have in this latter case that the degree of $R(x)$ is zero. Thus in *either* case, the remainder $R(x)$ is a constant. Denoting this constant by c, we can rewrite equation (2) as

$$f(x) = (x - r) \cdot q(x) + c$$

Now if we set $x = r$ in this identity, we obtain

$$f(r) = (r - r) \cdot q(r) + c$$

or

$$f(r) = c$$

We've now shown that $f(r) = c$. But by definition, c is the remainder $R(x)$. Thus $f(r) = R(x)$. This proves the Remainder Theorem.

Example 3 Let $f(x) = 2x^3 - 5x^2 + x - 6$. Use the Remainder Theorem to evaluate $f(3)$.

Solution According to the Remainder Theorem, $f(3)$ is the remainder when $f(x)$ is divided by $x - 3$. Using synthetic division, we have

$$\begin{array}{r|rrrr} 3 & 2 & -5 & 1 & -6 \\ & & 6 & 3 & 12 \\ \hline & 2 & 1 & 4 & 6 \end{array}$$

The remainder is 6 and therefore $f(3) = 6$.

From our experience with quadratic equations we know that there is a close connection between factoring a quadratic polynomial $f(x)$ and solving the polynomial equation $f(x) = 0$. The Factor Theorem (in the box that follows) states this relationship between roots and factors in a precise form. Furthermore, the Factor Theorem tells us that this relationship holds for polynomials of all degrees, not just quadratics.

The Factor Theorem

Let $f(x)$ be a polynomial. If $f(r) = 0$, then $x - r$ is a factor of $f(x)$. Conversely, if $x - r$ is a factor of $f(x)$, then $f(r) = 0$.

In terms of roots, the Factor Theorem may be summarized by saying that r is a root of the equation $f(x) = 0$ if and only if $x - r$ is a factor of $f(x)$. To prove the Factor Theorem, let us begin by assuming that $f(r) = 0$. We want to show that $x - r$ is a factor of $f(x)$. Now, according to the Remainder Theorem, if $f(x)$ is divided by $x - r$, the remainder is $f(r)$. So we can write

$$f(x) = (x - r)\cdot q(x) + f(r) \qquad \text{for some polynomial } q(x)$$

But since $f(r)$ is zero, this equation becomes

$$f(x) = (x - r)\cdot q(x)$$

This last equation tells us that $x - r$ is a factor of $f(x)$, as we wished to prove. Now conversely, assume that $x - r$ is a factor of $f(x)$. We want to show that $f(r) = 0$. Since $x - r$ is a factor of $f(x)$, we may write

$$f(x) = (x - r)\cdot q(x) \qquad \text{for some polynomial } q(x)$$

If we now let $x = r$ in this last equation, we obtain

$$f(r) = (r - r)\cdot q(r)$$

or

$$f(r) = 0 \qquad \text{as we wished to show}$$

The example that follows indicates how the Factor Theorem may be used in solving equations.

Example 4 Solve the equation $x^3 - 2x + 1 = 0$, given that one root is $x = 1$.

Solution Since $x = 1$ is a root, the Factor Theorem tells us that $x - 1$ is a factor of $x^3 - 2x + 1$. In other words

$$x^3 - 2x + 1 = (x - 1)\cdot q(x) \qquad \text{for some polynomial } q(x)$$

To determine this other factor $q(x)$, we divide $x^3 - 2x + 1$ by $x - 1$.

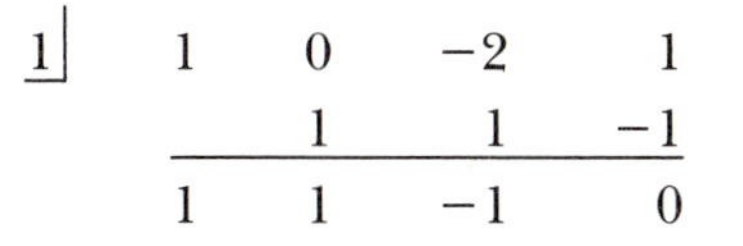

Thus $q(x) = x^2 + x - 1$ and we have the factorization

$$x^3 - 2x + 1 = (x - 1)(x^2 + x - 1)$$

Using this identity, the original equation becomes

$$(x - 1)(x^2 + x - 1) = 0$$

Now the problem is reduced to solving the linear equation $x - 1 = 0$ and the quadratic equation $x^2 + x - 1 = 0$. The linear equation yields $x = 1$, but we already knew that 1 was a root. So if we are to find any additional roots by this method, they must come from the equation $x^2 + x - 1 = 0$. Using the quadratic formula, we obtain

$$x = \frac{-1 \pm \sqrt{(1)^2 - 4(1)(-1)}}{2(1)} = \frac{-1 \pm \sqrt{5}}{2}$$

We now have three roots of the cubic equation $x^3 - 2x + 1 = 0$. These are

$$1 \qquad \frac{-1 + \sqrt{5}}{2} \qquad \frac{-1 - \sqrt{5}}{2}$$

As you'll see in the next section, a cubic equation can have at most three roots. So in this case we have determined all of the roots; that is, we've solved the equation.

Before going on to other examples, let's take a moment to summarize the technique used in Example 4. We want to solve a polynomial equation $f(x) = 0$, given that one root is $x = r$. Since r is a root, the Factor Theorem tells us that $x - r$ is a factor of $f(x)$. Then with the aid of synthetic division we obtain a factorization

$$f(x) = (x - r) \cdot q(x) \qquad \text{for some polynomial } q(x)$$

This gives rise to the two equations $x - r = 0$ and $q(x) = 0$. Since the first of these only reasserts that $x = r$ is a root, we try to solve the second equation $q(x) = 0$. We refer to the equation $q(x) = 0$ as the *reduced equation*. Example 4 shows you the idea behind this terminology; the degree of $q(x)$ is one less than that of $f(x)$. If, as in Example 4, the reduced equation happens to be a quadratic equation, then we can always determine the remaining roots by factoring or by the quadratic formula. In subsequent sections we will look at techniques that are helpful in cases where $q(x)$ is not quadratic.

Example 5 Solve the equation $x^4 + 2x^3 - 7x^2 - 20x - 12 = 0$, given that $x = 3$ and $x = -2$ are roots.

Solution Since $x = 3$ is a root, the Factor Theorem tells us that $x - 3$ is a factor of the polynomial $x^4 + 2x^3 - 7x^2 - 20x - 12$. That is,

$$x^4 + 2x^3 - 7x^2 - 20x - 12 = (x - 3) \cdot q(x) \qquad \text{for some polynomial } q(x)$$

As in Example 4, $q(x)$ can be found by synthetic division. We have

$$\begin{array}{r|rrrrr} 3 & 1 & 2 & -7 & -20 & -12 \\ & & 3 & 15 & 24 & 12 \\ \hline & 1 & 5 & 8 & 4 & 0 \end{array}$$

Thus $q(x) = x^3 + 5x^2 + 8x + 4$ and our original equation is equivalent to

$$(x - 3)(x^3 + 5x^2 + 8x + 4) = 0$$

Now $x = -2$ is a root of this equation. But $x = -2$ surely is not a root of the equation $x - 3 = 0$. Therefore $x = -2$ must be a root of the reduced equation $x^3 + 5x^2 + 8x + 4 = 0$. Again, the Factor Theorem is applicable. Since $x = -2$ is a root of the reduced equation, $x + 2$ must be a factor of $x^3 + 5x^2 + 8x + 4$. That is,

$$x^3 + 5x^2 + 8x + 4 = (x + 2)\cdot p(x) \qquad \text{for some polynomial } p(x)$$

We can use synthetic division to determine $p(x)$:

$$\begin{array}{r|rrrr} -2 & 1 & 5 & 8 & 4 \\ & & -2 & -6 & -4 \\ \hline & 1 & 3 & 2 & 0 \end{array}$$

Thus $p(x) = x^2 + 3x + 2$ and our reduced equation can be written

$$(x + 2)(x^2 + 3x + 2) = 0$$

This gives rise to a second reduced equation

$$x^2 + 3x + 2 = 0$$

In this case, the roots are readily obtained by factoring. We have

$$x^2 + 3x + 2 = 0$$

$$(x + 2)(x + 1) = 0$$

$$\begin{array}{r|l} x + 2 = 0 & x + 1 = 0 \\ x = -2 & x = -1 \end{array}$$

Now, of the two roots we've just found, the -2 happens to be one of the roots that we were initially given in the statement of the problem. The -1, on the other hand, is a distinct additional root. In summary then, we have three distinct roots: 3, -2, and -1. The root -2 has multiplicity two. As you'll see in the next section, a fourth-degree equation can have at most four (not necessarily distinct) roots. So in the case at hand, we have found all of the roots of the given equation.

Example 6 Find a polynomial equation for which -1, 4, and 5 are roots.

Solution If $f(x)$ is any polynomial containing the factors $(x + 1)$, $(x - 4)$, and $(x - 5)$, then the equation $f(x) = 0$ will certainly be satisfied when $x = -1$, $x = 4$, or $x = 5$. The simplest polynomial equation in this case then is

$$(x + 1)(x - 4)(x - 5) = 0$$

or

$$(x + 1)(x^2 - 9x + 20) = 0$$

After carrying out the indicated multiplication, we find that this last equation can be written $x^3 - 8x^2 + 11x + 20 = 0$. This is a polynomial equation that satisfies the required conditions.

Exercise Set 10.2

In Exercises 1–6, determine whether the given value for the variable is a root of the equation.

1. $12x - 8 = 112$; $x = 10$
2. $12x^2 - x - 20 = 0$; $x = \frac{5}{4}$
3. $x^2 - 2x - 4 = 0$; $x = 1 - \sqrt{5}$
4. $1 - x + x^2 - x^3 = 0$; $x = -1$
5. $2x^2 - 3x + 1 = 0$; $x = \frac{1}{2}$
6. $(x - 1)(x - 2)(x - 3) = 0$; $x = 4$

For Exercises 7–18, determine whether the given value is a zero of the function.

7. $f(x) = 3x - 2$; $x = \frac{2}{3}$
8. $g(x) = 1 + x^2$; $x = -1$
9. $h(x) = 5x^3 - x^2 + 2x + 8$; $x = -1$
10. $F(x) = -2x^5 + 3x^4 + 8x^3$; $x = 0$
11. $f(t) = 1 + 2t + t^3 - t^5$; $t = 2$
12. $f(t) = 1 + 2t + t^3 - t^5$; $t = \sqrt{2}$
13. $f(x) = 2x^3 - 3x + 1$; **(a)** $x = \frac{1}{2}(\sqrt{3} - 1)$ **(b)** $x = \frac{1}{2}(\sqrt{3} + 1)$
14. $g(x) = x^4 + 8x^3 + 9x^2 - 8x - 10$; **(a)** $x = 1$ **(b)** $x = \sqrt{6} - 4$ **(c)** $x = \sqrt{6} + 4$
15. $f(x) = \frac{1}{2}x^2 + bx + c$; $x = -b - \sqrt{b^2 - 2c}$

*16. $Q(x) = ax^2 + bx + c$; $x = (-b + \sqrt{b^2 - 4ac})(2a)^{-1}$
Hint: Look before you leap!

*17. $F(x) = 2x^4 + 4x + 1$; $x = \frac{1}{2}(-\sqrt{2} + \sqrt{2\sqrt{2} - 2})$

18. $f(x) = x^3 - 3x^2 + 3x - 3$; **(a)** $x = \sqrt[3]{2} - 1$ **(b)** $x = \sqrt[3]{2} + 1$
19. List the distinct roots of each of the following equations. In the case of a repeated root, give its multiplicity.
    (a) $(x - 1)(x - 2)^3(x - 3) = 0$
    (b) $(x - 1)(x - 1)(x - 1) = 0$
    (c) $(x - 5)^6(x + 1)^4 = 0$
    (d) $x^5(x - 1) = 0$
20. In this exercise we verify that the Remainder Theorem is valid for the cubic polynomial $g(x) = ax^3 + bx^2 + cx + d$.
    (a) Compute $g(r)$.
    (b) Using synthetic division, divide $g(x)$ by $x - r$. Check that the remainder you obtain is the same as the answer in part (a).

In Exercises 21–28, use the Remainder Theorem (as in Example 3) to evaluate f(x) *for the given value of* x.

21. $f(x) = 4x^3 - 6x^2 + x - 5$; $x = -3$
22. $f(x) = 2x^3 - x - 4$; $x = 4$
23. $f(x) = 6x^4 + 5x^3 - 8x^2 - 10x - 3$; $x = \frac{1}{2}$
24. $f(x) = x^5 - x^4 - x^3 - x^2 - x - 1$; $x = -2$
25. $f(x) = x^2 + 3x - 4$; $x = -\sqrt{2}$
26. $f(x) = -3x^3 + 8x^2 - 12$; $x = 5$
27. $f(x) = \frac{1}{2}x^3 - 5x^2 - 13x - 10$; $x = 12$
28. $f(x) = x^7 - 7x^6 + 5x^4 + 1$; $x = -3$

In each of Exercises 29–42, you are given a polynomial equation and one or more roots. Solve each equation using the method shown in Examples 4 and 5. To help decide if you have found all the roots in each case, you may rely on the following theorem, discussed in the next section: A polynomial equation of degree n *has at most* n *(not necessarily distinct) roots.*

29. $x^3 - 4x^2 - 9x + 36 = 0$; -3 is a root.
30. $x^3 + 7x^2 + 11x + 5 = 0$; -1 is a root.
31. $x^3 + x^2 - 7x + 5 = 0$; 1 is a root.
32. $x^3 + 8x^2 - 3x - 24 = 0$; -8 is a root.
33. $3x^3 - 5x^2 - 16x + 12 = 0$; -2 is a root.
34. $2x^3 - 5x^2 - 46x + 24 = 0$; 6 is a root.
35. $2x^3 + x^2 - 5x - 3 = 0$; $-\frac{3}{2}$ is a root.
36. $6x^4 - 19x^3 - 25x^2 + 18x + 8 = 0$; 4 and $-\frac{1}{3}$ are roots.
37. $x^4 - 15x^3 + 75x^2 - 125x = 0$; 5 is a root.
38. $2x^3 + 5x^2 - 8x - 20 = 0$; 2 is a root.
39. $x^4 + 2x^3 - 23x^2 - 24x + 144 = 0$; -4 and 3 are roots.
40. $6x^5 + 5x^4 - 29x^3 - 25x^2 - 5x = 0$; $\sqrt{5}$ and $-\frac{1}{3}$ are roots.
41. $x^3 + 7x^2 - 19x - 9 = 0$; -9 is a root.
42. $4x^5 - 15x^4 + 8x^3 + 19x^2 - 12x - 4 = 0$; 1, 2, and -1 are roots.
43. Find a polynomial equation of degree 3 in which the coefficient of x^3 is 1 and in which 3, -4, and 5 are roots.
44. Find a polynomial equation of degree 3 in which the coefficients are integers and $\frac{1}{2}$, $\frac{2}{5}$, and $-\frac{3}{4}$ are roots.
45. Find a cubic equation of the form $x^3 + bx^2 + cx + d = 0$, in which -1 and -6 are roots and -1 is a root of multiplicity two.
46. Find a fourth-degree equation in which the coefficient of x^4 is 1 and for which 0, -1, and 1 are roots and the multiplicity of 0 is two.
47. Find c and d such that 1 and 2 are roots of the equation $x^3 - 4x^2 + cx + d = 0$.
48. Let $g(x) = ax^2 + bx + c$. Show that when $g(x)$ is divided by $x + r$, the remainder is $g(-r)$.

49. Determine values for a and b such that $x - 1$ is a factor of both $x^3 + x^2 + ax + b$ and $x^3 - x^2 - ax + b$.

50. Determine a quadratic equation with the given roots.

(a) $\frac{a}{b}, -\frac{b}{a}$

(b) $-a + 2\sqrt{2b}, -a - 2\sqrt{2b}$

***51.** One root of the equation $x^2 + bx + 1 = 0$ is twice the other. Find b. (Two answers.)

***52.** Determine a value for a such that one root of the equation $ax^2 + x - 1 = 0$ is five times the other.

C *In Exercises 53–56, use the Remainder Theorem (as in Example 3) to evaluate* f(x) *for the given value of* x. *Use a calculator, and round off your answers to two decimal places.*

53. $f(x) = x^3 - 3x^2 + 12x + 9$; $x = 1.16$

54. $f(x) = x^3 - 2x - 5$; **(a)** $x = 2.09$ **(b)** $x = 2.094$ **(c)** $x = 2.0945$

55. $f(x) = x^3 - 5x - 2$; **(a)** $x = 2.41$ **(b)** $x = 2.42$

56. $f(x) = x^4 - 2x^3 - 5x^2 + 10x - 3$; **(a)** $x = -2.3$ **(b)** $x = -2.302$ **(c)** $x = -2.30277$

10.3 The Fundamental Theorem of Algebra

Every equation of algebra has as many solutions as the exponent of the highest term indicates.

Albert Girard, 1629, as quoted in *The History of Mathematics, An Introduction* by David M. Burton (Boston: Allyn and Bacon, 1985)

Does every polynomial equation (of degree at least 1) have a root? Or, to put the question another way, is it possible to write a polynomial equation that has no solution? Certainly if we consider only real roots (that is, roots that are real numbers), then it's easy to specify an equation with no real root:

$$x^2 = -1$$

This equation has no real root because the square of a real number is never negative. On the other hand, if we expand our base of operations from the real number system to the complex number system, then $x^2 = -1$ indeed has a root. In fact, both i and $-i$ are roots in this case.

As it turns out, the situation just described for the equation $x^2 = -1$ holds quite generally. That is, within the complex number system, every polynomial equation of degree at least 1 has at least one root. This is the substance of a remarkable theorem that was first proved by the great mathematician Carl Friedrich Gauss* in 1799. (Gauss was but 22 years old at the time.) Although there are many fundamental theorems in algebra, Gauss' result has come to be known as the *Fundamental Theorem of Algebra.*

The Fundamental Theorem of Algebra

Every polynomial equation of the form

$$a_nx^n + a_{n-1}x^{n-1} + \cdots + a_1x + a_0 = 0 \qquad n \geq 1,\ a_n \neq 0$$

has at least one root among the complex numbers. (This root may be a real number.)

*In the opinion of George F. Simmons in his text *Calculus with Analytic Geometry* (New York: McGraw-Hill, 1985), "Gauss was the greatest of all mathematicians and perhaps the most richly gifted genius of whom there is any record. This gigantic figure, towering at the beginning of the nineteenth century, separates the modern era in mathematics from all that went before."

Although for the most part in this section we consider only polynomials with real coefficients, Gauss' theorem still applies in cases in which some or all of the coefficients are nonreal complex numbers. The proof of Gauss' theorem is usually given in the post-calculus course Complex Variables.

The Fundamental Theorem of Algebra asserts that every polynomial equation of degree at least 1 has a root. There are two initial observations to be made here. First, notice that the theorem says nothing about actually finding the root. Second, notice that the theorem deals only with polynomial equations. Indeed, it's easy to specify a nonpolynomial equation that does not have a root. Such an equation is $1/x = 0$. The expression on the left-hand side of this equation can never be zero because the numerator is 1.

Example 1 Which of the following equations have at least one root?

(a) $x^3 - 17x^2 + 6x - 1 = 0$ **(b)** $\sqrt{2}x^{47} - \pi x^{25} + \sqrt{3} = 0$

(c) $x^2 - 2ix + (3 + i) = 0$

Solution All three equations are polynomial equations. So, according to the Fundamental Theorem of Algebra, each equation has at least one root.

In Chapter 2 we used factoring as a tool for solving various polynomial equations. The next theorem, a consequence of the Fundamental Theorem of Algebra, tells us that (in principle, at least) any polynomial of degree n can be factored into a product of n linear factors. In proving this result we'll need to use the Factor Theorem. If you reread the proof of that theorem in the previous section, you'll see that it makes no difference whether the number r appearing in the factor $x - r$ is a real number or a nonreal complex number. Thus the Factor Theorem is valid in both cases.

The Linear Factors Theorem

Let $f(x) = a_n x^n + a_{n-1}x^{n-1} + \cdots + a_1 x + a_0 \quad (n \geq 1, a_n \neq 0)$. Then $f(x)$ can be expressed as a product of n linear factors:

$$f(x) = a_n(x - r_1)(x - r_2) \cdots (x - r_n)$$

(The complex numbers r_k appearing in these factors are not necessarily all distinct, and some or all of the r_k may be real numbers.)

Proof of the Linear Factors Theorem According to the Fundamental Theorem of Algebra, the equation $f(x) = 0$ has a root. Call this root r_1. By the Factor Theorem, $x - r_1$ is a factor of $f(x)$ and we can write

$$f(x) = (x - r_1) \cdot Q_1(x)$$

for some polynomial $Q_1(x)$ of degree $n - 1$ and with leading coefficient a_n. If the degree of $Q_1(x)$ happens to be zero, we are done. On the other hand, if the degree of $Q_1(x)$ is at least 1, another application of the Fundamental Theorem of Algebra followed by the Factor Theorem gives us

$$Q_1(x) = (x - r_2) \cdot Q_2(x)$$

where the degree of $Q_2(x)$ is $n - 2$ and the leading coefficient of $Q_2(x)$ is a_n. We now have

$$f(x) = (x - r_1)(x - r_2) \cdot Q_2(x)$$

We continue this process until the quotient is $Q_n(x) = a_n$. As a result, we obtain

$$f(x) = (x - r_1)(x - r_2) \cdots (x - r_n)a_n$$

or

$$f(x) = a_n(x - r_1)(x - r_2) \cdots (x - r_n) \qquad \text{as we wished to show}$$

The Linear Factors Theorem tells us that any polynomial can be expressed as a product of linear factors. The theorem gives us no information, however, as to how those factors can actually be obtained in practice. The next example indicates one case in which the factors are readily obtained; this is the case with quadratic polynomials.

Example 2 Express each of the following second-degree polynomials in the form $a_n(x - r_1)(x - r_2)$.

(a) $3x^2 - 5x - 2$

(b) $x^2 - 4x + 5$

Solution

(a) A factorization for $3x^2 - 5x - 2$ can be found by simple trial and error. We have

$$3x^2 - 5x - 2 = (3x + 1)(x - 2)$$

Now write the factor $3x + 1$ as $3(x + \frac{1}{3})$. This in turn can be written as $3[x - (-\frac{1}{3})]$. The final factorization then is

$$3x^2 - 5x - 2 = 3[x - (-\tfrac{1}{3})](x - 2)$$

(b) From the Factor Theorem, or from our more elementary work with quadratic equations, we know that if r_1 and r_2 are the roots of the equation $x^2 - 4x + 5 = 0$, then $x - r_1$ and $x - r_2$ are the factors of $x^2 - 4x + 5$. That is, $x^2 - 4x + 5 = (x - r_1)(x - r_2)$. The values for r_1 and r_2 in this case are readily obtained by using the quadratic formula. As you can check, the results are

$$r_1 = 2 + i \qquad r_2 = 2 - i$$

The required factorization is therefore

$$x^2 - 4x + 5 = [x - (2 + i)][x - (2 - i)]$$

Using the Linear Factors Theorem, we can show (starting in the next paragraph) that every polynomial equation of degree n ($n \geq 1$) has exactly n roots. To help you follow the reasoning, we make two preliminary comments. First, we agree that a root of multiplicity k will be counted as k roots. (The concept of multiplicity was defined in Section 10.2.) For example, although the third-degree equation $(x - 1)(x - 4)^2 = 0$ has but two distinct roots, namely 1 and 4, it has three roots *if* we agree to count the repeated root 4 two times. The second preliminary comment concens the Zero-Product Property: $pq = 0$ if and only if $p = 0$ or $q = 0$. When we stated this in Section 2.2, we were working within the real number system. However, the property is also valid within the complex number system. Now let's state and prove our theorem.

Theorem

Every polynomial equation of degree $n \geq 1$ has exactly n roots, where a root of multiplicity k is counted k times.

Proof of Theorem Using the Linear Factors Theorem, we can write the nth degree polynomial equation as

$$f(x) = a_n(x - r_1)(x - r_2) \cdots (x - r_n) = 0 \qquad a_n \neq 0 \tag{1}$$

By the Factor Theorem, each of the numbers $r_1, r_2, \ldots, r_n$ is a root. It is possible that some of these numbers are equal; in other words, we may have repeated roots in this list. In any case though, if we agree to count a root of multiplicity k as k roots, then we obtain exactly n roots from the list $r_1, r_2, \ldots, r_n$. Furthermore, the equation $f(x) = 0$ can have no other roots, as we now show. Suppose that r is any number distinct from all the numbers $r_1, r_2, \ldots, r_n$. Replacing x with r in equation (1) yields

$$f(r) = a_n(r - r_1)(r - r_2) \cdots (r - r_n)$$

But the expression on the right-hand side of this last equation cannot be zero because none of the factors is zero. Thus $f(r)$ is not zero, and so r is not a root. This completes the proof of the theorem.

Example 3 Find a polynomial $f(x)$ with leading coefficient 1 such that the equation $f(x) = 0$ has the following roots and no others:

Root	Multiplicity
3	2
−2	1
0	2

What is the degree of this polynomial?

Solution $(x - 3)^2$, $(x + 2)$, and $(x - 0)^2$ all must appear as factors of $f(x)$, and furthermore no other linear factor may appear. The form of $f(x)$ is therefore

$$f(x) = a_n(x - 3)^2(x + 2)(x - 0)^2$$

Since the leading coefficient a_n is to be 1, we can rewrite this last equation as

$$f(x) = x^2(x - 3)^2(x + 2)$$

This is the required polynomial. The degree here is 5. This can be seen either by multiplying out the factors or by simply adding the multiplicities of the roots.

Example 4 Find a quadratic function f with zeros of 3 and 5 whose graph passes through the point $(2, -9)$.

Solution The general form of a quadratic function with 3 and 5 as zeros is $f(x) = a_n(x - 3)(x - 5)$. Since the graph passes through $(2, -9)$, we have

$$-9 = a_n(2 - 3)(2 - 5)$$

or

$$-9 = a_n(3)$$

or

$$-3 = a_n$$

The required function is therefore

$$f(x) = -3(x - 3)(x - 5)$$

If we wish, we can carry out the multiplication and rewrite this as

$$f(x) = -3x^2 + 24x - 45$$

The next example shows how we can use the factored form of a polynomial $f(x)$ to determine the relationships between the coefficients of the polynomial and the roots of the equation $f(x) = 0$.

Example 5 Let r_1 and r_2 be the roots of the equation $x^2 + bx + c = 0$. Show that

$$r_1 r_2 = c \quad \text{and} \quad r_1 + r_2 = -b$$

Solution Since r_1 and r_2 are the roots of the equation $x^2 + bx + c = 0$, we have the identity

$$x^2 + bx + c = (x - r_1)(x - r_2)$$

After multiplying out the right-hand side, this identity can be rewritten as

$$x^2 + bx + c = x^2 - (r_1 + r_2)x + r_1 r_2$$

By equating coefficients now, we readily obtain $r_1 r_2 = c$ and $r_1 + r_2 = -b$, as required.

The technique used in Example 5 can be used to obtain similar relationships between the roots and the coefficients for polynomial equations of any given degree. In Table 1, for instance, we show the relationships obtained in Example 5, along with the corresponding relationships that can be derived for a cubic equation. (Exercise 29 at the end of this section asks you to verify the results for the cubic equation.)

Table 1

Equation	Roots	Relationships Between Roots and Coefficients
$x^2 + bx + c = 0$	r_1, r_2	$r_1 + r_2 = -b$ $r_1 r_2 = c$
$x^3 + bx^2 + cx + d = 0$	r_1, r_2, r_3	$r_1 + r_2 + r_3 = -b$ $r_1 r_2 + r_2 r_3 + r_3 r_1 = c$ $r_1 r_2 r_3 = -d$

We conclude this section with some remarks concerning the solution of polynomial equations by formulas. You know that the roots of the quadratic equation $ax^2 + bx + c = 0$ are given by the formula

$$x = \frac{-b \pm \sqrt{b^2 - 4ac}}{2a}$$

The question is, are there similar formulas for the solutions of higher-degree equations? By "similar" here we mean a formula involving the coefficients and radicals. To answer this question, we look at a bit of history. As early as 1700 B.C., Babylonian mathematicians were able to solve quadratic equations. This is clear from the study of the clay tablets with cuneiform numerals that archeologists have found. The ancient Greeks were also able to solve quadratic equations. Like the Babylonians, the Greeks worked without the aid of algebra as we know it. The mathematicians of ancient Greece used geometric constructions to solve equations. Of course, since all quantities were interpreted geometrically, negative roots were never considered.

The general quadratic formula was known to the Moslem mathematicians sometime before A.D. 1000. For the next 500 years after that, mathematicians searched for, but did not discover, a formula to solve the general cubic equation. Indeed, in 1494, Lucas Pacioli stated in his text *Summa di Arithmetica* that the general cubic equation could not be solved by the algebraic techniques then available. All of this was to change however within the next several decades.

Around 1515, the Italian mathematician Scipione del Ferro solved the cubic equation $x^3 + px + q = 0$ using algebraic techniques. This essentially constituted a solution of the seemingly more general equation $x^3 + bx^2 + cx + d = 0$. The reason for this is that if we make the substitution $x = y - b/3$ in the latter equation, the result is a cubic equation with no y^2 term. By 1540, the Italian mathematician Ludovico Ferrari had solved the general fourth-degree equation. Actually, at that time in Renaissance Italy, there was considerable controversy as to exactly who discovered the various formulas first. Details of the dispute can be found in any text on the history of mathematics. But for our purposes here, the point is simply that, by the middle of the sixteenth century, all polynomial equations of degree 4 or less could be solved by the formulas that had been discovered. The common feature of these formulas was that they involved the coefficients, the four basic operations of arithmetic, and various radicals. For example, a formula for the solution of the equation $x^3 + px + q = 0$ is as follows:

$$x = \sqrt[3]{\frac{-q}{2} + \sqrt{\frac{q^2}{4} + \frac{p^3}{27}}} + \sqrt[3]{\frac{-q}{2} - \sqrt{\frac{q^2}{4} + \frac{p^3}{27}}}$$

To give yourself an idea of the practical difficulties inherent in computing with this formula, try using it to show that $x = -2$ is a root of the equation $x^3 + 4x + 16 = 0$.

For more than 200 years after the cubic and quartic (fourth-degree) equations had been solved, mathematicians looked for a formula that would yield the solutions of the general fifth-degree equation. The first breakthrough here, if it can be called that, occurred in 1770, when the French mathematician Joseph Louis Lagrange found a technique that served to unify and summarize all of the previous methods used for the equations of degrees 2, 3, and 4. However, Lagrange then showed that his technique could not work in the case of the general fifth-degree equation. While we will not describe the details of Lagrange's work here, it is worth pointing out that he relied on the types of relationships between roots and coefficients that we looked at in Table 1 of this section.

Finally, in 1828, the Norwegian mathematician Niels Henrik Abel proved that, for the general polynomial equation of degree 5 or higher, there could be no formula yielding the solutions in terms of the coefficients and radicals. This is not to say that such equations do not possess solutions. In fact they must, as we saw earlier in this section. It is just that we cannot in every case express the solutions in terms of the coefficients and radicals. By way of example, it can be shown that the equation $x^5 - 6x + 3 = 0$ has a real root between 0 and 1, but that this number cannot be expressed in terms of the coefficients and radicals. (We can however compute the root to as many decimal places as we wish, as you'll see in Section 10.4.) In 1830, the French mathematician Evariste Galois completed matters by giving conditions for determining exactly which polynomial equations can be solved in terms of coefficients and radicals.

Exercise Set 10.3

According to the Fundamental Theorem of Algebra, which of the equations in Exercises 1 and 2 have at least one root?

1. **(a)** $x^5 - 14x^4 + 8x + 53 = 0$
(b) $(4.17)x^3 + (2.06)x^2 + (0.01)x + 1.23 = 0$
(c) $ix^2 + (2 + 3i)x - 17 = 0$
(d) $x^{2.1} + 3x^{0.3} + 1 = 0$

2. **(a)** $\sqrt{3}x^{17} + \sqrt{2}x^{13} + \sqrt{5} = 0$
(b) $17x^{\sqrt{3}} + 13x^{\sqrt{2}} + \sqrt{5} = 0$
(c) $\dfrac{1}{x^2 + 1} = 0$
(d) $2^{3x} - 2^x - 1 = 0$

For Exercises 3–10, express each polynomial in the form $a_n(x - r_1)(x - r_2) \cdots (x - r_n)$.

3. $x^2 - 2x - 3$

4. $x^3 - 2x^2 - 3x$

5. $4x^2 + 23x - 6$

6. $6x^2 + x - 12$

7. $x^2 - 5$

8. $x^2 + 5$

9. $x^2 - 10x + 26$

10. $x^3 + 2x^2 - 3x - 6$

In Exercises 11–16, find a polynomial f(x) *with leading coefficient* 1 *such that the equation* f(x) = 0 *has the given roots and no others. If the degree of* f(x) *is 4 or more, express* f(x) *in factored form; otherwise, express* f(x) *in the form* $a_nx^n + a_{n-1}x^{n-1} + \cdots + a_1x + a_0$.

11.

Root	1	−3
Multiplicity	2	1

12.

Root	0	4
Multiplicity	2	1

13.

Root	2	−2	$2i$	$-2i$
Multiplicity	1	1	1	1

14.

Root	$2 + i$	$2 - i$
Multiplicity	1	1

15.

Root	$\sqrt{3}$	$-\sqrt{3}$	$4i$	$-4i$
Multiplicity	2	2	1	1

16.

Root	5	1	$1 - i$	$1 + i$
Multiplicity	2	3	1	1

For Exercises 17–20, express f(x) *in the form* $a_nx^n + a_{n-1}x^{n-1} + \cdots + a_1x + a_0$.

17. Find a quadratic function with zeros −4 and 9, whose graph passes through the point (3, 5).

18. Find a quadratic function that has a maximum value 2 and that has −2 and 4 as zeros.

19. Find a third-degree polynomial function with zeros −5, 2, and 3, whose graph passes through the point (0, 1).

20. Find a fourth-degree polynomial function with zeros $\sqrt{2}$, $-\sqrt{2}$, 1, and −1, whose graph passes through (2, −20).

For Exercises 21–27, find a quadratic equation with the given roots and no others. Write your answers in the form $Ax^2 + Bx + C = 0$. Suggestion: *Make use of Table 1.*

21. $r_1 = -i$, $r_2 = -\sqrt{3}$

22. $r_1 = 1 + i\sqrt{3}$, $r_2 = 1 - i\sqrt{3}$

23. $r_1 = 9$, $r_2 = -6$

24. $r_1 = 5$, $r_2 = \frac{3}{4}$

25. $r_1 = 1 + \sqrt{5}$, $r_2 = 1 - \sqrt{5}$

26. $r_1 = 6 - 5i$, $r_2 = 6 + 5i$

27. $r_1 = a + \sqrt{b}$, $r_2 = a - \sqrt{b}$ $(b > 0)$

***28.** Suppose that p and q are positive integers with $p > q$. Find a quadratic equation with integer coefficients whose roots are

$$\frac{\sqrt{p}}{\sqrt{p} \pm \sqrt{p-q}}$$

29. Let r_1, r_2, and r_3 be the roots of the equation $x^3 + bx^2 + cx + d = 0$. Use the method of Example 5 to verify the following relationships.

$$r_1 + r_2 + r_3 = -b$$
$$r_1r_2 + r_2r_3 + r_3r_1 = c$$
$$r_1r_2r_3 = -d$$

30. Let r_1, r_2, r_3, and r_4 be the roots of the equation $x^4 + bx^3 + cx^2 + dx + e = 0$. Use the method of Example 5 to prove the following facts.

$$r_1 + r_2 + r_3 + r_4 = -b$$
$$r_1r_2 + r_2r_3 + r_3r_4 + r_4r_1 + r_2r_4 + r_3r_1 = c$$
$$r_1r_2r_3 + r_2r_3r_4 + r_3r_4r_1 + r_4r_1r_2 = -d$$
$$r_1r_2r_3r_4 = e$$

***31.** Solve the equation $x^3 - 4x^2 - 9x + 36 = 0$ given that the sum of two of the roots is 0. *Suggestion:* Use Table 1.

***32.** Solve the equation $x^3 - 12x + 16 = 0$ given that two of the roots are equal. *Suggestion:* Use Table 1.

***33.** Let α and β be the roots of the equation $x^2 + bx + c = 0$.

(a) Show that $\alpha^2 + \beta^2 = b^2 - 2c$. *Hint:* Use Table 1 along with the identity $\alpha^2 + \beta^2 = (\alpha + \beta)^2 - 2\alpha\beta$.

(b) Show that $\dfrac{1}{\alpha^2} + \dfrac{1}{\beta^2} = \dfrac{b^2 - 2c}{c^2}$.

(c) Show that $\alpha^3 + \beta^3 = -b(b^2 - 3c)$.

34. Express the polynomial $x^4 + 64$ as a product of four linear factors. *Hint:* First add and subtract the quantity $16x^2$ so that you can use the difference-of-squares factoring.

35. True or False. Mark T if the statement is true without exception, otherwise F.

(a) Every equation has a root.

(b) Every polynomial equation of degree at least 1 has a root.

(c) Every polynomial equation of degree 4 has four distinct roots.

(d) No cubic equation can have a root of multiplicity four.

(e) The degree of the polynomial

$$x(x-1)(x-2)(x-3)$$

is 3.

(f) The degree of the polynomial $6(x+1)^2(x-5)^4$ is 6.

(g) Every polynomial of degree n, where $n \geq 1$, can be written in the form

$$(x - r_1)(x - r_2) \cdots (x - r_n)$$

(h) The sum of the roots of the equation $x^2 - px + q = 0$ is p.

(i) The product of the roots of the equation $2x^3 - x^2 + 3x - 1 = 0$ is 1. *Hint:* Use Table 1.

(j) Although a polynomial equation of degree n may have n distinct roots, the Fundamental Theorem of Algebra only tells us how to find one of the roots.

(k) Every polynomial equation of degree $n \geq 1$ has at least one real root.

(l) If all of the coefficients in a polynomial equation are real, then at least one of the roots must be real.

(m) Every cubic equation whose roots are $\sqrt{5}$, $\sqrt{6}$, and $\sqrt{7}$ can be written in the form

$$a_n(x - \sqrt{5})(x - \sqrt{6})(x - \sqrt{7}) = 0$$

10.4 Rational and Irrational Roots

As we saw in the previous section, not every polynomial equation has a real root. Furthermore, even if a polynomial equation does possess a real root, that root needn't necessarily be a rational number. (The equation $x^2 = 2$ provides a simple example.) If a polynomial equation does have a rational root, however, we can find that root by applying the *Rational Roots Theorem*, which we now state.

The Rational Roots Theorem

Consider the polynomial equation

$$a_nx^n + a_{n-1}x^{n-1} + \cdots + a_1x + a_0 = 0 \qquad n \geq 1,\ a_n \neq 0$$

and suppose that all of the coefficients are integers. Let p/q be a rational number, where p and q have no common factors other than ± 1. If p/q is a root of the equation, then p is a factor of a_0 and q is a factor of a_n.

A proof of the Rational Roots Theorem is outlined in Exercise 29 at the end of this section. For the moment, though, let's just see why the theorem is plausible. Suppose that the two rational numbers a/b and c/d are the roots of a certain quadratic equation. Then, from our experience with quadratics (or by the Linear Factors Theorem), we know that the equation can be written in the form

$$k\left(x - \frac{a}{b}\right)\left(x - \frac{c}{d}\right) = 0 \tag{1}$$

where k is a constant. Now, as Exercise 28 asks you to check, if we carry out the multiplication and clear of fractions, equation (1) becomes

$$(kbd)x^2 - (kad + kbc)x + (kac) = 0 \tag{2}$$

Observe here that a and c (the numerators of the two roots) are factors of the constant term (kac) in equation (2), just as the Rational Roots Theorem asserts. Furthermore, b and d (the denominators of the roots) are factors of the coefficient of the x^2 term in equation (2), again just as the theorem asserts.

The example that follows shows how the Rational Roots Theorem can be used to solve a polynomial equation.

Example 1 Find the rational roots (if any) of the equation $2x^3 - x^2 - 9x - 4 = 0$. Solve the equation.

Solution First we list the factors of a_0, the factors of a_n, and the possibilities for rational roots:

$$\text{factors of } a_0 = 4: \quad \pm 1, \pm 2, \pm 4$$

$$\text{factors of } a_3 = 2: \quad \pm 1, \pm 2$$

$$\text{possible rational roots:} \quad \pm\frac{1}{1}, \pm\frac{1}{2}, \pm\frac{2}{1}, \pm\frac{4}{1}$$

Now we can use synthetic division to test whether or not any of these possibilities is a root. (A zero remainder will tell us that we have a root.) As you can check, the first three possibilities (1, -1, and $\frac{1}{2}$) are not roots. However, using $-\frac{1}{2}$ we have

$$\begin{array}{r|rrrr} -\frac{1}{2} & 2 & -1 & -9 & -4 \\ & & -1 & 1 & 4 \\ \hline & 2 & -2 & -8 & 0 \end{array}$$

Thus $x = -\frac{1}{2}$ is a root. We could now continue to check the remaining possibilities in this same manner. However, at this point it is simpler to consider the reduced equation $2x^2 - 2x - 8 = 0$, or $x^2 - x - 4 = 0$. Since this is a quadratic equation, it can be solved directly. We have

$$x^2 - x - 4 = 0 \qquad x = \frac{-(-1) \pm \sqrt{(-1)^2 - 4(1)(-4)}}{2(1)}$$

$$x = \frac{1 \pm \sqrt{17}}{2}$$

We've now determined three distinct roots. Since the degree of the original equation is 3, there can be no other roots. We conclude that $x = -\frac{1}{2}$ is the only rational root. The three roots of the equation are $-\frac{1}{2}, \frac{1 + \sqrt{17}}{2}$, and $\frac{1 - \sqrt{17}}{2}$.

As Example 1 indicates, the number of possibilities for rational roots can be relatively large, even for rather simple equations. The next theorem we develop allows us to reduce the number of possibilities. We say that a real number B is an *upper bound* for the roots of an equation if every real root is less than or equal to B. Similarly, a real number b is a *lower bound* if every real root is greater than or equal to b. The following theorem tells us how synthetic division can be used in determining upper and lower bounds for roots.

The Upper and Lower Bound Theorem for Roots

Consider the polynomial equation

$$f(x) = a_n x^n + a_{n-1}x^{n-1} + \cdots + a_1 x + a_0 = 0$$

where the coefficients are all real and a_n is positive.

1. If we use synthetic division to divide $f(x)$ by $x - B$, where $B > 0$, and we obtain a third row containing no negative numbers, then B is an upper bound for the roots of $f(x) = 0$.
2. If we use synthetic division to divide $f(x)$ by $x + |b|$, where $b < 0$, and we obtain a third row in which the numbers are alternately positive and negative (or zero), then b is a lower bound for the roots of $f(x) = 0$.

We'll prove the first part of this theorem. A proof of the second part can be developed along similar lines. To prove the first part of the theorem, we use the Division Algorithm to write

$$f(x) = (x - B) \cdot Q(x) + R \tag{3}$$

The remainder R here is a constant that may be zero. To show that B is an upper bound, we must show that any number greater than B is not a root. Toward this end, let p be a number that is greater than B. Note that p must be positive since B is positive. Then with $x = p$, equation (3) becomes

$$f(p) = (p - B) \cdot Q(p) + R \tag{4}$$

We are going to show now that the right-hand side of equation (4) is positive. This will tell us that p is not a root. First, look at the factor $(p - B)$. This is positive since p is greater than B. Next consider $Q(p)$. By hypothesis, the coefficients of $Q(x)$ are nonnegative. Furthermore, the leading coefficient of $Q(x)$ is a_n, which is positive. Since p is also positive, it follows that $Q(p)$ must be positive. Finally, the number R is nonnegative because all of the numbers in the third line of the synthetic division of $f(x)$ by $x - B$ are nonnegative. It follows now that the right-hand side of equation (4) is positive. Consequently, $f(p)$ is not zero and p is not a root of the equation $f(x) = 0$. This is what we wished to show.

Example 2 Determine the rational roots, or show that none exist, for the equation

$$\frac{1}{4}x^4 - \frac{3}{4}x^3 + \frac{17}{4}x^2 + 4x + 5 = 0$$

Solution We'll use the Rational Roots Theorem along with the Upper and Lower Bound Theorem. First of all, in order to apply the Rational Roots Theorem, our equation must have integer coefficients. In view of this, we multiply both sides of the given equation by 4 to obtain

$$x^4 - 3x^3 + 17x^2 + 16x + 20 = 0$$

As in the previous example, we list the factors of a_0, the factors of a_n, and the possibilities for rational roots:

$$\begin{array}{rl} \text{factors of } a_0 = 20: & \pm 1, \pm 2, \pm 4, \pm 5, \pm 10, \pm 20 \\ \text{factors of } a_4 = 1: & \pm 1 \\ \text{possible rational roots:} & \pm 1, \pm 2, \pm 4, \pm 5, \pm 10, \pm 20 \end{array}$$

Our strategy here will be to first check for positive rational roots, beginning with 1 and working upward. The checks for $x = 1$, $x = 2$, and $x = 4$ are as follows.

$$\begin{array}{r|rrrrr} 1 & 1 & -3 & 17 & 16 & 20 \\ & & 1 & -2 & 15 & 31 \\ \hline & 1 & -2 & 15 & 31 & 51 \end{array} \qquad \begin{array}{r|rrrrr} 2 & 1 & -3 & 17 & 16 & 20 \\ & & 2 & -2 & 30 & 92 \\ \hline & 1 & -1 & 15 & 46 & 112 \end{array} \qquad \begin{array}{r|rrrrr} 4 & 1 & -3 & 17 & 16 & 20 \\ & & 4 & 4 & 84 & 400 \\ \hline & 1 & 1 & 21 & 100 & 420 \end{array}$$

As you can see, none of the remainders here is zero. However, notice that in the division corresponding to $x = 4$, all of the numbers appearing in the third row are nonnegative. It follows therefore that 4 is an upper bound for the roots of the given equation. In view of this, we needn't bother to check the remaining values $x = 5$, $x = 10$, and $x = 20$, since none of those can be roots. At this point we can conclude that the given equation has no positive rational roots.

Next we check for negative rational roots, beginning with -1 and working downward (if necessary). Checking $x = -1$, we have

$$\begin{array}{r|rrrrr} -1 & 1 & -3 & 17 & 16 & 20 \\ & & -1 & 4 & -21 & 5 \\ \hline & 1 & -4 & 21 & -5 & 25 \end{array}$$

Two conclusions can be drawn from this synthetic division. First, $x = -1$ is not a root of the equation. Second, -1 is a lower bound for the roots because the signs in the third row of the synthetic division alternate. This means that we needn't bother to check whether any of the numbers -2, -4, -5, -15, or -20 are roots; none of them can be roots, since they are all less than -1.

Let us summarize our results now. We've shown that the given equation has no positive rational roots and no negative rational roots. Furthermore, by inspection we see that zero is not a root of the equation. Thus the given equation possesses no rational roots.

We conclude this section by demonstrating a method for approximating irrational roots. The method depends on the following theorem.

The Location Theorem

Let $f(x)$ be a polynomial. If $f(a)$ and $f(b)$ have opposite signs, then the equation $f(x) = 0$ has at least one root between a and b.

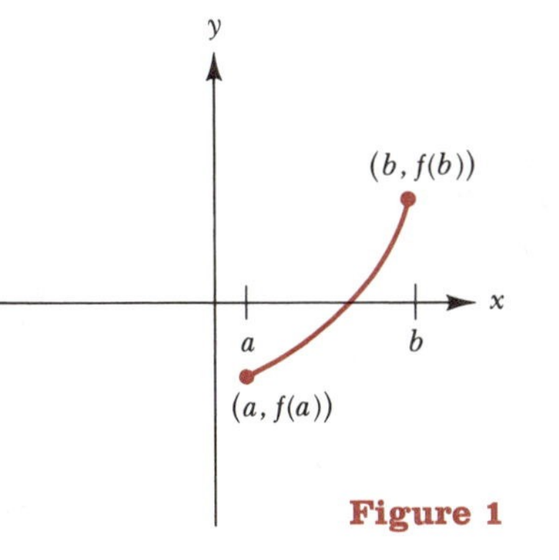

Figure 1

Figure 1 indicates why this theorem is plausible in view of our work graphing polynomial functions in Section 5.2. If the point $(a, f(a))$ lies below the x-axis, and $(b, f(b))$ lies above the x-axis, then it would certainly seem that the graph of f must cross the x-axis at some point x_0 between a and b. At this intercept we have $f(x_0) = 0$; that is, x_0 is a root of the equation $f(x) = 0$. (The Location Theorem is a special case of the so-called Intermediate Value Theorem; the proof is usually discussed in calculus courses.)

Our technique for approaching (or "locating") irrational roots uses the *method of successive approximations*. We'll demonstrate the method in Example 3.

Example 3 The following equation has exactly one positive root. Locate this root between successive hundredths.

$$f(x) = x^3 + 2x - 4 = 0$$

Solution First we need to find two numbers a and b such that $f(a)$ and $f(b)$ have opposite signs. By inspection (or by trial and error) we find that $f(1) = -1$ and $f(2) = 8$. Thus, according to the Location Theorem, the equation $f(x) = 0$ has a root in the interval $(1, 2)$.

Now that we have located the root between successive integers, we can locate it between successive tenths. We compute $f(1.0)$, $f(1.1)$, $f(1.2)$, and so on, up through $f(2.0)$ if necessary, until we find a sign change. As you can verify using synthetic division and a calculator, the results are as follows.

$$\begin{aligned} f(1.0) &= -1 \\ f(1.1) &= -0.469 \\ f(1.2) &= 0.128 \end{aligned} \Bigg\} \longleftarrow \text{sign change}$$

This shows that the root lies between 1.1 and 1.2.

Having located the root between successive tenths, we follow a similar process to locate the root between successive hundredths. Using synthetic division and a calculator, we obtain these results:

$$\begin{array}{ll} f(1.10) = -0.469 & f(1.15) \approx -0.179 \\ f(1.11) \approx -0.412 & f(1.16) \approx -0.119 \\ f(1.12) \approx -0.355 & f(1.17) \approx -0.058 \\ f(1.13) \approx -0.297 & f(1.18) \approx 0.003 \\ f(1.14) \approx -0.238 & \end{array} \Big\} \leftarrow \text{sign change}$$

This shows that the root lies between 1.17 and 1.18, so we've located the root between successive hundredths, as required.

The procedure described in Example 3 could be continued to yield closer and closer approximations for the required root. We note in passing that such calculations are easily handled with the aid of a programmable calculator or a computer. [See, for example, page 47 in the paperback text *Calculus by Calculator* by Maurice D. Weir (Englewood Cliffs, NJ: Prentice-Hall, 1982).]

Exercise Set 10.4

In Exercises 1–6, list the possibilities for rational roots.

1. $4x^3 - 9x^2 - 15x + 3 = 0$

2. $x^4 - x^3 + 10x^2 - 24 = 0$

3. $8x^5 - x^2 + 9 = 0$

4. $18x^4 - 10x^3 + x^2 - 4 = 0$

5. $\frac{2}{3}x^3 - x^2 - 5x + 2 = 0$

6. $\frac{1}{2}x^4 - 5x^3 + \frac{4}{3}x^2 + 8x - \frac{1}{3} = 0$

In Exercises 7–12, show that each equation has no rational roots.

7. $x^3 - 3x + 1 = 0$

8. $x^3 + 8x^2 - 1 = 0$

9. $x^3 + x^2 - x + 1 = 0$

10. $x^4 + 4x^3 + 4x^2 - 16 = 0$

11. $12x^4 - x^2 - 6 = 0$

12. $4x^5 - x^4 - x^3 - x^2 + x - 8 = 0$

For Exercises 13–25, find the rational roots of each equation and then solve the equation. (Use the Rational Roots Theorem and the Upper and Lower Bound Theorem, as in Example 2.)

13. $x^3 + 3x^2 - x - 3 = 0$

14. $2x^3 - 5x^2 - 3x + 9 = 0$

15. $4x^3 + x^2 - 20x - 5 = 0$

16. $3x^3 - 16x^2 + 17x - 4 = 0$

17. $9x^3 + 18x^2 + 11x + 2 = 0$

18. $4x^3 - 10x^2 - 25x + 4 = 0$

19. $x^4 + x^3 - 25x^2 - x + 24 = 0$

20. $10x^4 + 107x^3 + 301x^2 + 171x + 23 = 0$

21. $x^4 - 4x^3 + 6x^2 - 4x + 1 = 0$

22. $24x^3 - 46x^2 + 29x - 6 = 0$

23. $x^3 - \frac{5}{2}x^2 - 23x + 12 = 0$

24. $x^3 - \frac{17}{3}x^2 - \frac{10}{3}x + 8 = 0$

25. $2x^4 - \frac{9}{10}x^3 - \frac{29}{10}x^2 + \frac{27}{20}x - \frac{3}{20} = 0$

In Exercises 26 and 27, determine upper and lower bounds for the real roots of the equations. (Follow the method used within the solution of Example 2.)

26. **(a)** $x^3 + 2x^2 - 5x + 20 = 0$
(b) $x^5 - 3x^2 + 100 = 0$

27. **(a)** $5x^4 - 10x - 12 = 0$
(b) $3x^4 - 4x^3 + 5x^2 - 2x - 4 = 0$
(c) $2x^4 - 7x^3 - 5x^2 + 28x - 12 = 0$

28. For equation (1) in this section, multiply out the left-hand side and then clear the equation of fractions. Check that your result agrees with equation (2).

29. This exercise outlines a proof of the Rational Roots Theorem. At one point in the proof we'll need to rely on the following fact, which is proved in courses such as Number Theory.

Fact from Number Theory Suppose that A, B, and C are integers and that A is a factor of the number BC. Then if A has no factor in common with C (other than ± 1), A must be a factor of B.

(a) Let $A = 2$, $B = 8$, and $C = 5$. Verify that the fact from Number Theory is correct here.

(b) Let $A = 20$, $B = 8$, and $C = 5$. Note that A is a factor of BC but A is not a factor of B. Why doesn't this contradict the fact from Number Theory?

(c) Now we're ready to prove the Rational Roots Theorem. We begin with a polynomial equation with integer coefficients:

$$a_nx^n + a_{n-1}x^{n-1} + \cdots + a_1x + a_0 = 0 \qquad n \geq 1,\ a_n \neq 0$$

We assume that the rational number p/q is a root of the equation, and that p and q have no common factors other than 1. Why is it true now that

$$a_n\left(\frac{p}{q}\right)^n + a_{n-1}\left(\frac{p}{q}\right)^{n-1} + \cdots + a_1\left(\frac{p}{q}\right) + a_0 = 0$$

(d) Show that the last equation in part (c) can be written

$$p(a_np^{n-1} + a_{n-1}qp^{n-2} + \cdots + a_1q^{n-1}) = -a_0q^n$$

Since p is a factor of the left-hand side of this last equation, p must also be a factor of the right-hand side. That is, p must be a factor of a_0q^n. But since p and q have no common factors, neither do p and q^n. Our fact from Number Theory now tells us that p must be a factor of a_0, as we wished to show. (The proof that q is a factor of a_n is carried out in a similar manner.)

30. The Location Theorem asserts that the polynomial equation $f(x) = 0$ has a root in the open interval (a, b) whenever $f(a)$ and $f(b)$ have unlike signs. If $f(a)$ and $f(b)$ have the same sign, can the equation $f(x) = 0$ have a root between a and b? *Hint:* Look at the graph of $f(x) = x^2 - 2x + 1$ with $a = 0$ and $b = 2$.

C *In Exercises 31–36, each equation has exactly one positive root. In each case, locate the root between successive hundredths. Use a calculator.*

31. $x^3 + x - 1 = 0$

32. $x^3 - 2x - 5 = 0$

33. $x^5 - 200 = 0$

34. $x^3 - 3x^2 + 3x - 26 = 0$

35. $x^3 - 8x^2 + 21x - 22 = 0$

36. $2x^4 - x^3 - 12x^2 - 16x - 8 = 0$

C *In Exercises 37–40, each equation has exactly one negative root. In each case, use a calculator to locate the root between successive hundredths.*

37. $x^3 + x^2 - 2x + 1 = 0$

38. $x^5 + 100 = 0$

39. $x^3 + 2x^2 + 2x + 101 = 0$

40. $x^4 + 4x^3 - 6x^2 - 8x - 3 = 0$

For Exercises 41 and 42 you need to know that a prime number *is a positive integer greater than 1 with no factors other than itself and 1. Thus a list of the first few primes looks like 2, 3, 5, 7, 11, 13, 17, ...*

***41.** Find all prime numbers p for which the equation $x^2 + x - p = 0$ has a rational root.

***42.** Find all prime numbers p for which the equation $x^3 + x^2 + x - p = 0$ has at least one rational root. For each value of p that you find, find the corresponding *real* roots of the equation.

***43.** Find all integral values of b for which the equation $x^3 - b^2x^2 + 3bx - 4 = 0$ has a rational root.

C *Use a calculator for Exercises 44–46. Round off your answers to two decimal places.*

44. On the same set of axes sketch the graphs of $y = x^3$ and $y = 1 - 3x$. Find the x-coordinate of the point where the graphs intersect.

45. On the same set of axes sketch the graphs of $y = x^3$ and $y = x + 1$. Find the x-coordinate of the point where the graphs intersect.

46. Find the x-coordinate of the point where the curves $y = x^2 - 1$ and $y = \sqrt{x}$ meet.

***47.** Let $P(a, b)$ be a point on the first-quadrant portion of the curve $y = x^2$ such that the distance of P from the origin is equal to ab. (Assume that $a \neq 0$.)

C **(a)** By using the method of this section, find the value of a; round off your answer to two decimal places.

(b) Find the exact value of a.

10.5 Conjugate Roots and Descartes' Rule of Signs

As you know from earlier work involving quadratic equations with real coefficients, when nonreal complex roots occur, they occur in conjugate pairs. For instance, as you can check by means of the quadratic formula, the roots of the equation $x^2 - 2x + 5 = 0$ are $1 + 2i$ and $1 - 2i$. The following theorem tells us that the situation is the same for all polynomial equations with real coefficients.

The Conjugate Roots Theorem

Let $f(x)$ be a polynomial, all of whose coefficients are real numbers. Suppose that $a + bi$ is a root of the equation $f(x) = 0$, where a and b are real and $b \neq 0$. Then $a - bi$ is also a root of the equation.

To prove the Conjugate Roots Theorem, we use four of the properties of complex conjugates discussed in Chapter 1:

property 1: $\overline{z_1}\,\overline{z_2} = \overline{z_1 z_2}$

property 2: $\overline{z}^m = \overline{z^m}$

property 3: $\overline{r} = r$ for every real number r

property 4: $\overline{z_1} + \overline{z_2} = \overline{z_1 + z_2}$

Using these facts, the proof of the theorem runs as follows. We begin with a polynomial with real coefficients:

$$f(x) = a_n x^n + a_{n-1}x^{n-1} + \cdots + a_1 x + a_0$$

We must show that if $z = a + bi$ is a root of $f(x) = 0$, then $\overline{z} = a - bi$ is also a root. We have

$$\begin{aligned}
f(\overline{z}) &= a_n\overline{z}^n + a_{n-1}\overline{z}^{n-1} + \cdots + a_1\overline{z} + a_0 \\
&= \overline{a_n}\,\overline{z^n} + \overline{a_{n-1}}\,\overline{z^{n-1}} + \cdots + \overline{a_1}\,\overline{z} + \overline{a_0} && \text{properties 3 and 2} \\
&= \overline{a_n z^n} + \overline{a_{n-1}z^{n-1}} + \cdots + \overline{a_1 z} + \overline{a_0} && \text{property 1} \\
&= \overline{a_n z^n + a_{n-1}z^{n-1} + \cdots + a_1 z + a_0} && \text{property 4} \\
&= \overline{f(z)} = \overline{0} && f(z) = 0 \text{ since } z \text{ is a root} \\
&= 0 && \text{property 3}
\end{aligned}$$

We've now shown that $f(\overline{z}) = 0$. Thus $\overline{z}$ is a root, as we wished to show.

Although the Conjugate Roots Theorem concerns nonreal complex roots, it can nevertheless be used to obtain information about real roots, as the next two examples indicate.

Example 1 Show that the equation $x^3 - 2x^2 + x - 1 = 0$ has at least one irrational root.

Solution Allowing for multiple roots, the given equation has three roots. Therefore, since nonreal complex roots occur in pairs, the equation can have either two nonreal complex roots or no nonreal complex roots. In either case then, the remaining root or roots must be real. Furthermore, the equation has no rational roots; this is easily checked using the Rational Roots Theorem. It follows now that the equation must have at least one irrational root. (The reasoning here relies on the fact that each real number is either rational or irrational, but not both.)

Example 2 Solve the equation $f(x) = 2x^4 - 3x^3 + 12x^2 + 22x - 60 = 0$, given that one root is $1 + 3i$.

Solution Since all of the coefficients of $f(x)$ are real numbers, we know that the conjugate of $1 + 3i$ must also be a root. Thus $1 + 3i$ and $1 - 3i$ are roots, from which it follows that $[x - (1 + 3i)]$ and $[x - (1 - 3i)]$ are factors of $f(x)$. As you can check, the product of these two factors is $x^2 - 2x + 10$. Thus we must have

$$f(x) = (x^2 - 2x + 10) \cdot Q(x)$$

for some polynomial $Q(x)$. We compute $Q(x)$ using long division:

$$\begin{array}{r|l} & 2x^2 + x - 6 \\ \hline x^2 - 2x + 10 & 2x^4 - 3x^3 + 12x^2 + 22x - 60 \\ & \underline{2x^4 - 4x^3 + 20x^2} \\ & \qquad x^3 - 8x^2 + 22x \\ & \qquad \underline{x^3 - 2x^2 + 10x} \\ & \qquad\quad -6x^2 + 12x - 60 \\ & \qquad\quad \underline{-6x^2 + 12x - 60} \\ & \qquad\qquad\qquad\qquad 0 \end{array}$$

Thus $Q(x)$ is $2x^2 + x - 6$, and the original equation becomes

$$(x^2 - 2x + 10)(2x^2 + x - 6) = 0$$

Any additional roots now can be found by solving the equation

$$2x^2 + x - 6 = 0$$

We have

$$2x^2 + x - 6 = 0$$

$$(2x - 3)(x + 2) = 0$$

$$\begin{array}{r|l} 2x - 3 = 0 & x + 2 = 0 \\ 2x = 3 & x = -2 \\ x = \dfrac{3}{2} & \end{array}$$

Now we have four distinct roots of the original equation: $1 + 3i$, $1 - 3i$, $\frac{3}{2}$, and -2. Since the degree of the equation is 4, there can be no other roots.

There is a theorem, similar to the Conjugate Roots Theorem, that tells us about irrational roots of the form $a + b\sqrt{c}$. As background for this theorem, we look at two preliminary examples. First, consider the equation $x^2 - 2x - 5 = 0$. As you can check, the roots in this case are $1 + \sqrt{6}$ and $1 - \sqrt{6}$. However, it is not true in general that irrational roots such as these always occur in pairs. Consider as a second example the quadratic equation

$$(x + 2)(x - \sqrt{3}) = 0$$

or

$$x^2 + (2 - \sqrt{3})x - 2\sqrt{3} = 0$$

Here, one of the roots is $\sqrt{3}$, yet $-\sqrt{3}$ is not a root. This type of behavior can occur in polynomial equations where not all of the coefficients are rational. On the other hand, when the coefficients are all rational, we do have the following theorem. (See Exercise 41 at the end of this section for a proof.)

Theorem

Let $f(x)$ be a polynomial in which all the coefficients are rational. Suppose that $a + b\sqrt{c}$ is a root of the equation $f(x) = 0$, where a, b, and c are rational and $\sqrt{c}$ is irrational. Then $a - b\sqrt{c}$ is also a root of the equation.

Example 3 Find a quadratic equation with rational coefficients and a leading coefficient of 1, one of whose roots is $r_1 = 4 + 5\sqrt{3}$.

Solution If one root is $r_1 = 4 + 5\sqrt{3}$, the other is $r_2 = 4 - 5\sqrt{3}$. We denote the required equation by $x^2 + bx + c = 0$. Then, according to Table 1 in Section 10.3, we have

$$b = -(r_1 + r_2) = -[(4 + 5\sqrt{3}) + (4 - 5\sqrt{3})] = -8$$

and

$$c = r_1 r_2 = (4 + 5\sqrt{3})(4 - 5\sqrt{3}) = 16 - 75 = -59$$

The required equation is therefore $x^2 - 8x - 59 = 0$. This answer can also be obtained without using the table. Since the roots are $4 \pm 5\sqrt{3}$, we can write the required equation as $[x - (4 + 5\sqrt{3})][x - (4 - 5\sqrt{3})] = 0$. As you can now check by multiplying the two factors, this equation is equivalent to $x^2 - 8x - 59 = 0$, as obtained previously.

We conclude this section with a discussion of *Descartes' rule of signs.* This rule (published by Descartes in 1637) provides us with information about the types of roots an equation may have, even before we attempt to solve the equation. In order to state Descartes' rule of signs, we first explain what is meant by a variation in sign in a polynomial with real coefficients. Suppose that $f(x)$ is a polynomial with real coefficients, written in descending powers of x. For example, let $f(x) = 2x^3 - 4x^2 - 3x + 1$. Then we say that there is a *variation in sign* if two successive coefficients have opposite signs. In the case of $f(x) = 2x^3 - 4x^2 - 3x + 1$, there are two variations in sign, the first occurring as we go from 2 to -4 and the second occurring as we go from -3 to 1. In looking for variations in sign, we ignore terms with zero coefficients. Here are several more examples of how we count variations in sign:

Polynomial	Number of Variations in Sign
$x^2 + 4x$	0
$-3x^5 + x^2 + 1$	1
$x^3 + 3x^2 - x + 6$	2

We now state Descartes' rule of signs and look at some examples. Although the proof of this theorem is not difficult, it is lengthy and we shall omit it here.

Descartes' Rule of Signs

Let $f(x)$ be a polynomial, all of whose coefficients are real numbers, and consider the equation $f(x) = 0$. Then

(a) the number of positive roots either is equal to the number of variations in sign of $f(x)$ or is less than that by an even integer;

(b) the number of negative roots either is equal to the number of variations in sign of $f(-x)$ or is less than that by an even integer.

Example 4 Use Descartes' rule of signs to obtain information regarding the roots of the equation $x^3 + 8x + 5 = 0$.

Solution Let $f(x) = x^3 + 8x + 5$. Then, since there are no variations in sign for $f(x)$, we see from part (a) of Descartes' rule that the given equation has no positive roots. Next we compute $f(-x)$ to learn about the possibilities for negative roots. As you can quickly check, we have $f(-x) = -x^3 - 8x + 5$. So $f(-x)$ has one sign change, and consequently [by part (b) of Descartes' rule] the original equation has one negative root. Furthermore, notice that zero is not a root of the equation. Thus the equation has but one real root, a negative root. Since the equation has a total of three roots, we can conclude now that we have

one negative root and two nonreal complex roots

The two nonreal roots will be complex conjugates.

Example 5 Use Descartes' rule to obtain information regarding the roots of the equation $x^4 + 3x^2 - 7x - 5 = 0$.

Solution Let $f(x) = x^4 + 3x^2 - 7x - 5$. Then $f(x)$ has one variation in sign. So, according to part (a) of Descartes' rule, the equation has one positive root. That leaves us three roots still to account for, since the degree of the equation is 4. We have $f(-x) = x^4 + 3x^2 + 7x - 5$. Since $f(-x)$ has one sign change, we know from part (b) of Descartes' rule that the equation has one negative root. Noting now that zero is not a root, we conclude that the two remaining roots must be nonreal complex roots. In summary then, the equation has one positive root, one negative root, and two nonreal complex (conjugate) roots.

Exercise Set 10.5

In Exercises 1–16 an equation is given, followed by one or more roots of the equation. In each case, determine the remaining roots.

1. $x^2 - 14x + 53 = 0$; $x = 7 - 2i$

2. $x^2 - x - \frac{1535}{4} = 0$; $x = \frac{1}{2} + 8\sqrt{6}$

3. $x^3 - 13x^2 + 59x - 87 = 0$; $x = 5 + 2i$

4. $x^4 - 10x^3 + 30x^2 - 10x - 51 = 0$; $x = 4 + i$

5. $x^4 + 10x^3 + 38x^2 + 66x + 45 = 0$; $x = -2 + i$

6. $2x^3 + 11x^2 + 30x - 18 = 0$; $x = -3 - 3i$

7. $4x^3 - 47x^2 + 232x + 61 = 0$; $x = 6 - 5i$

8. $9x^4 + 18x^3 + 20x^2 - 32x - 64 = 0$; $x = -1 + \sqrt{3}i$

9. $4x^4 - 32x^3 + 81x^2 - 72x + 162 = 0$; $x = 4 + \sqrt{2}i$

10. $2x^4 - 17x^3 + 137x^2 - 57x - 65 = 0$; $x = 4 - 7i$

11. $x^4 - 22x^3 + 140x^2 - 128x - 416 = 0;\ x = 10 + 2i$

12. $4x^4 - 8x^3 + 24x^2 - 20x + 25 = 0;\quad x = \frac{1 + 3i}{2}$

13. $15x^3 - 16x^2 + 9x - 2 = 0;\quad x = \frac{1 + \sqrt{2}i}{3}$

14. $x^5 - 5x^4 + 30x^3 + 18x^2 + 92x - 136 = 0;$ $x = -1 + i\sqrt{3}, x = 3 - 5i$

15. $x^7 - 3x^6 - 4x^5 + 30x^4 + 27x^3 - 13x^2 - 64x + 26 = 0;\quad x = 3 - 2i, x = -1 + i, x = 1$

16. $x^6 - 2x^5 - 2x^4 + 2x^3 + 2x + 1 = 0; x = 1 + \sqrt{2}$

In Exercises 17–20, find a quadratic equation with rational coefficients, one of whose roots is the given number. Write your answer so that the coefficient of x^2 *is* 1. Suggestion: *Use the method shown in Example 3.*

17. $r_1 = 1 + \sqrt{6}$

18. $r_1 = 2 - \sqrt{3}$

19. $r_1 = \frac{2 + \sqrt{10}}{3}$

20. $r_1 = \frac{1}{2} + \frac{1}{4}\sqrt{5}$

In Exercises 21–36, use Descartes' rule of signs to obtain information regarding the roots of the equations.

21. $x^3 + 5 = 0$

22. $x^4 + x^2 + 1 = 0$

23. $2x^5 + 3x + 4 = 0$

24. $x^3 + 8x - 3 = 0$

25. $5x^4 + 2x - 7 = 0$

26. $x^3 - 4x^2 + x - 1 = 0$

27. $x^3 - 4x^2 - x - 1 = 0$

28. $x^8 + 4x^6 + 3x^4 + 2x^2 + 5 = 0$

29. $3x^8 + x^6 - 2x^2 - 4 = 0$

30. $12x^4 - 5x^3 - 7x^2 - 4 = 0$

31. $x^9 - 2 = 0$

32. $x^9 + 2 = 0$

33. $x^8 - 2 = 0$

34. $x^8 + 2 = 0$

35. $x^6 + x^2 - x - 1 = 0$

36. $x^7 + x^2 - x - 1 = 0$

37. Consider the equation $x^4 + cx^2 + dx - e = 0$, where c, d, and e are positive. Show that the equation has one positive, one negative, and two nonreal complex roots.

38. Consider the equation $x^n - 1 = 0$.

(a) Show that the equation has $n - 2$ nonreal complex roots when n is even.

(b) How many nonreal complex roots are there when n is odd?

39. Find the polynomial $f(x)$ of lowest degree, with rational coefficients and with leading coefficient 1, such that $\sqrt{3} + 2i$ is a root of the equation $f(x) = 0$.

40. Find a quadratic polynomial $f(x)$, with integer coefficients, such that $\frac{2 + \sqrt{3}}{2 - \sqrt{3}}$ is a root of the equation $f(x) = 0$. *Hint:* First rationalize the given root.

***41.** Let $f(x)$ be a polynomial, all of whose coefficients are rational. Suppose that $a + b\sqrt{c}$ is a root of $f(x) = 0$, where a, b, and c are rational and $\sqrt{c}$ is irrational. Complete the following steps to prove that $a - b\sqrt{c}$ is also a root of the equation $f(x) = 0$.

(a) If $b = 0$, we're done. Why?

(b) (From now on we'll assume that $b \neq 0$.) Let $d(x) = [x - (a + b\sqrt{c})][x - (a - b\sqrt{c})]$. Explain why $d(a + b\sqrt{c}) = 0$.

(c) Verify that $d(x) = (x - a)^2 - b^2c$. Thus $d(x)$ is a quadratic polynomial with rational coefficients.

(d) Now suppose that we use the long division process to divide the polynomial $f(x)$ by the quadratic polynomial $d(x)$. We'll obtain a quotient $Q(x)$ and a remainder. Since the degree of $d(x)$ is 2, our remainder will be of degree 1 or less. In other words, the general form of this remainder will be $Cx + D$. Furthermore, C and D will have to be rational, because all of the coefficients in $f(x)$ and in $d(x)$ are rational. In summary, we have the identity

$$f(x) = d(x) \cdot Q(x) + (Cx + D)$$

In this identity make the substitution $x = a + b\sqrt{c}$ and conclude that $C = D = 0$.

(e) Using the result in part (d), we have

$$f(x) = [x - (a + b\sqrt{c})][x - (a - b\sqrt{c})] \cdot Q(x)$$

Let $x = a - b\sqrt{c}$ in this last identity and conclude from the result that $a - b\sqrt{c}$ is a root of $f(x) = 0$, as required.

Chapter 10 Review Checklist

- □ Division Algorithm (page 497)
- □ Synthetic division (pages 498–499)
- □ Root or solution (page 501)
- □ Zero of a function (page 501)
- □ Multiplicity (of a root) (page 502)
- □ The Remainder Theorem (page 502)
- □ The Factor Theorem (page 504)
- □ Reduced equation (page 505)
- □ The Fundamental Theorem of Algebra (page 508)
- □ The Linear Factors Theorem (page 509)
- □ The Rational Roots Theorem (page 516)
- □ The Upper and Lower Bound Theorem (page 517)
- □ The Location Theorem (page 519)
- □ The Conjugate Roots Theorem (page 522)
- □ Variation in sign (page 524)
- □ Descartes' rule of signs (page 525)

Chapter 10 Review Exercises

In Exercises 1 and 2 you are given polynomials p(x) *and* d(x). *In each case, use the long division process to determine polynomials* q(x) *and* R(x) *such that*

$$p(x) = d(x) \cdot q(x) + R(x)$$

where either R(x) = 0 *or the degree of* R(x) *is less than the degree of* d(x).

1. $p(x) = x^4 + 3x^3 - x^2 - 5x + 1$; $d(x) = x + 2$

2. $p(x) = 4x^4 + 2x + 1$; $d(x) = 2x^2 + 1$

For Exercises 3–8, use synthetic division to find the quotients and the remainders.

3. $\dfrac{x^4 - 2x^2 + 8}{x - 3}$

4. $\dfrac{x^3 - 1}{x - 2}$

5. $\dfrac{2x^3 - 5x^2 - 6x - 3}{x + 4}$

6. $\dfrac{x^3 + x - 3\sqrt{2}}{x - \sqrt{2}}$

7. $\dfrac{5x^2 - 19x - 4}{x + 0.2}$

8. $\dfrac{x^3 - 3a^2x^2 - 4a^4x + 9a^6}{x - a^2}$

In Exercises 9–16, use synthetic division and the Remainder Theorem to find the indicated values of the functions.

9. $f(x) = x^5 - 10x + 4$; $f(10)$

10. $f(x) = x^4 + 2x^3 - x$; $f(-2)$

11. $f(x) = x^3 - 10x^2 + x - 1$; $f(\frac{1}{10})$

12. $f(x) = x^4 - 2a^2x^2 + 3a^3x - a^4$; $f(-a)$

13. $f(x) = x^3 + 3x^2 + 3x + 1$; $f(a - 1)$

14. $f(x) = x^3 - 1$; $f(1.1)$

C **15.** $f(x) = x^4 + 4x^3 - 6x^2 - 8x - 2$
- **(a)** $f(-0.3)$ (Round off the result to two decimal places.)
- **(b)** $f(-0.39)$ (Round off the result to three decimal places.)
- **(c)** $f(-0.394)$ (Round off the result to five decimal places.)

C **16.** $f(-4.907)$, where f is the function in Exercise 15. (Round off the result to three decimal places.)

17. Find a value for a such that 3 is a root of the equation $x^3 - 4x^2 - ax - 6 = 0$.

18. For which values of b will -1 be a root of the equation $x^3 + 2b^2x^2 + x - 48 = 0$?

19. For which values of a will $x - 1$ be a factor of $a^2x^3 + 3ax^2 + 2$?

20. Use synthetic division to verify that $\sqrt{2} - 1$ is a root of the equation $x^6 + 14x^3 - 1 = 0$.

21. Let $f(x) = ax^3 + bx^2 + cx + d$ and suppose that r is a root of the equation $f(x) = 0$.
- **(a)** Show that $r - h$ is a root of the equation $f(x + h) = 0$.
- **(b)** Show that $-r$ is a root of the equation $f(-x) = 0$.
- **(c)** Show that kr is a root of the equation $f(x/k) = 0$.

22. Suppose that r is a root of the equation $a_2x^2 + a_1x + a_0 = 0$. Show that mr is a root of the equation $a_2x^2 + ma_1x + m^2a_0 = 0$.

In Exercises 23–28, list the possibilities for the rational roots of the equations.

23. $x^5 - 12x^3 + x - 18 = 0$

24. $x^5 - 12x^3 + x - 17 = 0$

25. $2x^4 - 125x^3 + 3x^2 - 8 = 0$

26. $\frac{3}{5}x^3 - 8x^2 - \frac{1}{2}x + \frac{3}{2} = 0$

27. $x^3 + x - p = 0$ (where p is a prime number)

28. $x^3 + x - pq = 0$ (where both p and q are prime numbers)

For Exercises 29–36, each equation has at least one rational root. Solve the equations. Suggestion: *Use the Upper and Lower Bound Theorem to eliminate some of the possibilities for rational roots.*

29. $2x^3 + x^2 - 7x - 6 = 0$

30. $x^3 + 6x^2 - 8x - 7 = 0$

31. $2x^3 - x^2 - 14x + 10 = 0$

32. $2x^3 + 12x^2 + 13x + 15 = 0$

33. $\frac{3}{2}x^3 + \frac{1}{2}x^2 + \frac{1}{2}x - 1 = 0$

34. $x^4 - 2x^3 - 13x^2 + 38x - 24 = 0$

35. $x^5 + x^4 - 14x^3 - 14x^2 + 49x + 49 = 0$

36. $8x^5 + 12x^4 + 14x^3 + 13x^2 + 6x + 1 = 0$

37. Solve the equation $x^3 - 9x^2 + 24x - 20 = 0$ using the fact that one of the roots has multiplicity two.

38. One root of the equation $x^2 + kx + 2k = 0$ $(k \neq 0)$ is twice the other. Find k and find the roots of the equation.

39. State each of the following theorems:
- **(a)** The Division Algorithm
- **(b)** The Remainder Theorem
- **(c)** The Factor Theorem
- **(d)** The Fundamental Theorem of Algebra

40. Find a quadratic equation with roots $a - \sqrt{a^2 - 1}$ and $a + \sqrt{a^2 - 1}$ $(a > 1)$.

In Exercises 41–44, write the polynomials in the form $a_n(x - r_1)(x - r_2) \cdots (x - r_n)$.

41. $6x^2 + 7x - 20$

42. $x^2 + x - 1$

43. $x^4 - 4x^3 + 5x - 20$ *Hint:* Use one of the factoring techniques in Section 1.8.

44. $x^4 - 4x^2 - 5$

In Exercises 45–48 equations are given, followed by one or more roots. Solve the equations.

45. $x^3 - 7x^2 + 25x - 39 = 0$; $x = 2 - 3i$

46. $x^3 + 6x^2 - 24x + 160 = 0$; $x = 2 + 2i\sqrt{3}$

47. $x^4 - 2x^3 - 4x^2 + 14x - 21 = 0$; $x = 1 + i\sqrt{2}$

48. $x^5 + x^4 - x^3 + x^2 + x - 1$; $x = \frac{1}{2}(1 + i\sqrt{3})$ and $x = \frac{1}{2}(-1 - \sqrt{5})$

For Exercises 49–54, use Descartes' rule of signs to obtain information regarding the roots of the equations.

49. $x^3 + 8x - 7 = 0$

50. $3x^4 + x^2 + 4x - 2 = 0$

51. $x^3 + 3x + 1 = 0$

52. $2x^6 + 3x^2 + 6 = 0$

53. $x^4 - 10 = 0$

54. $x^4 + 5x^2 - x + 2 = 0$

55. Consider the equation $x^3 + x^2 + x + 1 = 0$.

(a) Use Descartes' rule to show that the equation has either one or three negative roots.

(b) Show now that the equation cannot have three negative roots. *Hint:* Multiply both sides of the equation by $x - 1$. Then simplify the left-hand side and reapply Descartes' rule to the new equation.

(c) Actually, the original equation can be solved using only the basic algebraic techniques discussed in Chapters 1 and 2. Solve the equation in this manner.

56. Use Descartes' rule to show that the equation $x^3 - x^2 + 3x + 2 = 0$ has no positive roots. *Hint:* Multiply both sides of the equation by $x + 1$ and apply Descartes' rule to the resulting equation.

C **57.** Let P be the point in the first quadrant where the curve $y = x^3$ intersects the circle $x^2 + y^2 = 1$. Locate the x-coordinate of P within successive hundredths.

C **58.** Let P be the point in the first quadrant where the parabola $y = 4 - x^2$ intersects the curve $y = x^3$. Locate the x-coordinate of P within successive hundredths.

59. Consider the equation $x^3 - 36x - 84 = 0$.

(a) Use Descartes' rule to check that this equation has exactly one positive root.

(b) Use the Upper and Lower Bound Theorem to show that 7 is an upper bound for the positive root.

C **(c)** Locate the positive root within successive hundredths.

60. Consider the equation $x^3 - 3x + 1 = 0$.

(a) Use Descartes' rule to check that this equation has exactly one negative root.

(b) Use the Upper and Lower Bound Theorem to show that -2 is a lower bound for the negative root.

C **(c)** Locate the negative root within successive hundredths.

In Exercises 61–64 find polynomial equations, with integer coefficients, having the given values as roots.

61. $4 - \sqrt{5}$

62. $a + b$ and $a - b$ (a and b are integers.)

63. $6 - 2i$ and $\sqrt{5}$

64. $\dfrac{5 + \sqrt{6}}{5 - \sqrt{6}}$ *Hint:* First rationalize the expression.

65. Find a fourth-degree polynomial equation, with integer coefficients, having the value $x = 1 + \sqrt{2} + \sqrt{3}$ as a root. *Hint:* Begin by writing the given relationship as $x - 1 = \sqrt{2} + \sqrt{3}$; then square both sides.

66. Find a cubic equation, with integer coefficients, having the value $x = 1 + \sqrt[3]{2}$ as a root. Is the value $1 - \sqrt[3]{2}$ also a root of the equation?

Chapter 10 Test

1. Let $f(x) = 6x^4 - 5x^3 + 7x^2 - 2x - 2$. Make use of the Remainder Theorem and synthetic division to compute $f(\frac{1}{2})$.

2. Solve the equation $x^3 + x^2 - 11x - 15 = 0$ given that one of the roots is -3.

3. List the possibilities for the rational roots of the equation $2x^5 - 4x^3 + x - 6 = 0$.

4. Find a quadratic function with zeros 1 and -8, and whose graph has a y-intercept of -24.

5. Use synthetic division to divide $4x^3 + x^2 - 8x + 3$ by $x + 1$.

6. **(a)** State the Factor Theorem.

(b) State the Fundamental Theorem of Algebra.

7. **(a)** The equation $x^3 - 2x^2 - 1 = 0$ has just one positive root. Use the Upper and Lower Bound Theorem to determine the smallest integer that is an upper bound for that root.

(b) Locate the root between successive tenths. (Use a calculator.)

8. Solve the equation $x^5 - 6x^4 + 11x^3 + 16x^2 - 50x + 52 = 0$ given that two of the roots are $1 + i$ and $3 - 2i$.

9. Let $p(x) = x^4 + 2x^3 - x + 6$ and $d(x) = x^2 + 1$. Use the long division process to find polynomials $q(x)$ and $R(x)$ such that $p(x) = d(x) \cdot q(x) + R(x)$.

10. Express the polynomial $2x^2 - 6x + 5$ in the factored form $a_n(x - r_1)(x - r_2)$.

11. Consider the equation $x^4 - x^3 + 24 = 0$.

(a) List the possibilities for rational roots.

(b) Use the Upper and Lower Bound Theorem to show that 2 is an upper bound for the roots.

(c) In view of parts (a) and (b), what possibilities now remain for positive rational roots?

(d) Which (if any) of the possibilities in part (c) are actually roots?

12. (a) Find the rational roots of the equation $2x^3 - x^2 - x - 3 = 0$.

(b) Find all solutions of the equation in part (a).

13. Use Descartes' rule of signs to obtain information regarding the roots of the equation $3x^4 + x^2 - 5x - 1 = 0$.

14. Find a cubic polynomial $f(x)$ with integer coefficients such that $1 - 3i$ and -2 are roots of the equation $f(x) = 0$.

15. Find a polynomial $f(x)$ with leading coefficient 1 such that the equation $f(x) = 0$ has the following roots and no others. Write your answer in the form

$$a_n(x - r_1)(x - r_2) \cdots (x - r_n)$$

Root	Multiplicity
2	1
$3i$	3
$1 + \sqrt{2}$	2

Chapter 11 Additional Topics

Introduction A strong background in algebra is an important prerequisite for courses in calculus and in probability and statistics. In this final chapter we develop several additional topics that serve as background in those areas of study. We begin in Section 11.1 with the Principle of Mathematical Induction. This provides a framework for proving statements about the natural numbers. In Section 11.2 we discuss the Binomial Theorem, which is used to analyze and expand expressions of the form $(a + b)^n$. As you'll see, the proof of the Binomial Theorem uses mathematical induction. Section 11.3 introduces the related (but distinct) concepts of sequences and series. Then in the next two sections (Sections 11.4 and 11.5), we study arithmetic and geometric sequences and series. The last topic introduced in Section 11.5 concerns the sum of an infinite geometric series. In a sense, this is an appropriate topic for our last chapter, for it is closely related to the idea of a *limit,* which is the starting point for calculus. The next two sections of this chapter introduce the important concepts of permutations, combinations, and probability. Finally, in the last section of the chapter, we explain the trigonometric form for complex numbers. This brings us to DeMoivre's Theorem and the nth roots of a complex number.

11.1 Mathematical Induction

Mathematical induction is not a method of discovery but a technique of proving rigorously what has already been discovered.

David M. Burton in his text *The History of Mathematics, An Introduction* (Boston: Allyn and Bacon, 1985)

Is mathematics an experimental science? The answer to this question is both "yes" and "no" as the following example illustrates. Consider the problem of determining a formula for the sum of the first n odd natural numbers:

$$1 + 3 + 5 + \cdots + (2n - 1)$$

We begin by doing some calculations in the hope that this may shed some light on the problem. Table 1 shows the results of calculating the sum of the first n odd natural numbers for values of n ranging from 1 to 5. Upon inspecting the table, we observe that each sum in the right-hand column is the square of the corresponding entry in the left-hand column. For instance, for $n = 5$ we see that

$$\underbrace{1 + 3 + 5 + 7 + 9}_{\text{five terms}} = 5^2$$

n	$1 + 3 + 5 + \cdots + (2n - 1)$	
1	1	$= 1$
2	$1 + 3$	$= 4$
3	$1 + 3 + 5$	$= 9$
4	$1 + 3 + 5 + 7$	$= 16$
5	$1 + 3 + 5 + 7 + 9$	$= 25$

Table 1

Let us try the next case beyond Table 1 and see if the pattern persists. That is, we want to know if it is true that

$$\underbrace{1 + 3 + 5 + 7 + 9 + 11}_{\text{six terms}} = 6^2$$

As you can easily check, this last equation is true. Thus, based on the experimental or empirical evidence, we are led to make the following conjecture.

Conjecture The sum of the first n odd natural numbers is n^2. That is, $1 + 3 + 5 + \cdots + (2n - 1) = n^2$, for each natural number n.

At this point, the "law" we've discovered is indeed really only a conjecture. After all, we've only checked it for values of n ranging from 1 to 6. It is conceivable at this point (although we may feel it is unlikely) that the conjecture is false for some or for many values of n. For the conjecture to be useful, we must be able to prove it holds without exception for all natural numbers n. In fact we will subsequently prove that this conjecture is valid. However, before explaining the method of proof to be used, let us look at one more example.

n	2^n	$(n + 1)^2$
1	2	4
2	4	9
3	8	16
4	16	25
5	32	36

Table 2

Again, let n denote a natural number. We pose the following question. Which quantity is the larger, 2^n, or $(n + 1)^2$? As before, we begin by doing some calculations. This is the experimental stage of our work. According to Table 2, the quantity $(n + 1)^2$ is larger than 2^n for each value of n up through $n = 5$. Thus we consider the following conjecture.

Conjecture $(n + 1)^2 > 2^n$ for all natural numbers n.

Again, we note that this is only a conjecture at this point. Indeed, if you try the next case beyond Table 2, you will see that the pattern does not persist. That is, when $n = 6$ we find that 2^n is 64, while $(n + 1)^2$ is only 49. So in this example, the conjecture is not true in general; we have found a value of n for which it fails.

The preceding examples indicate that experimentation does have a place in mathematics, but that we must be careful with the results of such experimentation. Where experimentation leads to a conjecture, proof is required before the conjecture can be viewed as a valid law. For the remainder of this section we shall discuss one such method of proof, that of *mathematical induction*.

In order to state the Principle of Mathematical Induction, we first introduce some notation. Suppose that for each natural number n we have a statement P_n to be proved. Consider, for instance, the first conjecture we mentioned:

$$1 + 3 + 5 + \cdots + (2n - 1) = n^2$$

If we denote this statement by P_n, then we have for example that

P_1 is the statement that $1 = 1^2$
P_2 is the statement that $1 + 3 = 2^2$
P_3 is the statement that $1 + 3 + 5 = 3^2$

With this notation we can now state the Principle of Mathematical Induction.

Principle of Mathematical Induction

Suppose that for each natural number n we have a statement P_n. Furthermore, suppose that the following two conditions hold.

1. P_1 is true.
2. For each natural number k, if P_k is true then P_{k+1} is true.

Then all of the statements are true; that is, P_n is true for all natural numbers n.

The idea behind mathematical induction is a simple one. Think of each statement P_n as the rung of a ladder to be climbed. Then there is the following analogy.

Mathematical Induction			Ladder Analogy		
Hypotheses	1.	P_1 is true.	Hypotheses	1′.	You can reach the first rung.
	2.	If P_k is true then P_{k+1} is true, for any k.		2′.	If you are on the kth rung you can reach the $(k+1)$st rung, for any k.
Conclusion	3.	P_n is true for all n.	Conclusion	3′.	You can climb the entire ladder.

According to the Principle of Mathematical Induction, we can prove that a statement or formula P_n is true for all n if we carry out the following two steps.

Step 1: Show that P_1 is true.

Step 2: Assume that P_k is true and on the basis of this assumption show that P_{k+1} is true.

In Step 2, the assumption that P_k is true is referred to as the *induction hypothesis.* Let us now turn to some examples of proof by mathematical induction.

Example 1 Use mathematical induction to prove that

$$1 + 3 + 5 + \cdots + (2n - 1) = n^2$$

for all natural numbers n.

Solution Let P_n denote the statement that $1 + 3 + 5 + \cdots + (2n - 1) = n^2$. Then we want to show that P_n is true for all natural numbers n.

Step 1: We must check that P_1 is true. But P_1 is just the statement that $1 = 1^2$, which is true.

Step 2: Assuming that P_k is true, we must show that P_{k+1} is true. Thus we assume that

$$1 + 3 + 5 + \cdots + (2k - 1) = k^2 \tag{1}$$

That is the induction hypothesis. We must show that

$$1 + 3 + 5 + \cdots + (2k - 1) + [2(k + 1) - 1] = (k + 1)^2 \tag{2}$$

To derive equation (2) from equation (1) now, we add the quantity $[2(k+1)-1]$ to both sides of equation (1). (The motivation for this stems from the observation that the left-hand sides of equations (1) and (2) differ only by the quantity $[2(k+1)-1]$.) We obtain

$$1 + 3 + 5 + \cdots + (2k-1) + [2(k+1)-1] = k^2 + [2(k+1)-1]$$

or

$$1 + 3 + 5 + \cdots + (2k-1) + [2(k+1)-1] = k^2 + 2k + 1$$

or

$$1 + 3 + 5 + \cdots + (2k-1) + [2(k+1)-1] = (k+1)^2$$

This last equation is what we wished to show.

Having now carried out Steps 1 and 2, we conclude by the Principle of Mathematical Induction that P_n is true for all natural numbers n.

Example 2 Use mathematical induction to prove that

$$2^3 + 4^3 + 6^3 + \cdots + (2n)^3 = 2n^2(n+1)^2$$

for all natural numbers n.

Solution Let P_n denote the statement that

$$2^3 + 4^3 + 6^3 + \cdots + (2n)^3 = 2n^2(n+1)^2$$

Then we want to show that P_n is true for all natural numbers n.

Step 1: We must check that P_1 is true. P_1 is the statement that

$$2^3 = 2(1^2)(1+1)^2$$

or

$$8 = 8$$

Thus P_1 is true.

Step 2: Assuming that P_k is true, we must show that P_{k+1} is true. Thus we assume that

$$2^3 + 4^3 + 6^3 + \cdots + (2k)^3 = 2k^2(k+1)^2 \tag{3}$$

We must show that

$$2^3 + 4^3 + 6^3 + \cdots + (2k)^3 + [2(k+1)]^3 = 2(k+1)^2(k+2)^2 \tag{4}$$

Adding $[2(k+1)]^3$ to both sides of equation (3) yields

$$\begin{aligned} 2^3 + 4^3 + 6^3 + \cdots + (2k)^3 + [2(k+1)]^3 &= 2k^2(k+1)^2 + [2(k+1)]^3 \\ &= 2k^2(k+1)^2 + 8(k+1)^3 \\ &= 2(k+1)^2[k^2 + 4(k+1)] \\ &= 2(k+1)^2(k^2 + 4k + 4) \\ &= 2(k+1)^2(k+2)^2 \end{aligned}$$

We have now derived equation (4) from equation (3), as we wished to do.

Having carried out Steps 1 and 2, we conclude by the Principle of Mathematical Induction that P_n is true for all natural numbers n.

Example 3 As indicated in Table 3, 3 is a factor of $2^{2n} - 1$ when $n = 1, 2, 3$, and 4. Use mathematical induction to show that 3 is a factor of $2^{2n} - 1$ for all natural numbers n.

n	$2^{2n} - 1$
1	$3\ (= 3 \cdot 1)$
2	$15\ (= 3 \cdot 5)$
3	$63\ (= 3 \cdot 21)$
4	$255\ (= 3 \cdot 85)$

Table 3

Solution Let P_n denote the statement that 3 is a factor of $2^{2n} - 1$. We want to show that P_n is true for all natural numbers n.

Step 1: We must check that P_1 is true. But P_1 in this case is just the statement that 3 is a factor of $2^{2(1)} - 1$; that is, 3 is a factor of 3, which is surely true.

Step 2: Assuming that P_k is true, we must show the P_{k+1} is true. Thus we assume that

$$3 \text{ is a factor of } 2^{2k} - 1 \tag{5}$$

and we must show that

$$3 \text{ is a factor of } 2^{2(k+1)} - 1 \tag{6}$$

The strategy here will be to rewrite the expression $2^{2(k+1)} - 1$ in such a way that the induction hypothesis [statement (5)] can be applied. We have

$$\begin{aligned} 2^{2(k+1)} - 1 &= 2^{2k+2} - 1 \\ &= 2^2 \cdot 2^{2k} - 1 \\ &= 4 \cdot 2^{2k} - 4 + 3 \\ &= 4(2^{2k} - 1) + 3 \end{aligned} \tag{7}$$

Now observe that 3 is a factor of the right-hand side of equation (7) (because 3 is a factor of $2^{2k} - 1$, according to the induction hypothesis). Consequently, 3 must be a factor of the left-hand side of equation (7), as we wished to show.

Having now completed Steps 1 and 2, we conclude by the Principle of Mathematical Induction that P_n is true for all natural numbers n. In other words, 3 is a factor of $2^{2n} - 1$ for all natural numbers n.

There are instances in which a given statement P_n is false for certain initial values of n, but true thereafter. An example of this is provided by the statement

$$2^n > (n + 1)^2 \qquad \text{for all natural numbers } n \geq 6$$

We can adapt the Principle of Mathematical Induction to prove this statement by beginning in Step 1 with a consideration of P_6 rather than P_1. This is done in Example 4.

Example 4 Use mathematical induction to prove that

$$2^n > (n + 1)^2$$

for all natural numbers $n \geq 6$.

Solution *Step 1:* We must first check that P_6 is true. But P_6 is simply the assertion that

$$2^6 > (6 + 1)^2$$

or

$$64 > 49$$

Thus P_6 is true.

Step 2: Assuming that P_k is true, where $k \geq 6$, we must show that P_{k+1} is true. Thus we assume that

$$2^k > (k + 1)^2 \qquad \text{where } k \geq 6 \tag{8}$$

We must show that

$$2^{k+1} > (k + 2)^2 \tag{9}$$

Multiplying both sides of inequality (8) by 2 gives us

$$2(2^k) > 2(k + 1)^2 = 2k^2 + 4k + 2$$

This may be rewritten as

$$2^{k+1} > k^2 + 4k + (k^2 + 2)$$

However, since $k \geq 6$, it is certainly true that

$$k^2 + 2 > 4$$

We therefore have that

$$2^{k+1} > k^2 + 4k + 4$$

or

$$2^{k+1} > (k + 2)^2 \qquad \text{as we wished to show}$$

Having now completed Steps 1 and 2, we conclude that P_n is true for all natural numbers $n \geq 6$.

Exercise Set 11.1

In Exercises 1–18, use the Principle of Mathematical Induction to show that the statements are true for all natural numbers.

1. $1 + 2 + 3 + \cdots + n = \dfrac{n(n + 1)}{2}$
2. $2 + 4 + 6 + \cdots + 2n = n(n + 1)$
3. $1 + 4 + 7 + \cdots + (3n - 2) = \dfrac{n(3n - 1)}{2}$
4. $5 + 9 + 13 + \cdots + (4n + 1) = n(2n + 3)$
5. $1^2 + 2^2 + 3^2 + \cdots + n^2 = \dfrac{n(n + 1)(2n + 1)}{6}$
6. $2^2 + 4^2 + 6^2 + \cdots + (2n)^2 = \dfrac{2n(n + 1)(2n + 1)}{3}$
7. $1^2 + 3^2 + 5^2 + \cdots + (2n - 1)^2 = \dfrac{n(2n - 1)(2n + 1)}{3}$
8. $2 + 2^2 + 2^3 + \cdots + 2^n = 2^{n+1} - 2$
9. $3 + 3^2 + 3^3 + \cdots + 3^n = \frac{1}{2}(3^{n+1} - 3)$
10. $e^x + e^{2x} + \cdots + e^{nx} = \dfrac{e^{(n+1)x} - e^x}{e^x - 1} \quad (x \neq 0)$
11. $1^3 + 2^3 + 3^3 + \cdots + n^3 = \left[\dfrac{n(n + 1)}{2}\right]^2$
12. $2^3 + 4^3 + 6^3 + \cdots + (2n)^3 = 2n^2(n + 1)^2$
13. $1^3 + 3^3 + 5^3 + \cdots + (2n - 1)^3 = n^2(2n^2 - 1)$
14. $1 \times 2 + 3 \times 4 + 5 \times 6 + \cdots + (2n - 1)(2n) = \dfrac{n(n + 1)(4n - 1)}{3}$
15. $1 \times 3 + 3 \times 5 + 5 \times 7 + \cdots + (2n - 1)(2n + 1) = \dfrac{n(4n^2 + 6n - 1)}{3}$

16. $\dfrac{1}{1 \times 3} + \dfrac{1}{2 \times 4} + \dfrac{1}{3 \times 5} + \cdots + \dfrac{1}{n(n+2)}$
$= \dfrac{n(3n+5)}{4(n+1)(n+2)}$

17. $1 + \dfrac{3}{2} + \dfrac{5}{2^2} + \dfrac{7}{2^3} + \cdots + \dfrac{2n-1}{2^{n-1}} = 6 - \dfrac{2n+3}{2^{n-1}}$

18. $1 + 2 \times 2 + 3 \times 2^2 + 4 \times 2^3 + \cdots + n \times 2^{n-1}$
$= (n-1)2^n + 1$

19. Show that $n \le 2^{n-1}$ for all natural numbers n.

20. Show that 3 is a factor of $n^3 + 2n$ for all natural numbers n.

21. Show that $n^2 + 4 < (n+1)^2$ for all natural numbers $n \ge 2$.

22. Show that $n^3 > (n+1)^2$ for all natural numbers $n \ge 3$.

23. Let $f(n) = \dfrac{1}{1 \times 2} + \dfrac{1}{2 \times 3} + \dfrac{1}{3 \times 4} + \cdots + \dfrac{1}{n(n+1)}$.

(a) Complete the following table.

n	1	2	3	4	5
$f(n)$					

(b) On the basis of the results in the table, what would you guess to be the value of $f(6)$? Compute $f(6)$ to see if this is correct.

(c) Make a conjecture about the value of $f(n)$ and prove it using mathematical induction.

24. Let $f(n) = \dfrac{1}{1 \times 3} + \dfrac{1}{3 \times 5} + \dfrac{1}{5 \times 7} + \cdots + \dfrac{1}{(2n-1)(2n+1)}$.

(a) Complete the following table.

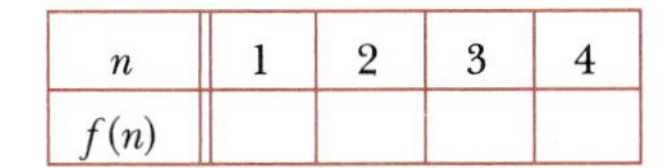

n	1	2	3	4
$f(n)$				

(b) On the basis of the results in the table, what would you guess to be the value of $f(5)$? Compute $f(5)$ to see if your guess is correct.

(c) Make a conjecture about the value of $f(n)$ and prove it using mathematical induction.

25. Suppose that a function f satisfies the following conditions:

$$f(1) = 1$$

and

$$f(n) = f(n-1) + 2\sqrt{f(n-1)} + 1 \quad \text{for } n \ge 2$$

(a) Complete the table.

n	1	2	3	4	5
$f(n)$					

(b) On the basis of the results in the table, what would you guess to be the value of $f(6)$? Compute $f(6)$ to see if your guess is correct.

(c) Make a conjecture about the value of $f(n)$ when n is a natural number, and prove the conjecture using mathematical induction.

26. This exercise demonstrates the necessity of carrying out both Step 1 and Step 2 before an induction proof can be considered valid.

(a) Let P_n denote the statement that $n^2 + 1$ is even. Check that P_1 is true. Then give an example showing that P_n is not true for all n.

(b) Let Q_n denote the statement that $n^2 + n$ is odd. Show that Step 2 of an induction proof can be completed in this case but Step 1 cannot.

27. A *prime number* is a natural number that has no factors other than itself and 1. For technical reasons, 1 is not considered a prime. Thus the list of the first few primes looks as follows:

$$2, 3, 5, 7, 11, 13, 17, \ldots$$

Let P_n be the statement that $n^2 + n + 11$ is prime. Check that P_n is true for all values of n less than 10. Check that P_{10} is false.

28. Prove that $1 + 2x + 3x^2 + \cdots + nx^{n-1} = \dfrac{1-x^n}{(1-x)^2} - \dfrac{nx^n}{1-x}$, $(x \ne 1)$, for all natural numbers n.

29. If $r \ne 1$, show that $1 + r + r^2 + \cdots + r^{n-1} = \dfrac{r^n - 1}{r - 1}$, for all natural numbers n.

30. Use mathematical induction to show that $x^n - 1 = (x-1)(1 + x + x^2 + \cdots + x^{n-1})$, for all natural numbers n.

***31.** Prove that 5 is a factor of $n^5 - n$ for all natural numbers $n \ge 2$.

***32.** Prove that 4 is a factor of $5^n + 3$ for all natural numbers n.

33. Use mathematical induction to show that $x - y$ is a factor of $x^n - y^n$ for all natural numbers n. *Suggestion for Step 2:* Verify and then use the fact that $x^{k+1} - y^{k+1} = x^k(x-y) + (x^k - y^k)y$.

In Exercises 34 and 35, use mathematical induction to prove that the formulas hold for all natural numbers n.

34. $\log_{10}(a_1 a_2 \cdots a_n)$
$= \log_{10} a_1 + \log_{10} a_2 + \cdots + \log_{10} a_n$

35. $(1 + p)^n \ge 1 + np$, where $p > -1$

11.2 The Binomial Theorem

A mathematician, like a painter or a poet, is a maker of patterns.
G. H. Hardy (1877–1947)

If you look back at Section 1.7 you'll see that two of the special products listed there are

$$(a + b)^2 = a^2 + 2ab + b^2$$

and

$$(a + b)^3 = a^3 + 3a^2b + 3ab^2 + b^3$$

Our goal now is to develop a general formula, known as the *Binomial Theorem*, for expanding any product of the form $(a + b)^n$, when n is a natural number.

We begin by looking for patterns in the expansion of $(a + b)^n$. To do this, let's list the expansions of $(a + b)^n$ for $n = 1, 2, 3, 4$, and 5. (Exercises 1 and 2 ask you to verify these results simply by repeated multiplication.)

$$(a + b)^1 = a + b$$
$$(a + b)^2 = a^2 + 2ab + b^2$$
$$(a + b)^3 = a^3 + 3a^2b + 3ab^2 + b^3$$
$$(a + b)^4 = a^4 + 4a^3b + 6a^2b^2 + 4ab^3 + b^4$$
$$(a + b)^5 = a^5 + 5a^4b + 10a^3b^2 + 10a^2b^3 + 5ab^4 + b^5$$

After surveying these results, we note the following patterns.

Property Summary

Patterns Observed in $(a + b)^n$ for $n = 1, 2, 3, 4, 5$

General Statement	Example
There are $n + 1$ terms.	There are $4(= 3 + 1)$ terms in the expansion of $(a + b)^3$.
The expansion begins with a^n and ends with b^n.	$(a + b)^3$ begins with a^3 and ends with b^3.
The sum of the exponents in each term is n.	The sum of the exponents in each term of $(a + b)^3$ is 3.
The exponents of a decrease by 1 from term to term.	$(a + b)^3 = a^{③} + 3a^{②}b + 3a^{①}b^2 + a^{⓪}b^3$
The exponents of b increase by 1 from term to term.	$(a + b)^3 = a^3b^{⓪} + 3a^2b^{①} + 3ab^{②} + b^{③}$
When n is even, the coefficients are symmetric about the middle term.	The sequence of coefficients for $(a + b)^4$ is 1, 4, 6, 4, 1.
When n is odd, the coefficients are symmetric about the two middle terms.	The sequence of coefficients for $(a + b)^5$ is 1, 5, 10, 10, 5, 1.

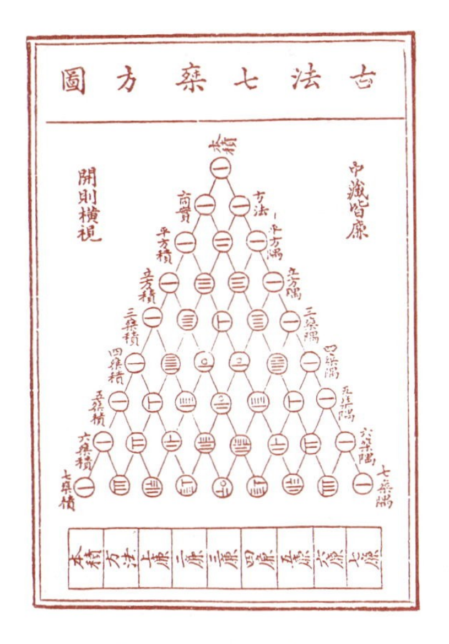

"Pascal's" Triangle
—by Chu-Shi-kie, A.D. 1303

It is a fact that the patterns we've listed above for $(a + b)^n$ persist for all natural numbers n. (This will follow from the Binomial Theorem, which is proved at the end of this section.) Thus, for example, the form of $(a + b)^6$ must be as follows:

$$(a + b)^6 = a^6 + \underset{\cdot}{?}a^5b + \underset{\cdot}{?}a^4b^2 + \underset{\cdot}{?}a^3b^3 + \underset{\cdot}{?}a^2b^4 + \underset{\cdot}{?}ab^5 + b^6$$

The problem now is to find the proper coefficient for each term. To do this we need to discover additional patterns in the expansion of $(a + b)^n$.

We've previously written out the expansions of $(a + b)^n$ for values of n ranging from 1 to 5. Let us now write only the coefficients appearing in those expansions. The resulting triangular array of numbers is known as *Pascal's triangle.** [For reasons of symmetry we begin with $(a + b)^0$ rather than $(a + b)^1$.]

$(a + b)^0$ 1
$(a + b)^1$ 1 1
$(a + b)^2$ 1 2 1
$(a + b)^3$ 1 3 3 1
$(a + b)^4$ 1 4 6 4 1
$(a + b)^5$ 1 5 10 10 5 1

The key observation regarding Pascal's triangle is this: *each entry in the array (other than the 1's along the sides) is the sum of the two numbers diagonally above it.* For instance, the 6 that appears in the fifth row is the sum of the two 3's diagonally above it. Using this observation, we can form as many additional rows as we please. The coefficients for $(a + b)^n$ will then appear in the $(n + 1)$st row of the array.† For instance, to obtain the row corresponding to $(a + b)^6$ we have

sixth row, $(a + b)^5$:

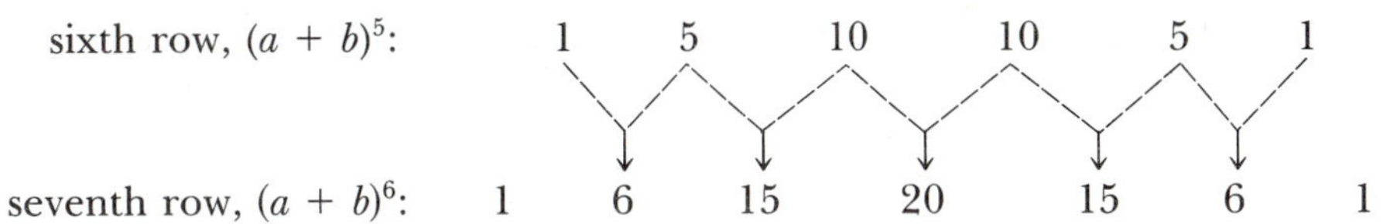

seventh row, $(a + b)^6$:

Thus the sequence of coefficients for $(a + b)^6$ is 1, 6, 15, 20, 15, 6, 1. This answers the question raised earlier about the expansion of $(a + b)^6$; we have

$$(a + b)^6 = a^6 + 6a^5b + 15a^4b^2 + 20a^3b^3 + 15a^2b^4 + 6ab^5 + b^6$$

For analytical work or for larger values of the exponent n, it is inefficient to rely on Pascal's triangle. For this reason, we point out another pattern in the expansion of $(a + b)^n$.

> In the expansion of $(a + b)^n$, the coefficient of any term after the first can be generated as follows. In the *previous* term, multiply the coefficient by the exponent of a and then divide by the number of that previous term.

*The array is named after Blaise Pascal, a 17th-century French mathematician and philosopher. However, as the illustration above indicates, the Pascal Triangle was known to Chinese mathematicians centuries earlier.

†That these numbers actually are the appropriate coefficients follows from the Binomial Theorem, which is proved at the end of this section.

To see how this observation is used, let's compute the second, third, and fourth coefficients in the expansion of $(a + b)^6$. To compute the coefficient of the second term we go back to the first term, which is a^6. We have

$$\text{coefficient of second term} = \frac{1 \cdot 6}{1} = 6$$

(1 = coefficient of first term; 6 = exponent of a in first term; denominator 1 = number of first term)

Thus the second term is $6a^5b$, and consequently we have

$$\text{coefficient of third term} = \frac{6 \cdot 5}{2} = 15$$

(6 = coefficient of second term; 5 = exponent of a in second term; denominator 2 = number of second term)

Continuing now with this method, you should check for yourself that the coefficient of the fourth term in the expansion of $(a + b)^6$ is 20. *Note:* We've found that the first four coefficients are 1, 6, 15, and 20. By symmetry now, it follows that the complete sequence of coefficients for this expansion is 1, 6, 15, 20, 15, 6, 1. No additional calculation for the coefficients is necessary.

Example 1 Expand $(2x - y^2)^7$.

Solution First we write the expansion of $(a + b)^7$ using the method explained just prior to this example, or using Pascal's triangle. As you should check for yourself, the expansion is

$$(a + b)^7 = a^7 + 7a^6b + 21a^5b^2 + 35a^4b^3 + 35a^3b^4 + 21a^2b^5 + 7ab^6 + b^7$$

Now we make the substitutions $a = 2x$ and $b = -y^2$. This yields

$$\begin{aligned}[2x + (-y^2)]^7 &= (2x)^7 + 7(2x)^6(-y^2) + 21(2x)^5(-y^2)^2 + 35(2x)^4(-y^2)^3 \\ &\quad + 35(2x)^3(-y^2)^4 + 21(2x)^2(-y^2)^5 + 7(2x)(-y^2)^6 \\ &\quad + (-y^2)^7 \\ &= 128x^7 - 448x^6y^2 + 672x^5y^4 - 560x^4y^6 + 280x^3y^8 \\ &\quad - 84x^2y^{10} + 14xy^{12} - y^{14}\end{aligned}$$

This is the required expansion. Notice how the signs alternate in the final answer; this is characteristic of all expansions of the form $(a - b)^n$.

Now let us turn to a consideration of the Binomial Theorem. First, we introduce two notations that are commonly used not only in connection with the Binomial Theorem, but in many other parts of mathematics as well.

Definition of $n!$
(Read "n Factorial")

$n! = 1 \cdot 2 \cdot 3 \cdot \cdots \cdot n$
where n is a natural number

$0! = 1$

Examples

$3! = 1 \times 2 \times 3 = 6$

$$\frac{6!}{4!} = \frac{6 \times 5 \times 4 \times 3 \times 2 \times 1}{4 \times 3 \times 2 \times 1} = 6 \times 5 = 30$$

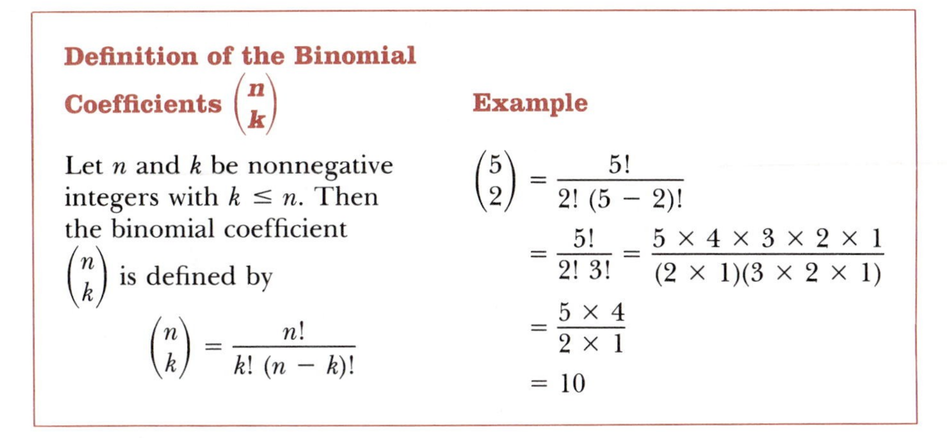

Definition of the Binomial Coefficients $\binom{n}{k}$

Let n and k be nonnegative integers with $k \le n$. Then the binomial coefficient $\binom{n}{k}$ is defined by

$$\binom{n}{k} = \frac{n!}{k!\,(n-k)!}$$

Example

$$\binom{5}{2} = \frac{5!}{2!\,(5-2)!} = \frac{5!}{2!\,3!} = \frac{5 \times 4 \times 3 \times 2 \times 1}{(2 \times 1)(3 \times 2 \times 1)} = \frac{5 \times 4}{2 \times 1} = 10$$

The binomial coefficients are so named because they are indeed the coefficients in the expansion of $(a + b)^n$. More precisely, the relationship is this:

> The coefficients in the expansion of $(a + b)^n$ are the $n + 1$ numbers $\binom{n}{0}, \binom{n}{1}, \binom{n}{2}, \ldots, \binom{n}{n}$.

We will see subsequently why this statement is true. For now, however, let us look at an example. Consider the binomial coefficients $\binom{3}{0}, \binom{3}{1}, \binom{3}{2}$, and $\binom{3}{3}$. According to our statement, these four quantities should be the coefficients in the expansion of $(a + b)^3$. Let us check.

$$\binom{3}{0} = \frac{3!}{0!\,(3-0)!} = \frac{3!}{1(3!)} = 1$$

$$\binom{3}{1} = \frac{3!}{1!\,(3-1)!} = \frac{3 \times 2 \times 1}{1(2 \times 1)} = 3$$

$$\binom{3}{2} = \frac{3!}{2!\,(3-2)!} = \frac{3 \times 2 \times 1}{(2 \times 1)1} = 3$$

$$\binom{3}{3} = \frac{3!}{3!\,(3-3)!} = \frac{3!}{3!\,0!} = 1$$

The values of $\binom{3}{0}, \binom{3}{1}, \binom{3}{2}$, and $\binom{3}{3}$ are thus 1, 3, 3, and 1, respectively. But these last four numbers are indeed the coefficients in the expansion of $(a + b)^3$, as we wished to check.

We are now in a position to state the Binomial Theorem. After doing this we will look at several applications of the theorem. Finally, at the end of this section, we'll use mathematical induction to prove the theorem. In the statement of the theorem that follows, we are assuming that the exponent n is a natural number.

Binomial Theorem

$$(a + b)^n = \binom{n}{0}a^n + \binom{n}{1}a^{n-1}b + \binom{n}{2}a^{n-2}b^2 + \cdots + \binom{n}{n-1}ab^{n-1} + \binom{n}{n}b^n$$

One of the uses of the Binomial Theorem is in identifying specific terms in an expansion without computing the entire expansion. This is particularly helpful when the exponent n is relatively large. Looking back at the statement of the Binomial Theorem, there are three observations we can make. First, the coefficient of the rth term is $\binom{n}{r-1}$. For instance, the coefficient of the third term is $\binom{n}{3-1} = \binom{n}{2}$. The second observation is that the exponent for a in the rth term is $n - (r - 1)$. For instance, the exponent for a in the third term is $n - (3 - 1) = n - 2$. Finally, we observe that the exponent for b in the rth term is $r - 1$, the same quantity that appears in the lower position of the corresponding binomial coefficient. For instance, the exponent for b in the third term is $r - 1 = 3 - 1 = 2$. We summarize these three observations now with the following statement:

> The rth term in the expansion of $(a + b)^n$ is
>
> $$\binom{n}{r-1} a^{n-r+1} b^{r-1}$$

The next three examples make use of this fact.

Example 2 Find the 15th term in the expansion of $\left(x^2 - \frac{1}{x}\right)^{18}$.

Solution Using the values $r = 15$, $n = 18$, $a = x^2$, and $b = -1/x$, we have

$$\begin{aligned}
\binom{n}{r-1} a^{n-r+1} b^{r-1} &= \binom{18}{15-1} (x^2)^{18-15+1} \left(\frac{-1}{x}\right)^{15-1} \\
&= \binom{18}{14} x^8 \cdot \frac{1}{x^{14}} \\
&= \frac{18 \times 17 \times 16 \times 15 \times (14!)}{14!(4 \times 3 \times 2 \times 1)} x^{-6} \\
&= \frac{18 \times 17 \times 16 \times 15}{4 \times 3 \times 2 \times 1} x^{-6} \\
&= 3060x^{-6} \qquad \text{as required}
\end{aligned}$$

Example 3 Find the coefficient of the term containing x^4 in the expansion of $(x + y^2)^{30}$.

Solution Again we use the fact that the rth term in the expansion of $(a + b)^n$ is $\binom{n}{r-1} a^{n-r+1} b^{r-1}$. In this case n is 30 and x plays the role of a. The exponent for x is then $n - r + 1$ or $30 - r + 1$. To see when this exponent is 4, we write

$$\begin{aligned}
30 - r + 1 &= 4 \\
-r &= -27 \\
r &= 27
\end{aligned}$$

The required coefficient is therefore $\binom{30}{27-1}$. We then have

$$\binom{30}{26} = \frac{30!}{26!\,(30-26)!}$$
$$= \frac{30 \times 29 \times 28 \times 27}{4 \times 3 \times 2 \times 1}$$

After carrying out the indicated arithmetic, we find that $\binom{30}{26} = 27{,}405$. This is the required coefficient.

Example 4 Find the coefficient of the term containing a^9 in the expansion of $(a + 2\sqrt{a})^{10}$.

Solution The rth term in this expansion will be

$$\binom{10}{r-1} a^{10-r+1}(2\sqrt{a})^{r-1}$$

We can rewrite this as

$$\binom{10}{r-1} a^{10-r+1}(2^{r-1})(a^{1/2})^{r-1}$$

or

$$2^{r-1}\binom{10}{r-1} a^{10-r+1+\frac{r-1}{2}}$$

This shows that the general form of the coefficient we wish to find is $2^{r-1}\binom{10}{r-1}$. We need to determine r now when the exponent of a is 9. Thus we require that

$$10 - r + 1 + \frac{r-1}{2} = 9$$

or

$$-r + \frac{r-1}{2} = -2$$

To solve this equation for r, multiply both sides by 2 to obtain

$$-2r + r - 1 = -4$$
$$-r = -3$$
$$r = 3$$

The required coefficient is now obtained by substituting $r = 3$ in the expression $2^{r-1}\binom{10}{r-1}$. We have

$$2^{r-1}\binom{10}{r-1} = 2^2\binom{10}{2}$$
$$= \frac{4 \times 10!}{2!\,(10-2)!} = 2 \times 10 \times 9$$
$$= 180 \qquad \text{as required}$$

There are three simple identities involving the binomial coefficients that will simplify our proof of the Binomial Theorem. For ease of reference let us label these identities 1, 2, and 3.

Identity 1: $\binom{r}{0} = 1$ for all nonnegative integers r

Identity 2: $\binom{r}{r} = 1$ for all nonnegative integers r

Identity 3: $\binom{k}{r} + \binom{k}{r-1} = \binom{k+1}{r}$ for all natural numbers k and r with $r \leq k$

All three of these identities can be proved directly from the definitions of the binomial coefficients, without the need for mathematical induction. The proofs of the first two are straightforward and we omit them here. The proof of Identity 3 runs as follows.

$$\binom{k}{r} + \binom{k}{r-1} = \frac{k!}{r!\,(k-r)!} + \frac{k!}{(r-1)!\,(k-r+1)!}$$

or

$$\binom{k}{r} + \binom{k}{r-1} = \frac{k!}{r(r-1)!\,(k-r)!} + \frac{k!}{(r-1)!\,(k-r+1)(k-r)!}$$

Now the common denominator on the right-hand side of the last equation is $r(r-1)!\,(k-r+1)(k-r)!$ Thus we have

$$\begin{aligned}\binom{k}{r} + \binom{k}{r-1} &= \frac{k!\,(k-r+1)}{r(r-1)!\,(k-r+1)(k-r)!} + \frac{k!\,r}{r(r-1)!\,(k-r+1)(k-r)!}\\ &= \frac{k!\,(k-r+1) + k!\,r}{r(r-1)!\,(k-r+1)(k-r)!}\\ &= \frac{k!\,(k-r+1+r)}{r(r-1)!\,(k-r+1)(k-r)!} = \frac{k!\,(k+1)}{r!\,(k-r+1)!}\\ &= \frac{(k+1)!}{r!\,[(k+1)-r]!}\\ &= \binom{k+1}{r} \qquad \text{as we wished to show}\end{aligned}$$

Taken together, the three identities we've listed show why the $(n+1)$st row of Pascal's triangle consists of the numbers $\binom{n}{0}, \binom{n}{1}, \binom{n}{2}, \ldots, \binom{n}{n}$. Identities 1 and 2 tell us that this row of numbers begins and ends with 1. Identity 3 is then just a statement of the fact that each entry in the row, other than the initial and final 1, is generated by adding the two entries diagonally above it.

We conclude this section by using mathematical induction to prove the Binomial Theorem. The statement P_n that we wish to prove for all natural numbers n is this:

$$(a+b)^n = \binom{n}{0}a^n + \binom{n}{1}a^{n-1}b + \binom{n}{2}a^{n-2}b^2 + \cdots + \binom{n}{n-1}ab^{n-1} + \binom{n}{n}b^n$$

First, we check that P_1 is true. P_1 asserts that

$$(a+b)^1 = \binom{1}{0}a^1 + \binom{1}{1}a^0 b$$

However, in view of Identities 1 and 2 this last equation becomes

$$(a+b)^1 = 1\cdot a + 1\cdot b \qquad \text{which is surely true}$$

Now let us assume that P_k is true and on the basis of this assumption show that P_{k+1} is true. P_k is the statement that

$$(a+b)^k = \binom{k}{0}a^k + \binom{k}{1}a^{k-1}b + \binom{k}{2}a^{k-2}b^2 + \cdots + \binom{k}{k-1}ab^{k-1} + \binom{k}{k}b^k$$

Multiplying both sides of this equation by the quantity $(a+b)$ yields

$$\begin{aligned}
(a+b)^{k+1} &= (a+b)\left[\binom{k}{0}a^k + \binom{k}{1}a^{k-1}b + \binom{k}{2}a^{k-2}b^2 + \cdots + \binom{k}{k-1}ab^{k-1} + \binom{k}{k}b^k\right]\\
&= a\left[\binom{k}{0}a^k + \binom{k}{1}a^{k-1}b + \binom{k}{2}a^{k-2}b^2 + \cdots + \binom{k}{k-1}ab^{k-1} + \binom{k}{k}b^k\right]\\
&\quad + b\left[\binom{k}{0}a^k + \binom{k}{1}a^{k-1}b + \binom{k}{2}a^{k-2}b^2 + \cdots + \binom{k}{k-1}ab^{k-1} + \binom{k}{k}b^k\right]\\
&= \binom{k}{0}a^{k+1} + \binom{k}{1}a^k b + \binom{k}{2}a^{k-1}b^2 + \cdots + \binom{k}{k-1}a^2b^{k-1} + \binom{k}{k}ab^k\\
&\quad + \binom{k}{0}a^k b + \binom{k}{1}a^{k-1}b^2 + \binom{k}{2}a^{k-2}b^3 + \cdots + \binom{k}{k-1}ab^k + \binom{k}{k}b^{k+1}\\
&= \binom{k}{0}a^{k+1} + \left[\binom{k}{1} + \binom{k}{0}\right]a^k b + \left[\binom{k}{2} + \binom{k}{1}\right]a^{k-1}b^2 + \cdots + \left[\binom{k}{k} + \binom{k}{k-1}\right]ab^k + \binom{k}{k}b^{k+1}
\end{aligned}$$

We can now make some substitutions on the right-hand side of this last equation. The initial binomial coefficient $\binom{k}{0}$ may be replaced by $\binom{k+1}{0}$, for both are equal to 1 according to Identity 1. Similarly, the binomial coefficient $\binom{k}{k}$ appearing at the end of the equation may be replaced by $\binom{k+1}{k+1}$, since both are equal to 1 according to Identity 2. Finally, we may use Identity 3 to simplify each of the sums in the brackets. We obtain

$$(a+b)^{k+1} = \binom{k+1}{0}a^{k+1} + \binom{k+1}{1}a^k b + \binom{k+1}{2}a^{k-1}b^2 + \cdots + \binom{k+1}{k}ab^k + \binom{k+1}{k+1}b^{k+1}$$

But this last equation is just the statement P_{k+1}; that is, we have derived P_{k+1} from P_k, as we wished to do. The induction proof is now complete.

Exercise Set 11.2

For Exercises 1 and 2, verify each statement directly, without using the techniques developed in this section.

1. (a) $(a+b)^2 = a^2 + 2ab + b^2$

(b) $(a+b)^3 = a^3 + 3a^2b + 3ab^2 + b^3$
Hint: $(a+b)^3 = (a+b)(a+b)^2$

2. (a) $(a+b)^4 = a^4 + 4a^3b + 6a^2b^2 + 4ab^3 + b^4$
Hint: Use the result in Exercise 1(b) and the fact that $(a+b)^4 = (a+b)(a+b)^3$.

(b) $(a+b)^5 = a^5 + 5a^4b + 10a^3b^2 + 10a^2b^3 + 5ab^4 + b^5$

In Exercises 3–28, carry out the indicated expansions.

3. $(a+b)^9$ **4.** $(a-b)^9$ **5.** $(2A+B)^3$

6. $(1+2x)^6$ **7.** $(1-2x)^6$ **8.** $(3x^2-y)^5$

9. $(\sqrt{x}+\sqrt{y})^4$ **10.** $(\sqrt{x}-\sqrt{y})^4$ **11.** $(x^2+y^2)^5$

12. $(5A-B^2)^3$ **13.** $\left(1-\frac{1}{x}\right)^6$ **14.** $(3x+y^2)^4$

15. $\left(\frac{x}{2}-\frac{y}{3}\right)^3$ **16.** $(1-z^2)^7$ **17.** $(ab^2+c)^7$

18. $\left(x-\frac{1}{x}\right)^8$ **19.** $(x+\sqrt{2})^8$ **20.** $(4A-\frac{1}{2})^5$

21. $(\sqrt{2}-1)^3$ **22.** $(1+\sqrt{5})^4$ **23.** $(\sqrt{2}+\sqrt{3})^5$

24. $(\frac{1}{2}-2a)^6$ **25.** $(2\sqrt[3]{2}-\sqrt[3]{4})^3$

***26.** $(x+y+1)^4$ *Suggestion:* Rewrite the expression as $[(x+y)+1]^4$.

***27.** $(x^2-2x-1)^5$ *Suggestion:* Rewrite the expression as $[x^2-(2x+1)]^5$.

***28.** $\left(x^2-2x-\frac{1}{x}\right)^6$

For Exercises 29–38, evaluate or simplify each expression.

29. $5!$

30. **(a)** $3!+2!$ **(b)** $(3+2)!$

31. $\binom{7}{3}\binom{3}{2}$

32. $\frac{20!}{18!}$

33. **(a)** $\binom{5}{3}$ **(b)** $\binom{5}{4}$

34. **(a)** $\binom{7}{7}$ **(b)** $\binom{7}{0}$

35. $\frac{(n+2)!}{n!}$

36. $\frac{n[(n-2)!]}{(n+1)!}$

37. $\binom{6}{4}+\binom{6}{3}-\binom{7}{4}$

38. $(3!)!+(3!)^2$

39. Find the 15th term in the expansion of $(a+b)^{16}$.

40. Find the third term in the expansion of $(a-b)^{30}$.

41. Find the 100th term in the expansion of $(1+x)^{100}$.

42. Find the 23rd term in the expansion of $\left(x-\frac{1}{x^2}\right)^{25}$.

43. Find the coefficient of the term containing a^4 in the expansion of $(\sqrt{a}-\sqrt{x})^{10}$.

44. Find the coefficient of the term containing a^4 in the expansion of $(3a-5x)^{12}$.

45. Find the coefficient of the term containing y^8 in the expansion of $\left(\frac{x}{2}-4y\right)^9$.

46. Find the coefficient of the term containing x^6 in the expansion of $\left(x^2+\frac{1}{x}\right)^{12}$.

47. Find the coefficient of the term containing x^3 in the expansion of $(1-\sqrt{x})^8$.

48. Find the coefficient of the term containing a^8 in the expansion of $\left(a-\frac{2}{\sqrt{a}}\right)^{14}$.

49. Find the term that does not contain A in the expansion of $\left(\frac{1}{A}+3A^2\right)^{12}$.

50. Find the coefficient of B^{-10} in the expansion of $\left(\frac{B^2}{2}-\frac{3}{B^3}\right)^{10}$.

***51.** Show that the coefficient of x^n in the expansion of $(1+x)^{2n}$ is $(2n)!/(n!)^2$.

***52.** Find n so that the coefficients of the 11th and 13th terms in $(1+x)^n$ are the same.

53. **(a)** Complete the following table.

k	0	1	2	3	4	5	6	7	8
$\binom{8}{k}$									

(b) Use the results in part (a) to verify that $\binom{8}{0}+\binom{8}{1}+\binom{8}{2}+\cdots+\binom{8}{8}=2^8$.

(c) By taking $a=b=1$ in the expansion of $(a+b)^n$, show that

$$\binom{n}{0}+\binom{n}{1}+\binom{n}{2}+\cdots+\binom{n}{n}=2^n$$

***54.** Two real numbers A and B are defined as follows:

$$A=\sqrt[99]{99!} \quad B=\sqrt[100]{100!}$$

Which number is larger, A or B? *Hint:* Compare A^{9900} and B^{9900}.

11.3 Introduction to Sequences and Series

This section and the next two sections in this chapter deal with *numerical sequences.* We'll begin with a somewhat informal definition of this concept. Then after looking at some examples and terminology, we will present a more formal definition. A *numerical sequence* is an ordered list of numbers. Here are four examples.

Example A: $1, \sqrt{2}, 10$

Example B: $2, 4, 6, 8, \ldots$

Example C: $1, \frac{1}{2}, \frac{1}{4}, \frac{1}{8}, \ldots$

Example D: $1, 1, 1, 1, \ldots$

The individual entries in a numerical sequence are called the *terms* of the sequence. In this chapter, the terms in each sequence will always be real numbers, so for convenience we shall drop the adjective *numerical* and refer simply to *sequences.* (It is worth pointing out, however, that in more advanced courses sequences are studied in which the individual terms are functions.) Any sequence possessing only a finite number of terms is called a *finite sequence.* Thus the sequence in Example A is a finite sequence. On the other hand, Examples B, C, and D are examples of what we call *infinite sequences;* each contains infinitely many terms. As Example D indicates, it is not necessary that all of the terms in a sequence be distinct. From now on in this chapter, all of the sequences discussed will be infinite sequences.

In Examples B, C, and D, the three dots are read "and so on." In using this notation, we are assuming that it is clear what the subsequent terms of the sequence are. Toward this end, we often specify a formula for the nth term in a sequence. Example B in this case would appear this way:

$$2, 4, 6, 8, \ldots, 2n, \ldots$$

A letter with subscripts is often used to denote the various terms in a sequence. For instance, if we denote the sequence in Example B by $a_1, a_2, a_3, \ldots$ then we have $a_1 = 2$, $a_2 = 4$, $a_3 = 6$, and in general $a_n = 2n$.

Example 1 Consider the sequence $a_1, a_2, a_3, \ldots$ in which the nth term a_n is given by

$$a_n = \frac{n}{n+1}$$

Compute the first three terms of the sequence as well as the 1000th term.

Solution To obtain the first term replace n by 1 in the given formula. This yields

$$a_1 = \frac{1}{1+1} = \frac{1}{2}$$

The other terms are obtained similarly. We have

$$a_2 = \frac{2}{2+1} = \frac{2}{3}$$

$$a_3 = \frac{3}{3+1} = \frac{3}{4}$$

$$a_{1000} = \frac{1000}{1000+1} = \frac{1000}{1001}$$

In the example just concluded we were given an explicit formula for the nth term of the sequence. The next example shows a different way of specifying a sequence, in which each term except the first is defined in terms of the previous term. This is an example of a *recursive definition.* (Recursive definitions are particularly useful in computer programming.)

Example 2 Compute the first three terms of the sequence $b_1, b_2, b_3, \ldots$ defined recursively by

$$b_1 = 4$$
$$b_n = 2(b_{n-1} - 1) \qquad \text{for } n \geq 2$$

Solution We are given the first term: $b_1 = 4$. To find b_2, replace n by 2 in the formula $b_n = 2(b_{n-1} - 1)$ to obtain

$$b_2 = 2(b_1 - 1) = 2(4 - 1) = 6$$

Thus $b_2 = 6$. Next we use this value of b_2 along with the formula

$$b_n = 2(b_{n-1} - 1)$$

to obtain b_3. Replacing n by 3 in this formula yields

$$b_3 = 2(b_2 - 1) = 2(6 - 1) = 10$$

We have now found the first three terms of the sequence: $b_1 = 4$, $b_2 = 6$, and $b_3 = 10$.

If you think about the central idea behind the notion of a sequence, you can see that a function is involved: For each input n we have an output a_n. For this reason the formal definition of a sequence is phrased as follows.

Definition

A *sequence* is a function whose domain is the set of natural numbers.

If we denote the function by f for the moment, then $f(1)$ would denote what we have been calling a_1, the first term of the sequence. Similarly, $f(2) = a_2$, $f(3) = a_3$, and so on. Notice that this definition does not specify that the range of the function consist of real numbers. This paves the way in more advanced mathematics courses for considering types of sequences other than the numerical ones with which we are working. We shall not pursue these ideas here, however. Instead we return to our main line of development.

We will often be interested in the sum of certain terms of a sequence. Consider, for example, the sequence

$$10, 20, 30, 40, \ldots$$

in which the nth term is $10n$. The sum of the first four terms in this sequence is

$$10 + 20 + 30 + 40 = 100$$

More generally, the sum of the first n terms in this sequence is indicated by writing

$$10 + 20 + 30 + \cdots + 10n$$

The above expression is an example of a *finite series,* which simply means a sum of a finite number of terms.

The sum of the first n terms of the sequence $a_1, a_2, a_3, \ldots$ may be indicated by writing

$$a_1 + a_2 + a_3 + \cdots + a_n$$

There is another way to indicate this sum using the so-called *sigma notation,* which we now introduce. The capital Greek letter sigma is written Σ. We define the notation $\sum_{k=1}^{n} a_k$ by the equation

$$\sum_{k=1}^{n} a_k = a_1 + a_2 + a_3 + \cdots + a_n$$

For example, $\sum_{k=1}^{3} a_k$ stands for the sum $a_1 + a_2 + a_3$, the idea in this case being to replace the subscript k successively by 1, 2, and 3 and add the results. As another example, perhaps more concrete, let us evaluate the expression $\sum_{k=1}^{4} k^2$. We have

$$\begin{aligned}\sum_{k=1}^{4} k^2 &= 1^2 + 2^2 + 3^2 + 4^2 \\ &= 1 + 4 + 9 + 16 \\ &= 30\end{aligned}$$

There is nothing special about the choice of the letter k in the expression $\sum_{k=1}^{4} k^2$. For instance, we could equally well write

$$\begin{aligned}\sum_{j=1}^{4} j^2 &= 1^2 + 2^2 + 3^2 + 4^2 \\ &= 30 \qquad \text{as before}\end{aligned}$$

The letter k in the expression $\sum_{k=1}^{4} k^2$ is called the *index of summation.* Similarly, the letter j appearing in $\sum_{j=1}^{4} j^2$ is the index of summation in that case. As we have seen, the choice of the letter used for the index of summation has no effect upon the value of the indicated sum. For this reason, the index of summation is referred to as a *dummy variable.* The next two examples provide further practice with the sigma notation.

Example 3 Express each of the following sums without sigma notation.

(a) $\sum_{k=1}^{3} (3k - 2)^2$ **(b)** $\sum_{i=1}^{4} ix^{i-1}$ **(c)** $\sum_{j=1}^{5} (a_{j+1} - a_j)$

Solution **(a)** The notation $\sum_{k=1}^{3} (3k - 2)^2$ directs us to replace k successively by 1, 2, and 3 in the expression $(3k - 2)^2$ and add the results. We thus obtain

$$\sum_{k=1}^{3} (3k - 2)^2 = 1^2 + 4^2 + 7^2 = 66$$

(b) The notation $\sum_{i=1}^{4} ix^{i-1}$ directs us to replace i successively by 1, 2, 3, and 4 in the expression ix^{i-1} and add the results. We have

$$\begin{aligned}\sum_{i=1}^{4} ix^{i-1} &= 1x^0 + 2x^1 + 3x^2 + 4x^3 \\ &= 1 + 2x + 3x^2 + 4x^3\end{aligned}$$

(c) To expand $\sum_{j=1}^{5} (a_{j+1} - a_j)$ we replace j successively by 1, 2, 3, 4, and 5 in the expression $(a_{j+1} - a_j)$ and add. We obtain

$$\sum_{j=1}^{5} (a_{j+1} - a_j) = (a_2 - a_1) + (a_3 - a_2) + (a_4 - a_3) + (a_5 - a_4) + (a_6 - a_5)$$

$$= a_2 - a_1 + a_3 - a_2 + a_4 - a_3 + a_5 - a_4 + a_6 - a_5$$

Now combining like terms, we have

$$\sum_{j=1}^{5} (a_{j+1} - a_j) = a_6 - a_1$$

Sums such as $\sum_{j=1}^{5} (a_{j+1} - a_j)$ are known as *collapsing* or *telescoping sums.*

Example 4 Use sigma notation to rewrite each sum.

(a) $\frac{x}{1!} + \frac{x^2}{2!} + \frac{x^3}{3!} + \cdots + \frac{x^{12}}{12!}$

(b) $\frac{x}{2!} + \frac{x^2}{3!} + \frac{x^3}{4!} + \cdots + \frac{x^n}{(n+1)!}$

Solution **(a)** Since the exponents on x run from 1 to 12, we choose a dummy variable, say k, running from 1 to 12. Also, we notice that if the numerator of a given term in the sum is x^k, then the corresponding denominator is $k!$ Consequently, the sum can be written

$$\frac{x}{1!} + \frac{x^2}{2!} + \frac{x^3}{3!} + \cdots + \frac{x^{12}}{12!} = \sum_{k=1}^{12} \frac{x^k}{k!}$$

(b) Since the exponents on x run from 1 to n, we choose a dummy variable, say k, running from 1 to n. (Note that both of the letters n and x would be inappropriate here as dummy variables.) Also, we notice that if the numerator of a given term in the sum is x^k, then the corresponding denominator is $(k+1)!$. Thus, the given sum can be written

$$\frac{x}{2!} + \frac{x^2}{3!} + \frac{x^3}{4!} + \cdots + \frac{x^n}{(n+1)!} = \sum_{k=1}^{n} \frac{x^k}{(k+1)!}$$

Exercise Set 11.3

In Exercises 1–20, compute the first five terms in each sequence. (In Exercises 1–12 where the nth term is given by an explicit formula, for instance $a_n = \frac{n}{n+1}$, assume the equation holds for all natural numbers n.)

1. $a_n = \frac{n}{n+1}$

2. $a_n = n^3$

3. $a_n = (n-1)^2$

4. $b_n = n!$

5. $b_n = \left(1 + \frac{1}{n}\right)^n$

6. $b_n = \frac{n^n}{n!}$

7. $u_n = (-1)^n$

8. $a_n = (-1)^n \sqrt{n^n}$

9. $a_n = (-1)^{n+1}/n!$

10. $S_n = \frac{1}{(n+1)!}$

11. $a_n = 3n$

12. $b_n = 3$

13. $a_1 = 1$
$a_n = (1 + a_{n-1})^2 \quad n \geq 2$

14. $a_1 = 2$
$a_n = \sqrt{a_{n-1}^2 + 1} \quad n \geq 2$

15. $a_1 = 2$
$a_2 = 2$
$a_n = a_{n-1}a_{n-2} \quad n \geq 3$

16. $F_1 = 1$
$F_2 = 1$
$F_n = F_{n-1} + F_{n-2} \quad n \geq 3$

17. $a_1 = 1$
$a_{n+1} = na_n \quad n \geq 1$

18. $a_1 = 1$
$a_2 = 2$
$a_n = a_{n-1}/a_{n-2} \quad n \geq 3$

19. $a_1 = 0$
$a_n = 2^{a_{n-1}} \quad n \geq 2$

20. $a_1 = 0$
$a_2 = 1$
$a_n = \frac{a_{n-1} + a_{n-2}}{2} \quad n \geq 3$

In Exercises 21–26, find the sum of the first five terms in the sequence whose n*th term is as given.*

21. $a_n = 2^n$

22. $b_n = 2^{-n}$

23. $a_n = n^2 - n$

24. $b_n = (n - 1)!$

25. $a_n = (-1)^n/n!$

26. $a_1 = 1$

$$a_n = \frac{1}{n-1} - \frac{1}{n+1} \qquad n \geq 2$$

27. Find the sum of the first five terms of the sequence

$$a_1 = 1$$
$$a_2 = 2$$
$$a_n = a_{n-1}^2 + a_{n-2}^2 \qquad n \geq 3$$

28. Find the sum of the first four terms of the sequence

$$a_1 = 2$$
$$a_n = \frac{n}{a_{n-1}} \qquad n \geq 2$$

29. Find the sum of the first four terms of the sequence

$$a_1 = 2$$
$$a_n = (a_{n-1})^2 \qquad n \geq 2$$

In Exercises 30–41, express each of the sums without using sigma notation. Simplify your answers where possible.

30. $\sum_{k=1}^{3} (k - 1)$

31. $\sum_{k=1}^{5} k$

32. $\sum_{k=4}^{5} k^2$

33. $\sum_{k=2}^{6} (1 - 2k)$

34. $\sum_{n=1}^{3} x^n$

35. $\sum_{n=1}^{3} (n - 1)x^{n-2}$

36. $\sum_{n=1}^{4} \frac{1}{n}$

37. $\sum_{n=0}^{4} 3^n$

38. $\sum_{j=1}^{9} \log_{10} \frac{j}{j+1}$

39. $\sum_{j=2}^{5} \log_{10} j$

40. $\sum_{j=1}^{6} \left(\frac{1}{j} - \frac{1}{j+1}\right)$

41. $\sum_{j=1}^{5} (x^{j+1} - x^j)$

In Exercises 42–51, rewrite the sums using sigma notation.

42. $5 + 5^2 + 5^3 + 5^4$

43. $5 + 5^2 + 5^3 + \cdots + 5^n$

44. $x + x^2 + x^3 + x^4 + x^5 + x^6$

45. $x + 2x^2 + 3x^3 + 4x^4 + 5x^5 + 6x^6$

46. $\frac{1}{1} + \frac{1}{2} + \frac{1}{3} + \cdots + \frac{1}{12}$

47. $\frac{1}{1} + \frac{1}{2} + \frac{1}{3} + \cdots + \frac{1}{n}$

48. $2 - 2^2 + 2^3 - 2^4 + 2^5$

49. $\binom{10}{3} + \binom{10}{4} + \binom{10}{5} + \cdots + \binom{10}{10}$

50. $1 - 2 + 3 - 4 + 5$

51. $\frac{1}{2} - \frac{1}{4} + \frac{1}{8} - \frac{1}{16} + \frac{1}{32} - \frac{1}{64} + \frac{1}{128}$

***52.** The *Fibonacci numbers* $F_1, F_2, F_3, \ldots$ are defined as follows:

$$F_1 = 1$$
$$F_2 = 1$$
$$F_{n+2} = F_n + F_{n+1} \qquad \text{for } n \geq 1$$

(a) Verify that the first 10 Fibonacci numbers are as follows:

F_1	F_2	F_3	F_4	F_5	F_6	F_7	F_8	F_9	F_{10}
1	1	2	3	5	8	13	21	34	55

(b) Complete the following table.

n	$F_1 + F_2 + F_3 + \cdots + F_n$	$F_{n+2} - 1$
1		
2		
3		
4		
5		

(c) Use mathematical induction to prove that

$$F_1 + F_2 + F_3 + \cdots + F_n = F_{n+2} - 1$$

for all natural numbers n

(d) Use mathematical induction to prove that

$$F_n \geq n \qquad \text{for all natural numbers } n \geq 5$$

(e) Use mathematical induction to show that

$$F_1^2 + F_2^2 + F_3^2 + \cdots + F_n^2 = F_nF_{n+1}$$

for all natural numbers n

(f) Use mathematical induction to show that

$$F_{n+1}^2 = F_nF_{n+2} + (-1)^n$$

for all natural numbers n

Suggestion for Step 2: Add $F_{k+1}F_{k+2}$ to both sides of the equation in the induction hypothesis. Then factor F_{k+1} from the left-hand side and factor F_{k+2} from the first two terms on the right-hand side.

(g) This part of the exercise requires a knowledge of matrix multiplication. Show that

$$\begin{pmatrix} 1 & 1 \\ 1 & 0 \end{pmatrix}^n = \begin{pmatrix} F_{n+1} & F_n \\ F_n & F_{n-1} \end{pmatrix} \qquad \text{for } n \geq 2$$

11.4 Arithmetic Sequences and Series

One of the most natural ways to generate a sequence is to begin with a fixed number a and then repeatedly add a fixed constant d. This yields the sequence

$$a, a + d, a + 2d, a + 3d, \ldots$$

This sequence is called an *arithmetic sequence* or *arithmetic progression.* Notice that the difference between any two consecutive terms is the constant d. We call d the *common difference.* Here are several examples of arithmetic sequences.

Example A:	$1, 2, 3, \ldots$
Example B:	$3, 7, 11, 15, \ldots$
Example C:	$10, 5, 0, -5, \ldots$

In Example A, the first term is $a = 1$ and the common difference is $d = 1$. For Example B, we have $a = 3$. The value of d in this example is found by subtracting any two consecutive terms. Thus $d = 4$. Finally, in Example C we have $a = 10$ and $d = -5$. Notice that when the common difference is negative, the terms of the sequence decrease.

We wish to develop a formula for the nth term a_n in the arithmetic sequence

$$a, a + d, a + 2d, \ldots$$

To do this, notice that

$$\begin{aligned} &\text{the first term is } && a + 0d \\ &\text{the second term is } && a + 1d \\ &\text{the third term is } && a + 2d \end{aligned}$$

On the basis of these observations, it appears that the nth term a_n is given by $a_n = a + (n - 1)d$. Let us use mathematical induction to show that this formula is indeed valid for all natural numbers n. We let P_n denote the statement that $a_n = a + (n - 1)d$. Then P_1 asserts that

$$a_1 = a + (1 - 1)d$$

or

$$a_1 = a$$

This last equation is true since a denotes the first term. Next we assume that P_k is true and on the basis of this assumption we must show that P_{k+1} is true. Now P_k asserts that

$$a_k = a + (k - 1)d$$

Adding d to both sides of this equation yields

$$a_k + d = a + (k - 1)d + d$$

The left-hand side of this last equation is, by definition, a_{k+1}. The right-hand side simplifies to $a + kd$. Thus we have

$$a_{k+1} = a + kd$$

But this is just the statement P_{k+1}. We've therefore shown that P_{k+1} follows from P_k. This completes the induction proof. We summarize our result as follows.

nth Term of an Arithmetic Sequence

The nth term of an arithmetic sequence $a, a + d, a + 2d, \ldots$ is given by

$$a_n = a + (n - 1)d$$

Example 1 Determine the 100th term of the arithmetic sequence 7, 10, 13, 16, ...

Solution The first term is $a = 7$ and the common difference is $d = 3$. Substituting these values in the formula $a_n = a + (n - 1)d$ yields

$$\begin{aligned} a_n &= 7 + (n - 1)3 \\ &= 3n + 4 \end{aligned}$$

To find the 100th term now, we replace n by 100 in this last equation to obtain

$$\begin{aligned} a_{100} &= 3(100) + 4 \\ &= 304 \qquad \text{as required} \end{aligned}$$

Example 2 Determine the arithmetic sequence in which the second term is -2 and the eighth term is 40.

Solution We are given that the second term is -2. Using this information in the formula $a_n = a + (n - 1)d$, we have

$$\begin{aligned} -2 &= a + (2 - 1)d \\ &= a + d \end{aligned}$$

This gives us one equation in two unknowns. We are also given that the eighth term is 40. Therefore

$$\begin{aligned} 40 &= a + (8 - 1)d \\ &= a + 7d \end{aligned}$$

We now have a system of two equations in two unknowns:

$$\begin{cases} -2 = a + d \\ 40 = a + 7d \end{cases}$$

Subtracting the first equation from the second gives us

$$\begin{aligned} 42 &= 6d \\ 7 &= d \end{aligned}$$

To find a, we replace d by 7 in the first equation of the system. This yields

$$\begin{aligned} -2 &= a + 7 \\ -9 &= a \end{aligned}$$

We've now determined the sequence, since we know the first term is -9 and the common difference is 7. The first few terms of the sequence are $-9, -2, 5, 12, \ldots$.

Next we would like to derive a formula for the sum of the first n terms of an arithmetic sequence. Such a sum is referred to as an *arithmetic series*. If we use S_n to denote the required sum, we have

$$S_n = a + (a + d) + \cdots + [a + (n - 2)d] + [a + (n - 1)d] \quad (1)$$

Now of course we must obtain the same sum if we add the terms from right to left rather than left to right. That is, we must have

$$S_n = [a + (n - 1)d] + [a + (n - 2)d] + \cdots + (a + d) + a \quad (2)$$

Let us now add the two equations we have. Adding the left-hand sides is easy; we obtain $2S_n$. Now we add the corresponding terms on the right-hand sides. For the first terms we have

$$\underset{\substack{\uparrow \\ \text{first term} \\ \text{in equation (1)}}}{a} + \underbrace{[a + (n - 1)d]}_{\substack{\text{first term} \\ \text{in equation (2)}}} = 2a + (n - 1)d$$

Next we add the second terms.

$$\underset{\substack{\uparrow \\ \text{second term} \\ \text{in equation (1)}}}{(a + d)} + \underbrace{[a + (n - 2)d]}_{\substack{\text{second term} \\ \text{in equation (2)}}} = 2a + d + (n - 2)d = 2a + (n - 1)d$$

Notice that the sum of the second terms is again $2a + (n - 1)d$, the same quantity we arrived at with the first terms. As you can check, this pattern continues all the way through to the last terms. For instance

$$\underbrace{[a + (n - 1)d]}_{\substack{\text{last term} \\ \text{in equation (1)}}} + \underset{\substack{\uparrow \\ \text{last term} \\ \text{in equation (2)}}}{a} = 2a + (n - 1)d$$

We conclude from these observations that by adding the right-hand sides of equations (1) and (2) we obtain the quantity $[2a + (n - 1)d]$ n times. Therefore

$$2S_n = n[2a + (n - 1)d]$$

or

$$S_n = \frac{n}{2}[2a + (n - 1)d]$$

This gives us the desired formula for the sum of the first n terms in an arithmetic sequence. There is an alternate version of this formula, which now follows rather quickly. We have

$$\begin{aligned} S_n &= \frac{n}{2}[2a + (n - 1)d] \\ &= \frac{n}{2}\{a + [a + (n - 1)d]\} \\ &= \frac{n}{2}(a + a_n) \\ &= n\left(\frac{a + a_n}{2}\right) \end{aligned}$$

This last form of the formula is easy to remember. It says that the sum of an arithmetic series is obtained by averaging the first and last terms and then multiplying this average by n, the number of terms. For reference, we summarize both formulas as follows.

Formulas for the Sum of an Arithmetic Series

$$S_n = \frac{n}{2}[2a + (n - 1)d]$$

$$S_n = n\left(\frac{a + a_n}{2}\right)$$

Example 3 Find the sum of the first 30 terms of the arithmetic sequence 2, 6, 10, 14, ...

Solution We have $a = 2$, $d = 4$, and $n = 30$. Substituting these values in the formula $S_n = \frac{n}{2}[2a + (n - 1)d]$ then yields

$$\begin{aligned} S_{30} &= \frac{30}{2}[2(2) + (30 - 1)4] \\ &= 15[4 + 29(4)] \\ &= 1800 \qquad \text{(Check the arithmetic.)} \end{aligned}$$

Example 4 In a certain arithmetic sequence, the first term is 6 and the 40th term is 71. Find the sum of the first 40 terms and also the common difference for the sequence.

Solution We have $a = 6$ and $a_{40} = 71$. Using this data in the formula

$$S_n = n\left(\frac{a + a_n}{2}\right)$$

yields

$$\begin{aligned} S_{40} &= \frac{40}{2}(6 + 71) = 20(77) \\ &= 1540 \end{aligned}$$

The sum of the first 40 terms is thus 1540. The value of d can now be found by using the formula $S_n = \frac{n}{2}[2a + (n - 1)d]$. We obtain

$$\begin{aligned} 1540 &= \frac{40}{2}[2(6) + (40 - 1)d] \\ 1540 &= 20(12 + 39d) \\ 1540 &= 240 + 780d \\ 1300 &= 780d \end{aligned}$$

or

$$d = \frac{1300}{780} = \frac{5}{3}.$$

The required value of d is therefore $\frac{5}{3}$.

Example 5 Show that $\sum_{k=1}^{50} (3k - 2)$ represents an arithmetic series and compute the sum.

Solution There are two different ways to see that $\sum_{k=1}^{50} (3k - 2)$ is an arithmetic series. One way is to simply write out the first few terms and look at the pattern. We have

$$\sum_{k=1}^{50} (3k - 2) = 1 + 4 + 7 + 10 + \cdots + 148$$

From this it is clear that we are indeed summing the terms in an arithmetic sequence in which $d = 3$ and $a = 1$.

Another, more formal way to show that $\sum_{k=1}^{50} (3k - 2)$ represents an arithmetic series is to prove that the difference between successive terms in the indicated sum is a constant. Now the form of a typical term in this sum is $(3k - 2)$. Thus the form of the next term must be $[3(k + 1) - 2]$. The difference between these is then

$$\begin{aligned}[3(k + 1) - 2] - (3k - 2) &= 3k + 3 - 2 - 3k + 2 \\ &= 3\end{aligned}$$

The difference therefore is constant, as we wished to show.

To evaluate $\sum_{k=1}^{50} (3k - 2)$ we may use either of the two formulas for the sum of an arithmetic series. Using the formula $S_n = \frac{n}{2}[2a + (n - 1)d]$ we obtain

$$\begin{aligned}S_{50} &= \frac{50}{2}[2(1) + 49(3)] \\ &= 25(149) \\ &= 3725\end{aligned}$$

Thus the required sum is 3725. You should check for yourself that the same value is obtained using the formula $S_n = n\left(\frac{a + a_n}{2}\right)$.

Exercise Set 11.4

1. Find the common difference d for each of the following arithmetic sequences.
   (a) $1, 3, 5, 7, \ldots$
   (b) $10, 6, 2, -2, \ldots$
   (c) $\frac{2}{3}, 1, \frac{4}{3}, \frac{5}{3}, \ldots$
   (d) $1, 1 + \sqrt{2}, 1 + 2\sqrt{2}, 1 + 3\sqrt{2}, \ldots$
2. Which of the following are arithmetic sequences?
   (a) $2, 4, 8, 16, \ldots$
   (b) $5, 9, 13, 17, \ldots$
   (c) $3, \frac{11}{5}, \frac{7}{5}, \frac{3}{5}, \ldots$
   (d) $-1, -1, -1, -1, \ldots$
   (e) $-1, 1, -1, 1, \ldots$

In Exercises 3–8, find the indicated term in each sequence.

3. a_{12}; $10, 21, 32, 43, \ldots$
4. a_{20}; $7, 2, -3, -8, \ldots$
5. a_{100}; $6, 11, 16, 21, \ldots$
6. a_{30}; $\frac{2}{5}, \frac{4}{5}, \frac{6}{5}, \frac{8}{5}, \ldots$
7. a_{1000}; $-1, 0, 1, 2, \ldots$
8. a_{15}; $42, 1, -40, -81, \ldots$
9. The fourth term in an arithmetic sequence is -6 and the 10th term is 5. Find the common difference and the first term.

10. The fifth term in an arithmetic sequence is $\frac{1}{2}$ and the 20th term is $\frac{7}{8}$. Find the first three terms of the sequence.

11. The 60th term in an arithmetic sequence is 105 and the common difference is 5. Find the first term.

12. Find the common difference in an arithmetic sequence in which $a_{10} - a_{20} = 70$.

13. Find the common difference in an arithmetic sequence in which $a_{15} - a_7 = -1$.

14. Find the sum of the first 16 terms in the sequence 2, 11, 20, 29, ...

15. Find the sum of the first 1000 terms in the sequence 1, 2, 3, 4, ...

16. Find the sum of the first 50 terms in an arithmetic series whose first term is -8 and whose 50th term is 139.

17. Find the sum: $\frac{\pi}{3} + \frac{2\pi}{3} + \pi + \frac{4\pi}{3} + \cdots + \frac{13\pi}{3}$.

18. Find the sum: $\frac{1}{e} + \frac{3}{e} + \frac{5}{e} + \cdots + \frac{21}{e}$.

19. Determine the first term of an arithmetic sequence in which the common difference is 5 and the sum of the first 38 terms is 3534.

20. The sum of the first 12 terms in an arithmetic sequence is 156. What is the sum of the first and 12th terms?

21. In a certain arithmetic sequence, the first term is 4 and the 16th term is -100. Find the sum of the first 16 terms and also the common difference for the sequence.

22. The fifth and 50th terms of an arithmetic sequence are 3 and 30, respectively. Find the sum of the first 10 terms.

23. The eighth term in an arithmetic sequence is 5 and the sum of the first 10 terms is 20. Find the common difference and the first term of the sequence.

In Exercises 24–26, find the sum of the indicated arithmetic series.

24. $\sum_{i=1}^{10}(2i - 1) = 1 + 3 + 5 + \cdots + 19$

25. $\sum_{k=1}^{20}(4k + 3)$

26. $\sum_{n=5}^{100}(2n - 1)$

27. The sum of three consecutive terms in an arithmetic sequence is 30, and their product is 360. Find the three terms. *Suggestion:* Let x denote the *middle* term and d the common difference.

28. The sum of three consecutive terms in an arithmetic sequence is 21, and the sum of their squares is 197. Find the three terms.

29. The sum of three consecutive terms in an arithmetic sequence is 6, and the sum of their cubes is 132. Find the three terms.

30. In a certain arithmetic sequence $a = -4$ and $d = 6$. If $S_n = 570$, find n.

31. Let $a_1 = \frac{1}{1 + \sqrt{2}}$, $a_2 = -1$, and $a_3 = \frac{1}{1 - \sqrt{2}}$.

(a) Show that $a_2 - a_1 = a_3 - a_2$.

(b) Find the sum of the first six terms in the arithmetic sequence

$$\frac{1}{1 + \sqrt{2}}, -1, \frac{1}{1 - \sqrt{2}}, \ldots$$

***32.** Let b denote a positive constant. Find the sum of the first n terms in the sequence

$$\frac{1}{1 + \sqrt{b}}, \frac{1}{1 - b}, \frac{1}{1 - \sqrt{b}}, \ldots$$

***33.** The sum of the first n terms in a certain arithmetic sequence is given by $S_n = 3n^2 - n$. Show that the rth term is given by $a_r = 6r - 4$.

***34.** Let $a_1, a_2, a_3, \ldots$ be an arithmetic sequence, and let S_k denote the sum of the first k terms. If $S_n/S_m = n^2/m^2$, show that

$$\frac{a_n}{a_m} = \frac{2n - 1}{2m - 1}$$

***35.** If the common difference in an arithmetic sequence is twice the first term, show that

$$\frac{S_n}{S_m} = \frac{n^2}{m^2}$$

***36.** The lengths of the sides of a right triangle form three consecutive terms in an arithmetic sequence. Show that the triangle is similar to the 3-4-5 right triangle.

***37.** Suppose that $1/a$, $1/b$, and $1/c$ are three consecutive terms in an arithmetic sequence.

(a) Show that $\frac{a}{c} = \frac{a - b}{b - c}$.

(b) Show that $b = \frac{2ac}{a + c}$.

***38.** a, b, and c are three positive numbers with $a > c$. If $1/a$, $1/b$, and $1/c$ are consecutive terms in an arithmetic sequence, show that

$$\ln(a + c) + \ln(a - 2b + c) = 2\ln(a - c)$$

11.5 Geometric Sequences and Series

By a *geometric sequence* or *geometric progression* we mean a sequence of the form

$$a, ar, ar^2, ar^3, \ldots \qquad \text{where } a \text{ and } r \text{ are constants}$$

As you can see, each term in a geometric sequence after the first is obtained by multiplying the previous term by r. The number r is called the *common ratio* since the ratio of any term to the previous one is always r. For instance, the ratio of the fourth term to the third is $ar^3/ar^2 = r$. Here are two examples of geometric sequences.

$$1, \frac{1}{2}, \frac{1}{4}, \frac{1}{8}, \ldots$$

$$10, -100, 1000, -10000, \ldots$$

In the first example we have $a = 1$ and $r = \frac{1}{2}$, while in the second we have $a = 10$ and $r = -10$.

Example 1 In a certain geometric sequence, the first term is 2, the third term is 3, and the common ratio is negative. Find the second term.

Solution Let x denote the second term so that the sequence begins

$$2, x, 3, \ldots$$

By definition the ratios $3/x$ and $x/2$ must be equal. Thus we have

$$\frac{3}{x} = \frac{x}{2}$$

$$x^2 = 6$$

$$x = \pm\sqrt{6}$$

Now the second term must be negative since the first term is positive and the common ratio is negative. Thus the second term is $x = -\sqrt{6}$, as required.

n	a_n
1	ar^0
2	ar^1
3	ar^2
4	ar^3
⋮	⋮

It is easy to see what the formula for the nth term of a geometric sequence ought to be by considering the table at the left. The table indicates that the exponent on r is one less than the value of n in each case. On the basis of this observation it appears that the nth term must be given by $a_n = ar^{n-1}$. Indeed, it can be shown by mathematical induction that this formula does hold for all natural numbers n. (Exercise 30 asks you to carry out the proof.) For reference we summarize this result as follows.

nth Term of a Geometric Sequence

The nth term of the geometric sequence $a, ar, ar^2, \ldots$ is given by

$$a_n = ar^{n-1}$$

Example 2 Find the seventh term in the geometric sequence 2, 6, 18, ...

Solution The common ratio r can be found by dividing the second term by the first. Thus $r = 3$. Now using $a = 2$, $r = 3$, and $n = 7$ in the formula $a_n = ar^{n-1}$, we have

$$a_7 = 2(3)^6 = 2(729) = 1458$$

The seventh term of the sequence is therefore 1458.

Now suppose that we begin with a geometric sequence $a, ar, ar^2, \ldots$ in which $r \neq 1$. If we add the first n terms and denote the sum by S_n, we have

$$S_n = a + ar + ar^2 + \cdots + ar^{n-2} + ar^{n-1} \quad (1)$$

This sum is called a *finite geometric series*. We would like to find a formula for S_n. To do this, multiply equation (1) by r to obtain

$$rS_n = ar + ar^2 + ar^3 + \cdots + ar^{n-1} + ar^n \quad (2)$$

Now subtract equation (2) from equation (1). This yields (after combining like terms)

$$S_n - rS_n = a - ar^n$$

Thus

$$S_n(1 - r) = a(1 - r^n)$$

or

$$S_n = \frac{a(1 - r^n)}{1 - r} \qquad (r \neq 1)$$

This is the formula for the sum of a finite geometric series. For reference we summarize this result in the box that follows.

Formula for the Sum of a Geometric Series

Let S_n denote the sum $a + ar + ar^2 + \cdots + ar^{n-1}$, and assume that $r \neq 1$. Then

$$S_n = \frac{a(1 - r^n)}{1 - r}$$

Example 3 Evaluate the sum $\frac{1}{2^1} + \frac{1}{2^2} + \frac{1}{2^3} + \cdots + \frac{1}{2^{10}}$.

Solution This is a finite geometric series with $a = \frac{1}{2}$, $r = \frac{1}{2}$, and $n = 10$. Using these values in the formula for S_n that we've just derived yields

$$\begin{aligned} S_{10} &= \frac{\frac{1}{2}[1 - (\frac{1}{2})^{10}]}{1 - \frac{1}{2}} \\ &= \frac{\frac{1}{2}[1 - \frac{1}{1024}]}{\frac{1}{2}} = 1 - \frac{1}{1024} \\ &= \frac{1023}{1024} \quad \text{as required} \end{aligned}$$

We would now like to attach a meaning to certain expressions of the form

$$a + ar + ar^2 + \cdots$$

Such an expression is called an *infinite geometric series*. The three dots indicate (intuitively, at least) that the additions are to be carried on indefinitely, without end. To see how to proceed here, let us look at some examples involving finite geometric series. In particular we will consider the series

$$\frac{1}{2^1} + \frac{1}{2^2} + \frac{1}{2^3} + \cdots + \frac{1}{2^n}$$

n	$S_n = \frac{1}{2} + \frac{1}{2^2} + \cdots + \frac{1}{2^n}$
1	0.5
2	0.75
5	0.96875
10	0.999023437. . .
15	0.999969482. . .
20	0.999999046. . .
25	0.999999970. . .

Table 1

for fixed values of n. The idea then is to look for a pattern as n grows ever larger. Let $S_1 = \frac{1}{2}$, $S_2 = \frac{1}{2^1} + \frac{1}{2^2}$, $S_3 = \frac{1}{2^1} + \frac{1}{2^2} + \frac{1}{2^3}$, and in general

$$S_n = \frac{1}{2^1} + \frac{1}{2^2} + \cdots + \frac{1}{2^n}$$

Then we can compute S_n for any given value of n by means of the formula for the sum of a finite geometric series. From Table 1, which displays the results of such calculations, it seems clear that as n grows larger and larger, the value of S_n grows ever closer to 1. More precisely (but leaving the details for calculus) it can be shown that the value of S_n can be made as close to 1 as we please, provided only that n is sufficiently large. For this reason we shall say that the *sum of the infinite geometric series* $\frac{1}{2^1} + \frac{1}{2^2} + \frac{1}{2^3} + \cdots$ is 1. That is,

$$\frac{1}{2^1} + \frac{1}{2^2} + \frac{1}{2^3} + \cdots = 1$$

We can motivate this last result in another way. First compute the sum of the finite geometric series $\frac{1}{2^1} + \frac{1}{2^2} + \cdots + \frac{1}{2^n}$. As you can check, the result is

$$S_n = 1 - \left(\frac{1}{2}\right)^n$$

Now as n grows larger and larger, the value of $(\frac{1}{2})^n$ gets closer and closer to zero. Thus as n grows ever larger, the value of S_n will more and more resemble $1 - 0$, or 1.

Now let us repeat the above reasoning to obtain a formula for the sum of the infinite geometric series

$$a + ar + ar^2 + \cdots \qquad \text{where} \quad |r| < 1$$

First consider the finite geometric series

$$a + ar + ar^2 + \cdots + ar^{n-1}$$

The sum S_n here is

$$S_n = \frac{a(1 - r^n)}{1 - r}$$

We want to know how S_n behaves as n grows ever larger. This is where the assumption $|r| < 1$ is crucial. Just as $(\frac{1}{2})^n$ approaches zero as n grows larger and larger, so will r^n approach zero as n grows larger and larger. Thus as n grows ever larger, the sum S_n will more and more resemble

$$\frac{a(1-0)}{1-r} = \frac{a}{1-r}$$

For this reason we say that the sum of the infinite geometric series is $\frac{a}{1-r}$. We will make free use of this result in the subsequent examples. However, a more rigorous development of infinite series properly belongs to calculus.

Formula for the Sum of an Infinite Geometric Series

Suppose that $|r| < 1$. Then the sum S of the infinite geometric series $a + ar + ar^2 + \cdots$ is given by

$$S = \frac{a}{1-r}$$

Example 4 Find the sum of the infinite geometric series $1 + \frac{2}{3} + \frac{4}{9} + \cdots$.

Solution In this case we have $a = 1$ and $r = \frac{2}{3}$. Thus

$$\begin{aligned} S &= \frac{a}{1-r} \\ &= \frac{1}{1-\frac{2}{3}} = \frac{1}{\frac{1}{3}} = 3 \end{aligned}$$

The sum of the series is 3.

Example 5 Find a fraction equivalent to the repeating decimal $0.2\overline{35}$.

Solution Let $S = 0.2\overline{35}$. Then we have

$$S = 0.2353535\ldots$$

or

$$S = \frac{2}{10} + \frac{35}{1000} + \frac{35}{100{,}000} + \frac{35}{10{,}000{,}000} + \cdots$$

Now the expression following $\frac{2}{10}$ on the right-hand side of this last equation is an infinite geometric series in which $a = \frac{35}{1000}$ and $r = \frac{1}{100}$. Thus

$$\begin{aligned} S &= \frac{2}{10} + \frac{a}{1-r} \\ &= \frac{2}{10} + \frac{\frac{35}{1000}}{1-\frac{1}{100}} \\ &= \frac{2}{10} + \frac{\frac{35}{1000}}{\frac{99}{100}} = \frac{2}{10} + \left(\frac{35}{1000} \times \frac{100}{99}\right) \\ &= \frac{2}{10} + \frac{35}{990} = \frac{2(99) + 35}{990} = \frac{233}{990} \end{aligned}$$

The given decimal is therefore equivalent to $\frac{233}{990}$.

Exercise Set 11.5

1. Find the second term in a geometric sequence in which the first term is 9, the third term is 4, and the common ratio is positive.
2. Find the fifth term in a geometric sequence in which the fourth term is 4, the sixth term is 6, and the common ratio is negative.
3. The product of the first three terms in a geometric sequence is 8000. If the first term is 4, find the second and third terms.

In Exercises 4–8, find the indicated term for the geometric sequence given.

4. a_7; $9, 81, 729, \ldots$
5. a_{100}; $-1, 1, -1, 1, \ldots$
6. a_9; $\frac{1}{2}, \frac{1}{4}, \frac{1}{8}, \ldots$
7. a_8; $\frac{2}{3}, \frac{4}{9}, \frac{8}{27}, \ldots$
8. a_6; $1, -\sqrt{2}, 2, \ldots$
9. Find the common ratio in a geometric sequence in which the first term is 1 and the seventh term is 4096.
10. Find the first term in a geometric sequence in which the common ratio is $\frac{4}{3}$ and the 10th term is $\frac{16}{9}$.
11. Find the sum of the first 10 terms of the sequence 7, 14, 28, ...
12. Find the sum of the first five terms of the sequence $-\frac{1}{2}, \frac{3}{10}, -\frac{9}{50}, \ldots$.
13. Find the sum: $1 + \sqrt{2} + 2 + \cdots + 32$.
14. Find the sum of the first 12 terms in the sequence $-4, -2, -1, \ldots$.

In Exercises 15–17, evaluate each sum.

15. $\sum_{k=1}^{6} \left(\frac{3}{2}\right)^k$
16. $\sum_{k=1}^{6} \left(\frac{2}{3}\right)^{k+1}$
17. $\sum_{k=2}^{6} \left(\frac{1}{10}\right)^k$

In Exercises 18–22, determine the sum of each infinite geometric series.

18. $\frac{1}{4} + \frac{1}{4^2} + \frac{1}{4^3} + \cdots$
19. $\frac{2}{3} - \frac{4}{9} + \frac{8}{27} - \cdots$
20. $\frac{9}{10} + \frac{9}{100} + \frac{9}{1000} + \cdots$
21. $1 + \frac{1}{1.01} + \frac{1}{(1.01)^2} + \cdots$
22. $-1 - \frac{1}{\sqrt{2}} - \frac{1}{2} - \cdots$

In Exercises 23–27, express each repeating decimal as a fraction.

23. $0.555\ldots$
24. $0.\overline{47}$
25. $0.1\overline{23}$
26. $0.050505\ldots$
27. $0.\overline{432}$
28. The lengths of the sides in a right triangle form three consecutive terms of a geometric sequence. Find the common ratio of the sequence. (There are two distinct answers.)
29. The product of three consecutive terms in a geometric sequence is -1000, and their sum is 15. Find the common ratio. There are two answers. *Suggestion:* Denote the terms by a/r, a, and ar.
30. Use mathematical induction to prove that the nth term of the geometric sequence $a, ar, ar^2, \ldots$ is ar^{n-1}.
31. Show that the sum of the infinite geometric series
$$\frac{\sqrt{3}}{\sqrt{3}+1} + \frac{\sqrt{3}}{\sqrt{3}+3} + \cdots$$
is $\frac{3}{2}$.

*32. Let A_1 denote the area of an equilateral triangle, each side of which is one unit long. A second equilateral triangle is formed by joining the midpoints of the sides of the first triangle. Let A_2 denote the area of this triangle. This process is then repeated to form a third triangle with area A_3, and so on. Find the sum of the areas $A_1 + A_2 + A_3 + \cdots$.

*33. Let $a_1, a_2, a_3, \cdots$ be a geometric sequence with $r \neq 1$. Let $S = a_1 + a_2 + a_3 + \cdots + a_n$ and let $T = \frac{1}{a_1} + \frac{1}{a_2} + \cdots + \frac{1}{a_n}$. Show that $\frac{S}{T} = a_1 a_n$.

*34. a, b, and c are three consecutive terms in a geometric sequence. Show that $\frac{1}{a+b}, \frac{1}{2b}$, and $\frac{1}{c+b}$ are three consecutive terms in an arithmetic sequence.

11.6 Permutations and Combinations

Suppose that you flip a coin twice, keeping track of the result each time. How many possible outcomes are there for this experiment? Using H and T to denote heads and tails, respectively, we can list the possible outcomes this way:

$$HH \quad HT \quad TH \quad TT$$

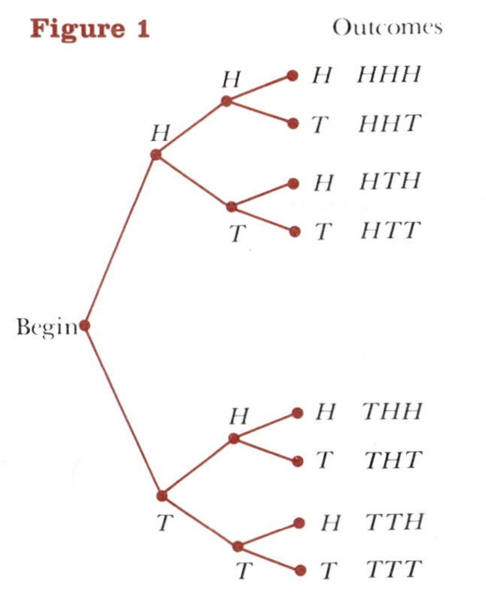

Figure 1

For instance, HH describes the experiment in which we obtain two heads in succession. Looking at this list then, we can say that there are four possible outcomes for the experiment.

Now let's ask the same question for an experiment in which we flip a coin three times in succession. How many possible outcomes are there? One way to organize the counting procedure uses the *tree diagram* in Figure 1. The tree diagram shows us every possibility at each stage in the experiment. Each path through the tree (from left to right) corresponds to a possible outcome for the experiment. For instance, the topmost path corresponds to the case in which we obtain three heads in succession (HHH). As you can see in Figure 1, there are eight possible outcomes for our experiment, corresponding to the eight branches at the right-hand side of the tree.

Example 1 There are four roads from town A to town B and three roads from town B to town C, as indicated schematically in Figure 2. Draw a tree diagram to find out how many routes there are from A to C via B.

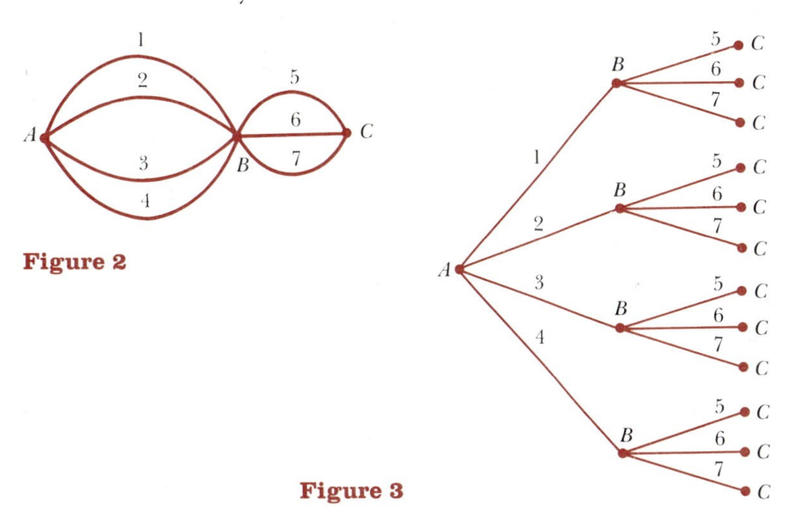

Figure 2

Figure 3

Solution Our tree diagram in Figure 3 has 12 branches in the right-hand column. (Count them!) So there are 12 routes from A to C via B.

Rather than actually counting all 12 branches in Figure 3, we could have reasoned as follows. There are four possible ways to go from A to B, and for each of these ways, there are then three ways to go from B to C. Thus there are 4 *times* 3, or 12, ways to go from A to C. The *Multiplication Principle*, which we now state, formalizes this idea.

The Multiplication Principle

Suppose that a task involves two consecutive operations. If the first operation can be performed in M ways, and the second operation can be performed in N ways, then the entire task can be performed in MN ways.

Example 2 How many different two-digit numbers can be formed using the digits 1, 3, 5, 7, and 9?

(a) If no digit is repeated (within a given number)?

(b) If repetitions are allowed?

Solution

(a) We can think of the task of forming a two-digit number as composed of two consecutive operations: First choose a tens' digit, then choose a units' digit. There are five choices for the tens' digit. Then, after choosing a tens' digit (and assuming that no repetitions are allowed), there will be four choices left for the units' digit. It follows now from the Multiplication Principle that there are 5×4 or 20 possible choices. That is, there are 20 different two-digit numbers that can be formed.

(b) With repetitions allowed, there are five choices for the tens' digit and five choices for the units' digit. Therefore (using the Multiplication Principle again) there are 5×5 or 25 different two-digit numbers that can be formed.

The Multiplication Principle can be extended to tasks involving three or more consecutive operations. This idea is used in the next example.

Example 3 You buy three books at a book sale. One by Hemingway, one by Joyce, and one by Fitzgerald. How many ways are there to arrange these on a (single) bookshelf?

Solution There are three positions we have to fill on the shelf: left, middle, and right. Starting with the left position, there are three choices. Then, once we make a choice for that left position, there are two choices left for the middle position. And finally, after choosing a book for that middle position, we'll have but one choice for the right position. Thus, according to the Multiplication Principle, there are $3 \times 2 \times 1$ or six ways to arrange the three books.

Let's list the six possible arrangements from Example 3. Using H for Hemingway, J for Joyce, and F for Fitzgerald, the six arrangements are:

$$\begin{array}{ccc} HJF & JHF & FJH \\ HFJ & JFH & FHJ \end{array}$$

Each of the six arrangements here is called a *permutation* of the set $\{H, J, F\}$. In general, by a *permutation* of a set of n distinct objects we just mean a listing or arrangement of those objects in a definite order. For instance, if we take the set $\{a, b, c, d\}$, then two (of the many possible) permutations of this set are *abcd* and *dcba*.

Example 4 How many permutations are there of the set $\{a, b, c, d\}$?

Solution To form a permutation, we have to assign a letter to each of four positions:

____ ____ ____ ____

We have four choices for the first position. Then, after filling that position, we'll have three letters left from which to choose for the second position. After carrying that out, there will be only two letters from which to choose for the third position. And finally, with the first three positions filled, there will be but one letter left for the fourth position. In summary we have

____ ____ ____ ____

4 choices ↗ 3 choices ↑ 2 choices ↖ 1 choice

Therefore (by the Multiplication Principle) the total number of permutations or arrangements is $4 \times 3 \times 2 \times 1$, or 24. Notice that this is $4!$. (Recall from Section 11.2 that $n!$ stands for the product of the first n natural numbers.)

In the example just concluded we saw that there were $4!$ permutations of a set with four elements. By following the same kind of reasoning (and using mathematical induction), it can be shown that there are $n!$ permutations of a set with n elements. For reference we state this fact in the box that follows.

> The number of permutations of a set with n elements is $n!$.

Example 5 A regular deck of playing cards contain 52 cards. How many arrangements (permutations) of the deck are there?

Solution The number of permutations of a set with 52 elements is $52!$. With the aid of a calculator we find

$$52! \approx 8 \times 10^{67}$$

We've said that a permutation of a set with n elements is a listing or arrangement of all n of those elements in some definite order. Sometimes, though, we want to consider arrangements that use less than all n of the elements. For example, let's take the set $\{H, J, F\}$ that we used earlier, and this time list the possible arrangements using only two elements at a time. We have

$$\begin{array}{ccc} HJ & JH & FH \\ HF & JF & FJ \end{array}$$

Each arrangement here is called a *permutation of three objects taken two at a time.* The number of such permutations is denoted by $P(3, 2)$. So we have $P(3, 2) = 6$, since there are six arrangements. More generally, we have the following definition.

> **Definition**
>
> By a *permutation of n objects taken r at a time,* we mean a listing or an arrangement of r of the objects in a definite order, where $r \leq n$. The number of such arrangements is denoted by
>
> $$P(n, r)$$

Example 6 Using the set $\{a, b, c, d\}$, list the permutations of these four letters taken two at a time. Evaluate $P(4, 2)$.

Solution

Using a and b:	ab, ba	Using b and c:	bc, cb
Using a and c:	ac, ca	Using b and d:	bd, db
Using a and d:	ad, da	Using c and d:	cd, dc

We've listed 12 permutations of the four given letters taken two at a time. (Convince yourself that we've listed all of the possibilities.) So $P(4, 2) = 12$.

We could have evaluated $P(4, 2)$ in Example 6 by reasoning as follows. To form a permutation of four letters taken two at a time, we have to fill two positions:

There are four choices for the first letter. Then, after making a choice, we're left with three choices for the second letter. Therefore (according to the Multiplication Principle) there are 4×3 or 12 possible choices, in agreement with the result in Example 6. By using this type of reasoning, we can obtain the following general formula for the number of permutations of n objects taken r at a time.

$$P(n, r) = n(n-1)\cdots(n-r+1)$$

Example 7 **(a)** Evaluate $P(8, 5)$, the number of permutations of eight objects taken five at a time.

(b) Evaluate $P(8, 8)$.

Solution **(a)** In the formula for $P(n, r)$, let $n = 8$ and $r = 5$. This yields

$$\begin{aligned} P(8, 5) &= 8(7)\cdots(8-5+1) \\ &= 8(7)(6)(5)(4) = 6720 \end{aligned}$$

So there are 6720 permutations of eight objects taken five at a time.

(b)

$$\begin{aligned} P(8, 8) &= 8(7)\cdots(8-8+1) \\ &= 8(7)\cdots(1) = 8! = 40{,}320 \end{aligned}$$

So there are 8! or 40,320 permutations of eight objects taken eight at a time. (Note that a permutation of eight objects taken eight at a time is just a permutation of those eight objects.)

Example 8 From a set of 10 pictures, five are to be selected and hung in a row. How many arrangements are possible?

Solution We want the number of permutations of 10 objects taken five at a time. This number is

$$\begin{aligned} P(10, 5) &= (10)(9)\cdots(10-5+1) \\ &= (10)(9)(8)(7)(6) \\ &= 30{,}240 \end{aligned}$$

There are, therefore, 30,240 possible arrangements.

Example 9 How many possible arrangements are there of the letters $\{a, b, c, d, e\}$ in which a and b are next to each other?

Solution First think of a and b as a single object and count the number of arrangements of the *four* objects (ab), c, d, and e. This number is

$$\begin{aligned} P(4, 4) &= 4! \qquad \text{(Why?)} \\ &= 4(3)(2)(1) = 24 \end{aligned}$$

Now, in each of these 24 arrangements, the letters a and b can appear in one of two ways, either ab or ba. Consequently (by the Multiplication Principle), the number of possible arrangements of $\{a, b, c, d, e\}$ in which a and b are adjacent is 24×2, or 48.

Example 10 How many possible arrangements are there of the letters $\{a, b, c, d, e\}$ in which a and b are not next to each other?

Solution In this type of counting problem, the direct approach is not the simplest. Instead, we reason as follows. The number of arrangements in which a and b are not adjacent equals the total number of possible arrangements of the five letters, minus the number of arrangements of the five letters in which a and b *are* adjacent. So (using the result in Example 9), the required number of arrangements is

$$P(5, 5) - 48 = 5(4)(3)(2)(1) - 48$$
$$= 120 - 48 = 72$$

Therefore there are 72 ways to arrange the five given letters such that a and b are not adjacent.

In working with permutations we've seen that the order in which the objects are listed is an essential factor. Now we want to introduce the idea of a *combination*, in which the order of the objects is not a consideration. Our work will be based on the following definition.

Definition

Suppose that r objects are selected from a set of n objects, where $r \leq n$. Then this selection or subset of r objects is called a *combination of the n objects taken r at a time*. The number of such combinations is denoted by

$$C(n, r)$$

To illustrate this definition, let's take the set $\{H, J, F\}$. As we saw earlier, there are six permutations of these three letters taken two at a time:

Permutations involving H and J:	HJ, JH
Permutations involving J and F:	JF, FJ
Permutations involving H and F:	HF, FH

Now the two permutations HJ and JH use exactly the same letters. So, corresponding to these two permutations, we have the single combination or set of letters $\{H, J\}$. Also, whether we write this set as $\{H, J\}$ or $\{J, H\}$, it's still the same combination of letters. Similarly, corresponding to the two permutations JF and FJ we have the single combination $\{J, F\}$. And finally, corresponding to the two permutations HF and FH, we have the combination $\{H, F\}$. Thus, although we have six permutations of the three letters taken two at a time, there are only three combinations of the letters taken two at a time. We can summarize these results by writing

$$P(3, 2) = 6 \quad \text{and} \quad C(3, 2) = 3$$

Example 11 Using the set $\{1, 2, 3, 4\}$, list both the combinations and the permutations of these four numbers taken three at a time. Evaluate $C(4, 3)$ and $P(4, 3)$.

Solution

Combination	*Permutations Generated from the Combination*					
$\{1, 2, 3\}$	123	132	213	231	312	321
$\{1, 2, 4\}$	124	142	241	214	412	421
$\{1, 3, 4\}$	134	143	341	314	413	431
$\{2, 3, 4\}$	234	243	342	324	423	432

We've found four combinations. (Convince yourself that there are no other distinct combinations using three of the four given digits.) Thus $C(4, 3) = 4$. Next, let's evaluate $P(4, 3)$. Assuming that the listing we've made is complete, a direct count shows that $P(4, 3) = 24$. You can check this result for yourself using the formula for $P(n, r)$ that was given earlier. Actually, there is yet another way to calculate $P(4, 3)$, and we'll show it here because it will help you to follow the reasoning in the next paragraph. Each of our combinations contains three numbers. So, corresponding to each combination there are 3! or six permutations. Thus, by the Multiplication Principle, we have $P(4, 3) = 6 \times C(4, 3) = 6 \times 4 = 24$.

We can derive a formula for $C(n, r)$ as follows. First of all, the number of permutations of n objects taken r at a time is

$$P(n, r) = n(n - 1) \cdots (n - r + 1)$$

We can think of the task of forming any one of these permutations as composed of two consecutive operations:

1. Select r of the n objects.
2. Then arrange these r objects in some definite order.

Since operation 1 can be performed in $C(n, r)$ ways, and then operation 2 can be performed in $r!$ ways, the Multiplication Principle tells us that

$$P(n, r) = C(n, r) \cdot r!$$

Therefore

$$C(n, r) = \frac{P(n, r)}{r!} = \frac{n(n - 1) \cdots (n - r + 1)}{r!}$$

We can simplify this last fraction. The technique is to multiply both the numerator and the denominator by the quantity $(n - r)!$. This gives us

$$C(n, r) = \frac{n(n - 1) \cdots (n - r + 1)[(n - r)!]}{r![(n - r)!]}$$

The numerator of this last fraction is just $n!$, so we have

$$C(n, r) = \frac{n!}{r!(n - r)!}$$

Do you recognize the fraction on the right-hand side of this last equation? It's the binomial coefficient $\binom{n}{r}$ that we defined in Section 11.2. In summary then, our formula for $C(n, r)$ is as follows.

$$C(n, r) = \binom{n}{r} = \frac{n!}{r!(n-r)!}$$

Example 12 There are 100 senators in the United States Senate. How many four-member committees are possible?

Solution Since we are interested in the members of a committee rather than the order in which they are chosen or seated, we use combinations rather than permutations. We want the number of combinations of the 100 senators taken 4 at a time. This number is

$$\begin{aligned} C(100, 4) = \binom{100}{4} &= \frac{100!}{4!(100-4)!} \\ &= \frac{100(99)(98)(97)(96!)}{4!(96!)} \\ &= \frac{100(99)(98)(97)}{4!} \\ &= 3{,}921{,}225 \qquad \text{(using a calculator)} \end{aligned}$$

So there are 3,921,225 possible four-member committees.

Example 13 Figure 4 shows a decagon (a ten-sided figure) with all its diagonals. How many diagonals are there?

Solution Each pair of vertices determines either a diagonal or a side of the decagon. So the total number of diagonals and sides is the same as the number of pairs of vertices. This latter number is $C(10, 2)$, the number of combinations of the ten vertices taken two at a time. Evaluating $C(10, 2)$, we have

$$C(10, 2) = \binom{10}{2} = \frac{10!}{2!(10-2)!} = \frac{10(9)(8!)}{2(8!)} = 45$$

Thus the total number of sides and diagonals is 45. Since 10 of these are sides, we're left with 45–10, or 35 diagonals. (For a different approach to this problem, see Exercise 53.)

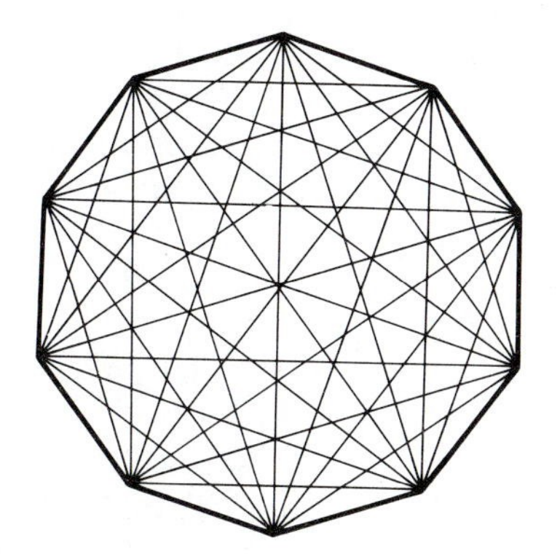

Figure 4

Example 14 On a certain committee there are eight Republicans and seven Democrats. How many possibilities are there for a five-member subcommittee that consists of three Republicans and two Democrats?

Solution Think of the task of forming a five-member subcommittee as composed of two consecutive operations: First choose three Republicans; then choose two Democrats. The number of ways to choose three Republicans is $C(8, 3)$, and the number of ways to choose two Democrats is $C(7, 2)$. Therefore, according to the Multiplication Principle, the total number of subcommittees here is

$$\begin{aligned} C(8, 3) \cdot C(7, 2) &= \binom{8}{3} \cdot \binom{7}{2} \\ &= \frac{8!}{3!(8-3)!} \cdot \frac{7!}{2!(7-2)!} \\ &= \frac{8(7)(6)}{3(2)} \cdot \frac{7(6)}{2} = 1176 \qquad \text{(Check the arithmetic.)} \end{aligned}$$

So there are 1176 possible subcommittees consisting of three Republicans and two Democrats.

Exercise Set 11.6

1. There are three roads from town A to town B and two roads from town B to town C. By drawing a tree diagram, find out how many routes there are from A to C via B. (Check your answer using the Multiplication Principle.)
2. There are five roads from town A to town B and three roads from town B to town C. Draw a tree diagram to find out how many routes there are from A to C via B. (Check your answer using the Multiplication Principle.)
3. There are three roads from town A to town B, two roads from town B to town C, and two roads from town C to town D. Draw a tree diagram to find out how many routes there are from town A to town D via B and C.
4. At the beginning of this section we noted that if a coin is flipped twice, there are four possible outcomes. Draw a tree diagram illustrating this.
5. How many different two-digit numbers can be formed using the digits 2, 4, 6, and 8 if no digit is repeated (within a given number)?
6. Follow Exercise 5, but allow repetitions.
7. How many three-digit numbers can be formed using the digits 1, 2, 3, 4, and 5 if no digit can be repeated (within a given number)?
8. How many two-digit numbers that are even can be formed using the digits 2, 4, 5, 7, and 9 if no digit is repeated (within a given number)?
9. How many three-digit numbers that are less than 600 can be formed using the digits 3, 4, 5, 6, 7, and 8 if repetitions are allowed?
10. In a certain state, the license plates contain three digits followed by three letters. (Repetitions of digits or letters on a given plate are allowed.) How many different license plates are possible?
11. Follow Exercise 10, but assume that repetitions are not allowed.
12. In how many ways can five books be arranged on a bookshelf?
13. In how many ways can 10 books be arranged on a bookshelf?
14. How many permutations are there of the set $\{a, e, i, o, u\}$?
15. If the number of permutations of a certain set of letters is 24, how many letters must there be in the set?
16. Five framed photographs are going to be hung on the wall in one row. If one particular photograph must always be in the middle, how many arrangements are possible?

For Exercises 17–20, evaluate each expression.

17. $P(8, 1)$
18. $P(8, 2)$
19. $P(6, 3)$
20. $P(11, 3)$
21. Show that $P(n, n-1) = n!$.

22. Show that $P(n, 1) = n$.

23. Which is larger, $P(100, 50)$ or $P(101, 51)$?

24. Which is larger, $P(100, 50)$ or $P(101, 52)$?

25. How many possible arrangements are there of the six letters $\{a, b, c, d, e, f\}$ in which b and c are next to each other? (Use the method shown in Example 9.)

26. How many arrangements are there of the six letters $\{a, b, c, d, e, f\}$ in which b and c are not next to each other? (Use the method shown in Example 10.)

27. How many four-digit numbers can be formed using the digits $\{2, 7, 1, 8\}$ such that 7 and 1 are next to each other? (No repetitions.)

28. How many four-digit numbers can be formed using the digits $\{2, 7, 1, 8\}$ such that 7 and 1 are not next to each other? (No repetitions.)

29. How many three-digit numbers can be formed from the ten digits $\{0, 1, 2, \ldots, 9\}$
   (a) if repetitions are allowed? (Don't forget, the first digit cannot be 0.)
   (b) if repetitions are not allowed?
   (c) if repetitions are not allowed and the numbers formed are even?

30. How many three-digit numbers can be formed using the digits $\{1, 2, 3, 4, 5, 6\}$
   (a) if repetitions are not allowed and each number ends with 32?
   (b) if repetitions are not allowed and the number formed in each case is odd?
   (c) if repetitions are not allowed and each number formed is divisible by 5?

31. How many three-letter code words can be formed using the letters in the word *equations*
   (a) if repetitions are allowed?
   (b) if repetitions are not allowed?
   (c) if repetitions are not allowed and each code word begins with q?
   (d) if repetitions are not allowed and each code word begins with a vowel and ends with a consonant?

32. Using the set $\{a, b, c, d\}$, list both the combinations and the permutations of these four letters taken two at a time. Evaluate $C(4, 2)$ and $P(4, 2)$.

For Exercises 33–38, evaluate each expression.

33. **(a)** $C(12, 11)$ **(b)** $C(12, 1)$

34. **(a)** $C(15, 14)$ **(b)** $C(15, 1)$

35. **(a)** $C(8, 3)$ **(b)** $C(8, 5)$

36. **(a)** $C(10, 8)$ **(b)** $C(10, 2)$

37. **(a)** $C(12, 5)$ **(b)** $C(12, 7)$

38. **(a)** $C(15, 13)$ **(b)** $C(15, 2)$

39. Simplify $C(n, n - 2)$.

40. Simplify $C(n, 1)$.

41. There are eight people on a certain committee. How many three-member subcommittees are possible?

42. How many five-member subcommittees are possible from a committee with 10 members?

43. There are 10 people in a certain club.
   (a) How many different six-member committees are possible?
   (b) Of the 10 people in the club, suppose that four are men and six are women. How many six-member committees are possible if each committee must be composed of three men and three women?

44. Suppose that you are one of the 12 people in a certain club.
   (a) How many five-member committees can be formed that include you?
   (b) How many five-member committees can be formed that exclude you?

45. Use the method shown in Example 13 to find out how many diagonals can be drawn in a hexagon. (A hexagon is a six-sided polygon.)

46. Use the method shown in Example 13 to count the number of diagonals that can be drawn in a regular 14-sided polygon.

47. Suppose that we are given eight points in a plane, and no three of the points lie on the same straight line. How many different triangles can be formed in which the vertices are chosen from these eight points?

48. Suppose that we are given 12 points lying on a certain circle. How many triangles can be formed if the vertices are chosen from the 12 points?

49. You have one penny, one nickel, one dime, and one quarter. How many different sums of money can you form using one or more of these coins? *Hint:* You need to compute $C(4, 1) + C(4, 2) + C(4, 3) + C(4, 4)$.

50. You have a \$1 bill, a \$5 bill, a \$10 bill, a \$20 bill, and a \$100 bill. How many different sums of money can you form using one or more of these bills? (Study the hint in the previous exercise.)

51. A group consists of nine Republicans and eight Democrats. How many different eight-member committees can be formed that contain
   (a) six Republicans and two Democrats?
   (b) at least six Republicans?

52. **(a)** From a group of 12 people, a committee with eight members will be formed. Then, from these eight, a subcommittee with three people will be chosen. How many ways are there to carry out these two consecutive operations?
   (b) From a group of 12 people, a committee of three will be chosen. Later, five additional people are picked to increase the committee size to eight. How many ways are there to carry out these two consecutive operations?

(c) Use parts (a) and (b) to explain why $\binom{12}{8}\binom{8}{3} = \binom{12}{3}\binom{9}{5}$.

(d) Verify the equality in part (c) by direct calculation (involving ratios of factorials).

(e) Show that $\binom{n}{k}\binom{k}{r} = \binom{n}{r}\binom{n-r}{k-r}$ where $n \geq k \geq r$. *Suggestion:* You can do this either by direct calculation or by repeating the reasoning used in parts (a), (b), and (c) with the quantities 12, 8, and 3 replaced by n, k, and r, respectively.

53. In Example 13 we computed the number of diagonals in a decagon by considering combinations. Here is a different way to approach the problem. First observe that there are seven diagonals emanating from each of the 10 vertices. Why does this imply now that the total number of diagonals is $\frac{7 \cdot 10}{2} = 35$?

11.7 Introduction to Probability

Laplace, a famous contributor to this branch of mathematics, once wrote that "the most important questions of life are, for the most part, really only questions of probability." At one time mathematicians had serious misgivings about probability theory; the logical foundations seemed flimsy. But after they learned how to resolve some confusing paradoxes, probability theory developed rapidly until now it is one of the chief areas of contact between mathematics and the world as a whole.

From *The Mathematical Sciences*, edited by the Committee on Support of Research in the Mathematical Sciences with the collaboration of George Boehm (Cambridge, MA: The M.I.T. Press, 1969).

In order to define the mathematical notion of probability, we first introduce the idea of a *sample space* for an experiment. As a particularly simple first example, we can think of flipping a coin as an experiment in which there are two possible *outcomes:* heads (H) or tails (T). The set $\{H, T\}$ consisting of all possible outcomes for this experiment is called a *sample space.* Intuitively at least, we believe that each outcome in this sample space is equally likely to occur.

As another example of this terminology, let's suppose that an experiment consists of tossing a single die and then noting the number of dots showing on the top face. Then a sample space for this experiment is the set

$$\{1, 2, 3, 4, 5, 6\}$$

Here, as in the previous example, our sample space consists of all possible outcomes for the experiment. And again, intuitively at least, we believe that each outcome is equally likely to occur. In this section, all the sample spaces that we consider are finite; that is, they contain finitely many outcomes.

Sometimes we wish to focus on a particular outcome, or a particular set of outcomes, in an experiment. For instance, in tossing the die we might be interested in whether the die shows an even number of dots. In this context, we refer to the set $\{2, 4, 6\}$ as an *event* in the sample space $\{1, 2, 3, 4, 5, 6\}$. More generally, any subset of a given sample space is referred to as an event. We shall use the notation $n(E)$ to denote the number of outcomes in an event E. For example, if E is the event "obtaining an even number" that we've just discussed, then $n(E) = 3$, because there are three outcomes in the set $E = \{2, 4, 6\}$.

Example 1 Use set notation to specify each of the following events in the die-tossing experiment using the sample space {1, 2, 3, 4, 5, 6}.

Event A: obtaining an odd number
Event B: obtaining a number less than 3
Event C: obtaining a 6
Event D: obtaining a number less than 7
Event E: obtaining a 7

Solution

Event A: {1, 3, 5}
Event B: {1, 2}
Event C: {6}
Event D: {1, 2, 3, 4, 5, 6}
Event E: $\emptyset$

Two of the events we've listed here deserve special mention. First, notice that Event D is the entire sample space. This is simply because every outcome in the sample space is a number less than 7. At the other extreme we have Event E, "obtaining a 7." There are no outcomes in the sample space corresponding to this event. So (remembering that an event is a subset of the sample space), we say that Event E is the so-called *empty set*, that is, the set with no elements. By convention, the empty set is denoted by the symbol $\emptyset$.

Example 2 Using the events A, B, C, D, and E in Example 1, compute $n(A)$, $n(B)$, $n(C)$, $n(D)$, and $n(E)$.

Solution We need only count the number of outcomes in each set. The results are

$$n(A) = 3 \qquad n(D) = 6$$
$$n(B) = 2 \qquad n(E) = 0$$
$$n(C) = 1$$

For each event E in a finite sample space we are going to define a number $P(E)$ called the *probability* of the event. Informally speaking, the number $P(E)$ is a rating of how likely it is that the event E will occur in any trial of the experiment. As you'll see, the number $P(E)$ always satisfies the condition $0 \le P(E) \le 1$.

Definition: Probability of an Event

Let S be a finite sample space in which each outcome is equally likely to occur. If E is an event in S, then the probability of E, denoted by $P(E)$, is defined by

$$P(E) = \frac{n(E)}{n(S)} = \frac{\text{number of outcomes in } E}{\text{number of outcomes in } S}$$

Example 3 Find the probability for each of the events A, B, C, D, and E discussed in Examples 1 and 2.

Solution

$$P(A) = P(\text{obtaining an odd number}) = \frac{n(A)}{n(S)} = \frac{3}{6} = \frac{1}{2}$$

$$P(B) = P(\text{obtaining a number less than 3}) = \frac{n(B)}{n(S)} = \frac{2}{6} = \frac{1}{3}$$

$$P(C) = P(\text{obtaining a 6}) = \frac{n(C)}{n(S)} = \frac{1}{6}$$

$$P(D) = P(\text{obtaining less than 7}) = \frac{n(D)}{n(S)} = \frac{6}{6} = 1$$

$$P(E) = P(\text{obtaining a 7}) = \frac{n(E)}{n(S)} = \frac{0}{6} = 0$$

Notice that each of the probabilities that we computed in Example 3 is a number between 0 and 1, as mentioned previously. Also note that a probability of 0 is assigned to the "impossible" event, whereas a probability of 1 is assigned to the "certain" event.

Example 4 Two (distinct) letters are selected at random from the set $\{a, b, c, d, e\}$. What is the probability that one of the letters is c?

Solution Let the sample space S consist of all combinations of the five given letters taken two at a time. For instance, *ab, ac, ad,* and *ae* are four of the outcomes in this sample space. We needn't list every outcome in S in order to find $n(S)$; for, from our work in the previous section, we know that

$$n(S) = C(5, 2) = \frac{5!}{2!\,3!} = 10$$

Now let E be the event "c is selected." We can specify E as follows:

$$E = \{ac, bc, dc, ec\}$$

Thus $n(E) = 4$, and consequently

$$P(E) = \frac{n(E)}{n(S)} = \frac{4}{10} = \frac{2}{5}$$

So the probability that one of the two letters selected is c is $\frac{2}{5}$. This is also expressed by saying that there is a 40% chance that one of the two letters is c.

Example 5 Suppose that five faulty transistors are accidentally packaged with 10 reliable transistors. What is the probability that three transistors chosen at random from this package all will be reliable?

Solution Let the sample space S consist of the combinations of the 15 transistors taken three at a time. Then we have

$$n(S) = C(15, 3) = \frac{15!}{3!\,12!} = 455$$

Next let E denote the event "choose three reliable transistors." Since these three must be chosen from the 10 reliable ones, we have

$$n(E) = C(10, 3) = \frac{10!}{3!\,7!} = 120$$

Thus

$$P(E) = \frac{n(E)}{n(S)} = \frac{120}{455} = \frac{24}{91} \qquad \text{as required}$$

To continue our development of probability a bit further, it will be convenient to introduce (or review, if you've seen this before) two basic ideas from set theory. Suppose that A and B are two sets. Then the *union* of these two sets, written $A \cup B$, is defined to be the set of elements that are in A *or* in B (or in both). For example, if

$$A = \{2, 3, 4, 5\} \quad \text{and} \quad B = \{4, 5, 6, 7\}$$

then

$$A \cup B = \{2, 3, 4, 5, 6, 7\}$$

The *intersection* of two sets A and B, written $A \cap B$, is defined as the set of elements that belong to both A *and* B. For example, using A and B as defined above, we have

$$A \cap B = \{4, 5\}$$

because 4 and 5 are the only elements common to both sets. As another example, if $E = \{1, 2\}$ and $F = \{3, 4\}$, then $E \cap F = \emptyset$, because there are no elements common to the two sets. Two sets that have no elements in common are said to be *disjoint.*

If A and B are events (remember, that means subsets of a sample space), then $A \cup B$ is the event "A or B," whereas $A \cap B$ is the event "A and B." Here's an example. Let S be the sample space $\{1, 2, 3, 4, 5, 6\}$ that we've used for the die-tossing experiment. And let the events A and B be as follows:

A: obtaining an even number

B: obtaining a number greater than four

Then we have

$$\begin{aligned} A \cup B &= \text{“obtaining an even number”} \\ &\quad \textit{or} \text{ “obtaining a number greater than four”} \\ &= \{2, 4, 6\} \cup \{5, 6\} \\ &= \{2, 4, 5, 6\} \end{aligned}$$

whereas

$$\begin{aligned} A \cap B &= \text{“obtaining an even number”} \\ &\quad \textit{and} \text{ “obtaining a number greater than four”} \\ &= \{2, 4, 6\} \cap \{5, 6\} \\ &= \{6\} \end{aligned}$$

Example 6 A coin is flipped three times and the result is recorded each time. Let A, B, and C denote the following events.

A: obtaining exactly two heads

B: obtaining no heads

C: obtaining more than one tail

Compute the following probabilities:

(a) $P(A)$, $P(B)$, and $P(C)$

(b) $P(A \text{ or } B)$

(c) $P(B \text{ or } C)$

Solution Our sample space S will be as follows. (See Figure 1 in Section 11.6 if you're not certain how we've obtained the eight outcomes in S.)

$$S = \{HHH, HHT, HTH, HTT, THH, THT, TTH, TTT\}$$

(a) Using the definitions given for A, B, and C, we have

$$A = \{HHT, HTH, THH\}$$
$$B = \{TTT\}$$
$$C = \{HTT, THT, TTH, TTT\}$$

Thus

$$P(A) = \frac{n(A)}{n(S)} = \frac{3}{8}$$
$$P(B) = \frac{n(B)}{n(S)} = \frac{1}{8}$$
$$P(C) = \frac{n(C)}{n(S)} = \frac{4}{8} = \frac{1}{2}$$

(b) The event "A or B" is $A \cup B$, so we have

$$A \cup B = \{HHT, HTH, THH, TTT\}$$

and consequently

$$P(A \cup B) = \frac{n(A \cup B)}{n(S)} = \frac{4}{8} = \frac{1}{2}$$

(c) Likewise, the event "B or C" is $B \cup C$, so we have

$$B \cup C = \{TTT, HTT, THT, TTH\}$$

and consequently

$$P(B \cup C) = \frac{n(B \cup C)}{n(S)} = \frac{4}{8} = \frac{1}{2} \qquad \text{as required}$$

We can use the results in Example 6 to introduce a rule that simplifies certain probability computations. If you review the results in Example 6, you'll see that $P(A \cup B) = \frac{1}{2}$, $P(A) = \frac{3}{8}$, and $P(B) = \frac{1}{8}$. Thus we have

$$P(A \cup B) = P(A) + P(B) \qquad \left(\text{because } \frac{1}{2} = \frac{3}{8} + \frac{1}{8}\right)$$

That is,

$$P(A \text{ or } B) = P(A) + P(B) \tag{1}$$

So (in this case at least), we find that the probability of the event "A or B" is the sum of the individual probabilities of A and B. Furthermore, it can be shown that equation (1) remains valid for any two events (in a given sample space) that have no outcomes in common. *Terminology:* Two events with no outcomes in common are said to be *mutually exclusive*. For instance, the events A and B in Example 6 are mutually exclusive. (Why?)

The remarks in the previous paragraph can be summarized this way: If E and F are mutually exclusive, then $P(E \text{ or } F) = P(E) + P(F)$. This formula is *not* valid, however, in cases in which the two events fail to be mutually exclusive. To see an example of this, let's take the events B and C in Example 6. In that example we found that $P(B \cup C) = \frac{1}{2}$, $P(B) = \frac{1}{8}$, and $P(C) = \frac{1}{2}$. Consequently we have

$$P(B \cup C) \neq P(B) + P(C) \qquad \left(\text{because } \frac{1}{2} \neq \frac{1}{8} + \frac{1}{2}\right)$$

The difficulty here, as we've said, has to do with the fact that B and C are not mutually exclusive. As you can check, B and C do have an outcome in common, namely TTT. Because of this, when we compute $P(B \cup C)$, the outcome TTT is counted but once; but when we compute $P(B) + P(C)$, the outcome TTT is counted twice (once with B and once again with C). This is essentially the reason that we obtain different values for the quantities $P(B \cup C)$ and $P(B) + P(C)$. By following this same type of reasoning (in which we keep track of how many times an outcome is counted) the following general formula can be derived. Notice that the formula agrees with equation (1) in the case when $A \cap B = \emptyset$ (that is, when A and B are mutually exclusive).

The Addition Theorem for Probabilities

If A and B are events in a sample space S, then

$$P(A \cup B) = P(A) + P(B) - P(A \cap B)$$

Example 7 Three balls are drawn at random from a bag containing eight red and seven yellow balls. Compute the probability for each of the following events:

(a) Drawing three red balls

(b) Drawing three yellow balls

(c) Drawing either three red or three yellow balls

Solution **(a)** The total number of possible outcomes in the experiment is $C(15, 3)$, the number of ways of choosing three balls from a set of 15. We have

$$C(15, 3) = \frac{15!}{3!\,12!} = \frac{15 \cdot 14 \cdot 13}{3 \cdot 2} = 455$$

Next, we need to know how many outcomes are in the event "drawing 3 red balls." This number is

$$C(8, 3) = \frac{8!}{3!\,5!} = \frac{8 \cdot 7 \cdot 6}{3 \cdot 2} = 56$$

Consequently we have

$$P(\text{drawing 3 red balls}) = \frac{56}{455} = \frac{8}{65}$$

(b) Following the reasoning used in part (a), we find

$$P(\text{drawing 3 yellow balls}) = \frac{C(7, 3)}{C(15, 3)} = \frac{35}{455} = \frac{1}{13}$$ (Check the arithmetic.)

(c) Let A and B denote the events described in parts (a) and (b), respectively. Then $P(A \cup B)$ is the event "drawing either three red or three yellow balls," and we have

$$\begin{aligned} P(A \cup B) &= P(A) + P(B) - P(A \cap B) \\ &= \frac{8}{65} + \frac{1}{13} - P(\emptyset) \\ &= \frac{8}{65} + \frac{5}{65} - 0 = \frac{13}{65} = \frac{1}{5} \end{aligned}$$

So the probability of drawing either three red or three yellow balls is $\frac{1}{5}$.

Example 8 A card is selected from a well-shuffled deck of 52 cards. What is the probability that it is a jack or a heart?

Solution In this experiment there are 52 possible outcomes. Let A denote the event "selecting a jack," and let B denote the event "selecting a heart." Then $P(A \cup B)$ is the probability of the event "selecting a jack or a heart," and we have

$$\begin{aligned} P(A \cup B) &= P(A) + P(B) - P(A \cap B) \\ &= \frac{4}{52} + \frac{13}{52} - \frac{1}{52} \qquad \text{(Why?)} \\ &= \frac{16}{52} = \frac{4}{13} \qquad \text{as required} \end{aligned}$$

Exercise Set 11.7

Exercises 1–20 refer to an experiment that consists of tossing a single die and then noting the number of dots showing on the top face. As in the text, assume that the sample space for this experiment is

$$S = \{1, 2, 3, 4, 5, 6\}$$

Let A, B, C, D, E, F, *and* G *be the following events.*

A: *obtaining an even number*
B: *obtaining a number greater than 3*
C: *obtaining a 1*
D: *obtaining a number greater than 1*
E: *obtaining an odd number*
F: *obtaining a number less than 8*
G: *obtaining 8*

1. Use set notation to specify the following events.
(a) A **(b)** B **(c)** $A \cup B$ **(d)** $A \cap B$

2. Use set notation to specify the following events.
(a) C **(b)** D **(c)** $C \cup D$ **(d)** $C \cap D$

3. Use set notation to specify the following events.
(a) E **(b)** F **(c)** $E \cup A$ **(d)** $E \cap A$

4. Use set notation to specify the following events.
(a) G **(b)** $G \cup B$ **(c)** $G \cap B$ **(d)** $G \cup F$

5. Are events A and B mutually exclusive?

6. Are events A and C mutually exclusive?

7. Are events C and D mutually exclusive?

8. Are events B and E mutually exclusive?

For Exercises 9–20, compute each of the indicated quantities.

9. **(a)** $n(A)$ **(b)** $n(B)$ **(c)** $n(A \cup B)$

10. **(a)** $n(C)$ **(b)** $n(D)$ **(c)** $n(C \cup D)$

11. **(a)** $n(E)$ **(b)** $n(F)$ **(c)** $n(G)$

12. $n(B \cap E)$

13. **(a)** $P(A)$ **(b)** $P(B)$

14. **(a)** $P(C)$ **(b)** $P(D)$

15. **(a)** $P(E)$ **(b)** $P(F)$

16. $P(G)$

17. **(a)** $P(A \cup B)$ **(b)** $P(A) + P(B)$

18. **(a)** $P(B \cup C)$ **(b)** $P(B) + P(C)$

19. **(a)** $P(B \cup E)$ **(b)** $P(B) + P(E)$

20. **(a)** $P(D) + P(E) - P(D \cap E)$ **(b)** $P(D \cup E)$

Exercises 21–40 refer to the experiment described in Example 6 in which a coin is flipped three times and the result is recorded each time. As in Example 6, let the sample space S be

S = {HHH, HHT, HTH, HTT, THH, THT, TTH, TTT}

Let A, B, C, D, E, *and* F *be the following events.*

A: *obtaining exactly two tails*
B: *obtaining two heads on the first two tosses*
C: *obtaining heads on the first toss*
D: *obtaining no heads*
E: *obtaining tails on both the second and third tosses*
F: *obtaining tails on the first toss*

Compute each of the following probabilities.

21. $P(A)$ **22.** $P(B)$ **23.** $P(C)$

24. $P(D)$ **25.** $P(E)$ **26.** $P(F)$

27. $P(A \cup B)$ **28.** $P(A \cap B)$ **29.** $P(A \cup C)$

30. $P(A \cap C)$ **31.** $P(C \cup F)$ **32.** $P(C \cap F)$

33. $P(E \cup F)$ **34.** $P(E \cap F)$ **35.** $P(B \cup C)$

36. $P(B \cap C)$ **37.** $P(D \cap A)$ **38.** $P(D \cup A)$

39. $P(D \cup E)$ **40.** $P(D \cap E)$

41. Two letters are selected at random (and without replacement) from the set $\{a, b, c, d, e, f\}$. What is the probability that one of the letters is d?

42. The numbers 1, 2, 3, and 4 are arranged in random order to form a four-digit number. What is the probability that the number formed is greater than 4000?

43. Suppose that two faulty transistors are accidentally packaged with a dozen reliable ones.
(a) If one transistor is chosen at random from the package, what is the probability that it will be reliable?
(b) If five transistors are chosen at random from the package, what is the probability that they are all reliable?

44. Suppose that five cartons of spoiled milk are inadvertently placed on a grocery shelf along with 15 cartons of fresh milk.
(a) If three cartons are selected at random, what is the probability that they all contain unspoiled milk?
(b) What is the probability that three cartons selected at random all contain spoiled milk?

In Exercises 45–48, use the Addition Theorem to compute the required probabilities.

45. If $P(A) = \frac{1}{3}$, $P(B) = \frac{1}{4}$, and $A \cap B = \emptyset$, compute $P(A \cup B)$.

46. If $P(A) = \frac{2}{5}$, $P(B) = \frac{3}{10}$, and $P(A \cap B) = \frac{1}{5}$, compute $P(A \cup B)$.

47. If $P(A \cup B) = \frac{7}{9}$, $P(A) = \frac{5}{9}$, and $P(B) = \frac{4}{9}$, compute $P(A \cap B)$.

48. If $P(A) = \frac{1}{3}$, $P(A \cup B) = \frac{2}{3}$, and $P(A \cap B) = \frac{1}{6}$, compute $P(B)$.

49. Two balls are drawn at random from a bag containing six red and 10 black balls. Compute the probability for each of the following events.
(a) Drawing two red balls
(b) Drawing two black balls
(c) Drawing either two red or two black balls

50. A card is selected at random from a deck of 52 cards. Compute the probability of each of the following events.
(a) Selecting a 4 or 2
(b) Selecting a red card or an ace
(c) Selecting an ace or a picture card

51. Your mathematics instructor claims that the probabilities of your getting A's on the first exam, on the second exam, and on both exams are $\frac{1}{2}$, $\frac{3}{10}$, and $\frac{1}{10}$, respectively. According to the figures, what is the probability that you will get an A on at least one exam? *Hint:* Let G and H denote the events "A on first exam" and "A on second exam," respectively. Then you want to compute $P(G \text{ or } H)$.

52. Two cards are drawn at random from a regular 52-card deck. Find the probability that either both are black or both are twos.

53. Five cards are drawn at random from a regular deck of 52 cards. Compute the probability that all five cards are clubs.

54. An experiment consists of selecting at random one of the 100 points in the x-y plane shown in the figure on page 580. Compute the probability for each of the following events.
(a) The point lies on the line $y = x$.
(b) The point lies on the line $y = 2x$.
(c) The point lies on the circle $x^2 + y^2 = 25$.
(d) The point lies within (but not on) the circle $x^2 + y^2 = 25$.

(e) The point lies on or within the circle $x^2 + y^2 = 25$.

(f) The point lies outside the circle $x^2 + y^2 = 25$.

(g) The point lies inside the circle $x^2 + y^2 = 25$ but not on the graph of $y = |x - 3|$.

(h) The point lies either on the circle $x^2 + y^2 = 25$ or on the line $y = x$.

(i) The point lies on the x-axis or on the circle $x^2 + y^2 = 25$.

(j) The sum of the x- and y-coordinates of the point is 18.

(k) The sum of the x- and y-coordinates of the point is 21.

(l) The y-coordinate of the point is greater than the square of the x-coordinate.

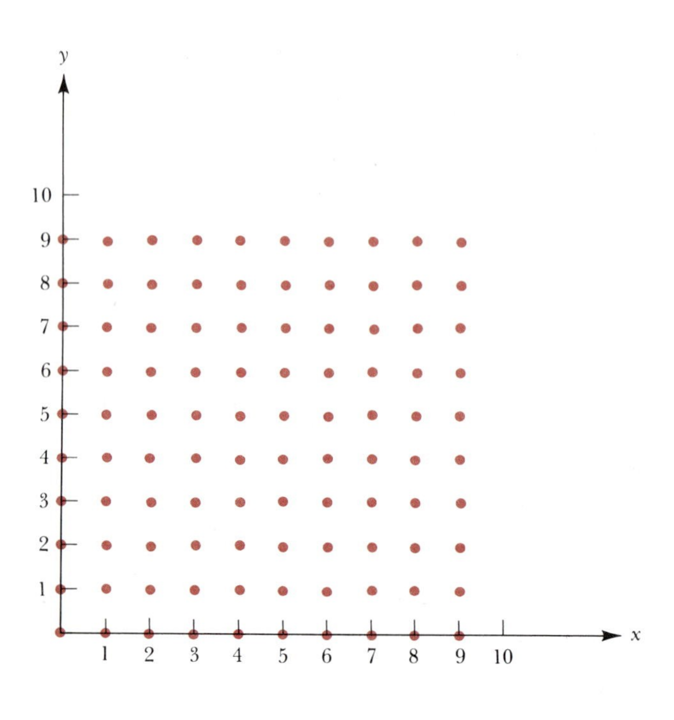

11.8 Trigonometric Form for Complex Numbers

Go to Mr. DeMoivre; he knows these things better than I do.

Isaac Newton (according to Boyer's *A History of Mathematics*, New York: John Wiley & Sons, Inc., 1968).

In this section we explore one of the many important connections between trigonometry and the complex number system. We begin with an observation that was first made in the year 1797 by the Norwegian surveyor and mathematician Caspar Wessel. Wessel realized, essentially, that the complex numbers could be visualized as points in the x-y plane, the complex number $a + bi$ being identified with the point (a, b). In this context, we often refer to the x-y plane as the *complex plane*, and we refer to the complex numbers as points in this plane.

Example 1 Plot the point $2 + 3i$ in the complex plane.

Solution The complex number $2 + 3i$ is identified with the point $(2, 3)$. See Figure 1.

Figure 1 The complex number $2 + 3i$ is identified with the point $(2, 3)$.

As indicated in Figure 2, the distance from the origin to the point $a + bi$ is denoted by r. We call the distance r the *modulus* of the complex number $a + bi$. The angle θ in Figure 2 (measured counterclockwise from the positive x-axis) is referred to as the *argument* of the complex number $a + bi$.

From Figure 2 we have the following three equations relating the quantities a, b, r, and θ. (Although Figure 2 shows $a + bi$ in the first quadrant, the equations remain valid for the other quadrants as well.)

$$r = \sqrt{a^2 + b^2} \quad (1)$$

$$a = r \cos \theta \quad (2)$$

$$b = r \sin \theta \quad (3)$$

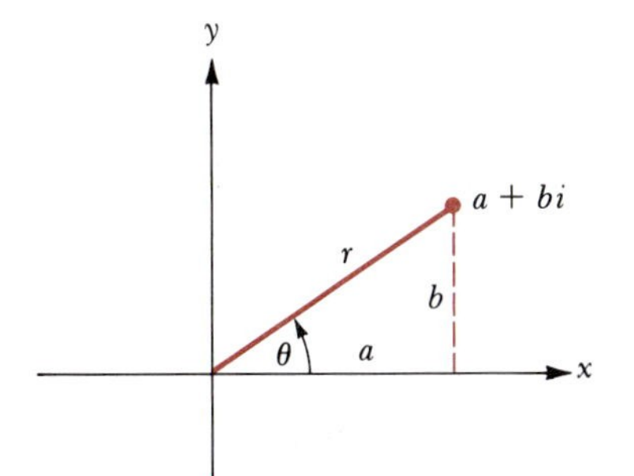

Figure 2

Now if we have a complex number $z = a + bi$, we can use equations (2) and (3) to write

$$z = (r \cos \theta) + (r \sin \theta)i = r(\cos \theta + i \sin \theta)$$

That is,

$$z = r(\cos \theta + i \sin \theta)$$

The expression that appears on the right-hand side of this last equation is called the *trigonometric* (or *polar*) *form* of the complex number z. In contrast to this, the expression $a + bi$ is referred to as the *rectangular form* of the complex number z.

Example 2 Express the complex number $z = 3[\cos(\pi/3) + i \sin(\pi/3)]$ in rectangular form.

Solution

$$z = 3\left(\cos \frac{\pi}{3} + i \sin \frac{\pi}{3}\right)$$
$$= 3\left(\frac{1}{2} + i\frac{\sqrt{3}}{2}\right) = \frac{3}{2} + \frac{3\sqrt{3}}{2}i$$

The rectangular form is therefore $\frac{3}{2} + (3\sqrt{3}/2)i$.

Example 3 Find the trigonometric form for the complex number $-\sqrt{2} + i\sqrt{2}$.

Solution We're asked to write the number in the form $r(\cos \theta + i \sin \theta)$. So we need to find r and θ. Using equation (1) and the values $a = -\sqrt{2}$, $b = \sqrt{2}$, we have

$$r = \sqrt{(-\sqrt{2})^2 + (\sqrt{2})^2} = \sqrt{2 + 2} = \sqrt{4} = 2$$

Now that we know r, we can use equations (2) and (3) to determine θ. From equation (2) we obtain

$$\cos \theta = \frac{a}{r} = \frac{-\sqrt{2}}{2} \tag{4}$$

Similarly, equation (3) gives us

$$\sin \theta = \frac{b}{r} = \frac{\sqrt{2}}{2} \tag{5}$$

One angle satisfying both equations (4) and (5) is $\theta = 3\pi/4$. (There are other angles; we'll return to this point in a moment.) In summary, then, we have $r = 2$ and $\theta = 3\pi/4$, so the required trigonometric form is

$$2\left(\cos \frac{3\pi}{4} + i \sin \frac{3\pi}{4}\right)$$

In the example just completed we noted that $\theta = 3\pi/4$ was but one angle satisfying the conditions $\cos \theta = -\sqrt{2}/2$ and $\sin \theta = \sqrt{2}/2$. Another such angle would be $(3\pi/4) + 2\pi$. Indeed, any angle of the form $(3\pi/4) + 2\pi k$, where k is an integer, would do just as well. The upshot of this is that θ, the argument of a complex number, is not uniquely determined. In Example 3 we followed a common convention in converting to trigonometric form:

we picked θ in the interval $0 \leq \theta < 2\pi$. Furthermore, although it won't cause us any difficulties in this section, you might also note that the argument θ is undefined for the complex number $0 + 0i$. (Why?)

We're ready now to derive a formula that makes it very easy to multiply two complex numbers in trigonometric form. Suppose that the two complex numbers are

$$r(\cos \alpha + i \sin \alpha) \quad \text{and} \quad R(\cos \beta + i \sin \beta)$$

Then their product is

$$\begin{aligned} rR[(\cos \alpha + i \sin \alpha)(\cos \beta + i \sin \beta)] \\ &= rR[(\cos \alpha \cos \beta - \sin \alpha \sin \beta) + i(\sin \alpha \cos \beta + \cos \alpha \sin \beta)] \\ &= rR[\cos(\alpha + \beta) + i \sin(\alpha + \beta)] \end{aligned}$$

We've used the addition formulas from Section 8.3.

Notice that the modulus of the product here is rR, that is, the product of the two original moduli. Also, the argument is $\alpha + \beta$, which is the sum of the two original arguments. So, to multiply two complex numbers, just multiply the moduli and add the arguments. There is a similar rule for obtaining the quotient of two complex numbers: Divide their moduli and subtract the arguments. This is stated more precisely in the box that follows, along with the multiplication rule. (For a proof of the division rule, see Exercise 71.)

> Let $z = r(\cos \alpha + i \sin \alpha)$ and $w = R(\cos \beta + i \sin \beta)$. Then
>
> $$zw = rR[\cos(\alpha + \beta) + i \sin(\alpha + \beta)]$$
>
> Also, if $R \neq 0$, then
>
> $$\frac{z}{w} = \frac{r}{R}[\cos(\alpha - \beta) + i \sin(\alpha - \beta)]$$

Example 4 Let $z = 8[\cos(5\pi/3) + i \sin(5\pi/3)]$ and $w = 4[\cos(2\pi/3) + i \sin(2\pi/3)]$. Compute **(a)** zw; **(b)** z/w. Express each answer in both trigonometric and rectangular form.

Solution **(a)**

$$\begin{aligned} zw &= (8)(4)\left[\cos\left(\frac{5\pi}{3} + \frac{2\pi}{3}\right) + i \sin\left(\frac{5\pi}{3} + \frac{2\pi}{3}\right)\right] \\ &= 32\left(\cos \frac{7\pi}{3} + i \sin \frac{7\pi}{3}\right) = 32\left(\cos \frac{\pi}{3} + i \sin \frac{\pi}{3}\right) \end{aligned}$$

This is the trigonometric form.

$$= 32\left(\frac{1}{2} + i\frac{\sqrt{3}}{2}\right) = 16 + 16\sqrt{3}\,i$$

This is the rectangular form.

(b)

$$\frac{z}{w} = \frac{8}{4}\left[\cos\left(\frac{5\pi}{3} - \frac{2\pi}{3}\right) + i \sin\left(\frac{5\pi}{3} - \frac{2\pi}{3}\right)\right]$$

$$= 2(\cos \pi + i \sin \pi)$$

This is the trigonometric form.

$$= 2(-1 + i \cdot 0)$$

$$= -2$$

This is the rectangular form.

Example 5 Compute z^2 where $z = r(\cos\theta + i\sin\theta)$.

Solution

$$\begin{aligned} z^2 &= [r(\cos\theta + i\sin\theta)][r\cos\theta + i\sin\theta] \\ &= r^2[\cos(\theta + \theta) + i\sin(\theta + \theta)] \\ &= r^2(\cos 2\theta + i\sin 2\theta) \end{aligned}$$

The result in Example 5 is a particular case of an important theorem attributed to Abraham DeMoivre (1667–1754). In the box that follows we state DeMoivre's Theorem. (Exercise 72 asks you to prove the theorem using mathematical induction.)

> **DeMoivre's Theorem**
>
> Let n be a natural number. Then
>
> $$[r(\cos\theta + i\sin\theta)]^n = r^n(\cos n\theta + i\sin n\theta)$$

Example 6 Use DeMoivre's Theorem to compute $(-\sqrt{2} + i\sqrt{2})^5$. Express the answer in rectangular form.

Solution In Example 3 we saw that the trigonometric form of $-\sqrt{2} + i\sqrt{2}$ is given by

$$-\sqrt{2} + i\sqrt{2} = 2\left(\cos\frac{3\pi}{4} + i\sin\frac{3\pi}{4}\right)$$

Therefore

$$\begin{aligned} (-\sqrt{2} + i\sqrt{2})^5 &= 2^5\left(\cos\frac{15\pi}{4} + i\sin\frac{15\pi}{4}\right) \\ &= 32\left[\frac{\sqrt{2}}{2} + i\left(-\frac{\sqrt{2}}{2}\right)\right] \\ &= 16\sqrt{2} - 16\sqrt{2}i \qquad \text{as required} \end{aligned}$$

The next two examples show how DeMoivre's Theorem is used in computing roots. If n is a natural number and $z^n = w$, then we say that z is an nth *root* of w. The work in the examples also relies on the following observation about equality between nonzero complex numbers in trigonometric form. If $r(\cos\theta + i\sin\theta) = R(\cos A + i\sin A)$, then $r = R$ and $\theta = A + 2\pi k$, where k is an integer.

Example 7 Find the cube roots of $8i$.

Solution First express $8i$ in trigonometric form. As you can see from Figure 3 on the following page, we have

$$8i = 8\left(\cos\frac{\pi}{2} + i\sin\frac{\pi}{2}\right)$$

Now let $z = r(\cos\theta + i\sin\theta)$ denote a cube root of $8i$. Then the equation $z^3 = 8i$ becomes

$$r^3(\cos 3\theta + i\sin 3\theta) = 8\left(\cos\frac{\pi}{2} + i\sin\frac{\pi}{2}\right) \tag{6}$$

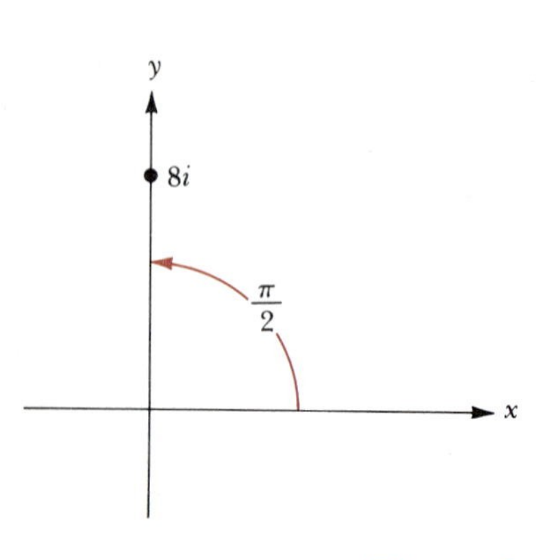

Figure 3

From equation (6) we conclude that $r^3 = 8$ and consequently $r = 2$. Also from equation (6) we have

$$3\theta = \frac{\pi}{2} + 2\pi k$$

or

$$\theta = \frac{\pi}{6} + \frac{2\pi k}{3} \qquad \text{dividing by 3} \tag{7}$$

Using $k = 0$, equation (7) yields $\theta = \pi/6$. Thus one of the cube roots of $8i$ is

$$2\left(\cos\frac{\pi}{6} + i\sin\frac{\pi}{6}\right) = 2\left(\frac{\sqrt{3}}{2} + i\cdot\frac{1}{2}\right) = \sqrt{3} + i$$

Next use $k = 1$ in equation (7). As you can check, this yields $\theta = 5\pi/6$. So another cube root of $8i$ is

$$2\left(\cos\frac{5\pi}{6} + i\sin\frac{5\pi}{6}\right) = 2\left(\frac{-\sqrt{3}}{2} + i\cdot\frac{1}{2}\right) = -\sqrt{3} + i$$

Similarly, using $k = 2$ in equation (7) yields $\theta = 3\pi/2$. (Verify this.) Consequently, a third cube root is

$$2\left(\cos\frac{3\pi}{2} + i\sin\frac{3\pi}{2}\right) = 2[0 + i(-1)] = -2i$$

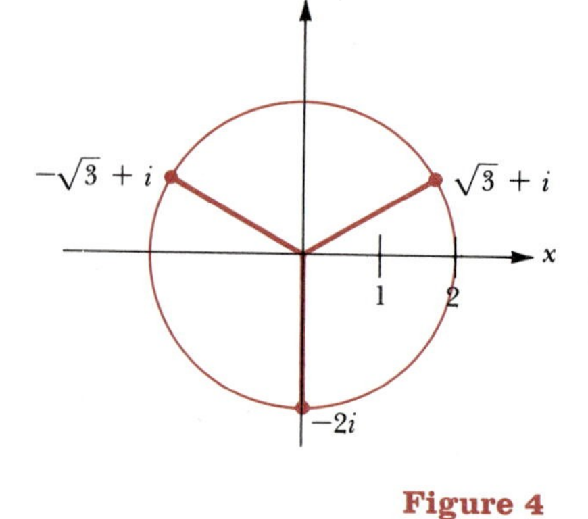

Figure 4

We've now found three distinct cube roots of $8i$. If you continue the process, using $k = 3$ for example, you'll find that no additional roots are obtained in this manner. We conclude from this that there are exactly three cube roots of $8i$. In Figure 4 we've plotted these cube roots. Notice that the points lie equally spaced on a circle of radius 2 about the origin.

Using DeMoivre's Theorem in Example 7, we found that the number $8i$ had three distinct cube roots. Along these same lines, it's true in general that any nonzero number $a + bi$ possesses exactly n distinct nth roots. The preceding sentence remains true even if $b = 0$. In the next example, for instance, we compute the five fifth roots of 2.

Example 8 Compute the five fifth roots of 2.

Solution We'll follow the procedure used in Example 7. In trigonometric form, the number 2 becomes $2(\cos 0 + i \sin 0)$. Now let $z = r(\cos\theta + i\sin\theta)$ denote a fifth root of 2. Then the equation $z^5 = 2$ becomes

$$r^5(\cos 5\theta + i\sin 5\theta) = 2(\cos 0 + i\sin 0) \tag{8}$$

From equation (8) we see that $r^5 = 2$ and consequently $r = 2^{1/5}$. Also from equation (8) we have

$$5\theta = 0 + 2\pi k$$

or

$$\theta = \frac{2\pi k}{5}$$

Using the values $k = 0, 1, 2, 3$, and 4 in succession, we obtain the following results:

k	0	1	2	3	4
θ	0	$2\pi/5$	$4\pi/5$	$6\pi/5$	$8\pi/5$

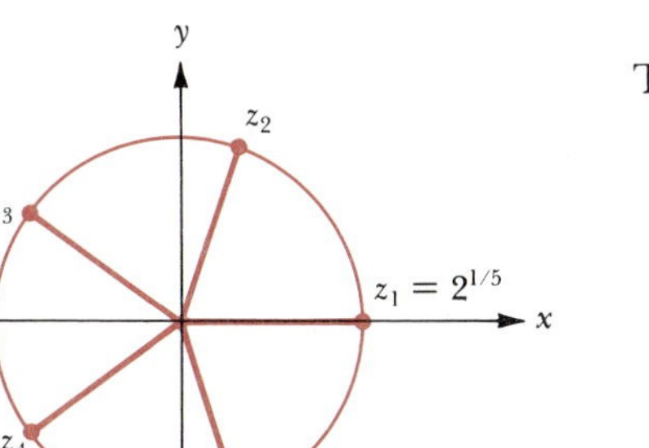

Figure 5

The five fifth roots of 2 are therefore

$$z_1 = 2^{1/5}(\cos 0 + i \sin 0) = 2^{1/5} \qquad z_4 = 2^{1/5}\left(\cos \frac{6\pi}{5} + i \sin \frac{6\pi}{5}\right)$$

$$z_2 = 2^{1/5}\left(\cos \frac{2\pi}{5} + i \sin \frac{2\pi}{5}\right) \qquad z_5 = 2^{1/5}\left(\cos \frac{8\pi}{5} + i \sin \frac{8\pi}{5}\right)$$

$$z_3 = 2^{1/5}\left(\cos \frac{4\pi}{5} + i \sin \frac{4\pi}{5}\right)$$

In Figure 5 we've plotted these five fifth roots. Notice that the points are equally spaced on a circle of radius $2^{1/5}$ about the origin.

Exercise Set 11.8

In Exercises 1–8, plot each point in the complex plane.

1. $4 + 2i$
2. $4 - 2i$
3. $-5 + i$
4. $-3 - 5i$
5. $1 - 4i$
6. i
7. $-i$
8. $1\ (= 1 + 0i)$

For Exercises 9–18, convert each complex number to rectangular form.

9. $2\left(\cos \frac{\pi}{4} + i \sin \frac{\pi}{4}\right)$
10. $6\left(\cos \frac{5\pi}{3} + i \sin \frac{5\pi}{3}\right)$

11. $4\left(\cos \frac{5\pi}{6} + i \sin \frac{5\pi}{6}\right)$

12. $3\left(\cos \frac{3\pi}{2} + i \sin \frac{3\pi}{2}\right)$

13. $\sqrt{2}(\cos 225° + i \sin 225°)$

14. $\frac{1}{2}(\cos 240° + i \sin 240°)$

15. $\sqrt{3}\left(\cos \frac{\pi}{2} + i \sin \frac{\pi}{2}\right)$

16. $5(\cos \pi + i \sin \pi)$

17. $4(\cos 75° + i \sin 75°)$ *Hint:* Use the addition formulas from Section 8.3 to evaluate $\cos 75°$ and $\sin 75°$.

18. $2\left(\cos \frac{\pi}{8} + i \sin \frac{\pi}{8}\right)$ *Hint:* Use the half-angle formulas from Section 8.4 to evaluate $\cos \frac{\pi}{8}$ and $\sin \frac{\pi}{8}$.

In Exercises 19–28, convert from rectangular to trigonometric form. (In each case choose an argument θ such that $0 \le \theta < 2\pi$.)

19. $\frac{\sqrt{3}}{2} + \frac{1}{2}i$
20. $\sqrt{2} + \sqrt{2}i$
21. $-1 + \sqrt{3}i$
22. -4
23. $-2\sqrt{3} - 2i$
24. $-3\sqrt{2} - 3\sqrt{2}i$
25. $-6i$
26. $-4 - 4\sqrt{3}i$
27. $\frac{\sqrt{3}}{4} - \frac{1}{4}i$
28. 16

For Exercises 29–54, carry out the indicated operations. Express your results in rectangular form for those cases in which the trigonometric functions are readily evaluated without tables or a calculator.

29. $2(\cos 22° + i \sin 22°) \times 3(\cos 38° + i \sin 38°)$

30. $4(\cos 5° + i \sin 5°) \times 6(\cos 130° + i \sin 130°)$

31. $\sqrt{2}\left(\cos \frac{\pi}{3} + i \sin \frac{\pi}{3}\right) \times \sqrt{2}\left(\cos \frac{4\pi}{3} + i \sin \frac{4\pi}{3}\right)$

32. $\left(\cos \frac{\pi}{5} + i \sin \frac{\pi}{5}\right) \times \left(\cos \frac{\pi}{20} + i \sin \frac{\pi}{20}\right)$

33. $3\left(\cos \frac{\pi}{7} + i \sin \frac{\pi}{7}\right) \times \sqrt{2}\left(\cos \frac{\pi}{7} + i \sin \frac{\pi}{7}\right)$

34. $\sqrt{3}(\cos 3° + i \sin 3°) \times \sqrt{3}(\cos 38° + i \sin 38°)$

35. $6(\cos 50° + i \sin 50°) \div 2(\cos 5° + i \sin 5°)$

36. $\sqrt{3}(\cos 140° + i \sin 140°) \div 3(\cos 5° + i \sin 5°)$

37. $2^{4/3}\left(\cos\frac{5\pi}{12} + i\sin\frac{5\pi}{12}\right) \div 2^{1/3}\left(\cos\frac{\pi}{4} + i\sin\frac{\pi}{4}\right)$

38. $\sqrt{6}\left(\cos\frac{16\pi}{9} + i\sin\frac{16\pi}{9}\right) \div \sqrt{2}\left(\cos\frac{\pi}{9} + i\sin\frac{\pi}{9}\right)$

39. $\left(\cos\frac{2\pi}{5} + i\sin\frac{2\pi}{5}\right) \div \left(\cos\frac{2\pi}{5} + i\sin\frac{2\pi}{5}\right)$

40. $\left(\cos\frac{2\pi}{5} + i\sin\frac{2\pi}{5}\right) \div \left(\cos\frac{\pi}{10} + i\sin\frac{\pi}{10}\right)$

41. $\left[3\left(\cos\frac{\pi}{3} + i\sin\frac{\pi}{3}\right)\right]^5$

42. $\left[\sqrt{2}\left(\cos\frac{5\pi}{6} + i\sin\frac{5\pi}{6}\right)\right]^4$

43. $\left[\frac{1}{2}\left(\cos\frac{\pi}{24} + i\sin\frac{\pi}{24}\right)\right]^6$

44. $[\sqrt{3}(\cos 70° + i\sin 70°)]^3$

45. $[2^{1/5}(\cos 63° + i\sin 63°)]^{10}$

46. $\left[2\left(\cos\frac{\pi}{5} + i\sin\frac{\pi}{5}\right)\right]^3$

47. $2(\cos 200° + i\sin 200°) \times \sqrt{2}(\cos 20° + i\sin 20°) \times \frac{1}{2}(\cos 5° + i\sin 5°)$

48. $\left[\frac{\cos(\pi/8) + i\sin(\pi/8)}{\cos(-\pi/8) + i\sin(-\pi/8)}\right]^5$

49. $\left(\frac{1}{2} - \frac{\sqrt{3}}{2}i\right)^5$ *Hint:* First convert to trigonometric form.

50. $(1 - i)^3$

51. $(-2 - 2i)^5$

52. $\left(-\frac{1}{2} + \frac{\sqrt{3}}{2}i\right)^6$

53. $(-2\sqrt{3} - 2i)^4$

54. $(1 + i)^{16}$

In Exercises 55–64, use DeMoivre's Theorem to find the indicated roots. Express the results in rectangular form.

55. Cube roots of $-27i$

56. Cube roots of 2

57. Eighth roots of 1

58. Fourth roots of i *Hint:* Use the half-angle formulas from Section 8.4.

59. Square roots of i

60. Fourth roots of $8 - 8\sqrt{3}i$ *Hint:* Use the addition formulas or the half-angle formulas.

61. Cube roots of 64

62. Square roots of $-\frac{1}{2} - \frac{\sqrt{3}}{2}i$

63. Sixth roots of 729

***64.** Square roots of $7 + 24i$ *Hint:* You'll need to use the half-angle formulas from Section 8.4.

65. **(a)** Compute the three cube roots of 1.
(b) Let z_1, z_2, and z_3 denote the three cube roots of 1. Verify that $z_1 + z_2 + z_3 = 0$ and $z_1z_2 + z_2z_3 + z_3z_1 = 0$.

66. **(a)** Compute the four fourth roots of 1.
(b) Verify that the sum of these four fourth roots is 0.

67. Evaluate $\left(\frac{-1 + i\sqrt{3}}{2}\right)^5 + \left(\frac{-1 - i\sqrt{3}}{2}\right)^5$. *Hint:* Use DeMoivre's Theorem.

68. Show that $\left(\frac{-1 + i\sqrt{3}}{2}\right)^6 + \left(\frac{-1 - i\sqrt{3}}{2}\right)^6 = 2$.

69. Compute $(\cos\theta + i\sin\theta)(\cos\theta - i\sin\theta)$.

70. In the identity $(\cos\theta + i\sin\theta)^2 = \cos 2\theta + i\sin 2\theta$, carry out the actual multiplication of the left-hand side of the equation. Then equate the corresponding real parts and the corresponding imaginary parts from each side of the equation that results. What do you obtain?

71. Show that $\frac{r(\cos\alpha + i\sin\alpha)}{R(\cos\beta + i\sin\beta)} = \frac{r}{R}[\cos(\alpha - \beta) + i\sin(\alpha - \beta)]$. *Suggestion:* Begin with the quantity on the left side and multiply it by $\frac{\cos\beta - i\sin\beta}{\cos\beta - i\sin\beta}$.

72. Use mathematical induction to prove that the equation

$$[r(\cos\theta + i\sin\theta)]^n = r^n(\cos n\theta + i\sin n\theta)$$

is valid for all positive integers n.

73. Show that $1 + \cos\theta + i\sin\theta = 2\cos\left(\frac{\theta}{2}\right)\left[\cos\left(\frac{\theta}{2}\right) + i\sin\left(\frac{\theta}{2}\right)\right]$.

74. If $z = r(\cos\theta + i\sin\theta)$, show that $1/z = (1/r)(\cos\theta - i\sin\theta)$. *Hint:* $\frac{1}{z} = \frac{1(\cos 0 + i\sin 0)}{r(\cos\theta + i\sin\theta)}$.

***75.** Show that $\frac{1 + \sin\theta + i\cos\theta}{1 + \sin\theta - i\cos\theta} = \sin\theta + i\cos\theta$. (Assume $\theta \neq \frac{3\pi}{2} + 2\pi k$, where k is an integer.) *Hint:* Work with the left-hand side; first "rationalize" the denominator by multiplying by $\frac{(1 + \sin\theta) + i\cos\theta}{(1 + \sin\theta) + i\cos\theta}$.

C *Use a calculator to complete Exercises 76–79.*

76. Compute $(9 + 9i)^6$.

77. Compute $(7 - 7i)^8$.

78. Compute the cube roots of $1 + 2i$. Express the answers in rectangular form, with the real and imaginary parts rounded off to two decimal places.

79. Compute the fifth roots of i. Express the answers in rectangular form, with the real and imaginary parts rounded off to two decimal places.

Chapter 11 Review Checklist

- □ Principle of Mathematical Induction (page 533)
- □ Pascal's triangle (page 539)
- □ $n!$ (page 540)
- □ $\binom{n}{k}$ (pages 541, 568)
- □ Binomial Theorem (page 541)
- □ Sequence (page 548)
- □ $\sum_{k=1}^{n} a_k$ (page 549)
- □ Arithmetic sequence (page 552)
- □ Formulas associated with arithmetic sequences
  - nth term: $a_n = a + (n-1)d$ (page 553)
  - Sum of n terms: $S_n = (n/2)[2a + (n-1)d] = n(a + a_n)/2$ (page 555)
- □ Geometric sequence (page 558)
- □ Formulas associated with geometric sequences
  - nth term: $a_n = ar^{n-1}$ (page 558)
  - Sum of n terms: $S_n = \dfrac{a(1-r^n)}{1-r}$ (page 558)
  - Sum of an infinite geometric series: $S = \dfrac{a}{1-r}$, where $|r| < 1$ (page 561)
- □ Tree diagram (page 563)
- □ Multiplication Principle (page 563)
- □ Permutation (page 564)
- □ $P(n, r)$ (definition, page 565; formula, page 566)
- □ $C(n, r)$ (definition, page 567; formula, page 569)
- □ Sample space, outcome, and event (page 572)
- □ $P(E)$, the probability of an event E (page 573)
- □ Mutually exclusive events (page 577)
- □ The Addition Theorem for Probabilities (page 577)
- □ Complex plane (page 580)
- □ Modulus and argument (page 580)
- □ Trigonometric (or polar) form (page 581)
- □ Rectangular form (page 581)
- □ Rules for multiplying and dividing numbers in trigonometric form (page 582)
- □ DeMoivre's Theorem (page 583)
- □ nth roots of a complex number (definition: page 583; examples: pages 583–585)

Chapter 11 Review Exercises

In Exercises 1–10, use the Principle of Mathematical Induction to show that the statements are true for all natural numbers.

1. $5 + 10 + 15 + \cdots + 5n = \frac{5}{2}n(n+1)$
2. $10 + 10^2 + 10^3 + \cdots + 10^n = \frac{10}{9}(10^n - 1)$
3. $1\cdot 2 + 2\cdot 3 + 3\cdot 4 + \cdots + n(n+1) = \frac{1}{3}n(n+1)(n+2)$
4. $\dfrac{1}{2} + \dfrac{2}{2^2} + \dfrac{3}{2^3} + \cdots + \dfrac{n}{2^n} = 2 - \dfrac{2+n}{2^n}$
5. $1 + 3\cdot 2 + 5\cdot 2^2 + 7\cdot 2^3 + \cdots + (2n-1)\cdot 2^{n-1} = 3 + (2n-3)\cdot 2^n$
6. $\dfrac{1}{1\cdot 4} + \dfrac{1}{4\cdot 7} + \dfrac{1}{7\cdot 10} + \cdots + \dfrac{1}{(3n-2)(3n+1)} = \dfrac{n}{3n+1}$
7. $1 + 2^2\cdot 2 + 3^2\cdot 2^2 + 4^2\cdot 2^3 + \cdots + n^2\cdot 2^{n-1} = (n^2 - 2n + 3)2^n - 3$
8. 9 is a factor of $n^3 + (n+1)^3 + (n+2)^3$.
9. 3 is a factor of $7^n - 1$.
10. 8 is a factor of $9^n - 1$.

For Exercises 11–20, expand the given expressions.

11. $(3a + b^2)^4$
12. $(5a - 2b)^3$
13. $(x + \sqrt{x})^4$
14. $(1 - \sqrt{3})^6$
15. $(x^2 - 2y^2)^5$
16. $\left(\dfrac{1}{a} + \dfrac{2}{b}\right)^3$
17. $\left(1 + \dfrac{1}{x}\right)^5$
18. $\left(x^3 + \dfrac{1}{x^2}\right)^6$
19. $(a\sqrt{b} - b\sqrt{a})^4$
20. $(x^{-2} + y^{5/2})^8$
21. Find the fifth term in the expansion of $(3x + y^2)^5$.
22. Find the eighth term in the expansion of $(2x - y)^9$.
23. Find the coefficient of the term containing a^5 in the expansion of $(a - 2b)^7$.

24. Find the coefficient of the term containing b^8 in the expansion of $\left(2a - \frac{b}{3}\right)^{10}$.

25. Find the coefficient of the term containing x^3 in the expansion of $(1 + \sqrt{x})^8$.

26. Expand $(1 + \sqrt{x} + x)^6$. *Suggestion:* Rewrite the expression as $[(1 + \sqrt{x}) + x]^6$.

In Exercises 27–32, verify each assertion by computing the indicated binomial coefficients.

27. $\binom{2}{0}^2 + \binom{2}{1}^2 + \binom{2}{2}^2 = \binom{4}{2}$

28. $\binom{3}{0}^2 + \binom{3}{1}^2 + \binom{3}{2}^2 + \binom{3}{3}^2 = \binom{6}{3}$

29. $\binom{4}{0}^2 + \binom{4}{1}^2 + \binom{4}{2}^2 + \binom{4}{3}^2 + \binom{4}{4}^2 = \binom{8}{4}$

30. $\binom{2}{0} + \binom{2}{1} + \binom{2}{2} = 2^2$

31. $\binom{3}{0} + \binom{3}{1} + \binom{3}{2} + \binom{3}{3} = 2^3$

32. $\binom{4}{0} + \binom{4}{1} + \binom{4}{2} + \binom{4}{3} + \binom{4}{4} = 2^4$

In Exercises 33–38, compute the first five terms in each sequence. (In Exercises 33–35, in which the n*th term is defined by a formula, assume that the formula holds for all natural numbers* n.*)*

33. $a_n = \dfrac{2n}{n + 1}$

34. $a_n = \dfrac{3n - 2}{3n + 2}$

35. $a_n = (-1)^n\left(1 - \dfrac{1}{n + 1}\right)^n$

36. $a_0 = 4$
$a_n = 2a_{n-1}, \quad n \geq 1$

37. $a_0 = -3$
$a_n = 4a_{n-1}, \quad n \geq 1$

38. $a_0 = 1$
$a_1 = 2$
$a_n = 3a_{n-1} + 2a_{n-2}, \; n \geq 2$

In Exercises 39 and 40, express each sum without using sigma notation. Simplify each answer.

39. $\sum_{k=1}^{5} (2k + 3)$

40. $\sum_{k=0}^{8} \left(\dfrac{1}{k + 1} - \dfrac{1}{k + 2}\right)$

In Exercises 41 and 42, rewrite each sum using sigma notation.

41. $5/3 + 5/3^2 + 5/3^3 + 5/3^4 + 5/3^5$

42. $1/2 - 2/2^2 + 3/2^3 - 4/2^4 + 5/2^5 - 6/2^6$

For Exercises 43–46, find the indicated term in each sequence.

43. a_{14}; $5, 9, 13, 17, \ldots$

44. a_{20}; $5, \frac{9}{2}, 4, \frac{7}{2}, \ldots$

45. a_{12}; $10, 5, \frac{5}{2}, \frac{5}{4}, \ldots$

46. a_{10}; $\sqrt{2} + 1, 1, \sqrt{2} - 1, 3 - 2\sqrt{2}, \ldots$

47. Find the sum of the first 100 terms in the sequence $-5, -2, 1, 4, \ldots$.

48. Find the sum of the first 45 terms in the sequence $10, \frac{29}{3}, \frac{28}{3}, 9, \ldots$.

49. Find the sum of the first 10 terms in the sequence $7, 70, 700, \ldots$.

50. Find the sum of the first 12 terms in the sequence $\frac{1}{3}, -\frac{2}{9}, \frac{4}{27}, -\frac{8}{81}, \ldots$.

In Exercises 51–54, find the sum of each infinite geometric series.

51. $\frac{3}{5} + \frac{3}{25} + \frac{3}{125} + \cdots$

52. $\frac{7}{10} + \frac{7}{100} + \frac{7}{1000} + \cdots$

53. $\frac{1}{9} - \frac{1}{81} + \frac{1}{729} - \cdots$

54. $1 + \dfrac{1}{1 + \sqrt{2}} + \dfrac{1}{(1 + \sqrt{2})^2} + \cdots$

55. Find a fraction equivalent to $0.\overline{45}$.

56. Find a fraction equivalent to $0.2\overline{13}$.

For Exercises 57–60, verify each equation using the formula given in the text for the sum of an arithmetic series:

$$S_n = \frac{n}{2}[2a + (n - 1)d]$$

(The formulas given in Exercises 57–60 appear in Elements of Algebra *by Leonhard Euler, first published in 1770.)*

57. $1 + 2 + 3 + \cdots + n = n + \dfrac{n(n - 1)}{2}$

58. $1 + 3 + 5 + \cdots \text{(to } n \text{ terms)} = n + \dfrac{2n(n - 1)}{2}$

59. $1 + 4 + 7 + \cdots \text{(to } n \text{ terms)} = n + \dfrac{3n(n - 1)}{2}$

60. $1 + 5 + 9 + \cdots \text{(to } n \text{ terms)} = n + \dfrac{4n(n - 1)}{2}$

C **61.** In this exercise you will use the following (remarkably simple) formula for approximating sums of powers of integers:

$$1^k + 2^k + 3^k + \cdots + n^k \approx \frac{(n + \frac{1}{2})^{k+1}}{k + 1} \qquad (1)$$

[This formula appears in the article "Sums of Powers of Integers," by B. L. Burrows and R. F. Talbot, published in the *American Mathematical Monthly*, vol. 91 (1984), p. 394.]

(a) Use formula (1) to estimate the sum $1^2 + 2^2 + 3^2 + \cdots + 50^2$. Round off your answer to the nearest integer.

(b) Compute the exact value of the sum in part (a) using the formula $\sum_{k=1}^{n} k^2 = \frac{n(n+1)(2n+1)}{6}$. (This formula can be proved using mathematical induction.) Then compute the percent error for the approximation obtained in part (a). the percent error is given by

$$\frac{|(\text{actual value}) - (\text{approximate value})|}{(\text{actual value})} \times 100$$

(c) Use formula (1) to estimate the sum $1^4 + 2^4 + 3^4 + \cdots + 200^4$. Round off your answer to 6 significant digits.

(d) The following formula for the sum $1^4 + 2^4 + \cdots + n^4$ can be proved using mathematical induction:

$$\sum_{k=1}^{n} k^4 = \frac{n(n+1)(2n+1)(3n^2+3n-1)}{30}.$$

Use this formula to compute the sum in part (c). Round off your answer to 6 significant digits. Then use this result to compute the percent error for the approximation in part (c).

62. There are three roads from town A to town B and six roads from town B to town C. How many routes are there from A to C via B?

63. How many different two-digit numbers can be formed using the digits 2, 4, 6, or 8? (Repetitions are allowed.)

64. How many permutations are there of the set $\{A, B, C, D, E\}$?

65. Using the set $\{a, b, c, d\}$, list the permutations of these four letters taken three at a time. Evaluate $P(4, 3)$.

66. Follow Exercise 65, using combinations rather than permutations, and evaluate $C(4, 3)$.

67. From a collection of 12 pictures, eight are to be selected and hung in a row. How many arrangements are possible?

68. How many possible arrangements are there of the letters a, b, c, d, e, f in which a, b, and c appear consecutively (but not necessarily in that order)?

69. How many possible arrangements are there of the letters a, b, c, d, e, f in which e and f are not next to each other?

70. How many four-member committees are possible in a club with 15 members?

71. The figure shows a 12-sided figure with all its diagonals. How many diagonals are there?

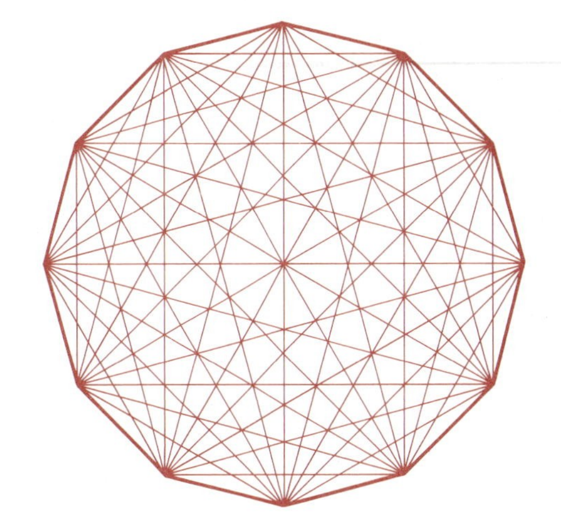

72. On a certain committee there are 9 males and 10 females. How many possibilities are there for a four-member subcommittee consisting of two females and two males?

73. Given 10 points in a plane, no three of which lie on the same line, find the number of different triangles that can be formed when the vertices are selected from the given points.

C **74.** According to *Stirling's Formula* [named after James Stirling (1692–1770)], the quantity $n!$ can be approximated as follows:

$$n! \approx \sqrt{2\pi n}\left(\frac{n}{e}\right)^n$$

In this formula, e is the constant 2.718 . . . (discussed in Section 6.1). Use a calculator to complete the following table. Round off your answers to 5 significant digits. As you will see, the numbers in the right-hand column approach 1 as n increases. This shows that, in a certain sense, the approximation improves as n increases.

n	$n!$	$\sqrt{2\pi n}\,(n/e)^n$	$\frac{n!}{\sqrt{2\pi n}\,(n/e)^n}$
10			
20			
30			
40			
50			
60			
65			

75. An integer between 1 and 15, inclusive, is selected at random. Compute the probabilities for the following events.
- **(a)** Selecting an even number.
- **(b)** Selecting an odd number.
- **(c)** Selecting either a number greater than 10 or an even number.
- **(d)** Selecting either a multiple of 2 or a multiple of 3.

76. **(a)** How many three-digit numbers are there with no repeated digits? (The first digit cannot be zero.)
(b) A three-digit number is selected at random. What is the probability that it has no repeated digits?

77. **(a)** How many three-digits numbers are there with one or more repeated digits?
(b) What is the probability that a three-digit number selected at random will contain one or more repeated digits?

For Exercises 78–88, consider an experiment in which two dice are tossed. Assume that one die is red and the other is white. Then the outcomes can be recorded as ordered pairs. Let us agree that the first element in any ordered pair corresponds to the number of dots showing on the red die and the second element corresponds to the number of dots showing on the white die. For example, (5, 4) denotes the outcome "5 on red and 4 on white." Compute the probabilities for the following events.

78. Obtaining a sum of 2

79. Obtaining a sum of 3

80. Obtaining a sum of 5

81. Obtaining a sum of 7

82. Obtaining a sum of 8

83. Obtaining a sum of either 7 or 11

84. Obtaining "doubles" (i.e., the same number of dots on both dice)

85. Obtaining a sum of at least 10

86. Obtaining a sum that is either even or less than 5

87. Obtaining any sum other than 7 and 8

88. Obtaining either doubles or a sum greater than 9

For Exercises 89–92, convert each complex number to rectangular form.

89. $3\left(\cos\frac{\pi}{3} + i\sin\frac{\pi}{3}\right)$

90. $\cos\frac{\pi}{6} + i\sin\frac{\pi}{6}$

91. $2^{1/4}\left(\cos\frac{7\pi}{4} + i\sin\frac{7\pi}{4}\right)$

92. $5\left[\cos\left(-\frac{\pi}{4}\right) + i\sin\left(-\frac{\pi}{4}\right)\right]$

In Exercises 93–96, express the complex numbers in trigonometric form.

93. $\frac{1}{2} + \frac{\sqrt{3}}{2}i$

94. $3i$

95. $-3\sqrt{2} - 3\sqrt{2}i$

96. $2\sqrt{3} - 2i$

For Exercises 97–106, carry out the indicated operations. Express your results in rectangular form for those cases in which the trigonometric functions are readily evaluated without tables or a calculator.

97. $5\left(\cos\frac{\pi}{7} + i\sin\frac{\pi}{7}\right) \times 2\left(\cos\frac{3\pi}{28} + i\sin\frac{3\pi}{28}\right)$

98. $4\left(\cos\frac{\pi}{12} + i\sin\frac{\pi}{12}\right) \times 3\left(\cos\frac{\pi}{12} + i\sin\frac{\pi}{12}\right)$

99. $8\left(\cos\frac{\pi}{12} + i\sin\frac{\pi}{12}\right) \div 4\left(\cos\frac{\pi}{3} + i\sin\frac{\pi}{3}\right)$

100. $4(\cos 32^\circ + i\sin 32^\circ) \div 2^{1/2}(\cos 2^\circ + i\sin 2^\circ)$

101. $\left(\cos\frac{\pi}{9} + i\sin\frac{\pi}{9}\right) \times 3\left(\cos\frac{4\pi}{9} + i\sin\frac{4\pi}{9}\right)$

102. $5(\cos 3^\circ + i\sin 3^\circ) \div 4(\cos 5^\circ + i\sin 5^\circ)$

103. $\left[3^{1/4}\left(\cos\frac{\pi}{36} + i\sin\frac{\pi}{36}\right)\right]^{12}$

104. $\left[2\left(\cos\frac{2\pi}{15} + i\sin\frac{2\pi}{15}\right)\right]^5$

105. $(\sqrt{3} + i)^{10}$

106. $(\sqrt{2} - \sqrt{2}i)^{15}$

In Exercises 107–110, use DeMoivre's Theorem to find the indicated roots. Express your results in rectangular form.

107. Sixth roots of 1

108. Cube roots of $-64i$

109. Square roots of $\sqrt{2} - \sqrt{2}i$

110. Fourth roots of $1 + \sqrt{3}i$

C **111.** Find the five fifth roots of $1 + i$. Express the roots in rectangular form. (Round off each decimal to two places in the final answer.)

112. If $z = r(\cos\theta + i\sin\theta)$, show that $r[\cos(\theta + \pi) + i\sin(\theta + \pi)] = -z$.

113. Trigonometric identities for $\cos n\theta$ and for $\sin n\theta$ can be derived using DeMoivre's Theorem. For example, according to DeMoivre's Theorem we have $(\cos\theta + i\sin\theta)^3 = \cos 3\theta + i\sin 3\theta$. Now expand the expression on the left-hand side of this last equation, and then equate the corresponding real parts and imaginary parts in the equation that results. What identities do you obtain?

Chapter 11 Test

1. Use the Principle of Mathematical Induction to show that the following formula is valid for all natural numbers n.

$$1^2 + 2^2 + 3^2 + \cdots + n^2 = \frac{n(n + 1)(2n + 1)}{6}$$

2. Express each of the following sums without using sigma notation, and then evaluate each sum.
   (a) $\sum_{k=0}^{2} (10k - 1)$
   (b) $\sum_{k=1}^{3} (-1)^k k^2$

3. **(a)** Write the formula for the sum S_n of a finite geometric series.
   (b) Evaluate the sum $\frac{3}{2} + \frac{3^2}{2^2} + \frac{3^3}{2^3} + \cdots + \frac{3^{10}}{2^{10}}$.

4. **(a)** Determine the coefficient of the term containing a^3 in the expansion of $(a - 2b^3)^{11}$.
   (b) Find the fifth term of the expansion in part (a).

5. Expand the expression $(3x^2 + y^3)^5$.

6. Determine the sum of the first 12 terms of an arithmetic sequence in which the first term is 8 and the 12th term is $\frac{43}{2}$.

7. Find the sum of the infinite geometric series $\frac{7}{10} + \frac{7}{100} + \frac{7}{1000} + \cdots$.

8. A sequence is defined recursively as follows: $a_1 = 1$, $a_2 = 1$, and $a_n = (a_{n-1})^2 + a_{n-2}$ for $n \geq 3$. Determine the fourth and the fifth terms in this sequence.

9. In a certain geometric sequence, the third term is 4 and the fifth term is 10. Find the sixth term, given that the common ratio is negative.

10. What is the 20th term in the arithmetic sequence $-61, -46, -31, \ldots$?

11. How many different three-digit numbers can be formed using the digits 2, 4, 6, 8, and 9 if no repetitions are allowed (within a given number)?

12. From a set of nine pictures, five are to be selected and hung in a row. How many arrangements are possible?

13. There are seven women and six men on a certain committee. How many possibilities are there for a four-member subcommittee consisting of two women and two men?

14. Suppose that three faulty batteries are accidentally shipped along with nine reliable batteries. What is the probability that two batteries selected at random from this shipment will both be reliable?

15. Two balls are selected at random from a bag containing five red and seven green balls. Compute the probability for each of the following events.
   (a) Drawing two red balls
   (b) Drawing two green balls
   (c) Drawing either two red or two green balls

16. Find the rectangular form for the complex number $z = 2\left(\cos \frac{2\pi}{3} + i \sin \frac{2\pi}{3}\right)$.

17. Find the trigonometric form of the complex number $\sqrt{2} - \sqrt{2}i$.

18. Complete the statement: According to DeMoivre's Theorem, if n is a natural number, then $[r(\cos \theta + i \sin \theta)]^n =$ ______.

19. Let $z = 3\left(\cos \frac{2\pi}{9} + i \sin \frac{2\pi}{9}\right)$ and $w = 5\left(\cos \frac{\pi}{9} + i \sin \frac{\pi}{9}\right)$. Compute the product zw and express the answer in rectangular form.

20. Compute the cube roots of $64i$.

Appendix

A.1 Significant Digits and Calculators

Many of the numbers that we use in scientific work and in daily life are approximations. In some cases the approximations arise because the numbers were obtained through measurements or experiments. Consider, for example, the following statement from an astronomy textbook:

The diameter of the Moon is 3476 km.

We interpret this statement as meaning that the actual diameter D is closer to 3476 km than it is to either 3475 km or 3477 km. In other words,

$$3475.5 \text{ km} \leq D \leq 3476.5 \text{ km}$$

The interval [3475.5, 3476.5] in this example provides information about the accuracy of the measurement. Another way to indicate accuracy in an approximation is by specifying the number of *significant digits* it contains. The measurement 3476 km has four significant digits. In general, the number of significant digits in a given number is found as follows.

Significant Digits

The number of significant digits in a given number is determined by counting the digits from left to right, beginning with the left-most nonzero digit.

Examples

Number	*Number of Significant Digits*
1.43	3
0.52	2
0.05	1
4837	4
4837.0	5

Numbers obtained through measurements are not the only source of approximations in scientific work. For example, to five significant digits, we have the following approximation for the irrational number π:

$$\pi \approx 3.1416$$

This statement tells us that π is closer to 3.1416 than it is to either 3.1415 or 3.1417. In other words,

$$3.14155 \leq \pi \leq 3.14165$$

Table 1 provides some additional examples of the ideas we've introduced.

Table 1

Number	Number of Significant Digits	Range of Measurement
37	2	[36.5, 37.5]
37.0	3	[36.95, 37.05]
268.1	4	[268.05, 268.15]
1.036	4	[1.0355, 1.0365]
0.036	2	[0.0355, 0.0365]

There is an ambiguity involving zero that can arise in counting significant digits. Suppose that someone measures the width w of a rectangle and reports the result as 30 cm. How many significant digits are there? If the value 30 cm was obtained by measuring to the nearest 10 cm, then only the digit 3 is significant, and we can conclude only that the width w lies in the range 25 cm $\leq w \leq$ 35 cm. On the other hand, if the 30 cm was obtained by measuring to the nearest 1 cm, then both the digits 3 and 0 are significant and we have 29.5 cm $\leq w \leq$ 30.5 cm.

By using *scientific notation* we can avoid the type of ambiguity discussed in the previous paragraph. As explained in Section 1.4, a number written in the form

$$b \times 10^n \qquad \text{where } 1 \leq b < 10 \text{ and } n \text{ is an integer}$$

is said to be expressed in scientific notation. For the example in the previous paragraph, then, we would write

$$w = 3 \times 10^1 \text{ cm} \qquad \text{if the measurement were to the nearest 10 cm}$$

and

$$w = 3.0 \times 10^1 \text{ cm} \qquad \text{if the measurement were to the nearest 1 cm}$$

As the figures in Table 2 indicate, for a number $b \times 10^n$ in scientific notation, the number of significant digits is just the number of digits in b. (This is one of the advantages in using scientific notation; the number of significant digits, and hence the accuracy of the measurement, is readily apparent.)

Table 2

Measurement	Number of Significant Digits	Range of Measurement
Mass of the earth:		
6×10^{27} gm	1	[5.5×10^{27} gm, 6.5×10^{27} gm]
6.0×10^{27} gm	2	[5.95×10^{27} gm, 6.05×10^{27} gm]
5.974×10^{27} gm	4	[5.9735×10^{27} gm, 5.9745×10^{27} gm]
Mass of a proton:		
1.67×10^{-24} gm	3	[1.665×10^{-24} gm, 1.675×10^{-24} gm]

There are a number of exercises in this text in which a calculator either is required or is extremely useful. Although you could work many of these exercises using tables instead of a calculator, this author recommends that you purchase a scientific calculator. *Scientific calculator* is a generic term describing a calculator with (at least) the following features or functions beyond the usual arithmetic functions:

1. Memory
2. Scientific notation
3. Powers and roots
4. Logarithms (base ten and base e)
5. Trigonometric functions
6. Inverse functions

Because of the variety and differences in the calculators that are available, no specific instructions are provided in this appendix for operating a calculator. When you buy a calculator, read the owner's manual carefully and work through some of the examples in it. In general, learn to use the memory capabilities of the calculator so that, as far as possible, you don't need to write down the results of the intermediate steps in a given calculation.

Many of the numerical exercises in the text ask that you round off the answers to a specified number of decimal places. Our rules for rounding off are as follows.

Rules for Rounding Off a Number (With More Than n Decimal Places) to n Decimal Places

1. If the digit in the $(n + 1)$st decimal place is greater than 5, increase the digit in the nth place by 1. If the digit in the $(n + 1)$st place is less than 5, leave the nth digit unchanged.
2. If the digit in the $(n + 1)$st decimal place is 5 and there is at least one nonzero digit to the right of this 5, increase the digit in the nth decimal place by 1.
3. If the digit in the $(n + 1)$st decimal place is 5 and there are no nonzero digits to the right of this 5, then increase the digit in the nth decimal place by 1 only if this results in an even digit.

The examples in Table 3 illustrate the use of these rules.

Table 3

Number	Rounded to One Decimal Place	Rounded to Three Decimal Places
4.3742	4.4	4.374
2.0515	2.1	2.052
2.9925	3.0	2.992

These same rules can be adapted for rounding off a result to a specified number of significant digits. As examples of this, we have

2347 rounded off to two significant digits is 2300 $= 2.3 \times 10^3$
2347 rounded off to three significant digits is 2350 $= 2.35 \times 10^3$
975 rounded off to two significant digits is 980 $= 9.8 \times 10^2$
0.985 rounded off to two significant digits is 0.98 $= 9.8 \times 10^{-1}$

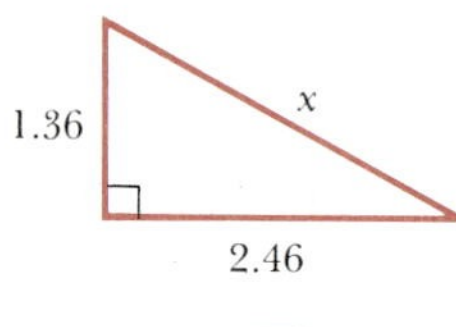

Figure 1

In calculator exercises that ask you to round off your answers, it's important that you postpone rounding until the final calculation is carried out. For example, suppose that you are required to determine the hypotenuse x of the right triangle in Figure 1 to two significant digits. Using the Pythagorean Theorem, we have

$$\begin{aligned} x &= \sqrt{(1.36)^2 + (2.46)^2} \\ &= 2.8 \quad \text{(using a calculator and rounding off the final result to two significant digits)} \end{aligned}$$

On the other hand, if we first round off each of the given lengths to two significant digits, we obtain

$$\begin{aligned} x &= \sqrt{(1.4)^2 + (2.5)^2} \\ &= 2.9 \quad \text{(to two significant digits)} \end{aligned}$$

This last result is inappropriate, and we can see why as follows. As Table 4 shows, the maximum possible values for the sides are 1.365 and 2.465, respectively.

Table 4

Number	Range of Measurement
1.36	[1.355, 1.365]
2.46	[2.455, 2.465]

Thus, the maximum possible value for the hypotenuse must be

$$\begin{aligned} \sqrt{(1.365)^2 + (2.465)^2} &= 2.817\ldots \quad \text{(calculator display)} \\ &= 2.8 \quad \text{(to two significant digits)} \end{aligned}$$

This shows that the value 2.9 is indeed inappropriate, as we stated previously.

An error often made by people working with calculators and approximations is to report a final answer with a greater degree of accuracy than the data warrant. Consider, for example, the right triangle in Figure 2. Using the Pythagorean Theorem and a calculator with an eight-digit display, we obtain

Figure 2

$$h = 3.6055513 \text{ cm}$$

This value for h is inappropriate, since common sense tells us that the answer should be no more accurate than the data used to obtain that answer. In particular, since the given sides of the triangle apparently were measured only to the nearest tenth of a centimeter, we certainly should not expect any improvement in accuracy for the resulting value of the hypotenuse. An appropriate form for the value of h here would be $h = 3.6$ cm. In general, for calculator exercises in this text that do not specify a required number of decimal places or significant digits in the final results, you should use the following guidelines.

Guidelines for Computing with Approximations

1. *For adding and subtracting:* Round off the final result so that it contains only as many decimal places as there are in the data with the fewest decimal places.
2. *For multiplying and dividing:* Round off the final result so that it contains only as many significant digits as there are in the data with the fewest significant digits.
3. *For powers and roots:* In computing a power or a root of a real number b, round off the result so that it contains as many significant digits as there are in b.

A.2 Some Proofs for Section 1.2

Some of the fundamental algebraic properties of the real number system are listed in Table 1 in Section 1.2. In this appendix we use those properties to prove the following theorem.

Theorem

Let a and b be real numbers. Then

(a) $a \cdot 0 = 0$

(b) $-a = (-1)a$

(c) $-(-a) = a$

(d) $a(-b) = -ab$

(e) $(-a)(-b) = ab$

Proof of part (a)

$a \cdot 0 = a \cdot (0 + 0)$	additive identity property
$= a \cdot 0 + a \cdot 0$	distributive property

Now since $a \cdot 0$ is a real number, it has an additive inverse $-(a \cdot 0)$. Adding this to both sides of the last equation, we obtain

$a \cdot 0 + [-(a \cdot 0)] = (a \cdot 0 + a \cdot 0) + [-(a \cdot 0)]$	
$a \cdot 0 + [-(a \cdot 0)] = a \cdot 0 + \{a \cdot 0 + [-(a \cdot 0)]\}$	associative property of addition
$0 = a \cdot 0 + 0$	additive inverse property
$0 = a \cdot 0$	additive identity property

Thus $a \cdot 0 = 0$, as we wished to show.

Proof of part (b)

$0 = 0 \cdot a$	using part (a) and the commutative property of multiplication
$= [1 + (-1)]a$	additive inverse property
$= 1 \cdot a + (-1)a$	distributive property
$= a + (-1)a$	multiplicative identity property

Now, by adding $-a$ to both sides of this last equation, we obtain

$$-a + 0 = -a + [a + (-1)a]$$

$-a = (-a + a) + (-1)a$	additive identity property and associative property of addition
$= 0 + (-1)a$	additive inverse property
$= (-1)a$	additive identity property

This last equation asserts that $-a = (-1)a$, as we wished to show.

Proof of part (c)

$-(-a) + (-a) = 0$	additive inverse property

By adding a to both sides of this last equation, we obtain

$$[-(-a) + (-a)] + a = 0 + a$$

$-(-a) + (-a + a) = a$	associative property of addition and additive identity property
$-(-a) + 0 = a$	additive inverse property
$-(-a) = a$	additive identity property

This last equation states that $-(-a) = a$, as we wished to show.

Proof of part (d)

$a(-b) = a[(-1)b]$	using part (b)
$= [a(-1)]b$	associative property of multiplication
$= [(-1)a]b$	commutative property of multiplication
$= (-1)(ab)$	associative property of multiplication
$= -(ab)$	using part (b)

Thus $a(-b) = -ab$, as we wished to show.

Proof of part (e)

$(-a)(-b) = -[(-a)b]$	using part (d)
$= -[b(-a)]$	commutative property of multiplication
$= -[-(ba)]$	using part (d)
$= ba$	using part (c)
$= ab$	commutative property of multiplication

We've now shown that $(-a)(-b) = ab$, as required.

A.3 $\sqrt{2}$ Is Irrational

The proof is by reductio ad absurdum, *and* reductio ad absurdum, *which Euclid loved so much, is one of a mathematician's finest weapons.*

G. H. Hardy (1877–1947)

We will use an *indirect proof* to show that the square root of two is an irrational number. The strategy is as follows.

1. We suppose that $\sqrt{2}$ is a rational number.
2. Using (1) and the usual rules of logic and algebra, we derive a contradiction.
3. On the basis of the contradiction in (2), we conclude that the supposition in (1) is untenable; that is, we conclude that $\sqrt{2}$ is irrational.

In carrying out the proof, we'll assume as known the following three statements.

- If x is an even natural number, then $x = 2k$, for some natural number k.
- Any rational number can be written in the form a/b, where the integers a and b have no common integral factors other than ± 1. (In other words, any fraction can be reduced to lowest terms.)
- If x is a natural number and x^2 is even, then x is even.

Our indirect proof now proceeds as follows. Suppose that $\sqrt{2}$ were a rational number. Then we would be able to write

$$\sqrt{2} = \frac{a}{b} \quad \text{where } a \text{ and } b \text{ are natural numbers with no common factor other than 1} \tag{1}$$

Since both sides of equation (1) are positive, we can square both sides to obtain the equivalent equation

$$2 = \frac{a^2}{b^2}$$

or

$$2b^2 = a^2 \tag{2}$$

Since the left-hand side of equation (2) is an even number, the right-hand side must be even. But if a^2 is even, then a is even, and so

$$a = 2k, \text{ for some natural number } k$$

Using this last equation to substitute for a in equation (2), we have

$$2b^2 = (2k)^2 = 4k^2$$

or

$$b^2 = 2k^2$$

Hence (reasoning as before) b^2 is even, and therefore b is even. But then we have that both b and a are even, contrary to our hypothesis that b and a have no common factor other than 1. We conclude from this that equation (1) cannot hold; that is, there is no rational number a/b such that $\sqrt{2} = a/b$. Thus $\sqrt{2}$ is irrational, as we wished to prove.

Tables

Table 1: Squares and Square Roots

No.	Sq.	Sq. Rt.	No.	Sq.	Sq. Rt.
1	1	1.000	51	2,601	7.141
2	4	1.414	52	2,704	7.211
3	9	1.732	53	2,809	7.280
4	16	2.000	54	2,916	7.348
5	25	2.236	55	3,025	7.416
6	36	2.449	56	3,136	7.483
7	49	2.646	57	3,249	7.550
8	64	2.828	58	3,364	7.616
9	81	3.000	59	3,481	7.681
10	100	3.162	60	3,600	7.746
11	121	3.317	61	3,721	7.810
12	144	3.464	62	3,844	7.874
13	169	3.606	63	3,969	7.937
14	196	3.742	64	4,096	8.000
15	225	3.873	65	4,225	8.062
16	256	4.000	66	4,356	8.124
17	289	4.123	67	4,489	8.185
18	324	4.243	68	4,624	8.246
19	361	4.359	69	4,761	8.307
20	400	4.472	70	4,900	8.367
21	441	4.583	71	5,041	8.426
22	484	4.690	72	5,184	8.485
23	529	4.796	73	5,329	8.544
24	576	4.899	74	5,476	8.602
25	625	5.000	75	5,625	8.660
26	676	5.099	76	5,776	8.718
27	729	5.196	77	5,929	8.775
28	784	5.292	78	6,084	8.832
29	841	5.385	79	6,241	8.888
30	900	5.477	80	6,400	8.944
31	961	5.568	81	6,561	9.000
32	1,024	5.657	82	6,724	9.055
33	1,089	5.745	83	6,889	9.110
34	1,156	5.831	84	7,056	9.165
35	1,225	5.916	85	7,225	9.220
36	1,296	6.000	86	7,396	9.274
37	1,369	6.083	87	7,569	9.327
38	1,444	6.164	88	7,744	9.381
39	1,521	6.245	89	7,921	9.434
40	1,600	6.325	90	8,100	9.487
41	1,681	6.403	91	8,281	9.539
42	1,764	6.481	92	8,464	9.592
43	1,849	6.557	93	8,649	9.644
44	1,936	6.633	94	8,836	9.695
45	2,025	6.708	95	9,025	9.747
46	2,116	6.782	96	9,216	9.798
47	2,209	6.856	97	9,409	9.849
48	2,304	6.928	98	9,604	9.899
49	2,401	7.000	99	9,801	9.950
50	2,500	7.071	100	10,000	10.000

Table 2: Exponential Functions

x	e^x	e^{-x}	x	e^x	e^{-x}
0.00	1.0000	1.0000	1.5	4.4817	0.2231
0.01	1.0101	0.9901	1.6	4.9530	0.2019
0.02	1.0202	0.9802	1.7	5.4739	0.1827
0.03	1.0305	0.9704	1.8	6.0496	0.1653
0.04	1.0408	0.9608	1.9	6.6859	0.1496
0.05	1.0513	0.9512	2.0	7.3891	0.1353
0.06	1.0618	0.9418	2.1	8.1662	0.1225
0.07	1.0725	0.9324	2.2	9.0250	0.1108
0.08	1.0833	0.9231	2.3	9.9742	0.1003
0.09	1.0942	0.9139	2.4	11.023	0.0907
0.10	1.1052	0.9048	2.5	12.182	0.0821
0.11	1.1163	0.8958	2.6	13.464	0.0743
0.12	1.1275	0.8869	2.7	14.880	0.0672
0.13	1.1388	0.8781	2.8	16.445	0.0608
0.14	1.1503	0.8694	2.9	18.174	0.0550
0.15	1.1618	0.8607	3.0	20.086	0.0498
0.16	1.1735	0.8521	3.1	22.198	0.0450
0.17	1.1853	0.8437	3.2	24.533	0.0408
0.18	1.1972	0.8353	3.3	27.113	0.0369
0.19	1.2092	0.8270	3.4	29.964	0.0334
0.20	1.2214	0.8187	3.5	33.115	0.0302
0.21	1.2337	0.8106	3.6	36.598	0.0273
0.22	1.2461	0.8025	3.7	40.447	0.0247
0.23	1.2586	0.7945	3.8	44.701	0.0224
0.24	1.2712	0.7866	3.9	49.402	0.0202
0.25	1.2840	0.7788	4.0	54.598	0.0183
0.30	1.3499	0.7408	4.1	60.340	0.0166
0.35	1.4191	0.7047	4.2	66.686	0.0150
0.40	1.4918	0.6703	4.3	73.700	0.0136
0.45	1.5683	0.6376	4.4	81.451	0.0123
0.50	1.6487	0.6065	4.5	90.017	0.0111
0.55	1.7333	0.5769	4.6	99.484	0.0101
0.60	1.8221	0.5488	4.7	109.95	0.0091
0.65	1.9155	0.5220	4.8	121.51	0.0082
0.70	2.0138	0.4966	4.9	134.29	0.0074
0.75	2.1170	0.4724	5.0	148.41	0.0067
0.80	2.2255	0.4493	5.5	244.69	0.0041
0.85	2.3396	0.4274	6.0	403.43	0.0025
0.90	2.4596	0.4066	6.5	665.14	0.0015
0.95	2.5857	0.3867	7.0	1096.6	0.0009
1.0	2.7183	0.3679	7.5	1808.0	0.0006
1.1	3.0042	0.3329	8.0	2981.0	0.0003
1.2	3.3201	0.3012	8.5	4914.8	0.0002
1.3	3.6693	0.2725	9.0	8103.1	0.0001
1.4	4.0552	0.2466	10.0	22026	0.00005

Table 3: Logarithms to the Base Ten

x	0	1	2	3	4	5	6	7	8	9
1.0	.0000	.0043	.0086	.0128	.0170	.0212	.0253	.0294	.0334	.0374
1.1	.0414	.0453	.0492	.0531	.0569	.0607	.0645	.0682	.0719	.0755
1.2	.0792	.0828	.0864	.0899	.0934	.0969	.1004	.1038	.1072	.1106
1.3	.1139	.1173	.1206	.1239	.1271	.1303	.1335	.1367	.1399	.1430
1.4	.1461	.1492	.1523	.1553	.1584	.1614	.1644	.1673	.1703	.1732
1.5	.1761	.1790	.1818	.1847	.1875	.1903	.1931	.1959	.1987	.2014
1.6	.2041	.2068	.2095	.2122	.2148	.2175	.2201	.2227	.2253	.2279
1.7	.2304	.2330	.2355	.2380	.2405	.2430	.2455	.2480	.2504	.2529
1.8	.2553	.2577	.2601	.2625	.2648	.2672	.2695	.2718	.2742	.2765
1.9	.2788	.2810	.2833	.2856	.2878	.2900	.2923	.2945	.2967	.2989
2.0	.3010	.3032	.3054	.3075	.3096	.3118	.3139	.3160	.3181	.3201
2.1	.3222	.3243	.3263	.3284	.3304	.3324	.3345	.3365	.3385	.3404
2.2	.3424	.3444	.3464	.3483	.3502	.3522	.3541	.3560	.3579	.3598
2.3	.3617	.3636	.3655	.3674	.3692	.3711	.3729	.3747	.3766	.3784
2.4	.3802	.3820	.3838	.3856	.3874	.3892	.3909	.3927	.3945	.3962
2.5	.3979	.3997	.4014	.4031	.4048	.4065	.4082	.4099	.4116	.4133
2.6	.4150	.4166	.4183	.4200	.4216	.4232	.4249	.4265	.4281	.4298
2.7	.4314	.4330	.4346	.4362	.4378	.4393	.4409	.4425	.4440	.4456
2.8	.4472	.4487	.4502	.4518	.4533	.4548	.4564	.4579	.4594	.4609
2.9	.4624	.4639	.4654	.4669	.4683	.4698	.4713	.4728	.4742	.4757
3.0	.4771	.4786	.4800	.4814	.4829	.4843	.4857	.4871	.4886	.4900
3.1	.4914	.4928	.4942	.4955	.4969	.4983	.4997	.5011	.5024	.5038
3.2	.5051	.5065	.5079	.5092	.5105	.5119	.5132	.5145	.5159	.5172
3.3	.5185	.5198	.5211	.5224	.5237	.5250	.5263	.5276	.5289	.5302
3.4	.5315	.5328	.5340	.5353	.5366	.5378	.5391	.5403	.5416	.5428
3.5	.5441	.5453	.5465	.5478	.5490	.5502	.5514	.5527	.5539	.5551
3.6	.5563	.5575	.5587	.5599	.5611	.5623	.5635	.5647	.5658	.5670
3.7	.5682	.5694	.5705	.5717	.5729	.5740	.5752	.5763	.5775	.5786
3.8	.5798	.5809	.5821	.5832	.5843	.5855	.5866	.5877	.5888	.5899
3.9	.5911	.5922	.5933	.5944	.5955	.5966	.5977	.5988	.5999	.6010
4.0	.6021	.6031	.6042	.6053	.6064	.6075	.6085	.6096	.6107	.6117
4.1	.6128	.6138	.6149	.6160	.6170	.6180	.6191	.6201	.6212	.6222
4.2	.6232	.6243	.6253	.6263	.6274	.6284	.6294	.6304	.6314	.6325
4.3	.6335	.6345	.6355	.6365	.6375	.6385	.6395	.6405	.6415	.6425
4.4	.6435	.6444	.6454	.6464	.6474	.6484	.6493	.6503	.6513	.6522
4.5	.6532	.6542	.6551	.6561	.6571	.6580	.6590	.6599	.6609	.6618
4.6	.6628	.6637	.6646	.6656	.6665	.6675	.6684	.6693	.6702	.6712
4.7	.6721	.6730	.6739	.6749	.6758	.6767	.6776	.6785	.6794	.6803
4.8	.6812	.6821	.6830	.6839	.6848	.6857	.6866	.6875	.6884	.6893
4.9	.6902	.6911	.6920	.6928	.6937	.6946	.6955	.6964	.6972	.6981
5.0	.6990	.6998	.7007	.7016	.7024	.7033	.7042	.7050	.7059	.7067
5.1	.7076	.7084	.7093	.7101	.7110	.7118	.7126	.7135	.7143	.7152
5.2	.7160	.7168	.7177	.7185	.7193	.7202	.7210	.7218	.7226	.7235
5.3	.7243	.7251	.7259	.7267	.7275	.7284	.7292	.7300	.7308	.7316
5.4	.7324	.7332	.7340	.7348	.7356	.7364	.7372	.7380	.7388	.7396
x	0	1	2	3	4	5	6	7	8	9

Table 3: Logarithms to the Base Ten (*continued*)

x	0	1	2	3	4	5	6	7	8	9
5.5	.7404	.7412	.7419	.7427	.7435	.7443	.7451	.7459	.7466	.7474
5.6	.7482	.7490	.7497	.7505	.7513	.7520	.7528	.7536	.7543	.7551
5.7	.7559	.7566	.7574	.7582	.7589	.7597	.7604	.7612	.7619	.7627
5.8	.7634	.7642	.7649	.7657	.7664	.7672	.7679	.7686	.7694	.7701
5.9	.7709	.7716	.7723	.7731	.7738	.7745	.7752	.7760	.7767	.7774
6.0	.7782	.7789	.7796	.7803	.7810	.7818	.7825	.7832	.7839	.7846
6.1	.7853	.7860	.7868	.7875	.7882	.7889	.7896	.7903	.7910	.7917
6.2	.7924	.7931	.7938	.7945	.7952	.7959	.7966	.7973	.7980	.7987
6.3	.7993	.8000	.8007	.8014	.8021	.8028	.8035	.8041	.8048	.8055
6.4	.8062	.8069	.8075	.8082	.8089	.8096	.8102	.8109	.8116	.8122
6.5	.8129	.8136	.8142	.8149	.8156	.8162	.8169	.8176	.8182	.8189
6.6	.8195	.8202	.8209	.8215	.8222	.8228	.8235	.8241	.8248	.8254
6.7	.8261	.8267	.8274	.8280	.8287	.8293	.8299	.8306	.8312	.8319
6.8	.8325	.8331	.8338	.8344	.8351	.8357	.8363	.8370	.8376	.8382
6.9	.8388	.8395	.8401	.8407	.8414	.8420	.8426	.8432	.8439	.8445
7.0	.8451	.8457	.8463	.8470	.8476	.8482	.8488	.8494	.8500	.8506
7.1	.8513	.8519	.8525	.8531	.8537	.8543	.8549	.8555	.8561	.8567
7.2	.8573	.8579	.8585	.8591	.8597	.8603	.8609	.8615	.8621	.8627
7.3	.8633	.8639	.8645	.8651	.8657	.8663	.8669	.8675	.8681	.8686
7.4	.8692	.8698	.8704	.8710	.8716	.8722	.8727	.8733	.8739	.8745
7.5	.8751	.8756	.8762	.8768	.8774	.8779	.8785	.8791	.8797	.8802
7.6	.8808	.8814	.8820	.8825	.8831	.8837	.8842	.8848	.8854	.8859
7.7	.8865	.8871	.8876	.8882	.8887	.8893	.8899	.8904	.8910	.8915
7.8	.8921	.8927	.8932	.8938	.8943	.8949	.8954	.8960	.8965	.8971
7.9	.8976	.8982	.8987	.8993	.8998	.9004	.9009	.9015	.9020	.9025
8.0	.9031	.9036	.9042	.9047	.9053	.9058	.9063	.9069	.9074	.9079
8.1	.9085	.9090	.9096	.9101	.9106	.9112	.9117	.9122	.9128	.9133
8.2	.9138	.9143	.9149	.9154	.9159	.9165	.9170	.9175	.9180	.9186
8.3	.9191	.9196	.9201	.9206	.9212	.9217	.9222	.9227	.9232	.9238
8.4	.9243	.9248	.9253	.9258	.9263	.9269	.9274	.9279	.9284	.9289
8.5	.9294	.9299	.9304	.9309	.9315	.9320	.9325	.9330	.9335	.9340
8.6	.9345	.9350	.9355	.9360	.9365	.9370	.9375	.9380	.9385	.9390
8.7	.9395	.9400	.9405	.9410	.9415	.9420	.9425	.9430	.9435	.9440
8.8	.9445	.9450	.9455	.9460	.9465	.9469	.9474	.9479	.9484	.9489
8.9	.9494	.9499	.9504	.9509	.9513	.9518	.9523	.9528	.9533	.9538
9.0	.9542	.9547	.9552	.9557	.9562	.9566	.9571	.9576	.9581	.9586
9.1	.9590	.9595	.9600	.9605	.9609	.9614	.9619	.9624	.9628	.9633
9.2	.9638	.9643	.9647	.9652	.9657	.9661	.9666	.9671	.9675	.9680
9.3	.9685	.9689	.9694	.9699	.9703	.9708	.9713	.9717	.9722	.9727
9.4	.9731	.9736	.9741	.9745	.9750	.9754	.9759	.9763	.9768	.9773
9.5	.9777	.9782	.9786	.9791	.9795	.9800	.9805	.9809	.9814	.9818
9.6	.9823	.9827	.9832	.9836	.9841	.9845	.9850	.9854	.9859	.9863
9.7	.9868	.9872	.9877	.9881	.9886	.9890	.9894	.9899	.9903	.9908
9.8	.9912	.9917	.9921	.9926	.9930	.9934	.9939	.9943	.9948	.9952
9.9	.9956	.9961	.9965	.9969	.9974	.9978	.9983	.9987	.9991	.9996

x 0 1 2 3 4 5 6 7 8 9

Table 4: Natural Logarithms

n	ln *n*	*n*	ln *n*	*n*	ln *n*
		4.5	1.5041	9.0	2.1972
0.1	−2.3026	4.6	1.5261	9.1	2.2083
0.2	−1.6094	4.7	1.5476	9.2	2.2192
0.3	−1.2040	4.8	1.5686	9.3	2.2300
0.4	−0.9163	4.9	1.5892	9.4	2.2407
0.5	−0.6931	5.0	1.6094	9.5	2.2513
0.6	−0.5108	5.1	1.6292	9.6	2.2618
0.7	−0.3567	5.2	1.6487	9.7	2.2721
0.8	−0.2231	5.3	1.6677	9.8	2.2824
0.9	−0.1054	5.4	1.6864	9.9	2.2925
1.0	0.0000	5.5	1.7047	10	2.3026
1.1	0.0953	5.6	1.7228	11	2.3979
1.2	0.1823	5.7	1.7405	12	2.4849
1.3	0.2624	5.8	1.7579	13	2.5649
1.4	0.3365	5.9	1.7750	14	2.6391
1.5	0.4055	6.0	1.7918	15	2.7081
1.6	0.4700	6.1	1.8083	16	2.7726
1.7	0.5306	6.2	1.8245	17	2.8332
1.8	0.5878	6.3	1.8405	18	2.8904
1.9	0.6419	6.4	1.8563	19	2.9444
2.0	0.6931	6.5	1.8718	20	2.9957
2.1	0.7419	6.6	1.8871	25	3.2189
2.2	0.7885	6.7	1.9021	30	3.4012
2.3	0.8329	6.8	1.9169	35	3.5553
2.4	0.8755	6.9	1.9315	40	3.6889
2.5	0.9163	7.0	1.9459	45	3.8067
2.6	0.9555	7.1	1.9601	50	3.9120
2.7	0.9933	7.2	1.9741	55	4.0073
2.8	1.0296	7.3	1.9879	60	4.0943
2.9	1.0647	7.4	2.0015	65	4.1744
3.0	1.0986	7.5	2.0149	70	4.2485
3.1	1.1314	7.6	2.0281	75	4.3175
3.2	1.1632	7.7	2.0142	80	4.3820
3.3	1.1939	7.8	2.0541	85	4.4427
3.4	1.2238	7.9	2.0669	90	4.4998
3.5	1.2528	8.0	2.0794	95	4.5539
3.6	1.2809	8.1	2.0919	100	4.6052
3.7	1.3083	8.2	2.1041		
3.8	1.3350	8.3	2.1163		
3.9	1.3610	8.4	2.1282		
4.0	1.3863	8.5	2.1401		
4.1	1.4110	8.6	2.1518		
4.2	1.4351	8.7	2.1633		
4.3	1.4586	8.8	2.1748		
4.4	1.4816	8.9	2.1861		

Table 5: Values of Trigonometric Functions

Angle θ									
Degrees	Radians	sin θ	csc θ	tan θ	cot θ	sec θ	cos θ		
0° 00′	.0000	.0000	No value	.0000	No value	1.000	1.0000	1.5708	90° 00′
10	029	029	343.8	029	343.8	000	000	679	50
20	058	058	171.9	058	171.9	000	000	650	40
30	087	087	114.6	087	114.6	000	1.0000	621	30
40	116	116	85.95	116	85.94	000	.9999	592	20
50	145	145	68.76	145	68.75	000	999	563	10
1° 00′	.0175	.0175	57.30	.0175	57.29	1.000	.9998	1.5533	89° 00′
10	204	204	49.11	204	49.10	000	998	504	50
20	233	233	42.98	233	42.96	000	997	475	40
30	262	262	38.20	262	38.19	000	997	446	30
40	291	291	34.38	291	34.37	000	996	417	20
50	320	320	31.26	320	31.24	001	995	388	10
2° 00′	.0349	.0349	28.65	.0349	28.64	1.001	.9994	1.5359	88° 00′
10	378	378	26.45	378	26.43	001	993	330	50
20	407	407	24.56	407	24.54	001	992	301	40
30	436	436	22.93	437	22.90	001	990	272	30
40	465	465	21.49	466	21.47	001	989	243	20
50	495	494	20.23	495	20.21	001	988	213	10
3° 00′	.0524	.0523	19.11	.0524	19.08	1.001	.9986	1.5184	87° 00′
10	553	552	18.10	553	18.07	002	985	155	50
20	582	581	17.20	582	17.17	002	983	126	40
30	611	610	16.38	612	16.35	002	981	097	30
40	640	640	15.64	641	15.60	002	980	068	20
50	669	669	14.96	670	14.92	002	978	039	10
4° 00′	.0698	.0698	14.34	.0699	14.30	1.002	.9976	1.5010	86° 00′
10	727	727	13.76	729	13.73	003	974	981	50
20	756	756	13.23	758	13.20	003	971	952	40
30	785	785	12.75	787	12.71	003	969	923	30
40	814	814	12.29	816	12.25	003	967	893	20
50	844	843	11.87	846	11.83	004	964	864	10
5° 00′	.0873	.0872	11.47	.0875	11.43	1.004	.9962	1.4835	85° 00′
10	902	901	11.10	904	11.06	004	959	806	50
20	931	929	10.76	934	10.71	004	957	777	40
30	960	958	10.43	963	10.39	005	954	748	30
40	.0989	.0987	10.13	.0992	10.08	005	951	719	20
50	.1018	.1016	9.839	.1022	9.788	005	948	690	10
6° 00′	.1047	.1045	9.567	.1051	9.514	1.006	.9945	1.4661	84° 00′
10	076	074	9.309	080	9.255	006	942	632	50
20	105	103	9.065	110	9.010	006	939	603	40
30	134	132	8.834	139	8.777	006	936	573	30
40	164	161	8.614	169	8.556	007	932	544	20
50	193	190	8.405	198	8.345	007	929	515	10
7° 00′	.1222	.1219	8.206	.1228	8.144	1.008	.9925	1.4486	83° 00′
10	251	248	8.016	257	7.953	008	922	457	50
20	280	276	7.834	287	7.770	008	918	428	40
30	309	305	7.661	317	7.596	009	914	399	30
40	338	334	7.496	346	7.429	009	911	370	20
50	367	363	7.337	376	7.269	009	907	341	10
8° 00′	.1396	.1392	7.185	.1405	7.115	1.010	.9903	1.4312	82° 00′
10	425	421	7.040	435	6.968	010	899	283	50
20	454	449	6.900	465	827	011	894	254	40
30	484	478	765	495	691	011	890	224	30
40	513	507	636	524	561	012	886	195	20
50	542	536	512	554	435	012	881	166	10
9° 00′	.1571	.1564	6.392	.1584	6.314	1.012	.9877	1.4137	81° 00′
		cos θ	sec θ	cot θ	tan θ	csc θ	sin θ	Radians	Degrees
									Angle θ

Table 5: Values of Trigonometric Functions (*continued*)

Angle θ									
Degrees	Radians	sin θ	csc θ	tan θ	cot θ	sec θ	cos θ		
9° 00′	.1571	.1564	6.392	.1584	6.314	1.012	.9877	1.4137	81° 00′
10	600	593	277	614	197	013	872	108	50
20	629	622	166	644	6.084	013	868	079	40
30	658	650	6.059	673	5.976	014	863	050	30
40	687	679	5.955	703	871	014	858	1.4021	20
50	716	708	855	733	769	015	853	1.3992	10
10° 00′	.1745	.1736	5.759	.1763	5.671	1.015	.9848	1.3963	80° 00′
10	774	765	665	793	576	016	843	934	50
20	804	794	575	823	485	016	838	904	40
30	833	822	487	853	396	017	833	875	30
40	862	851	403	883	309	018	827	846	20
50	891	880	320	914	226	018	822	817	10
11° 00′	.1920	.1908	5.241	.1944	5.145	1.019	.9816	1.3788	79° 00′
10	949	937	164	.1974	5.066	019	811	759	50
20	.1978	965	089	.2004	4.989	020	805	730	40
30	.2007	.1994	5.016	035	915	020	799	701	30
40	036	.2022	4.945	065	843	021	793	672	20
50	065	051	876	095	773	022	787	643	10
12° 00′	.2094	.2079	4.810	.2126	4.705	1.022	.9781	1.3614	78° 00′
10	123	108	745	156	638	023	775	584	50
20	153	136	682	186	574	024	769	555	40
30	182	164	620	217	511	024	763	526	30
40	211	193	560	247	449	025	757	497	20
50	240	221	502	278	390	026	750	468	10
13° 00′	.2269	.2250	4.445	.2309	4.331	1.026	.9744	1.3439	77° 00′
10	298	278	390	339	275	027	737	410	50
20	327	306	336	370	219	028	730	381	40
30	356	334	284	401	165	028	724	352	30
40	385	363	232	432	113	029	717	323	20
50	414	391	182	462	061	030	710	294	10
14° 00′	.2443	.2419	4.134	.2493	4.011	1.031	.9703	1.3265	76° 00′
10	473	447	086	524	3.962	031	696	235	50
20	502	476	4.039	555	914	032	689	206	40
30	531	504	3.994	586	867	033	681	177	30
40	560	532	950	617	821	034	674	148	20
50	589	560	906	648	776	034	667	119	10
15° 00′	.2618	.2588	3.864	.2679	3.732	1.035	.9659	1.3090	75° 00′
10	647	616	822	711	689	036	652	061	50
20	676	644	782	742	647	037	644	032	40
30	705	672	742	773	606	038	636	1.3003	30
40	734	700	703	805	566	039	628	1.2974	20
50	763	728	665	836	526	039	621	945	10
16° 00′	.2793	.2756	3.628	.2867	3.487	1.040	.9613	1.2915	74° 00′
10	822	784	592	899	450	041	605	886	50
20	851	812	556	931	412	042	596	857	40
30	880	840	521	962	376	043	588	828	30
40	909	868	487	.2944	340	044	580	799	20
50	938	896	453	.3026	305	045	572	770	10
17° 00′	.2967	.2924	3.420	.3057	3.271	1.046	.9563	1.2741	73° 00′
10	.2996	952	388	089	237	047	555	712	50
20	.3025	.2979	357	121	204	048	546	683	40
30	054	.3007	326	153	172	048	537	654	30
40	083	035	295	185	140	049	528	625	20
50	113	062	265	217	108	050	520	595	10
18° 00′	.3142	.3090	3.236	.3249	3.078	1.051	.9511	1.2566	72° 00′
		cos θ	sec θ	cot θ	tan θ	csc θ	sin θ	Radians	Degrees
									Angle θ

Table 5: Values of Trigonometric Functions (*continued*)

Angle θ									
Degrees	**Radians**	**sin θ**	**csc θ**	**tan θ**	**cot θ**	**sec θ**	**cos θ**		
18° 00′	.3142	.3090	3.236	.3249	3.078	1.051	.9511	1.2566	72° 00′
10	171	118	207	281	047	052	502	537	50
20	200	145	179	314	3.018	053	492	508	40
30	229	173	152	346	2.989	054	483	479	30
40	258	201	124	378	960	056	474	450	20
50	287	228	098	411	932	057	465	421	10
19° 00′	.3316	.3256	3.072	.3443	2.904	1.058	.9455	1.2392	71° 00′
10	345	283	046	476	877	059	446	363	50
20	374	311	3.021	508	850	060	436	334	40
30	403	338	2.996	541	824	061	426	305	30
40	432	365	971	574	798	062	417	275	20
50	462	393	947	607	773	063	407	246	10
20° 00′	.3491	.3420	2.924	.3640	2.747	1.064	.9397	1.2217	70° 00′
10	520	448	901	673	723	065	387	188	50
20	549	475	878	706	699	066	377	159	40
30	578	502	855	739	675	068	367	130	30
40	607	529	833	772	651	069	356	101	20
50	636	557	812	805	628	070	346	072	10
21° 00′	.3665	.3584	2.790	.3839	2.605	1.071	.9336	1.2043	69° 00′
10	694	611	769	872	583	072	325	1.2014	50
20	723	638	749	906	560	074	315	1.1985	40
30	752	665	729	939	539	075	304	956	30
40	782	692	709	.3973	517	076	293	926	20
50	811	719	689	.4006	496	077	283	897	10
22° 00′	.3840	.3746	2.669	.4040	2.475	1.079	.9272	1.1868	68° 00′
10	869	773	650	074	455	080	261	839	50
20	898	800	632	108	434	081	250	810	40
30	927	827	613	142	414	082	239	781	30
40	956	854	595	176	394	084	228	752	20
50	985	881	577	210	375	085	216	723	10
23° 00′	.4014	.3907	2.559	.4245	2.356	1.086	.9205	1.1694	67° 00′
10	043	934	542	279	337	088	194	665	50
20	072	961	525	314	318	089	182	636	40
30	102	.3987	508	348	300	090	171	606	30
40	131	.4014	491	383	282	092	159	577	20
50	160	041	475	417	264	093	147	548	10
24° 00′	.4189	.4067	2.459	.4452	2.246	1.095	.9135	1.1519	66° 00′
10	218	094	443	487	229	096	124	490	50
20	247	120	427	522	211	097	112	461	40
30	276	147	411	557	194	099	100	432	30
40	305	173	396	592	177	100	088	403	20
50	334	200	381	628	161	102	075	374	10
25° 00′	.4363	.4226	2.366	.4663	2.145	1.103	.9063	1.1345	65° 00′
10	392	253	352	699	128	105	051	316	50
20	422	279	337	734	112	106	038	286	40
30	451	305	323	770	097	108	026	257	30
40	480	331	309	806	081	109	013	228	20
50	509	358	295	841	066	111	.9001	199	10
26° 00′	.4538	.4384	2.281	.4877	2.050	1.113	.8988	1.1170	64° 00′
10	567	410	268	913	035	114	975	141	50
20	596	436	254	950	020	116	962	112	40
30	625	462	241	.4986	2.006	117	949	083	30
40	654	488	228	.5022	1.991	119	936	054	20
50	683	514	215	059	977	121	923	1.1025	10
27° 00′	.4712	.4540	2.203	.5095	1.963	1.122	.8910	1.0996	63° 00′
		cos θ	**sec θ**	**cot θ**	**tan θ**	**csc θ**	**sin θ**	**Radians**	**Degrees**
									Angle θ

Table 5: Values of Trigonometric Functions (*continued*)

Angle θ Degrees	Radians	sin θ	csc θ	tan θ	cot θ	sec θ	cos θ		
27° 00′	.4712	.4540	2.203	.5095	1.963	1.122	.8910	1.0996	63° 00′
10	741	566	190	132	949	124	897	966	50
20	771	592	178	169	935	126	884	937	40
30	800	617	166	206	921	127	870	908	30
40	829	643	154	243	907	129	857	879	20
50	858	669	142	280	894	131	843	850	10
28° 00′	.4887	.4695	2.130	.5317	1.881	1.133	.8829	1.0821	62° 00′
10	916	720	118	354	868	134	816	792	50
20	945	746	107	392	855	136	802	763	40
30	.4974	772	096	430	842	138	788	734	30
40	.5003	797	085	467	829	140	774	705	20
50	032	823	074	505	816	142	760	676	10
29° 00′	.5061	.4848	2.063	.5543	1.804	1.143	.8746	1.0647	61° 00′
10	091	874	052	581	792	145	732	617	50
20	120	899	041	619	780	147	718	588	40
30	149	924	031	658	767	149	704	559	30
40	178	950	020	696	756	151	689	530	20
50	207	.4975	010	735	744	153	675	501	10
30° 00′	.5236	.5000	2.000	.5774	1.732	1.155	.8660	1.0472	60° 00′
10	265	025	1.990	812	720	157	646	443	50
20	294	050	980	851	709	159	631	414	40
30	323	075	970	890	698	161	616	385	30
40	352	100	961	930	686	163	601	356	20
50	381	125	951	.5969	675	165	587	327	10
31° 00′	.5411	.5150	1.942	.6009	1.664	1.167	.8572	1.0297	59° 00′
10	440	175	932	048	653	169	557	268	50
20	469	200	923	088	643	171	542	239	40
30	498	225	914	128	632	173	526	210	30
40	527	250	905	168	621	175	511	181	20
50	556	275	896	208	611	177	496	152	10
32° 00′	.5585	.5299	1.887	.6249	1.600	1.179	.8480	1.0123	58° 00′
10	614	324	878	289	590	181	465	094	50
20	643	348	870	330	580	184	450	065	40
30	672	373	861	371	570	186	434	036	30
40	701	398	853	412	560	188	418	1.0007	20
50	730	422	844	453	550	190	403	.9977	10
33° 00′	.5760	.5446	1.836	.6494	1.540	1.192	.8387	.9948	57° 00′
10	789	471	828	536	530	195	371	919	50
20	818	495	820	577	520	197	355	890	40
30	847	519	812	619	511	199	339	861	30
40	876	544	804	661	501	202	323	832	20
50	905	568	796	703	492	204	307	803	10
34° 00′	.5934	.5592	1.788	.6745	1.483	1.206	.8290	.9774	56° 00′
10	963	616	781	787	473	209	274	745	50
20	.5992	640	773	830	464	211	258	716	40
30	.6021	644	766	873	455	213	241	687	30
40	050	688	758	916	446	216	225	657	20
50	080	712	751	.6959	437	218	208	628	10
35° 00′	.6109	.5736	1.743	.7002	1.428	1.221	.8192	.9599	55° 00′
10	138	760	736	046	419	223	175	570	50
20	167	783	729	089	411	226	158	541	40
30	196	807	722	133	402	228	141	512	30
40	225	831	715	177	393	231	124	483	20
50	254	854	708	221	385	233	107	454	10
36° 00′	.6283	.5878	1.701	.7265	1.376	1.236	.8090	.9425	54° 00′
		cos θ	sec θ	cot θ	tan θ	csc θ	sin θ	Radians	Degrees (Angle θ)

Table 5: Values of Trigonometric Functions (*continued*)

Angle θ Degrees	Radians	sin θ	csc θ	tan θ	cot θ	sec θ	cos θ		
36° 00′	.6283	.5878	1.701	.7265	1.376	1.236	.8090	.9425	54° 00′
10	312	901	695	310	368	239	073	396	50
20	341	925	688	355	360	241	056	367	40
30	370	948	681	400	351	244	039	338	30
40	400	972	675	445	343	247	021	308	20
50	429	.5995	668	490	335	249	.8004	279	10
37° 00′	.6458	.6018	1.662	.7536	1.327	1.252	.7986	.9250	53° 00′
10	487	041	655	581	319	255	969	221	50
20	516	065	649	627	311	258	951	192	40
30	545	088	643	673	303	260	934	163	30
40	574	111	636	720	295	263	916	134	20
50	603	134	630	766	288	266	898	105	10
38° 00′	.6632	.6157	1.624	.7813	1.280	1.269	.7880	.9076	52° 00′
10	661	180	618	860	272	272	862	047	50
20	690	202	612	907	265	275	844	.9018	40
30	720	225	606	.7954	257	278	826	.8988	30
40	749	248	601	.8002	250	281	808	959	20
50	778	271	595	050	242	284	790	930	10
39° 00′	.6807	.6293	1.589	.8098	1.235	1.287	.7771	.8901	51° 00′
10	836	316	583	146	228	290	753	872	50
20	865	338	578	195	220	293	735	843	40
30	894	361	572	243	213	296	716	814	30
40	923	383	567	292	206	299	698	785	20
50	952	406	561	342	199	302	679	756	10
40° 00′	.6981	.6428	1.556	.8391	1.192	1.305	.7660	.8727	50° 00′
10	.7010	450	550	441	185	309	642	698	50
20	039	472	545	491	178	312	623	668	40
30	069	494	540	541	171	315	604	639	30
40	098	517	535	591	164	318	585	610	20
50	127	539	529	642	157	322	566	581	10
41° 00′	.7156	.6561	1.524	.8693	1.150	1.325	.7547	.8552	49° 00′
10	185	583	519	744	144	328	528	523	50
20	214	604	514	796	137	332	509	494	40
30	243	626	509	847	130	335	490	465	30
40	272	648	504	899	124	339	470	436	20
50	301	670	499	.8952	117	342	451	407	10
42° 00′	.7330	.6691	1.494	.9004	1.111	1.346	.7431	.8378	48° 00′
10	359	713	490	057	104	349	412	348	50
20	389	734	485	110	098	353	392	319	40
30	418	756	480	163	091	356	373	290	30
40	447	777	476	217	085	360	353	261	20
50	476	799	471	271	079	364	333	232	10
43° 00′	.7505	.6820	1.466	.9325	1.072	1.367	.7314	.8203	47° 00′
10	534	841	462	380	066	371	294	174	50
20	563	862	457	435	060	375	274	145	40
30	592	884	453	490	054	379	254	116	30
40	621	905	448	545	048	382	234	087	20
50	650	926	444	601	042	386	214	058	10
44° 00′	.7679	.6947	1.440	.9657	1.036	1.390	.7193	.8029	46° 00′
10	709	967	435	713	030	394	173	.7999	50
20	738	.6988	431	770	024	398	153	970	40
30	767	.7009	427	827	018	402	133	941	30
40	796	030	423	884	012	406	112	912	20
50	825	050	418	.9942	006	410	092	883	10
45° 00′	.7854	.7071	1.414	1.000	1.000	1.414	.7071	.7854	45° 00′
		cos θ	sec θ	cot θ	tan θ	csc θ	sin θ	Radians	Degrees
									Angle θ

Answers to Selected Exercises

Exercise Set 1.1

1. x^5y^2 **3.** $(x + 1)^3$ **5.** $4x$ **7.** $3a + 2b$ **9.** $(2a + 1)^3(2b + 1)^2$ **11.** 22 **13.** 25 **15.** 36 **17.** 21 **19.** 42 **21.** 2 **23.** 6 **25.** 1000 **27.** $x + 2$ ($2 + x$ is also correct.) **29.** $4(a + b^2)$ **31.** $x^2 + y^2$ **33.** $x + 2x^2$ **35.** $\frac{x + y + z}{3}$ **37.** $\frac{x^2 + y^2 + z^2}{3}$ **39.** $2xy - 1$ **41.** area: $9\pi y^2$; circumference: $6\pi y$ **43.** $V = 36\pi x^3$ **45.** (a) $V = \pi r^3$; (b) $V = 64\pi\ \text{cm}^3$ **47.** (a) $V = 12w^3$; (b) $S = w^2$ **49.** $C = \pi s$ **51.** 10.8 cm^3 **53.** 86.8 cm^3 **55.** (a) 1.12

Exercise Set 1.2

1. 832 **3.** 4410 **5.** (a) $A^2 + 2AB + B^2$; (b) $A^2 - B^2$; (c) $A^2 - 2AB + B^2$ **7.** (a) $x^2 + 4xy + 4y^2$; (b) $x^2 - 4y^2$; (c) $x^2 - 4xy + y^2$ **9.** (a) $Ax^2 - Ax + A$; (b) $x^3 + 1$
11. commutative property of addition
13. additive inverse property
15. associative property of addition
17. associative property of multiplication
19. identity property for addition
21. closure (addition)
23. distributive property
25. -23 **27.** 2 **29.** -1 **31.** 2 **33.** $\frac{31}{15}$ **35.** $\frac{-5}{x}$ **37.** $\frac{2y^2}{x + y}$ **39.** $\frac{-49}{36}$ **41.** $\frac{6}{35}$ **43.** $\frac{3}{5}$ **45.** $\frac{22}{37}$ **47.** -1 **49.** $\frac{5}{7}$ **51.** 16 **53.** 0 **55.** 1 **57.** 145 **63.** $y = -1.6$ (to two significant digits)

65. (a)

x	1	1.250	1.500	1.750	2.000	2.250
y	0	.2880	.3704	.3848	.3750	.3566

(b) $x = 1.750$; (c) $y = .3849$

Exercise Set 1.3

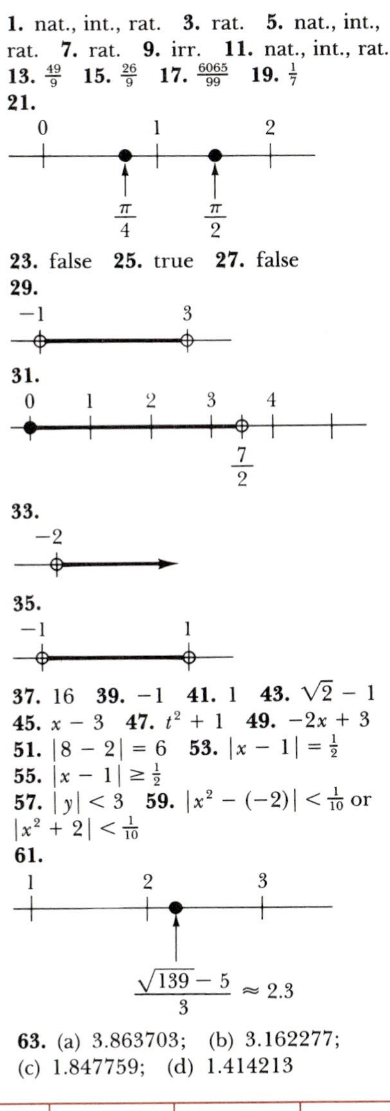

1. nat., int., rat. **3.** rat. **5.** nat., int., rat. **7.** rat. **9.** irr. **11.** nat., int., rat. **13.** $\frac{49}{9}$ **15.** $\frac{26}{9}$ **17.** $\frac{6065}{99}$ **19.** $\frac{1}{7}$
21.

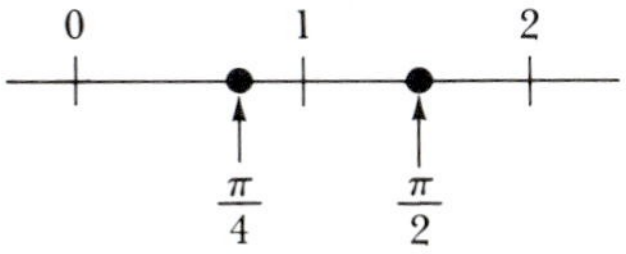

23. false **25.** true **27.** false
29.

31.

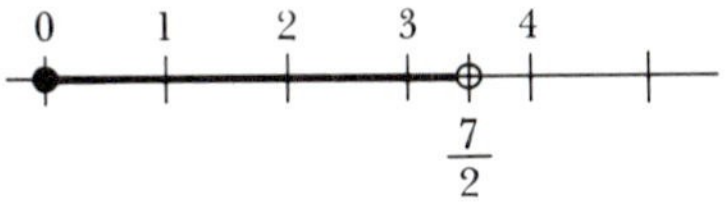

33.

35.

37. 16 **39.** -1 **41.** 1 **43.** $\sqrt{2} - 1$ **45.** $x - 3$ **47.** $t^2 + 1$ **49.** $-2x + 3$ **51.** $|8 - 2| = 6$ **53.** $|x - 1| = \frac{1}{2}$ **55.** $|x - 1| \geq \frac{1}{2}$ **57.** $|y| < 3$ **59.** $|x^2 - (-2)| < \frac{1}{10}$ or $|x^2 + 2| < \frac{1}{10}$
61.

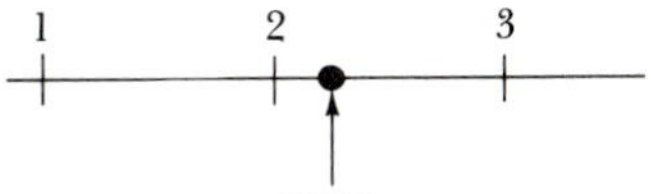

$\frac{\sqrt{139} - 5}{3} \approx 2.3$

63. (a) 3.863703; (b) 3.162277; (c) 1.847759; (d) 1.414213

Exercise Set 1.4

1. -10 **3.** $\frac{2}{3}$ **5.** 2 **7.** $(x + y)^2$ **9.** $(x + y)^2 + 3$ **11.** $[\frac{1}{2}(x^2 - 2y^3)]^2$ **13.** $|x - 1|^3$ **15.** a^{15} **17.** $(x + 1)^{10}$ **19.** 100 **21.** a^6 **23.** $\frac{x^4}{y^5}$ **25.** $16x^6$ **27.** 1 **29.** 1 **31.** $\frac{11}{100}$ **33.** 5 **35.** 1 **37.** $\frac{1}{a^6b^3}$ **39.** $\frac{a^4b^2}{c^6}$ **41.** $\frac{y^{12}}{x^6z^{12}}$ **43.** $-\frac{1}{x^4y^{10}}$ **45.** $\frac{1}{16x}$ **47.** y^6 **49.** 576 **51.** 12 **53.** 9.29×10^7 **55.** 6.68×10^4 **57.** 2.5×10^{19} **59.** (a) 8.6688×10^1; (b) 1.0604772×10^4; (c) 8.9424×10^4 **61.** (a) 1×10^{-9}; (b) 1×10^{-18} (c) 1×10^{-24} **63.** -1.602×10^{-19} **65.** $\frac{1}{x^c}$ **67.** 9^{10} **69.** 2.3×10^2
71. 8.97×10^2
73. In each case except for Pluto, the value of $\frac{a^3}{T^2}$ is 2.51×10^{19}. For Pluto, the ratio is 2.49×10^{19}.

Exercise Set 1.5

1. false **3.** false **5.** false **7.** $3\sqrt{2}$ **9.** $\frac{5}{2}$ **11.** $-2\sqrt{3}$ **13.** $3\sqrt{3}$ **15.** $\frac{7}{5}$ **17.** $6x^2$ **19.** $ab\sqrt{ab}$ **21.** $6ab^2c^2\sqrt{2ac}$ **23.** $8ab\sqrt{3ab} + 2\sqrt{3ab}$ **25.** $\frac{4\sqrt{7}}{7}$ **27.** $\frac{\sqrt{2}}{4}$ **29.** $\frac{1 - \sqrt{5}}{-4}$ $\left(\text{or } \frac{\sqrt{5} - 1}{4}\right)$ **31.** $-2 - \sqrt{3}$ **33.** $\frac{61\sqrt{5}}{5}$ **35.** $\frac{1}{\sqrt{x} + \sqrt{5}}$ **37.** $\frac{1}{\sqrt{2 + h} + \sqrt{2}}$ **39.** 13 **41.** $\sqrt{2}$ **43.** 1 **45.** $\sqrt{6}$ **47.** $\sqrt{7 + 2\sqrt{3}}$ **55.** $\frac{a + b}{a - b}$
59. (a) To six decimal places, both values are 0.646083.
60. For the first equation, the common value of both sides to six decimal places is 1.414213. For the second equation,

the value of both sides to six decimal places is 3.141592. (Actually, the values in the second equation agree to eight places; this approximation for π was discovered by Ramanujan.)

a	b	$\sqrt{ab}$	$\frac{a+b}{2}$	$\sqrt{\frac{a^2+b^2}{2}}$	Largest	Smallest
1	2	1.4142	1.5000	1.5811	R.M.	G.M.
1	3	1.7320	2	2.2361	R.M.	G.M.
1	4	2	2.5000	2.9155	R.M.	G.M.
2	3	2.4495	2.5000	2.5495	R.M.	G.M.
3	4	3.4641	3.5000	3.5355	R.M.	G.M.
9	10	9.4868	9.5000	9.5131	R.M.	G.M.
99	100	99.4987	99.5000	99.5012	R.M.	G.M.
999	1000	999.4999	999.5000	999.5001	R.M.	G.M.

Exercise Set 1.6

1. -4 **3.** $\frac{1}{5}$ **5.** undefined **7.** $\frac{3}{2}$ **9.** 2 **11.** $\sqrt[3]{2}$ **13.** $5\sqrt[5]{2}$ **15.** $-14\sqrt{6}$ **17.** $2\sqrt{2}$ **19.** $2ab\sqrt[4]{b}$ **21.** $3ab\sqrt{2a}$ **23.** $\frac{2a^4\sqrt[3]{2b^2}}{c^3}$ **25.** $\frac{a^2\sqrt[6]{5}}{b}$ **27.** $\frac{5x^2y}{z^3}$ **29.** $\sqrt[3]{4}$ **31.** $\frac{\sqrt[3]{2ab}}{2ab^3}$ **33.** $\frac{\sqrt[5]{162b^4}}{3b}$ **35.** $\frac{1}{4}$ **37.** 4 **39.** 10 **41.** -5 **43.** $\frac{6643}{81}$ **45.** $3^{1/5}a^{4/5}$ **47.** $(x^2+1)^2$ **49.** $\frac{1}{(x^2+1)^2}$ **51.** $x^{-1/2}+x^{1/2}$ **53.** $x^{1/4}$ **55.** $2^{1/6}$ **57.** $\frac{x^{1/10}y^{1/15}}{z^{1/10}}$ **59.** $\sqrt[35]{x^{21}y^{20}}$ **61.** (a) With $x = 2$ and $m = 1$ we have $2 = \frac{1}{2}$, which is false. (b) With $x = 2$ we have $2^{-1/2} = \frac{1}{4}$, or $1/\sqrt{2} = \frac{1}{4}$, which is false. **65.** (a) $2^{3/2}$; (b) $5^{1/2}$; (c) $2^{1/2}$; (d) $(\frac{1}{2})^{1/3}$; (e) $10^{1/10}$ **67.** $9^{10/9}$

69.

a	b	c	$(abc)^{1/3}$	$\frac{a+b+c}{3}$	Larger?
1	1	6	1.82	2.67	A.M.
1	2	5	2.15	2.67	A.M.
1	3	4	2.29	2.67	A.M.
2	2	4	2.52	2.67	A.M.
2	3	3	2.62	2.67	A.M.
$\frac{8}{3}$	$\frac{8}{3}$	$\frac{8}{3}$	2.67	2.67	same
12	1	1	2.29	4.67	A.M.
6	2	1	2.29	3.00	A.M.
4	3	1	2.29	2.67	A.M.
3	2	2	2.29	2.33	A.M.
$\sqrt[3]{12}$	$\sqrt[3]{12}$	$\sqrt[3]{12}$	2.29	2.29	same

71. (a) first 6 decimal places: 3.236068; (b) first 6 decimal places: 0.312050; (c) first 6 decimal places: 1.000000

Exercise Set 1.7

1. all real numbers **3.** all real numbers **5.** all nonnegative real numbers **7.** degree: 3; coefficients: 4, -2, -6, -1 **9.** degree: 1; coefficient: 5 **11.** $20x^2 + 2x + 1$ **13.** $-2x - 2$ **15.** $-4x^2 - 12x - 8$ **17.** $4x^2 - 19x + 10$ **19.** $-ax^2 + 4bx$ **21.** $2x^3 - 8x^2 - 10x$ **23.** $x^2 - 3x + 2$ **25.** $2x^2 + 6x + 4$ **27.** $x^4 + 4x^3 + 3x^2$ **29.** $4x^2y^2 - 8xy + 3$ **31.** $x^2 + 2 + \frac{1}{x^2}$ **33.** $a + b + 4\sqrt{a+b} + 3$ **35.** $x - 3x^{1/2} + 2$ **37.** $y^3 - y^2 - 11y - 10$ **39.** $x^2 - 4y^2 + 4yz - z^2$ **41.** $x^2 - y^2$ **43.** $a^4 - 9y^2$ **45.** $x^2 + 2xy + y^2 - 4$ **47.** $x^2 - 6x + 9$ **49.** $x^4 + 2x^2 + 1$ **51.** $x^4 + 2x^2a^2 + a^4$ **53.** $x^3 + 6x^2y + 12xy^2 + 8y^3$ **55.** $a^3 + 3a^2 + 3a + 1$ **57.** $x^3 - y^3$ **59.** $x^3 + 1$ **61.** $x - y$ **63.** quotient: $x + 7$; remainder: 6 **65.** quotient: $x^2 - 3$; remainder: 14 **67.** quotient: $x^5 + 2x^4 + 4x^3 + 8x^2 + 16x + 32$; remainder: 0 **69.** quotient: a; remainder: $b - ac$ **77.** 0 **79.** $k = -7$ **81.** (a) $ar^2 + br + c$; (b) $ar^3 + br^2 + cr + d$; (c) $ar^4 + br^3 + cr^2 + dr + e$

83. (a)

x	$\frac{x^3-8}{x-2}$
1.9	11.4100
1.99	11.9401
1.999	11.9940
1.9999	11.9994
1.99999	11.9999

x	$\frac{x^3-8}{x-2}$
2.1	12.6100
2.01	12.0601
2.001	12.0060
2.0001	12.0006
2.00001	12.0000

(b) The target value appears to be 12.

85.

x	$1+\frac{x}{2}$	$\sqrt{1+x}$
0.1	1.05	1.048808
0.01	1.005	1.004987
0.001	1.005	1.000499

Exercise Set 1.8

1. (a) $(x - 8)(x + 8)$; (b) $7x^2(x^2 + 2)$; (c) $z(11 - z)(11 + z)$; (d) $(ab - c)(ab + c)$ **3.** (a) $(x + 3)(x - 1)$; (b) $(x - 3)(x + 1)$; (c) irreducible; (d) $(-x + 3)(x + 1)$ **5.** (a) $(x + 1)(x^2 - x + 1)$; (b) $(x + 6)(x^2 - 6x + 36)$;

5. (c) $8(5 - x^2)(25 + 5x^2 + x^4)$; (d) $(4ax - 5)(16a^2x^2 + 20ax + 25)$ **7.** $2x(1 - x)(1 + x)$ **9.** $x^3(10 - x)(10 + x)$ **11.** $x^2(2x - 3)(x + 3)$ **13.** $x(2x - 5)^2$ **15.** $(xz + t)(xz + y)$ **17.** $(a^2 + b^2)(t^2 - c)$ **19.** $x(x - 18)(x + 5)$ **21.** $(4x + 3y)(x - 8y)$ **23.** irreducible **25.** $(1 - x - y)(1 + x + y)$ **27.** $(x - 1)(x + 1)(x^2 + 1)(x^4 + 1)$ **29.** $(x + 1)^3$ **31.** $(3x + 4)^3$ **33.** $(2a - b)(4a^2 + 2ab + b^2)(2a + b) \times (4a^2 - 2ab + b^2)$ **35.** $(x - 3)(x + 3)(x - 4)(x + 4)$ **37.** irreducible **39.** $x(x + 17)(x - 15)$ **41.** $(x + a)(x^2 - xa + a^2 + 1)$ **43.** $(x - y - a)(x - y + a)$ **45.** $x(7x - 3)(3x + 13)$ **47.** $(5 - 2x + 3y)(5 + 2x - 3y)$ **49.** $(ax + b)(x + 1)$ **51.** $(x^2 - 4x + 8)(x^2 + 4x + 8)$ **53.** $2(x - t)(x + y + z + t)$ **55.** $3(b - a)(c - b)(c - a)$ **57.** 1999 **59.** 271 **61.** $-x(x + 1)^{1/2}$ **63.** $x(x + 1)^{-3/2}$ **65.** $\dfrac{2x(x - 1)}{(x - 2)^4}$ **67.** $[(4x + 5)x - 1]x + 10$ **69.** $\{[(5x + 2)x - 7]x + 8\} + 2$ **71.** $[(ax + b)x + c]x + d$ **73.** 24 **75.** -77 **77.** -16.1875 **79.** 17,689,213 **81.** Answers will vary, depending on the brand of calculator. An HP 11c yields 10^{-9}. (Doing the calculations by hand shows that the exact answer is zero.)

Exercise Set 1.9

1. $x - 3$ **3.** $\dfrac{1}{(x - 2)(x^2 + 4)}$ **5.** $\dfrac{1}{x - 2}$ **7.** $\dfrac{3b}{2a}$ **9.** $\dfrac{a^2 + 1}{a - 1}$ **11.** $\dfrac{x^2 + xy + y^2}{(x - y)^2}$ **13.** 2 **15.** $\dfrac{x^2 + x}{(x + 4)(x - 2)}$ **17.** $\dfrac{x^2 - xy + y^2}{(x - y)(x + y)}$ **19.** $\dfrac{x - 3y}{x + 3y}$ **21.** $\dfrac{4x - 2}{x^2}$ **23.** $\dfrac{36 - a^2}{6a}$ **25.** $\dfrac{4x + 11}{(x + 3)(x + 2)}$ **27.** $\dfrac{3x^2 + 6x - 6}{(x - 2)(x + 2)}$ **29.** $\dfrac{6ax^2 - 4ax + a}{(x - 1)^3}$ **31.** $\dfrac{2x^2 - 3}{(x - 3)(x + 3)(x - 2)}$ **33.** $\dfrac{8}{x - 5}$ **35.** $\dfrac{2a}{a + b}$ **37.** $\dfrac{-9}{(x + 5)(x - 4)^2}$

39. $\dfrac{-p^2 - pq - q^2}{(2p + q)(p - 5q)p}$ **41.** 0 **43.** 0 **45.** 0 **47.** $\dfrac{x^2 + x - a - 1}{(x + a)(x - a)}$ **49.** $\dfrac{px + q}{(x - a)(x - b)}$ **51.** $\dfrac{2}{c - a}$ **53.** $\dfrac{x + 4}{3 - 2x}$ **55.** $a - 1$ **57.** $-\dfrac{1}{2(2 + h)}$ **59.** $\dfrac{a + x}{x^2a^2}$ **61.** $\dfrac{a^2 - b^2}{2}$ **63.** $\left(\dfrac{a}{b}\right)^{a+b}$ **65.** -1 **67.** (a) The sum and product both equal $\frac{1}{42}$. **69.** 1

Exercise Set 1.10

1.

i^2	i^3	i^4	i^5	i^6	i^7	i^8
-1	$-i$	1	i	-1	$-i$	1

3. (a) -6; (b) $\sqrt{2}$; (c) 5; (d) 0 **5.** $a = 3, b = 4$ **7.** $-10 + 13i$ **9.** $-1 + 20i$ **11.** 4 **13.** $8 + 9i$ **15.** $-19 + 30i$ **17.** $52 + 117i$ **19.** $16 + 36i$ **21.** $24i$ **23.** $-539 - 1140i$ **25.** $-119 - 120i$ **27.** $\frac{-5}{13} - \frac{12}{13}i$ **29.** $\frac{35}{13} + \frac{6}{13}i$ **31.** $\frac{-19}{1261} - \frac{30}{1261}i$ **33.** $\frac{2}{41} - \frac{23}{41}i$ **35.** $-i$ **37.** $\bar{z} = 3 - 5i$; $\bar{\bar{z}} = 3 + 5i$ **39.** $44 + 52i$ **41.** (a) $2\sqrt{5}$; (b) $\sqrt{34}$; (c) $2\sqrt{170}$ **43.** $-i$ **45.** $-28\sqrt{2}i$ **47.** i **49.** real part: $\dfrac{2a^2 - 2b^2}{a^2 + b^2}$; imaginary part: 0

Chapter 1 Review Exercises

1. $(a - 4b)(a + 4b)$ **3.** $(1 - a)(7 + 4a + a^2)$ **5.** $x(ax + b)^2$ **7.** $(4x + 1)(2x + 1)$ **9.** $a^4x^4(1 - xa)(1 + xa)(1 + x^2a^2)$ **11.** $(2 + a)^3$ **13.** $z^3(4xy + 1)(xy - 1)$ **15.** $(1 - x)(1 + x + x^2)(1 + x) \times (1 - x + x^2)$ **17.** $(ax + b - 2abx)(ax + b + 2abx)$ **19.** $(a - b)(a + b + c + ab)$ **21.** 8 **23.** 1 **25.** $\frac{1}{16}$ **27.** $\frac{1}{9}$ **29.** ab^3c^4 **31.** $2a^2b^4$ **33.** $5\sqrt[3]{2}$ **35.** $5ab\sqrt{6b}$ **37.** 1.4×10^{-3} **39.** 1.2001×10^1 **41.** $\dfrac{x - 2}{(x - 1)^2}$ **43.** $\dfrac{1}{x}$ **45.** $\dfrac{-x^2 - 2x + 1}{(x - 1)(x + 2)}$ **47.** $\dfrac{a^2 + ax^2 - a^3}{x^4 - 2a^2x^2 + a^4 - ax^2 + a^3}$ **49.** $\dfrac{x^2 + x - 1}{-x^2 + x + 1}$ **51.** $2\sqrt{2}$ **53.** $-\sqrt{5} - \sqrt{6}$ **55.** $\dfrac{a^2 + \sqrt{a^4 - x^4}}{x^2}$ **57.** 2 **59.** $\frac{7}{8}$ **61.** 15 **63.** -37 **65.** -12 **67.** $x^3 + 2x^2 - x - 2$

69. $3x + 2$ **71.** quotient: $x^2 - x$; remainder: 4 **73.** quotient: $x^2 - 2$; remainder: 3 **75.** $33 + 9i$ **77.** $3 - 5i$ **79.** $\dfrac{5 + i}{2}$ **81.** $\dfrac{1 - i\sqrt{3}}{2}$ **83.** 1 **85.** $1 - 5i$ **87.** $-(4\sqrt{2})i$ **89.** 5 **91.** 6 **93.** $-\sqrt{2} + 2i$ **95.** $\dfrac{8 - i}{5}$ **97.** distributive (three times); commutative (multiplication) **99.** distributive **101.** 26 **103.** $2\sqrt{3}$ **105.** $x^2 + 1$ **107.** $|x - a| < \frac{1}{2}$ **109.** $|x - (-1)| = 5$, or $|x + 1| = 5$ **111.** $\sqrt{6} - 2$ **113.** $x^4 + x^2 + 1$ **115.** 1

117.

2 3 4 5 6

119.

−1 0 1 2

121. $81x^2 - y^2$ **123.** $4a^2b^2 - 12ab + 9$ **125.** $a^4x^4 - 2a^2b^2x^2 + b^4$ **127.** $x^6 - 6x^4 + 12x^2 - 8$ **129.** $x^{3n} + y^{3n}$

Exercise Set 2.1

1. -1 **3.** 3 **5.** -11 **7.** 12 **9.** 3 **11.** -1 **13.** -6 **15.** -7 **17.** -3 **19.** -15 **21.** 9 **23.** $-\frac{13}{7}$ **25.** 8 **27.** all real numbers except ± 5 **29.** no solution **31.** no solution **33.** 3 **35.** 6 and 4 **37.** 5 and $-\frac{10}{3}$ **39.** -3 **41.** $\dfrac{b + 1}{a}$ **43.** 1 **45.** $\dfrac{1}{a + b}$ **47.** $\dfrac{ab}{b + c}$ **49.** $\dfrac{a^2 - b^2}{2a}$ **51.** $\dfrac{1}{b + a}$ **53.** $\dfrac{pq}{p + q}$ **55.** $\dfrac{y - b}{m}$ **57.** $a + b$ **59.** $\dfrac{2pq}{p + q}$ **61.** (a) $(a + b + 1)^2$; (b) $a + b + 1$ **63.** $\dfrac{-Ax - C}{B}$ **65.** $2x - x_2$ **67.** $\dfrac{S - 2LH}{2L + 2H}$ **69.** $\dfrac{S - 2\pi r^2}{2\pi r}$ **71.** $\dfrac{S - a}{S - l}$ **73.** $c = 2s - a - b$ **75.** $\dfrac{6A - (b - a)f_0 - (b - a)f_2}{4(b - a)}$ $\left(\text{or } \dfrac{3A}{2(b - a)} - \dfrac{f_0 + f_2}{4}\right)$ **77.** 6.80 **79.** 0.04 **81.** 4.45 and -4.17

Exercise Set 2.2

1. yes **3.** no **5.** no **7.** 6, -1
9. ± 10 **11.** $\frac{6}{5}$ (double root) **13.** $-\frac{1}{5}, \frac{3}{2}$
15. 1, -3 **17.** $-\frac{1}{2}$, 8 **19.** $\frac{1}{2}$, 6
21. 12, -13 **23.** ± 7 **25.** $\pm i$ **27.** $\pm\frac{1}{2}$
29. ± 8 **31.** $2 \pm \sqrt{14}$ **33.** $\dfrac{5 \pm \sqrt{21}}{2}$
35. 0, -8 **37.** $\dfrac{-3 \pm \sqrt{11}}{2}$
39. $\dfrac{-9 \pm 2\sqrt{21}}{2}$ **41.** $-\dfrac{1}{4}, -\dfrac{5}{2}$ **43.** 1, $-\dfrac{1}{2}$ **45.** $\dfrac{-1 \pm \sqrt{5}}{2}$ **47.** $-1, -9$
49. 3, -8 **51.** $\dfrac{1 \pm \sqrt{21}}{2}$
53. $\dfrac{-3 \pm \sqrt{41}}{4}$ **55.** $\pm\sqrt{17}$ **57.** $-\dfrac{1}{6}$, $-\dfrac{5}{2}$ **59.** $\dfrac{1 \pm \sqrt{41}}{4}$ **61.** $\dfrac{-6 \pm \sqrt{42}}{-6}$ $\left(\text{or } \dfrac{6 \pm \sqrt{42}}{6}\right)$ **63.** two real roots
65. two real roots **67.** one distinct real root **69.** two real roots **71.** 36
73. $\pm 2\sqrt{5}$ **75.** $\pm\frac{1}{5}$ **77.** $\frac{2}{3}$, 1
79. $\frac{1}{2}$, -1 **81.** $\dfrac{-2 \pm \sqrt{13}}{3}$ **83.** $\pm 2\sqrt{6}$
85. $\dfrac{1 \pm \sqrt{5}}{2}$ **87.** $\dfrac{3 \pm \sqrt{15}}{3}$ **89.** $\dfrac{\sqrt{5}}{2}$, $\dfrac{-2\sqrt{5}}{5}$ **91.** $\dfrac{5}{2}$, -4 **93.** $\dfrac{1}{3}, \dfrac{1}{2}$ **95.** 1
97. $-3 \pm 2\sqrt{2}$ **99.** all real numbers except -1 and -3 **101.** $\frac{1}{y}, \frac{1}{2y}$ **103.** 0, $p + q$ **105.** $\dfrac{4a}{3}, -\dfrac{5a}{4}$ **107.** 1, $\dfrac{a+b-c}{a+b+c}$
109. $\sqrt{\dfrac{b}{a}}\left(= \dfrac{\sqrt{ab}}{a}\right)$, $\sqrt{\dfrac{a}{b}}\left(= \dfrac{\sqrt{ab}}{b}\right)$
111. 4.75, 0.01 **113.** $-1.47, -1.53$

Exercise Set 2.3

1. 12 and -5 **3.** 13 and 4 **5.** 21
7. 8 **9.** 5 **11.** 6 and 12 **13.** 78
15. 44 **17.** 5 cm, 12 cm, 13 cm
19. $2\frac{1}{2}$ in. **21.** 20 cm **23.** 4.8 in. and 5.2 in. **25.** 10 cm, 30 cm, 30 cm
27. 8 yr **29.** 10 yr **31.** 6.6%
33. rate for \$8000 is 7.5%; rate for \$9000 is 8.5% **35.** \$11,727
37. 80 cm³ of 10% solution; 120 cm³ of 35% solution **39.** 15 lb **41.** 8 tons
43. 12 cm **45.** 17 cm **47.** $\dfrac{3 - \sqrt{5}}{2}$
49. $\dfrac{1 + \sqrt{5}}{2}$ **51.** (a) 6 sec; (b) 1 sec and 5 sec **53.** $12\frac{3}{8}$ min (about 12 min and 22 sec) **55.** 55 mph **57.** 350 mph
59. 15 mi **61.** 16 in. by 12 in. **63.** 26
65. $\dfrac{2v_1v_2}{v_1 + v_2}$ **67.** $\dfrac{100(d + D)}{dr + DR}$ yr
69. $\sqrt{\dfrac{b + \pi r^2}{\pi}} - r$ **71.** $\dfrac{b(c - a)}{c}$ lb
73. $\dfrac{L}{4}$ in. and $\dfrac{3L}{4}$ in. **75.** \$5117
77. 9.76 ft

Exercise Set 2.4

1. 0, 25 **3.** 10 **5.** 0, ± 4 **7.** 1, ± 15
9. 0, $\frac{5}{2}$ ($\frac{5}{2}$ is a double root.) **11.** 0, -3, 1
13. 0, $\frac{1}{4}$, -6 **15.** $\pm\sqrt{3}$ **17.** ± 1
19. $\dfrac{\pm\sqrt{6}}{3}$ **21.** 0, $\pm\sqrt{6}$, ± 2
23. $\pm\sqrt{\frac{1}{2}(\sqrt{5} - 1)}$
25. $\pm\sqrt{\frac{1}{2}(-3 + \sqrt{17}}$ **27.** -2, 1
29. $\frac{1}{3}, \frac{1}{4}$ **31.** $\frac{3}{5}$, 4 **33.** $\pm\frac{1}{3}$ **35.** 24
37. -6, 1 **39.** 7 **41.** 27 **43.** ± 3
45. $1 + \sqrt[3]{7}$ **47.** -4 **49.** $\pm 1, \pm\frac{1}{27}$
51. 0, 1 **53.** $-4, -\frac{20}{9}$ **55.** no solutions
57. 2, -1 **59.** 4, 20 **61.** -1 **63.** 5
65. 1 **67.** $a + b$ **69.** $\dfrac{a^2b^2}{(2a + b)^2}$
71. $\dfrac{1 \pm \sqrt{21}}{2}$ **73.** $-\dfrac{a}{15}$ **75.** 0, $\dfrac{1 - \sqrt{5}}{2}$, -1 **77.** ± 1.80 **79.** 0.47
81. 23.42 cm

Exercise Set 2.5

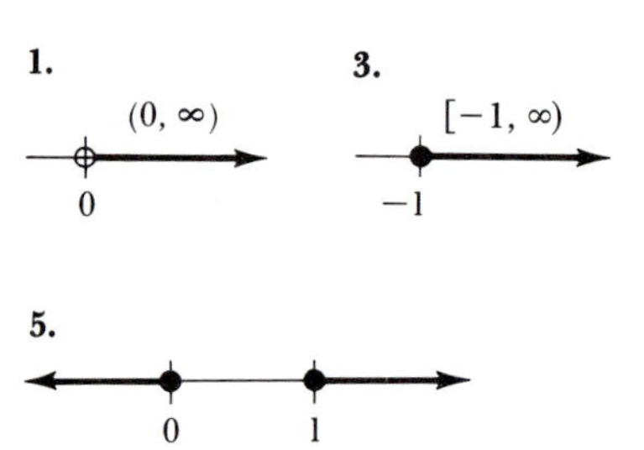

7. $(-\infty, -1)$ **9.** $[\frac{1}{3}, \infty)$ **11.** $(9, \infty)$
13. $[-\frac{10}{3}, \infty)$ **15.** $(-\infty, 1)$ **17.** $[0, \infty)$
19. $(-\infty, -3)$ **21.** $(-\infty, -5]$
23. $(-\infty, \frac{5}{2})$ **25.** $[-\frac{17}{23}, \infty)$ **27.** $[\frac{42}{5}, \infty)$
29. $[4, 6]$ **31.** $(-22, -10)$ **33.** $[-\frac{1}{2}, 1]$
35. $(3.98, 3.998)$ **37.** $(-2, -\frac{17}{10})$
39. \$2000 **41.** $-13 \le F \le 887$
43. $\frac{5}{16} \le t \le \frac{5}{8}$ **45.** periods longer than 10 mo **47.** $(-\frac{1}{2}, \frac{1}{2})$
49. $(-\infty, 0) \cup (0, \infty)$ **51.** $(1, 3)$
53. $[-4, 6]$ **55.** $(-\infty, -\frac{22}{3}) \cup [4, \infty)$
57. $(a - b, a + b)$ **59.** $(-11, 1)$
61. $(-0.505, -0.495)$
63. (a) $\left(-\dfrac{121}{120}, -\dfrac{119}{120}\right)$; (b) $\left(\dfrac{-12 - \epsilon}{12}, \dfrac{-12 + \epsilon}{12}\right)$

65. (a)

x	y	$\frac{x}{y} + \frac{y}{x}$	$\frac{x}{y} + \frac{y}{x} \ge 2$
1	1	2	true
2	3	2.1667	true
3	5	2.2667	true
4	7	2.3214	true
5	9	2.3556	true
9	10	2.0111	true
49	50	2.0004	true
99	100	2.0001	true

67. (a)

n	$\sqrt{n+1} + \sqrt{n-1}$
1	1.41
2	2.73
3	3.41
4	3.97
5	4.45
10	6.32
5000	141.42
10,000	200.00

(b) $n = 2500$

Exercise Set 2.6

1. $(-3, 2)$ **3.** $(-\infty, 2) \cup (9, \infty)$
5. $(-\infty, 4] \cup [5, \infty)$
7. $(-\infty, -4] \cup [4, \infty)$ **9.** no solution
11. $(-7, -6) \cup (0, \infty)$
13. $[-15, 0] \cup [15, \infty)$ **15.** $(-\infty, \infty)$
17. $(-\infty, -\frac{3}{4}) \cup (-\frac{2}{3}, 0)$
19. $\left(-\infty, \dfrac{-1 - \sqrt{5}}{2}\right) \cup \left(\dfrac{-1 + \sqrt{5}}{2}, \infty\right)$
21. $[4 - \sqrt{14}, 4 + \sqrt{14}]$
23. $[-4, -3] \cup [1, \infty)$
25. $(-\infty, -6) \cup (-5, -4)$
27. $(-\infty, -\frac{1}{3}) \cup (\frac{1}{3}, 2) \cup (2, \infty)$
29. $(-\infty, -\frac{2}{3}]$ **31.** $[-2, 2] \cup [\sqrt{5}, \infty)$
33. $\left(-\infty, \dfrac{3 - \sqrt{11}}{2}\right) \cup \left(\dfrac{3 + \sqrt{11}}{2}, 4\right)$
35. $(-2, -1) \cup (1, \infty)$ **37.** $(-1, 1]$
39. $(-\infty, \frac{3}{2}) \cup [2, \infty)$
41. $(-\infty, -1) \cup (0, 9)$
43. $(-\frac{7}{2}, -\frac{4}{3}) \cup (-1, 0) \cup (1, \infty)$
45. $(-5, \infty)$ **47.** $[-1, 1) \cup (2, 4]$
49. $(-1, 2 - \sqrt{5}) \cup (1, 2 + \sqrt{5})$
51. $(-\frac{3}{2}, 0)$ **53.** $(-\infty, -1) \cup (0, \infty)$
55. $b \le -2$ or $b \ge 2$
57. $a < 0$ or $a > \frac{1}{2}$ **59.** $\frac{1}{5} < x < \frac{4}{5}$
61. $0 < r < 2$

63. (a)

x	$x^2 + 1000$	$2x^2 + x$	$\dfrac{x^2 + 1000}{2x^2 + x}$
1	1001	3	333.67
2	1004	10	100.40
5	1025	55	18.64
10	1100	210	5.24
100	11,000	20,100	0.55
200	41,000	80,200	0.51
10,000	100,001,000	200,010,000	0.50

(b) $-\frac{1}{2} < x < 0$ or $x > 2000$; (c) 2001 **65.** $[-1.5, 1.6]$

Chapter 2 Review Exercises

1. true **3.** false **5.** false **7.** true **9.** false **11.** false **13.** $\frac{1}{3}$ **15.** $-\frac{37}{11}$ **17.** $-\frac{11}{3}$ **19.** 5 **21.** 8 **23.** -2 and -6 **25.** 1 and $-\frac{1}{3}$ **27.** -9 and 11 **29.** $\frac{1}{3}$, $-\frac{1}{2}$ **31.** -6 and 4 **33.** $\frac{14}{11}$ and -1 **35.** 10 and -1 **37.** $\dfrac{-3 \pm \sqrt{3}}{2}$ **39.** $\frac{2}{3}$, 3 **41.** $-2 \pm \sqrt{5}$ **43.** $\dfrac{-1 \pm \sqrt{3}}{2}$ **45.** $\pm\sqrt{3}$ **47.** ± 2 and $\pm\sqrt{3}$ **49.** 0, $\pm\sqrt{1 + \sqrt{3}}$ **51.** -6 and -8 **53.** ± 1 and $\pm\frac{27}{8}$ **55.** 6561 and 256 **57.** -7 **59.** 169 **61.** $\frac{1}{4}$ and $\frac{37}{36}$ **63.** 2 **65.** 16 **67.** $\frac{1}{2}$ **69.** 10 **71.** 2 **73.** $\dfrac{d - b}{a - c}$ **75.** $a + b$ **77.** $\dfrac{2ab}{b + a}$ **79.** $-8b$ and $4b$ **81.** $\dfrac{1}{2y}$ (double root) **83.** $\dfrac{a + b}{ab}$ and $-\dfrac{2}{a}$ **85.** $-a$ and $-b$ **87.** $\left(\dfrac{a - b}{a + b}\right)^2$ and $\left(\dfrac{a + b}{a - b}\right)^2$ **89.** $[2, \infty)$ **91.** $(-\infty, 3]$ **93.** $(\frac{1}{2}, 1)$ **95.** $(-7, -5)$ **97.** $[6, \infty)$ **99.** $(1, 3)$ **101.** $(\frac{1}{5}, 1)$ **103.** $[9, 12]$ **105.** $(-\infty, -\frac{8}{3}) \cup (\frac{1}{2}, \infty)$ **107.** $(3 - \sqrt{10}, 3 + \sqrt{10})$ **109.** $(-\infty, -12) \cup (1, 8)$ **111.** $(-\infty, -5] \cup [5, \infty)$ **113.** $(-\infty, -12) \cup (5, \infty)$ **115.** $[-3, 3) \cup (3, 8]$ **117.** $(-\frac{5}{2}, 4)$ **119.** $(-\infty, 1) \cup [\frac{14}{11}, 2) \cup [5, \infty)$ **121.** $(-\infty, -\sqrt{A}) \cup (-\sqrt{B}, 0) \cup (0, \sqrt{B}) \cup (\sqrt{A}, \infty)$, where $A = \frac{1}{2}(3 + \sqrt{5})$ and $B = \frac{1}{2}(3 - \sqrt{5})$ **123.** $k \le \frac{9}{5}$ **125.** $k < 0$ or $k \ge 1$ **127.** 2750 lb **129.** 523, 525, 527, 529 **131.** \$2000 at 10%; \$4000 at 12% **133.** $\dfrac{Bt + d}{A}$ hr **135.** $-1 + \sqrt{2}$ cm **137.** $x > 16$ cm **139.** 21; 22; 23 **141.** 6 and -5 **143.** 12 cm by 4 cm by 1 cm **147.** the ball thrown from 100 ft **149.** $\dfrac{\sqrt{2}}{2}$ cm

Exercise Set 3.1

1.

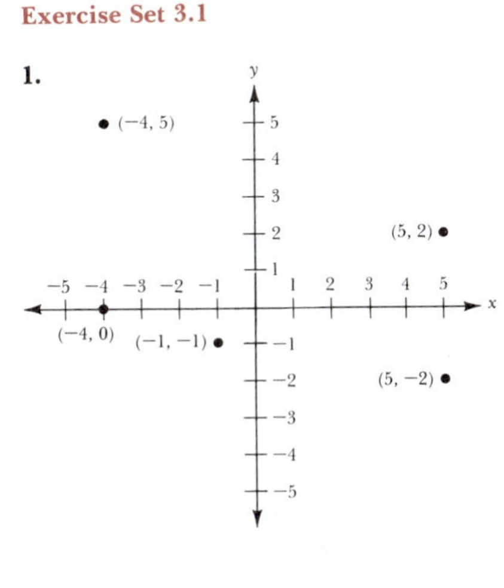

3. (a) (b) 6

5. (a) $(1, -4)$; (b) $(12, 4)$ **7.** $(11, -8)$ **9.** (a)

(b) AC: $(0, \frac{7}{2})$; BD: $(0, \frac{7}{2})$; (c) They are equal. **11.** (a) $\sqrt{65}$; (b) $\sqrt{58}$ **13.** (a) 5; (b) $\sqrt{65}$ **15.** (a) Two sides are $\sqrt{53}$; (b) Two sides are 5; (c) Two sides are $\sqrt{65}$. **17.** (a) 102, 13; (b) 85.5 **19.** (a) $(4, \frac{1}{2})$ (b) $(-6, 7)$ **23.** (a) $-2 \pm 4\sqrt{2}$; (b) no **33.** (a) 8; (b) $\frac{21}{2}$

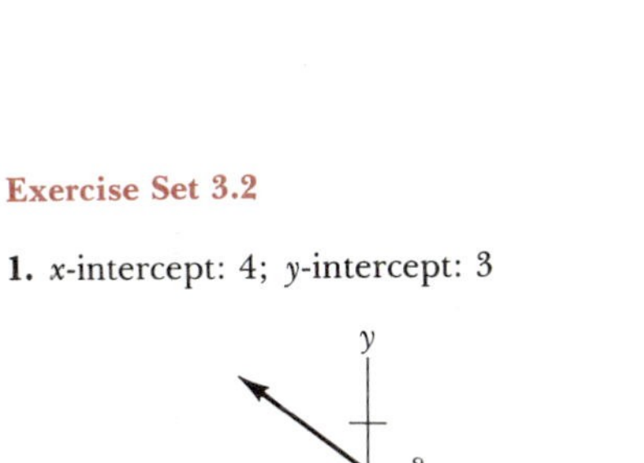

Exercise Set 3.2

1. x-intercept: 4; y-intercept: 3

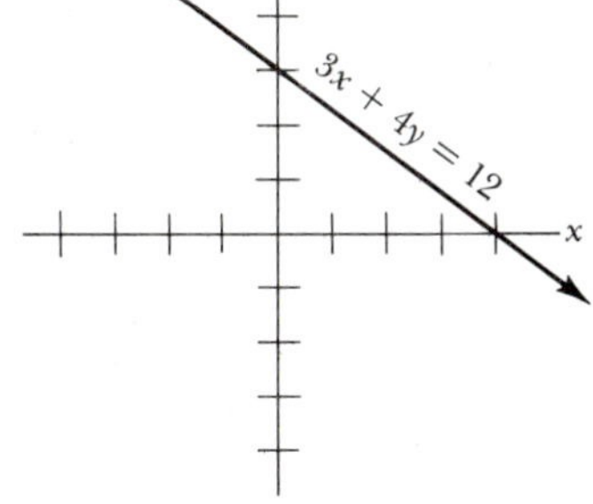

3. x-intercept: 2; y-intercept: -4

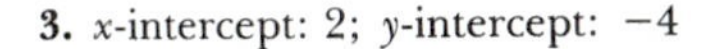

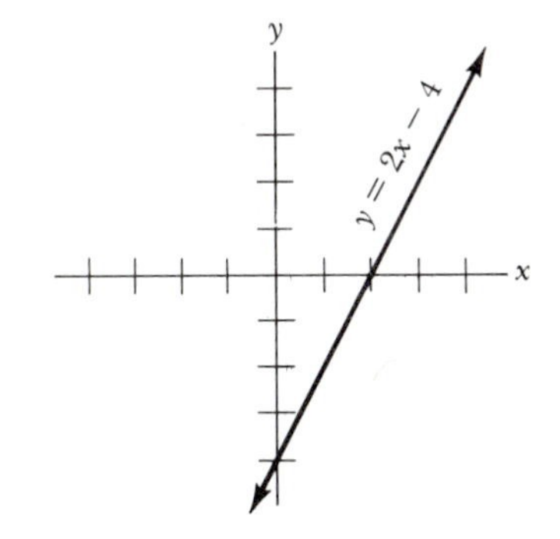

5. x-intercept: 1; y-intercept: 1

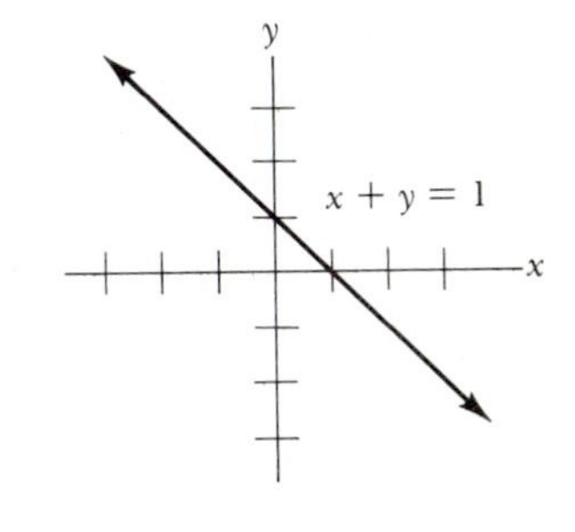

7. x-intercepts: -1, 6; y-intercept: -6
9. x-intercepts: $\dfrac{-1 \pm \sqrt{5}}{2}$; y-intercept: -1
11. x-intercept: 2; y-intercept: 3
13. x-intercept: $\frac{10}{3}$; y-intercept: -2
15. x-intercept: 2; y-intercept: -8
17. y-axis **19.** y-axis **21.** x-axis, y-axis, origin **23.** none **25.** none
27. x-intercept: 0; y-intercept: 0; symmetry: y-axis

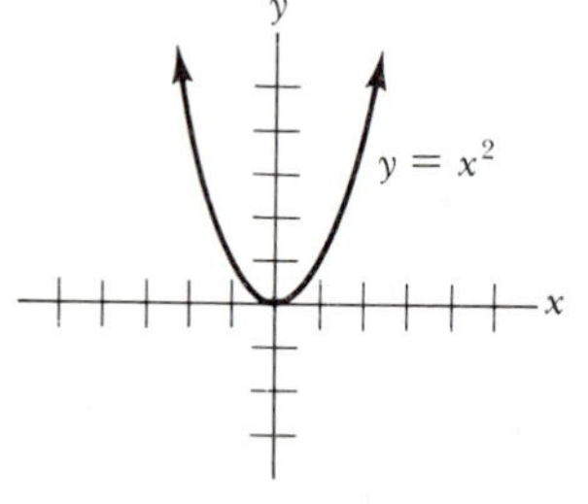

29. symmetry: origin

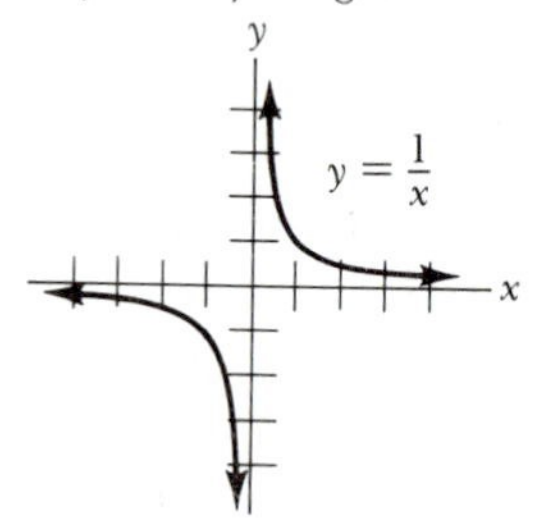

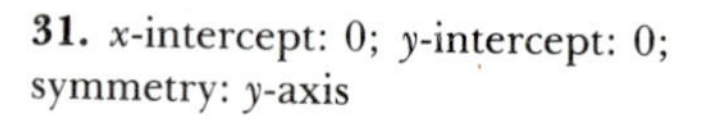
31. x-intercept: 0; y-intercept: 0; symmetry: y-axis

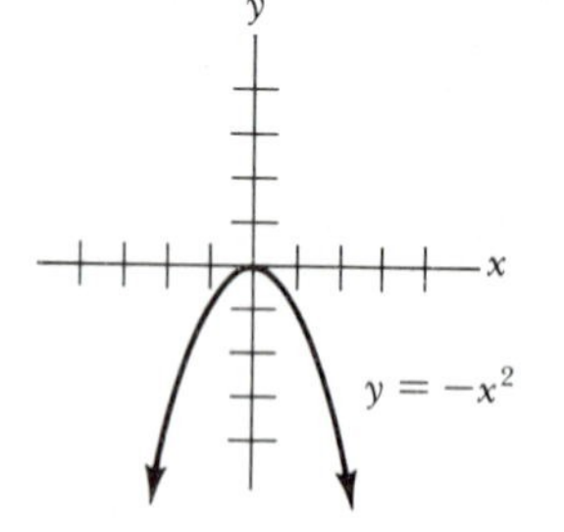

33. symmetry: y-axis

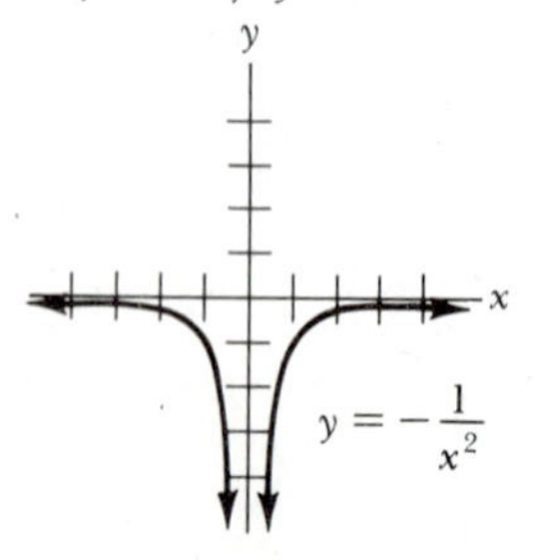

35. x-intercept: 0; y-intercept: 0; symmetry: y-axis

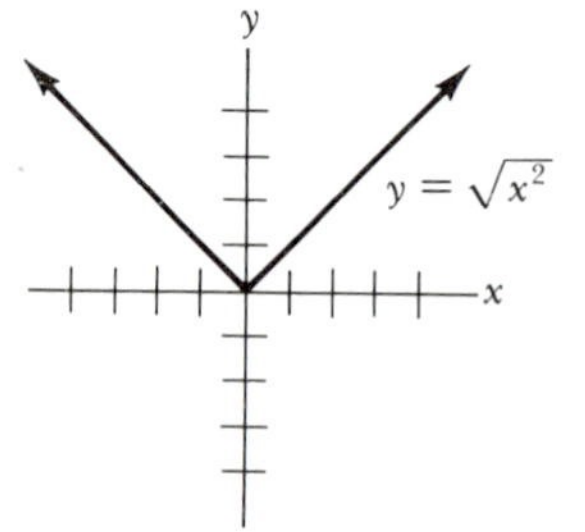

37. x-intercept: 1; y-intercept: 1

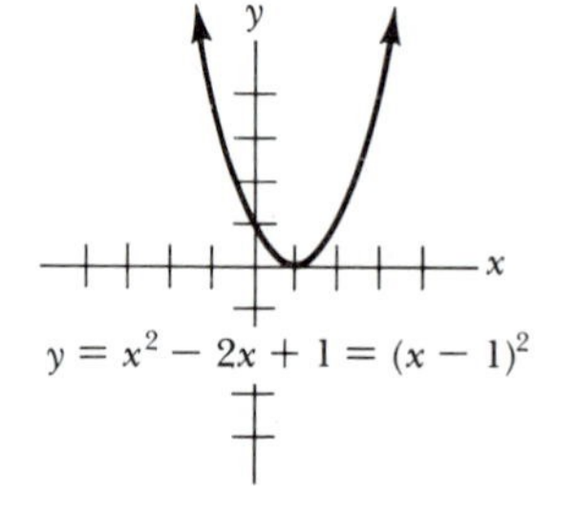

39. y-intercept: 1

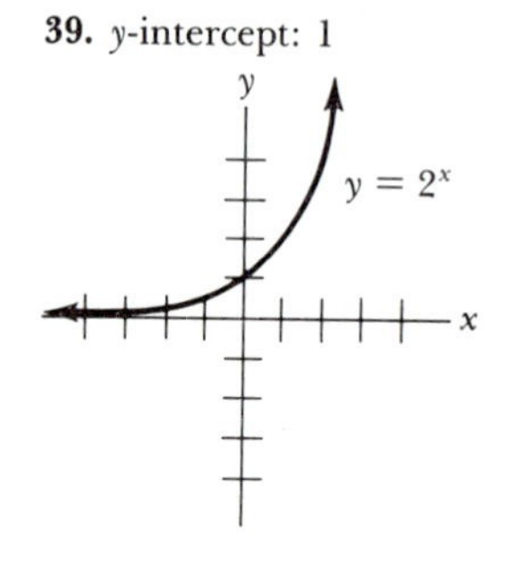

41. x-intercepts: $\dfrac{-1 \pm \sqrt{33}}{4}$; y-intercept: -4

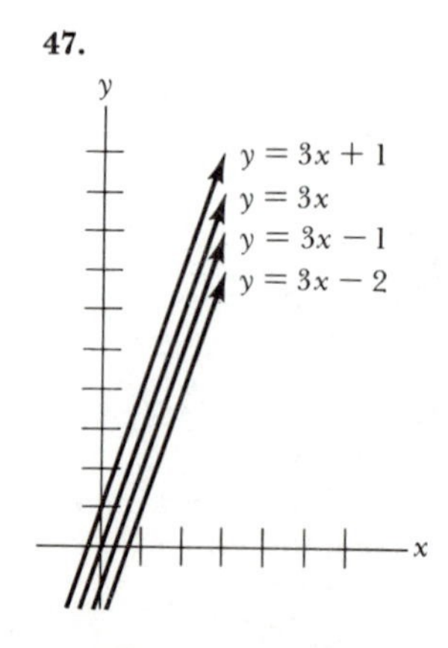

43. (a) x-intercept: 2; y-intercept: -6

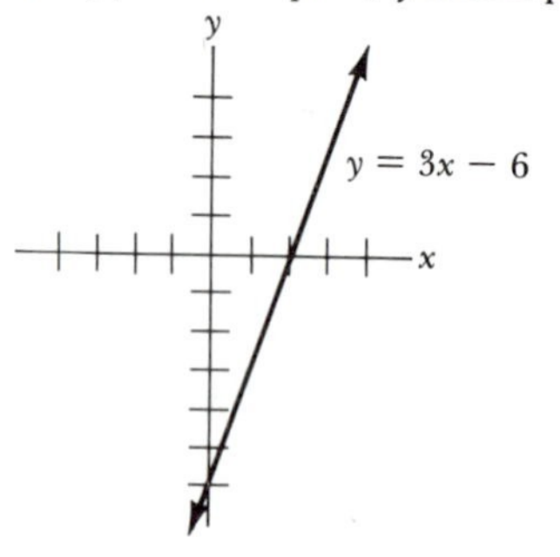

(b) x-intercept: 2; y-intercept: 6

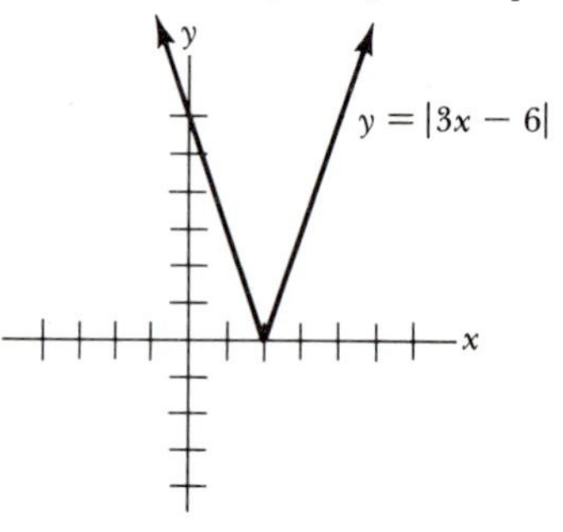

45. x-intercept: -3; y-intercept: 3

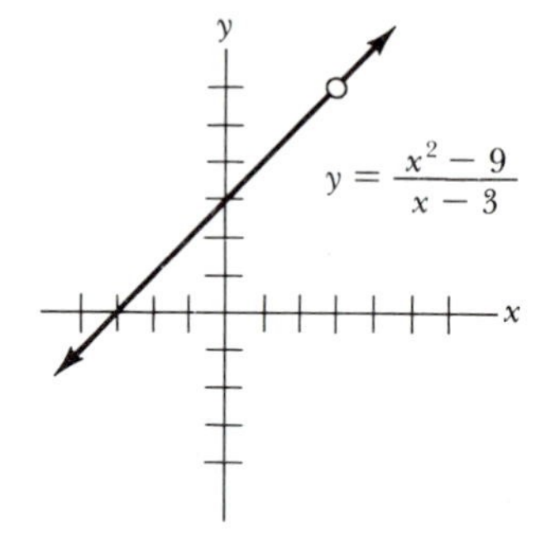

47.

49. (a)

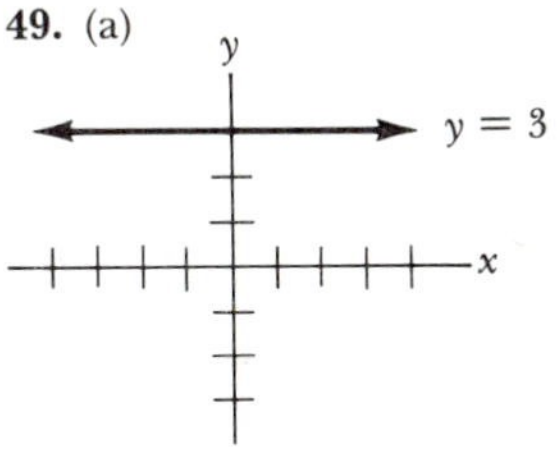

(b)

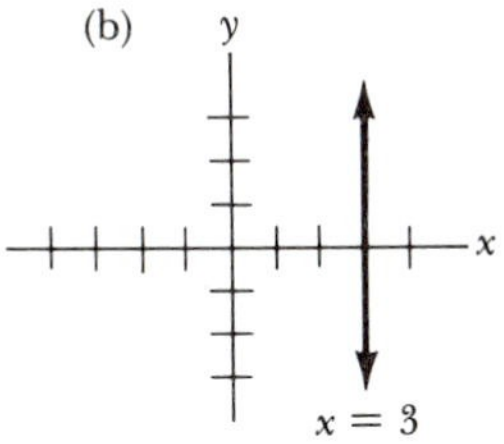

(c)

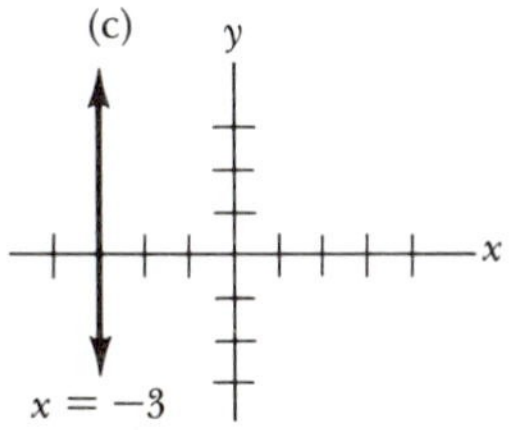

(d)

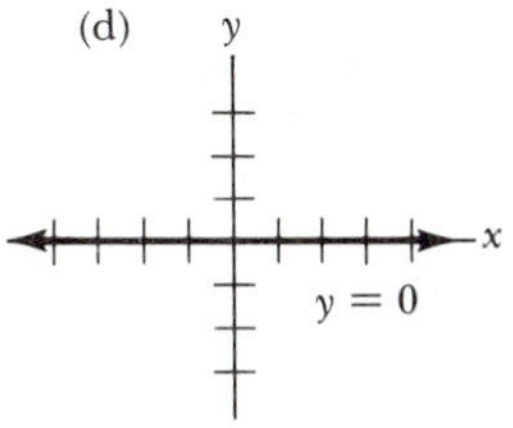

(e)

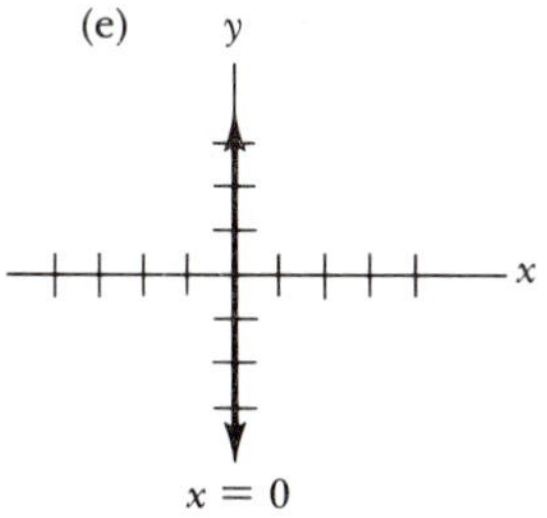

51. (a)

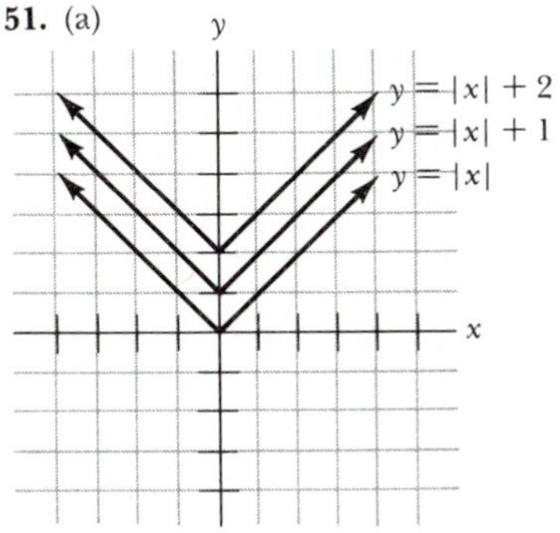

(b)

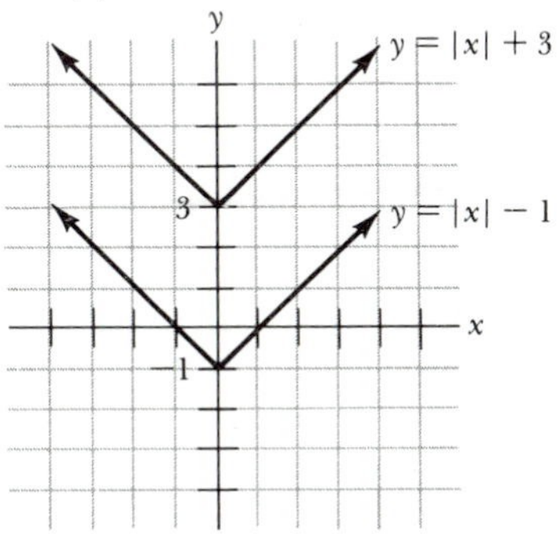

53. (a)

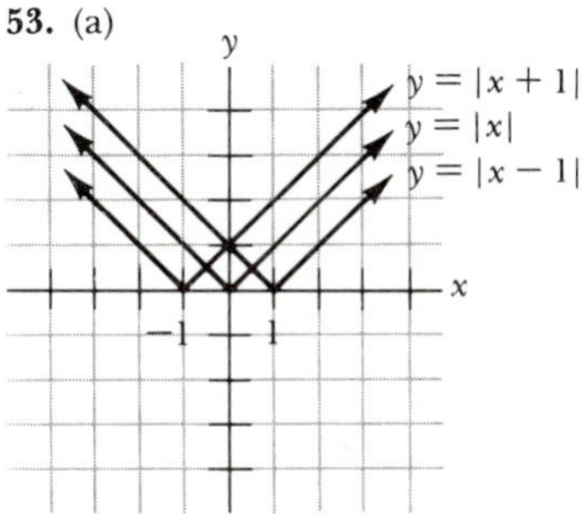

(b)

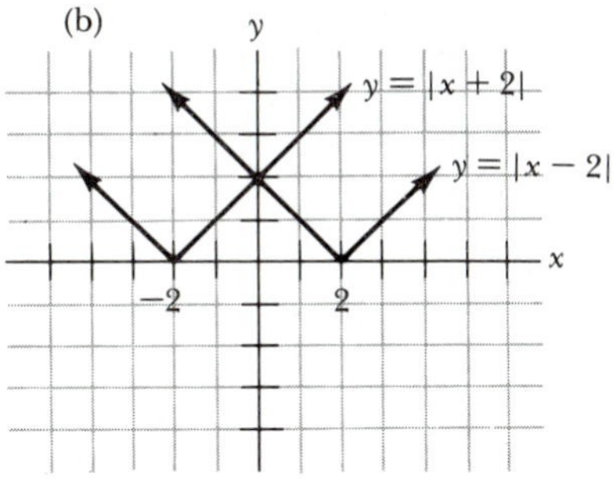

55. (a)

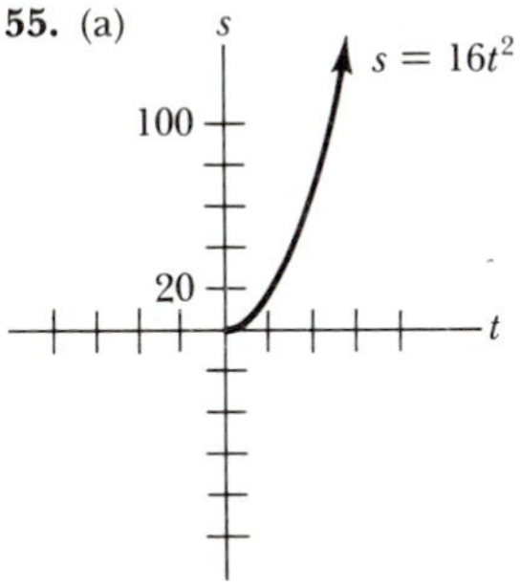

(b) $0 \leq s \leq 16$; (c) $16 \leq s \leq 64$

57.

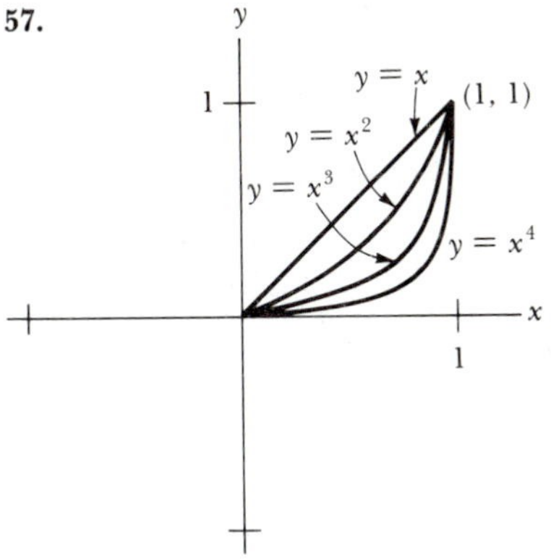

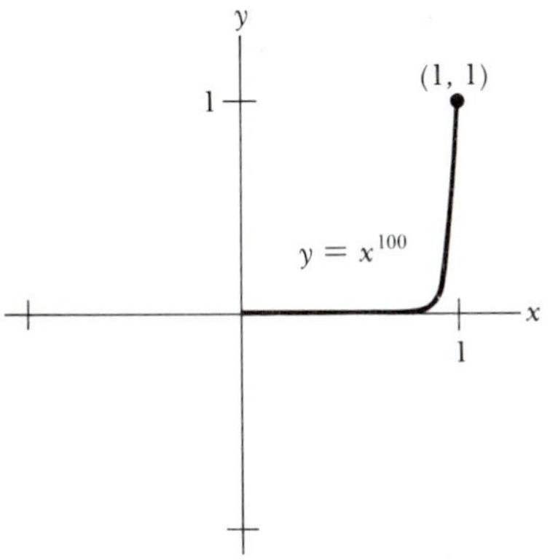

59. (a) 1000; (b) 1.7 hr; (c) 4 hr; (d) $t = 4$ and $t = 5$

61. (a) $(x - 1)^2 + (y - 3)^2 = 1$

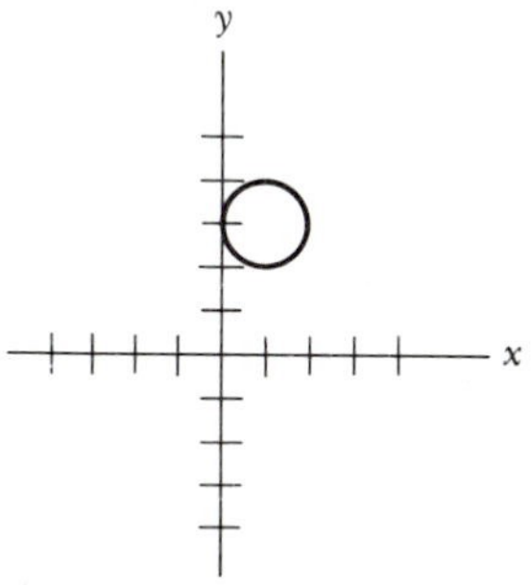

(b) $(x + 1)^2 + (y - 3)^2 = 4$

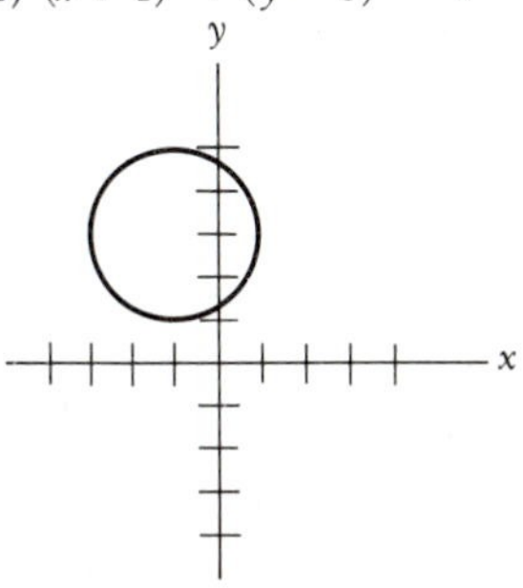

(c) $x^2 + y^2 = 9$

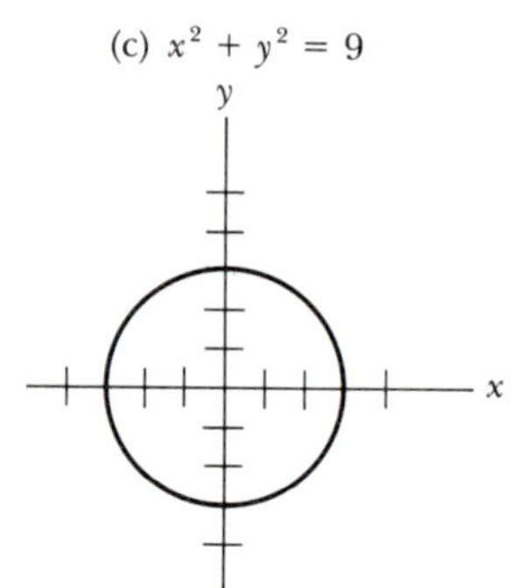

63. $(x - 3)^2 + (y - 2)^2 = 169$
65. center: $(-4, 3)$; radius: 1
67. center: $(-3, \frac{1}{3})$; radius: $\sqrt{2}$; x-intercepts: $-3 \pm \dfrac{\sqrt{17}}{3}$ **69.** center: $(-\frac{3}{2}, \frac{5}{2})$; radius: 3; x-intercepts: $\dfrac{-3 \pm \sqrt{11}}{2}$; y-intercepts: $\dfrac{5 \pm 3\sqrt{3}}{2}$
71. (a) $(x - 3)^2 + (y - 5)^2 = 25$; (b) $(x - 3)^2 + (y - 5)^2 = 9$; (c) $(x - 3)^2 + (y - 5)^2 = 34$
73. 24π sq units
75. $(x + \frac{1}{2})^2 + (y - 2)^2 = \frac{169}{4}$
77.

79.

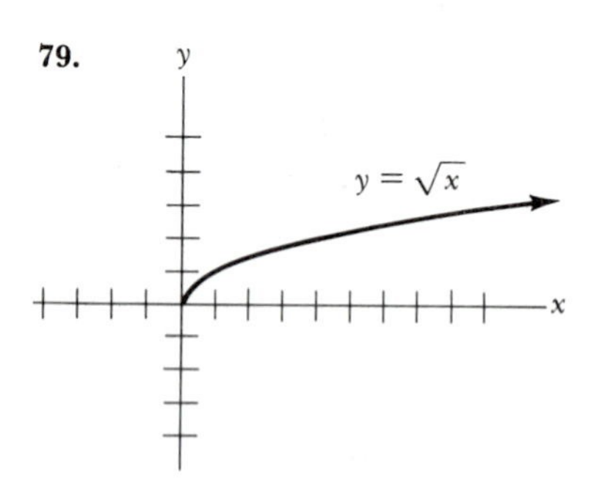

81. x-intercept: 0; y-intercept: 0; symmetry: y-axis

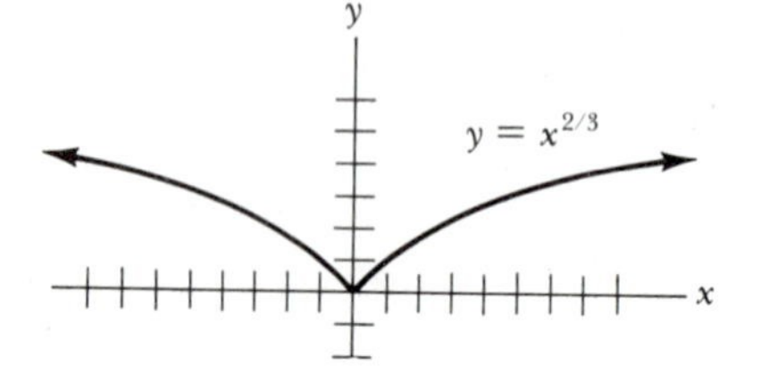

83.

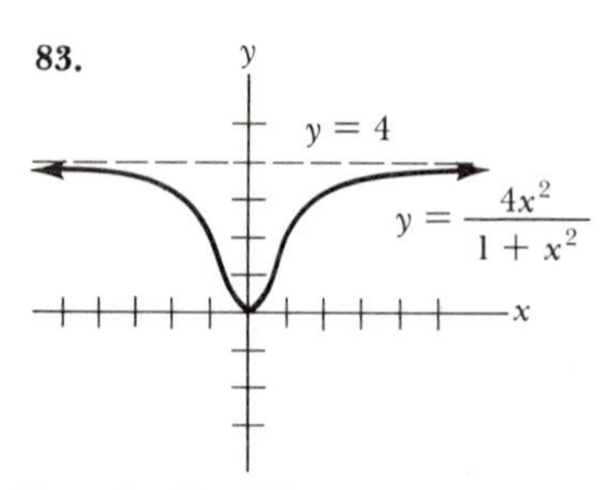

Exercise Set 3.3

1. (a) -2; (b) 3; (c) $-\frac{7}{3}$; (d) 3; (e) $\frac{9}{7}$; (f) $-\frac{1}{2}$; (g) $\frac{27}{20}$; (h) -1
3. (a) -2; (b) -2; (c) -2 **5.** (a) 2; (b) $-\frac{4}{5}$; (c) 1; (d) $\frac{13}{3}$; (e) 0
7. L_1: $\frac{1}{3}$; L_2: $-\frac{1}{2}$; L_3: 3
11. (a) $-\frac{3}{2}$; (b) $-\frac{3}{2}$
13. (a) slope: 1

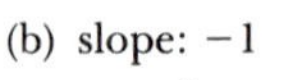

(b) slope: -1

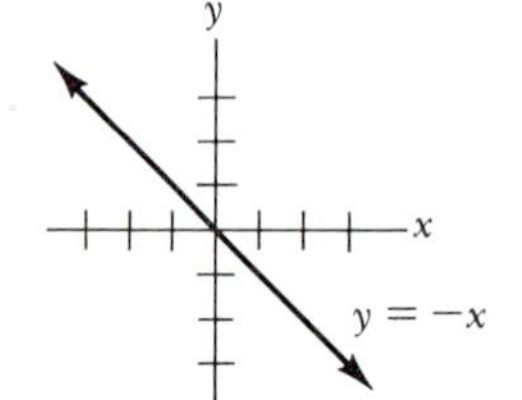

(c) slope: 5

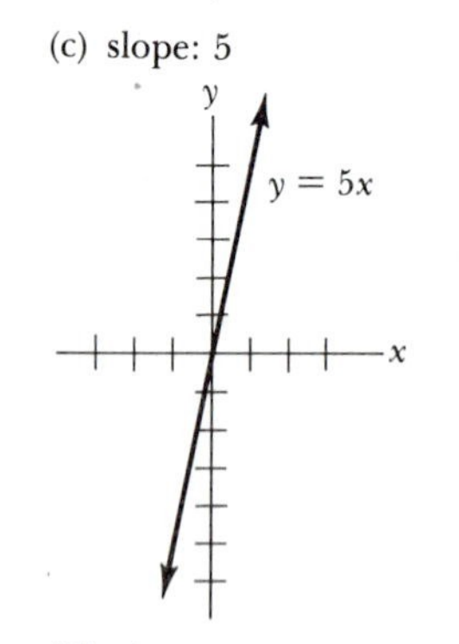

(d) slope: -5

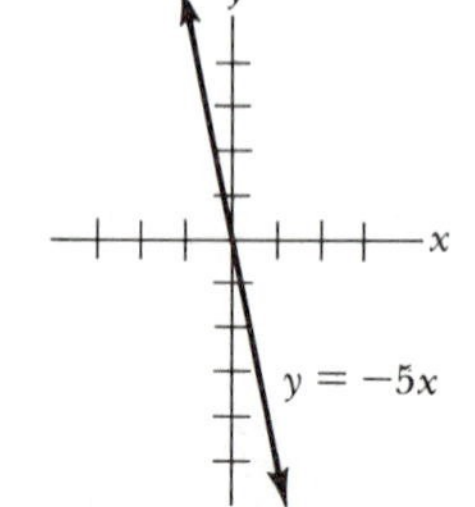

17. slope: 5

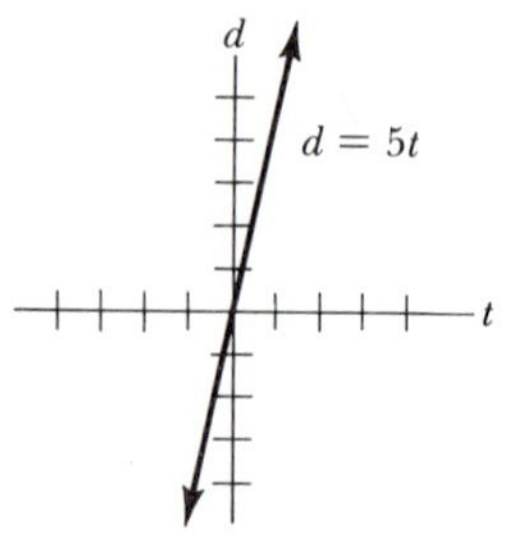

19. (a) $\frac{4}{5}$ ft/sec; (b) 0 cm/sec; (c) 8 mi/hr
21. slope: 0.5; marginal cost: $0.50 per album

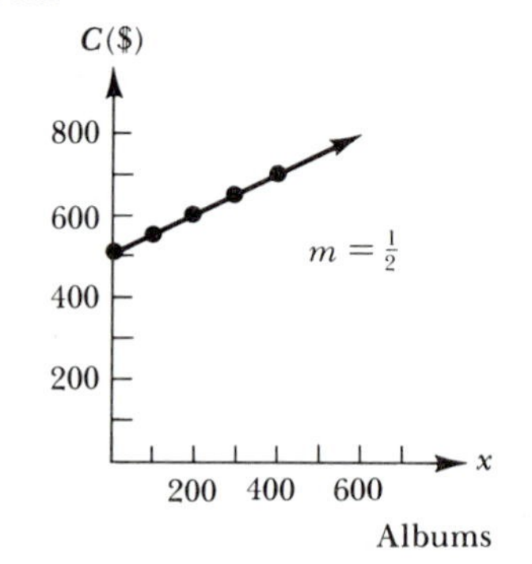

27. slope: 31 mi/gal, which is the car's mileage **29.** $(6, 36)$ **31.** $(8, \frac{1}{8})$
33. (a) $\dfrac{\sqrt{3}}{3}$; (b) $\sqrt{3}$; (c) 1
35. slope: 27. $y = x^3$ has a steeper tangent line.

Exercise Set 3.4

1. (a) $y = -5x - 9$; (b) $y = 4x - 20$; (c) $y = \frac{1}{3}x + \frac{4}{3}$; (d) $y = -x + 1$
3. (a) $y = 2x$; (b) $y = -2x - 4$; (c) $y = \frac{1}{7}x - \frac{11}{7}$; (d) $y = 9$; (e) $y = -2x + \frac{9}{2}$; (f) $y = -x + 25$
5. (a) $y = -4x + 7$; (b) $y = 2x + \frac{3}{2}$; (c) $y = -\frac{4}{3}x + 14$; (d) $y = 14$; (e) $y = 14x$
7. (a) $y = 4x + 11$

(b) $y = \frac{1}{2}x - \frac{5}{4}$

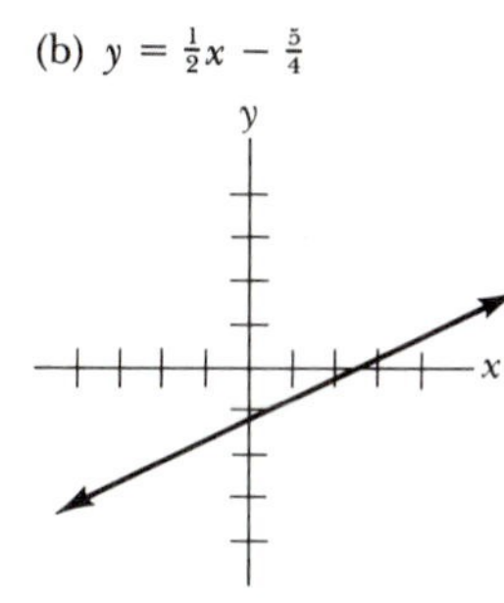

(c) $y = -\frac{5}{6}x + 5$

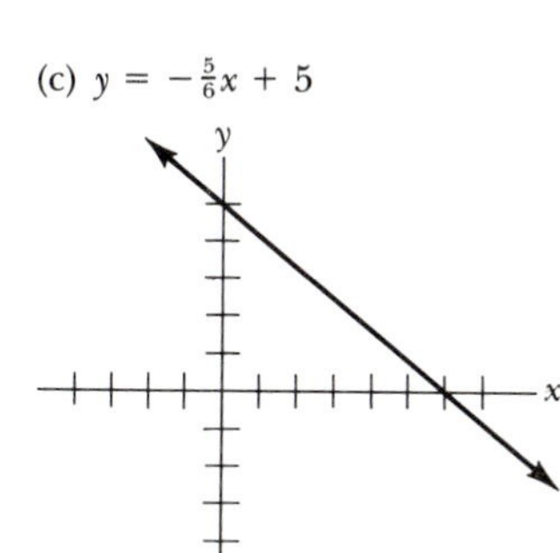

(d) $y = \frac{3}{4}x + \frac{3}{2}$

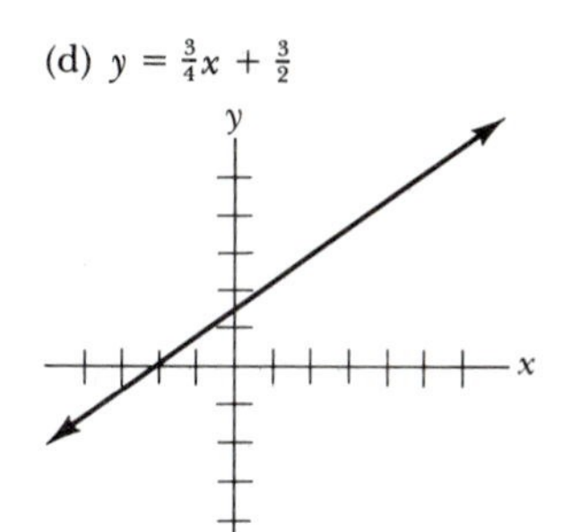

(e) $y = 4x - 2$

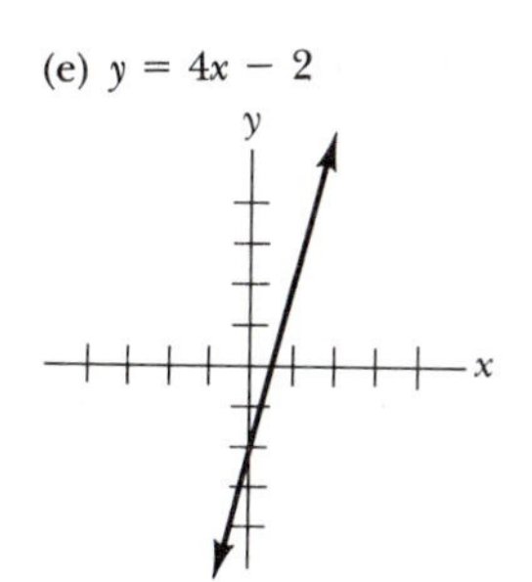

(f) $y = \frac{2}{7}x$

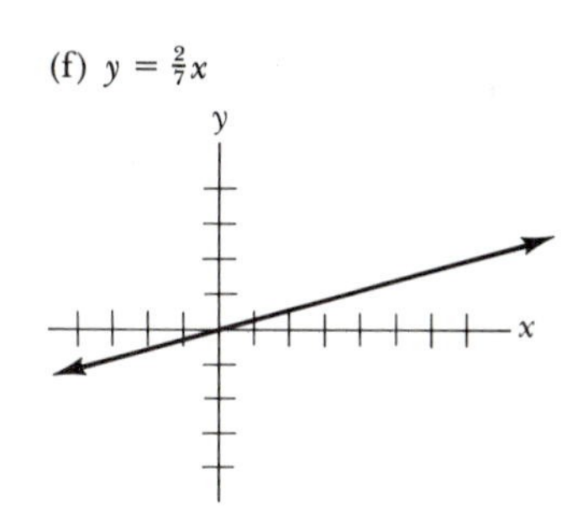

(g) $y = -\frac{11}{6}x + 8$

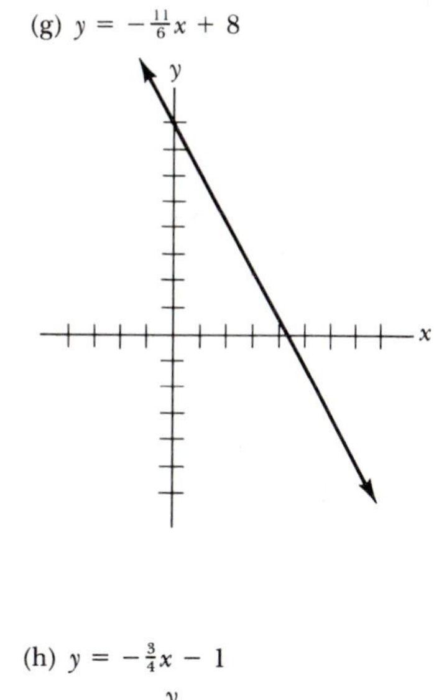

(h) $y = -\frac{3}{4}x - 1$

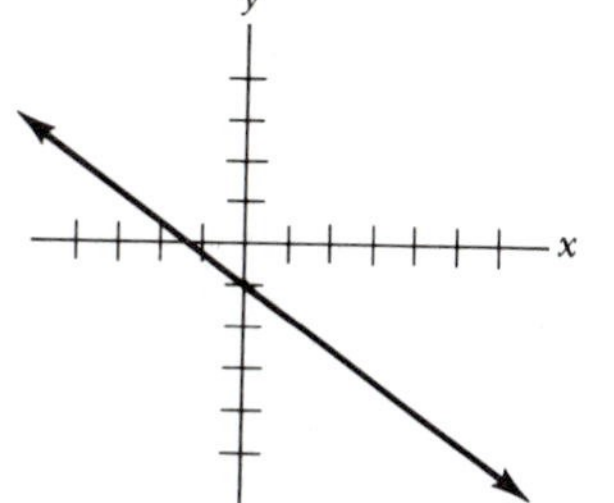

(i) $y = \frac{4}{13}x + \frac{2}{13}$

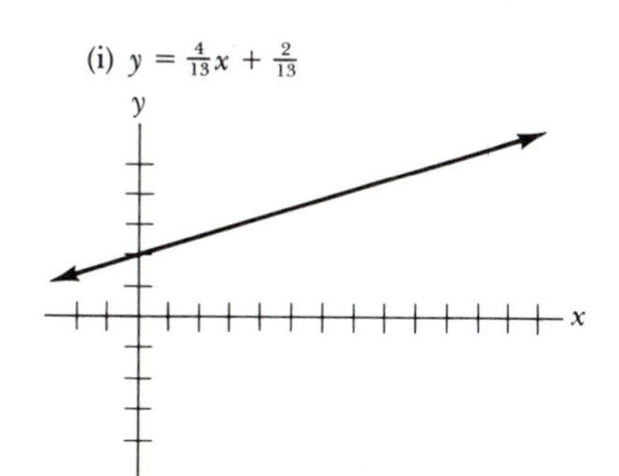

(j) $y = \dfrac{\sqrt{2}}{6}x + \sqrt{2}$

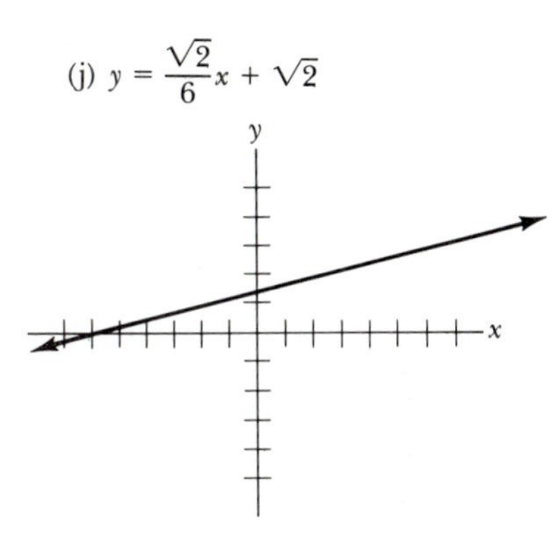

9. $y = -x - 4$

$x + y = 0$

$y = -x - 4$

11. $\frac{26}{5}$ **13.** (a) $y = -\frac{7}{3}x + 7$; (b) $y = -2x - 2$; (c) $y = \frac{4}{15}x - \frac{1}{3}$; (d) $y = -x + \pi$ **15.** $\frac{1369}{28} \approx 48.9$ **17.** (a) \$530; (b) \$538; (c) \$8 per unit **19.** (a) 5 ft/sec; (b) 75 ft; (c) 80 ft **21.** perpendicular **23.** $y = \frac{2}{5}x + \frac{12}{5}$

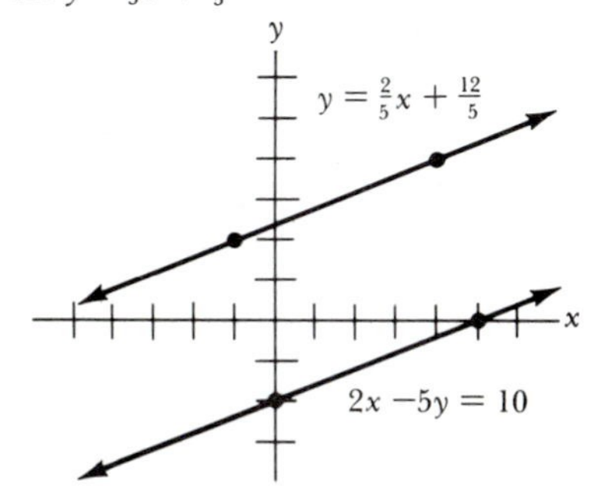

25. $y = -\frac{5}{4}x$ **27.** $y = \frac{15}{1}$ **29.** $y = 3x + 13$

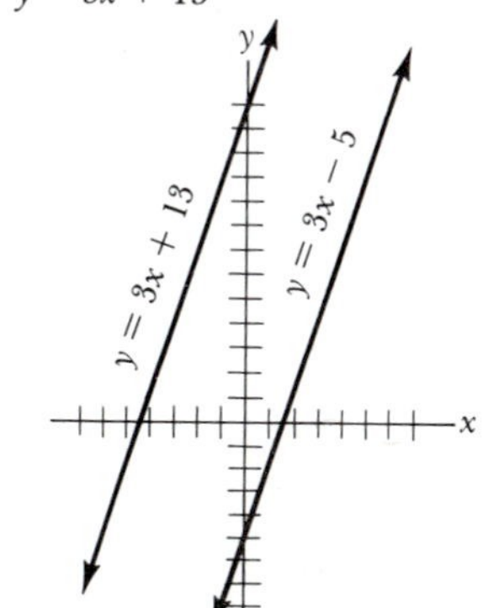

31. (a)

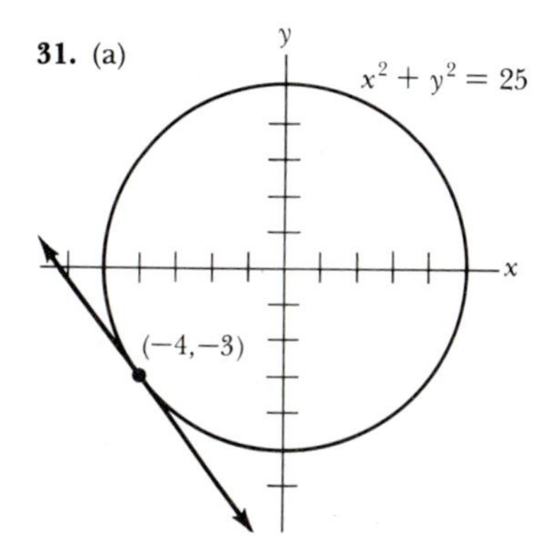

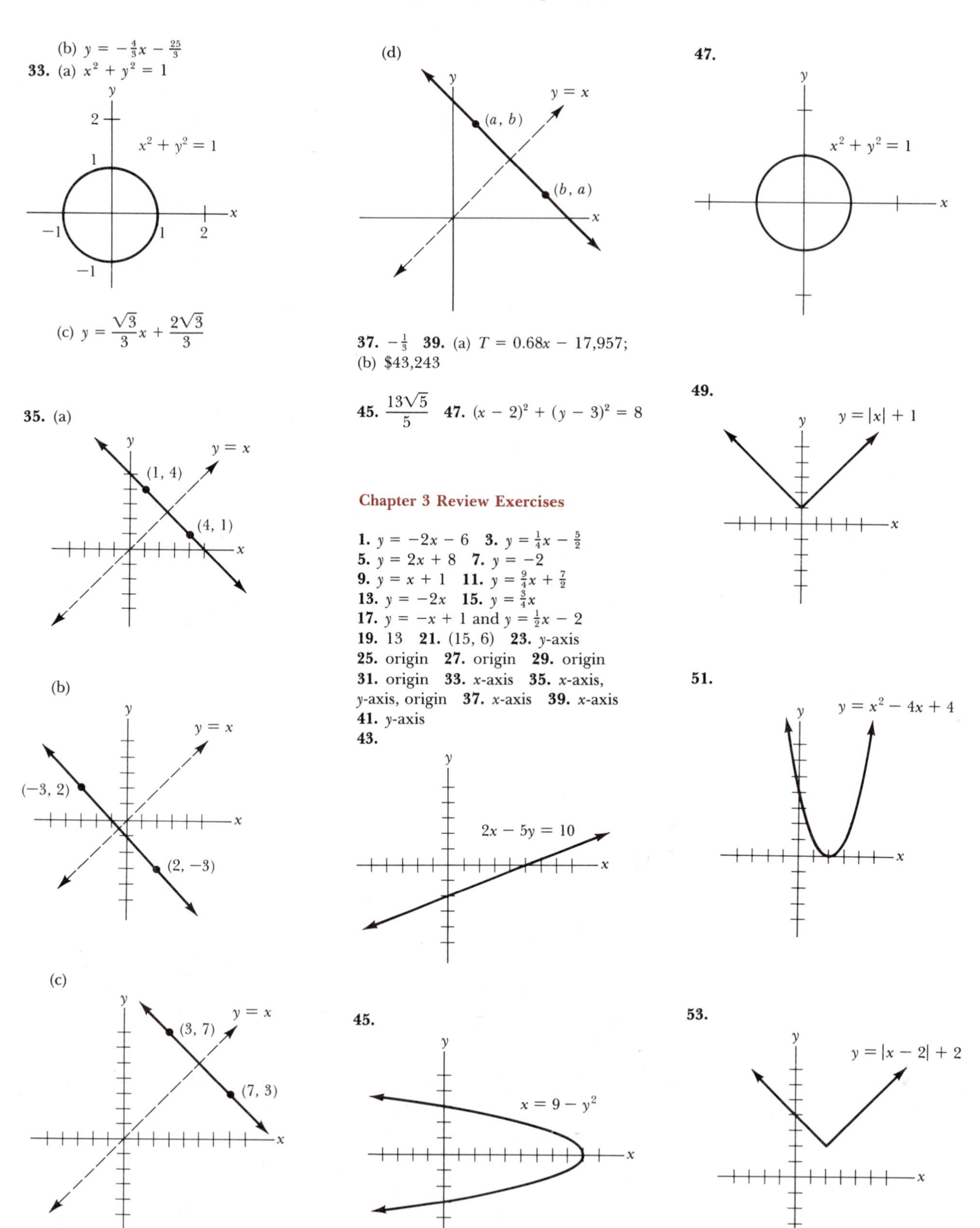

(b) $y = -\frac{4}{3}x - \frac{25}{3}$

33. (a) $x^2 + y^2 = 1$

(c) $y = \frac{\sqrt{3}}{3}x + \frac{2\sqrt{3}}{3}$

35. (a)

(b)

(c)

(d)

37. $-\frac{1}{3}$ **39.** (a) $T = 0.68x - 17{,}957$; (b) \$43,243

45. $\frac{13\sqrt{5}}{5}$ **47.** $(x - 2)^2 + (y - 3)^2 = 8$

Chapter 3 Review Exercises

1. $y = -2x - 6$ **3.** $y = \frac{1}{4}x - \frac{5}{2}$
5. $y = 2x + 8$ **7.** $y = -2$
9. $y = x + 1$ **11.** $y = \frac{9}{4}x + \frac{7}{2}$
13. $y = -2x$ **15.** $y = \frac{3}{4}x$
17. $y = -x + 1$ and $y = \frac{1}{2}x - 2$
19. 13 **21.** (15, 6) **23.** y-axis
25. origin **27.** origin **29.** origin
31. origin **33.** x-axis **35.** x-axis, y-axis, origin **37.** x-axis **39.** x-axis
41. y-axis
43.

45.

47.

49.

51.

53.

55.

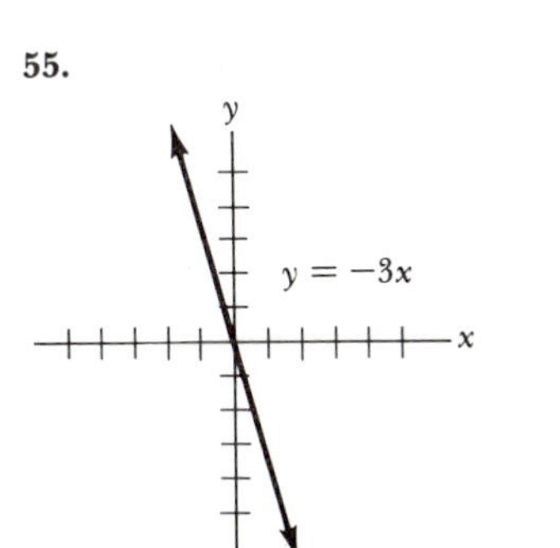

57.

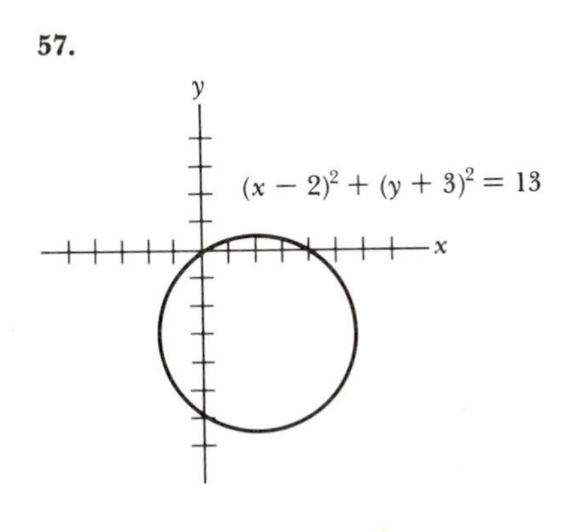

59.

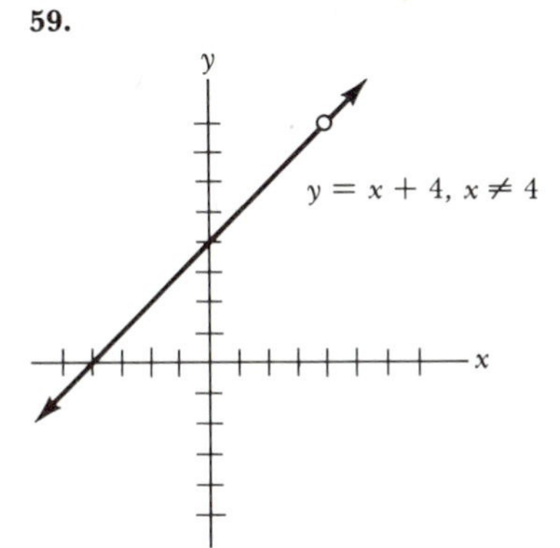

61.

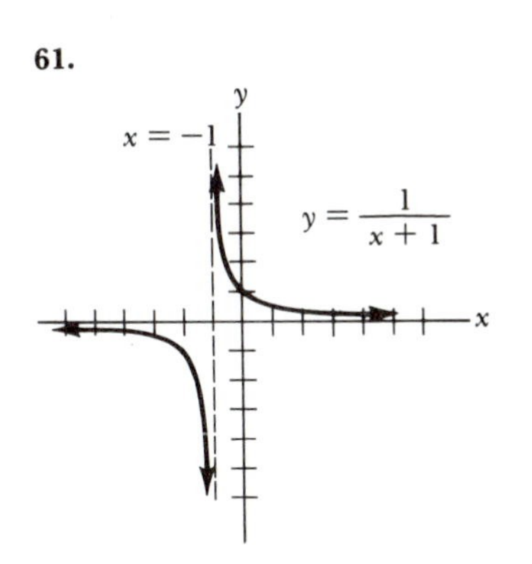

63. $t = 19$ **65.** $x^2 - 2x + 4$
67. $\overline{MA} = \overline{MB} = \overline{MC} = \sqrt{b^2 + c^2}$
69. (a) $\sqrt{130}$; (b) $-\frac{11}{3}$ (c) $(\frac{7}{2}, -\frac{1}{2})$

71. (a) 2; (b) $\dfrac{\sqrt{3}}{3}$; (c) (0, 0)
73. $\frac{32}{3}$ **75.** (a) \$12 per tire; (b) \$20,500; (c) \$20,512
79. $2h^2 + 2k^2 - 2r^2$

Exercise Set 4.1

1. (a) all reals; (b) all reals except 3; (c) $[0, \infty)$; (d) all reals **3.** (a) all reals; (b) all reals except 0; (c) all reals; (d) all reals except -4 **5.** all reals except 0, -2, 2 **7.** (a) $(-\infty, -1] \cup [5, \infty)$; (b) $(-\infty, -2] \cup (4, \infty)$ **9.** all reals except 1 **11.** all reals except -1
13. $[1, \infty)$ **15.** f, g, F, H **17.** (a) f: $\{1, 2, 3\}$; g: $\{2, 3\}$; F: $\{1\}$; H: $\{1, 2\}$; (b) g: $\{i, j\}$; F: $\{i, j\}$; G: $\{k\}$
19. (a) $y = (x - 3)^2$; (b) $y = x^2 - 3$; (c) $y = (3x)^2$; (d) $y = 3x^2$ **21.** (a) -1; (b) 1; (c) 5; (d) $-\frac{5}{4}$; (e) $z^2 - 3z + 1$; (f) $x^2 - x - 1$; (g) $a^2 - a - 1$; (h) $x^2 + 3x + 1$; (i) 1; (j) $4 - 3\sqrt{3}$; (k) $1 - \sqrt{2}$; (l) 2 **23.** (a) $12x^2$; (b) $6x^2$; (c) $3x^4$; (d) $9x^4$; (e) $\dfrac{3x^2}{4}$; (f) $\dfrac{3x^2}{2}$ **25.** (a) 1; (b) -7; (c) -3; (d) $-\frac{7}{18}$; (e) $-7 + 4\sqrt{3}$; (f) $1 - 2x^4$; (g) $-2x^2 - 4x - 1$; (h) $1 - 2x^2 - 4xh - 2h^2$; (i) $-4xh - 2h^2$; (j) $-4x - 2h$
27. (a) domain = range = all reals except 2; (b) $\frac{1}{2}$; (c) 0; (d) 1; (e) $\dfrac{2x^2 - 1}{x^2 - 2}$; (f) $\dfrac{2 - x}{1 - 2x}$; (g) $\dfrac{2a - 1}{a - 2}$; (h) $\dfrac{2x - 3}{x - 3}$
29. (a) $d(1) = 80$, $d(\frac{3}{2}) = 108$, $d(2) = 128$, $d(t_0) = -16t_0^2 + 96t_0$; (b) 0, 6; (c) 5.99, 0.01 **31.** $g(3) = 1$, $g(x + 4) = |x|$ **33.** (a) -3; (b) $2x + h$; (c) $4x + 2h - 3$; (d) $3x^2 + 3xh + h^2$; (e) 0 **35.** $-\frac{1}{4}$ **37.** (a) $\frac{11}{5}$; (b) 2, -2; (c) 0, 1; (d) $\frac{3}{2}$, -1 (e) $\dfrac{\sqrt{5}}{5}$, $\dfrac{-\sqrt{5}}{5}$; (f) no reals **39.** (a) $h + 5$; (b) $t + 3$; (c) $t + t_0 + 1$ **41.** $\dfrac{1 - ax}{1 + ax}$
47. $\dfrac{19}{(2x + 2h + 1)(2x + 1)}$ **49.** $\dfrac{2u - 4}{u + 12}$
53. $-\dfrac{1}{3}$ **57.** $\dfrac{1}{\sqrt{x + h} + \sqrt{x}}$ **61.** (a) x; (b) $\frac{22}{7}$ **65.** $P(1) = 43$, $P(2) = 47$, $P(3) = 53$, $P(4) = 61$. 40 and 41 are not prime. **67.** 0 **69.** (a) -0.041; (b) -0.040 **71.** (a) \$125.51; (b) \$363.76 **73.** (a) $g(2) = 1.4142$, $g(3) = 1.4422$, $g(4) = 1.4142$, $g(5) = 1.3797$, $g(6) = 1.3480$, $g(7) = 1.3205$, $g(8) = 1.2968$; (b) 15

Exercise Set 4.2

1. (a) $P(x) = 2x + 2\sqrt{144 - x^2}$; (b) $A(x) = x\sqrt{144 - x^2}$
3. (a) $D(x) = \sqrt{x^4 + 3x^2 + 1}$; (b) $m(x) = \dfrac{x^2 + 1}{x}$ **5.** (a) $A(y) = \dfrac{\pi y^2}{4}$; (b) $A(y) = \dfrac{\pi^2 y^2}{16}$ **7.** (a) $P(x) = 16x - x^2$; (b) $S(x) = 2x^2 - 32x + 256$; (c) $D(x) = (16 - x)^3 - x^3$; (d) $A(x) = 8$
9. $R(x) = -\frac{1}{4}x^2 + 8x$
11. (a) $P(1) = 17.875$, $P(2) = 19.492$, $P(3) = 20.832$, $P(4) = 21.856$, $P(5) = 22.490$, $P(6) = 22.583$, $P(7) = 21.746$; (b) $P(1) < P(2) < P(3) < P(7) < P(4) < P(5) < P(6)$
13. (a) $h(s) = \sqrt{3}s$; (b) $A(s) = \sqrt{3}s^2$; (c) $4\sqrt{3}$ cm; (d) $\dfrac{25\sqrt{3}}{4}$ sq in.
15. $V(r) = 2\pi r^3$ **17.** (a) $h(r) = \dfrac{12}{r^2}$; (b) $S(r) = 2\pi r^2 + \dfrac{24\pi}{r}$
19. $V(S) = \dfrac{S\sqrt{\pi S}}{6\pi}$
21. $A(x) = \dfrac{1}{2}x\sqrt{400 - x^2}$
23. $d(x) = \dfrac{(x + 4)\sqrt{x^2 + 25}}{x}$
25. (a) $\dfrac{a^2 + 1}{a}$
27. (a) $V(r) = \dfrac{\sqrt{3}}{3}\pi r^3$; (b) $S(r) = 2\pi r^2$
29. (a) $r(h) = \dfrac{3h}{\sqrt{h^2 - 9}}$; (b) $h(r) = \dfrac{3r}{\sqrt{r^2 - 9}}$
31. $A(x) = \dfrac{4x^2 + \pi(14 - x)^2}{16\pi}$
33. $A(r) = \dfrac{r(1 - 4\pi r)}{4}$ **35.** $A(x) = \dfrac{\pi x^2}{3}$
37. $A(h) = \pi R^2 - h\sqrt{2Rh - h^2}$
39. $V(x) = 4x^3 - 28x^2 + 48x$
41. $A(x) = 500x - 2x^2$
43. $A(r) = 32r - 2r^2 - \dfrac{\pi r^2}{2}$
45. (a) $y(s) = \dfrac{3s}{\sqrt{1 - s^2}}$; (b) $s(y) = \dfrac{y}{\sqrt{y^2 + 9}}$; (c) $z(s) = \dfrac{3}{\sqrt{1 - s^2}}$; (d) $s(z) = \dfrac{\sqrt{z^2 - 9}}{z}$

47. $A(m) = \dfrac{2m^2 - 8m + 8}{m^2 - 4m}$
49. Table 1: $x = 1$; Table 2: $x = 0.50$; Table 3: $x = 0.500$

Exercise Set 4.3

1. (a) positive; (b) $f(-2) = 4, f(1) = 1, f(2) = 2, f(3) = 0$; (c) $f(2)$; (d) -3; (e) 3; (f) domain = range = $[-2, 4]$
3. (a) $f(3) = 3, f(\pi) = 3, f(\frac{39}{10}) = 3, f(-\frac{3}{2}) = -2$; (b) the set of all integers
5. (a) false; (b) true; (c) false; (d) false; (e) true **7.** $f(x)$: max = 2, min = -3; $g(x)$: max = 3, min = -1; $h(x)$: max = 4, min = -2 **9.** $\frac{54}{7}$ **11.** 7
13. $(1, T(1))$ and $(4, T(4))$
15. (a) (b)

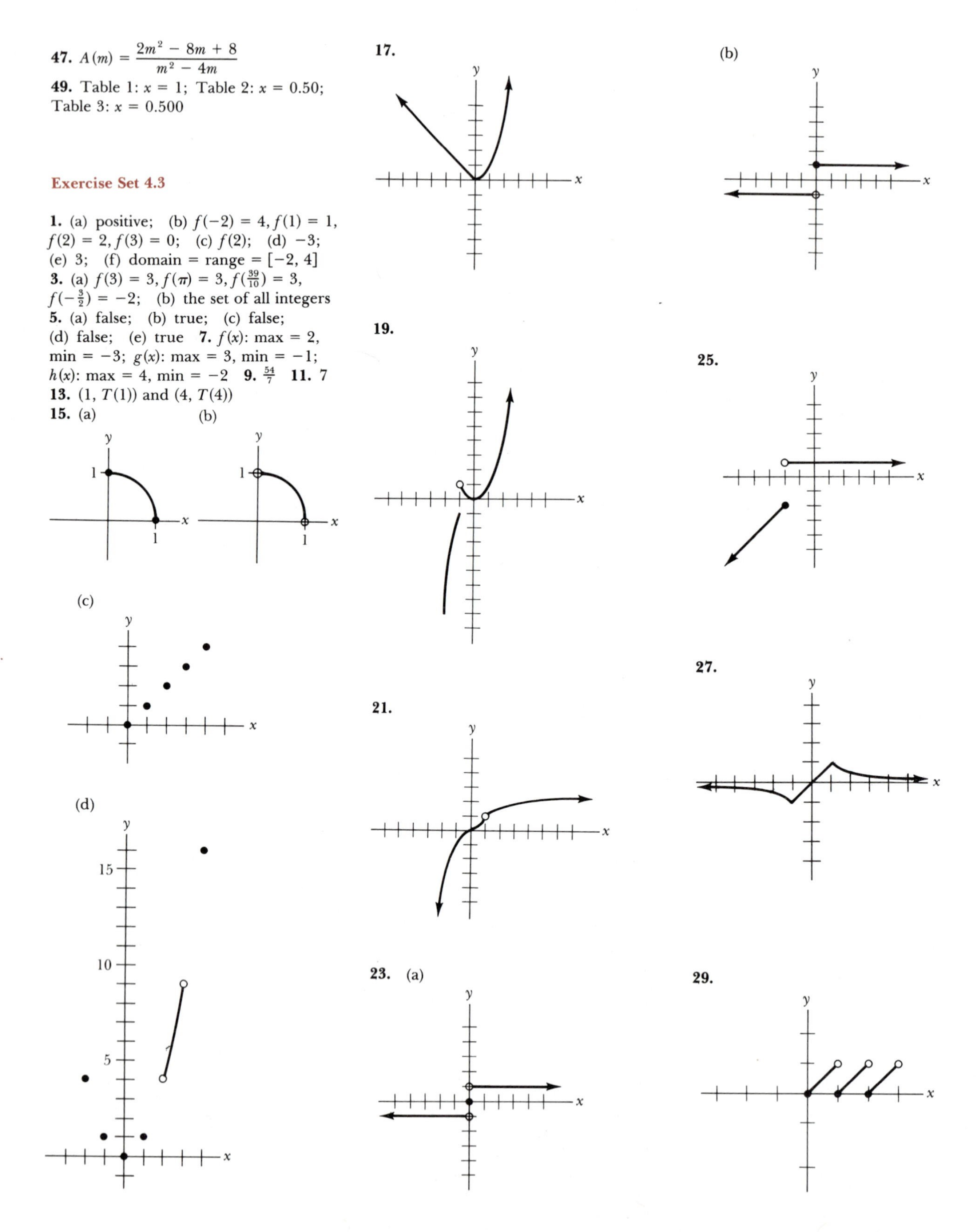

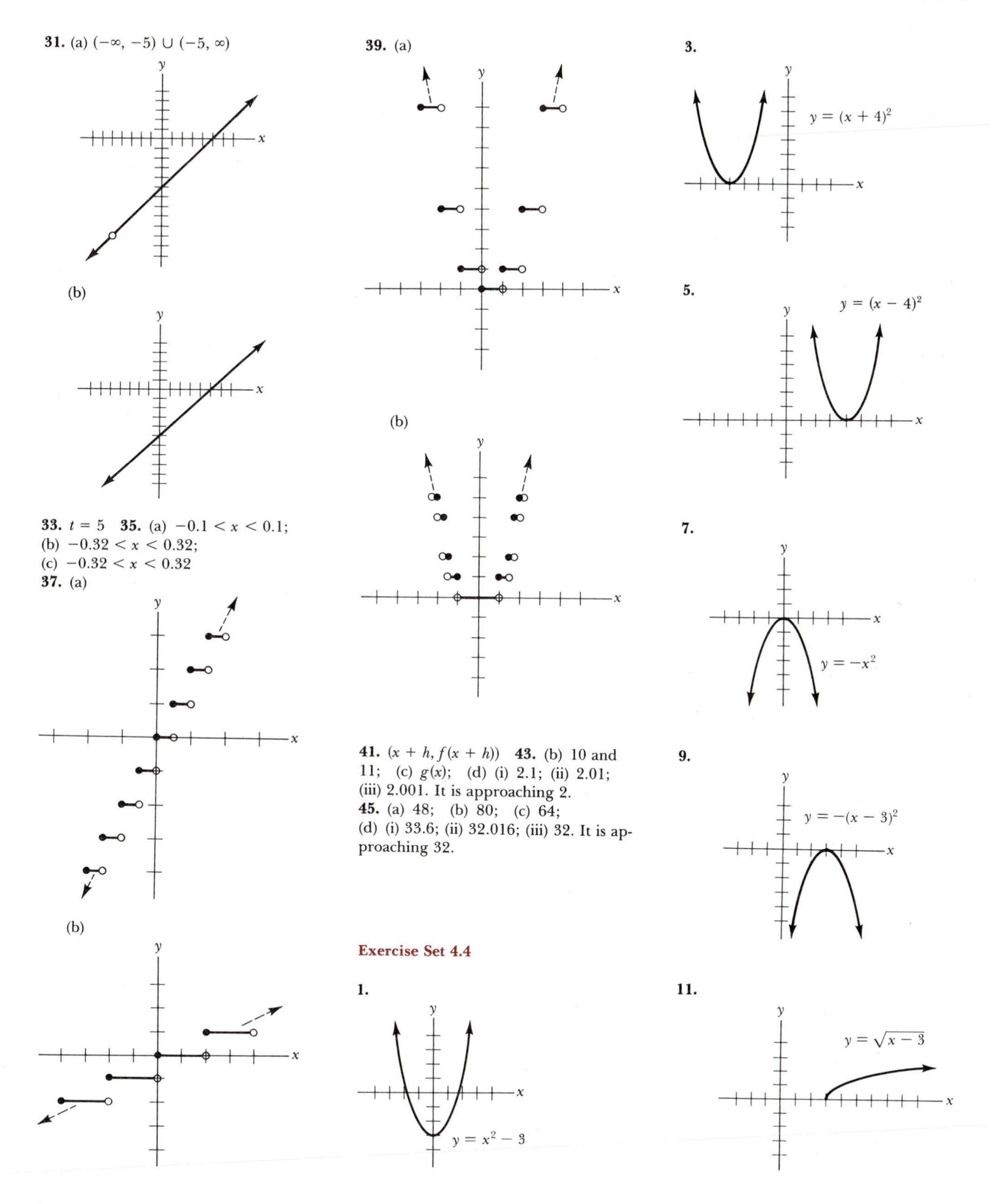

31. (a) $(-\infty, -5) \cup (-5, \infty)$

(b)

33. $t = 5$ **35.** (a) $-0.1 < x < 0.1$; (b) $-0.32 < x < 0.32$; (c) $-0.32 < x < 0.32$

37. (a)

(b)

39. (a)

(b)

41. $(x + h, f(x + h))$ **43.** (b) 10 and 11; (c) $g(x)$; (d) (i) 2.1; (ii) 2.01; (iii) 2.001. It is approaching 2.

45. (a) 48; (b) 80; (c) 64; (d) (i) 33.6; (ii) 32.016; (iii) 32. It is approaching 32.

Exercise Set 4.4

1.

3.

5.

7.

9.

11.

13.

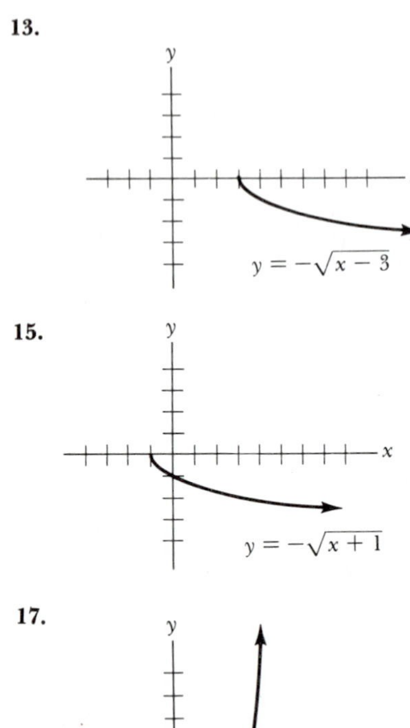

15.

17.

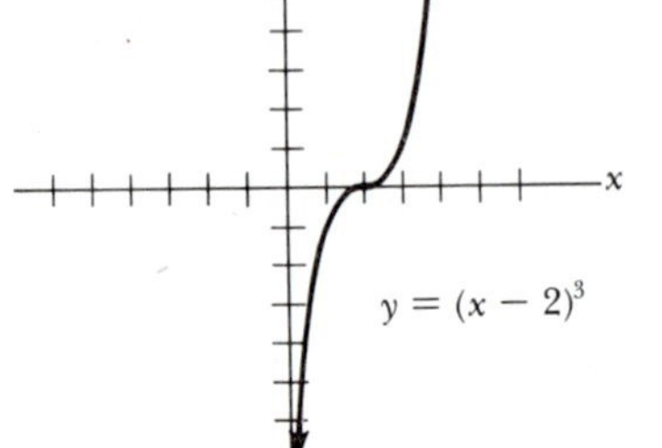

19.

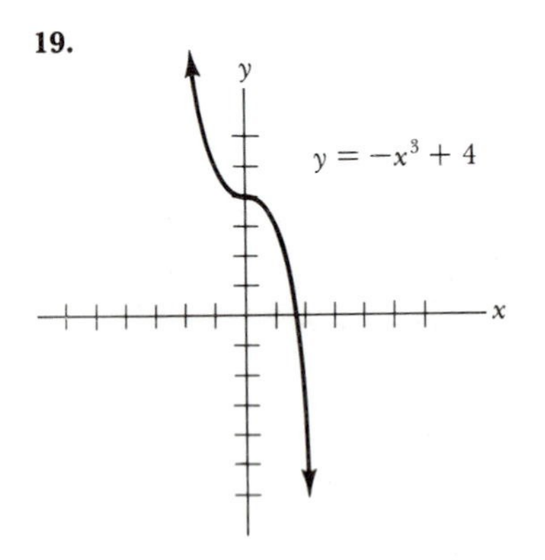

21. (a)

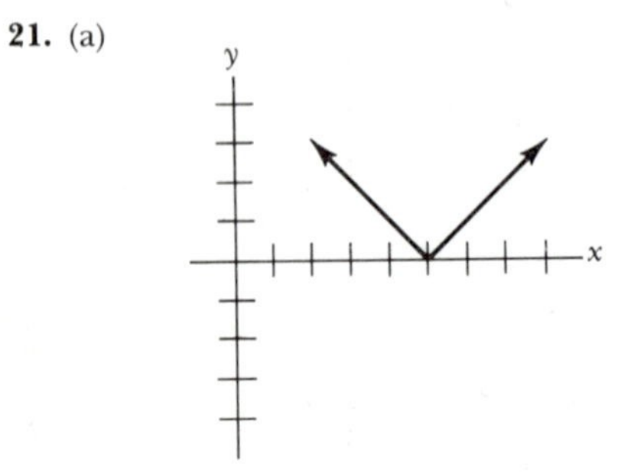

(b)

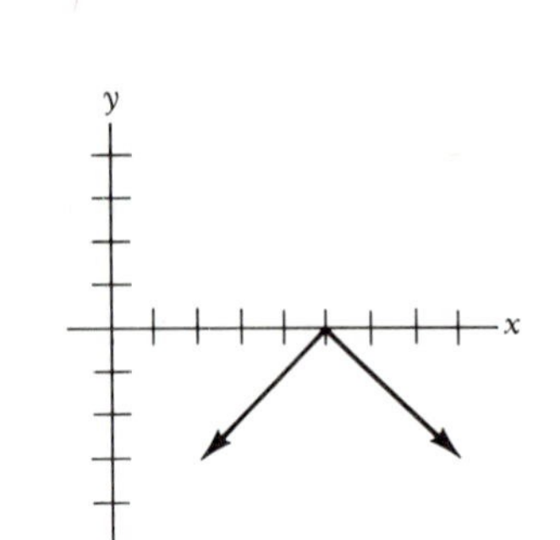

(c)

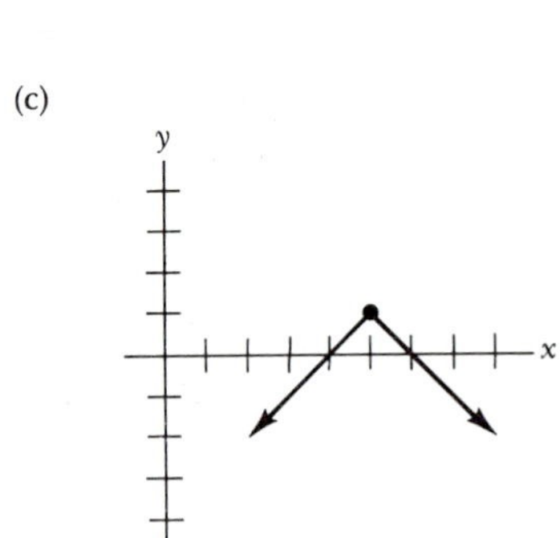

(d)

23. (a)

(b)

(c)

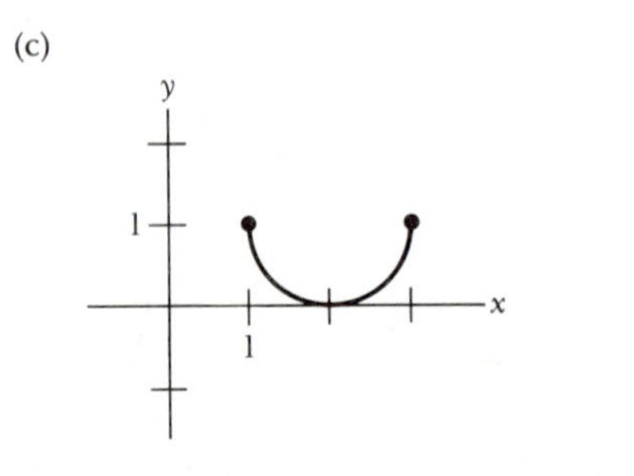

25. (a)

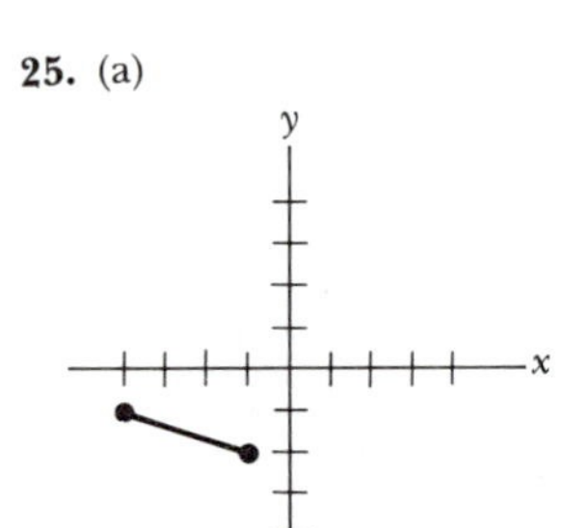

(b)

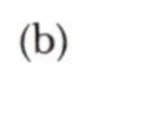

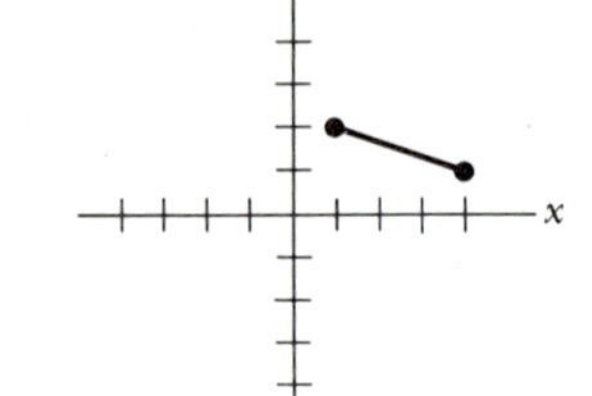

(c)

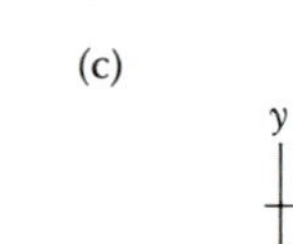

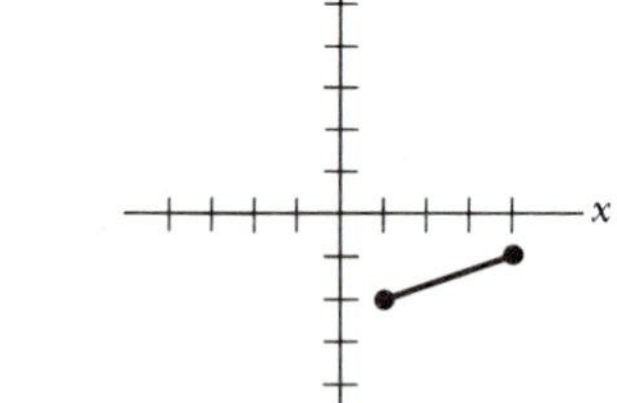

27. (a)

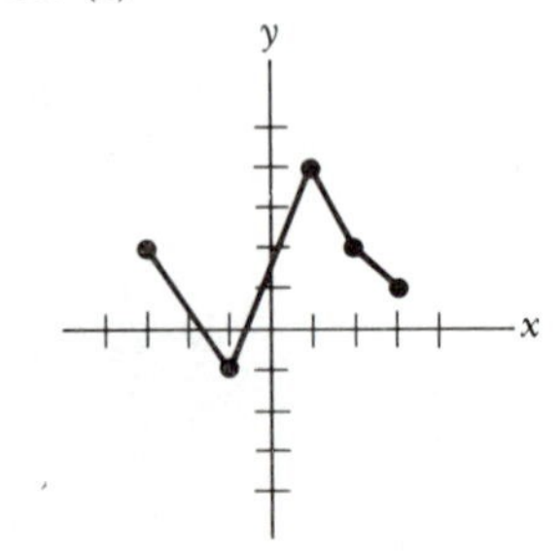

(b)

(c)

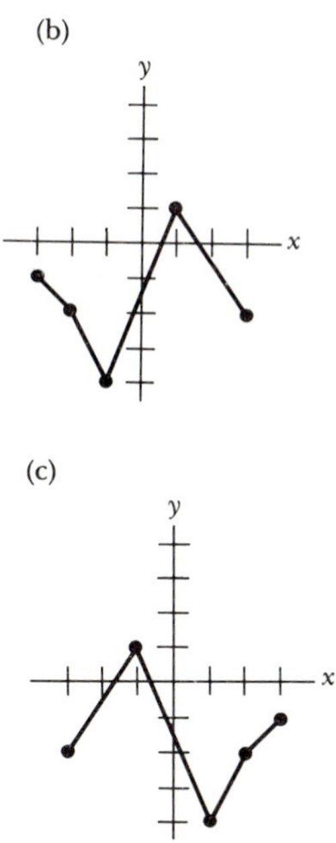

29.

31.

33. (a) y-intercept: 4

(b) x-intercepts: $\pm\sqrt{3}$; y-intercept: -3

(c) x-intercept: -1; y-intercept: 1

(d) y-intercept: 19

(e) x-intercepts: $4 \pm \sqrt{2}$; y-intercept: -14

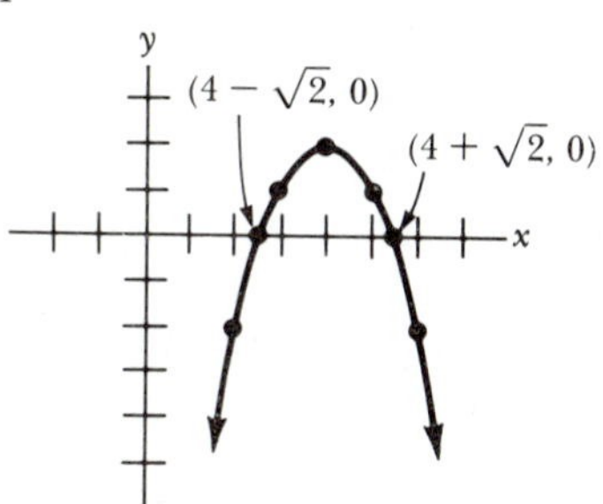

(f) x-intercept: 100; y-intercept: 10,000

Exercise Set 4.5

1. (a) $x^2 - x - 7$; (b) $-x^2 + 5x + 5$; (c) 5 **3.** (a) $x^2 - 2x - 8$; (b) $-x^2 + 2x + 8$ **5.** (a) $4x - 2$; (b) $4x - 2$; (c) -4 **7.** (a) $\dfrac{-x^4 + 22x^2 - 4x - 80}{2x^3 - x^2 - 18x + 9}$; (b) $-\dfrac{80}{9}$ **9.** (a) $-x^5 + 9x^3 + 2x^2 - 18$; (b) $-x^5 + 9x^3 + 2x^2 - 18$; (c) -24 **11.** (a) $-6x - 14$; (b) -74; (c) $-6x - 7$ (d) -67 **13.** (a) $-x$, 2, $2 - x$, 4; (b) $4x^4$, 64, $2x^4$, 32; (c) x^6, 64, x^6, 64; (d) $9x^2 - 3x - 6$, 36, $-3x^2 + 9x + 14$, -16; (e) $\dfrac{1 - x^4}{3}$, -5, $1 - \dfrac{x^4}{81}$, $\dfrac{65}{81}$; (f) 2^{x^2+1}, 32, 2^{2x+1}, $\dfrac{17}{16}$; (g) $3x^5 - 4x^2$, -112, $3x^5 - 4x^2$, -112; (h) x, -2, x, -2 **15.** (a) $\dfrac{-x + 7}{6x}$; (b) $\dfrac{-t + 7}{6t}$; (c) $\dfrac{5}{12}$; (d) $\dfrac{1 - 6x}{7}$; (e) $\dfrac{1 - 6y}{7}$; (f) $-\dfrac{11}{7}$ **17.** (a) 7; (b) x; (c) yes! **19.** (a) 1; (b) -3; (c) -1; (d) 2; (e) 2; (f) -3 **21.** (a) $4x^3 - 3x^2 + 6x - 1$, $4x^3 - 3x^2 + 6x - 1$; (b) $ax^2 + bx + c$, $ax^2 + bx + c$; (c) $(f \circ I)(x) = (I \circ f)(x) = f(x)$ **23.** $(f \circ g)(0) = 1$, $(f \circ g)(1) = 3$, $(f \circ g)(2) = 2$, $(f \circ g)(3)$ is undefined, $(f \circ g)(4) = 2$, $(g \circ f)(-1) = 0$, $(g \circ f)(0) = 0$, $(g \circ f)(1) = 3$, $(g \circ f)(2) = 4$, $(g \circ f)(3) = 2$, $(g \circ f)(4)$ is undefined **25.** (a) $6x - 7$

(b) $6x - 1$

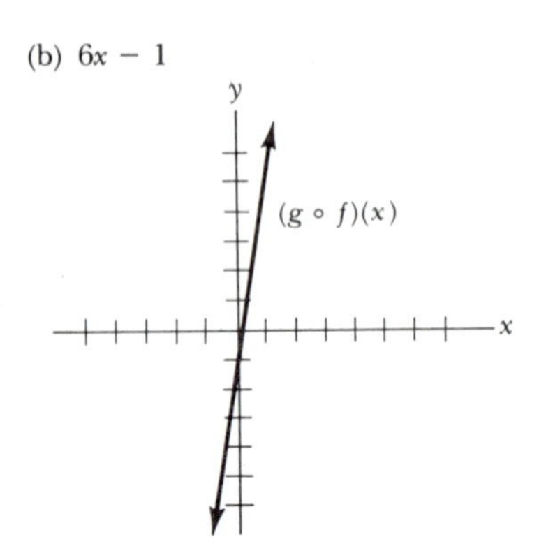

27. (a) domain: $[0, \infty)$; range: $[-3, \infty)$

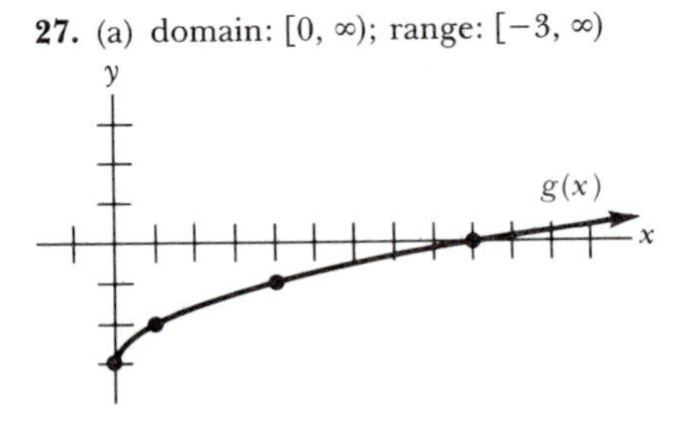

(b) domain: all real numbers; range: all real numbers

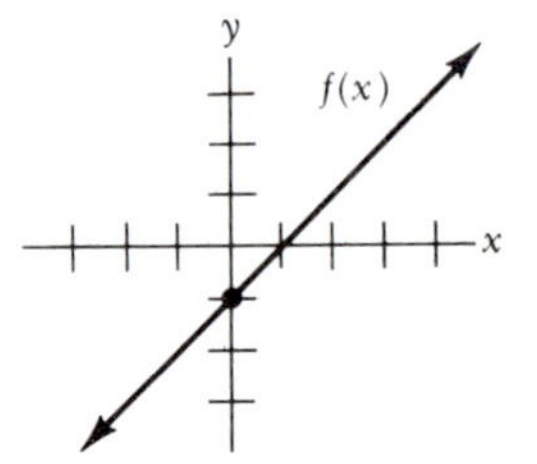

(c) $\sqrt{x} - 4$. domain: $[0, \infty)$; range: $[-4, \infty)$

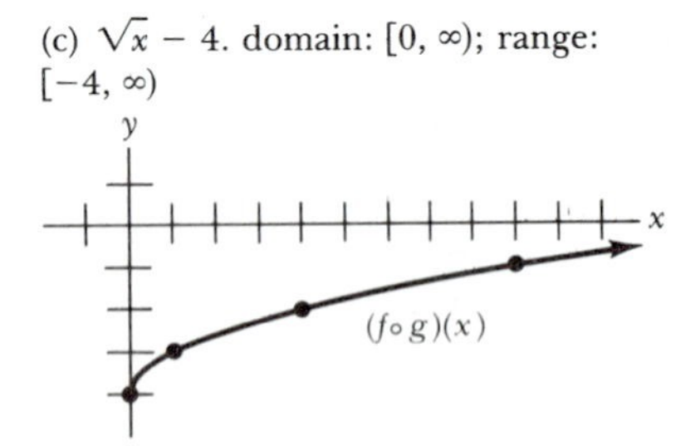

(d) $\sqrt{x - 1} - 3$. domain: $[1, \infty)$; range: $[-3, \infty)$

(e)

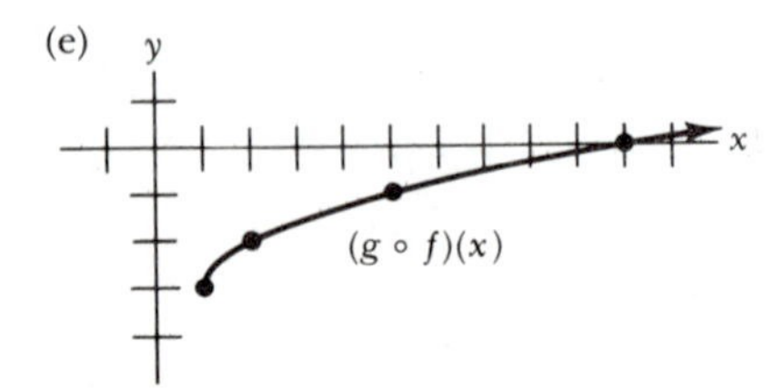

29. $V[g(t)] = \dfrac{\pi(t + 4)^3}{6}$; 2 sec
31. (a) $100(-2t^2 + 8t + 10)$; (b) 1000, 1800, 0 **33.** $f(x) = x^3$, $g(x) = 1 + x^2$ **35.** (a) $f(x) = \sqrt[3]{x}$, $g(x) = 3x + 4$; (b) $f(x) = |x|$, $g(x) = 2x - 3$; (c) $f(x) = x^5$, $g(x) = ax + b$; (d) $f(x) = \dfrac{1}{x}$, $g(x) = \sqrt{x}$
37. (a) $(b \circ c)(x)$; (b) $(a \circ d)(x)$; (c) $(c \circ d)(x)$; (d) $(c \circ b)(x)$; (e) $(c \circ a)(x)$; (f) $(a \circ c)(x)$; (g) $(b \circ d)(x)$ or $(d \circ b)(x)$
39. $5x - 4$ **41.** (a) x; (b) $\frac{113}{355}$
43. (a) $2x + 2a - 2$; (b) $4x + 4a - 4$
45. (a) $\dfrac{x^2}{2} + 1$; (b) $\dfrac{(x + 1)^2}{2}$; (c) $\dfrac{x^2}{4} + 1$; (d) $\dfrac{(x + 1)^2}{4}$; (e) $\dfrac{x^2 + 1}{2}$
47. (a) $1 - 9x^2$; (b) $3 - 3x^2$; (c) $1 - 6x + 9x^2$; (d) $3 - 6x + 3x^2$
49. (a) $48x^5 + 240x^3 + 300x$; (b) -588; (c) $48t^5 + 240t^3 + 300t$

Exercise Set 4.6

3. (a) 4; (b) -1; (c) $\sqrt{2}$; (d) $t + 1$; (e) 1, 0; (f) $-2, -1$ **5.** (a) $\dfrac{x + 1}{3}$;
(c)

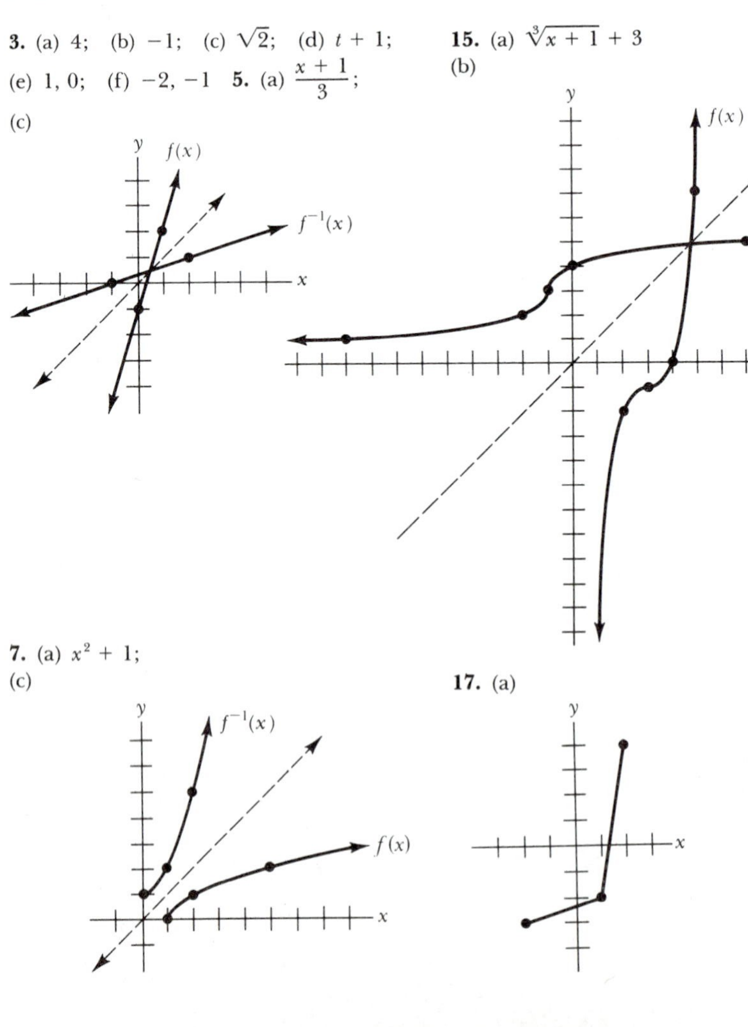

7. (a) $x^2 + 1$;
(c)

9. (a) domain: $(-\infty, 3) \cup (3, \infty)$; range: $(-\infty, 1) \cup (1, \infty)$; (b) $\dfrac{3x + 2}{x - 1}$; (c) domain: $(-\infty, 1) \cup (1, \infty)$; range: $(-\infty, 3) \cup (3, \infty)$

11. $\sqrt[3]{\dfrac{x - 1}{2}}$

13.

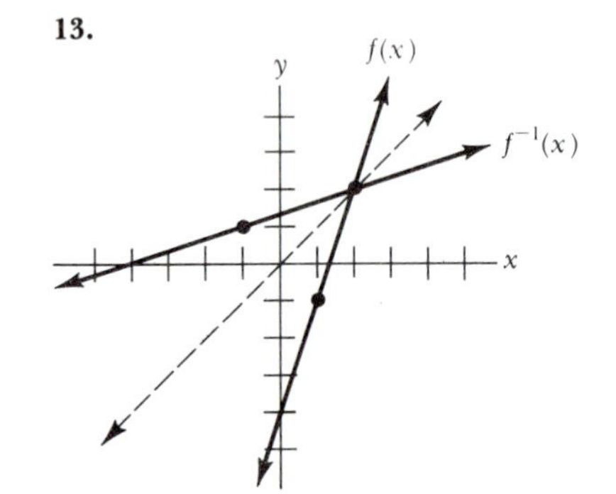

15. (a) $\sqrt[3]{x + 1} + 3$
(b)

17. (a)

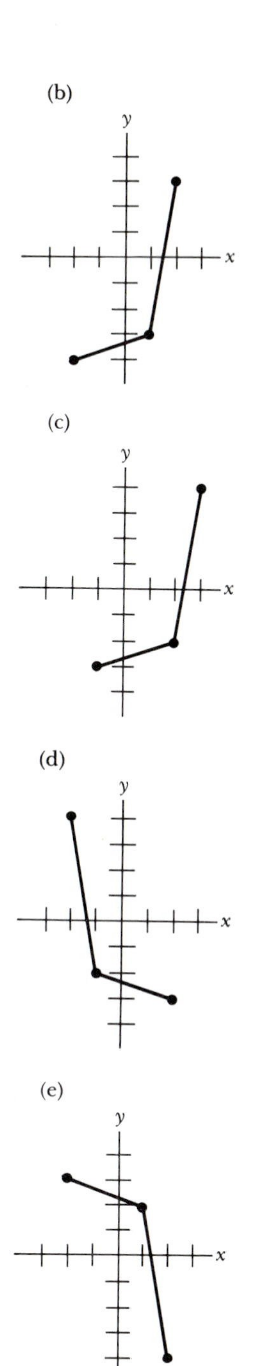

21. (a) x^2, $[0, \infty)$; (b) (i) f; (ii) f^{-1}; (iii) f; (iv) f^{-1}; (v) f; (vi) f^{-1}; (vii) f^{-1}; (viii) f **23.** (a) $\dfrac{b - dx}{cx - a}$; (b) $\dfrac{(a^2 + bc)x + (ab + bd)}{(ac + cd)x + (bc + d^2)}$; (c) $\dfrac{ab + bd}{bc + d^2}$; (d) x; (e) 0 **27.** no **29.** yes **31.** yes **33.** no **35.** no **37.** yes **39.** $m = \frac{2}{3}$, $b = 1$ **43.** yes

Exercise Set 4.7

1. (a) $y = kx$, (b) $A = \dfrac{k}{B}$
3. (a) $x = kuv^2$; (b) $z = kA^2B^3$
5. (a) $F = \dfrac{k}{r^2}$; (b) $V^2 = k(U^2 + T^2)$
7. (a) -2; (b) $A = -\dfrac{2}{B}$; (c) $-\dfrac{8}{5}$
9. -24 **11.** multiplied by 6
13. multiplied by 8 **15.** (a) $S = 4\pi r^2$; (b) 12π cm^2 **17.** multiplied by 48
19. 122.5 m **21.** (a) $V = \dfrac{2.05}{P}$; (b) 2.05 liters **23.** (a) multiplied by 6; (b) multiplied by $\sqrt{2}$ **25.** 110.6 lb
33. (a) multiplied by $\frac{9}{4}$; (b) 70.71 mph
41. $4000(\sqrt{2} - 1)$ **43.** 3,966,064,430 light years

Chapter 4 Review Exercises

1. x-intercepts: 2, -2; y-intercept: 4

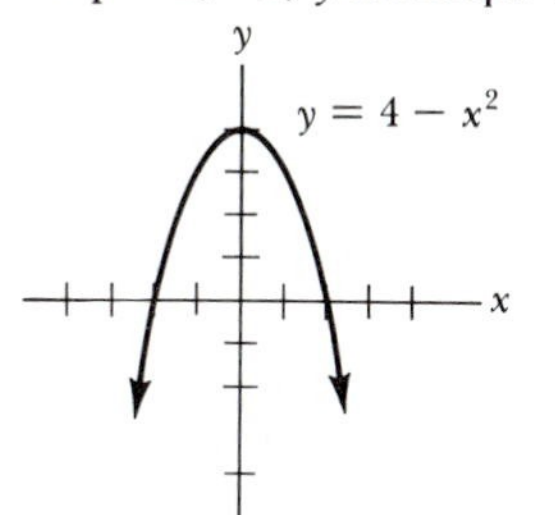

3. x-intercept: -1

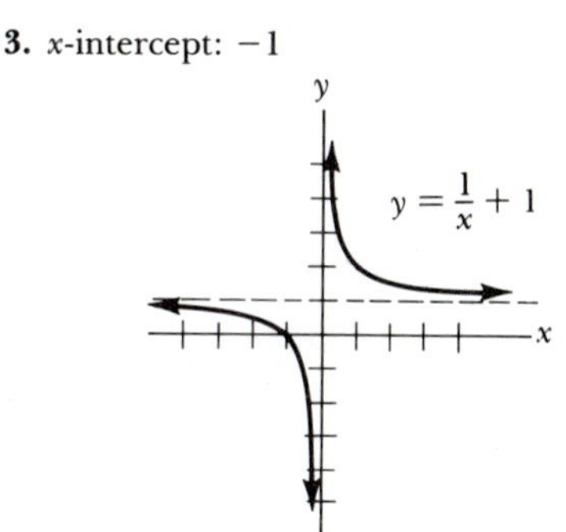

5. x-intercept: -2; y-intercept: 2

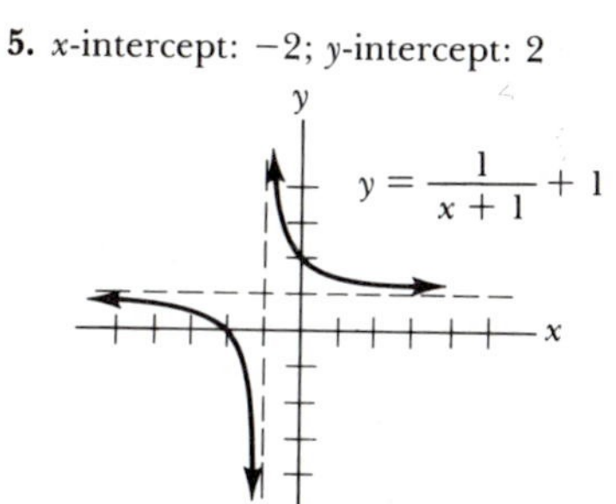

7. x-intercept: 4

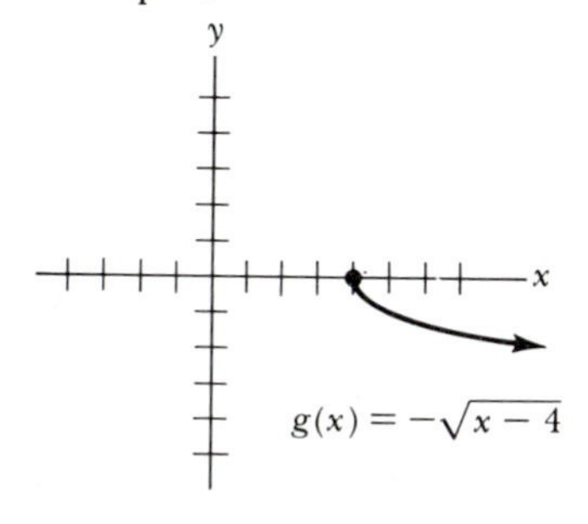

9. y-intercept: 2

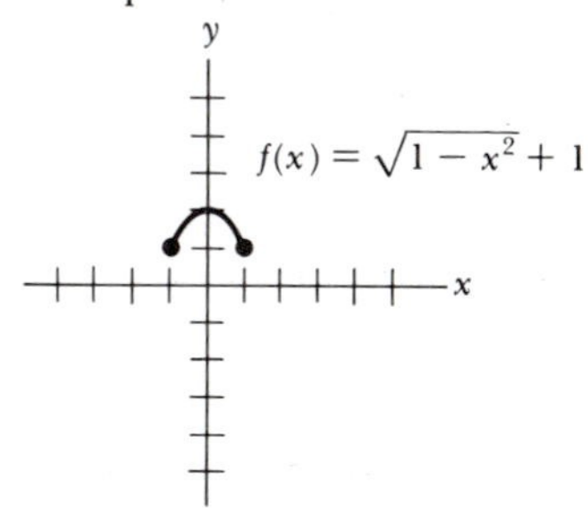

11. x-intercept: -16; y-intercept: 4

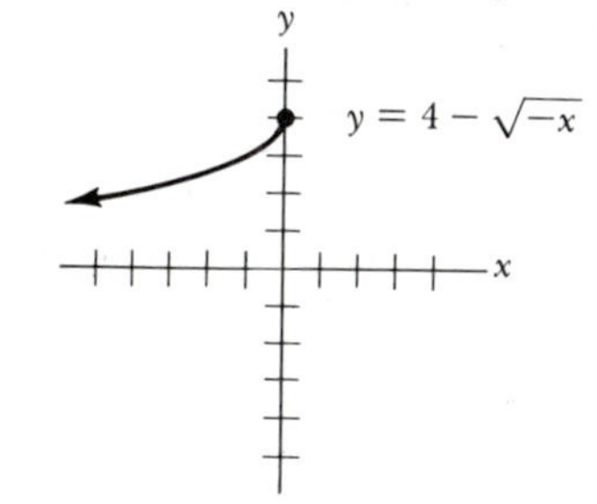

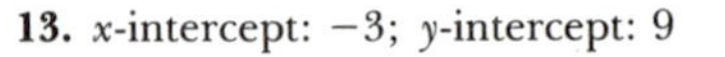

13. x-intercept: -3; y-intercept: 9

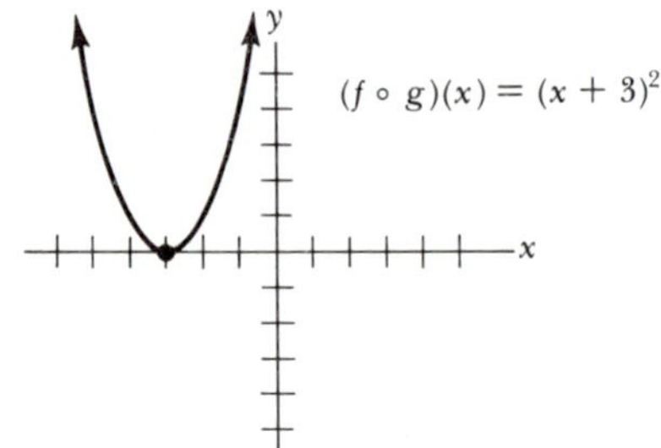

15. x-intercept: -1; y-intercept: 1

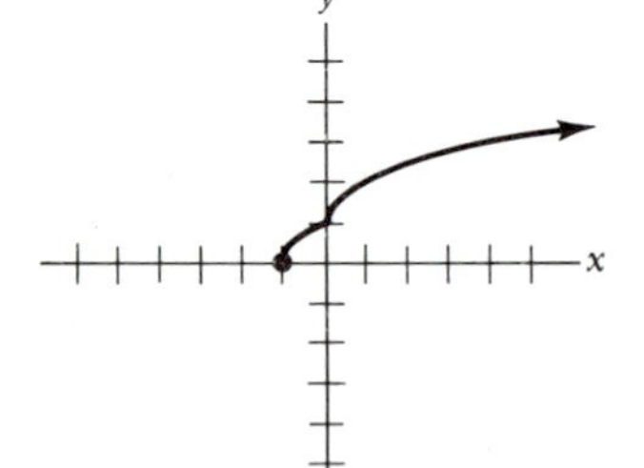

17. x-intercepts: 1, 3; y-intercept: 1

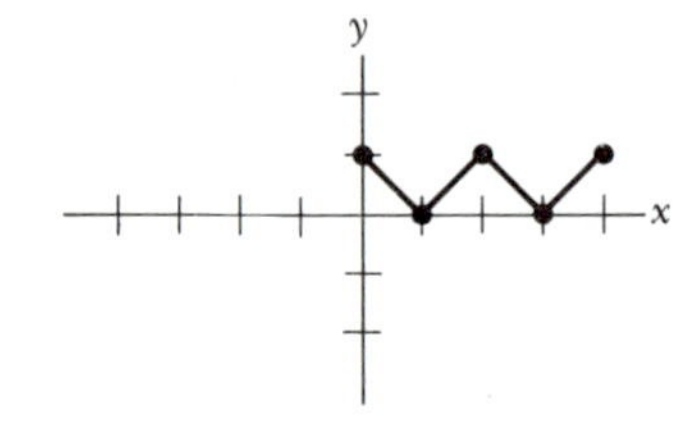

19. x-intercepts: 1, -1; y-intercepts: 1, -1

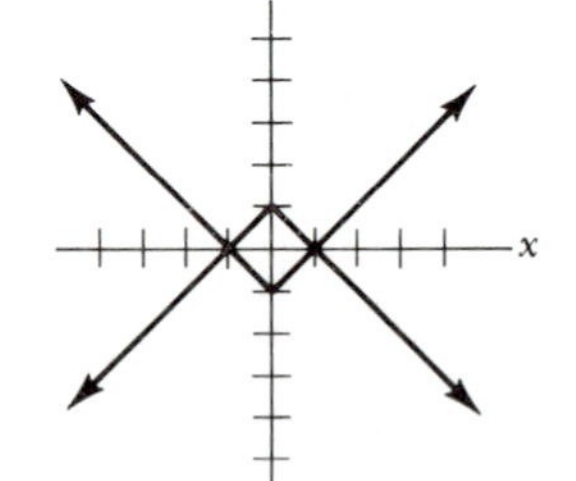

21. x-intercept: $\sqrt[3]{2}$; y-intercept: -2

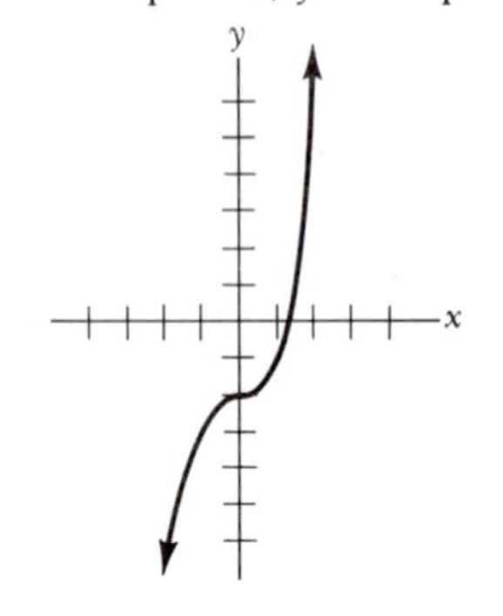

23.

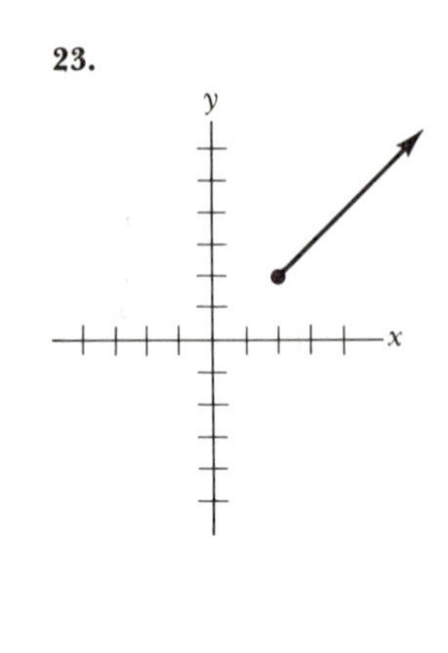

25. $-\frac{1}{2}$ **27.** $(0, -1)$ **29.** all real numbers **31.** $x \neq \frac{1}{3}, -\frac{3}{2}$ **33.** all real numbers **35.** $[-\sqrt{5}, \sqrt{5}]$ **37.** all noninteger real numbers **39.** all real numbers except 2 **41.** all real numbers except $\frac{2}{5}$ **43.** all real numbers except -1 **45.** $(g \circ f)(x)$ **47.** $(g \circ G)(x)$ **49.** $(F \circ g \circ g)(x)$ **51.** $(f \circ G \circ g \circ g \circ g)(x)$ **53.** $2 + \sqrt{2}$ **55.** $t^2 - t$ **57.** $1 - 4x$ **59.** $1 - 2x - 2h$ **61.** $x^2 + x - 1$ **63.** $2xh + h^2 - h$ **65.** $x - 1$ **67.** $x^4 - 2x^3 + x$ **69.** -11 **71.** $-6x^2 + 4x + 1$ **73.** -2 **75.** $\dfrac{7}{(x+4)(a+4)}$ **77.** x **79.** $\dfrac{13}{3}$ **81.** $\dfrac{1-x}{2}$ **83.** $-\dfrac{1}{2}$ **85.** $x + 1$ **87.** domain $= [-6, 8]$; range $= [-2, 4]$ **89.** $f(-\frac{5}{2})$ **91.** 1 **93.** $[-6, 5]$ **95.** no **97.** -2 **99.** $[0, 4]$ **101.** 5 **103.** (a) 2; (b) 3; (c) 2; (d) 3 **105.** -4 **107.** $[3, 5]$ **109.** (a) $(-6, 5)$; (b) $(0, 0)$; (c) $(7, 5)$; (d) $(0, 5)$; (e) $(10, 0)$; (f) $(-10, 0)$ **111.** $\frac{9}{16}$ **113.** 5.41% **115.** 3 **117.** (a) $\dfrac{1}{x^2}$; (b) no **119.** $A(r) = 4r^2$ **121.** (a) $S(x) = 2x^2 - 16x + 64$ (b) $C(x) = 24x^2 - 192x + 512$ **123.** (a) 8 units, \$16: (b) 4 units, \$16 **125.** $\dfrac{-9m^2 + 24m - 16}{2m}$ **127.** $f^{-1}(x) = 3 + (2 - x)^2$

$y = x$ $\quad f^{-1}(x)$ $\quad f(x)$

Exercise Set 5.1

1. $f(x) = \frac{2}{3}x + \frac{2}{3}$ **3.** $g(x) = \sqrt{2}\,x$ **5.** $f(x) = x - \frac{7}{2}$ **7.** $f(x) = \sqrt{3}$ **9.** yes **11.** $V(t) = -2375t + 20{,}000$ **13.** (a) $V(t) = -12{,}000t + 60{,}000$ (b)

End of Year	Yearly Depreciation	Accumulated Depreciation	Value V
0	0	0	60,000
1	12,000	12,000	48,000
2	12,000	24,000	36,000
3	12,000	36,000	24,000
4	12,000	48,000	12,000
5	12,000	60,000	0

15. vertex: $(-2, 0)$; axis: $x = -2$; minimum: 0; x-intercept: -2; y-intercept: 4

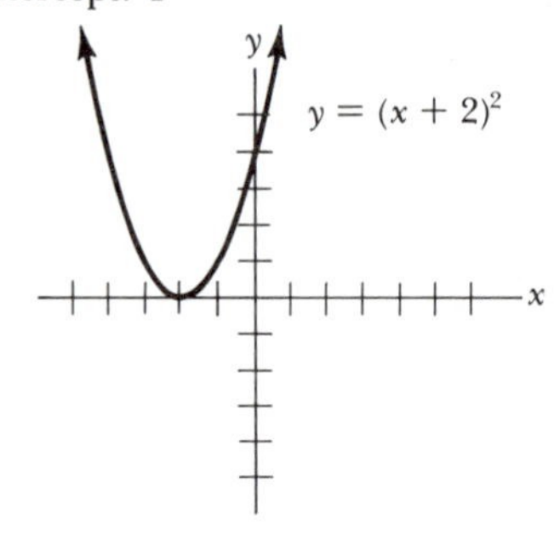

17. vertex: $(-2, 0)$; axis: $x = -2$; minimum: 0; x-intercept: -2; y-intercept: 8

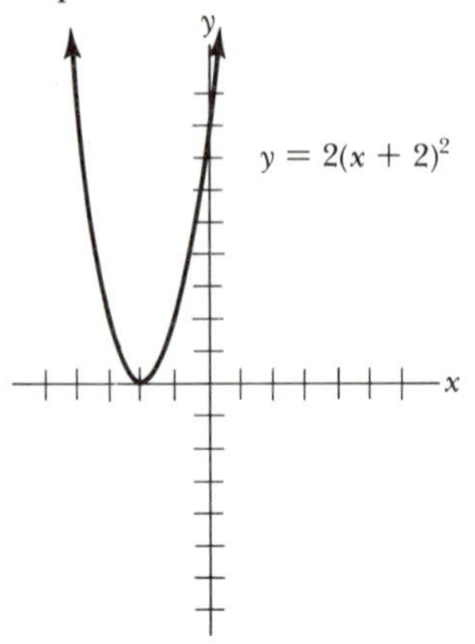

19. vertex: $(-2, 4)$; axis: $x = -2$; maximum: 4; x-intercepts: $-2 \pm \sqrt{2}$; y-intercept: -4

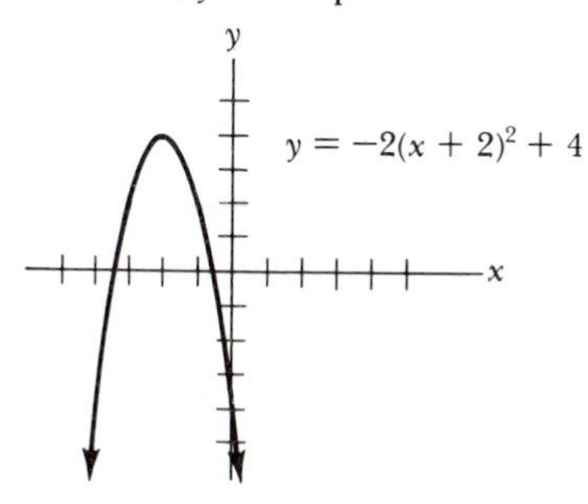

21. vertex: $(2, -4)$; axis: $x = 2$; minimum: -4; x-intercepts: 0, 4; y-intercept: 0

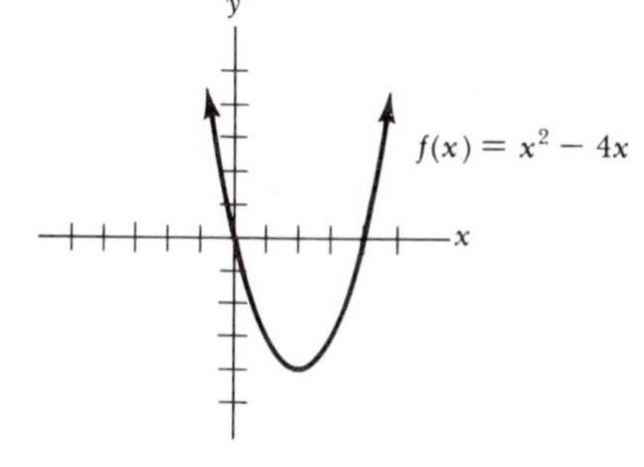

23. vertex: $(0, 1)$; axis: $x = 0$; maximum: 1; x-intercepts: ± 1; y-intercept: 1

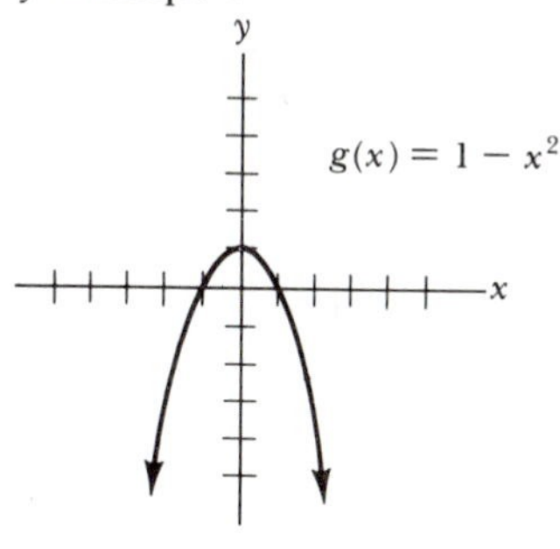

25. vertex: $(1, -4)$; axis: $x = 1$; minimum: -4; x-intercepts: 3, -1; y-intercept: -3

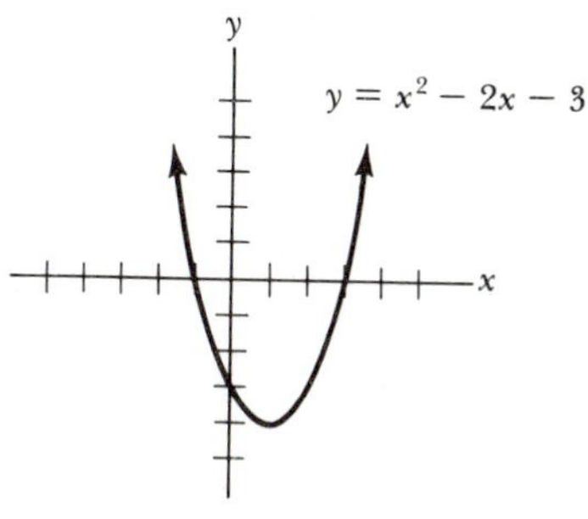

27. vertex: $(3, 11)$; axis: $x = 3$; maximum: 11; x-intercepts: $3 \pm \sqrt{11}$; y-intercept: 2

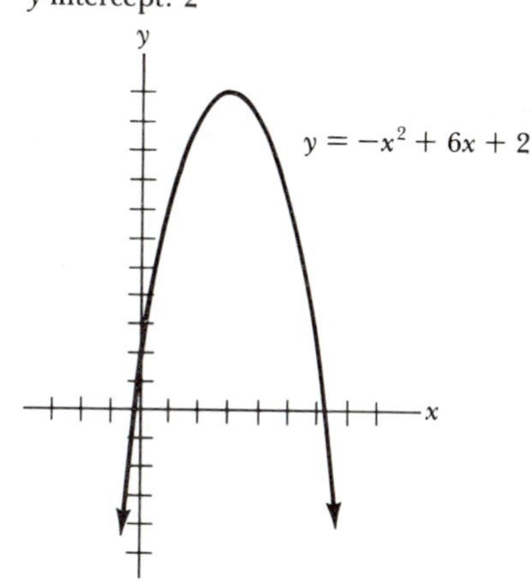

29. (a) minimum: -13; (b) maximum: $\frac{17}{8}$; (c) maximum: 1000 **31.** 5 units **33.** (a) linear; (b) quadratic; (c) quadratic; (d) quadratic; (e) quadratic; (f) neither **35.** (a) $y = \frac{1}{4}(x - 3)^2 - 1$; (b) $y = (x - 1)^2$ **37.** $(x + 3)^2 + (y + 4)^2 = 25$ **39.** $\frac{25}{4}$ **41.** $\frac{1}{2}$ **43.** $\frac{25}{4} \times \frac{25}{4}$ **45.** 1250 in^2 **47.** (a) 18; (b) $\frac{23}{4}$; (c) $\frac{47}{8}$; (d) $\frac{95}{16}$ **49.** (a) 16 ft, 12 ft; (b) 16 ft; (c) $\frac{1}{4}$ sec, $\frac{7}{4}$ sec **51.** $\frac{\sqrt{7}}{2}$ **53.** (a) $\frac{1}{2}$; (b) $\frac{1}{4}$ **55.** 125 ft × 250 ft **57.** 40 **59.** 60 units, \$900, \$15 **61.** 16.95 in. **63.** 3 P.M., no **65.** (a) $\frac{36}{13}$; (b) $\frac{6\sqrt{13}}{13}$ **67.** (a) $\frac{225}{2}$; (b) $\frac{c^2}{2}$ **69.** $\frac{1}{2}$ **71.** $\frac{49}{12}$ **73.** $2R^2$ **75.** $\frac{1}{16\pi}$ mi **79.** $\frac{1}{4}P \times \frac{1}{4}P$ **83.** -16 **85.** (b) 2 **89.** -27.87 **91.** $V(t) = -4050t + 26{,}450$

Exercise Set 5.2

1. yes **3.** yes **5.** no **7.** y-intercept: 5

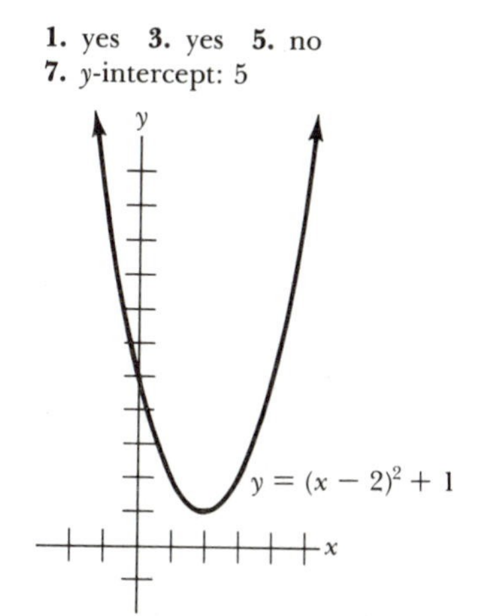

9. x-intercept: 1; y-intercept: -1

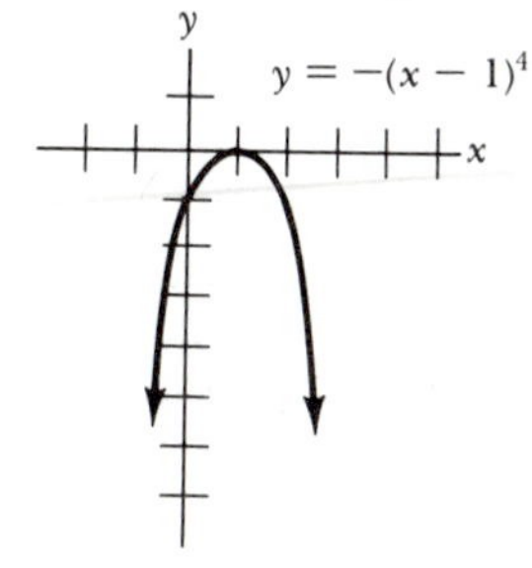

11. x-intercept: $4 + \sqrt[3]{2}$; y-intercept: -66

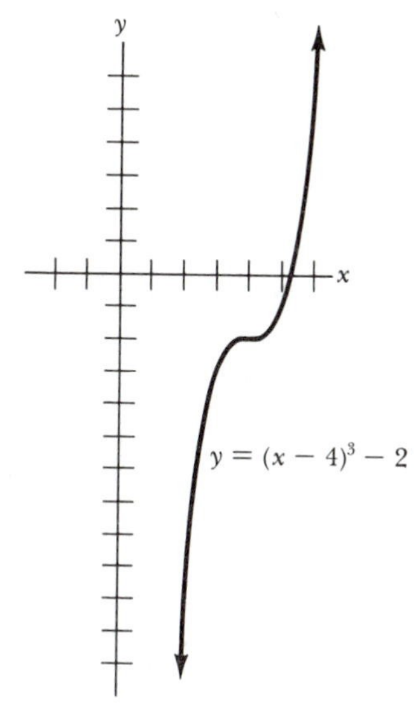

13. x-intercept: -5; y-intercept: -1250

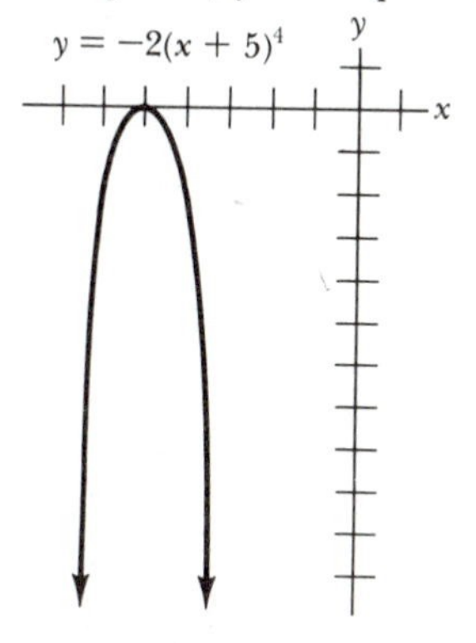

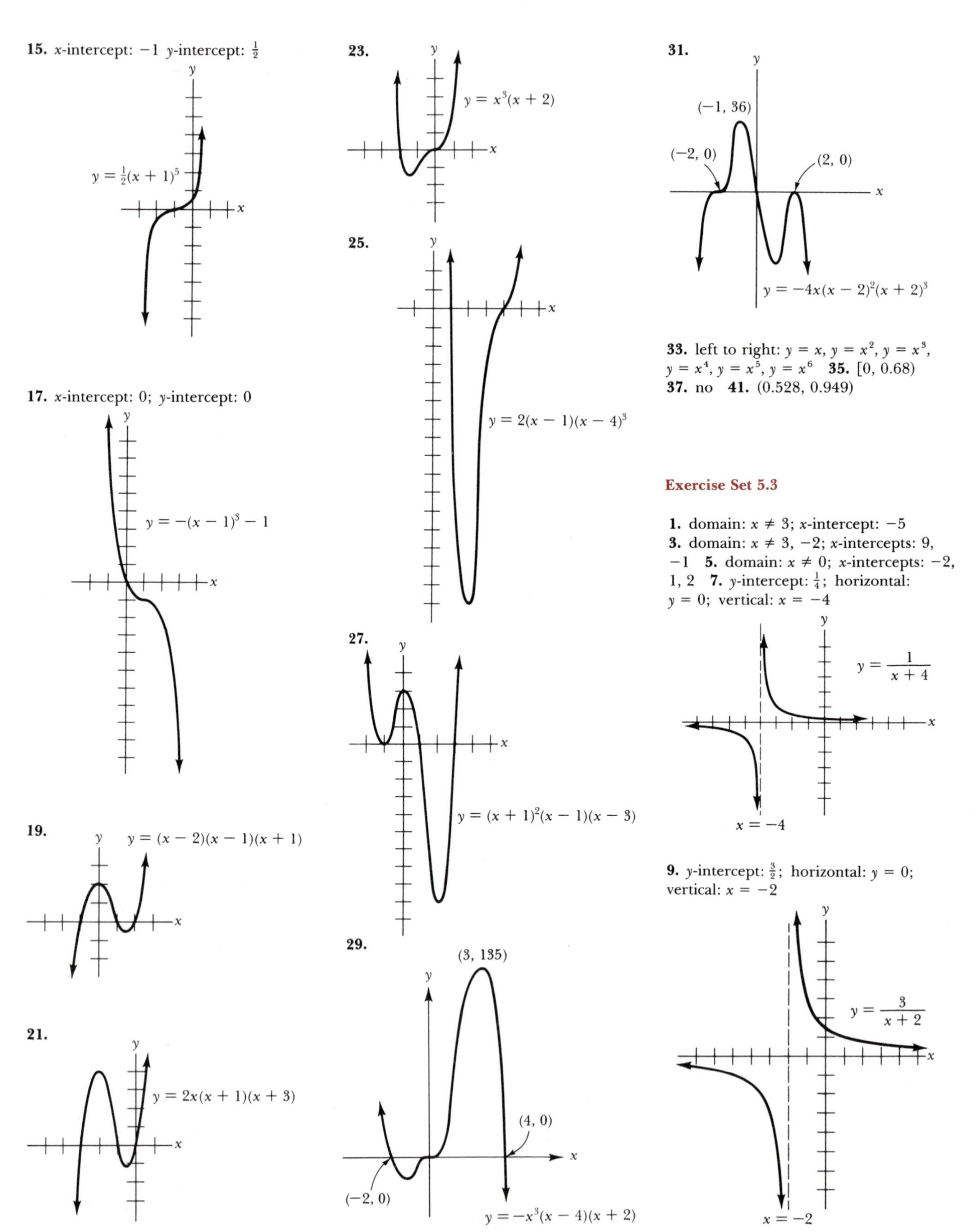
15. x-intercept: −1 y-intercept: 1/2
y = 1/2(x + 1)^5
17. x-intercept: 0; y-intercept: 0
y = −(x − 1)^3 − 1
19.
y = (x − 2)(x − 1)(x + 1)
21.
y = 2x(x + 1)(x + 3)
23.
y = x^3(x + 2)
25.
y = 2(x − 1)(x − 4)^3
27.
y = (x + 1)^2(x − 1)(x − 3)
29.
(3, 135)
(4, 0)
(−2, 0)
y = −x^3(x − 4)(x + 2)
31.
(−1, 36)
(−2, 0)
(2, 0)
y = −4x(x − 2)^2(x + 2)^3
33. left to right: y = x, y = x^2, y = x^3, y = x^4, y = x^5, y = x^6 35. [0, 0.68)
37. no 41. (0.528, 0.949)
Exercise Set 5.3
1. domain: x ≠ 3; x-intercept: −5
3. domain: x ≠ 3, −2; x-intercepts: 9, −1 5. domain: x ≠ 0; x-intercepts: −2, 1, 2 7. y-intercept: 1/4; horizontal: y = 0; vertical: x = −4
y = 1/(x + 4)
x = −4
9. y-intercept: 3/2; horizontal: y = 0; vertical: x = −2
y = 3/(x + 2)
x = −2
x
y

11. y-intercept: $\frac{2}{3}$; horizontal: $y = 0$; vertical: $x = 3$

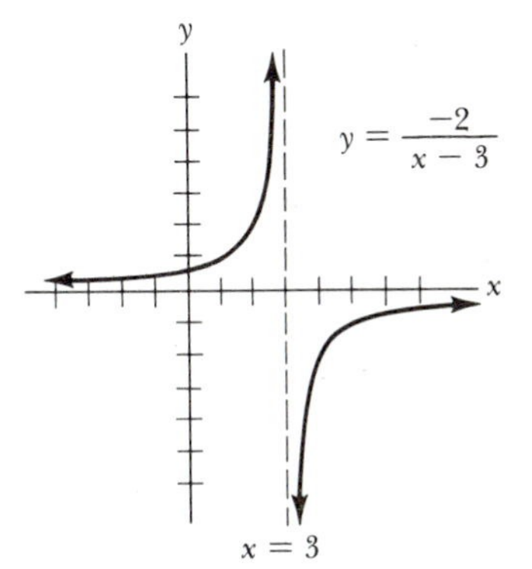

13.
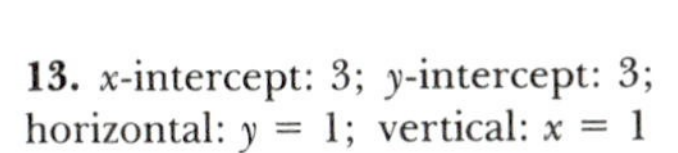
x-intercept: 3; y-intercept: 3; horizontal: $y = 1$; vertical: $x = 1$

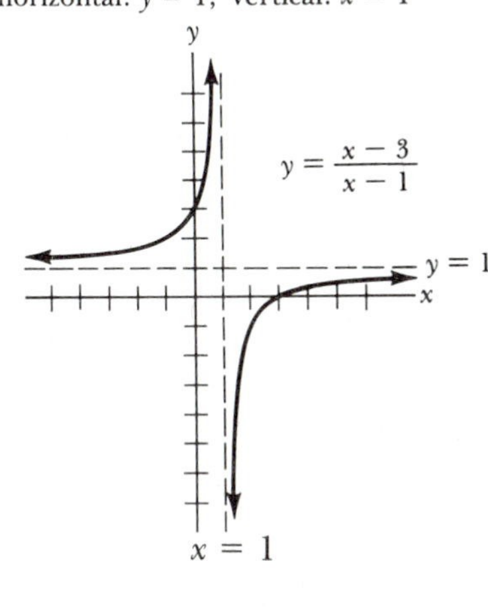

15.
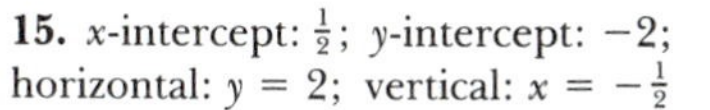
x-intercept: $\frac{1}{2}$; y-intercept: -2; horizontal: $y = 2$; vertical: $x = -\frac{1}{2}$

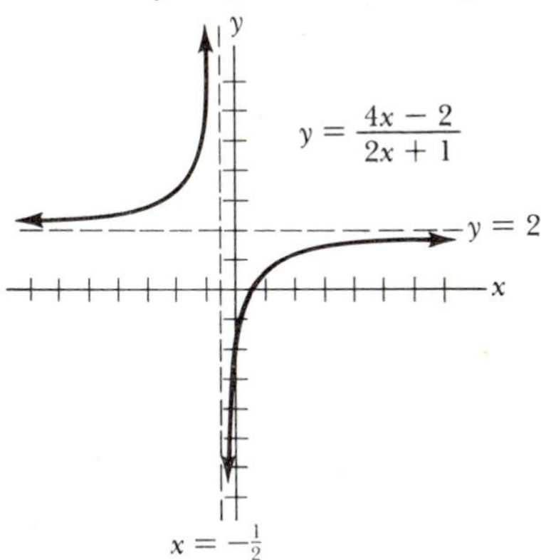

17. y-intercept: $\frac{1}{4}$; horizontal: $y = 0$; vertical: $x = 2$

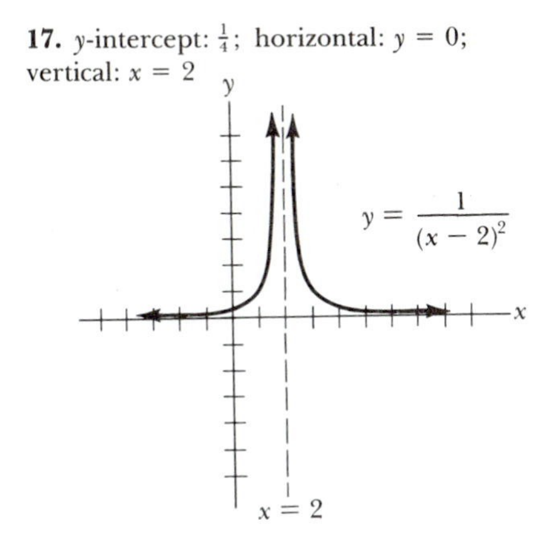

19.
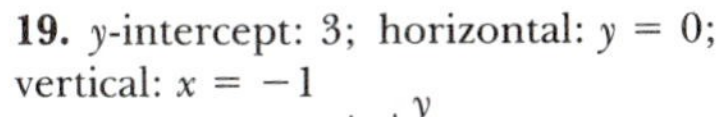
y-intercept: 3; horizontal: $y = 0$; vertical: $x = -1$

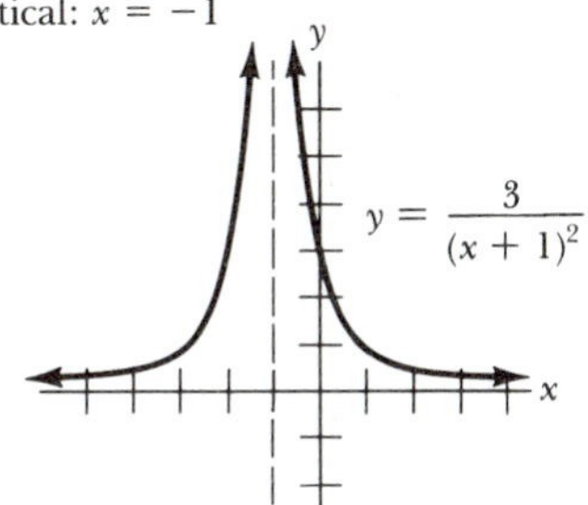

21. y-intercept: $\frac{1}{8}$; horizontal: $y = 0$; vertical: $x = -2$

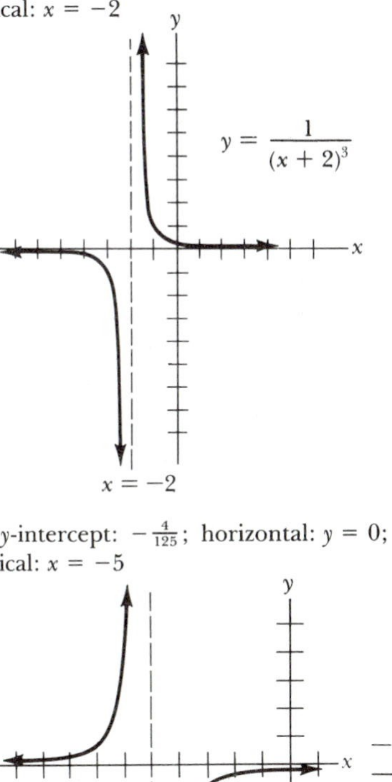

23. y-intercept: $-\frac{4}{125}$; horizontal: $y = 0$; vertical: $x = -5$

25. x-intercept: 0; y-intercept: 0; horizontal: $y = 0$; vertical: $x = -2$, $x = 2$

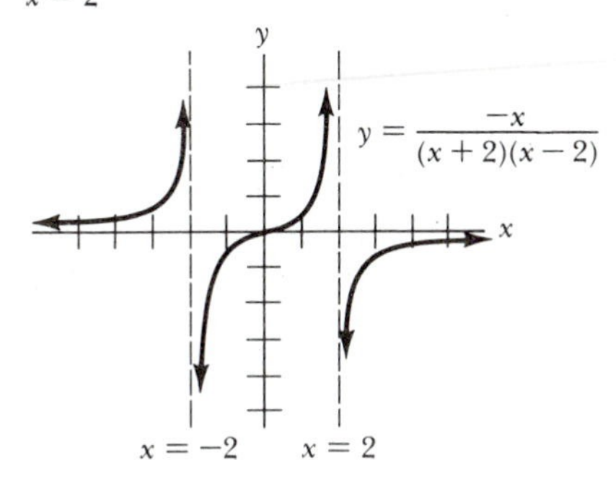

27. x-intercept: 0; y-intercept: 0; horizontal: $y = 0$; vertical: $x = -3$, $x = 1$

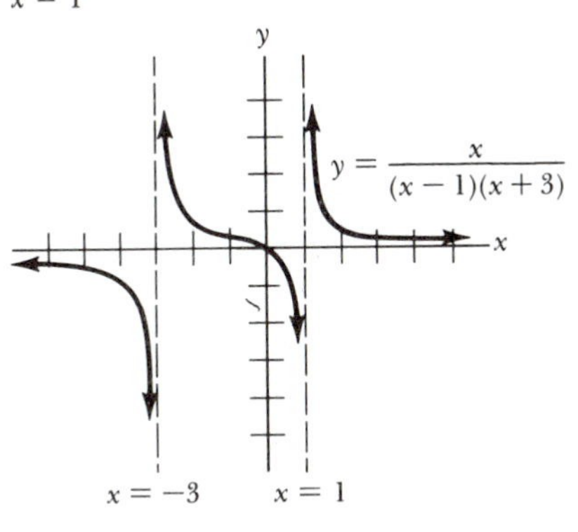

29. x-intercepts: -2, 4; y-intercept: $-\frac{8}{3}$; horizontal: $y = 1$; vertical: $x = 1$, $x = 3$

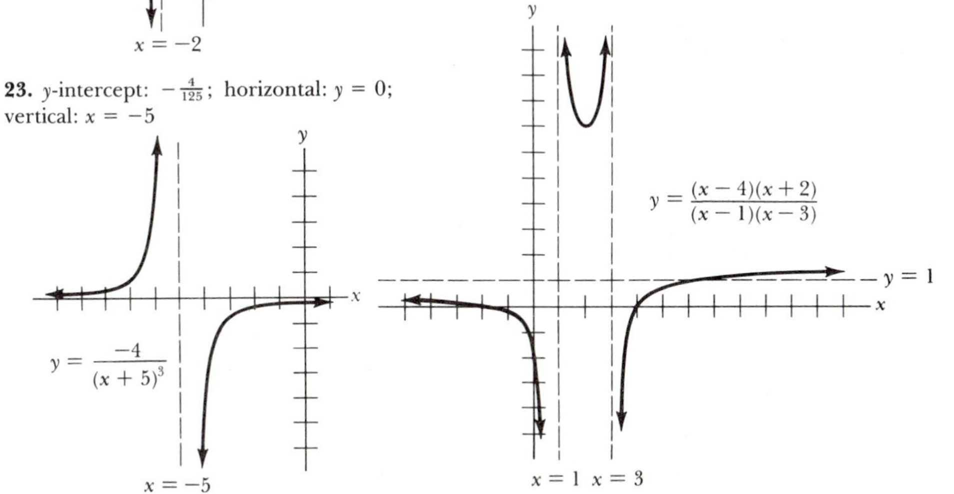

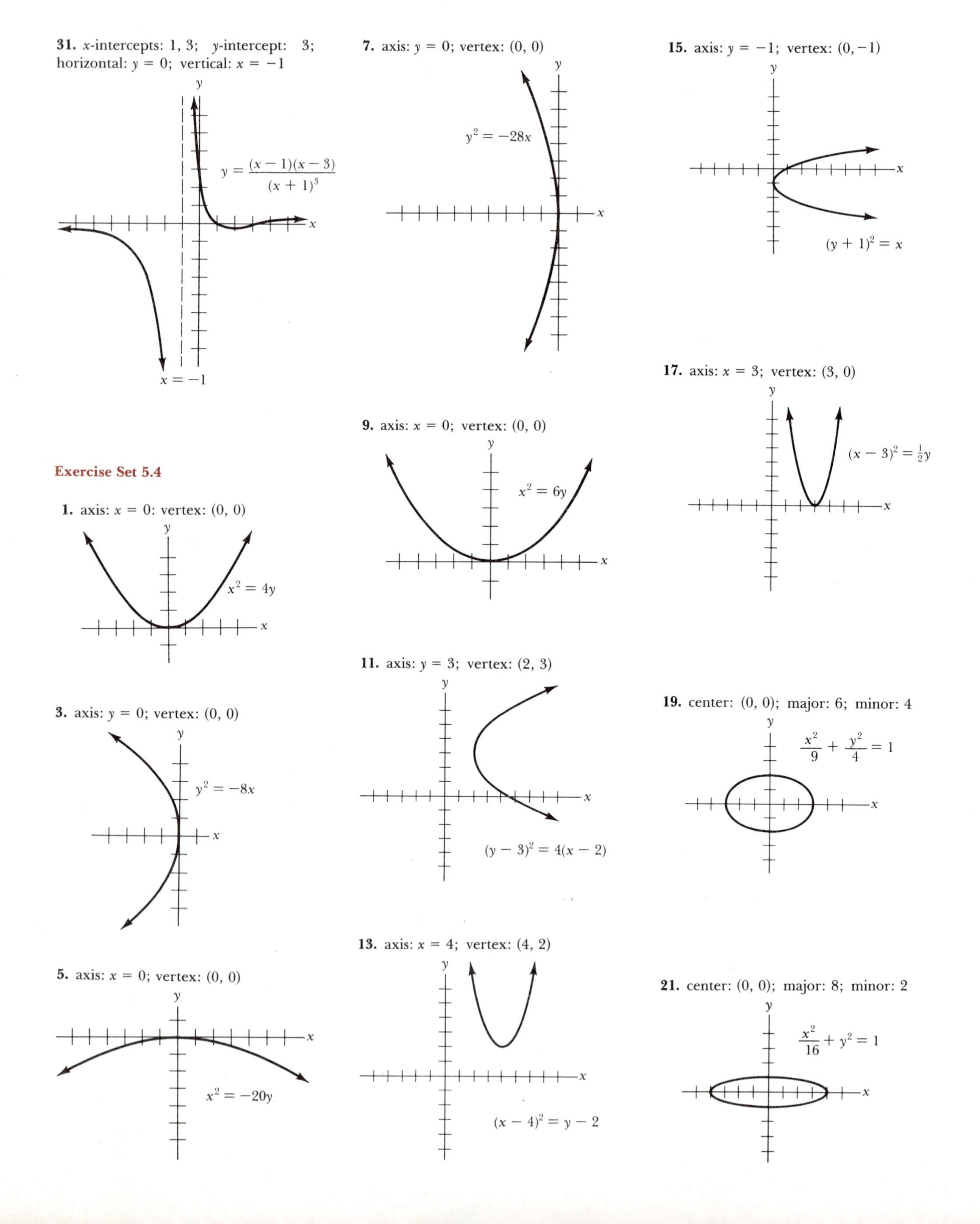
31. x-intercepts: 1, 3; y-intercept: 3; horizontal: y = 0; vertical: x = −1
y = (x − 1)(x − 3) / (x + 1)^3
x = −1
Exercise Set 5.4
1. axis: x = 0: vertex: (0, 0)
x^2 = 4y
3. axis: y = 0; vertex: (0, 0)
y^2 = −8x
5. axis: x = 0; vertex: (0, 0)
x^2 = −20y
7. axis: y = 0; vertex: (0, 0)
y^2 = −28x
9. axis: x = 0; vertex: (0, 0)
x^2 = 6y
11. axis: y = 3; vertex: (2, 3)
(y − 3)^2 = 4(x − 2)
13. axis: x = 4; vertex: (4, 2)
(x − 4)^2 = y − 2
15. axis: y = −1; vertex: (0, −1)
(y + 1)^2 = x
17. axis: x = 3; vertex: (3, 0)
(x − 3)^2 = 1/2 y
19. center: (0, 0); major: 6; minor: 4
x^2/9 + y^2/4 = 1
21. center: (0, 0); major: 8; minor: 2
x^2/16 + y^2 = 1

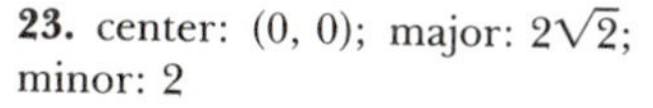

23. center: (0, 0); major: $2\sqrt{2}$; minor: 2

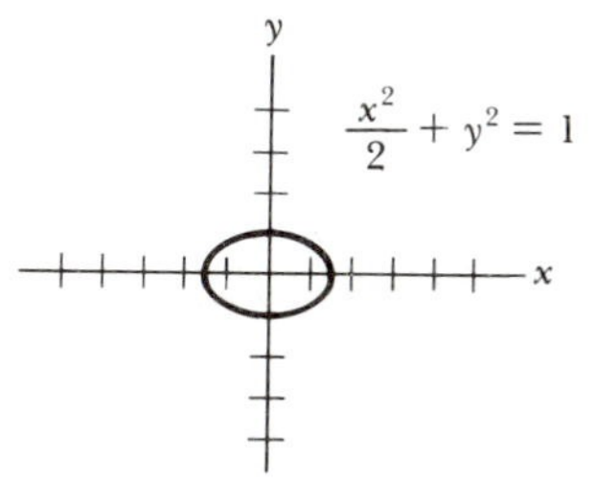

25. center: (0, 0); major: 8; minor: 6

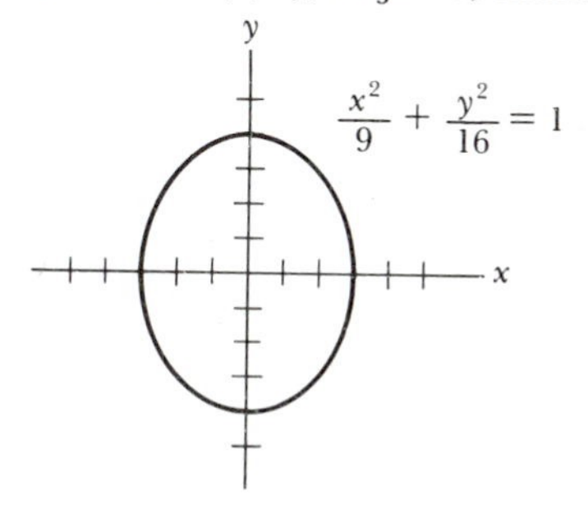

27. center: (0, 0); major: $\frac{2\sqrt{15}}{3}$; minor: $\frac{2\sqrt{3}}{3}$

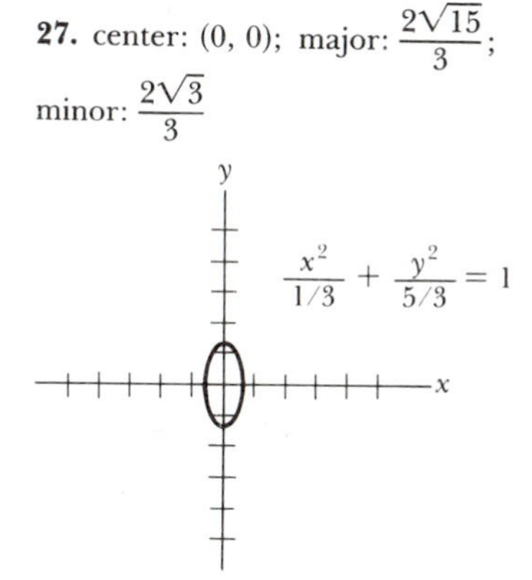

29. center: (5, −1); major: 10; minor: 6

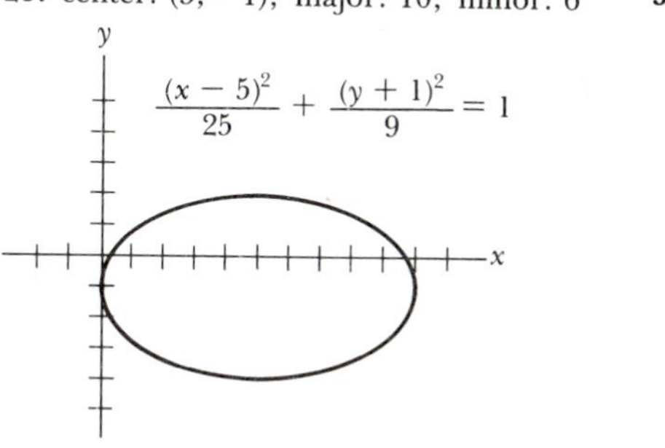

31. center: (1, 2); major: 4; minor: 2

$(x-1)^2 + \frac{(y-2)^2}{4} = 1$

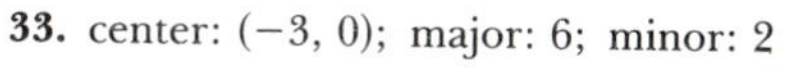

33. center: (−3, 0); major: 6; minor: 2

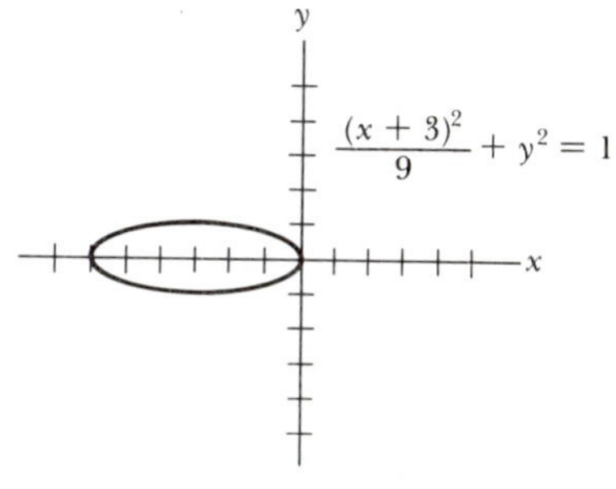

35. center: (1, −2); major: 4; minor: $2\sqrt{3}$

$\frac{(x-1)^2}{4} + \frac{(y+2)^2}{3} = 1$

37. the point (4, 6)

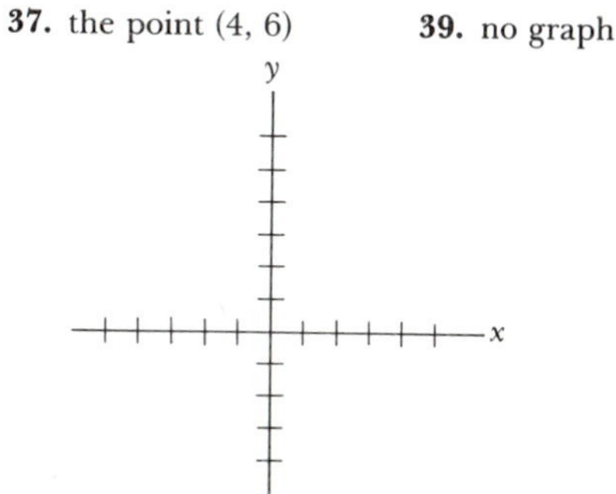

39. no graph

41. $2\sqrt{30}\,\pi$

43.

$\frac{x^2}{9} + \frac{y^2}{4} = 1$

45. C_2

Exercise Set 5.5

1. center: (0, 0); vertices: (2, 0), (−2, 0); transverse: 4; asymptotes: $y = \pm\frac{1}{2}x$

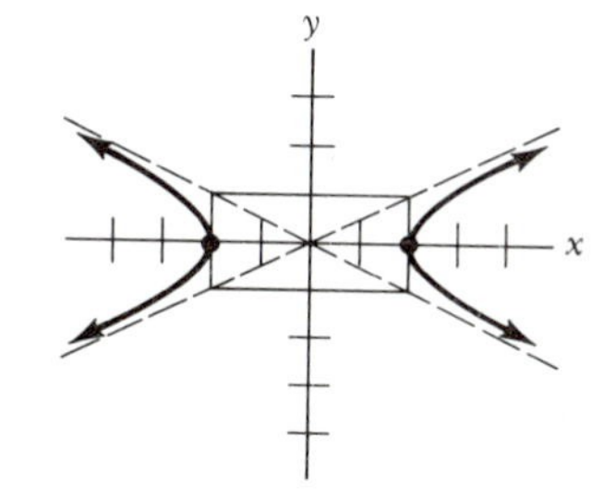

3. center: (0, 0); vertices: (0, 2), (0, −2); transverse: 4; asymptotes: $y = \pm 2x$

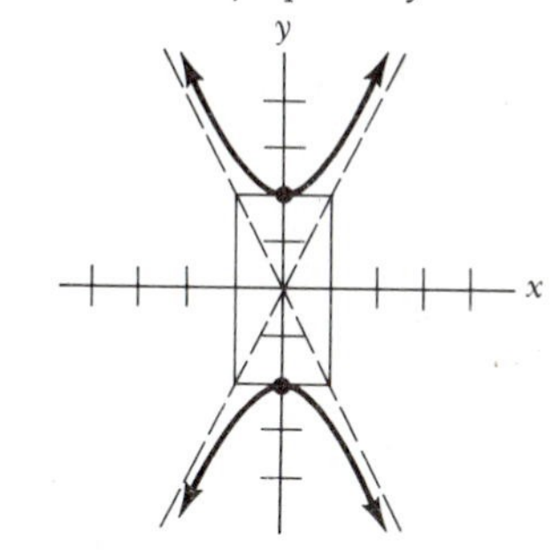

5. center: (0, 0); vertices: (5, 0), (−5, 0); transverse: 10; asymptotes: $y = \pm\frac{4}{5}x$

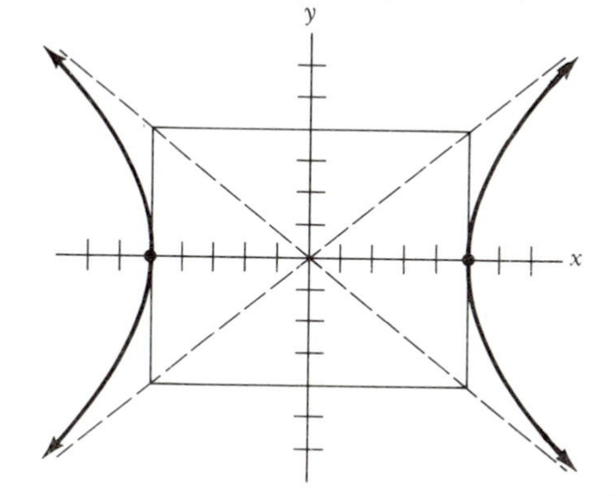

7. center: (0, 0); vertices: $(0, \frac{\sqrt{2}}{2})$, $(0, -\frac{\sqrt{2}}{2})$; transverse: $\sqrt{2}$; asymptotes: $y = \pm\sqrt{\frac{3}{2}}\,x$

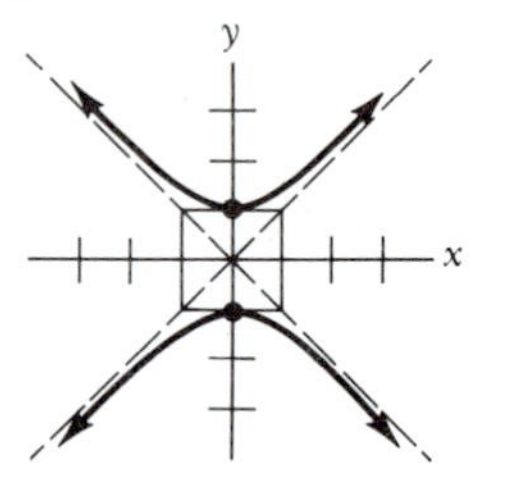

9. center: (0, 0); vertices: (0, 5), (0, −5); transverse: 10; asymptotes: $y = \pm\frac{5}{2}x$

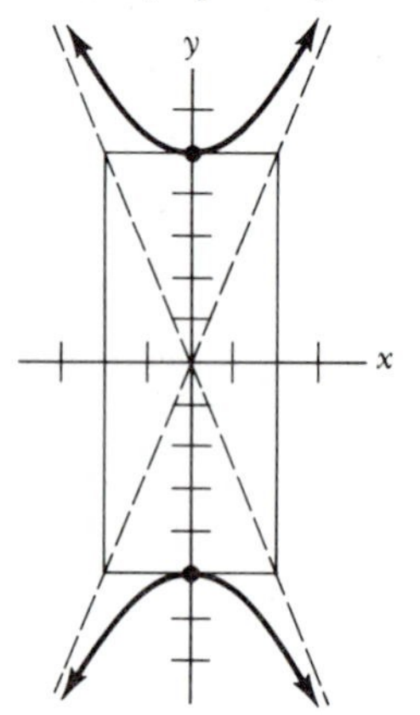

11. center: (5, −1); vertices: (10, −1), (0, −1); transverse: 10; asymptotes: $y = \frac{3}{5}x - 4$, $y = -\frac{3}{5}x + 2$

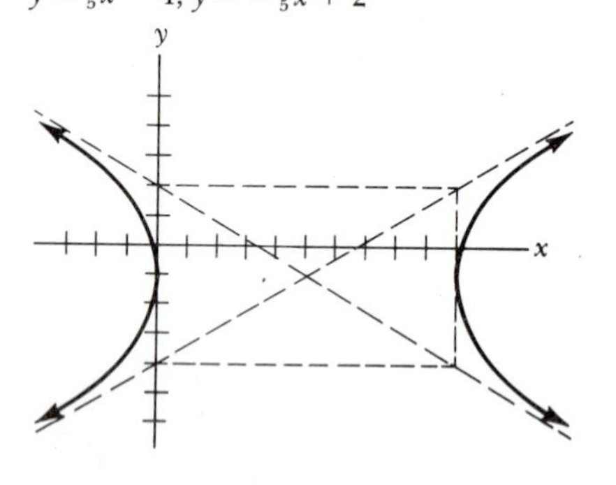

13. center: (1, 2); vertices: (1, 4), (1, 0); transverse: 4; asymptotes: $y = 2x$, $y = -2x + 4$

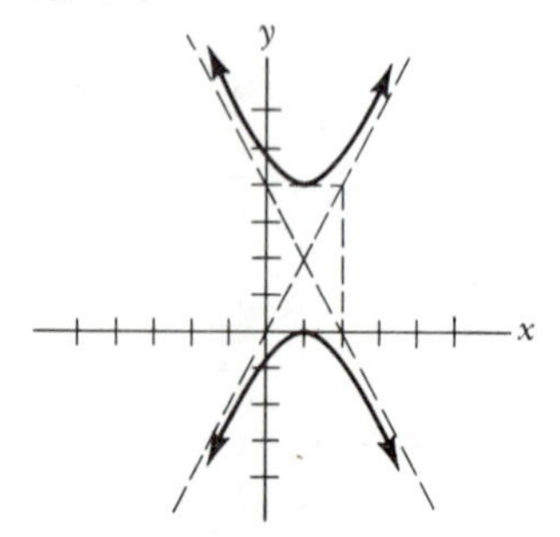

15. center: (−3, 4); vertices: (1, 4), (−7, 4); transverse: 8; asymptotes: $y = x + 7$, $y = -x + 1$

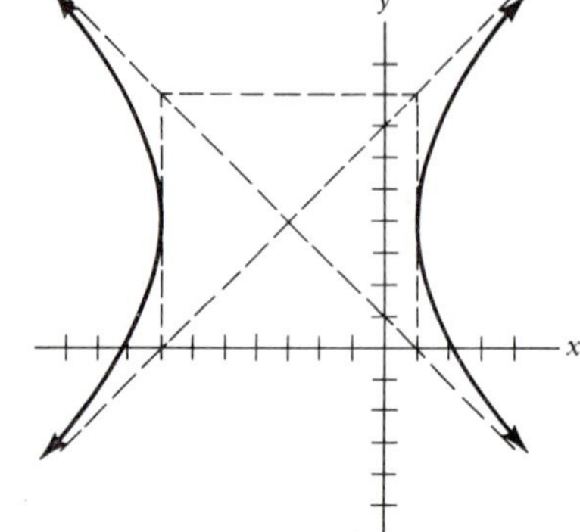

17. center: (0, 1); vertices: (2, 1), (−2, 1); transverse: 4; asymptotes: $y = x + 1$, $y = -x + 1$

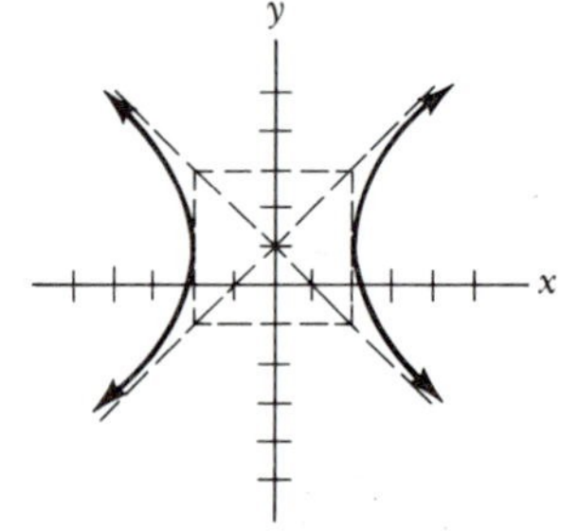

19. center: (2, 1); vertices: (−1, 1), (5, 1); transverse: 6; asymptotes: $y = x - 1$, $y = -x + 3$

21. center: (0, −4); vertices: (0, −9), (0, 1); transverse: 10; asymptotes: $y = 5x - 4$, $y = -5x - 4$

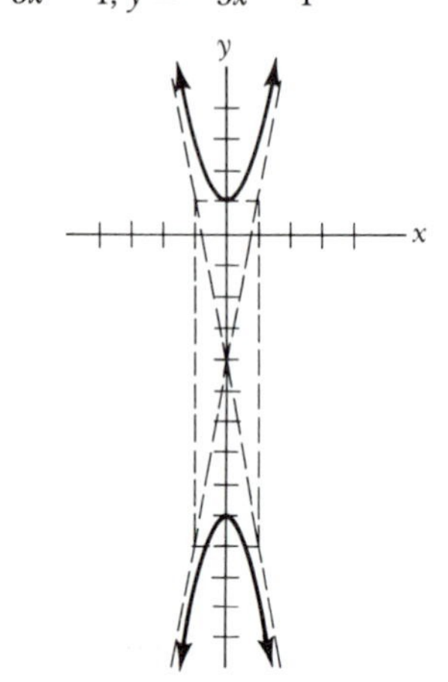

23. asymptotes: $y = x + 3$, $y = -x - 4$

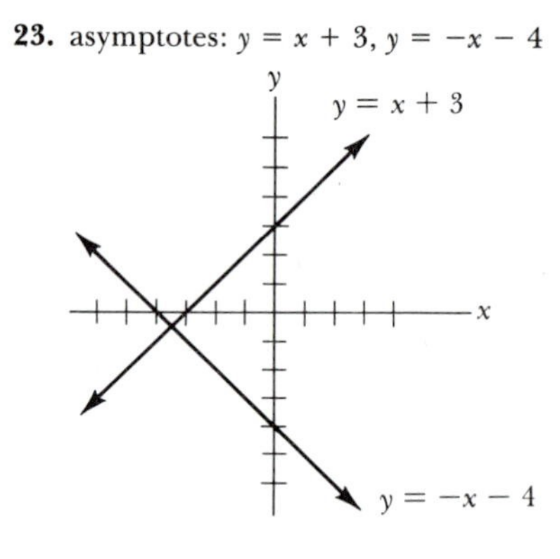

27. $\overline{PQ} = 0.1, 0.02, 0.01, 0.002, 0.001, 0.0001$

Chapter 5 Review Exercises

1. $f(x) = x + 2$ **3.** $f(x) = \frac{3}{8}x - \frac{5}{2}$
5. $f(x) = -\frac{3}{4}x + \frac{11}{4}$
7. vertex: (−1, −4); x-intercepts: 1, −3; y-intercept: −3

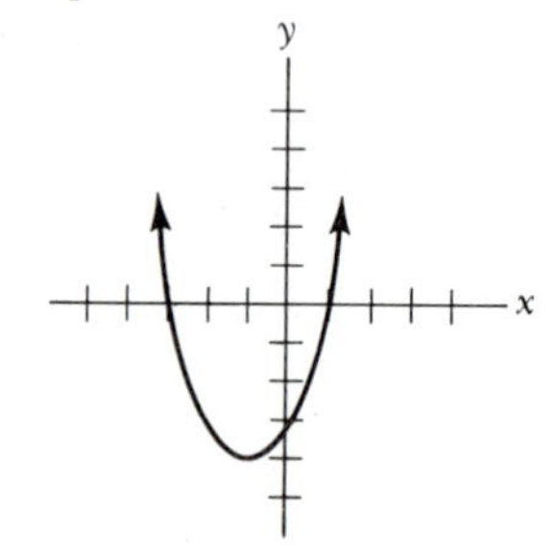

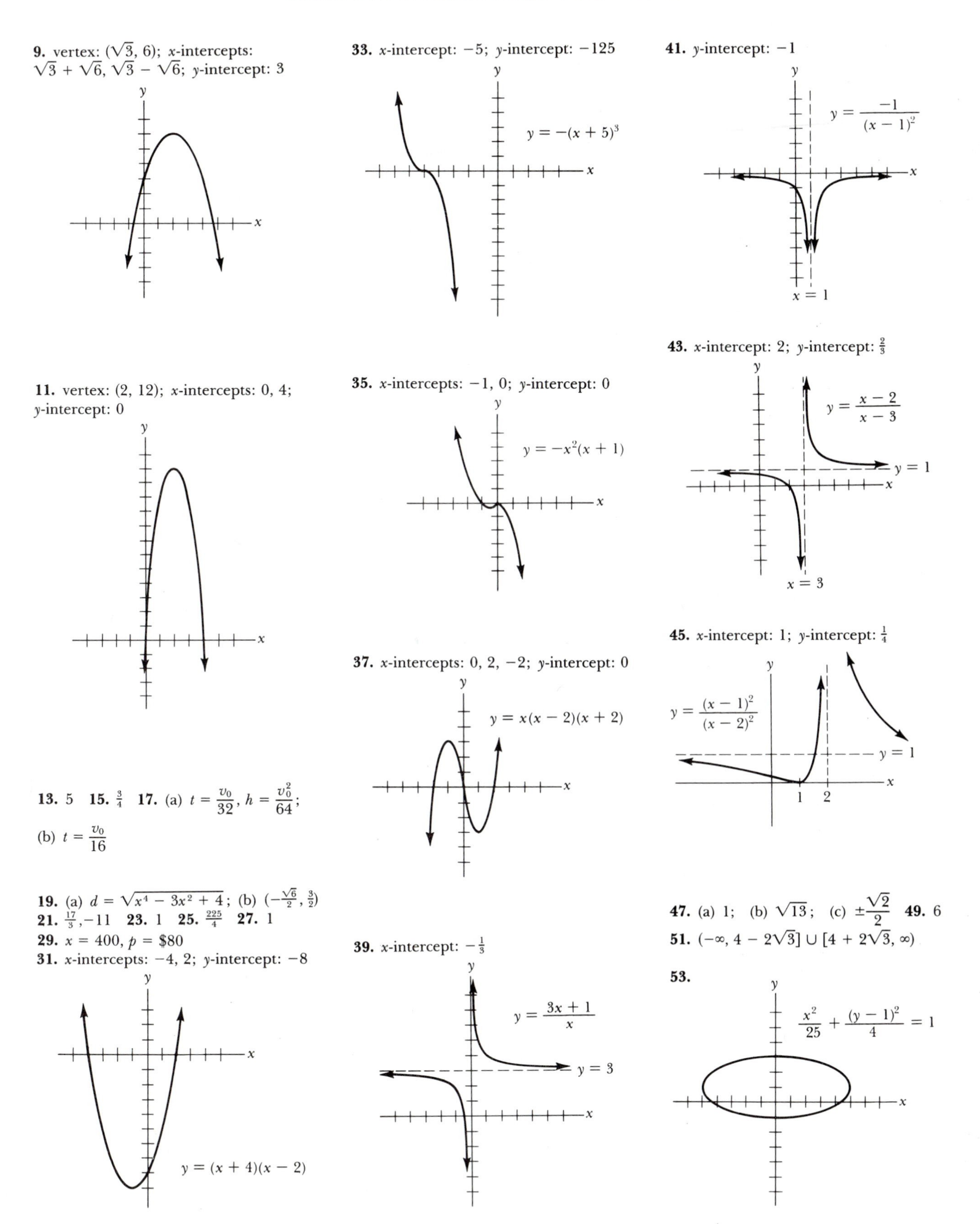

9. vertex: $(\sqrt{3}, 6)$; x-intercepts: $\sqrt{3} + \sqrt{6}, \sqrt{3} - \sqrt{6}$; y-intercept: 3

11. vertex: (2, 12); x-intercepts: 0, 4; y-intercept: 0

13. 5 **15.** $\frac{3}{4}$ **17.** (a) $t = \frac{v_0}{32}, h = \frac{v_0^2}{64}$; (b) $t = \frac{v_0}{16}$

19. (a) $d = \sqrt{x^4 - 3x^2 + 4}$; (b) $(-\frac{\sqrt{6}}{2}, \frac{3}{2})$
21. $\frac{17}{3}, -11$ **23.** 1 **25.** $\frac{225}{4}$ **27.** 1
29. $x = 400$, $p = \$80$
31. x-intercepts: -4, 2; y-intercept: -8

33. x-intercept: -5; y-intercept: -125

35. x-intercepts: -1, 0; y-intercept: 0

37. x-intercepts: 0, 2, -2; y-intercept: 0

39. x-intercept: $-\frac{1}{3}$

41. y-intercept: -1

43. x-intercept: 2; y-intercept: $\frac{2}{3}$

45. x-intercept: 1; y-intercept: $\frac{1}{4}$

47. (a) 1; (b) $\sqrt{13}$; (c) $\pm\frac{\sqrt{2}}{2}$ **49.** 6
51. $(-\infty, 4 - 2\sqrt{3}] \cup [4 + 2\sqrt{3}, \infty)$

53.

55.

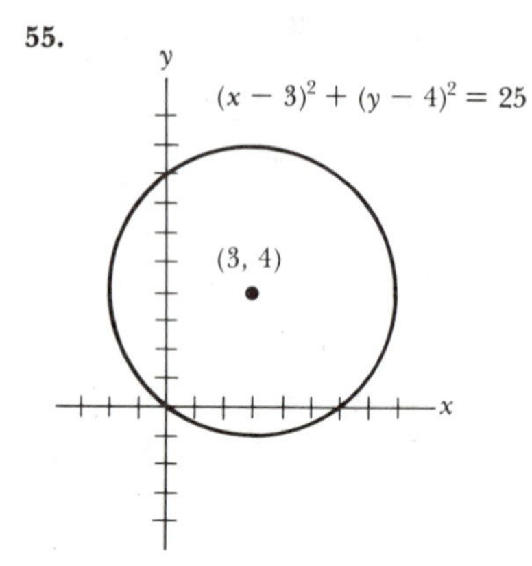

57.

y

$2(y + 3)^2 = x + 1$

x

59.

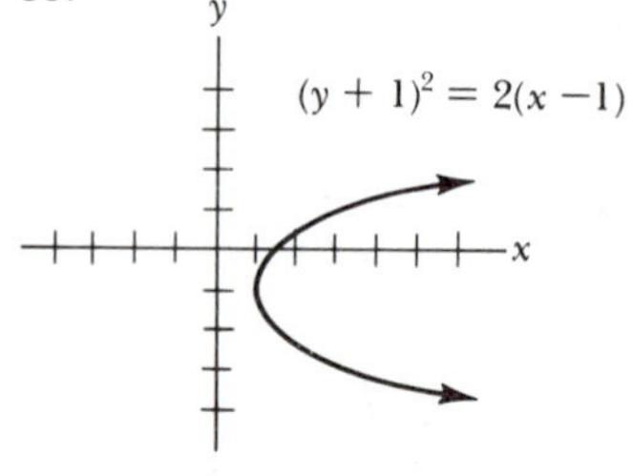

61. side $= x\sqrt{1 + x^2}$; area $= x^3$

Exercise Set 6.1

1. (a) 10^6; (b) 10^{15}; (c) 2×10^6; (d) $\frac{1}{2} \times 10^{18}$

3.

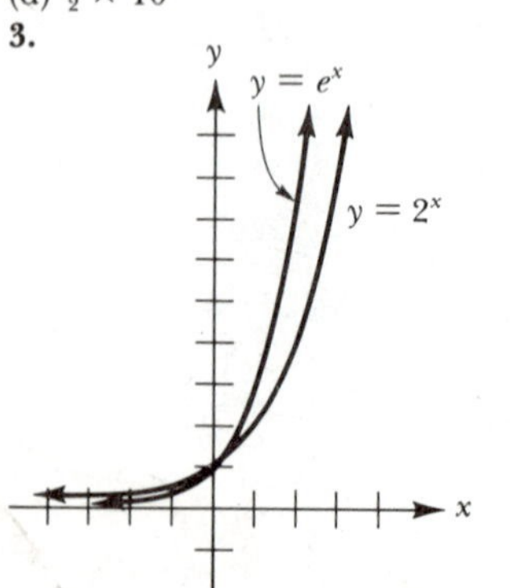

5.

x	x^2	2^x	Larger?
0	0	1	2^x
1	1	2	2^x
2	4	4	=
3	9	8	x^2
4	16	16	=
5	25	32	2^x
6	36	64	2^x
10	100	1024	2^x

2^x seems to be larger.

7. (a) domain: all reals; range: $y < 1$; x-intercept: 0; y-intercept: 0; asymptote: $y = 1$

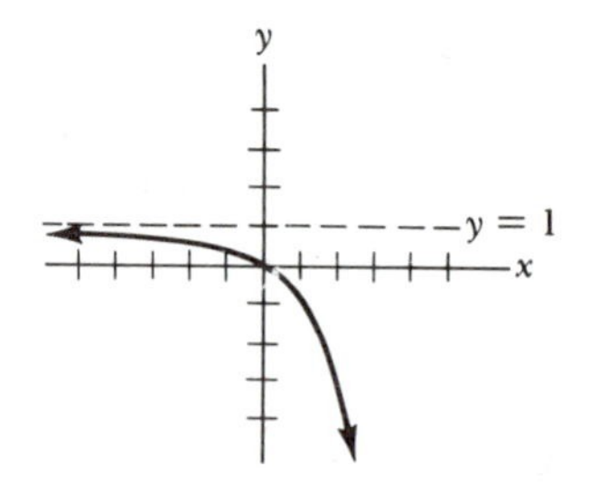

(b) domain: all reals; range: $y < 2$; x-intercept: 1; y-intercept: 1; asymptote: $y = 2$

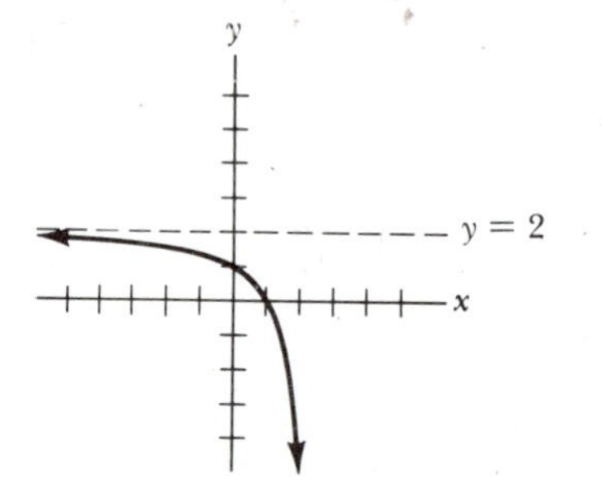

9. (a) domain: all reals; range: $y < 1$; x-intercept: 0; y-intercept: 0; asymptote: $y = 1$

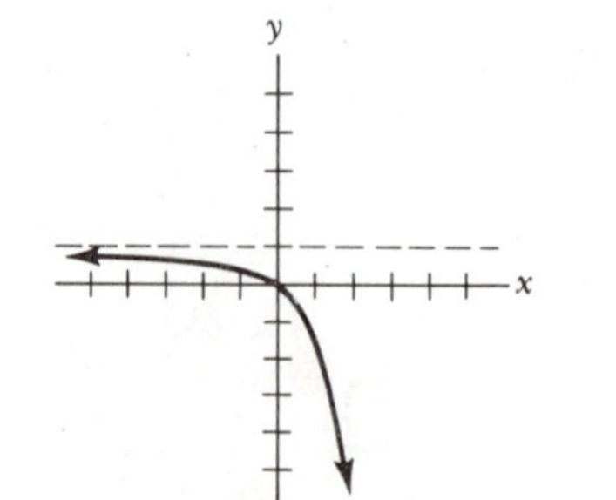

(b) domain: all reals; range: $y < e$; x-intercept: 1; y-intercept: $e - 1$; asymptote: $y = e$

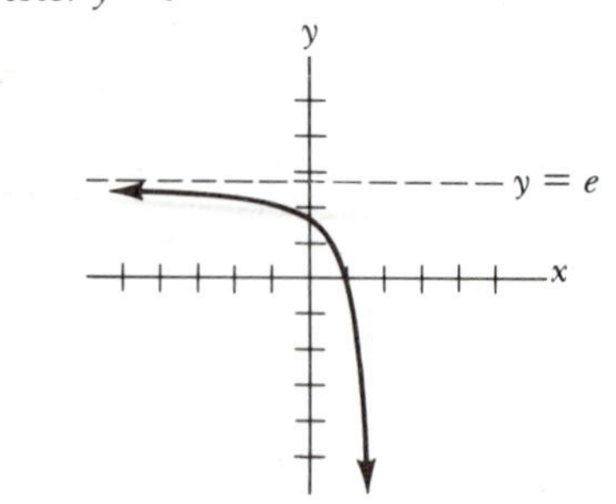

11. (a) domain: all reals; range: $y > 0$; y-intercept: 3; asymptote: $y = 0$

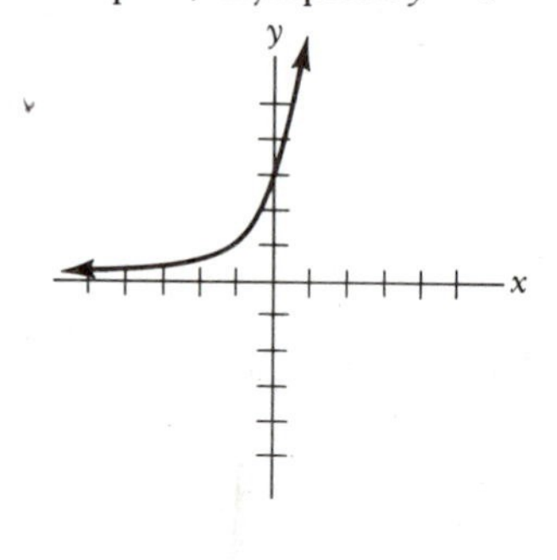

(b) domain: all reals; range: $y > 1$; y-intercept: 4; asymptote: $y = 1$

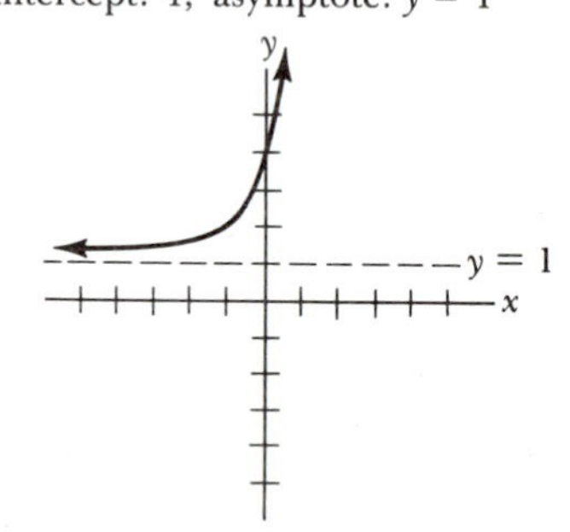

(c) domain: all reals; range: $y < 0$; y-intercept: -3; asymptote: $y = 0$

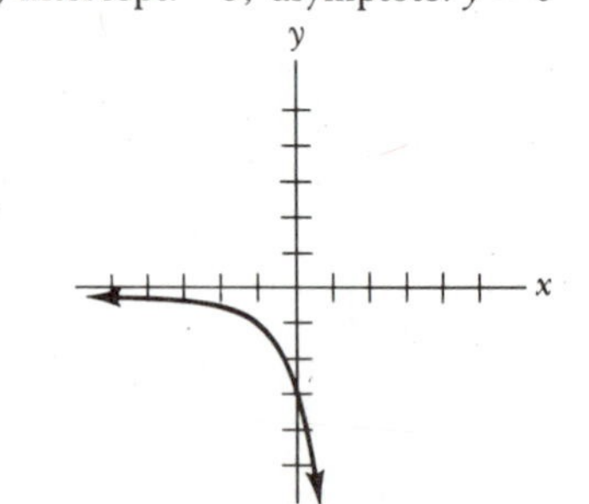

13. (a) domain: all reals; range: $y > 0$; y-intercept: e; asymptote: $y = 0$
(b) domain: all reals; range: $y < 0$; y-intercept: $-e$; asymptote: $y = 0$

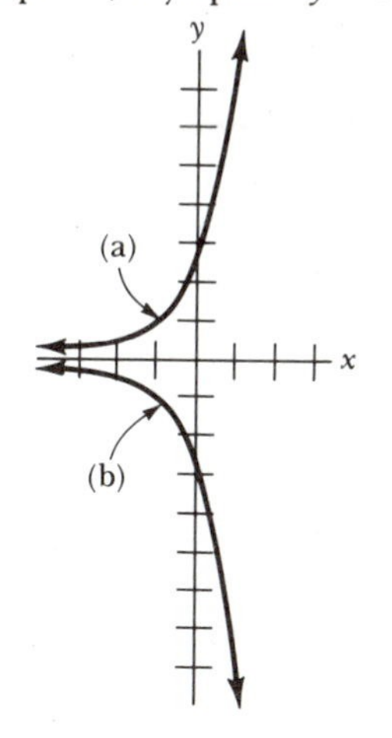

15. (a) domain: all reals; range: $y > 0$; y-intercept: 1; asymptote: $y = 0$
(b) domain: all reals; range: $y > 0$; y-intercept: 1; asymptote: $y = 0$

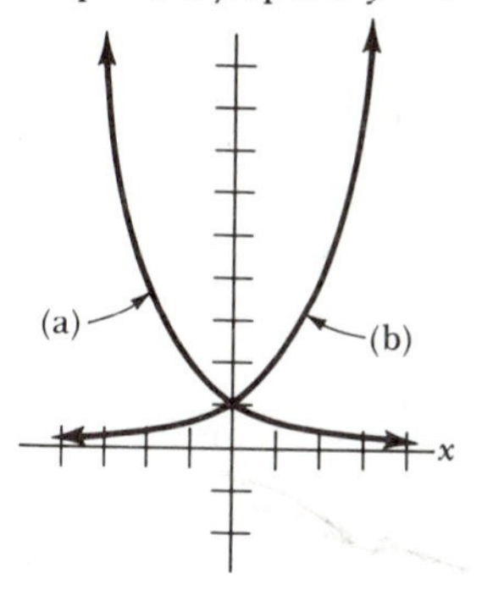

17. (a) ± 4; (b) $1, \frac{3}{2}$; **19.** 5
21. (a) 31,560; (b) 67,500 **23.** (a) 4; (b) 2; (c) 1; (d) $\frac{1}{2}$ (e) $\frac{1}{128}$
25. (a) 9.76 g; (b) 7.81 g;
27. $9.31 \times 10^{-8}\%$
31.

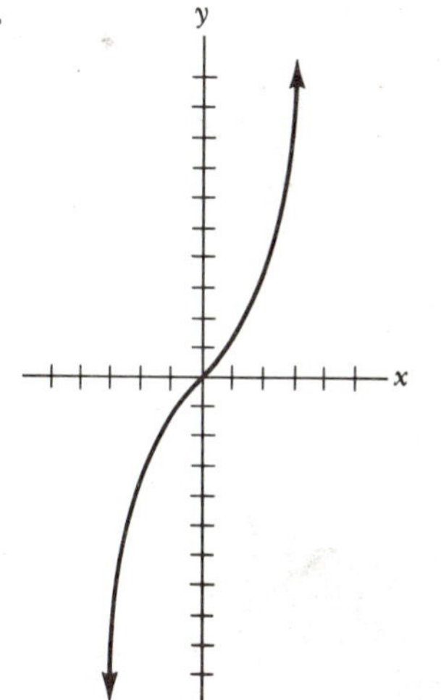

33. (a) $y = x + 1$; (b) See table.
(c)

x	$x + 1$	e^x
1	2	2.71828
0.5	1.5	1.64872
0.1	1.1	1.10517
0.01	1.01	1.01005
0.001	1.001	1.001005
0.0001	1.0001	1.000100005

35. (a)

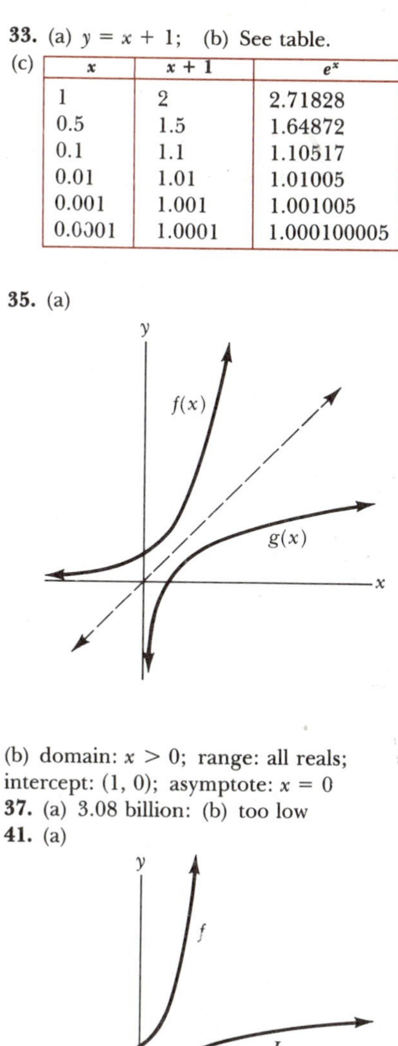

(b) domain: $x > 0$; range: all reals; intercept: (1, 0); asymptote: $x = 0$
37. (a) 3.08 billion: (b) too low
41. (a)

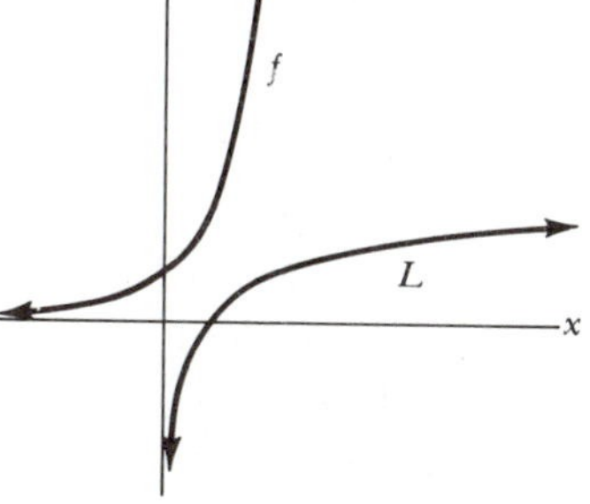

(b) domain: $x > 0$; range: all reals; intercept: (1, 0); asymptote: $x = 0$
(c) (i) intercept: (1, 0); asymptote: $x = 0$

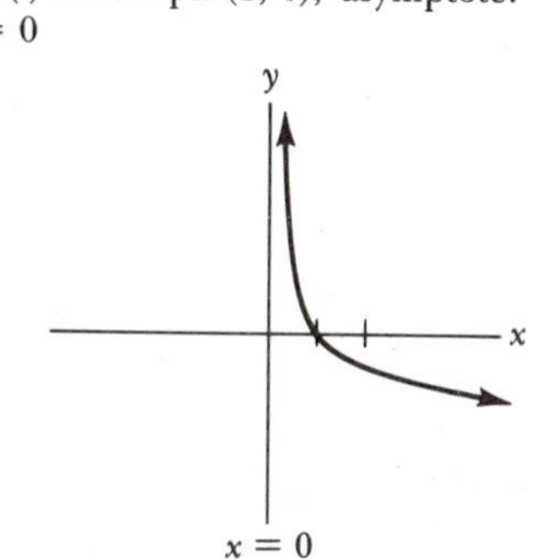

(ii) intercept: $(-1, 0)$; asymptote: $x = 0$

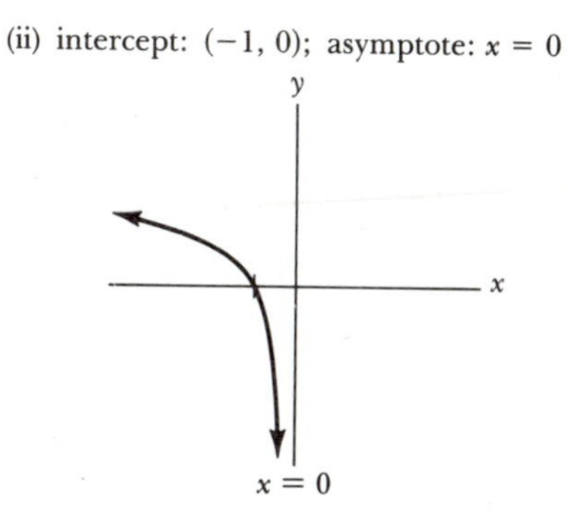

(iii) intercept: (2, 0); asymptote: $x = 1$

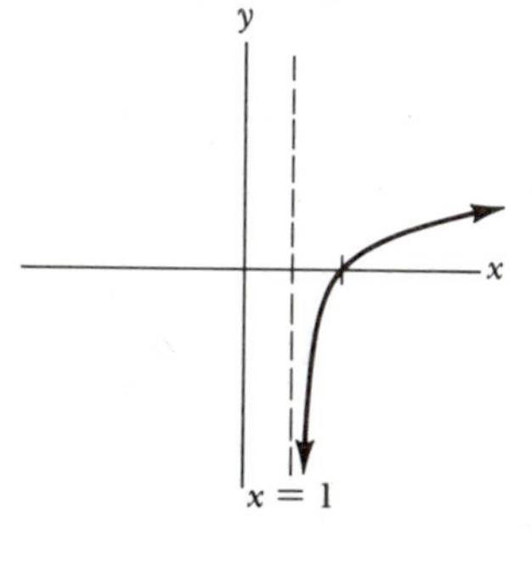

Exercise Set 6.2

1. (a) no; (b) yes; (c) yes
3. (a) $\frac{4x + 1}{2 - 3x}$; (b) $\frac{3x + 4}{2x - 1}$;
(c) $\frac{1}{2}$; (d) -4 **5.** $(-1, 3)$ and $(6, -1)$
7. (a) $\log_3 9 = 2$; (b) $\log_{10} 1000 = 3$; (c) $\log_7 343 = 3$; (d) $\log_2 \sqrt{2} = \frac{1}{2}$
9. (a) $2^5 = 32$; (b) $10^0 = 1$; (c) $e^{1/2} = \sqrt{e}$; (d) $3^{-4} = \frac{1}{81}$; (e) $t^v = u$ **11.** $\log_5 30$
13. (a) $\frac{3}{2}$; (b) $-\frac{5}{2}$; (c) $\frac{3}{2}$
15. (a) $x = 2$; (b) $x = \frac{1}{5}$ **17.** (a) $(0, \infty)$; (b) $(-\infty, \frac{3}{4})$; (c) $x \neq 0$; (d) $(0, \infty)$; (e) $(-\infty, -5) \cup (5, \infty)$ **19.** domain: $(0, e) \cup (e, \infty)$; range: $(-\infty, 0) \cup (0, \infty)$

21. (a) domain: $(-1, \infty)$; range: all reals; x-intercept: 0; y-intercept: 0; asymptote: $x = -1$

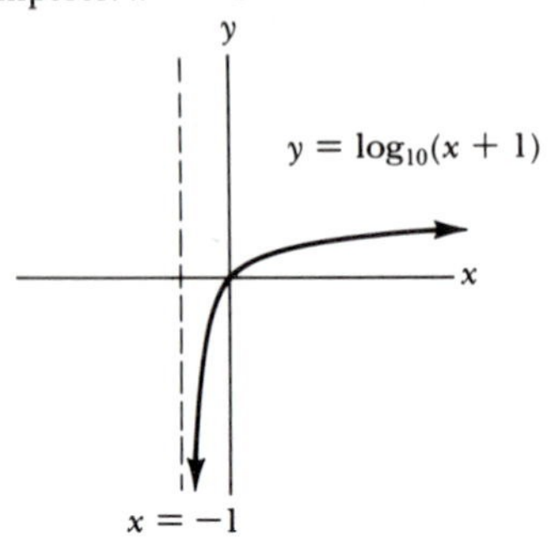

(b) domain: $(-1, \infty)$; range: all reals; x-intercept: 0; y-intercept: 0; asymptote: $x = -1$

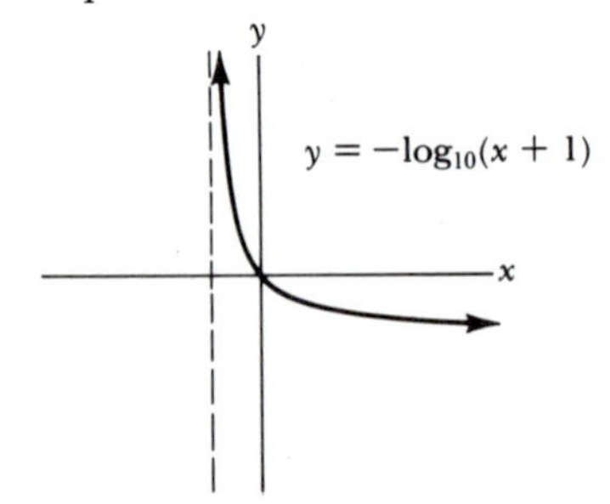

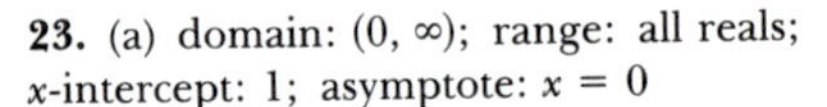

23. (a) domain: $(0, \infty)$; range: all reals; x-intercept: 1; asymptote: $x = 0$

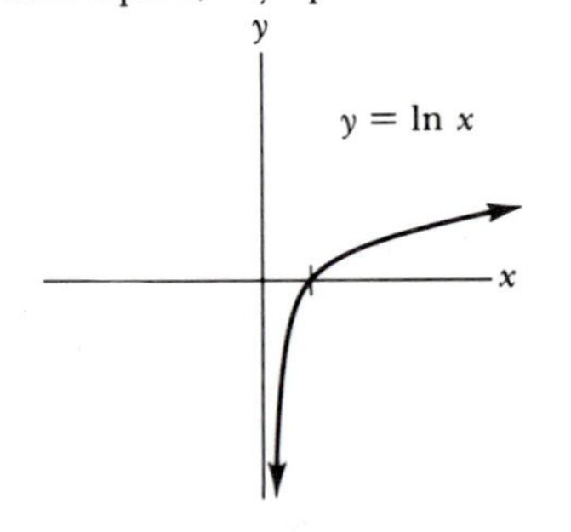

(b) domain: $(-\infty, 0)$; range: all reals; x-intercept: -1; asymptote: $x = 0$

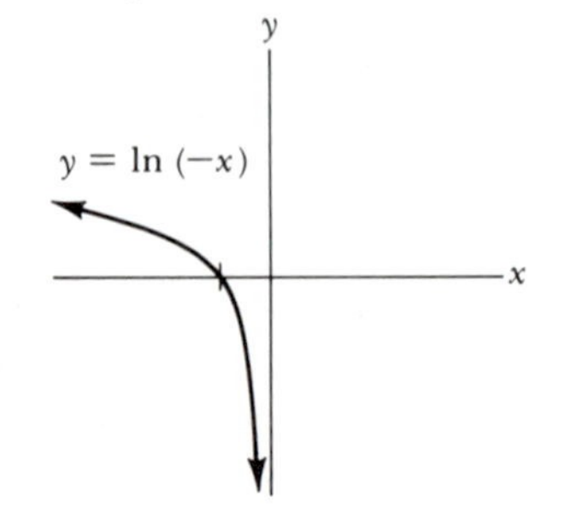

(c) domain: $(-\infty, 0)$; range: all reals; x-intercept: $-e$; asymptote: $x = 0$

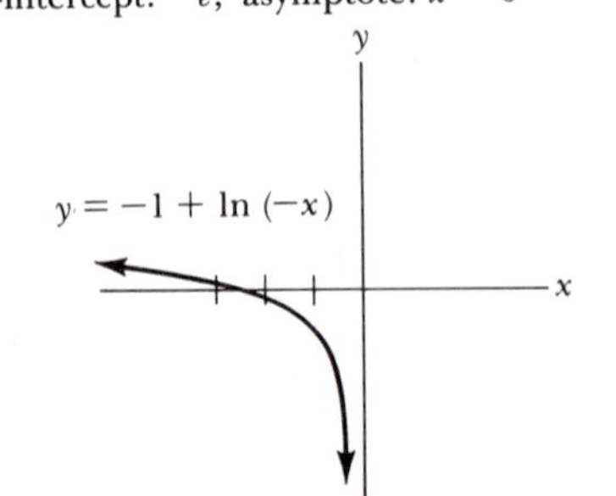

25. (a) 4; (b) -1; (c) $\frac{1}{2}$
27. (a) $x = \pm\sqrt{\log_{10} 40}$; (b) $x = \log_{10} \sqrt{40}$ (c) part (a)
29. $2^{5x} > 0$ **31.** $t = \dfrac{\sqrt{1 + \ln 6}}{2}$
33. (a) $k = -1.54 \times 10^{-10}$; (b) 99.9999846% **35.** (a) $k = \ln \frac{1}{2}$; (b) 3.32 yr **37.** (a) 0.51 hr; (b) 2.89 hr
39. $f^{-1}(x) = -1 + \ln x$; x-intercept: e; asymptote: $x = 0$

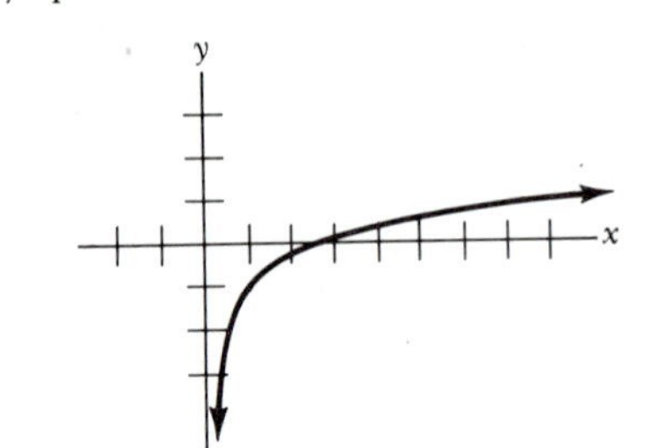

41.

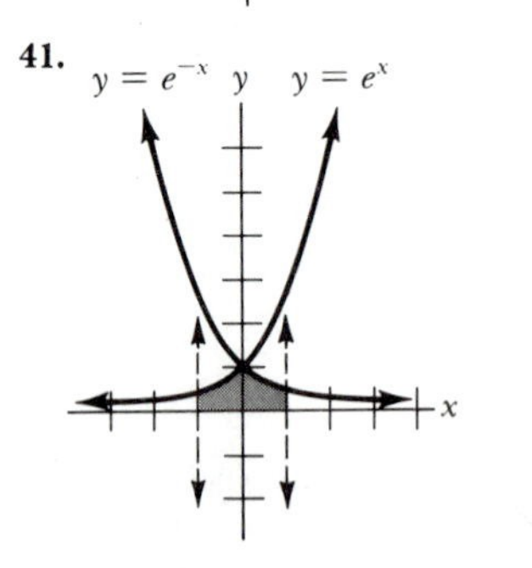

43. (b) $k = -0.182$; (c) 25 days
45. $x = \ln 6$ **47.** 10^{30}
49. $t = \dfrac{\ln(A_0/B_0)}{k_2 - k_1}$
51. (a) $(1, \infty)$; (b) $f^{-1}(x) = e^{(e^x)}$

Exercise Set 6.3

1. 1 **3.** $\frac{1}{2}$ **5.** 4 **7.** 0 **9.** 4
11. $\log_{10} 60$ **13.** $\log_5 20$ **15.** (a) $\ln 6$; (b) $\ln\left(\dfrac{3}{16384}\right)$ **17.** $\log_b\left[\dfrac{4(1 + x)^3}{(1 - x)^{3/2}}\right]$
19. $\log_{10}\left[\dfrac{27\sqrt{x + 1}}{(x^2 + 1)^6}\right]$
21. (a) $2 \log_{10} x - \log_{10}(1 + x^2)$; (b) $2 \ln x - \frac{1}{2} \ln(1 + x^2)$
23. (a) $\frac{1}{2} \log_{10}(3 + x) + \frac{1}{2} \log_{10}(3 - x)$; (b) $\frac{1}{2} \ln(2 + x) + \frac{1}{2} \ln(2 - x) - \ln(x - 1) - \frac{3}{2} \ln(x + 1)$ **25.** (a) $\frac{1}{2} \log_b x - \frac{1}{2}$; (b) $\ln(1 + x^2) + \ln(1 + x^4) + \ln(1 + x^6)$
27. (a) $a + 2b + 3c$; (b) $\frac{1}{2}(1 + a + b + c)$; (c) $1 + a - \frac{1}{2}b - \frac{1}{2}c$; (d) $2 + 2a - 4b - \frac{1}{3}c$; (e) $5a + 5b - c$
29. $x = \dfrac{\ln 5 - \ln 2 + 1}{2}$
31. $t = \ln 2 - \ln 3 - 1$ **33.** $x = \dfrac{\ln 9}{\ln 2}$
35. $x = \dfrac{\ln 10}{\ln 5 - \ln 2}$ **37.** $x = \dfrac{1}{2}$
39. $x = 3$ **41.** $x = \frac{203}{99}$ **43.** $x = 3$
45. $x = 7$ **47.** (a) $x = \dfrac{10^y}{3(10^y) - 1}$; (b) $x = \dfrac{1 - y}{2}$ **49.** (a) (i) $\dfrac{\log_{10} 5}{\log_{10} 2}$; (ii) $\dfrac{1}{\log_{10} 5}$; (iii) $\dfrac{\log_{10} 3}{\log_{10} e}$; (iv) $\dfrac{\log_{10} 2}{\log_{10} b}$; (v) $\dfrac{\log_{10} b}{\log_{10} 2}$; (b) (i) $\dfrac{\ln 6}{\ln 10}$; (ii) $\dfrac{\ln 10}{\ln 2}$; (iii) $\dfrac{\ln[(\ln x)/(\ln 10)]}{\ln 10}$ **51.** (a) true; (b) true; (c) true; (d) false; (e) true; (f) false; (g) true; (h) false; (i) true; (j) false; (k) false; (l) true; (m) true

57. x^3 **59.** $\frac{1}{2}$ **61.** $\beta e^{\alpha/3}$ **65.** $\log_2 5$
79. 2345.6
81.

x	100	1000	10^6
y	0.4234	0.6589	0.9654

x	10^{20}	10^{50}	10^{99}
y	1.3428	1.5573	1.6918

Exercise Set 6.4

1. \$1009.98 **3.** 8.45% **5.** \$767.27
7. (a) \$3869.68; (b) \$4006.39 **9.** 13 quarters (3 yr, 1 quarter) **11.** \$3487.50
13. \$2610.23 **15.** 5.83% **17.** first
19. (a) 14 yr; (b) 13.86 yr; (c) 1.01%

21. $\$2.65 \times 10^{13}$ **23.** (a) 14 yr
(b)

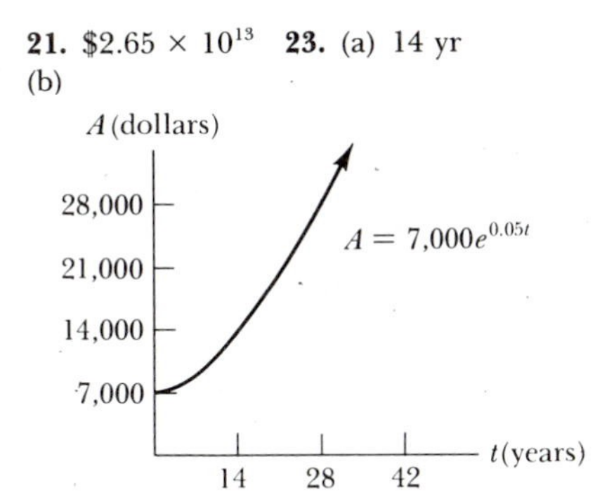

25.

	1975 Population	% Population	Relative Growth Rate	2000 Population	% Population
World	4.090	100	1.8	6.414	100
More Developed Regions	1.131	27.7	0.6	1.314	20.5
Less Developed Regions	2.959	72.3	2.1	5.002	78.0

27.

	1975 Population	Relative Growth Rate	2000 Population	% Increase
Low	4.043	1.5	5.883	45.5
Medium	4.090	1.8	6.414	56.8
High	4.134	2.0	6.816	64.9

29. (a) $k = 0.02$; (b) 170,853,155; (c) less **31.** (a) 0.68%; (b) 0.96%; (c) 19,751,512; (d) higher
33. (a)

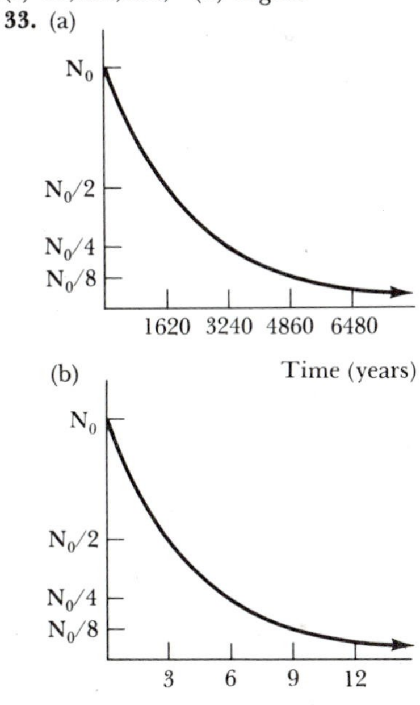

35. (a) $k = -0.0533$; (b) 58.7%; 0.5% **37.** (a) $k = -0.0248$; (b) 279 yr; (c) 280 yr **39.** (a) 2020; (b) 2035; (c) 2019 **41.** (a) 200 yr (b) 132 yr **43.** (a) $E = 10^{11.4} \times 10^{1.5m}$; (b) 44.7 **45.** $k = -1.4748 \times 10^{-11}$ **47.** 4.181 billion years old **49.** 15,500 years old **51.** 11,500 years

Chapter 6 Review Exercises

1. asymptote: $y = 0$; y-intercept: 1

3. asymptote: $x = 0$; x-intercept: 1

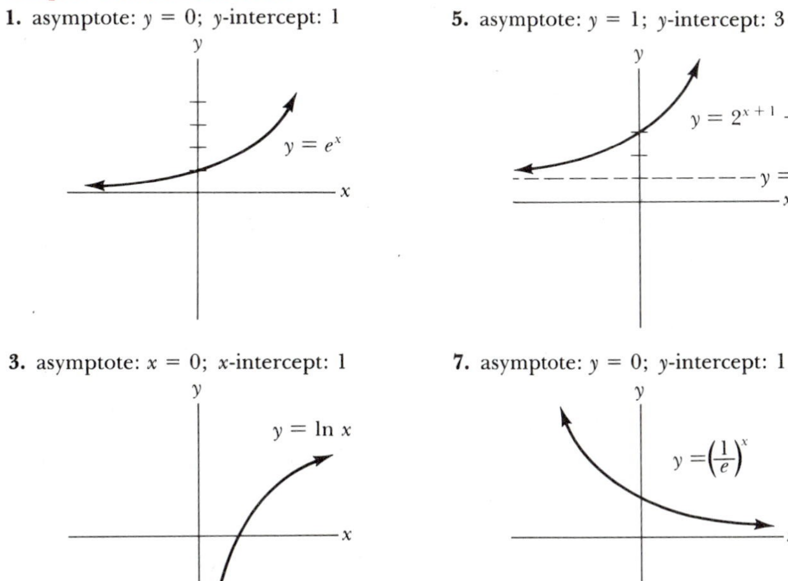

5. asymptote: $y = 1$; y-intercept: 3

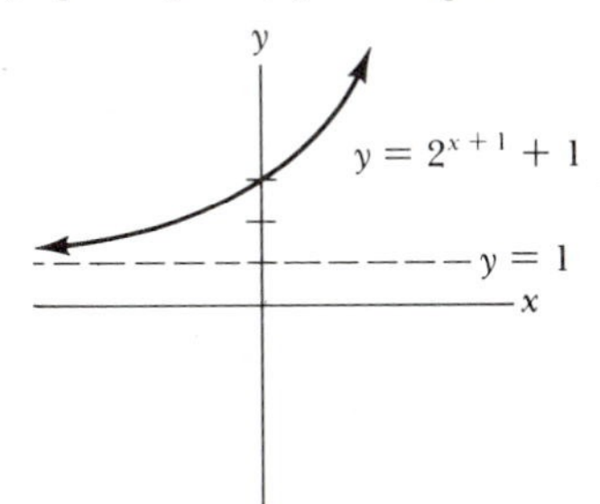

7. asymptote: $y = 0$; y-intercept: 1

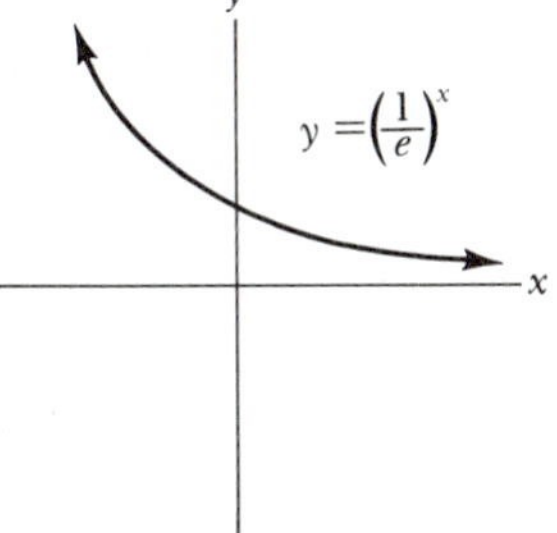

9. asymptote: $y = 1$; y-intercept: $e + 1$

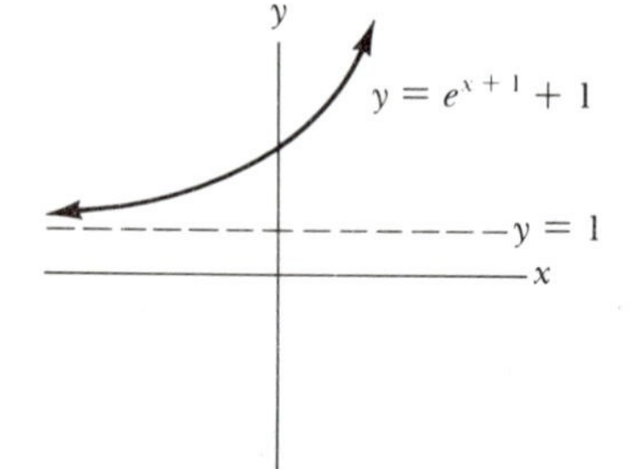

11. x-intercept: 0; y-intercept: 0

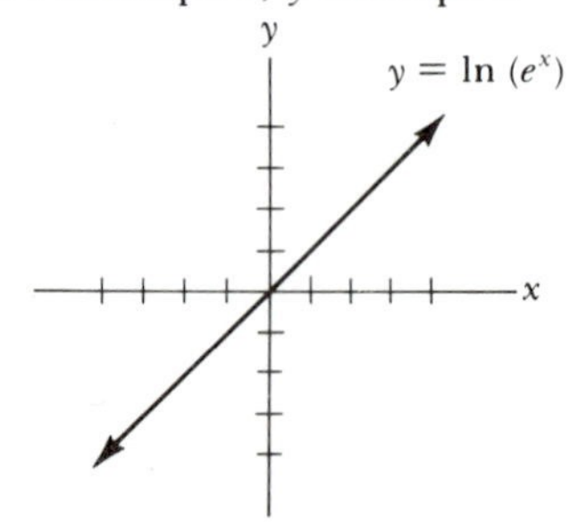

13. $x = 4$ **15.** $x = 3$ **17.** $x = \frac{2}{3}$ **19.** $x = \sqrt[3]{3}$ **21.** $x = \frac{1}{10}$ **23.** $x = \frac{200}{99}$ **25.** $x = 2$ **27.** all $x > 0$ **29.** $x = 1$ **31.** $\frac{1}{2}$ **33.** $\frac{1}{5}$ **35.** -1 **37.** 16 **39.** 4 **41.** 2 **43.** 2 **45.** $\frac{9}{14}$ **47.** $2a + 3b + \frac{1}{2}c$ **49.** $8a + 4b$ **51.** 2 and 3 **53.** 2 and 3 **55.** -3 and -2 **57.** (a) third; (b) $x = \dfrac{-2e}{e+1}$ **59.** $k = \dfrac{\ln(\frac{1}{2})}{T}$ **61.** 6.25% **63.** $t = \dfrac{d \ln(c/b)}{\ln(\frac{1}{2})}$ **65.** $\log_{10} 2$ **67.** ln 10 **69.** $\ln(x^a y^b)$ **71.** $\frac{1}{2}\ln(x-3) + \frac{1}{2}\ln(x+4)$ **73.** $3\log_{10} x - \frac{1}{2}\log_{10}(x+1)$ **75.** $\frac{1}{3}\log_{10} x - \frac{2}{3}$ **77.** $3\ln(1+2e) - 3\ln(1-2e)$ **79.** $t = \dfrac{\ln 2}{\ln(1 + R/100)}$ **81.** 9.92% **83.** $t = \dfrac{100 \ln 2}{R}$ **85.** $t = \dfrac{\ln n}{4\ln(1 + R/400)}$ **87.** (a) $(0, \infty)$; (b) $[1, \infty)$ **89.** (a) $x \neq 5, -3$ (b) $(-\infty, -3) \cup (5, \infty)$ **91.** $(-\infty, -1) \cup (1, \infty)$ **93.** -0.7

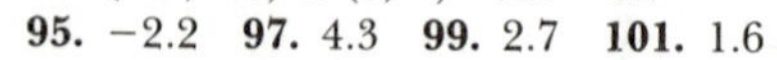

95. -2.2 **97.** 4.3 **99.** 2.7 **101.** 1.6

Exercise Set 7.1

1. (b) $\frac{15}{17}, \frac{15}{17}$; (c) $\frac{8}{17}, \frac{8}{17}$; (d) $\frac{15}{8}, \frac{17}{15}, \frac{17}{8}, \frac{8}{15}$; (e) $\dfrac{8}{15}, \dfrac{17}{8}, \dfrac{17}{15}, \dfrac{15}{8}$ **3.** (a) $\dfrac{2\sqrt{29}}{29}, \dfrac{5\sqrt{29}}{29}$; (b) 1; (c) $\dfrac{2}{5}, \dfrac{2}{5}$; (d) $\dfrac{\sqrt{29}}{2}, \dfrac{\sqrt{29}}{5}, \dfrac{5}{2}$ **5.** 6, $6\sqrt{2}$ cm **7.** $\dfrac{2}{3}$ **9.** $\dfrac{29}{21}$ **13.** 554 ft **15.** 34 million mi **17.** (a) $\frac{3}{2}$ sq in.; (b) 11.3 sq cm; (c) 21.2 sq ft **19.** 7.05 in. **21.** 10,657 ft **23.** $\cos \alpha$, $\sin \alpha$, $\tan \alpha$ **25.** 136.3 m **27.** $\sqrt{2}$, 45°; $\sqrt{3}$, 35°; 2, 30°; $\sqrt{5}$, 27°; $\sqrt{6}$, 24° **29.** $\sin \beta = \dfrac{a}{\sqrt{1 + a^2}}$; $\cos \beta = \dfrac{1}{\sqrt{1 + a^2}}$; $\tan \beta = a$; $\csc \beta = \dfrac{\sqrt{1 + a^2}}{a}$; $\cot \beta = \dfrac{1}{a}$ **31.** $4 \sec \theta + 5 \csc \theta$ **33.** (a) $\sin \theta$; (b) $\cos \theta$; (c) $\tan \theta$; (d) $\sec \theta$; (e) $\csc \theta$; (f) $\cot \theta$ **39.** (b)

n	5	10	50	100
A_n	2.38	2.94	3.1333	3.1395

n	1000	5000	10,000
A_n	3.141572	3.1415918	3.1415924

41. approximation: 0.8660; actual: 0.8678 **43.** (b) 1076 mi

Exercise Set 7.2

1. (a) $1 - 2\cos\theta + \cos^2\theta$; (b) $1 + 2\sin\theta\cos\theta$; (c) $\dfrac{\cos\theta\sin\theta + 1}{\sin\theta}$; (d) $2 + \tan x + \cot x$; (e) $2 + \sin^2 B \cos^2 B$ **3.** (a) $(\tan\theta - 6)(\tan\theta + 1)$; (b) $(\sin B + \cos B)(\sin B - \cos B)$; (c) $(\cos A + 1)^2$; (d) $3\sin\theta(\cos^2\theta + 2)$; (e) $(\csc\alpha - 3)(\csc\alpha + 1)$; (f) $(3\sin\theta + 2)(2\sin\theta + 1)$ **5.** -1 **7.** (a) $\csc\theta$; (b) $\tan\theta$; (c) $\cos^2 B$ **11.** $\frac{12}{13}, \frac{12}{5}$ **13.** $\sin\theta = \frac{3}{5}$, $\sec\theta = \frac{5}{4}$, $\csc\theta = \frac{5}{3}$, $\tan\theta = \frac{3}{4}$, $\cot\theta = \dfrac{4}{3}$ **15.** (a) $\dfrac{\sqrt{10 + 2\sqrt{5}}}{4}$; (b) $\dfrac{1}{4}(\sqrt{5} - 1)$ **41.** (h) $\dfrac{\sqrt{6} - \sqrt{2}}{4}$

Exercise Set 7.3

1. (a)

(b)

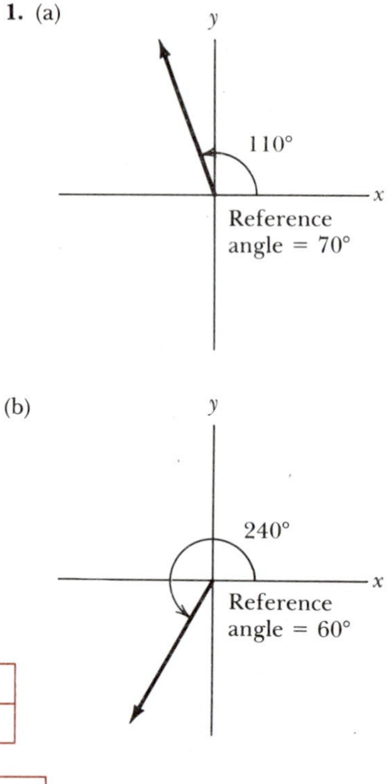

(c)

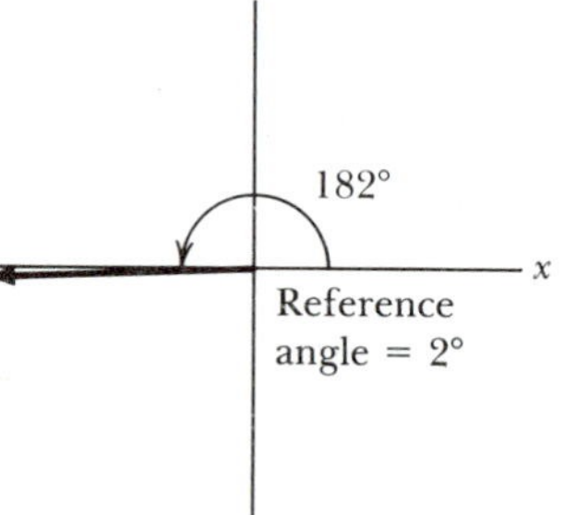

(d) Reference angle = 60°; 300°

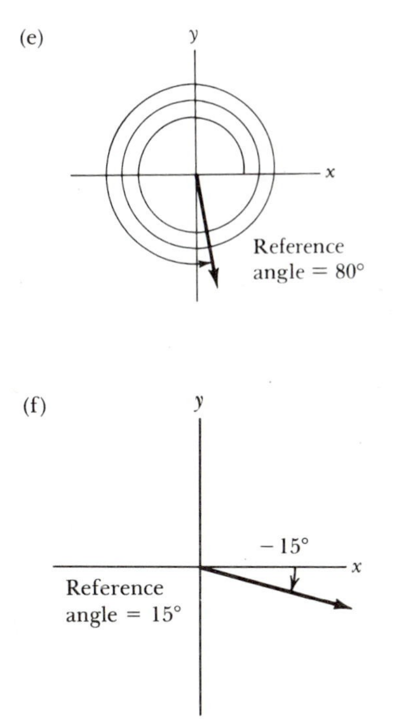

3. See Table 1. **5.** $-\frac{\sqrt{5}}{2}$

7. (a) $-\frac{\sqrt{2}}{2}$; (b) $\frac{1}{2}$; (c) $-\frac{\sqrt{3}}{3}$; (d) $-\frac{\sqrt{3}}{2}$; (e) $-\frac{2\sqrt{3}}{3}$; (f) $-\sqrt{2}$; (g) -1; (h) $-\frac{\sqrt{3}}{3}$; (i) $\frac{-\sqrt{3}}{3}$

9. $\sin\theta = \frac{-\sqrt{2}}{2}$, $\cos\theta = \frac{\sqrt{2}}{2}$, $\tan\theta = -1$, $\cot\theta = -1$, $\sec\theta = \sqrt{2}$, $\csc\theta = -\sqrt{2}$

11. (a)

θ	0°	90°	180°	270°	360°	450°	540°	630°	720°
$\sin\theta$	0	1	0	−1	0	1	0	−1	0
$\cos\theta$	1	0	−1	0	1	0	−1	0	1

(b)

θ	30°	60°	90°	120°	150°	180°	210°	240°	270°	300°	330°	360°
$\sin\theta$	$\frac{1}{2}$	$\frac{\sqrt{3}}{2}$	1	$\frac{\sqrt{3}}{2}$	$\frac{1}{2}$	0	$-\frac{1}{2}$	$-\frac{\sqrt{3}}{2}$	−1	$\frac{-\sqrt{3}}{2}$	$-\frac{1}{2}$	0

13. (a)

	Quadrant I	Quadrant II	Quadrant III	Quadrant IV
$\sin\theta$	positive	positive	negative	negative
$\cos\theta$	positive	negative	negative	positive
$\tan\theta$	positive	negative	positive	negative

15. $\frac{-4}{5}$ **17.** $\frac{-4\sqrt{15}}{15}$ **19.** 120°, 240°, −120° (many answers are possible)

21. $\frac{3\sqrt{3}}{4}$ sq units **23.** (a) $\frac{1}{4}(\sqrt{2} - \sqrt{6})$; (b) $-\frac{1}{4}\sqrt{10 + 2\sqrt{5}}$; (c) $\frac{\sqrt{2} - \sqrt{6}}{2\sqrt{2 + \sqrt{3}}}$; (d) $\frac{1}{4}(\sqrt{2} - \sqrt{6})$; (e) $\frac{1}{4}\sqrt{10 + 2\sqrt{5}}$; (f) $-\frac{1}{4}\sqrt{10 + 2\sqrt{5}}$; (g) $\frac{1}{4}(\sqrt{6} - \sqrt{2})$; (h) $\frac{1}{4}\sqrt{10 + 2\sqrt{5}}$; (i) $-\sqrt{6} - \sqrt{2}$; (j) $\frac{-\sqrt{50 - 10\sqrt{5}}}{5}$ (k) 1; (l) 1

27. (a) 30° (many answers are possible); (b) 225° (many answers are possible)

29. (a) $P(\cos\theta, \sin\theta)$; $Q(\cos\phi, \sin\phi)$

Exercise Set 7.4

1. (a) $4\sqrt{6}$ cm; (b) $4\sqrt{2}$ in.; (c) $3\sqrt{2}$ cm **3.** $b = \frac{\sqrt{3}}{3}x$, $c = \frac{\sqrt{3}}{3}x$

5. 19.15 sq in. **7.** (a) $A = 63°$, $C = 51°$, $c = 25.6$; (b) $B = 14°$, $C = 41°$, $b = 15$

9. (c) $C = 105°$, $c = 1.93$; (d) $C = 15°$, $c = 0.52$ **11.** (a) 21 cm; (b) 5.4 in.

13. 5.88 **15.** $A = 28.96°$, $B = 46.57°$, $c = 104.47$ **17.** 3.90, 9.42

19. 159.6 ft **29.** (c) $\frac{\sqrt{2} + \sqrt{6}}{4}$; (d) $\frac{\sqrt{2} + \sqrt{6}}{4}$

Chapter 7 Review Exercises

1. $a = \frac{1}{2}$, $c = \frac{\sqrt{3}}{2}$ **3.** $b = \frac{35}{2}$

5. $\sin A = \frac{\sqrt{55}}{8}$, $\cot A = \frac{3\sqrt{55}}{55}$

7. $a = \sqrt{41 + 20\sqrt{3}}$, area = 5

9. $\cos A = -\frac{1}{15}$, $\sin A = \frac{4\sqrt{14}}{15}$

11. $\frac{1}{5}$ **13.** $c = 3$ **15.** $\sin A = \frac{\sqrt{2}}{8}$, $\cos A = \frac{\sqrt{62}}{8}$, $B = 34.8°$ **17.** $c = 6$

19. $\frac{\sqrt{2}}{2}$ **21.** $\sqrt{3}$ **23.** −2 **25.** −1

27. $\frac{\sqrt{2}}{2}$ **29.** 1 **31.** $\frac{-2\sqrt{3}}{3}$ **33.** 2

35. (a) no; (b) 1 **37.** $\sin\theta = \frac{4}{5}$, $\tan\theta = \frac{4}{3}$ **39.** $\tan\theta = \frac{-24}{7}$

41. $\cot\theta = \frac{5}{12}$

43. $\tan(90° - \theta) = \frac{3\sqrt{7}}{7}$

45. $\cos(180° - \theta) = \frac{-7}{9}$

47. $\sin\theta - \cos\theta$ **49.** $\sin A \cos A$

51. $\sin^2 A$ **53.** $\cos A + \sin A$

55. $\sin A \cos A$ **57.** $2 \sin A \cos A$

59. $\sqrt{1 - a^2}$ **61.** $\sqrt{1 - a^2}$ **63.** $-a$

65. $-a$ **67.** $-\sqrt{1 - a^2}$

69. $\sqrt{68 - 16\sqrt{3}}$

71. 3600 tan 35° cm² **107.** $\frac{2\sqrt{ab}}{a + b}$

113. 12 cm **115.** 13.25 cm
117. 4.04 cm **119.** 9.53 cm

121.

a	$a \cdot 90°$	$\sin(a \cdot 90°)$	$\frac{11a - 3a^3}{7 + a^2}$
0.0	0°	0	0
0.1	9°	0.156	0.153
0.2	18°	0.309	0.295
0.3	27°	0.454	0.427
0.4	36°	0.588	0.547
0.5	45°	0.707	0.655
0.6	54°	0.809	0.750
0.7	63°	0.891	0.832
0.8	72°	0.951	0.901
0.9	81°	0.988	0.956
1.0	90°	1	1

Exercise Set 8.1

1. (a) $\frac{\pi}{3}$; (b) $\frac{5\pi}{4}$; (c) $\frac{\pi}{5}$; (d) $\frac{5\pi}{2}$; (e) 0 **3.** (a) $\frac{7\pi}{36}$; (b) $\frac{\pi}{8}$; (c) $\frac{\pi}{90}$; (d) $\frac{5\pi}{9}$ **5.** smaller

7.

θ	0	$\frac{\pi}{2}$	π	$\frac{3\pi}{2}$	2π
$\cos\theta$	1	0	−1	0	1

9. (a) $\frac{\pi}{2}$ cm; (b) $\frac{5\pi}{6}$ in.; (c) 4π ft

11. 1 radian **13.** $\frac{2\pi}{5}$ radians

15. $A = \theta + \sin\theta$ **17.** (a) $\frac{2\pi}{3}$ radians; (b) $\frac{\pi}{6}$ radians **19.** 135π ft

21. (b) $\theta = \frac{\pi}{2}$

Exercise Set 8.2

1.

θ	$\sin\theta$	$\cos\theta$	$\tan\theta$	$\csc\theta$	$\sec\theta$	$\cot\theta$
0	0	1	0	undef.	1	undef.
$\frac{\pi}{6}$	$\frac{1}{2}$	$\frac{\sqrt{3}}{2}$	$\frac{\sqrt{3}}{3}$	2	$\frac{2\sqrt{3}}{3}$	$\sqrt{3}$
$\frac{\pi}{4}$	$\frac{\sqrt{2}}{2}$	$\frac{\sqrt{2}}{2}$	1	$\sqrt{2}$	$\sqrt{2}$	1
$\frac{\pi}{3}$	$\frac{\sqrt{3}}{2}$	$\frac{1}{2}$	$\sqrt{3}$	$\frac{2\sqrt{3}}{3}$	2	$\frac{\sqrt{3}}{3}$
$\frac{\pi}{2}$	1	0	undef.	1	undef.	0
$\frac{2\pi}{3}$	$\frac{\sqrt{3}}{2}$	$-\frac{1}{2}$	$-\sqrt{3}$	$\frac{2\sqrt{3}}{3}$	−2	$-\frac{\sqrt{3}}{3}$
$\frac{3\pi}{4}$	$\frac{\sqrt{2}}{2}$	$-\frac{\sqrt{2}}{2}$	−1	$\sqrt{2}$	$-\sqrt{2}$	−1
$\frac{5\pi}{6}$	$\frac{1}{2}$	$-\frac{\sqrt{3}}{2}$	$-\frac{\sqrt{3}}{3}$	2	$-\frac{2\sqrt{3}}{3}$	$-\sqrt{3}$
π	0	−1	0	undef.	−1	undef.

3. $\cos\theta = \frac{-4}{5}$, $\tan\theta = \frac{3}{4}$

5. $\tan t = \frac{-\sqrt{39}}{13}$ **7.** $\sec\alpha = \frac{13}{5}$, $\cos\alpha = \frac{5}{13}$, $\sin\alpha = \frac{12}{13}$

13. $\frac{\sqrt{7}\cos\theta}{7}$ **15.** (a) $-\frac{1 + 2\sqrt{2}}{3}$; (b) 1 **17.** (a) $\frac{\sqrt{2}}{2}$; (b) $\frac{\sqrt{3}}{2}$; (c) 1

19. (a) $\sin\theta = \frac{15}{17}$, $\cos\theta = \frac{-8}{17}$, $\tan\theta = \frac{-15}{8}$, $\cot\theta = \frac{-8}{15}$, $\csc\theta = \frac{17}{15}$, $\sec\theta = \frac{-17}{8}$; (b) $\sin(-\theta) = \frac{-15}{17}$, $\cos(-\theta) = \frac{-8}{17}$, $\tan(-\theta) = \frac{15}{8}$, $\cot(-\theta) = \frac{8}{15}$, $\csc(-\theta) = \frac{-17}{15}$, $\sec(-\theta) = \frac{-17}{8}$ **23.** (a) $\sin\theta = \frac{2\sqrt{5}}{5}$, $\cos\theta = \frac{\sqrt{5}}{5}$; (b) $9Y^2 - X^2$

25.

θ	$\theta - \frac{\theta^3}{6}$	$\sin\theta$
.1	.099833 · · ·	.099833 · · ·
.2	.198666 · · ·	.198669 · · ·
.3	.295500 · · ·	.295520 · · ·
1.0	.833333 · · ·	.841470 · · ·

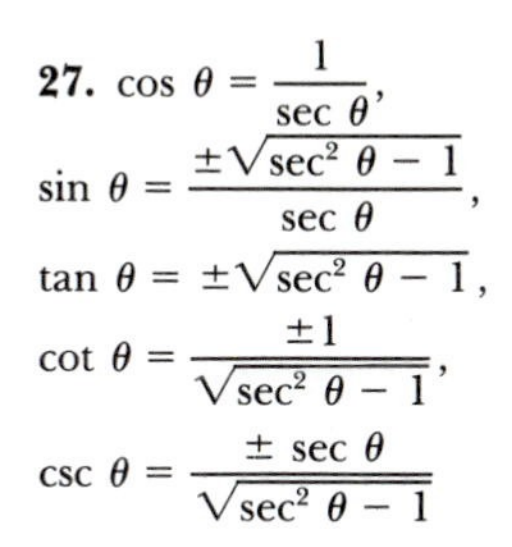

27. $\cos\theta = \dfrac{1}{\sec\theta}$,

$\sin\theta = \dfrac{\pm\sqrt{\sec^2\theta - 1}}{\sec\theta}$,

$\tan\theta = \pm\sqrt{\sec^2\theta - 1}$,

$\cot\theta = \dfrac{\pm 1}{\sqrt{\sec^2\theta - 1}}$,

$\csc\theta = \dfrac{\pm\sec\theta}{\sqrt{\sec^2\theta - 1}}$

29. (a)

θ	$\sin\theta$	Which Is Larger, θ or $\sin\theta$?
.1	.0998	θ
.2	.1987	θ
.3	.2955	θ
.4	.3894	θ
.5	.4794	θ

Exercise Set 8.3

1. (a) $\cos\theta$; (b) $-\cos\theta$; (c) $\sin\theta$; (d) $-\sin\theta$ **3.** $\dfrac{\sqrt{6}-\sqrt{2}}{4}$ **5.** $\dfrac{\sqrt{6}+\sqrt{2}}{4}$

7. (a) $\dfrac{56}{65}$; (b) $\dfrac{16}{65}$; (c) $\dfrac{33}{65}$; (d) $\dfrac{-63}{65}$; (e) $\dfrac{56}{33}$; (f) $\dfrac{-16}{63}$ **9.** (a) $\dfrac{-5}{13}$; (b) $\dfrac{119}{169}$

11. $\sin(\theta + \beta) = \dfrac{2\sqrt{3} - 3}{2\sqrt{13}}$, $\cos(\beta - \theta) = \dfrac{2 - 3\sqrt{3}}{2\sqrt{13}}$ **13.** $\dfrac{\pi}{4}$

15. $\cos\theta$

Exercise Set 8.4

1. (a) $\dfrac{-17}{6}$; (b) $\dfrac{-1}{18}$; (c) $\dfrac{-6}{17}$; (d) $\dfrac{-7}{24}$

3. (a) $\dfrac{24}{7}$; (b) $\dfrac{-24}{7}$; (c) $\dfrac{-12}{5}$; (d) $\dfrac{12}{5}$

5. (a) $\dfrac{3\sqrt{10}}{10}$; (b) $\sqrt{\dfrac{\sqrt{13}+2}{2\sqrt{13}}}$

(c) $\dfrac{3\sqrt{\sqrt{13}+2} + 4\sqrt{\sqrt{13}-2}}{5\sqrt{2\sqrt{13}}}$;

(d) $\dfrac{8\sqrt{65}}{65}$ **7.** (a) $\dfrac{1}{3}$; (b) 3

13. (a) $\sqrt{\dfrac{z-1}{2z}}$; (b) $\dfrac{2\sqrt{z^2-1}}{z^2}$;

(c) $\dfrac{2-z^2}{z^2}$ **15.** (a) $2\sin 8\theta + 2\sin 4\theta$;

(b) $\cos 2A - \cos 4A$;

(c) $\frac{1}{2}\cos 6\theta + \frac{1}{2}\cos 2\theta$

Exercise Set 8.5

1. (a) period: 10; amplitude: 6;
(b) period: 6; amplitude: $\frac{3}{2}$
(c) period: 4; amplitude: 5;
(d) period: $\dfrac{2\pi}{5}$; amplitude: 2

3. (a)

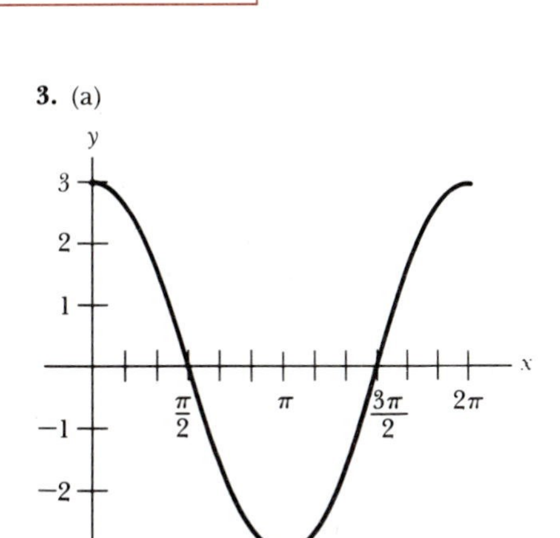

amplitude: 3
period: 2 π
x-intercepts: $\dfrac{\pi}{2}$ and $\dfrac{3\pi}{2}$
highs: $(0, 3)$, $(2\pi, 3)$
low: $(\pi, -3)$

(b)

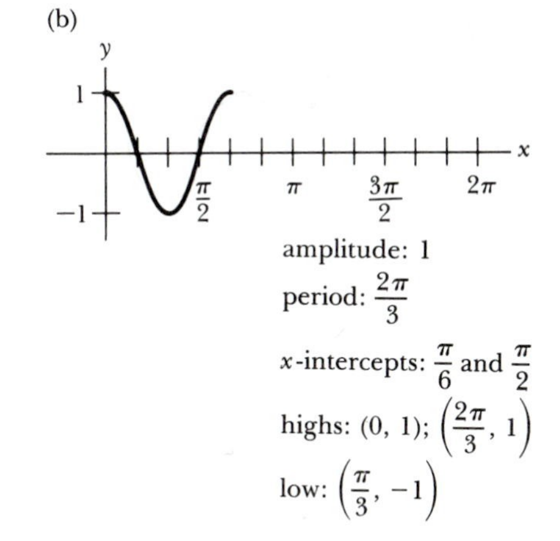

amplitude: 1
period: $\dfrac{2\pi}{3}$
x-intercepts: $\dfrac{\pi}{6}$ and $\dfrac{\pi}{2}$
highs: $(0, 1)$; $\left(\dfrac{2\pi}{3}, 1\right)$
low: $\left(\dfrac{\pi}{3}, -1\right)$

(c)

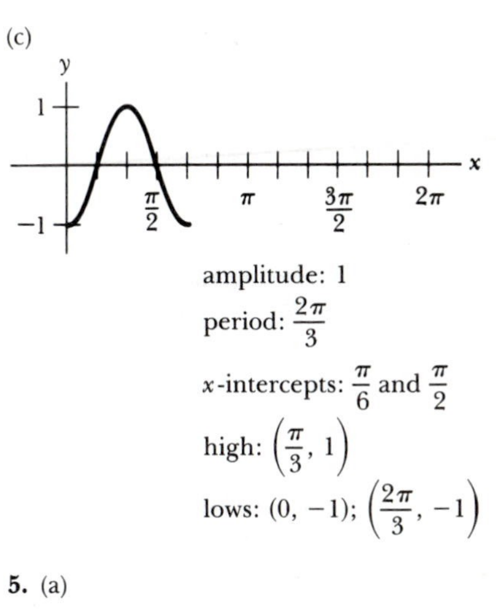

amplitude: 1
period: $\dfrac{2\pi}{3}$
x-intercepts: $\dfrac{\pi}{6}$ and $\dfrac{\pi}{2}$
high: $\left(\dfrac{\pi}{3}, 1\right)$
lows: $(0, -1)$; $\left(\dfrac{2\pi}{3}, -1\right)$

5. (a)

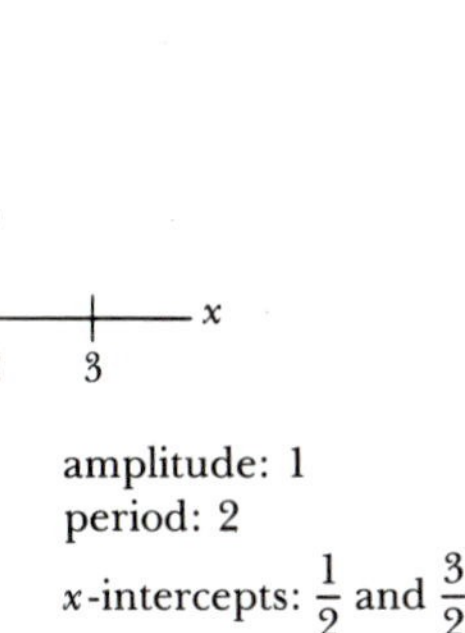

amplitude: 1
period: 2
x-intercepts: $\dfrac{1}{2}$ and $\dfrac{3}{2}$
highs: $(0, 1)$; $(2, 1)$
low: $(1, -1)$

(b)

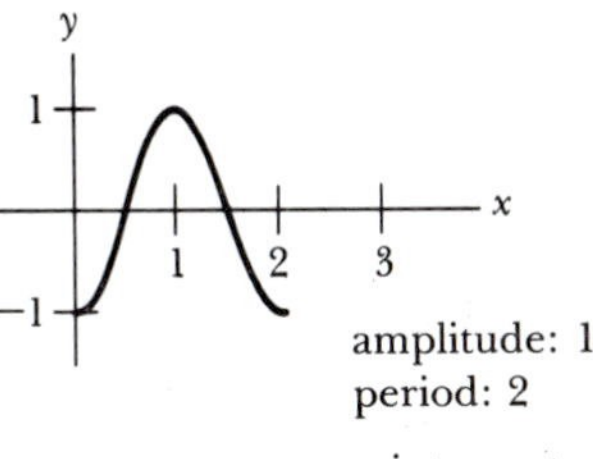

amplitude: 1
period: 2
x-intercepts: $\dfrac{1}{2}$ and $\dfrac{3}{2}$
high: $(1, 1)$
lows: $(0, -1)$; $(2, -1)$

(c)

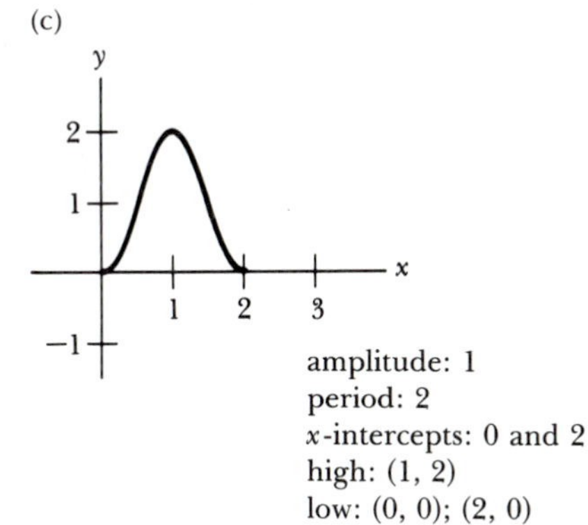

amplitude: 1
period: 2
x-intercepts: 0 and 2
high: $(1, 2)$
low: $(0, 0)$; $(2, 0)$

7. (a)

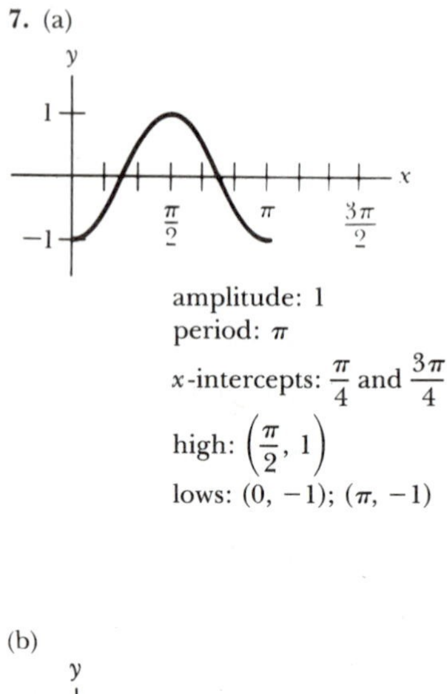

amplitude: 1
period: π
x-intercepts: $\frac{\pi}{4}$ and $\frac{3\pi}{4}$
high: $\left(\frac{\pi}{2}, 1\right)$
lows: $(0, -1)$; $(\pi, -1)$

(b)

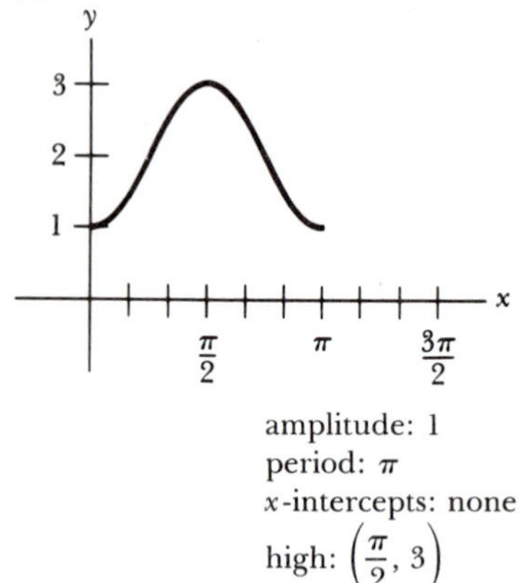

amplitude: 1
period: π
x-intercepts: none
high: $\left(\frac{\pi}{2}, 3\right)$
lows: $(0, 1)$; $(\pi, 1)$

9. (a)

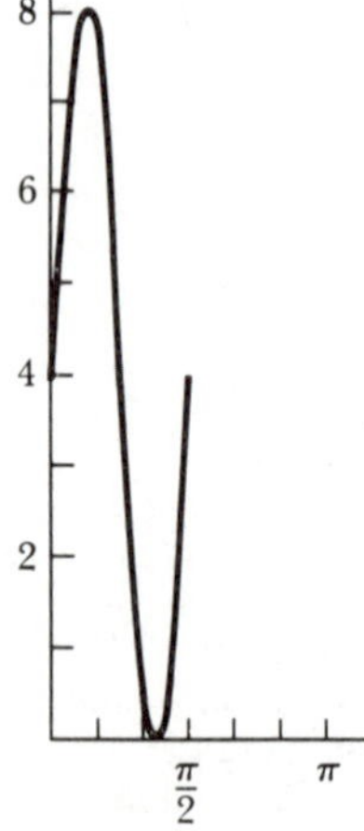

amplitude: 4
period: $\frac{\pi}{2}$
x-intercept: $\frac{3\pi}{8}$
high: $\left(\frac{\pi}{8}, 8\right)$
low: $\left(\frac{3}{8}\pi, 0\right)$

(b)

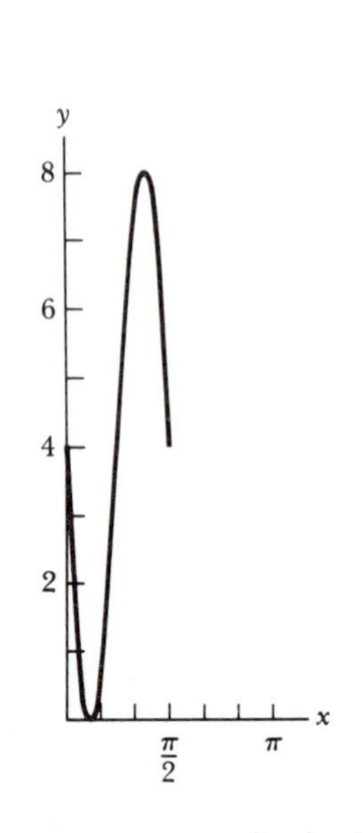

amplitude: 4
period: $\frac{\pi}{2}$
x-intercept: $\frac{\pi}{8}$
high: $\left(\frac{3\pi}{8}, 8\right)$
low: $\left(\frac{\pi}{8}, 0\right)$

11.

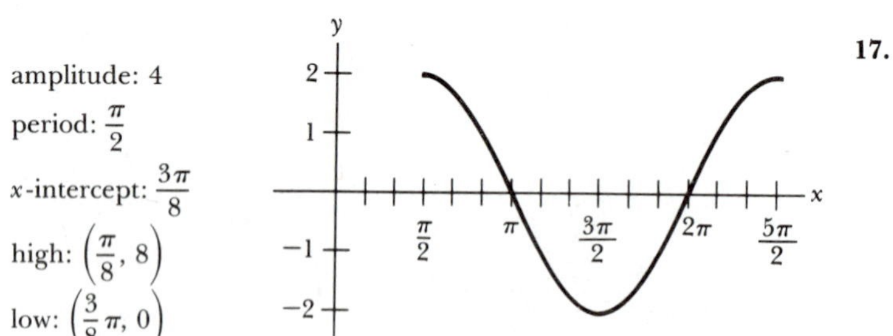

amplitude: 2
period: 2π
phase shift: $\frac{\pi}{2}$
x-intercepts: π and 2π
highs: $\left(\frac{\pi}{2}, 2\right)$; $\left(\frac{5\pi}{2}, 2\right)$
low: $\left(\frac{3}{2}\pi, -2\right)$

13.

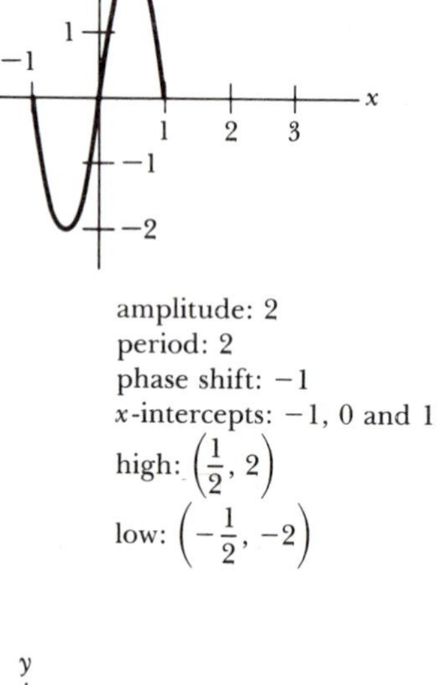

amplitude: 2
period: 2
phase shift: -1
x-intercepts: -1, 0 and 1
high: $\left(\frac{1}{2}, 2\right)$
low: $\left(-\frac{1}{2}, -2\right)$

15.

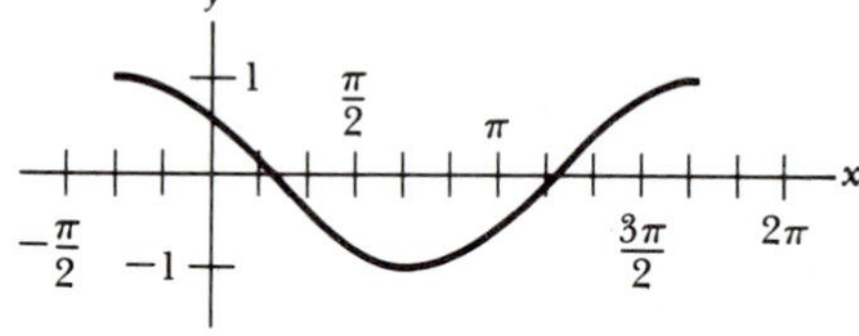

amplitude: 1

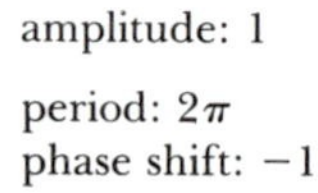

period: 2π
phase shift: -1

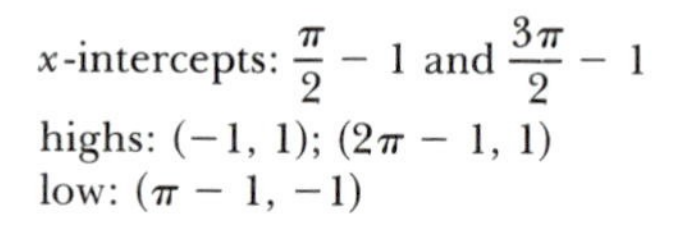

x-intercepts: $\frac{\pi}{2} - 1$ and $\frac{3\pi}{2} - 1$
highs: $(-1, 1)$; $(2\pi - 1, 1)$
low: $(\pi - 1, -1)$

17.

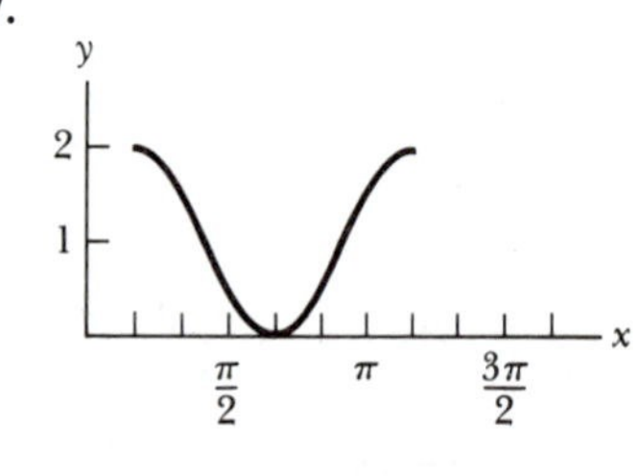

amplitude: 1
period: π
phase shift: $\frac{\pi}{6}$
x-intercept: $\frac{2\pi}{3}$
highs: $\left(\frac{\pi}{6}, 2\right)$; $\left(\frac{7\pi}{6}, 2\right)$
low: $\left(\frac{2}{3}\pi, 0\right)$

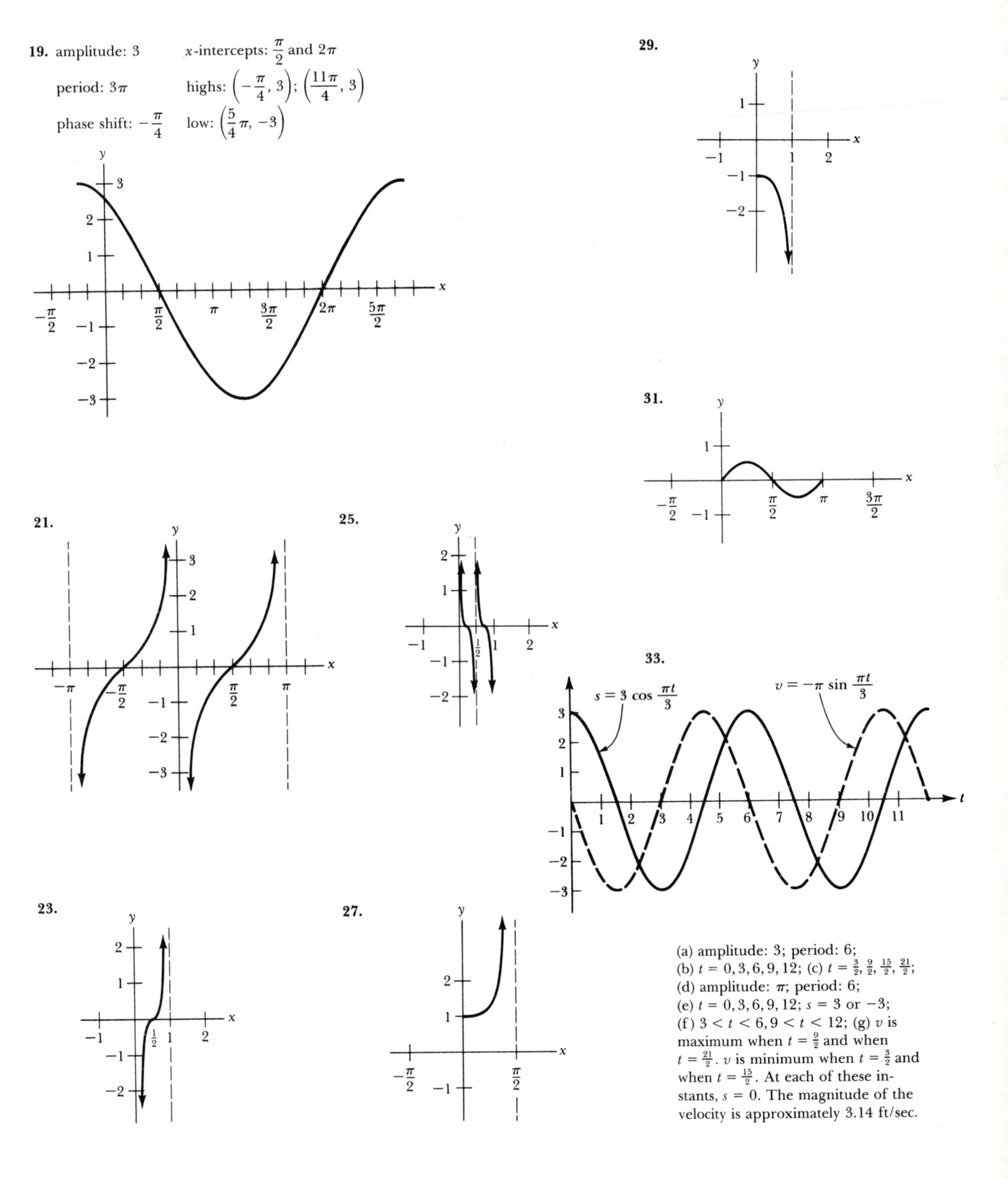

19. amplitude: 3 x-intercepts: $\frac{\pi}{2}$ and 2π

period: 3π highs: $\left(-\frac{\pi}{4}, 3\right)$; $\left(\frac{11\pi}{4}, 3\right)$

phase shift: $-\frac{\pi}{4}$ low: $\left(\frac{5}{4}\pi, -3\right)$

21.

23.

25.

27.

29.

31.

33. $s = 3\cos\frac{\pi t}{3}$ $v = -\pi \sin\frac{\pi t}{3}$

(a) amplitude: 3; period: 6;
(b) $t = 0, 3, 6, 9, 12$; (c) $t = \frac{3}{2}, \frac{9}{2}, \frac{15}{2}, \frac{21}{2}$;
(d) amplitude: π; period: 6;
(e) $t = 0, 3, 6, 9, 12$; $s = 3$ or -3;
(f) $3 < t < 6, 9 < t < 12$; (g) v is maximum when $t = \frac{9}{2}$ and when $t = \frac{21}{2}$. v is minimum when $t = \frac{3}{2}$ and when $t = \frac{15}{2}$. At each of these instants, $s = 0$. The magnitude of the velocity is approximately 3.14 ft/sec.

35.

37.

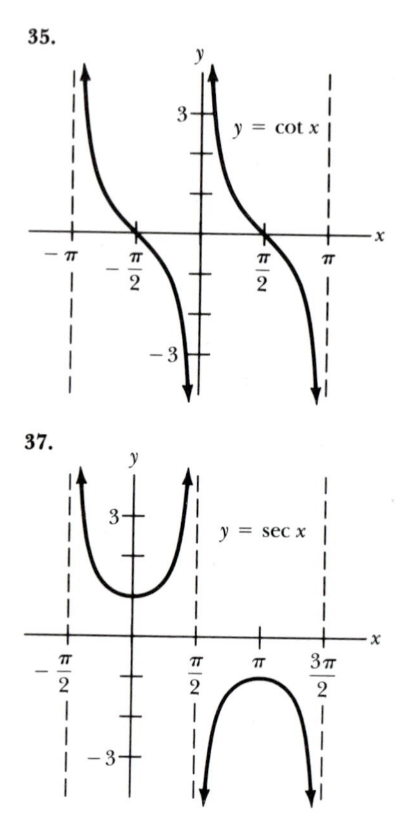

39. (a) amplitude: 2; period: π; (b) -2

Exercise Set 8.6

1. yes **3.** no **5.** no **7.** $\theta = \pi + 2\pi k$; k = any integer **9.** $x = \frac{\pi}{6} + 2\pi k$; $x = \frac{5\pi}{6} + 2\pi k$; $x = \frac{\pi}{2} + 2\pi k$; k = any integer **11.** no solution **13.** $\theta = \pi + 2\pi k$; k = any integer **15.** $x = \frac{\pi}{3} + \pi k$; $x = \frac{2\pi}{3} + \pi k$; k = any integer **17.** $\alpha = \frac{\pi}{6} + 2\pi k$; k = any integer **19.** 67.5°, 157.5°, 247.5°, 337.5° **21.** 60°, 300° **23.** 30°, 150°, 210°, 330° **25.** 90°, 270° **27.** no solution **29.** 90°, 199.5°, 340.5° **31.** no solution **33.** 0°, 126.9° **35.** $\frac{\pi}{12}, \frac{\pi}{6}, \frac{7\pi}{12}, \frac{2\pi}{3}, \frac{13\pi}{12}, \frac{7\pi}{6}, \frac{19\pi}{12}, \frac{5\pi}{3}$ **37.** $0, \frac{\pi}{8}, \frac{3\pi}{8}, \frac{5\pi}{8}, \frac{7\pi}{8}, \pi, \frac{9\pi}{8}, \frac{11\pi}{8}, \frac{13\pi}{8}, \frac{15\pi}{8}, 2\pi$ **39.** $\frac{\pi}{10}, \frac{\pi}{2}, \frac{9\pi}{10}, \frac{13\pi}{10}, \frac{17\pi}{10}$ **41.** $\frac{\pi}{3}, \frac{4\pi}{3}$ **43.** 60.45°

Exercise Set 8.7

1. $\frac{\pi}{3}$ **3.** $\frac{\pi}{3}$ **5.** $\frac{-\pi}{6}$ **7.** $\frac{\pi}{4}$ **9.** undefined **11.** $\frac{1}{4}$ **13.** $\frac{3}{4}$ **15.** $\frac{-\pi}{7}$ **17.** $\frac{\pi}{2}$ **19.** 0 **21.** $\frac{4}{3}$ **23.** $\frac{\sqrt{2}}{2}$ **25.** $\frac{12}{5}$ **27.** $\frac{1}{2}$ **29.** $\frac{2\sqrt{2}}{3}$ **31.** (a) 0.84; (b) 0.84; (c) 1.26; (d) 0.90 **33.** $\sqrt{2}$ **37.** $\frac{119}{169}$ **47.** $\frac{\pi}{2}$

Chapter 8 Review Exercises

1. 2π **3.** $\frac{\pi}{180}$ **5.** $\frac{\pi}{24}$ **7.** $\frac{1}{180}$ **9.** 135° **11.** 900° **13.** 150° **15.** $\frac{360°}{\pi}$ **17.** $s = 2\pi$ cm, $A = 16\pi$ cm^2 **19.** $\frac{1}{2}$ cm^2 **21.** $\frac{20}{\pi}$ cm **23.** $r = 3$ cm, $\theta = 4$ **25.** -1 **27.** $\frac{2\sqrt{3}}{3}$ **29.** $-\sqrt{3}$ **31.** $\frac{1}{2}$ **33.** 1 **35.** -2 **37.** $\frac{7\pi}{18}$ **39.** 0.841 **41.** -1.000 **43.** 0.012 **49.** $5 \cos\theta$ **51.** $10 \tan\theta$ **53.** $\frac{\sqrt{5}}{5}\cos\theta$ **55.** $\frac{\sqrt{10}}{10}$ **57.** $\frac{527}{625}$ **59.** $\frac{2\sqrt{29}}{29}$ **61.** $\frac{\sqrt{2}}{10}$ **63.** $\frac{\pi - 2}{2}$ cm^2 **75.** $\frac{-16}{65}$ **77.** $\sin x$ **79.** 1 **81.** 0 **83.** $\cos x$ **85.** -3 **87.** $\frac{1104}{1105}$ **89.** 1 **91.** $y = 4 \sin x$ **93.** $y = 2 \sin 4\left(x - \frac{\pi}{8}\right)$ **95.** $\sqrt{2}$ **97.** x-intercepts: $\frac{1}{2}, \frac{5}{2}, \frac{9}{2}$

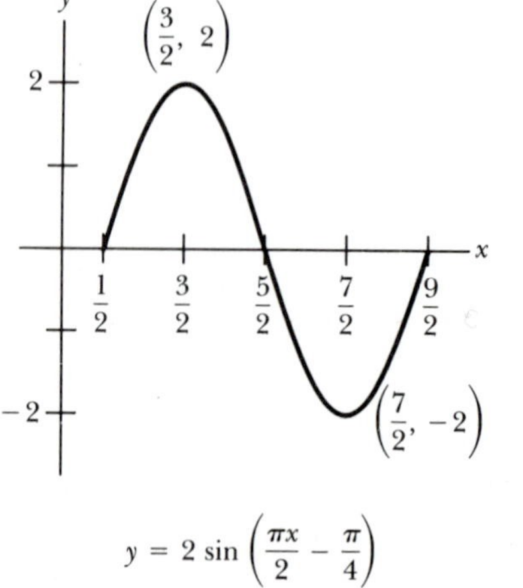

$y = 2 \sin\left(\frac{\pi x}{2} - \frac{\pi}{4}\right)$

99. x-intercepts: $-\frac{1}{2}, \frac{5}{2}$

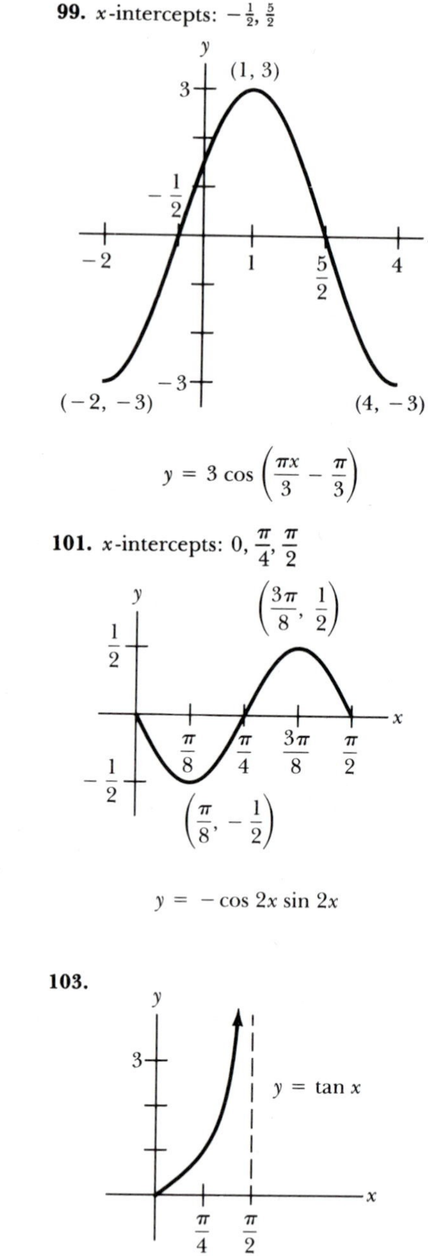

$y = 3 \cos\left(\frac{\pi x}{3} - \frac{\pi}{3}\right)$

101. x-intercepts: $0, \frac{\pi}{4}, \frac{\pi}{2}$

$y = -\cos 2x \sin 2x$

103.

105.

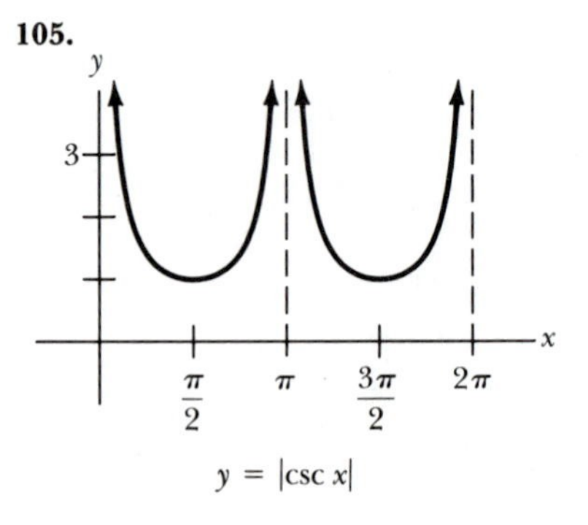

$y = |\csc x|$

107.

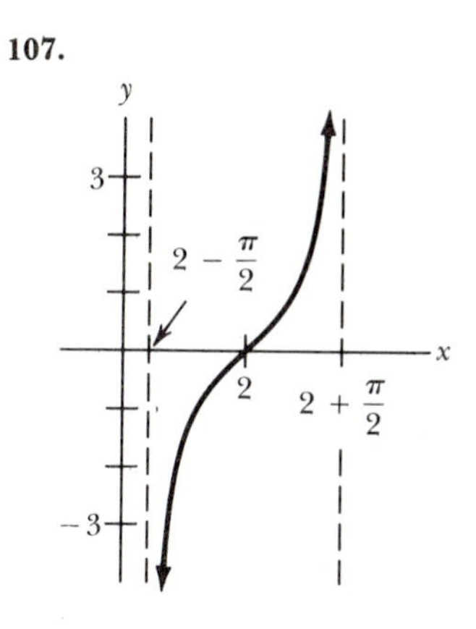

$y = \tan(x - 2)$

151. $\frac{\pi}{4}, \frac{\pi}{2}, \frac{5\pi}{4}, \frac{3\pi}{2}$ **153.** $\frac{\pi}{9}, \frac{2\pi}{9}, \frac{7\pi}{9}, \frac{8\pi}{9}, \frac{13\pi}{9}, \frac{14\pi}{9}$ **155.** $0, \frac{2\pi}{3}, \pi, \frac{4\pi}{3}$ **157.** $\frac{\pi}{6}, \frac{5\pi}{6}, \frac{3\pi}{2}$ **159.** $\frac{\pi}{3}, \pi, \frac{5\pi}{3}$ **161.** $\frac{\pi}{2}, \frac{3\pi}{2}$ **163.** $\frac{\pi}{3}, \frac{\pi}{2}, \frac{2\pi}{3}, \frac{4\pi}{3}, \frac{3\pi}{2}, \frac{5\pi}{3}$ **165.** $\frac{2\pi}{45}, \frac{14\pi}{15}$ **167.** $\frac{\pi}{6}, \frac{\pi}{3}, \frac{2\pi}{3}, \frac{5\pi}{6}, \frac{7\pi}{6}, \frac{4\pi}{3}, \frac{5\pi}{3}, \frac{11\pi}{6}$ **169.** $\frac{\pi}{2}, \pi$ **173.** $\frac{\sqrt{3}}{2}$ **175.** $\frac{\pi}{6}$ **177.** $\frac{\pi}{6}$ **179.** $\frac{\pi}{3}$ **181.** $\frac{2\pi}{3}$ **183.** 0 **185.** $\frac{2}{7}$ **187.** $\frac{-\sqrt{2}}{2}$ **189.** $\frac{3\sqrt{2}}{2}$ **191.** $\frac{\sqrt{3}}{2}$ **193.** $\frac{17}{7}$ **195.** $\frac{-4}{3}$ **197.** $\frac{3\sqrt{10}}{10}$ **199.** $\frac{\sqrt{2}}{2}$

Exercise Set 9.1

1. (a) yes; (b) no; (c) yes; (d) yes **3.** yes **5.** no **7.** yes **9.** $(-6, \frac{1}{2})$ **11.** $(-\frac{5}{2}, \frac{13}{2})$ **13.** $(\frac{1}{5}, \frac{7}{10})$ **15.** $(-\frac{60}{13}, \frac{60}{13})$ **17.** $(\sqrt{6}, 1)$ **19.** $(-\frac{2}{9}, \frac{23}{27})$ **21.** $(-\frac{283}{242}, -\frac{3}{121})$ **23.** $(-\frac{226}{25}, -\frac{939}{50})$ **25.** $(\frac{5}{8}, 0)$ **27.** $\left(\frac{4\sqrt{3} - \sqrt{10} - 5\sqrt{2}}{4}, \frac{-3 - \sqrt{5}}{2}\right)$ **29.** $b = 3, c = 4$ **31.** $A = -28$, $B = -22$ **33.** $\frac{49}{20}$ **35.** 10%: 80 cm³; 35%: 120 cm³ **37.** 8 lb of each **39.** $\left(\frac{ab}{a + b}, \frac{ab}{a + b}\right), a \neq -b$ **41.** $\left(\frac{a + b}{ab}, \frac{-1}{ab}\right), ab \neq 0$ **43.** $(-\frac{5}{19}, -\frac{15}{58})$ **45.** $(w, z) = (5, 2)$ **47.** $(-\frac{1}{11}, \frac{1}{9})$ **49.** $(\frac{5}{2}, \frac{1}{2})$ **51.** 59 **53.** $l = 12$ in., $w = 5$ in. **55.** (a) (60, 40); (b) (60, 40) **57.** $(0, a)$ **59.** $(\frac{1}{b}, \frac{1}{a})$ **61.** $(-ab, a+b)$, **63.** $\left(\frac{ce - bf}{ae - bd}, \frac{af - cd}{ae - bd}\right)$

Exercise Set 9.2

1. $(-3, -2, -1)$ **3.** $(-\frac{1}{60}, -\frac{2}{15}, \frac{3}{5})$ **5.** $(\frac{25}{36}, \frac{5}{9}, -\frac{1}{3})$ **7.** $(x, \frac{x}{8}, 0)$, x is any real number **9.** $(-1, 0, 1, -5)$ **11.** $(\frac{11}{3}, \frac{8}{3}, \frac{17}{3})$ **13.** $(1, 0, 1)$ **15.** $(2, 3, 1)$ **17.** $(1, 1, -1)$ **19.** $(-\frac{1}{2}, \frac{5}{2}, \frac{5}{2})$ **21.** $\left(\frac{11 - 5z}{7}, \frac{-3z - 6}{7}, z\right)$, z is any real number **23.** $(4, 1, -3, 2)$ **25.** $\left(\frac{17 - 17z}{5}, \frac{8z - 3}{5}, z\right)$, z is any real number **27.** $\left(\frac{12 + 10w}{11}, \frac{146 + 19w}{55}, \frac{159 + 61w}{55}, w\right)$, w is any real number **29.** $\left(-\frac{5z}{12}, \frac{2z + 3}{3}, z\right)$, z is any real number **31.** $(1, 1, -1)$ **33.** $\left(\frac{b + c}{2}, \frac{a - c}{2}, \frac{a - b}{2}\right)$ **35.** $A = \frac{1}{16}$, $B = -\frac{1}{16}, C = -\frac{1}{4}$ **37.** $A = \frac{3}{64}$, $B = -\frac{3}{64}, C = -\frac{3}{8}$ **39.** $A = \frac{1}{3}, B = \frac{1}{3}$, $C = -\frac{1}{3}, D = \frac{1}{3}$ **41.** $A = 4, B = 4$ **43.** $A = \frac{1}{2}, B = \frac{1}{2}, C = -\frac{1}{2}, D = \frac{1}{2}$ **47.** $a = -1, b = -4, c = -5$ **49.** $(x - 3)^2 + (y + 4)^2 = 25$; center: $(3, -4)$; radius: 5 **51.** $-1, 3$ **53.** $\left(\frac{a(b + c - a)}{2}, \frac{b(a + c - b)}{2}, \frac{c(a + b - c)}{2}\right)$ **55.** $\left(\frac{a + b + c - 2d}{3}, \frac{a + b + d - 2c}{3}, \frac{a + c + d - 2b}{3}, \frac{b + c + d - 2a}{3}\right)$ **57.** (a) $\left(\frac{a(\lambda + 1) - b - c}{(\lambda - 1)(\lambda + 2)}, \frac{b(\lambda + 1) - a - c}{(\lambda - 1)(\lambda + 2)}, \frac{c(\lambda + 1) - a - b}{(\lambda - 1)(\lambda + 2)}\right)$ (b) $\left(\frac{3z + 2c + b}{3}, \frac{3z + c - b}{3}, z\right)$, z is any real number (c) $(a - z - y, y, z)$, y and z are any real numbers **59.** $A = \frac{1}{2a}, B = -\frac{1}{2a}$ **61.** $A = \frac{ap + q}{2a}$, $B = \frac{ap - q}{2a}$ **63.** 60 mi

Exercise Set 9.3

1. (a) 2×3; (b) 3×2 **3.** 5×4 **5.** coefficient: $\begin{pmatrix} 5 & -1 & 1 \\ 0 & 4 & 2 \\ 3 & 1 & 1 \end{pmatrix}$; augmented: $\left(\begin{array}{ccc|c} 5 & -1 & 1 & 0 \\ 0 & 4 & 2 & 1 \\ 3 & 1 & 1 & -1 \end{array}\right)$ **7.** coefficient: $\begin{pmatrix} 8 & -8 & 0 \\ 1 & -1 & 1 \end{pmatrix}$; augmented: $\left(\begin{array}{ccc|c} 8 & -8 & 0 & 5 \\ 1 & -1 & 1 & 1 \end{array}\right)$ **9.** $(-1, -2, 3)$ **11.** $(-5, 1, 3)$ **13.** $(6, -12, 18)$ **15.** $(8, 9, -1)$ **17.** $\left(\frac{9z + 5}{19}, \frac{31z - 6}{19}, z\right)$, z is any real number **19.** $(2, -1, 0, 3)$ **21.** no solution **23.** $\begin{pmatrix} 3 & 2 \\ 2 & 4 \end{pmatrix}$ **25.** $\begin{pmatrix} 6 & 4 \\ 4 & 8 \end{pmatrix}$ **27.** $\begin{pmatrix} 11 & -2 \\ 11 & 1 \end{pmatrix}$ **29.** $\begin{pmatrix} 2 & 3 \\ -1 & 4 \end{pmatrix}$ **31.** undefined **33.** $\begin{pmatrix} 10 & -2 \\ -8 & 0 \\ 4 & 6 \end{pmatrix}$ **35.** $\begin{pmatrix} 2 & 4 & 11 \\ -12 & 16 & 19 \\ 14 & 12 & 43 \end{pmatrix}$ **37.** $\begin{pmatrix} -9 & 10 & 10 \\ 4 & -8 & -12 \\ 10 & 4 & 21 \end{pmatrix}$ **39.** undefined **41.** $\begin{pmatrix} 0 & 0 & 0 \\ 0 & 0 & 0 \\ 0 & 0 & 0 \end{pmatrix}$ **43.** $\begin{pmatrix} 4 & 2 \\ 2 & 5 \end{pmatrix}$ **45.** undefined **47.** $\begin{pmatrix} 1 & 18 \\ -6 & 13 \end{pmatrix}$ **49.** $\begin{pmatrix} -16 & 75 \\ -25 & 34 \end{pmatrix}$ **51.** $\begin{pmatrix} \frac{1}{2} & \frac{3}{2} & 0 \\ \frac{13}{2} & \frac{13}{2} & -\frac{1}{2} \end{pmatrix}$ **53.** $\begin{pmatrix} \frac{11}{2} & \frac{3}{2} & \frac{7}{2} \\ -\frac{7}{2} & 6 & \frac{15}{2} \end{pmatrix}$ **55.** (a) $\begin{pmatrix} -13 & 1 & 40 \\ 43 & 17 & 0 \\ 89 & 61 & 60 \end{pmatrix}$; (b) $\begin{pmatrix} -13 & 1 & 40 \\ 43 & 17 & 0 \\ 89 & 61 & 60 \end{pmatrix}$; (c) $\begin{pmatrix} -52 & -82 & 61 \\ 87 & 141 & 0 \\ 216 & 318 & 165 \end{pmatrix}$; (d) $\begin{pmatrix} -52 & -82 & 61 \\ 87 & 141 & 0 \\ 216 & 318 & 165 \end{pmatrix}$ **57.** (a) $\begin{pmatrix} 16 & 20 \\ 24 & 28 \end{pmatrix}$; (b) $\begin{pmatrix} 18 & 26 \\ 18 & 26 \end{pmatrix}$; (c) $\begin{pmatrix} 14 & 14 \\ 30 & 30 \end{pmatrix}$; (d) $\begin{pmatrix} 18 & 26 \\ 18 & 26 \end{pmatrix}$ **59.** (a) $\begin{pmatrix} x \\ -y \end{pmatrix}$; (b) $\begin{pmatrix} -x \\ y \end{pmatrix}$; (c) $\begin{pmatrix} -x \\ -y \end{pmatrix}$ **63.** (c) $\begin{pmatrix} 4 & 20 \\ \frac{11}{2} & 27 \end{pmatrix}$

Exercise Set 9.4

1. (a) 29; (b) -29 **3.** (a) 0; (b) 0 **5.** -1 **7.** -60 **9.** 9 **11.** (a) 314; (b) 674; (c) part (b) **13.** (a) 0; (b) 0; (c) 0; (d) 0 **15.** 0 **17.** -3 **19.** 6848 **21.** 17120 **23.** $(y - x)(z - x)(z - y)$ **25.** xy **27.** $y = 3x - 27$ **29.** $y = \frac{\sqrt{3}}{3}x$ **31.** -300 **33.** 7840 **35.** (a) $(0, 0, 0)$; (b) -1; (c) $(0, 0, 0)$ **37.** $(-1, 2, -2)$ **39.** $(1, -3, -2)$ **41.** $(2 - w, 2 + 2w, w)$, w is any real number **43.** no solution **45.** $(2, -1, 3, 1)$ **47.** $x = 1, -5$ **59.** (a) 1; (b) 1 **63.** $\left(\frac{k(k-b)(k-c)}{a(a-b)(a-c)}, \frac{k(k-a)(k-c)}{b(b-a)(b-c)}, \frac{k(k-a)(k-b)}{c(c-a)(c-b)}\right)$

Exercise Set 9.5

1. $(0, 0), (3, 0)$ **3.** $(2\sqrt{6}, 1), (-2\sqrt{6}, 1)$ **5.** $(-1, -1)$ **7.** $(2, 3), (2, -3), (-2, 3), (-2, -3)$ **9.** $(1, 0), (-1, 0)$ **11.** $\left(\frac{-1 + \sqrt{65}}{8}, \frac{1 + \sqrt{65}}{2}\right)$, $\left(\frac{-1 - \sqrt{65}}{8}, \frac{1 - \sqrt{65}}{2}\right)$ **13.** $\left(\frac{\sqrt{17}}{17}, 1\right)$, $\left(\frac{\sqrt{17}}{17}, -1\right)$, $\left(-\frac{\sqrt{17}}{17}, 1\right)$, $\left(-\frac{\sqrt{17}}{17}, -1\right)$ **15.** $(1, 0), (4, -\sqrt{3})$ **17.** $(2, 4)$ **19.** $(100, 1000), (100, \frac{1}{1000}), (\frac{1}{100}, 1000), (\frac{1}{100}, \frac{1}{1000})$ **21.** $\left(\frac{2 \ln 2 - 5 \ln 3}{\ln 2 - \ln 3}, \frac{3 \ln 2}{\ln 2 - \ln 3}\right)$ **25.** $(\frac{1}{a}, \frac{1}{b})$ **27.** $(9, 14), (14, 9)$ **29.** width: $\frac{p - \sqrt{2d^2 - p^2}}{2}$; length: $\frac{p + \sqrt{2d^2 - p^2}}{2}$ **31.** $(5, 4), (5, -4)$ **33.** Let $A = \sqrt{p^2 + q^2 + r^2}$. Solutions are: $\left(\frac{p^2}{A}, \frac{q^2}{A}, \frac{r^2}{A}\right)$; $\left(\frac{-p^2}{A}, \frac{-q^2}{A}, \frac{-r^2}{A}\right)$ **35.** 9 cm, 40 cm **37.** width: 3 cm; length: 20 cm **39.** $(1, 2), (-1, -2), (2, 1), (-2, -1)$ **41.** $\left(\frac{1 + \sqrt{13}}{2}, \frac{-1 + \sqrt{13}}{2}\right)$, $\left(\frac{-1 + \sqrt{13}}{2}, \frac{1 + \sqrt{13}}{2}\right)$, $\left(\frac{1 - \sqrt{13}}{2}, \frac{-1 - \sqrt{13}}{2}\right)$, $\left(\frac{-1 - \sqrt{13}}{2}, \frac{1 - \sqrt{13}}{2}\right)$ **43.** $(2, 3), (-2, -3), (1, 2), (-1, -2)$ **45.** $(e^{9/2}, e^3)$

Exercise Set 9.6

1. (a) no; (b) yes **3.** (a) yes; (b) no

5.

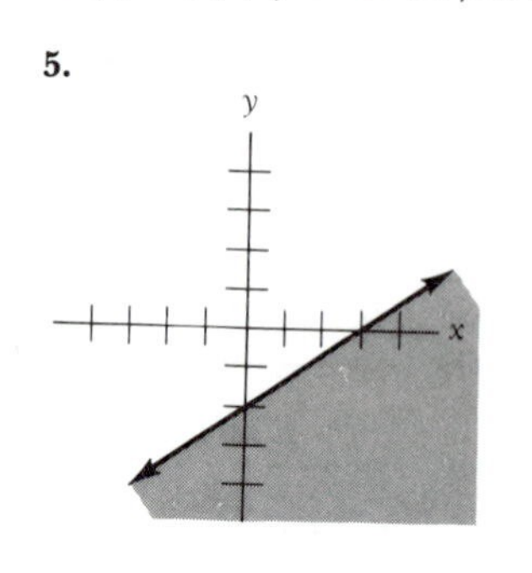

7.

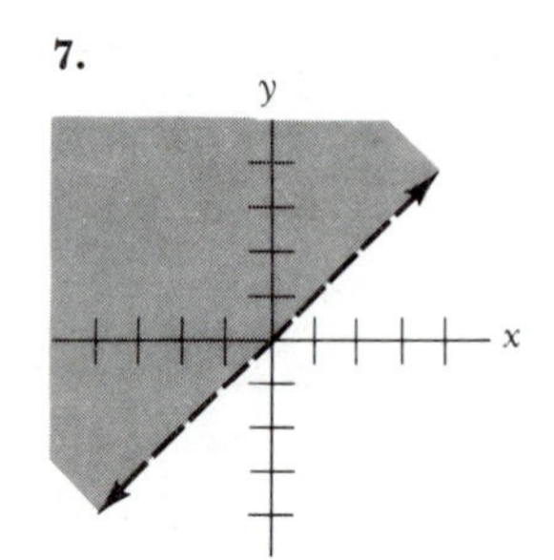

9.

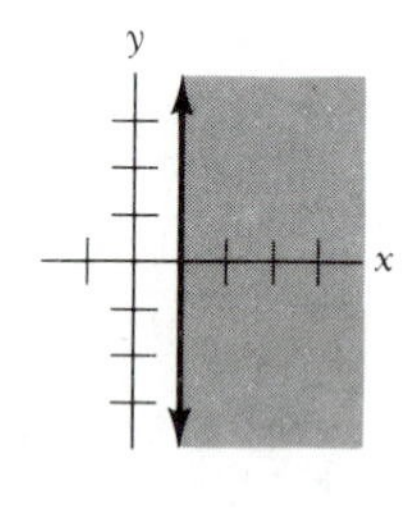

11.

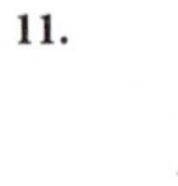

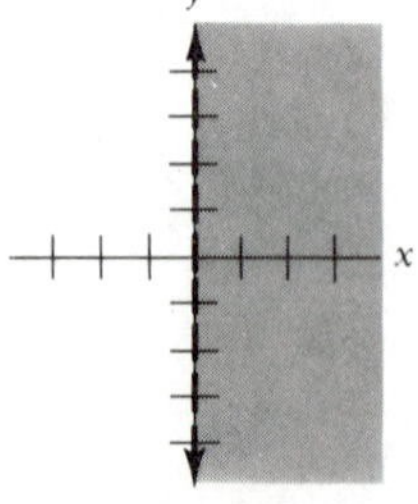

13.

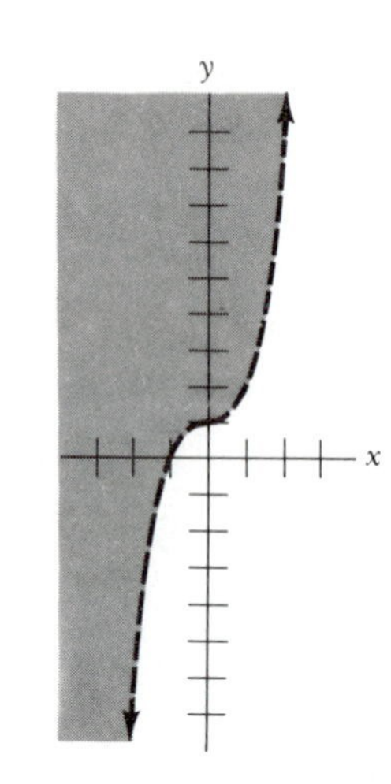

15.

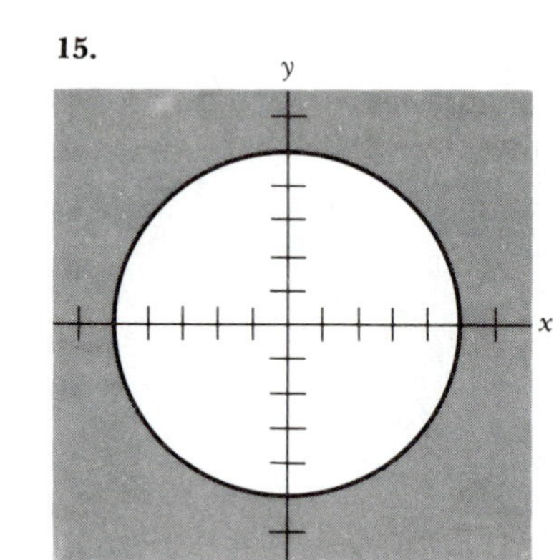

17.

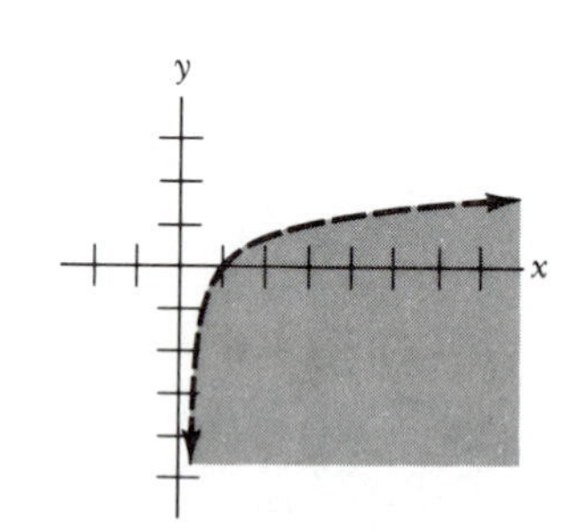

19.

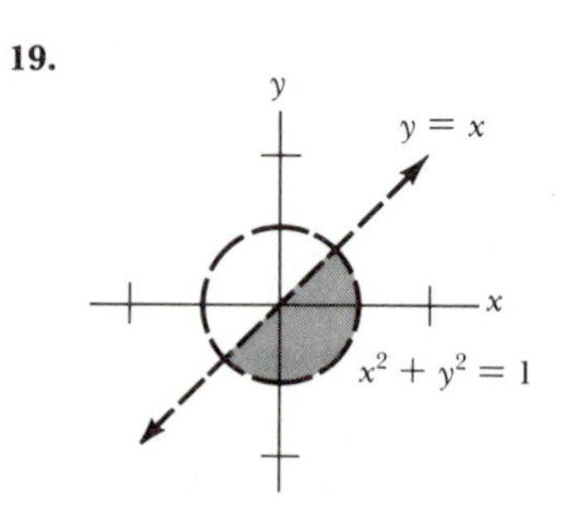

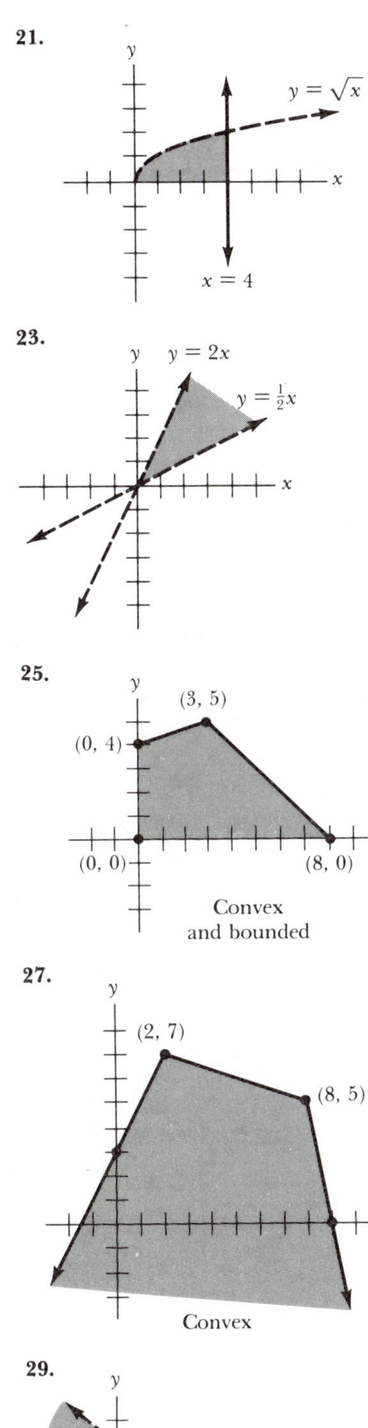

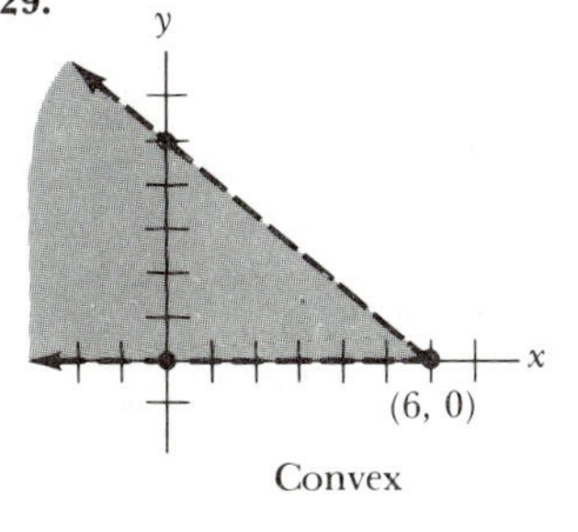

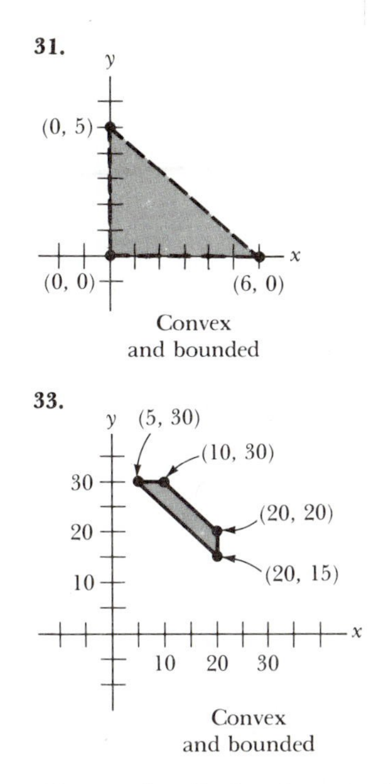

35. max: 9 at (3, 4); min: −5 at (5, 0) **37.** max: 4 at (1, 2); min: −4 at (−1, −2) **39.** max: 113 at (6, 6); min: 27 at (2, 1) **41.** max: 188 at (6, 7); min: 100 at (0, 0)
43. (a) max: 819 at (1, 8); min: 0 at (0, 0); (b) max: 826 at (6, 7); min: 0 at (0, 0) **45.** (a) max: 12 at (9, 3); (b) (0, −100); (c) (0, −1000); (d) $(0, -10^6)$
47. W_1 to R_1: 55 televisions; W_2 to R_2: 75 televisions; cost: \$965 **49.** A: 800 units; B: 350 units; profit: \$760
51. \$23,000 when 100 acres of cherry tomatoes and 500 acres of regular tomatoes are planted **53.** A: \$1000; B: \$4000; C: \$5000; profit: \$540
55. A: \$1500; B: \$7500; C: \$1000; return: \$645 **57.** \$1100 when 50 small and 25 large are produced

Chapter 9 Review Exercises

1. (3, −5) **3.** (−1, 4) **5.** (−3, 15)
7. $\left(-\frac{18}{5}, \frac{8}{5}\right)$ **9.** $\left(\frac{2}{3}, -\frac{1}{5}\right)$ **11.** (−12, −8)
13. $\left(\frac{1}{3}, -\frac{1}{4}\right)$
15. $\left(\frac{-1}{a^2 - 3a + 1}, \frac{1 - a}{a^2 - 3a + 1}\right)$, $a^2 - 3a + 1 \neq 0$ **17.** $(a^2, 1 - a^2)$
19. $(a^2 - b^2, a^2 + b^2)$
21. $\left(\frac{pq(p + q)}{p^2 + q^2}, \frac{p^3 - q^3}{p^2 + q^2}\right)$, $(p, q) \neq (0, 0)$
23. $\left(\frac{a}{a - b}, \frac{b}{a + b}\right)$, $ab \neq 0$, $9a \neq 8b$
25. (2, 3, 4) **27.** (−1, −2, 0) **29.** no solution **31.** $(x, 6 - 2x, -1)$, x is any real number **33.** no solution
35. $(2b - z, a - b, z)$, z is any real number **37.** (4, 3, −1, 2) **39.** $A = \frac{1}{20}$, $B = -\frac{1}{20}$ **41.** $A = 2$, $B = -2$
43. $A = -\frac{5}{4}$, $B = \frac{5}{4}$ **45.** $A = \frac{1}{16}$, $B = -\frac{1}{16}$, $C = -\frac{1}{4}$ **47.** $A = 3$, $B = 1$, $C = 0$ **49.** $A = \frac{1}{48}$, $B = -\frac{1}{48}$, $C = \frac{1}{6}$
51. $A = 0$, $B = 1$, $C = a$ **53.** $A = b$, $B = -a$ **55.** 34 **57.** 0 **59.** 0 **61.** 24
63. $3abc - a^3 - b^3 - c^3$
65. $a^2 + b^2 + c^2 + d^2 + x$ **67.** 31 and 21 **69.** (a) $\left(\frac{2}{3}, 2\right)$; (b) $\left(\frac{2}{3}, 2\right)$; (c) $\left(\frac{2}{3}, 2\right)$; (d) They all equal 2.
73. $\left(x + \frac{17}{6}\right)^2 + \left(y + \frac{8}{3}\right)^2 = \frac{245}{36}$
75. $\begin{pmatrix} 10 & -2 \\ 4 & 26 \end{pmatrix}$ **77.** $\begin{pmatrix} 8 & 4 \\ 4 & 32 \end{pmatrix}$
79. $\begin{pmatrix} 4 & -13 \\ 7 & 41 \end{pmatrix}$ **81.** $\begin{pmatrix} -3 & -14 \\ -4 & 3 \end{pmatrix}$
83. $\begin{pmatrix} 1 & 1 \\ 1 & 7 \end{pmatrix}$ **85.** $\begin{pmatrix} 1 & -11 \\ 6 & 36 \end{pmatrix}$
87. $\begin{pmatrix} 4 & 3 \\ 10 & 33 \end{pmatrix}$ **89.** $\begin{pmatrix} -42 & 58 \\ 5 & 20 \end{pmatrix}$
91. undefined **93.** undefined
95. $\begin{pmatrix} 4 & -1 \\ 2 & 12 \end{pmatrix}$ **97.** $\begin{pmatrix} -4 & 13 \\ -7 & -41 \end{pmatrix}$
99. $\begin{pmatrix} 13 & -1 \\ 5 & 35 \end{pmatrix}$ **101.** $\begin{pmatrix} 2 & -14 \\ -2 & -4 \\ -20 & 2 \end{pmatrix}$
103. $\begin{pmatrix} 0 & 0 \\ 0 & 0 \\ 0 & 0 \end{pmatrix}$ **105.** $\begin{pmatrix} 2 & -26 \\ -2 & -42 \end{pmatrix}$
109. $X = \begin{pmatrix} 6 & 0 & 2 \\ 5 & 5 & 7 \end{pmatrix}$
111. $\begin{pmatrix} \frac{3}{2} & -\frac{1}{2} & \frac{7}{2} \\ -\frac{11}{2} & -\frac{7}{2} & -4 \end{pmatrix}$
113. $\left(-\frac{3}{5}, -\frac{13}{20}, \frac{31}{20}\right)$ **115.** no solution
117. $(-5 - 4y, y, -8 - 5y)$, y is any real number **119.** (4, 1, −1, 3) **121.** (0, 0) and (6, 36) **123.** (3, 0) and (−3, 0)
125. $\left(\frac{5\sqrt{2}}{2}, \frac{\sqrt{14}}{2}\right)$, $\left(-\frac{5\sqrt{2}}{2}, \frac{\sqrt{14}}{2}\right)$, $\left(\frac{5\sqrt{2}}{2}, -\frac{\sqrt{14}}{2}\right)$, and $\left(-\frac{5\sqrt{2}}{2}, -\frac{\sqrt{14}}{2}\right)$
127. $\left(\frac{-1 + \sqrt{5}}{2}, \frac{\sqrt{-2 + 2\sqrt{5}}}{2}\right)$
129. $\left(\frac{\sqrt{-2 + 2\sqrt{17}}}{4}, \frac{-1 + \sqrt{17}}{4}\right)$ and $\left(-\frac{\sqrt{-2 + 2\sqrt{17}}}{4}, \frac{-1 + \sqrt{17}}{4}\right)$
131. $(\sqrt{2}, 3\sqrt{2})$, $(-\sqrt{2}, -3\sqrt{2})$, $\left(\frac{7\sqrt{22}}{33}, \frac{31\sqrt{22}}{33}\right)$, and $\left(-\frac{7\sqrt{22}}{33}, -\frac{31\sqrt{22}}{33}\right)$ **133.** (5, 2), (5, −4), (1, 2), and (1, −4) **135.** $\frac{as}{a + b}$

and $\frac{bs}{a+b}$ **137.** 24, 60, and 120
139. $\frac{\sqrt{n+2m}+\sqrt{n-2m}}{2}$ and $\frac{\sqrt{n+2m}-\sqrt{n-2m}}{2}$; $\frac{\sqrt{n-2m}-\sqrt{n+2m}}{2}$ and $\frac{-\sqrt{n-2m}-\sqrt{n+2m}}{2}$

141.

$y = 4x$ $y = x$ $x^2 + y^2 = 1$

143.

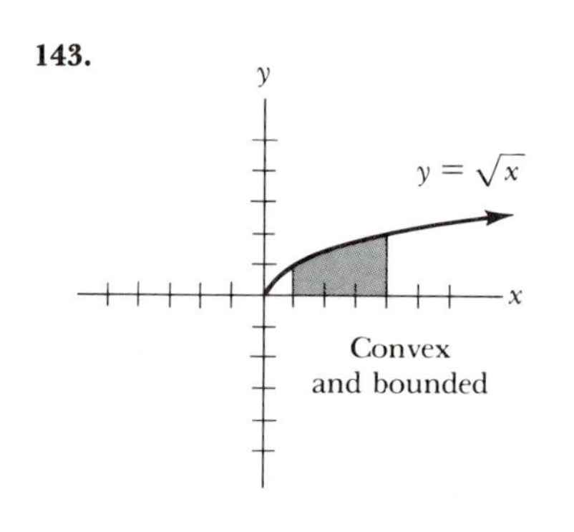

145.

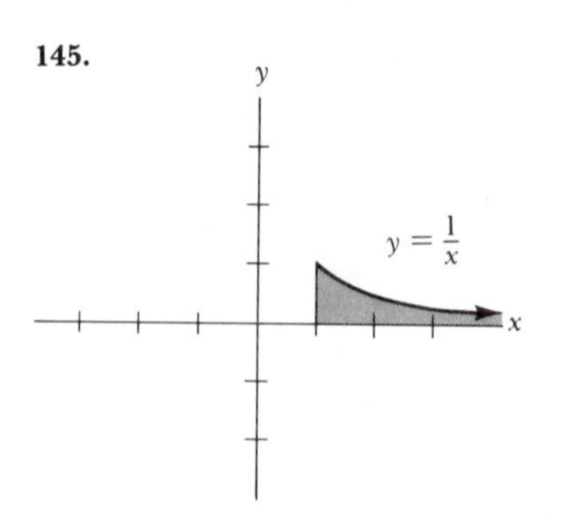

147. max: 16 at (4, 2); min: 0 at (0, 0) **149.** max: 384 at (9, 10); min: 49 at (0, 0) **151.** $34,000: 17 daytime and 0 evening

Exercise Set 10.1

1. quotient: $x - 1$; remainder: -7 **3.** quotient: $4x - 5$; remainder: 0 **5.** quotient: $6x^2 + 19x + 78$; remainder: 313 **7.** quotient: $x^2 + 2x + 4$; remainder: 7 **9.** quotient: $x^4 - 2x^3 + 4x^2 - 8x + 16$; remainder: -33 **11.** quotient: $x^3 - 10x^2 + 40x - 160$; remainder: 642 **13.** quotient: $x^2 + 6x + 57$; remainder: 576 **15.** quotient: $x^2 - 6x + 30$; remainder: -150 **17.** quotient: $54x^2 + 9x - 21$; remainder: 0 **19.** quotient: $x^3 + 3x^2 + 12x + 36$; remainder: 120 **21.** quotient: $ax + ar + b$; remainder: $ar^2 + br + c$ **23.** quotient: $x - \frac{4}{3}$; remainder: $-\frac{13}{9}$ **25.** $k = 4$

Exercise Set 10.2

1. yes **3.** yes **5.** yes **7.** yes **9.** yes **11.** no **13.** (a) yes; (b) no **15.** yes **17.** yes **19.** (a) 1, 2 (multiplicity 3), 3; (b) 1 (multiplicity 3); (c) 5 (multiplicity 6), -1 (multiplicity 4); (d) 0 (multiplicity 5), 1 **21.** -170 **23.** -9 **25.** $-3\sqrt{2} - 2$ **27.** -22 **29.** $-3, 4, 3$ **31.** $1, -1 + \sqrt{6}, -1 - \sqrt{6}$ **33.** $-2, \frac{2}{3}, 3$ **35.** $-\frac{3}{2}, \frac{1+\sqrt{5}}{2}, \frac{1-\sqrt{5}}{2}$ **37.** 0, 5 **39.** $-4, 3$ **41.** $-9, 1 + \sqrt{2}, 1 - \sqrt{2}$ **43.** $x^3 - 4x^2 - 17x + 60 = 0$ **45.** $x^3 + 8x^2 + 13x + 6 = 0$ **47.** $c = 5, d = -2$ **49.** $a = -1, b = -1$ **51.** $b = -\frac{3\sqrt{2}}{2}$ or $b = \frac{3\sqrt{2}}{2}$ **53.** 20.44 **55.** (a) -0.05; (b) 0.07

Exercise Set 10.3

1. (a) yes; (b) yes; (c) yes; (d) no **3.** $(x - (-1))(x - 3)$ **5.** $4(x - \frac{1}{4})(x - (-6))$ **7.** $(x - (-\sqrt{5}))(x - \sqrt{5})$ **9.** $(x - (5 + i))(x - (5 - i))$ **11.** $x^3 + x^2 - 5x + 3$ **13.** $(x - 2)(x - (-2))(x - 2i)(x - (-2i))$ **15.** $(x - \sqrt{3})^2(x - (-\sqrt{3}))^2(x - 4i) \times (x - (-4i))$ **17.** $f(x) = -\frac{5}{42}x^2 + \frac{25}{42}x + \frac{30}{7}$ **19.** $f(x) = \frac{1}{30}x^3 - \frac{19}{30}x + 1$ **21.** $x^2 + (i + \sqrt{3})x + i\sqrt{3} = 0$ **23.** $x^2 - 3x - 54 = 0$ **25.** $x^2 - 2x - 4 = 0$ **27.** $x^2 - 2ax + a^2 - b = 0$ **31.** $x = 4, 3, -3$ **35.** (a) false; (b) true; (c) false; (d) true; (e) false; (f) true; (g) true; (h) true; (i) true; (j) false; (k) false; (l) false; (m) true

Exercise Set 10.4

1. $\pm 1, \pm\frac{1}{2}, \pm\frac{1}{4}, \pm 3, \pm\frac{3}{2}, \pm\frac{3}{4}$ **3.** $\pm 1, \pm\frac{1}{2}, \pm\frac{1}{4}, \pm\frac{1}{8}, \pm 3, \pm\frac{3}{2}, \pm\frac{3}{4}, \pm\frac{3}{8}, \pm 9, \pm\frac{9}{2}, \pm\frac{9}{4}, \pm\frac{9}{8}$ **5.** $\pm 1, \pm\frac{1}{2}, \pm 2, \pm 3, \pm\frac{3}{2}, \pm 6$ **13.** $x = 1, -1, -3$ **15.** $x = -\frac{1}{4}, -\sqrt{5}, \sqrt{5}$ **17.** $x = -1, -\frac{2}{3}, -\frac{1}{3}$ **19.** $x = 1, -1, \frac{-1+\sqrt{97}}{2}, \frac{-1-\sqrt{97}}{2}$ **21.** $x = 1$ **23.** $x = \frac{1}{2}, 6, -4$ **25.** $x = \frac{1}{4}, \frac{1}{5}, -\frac{\sqrt{6}}{2}, \frac{\sqrt{6}}{2}$ **27.** (a) upper: 2; lower: -1; (b) upper: $\frac{4}{3}$; lower: $-\frac{2}{3}$; (c) upper: 6; lower: -2 **31.** $0.68 < x < 0.69$ **33.** $2.88 < x < 2.89$ **35.** $4.31 < x < 4.32$ **37.** $-2.15 < x < -2.14$ **39.** $-5.27 < x < -5.26$ **41.** $p = 2$ **43.** $b = 2$ **45.** $x = 1.32$

$y = x^3$ $y = x + 1$

47. (a) $a = 1.27$; (b) $a = \frac{\sqrt{2 + 2\sqrt{5}}}{2} = 1.272\ldots$

Exercise Set 10.5

1. $7 + 2i$ **3.** $5 - 2i, 3$ **5.** $-2 - i, -3$ **7.** $6 + 5i, -\frac{1}{4}$ **9.** $4 - \sqrt{2}i, -\frac{3i}{2}, \frac{3i}{2}$ **11.** $10 - 2i, 1 + \sqrt{5}, 1 - \sqrt{5}$ **13.** $\frac{1 - \sqrt{2}i}{3}, \frac{2}{5}$ **15.** $3 + 2i, -1 - i, -1 + \sqrt{2}, -1 - \sqrt{2}$ **17.** $x^2 - 2x - 5 = 0$ **19.** $x^2 - \frac{4}{3}x - \frac{2}{3} = 0$ **21.** 2 complex and 1 negative real **23.** 4 complex and 1 negative real **25.** 2 complex, 1 positive real, and 1 negative real **27.** 1 positive real and 2 negative real, or 1 positive real and 2 complex **29.** 1 positive real, 1 negative real, and 6 complex **31.** 1 positive real and 8 complex

33. 1 positive real, 1 negative real, and 6 complex **35.** 1 positive real, 1 negative real, and 4 complex **37.** 1 positive real, 1 negative real, and 2 complex **39.** $f(x) = x^4 + 2x^2 + 49$

Chapter 10 Review Exercises

1. $q(x) = x^3 + x^2 - 3x + 1$; $R(x) = -1$ **3.** quotient: $x^3 + 3x^2 + 7x + 21$; remainder: 71 **5.** quotient: $2x^2 - 13x + 46$; remainder: -187 **7.** quotient: $5x - 20$; remainder: 0 **9.** 99,904 **11.** $-\frac{999}{1000}$ **13.** a^3 **15.** (a) -0.24; (b) -0.007; (c) 0.00003 **17.** $a = -5$ **19.** $a = -1, -2$ **23.** $\pm1, \pm2, \pm3, \pm6, \pm9, \pm18$ **25.** $\pm1, \pm\frac{1}{2}, \pm2, \pm4, \pm8$ **27.** $\pm1, \pm p$ **29.** $x = 2, -\frac{3}{2}, -1$ **31.** $x = \frac{5}{2}, -1 + \sqrt{3}, -1 - \sqrt{3}$ **33.** $x = \frac{2}{3}, \dfrac{-1 + i\sqrt{3}}{2}, \dfrac{-1 - i\sqrt{3}}{2}$ **35.** $x = -1, -\sqrt{7}, \sqrt{7}$ **37.** $x = 2, 5$ **39.** (a) See beginning of Section 10.1; (b) See Section 10.2; (c) See Section 10.2; (d) See beginning of Section 10.3. **41.** $6(x - \frac{1}{3})(x - (-\frac{5}{2}))$ **43.** $(x - 4)(x - (-\sqrt[3]{5})) \times \left(x - \left(\dfrac{\sqrt[3]{5} + i\sqrt{3\sqrt[3]{25}}}{2}\right)\right) \times \left(x - \left(\dfrac{\sqrt[3]{5} - i\sqrt{3\sqrt[3]{25}}}{2}\right)\right)$ **45.** $x = 2 - 3i, 2 + 3i, 3$ **47.** $x = 1 + i\sqrt{2}, 1 - i\sqrt{2}, \sqrt{7}, -\sqrt{7}$ **49.** 1 positive real and 2 complex **51.** 1 negative real and 2 complex **53.** 1 positive real, 1 negative real, and 2 complex **55.** (c) $x = -1, i, -i$ **57.** $0.82 < x < 0.83$ **59.** (c) $6.93 < x < 6.94$ **61.** $x^2 - 8x + 11 = 0$ **63.** $x^4 - 12x^3 + 35x^2 + 60x - 200 = 0$ **65.** $x^4 - 4x^3 - 4x^2 + 16x - 8 = 0$

Exercise Set 11.1

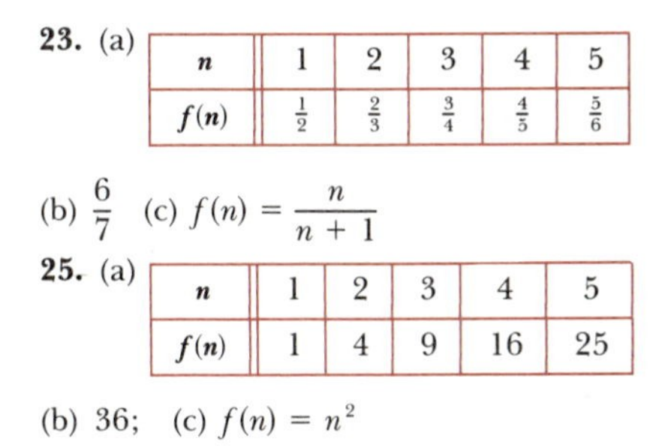

23. (a)

n	1	2	3	4	5
$f(n)$	$\frac{1}{2}$	$\frac{2}{3}$	$\frac{3}{4}$	$\frac{4}{5}$	$\frac{5}{6}$

(b) $\frac{6}{7}$ (c) $f(n) = \dfrac{n}{n+1}$

25. (a)

n	1	2	3	4	5
$f(n)$	1	4	9	16	25

(b) 36; (c) $f(n) = n^2$

Exercise Set 11.2

3. $a^9 + 9a^8b + 36a^7b^2 + 84a^6b^3 + 126a^5b^4 + 126a^4b^5 + 84a^3b^6 + 36a^2b^7 + 9ab^8 + b^9$ **5.** $8A^3 + 12A^2B + 6AB^2 + B^3$ **7.** $1 - 12x + 60x^2 - 160x^3 + 240x^4 - 192x^5 + 64x^6$ **9.** $x^2 + 4x\sqrt{xy} + 6xy + 4y\sqrt{xy} + y^2$ **11.** $x^{10} + 5x^8y^2 + 10x^6y^4 + 10x^4y^6 + 5x^2y^8 + y^{10}$ **13.** $1 - \dfrac{6}{x} + \dfrac{15}{x^2} - \dfrac{20}{x^3} + \dfrac{15}{x^4} - \dfrac{6}{x^5} + \dfrac{1}{x^6}$ **15.** $\dfrac{x^3}{8} - \dfrac{x^2y}{4} + \dfrac{xy^2}{6} - \dfrac{y^3}{27}$ **17.** $a^7b^{14} + 7a^6b^{12}c + 21a^5b^{10}c^2 + 35a^4b^8c^3 + 35a^3b^6c^4 + 21a^2b^4c^5 + 7ab^2c^6 + c^7$ **19.** $x^8 + 8\sqrt{2}x^7 + 56x^6 + 112\sqrt{2}x^5 + 280x^4 + 224\sqrt{2}x^3 + 224x^2 + 64\sqrt{2}x + 16$ **21.** $5\sqrt{2} - 7$ **23.** $89\sqrt{3} + 109\sqrt{2}$ **25.** $12 - 24\sqrt[3]{2} + 12\sqrt[3]{4}$ **27.** $x^{10} - 10x^9 + 35x^8 - 40x^7 - 30x^6 + 68x^5 + 30x^4 - 40x^3 - 35x^2 - 10x - 1$ **29.** 120 **31.** 105 **33.** (a) 10; (b) 5 **35.** $n^2 + 3n + 2$ **37.** 0 **39.** $120a^2b^{14}$ **41.** $100x^{99}$ **43.** 45 **45.** 294,912 **47.** 28 **49.** 40,095 **53.** (a)

k	0	1	2	3	4	5	6	7	8
$\binom{8}{k}$	1	8	28	56	70	56	28	8	1

(b) 1 **19.** (a) $\frac{5}{6}$; (b) 1 **21.** $\frac{3}{8}$ **23.** $\frac{1}{2}$ **25.** $\frac{1}{4}$ **27.** $\frac{5}{8}$ **29.** $\frac{3}{4}$ **31.** 1 **33.** $\frac{5}{8}$ **35.** $\frac{1}{2}$ **37.** 0 **39.** $\frac{1}{4}$ **41.** $\frac{1}{3}$ **43.** (a) $\frac{6}{7}$; (b) $\frac{36}{91}$ **45.** $\frac{7}{12}$ **47.** $\frac{2}{9}$ **49.** (a) $\frac{1}{8}$; (b) $\frac{3}{8}$; (c) $\frac{1}{2}$ **51.** $\frac{7}{10}$ **53.** $\frac{33}{66640}$

Exercise Set 11.3

1. $\frac{1}{2}, \frac{2}{3}, \frac{3}{4}, \frac{4}{5}, \frac{5}{6}$ **3.** 0, 1, 4, 9, 16 **5.** $2, \frac{9}{4}, \frac{64}{27}, \frac{625}{256}, \frac{7776}{3125}$ **7.** $-1, 1, -1, 1, -1$ **9.** $1, -\frac{1}{2}, \frac{1}{6}, -\frac{1}{24}, \frac{1}{120}$ **11.** 3, 6, 9, 12, 15 **13.** 1, 4, 25, 676, 458,329 **15.** 2, 2, 4, 8, 32 **17.** 1, 1, 2, 6, 24 **19.** 0, 1, 2, 4, 16 **21.** 62 **23.** 40 **25.** $-\frac{19}{30}$ **27.** 903 **29.** 278 **31.** 15 **33.** -35 **35.** $1 + 2x$ **37.** 121 **39.** $\log_{10}(120)$ **41.** $x^6 - x$ **43.** $\sum_{j=1}^{n} 5^j$ **45.** $\sum_{j=1}^{6} jx^j$ **47.** $\sum_{k=1}^{n} \frac{1}{k}$ **49.** $\sum_{j=3}^{10} \binom{10}{j}$ **51.** $\sum_{j=1}^{7} (-1)^{j+1}\frac{1}{2^j}$

Exercise Set 11.4

1. (a) 2; (b) -4; (c) $\frac{1}{3}$; (d) $\sqrt{2}$ **3.** 131 **5.** 501 **7.** 998 **9.** $a = -\frac{23}{2}$, $d = \frac{11}{6}$ **11.** -190 **13.** $-\frac{1}{8}$ **15.** 500,500 **17.** $\dfrac{91\pi}{3}$ **19.** $\dfrac{1}{2}$ **21.** $-\frac{104}{15}$ **23.** $d = \frac{6}{5}$, $a = -\frac{17}{5}$ **25.** 900 **27.** 2, 10, 18 or 18, 10, 2 **29.** -1, 2, 5 or 5, 2, -1 **31.** (b) $-3(2 + 3\sqrt{2})$

Exercise Set 11.5

1. 6 **3.** 20, 100 **5.** 1 **7.** $\frac{256}{6561}$ **9.** 4 **11.** 7161 **13.** $63 + 31\sqrt{2}$ **15.** $\frac{1995}{64}$ **17.** 0.011111 **19.** $\frac{2}{5}$ **21.** 101 **23.** $\frac{5}{9}$ **25.** $\frac{61}{495}$ **27.** $\frac{16}{37}$ **29.** $-2, -\frac{1}{2}$

Exercise Set 11.6

1. 6 **3.** 12 **5.** 12 **7.** 60 **9.** 108 **11.** 11,232,000 **13.** 3,628,800 **15.** 4 **17.** 8 **19.** 120 **23.** $P(101, 51)$ **25.** 240 **27.** 12 **29.** (a) 900; (b) 648; (c) 328 **31.** (a) 729; (b) 504; (c) 56; (d) 140 **33.** (a) 12; (b) 12 **35.** (a) 56; (b) 56 **37.** (a) 792; (b) 792 **39.** $\dfrac{n(n-1)}{2}$ **41.** 56 **43.** (a) 210; (b) 80 **45.** 9 **47.** 56 **49.** 15 **51.** (a) 2352; (b) 2649

Exercise Set 11.7

1. (a) {2, 4, 6}; (b) {4, 5, 6}; (c) {2, 4, 5, 6}; (d) {4, 6} **3.** (a) {1, 3, 5}; (b) {1, 2, 3, 4, 5, 6}; (c) {1, 2, 3, 4, 5, 6}; (d) ø **5.** no **7.** yes **9.** (a) 3; (b) 3; (c) 4 **11.** (a) 3; (b) 6; (c) 0 **13.** (a) $\frac{1}{2}$; (b) $\frac{1}{2}$ **15.** (a) $\frac{1}{2}$; (b) 1 **17.** (a) $\frac{2}{3}$; (b) 1 **19.** (a) $\frac{5}{6}$; (b) 1 **21.** $\frac{3}{8}$ **23.** $\frac{1}{2}$ **25.** $\frac{1}{4}$ **27.** $\frac{5}{8}$ **29.** $\frac{3}{4}$ **31.** 1 **33.** $\frac{5}{8}$ **35.** $\frac{1}{2}$ **37.** 0 **39.** $\frac{1}{4}$ **41.** $\frac{1}{3}$ **43.** (a) $\frac{6}{7}$; (b) $\frac{36}{91}$ **45.** $\frac{7}{12}$ **47.** $\frac{2}{9}$ **49.** (a) $\frac{1}{8}$; (b) $\frac{3}{8}$; (c) $\frac{1}{2}$ **51.** $\frac{7}{10}$ **53.** $\frac{33}{66640}$

Exercise Set 11.8

1, 3, 5, 7

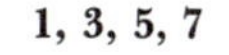

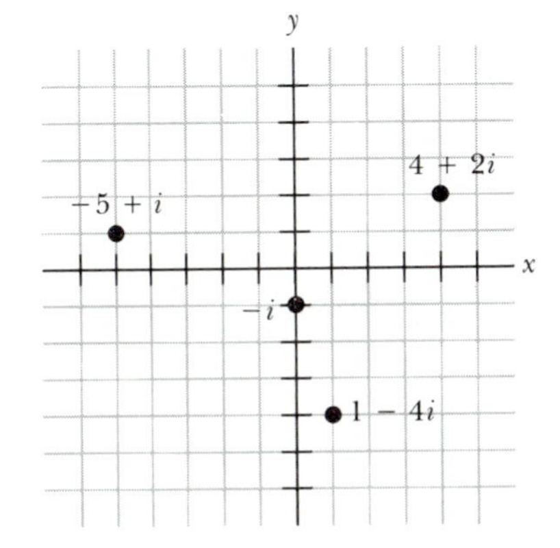

9. $\sqrt{2} + \sqrt{2}i$ **11.** $-2\sqrt{3} + 2i$ **13.** $-1 - i$ **15.** $\sqrt{3}i$ **17.** $(\sqrt{6} - \sqrt{2}) + (\sqrt{6} + \sqrt{2})i$ **19.** $\cos\dfrac{\pi}{6} + i\sin\dfrac{\pi}{6}$

21. $2\left(\cos\frac{2\pi}{3} + i\sin\frac{2\pi}{3}\right)$
23. $4\left(\cos\frac{7\pi}{6} + i\sin\frac{7\pi}{6}\right)$
25. $6\left(\cos\frac{3\pi}{2} + i\sin\frac{3\pi}{2}\right)$
27. $\frac{1}{2}\left(\cos\frac{11\pi}{6} + i\sin\frac{11\pi}{6}\right)$
29. $3 + 3\sqrt{3}i$ **31.** $1 - \sqrt{3}i$
33. $3\sqrt{2}\left(\cos\frac{2\pi}{7} + i\sin\frac{2\pi}{7}\right)$
35. $\frac{3\sqrt{2}}{2} + \frac{3\sqrt{2}}{2}i$ **37.** $\sqrt{3} + i$
39. 1 **41.** $\frac{243}{2} - \frac{243\sqrt{3}}{2}i$
43. $\frac{\sqrt{2}}{128} + \frac{\sqrt{2}}{128}i$ **45.** $-4i$ **47.** $-1 - i$
49. $\frac{1}{2} + \frac{\sqrt{3}}{2}i$ **51.** $128 + 128i$
53. $-128 + 128\sqrt{3}i$ **55.** $3i$, $\frac{-3\sqrt{3}}{2} - \frac{3}{2}i$, $\frac{3\sqrt{3}}{2} - \frac{3}{2}i$ **57.** ± 1, $\pm i$, $\frac{\sqrt{2}}{2} \pm \frac{\sqrt{2}}{2}i$, $\frac{-\sqrt{2}}{2} \pm \frac{\sqrt{2}}{2}i$
59. $\frac{\sqrt{2}}{2} + \frac{\sqrt{2}}{2}i$, $\frac{-\sqrt{2}}{2} - \frac{\sqrt{2}}{2}i$
61. 4, $-2 + 2\sqrt{3}i$, $-2 - 2\sqrt{3}i$
63. ± 3, $\frac{3}{2} \pm \frac{3\sqrt{3}}{2}i$, $-\frac{3}{2} \pm \frac{3\sqrt{3}}{2}i$
65. (a) 1, $-\frac{1}{2} \pm \frac{\sqrt{3}}{2}i$ **67.** -1
69. 1 **77.** 92,236,816
79. $0.95 + 0.31i$, i, $-0.95 + 0.31i$, $-0.59 - 0.81i$, $0.59 - 0.81i$

Chapter 11 Review Exercises

11. $81a^4 + 108a^3b^2 + 54a^2b^4 + 12ab^6 + b^8$ **13.** $x^4 + 4x^3\sqrt{x} + 6x^3 + 4x^2\sqrt{x} + x^2$ **15.** $x^{10} - 10x^8y^2 + 40x^6y^4 - 80x^4y^6 + 80x^2y^8 - 32y^{10}$
17. $1 + \frac{5}{x} + \frac{10}{x^2} + \frac{10}{x^3} + \frac{5}{x^4} + \frac{1}{x^5}$
19. $a^4b^2 - 4a^3b^2\sqrt{ab} + 6a^3b^3 - 4a^2b^3\sqrt{ab} + a^2b^4$ **21.** $15xy^8$ **23.** 84
25. 28 **33.** $1, \frac{4}{3}, \frac{3}{2}, \frac{8}{5}, \frac{5}{3}$ **35.** $-\frac{1}{2}, \frac{4}{9}, -\frac{27}{64}, \frac{256}{625}, -\frac{3125}{7776}$ **37.** $-3, -12, -48, -192, -768$ **39.** 45 **41.** $\sum_{k=1}^{5} \frac{5}{3^k}$ **43.** 57
45. $\frac{5}{1024}$ **47.** 14,350
49. 7,777,777,777 **51.** $\frac{3}{4}$ **53.** $\frac{1}{10}$
55. $\frac{5}{11}$ **61.** (a) 42,929; (b) 42,925; 0.00932% error; (c) 6.48040×10^{10}; (d) 2×10^{-13}% error **63.** 16 **65.** 24 permutations
67. 19,958,400 **69.** 480 **71.** 54
73. 120 **75.** (a) $\frac{7}{15}$; (b) $\frac{8}{15}$; (c) $\frac{2}{3}$; (d) $\frac{2}{3}$ **77.** (a) 252; (b) $\frac{7}{25}$ **79.** $\frac{1}{18}$ **81.** $\frac{1}{6}$
83. $\frac{2}{3}$ **85.** $\frac{1}{6}$ **87.** $\frac{25}{36}$ **89.** $\frac{3}{2} + \frac{3\sqrt{3}}{2}i$
91. $2^{-1/4} - 2^{-1/4}i$
93. $1\left(\cos\frac{\pi}{3} + i\sin\frac{\pi}{3}\right)$
95. $6\left(\cos\frac{5\pi}{4} + i\sin\frac{5\pi}{4}\right)$
97. $5\sqrt{2} + 5\sqrt{2}i$ **99.** $\sqrt{2} - \sqrt{2}i$
101. $3\left(\cos\frac{5\pi}{9} + i\sin\frac{5\pi}{9}\right)$
103. $\frac{27}{2} + \frac{27\sqrt{3}}{2}i$ **105.** $512 - 512\sqrt{3}i$
107. 1, $\frac{1}{2} + \frac{\sqrt{3}}{2}i$, $-\frac{1}{2} + \frac{\sqrt{3}}{2}i$, -1, $-\frac{1}{2} - \frac{\sqrt{3}}{2}i$, $\frac{1}{2} - \frac{\sqrt{3}}{2}i$
109. $-\frac{\sqrt{4 + 2\sqrt{2}}}{2} + \frac{\sqrt{4 - 2\sqrt{2}}}{2}i$, $\frac{\sqrt{4 + 2\sqrt{2}}}{2} - \frac{\sqrt{4 - 2\sqrt{2}}}{2}i$
111. $1.06 + 0.17i$, $0.17 + 1.06i$, $-0.95 + 0.49i$, $-0.76 - 0.76i$, $0.49 - 0.95i$
113. $\cos 3\theta = \cos^3\theta - 3\cos\theta\sin^2\theta$, $\sin 3\theta = 3\cos^2\theta\sin\theta - \sin^3\theta$

Index